1.0366

					VIII A
					2 **He** 4.0026

III A	IV A	V A	VI A	VII A	
5 **B** 10.81	6 **C** 12.011	7 **N** 14.007	8 **O** 15.999	9 **F** 18.998	10 **Ne** 20.179

	I B	II B	13 **Al** 26.98	14 **Si** 28.09	15 **P** 30.974	16 **S** 32.06	17 **Cl** 35.453	18 **Ar** 39.948
28 **Ni** 58.71	29 **Cu** 63.55	30 **Zn** 65.37	31 **Ga** 69.72	32 **Ge** 72.59	33 **As** 74.92	34 **Se** 78.96	35 **Br** 79.904	36 **Kr** 83.80
46 **Pd** 106.4	47 **Ag** 107.87	48 **Cd** 112.40	49 **In** 114.82	50 **Sn** 118.69	51 **Sb** 121.75	52 **Te** 127.60	53 **I** 126.90	54 **Xe** 131.30
78 **Pt** 195.09	79 **Au** 196.97	80 **Hg** 200.59	81 **Tl** 204.37	82 **Pb** 207.2	83 **Bi** 208.98	84 **Po** (210)	85 **At** (210)	86 **Rn** (222)

Colored symbols denote elements essential to living organisms.

64 **Gd** 157.25	65 **Tb** 158.92	66 **Dy** 162.50	67 **Ho** 164.93	68 **Er** 167.26	69 **Tm** 168.93	70 **Yb** 173.04	71 **Lu** 174.97
96 **Cm** (247)	97 **Bk** (247)	98 **Cf** (251)	99 **Es** (254)	100 **Fm** (253)	101 **Md** (256)	102 **No** (254)	103 **Lw** (257)

Chemistry:
AN ECOLOGICAL APPROACH

Harper's Chemistry Series
Stuart A. Rice, EDITOR

Chemistry:
AN ECOLOGICAL APPROACH

Roger G. Gymer

Harper & Row, Publishers
New York, Evanston, San Francisco, London

About the Author

Roger Gymer received his education in chemistry at Western Reserve University, the University of Minnesota, and Case Institute of Technology (now Case Western Reserve University). After receiving the Ph.D. from Case in 1961, he taught at California State University in Fresno, the College of the Virgin Islands, Case Institute of Technology, and Fort Lewis College (Colorado); and he also did research with the National Aeronautics and Space Administration. He was associated with the Advisory Council on College Chemistry at Stanford University prior to devoting himself to writing and consulting for environmental impact studies. Now an Adjunct Professor of Chemistry at California State University in Fresno, Dr. Gymer has had a long-time interest in conservation and is a member of the Wilderness Society and the Sierra Club.

To Anne

Contents

Credits for photographs
Part One: NASA; Part Two: T. C. Naughton; Part Three: Michel Cosson; Part Four: Koch, Ralpho-
Guillumette; Part Five: Rebecca Sacks; Part Six: United States Department of Agriculture.

Preface

This is an introductory chemistry text intended primarily for students studying subjects related to chemistry and for students of chemistry who wish to relate chemistry to other disciplines. Its intention is to present, without superficiality, the basic principles of modern chemistry within a loose framework of man's interactions with the natural environment.

The environment of any living organism, such as man, includes all other forms of life on earth as well as his physical environment: the earth, air, and water. Ecology, defined for the present as the study of how the household of nature is kept in order, attempts to point out and explain the interrelations of living organisms with their total environment. A good many of these relations are chemical in origin, and it is with this in mind that most of the ideas of basic chemistry are introduced.

Behind this approach to chemistry is the realization that man is increasingly acknowledged to be the principal contributor to the deterioration of his environment and the decline in the quality of life. Population growth, urbanization, and technological manipulation of the natural environment have resulted in such destruction of our natural resources and pollution of our living space that some fear that man has placed himself, as he has other forms of life, on the list of endangered species. This is a sad situation, but it is not yet hopeless. There are ways to prevent or reduce man's impact on his surroundings, and there are ways to cure or reverse the backlash of a wounded environment on man. But if man is to evaluate intelligently the present and future consequences of his actions on the environment, he will need a deeper understanding of the scientific foundations of his technology and of the interaction of this technology with his surroundings. This viewpoint is developed further in the Epilogue.

The presentation here of the chemical background of environmental problems steers a middle course between the detailed treatments of specialists and the popular, but oversimplified and often misleading, discussions found in the news media. It is my hope that this book will help the student do two things: understand the principles of modern chemistry and understand how these principles may contribute to an intelligent approach to conservation—the wise use of our natural resources and the protection of our environment.

Although all the topics usually found in an introductory chemistry text are included here, emphasis has been placed on an interdisciplinary attack on chemical areas having to do with life and its well-being, rather than on the more inclusive "environmental" subject matter. This emphasis does not mean that certain major chemical topics have been excluded if they seem to be less directly related to life or ecology. Indeed, some of these topics have been more fully, but qualitatively, developed than usual, partly in order to avoid some of the hazy conceptions caused by too concise a treatment. It is intended that the instructor use his own discretion on how much class time to devote to these sections. Extensive cross-referencing permits some of the development to be skipped, if desired, and referred to later.

Because of the approach taken, it is inevitable that many ideas from other fields such as molecular biology and biochemistry as well as ecology must be used. Nevertheless, no comprehensive discussion of any of these areas is attempted. On the other hand, no attempt is made to exclude ideas from any discipline if they will serve to enhance our understanding and treatment of our natural environment provided our principal focus is on basic chemistry. For not only are man and all parts of his environment related; but, in today's situation, all disciplines—the sciences and the humanities—are related and integral in terms of protecting life. As a result, some of the topics treated here either are not covered in the usual introductory chemistry course or are passed over quite rapidly. That they have been included or developed more fully here reflects (1) the intention to base the chemistry on the sub-microscopic level of electrons, atoms, and molecules in order to show how subtle events at this level may be propagated up through levels of biological organization to affect organisms and ecosystems, and (2) the necessity to provide some background material in interdisciplinary areas.

In traditional chemistry texts all material dealing with a given chemical topic, such as solution properties or chemical energetics, is usually completely developed in one chapter. Subsequent chapters may use some of this material to develop another fairly self-contained topic. Such an orderly, logical path from one chemical concept or topic to the next—an approach that may be useful for illustrating the structure of chemistry—is, with the exception of some material in the first half of the book, not used here. As far as possible, chemical ideas are introduced as needed to explain ecological and environmental relationships, and the ideas are developed only to the point needed to understand these relationships at a basic level. In doing this, chemical principles have been stressed, and the need for the reader to memorize disconnected descriptive material has been minimized.

The abandonment of the chemical topics approach for an ecological approach does not mean that the text is based on a sophisticated ecological framework that stresses cycles of material and the flow of energy in nature. If anything, the text is patterned loosely on the basic environmental needs of life and on certain characteristics of life. The title phrase "an ecological approach" here connotes more a perspective or philosophy than a concern with ecology as a body of scientific knowledge. I am, of course, aware that some exceptions can be taken to the phrase, but I feel that it describes the philosophy, if not the framework, of the book better than any other.

Chemistry texts have almost always presented their material in an objective manner without the introduction of value judgments. While basic science is indeed neutral in this respect, the uses of science in technology and the problems associated with areas such as urbanization and population are open to such decisions. I have exposed my leanings on several of these areas, although not as much as I had originally intended. The student is encouraged to make his own value judgments as he proceeds through this book.

I encourage all readers to send me their suggestions, criticisms, and corrections to any errors.

ROGER G. GYMER

Three Rivers, California

Chemistry:
AN ECOLOGICAL APPROACH

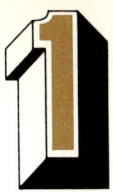

Chemistry and the Biosphere

*We travel together, passengers on a little space ship
. . . dependent on its vulnerable reserves of air and
soil; . . . all committed for our safety to its security
and peace; . . . preserved from annihilation only by
the care, the work and, I will say, the love we give
our fragile craft.*

Adlai Stevenson

As man prepares to send human beings on long journeys through outer space aboard spaceships, he must give considerable thought to the care and feeding of the passengers on these ships. Aerospace scientists realize that the environment aboard these ships must be finely controlled and regulated if the passengers are to survive long trips. The study of life-support systems in the isolated or closed environment that exists in a spaceship on a long journey must concern itself with the wise use and protection of the basic needs of life— principally food, water, oxygen, and energy. Obviously, in an environment cut off from the earth, wise use of the resources stored in the ship at its starting point is important, because most of these resources cannot be renewed. So carbon dioxide produced by respiration must somehow be removed from the atmosphere carried along in the ship; water, in the absence of augmenting sources, must be made or regenerated from urine and perspiration. Food must be conserved or grown, and wastes must be ejected or recycled and used again.

The quotation above indicates that those of us who remain on earth need to think along the same lines so far as the planet Earth is concerned. The growth of population, combined with increasing technological advances and industrialization, have resulted in a staggering rise in the rate of use of the resources of the earth. We are burning up oxygen from the atmosphere, replacing it with carbon dioxide and a host of much more toxic compounds. We are using up nonrenewable resources, such as fossil fuels, as well as renewable resources, such as air and water. Because we think that new kinds of resources will eventually be found or that the supply of certain materials is endless, we have given little thought to the idea that our "vulnerable resources of air and soil" might be damaged beyond repair or wasted for all time. And until recently, we have not worried much about the fact that damage to one part of our environment, whether it be living or nonliving, can affect other parts of our environment in subtle and unknown ways. At one time, our natural environment could renew certain resources at a rate nearly equal to the rate at which man used them. Now we are using many natural resources at a rate greatly exceeding the restorative processes in nature. At one time, our environment, again by natural means, could assimilate and dispose of the wastes of man and his civilization. But now the wastes produced by man are being poured back into our environment in ever-increasing amounts and variety. Where nature retains the ability to decompose wastes, we are exceeding the natural assimilative and restorative powers of nature. The results are well known: polluted air and water, destruction of wildlife, disappearance or dispersal of natural resources, and an overall degradation and destruction of hospitable and livable surroundings.

If we do regard the earth as a huge spaceship, we must view it as a complex system of many parts which are all interconnected to some extent. To begin our discussion, we can group the many components of this earth system into three parts: man, all other life, and the nonliving or physical surroundings. Man is not singled out because he is especially unique from other forms of life, but because his impact on the total system exceeds that of all other life forms. These three major components can be drawn as in Figure 1.1, where the arrows indicate a wide variety of interactions and interrelationships

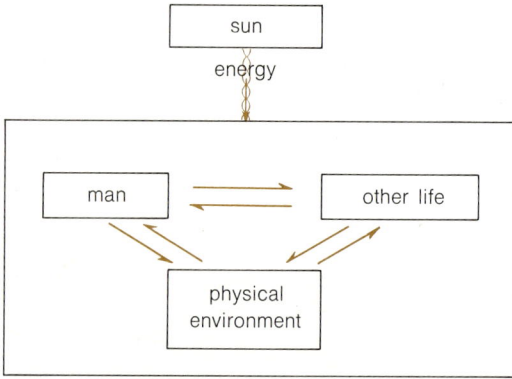

Figure 1.1. Interacting subsystems in the biosphere.

between and among the subsystems. We could, if we wanted, simplify the diagram even more by merging the man and other life subsystems into one subsystem including all life on earth.

If our attention is limited to that part of earth's physical environment in which life, as we know it, exists, then the three subsystems in the large rectangle of Figure 1.1 constitute what is called the *biosphere*, that portion of the earth inhabited by life. This is a relatively thin zone near the surface of the earth, for life does not exist to any appreciable depth in the soil nor very high up in the atmosphere. Within the biosphere occur all the dynamic, interlocking processes of the living and nonliving world. The power to drive this intricately balanced and complex "machine" is provided by the solar energy entering the biosphere.

Within a portion of the biosphere and at a particular place, there may be stronger links among some life forms than among others. These associated life forms will then have strong connections to their common environment. If we think of drawing a boundary around these strongly associated organisms and their common environment, we have created a smaller unit of the biosphere. This unit is called an *ecosystem*. The unit is composed of the nonliving or physical environment and the collection of living organisms that are particularly dependent on each other and on the physical environment. The study of the structure of the ecosystem and of the relationships between the organisms and their total environment—relations that give rise to the structure of the ecosystem—is the science of *ecology*. The word "ecology" is derived from the Greek *oikos*, meaning "home," "house," or "place to live." Literally, then, ecology is the study of the household (of nature). In the respect that ecology is the study of a natural system, where we describe a system as a number of parts making up a complex whole, ecology is really just an application of the use of the powerful systems analysis approach. All systems, whether ecosystems, a human body, a large factory, or a nation's economy, have in common an internal interconnectedness. Systems analysis means the consideration of that complex interrelated whole and not just of the individual parts.

We are going to make use of only a small part of the wide field of ecology, and we shall not cover all the concepts and terminology of the field. Ecology, in the beginning, was a branch of descriptive biology. Today it is a multi-disciplinary field that is becoming more quantitative. Again, we shall not be concerned so much with ecology as a field of science or even a grouping of disciplines as we shall with ecology as a way of thought or as a way of viewing some of our environmental relationships.

Chemistry can be defined as the integrated study of the preparation, properties, and reactions of the chemical elements and their compounds, and of the systems they form. You can perhaps already see that many of the relationships between organisms and their environment will be chemical in nature, because organisms and their environments can be considered to be interacting chemical systems. To get a little bit ahead of ourselves, many of modern man's technological processes spew chemical wastes into the environment. Subsequent reaction of these chemicals with either the living or the nonliving (biotic or abiotic) parts of the environment may lead to secondary chemical effects in man or in other organisms of the ecosystem on which man depends. What we wish to do is study the basic chemical processes, generally at a microscopic or molecular level, and see how the chemistry leads to or is relevant to changes in the structure of the ecosystem. This will be discussed further after we look at a simple example of an ecosystem, discuss material cycles and energy flow, and learn something about levels of organization in nature.

A Simple Example of an Ecosystem: A Fresh-Water Lake

To get some idea of the structure of an ecosystem and what goes on in it, a fresh-water lake will be used as an example. Usually, the boundaries of this ecosystem can be taken as the shores and the surface of the lake, but this choice is somewhat arbitrary, just as it is when any ecosystem is delineated.

Within any ecosystem there are at least two, and usually three, kinds of living organisms. One, like the algae in a lake, grows and *produces* more complex substances from the relatively simple chemical materials in the lake. Algae, small floating green plants, do this by photosynthesis, using carbon dioxide, water, and various nutrients in the lake, such as the nitrates and phosphates commonly found in agricultural fertilizers. In addition to producing the more complex carbohydrates, proteins, and other molecules, the algae also give off oxygen gas.

The second kind of living organism usually present consumes the algae and may, in turn, be eaten by other *consumer* organisms. In the lake, a variety of insects and fish perform the consumer role.

When the producers and consumers die, their bodies are acted upon by *decomposer* organisms such as bacteria and fungi. This decomposition eventually breaks down the complex chemical substances in the bodies to the simple substances that can again be used by the producers for their growth. Thus, a material cycle begins. Organisms, eating and being eaten by other organisms, are part of an often complex food chain or web whereby materials and the energy needed for life are transferred from one participant to another and to the nonliving, or abiotic, environment. Most substances, or their constituent chemical elements, undergo a series of chemical transformations as they traverse their cycles from the abiotic environment, through the organisms in the food chain, and back to the abiotic environment. Whether these chemical reactions will be beneficial or detrimental to life depends on the chemical properties of the substances themselves, the chemical properties of other materials in the environment, and the biochemical makeup of the living organisms present in the environment. Also, in general, the chemical reactions are accompanied by the release or the absorption of energy, which may be used to drive the cycles and to power the living organisms. In essence, the cyclical movement of materials and the flow of energy through the different parts of the ecosystem both determine and are determined by the relations of organism to organism and of organisms to their physical environment.

Each kind of organism depends on the existence of certain substances in its surroundings as well as on certain environmental factors such as temperature and light. Because the growth of a consumer depends on the growth of a producer, the consumer is indirectly dependent on conditions that may directly concern only the producer. In the lake, the survival and reproduction of the fish depend on the availability of sunlight and carbon dioxide to the algae. Not only are the bodies of the algae necessary as food for the fish, but the oxygen produced by photosynthesis is essential to the life of the consumer organisms and to the decomposer organisms. On the other hand, if there is too great a growth of algae in the lake, the large mass of decaying algae will use up the dissolved oxygen in the water, and the fish will die. This situation could come about by the addition of large amounts of nitrate and phosphate from sewage. Ordinarily, most nutrients like these are supplied to the lake in the natural water runoff from the surrounding land. There is still a material cycle, however. If one wished to consider this natural drainage contribution more fully, then the boundaries of the lake ecosystem could be extended beyond the shores. And if the effect of sewage becomes a major factor, man should be included in the community. The extent of an ecosystem thus depends on the strengths of possible interactions and on the length of time the ecosystem may be under consideration.

Increased water temperature, perhaps due to disposal of waste or cooling water from factories or power plants, would also affect the amount of oxygen dissolved in the water. If we chose, we could make a list of the many chemical reactions involved in the movement of oxygen and other materials through the lake ecosystem, but it will become clear that there are a number of interrelations, many of them basically chemical in origin, that can affect the operation of an ecosystem and thereby its makeup or structure.

In considering the structure of an ecosystem, ecologists often talk about the *population* of one kind of organism, say a lake trout, in the ecosystem. The ecology of populations encompasses the numbers of each particular organism, its birth and death rates, age distribution, population density, and other characteristics of each group of interacting like species. But, as we have seen, the trout are also intimately associated with other kinds of organisms. The term *community* refers to the cohesive unit composed of all interacting organisms in the ecosystem. Questions about food chains (what eats what), the change in species composition with time, and the like are included in community ecology. It is said that the study of populations and communities is the true province of ecology, considered as a field of biology. However, we hope to show how at least some of the features of populations and communities are affected, if not caused, by some physical and chemical factors at lower levels of organization.

By levels of organization we mean the ways that living things can come together to form units of large size and of generally greater complexity. A spectrum of biological organization is shown in Figure 1.2.

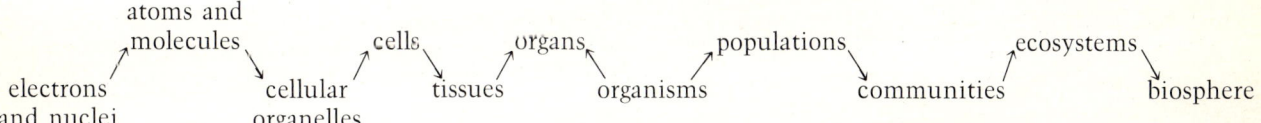

Figure 1.2. Levels of organization of biological systems.

Events at lower levels of organization often determine the structure and stability of succeeding levels. Also, in any particular level, the forces that maintain a balanced organization are often related to factors basic to the levels both above and below that level.

Everyone has heard the phrase "the balance of nature," but it has been most commonly used to describe what happens on the community level when the relation between a consumer (predator) species and a species on which it feeds (prey) is disturbed, perhaps when a lowered predator population causes an unfavorable population growth of the prey species. Under natural conditions, such disturbances are adjusted and balance reestablished by self-regulating mechanisms in the community. There are other types of balancing mechanisms at all levels. In ecology, such mechanisms are called *homeostatic mechanisms*, and the regulating process by which a system maintains balance or stability is called *homeostasis*. The environmental problems we have today are, at heart, the inability of the natural homeostatic mechanisms to operate, or to operate rapidly enough, after man's activities have thrown the environment out of kilter. While it is true that no molecular approach may reveal the causes of the predator–prey population effects, many of the balancing or control mechanisms operate at the atomic and molecular levels.

Man, either through ignorance or shortsightedness, often interferes with the balance of nature; and nature, in turn, often reacts back on man. The effect of a disturbance at one level of organization may be transmitted along the organizational ladder, often by subtle or delayed actions, because of the interrelations that exist between levels. Because man is linked with his natural environment by an enormous number of connections, some very apparent, some very subtle, he has the capacity of harming his environment and himself. His environment will generally suffer some tangible physical damage, but man can be harmed in numerous other ways.

Recognition of how effects, chemical or otherwise, may originate at any level and of how the consequences are propagated up or down the ladder of organization is only part of the problem. If conservation can be said to be the study of the relationships between the human community and the planetary environment, and if the human community includes both man and his institutions, then the relations between man and his institutions must be studied. This study, sometimes called human ecology, will be discussed at the end of the book in Chapter 21.

At this point, let's get on our organizational ladder and begin climbing. We shall look first at our environment at the atomic and molecular level to try to get some idea of why our present natural world may be especially fit, even unique, for life. We shall consider such fitness to result from, and be defined in terms of, the way the physical and chemical behavior of the primary constituents of the environment meet the conditions required by present-day living organisms.

References

Cole, L. C., "The Ecosphere," *Scientific American*, **198,** 83 (April 1958).

Grobstein, C., "The Strategy of Life," Freeman, San Francisco, 1965.
> A discussion of levels of organization by a biologist.

Handler, P., Ed., "Biology and the Future of Man," Oxford, New York, 1970, Chapter 11, pp. 431–473.

Hutchinson, G. E., "The Biosphere," *Scientific American*, **223,** 45 (September 1970).
> This entire issue of *Scientific American* is devoted to the biosphere, emphasizing material cycles and energy flow.

Odum, E. P., "Fundamentals of Ecology," 3rd ed, Saunders, Philadelphia, 1971.

Woodwell, G. M., "Toxic Substances and Ecological Cycles," *Scientific American*, **216,** 24 (March 1967).
> Most of this article is about the concentration of toxic substances, especially the insecticide DDT, in food chains. However, it brings up a number of concepts that will be useful at this early point in the text and that will give previews of some of the subjects to be discussed in later chapters. The article offers a good perspective of the ecological side of ecological chemistry. Recommended reading—now and later.

ONE

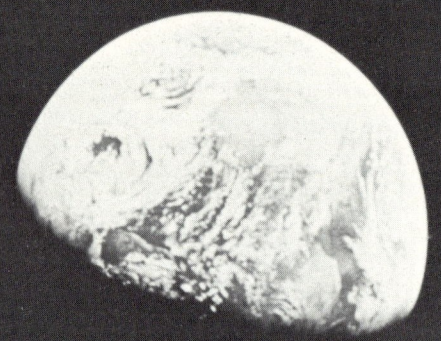

The Chemical Fitness of the Environment

The two major topics we shall be considering throughout this book are life and the environment in which life exists. Neither life nor the environment are constant, never-changing parts of the history of the earth. We know, for instance, that the climate at specific points on the earth's surface is much different today from what it was millions of years ago. We also know that the forms of animal and plant species have changed over the course of geological time. Living organisms and their environment can both change slowly and often do so at the same time. A good example of this is the penguin in Antarctica. The climate in Antarctica was once tropical, or at least temperate. The penguins were then true birds of the forest. As the climate of the Antarctic changed, the wings of the penguins evolved into flippers for use in swimming rather than flying.

For purposes of discussion, let us allow only one variable, life form or environment, to change while regarding the other as static. First, consider that the environment remains the same for an extended period of time. Living organisms then become adapted to this environment through behavioral or evolutionary changes or both—or they try to move to another, more favorable, environment. Those organisms that adapt to their external physical or biological environment in this way can survive, reproduce, and flourish. Those that do not or cannot adjust to changing conditions are eventually doomed to extinction as a species. Those species that are able to adjust to the environment will continue to propagate, and those survive will be passed down from one generation to the next. The fittest will survive. This is basically the theory of natural selection and biological evolution enunciated by the English naturalist Charles Darwin (1809–1882) in 1859. (Later we shall discuss a "chemical evolution," which precedes biological evolution. A natural selection also acts in chemical evolution.) Today, human technology is contributing to changes in environment that are so rapid that many species cannot adapt quickly enough.

But now, consider what might happen if living organisms remain static in the form that we call life today. Does the environment change to adapt to life as we know it? Not likely. There is no such thing as adaptation or natural selection of the environment in its relation to life, although in many ways the existence of life can change the environment (see Chapters 18 and 20).

What we can consider, however, is not *adaptation* but the *fitness* of the environment for life as we know it. That is, what makes our present environment especially fit for the existence of life? This fitness predates adaptation.

This entire question was considered in some detail in 1913 by Lawrence J. Henderson (American biological chemist, 1878–1942) in his classic book, "The Fitness of the Environment." He reasoned that the environment must have certain qualities in order to support and give rise to certain major characteristics of life. The qualities of such a fit environment arise in large part from the physical and chemical characteristics of this environment and the way in which the resulting properties provide the "fittest possible abode of life." Henderson showed, on the basis of the chemical knowledge then available, that the present chemical composition of the environment, as well as of life, seemed to be particularly suitable for the existence and functioning of life. Briefly, the chemical elements and their compounds found in living organisms, as well as in their external physical environment, were shown to be particularly fit, and often unique, to the performance of their function in sustaining living processes or as a habitat for life. These properties are now known to depend in large part on the microscopic or electronic, atomic, and molecular makeup of matter.

What was demonstrated in 1913 is even more powerfully demonstrable today. We know, for example, that the structure of the hydrogen atom and of the oxygen atom are responsible for the structure, both geometric and electronic, of the water molecule. These are, in turn, mainly responsible for the unique behavior and properties of water, an essential ingredient of living organisms and of the life-supporting external environment.

Details will be given when we have considered some basic knowledge of atoms and molecules: what they are, what they are made of, how atoms combine to form molecules, and how molecules interact. Our first task, then, is to investigate extensively the nature of atoms and molecules. We shall try to do this in as general a way as possible, using as specific examples only those substances, out of a vast number of possible substances, of particular importance for life.

Atoms and Electrons

The primal elements must needs be made
Of stuff immortal, whereinto all things
At their last hour can be dissolved, that so
Their matter for the birth of other things
May be supplied. These primal bodies, then,
Are solid, single, indivisible;
Nor could they else throughout the eternal roll
Of endless time now gone have been preserved,
And of their substance build all things anew.

Lucretius—*De Rerum Natura*
(*On the Nature of Things*)

The concept of atoms and the development of atomic theory have a long history, one sprinkled with the names of many illustrious philosophers and scientists. In more recent times, our picture of the atom has gone from the view of charged particles interacting according to the laws of classical (Newtonian) physics to a view that appears to be at odds with our experience with ordinary objects. Both of these ways of looking at an atom, the classical and the quantum theory of atomic structure, will be considered in this chapter.

Classical Atomic Theory

RONS, PROTONS,
AND NEUTRONS

Atoms are the fundamental building blocks of matter. At one time they were thought to be indivisible things that could not be broken down into any component parts, but the discovery of the phenomenon of natural or spontaneous radioactive decay (see Chapter 20) disposed of this belief. Atoms have been shown to be composed of at least three elementary particles: the electron, the proton, and the neutron. The histories of the discoveries of these three particles can be found in any standard freshman chemistry or physics text. We shall assume the existence of these three particles and shall concern ourselves only with their properties and with their place in the atom.

A proton has a positive electric charge whose magnitude is arbitrarily defined as one, or unity. The electron carries an opposite electric charge of equal magnitude, -1. A neutron, as the name implies, is electrically neutral. In an electrically neutral atom, the number of protons must equal the number of electrons. The charges of these particles, as well as their masses, are summarized in Table 2.1.

Table 2.1
Charge and mass of the fundamental particles[a]

particle	symbols	charge relative to proton	mass relative to proton	mass relative to ^{12}C, amu
electron	$_{-1}^{0}e$, β^-, e^-	-1	1/1836	0.000549
proton	$_{1}^{1}p$, p	$+1$	1	1.00728
neutron	$_{0}^{1}n$, n	0	1.001	1.00867

[a] One atomic mass unit (amu) is 1.66×10^{-24} g. The left superscripts on symbols refer to the approximate mass of the particle relative to the proton mass. Left subscripts indicate relative charge, as do any signs at the upper right.

The masses in Table 2.1 are given in terms of (1) mass relative to that of the proton and (2) mass relative to that of the ^{12}C atom, which is considered to have a mass of exactly 12 atomic mass units (amu). These units will be explained more fully shortly, and they are included here primarily to give some idea of the very small mass of the electron relative to that of the proton and neutron.

2.2 GROSS STRUCTURE OF THE ATOM

Knowledge of the existence of the three component particles of the atom tells nothing of the gross structure of an atom. How do these three particles fit into an atom? How big is an atom? To answer these important questions, we can go back in history to a classic experiment done in 1911 by Lord Rutherford (Ernest Rutherford, British physicist, 1871–1937; Nobel prize in chemistry, 1908). This experiment led to the nuclear model of the atom.

In this experiment, positively charged alpha particles (alpha particles are made up of 2 protons and 2 neutrons, thus carrying a $+2$ charge) were shot at a very thin foil of gold metal. Most of the alpha particles passed right through the foil with little or no deflection from their original path. A very few (about 1 in 10,000) were considerably deflected from their original path. The situation is shown in Figure 2.1.

Rutherford showed that the results of this experiment could be explained only if the metal foil was composed of atoms made up of a very small, positively charged *nucleus* surrounded at comparatively large distances by very light, negatively charged electrons. Since like electric charges repel one other, the kinds of deflections observed for the incoming alpha particle explainable on the basis of a collision with, or a near approach to,

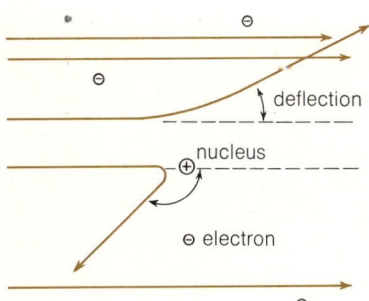

Figure 2.1. Scattering of alpha particles by atomic nuclei. The closeness of the electrons to the nucleus is exaggerated in this figure.

more highly charged and massive body. This body was called the nucleus. To explain those very large deflections where an alpha particle in effect bounced off the metal foil, it was necessary to assume that all the positive charge and essentially all the mass of an atom was concentrated in the nucleus. From the fact that so few of the projectiles were deflected, it was deduced that the nucleus was extremely small and that there were comparatively tremendous distances between adjacent nuclei in the foil.

An alpha particle has a mass 7300 times greater than an electron; so, despite their unlike charges, there would be no measurable deflection of an alpha particle even on direct collision with isolated electrons. Thus, the observed deflections on which Rutherford based his deductions gave no information about the position of the electrons in the atom except to show that they were located outside the nucleus. The main results from this scattering experiment were then:

1. Atoms are composed of very small, positively charged nuclei. The protons therefore reside in the nucleus.

2. The nucleus is very dense, containing approximately 99.9 percent of the mass of an atom. (It was later discovered that the neutrons as well as the protons reside in the nucleus.)

3. There are large spaces between adjacent nuclei.

4. The electrons reside outside the nucleus at relatively tremendous distances from the nucleus. An atom is, therefore, mostly empty space.

5. Matter, which is made up of these nuclear atoms, may also be said to be mostly empty space, because the nuclei of adjacent atoms are separated by the space through which the electrons move.

We may, for the present, imagine the nucleus of a typical atom as being a sphere about the size of a golf ball (about $1\frac{1}{2}$ in. in diameter) and an electron as being a fly circling the nucleus at a distance of about $1\frac{1}{2}$ miles. If the nucleus were the size of a pencil eraser, its density would cause it to weigh 300 million tons.

The nuclear model is often called the planetary or solar-system model of the atom because of its apparent similarity to the structure of our solar system, with the sun being analogous to the nucleus and the planets to the electrons. In fact, the size and distance comparisons are quite similar. The diameter of an atom (to the outermost electrons) is of the order of 10^{-8} cm, while the diameter of a nucleus is about 10^{-13} cm. The ratio of the diameter of an atom to that of its nucleus is therefore about $10^{-8}/10^{-13} = 10^5$ (or 100,000). (See Appendix A.1 for a discussion of exponents.) The corresponding ratio of the largest planetary orbit to the diameter of the sun is about 10^4 (10,000). These comparisons serve only to give some idea of the spatial extent of an atom. You should not place too much credence in the picture of an electron as a particle with a point charge rotating about the nucleus; this aspect of the nuclear model has been substantially modified.

A more detailed analysis of the results of such scattering experiments yields information on the number of protons in the nucleus of the scattering atom. This analysis is based on the forces of electric repulsion to be expected between the positive alpha projectile and the bombarded nucleus. The results show that the charge on the nucleus is some whole number of positive proton charges. From experiments using different metal foils, the number of protons in the nuclei of various atoms can be determined.

We now know where the proton and the electrons are. What about the neutrons? At the time of Rutherford's experiment, the weights or masses of various atoms were known relative to the weight of oxygen atoms. Natural oxygen as obtained from the air or from a chemical reaction was considered by chemists to have an "average" atom that was arbitrarily assigned a weight of exactly 16.0000 amu. Other atoms were assigned masses relative to oxygen so that, for instance, an atom that weighed twice as much as an oxygen atom would be assigned an atomic mass of 32.0000 amu.

Because the electrons in an atom make an almost negligible contribution to the mass of an atom except in the heaviest atoms, almost all of the mass is concentrated in the nucleus. It was found that the total mass of the protons in a particular nucleus was generally much lower than the known mass of a particular atom. The difference or discrepancy in mass for a given atom was made up by assigning a number of neutrons to the nucleus sufficient to make up the deficiency in mass.

2.3 ELEMENTS   The number of protons in the nucleus of an atom, the nuclear charge, is called the *atomic number* of that atom and is given the symbol Z. In the neutral atom, the nuclear charge controls the number of electrons outside the nucleus; as you will soon see, these electrons control all the chemical properties of the atom. The nuclear charge or atomic number Z is thus an extremely important number, since all atoms with the same atomic number behave very much alike chemically. Since atoms having the same chemical behavior are generally said to define a chemical element, the precise definition is as follows: all atoms having the same atomic number are atoms of the same *element*. Listed in Table 2.2 are the names and symbols of the elements having from 1 to 20 protons in their nuclei. Also listed are rounded-off atomic mass values. Starting with hydrogen, we see that there are no neutrons in the nucleus of hydrogen with atomic mass of 1 amu. Helium has 2 neutrons in its nucleus along with 2 protons. Lithium has 3 neutrons and 3 protons; beryllium has 4 protons and 5 neutrons, and so on.

2.4 MASS NUMBER   The sum of the number of protons and neutrons in the nucleus of an atom is called the *mass number A*. The mass number is always a whole number. The number of neutrons in the nucleus is $A - Z$. It is customary to denote a particular atom X by the symbol $^A_Z X$.

It is possible for atoms of the same element to have slightly different numbers of neutrons in their nuclei; that is, they have different mass numbers but identical atomic numbers. Such atoms are called *isotopes*. Many elements have several isotopes. In naturally occurring oxygen, for instance, there are atoms having three different mass numbers, 16, 17, and 18. All have 8 protons in their nuclei. The 3 isotopes of oxygen and their nuclear constitution are

$^{16}_{8}O$ 8 protons, 8 neutrons

$^{17}_{8}O$ 8 protons, 9 neutrons

$^{18}_{8}O$ 8 protons, 10 neutrons

Isotopes are commonly denoted by saying "oxygen-16," "oxygen-17," and so on. Only for hydrogen are the isotopes given distinctive names and symbols:

$^{1}_{1}H$ hydrogen

$^{2}_{1}H$ or $^{2}_{1}D$ deuterium (heavy hydrogen)

$^{3}_{1}H$ or $^{3}_{1}T$ tritium

Table 2.2 The first 20 elements	element	symbol	atomic number Z	atomic mass, amu, rounded off to the nearest whole number
	hydrogen	H	1	1
	helium	He	2	4
	lithium	Li	3	6
	beryllium	Be	4	9
	boron	B	5	11
	carbon	C	6	12
	nitrogen	N	7	14
	oxygen	O	8	16
	fluorine	F	9	19
	neon	Ne	10	20
	sodium	Na	11	23
	magnesium	Mg	12	24
	aluminum	Al	13	27
	silicon	Si	14	28
	phosphorus	P	15	31
	sulfur	S	16	32
	chlorine	Cl	17	35
	argon	Ar	18	40
	potassium	K	19	39
	calcium	Ca	20	40

Not all elements have more than one naturally occurring isotope. Fluorine, for example, has only one kind of nucleus, one that contains 9 protons and 10 neutrons.

2.5 ATOMIC MASSES

Atoms can be converted into charged particles called *ions* by the removal of one or more electrons. When a stream of moving ions is subjected to electric and magnetic forces in an instrument known as a mass spectrograph, ions of different mass may be separated from one another. The masses of the ions may be determined relative to the mass of some reference ion or atom that is arbitrarily assigned an atomic mass. Since 1961, the carbon-12 isotope, ^{12}C, has been used to assign all other atomic masses. Its mass is arbitrarily assigned as exactly 12 amu. The masses of the elementary particles are also defined with respect to ^{12}C. The *atomic mass unit*, amu, is therefore defined as one-twelfth the mass of the lightest naturally occurring isotope of carbon. The masses of various atoms (nuclei + electrons) are given in Table 2.3. Note that, unlike mass numbers A, the atomic masses are not whole numbers except for ^{12}C.

2.6 ATOMIC WEIGHTS

You can see from Table 2.3 that many elements have two or more naturally occurring isotopes. (Some elements also have artificial isotopes, which do not occur in nature.) The relative proportions of the different isotopes of an element in the natural mixture can easily be determined with the mass spectrograph. Table 2.3 includes the relative abundances of the listed isotopes in the natural mixture.

Table 2.3 Isotopes of some elements element		mass of isotope, amu	relative abundance, %
hydrogen	$_1^1H$	1.008	99.98
	$_1^2H$	2.014	0.02
	$_1^3H$	3.016	0
carbon	$_6^{12}C$	exactly 12 (standard)	98.89
	$_6^{13}C$	13.003	1.11
nitrogen	$_7^{14}N$	14.003	99.63
	$_7^{15}N$	15.000	0.37
oxygen	$_8^{16}O$	15.995	99.76
	$_8^{17}O$	16.999	0.04
	$_8^{18}O$	17.999	0.20
fluorine	$_9^{19}F$	18.998	100
chlorine	$_{17}^{35}Cl$	34.969	75.53
	$_{17}^{37}Cl$	36.966	24.47
uranium	$_{92}^{235}U$	235.044	0.71
	$_{92}^{238}U$	238.051	99.28

The elements with which the chemist, biologist, or geologist work are almost always the naturally occurring elements or are derived by chemical reaction from the naturally occurring compounds of the element. This means, for instance, that when chlorine is involved in any chemical reaction in nature, the chlorine is actually a mixture of two isotopes, one of mass number 35 and one of 37. The atomic masses of these two isotopes are, respectively, 34.97 amu and 36.97 amu.

$_{17}^{35}Cl$ 34.97 amu   75.5% relative abundance

$_{17}^{37}Cl$ 36.97 amu   24.5% relative abundance

For mass calculations involving chemical reactions, we must use an average value for the mass of naturally occurring chlorine. We use an average weighted on the basis of the relative abundances of the isotopes in chlorine. For instance, in 1000 atoms of natural chlorine, 755 of the atoms weigh 34 97 amu or a total of 755×34.97 amu. The other 245 atoms weigh 36.97 amu apiece for a total of 245×36.97 amu. The *average* mass of one atom of natural chlorine (or the mass of an average atom) is obtained by dividing the total mass of 1000 atoms by 1000:

$$\frac{(755\ \text{atoms})(34.97\ \text{amu atom}^{-1}) + (245\ \text{atoms})(36.97\ \text{amu atom}^{-1})}{1000\ \text{atoms}}$$

$$= 35.5\ \text{amu, average}$$

(See Appendixes A.2 and A.4 for the use of significant figures and units in calculations.) This average, called the *atomic weight* by chemists, is the value found in Table 2.4.

The atomic weights so derived find extensive use in chemistry, unlike

Table 2.4 Atomic weights[a] element	symbol	atomic number	atomic weight (relative to ^{12}C = 12.00000)
actinium	Ac	89	(227)
aluminum	Al	13	26.9815
americium	Am	95	(243)
antimony	Sb	51	121.75
argon	Ar	18	39.948
arsenic	As	33	74.9216
astatine	At	85	(210)
barium	Ba	56	137.34
berkelium	Bk	97	(247)
beryllium	Be	4	9.01218
bismuth	Bi	83	208.9806
boron	B	5	10.81
bromine	Br	35	79.904
cadmium	Cd	48	112.40
calcium	Ca	20	40.08
californium	Cf	98	(251)
carbon	C	6	12.011
cerium	Ce	58	140.12
cesium	Cs	55	132.9055
chlorine	Cl	17	35.453
chromium	Cr	24	51.996
cobalt	Co	27	58.9332
copper	Cu	29	63.546
curium	Cm	96	(247)
dysprosium	Dy	66	162.50
einsteinium	Es	99	(254)
erbium	Er	68	167.26
europium	Eu	63	151.96
fermium	Fm	100	(253)
fluorine	F	9	18.9984
francium	Fr	87	(223)
gadolinium	Gd	64	157.25
gallium	Ga	31	69.72
germanium	Ge	32	72.59
gold	Au	79	196.9665
hafnium	Hf	72	178.49
helium	He	2	4.00260
holmium	Ho	67	164.9303
hydrogen	H	1	1.0080
indium	In	49	114.82
iodine	I	53	126.9045
iridium	Ir	77	192.22
iron	Fe	26	55.847
krypton	Kr	36	83.80
lanthanum	La	57	138.9055
lawrencium	Lw	103	(257)

element	symbol	atomic number	atomic weight (relative to $^{12}C = 12.00000$)
lead	Pb	82	207.2
lithium	Li	3	6.941
lutetium	Lu	71	174.97
magnesium	Mg	12	24.305
manganese	Mn	25	54.9380
mendelevium	Md	101	(256)
mercury	Hg	80	200.59
molybdenum	Mo	42	95.94
neodymium	Nd	60	144.24
neon	Ne	10	20.179
neptunium	Np	93	237.0482
nickel	Ni	28	58.71
niobium	Nb	41	92.9064
nitrogen	N	7	14.0067
nobelium	No	102	(254)
osmium	Os	76	190.2
oxygen	O	8	15.9994
palladium	Pd	46	106.4
phosphorus	P	15	30.9738
platinum	Pt	78	195.09
plutonium	Pu	94	(244)
polonium	Po	84	(209)
potassium	K	19	39.102
praseodymium	Pr	59	140.9077
promethium	Pm	61	(145)
protactinium	Pa	91	231.0359
radium	Ra	88	226.0254
radon	Rn	86	(222)
rhenium	Re	75	186.2
rhodium	Rh	45	102.9055
rubidium	Rb	37	85.467
ruthenium	Ru	44	101.07
samarium	Sm	62	150.4
scandium	Sc	21	44.9559
selenium	Se	34	78.96
silicon	Si	14	28.086
silver	Ag	47	107.868
sodium	Na	11	22.9898
strontium	Sr	38	87.62
sulfur	S	16	32.06
tantalum	Ta	73	180.9479
technetium	Tc	43	98.9062
tellurium	Te	52	127.60
terbium	Tb	65	158.9254
thallium	Tl	81	204.37
thorium	Th	90	232.0381

Table 2.4 Atomic weights[a] (Continued)	element	symbol	atomic number	atomic weight (relative to $^{12}C = 12.00000$)
	thulium	Tm	69	168.9342
	tin	Sn	50	118.69
	titanium	Ti	22	47.90
	tungsten	W	74	183.85
	uranium	U	92	238.029
	vanadium	V	23	50.941
	xenon	Xe	54	131.30
	ytterbium	Yb	70	173.04
	yttrium	Y	39	88.9059
	zinc	Zn	30	65.37
	zirconium	Zr	40	91.22

[a] Atomic weight values in parentheses are the mass numbers of the most stable or best-known isotopes. Elements 104 and 105, which have not yet received official names, are not listed.

the atomic or isotopic masses for a particular isotope. Only when we are dealing with changes in the nucleus do we have to worry about individual atomic masses. The reasons for this are two: (1) we are almost always dealing with a natural mixture of isotopes, and (2) chemical reactions involve only the extranuclear electrons and are not affected to any appreciable extent by the smaller differences in mass of isotopes of the same element. The major exceptions to the second point occur when isotopes of the lighter elements are involved, since the relative differences in the masses of $^{1}_{1}H$ and $^{2}_{1}H$, for example, are large.

2.7 GRAM-ATOMS, MOLES, AND AVOGADRO'S NUMBER

Technically, atomic masses and the atomic weight values given in Table 2.4 have no units, amu or otherwise, associated with them. Atomic weights are pure numbers without dimensions, because they represent the weight of an atom of an element divided by one-twelfth the weight of a ^{12}C atom. As long as the units used for the weights of the element and for ^{12}C are the same, the quotient has no units, because all units cancel in the division (see Appendix A.4). Just the same, chemists generally concern themselves with chemical reactions where the quantities of reactants and their products are measured in terms of grams or some other suitably large unit, rather than in terms of individual atoms. For this purpose, it is convenient to define a quantity of an element equal to its atomic weight expressed in grams. The atomic weight expressed in grams is called a gram-atomic weight or a *gram-atom* (g-atom) of the element. A gram-atom of oxygen atoms weighs 15.999 g; a gram-atom of hydrogen weighs 1.008 g; a gram-atom of helium weighs 4.003 g.

The number of gram-atoms in any portion of elemental matter is obtained by dividing the weight of the portion by the gram-atomic weight. In 36.46 g of magnesium, for example, there are 36.46 g/24.30 g(g-atom)$^{-1}$ or 1.500 g-atoms of magnesium.

The number of atoms in a gram-atom of *any* element is an important number in chemistry. This number can be calculated using helium as an example. The mass (or weight) of an average helium atom is 4.00 amu atom^{-1} (to three significant figures). One amu is 1.66×10^{-24} g. Therefore, the mass of 1 helium atom is

$$(4.00 \text{ amu atom}^{-1})(1.66 \times 10^{-24} \text{ g amu}^{-1})$$

The number of helium atoms in 1 g-atom of helium is

$$\frac{4.00 \text{ g(g-atom)}^{-1}}{(4.00 \text{ amu atom}^{-1})(1.66 \times 10^{-24} \text{ g amu}^{-1})}$$
$$= 6.02 \times 10^{23} \text{ atom(g-atom)}^{-1}$$

The number 6.02×10^{23} is called *Avogadro's number*, N_A. The same number would have been obtained for the number of atoms in 12.011 g of carbon, 200.5 g of mercury, or 1 g-atom of any element.

The significance and application of Avogadro's number is not limited to atoms. We shall also be dealing with combinations of atoms called molecules, such as H_2 and H_2O, as well as with ions and combinations of ions. In the general case, we shall refer to 6.02×10^{23} units of a substance as being 1 *mole* (abbreviated mol) of that substance. Several examples are in order.

One mole of hydrogen atoms (H) is 6.02×10^{23} hydrogen atoms. It is also 1 g-atom of hydrogen atoms.

One mole of hydrogen gas, which is made up of hydrogen molecules (H_2), contains 6.02×10^{23} hydrogen *molecules* (or 2 g-atoms of hydrogen *atoms*).

One mole of solid sodium chloride (NaCl) contains 6.02×10^{23} NaCl *formula units* (NaCl is not considered to be a molecule except in the gaseous state). Each mole of NaCl also contains 1 mol each of sodium ion, Na^+, and chloride ion, Cl^-.

For a molecular combination of two or more atoms, a *molecular weight* equals the sum of the atomic weights of the atoms making up the molecule. When, as is usually the case, gram-atomic weights are used, the resulting molecular weight is often called the *gram-molecular weight*. In this book, the term molecular weight is assumed to mean gram-molecular weight.

EXAMPLES The molecular weight of H_2 is $2(1.008) = 2.016$ g.
The molecular weight of water, H_2O, is $2(1.008) + 15.999 = 18.015$ g.

For substances like NaCl (Na^+Cl^-), which really have no true molecules, the term "gram-formula weight" would be technically correct, but to avoid too many names we shall consider 58.4 g of NaCl to be its (gram) molecular weight.

In analogy with gram-atomic weights, a gram-molecular weight of any substance contains 6.02×10^{23} units (atoms, molecules, ions, formula units) of that substance. Therefore, if a *mole* is defined fundamentally as 6.02×10^{23} units, then we can also say that a (gram) molecular weight of any substance is commonly called a mole.

EXAMPLES
$$1.008 \text{ g of H} = 1 \text{ g-atom of H } atoms$$
$$= 1 \text{ mol of H } atoms$$
$$2.016 \text{ g of } H_2 = 1 \text{ mol of } H_2 \text{ } molecules$$
$$= 2 \text{ g-atoms of H } atoms$$
$$= 2 \text{ mol of H } atoms$$

$$18.015 \text{ g of } H_2O = 1 \text{ mol of } H_2O$$
$$= 6.02 \times 10^{23} \ H_2O \text{ molecules}$$
$$29.2 \text{ g of NaCl} = 0.500 \text{ mol of NaCl}$$
$$= 3.01 \times 10^{23} \text{ NaCl formula units}$$
$$= 0.500 \text{ mol of Na}^+ \text{ ions}$$
$$= 0.500 \text{ mol of Cl}^- \text{ ions}$$

2.8 THE PERIODIC LAW

There are now (1971) 105 elements (assuming that experimental evidence for element 105 is confirmed) known, as listed in Table 2.4. In 1869, there were only about 66 known elements. In that year, the Russian chemist Dmitri Mendeleev (1834–1907) arranged these elements in order of increasing atomic weight and found that similar properties recur at intervals or periodically. That is, *the physical and chemical properties of the elements are periodic functions of their atomic weights.* This is a statement of the *periodic law.* A simple form of a modern table is shown in Figure 2.2, but the elements are now arranged in order of their atomic numbers. Only rarely, however, does the order based on atomic number differ from the order based on the earlier atomic weights. Elements in the same vertical columns have similar, but not identical, properties.

2.9 NEUTRONS AND OTHER NUCLEAR PARTICLES

A nucleus containing two or more positively charged protons should, according to classical electrostatics, be unstable since the protons should repel each other. In some way, the neutrons in the nucleus interact with the protons to prevent this from happening. The neutrons function as some sort of nuclear glue. The stability of nuclei is related to the ratio of protons to neutrons in the nucleus, a higher ratio of neutrons to protons being needed to give a stable nucleus as the number of protons increases. This will be discussed in more detail in Chapter 20.

Since the early 1930's, when the nucleus of the atom was first broken apart, many elementary particles in addition to the electron, proton, and neutron have been discovered in the nucleus. The determination of the function of the many particles making up the "nuclear zoo" is one of the major goals of nuclear physics, because interactions among these particles are believed to give rise to the awesome forces holding nuclei together. However, these particles need not concern us here, because the three particles already named are essentially all we need to understand the chemical properties of the elements.

2.10 ELEMENTS IN LIVING ORGANISMS

When we ask why the atoms—or elements—that compose living organisms are the ones they are, we are really talking about the *fitness* of these atoms for life as we know it. What properties of these elements make them particularly fit or suitable, above all others in the periodic table, for the role they must play?

By and large, only 20 of the 105 known elements play an essential role in life. Many of these elements are present in living organisms in extremely small amounts, less than 0.01 percent of all the atoms in an organism, but they are absolutely essential. In terms of percent composition, by weight or by atoms, the four elements carbon, hydrogen, nitrogen, and oxygen are the most abundant, followed by calcium, phosphorus, chlorine, and sulfur, as shown in Table 2.5, where values are given in weight percent. Why are these

1	2	3	4	5	6	7	8	9	10	11	12	13	14	15	16	17	18
1 H 1.01																	2 He 4.00
3 Li 6.94	4 Be 9.01											5 B 10.8	6 C 12.0	7 N 14.0	8 O 16.0	9 F 19.0	10 Ne 20.2
11 Na 23.0	12 Mg 24.3											13 Al 27.0	14 Si 28.1	15 P 31.0	16 S 32.1	17 Cl 35.5	18 Ar 39.9
19 K 39.1	20 Ca 40.1	21 Sc 45.0	22 Ti 47.9	23 V 50.9	24 Cr 52.0	25 Mn 54.9	26 Fe 55.8	27 Co 58.9	28 Ni 58.7	29 Cu 63.5	30 Zn 65.4	31 Ga 69.7	32 Ge 72.6	33 As 74.9	34 Se 79.0	35 Br 79.9	36 Kr 83.8
37 Rb 85.5	38 Sr 87.6	39 Y 88.9	40 Zr 91.2	41 Nb 92.9	42 Mo 95.9	43 Tc (99)	44 Ru 101.	45 Rh 103.	46 Pd 106.	47 Ag 108.	48 Cd 112.	49 In 115.	50 Sn 119.	51 Sb 122.	52 Te 128.	53 I 127.	54 Xe 131.
55 Cs 133.	56 Ba 137.	57 La 139.	72 Hf 178.	73 Ta 181.	74 W 184.	75 Re 186.	76 Os 190.	77 Ir 192.	78 Pt 195.	79 Au 197.	80 Hg 201.	81 Tl 204.	82 Pb 207.	83 Bi 209.	84 Po (210)	85 At (210)	86 Rn (222)
87 Fr (223)	88 Ra (226)	89 Ac (227)	104	105													

58 Ce 140.	59 Pr 141.	60 Nd 144.	61 Pm (147)	62 Sm 150.	63 Eu 152.	64 Gd 157.	65 Tb 159.	66 Dy 162.	67 Ho 165.	68 Er 167.	69 Tm 169.	70 Yb 173.	71 Lu 175.
90 Th 232.	91 Pa (231)	92 U 238.	93 Np (237)	94 Pu (242)	95 Am (243)	96 Cm (247)	97 Bk (247)	98 Cf (249)	99 Es (254)	100 Fm (253)	101 Md (256)	102 No (254)	103 Lw (257)

Figure 2.2. The periodic table of the elements. Elements commonly found in living systems are shaded. The number above the symbols is the atomic number; the number below the symbols is the rounded-off atomic weight. In addition to the essential elements shown here, recent studies show that fluorine (F), vanadium (V), chromium (Cr), selenium (Se), tin (Sn), and possibly aluminium (Al), nickel (Ni), and germanium (Ge) are essential to certain organisms.

elements, out of all those in the periodic table, so common in living organisms? Table 2.6 compares the abundance, in atom percent, of some of these elements in living organisms (primarily terrestrial vegetation) with their abundance in various regions of the earth. Although it is apparent that life selects the lighter elements and that the lighter elements also tend to be the most abundant on the earth (and in the universe), there are enough discrepancies to indicate that life did not simply select the most available elements. Table 2.6 shows that the abundance and distribution of the main biological elements differ considerably from that in their surroundings. On the basis of atom percent, life concentrates some elements from its surroundings and rejects others (see, however, the note to Table 2.5).

The real reason that the lighter elements are favored is based on the electronic structure of their atoms, because almost all of these elements are present in living matter, not as free or neutral atoms, but combined with other atoms to form molecules. Many of these molecules are made up of thousands

Table 2.5

Values of the percentage of body weight of the principal elements found in living organisms[a]

primary 1–60%		secondary 0.05–1%		microconstituents 0.05% or less
C	20%	Na	0.1%	B
H	10%	Mg	0.07%	Fe
N	2.5%	S	0.14%	Si
O	63%	Cl	0.16%	Mn
P	1.1%	K	0.11%	Cu
		Ca	0.05–2.45%	I
				Co
				Mo
				Zn

[a] On a percent-by-weight basis, the elements C, H, N, O, P, and S are considerably concentrated in living organisms compared to their average percentage in the earth's crust. The percents by weight in the crust are C, 0.02%; H, 0.14%; N, 0.002%; O, 46.6%; P, 0.11%; S, 0.026%.

of atoms of different elements. The fitness of these major elements must be connected somehow with the ways in which they can undergo changes to form entities that differ from free atoms. These changes are almost completely dependent on the electrons around the nucleus of an atom. It is the study of what happens to the extranuclear electrons that might be said to define the whole province of modern chemistry.

2.11 ELECTRONS AND INTERACTIONS OF ATOMS

One of the simplest ways an atom can change is by the loss or gain of one or more electrons. The atom then becomes an electrically charged body called an *ion*. A positively charged ion is formed when electrons are removed from the neutral atom, leaving an excess of nuclear protons. When electrons are added to a neutral atom, the result is, of course, a negatively charged ion, and the magnitude of the charge is equal to the number of electrons added. Since opposite charges attract, oppositely charged ions might come together and be held by simple electrostatic attraction, and we would have our first "molecule," the atoms of which would be held together by ionic or electrostatic bonds. In fact, something like this often happens, and we shall look briefly at some of the factors involved.

When an electron is removed from a neutral atom to form a singly charged positive ion, energy must be expended and work must be done by some outside agency. A negative electron is being pulled away from a positively charged ion, and work is done against the normal attractive forces between oppositely charged bodies, in much the same way that work must be done to stretch and separate the ends of a spring.

As in economics or finance, where one must take some notice of the profit and loss, or the income and outgo, of money, so in chemistry one must do the same thing in terms of energy currency. A process usually will not take place by itself, that is, spontaneously, if it takes more energy or work than can be gained at the end of the process. For the time being, we can say that physical and chemical processes will not occur by themselves if energy or work must be done by some outside agency to drive the reaction or process.

Table 2.6
Relative average abundance of
the elements in atom percent
(atoms per 100 atoms)

element	living organisms (plants)	earth's crust	hydrosphere	atmosphere
hydrogen	49.8	2.92	66.4	
oxygen	24.9	60.4	33	21
carbon	24.9	0.16	0.0014	0.03
nitrogen	0.27[a]			78.3
calcium	0.073	1.88	0.0006	
potassium	0.046	1.37	0.0006	
silicon	0.033	20.5		
magnesium	0.031	1.77	0.034	
phosphorus	0.030	0.08		
sulfur	0.017	0.04	0.017	
aluminum	0.016	6.2		
sodium		2.49	0.28	
iron		1.90		
titanium		0.27		
chlorine			0.33	
boron			0.0002	
argon				0.93
neon				0.0018

[a] The low value of N results because the biosphere is chiefly wood.
SOURCE: Edward S. Deevey, Jr., "Mineral Cycles," *Scientific American*, **223,** 149 (September 1970).

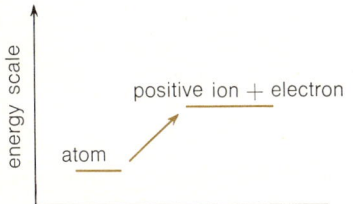

Figure 2.3. The removal of an electron from an atom is an "uphill" or endoergic process in which energy is absorbed.

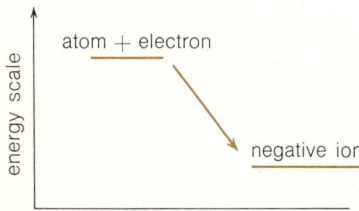

Figure 2.4. The addition of an electron to an atom may be a "downhill" or exoergic process resulting in the release of energy.

In a very general way, this result comes about because of the natural tendency of things to run "downhill," that is, to lose energy. (Energy is not the whole story. In the discussion of disorder and entropy in Chapter 16, we shall see that many processes are spontaneous even when there is no loss of energy, but this need not bother us now.) When we remove an electron from an isolated atom, the resulting system of the electron at infinity plus an ion has more energy than did the neutral atom:

$$\text{atom} + \text{energy} \longrightarrow \text{positive ion} + \text{electron}$$

(see Figure 2.3). You can see that the loss of an electron by a neutral atom is not likely to be a spontaneous process, taken by itself. In any event, things are such that the free electrons do not wander about by themselves after being removed from the atom. The electrons soon find themselves picked up by another atom, or perhaps a positive ion, because an electron loss cannot occur without an electron gain somewhere else. In contrast to the removal of an electron from an atom, the addition of an electron to an atom to give a negative ion is often a downhill process, energetically speaking. That is, with some atoms we have

$$\text{atom} + \text{electron} \longrightarrow \text{negative ion} + \text{energy}$$

(see Figure 2.4).

The *ionization energy* (IE) of an atom is the energy that must be supplied to remove an electron from the isolated gaseous atom. If one electron is removed at a time, we have a first IE and a second IE, and so on. Successive

IE's become larger in part because the positive charge, and therefore the attractive force, of the remaining gaseous positive ion for the electron becomes larger each time.

The energy gained when an electron is added to gaseous atoms is called the *electron affinity* (EA). The larger its value, the greater the tendency for an atom to attach an electron and become a gaseous negative ion. Using X as the symbol for some atom, we have

$$X(g) + IE \longrightarrow X^+(g) + e^- \qquad \text{(energy absorbed), endoergic}$$

and

$$X(g) + e^- \longrightarrow X^-(g) + EA$$
$$\text{(energy released or evolved), exoergic}$$

Processes in which energy is absorbed are called *endoergic* processes. "Downhill" reactions, or processes where energy is evolved, are *exoergic*. (The terms endothermic and exothermic are also commonly used. Their use in this text is restricted to cases where thermal energy, i.e. heat, is absorbed or evolved, respectively.) Exoergic processes tend to be spontaneous, that is, to happen by themselves without external help.

Sign Convention for Energy Changes

We could continue to write the energies involved in reactions as we have above; that is, as a reactant in the case of absorbed energy or as a product in the case of evolved energy. But this method will become clumsy later, so another method of writing the energy changes outside of the reaction equation will be briefly shown. First, a convention must be set up so that we can give an algebraic sign to energy changes.

By convention, energy evolved is given a negative sign, and energy absorbed is given a positive sign. This agrees with the downhill and uphill aspects. Alternatively, one can think, for instance, in a reaction that evolves energy, of a high-energy system going to a low-energy system by the subtraction (loss or evolution) of energy. Then,

$$X(g) \longrightarrow X^+(g) + e^- \qquad |IE|$$

and

$$X(g) + e^- \longrightarrow X^-(g) \qquad -|EA|$$

where the vertical lines denote the absolute value of the energy term, in whatever units desired.

Ionization energies and electron affinities are usually expressed in terms of electron volts. One electron volt (eV) is defined as the energy acquired by an electron or any other particle carrying a unit charge when it is accelerated through a potential difference of 1 volt. Tables 2.7 and 2.8 list the IE's and EA's for several elements.

With these data we might expect, in a very rough way, to determine if a pair of gaseous atoms may gain and lose electrons to form a pair of gaseous ions. Take, as an example, sodium atoms and chlorine atoms,

$$
\begin{array}{ll}
Na(g) \longrightarrow Na^+(g) + e^- & IE = 5.1 \text{ eV} \\
\underline{Cl(g) + e^- \longrightarrow Cl^-(g)} & \underline{EA = -3.8 \text{ eV}} \\
Na(g) + Cl(g) \longrightarrow Na^+(g) + Cl^-(g) & IE + EA = +1.3 \text{ eV} \\
& \qquad\qquad \text{(endoergic)}
\end{array}
$$

Table 2.7
Ionization energies, eV

H							He
13.6							24.6
Li	Be	B	C	N	O	F	Ne
5.4	9.3	8.3	11.3	14.5	13.6	17.4	21.6
Na	Mg	Al	Si	P	S	Cl	Ar
5.1	7.6	6.0	8.1	11.0	10.4	13.0	15.8
K						Br	Kr
4.3						11.8	14.0
Rb						I	Xe
4.2						10.5	12.1
Cs							
3.9							

where the two reactions and their energies have been added to obtain the overall change. We still do not have any kind of association between the sodium and chloride ions. The two ions are not bonded but are separated in space. The bringing together of two oppositely charged ions is an exoergic process that evolves energy. This is shown in Figure 2.5, which indicates the change in energy as two oppositely charged gaseous ions are brought closer to each other. Until the ions are very close to one another, the energy of the system decreases, and this decrease in energy represents energy released by the system. Finally, when the two ions are very close to one another, repulsive forces come into play between the like-charged nuclei of the two ions and between the electrons around each ion. The energy of repulsion is shown by the steeply rising curve. The net decrease in energy at any interionic distance is the sum of the attractive and repulsive curves at that point, as shown.

The energies of attraction and repulsion between the sodium and chloride ions must be added to (IE + EA) when these ions are brought together.

Table 2.8
Electron affinities, eV

H						
0.7						
Li	Be	B	C	N	O	F
0.5	0	0.3	1.1	0.2	1.5	3.6
Na	Mg	Al	Si	P	S	Cl
0.7	0	0.4	1.9	0.8	2.1	3.8
						Br
						3.5
						I
						3.2

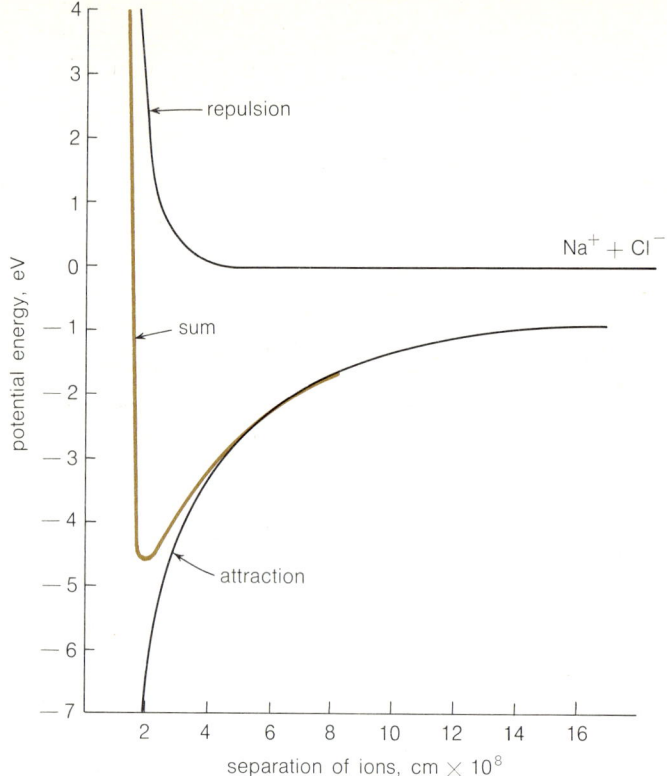

Figure 2.5. Energies of interaction of gaseous sodium and chloride ions as a function of distance between the ions. The minimum in the net potential energy curve represents the separation between the nuclei in the stable NaCl molecule. (Distinct NaCl molecules exist only in the gaseous state.)

Figure 2.5 indicates that at most distances the attractive energy overbalances the repulsive energy. Furthermore, the difference between the attractive and repulsive energies is great enough (in a negative or exoergic sense) that the complete energy change in going from gaseous atoms to a gaseous Na^+Cl^- ion pair, IE + EA + attractive energy + repulsive energy, is exoergic or negative.

$$
\begin{array}{ll}
Na(g) \longrightarrow Na^+(g) + e^- & \text{IE} \\
Cl(g) + e^- \longrightarrow Cl^-(g) & \text{EA} \\
\underline{Na^+(g) + Cl^-(g) \longrightarrow Na^+Cl^-(g)} & \underline{E_{\text{attr.}} + E_{\text{rep.}}} \\
Na(g) + Cl(g) \longrightarrow Na^+Cl^-(g) & \text{IE} + \text{EA} + E_{\text{attr.}} + E_{\text{rep.}}
\end{array}
$$

The result indicates that the interaction of a sodium and a chlorine atom to form a gaseous ion pair held together by electrostatic forces takes place spontaneously.

This simple picture and treatment of the combination of ions leaves much to be desired. Not only is the treatment incomplete for the coming together of the gaseous ions into the common ionic solids, but it fails to explain an enormous number of atomic combinations that we know exist and that are predominant in living organisms. Questions that must be answered, and cannot on the basis of this brief treatment, are how do 2 hydrogen or 2 chlorine atoms stick together to give normal diatomic hydrogen, H_2, or chlorine, Cl_2? Why is there no Cl_3? Why is it that some atoms only gain electrons to become negative ions while others only lose electrons? Why is there no Na^{2+}?

To answer these and other questions requires an excursion into the ideas and background of the elementary quantum theory of the atom. We shall begin by looking at some of the historical reasons that pointed up the need for a new type of treatment.

The Quantum Theory of Atomic Structure

2.12 THE NEED FOR A NEW THEORY

The hydrogen atom, being the simplest possible atomic system, is the starting point for any attempt to understand the electronic structure of atoms. The early treatment of the hydrogen atom was based on classical or Newtonian laws of motion, and some of the results of that treatment will be briefly noted here.

The simple electron–proton system of the hydrogen atom can be thought of as being composed of a stationary proton circled by the moving electron (Figure 2.6). The total energy E of such a system is made up of two types of energy, kinetic energy T and potential energy U, so that $E = T + U$.

Kinetic energy is the energy of motion of a body, in this case, that of the electron alone. The kinetic energy of a body of mass m moving with a velocity v is always $T = \frac{1}{2}mv^2$.

Potential energy is the energy that a body has by virtue of its position relative to some reference point. Unlike the kinetic energy, potential energy comes in various forms depending on the kinds of forces acting on the body. For instance, a body of mass m has a gravitational potential energy mgh, where g is the acceleration due to gravity, by virtue of its height h above the surface of the earth. The potential energy of the electron–proton system is the electrostatic potential energy of the negative electron under the attractive influence of the positive proton. This potential energy is given by the formula

$$U = -\frac{e^2}{r}$$

where e is the magnitude of the electronic charge and r is the distance between the proton and the electron. The potential energy of the system is more negative the closer the electron is to the positive proton. The movement of the electron from one position to another location farther from the proton results in a more positive U (a higher potential energy), and this means that work has been done on the electron–proton system. The lower the potential energy (the larger the magnitude of the negative U), the more stable the system will be.

For the hydrogen atom, the total electronic energy is

$$E = T + U = \frac{1}{2}mv^2 - \frac{e^2}{r}$$

According to the classical treatment of the simple model of the hydrogen atom, two situations should result. One is that the atom may have any electronic energy—the whole range of values from $-\infty$ (infinity) to zero. The other situation arises because an electron moving in a circular path is continually

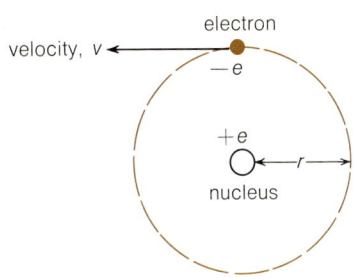

velocity, v ← electron $-e$

$+e$ ← r →

nucleus

Figure 2.6. Classical model of the hydrogen atom.

being accelerated. Acceleration can refer to a change of direction at constant speed as well as to a change of speed only. According to classical electric theory, any accelerating electrically charged particle must emit radiation and, by radiating, lose energy. The continual loss of energy by the electron should eventually cause it to spiral down into the nucleus. Such an atom would not be stable.

We know that hydrogen atoms are stable for comparatively long periods of time. We also know that when hydrogen atoms do radiate energy, only certain quantities of energy are emitted, not a whole range. To understand these two facts, and particularly the latter one, a knowledge of the nature of radiation is necessary, because our concepts of the nature of radiation and of matter are inseparably tied together.

2.13 ELECTROMAGNETIC RADIATION

The nature of radiation and of light, which is visible radiation, can be understood in terms of two different theories, the wave theory and the corpuscular or photon theory. Historically, the photon theory came first, having been put forward by the English mathematician and natural philosopher Sir Isaac Newton (1642–1727) in 1704. In its simplest form, it considers light to be made up of particles or corpuscles which travel in straight lines from the source of radiation. This theory was largely supplanted in the early 1800's by the wave theory of light, which, together with the electromagnetic theory of radiation, pictures light as electromagnetic waves composed of oscillating or vibrating electric and magnetic fields.

Before going further, the term "field" needs explanation. An electric field is said to exist in a region of space when there is a force on a charged body placed in that region. The force F on that charged body is related to the strength E of the electric field and to the magnitude of the charge q on the body by the equation $F = qE$. The direction of the electric field E is in the direction of the force on a positive charge inserted into the field. What this means is that the electric field represents a force field in space—a region where electrically charged particles such as protons or electrons *could* experience a force if they were there. However, the field exists even in the absence of any charged bodies or other matter. Therefore, when we talk about light waves or vibrating fields, we should not feel that anything material has to "wave." The vibrating electric and magnetic fields exist even in a vacuum, so that radiation can travel through a vacuum. The magnitude and direction of the electric field can vary at different locations in space and at different times.

In an electromagnetic wave, there are two force fields corresponding to an electric force and to a magnetic force. These fields are shown in Figure 2.7. Note that the electric field E is "in phase" with the magnetic field. That is, when the value of the electric force field is at its maximum, so is that of the magnetic field. We need only consider the electric field from now on and

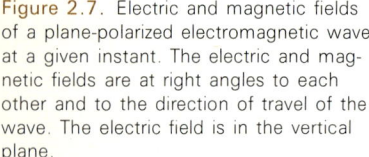
Figure 2.7. Electric and magnetic fields of a plane-polarized electromagnetic wave at a given instant. The electric and magnetic fields are at right angles to each other and to the direction of travel of the wave. The electric field is in the vertical plane.

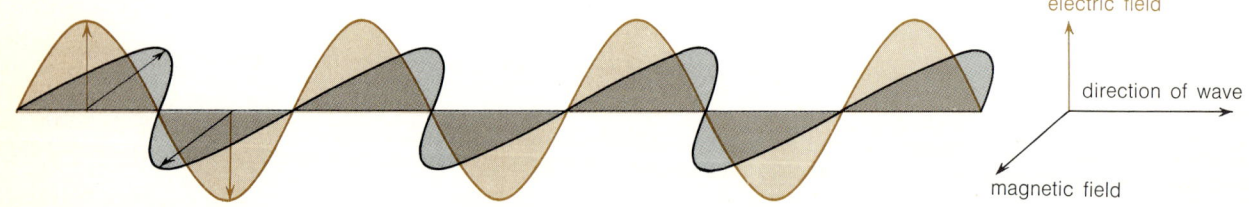

electric field

direction of wave

magnetic field

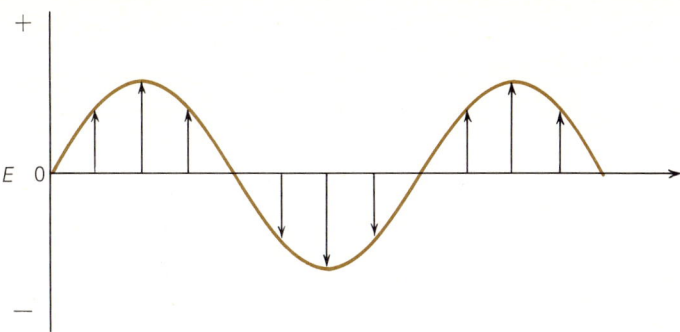

Figure 2.8. Variation with position of the electric field associated with plane-polarized radiation.

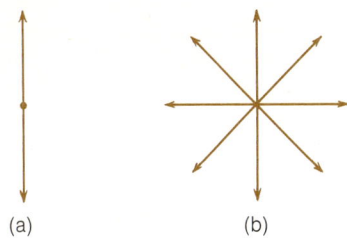

(a)   (b)

Figure 2.9. The electric field of plane-polarized light (a) vibrates in one plane, that of ordinary light (b) in many planes.

shall leave the accompanying magnetic field out of the picture, as shown in Figure 2.8. What is known as *plane-polarized* radiation, in which the electric (and magnetic) field vibrates in only one plane, will, for simplicity, be shown in our figures. Ordinary light, even of one pure color, is composed of electric and magnetic fields that vibrate in all planes. This is shown in Figure 2.9, where the direction of viewing is straight down the direction of propagation. You may think of the wave as moving from left to right, much as water waves do. As the wave passes a given point x, the strength of the electric field changes with time from a maximum, in a positive direction, through zero, to a maximum in the negative direction, as shown in Figure 2.10.

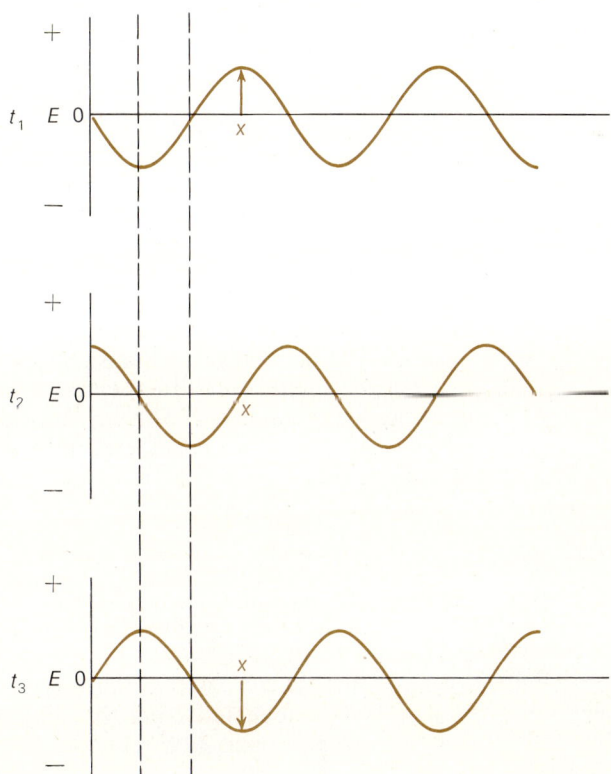

Figure 2.10. The electric field of radiation at successive times. The wave is moving to the right.

The electric field is a quantity known as *vector;* it has both magnitude and direction. A common example of a vector is velocity, **v.** To specify accurately the velocity of a moving object, two quantities are necessary: (1) the magnitude or speed of the object and (2) the direction in which the object is moving. Similarly, the complete effect of a force on an object can be given only if both the magnitude of the force and the direction of the force or push are specified. It is the same with the electric field. The magnitude here is specified by the length of the arrows; the direction is given by the arrowhead. Later, in discussing how radiation interacts with matter, it will be important to consider the variation in strength and direction of the electric field associated with a light wave.

2.14 WAVES Two of the most important parameters used to describe waves are the wavelength and the frequency. The wavelength λ (Greek, lambda) is the distance between adjacent crests, adjacent troughs, or any point and the next point where the wave starts to repeat itself, as shown in Figure 2.11. The frequency

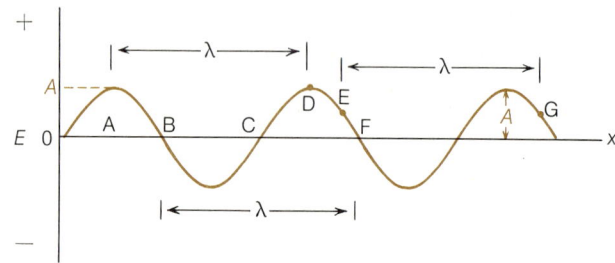

Figure 2.11. Wavelength λ of a wave. Any of the segments AD, BF, EG equals λ. The maximum displacement of the wave is its amplitude A.

ν (Greek, nu) of the wave is the number of wave crests or troughs that pass a given point per unit time. The wavelength and the frequency are related by

$$c = \lambda\nu$$

where, for electromagnetic radiation, c is the speed of light in a vacuum, 3×10^{10} cm sec^{-1}.

For the time being, the units we shall use for λ are centimeters or angstroms and the units for ν are then per sec or sec^{-1}. Frequency is usually given in cycles per second (cps) or hertz (Hz). Cycles is not a fundamental dimensional unit and is not considered in calculations (see Appendix A.4). You can check this:

$$\nu = \frac{c}{\lambda} = \frac{\text{cm sec}^{-1}}{\text{cm}} = \frac{1}{\text{sec}} = \text{sec}^{-1} \text{ or Hz}$$

An angstrom, Å, is 10^{-8} cm. This unit was chosen to make many wavelengths come out to be easily handled numbers. Another unit in common use is the nanometer (nm), which is 10^{-9} m or 10^{-7} cm.

As Figure 2.12 shows, various kinds of radiation have different wavelengths and frequencies. Note that the wavelength and the frequency are inversely proportional: as λ increases, ν decreases and vice versa. All types

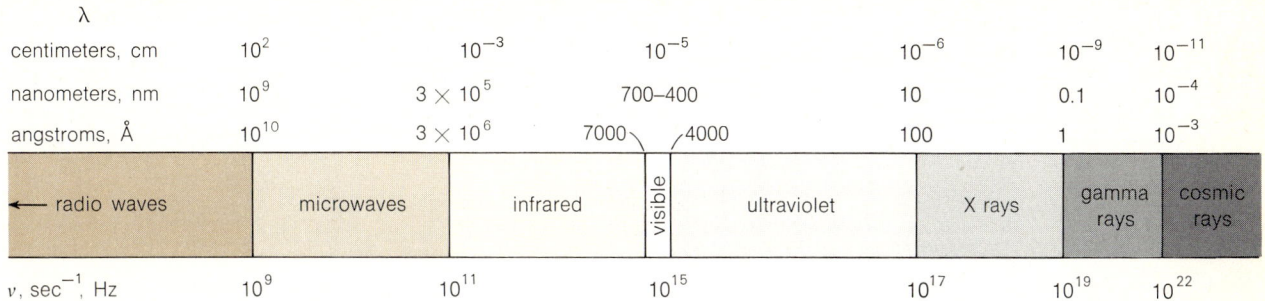

λ

centimeters, cm	10^2		10^{-3}		10^{-5}		10^{-6}		10^{-9}	10^{-11}
nanometers, nm	10^9		3×10^5		700–400		10		0.1	10^{-4}
angstroms, Å	10^{10}		3×10^6	7000	4000		100		1	10^{-3}

| ← radio waves | microwaves | infrared | visible | ultraviolet | X rays | gamma rays | cosmic rays |

ν, $\sec^{-1}$, Hz 10^9 10^{11} 10^{15} 10^{17} 10^{19} 10^{22}

Figure 2.12. The electromagnetic spectrum. The upper scales give the approximate wavelength ranges of the spectral regions in some common units. These units can be related via
1 Å = 10^{-8} cm = 0.1 nm;
1 nm = 10^{-9} m = 10^{-7} cm. The lower scale gives the frequency in cycles per second or hertz.

of electromagnetic radiation have the same speed in vacuum, 3×10^{10} cm $\sec^{-1}$. Visible light is only a small part of the electromagnetic spectrum and is composed of radiations that have wavelengths within the narrow range 4000–7000 Å (400–700 nm) or from violet to red. White light, as from the sun, is a mixture of light of all wavelengths, as nature shows in a rainbow where water droplets act as small prisms to disperse the white light into its components. Figure 2.13 shows a common prism and the spectrum of sunlight. Each color of visible light is associated with a particular wavelength region.

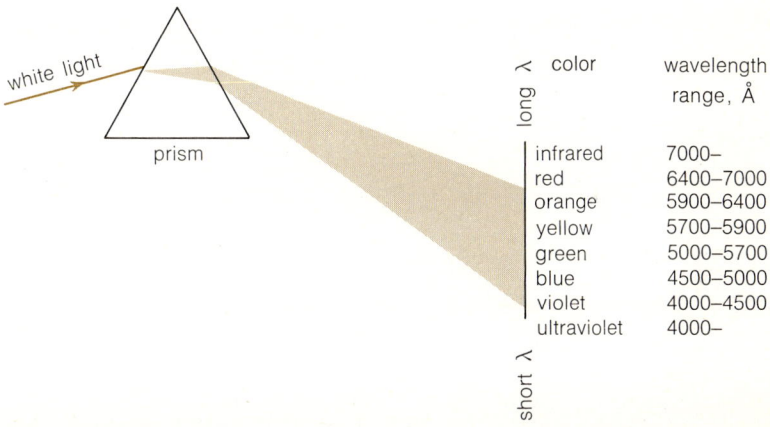

	color	wavelength range, Å
infrared	7000–
red	6400–7000
orange	5900–6400
yellow	5700–5900
green	5000–5700
blue	4500–5000
violet	4000–4500
ultraviolet	4000–

Figure 2.13. Dispersion of white light by a prism.

To return to the electric field associated with radiation, we must also consider briefly the magnitude of the electric field. The intensity of radiation, or the brightness of visible light, is related to the maximum value of the electric field strength. This maximum value is called the *amplitude* of the wave. At a crest of the wave, E has a maximum value A. The intensity of the radiation is proportional to the square of the absolute value of the amplitude, $|A|^2$.

As an electromagnetic wave passes a given point in space, the electric field direction alternates as the magnitude rises to a maximum, falls to zero, and attains a maximum with opposite field direction. Thus, a charged particle located at that point will experience varying forces, both in magnitude and in direction, as the wave passes.

**2.15 TWO IMPORTANT
CHARACTERISTICS OF
WAVES**

There are two properties shown by light and other electromagnetic radiation that can be explained only on the basis of the wave theory of radiation. The corpuscular or photon theory is incapable of describing how these phenomena arise. These two qualities, interference and diffraction, are closely related.

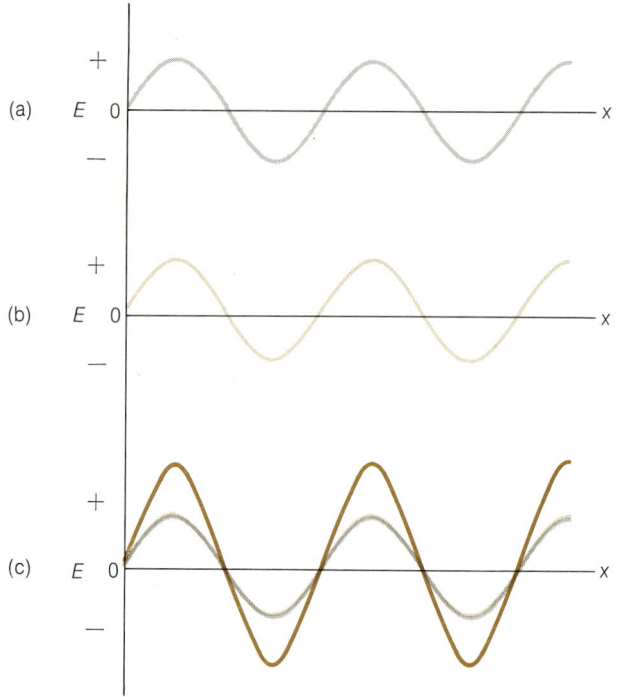

Figure 2.14. Constructive interference of two in-phase waves. The separate waves are shown in (a) and (b), their sum in (c).

Interference of Waves

Consider two waves having equal wavelengths. First assume that the two waves start from the same point in phase, that is, with the crest of one wave co-incident with the crest of the other. When these two waves are superimposed, a resultant wave is produced. The displacement (the electric field strength) at any point of the resultant wave is found by adding algebraically the displacements at that point of the individual waves (Figure 2.14). In this particular case, the displacement at any point and the amplitude of the resultant wave are twice as large as the corresponding quantities for either wave. This is an example of *constructive interference* or *reinforcement*.

Now consider that the two waves, again with the same amplitude (but only for simplicity) and the same wavelength, are completely out of phase, as in Figure 2.15; that is, a crest on one is coincident with a trough on the other. Adding displacements, again with respect to sign, gives a resultant wave having zero displacement at all points. This is an example of complete *destructive interference* or *cancellation*.

Interference is not limited to waves that have the same wavelength or amplitude, or to just two waves. There are all degrees of interference between complete reinforcement and complete cancellation.

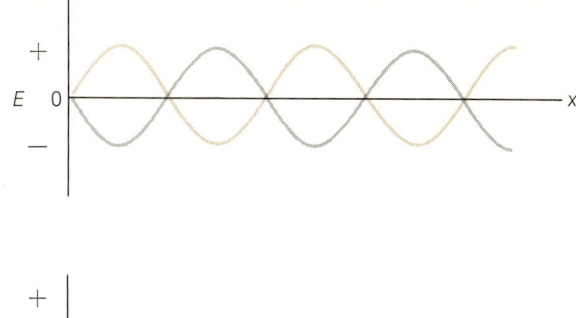

resultant wave
$E = 0$ everywhere

Figure 2.15. Complete destructive interference of two out-of-phase waves.

Diffraction of Radiation
Diffraction is another phenomenon connected with wave motion, and it is
closely related to interference. Diffraction shows that light does not travel
in straight lines but spreads out, or diffracts. In effect, light can go around
corners much as a group of spreading waves or ripples on a body of water
can get around an obstacle. This spreading of waves results in fuzzy edges
on shadows rather than the sharp edges expected if light were composed of
corpuscles moving in straight lines.

The diffraction effects we observe on a screen are due to interference
of waves that arrive out of phase with one another. To simplify the treatment
of diffraction phenomena, we shall introduce the idea of *wave fronts* for a
set of waves originating at some source. Wave fronts correspond to positions
of equal displacement of the waves from the source of light. Their construction
is shown in Figure 2.16 for a group of in-phase waves (this phase restriction
is not necessary to obtain diffraction).

A classic experiment in wave optics is Young's double-slit experiment.
Here two very small holes or two narrow, closely spaced slits are used. The
distance between holes or slits must be of the same order of magnitude as
the wavelength of the radiation if the effect is to be easily observed. The
pattern on the screen is shown in Figure 2.17. The interpretation of this
diffraction pattern is shown in Figure 2.18, where the wave fronts (points of
equal displacement—maximum upward here) are shown. Although only the
wave picture of light propagation can explain interference and diffraction,
there are certain light phenomena that only the corpuscular theory can easily
explain.

**2.16 PHOTON THEORY OF
RADIATION**
With the successes of the wave theory, the corpuscular theory had been
pushed into the background by the 1900's and had been all but forgotten.
However, beginning in 1900, a series of investigations on the interaction of
matter and radiation led both to a revival of the corpuscular theory and to
a fundamentally new way of looking at matter, the quantum theory. In fact,
the development of the modern corpuscular theory of radiation and of the

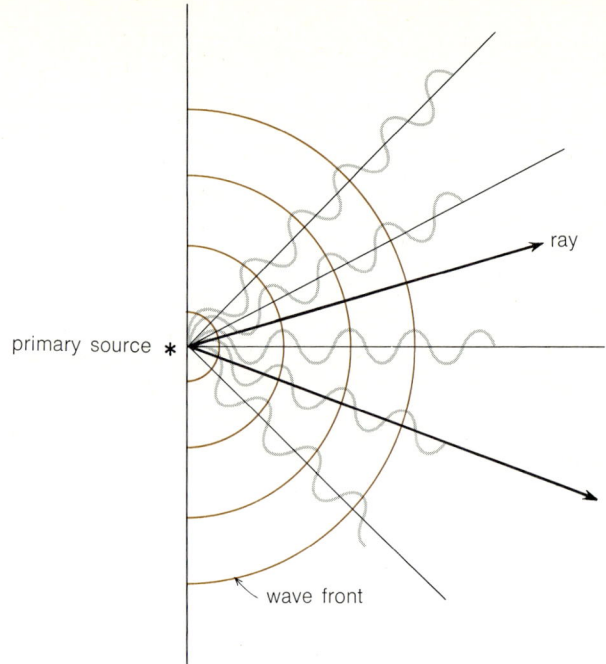

Figure 2.16. Construction of wave fronts. All waves emerging from the small hole (the secondary source) are in phase. Semicircles with the aperture at their center (hemispheres in three dimensions) are drawn passing through points of equal displacement on each wave. The semicircles (or hemispheres) represent the wave front. Lines drawn perpendicular to the fronts from the aperture are called *light rays*.

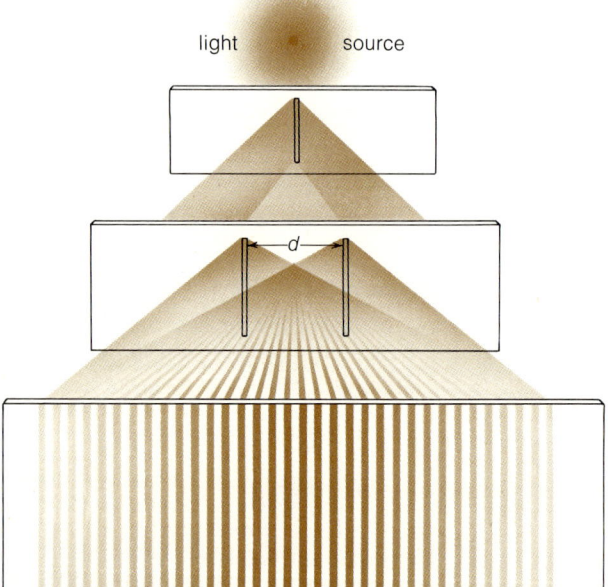

Figure 2.17. Interference fringes formed by the diffraction of light by two very narrow slits. The distance *d* between the two slits must be similar to the wavelength of the light used. A similar diffraction effect can be seen by looking at a light source through the narrow crack formed by two fingers held close to the eye.

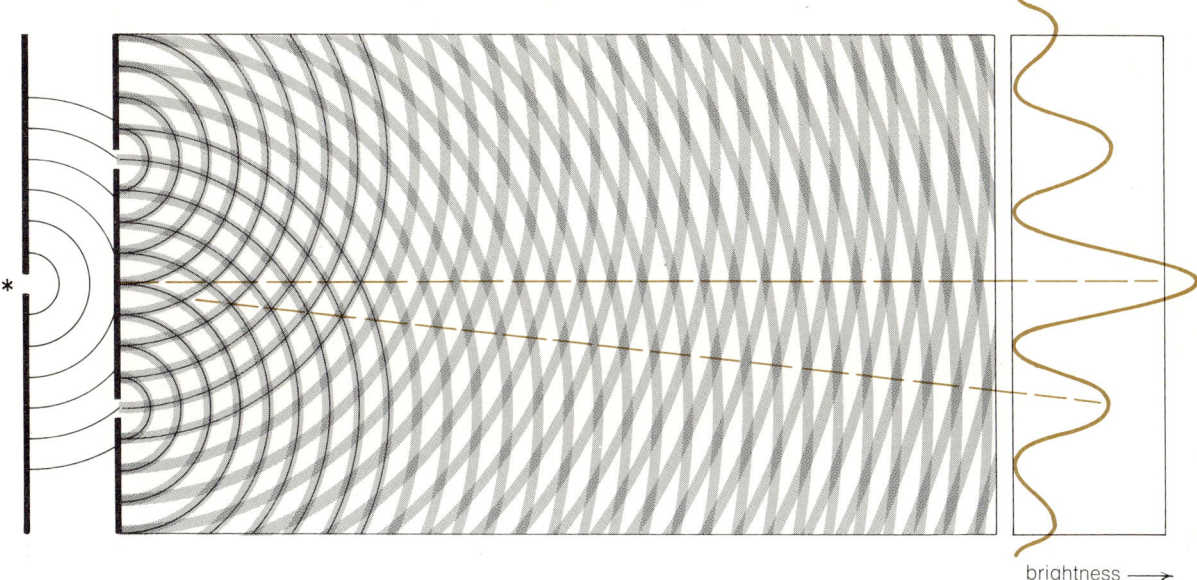

brightness ⟶

Figure 2.18. Interpretation of the diffraction pattern of Figure 2.17. Lines drawn through the points of intersection of the wave fronts (here corresponding to maximum positive displacement) intersect the screen at points of maximum light intensity, since these lines are the loci of points of maximum reinforcement. Other points correspond to varying degrees of destructive interference. Look along the figure from right to left at a grazing angle.

quantum theory of matter are so close that it is almost impossible to distinguish them. Here only a brief review of two of the investigations will be given—ones that are of great importance.

The first of these investigations was undertaken by Max Planck (German physicist, 1858–1947; Nobel prize in physics, 1918) in 1900 in an effort to derive theoretically an equation to describe (1) the variation of intensity with wavelength of the radiation emitted by heated matter at a given temperature, and (2) its change with temperature. The experimental results are shown in Figure 2.19. This figure illustrates the well-known occurrence of a heated body progressively emitting more and more radiation of short wavelength as its temperature increases, that is, as it goes from a colorless warm through red hot to white hot. Planck found, as others had, that the laws of classical physics were incapable of explaining the result.

To explain the results, he had to make a revolutionary and somewhat unpopular proposal, which led to the quantum theory of matter. He proposed that matter, or the parts of matter responsible for the absorption and emission of radiation, could exist only in certain well-defined or discrete energy states. In particular, if we think of some kind of vibrating system (a vibrating atom, molecule, or charge) as being responsible for radiation emission, and if a particular system vibrates at a mechanical frequency ν, then Planck said that this oscillator could have only one of a set of definite energies: $E = nh\nu$, where n may be 0, 1, 2, . . . , any whole number and h is a universal constant, now known as Planck's constant. This oscillator could not have, at any time, an energy of 1.50 $h\nu$, 2.75 $h\nu$, and so on. Another way to look at it is to draw a set of equally spaced levels representing the discrete energies (Figure 2.20). Then the energy of the vibrating system must lie on one of the lines and cannot lie between the lines.

According to Planck's postulate, the frequency of the radiation emitted

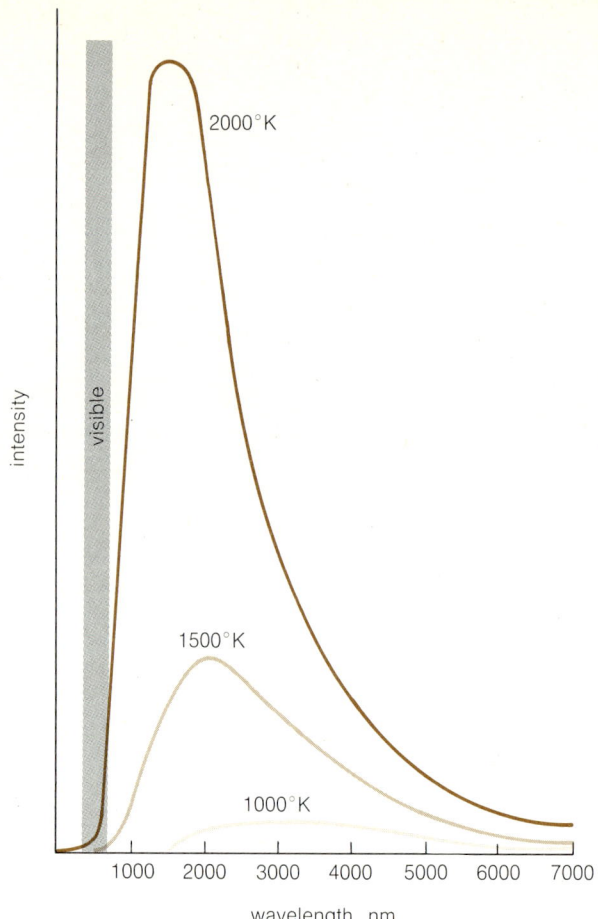

Figure 2.19. Emission of radiation by a heated body. The hotter the body, the shorter the wavelength of the radiation of greatest intensity. The visible region of the spectrum is indicated. A radiating body must reach a temperature of about 6000°K before its radiation resembles sunlight. Incandescent lamp filaments are about 3000°K, and their light has more long-wave orange components in it than sunlight.

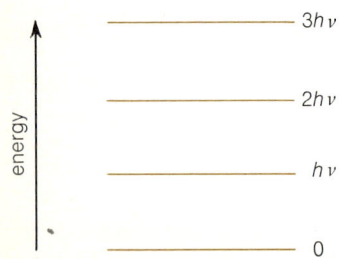

Figure 2.20. Allowed energy levels of quantized matter. Energies between the lines are not permitted.

by a given oscillator is the same as the mechanical frequency of that oscillator. The oscillator can absorb no energy less than the amount $h\nu$ required to increase its energy to the next highest level. Conversely, when emitting energy, an oscillator can lose energy only in steps of $h\nu$, corresponding to movement from one energy level to the next lower level; the frequency of the radiation emitted in this process is again ν_{mech}.

An analogy to the loss of energy by a simple mechanical system may help here. A ball rolling down a smooth incline appears to lose gravitational potential energy smoothly and continuously. This is the kind of gradual energy loss that classical theory predicted for electronic and atomic systems. However, a ball bouncing down a flight of stairs seems to lose energy in chunks (or steps, if you will) and not gradually. The height of each step determines how much energy the ball can lose at a time. This is analogous to the existence of definite energy levels in quantum theory.

Therefore, and this is the basis of the quantum theory, the absorption and emission of radiation by a given oscillator can take place only by transitions between definite energy levels of the oscillator. A result that Planck himself was not entirely convinced of at the time is that the radiation of frequency ν emitted should come out in definite chunks, or quanta, as the vibrating system goes from one energy level to a lower one. Based on this model of energy absorption and emission, Planck was able to derive the experimental results. Classical physics, which said that the energy of the oscillator could vary continuously (that is, between the lines in Figure 2.20) during the exchange of radiation had failed.

Planck's use of the equation $E = nh\nu$ in his theory had no theoretical justification at the time other than that it worked. No previous theory or basic principle led up to it. It could not be derived. Planck's results led to the conclusion that the radiation associated with discontinuous changes in energy levels of atomic or molecular oscillators was composed of quanta or packets of energy $h\nu$. In 1905, Albert Einstein (German-born, naturalized American physicist, 1879–1955; Nobel prize in physics, 1921) postulated that all radiation was quantized or composed of discrete energy packets just as was matter. He called attention to the photoelectric effect to illustrate his theory, and in doing so he gave support to the particle or corpuscular theory of light.

When light falls on the surface of certain metals, electrons are emitted from the surface. This is the *photoelectric effect*. Regardless of the intensity of the light, no electrons are emitted from the surface until the frequency of the incident light becomes larger than a certain value, but after this threshold frequency is reached, the number of electrons emitted depends only on the intensity of the light. These features of the photoelectric effect were easily explained by the photon theory of radiation put forth by Einstein. Einstein said that the energy associated with light existed as definite packets called quanta or photons. The energy ϵ carried by a photon is equal to $h\nu$, where ν is now the frequency of the radiation, not the mechanical frequency of vibration of some mechanical oscillator. In the photoelectric effect, a photon absorbed by the metal transfers its energy to an electron. If the photon energy, $\epsilon = h\nu$, is high enough, the electron can escape from the metal surface. The number of electrons emitted is proportional to the number of photons striking the surface, and this is proportional to the intensity of the light.

We now accept two complementary theories of light—the wave theory and the photon theory. Some phenomena such as diffraction can be explained only by the wave theory; other phenomena can best be explained by the photon theory. To have two theories for the same thing caused some consternation for a time, but today most people accept both theories as simultaneously right and not contradictory. The wave theory provides a statistical picture of the behavior of photons. That is, when we think about the intensity of radiation (the square of the amplitude of the wave), we are really thinking about the number of photons passing by or through a given area. We shall return to this point when discussing waves associated with matter.

2.17 ATOMIC SPECTRA What does all this material on light and radiation have to do with the structure of atoms and molecules? First, these investigations gave rise to the quantum theory: the idea that material systems can exist only in certain definite energy states characterized by an integral number n, a quantum number such that $E = nh\nu$. The concept of discrete energy levels is basic to much of modern atomic and molecular science. Second, they revived the corpuscular theory of light, rechristening it the photon theory and placing it alongside the older wave theory. They showed the *dual nature* of radiation. These ideas, the quantum nature of matter and the dual nature of radiation, helped to bring about the modern and far-reaching investigations of the structure of atoms and molecules.

We have already mentioned that the classical theory of the atom, a negative electron circling a positive nucleus, predicted that such a system would be unstable and would collapse in a very short time, with the emission of radiation of a great many wavelengths in a continuous spectrum. The actual spectrum of hydrogen and other atoms is not at all continuous.

When hydrogen gas is excited or energized by passing an electric current through a hydrogen-filled tube similar in design to the familiar neon sign tubes, radiation is emitted. When this radiation is dispersed by a prism, it is found to be composed of a number of definite wavelengths or "lines," as shown in Figure 2.21, where the visible lines are labeled.

Visible line spectra of other atoms also consist of discrete lines, but the spectra are more complex. Portions of the line spectra of other elements, such as the neon and argon used in signs, are more familiar than that of hydrogen. Some of these spectra are shown in Figure 2.21.

The characteristic line spectra emitted by the excited atoms of elements are an extremely useful way of determining if that element is present in a mixture. The unknown substance is placed between two carbon electrodes, and an electric arc is struck between the electrodes. The sample vaporizes, and the gaseous atoms and ions are energized and caused to emit their characteristic line spectra (most molecules are dissociated or broken up into atoms and ions). By dispersing the emitted radiation and determining the frequency of the lines, it is possible to tell which elements are present. The intensity of the lines also can be measured to tell how much of a given element is present. This technique of analysis is called *emission spectroscopy*.

2.18 THE BOHR THEORY

To explain the characteristic line spectrum of the hydrogen atom, the Danish physicist Niels Bohr (1885–1962; Nobel prize in physics, 1922) postulated that the hydrogen atom could exist only in certain discrete energy states, called stable or *stationary states*, and that the atom radiated energy of a definite frequency when it dropped from a higher energy state to a lower energy state. The energy of the atom, by definition, is due entirely to its electronic energy, so this is the same as saying that the electron can exist only in certain energy states, which are stable in the sense that the electron does not absorb or emit radiation while in these states, but only during a transition from one state to another. Since, for simplicity, Bohr assumed the path of the electron about the nucleus to be a closed circular orbit, this postulate went against the classical requirement that an accelerating electron must lose energy.

In addition to the postulate that the electron neither loses nor gains energy while in a stationary state (i.e., its energy is constant, fixed, or stationary), Bohr also made the following postulates:

1. The frequency of the radiation, ν_{rad}, emitted when the electron dropped from one energy level E_2 to a lower one E_1 closer to the nucleus is related to the energy difference ΔE between these levels by

$$\Delta E = E_2 - E_1 = h\nu_{rad}$$

2. The product of the momentum, mv, of the electron in an energy state and the circumference, $2\pi r$, of its circular path could only be multiples of

Figure 2.21. (Above) Dispersion of white light into a continuous spectrum (see also Figure 2.13). (Below) (1) Continuous spectrum of the white light from an incandescent solid (see also Figure 2.19). (2) Solar spectrum, showing dark lines resulting from the absorption or blockage (Section 6.4) of certain wavelengths by cooler elements in the sun's atmosphere. Below the sun's spectrum are shown the visible line spectra of some elements. The wavelengths of the lines in the hydrogen spectrum are 4102 A (violet), 4341 A (blue-violet), 4861 A (blue-green), and 6563 A (red-orange). (From Charles W. Keenan and Jesse H. Wood, "General College Chemistry," Fourth Edition, Harper & Row, New York, 1971, page opposite page 51.) ▶

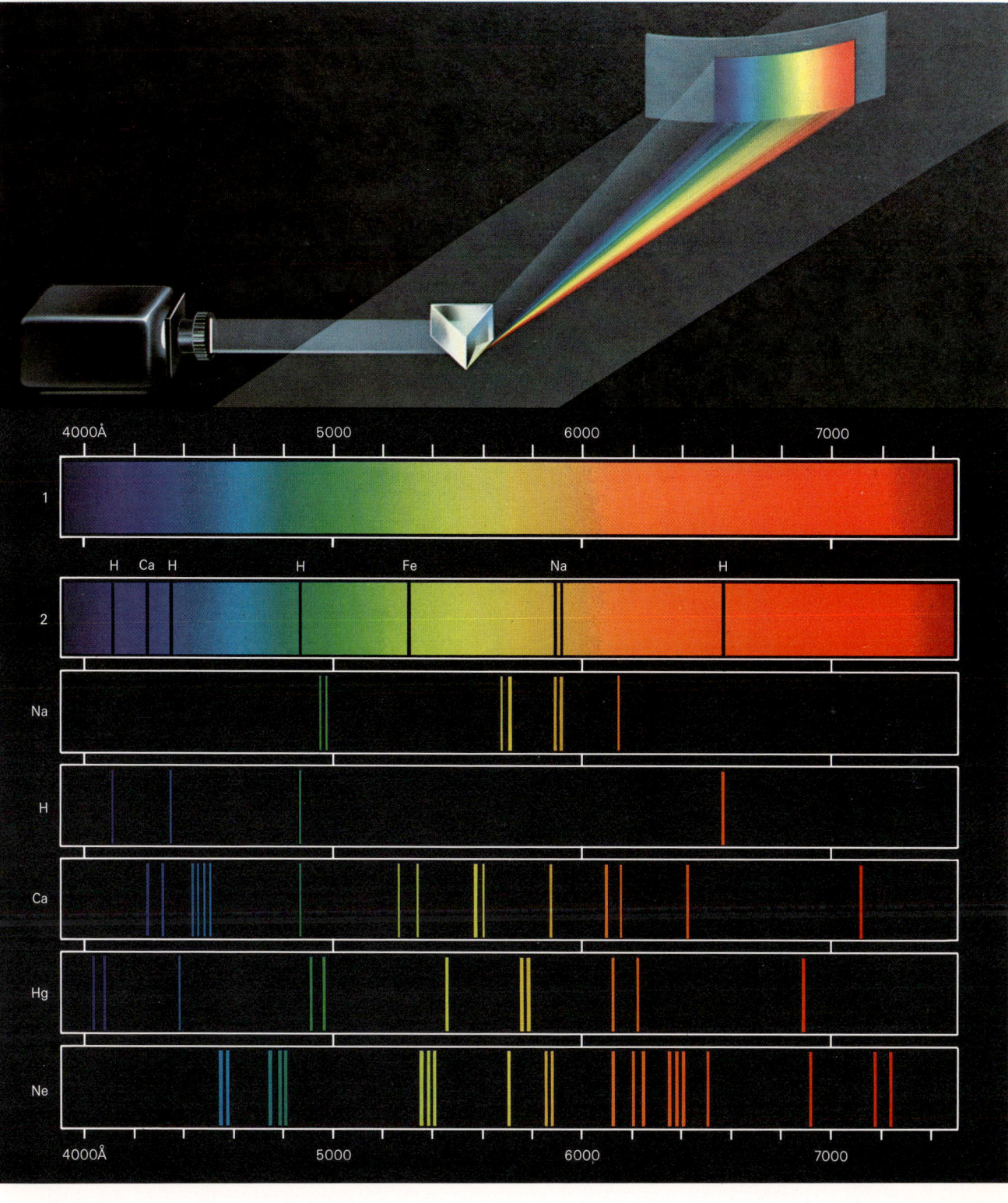

Planck's constant h, that is, $mv2\pi r = nh$, where n is a whole number 1, 2, . . . called the *quantum number* of the energy state.

Note that $\Delta E = h\nu_{\text{rad}}$ is not exactly the same as Planck's hypothesis. In Planck's original equation, ν referred to the frequency of vibration of a mechanical system. In Bohr's relation, ν refers to the frequency of light. Nor is $\Delta E = h\nu_{\text{rad}}$ the same as Einstein's equation for photon energy, $\epsilon = h\nu_{\text{rad}}$, since here ϵ does not refer to an energy difference or change, but to the energy of a quantum of radiation. Bohr's assumption is a combination of these.

Using these postulates, together with classical physical laws, Bohr was able to calculate the allowed, or discrete, energy levels for the hydrogen atom. The result was

$$E_n = -\frac{2\pi^2 e^4 m}{h^2 n^2}$$

From this the wavelengths of the spectral lines were calculated and found to agree with experiment. Figure 2.22 shows how different series of spectral lines in the hydrogen spectrum can result through electron transitions from higher energy levels (E_2, n_2) to lower levels (E_1, n_1).

The ability of the Bohr treatment to calculate the experimentally measured spectral frequencies of the hydrogen atom was a great triumph. However, it soon became apparent that the Bohr theory, even with refinements, failed completely when it came to reproducing the experimental line spectra of atoms and ions having more than one electron. The main fault, as it turned out, was the idea that the electron behaved as a small charged particle subject to the laws of Newtonian physics. Nevertheless, the Bohr theory, by its postulate of discrete energy states, marked the beginning of the quantum theory of atomic structure as opposed to the earlier quantum theory of radiation.

2.19 MATTER WAVES   Something new was needed. This new thing was supplied in 1924 when Louis de Broglie (French physicist, 1892–   ; Nobel prize in physics, 1929) reasoned that if radiation had a dual or wave–particle nature, so might matter. He suggested that matter, electrons in particular, might exhibit wave properties, and that the link between the wave and particle picture of matter was given by the relation

$$\lambda = \frac{h}{mv} = \frac{h}{p}$$

where

  m = mass of the electron
  v = velocity of the electron
  p = linear momentum, mv, of the electron
  h = Planck's constant = 6.6×10^{-27} erg sec = 6.6×10^{-27} g cm^2 sec^{-1}
  λ = wavelength associated with the electron

This equation is the bridge between the wave attributes of matter (λ) and the particle picture (p or mv).

Now, if electrons did have wave properties, they should show diffraction and interference phenomena just as light waves do. This sort of diffraction, electron diffraction, was soon found when an electron beam directed at a metal crystal gave, on reflection, a diffraction pattern similar to that observed with X-rays and crystals. As mentioned before, X-rays are part of the spectrum

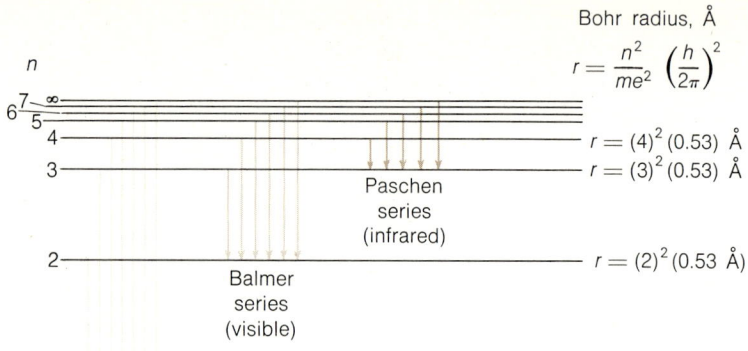

Figure 2.22. Spectral transitions in atomic hydrogen. Two other series for transitions to $n = 4$ and $n = 5$ exist in the infrared. As the magnitude of the electron energy change increases, the wavelength of the spectral line decreases and the frequency increases. Some lines of the Balmer series lie in the visible range (Figure 2.21). At the right of the diagram are shown the radii of the allowed electron "orbits" as calculated from the Bohr theory.

of electromagnetic radiation, and diffraction effects, as in Young's experiment, can be explained only by invoking the wave theory of electromagnetic radiation. Similarly, diffraction of an electron beam can be explained only by attributing wave properties to a beam of moving electrons, just as de Broglie suggested. The phenomenon of diffraction of waves by a crystal is explained in more detail in Chapter 13. All we need note now is that the closely spaced layers of atoms in a crystal act like the series of slits (or sources) in Young's experiment (Section 2.15) and give a similar diffraction pattern. As pointed out before, the diffraction pattern is noticed only if the distance between the slits, or primary sources, is of the order of magnitude of the wavelength of the radiation, or electrons. Therefore, to observe the diffraction pattern of electron waves, a set of slits, or planes of reflecting atoms very close together, is needed. The wavelength of electrons, and therefore the atomic spacing required, depends on the velocity of the electrons as shown in the de Broglie relation $\lambda = h/mv$.

EXAMPLE | The velocity of an electron accelerated through a potential difference of 54 volts is 4.4×10^8 cm sec^{-1}. With $m = 9.1 \times 10^{-28}$ g and $h = 6.6 \times 10^{-27}$ erg sec, the de Broglie wavelength of the electron is 1.6 Å. The distance between reflecting planes of atoms in a nickel crystal is 2.2 Å, so crystals like this are able to diffract electrons.

We can see from the de Broglie relation that diffraction effects with particles heavier than electrons, such as baseballs or airplanes, are not going to be observed in our everyday life. Take, for instance, a golf ball with a mass of 47 g and a velocity of $v = 3000$ cm sec^{-1} (about 70 mph). Then

$$\lambda = \frac{h}{mv} = \frac{6.6 \times 10^{-27} \text{ g cm}^2 \text{ sec}^{-1}}{(47 \text{ g})(3.0 \times 10^3 \text{ cm sec}^{-1})}$$

$$\approx 4.7 \times 10^{-26} \text{ cm}$$

There are no common materials, natural or man-made, having slits or holes with widths comparable to the small wavelength of a golf ball or even a grain of sand. Consequently, the diffraction of matter waves cannot be detected for such massive particles. To observe diffraction effects with the materials at our disposal, we must have longer wavelengths. Matter waves of sufficiently long wavelength will be associated only with light particles moving at low speeds. But since the particle velocities are rather large in many atomic and molecular systems, diffraction effects are usually detected only with particles of very small mass.

There is another way to look at either $\Delta E = h\nu$ or $\lambda = h/p$, and that is through the constant h. What would happen if our universe had a larger value for h? One result, from $\lambda = h/mv$, is that wave properties of matter would become more apparent. But the other equation, $\Delta E = h\nu$, tells us that the difference between allowed and discrete energy states of material systems is proportional to h. If h has a large value, then the transition between energy states of material systems should become more apparent. Perhaps, instead of a golf ball having a continuous range of velocities, it might turn out to have only a discrete set of allowed speeds, so that an accelerating ball would appear to move in a jerky manner, changing speed abruptly.

A way to think of the constant h is in terms of the total energy of a system and of the relative change in this energy as the system goes from one energy state to another. Systems of large mass, such as a golf ball, have a large initial store of energy. The loss of a few quanta, $h\nu$, of energy in going from a high-energy state to a low-energy state makes little difference. The analogy might be extended to our smallest unit of money, the penny. Consider a rich man and a poor man. The first has a large store of money (energy), while the latter has only a little money (energy). Suppose the first man has $100 and the latter has 10¢. If our "money quantum, h" is equal to 1¢ for both, who will notice the loss of one quantum of money the most? The poor man will, because he has lost 10 percent of his original money (energy), while the rich man has lost only 0.01 percent of his. The outward effect of losing or gaining a unit of money (energy) depends on how much was there to begin with. The common large-scale features of the world that we can see have a large store of energy, so the change in going from one energy state to another seems unimportant, and the system appears to go smoothly through a continuum of energies.

The wave theory of matter gives a statistical picture of matter in much the same way that the wave theory of radiation gives statistical information about photons. The following discussion of a Young's double-slit diffraction experiment points this out and can be applied both to diffraction of radiation photons and to electrons.

Suppose there is some method whereby we can shoot just 1 photon (electron) from a source *s* toward the set of slits shown in Figure 2.23. Where will this photon (electron) strike the screen at the right? We don't know for sure—it may strike anywhere. But we shall put a dot at the place where it does strike. Now let us do this a great many times for individually released photons (electrons) and make a mark each time. Soon we would end up with a bunch of dots as shown. The density of dots at a point gives a measure of the probability of a photon (electron) hitting the screen at that point. You can see that if you draw a line around the most dense portions that you get a diffraction pattern in terms of the intensity of radiation (or electron waves) that falls on a particular area of the screen. This intensity or density of dots per unit area at a point gives the probability of a single photon (or electron) going through the slits and ending up at that point on the screen. The complete diffraction pattern is not obtained when a single photon or electron is allowed through the slits, but only when a large number—a beam of particles—go through the slits. A single electron will not end up on the screen at a position predicted by its properties as a particle.

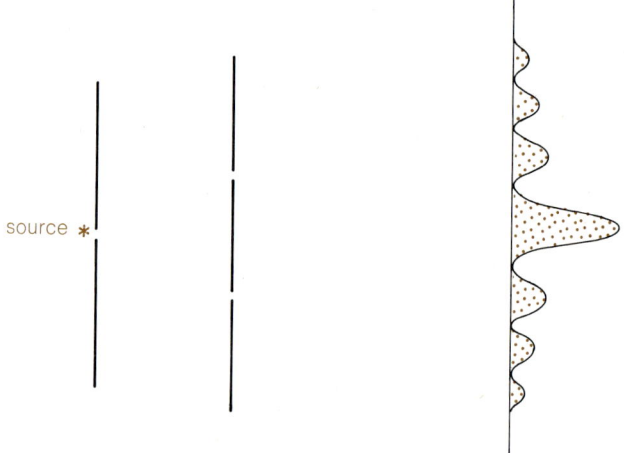

source ✱

Figure 2.23. Diffraction pattern formed by individual electrons or photons. The density of dots is proportional to the wave intensity at that point.

2.20 THE UNCERTAINTY PRINCIPLE

A great deal is made of the terms "probability," "density," or "probability density" when the location of an electron is being discussed. This is because it is impossible to know exactly where an electron is or exactly where it will be, especially if we want to know its energy very closely. This results because of a fundamental law of nature involving particles such as electrons. This law is called the *Heisenberg uncertainty principle* and is a consequence of the wave–particle duality of matter. The principle is named after Werner Heisenberg (1901–   ; Nobel prize in physics, 1932), the German physicist who was originally responsible for the general application of quantum theory to matter. This principle says that it is impossible to specify exactly both the location and the momentum (or energy) of an electron. In equation form,

$$\Delta p_x \, \Delta x \geq h$$

That is, the uncertainty in momentum $\Delta p = \Delta(mv)$ of an electron moving in the x direction multiplied by the uncertainty in its position Δx must be greater than ($>$) or equal to h. Most of the time we are interested in knowing the energy and therefore the momentum or velocity of an electron very accurately. To have this information, we must sacrifice some information about the location of the electron. So we must talk about where the electron might be, or the area in which it might be, rather than its exact position. Instead of saying that an electron with energy T is at some point x, we have to say that the probability that the electron is at x is perhaps 50 : 50, or that the probability per unit volume of finding the electron in a given region around the point x is, say, 80 percent (20 percent of the time it is not in that area).

The uncertainty in the position of an electron, considered as a point particle, reminds us of the wave properties of electrons, because there is a certain amount of difficulty in determining exactly where a wave is in space. Unless a wave has an extremely high amplitude at some point in space, we cannot say that the wave has any special location. It is for reasons such as this that we shall start to think of electrons as being represented by waves spread over space rather than as particles with a particular location at any given time.

2.21 WAVE FUNCTIONS

If the electron does have wave properties, then perhaps electrons could be described by a mathematical equation of the type used to study wave phenomena—sound, radiation, vibrating strings, etc.—particularly a form of wave called a *standing wave*. If one end of a rope is tied to a wall and you shake the other end, standing waves will be generated. At the first shake, a wave travels down the rope to the wall, but this wave is reflected or bounced off the wall and travels back along the rope to your hand. If the rope is continually shaken with the proper frequencies, the direct waves traveling to the wall and the reflected waves add up to a standing wave in which the points of no displacement, the *nodes*, do not move along the rope. The displacements of other parts of the rope merely vibrate about a given position as shown in Figure 2.24, which represents a time exposure of a slightly different system, a string fastened at both ends and set into motion by plucking.

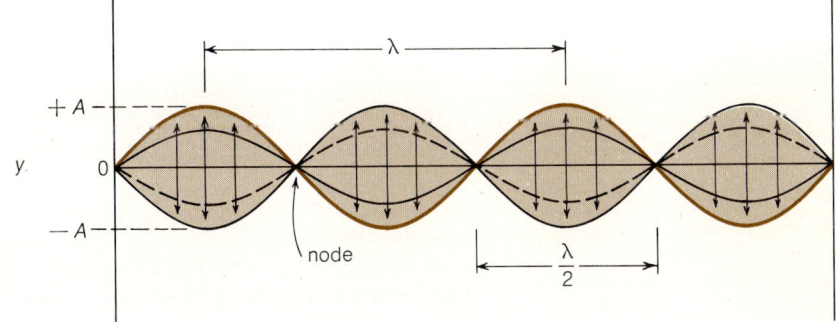

Figure 2.24. Standing wave in a plucked string. The heavy line corresponds to the instantaneous form of the wave at a position of maximum displacement. Other positions of the string lie within the shaded regions. Arrows show the vertical movement of points on the string. There is no motion at the nodes. The distance between adjacent nodes is $\lambda/2$.

Standing waves are most important in describing the motion of an electron in an atom or molecule, because the standing waves in a given system can have only a certain set of wavelengths or frequencies. This set of wavelengths represents the discrete energy levels, the stationary states, in which the energy of the system does not change with time. The trigonometric equation

$$y = A \sin \frac{2\pi x}{\lambda}$$

is a simple representation of such a standing wave at a given time. Here y is the vertical displacement of a point at any position x on the wave, A is the maximum displacement of the wave, and λ is the wavelength. In this special case, the standing wave keeps the same sinusoidal form at all times, and a fast exposure would show no movement, as shown by the heavy curve in Figure 2.24.

The wave equation used to describe electron standing waves is called the *Schrödinger equation*, named after the Austrian physicist Erwin Schrödinger, who received the Nobel prize for physics in 1933. It is a mathematical expression relating a quantity called ψ (Greek, psi, pronounced "sigh") to the total energy E of the electron. The ψ of the Schrödinger wave equation is analogous to the displacement y of a stationary wave and is called the *wave function*. As with the displacement of a vibrating string from the horizontal, ψ can be positive or negative.

The Schrödinger wave equation is really another form of the fundamental energy equation $E = T + U$, where $T = \frac{1}{2}mv^2$ and U is the potential energy of an electron in various force fields. There is no derivation of the Schrödinger equation from first principles, any more than there is a theoretical derivation of $F = ma$, the foundation of classical or Newtonian mechanics, although there are various ways at least to make the equation look plausible. Its sole justification is that it works!

The chemically important quantities in the wave equation are the energy E and the wave function ψ. We have already discussed the energy of electrons. If there is just one electron in the system, E is the energy of that electron and therefore the energy of the system. The energy is obviously important, because almost all changes in an atomic or molecular system are controlled by energy changes. Chemists are particularly interested in knowing the energies of atomic and molecular systems.

The other important term, ψ, is related to the location of the electron. It gives the form of the standing wave associated with the electron. In electromagnetic radiation, the square of the wave amplitude measures the intensity of the radiation, or the photon intensity or density at a certain point. So too with electrons, the value of ψ^2 at a specified position x, $\psi^2_{(x)}$, is a measure of the probability of finding the one electron within a unit of length centered at the point x.

In three dimensions, $\psi^2_{(x,y,z)}$ is the probability per unit volume or the probability density. If we take a very small region of space surrounding the point (x, y, z), the product of $\psi^2_{(x,y,z)}$ and the volume of that space is equal to the probability of finding the electron in that region. (The smaller the volume surrounding the point, the better will the calculated probability represent the exact value at that point.) Alternatively, with several electrons, $\psi^2_{(x,y,z)}$ is the density of electrons in that region. The one electron–several electron interpretation is the same as in a Young electron diffraction experiment. With just one electron, $\psi^2_{(x)}$ is the probability of having a single electron end up at a given point x on the screen. With many electrons, $\psi^2_{(x)}$ gives the density of electrons per unit area at a given place on the screen.

It is worthwhile to point out again that ψ or ψ^2 also gives information about the electron wave at various regions in space. In regions where ψ or ψ^2 is large, the wave amplitude is large and the electron, and its negative charge, has a higher probability of being located in those regions. The electron charge is not located entirely at any one point, but is spread over space. Therefore, we can begin to think of the electron as a diffuse cloud of negative charge occupying a region of space, even though we shall continue to use the word "electron" rather than "electron wave."

ψ or ψ^2 thus gives information about the location of an electron, or about the form of the standing wave of the electron. Chemists are almost, but not quite, as interested in this as in the energy of the electron. Because of the uncertainty principle, the more accurately the energy of the electron is known, or calculated, the more uncertain the position of the electron and the more uncertain is ψ or ψ^2.

The mathematical details of solving the Schrödinger equation for the energy of an electronic system, even with a simple ψ, will not concern us here. What we are going to do is find the energy of a very simple one-electron, one-dimensional system without using the wave equation. We shall make use of the physical picture of the problem to draw the standing waves that would satisfy the Schrödinger equation. The results of this simple but artificial problem will show us general features of electronic systems, features that will appear again and again.

2.22 PARTICLE IN A ONE-DIMENSIONAL BOX

The particle-in-a-box model, more than any other, provides us with a wealth of qualitative information about electron energy levels and probability densities. The "box" is not a material box, but one whose sides are formed by force fields that prevent the electron from escaping. The electron is, in the simplest case, restricted to one-dimensional motion back and forth along the horizontal or x axis. At the edges of the box (at $x = 0, l$), the potential is infinitely large so that the particle is definitely kept in the box. Within the box, let the potential energy U be zero, as shown in Figure 2.25.

Now, what about the waves of the confined electron? They are standing waves and will have the same form as the standing waves formed by a string stretched between two walls (Figure 2.24). We know that at the fastened ends the displacement of the string must always be zero, that is, $\psi_{(0)} = \psi_{(l)} = 0$. These are known as *boundary conditions*. All situations involving electrons confined to a region of space involve boundary conditions, and these boundary conditions are what give rise to the quantum numbers (in a natural and not arbitrary way as they were in the Bohr treatment) and discrete energy levels in a strict mathematical treatment. However, we shall supply the quantum numbers by making an analogy between the electron waves and standing waves in a plucked string. Such a string, as we have seen, must have no displacement at its ends. Therefore, its allowed wave forms ψ are as shown in Figure 2.26. These represent the possible forms of ψ suitable for the particular physical conditions here.

By inspection of these possible ψ's we can generalize that, for any box, there will be $n\lambda/2$ wavelengths in the box, where $n = 1, 2, \ldots, n$. Therefore, for a given ψ_n, the width of the box is given by

$$l = \frac{n\lambda}{2}$$

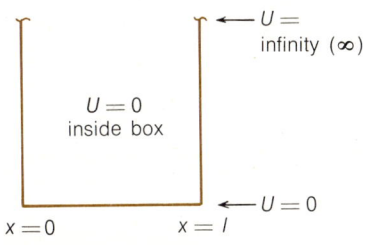

$U = \infty$ infinity (∞)

$U = 0$ inside box

$U = 0$

$x = 0$ $x = l$

Figure 2.25. A one-dimensional box with infinitely high sides.

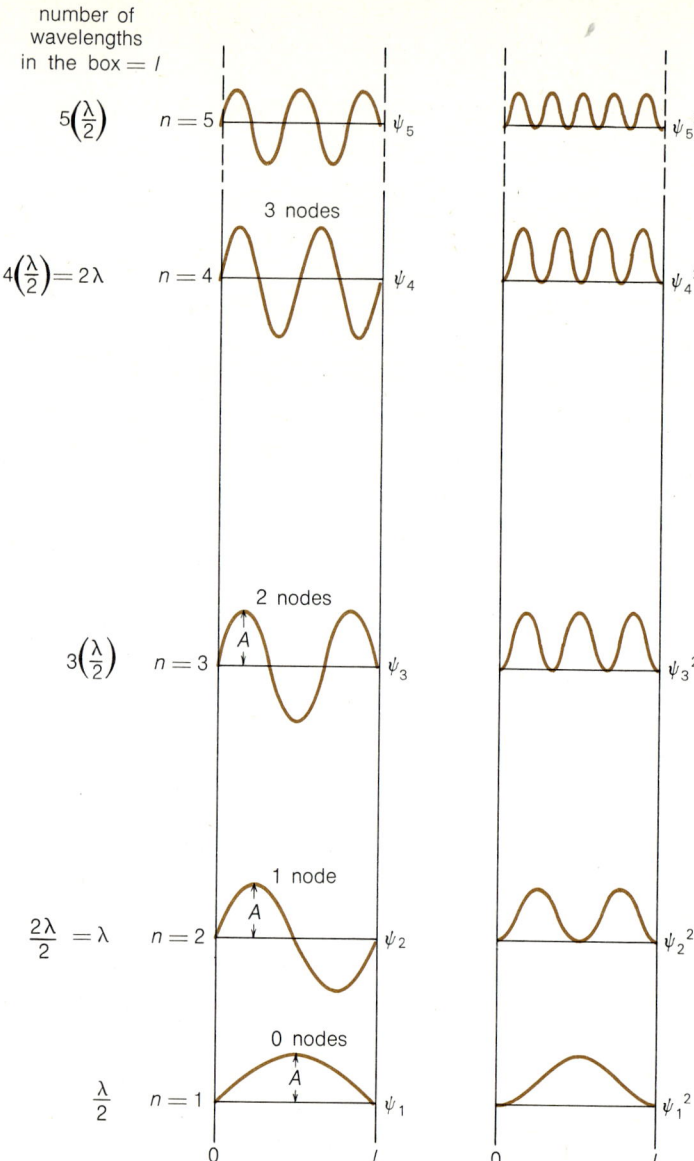

Figure 2.26. Wave functions ψ and probability densities ψ^2 for a particle in a one-dimensional box. The electron probability is not constant along the line of motion.

For any system, $E = T + U$ but $U = 0$ in the box, so

$$E = T = \frac{mv^2}{2} = \frac{p^2}{2m}$$

But, by the de Broglie relation, $p = h/\lambda$, so

$$E = \frac{h^2}{2m\lambda^2}$$

and substituting $\lambda = 2l/n$,

$$E = \frac{h^2}{8ml^2} n^2$$

These are our discrete or quantized energy levels. For each ψ_n there is an E_n.

$$\psi_n = A \sin \frac{n\pi}{l} x \qquad E_n = \frac{n^2 h^2}{8ml^2}$$

$$\psi_1 = A \sin \frac{\pi}{l} x \qquad E_1 = \frac{h^2}{8ml^2}$$

$$\psi_2 = A \sin \frac{2\pi}{l} x \qquad E_2 = \frac{4h^2}{8ml^2}$$

$$\psi_3 = A \sin \frac{3\pi}{l} x \qquad E_3 = \frac{9h^2}{8ml^2}$$

Figure 2.26 shows various ψ^2's as well as ψ's. Remember that $\psi_{(x)}^2$ is a measure of the probability of finding the electron at a certain position x. Points, aside from the edges of the box, where $\psi = \psi^2 = 0$ are called *nodes*. Any ψ represents a standing wave, as in a string plucked in the middle, one-fourth of the way from the end, and so forth. Several important generalizations concerning energy levels can be made from this simple model:

1. For the same value of n, the energy is inversely proportional to the mass of the particle and the width (size) of the box. Thus, as the particle becomes heavier or the box larger, the energy levels become more closely spaced, and it is only when the quantity ml^2 is of the same order as h^2 that quantized energy levels become important in experimental measurements.

2. Each ψ_n has $n - 1$ nodes. It is a general property of wave functions that the greater the number of nodes, the higher the energy of the corresponding state. (The greater $n - 1$, the shorter λ in a given box, and since $\lambda = h/p$, $p = h/\lambda$. A smaller λ means a large momentum p and a large energy E, because the energy is proportional to the square of the momentum.)

3. As the size of the box becomes larger (or the particle heavier), a greater number of energy levels are allowed within a fixed energy range (Figure 2.27).

4. As the box gets larger (Figure 2.27), the value of the lowest level E_1 becomes lower (lies closer to the bottom).

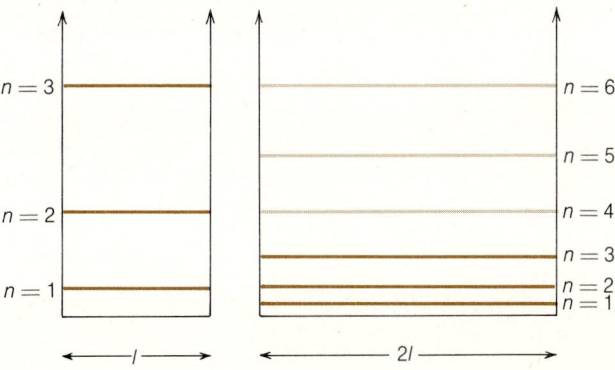

Figure 2.27. Effect of a larger box on energy levels. More energy levels are allowed, and the lowest permitted level lies lower in a larger box.

2.23 DEGENERATE ENERGY LEVELS

In the one-dimensional particle-in-a-box problem, all the wave functions ψ_n were different, and they all corresponded to energy values E_n that were different. In some instances, however, different ψ's can correspond to the same energy. Then the energy level is said to be *degenerate*. This situation would arise in a three-dimensional box where two or more sides are of equal length.

A particular energy level is said to be threefold or triply degenerate if three different wave functions give the same energy value. There are still three energy levels, but they are superimposed, or lie on top of each other, although Figure 2.28 shows them slightly separated for clarity. In some situations, for instance in a box with two equal sides, we could find doubly degenerate levels.

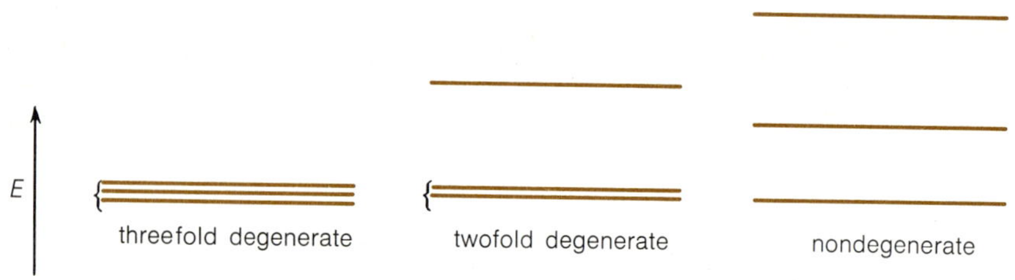

Figure 2.28. Degenerate energy levels. Degenerate levels are shown slightly separated.

Simple and artificial as it is, the particle-in-a-box model gives results that will help us explain a great many chemical problems, from the absorption of radiation by molecules to the direction and extent of chemical reaction. We shall refer to it often, but the "box" we shall talk about is an atom or a molecule within which electrons are confined to move.

2.24 THE HYDROGEN ATOM

Provided that you don't worry about "what is waving" in an electron wave, we are now ready to look at the energy levels and the wave functions of the electron in the simplest atom, the hydrogen atom. The wave mechanical treatment of this atom forms the basis of our ideas about the electronic configurations of all atoms and, as such, lays the groundwork for a study of how and why atoms bond together to form molecules.

The complete solution of the wave equation for the hydrogen atom yields both a set of wave functions ψ and the energies E corresponding to each of these wave functions just as in the particle-in-a-box problem. Each wave function is characterized by three quantum numbers.

In discussing the wave functions and, therefore, the probable position of the electron, it is often convenient to talk about (1) how far the electron is from the nucleus, and (2) in what direction the electron lies from the nucleus. The first relates loosely to the radius r of the electron "orbit," while the second relates to the angular orientation of the electron position. For this reason, it is convenient and proper to break the electronic wave function into a product of two parts, a radial part $\psi(r)$ and an angular part, $\psi(a)$.

$$\psi_H = \psi(r) \cdot \psi(a)$$

The radial part $\psi(r)$ is called the *radial wave function*, and its value is determined by the distance of the electron from the nucleus. The angular part $\psi(a)$ is the *angular wave function*, and its value depends only on the direction of the electron from the nucleus.

The energy E of the electron in a free hydrogen atom turns out to depend essentially only on the radial coordinate, r, of the electron. This energy is calculated to be

$$E_\text{H} = -\frac{2\pi^2 me^4}{n^2 h^2} \qquad n = 1, 2, 3, \ldots, \infty \text{ (infinity)}$$

(Again, negative energies are more stable.) This is the same energy as was originally calculated in the Bohr theory. However, the Schrödinger equation works for other atoms, while Bohr's theory does not.

You see that there is a quantum number, n, in this equation. It is called the *principal* quantum number. The energy is quantized because, for every value of n, there will be a value for E. There are two other quantum numbers, l and m, which arise because of the boundary conditions of the atomic system. The names of all the quantum numbers and their allowed values are

The principal quantum number $n = 1, 2, 3, \ldots, \infty$. The larger n is, the higher the energy E of the electron and the farther the electron is from the nucleus.

The secondary, or azimuthal, quantum number $l = 0, 1, \ldots, n - 1$. The quantum number l is related to the shape, and to a lesser extent the radius, of an "orbit" of the electron around the nucleus.

The magnetic quantum number $m = l, l - 1, \ldots, 0, \ldots, -l$. The quantum number m is related to the orientation of an "orbit" in space.

There is a fourth quantum number, the *spin quantum number s*, that enters independently. It has only two values, $\pm\frac{1}{2}$.

An electron in an atom is characterized by a set of four quantum numbers.

Electrons are designated by the letters s, p, d, and f according to the value of their secondary quantum number, l, as follows:

$l = 0$ s electron
$l = 1$ p electron
$l = 2$ d electron
$l = 3$ f electron

(The letters s, p, d, f originate from the terms "sharp," "principal," "diffuse," and "fundamental," whose initial letters were formerly used to designate groups of lines in atomic spectra.)

If an electron has $n = 1$, the other possible quantum numbers for that electron are

$l = 0, \ldots, n - 1 = 1 - 1 = 0$ thus, $l = 0$ only
$m = l$ to $-l$ thus, $m = 0$ only

Electron n and l values are usually denoted by the numerical value of the principal quantum number followed by the letter equivalent of the secondary quantum number. An electron with the wave function and energy corresponding to $n = 1$, and $l = 0$, is a 1s electron.

When $n = 2$, $l = 0, 1$ since $l = 0, \ldots, (n - 1)$. Therefore, we can have both 2s and 2p electrons.

But $m = l$ to $-l$. For the $l = 0$ electron (2s), $m = 0$ only. But for the $l = 1$ electron (2p), $m = 1, 0, -1$, so there can be three combinations with $n = 2$ and $l = 1$, or three 2p electrons.

Summary of $n = 2$:

$$
\begin{array}{llll}
n = 2 & l = 0 & m = 0 & 2s \\
 & & \left\{\begin{array}{l} m = 1 \\ m = 0 \\ m = -1 \end{array}\right. & \begin{array}{l} 2p \\ 2p \\ 2p \end{array} \\
n = 2 & l = 1 & &
\end{array}
$$

In general, for a given n value, there can be a total of n different l values and a total of n^2 different combinations of n, l, and m.

2.25 ATOMIC ORBITALS

The wave function of an electron in a hydrogen atom, $\psi_H = \psi_{(n,l,m)}$ has a different form for every set of quantum numbers n, l, m that the electron may have. Thus, the wave function for a 1s electron could be abbreviated $\psi_{(1,0,0)}$ or simply ψ_{1s}. The wave function for a 2s electron might be written $\psi_{(n,l,m)} = \psi_{(2,0,0)} = \psi_{2s}$; and that for a 2p electron can be one of the set $\psi_{(2,1,1)}$, $\psi_{(2,1,0)}$, $\psi_{(2,1,-1)}$. Each of these is a different wave function.

Any one atomic wave function describes an electron in an atom. One might say that it describes the path, or the orbit, of the electron. However, the word "orbit" implies a definite trajectory or path in much the same way that we think of the orbit of a spaceship going around the earth or moon. So we shall abandon the word "orbit," because this implies that the position of the electron is definite or certain and also gives the impression of a particulate electron. Instead, we shall say that each wave function describes an *orbital*, in this case an *atomic orbital* (A.O.). The word "orbital" will be used in two ways: to denote that region of space in which the electron described by the particular set of quantum numbers is most likely to be found, and to specify the energy of that electron.

We can think of the atomic orbital as being the mathematical form of the wave function. As before, the square of the wave function, ψ^2, evaluated at a particular point x, y, z in space is a measure of the probability per unit volume of finding an electron near that point. Different orbitals have different probabilities of electrons being found at the same point in space.

Yet another way of thinking about orbitals is to think of them as regions in space in which electrons with a certain set of quantum numbers are constrained to move. These regions or orbitals are a property of an atom even though there may be no electron in them, just as we say that a flight path or corridor for airplanes exists in a certain region of the sky even though there is no airplane in sight. We know that the probability of an airplane being in that region is much greater than in any other region. The flight path exists even in the absence of an airplane. In this respect, the word "orbital" can also stand for "energy level," since the allowed energy levels are there even if the electron is not.

2.26 SHAPES AND ENERGIES OF ATOMIC ORBITALS IN THE HYDROGEN ATOM

If an atomic orbital ψ (or ψ^2) is supposed to describe a region of space, that region of space must have some sort of shape. There are several ways to represent the shape of atomic orbitals. We shall look at some of these ways, and we shall make use of the fact that ψ^2, which can only be zero or positive, is proportional to the probability of finding an electron at a particular place.

Let us first look at how the probability (or as we shall see in a moment, the electron "charge cloud" density) varies as one travels out from the nucleus

($r = 0$) to an infinite distance from the nucleus ($r = \infty$). At various distances r from the nucleus, we can calculate the value of the term $4\pi r^2 \psi^2(r)$, where $\psi(r)$ is the radial part of ψ_H. This combination of terms, called the *radial distribution function*, gives the probability per unit distance of finding an electron in a spherical shell of very small thickness at a distance r from the nucleus. When the values of $4\pi r^2 \psi^2(r)$ and $\psi(r)$ are graphed versus r for the various orbitals 1s, 2s, 2p, 3p, 3d, the plots shown in Figure 2.29 are obtained.

These figures illustrate the following:

1. Places where the lines touch the horizontal axis (except at the ends of the lines) are called *radial nodes*. The probability of an electron being in a region this far from the nucleus is zero.

The 1s orbital has no radial node.
The 2s orbital has one radial node.
A 2p orbital has no radial node.
The 3s orbital has two radial nodes.
A 3p orbital has one radial node.
A 3d orbital has no radial node.

2. The position of maximum probability of a 1s electron is closer to the nucleus (in fact, it is just equal to the first Bohr radius calculated by Bohr, $r_0 = 0.529$ Å) than that of a 2s electron. The same comparison applies to 2s and 3s.

3. Just as not all orbitals have the same probability at a given point, so too, not all orbitals correspond to the same electron energy. The 1s orbital has a lower energy than a 2s orbital, because the average distance (or position of maximum probability) for an electron in a 1s ($n = 1$) orbital is closer to the nucleus than for a 2s ($n = 2$) electron. That is, a 1s electron is held more tightly by the nucleus than a 2s electron—or the ionization energy of an electron in a 1s orbital is greater than that of a 2s electron. In the one-electron hydrogen atom, the 2s orbital and the 2p orbitals have equal energies. Although the maximum for a 2p electron is closer to the nucleus, an electron in a 2s orbital shows some probability of venturing closer to the nucleus than a 2p electron. The small hump in the 2s radial distribution function shows this. Similarly, in hydrogen, the 3s, 3p, and 3d orbitals all have the same energy. In hydrogen, then, the orbital energies increase with increasing n:

$$(1s) < (2s, 2p) < (3s, 3p, 3d) < (4s, 4p, 4d, 4f)$$

(The symbol $<$ means "is less than.")

The plots in Figure 2.29 should not give the impression that the same view of electron distribution is seen in any direction from the nucleus. Imagine yourself standing on top of a proton at its north pole. For a 1s, 2s, 3s, . . . or any s electron, $\psi(r)$, $\psi^2(r)$, or $4\pi r^2 \psi^2(r)$ is spherically symmetric. That is, you would see the same electron distribution no matter in what direction you looked. For a 1s electron, the nucleus is the center of a spherical electron distribution with the same electron density gradient, or probability of finding an electron, in all directions. A 2s electron (or orbital) would have two spherical regions of finite electron density separated by a node. Figures 2.30(a) and 2.30(b) give cross sections of $\psi(r)$ and $\psi^2(r)$ for the 1s and 2s orbitals.

However, for p and d electrons, the electron probability distribution is

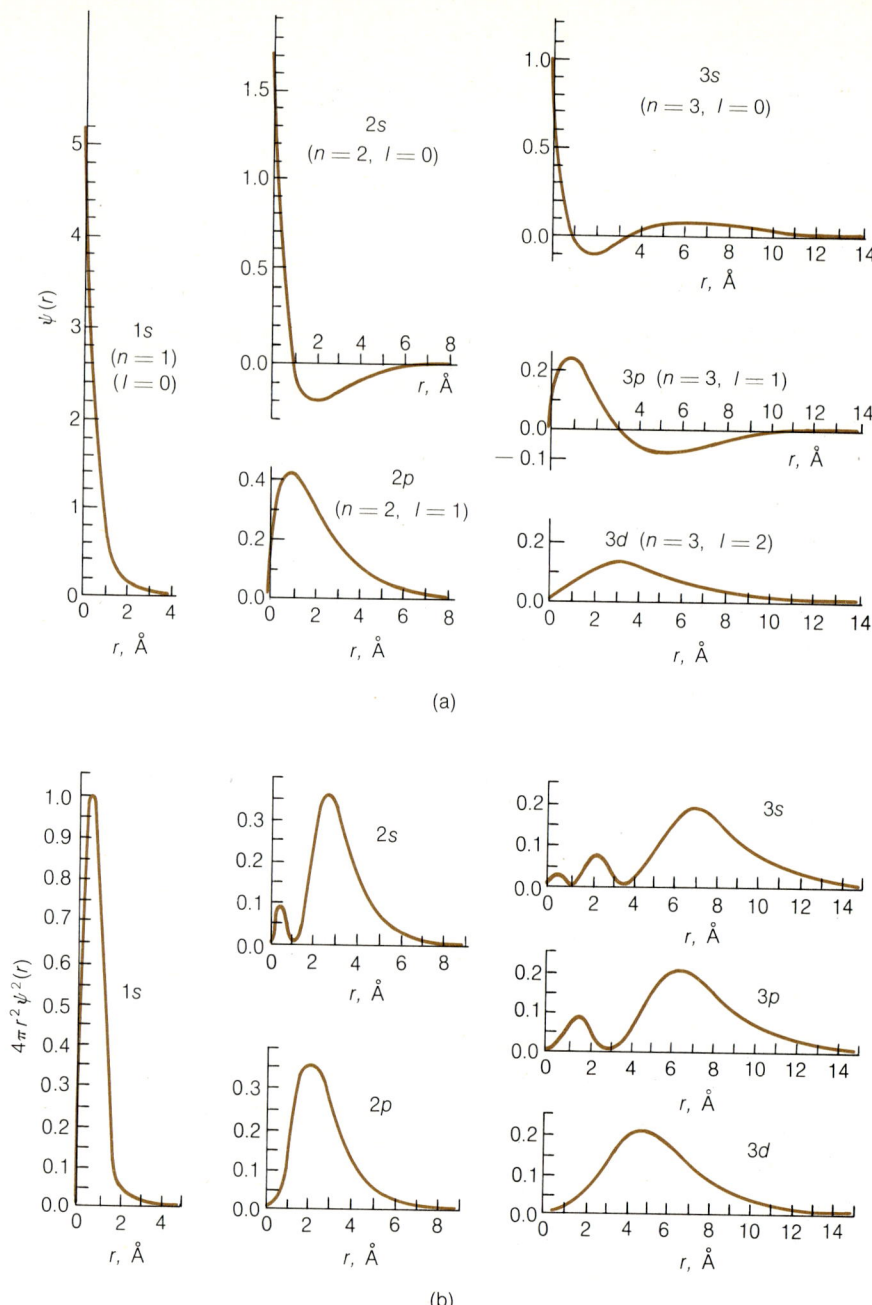

Figure 2.29. (a) Radial wave functions $\psi(r)$ and (b) radial distribution functions $4\pi r^2 \psi^2(r)$ for atomic orbitals in hydrogen.

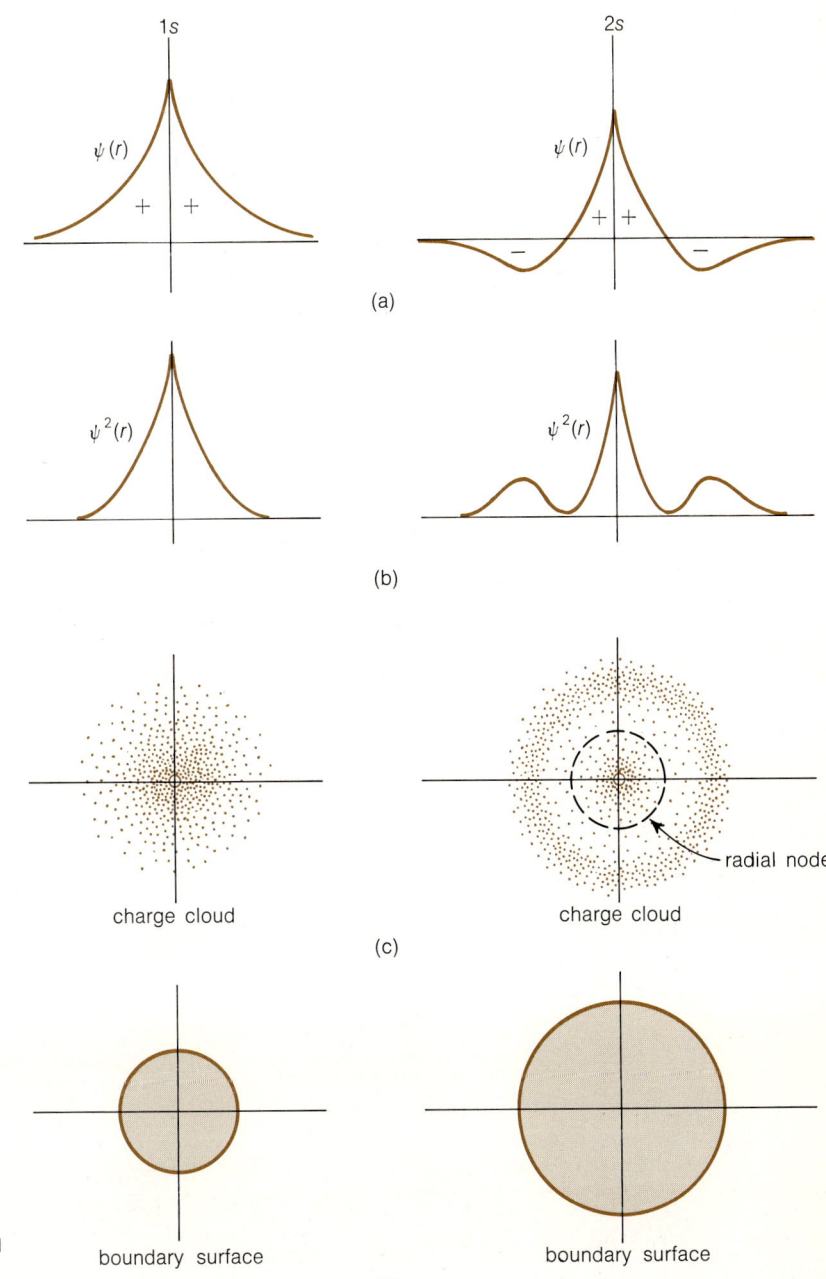

Figure 2.30. Representations of the spherically symmetric hydrogen $1s$ and $2s$ atomic orbitals. (a) Radial wave function $\psi(r)$. (b) Radial probability density $\psi^2(r)$. (c) Electron charge cloud. (d) Boundary surface diagrams. The plus and minus signs in (a) refer to the phase or displacement of the wave function.

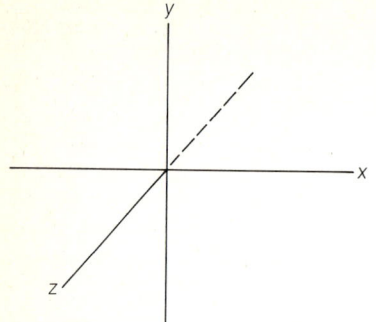

Figure 2.31. Rectangular coordinate axes.

not the same in every direction from the nucleus. Again, imagine yourself at the north pole of a proton. Let us select a particular 2p electron (or a 2p orbital containing an electron). Remember that there are three p-type orbitals, whether $2p$, $3p$, . . . or np. You may also imagine a set of rectangular coordinate axes on the proton, as in Figure 2.31. The three p orbitals for any given value of n are equivalent, or triply degenerate, with each p orbital along one of the axes. Let us consider the p orbital (p_x) directed along the x axis. Now, if you look to your right or left (east or west) along the x axis, you will see the hump or rim shown in the radial distribution function, and also in the radial probability density, $\psi^2(r)$, for a $2p_x$ electron. However, if you look along the

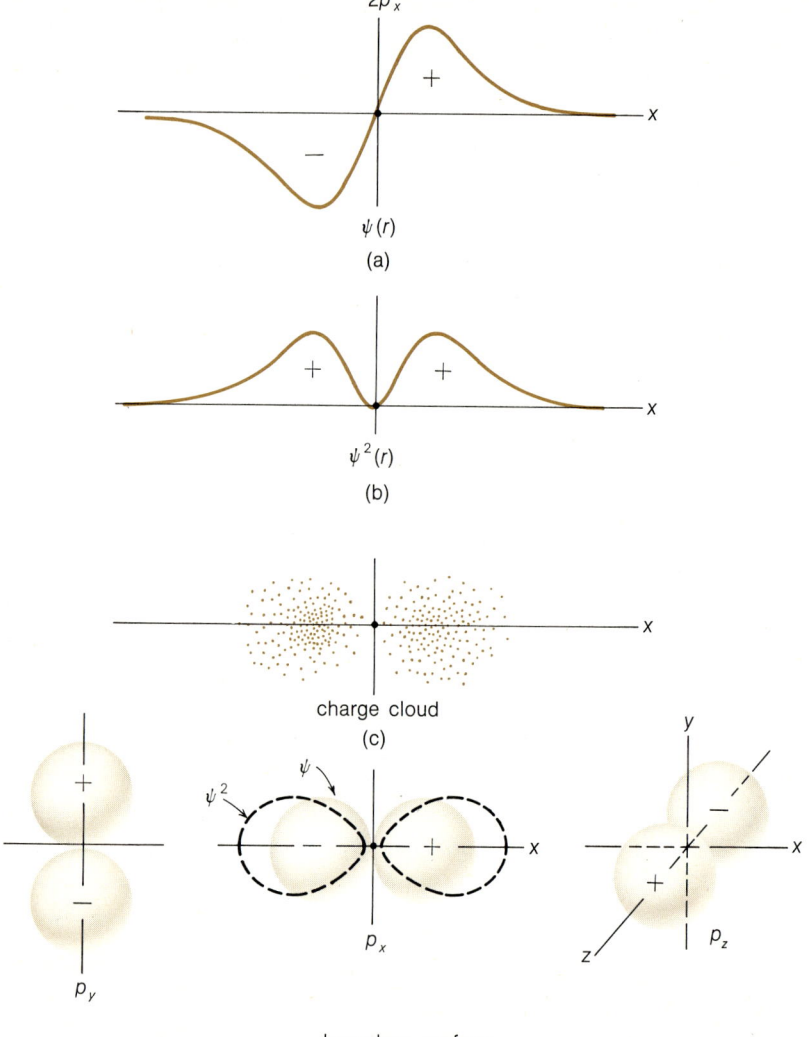

Figure 2.32. Representations of the p_x orbital. (a) Radial wave function $\psi(r)$. (b) Probability density $\psi^2(r)$. (c) Electron charge cloud. (d) Boundary surface diagrams for p_x, p_y, and p_z orbitals. The lobes of the boundary surface diagrams would be more elongated than spherical if the boundary surface of $\psi^2(r)$ were actually used; this is shown by the dotted outline. For ease in drawing, spherical lobes, which correspond more to the boundary of $\psi(r)$, are used here. The signs in (d) correspond to the phases in $\psi(r)$.

z axis, or up or down along the y axis, you will discover that the probability of finding $2p_x$ electrons in those directions is zero. This p orbital is not spherically symmetric; it has decided directional properties. The directional properties of the p- and d-type orbitals are shown in Figures 2.32 and 2.33.

2.27 CHARGE CLOUD OR ELECTRON DENSITY PICTURES OF ATOMIC ORBITALS

In Figures 2.30(c) and 2.32(c), the stippled portions of the orbitals indicate the areas where an electron will most probably be found. The denser the stippling, the greater the probability. To see how you might obtain such a diagram, imagine that you plot the position of one electron in an orbital over a long period of time, placing a dot at the point where you find the electron at any given instant. After placing several hundred dots, you would end up with something like these figures, where the density of dots is proportional to the chance of finding an electron. The electron "spends" most of its time in the darkest regions. The electron charge cloud picture also can be thought of as representing a time average, like a blurred time exposure of a rapidly moving object. Usually, a boundary is drawn around the region within which there is about a 90 percent chance of finding an electron, the shading is omitted, and we get the commonly used boundary surface diagram shown in Figures 2.30(d) and 2.32(d).

The 3d orbitals—there are five of them, with $m = 2, 1, 0, -1, -2$—have four lobes. The boundary surfaces of two of the 3d orbitals are shown in Figure 2.33.

We now have the shapes of the major orbitals s, p, d. We see that the s orbitals, whether 1s, 2s, 3s, . . . are all spherical and have no preferred directional properties. A p orbital is dumbbell shaped, with two directional lobes pointing in opposite directions. For the three p orbitals of any principal quantum number (except $n = 1$, for which there are no p orbitals), we may pick the directions along the axes of a rectangular coordinate system, that is, p_x, p_y, and p_z. There are five d orbitals of any given principal quantum number (except $n = 1, 2$), and they are directional also.

The lobes of an orbital, their size or extent, and their direction are important in establishing many aspects of bonding between atoms, because it is the interaction of orbitals, or the electrons in orbitals, on adjoining atoms that controls bond formation.

An orbital is said to have an angular node when a plane or planes can be passed through the orbital such that the value of the orbital is zero everywhere on that plane, as shown in Figure 2.34. The total number of nodes, both radial and angular, in an atomic orbital is $n - 1$. The number of radial or spherical nodal surfaces is $n - l - 1$. Therefore, $l =$ the number of angular nodes. In the one-electron hydrogen atom, all orbitals with the same total number of nodes have equal energies. As in the particle-in-a-box model, the more nodes the higher the energy.

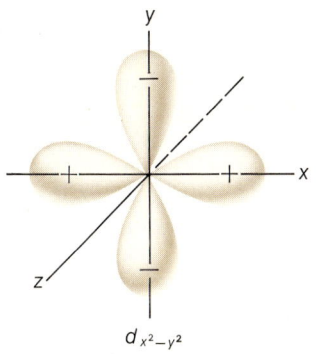

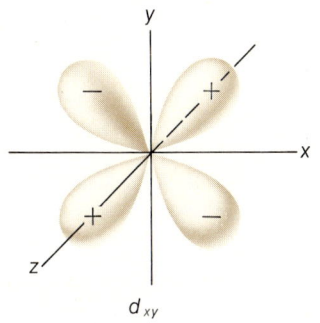

Figure 2.33. Boundary surfaces of two of the five d atomic orbitals. The other d orbitals are designated d_{yz}, d_{zx}, and d_{z^2}. Both orbitals lie in the xy plane.

2.28 ORBITAL ENERGIES IN MULTIELECTRON ATOMS

Once there are two or more electrons in an atom, the energy values of the orbitals are no longer dependent on the n value alone. One reason for this can be seen by looking at the radial distribution functions for the 1s, 2s, and 2p electrons in Figure 2.29. The 2s electron penetrates closer to the nucleus than a 2p electron and even gets inside the 1s hump to some extent. The 1s electron(s) already present screen the positive nuclear charge from the 2p electron more than they do for the 2s electron, because the 2s electron can

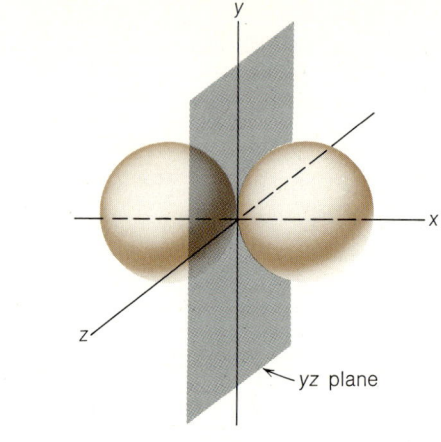

$l = 1$; 1 angular node

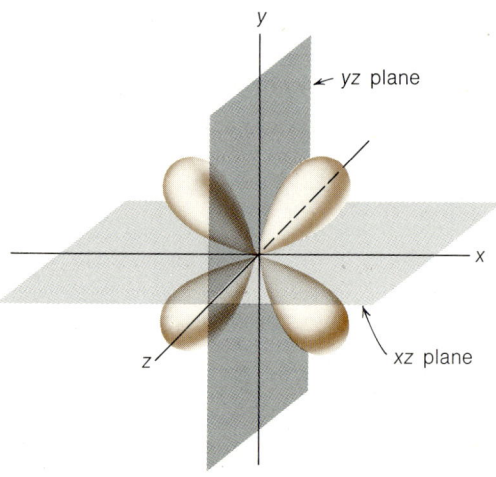

Figure 2.34. Nodal planes for orbitals with angular nodes.

$l = 2$; 2 angular nodes

penetrate the charge cloud of a 1s electron. Thus, the 2p electron is held less tightly and has a higher energy than a 2s electron.

We can find the order of orbital energies for atoms other than hydrogen by using a rule: the orbitals increase in energy in the order of $(n + l)$. If two $(n + l)$ values are equal, the orbital of lowest energy is that with the lowest n. The order and the relative spacing of the orbital energies is shown in Figure 2.35.

2.29 PUTTING ELECTRONS IN THE ORBITALS

We are now ready to determine the electronic configurations of atoms by inserting electrons into the available orbitals. Every time we add an electron to the set of atomic orbitals, we also add a proton to the nucleus to preserve neutrality. As might be expected, each added electron goes into the available

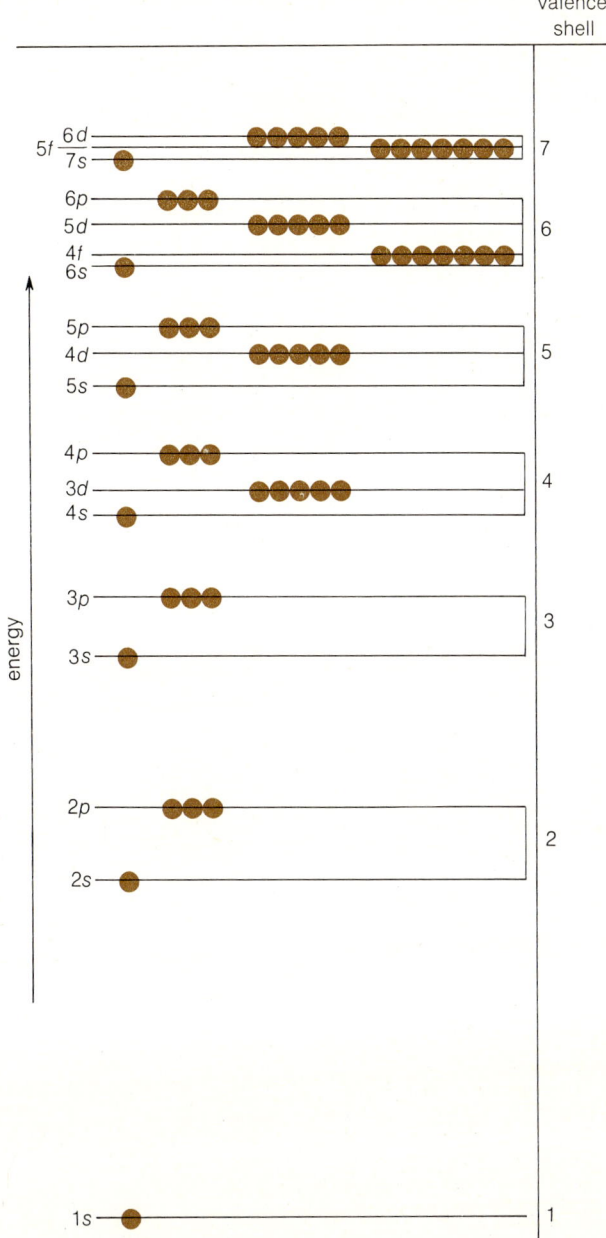

Figure 2.35. Energy levels for atomic orbitals in atoms with more than one electron. The circles indicate the compartments or suborbitals in each orbital level.

orbital with the lowest energy. But we still do not know how many electrons we can put into each orbital.

The answer to this question is given by the *Pauli exclusion principle* (named after the Austrian physicist Wolfgang Pauli, 1900– ; Nobel prize in physics, 1945), which states that no two electrons in the same atom may have the same set of quantum numbers. To use this principle, the spin quantum number s must also be considered. Two allowed values of s exist for each allowed value of the magnetic quantum number m that an electron may have. The s values are often denoted by vertical arrows, ↑ for $+\frac{1}{2}$, ↓ for $-\frac{1}{2}$.

An s orbital may hold a maximum of 2 electrons, each with opposite spin:

s

The set of p orbitals of any principal quantum number consists of three orbitals, each of which can hold a maximum of 2 electrons with opposite spin.

p

Similarly,

d

Of course, the orbitals can also be unoccupied or hold only 1 electron.

In summary, Table 2.9 shows how many $1s$, $2s$, $2p$, . . . electrons an atom may have.

Now let's add electrons (and protons) and build up atoms, as shown in Figure 2.36. The first 2 electrons go into the $1s$ orbital; the next 2 electrons

Table 2.9		allowed values of quantum numbers				
Quantum numbers and orbital occupation	orbital	n	l	m	s	total electrons in orbital
	$1s$	1	0	0	$\pm\frac{1}{2}$	2
	$2s$	2	0	0	$\pm\frac{1}{2}$	2
	$2p$	2	1	$+1$	$\pm\frac{1}{2}$	
				0	$\pm\frac{1}{2}$	6
				-1	$\pm\frac{1}{2}$	
	$3s$	3	0	0	$\pm\frac{1}{2}$	2
	$3p$	3	1	$+1$	$\pm\frac{1}{2}$	
				0	$\pm\frac{1}{2}$	6
				-1	$\pm\frac{1}{2}$	
	$3d$	3	2	2	$\pm\frac{1}{2}$	
				1	$\pm\frac{1}{2}$	
				0	$\pm\frac{1}{2}$	10
				-1	$\pm\frac{1}{2}$	
				-2	$\pm\frac{1}{2}$	
	$4s$					2
	$4p$					6
	$4d$					10
	$4f$					14

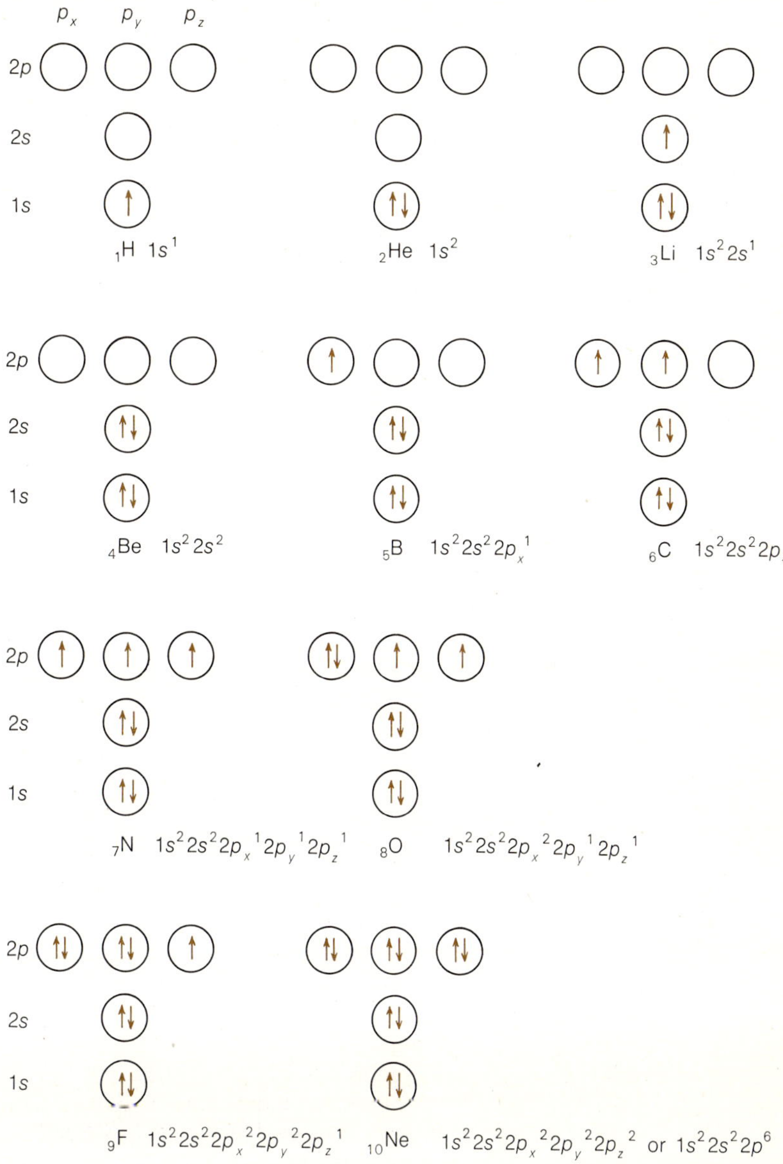

p_x p_y p_z

2p

2s

1s

$_1$H $1s^1$ $_2$He $1s^2$ $_3$Li $1s^2 2s^1$

2p

2s

1s

$_4$Be $1s^2 2s^2$ $_5$B $1s^2 2s^2 2p_x^1$ $_6$C $1s^2 2s^2 2p_x^1 2p_y^1$

2p

2s

1s

$_7$N $1s^2 2s^2 2p_x^1 2p_y^1 2p_z^1$ $_8$O $1s^2 2s^2 2p_x^2 2p_y^1 2p_z^1$

2p

2s

1s

$_9$F $1s^2 2s^2 2p_x^2 2p_y^2 2p_z^1$ $_{10}$Ne $1s^2 2s^2 2p_x^2 2p_y^2 2p_z^2$ or $1s^2 2s^2 2p^6$

Figure 2.36. Orbital filling diagrams.

go into the 2s orbital. The fifth electron goes into a 2p orbital as in boron.

When the sixth electron (and proton) enters the atom, a decision must be made. Does the sixth electron pair up in the same p orbital as the fifth,

or does it enter another p orbital of the degenerate set,

The answer to this is given by *Hund's rule of maximum multiplicity*, which says that electrons will go into the orbitals singly with parallel (same) spins, and that after singly occupying each one of a set of degenerate orbitals, further electrons will double up with paired spins. We can think of the unpaired electrons as staying away from each other by occupying different regions of space.

When giving the electron configuration, it is best at this point to write the separate p orbitals with their electrons. The number of unpaired electrons in an atom is important, and $2p_x^1 2p_y^1$, rather than $2p^2$, makes it more evident that $_6$C has 2 unpaired electrons.

The electron configuration notation for neon, $_{10}$Ne, is given in two ways. If you know that the capacity of the set of p orbitals is 6 electrons or three pairs, then it is all right to use $2p^6$. In neon, two "valence shells" are filled. A valence shell is defined as that set of orbitals whose energy levels are included between ns and np, counting all orbitals in between even if their principal quantum number is smaller than n. For instance,

$1s^2$ constitutes a filled first valence shell (sometimes called the K shell)

$2s^2 2p^6$ constitutes a filled second valence shell (sometimes called the L shell)

$3s^2 3p^6$ constitutes a filled third valence shell

$4s^2 3d^{10} 4p^6$ constitutes a filled fourth valence shell

$5s^2 4d^{10} 5p^6$ constitutes a filled fifth valence shell

$6s^2 4f^{14} 5d^{10} 6p^6$ constitutes a filled sixth valence shell.

Note that there is a comparatively large energy gap between valence shells (Figure 2.35). That is, it would take considerable energy to promote an electron from the top p level of one valence shell to the lowest s level of the next higher shell, especially when the n quantum number is small.

A set of equivalent d orbitals also fills in accordance with Hund's rules. Although it is possible to denote the set of five d orbitals by letters referring to their directions along rectangular axes, we shall merely write out five $3d$'s where necessary, so that, for instance, the electronic configuration of cobalt, $_{27}$Co, is $1s^2 2s^2 2p^6 3s^2 3p^6 4s^2 3d^2 3d^2 3d^1 3d^1 3d^1$.

Within the series of elements from potassium, $_{19}$K, to zinc, $_{30}$Zn, the fourth valence shell, which includes the set of $3d$ orbitals, is being filled. All the rules are followed except for apparent exceptions at $_{24}$Cr and $_{29}$Cu, which have incomplete $4s$ levels but a half-filled and fully filled $3d$ level, respectively. The same thing happens with elements of higher atomic number where filled or half-filled d or f levels are involved.

Electronic configurations for the entire periodic table to element number 104 are given in Table 2.10.

2.30 ELECTRONIC CONFIGURATIONS AND THE PERIODIC TABLE

Now that we have the orbital electronic configurations of the atoms of the chemical elements, we can return to the periodic table and see how this table reflects the electronic structure of the atoms. Remember that this table was originally set up without any knowledge of orbitals, quantum numbers, energy levels, unpaired electrons, and so on. All that was known were the atomic numbers and weights of the known elements and the fact that there were certain similarities in chemical and physical properties or behavior among these elements.

Table 2.10
Electronic configurations of the elements

Z	atom	electronic configuration	Z	atom	electronic configuration	Z	atom	electronic configuration
1	H	$1s^1$	36	Kr	$(Ar)4s^23d^{10}4p^6$	71	Lu	$(Xe)6s^24f^{14}5d^1$
2	He	$1s^2$	37	Rb	$(Kr)5s^1$	72	Hf	$(Xe)6s^24f^{14}5d^2$
3	Li	$(He)2s^1$	38	Sr	$(Kr)5s^2$	73	Ta	$(Xe)6s^24f^{14}5d^3$
4	Be	$(He)2s^2$	39	Y	$(Kr)5s^24d^1$	74	W	$(Xe)6s^24f^{14}5d^4$
5	B	$(He)2s^22p^1$	40	Zr	$(Kr)5s^24d^2$	75	Re	$(Xe)6s^24f^{14}5d^5$
6	C	$(He)2s^22p^2$	41	Nb	$(Kr)5s^14d^4$	76	Os	$(Xe)6s^24f^{14}5d^6$
7	N	$(He)2s^22p^3$	42	Mo	$(Kr)5s^14d^5$	77	Ir	$(Xe)6s^24f^{14}5d^7$
8	O	$(He)2s^22p^4$	43	Tc	$(Kr)5s^24d^5$	78	Pt	$(Xe)6s^14f^{14}5d^9$
9	F	$(He)2s^22p^5$	44	Ru	$(Kr)5s^14d^7$	79	Au	$(Xe)6s^14f^{14}5d^{10}$
10	Ne	$(He)2s^22p^6$	45	Rh	$(Kr)5s^14d^8$	80	Hg	$(Xe)6s^24f^{14}5d^{10}$
11	Na	$(Ne)3s^1$	46	Pd	$(Kr)4d^{10}$	81	Tl	$(Xe)6s^24f^{14}5d^{10}6p^1$
12	Mg	$(Ne)3s^2$	47	Ag	$(Kr)5s^14d^{10}$	82	Pb	$(Xe)6s^24f^{14}5d^{10}6p^2$
13	Al	$(Ne)3s^23p^1$	48	Cd	$(Kr)5s^24d^{10}$	83	Bi	$(Xe)6s^24f^{14}5d^{10}6p^3$
14	Si	$(Ne)3s^23p^2$	49	In	$(Kr)5s^24d^{10}5p^1$	84	Po	$(Xe)6s^24f^{14}5d^{10}6p^4$
15	P	$(Ne)3s^23p^3$	50	Sn	$(Kr)5s^24d^{10}5p^2$	85	At	$(Xe)6s^24f^{14}5d^{10}6p^5$
16	S	$(Ne)3s^23p^4$	51	Sb	$(Kr)5s^24d^{10}5p^3$	86	Rn	$(Xe)6s^24f^{14}5d^{10}6p^6$
17	Cl	$(Ne)3s^23p^5$	52	Te	$(Kr)5s^24d^{10}5p^4$	87	Fr	$(Rn)7s^1$
18	Ar	$(Ne)3s^23p^6$	53	I	$(Kr)5s^24d^{10}5p^5$	88	Ra	$(Rn)7s^2$
19	K	$(Ar)4s^1$	54	Xe	$(Kr)5s^24d^{10}5p^6$	89	Ac	$(Rn)7s^26d^1$
20	Ca	$(Ar)4s^2$	55	Cs	$(Xe)6s^1$	90	Th	$(Rn)7s^26d^2$
21	Sc	$(Ar)4s^23d^1$	56	Ba	$(Xe)6s^2$	91	Pa	$(Rn)7s^25f^26d^1$
22	Ti	$(Ar)4s^23d^2$	57	La	$(Xe)6s^25d^1$	92	U	$(Rn)7s^25f^36d^1$
23	V	$(Ar)4s^23d^3$	58	Ce	$(Xe)6s^24f^2$	93	Np	$(Rn)7s^25f^46d^1$
24	Cr	$(Ar)4s^13d^5$	59	Pr	$(Xe)6s^24f^3$	94	Pu	$(Rn)7s^25f^6$
25	Mn	$(Ar)4s^23d^5$	60	Nd	$(Xe)6s^24f^4$	95	Am	$(Rn)7s^25f^7$
26	Fe	$(Ar)4s^23d^6$	61	Pm	$(Xe)6s^24f^5$	96	Cm	$(Rn)7s^25f^76d^1$
27	Co	$(Ar)4s^23d^7$	62	Sm	$(Xe)6s^24f^6$	97	Bk	$(Rn)7s^25f^86d^1$
28	Ni	$(Ar)4s^23d^8$	63	Eu	$(Xe)6s^24f^7$	98	Cf	$(Rn)7s^25f^{10}$
29	Cu	$(Ar)4s^13d^{10}$	64	Gd	$(Xe)6s^24f^75d$	99	Es	$(Rn)7s^25f^{11}$
30	Zn	$(Ar)4s^23d^{10}$	65	Tb	$(Xe)6s^24f^9$	100	Fm	$(Rn)7s^25f^{12}$
31	Ga	$(Ar)4s^23d^{10}4p^1$	66	Dy	$(Xe)6s^24f^{10}$	101	Md	$(Rn)7s^25f^{13}$
32	Ge	$(Ar)4s^23d^{10}4p^2$	67	Ho	$(Xe)6s^24f^{11}$	102	No	$(Rn)7s^25f^{14}$
33	As	$(Ar)4s^23d^{10}4p^3$	68	Er	$(Xe)6s^24f^{12}$	103	Lw	$(Rn)7s^25f^{14}6d^1$
34	Se	$(Ar)4s^23d^{10}4p^4$	69	Tm	$(Xe)6s^24f^{13}$	104		$(Rn)7s^25f^{14}6d^2$
35	Br	$(Ar)4s^23d^{10}4p^5$	70	Yb	$(Xe)6s^24f^{14}$			

The periodic classification of the elements has had a long history of development, starting around the early 1800's. This development continues today, but is generally confined to trying to devise new shapes, both two and three dimensional, of the periodic table and to trying to iron out certain small irregularities and increase the ease of reading of the table. These are all small details. Today's commonly used form of the table is derived from Mendeleev's classification of 1869. This "long" form is given in Figure 2.37.

We shall examine first how the electronic configurations of the elements

IA																	0
1																	2
H	IIA											IIIA	IVA	VA	VIA	VIIA	He
1.01																	4.00
3	4											5	6	7	8	9	10
Li	Be											B	C	N	O	F	Ne
6.94	9.01											10.81	12.01	14.01	16.00	19.00	20.18
11	12	IIIB	IVB	VB	VIB	VIIB		VIII		IB	IIB	13	14	15	16	17	18
Na	Mg											Al	Si	P	S	Cl	Ar
22.99	24.31											26.98	28.09	30.97	32.06	35.45	39.95
19	20	21	22	23	24	25	26	27	28	29	30	31	32	33	34	35	36
K	Ca	Sc	Ti	V	Cr	Mn	Fe	Co	Ni	Cu	Zn	Ga	Ge	As	Se	Br	Kr
39.10	40.08	44.96	47.90	50.94	52.00	54.94	55.85	58.93	58.71	63.55	65.37	69.72	72.59	74.92	78.96	79.90	83.80
37	38	39	40	41	42	43	44	45	46	47	48	49	50	51	52	53	54
Rb	Sr	Y	Zr	Nb	Mo	Tc	Ru	Rh	Pd	Ag	Cd	In	Sn	Sb	Te	I	Xe
85.47	87.62	88.91	91.22	92.91	95.94	(99)	101.07	102.91	106.4	107.87	112.40	114.82	118.69	121.75	127.60	126.90	131.30
55	56	57	72	73	74	75	76	77	78	79	80	81	82	83	84	85	86
Cs	Ba	La	Hf	Ta	W	Re	Os	Ir	Pt	Au	Hg	Tl	Pb	Bi	Po	At	Rn
132.91	137.34	138.91	178.49	180.95	183.85	186.2	190.2	192.2	195.09	196.97	200.59	204.37	207.19	208.98	(210)	(210)	(222)
87	88	89	104	105													
Fr	Ra	Ac															
(223)	(226)	(227)															

58	59	60	61	62	63	64	65	66	67	68	69	70	71
Ce	Pr	Nd	Pm	Sm	Eu	Gd	Tb	Dy	Ho	Er	Tm	Yb	Lu
140.12	140.91	144.24	(147)	150.35	151.96	157.25	158.92	162.50	164.93	167.26	168.93	173.04	174.97

90	91	92	93	94	95	96	97	98	99	100	101	102	103
Th	Pa	U	Np	Pu	Am	Cm	Bk	Cf	Es	Fm	Md	No	Lw
232.04	(231)	238.03	(237)	(242)	(243)	(247)	(247)	(251)	(254)	(253)	(256)	(254)	(257)

Figure 2.37. The periodic table.

are reflected in the table and then look at some of the trends in selected properties of elements. In the long form of the periodic table, the elements are arranged in horizontal *periods* and in vertical columns or *groups*. The periods are of different length, having 2, 8, 8, 18, 18, 32, and (so far) 19 members. Roman numerals at the top of the groups are sometimes used to identify the groups.

Let us take first group IA, also called the alkali metals, and look at the electronic configuration of their valence or outermost shells. By reference to Table 2.10, you can see that their outermost electronic configuration is ns^1, a single unpaired electron in the nth valence shell. Thus,

$$\text{Li } 1s^2 2s^1 \quad \text{ or } \quad (\text{He}) 2s^1 \quad \text{ in period 2}$$
$$\text{Na } 1s^2 2s^2 2p^6 3s^1 \quad \text{ or } \quad (\text{Ne}) 3s^1 \quad \text{ in period 3}$$
$$\text{K } (\text{Ar}) 4s^1 \quad \text{ in period 4}$$

In group IIA, also called the alkaline earth elements, the common outermost electronic configuration is ns^2. Similarly, we have,

group IIIB	$ns^2(n-1)d^1np^0$	
group IVB	$ns^2(n-1)d^2np^0$	
group VB	$ns^2(n-1)d^3np^0$	
group VIB	$ns^1(n-1)d^5np^0$	
group VIIB	$ns^2(n-1)d^5np^0$	transition elements
group VIII	$ns^{0-2}(n-1)d^{6-10}np^0$ (see Table 2.10)	
group IB (coinage metals)	$ns^1(n-1)d^{10}np^0$	
group IIB	$ns^2(n-1)d^{10}np^0$	
group IIIA	ns^2np^1 or $ns^2(n-1)d^{10}np^1$	
group IVA	ns^2np^2 or $ns^2(n-1)d^{10}np^2$	
group VA	ns^2np^3 or $ns^2(n-1)d^{10}np^3$	
group VIA	ns^2np^4 or $ns^2(n-1)d^{10}np^4$	
group VIIA (halogens)	ns^2np^5 or $ns^2(n-1)d^{10}np^5$	
group 0 (inert gases)	ns^2np^6 or $ns^2(n-1)d^{10}np^6$	

You can see that the elements in each group have pretty much the same outermost electronic configuration. Since it is these outermost or valence electrons that are involved in bonding and in chemical reactions with other atoms, elements in the same group ought to have similar, but not identical, properties.

Elements in the same period have their outermost valence electrons in the same valence shell. Thus, in period 4 the fourth valence shell ($4s3d4p$) is being filled with electrons as one proceeds across the table until group 0, the inert gases, is reached. The length of a period is governed by the number of electrons needed to fill a valence shell. In group 0, the valence shell is filled with paired electrons. This configuration, $ns^2 \cdots np^6$, is an extremely stable, or low-energy, one. Atoms with the inert gas configuration are unusually unreactive or inert. This fact has important bearings on chemical combination.

The Roman group numbers have some relation to the number of electrons in the valence shell [if $(n-1)d$ and $(n-2)f$ electrons are ignored], but otherwise these group designations are not particularly important.

Hydrogen is a little hard to place in the periodic table, because it can behave like the alkali metals in group IA by losing its electron to form a positive ion or like the halogens in group VIIB by adding or sharing an electron. Here it is placed in group IA only. It is sometimes placed by itself above the table.

For the most part, we shall be concerned with the so-called "representative" elements. These are the elements in which the orbitals of maximum principal quantum number are incompletely filled while all levels of lesser n are filled. This covers anything from ns^1 through ns^2np^5. These elements are quite near the edges of the table and include the elements C, O, H, N, P, and S, which constitute the bulk of living matter.

The two lines of elements $_{58}$Ce to $_{71}$Lu and $_{90}$Th to $_{103}$Lw below the table fill the sixth and seventh periods after $_{57}$La and $_{89}$Ac, respectively. They do not fit into the table well, and their placement is a compromise. Other forms of the table have attempted to solve this position problem. These two series of elements are, as inspection of the table of electronic configurations shows, formed by entry of electrons into underlying 4f and 5f levels, and this accounts for the length of the sixth period and the potential length of the seventh period. These elements are called *inner transition* elements or *rare earths*. We shall not discuss them further except to say that their chemical properties are very similar to each other as a result of their similar outer electron configuration and their similar size.

2.31 TRENDS IN THE PERIODIC TABLE

The values of selected properties of atoms change as we go across the periods and down through the groups in the table. The properties in question are (1) atomic size, (2) ionization energy, and (3) electron affinity. The latter two were discussed in Section 2.11. A knowledge of the trends of these properties is helpful in understanding chemical bonding.

Atomic Size

The size of an atom is determined by the extent of the electron cloud surrounding it. In general, the size of this cloud is determined by the electron density distribution of the outermost or valence electrons; this may be estimated from the radial probability density diagrams of Figure 2.29. However, it is not possible to make a good estimate of the size of an atom from such plots, especially in the more complex atoms, because the innermost electrons tend to shield or screen valence electrons from the full influence of the nuclear charge.

Fortunately, there are numerous experimental techniques (X-ray, neutron and electron diffraction, spectroscopy, etc.) that permit determination of the radius of atoms corresponding to their "just touching" distances. For example, it is not difficult to measure the distance between two atoms, such as the Cl—Cl distance in Cl_2, and divide that distance in two to get the chlorine atom radius.

Table 2.11 gives values for some atomic radii. Going across the table from left to right in any period, the radius decreases with increasing atomic number. In any group, the radius increases going from top to bottom as atomic number increases. These trends can be summarized by arrows showing increasing radii, as in Figure 2.38.

Table 2.11
Atomic (covalent) radii, Å

H 0.37							He 0.4–0.6
Li 1.23	Be 0.89	B 0.80	C 0.77	N 0.7	O 0.74	F 0.72	Ne 0.70
Na 1.57	Mg 1.36	Al 1.25	Si 1.17	P 1.10	S 1.04	Cl 0.99	Ar 0.94
K 2.03	Ca 1.74	Ga 1.25				Br 1.14	Kr 1.09
Rb 2.16						I 1.33	Xe 1.30
Cs 2.35							Rn 1.4–1.5

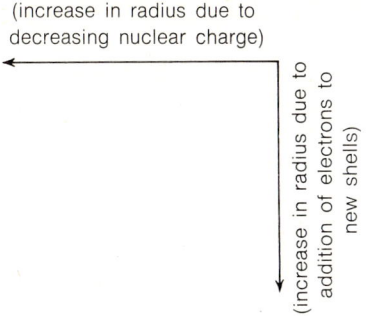

Figure 2.38. Periodic trends in atomic radii.

The size decreases through a given horizontal period because of increasing nuclear charge. While it is true that an extra valence electron is added to a level or orbital as one goes from one element to another, an extra proton is also added to the nucleus. Electrons that go into the same valence shell as one progresses through a period do not expand the electron charge cloud about the atom to any degree. In the absence of an increased nuclear charge, these electrons would all be at about the same distance from the nucleus. With the increased nuclear charge, they are drawn in to the nucleus, leading to a decrease in radius.

In going down through a group, the new electrons have gone into a higher valence shell, in which the location of the electron charge cloud is quite a bit farther from the nucleus than that of the electrons in the next lower valence shell. The increased nuclear charge is not sufficient to counteract the expansion due to occupation of higher valence shells.

Ionization Energies

The ionization energy of an atom has been defined as the energy necessary to remove an electron from an isolated gaseous atom to give a positively charged gaseous ion. More specifically, the first ionization energy refers to the energy needed to remove the outermost, or most loosely bound, valence electron from an atom. This energy is given in electron volts (eV), but only the relative values need concern us now.

First ionization energies for some elements were shown in Table 2.7. The *trends* in IE are shown in Figure 2.39. The ionization energy of the outermost electron depends on

1. nuclear charge, provided outer electronic configurations are similar,
2. atomic size,
3. screening effect of inner electrons, and
4. penetration of the outermost electron (ns penetrates more than np, np more than nd, etc.).

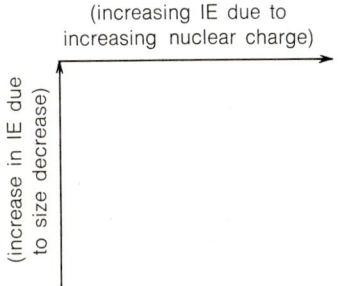

Figure 2.39. Periodic trends in ionization energies.

These factors do not all work in the same direction. As we go across any period, the ionization energy generally increases. This increasing difficulty of removing the electron is due primarily to increasing nuclear charge with the related decrease in size of the atom.

The decrease at B ($1s^2 2s^2 2p^1$) and at Al is due to the lower penetration of the p electron as compared to an s electron.

The decrease at O and S is really imaginary and is due to a larger than usual increase at N ($1s^2 2s^2 2p^3$) and P ($1s^2 2s^2 2p^6 3s^2 3p^3$) because of the increased stability of a half-filled set of p orbitals.

The decreasing ionization energy as one goes down a group is due to increasing atomic size together with increased shielding of the nucleus by inner electrons.

The important thing to see is that those atoms to the left of the periodic table and toward the bottom tend to lose their outermost electron most easily. This is characteristic of metals and so-called *electropositive* elements in general.

Electron Affinity

The same factors that determine the ionization energy of an atom also influence the electron affinity, the energy evolved when an electron is added to the neutral gaseous atom to form a gaseous negative ion. Some electron affinities were given in Table 2.8. Periodic trends are not too evident, but atoms with the greatest tendency to gain electrons and the highest electron affinities lie at the right of the table, close to the inert gases. Such atoms are often called highly *electronegative*. A fuller definition of electronegativity will appear in Chapter 4.

Free atoms are relatively rare in our environment. Combination of two or more atoms into molecules or other groupings bound together by various forces are much more common, and they determine in a more direct manner the properties of the environment. We now turn to the subject of chemical bonding and molecular structure to complete the groundwork for our entry into the environment of life.

2.32 SUMMARY

An atom is composed of a compact, positively charged nucleus surrounded by electrons. The nucleus is made up of protons and neutrons. In a neutral atom, the number of nuclear protons and the number of extranuclear electrons are equal. The number of protons in the nucleus, the nuclear charge, is called the atomic number Z of the atom. Each chemical element is characterized by a specific atomic number. The chemical behavior of an atom is determined by the number and arrangement of the extranuclear electrons, so that atoms of the same element have identical chemical properties.

The mass number A of an atom is equal to the sum of the numbers of protons and neutrons in the nucleus. The mass number is an integer and is useful in distinguishing isotopes, atoms of the same element having different numbers of neutrons in the nucleus. Unlike the mass number, the atomic (or isotopic) mass of an atom is not a whole number except for the atomic mass of the ^{12}C isotope. Carbon-12 is arbitrarily assigned an atomic mass of exactly 12 atomic mass units (amu), and the masses of all other atoms are given relative to that of the ^{12}C isotope. The atomic weight of an element is the average mass of an atom of the element in the naturally occurring mixture of isotopes of that element.

A gram-atom, or gram-atomic weight, of any element is the mass in grams numerically equal to its atomic weight. Similarly, the gram-molecular weight of any substance is a mass of the substance equal to the molecular weight expressed in grams. Avogadro's number, 6.02×10^{23}, is the number of atoms in 1 g-atom of any element, or the number of molecules in 1 gram-molecular weight of any substance or, in general, the number of formula units in 1 mol of any substance.

In general, chemical reactions or processes can take place spontaneously if the overall energy change for the process is exoergic, that is, if energy is released. Using arguments based on the energy changes when one atom loses an electron and another gains an electron, and on the energy change when the oppositely charged ions approach each other, it is possible to show that simple ion–ion electrostatic forces may hold some atoms (or ions) together. Of the 105 known elements, only a very few are involved in living organisms to any great extent. Most of those that do participate in living systems do not lose or gain electrons to form substances held together by ionic bonds. Because the chemical behavior of atoms is determined by the electrons around the nucleus, some other explanation of the atomic combinations prevalent in life must be sought in the behavior of electrons in atoms. This explanation is based on the quantum theory of the electronic structure of atoms.

Prior to 1900, there were a number of commonly observed phenomena that could not be explained with classical physical theory. Among these were the stability of atoms, the characteristic line spectra of atoms, the photoelectric effect, and the properties of the radiation emitted by heated bodies. The features of the latter phenomenon were finally interpreted by Max Planck, but only after making the bold and unprecedented assumption that the radiating constituents of a heated body were confined to a set of discrete or quantized energy levels. Radiation is not emitted continuously, but only in definite packets or quanta.

Later, Bohr succeeded in deriving the observed line spectrum of hydrogen by postulating that the hydrogen atom, or its electron, exists in certain energy states in which it does not radiate energy. Radiation occurs only when the atom makes a transition from one energy state to a lower energy state. Despite Bohr's success with the hydrogen atom, other atomic systems did not yield to Bohr's arbitrary assumptions.

Radiation is known to exhibit a dual wave–particle nature. According to the wave theory of light, radiation is composed of electromagnetic waves moving through space. Radiation is characterized by the wavelength, frequency, and speed of propagation of these waves. According to the particle or photon theory of radiation, the energy of light is carried in small packets called photons or light quanta of energy $h\nu$. These two theories of light are complementary. Some phenomena, such as interference and diffraction, can be explained only by the wave theory. Other phenomena are best explained in terms of the photon theory.

De Broglie reasoned that if light had a dual nature, so also might matter. It was soon shown experimentally that particles of matter, particularly very light particles like electrons, are endowed with wavelike properties, the wavelength of the matter wave being $\lambda = h/mv$. Upon pursuing the analogy between electron waves and ordinary waves, a wave equation was developed that related the form of the electron wave to the allowed energy levels of the electron. The wave function describing the electron wave is related to the probability of finding the electron in a given region of space. An electron, when regarded as a wave, is no longer localized at a point, but is more or less spread out through space.

In an atom, the wave function for an electron is called an atomic orbital. The atomic orbital specifies both the energy of the electron and the region occupied by the electron. Each atom has a series of atomic orbitals or set of allowed energy levels that may be occupied by electrons subject to restrictions on the permitted values of a set of four quantum numbers. In a multielectron atom, the electrons fill the allowed orbitals to determine the electronic configuration of the atom. These configurations control the chemical properties of the atom and are reflected in the periodic table.

Questions and Problems

1. Which of the following represent isotopes? X or Y may be the same or different; identify the element symbolized by X or Y.
   a. $^{54}_{26}X$ and $^{58}_{26}Y$
   b. $^{64}_{30}X$ and $^{64}_{28}Y$
   c. $^{31}_{15}X$ and $^{27}_{13}Y$

2. What is the mass in grams of the following?
   a. one ^{35}Cl atom
   b. an average Cl atom in natural chlorine
   c. an electron
   d. 3.01×10^{23} molecules of natural Cl_2
   e. 3.01×10^{23} atoms of ^{35}Cl

3. How many molecules are there in the following?
   a. 1.5 mol of H_2 gas
   b. 1.5 mol of glucose, $C_6H_{12}O_6$
   c. 8 g of oxygen, O_2
   d. 8 g of methane, CH_4
4. How many moles are represented by the following amounts of material? Specify the particle unit in each case.
   a. 85.16 g of ammonia, NH_3
   b. 9.03×10^{21} atoms of nitrogen
   c. 1.5×10^{22} molecules of nitrogen, N_2
   d. 55.5 g of $CaCl_2$ (Ca^{2+} $2Cl^-$)
5. As practice in unit conversions, calculate the velocity of light c in miles per hour. Take $c = 3.00 \times 10^{10}$ cm sec^{-1}. There are 5280 ft in 1 mile and 2.54 cm in 1 in. Observe significant figures (See Appendixes A.2 and A.4).
6. A sample of calcium carbonate, $CaCO_3$, has a mass of 25 g.
   a. How many gram-atoms of calcium, carbon, and oxygen are in the sample (see Appendix B.1)?
   b. How much of each element, in grams, is contained in the 25-g sample?
7. If the symbol or formula of an element, molecule, or compound represents 1 mol or gram-formula weight of that substance, interpret the following equation for a chemical reaction in terms of
   a. moles
   b. grams of reactants and products

$$2S + 3O_2 \longrightarrow 2SO_3$$

   c. How many grams of sulfur trioxide can be prepared from 8.016 g of sulfur? (See Appendix B.2 for a discussion of calculations based on chemical equations.)
   d. How many grams of oxygen are required to react with the sulfur?
8. The spectral region where the human eye is most sensitive or the region of maximum visibility is centered at about 5560 Å. Would sodium vapor, mercury vapor, or neon lights be best for street lamps? (See Figure 2.21.)

9. Consider Bohr's postulates,

$$mv2\pi r = nh$$

and

$$\Delta E = E_2 - E_1 = h\nu$$

together with the equality of the attractive force by the proton nucleus on the electron, that is, e^2/r^2, to the so-called centripetal force required to keep any object in a circular orbit of radius r, mv^2/r, and

$$E_{electron} = \tfrac{1}{2}mv^2 + \left(-\frac{e^2}{r}\right)$$

   a. Show that the allowed radii of the electron orbits are

$$r = \frac{n^2}{me^2}\left(\frac{h}{2\pi}\right)^2$$

   b. Show that the allowed energies are

$$E_n = -\frac{2\pi^2 e^4 m}{h^2 n^2}$$

   c. Show that

$$\Delta E = E_{n_2} - E_{n_1} = h\nu$$

   leads to

$$\nu_H = \frac{2\pi^2 e^4 m}{h^3}\left(\frac{1}{n_1{}^2} - \frac{1}{n_2{}^2}\right)$$

   or

$$\frac{1}{\lambda} = R\left(\frac{1}{n_1{}^2} - \frac{1}{n_2{}^2}\right)$$

   where

$$R = \frac{2\pi^2 e^4 m}{h^3 c}$$

10. How would the uncertainty principle and its consequences be changed in a universe with a much larger value for h?
11. Consider the wave functions ψ_1, ψ_2, and ψ_3 for a particle in a one-dimensional box (Figure 2.26). For each of these wave functions, where are the most probable positions of the particle? Sketch particle density cloud figures for the ψ's to illustrate your answer.
12. In the result for the particle-in-a-box, why do we not consider E_n and ψ_n for $n = 0$ as acceptable values for discussion?
13. If the appropriate values for mass, charge, and orbital radius are used, some idea of the velocity of an electron in a classical atom can be obtained. A typical value for r, the orbital radius (and effectively the radius of the atom) is 1 Å $= 1 \times 10^{-8}$ cm. With $m = 9.1 \times 10^{-28}$ g and $e = 4.8 \times 10^{-10}$ electrostatic

units (esu), calculate the electron's velocity in miles per hour from the equality

$$\frac{mv^2}{r} = \frac{e^2}{r^2}$$

14. How many $l = 3$ or f orbitals are there? How many electrons will they hold? Show the combinations of quantum numbers for the f orbitals.
15. According to $\epsilon = h\nu = hc/\lambda$, radiation of various frequencies and wavelengths can also be characterized by the energy of its photons. With h in erg seconds, $c = 3 \times 10^{10}$ cm sec^{-1}, and λ in centimeters, ϵ is in units of ergs. Calculate the photon energy of
    a. infrared radiation with $\lambda = 5000$ nm
    b. visible radiation with $\lambda = 5000$ Å (green)
    c. ultraviolet radiation with $\lambda = 500$ Å
    Note that the photon energy of the radiation also corresponds to the difference in energy, ΔE, of the energy levels between which the emitting or absorbing system makes a transition.

16. Assume that the visible radiation of $\lambda = 5000$ Å emitted or absorbed in the previous problem is associated with an $n = 2$ to (or from) an $n = 1$ transition of an electron in a one-dimensional box. What is the width of the box?
17. Calculate the wavelength of the radiation associated with an electronic transition between the energy states $E_{n=2}$ and $E_{n=4}$ of the hydrogen atom. The energy calculation involves the electron charge e. The appropriate value of e is 4.80×10^{-10} electrostatic units (esu). An esu has the fundamental units of g$^{1/2}$ cm$^{3/2}$ sec^{-1}.
18. Give the complete electronic configurations of the following elements, showing all unpaired electrons: $_{13}$Al, $_{16}$S, $_{20}$Ca, $_{25}$Mn, $_{36}$Kr, $_{47}$Ag, $_{53}$I.
19. Inspect Figures 2.29, 2.30, and 2.32, and discuss similarities and differences in the plots of an atomic orbital when presented as $\psi(r)$, $4\pi r^2\psi^2(r)$, and $\psi^2(r)$.

Answers

2. a. 5.8×10^{-23} g
   b. 5.9×10^{-23} g
   c. 9.1×10^{-28} g
   d. 35.5 g
   e. 17.5 g
3. a. 9.0×10^{23}
   b. 9.0×10^{23}
   c. 1.5×10^{23}
   d. 3.0×10^{23}
4. a. 5.00 mol of NH_3 molecules
   b. 0.015 mol (or g-atom) of N atoms
   c. 0.025 mol of N_2 molecules
   d. 0.50 mol (or gram-formula weight) of $CaCl_2$ or 0.50 mol of Ca^{2+} ions and 1.0 mol of Cl$^-$ ions

5. 6.72×10^8 mph
6. a. 0.25 g-atom of Ca, 0.25 g-atom of C, 0.75 g-atom of O
   b. 10 g of Ca, 3 g of C, 12 g of O
7. b. 20 g of SO_3
   c. 12 g of O_2
13. about 3.6×10^6 mph (1.6×10^8 cm sec^{-1})
15. a. 4.97×10^{-13} erg
    b. 4.97×10^{-12} erg
    c. 4.97×10^{-11} erg
16. $l = 6$ Å
17. 4870 Å

References

FITNESS

Blum, H. F., "Time's Arrow and Evolution," 3rd ed., Princeton University Press, Princeton, N. J., 1968.
    Chapter VI deals specifically with fitness. The book as a whole concerns the processes of organic evolution at the molecular level and is well worth reading in conjunction with later chapters in this text.

Henderson, L. J., "The Fitness of the Environment," Macmillan, New York, 1913. (Republished by Beacon Press, Boston, 1958.)
    The subtitle of this book is "An Inquiry into the Biological Significance of the Properties of Matter."

Needham, A. E., "The Uniqueness of Biological Materials," Pergamon, Elmsford, N. Y., 1965.

> A descriptive account of the comparative chemical fitness of all elements and compounds of biological importance, with emphasis on the biological function of elements, ions, and compounds. Contains sections on the higher organization of living forms and on the uniqueness of life on earth.

CLASSICAL ATOMIC STRUCTURE

Ashford, Theodore A., Rutherford's Theory of the Nuclear Atom, in "The Mystery of Matter," Louise B. Young, Ed., Oxford University Press, New York, 1965, Part 2, pp. 84–95.

Feinberg, Gerald, "Ordinary Matter," *Scientific American*, **216,** 126 (May 1967).

LIGHT

Clayton, Roderick K., "Light and Living Matter, Vol. 1: The Physical Part," McGraw-Hill, New York, 1970.

Scientific American, **219** (September 1968).

> Entire issue devoted to light.

QUANTUM THEORY OF ATOMIC STRUCTURE

Christiansen, G. S., and Garrett, P. H., "Structure and Change: An Introduction to the Science of Matter," Freeman, San Francisco, 1960, Chapters 25, 26, 28, 29, 31.

Hoffman, Banesh, "The Strange Story of the Quantum, 2nd ed, Dover, New York, 1959.

Pimentel, George C., and Spratley, R. D., "Chemical Bonding Clarified Through Quantum Mechanics, Holden-Day, San Francisco, 1969, Chapters 1, 2.

Weisskopf, Victor, Atomic Structure and Quantum Theory, in "The Mystery of Matter," Louise B. Young, Ed., Oxford University Press, New York, 1965, Part 3, pp. 95–121.

OTHER

Kieffer, William F., "The Mole Concept in Chemistry," Van Nostrand Reinhold, New York, 1962.

Chemical Bonding and Molecules

Pangloss taught metaphysico-theologo-cosmolonigol-ogy. He proved admirably that there is no effect without a cause and that, in this best of all possible worlds, . . . "things cannot be otherwise; for, since everything is made for an end, everything is neces-sarily for the best end. Observe that noses were made to wear spectacles; and so we have spectacles. . . ."
Voltaire—*Candide*

3.1 ENERGY AND INERT GAS CONFIGURATIONS

We are now ready to tackle the formation of bonds between atoms in more detail than before. In doing this, we shall consider the energy levels of atomic orbitals as well as their geometric form.

Any time two or more atoms come together and form a stable combina-tion, the energy of this combination must be less than the combined energies of the separate atoms. This decrease in energy is the essence of chemical bond formation, a reflection of the normal tendency of things to seek, and stay in, a lower energy state. The simple energy picture is shown in Figure 3.1. To simplify the discussion, we shall consider interactions involving only two atoms, although the same general principles apply to cases of three or more atoms.

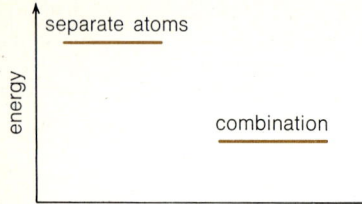

Figure 3.1. Chemical bonds form because of a decrease in potential energy as the atoms approach one another. There may be various pathways between the atoms and the combination, so no connecting arrow is drawn here.

The formation of ions has been mentioned briefly in Chapter 2. There it was pointed out that atoms with low ionization energies have a tendency to lose electrons to form positive ions, while atoms with high electron affinities tend to gain electrons and form negative ions. Now that we have information about orbital energies and valence shell energies, additional insights can be obtained into this process.

The first thing to notice is that the inert gases in group 0 of the periodic table have an electronic configuration that confers a special stability to these atoms. They tend neither to gain nor to lose electrons. This can be seen from the high values of their ionization energies (Table 2.7) and electron affinities (the latter are assumed to be zero for inert gases). But another way to look at their great stability is in terms of the orbital energy levels of the electrons in the inert gases. Figure 2.35 showed that the inert gases are characterized by completely filled valence shells. Because this configuration, with all of the electrons paired, is so stable, it would appear reasonable that other atoms would tend to approach the low-energy electronic configuration of the inert gases. To a large degree, elements behave as if they want to attain the inert gas configuration.

3.2 TRANSFER OF ELECTRONS AND IONIC BONDING

One obvious way an atom may attain the inert gas electron configuration is by the outright gain or loss of a sufficient number of electrons so that it ends up with the same number of electrons (but not protons or neutrons) as an inert gas.

Take lithium, for example. By the loss of 1 electron, lithium changes to the lithium ion, with one positive charge:

$$\text{Li } (1s^2 2s^1) \longrightarrow \text{Li}^+ (1s^2) + e^-$$

The Li^+ ion has the electronic configuration of helium. Other atoms having 1 to 3 electrons outside the closed-shell arrangement of a nearby inert gas may also lose electrons to attain an inert gas configuration.

$$\text{Na } (1s^2 2s^2 2p^6 3s^1) \longrightarrow \text{Na}^+ (1s^2 2s^2 2p^6) + e^-$$
$$\text{Na}^+ \text{ has the Ne configuration}$$

$$\text{Mg } (1s^2 2s^2 2p^6 3s^2) \longrightarrow \text{Mg}^{2+} (1s^2 2s^2 2p^6) + 2e^-$$
$$\text{Mg}^{2+} \text{ has the Ne configuration}$$

$$\text{Al } (1s^2 2s^2 2p^6 3s^2 3p^1) \longrightarrow \text{Al}^{3+} (1s^2 2s^2 2p^6) + 3e^-$$
$$\text{Al}^{3+} \text{ has the Ne configuration}$$

$$\text{K } (1s^2 2s^2 2p^6 3s^2 3p^6 4s^1) \longrightarrow \text{K}^+ (1s^2 2s^2 2p^6 3s^2 3p^6) + e^-$$
$$\text{K}^+ \text{ has the Ar configuration}$$

The formation of positive ions by the complete loss of electrons is limited to atoms with atomic numbers 1 to 3 greater than that of the nearest inert gas atom. That is, positive ions (cations) with a charge greater than $+3$ are rarely formed.

Some atoms having a nearly full valence shell can attain an inert gas configuration by gaining electrons. Examples are

$$\text{F } (1s^2 2s^2 2p^5) + e^- \longrightarrow \text{F}^- (1s^2 2s^2 2p^6) \quad \text{(Ne configuration)}$$
$$\text{Cl } (1s^2 2s^2 2p^6 3s^2 3p^5) + e^- \longrightarrow \text{Cl}^- \quad \text{(Ar configuration)}$$
$$\text{S } (1s^2 2s^2 2p^6 3s^2 3p^4) + 2e^- \longrightarrow \text{S}^{2-} \quad \text{(Ar configuration)}$$

Simple negative ions (anions) with a charge greater than -3 are rare.

The ions formed from neutral atoms have their own special properties. These properties are vastly different from those of the atom. The fact that the chemical behavior of the ion is not the same as the behavior of the neutral atom is sometimes ignored by many nonchemists, perhaps because the atom and ion are often incorrectly called by the same name (see Appendix C). Thus, one may hear the opponents of water fluoridation say that, because fluor*ine*, whether atomic (F) or molecular (F_2), is deadly, so is the fluor*ide* ion F^- that is put into water to reduce tooth decay in the young. There may be reasons to avoid too large a concentration of fluoride ion in drinking water, but this is not one of them. After all, hardly anyone worries about sprinkling sodium ions and chloride ions on their food, and each of these ions is derived from a dangerous neutral atom.

Another thing must be considered. When an atom loses an electron, this electron must go somewhere. So, in a simple way, for an atom to lose an electron another atom must gain that electron. The electron usually goes to an atom with a high electron affinity, an atom that will accept it with the evolution of energy. It is this evolved energy that can be thought of as pulling the electron away from the original atom. The joint formation of ions by two atoms should occur most readily between elements with low ionization energies and elements with high electron affinities. The energetics of this process were noted in Chapter 2, where it was pointed out that the electron affinity is usually not sufficient to counterbalance the ionization energy, and other factors must be considered. When sufficient energy is not available from the electron affinity of one atom to supply the ionization energy of the other atom, extra energy may become available because of the potential energy of attraction between ions of unlike charges.

To summarize, atoms close to the inert gases may achieve the stable inert gas configurations by the complete loss or gain of electrons. Even here, ion formation is usually subject to the availability of other sources of energy besides the electron affinity of one atom. However, the energy demands of high ionization energies and the energy gains attributable to low electron affinities are such that elements several places removed from the inert gases in atomic number cannot achieve inert gas arrangements by electron transfer.

Once the oppositely charged ions are formed, they attract each other by simple electrostatic interaction. This electrostatic attractive force is the bond that holds the ions together; it is the *ionic*, or *electrovalent*, bond. The number of electrons gained or lost by an atom, or the charge on the resulting ion, is called its *ionic valence* and can be positive or negative.

3.3 COVALENT BONDING   There is another way that atoms can attain the stability of the inert gas configuration; this involves a sharing of electrons between two or more atoms rather than a complete transfer of electrons as in purely ionic bonding. This type of electron sharing must be invoked to explain the existence of bonds between atoms that do not differ greatly in ionization energy and electron affinity. In particular, it is needed to describe the bond between identical atoms, as in H_2 or Cl_2. Such shared electron bonds are called *covalent bonds*.

Covalent bonds are extremely important in our discussion, because they are the bonds that hold all the structural materials of living organisms together. Carbon, hydrogen, nitrogen, oxygen, phosphorus, and sulfur are among the elements predominantly involved in this covalent bonding. Ions such as Na^+

and Cl^- play little part in structural molecules, but they are involved in the operations taking place within these structures.

The formation of a covalent chemical bond between two atoms comes about by the sharing of an electron pair between the two atoms. To gain insight into electron pair sharing, let us first look at the hydrogen molecule, H_2.

The electronic configuration of the hydrogen atom is $1s$. If, somehow, this atom could get one more electron in its $1s$ orbital it would have the helium configuration ($1s^2$). But if we want to form a bond with another hydrogen atom, there is no reason why one hydrogen atom should give up its one electron to an identical atom to form H^+H^-. By sharing their electrons, both atoms attain the helium configuration.

$$\left.\begin{array}{l} \text{H } 1s^1 \\ \text{H } 1s^1 \end{array}\right\} 1s^2 \text{ He}$$

A few more examples are shown in Figure 3.2. In F_2, by sharing the $2p_z$ electrons, each F atom can attain the Ne configuration. The same applies to Cl_2 except that $3p_z$ electrons are shared.

Because the electrons in a covalent bond are almost always shared in pairs, the covalent bond is often called an *electron pair bond*. This does not mean

$$\text{F } 1s^2 2s^2 2p_x{}^2 2p_y{}^2 \;\boxed{2p_z{}^1}$$
$$\text{F } 1s^2 2s^2 2p_x{}^2 2p_y{}^2 \;\boxed{2p_z{}^1}$$

(a) F—F. The dash represents a shared pair in the F_2 molecule.

$$\text{N } 1s^2 2s^2 \;\boxed{2p_x{}^1 \;|\; 2p_y{}^1 \;|\; 2p_z{}^1}$$
$$\text{N } 1s^2 2s^2 \;\boxed{2p_x{}^1 \;|\; 2p_y{}^1 \;|\; 2p_z{}^1}$$

(b) N≡N, a triple bond.

$$\begin{array}{ll} \text{N} & 1s^2 2s^2 \;\boxed{2p_x{}^1 \;|\; 2p_y{}^1 \;|\; 2p_z{}^1} \\ \text{H}_1 & \boxed{1s^1} \\ \text{H}_2 & \boxed{1s^1} \\ \text{H}_3 & \boxed{1s^1} \end{array}$$

(c) NH_3 or H—N—H with three single covalent bonds.
$|$
H

$$\text{P}\quad 1s^2 2s^2 2p^6 3s^2 3p_x 3p_y 3p_z \qquad 3s \text{ unpaired and promoted}$$

$$\begin{array}{l} \text{P* } 1s^2 2s^2 2p^6 \;\boxed{3s^1 \;|\; 3p_x{}^1 \;|\; 3p_y{}^1 \;|\; 3p_z{}^1 \;|\; 3d^1} \\ \text{5(Cl)} \qquad\qquad\quad \boxed{3p_z{}^1 \;|\; 3p_z{}^1 \;|\; 3p_z{}^1 \;|\; 3p_z{}^1 \;|\; 3p_z{}^1} \end{array}$$

(d) PCl_5, with five covalent bonds.

Figure 3.2. Covalent bond formation by electron sharing.

that only one pair of electrons may be involved between two atoms; more than one covalent bond may join two atoms. Consider the nitrogen atom:

$$\text{N } 1s^2 2s^2 2p_x{}^1 2p_y{}^1 2p_z{}^1$$

If, somehow, the 3 unpaired $2p$ electrons could pair up with 3 other electrons, this nitrogen atom could attain the inert gas configuration of neon. Let us see how this can happen, first with another nitrogen atom. With a second nitrogen atom, there are 3 shared electron pairs, as shown in Figure 3.2(b). In the N_2 molecule, there are three electron pair or covalent bonds; and this is an example of a multiple bond, in this case a triple bond. Other elements, especially carbon, are often joined by double bonds.

One nitrogen atom can also combine with three hydrogen atoms, as in Figure 3.2(c). The nitrogen would get its closed neon shell, and each hydrogen would get a helium configuration. Ammonia, NH_3, with three single covalent bonds, would be formed.

In all of these examples, each atom has 1 or more unpaired electrons; and the covalent bonds have been formed by the pairing of unpaired electrons, one from each atom. The number of unpaired electrons an atom has available for bonding may change if paired electrons can be unpaired without much addition of energy. This may occur if no change in the principal quantum number of the electrons involved is required. For example, nitrogen, N ($1s^2 2s^3 2p_x{}^1 2p_y{}^1 2p_z{}^1$), is limited to 3 unpaired electrons, because there are no more $n = 2$ orbitals to accommodate an unpaired $2s$ electron. However, phosphorus, P (Ne $3s^2 3p_x{}^1 3p_y{}^1 3p_z{}^1$), may have 3 (Ne $3s^2 3p_x{}^1 3p_y{}^1 3p_z{}^1$) or 5 (Ne $3s^1 3p_x{}^1 3p_y{}^1 3p_z{}^1 3d^1$) unpaired electrons because of the availability of $3d$ orbitals to accept an unpaired and promoted $3s$ electron. An example of a phosphorus compound with five covalent bonds is PCl_5 in Figure 3.2(d).

Covalent bonds are not always formed by a contribution of one electron from each atom. For instance, in the ammonium ion, $NH_4{}^+$, nitrogen shares four pairs of electrons with the hydrogens. It is said to have a covalence of 4 and is not limited to a covalence of 3, as in NH_3. In $NH_4{}^+$, one of the N—H bonds was formed by the sharing of the 2 paired nitrogen $2s$ electrons with a hydrogen ion. The formation of this bond, often called a coordinate covalent bond, will be discussed in Section 3.13.

3.4 THE RULE OF EIGHT   The ideas we have just discussed may be summarized by a rule known as the *rule of eight* (or octet rule), which states that when an atom combines with another it loses, gains, or shares electrons in number sufficient for it to obtain a closed, inert gas type of configuration of 8 electrons (except for atoms close to helium).

Examples of the Rule of Eight   *Ionic:* Na^+Cl^- The sodium atom transfers its $3s$ electron to a chlorine atom. The resulting sodium ion has the neon configuration, and the chloride ion has the argon configuration.

$$Na^+ \ 1s^2(2s^2 2p^6)$$
$$Cl^- \ 1s^2 2s^2 2p^6(3s^2 3p^6)$$

The parentheses enclose the outer configuration of 8 electrons.

Ionic: Mg^{2+} $2Cl^-$ Similarly, a magnesium atom may transfer its two $3s$ electrons to two chlorine atoms, one electron to each chlorine, to form ionic magnesium chloride, $MgCl_2$.

$$Mg^{2+} \; 1s^2(2s^22p^6)$$
$$2Cl^- \; 1s^22s^22p^6(3s^23p^6)$$

Illustrations of covalent bond formation where the rule of eight is followed are included in Figure 3.2. An exception to the rule, PCl_5, is also shown. Similar availability of d orbitals leads to many other exceptions.

3.5 LEWIS ELECTRON DOT STRUCTURES

The bonding in many molecules and ionic compounds can be simply depicted through the use of Lewis electron dot structures. The *valence electrons* of an atom are represented by dots placed around the symbol of the element. The symbol represents the nucleus plus all other nonvalence electrons.

Dot structures for some neutral atoms are

hydrogen	·H
helium	·He·
lithium	·Li
beryllium	·Be·
boron	·B·
carbon	·C·
nitrogen	:N·
oxygen	:O:
fluorine	:F·
sodium	·Na
magnesium	·Mg·

When an atom forms an ion by the gain or loss of electrons, dots are added or subtracted and the ionic charge is noted.

positive hydrogen ion	H^+
hydride ion	$[\cdot H \cdot]^-$
lithium ion	Li^+
fluoride ion	$[:\ddot{F}:]^-$
magnesium ion	Mg^{2+}

The dot structure for a couple of ionic compounds is

sodium chloride	$Na^+[:\ddot{Cl}:]^-$
calcium fluoride	$Ca^{2+}[:\ddot{F}:]_2{}^-$

or $[:\ddot{F}:]^- Ca^{2+}[:\ddot{F}:]^-$

For compounds with covalent bonds, the electrons of each atom may be shown by dots or by two different symbols as in the examples below.

H_2 H:H   or   H ×H

H_2O H:Ö: or   H ×Ö:
           H   H

HCl   H:C̈l: or   H ×C̈l:

N_2 :N:::N   or   :N×××N×

Cl_2 :C̈l:C̈l: or   :C̈l ×C̈l×

The representation using the ×'s is helpful in showing that some electrons were originally associated with one atom. However, all electrons are alike, and it is impossible to determine the origin of shared electrons.

All of the above examples illustrate the rule of eight (or the rule of two in the case of atoms that attain the helium configuration). Positive ions have no dots around them, because they are understood to have an inert gas configuration.

To test the rule of eight, count valence electrons around each atom. In counting electrons around a given covalently bonded atom, a shared pair is thought of as belonging exclusively to that atom:

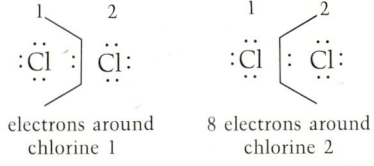

8 electrons around   8 electrons around
   chlorine 1   chlorine 2

Both chlorines satisfy the rule of eight.

      :N:::N: :N:::N:
$8e^-$ $8e^-$

Shared electron pairs are often represented by a dash, as was done earlier, and unshared pairs are often not written down. Examples are

H—C̈l: or   H—Cl

:N≡N: or   N≡N

The rule of eight is not without exceptions, as we have seen in the case of PCl_5. Other exceptions will be noted as more compounds are discussed. Despite the exceptions, the use of the rule of eight and of Lewis dot structures based on the rule will be found useful and reliable in the majority of compounds involving first- and second-period elements that we shall encounter.

3.6 THE WAVE MECHANICAL PICTURE OF BONDING

The previous discussion of covalent bonding does not really explain what happens to the electrons when a bond is formed, and it does not explain why such a bond holds two atoms together. Of course, one could say that since the two atoms get an inert gas configuration by sharing electrons, they are both in a lower or more stable energy state than that of the atoms taken separately. This is true, but it still explains little. To understand why the energy of the combination is lower, we must turn to wave mechanics and solve the Schrödinger equation for the problem of two atoms with their positive nuclei and all their electrons. We shall not use the Schrödinger equation here, but we shall use electron waves and, in particular, orbital wave functions of those atomic electrons involved in bond formation.

Let us return once more to the radial wave functions $\psi(r)$ for electrons in an atom. We shall focus our attention on the orbitals of $1s$ and $2p$ electrons for simplicity. In the redrawn plots of the radial wave functions $\psi(r)$ [not $4\pi r^2 \psi^2(r)$ or $\psi^2(r)$] shown in Figure 3.3, an entire cross section through the nucleus in the direction of greatest electron density of the orbitals is shown, along with the positive and negative signs indicating the direction of displacement or phase in the electron wave. Also shown are the boundary diagrams, again with signs. Remember, however, that when we are thinking, not about waves, but about the probability of finding an electron somewhere, we shall use $\psi^2(r)$, and since this quantity is always positive no signs need be specified.

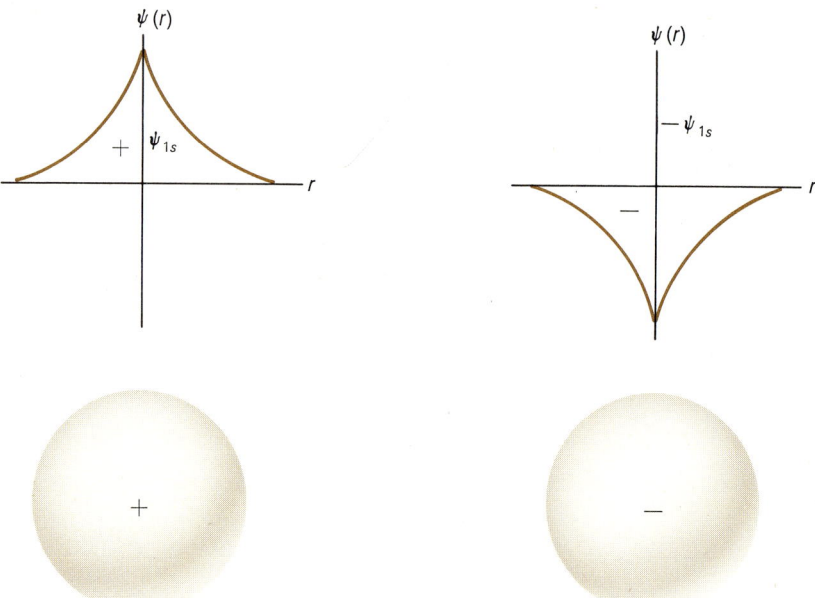

Figure 3.3. $1s$ atomic orbitals with opposite phases.

Now let us take two hydrogen atoms A and B, each with a $1s$ atomic orbital, and move these atoms closer and closer together. Two possibilities result, depending on whether the electron wave functions of the two orbitals have the same or the opposite phases. The two orbitals are shown in Figure 3.3.

Case 1. Same Phases At relatively large distances ($\sim$8 Å apart), the electron wave functions of the two orbitals do not overlap and we still have separate $1s$ orbitals. As the atoms come closer together, overlapping of the waves results in constructive interference [add the displacements $\psi_A(r) + \psi_B(r)$]. At the actual internuclear distance in H_2 ($r \sim 1$ Å), the situation is as shown in Figure 3.4. By squaring the resultant electron wave, you get an idea of the electron density distribution in the molecule [Figures 3.4(c) and 3.4(d)]. There is a pileup of electron density between the nuclei, and this is what holds the nuclei together. Why? There are at least two nonmathematical ways to answer this question. One way is based on arguments using plain classical electrostatic forces; the other way makes use of arguments stemming from the particle-in-a-box model.

In the first way, one could say that the presence of an electron charge density between the two positive nuclei decreases the repulsive electrostatic force between the nuclei by partially shielding them from each other. In

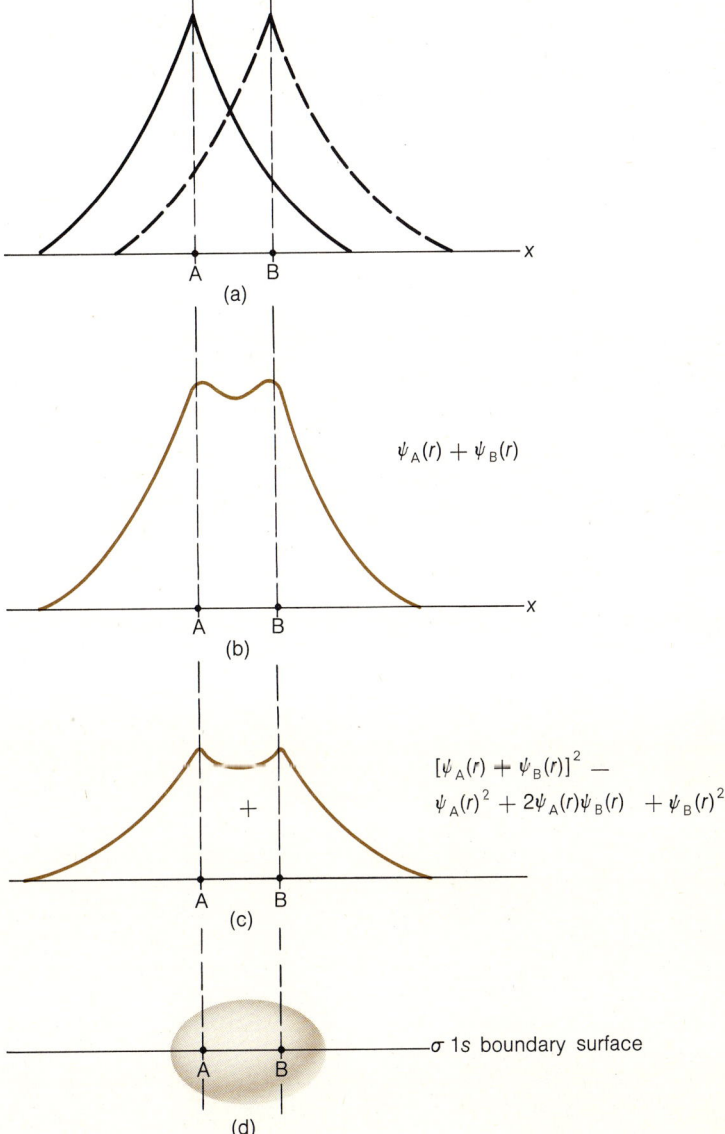

Figure 3.4. Constructive overlap of $1s$ atomic orbitals on atoms A and B. (a) Separate orbitals at the internuclear distance in H_2. (b) Resultant of the two orbitals. (c) Resultant probability density. (d) Boundary surface of (c). The term $2\psi_A(r)\psi_B(r)$ represents the additional electron density between the nuclei.

$\psi_A(r) + \psi_B(r)$

$[\psi_A(r) + \psi_B(r)]^2 -$
$\psi_A(r)^2 + 2\psi_A(r)\psi_B(r) + \psi_B(r)^2$

σ $1s$ boundary surface

addition, each electron now experiences attractive forces from two nuclei. To this extent, the covalent bond is merely a manifestation of the well-known electrostatic (Coulomb) forces and involves no mysterious wave mechanical concepts to explain the attractive force between the nuclei.

According to the particle-in-a-box model, an electron, originally confined to the region about a single hydrogen, now can move in a larger region encompassing both nuclei. So far as an electron is concerned, its box has become larger. Referring to the discussion of electron energy levels and box size, we see that the lowest, or ground state, energy level has a lower energy in a larger box. With hydrogen we are dealing with only 2 electrons. They can both go into the lowest energy level for the molecular box if they have opposite spins. Their total energy is lower in the molecule than in their separate atoms, even when electron and nuclear repulsion is considered (Figure 3.5). Therefore, the electronic energy of the molecule is lower than the original energy of two hydrogen atoms, and a stable molecule is formed. This explanation of the covalent bond is based on the fact that the electrons are no longer localized on one atom but are spread over two nuclei in the molecule.

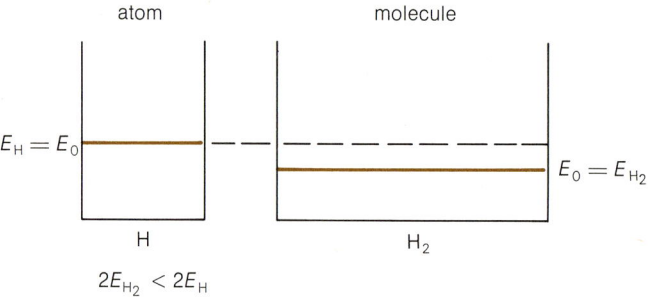

Figure 3.5. Electron energy levels in atoms and molecules. The total energy of two electrons in the molecule, $2E_{H_2}$, is lower than their total energy in the atoms, $2E_{H}$.

If the two hydrogen nuclei are forced closer together than the equilibrium distance, the electrons will once again become localized, and the energy of the combination will increase. In terms of the electrostatic interpretation, proton–proton repulsion will increase. The change of the potential energy with nuclear separation is shown by the lower curve in Figure 3.6, where the minimum occurs at the equilibrium separation.

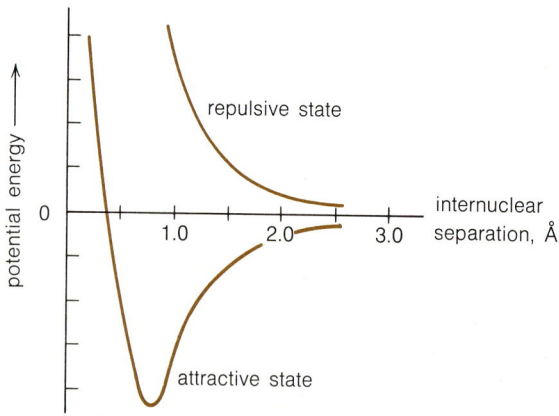

Figure 3.6. Potential energy of two hydrogen atoms as a function of their nuclear separation.

Case 2. Opposite Phases

Now, it is just as likely that the electron waves of the two $1s$ orbitals in the hydrogen atoms will have opposite phases. When the two atoms are brought closer together, destructive interference will take place, as shown in Figure 3.7. The result is a node or a region of zero electron density between the two nuclei. No stable or attractive bond results. The reason for the lack of a bond might again be roughly justified in two ways. On the basis of the particle-in-a-box model, the mutual cancellation of parts of the two atomic orbitals on each nucleus gives a smaller box and a higher energy for each electron. Alternatively, the region of zero electron density, the node, between the two nuclei results in less screening of the two positive nuclei from each other and more repulsion than in the previous case of the constructive interference. As Figure 3.6 shows, in this state there is no minimum at any internuclear separation to indicate the formation of a bond. At all distances, the two atoms repel each other.

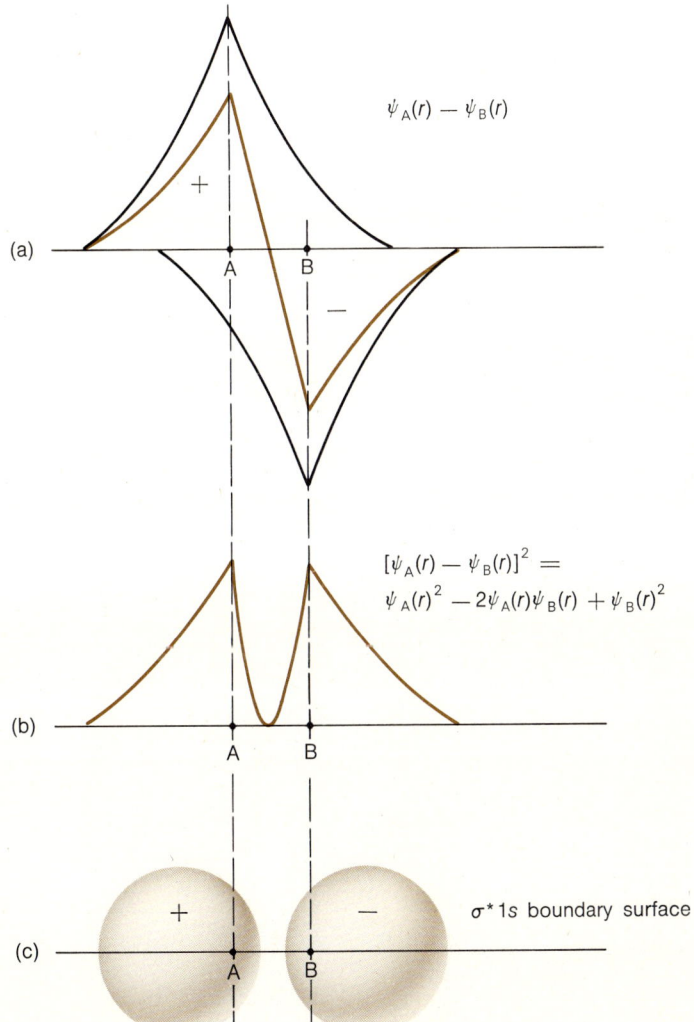

Figure 3.7. Destructive interference of $1s$ atomic orbitals. (a) The individual orbitals and their resultant. (b) Resultant probability density. (c) Boundary surface. The term $-2\psi_A(r)\psi_B(r)$ indicates the decrease in electron density between the nuclei.

(a) $\psi_A(r) - \psi_B(r)$

(b) $[\psi_A(r) - \psi_B(r)]^2 = \psi_A(r)^2 - 2\psi_A(r)\psi_B(r) + \psi_B(r)^2$

(c) $\sigma^* 1s$ boundary surface

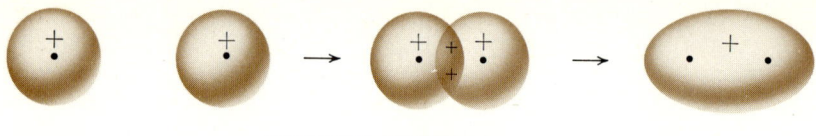

constructive interference

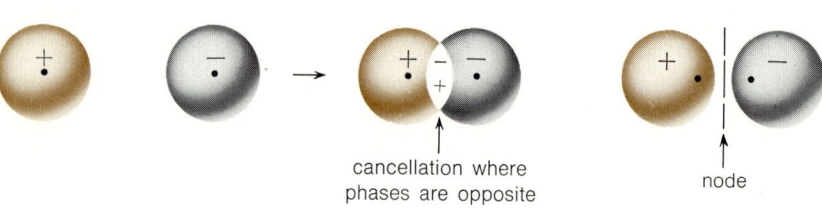

cancellation where
phases are opposite

node

destructive interference

Figure 3.8. Boundary surface overlap diagrams for constructive and destructive interference of $1s$ atomic orbitals. The end results are identical to the boundary surfaces shown in Figures 3.4(d) and 3.7(c).

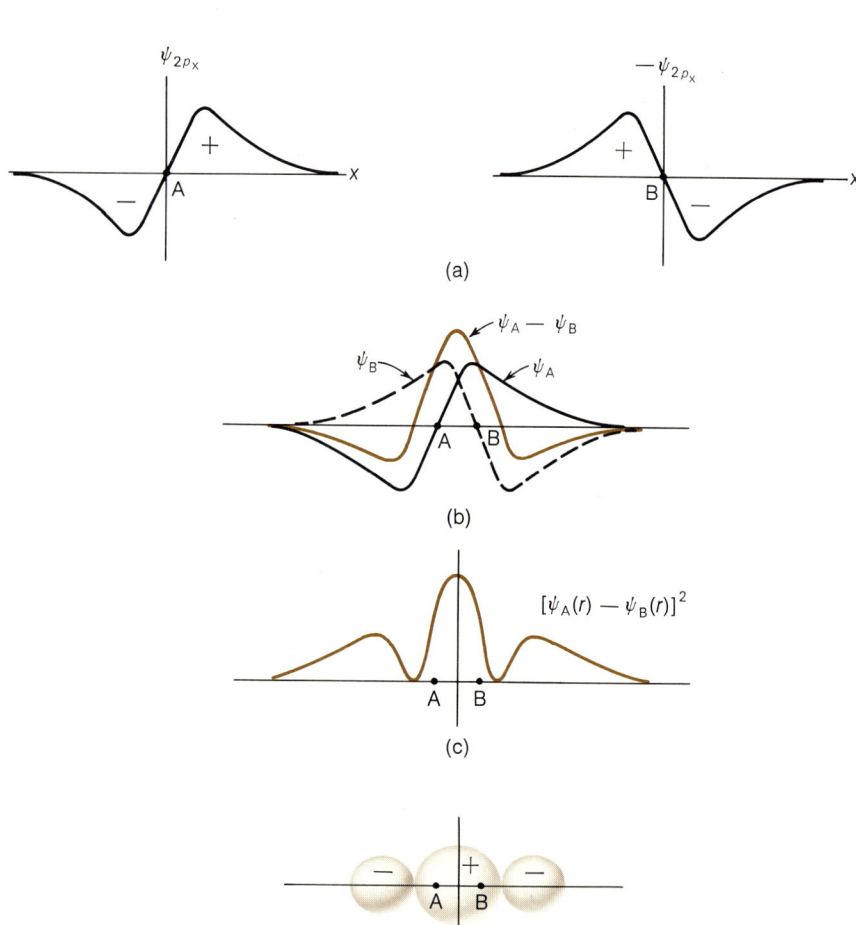

(a)

(b)

(c)

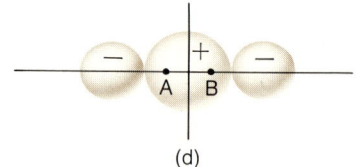

(d)

Figure 3.9. Constructive interference of $2p_x$ atomic orbitals. (a) The separated orbitals. (b) The orbitals at their internuclear separation and their resultant, $\psi_A(r) - \psi_B(r)$. (c) Resultant probability density. (d) Boundary surface.

In both of the situations we have described here, the attractive and repulsive states, we have been overlapping atomic orbitals. In the attractive state, there was constructive interference or reinforcement, and maximum overlap took place. In the repulsive state, with the phase differences, destructive interference or cancellation took place, and no net overlap resulted. The two cases can be shown simply (Figure 3.8) by using boundary overlap diagrams.

Figures 3.9 and 3.10 show what happens when two atoms with p_x atomic orbitals in their valence shell are brought together. The nuclei lie along the x axis. Boundary surface diagrams for the processes are shown in Figure 3.11. The constructive and destructive interference of two p_y (or two p_z) atomic

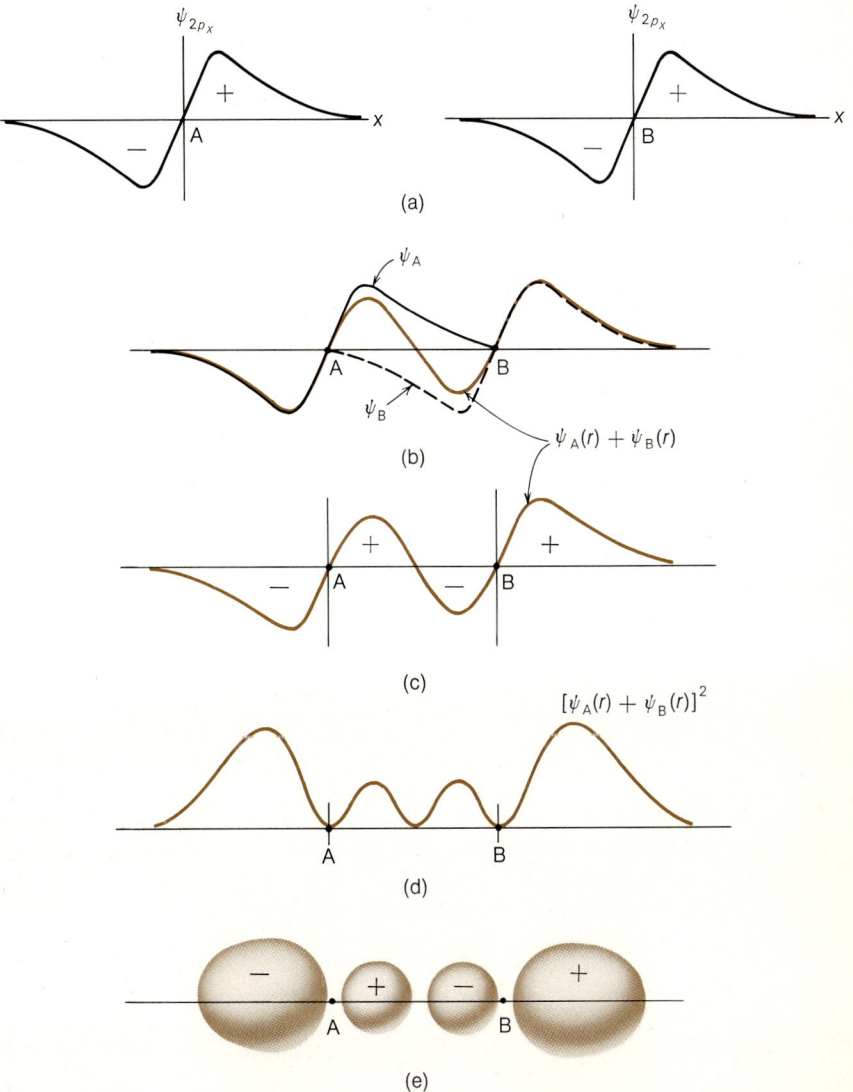

Figure 3.10. Destructive interference of $2p_x$ atomic orbitals. (a) Separated orbitals. (b) The orbitals at their internuclear separation and their resultant, $\psi_A(r) + \psi_B(r)$. (c) The resultant (for clarity). (d) Resultant probability density. (e) Boundary surface.

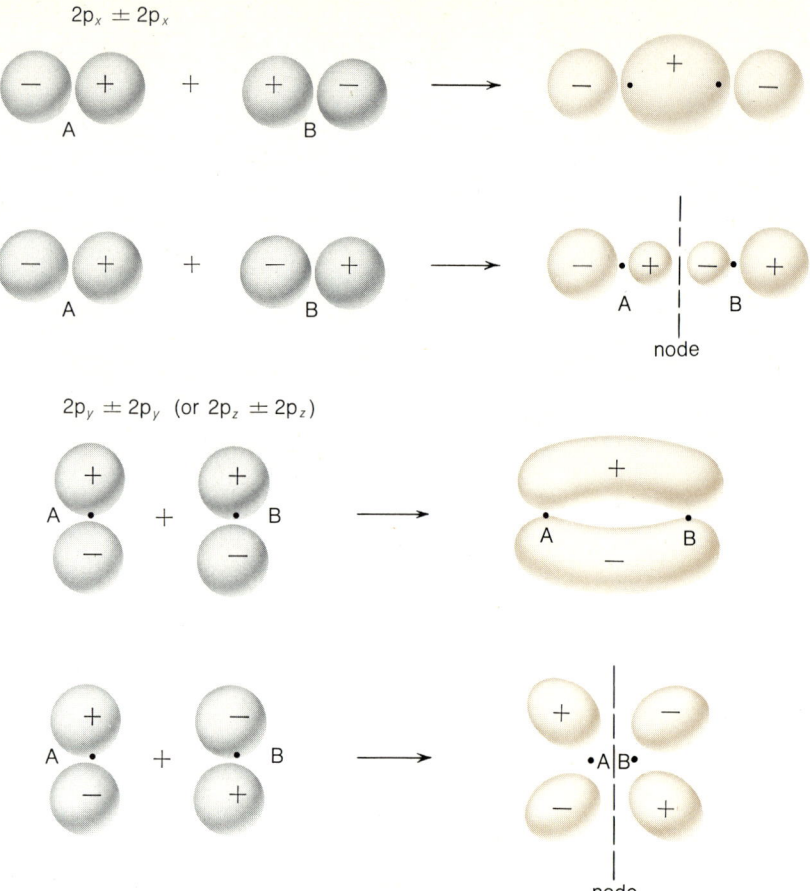

Figure 3.11. Boundary surface overlap diagrams for p atomic orbitals.

orbitals is also shown in the simple overlap diagrams in Figure 3.11. The y and z axes are perpendicular to the line joining the nuclei and to each other.

The unpaired electron in a fluorine atom, for instance, does not know if it is in a p_x, a p_y, or a p_z orbital. When two fluorine atoms come together to form a bond, they will align themselves so as to cause maximum overlap between their p orbitals. We have arbitrarily chosen to call their direction of alignment the x axis. Despite the fact that our electronic configurations of fluorine have again arbitrarily labeled the unpaired electron p_z, the bond formed will be of the type shown in Figure 3.9.

3.7 MOLECULAR ORBITALS   All of the diagrams that have been drawn show the formation of what are called *molecular orbitals* (M.O.'s). Just as atomic orbitals describe electrons in the neighborhood of one nucleus in atoms, molecular orbitals describe the state of electrons in molecules where there are two or more nuclei. So far only diatomic molecules where the electron or electrons are localized in the region of two nuclei have been considered, but we shall see that other molecular orbitals can encompass more than two nuclei. However, the molecular orbitals of the hydrogen molecule serve as the starting point for a discussion of more complex molecules, just as the atomic orbitals of the hydrogen atom provide a basis for more complex atoms.

In Figure 3.12 are collected representations of the charge clouds (with phase signs) of a number of the molecular orbitals for diatomics with identical nuclei (homonuclear diatomics). The star (*) denotes an *antibonding* molecular orbital, one with a node *between* the nuclei; that is, an orbital with a nodal plane perpendicular to the line (the x axis) joining the nuclei. The σ (Greek sigma) and π (Greek pi) designations refer to certain spatial symmetry properties of the orbitals. The σ orbitals have symmetry with respect to reflection (or rotation by 180° about the bond axis) through or across the internuclear axis. If we imagine a plane perpendicular to the paper and

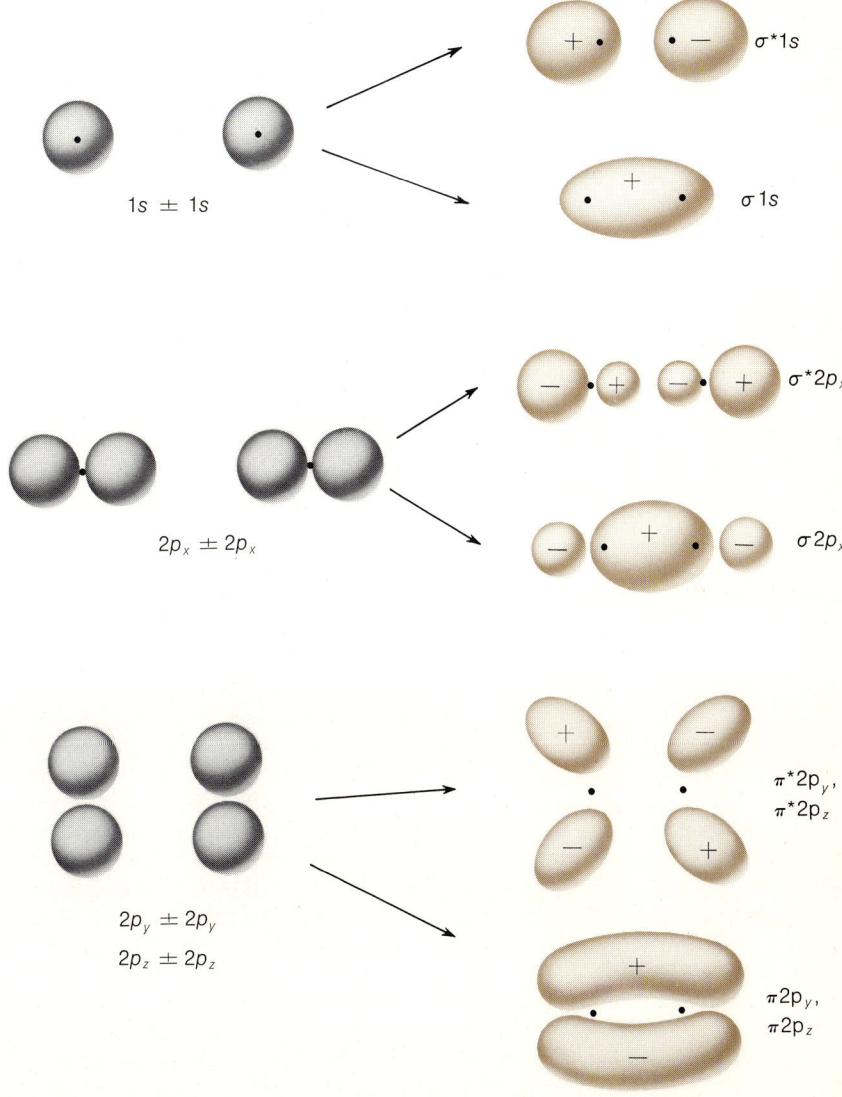

Figure 3.12. Molecular orbitals for homonuclear diatomic molecules. This figure summarizes Figures 3.8–3.11 and labels the resulting molecular orbitals.

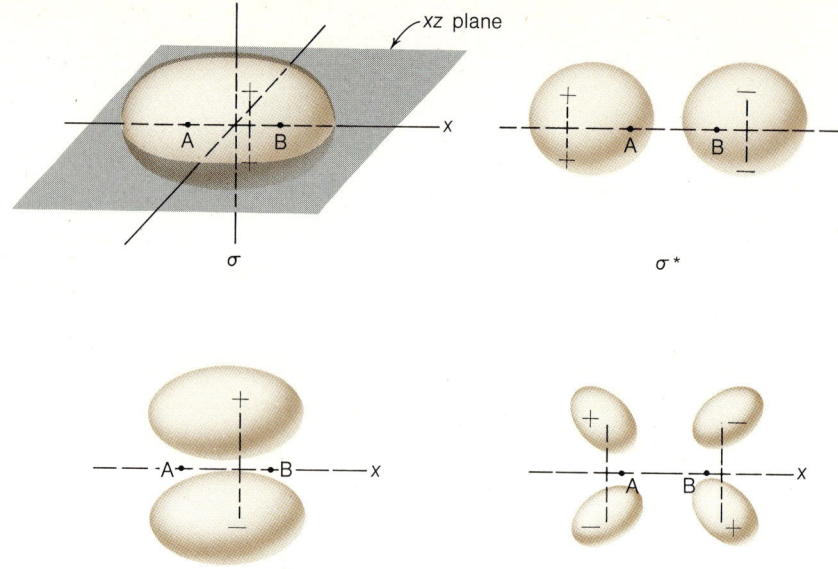

Figure 3.13. Symmetry properties of σ and π molecular orbitals. The π orbitals change sign on reflection across the internuclear axis (x).

containing the internuclear axis, two points on a line perpendicular to this plane and lying on opposite sides of the axis have the same sign and charge density (see Figure 3.13). The π orbitals change sign under the same operation. Also, in σ orbitals the electron charge density lies along the line between bonded nuclei, while π orbitals have their charge density off this axis. Atomic orbitals are often assigned the same notation, an s and a p_x atomic orbital being σ type, p_y and p_z being π type if the x axis is taken as the line between the nuclei.

3.8 ENERGIES OF MOLECULAR ORBITALS

Molecular orbitals can be ordered according to their energies in the same way as atomic orbitals. In increasing energy the order is, for the molecular orbitals for homonuclear diatomics where the x axis is chosen as the internuclear axis,

$$\sigma 1s < \sigma^* 1s < \sigma 2s < \sigma^* 2s < \pi 2p_y = \pi 2p_z < \sigma 2p_x < \pi^* 2p_y = \pi^* 2p_z < \sigma^* 2p_x$$

(The energies of the $\sigma 2p_x$ and the degenerate $\pi 2p_y$, $\pi 2p_z$ may be interchanged in some molecules, and many texts show the $\sigma 2p_x$ lower in energy. This need not concern us here.) The energies are roughly proportional to the number of nodes in the orbitals.

The two π orbitals are degenerate, as are the two π^* molecular orbitals. Each of the molecular orbitals can hold a maximum of 2 electrons, provided the spins of the electrons are opposed.

To determine the electronic structure of a homonuclear diatomic molecule, a building-up process similar to that used in determining the electronic configurations of atoms is used. That is, the two nuclei minus their electrons are imagined to be at their equilibrium distance. Electrons are then added one by one to the system, the electrons going into the lowest-energy molecular orbitals first. The Pauli exclusion principle and Hund's rule are followed.

When two identical atomic orbitals are combined or overlapped, we have seen that two molecular orbitals can be formed, one bonding and one anti-bonding. In general, when n atomic orbitals are combined, n molecular orbitals are possible. On an energy scale, the energies of the atomic orbitals and the molecular orbitals look like those in Figure 3.14.

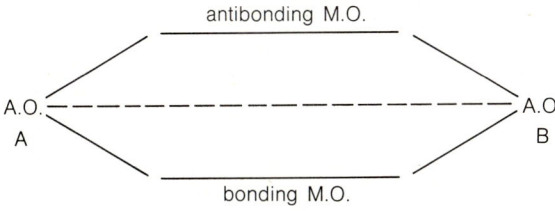

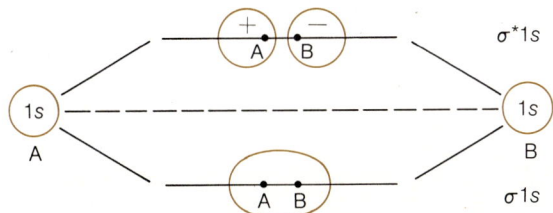

Figure 3.14. Energies of bonding and antibonding molecular orbitals.

The energy of an antibonding molecular orbital is farther above the original atomic orbital energies than the energy of the bonding molecular orbital is below. This means that an electron in an antibonding molecular orbital, such as σ^*1s, contributes more to repulsion of the two atoms than an electron in a bonding molecular orbital, such as $\sigma 1s$, contributes to attraction. An electron in an antibonding orbital is therefore more antibonding than a bonding electron is bonding.

With this in mind, we can successively add electrons to our scheme of molecular orbitals to build electron configurations and get some idea of the bonding or lack of it in some molecules.

$\quad$ H $1s^1$ $\quad$ H$_2$ $\quad$ 2 electrons

The molecular electronic configuration is $(\sigma 1s)^2$. Two bonding electrons mean one strong covalent bond.

$\quad$ He $1s^2$ $\quad$ He$_2$ $\quad$ 4 electrons total
$\quad\quad\quad\quad\quad$ He$_2$ $\quad$ $(\sigma 1s)^2(\sigma^*1s)^2$

The repulsive effect of the 2 antibonding electrons overbalances the 2 bonding electrons, and no such molecule exists. Note that this agrees with the argument that two He atoms will not bond because neither has unpaired electrons and too much energy would be required to unpair the two $1s$ electrons in each atom.

$\quad$ Li $1s^2 2s^1$ $\quad$ Li$_2$ $\quad$ 6 electrons total
$\quad\quad\quad\quad\quad$ Li$_2$ $\quad$ $(\sigma 1s)^2(\sigma^*1s)^2(\sigma 2s)^2$

The result is one covalent bond, which is weaker than that in H_2, because the net effect of the $\sigma 1s$ and $\sigma^* 1s$ is repulsion. The Li_2 molecule exists in the gaseous state.

Moving on to the oxygen molecule, O_2, with a total of 16 electrons, we find

O $1s^2 2s^2 2p_x^2 2p_y^1 2p_z^1$
O_2 $(\sigma 1s)^2 (\sigma^* 1s)^2 (\sigma 2s)^2 (\sigma^* 2s)^2 (\pi 2p_y)^2 (\pi 2p_z)^2 (\sigma 2p_x)^2 (\pi^* 2p_y)^1 (\pi^* 2p_z)^1$

The assignment of electrons is also shown in Figure 3.15. The unpaired electrons in the set of degenerate π^* molecular orbitals are due to the operation of Hund's rule.

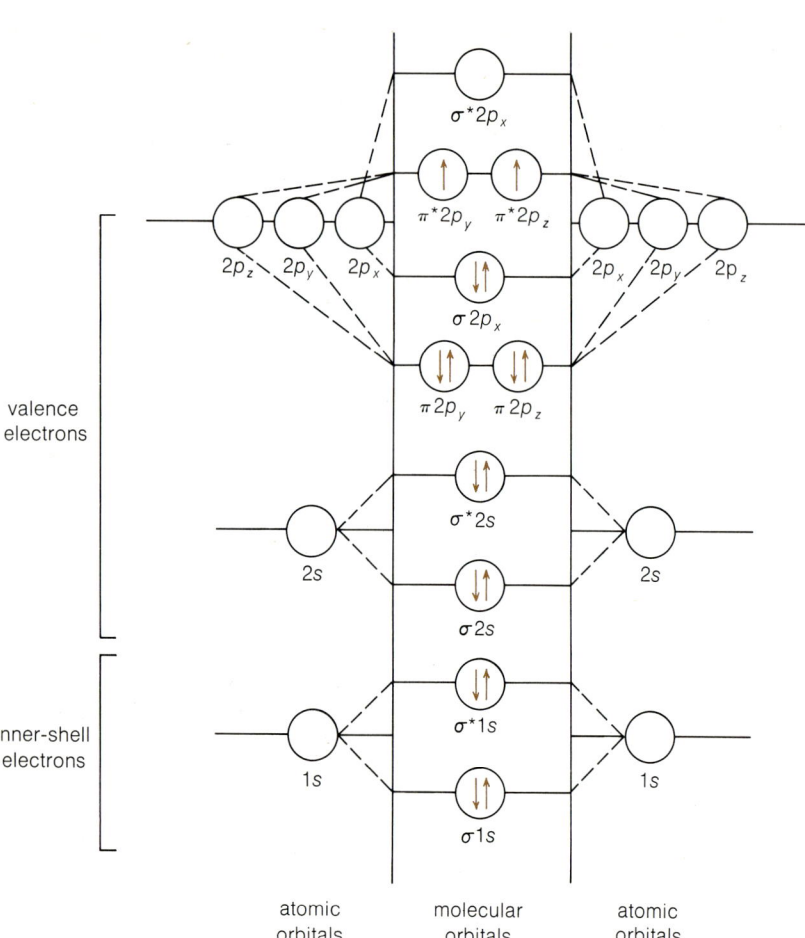

Figure 3.15. Electron filling of the molecular orbitals for the oxygen molecule. The $\sigma 2p$ and $\pi 2p$ orbitals have similar energies and may change places in other molecules.

The fluorine molecule, F_2, has the molecular orbital configuration

F_2 $(\sigma 1s)^2 (\sigma^* 1s)^2 (\sigma 2s)^2 (\sigma^* 2s)^2 (\pi 2p_y)^2 (\pi 2p_z)^2 (\sigma 2p_x)^2 (\pi^* 2p_y)^2 (\pi^* 2p_z)^2$

The Ne_2 molecule would have the configuration

Ne_2 $(\sigma 1s)^2 (\sigma^* 1s)^2 (\sigma 2s)^2 (\sigma^* 2s)^2 (\pi 2p_y)^2 (\pi 2p_z)^2 (\sigma 2p_x)^2 (\pi^* 2p_y)^2 (\pi^* 2p_z)^2 (\sigma^* 2p_x)^2$

As in He, we find an equal number of bonding and antibonding orbitals. The antibonding molecular orbitals cancel out the bonding molecular orbitals, so no Ne_2 molecule is formed.

3.9 BOND ORDER IN HOMONUCLEAR DIATOMICS

Multiple bonds have been mentioned briefly before. It is of interest to see how single, double, triple, and even fractional bonds come about in the molecular orbital scheme. The single covalent bond in H_2 and Li_2 has been referred to, as has the absence of a bond in He_2. Let us be more general and define a *bond order* as

    1. the number of *pairs* of electrons in bonding orbitals minus the number of *pairs* in antibonding orbitals, or

    2. one-half of the excess of bonding electrons over antibonding electrons.

Table 3.1
Bond orders

molecule	bond order	
H_2	1	single bond
He_2	0	no bond
Li_2	1	single bond
N_2	3	triple bond
O_2	2	double bond
F_2	1	single bond
Ne_2	0	no bond

In counting pairs or individual electrons, inner-shell electrons may be ignored. Bond orders in the molecules used as examples are given in Table 3.1. In terms of electron pairing, the bond in O_2 would have been expected to be a double bond with no unpaired electrons, as in the Lewis structure $\ddot{O}=\ddot{O}$. Experimental evidence of the magnetic behavior of oxygen, however, shows the bond to be as the molecular orbital treatment indicates, with two unpaired electrons. Possible dot structures of O_2 can be variously represented by

$$:\dot{\ddot{O}}{-}\dot{\ddot{O}}: \quad \text{or} \quad :O{\vdots\vdots}O: \quad \text{or} \quad \dot{\ddot{O}}{\equiv}\dot{\ddot{O}}$$

3.10 BONDING IN DIATOMICS WITH DIFFERENT ATOMS

Overlapping of atomic orbitals between dissimilar atoms occurs in the same way as in homonuclear diatomics. Of course, the overlapping atomic orbitals need no longer be identical, but to ensure proper overlap and produce a good bond, the two atomic orbitals must satisfy the following conditions:

    1. They must have similar orbital energies. This condition is usually met by the valence electrons (or orbitals) in different atoms.

    2. They must have the proper phase and symmetry properties to give an attractive bond by constructive interference.

Condition 2 is illustrated in Figure 3.16 with simple boundary diagrams, with the internuclear axis being the x axis. $1s + 2p_x$ illustrates constructive interference and a net overlap to give a bonding σ molecular orbital. $1s - 2p_x$ shows the formation of an antibonding σ^* molecular orbital because of the destructive interference of the s and p_x atomic orbitals. $1s \pm 2p_y$ (or $2p_z$) produces no net overlap, because the constructive interference in the upper region of overlap cancels the destructive interference in the lower region. The

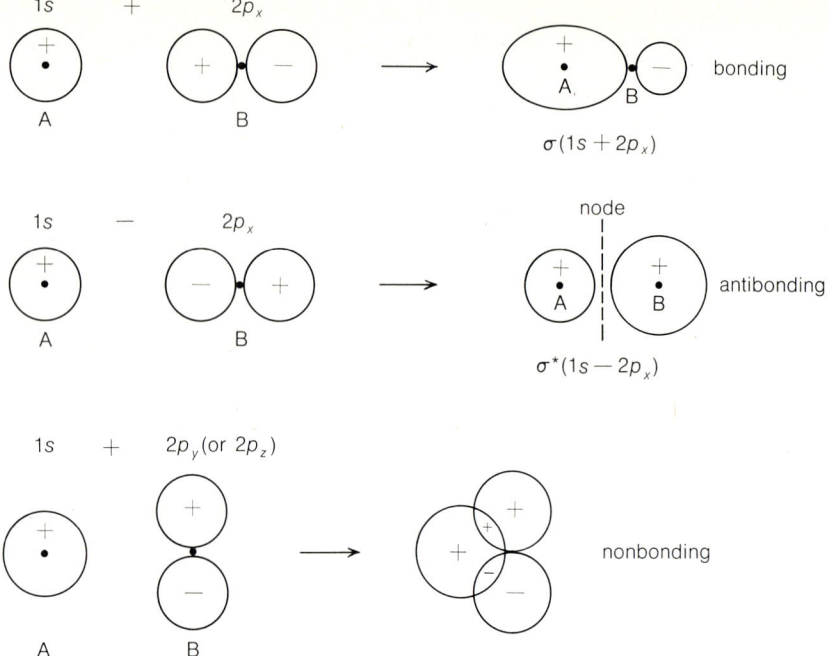

Figure 3.16. Orbital overlap for unlike atomic orbitals.

resulting molecular orbital is neither bonding (attractive) nor antibonding (repulsive). The result is called a nonbonding molecular orbital and does not contribute either positively or negatively to a bond between the two atoms. Nonbonding molecular orbitals lead to neither an increase nor a decrease in energy over that of the combined atomic orbitals. Similarly, a p_x atomic orbital on one atom would give no net overlap with a p_y or a p_z orbital.

As an example, a hydrogen atom with its $1s$ atomic orbital can overlap and bond with a fluorine atom with its $2p_x$ orbital:

$$\text{HF } 1s_F^2 2s_F^2 2p_{y_F}^2 2p_{z_F}^2 (\sigma 1s_H 2p_{x_F})^2$$

Again, it makes no difference to which p orbital we had previously assigned the unpaired electron in fluorine.

When different atoms are bonded, there may be a competition for the electron pair. In the extreme, one atom may draw the electron pair completely over to itself, with the formation of an ionic bond. The degree of competition for the electrons in a covalent bond leads to bonds with partial ionic and partial covalent character. Most bonds are somewhere in between the extremes of 100 percent ionic and 100 percent covalent, as will be discussed in Chapter 4.

3.11 EXCITED STATES

A distinct advantage of the molecular orbital picture of bonding is that it allows for excited states of molecules. These are higher-energy states in which the molecule is still intact but is in one of its more energetic quantum states. Molecules have a set of discrete electronic energy states just as atoms have. In an excited state, one or more of the electrons in the molecule has been promoted to one of the higher, previously unoccupied molecular orbitals.

As a very simple example, consider H_2. Its ground state configuration is $(\sigma 1s)^2$.

H_2 ◯ $\sigma^* 1s$

(↑↓) $\sigma 1s$

Two excited states of H_2 can be thought of:

H_2^* (↑) $\sigma^* 1s$

(↑) $\sigma 1s$

or

H_2^* (↓) $\sigma^* 1s$

(↑) $\sigma 1s$

$\left. \begin{array}{l} \end{array} \right\}$ H_2^* $(\sigma 1s)^1 (\sigma^* 1s)^1$

These two excited states of H_2 differ in electron spin. An even higher energy state would be

H_2^* $(\sigma 1s)^1 (\sigma 2s)^1$

Although we can write down many excited states, there is no guarantee that the excited state will be stable, in the sense that the excited molecule may exist for any length of time before it loses its excess energy in some way.

Excited states in more complex molecules, especially those with π electrons, help to explain how molecules interact with radiation and how energy is transferred from one molecule to another. They are necessary to the understanding of energy transfer among and along molecules in living organisms.

3.12 BONDING IN POLYATOMIC MOLECULES

Overlapping of atomic orbitals in polyatomic molecules and the formation of bonds follow the same procedure as in diatomic molecules, although two new features (hybridization and delocalization or resonance) will be introduced which are of great importance in biological molecules.

First, let us take water as a simple example of a polyatomic molecule. In our earlier electron pairing picture, we would say that a $1s$ electron from each of two hydrogen atoms paired with the 2 unpaired p electrons on oxygen:

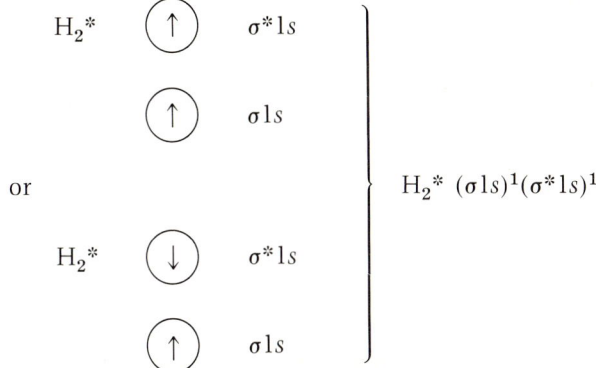

H
O $1s^2 2s^2 2p_z^2$ $\begin{pmatrix} 2p_y^{\ 1} \\ 1s^1 \end{pmatrix}$ $\begin{pmatrix} 1s^1 \\ 2p_x^{\ 1} \end{pmatrix}$
H

Note that the nuclei now lie along two perpendicular axes. The exact labeling of the axes is unimportant here.

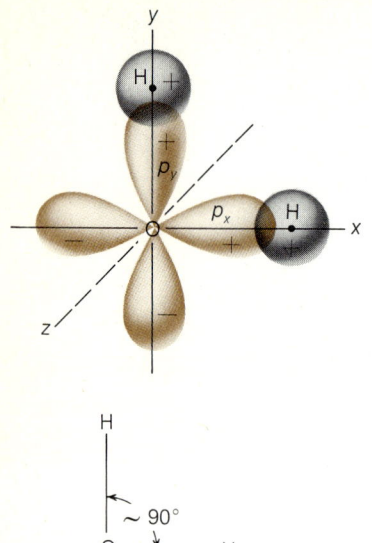

In terms of the molecular orbital picture, we say that the $1s$ orbitals of hydrogen overlap with the $2p$ orbitals of oxygen as shown in Figure 3.17. To obtain maximum overlap, the hydrogen atoms with their $1s$ orbitals must approach and bond along the perpendicular axes of the oxygen p orbitals.

If two localized σ-type ($\sigma 1s_H 2p_O$) molecular orbitals are formed between the two hydrogens and oxygen, the HOH angle would be expected to be 90°, since the two p atomic orbitals of oxygen are perpendicular to each other. We shall discuss bond angles and molecular geometry later (Section 3.14), but you can already see that the shape of a molecule is governed by the electronic configurations of its component atoms.

Other examples of simple polyatomic molecules with localized bonding electrons are shown with their orbital overlap diagrams in Figure 3.18. In each of these, the bonding atoms approach each other so as to get maximum overlap of their atomic orbitals.

Figure 3.17. Orbital overlap diagram for water. A 90° bond angle is predicted on the basis of simple atomic overlap. Elongated p orbitals (ψ^2) are used to reduce crowding in the figure.

Figure 3.18. Orbital overlap diagram for NH_3 and H_2O_2. The bond angles are those expected from simple atomic orbital overlap. The H—N—H bond angles and the H—O—O bond angles are 90°.

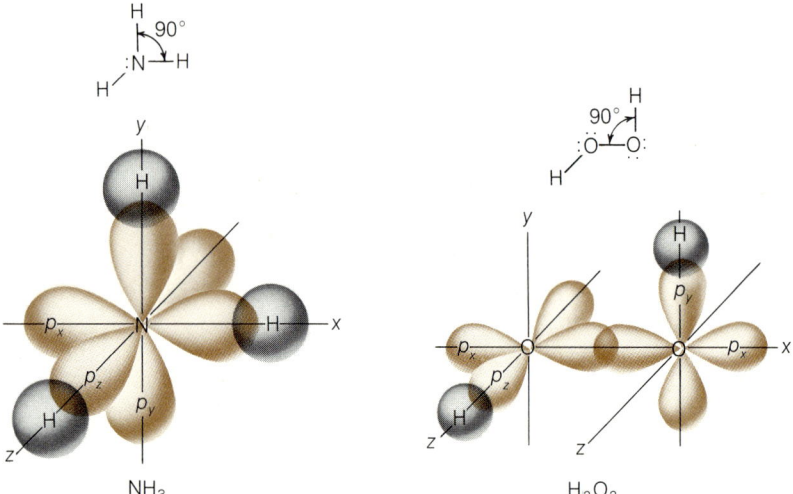

NH_3 H_2O_2

3.13 COORDINATE COVALENT BONDS

In the molecular orbital picture, we talk about orbitals on atoms overlapping to give molecular orbitals. Then we can think of putting electrons into the possible molecular orbitals in the same manner that we imagined we fed electrons into atomic orbitals (or energy levels) in Section 2.29. In H_2, H_2O, and many other molecules, each overlapping atomic orbital appears to contribute one electron to any one molecular orbital.

However, a perfectly respectable molecular orbital can be formed from the overlapping of appropriate atomic orbitals whether these orbitals contain atomic electrons or not. If one of two atomic orbitals originally held 2 paired electrons in one of the bonding atoms, while the other atomic orbital held none, then both of these electrons may fill the molecular orbital, and a *coordinate covalent* bond is formed.

Actually, it is immaterial from which atom the electrons come. Once the coordinate bond is formed, it is identical to any covalent bond. It is our use of Lewis electron dot structures that causes us to think of both electrons of the electron pair bond as coming from the same atom. In the ammonium ion, we consider one of the N—H bonds to be composed solely of a nitrogen unshared pair of electrons, as indicated in Figure 3.19.

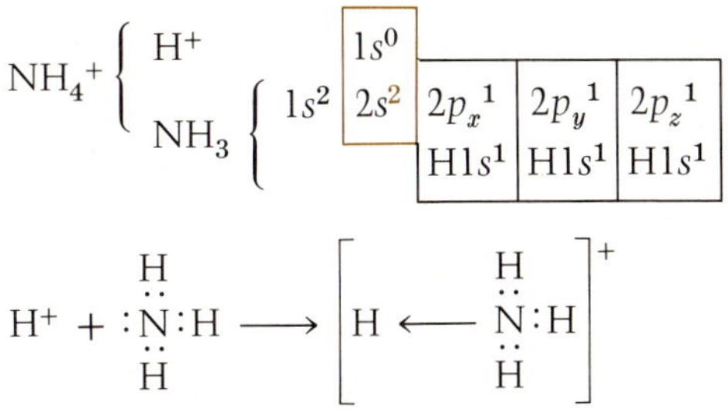

Figure 3.19. Coordinate covalent bonding. The $2s$ orbital (actually a hybrid sp^3 orbital) of N overlaps with the empty $1s$ orbital of H^+; both electrons in the resulting bond may be considered to come from the unshared pair in N. An arrow showing the direction of electron pair donation is often used. All bonds in NH_4^+ are identical.

$$NH_4^+ \begin{cases} H^+ & \begin{array}{|c|} \hline 1s^0 \\ \hline \end{array} \\ NH_3 & 1s^2 \begin{array}{|c|c|c|c|} \hline 2s^2 & 2p_x{}^1 & 2p_y{}^1 & 2p_z{}^1 \\ & H1s^1 & H1s^1 & H1s^1 \\ \hline \end{array} \end{cases}$$

$$H^+ + \overset{..}{\underset{H}{:N:}}H \longrightarrow \left[H \longleftarrow \overset{H}{\underset{H}{\overset{..}{N}:}}H \right]^+$$

In principle, a coordinate bond may be formed between any entity having an unshared pair of valence electrons and an entity having empty valence orbitals. The coordination compounds or complexes most commonly involve transition metal ions, such as Fe^{3+}, having available d orbitals. An example of a complex ion is given in Figure 3.20.

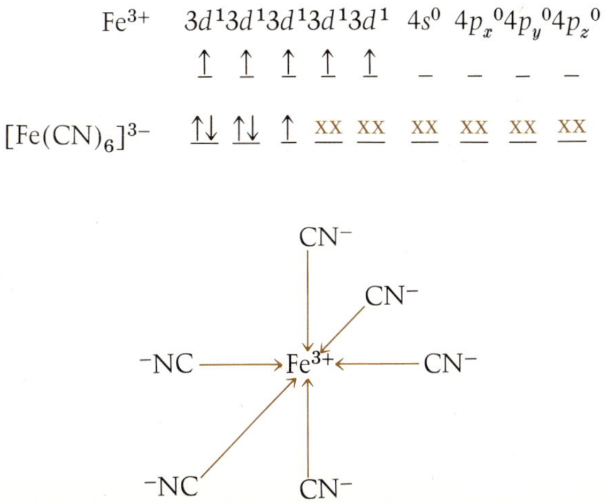

$$Fe^{3+} \quad 3d^1 3d^1 3d^1 3d^1 3d^1 \quad 4s^0 \quad 4p_x{}^0 4p_y{}^0 4p_z{}^0$$

Figure 3.20. The $[Fe(CN)_6]^{3-}$ complex ion. In this interpretation of bonding, six coordinate covalent bonds result from electron pairs donated to Fe^{3+} by the six cyanide ions, $(\overset{x}{\underset{x}{}}C\equiv N:)^-$, preceded by rearrangement of the iron $3d$ electrons. Complexes or complex ions are compounds containing a central atom or ion, usually a metal, surrounded by a cluster of ions or molecules. The name complex is usually restricted to compounds that are formed by direct reaction between the simple ions or molecules or that dissociate into their components. This definition includes the common hydrated ions or aquo complexes such as $Cu(H_2O)_4{}^{2+}$ and $Al(H_2O)_6{}^{3+}$, but excludes $CO_3{}^{2-}$, $SO_4{}^{2-}$, and the like.

$[Fe(CN)_6]^{3-}$

3.14 HYBRIDIZATION

There is one very serious omission in the molecular examples we have studied so far. The element carbon has not been considered at all. Carbon is one atom common to the huge number of organic molecules, so named because they form the structural materials of all life as we know it and they were once believed to be formed only by living things. The bonding and structure of these carbon-containing organic or biological molecules will play a large part in our later discussions. The consideration of carbon introduces us to new bonding factors that occur in some degree with all other atoms.

To start with, let us investigate the features of carbon bonding based on its electronic configuration,

$$C \quad 1s^2 2s^2 2p_x{}^1 2p_y{}^1$$

Here there are 2 unpaired electrons—two bonds? What about the simple and common molecule CH_4? Unpairing of the two $2s$ electrons might give

$$C^* \; 1s^2 2s^1 2p_x^{\;1} 2p_y^{\;1} 2p_z^{\;1}$$

allowing four possible bonds. What are the expected angles between such bonds? Three are at right angles to one another; the fourth (s-) has no particular angle to the others. But we know from experiment that the angles between the four C—H bonds in methane gas, CH_4, are 109°28′ and are all alike, as shown in Figure 3.21.

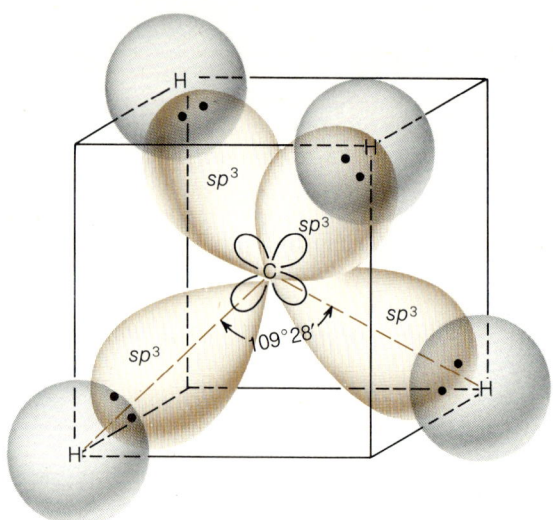

Figure 3.21. Tetrahedral or sp^3 hybrid orbitals in methane, CH_4.

How can the valence of carbon and the bond angles in CH_4 be explained? Remember that the bonded atoms in a molecule will tend to be arranged in a configuration that will give the lowest possible energy. To get a good strong bond between two atoms, it is necessary to have maximum overlap between the atomic orbitals of the two atoms.

Carbon can form equally strong bonds, with maximum overlap, with four hydrogen atoms if we allow it to undergo an internal process called *hybridization* on the approach of the four hydrogens.

Briefly, we think of the $2s$ and the three $2p$ atomic orbitals of the carbon atom as combining or mixing to give four *equivalent hybrid* orbitals. These hybrid orbitals are directed at angles of 109°28′ to each other. The carbon atom can be imagined to be at the center of a cube, with the orbitals directed at opposing corners as in Figure 3.21. The solid that connects these corners by planes is called a tetrahedron, so a carbon atom hybridized in this fashion is called a *tetrahedral* carbon atom or one having *tetrahedral hybrid* orbitals. Because one s orbital and three p orbitals are mixed or hybridized, tetrahedral hybrid orbitals are also called sp^3 hybrids. There are several advantages to mixing the carbon atomic orbitals in this way:

1. The addition of some s character to the p orbitals makes the lobes of each orbital fatter on one side, thereby increasing the amount of overlap with an atomic orbital on another atom.

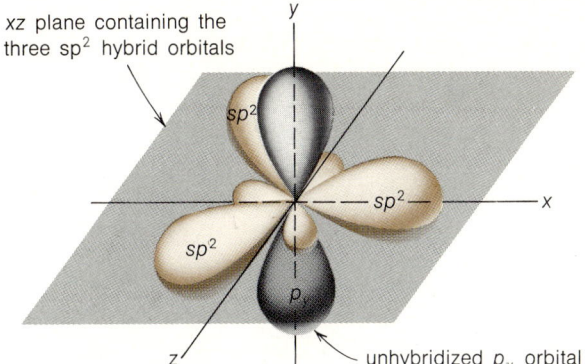
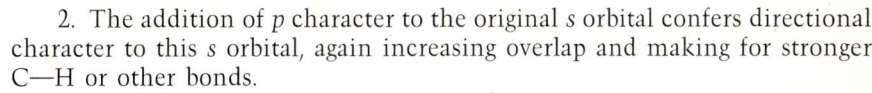

Figure 3.22. Trigonal or *sp²* hybrid orbitals. The *sp²* orbitals are in the same plane. The unhybridized *p* atomic orbital is perpendicular to this plane.

2. The addition of *p* character to the original *s* orbital confers directional character to this *s* orbital, again increasing overlap and making for stronger C—H or other bonds.

3. The 109°28' angle between the four equivalent hybrid orbitals results in the four electron pairs of the four potential bonds being as far away from each other as possible. Since electron pairs in filled orbitals repel each other particularly strongly, this is a favorable configuration of low energy. In fact, the angles that result between hybrid orbitals could be rationalized solely on the basis of the configurations that will keep the valence shell electron pairs in the σ-type hybrid orbitals plus the pairs in atomic orbitals not used in bonding as far apart as possible.

Not all *p* orbitals need be involved in this "mixing" process; one or two may be left out of the process. We can then have *sp²* planar *trigonal* hybrid orbitals, shown in Figure 3.22. Here one *p* orbital remains as pure *p* and does not enter into the mixing process. Trigonal orbitals lie in the same plane, and the angle between them is 120°. There are three equivalent trigonal orbitals. The remaining *p* orbital in carbon is perpendicular to the plane containing the trigonal hybrid orbitals. This is also shown in Figure 3.22.

Digonal or *sp* hybrid orbitals involve only one of the *p* orbitals. As before, we get as many hybrid orbitals as the number of atomic orbitals used in the mixing process. The two digonal orbitals have an angle of 180° between them. Because of the simplicity of drawing only one *s* and one *p* orbital, we can show how the mixing of an *s* atomic orbital and a *p* atomic orbital on one atom results in a hybrid orbital. Figure 3.23 shows how the addition of the $2s$ and $2p_x$ on the same atom gives one hybrid digonal orbital, while the subtraction, $2s - 2p_x$, can give the other hybrid. The two opposed *sp* hybrid orbitals lie in the same plane, with the two unused *p* orbitals each perpendicular to the line of the *sp* hybrids. This is shown in Figure 3.24.

So far as the carbon atom is concerned, these three types of hybridization are all we need to discuss its bonding and the structures of its compounds.

In carbon compounds where all bonds to carbon are single bonds, the carbon atom is hybridized in tetrahedral or *sp³* fashion. Consider propane, C_3H_8, as an example. All the C—H and C—C bonds are σ type. A space representation of propane is shown in Figure 3.25 along with the simpler structural formula. Another example is ethyl alcohol, or ethanol, C_2H_5OH, also shown in Figure 3.25.

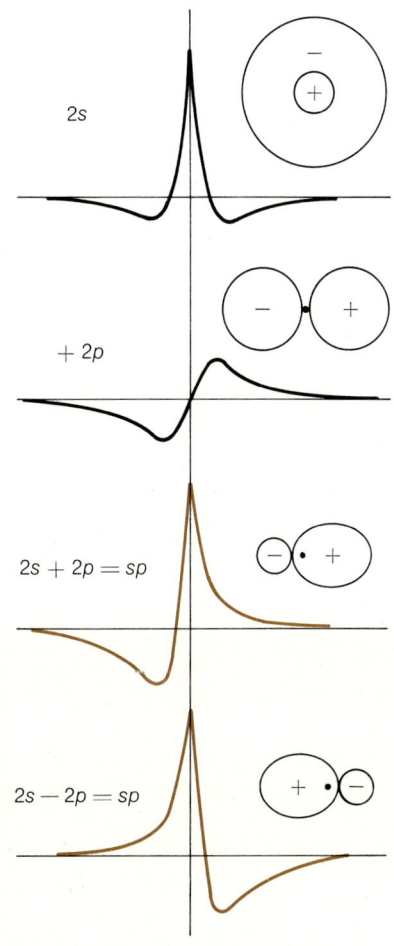

Figure 3.23. Mixing of $2s$ and $2p$ atomic orbitals to form digonal or *sp* hybrid orbitals. Boundary surfaces are shown at the right.

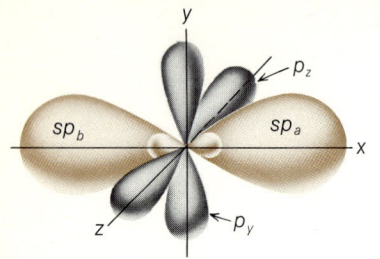

Figure 3.24. Digonal or *sp* hybrid orbitals. The two unhybridized *p* orbitals are at right angles to the *sp* hybrids and to one another.

Whenever there is a carbon with one double bond between it and another atom, that carbon atom has trigonal, sp^2, hybrid orbitals. Molecular overlap diagrams for ethene (ethylene), $H_2C{=}CH_2$, and for propanal (propionaldehyde)

$$H_3C{-}CH_2{-}C\overset{\textstyle O}{\underset{\textstyle H}{\diagdown}}$$

are shown in Figure 3.26

A carbon atom bonded to an adjacent atom by a triple bond or to two adjacent atoms by double bonds has digonal or *sp* hybridization. Examples are ethyne (acetylene), $HC{\equiv}CH$, and propadiene (allene), $H_2C{=}C{=}CH_2$, where the middle C has *sp* hybridization and the two end carbons have sp^2. The overlap diagrams are sketched in Figure 3.27.

Hybridization is not confined to the carbon atom and may occur with

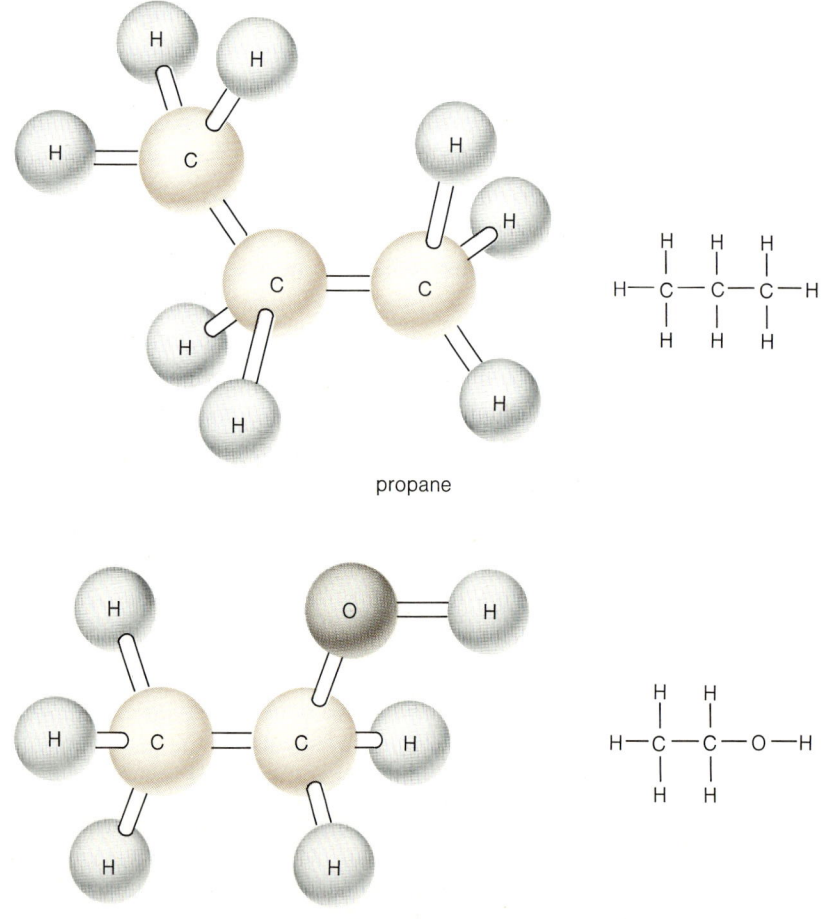

propane

ethyl alcohol

Figure 3.25. Space representations and simple structural formulas for propane and ethyl alcohol.

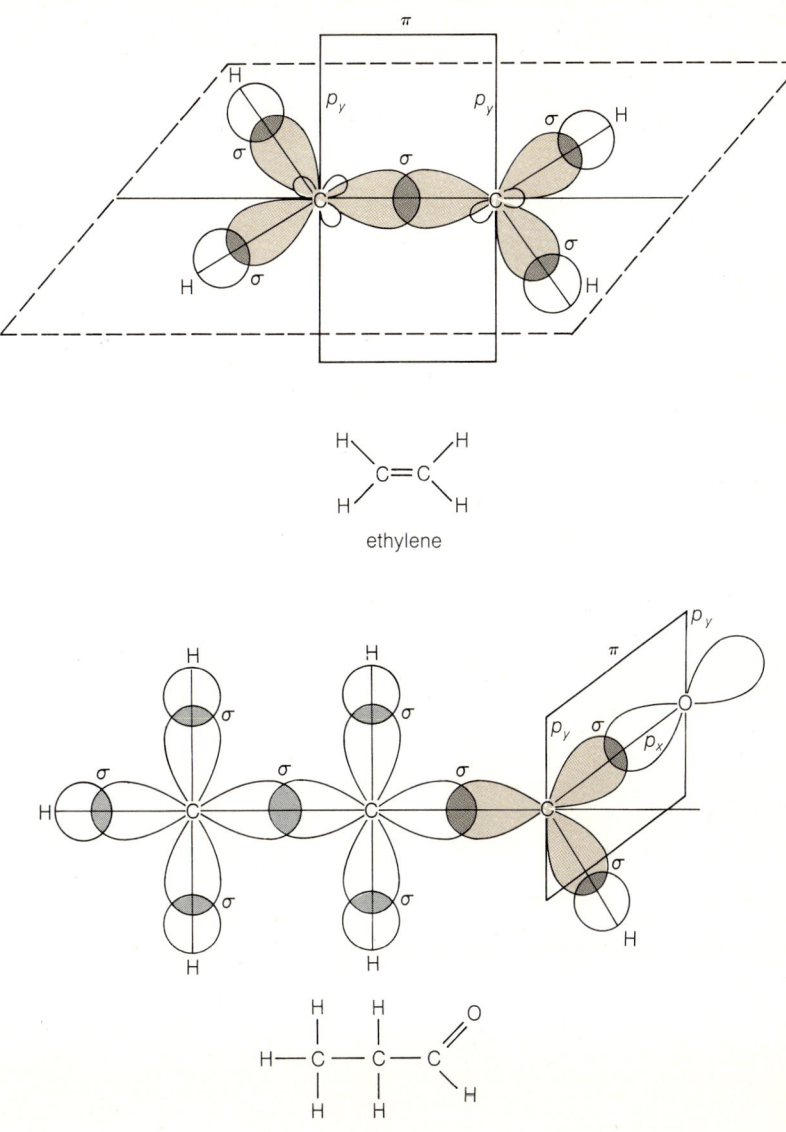

ethylene

Figure 3.26. Simplified overlap diagram for molecules with sp^2 hybrid orbitals (color). To prevent crowding, only a single line is drawn to indicate overlap between pure p orbitals. Tetrahedral bond angles due to sp^3 orbitals (gray) are not shown.

propanal

any atom where s, p, or s, p, and d orbital energy levels lie close together, whether there are electrons present in these levels or not. Table 3.2 lists some other types of hybridization.

3.15 ELECTRON DELOCALIZATION AND RESONANCE

The term "delocalization" has already been used to indicate the fact that an electron in a bonding covalent orbital in a diatomic molecule is no longer restricted to motion about one nucleus as in the free atom, but has a larger volume available to it, the bonding molecular orbital that encompasses both nuclei. We shall now see that in certain polyatomic molecules, specifically those involving multiple bonds, there is an opportunity for the delocalization

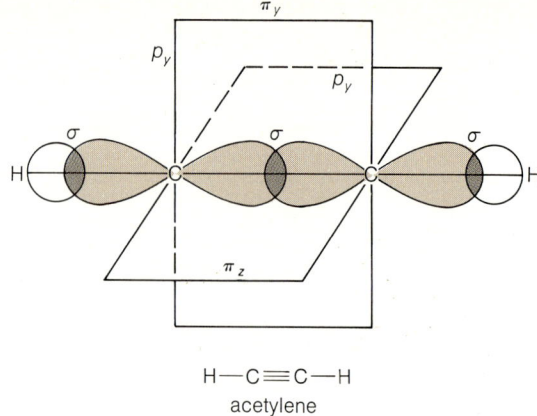

H—C≡C—H
acetylene

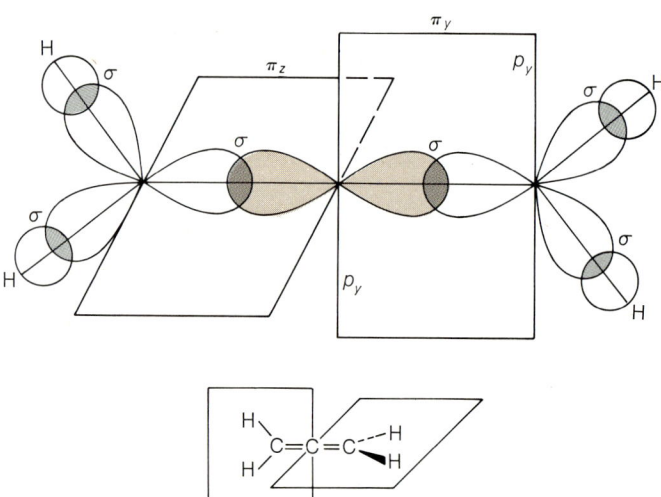

Figure 3.27. Simplified overlap diagram for molecules with *sp* hybrid orbitals (color). All *sp²* orbitals are shown in gray.

allene

Table 3.2 Types of hybridization	type	number of equivalent hybrid orbitals	angle between hybrid orbitals	spatial arrangement of bonds
	sp	2	180°	linear
	dp	2	180°	linear
	sp^2	3	120°	trigonal planar
	dp^2	3	60°	trigonal pyramidal
	sp^3	4	109°28′	tetrahedral
	dsp^2	4	90°	tetragonal planar
	d^2sp^3	6	90°	octahedral
	d^4sp	6	73°09′ and 133°37′	trigonal prism

of certain electrons over more than two nuclei. According to the particle-in-a-box model, this additional delocalization results in a lower ground state for the electronic energy of the molecule or, what is the same thing, increased stability of the molecule.

The classic example of electron delocalization is the benzene molecule, C_6H_6. However, this is a rather large molecule for our present purposes, and we shall illustrate electron delocalization with a simpler substance, the carbonate ion, CO_3^{2-}. Electron delocalization in benzene will be treated in Section 15.5.

The bond structure of the carbonate ion is

$$\left[\begin{array}{c} :O: \\ \| \\ C \\ :O. \quad .O: \end{array} \right]^{2-}$$

The carbon atom is trigonally hybridized. Each oxygen is bonded to the carbon by a σ-type bond formed by the overlap of a carbon sp^2 hybrid orbital and a p atomic orbital from each oxygen. The carbon atom and all oxygen atoms are in the same plane. Trigonal hybridization leaves one carbon p atomic orbital alone, and this p orbital is perpendicular to the plane of the carbon and oxygen atoms. Let us call this the p_y orbital. This carbon p_y orbital can overlap with an oxygen p_y orbital to form a π bond. But which oxygen will be chosen? Any of the three oxygens may be used, and the possible O_{p_y}–C_{p_y} overlap schemes are shown in Figure 3.28, with the assumption that the signs (phases) of all p_y orbitals are as shown. It might appear that the carbonate ion could exist in one form having a double bond between C and O_1; in another form with a double bond between C and O_2; and in yet a third form with the π bond between C and O_3. The bond structures would be

$$\left[\begin{array}{c} O_2 \\ \diagdown \\ C{=}O_1 \\ \diagup \\ O_3 \end{array} \right]^{2-} \quad \left[\begin{array}{c} O \\ \diagdown\diagdown \\ C{-}O \\ \diagup \\ O \end{array} \right]^{2-} \quad \left[\begin{array}{c} O \\ \diagdown \\ C{-}O \\ \diagup\diagup \\ O \end{array} \right]^{2-}$$

$$\qquad \text{I} \qquad\qquad\qquad \text{II} \qquad\qquad\qquad \text{III}$$

These structures may be thought of in two ways: (1) as the result of different pairing schemes of the p_y electrons in the carbon and oxygen atoms, or (2) as the result of different overlap choices of adjacent carbon and oxygen p_y orbitals to form localized π molecular orbitals, into each of which are inserted 2 electrons with opposite spins.

But whichever way distinct bond structures like these are thought to arise, none of them has any reality. They are fictitious pictures. There is no reason why the p_y electron or atomic orbital on the carbon atom should discriminate among the p_y orbitals on the three oxygen atoms. What happens is (with the p_y orbitals in the orientation shown in Figure 3.28) that the carbon p_y overlaps with all of the oxygen p_y orbitals to form one large π molecular orbital encompassing all four atoms. This molecular orbital, shown in Figure 3.29, holds 2 paired electrons.

The important thing here is that, because of the spreading out or delocalization of the p_y (or π-forming) electrons, the π electron box has more room.

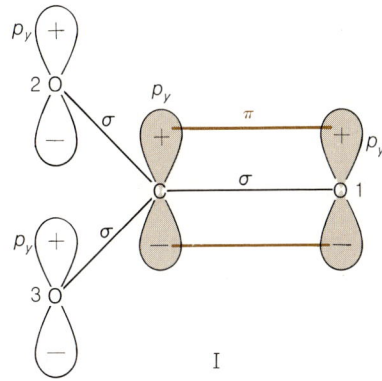

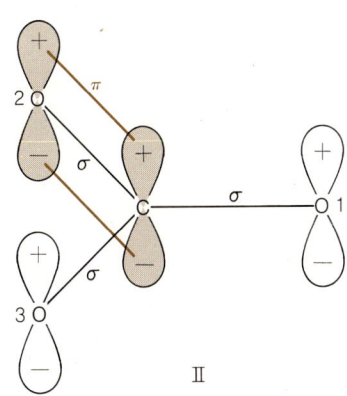

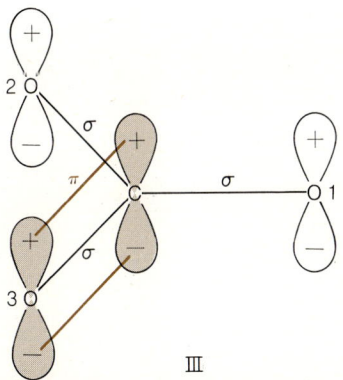

Figure 3.28. Pi orbital overlap in three equivalent contributing bond structures of the carbonate ion.

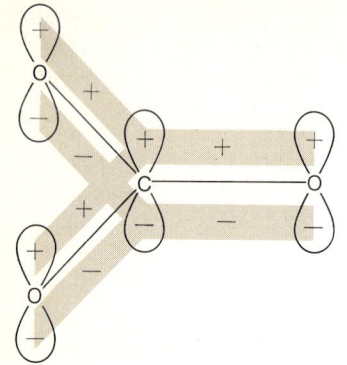

Figure 3.29. A π molecular orbital in the carbonate ion. Other molecular orbitals may be obtained by varying the orientation of the p atomic orbitals. The π orbital shown here is the one of lowest energy.

And according to the particle-in-a-box model, the energy of the carbonate ion is lower and the ion more stable than if it had pure double and single C—O bonds as shown in the three (imaginary) bond structures. In the carbonate ion, the bond order of each C—O bond can be said to be about $1\frac{1}{3}$, somewhere between a pure single and a pure double bond. Physical measurements show that all the C—O bonds are equivalent and have a bond length of 1.31 Å.

Molecules and ions such as CO_3^{2-}, for which different (Lewis-type) bond structures can be drawn, are said to have *resonance* structures, and the actual form of the molecule is a *resonance hybrid*. The meaning of the term hybrid can be appreciated by thinking about another hybrid, the mule, which is the offspring of a horse and a donkey. A mule is a mule. It is not a horse half of the time and a donkey the other half. It has some of the characteristics of its dissimilar parents, but it is a distinct animal. In fact, animal hybrids are usually more hardy than either of their parent species. In the same way, the actual carbonate ion, the resonance hybrid, is a distinct ion and is more stable or resistant to reaction than any of its imaginary so-called contributing structures I, II, or III. Resonance, or electron delocalization, is indicated by the double-headed arrow used in Figure 3.30, where other examples of molecules and ions showing electron delocalization are given.

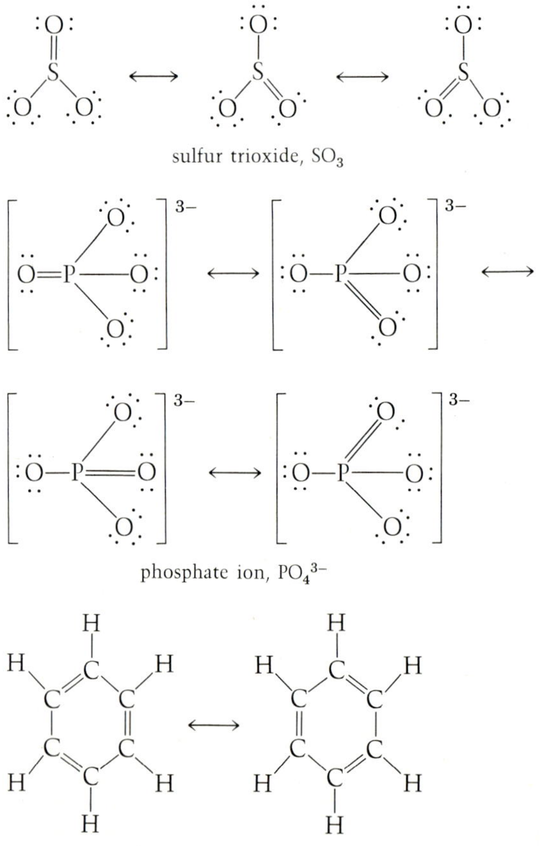

sulfur trioxide, SO_3

phosphate ion, PO_4^{3-}

Figure 3.30. Contributing structures for some molecules and ions involving resonance (electron delocalization).

benzene, C_6H_6

It is important to keep in mind that the contributing bond structures drawn to picture a resonance hybrid must be identical in the positions of the atomic nuclei and in the number of electron pairs, and have similar energy (stability). In a way, these and other stipulations for possible contributing structures illustrate the well-known fact that hybrids can result only from similar structures, but no hybrid will emerge from resonance between dissimilar structures. Thus, a horse and a donkey can produce hybrid offspring, but a horse and a monkey cannot, except in mythology.

Resonance occurs most commonly in molecules containing multiple bonds, that is, so-called unsaturated compounds. This is especially true in molecules in which the bond structures imply alternating double and single bonds as in benzene. Compounds of these types will be important to us because delocalization of π electrons is a phenomenon common to many of those molecules and structures found to be essential to living systems.

3.16 BOND ENERGY The strength of bonds has been mentioned in a qualitative way, but we now need a quantitative measure of the strength of the bond between two atoms. The energy that must be supplied to break a bond and pull the atoms apart is a good measure of how tightly two atoms are held together by a bond.

In simple diatomic molecules, the energy associated with a bond is easy to define. The *bond energy*, represented by the symbol ΔH, is the energy required to dissociate the *gaseous* molecule in its ground state into two *gaseous* atoms in their ground states. The energy may be measured in units of kilocalories (kcal), where 1 kcal is the amount of energy required to raise the temperature of 1000 g of water by 1°C.

The bond energies listed below are those required to break 1 mol (6.03×10^{23}) of bonds at a reference temperature of 298°K (25°C).

(H—H)	$H_2(g) \longrightarrow 2H(g)$	$\Delta H = 104$ kcal/mol
(O=O)	$O_2(g) \longrightarrow 2O(g)$	$\Delta H = 118$ kcal/mol
(Cl—Cl)	$Cl_2(g) \longrightarrow 2Cl(g)$	$\Delta H = 58$ kcal/mol
(N≡N)	$N_2(g) \longrightarrow 2N(g)$	$\Delta H = 226$ kcal/mol
(H—Cl)	$HCl(g) \longrightarrow H(g) + Cl(g)$	$\Delta H = 103$ kcal/mol

The bond energies of bonds in polyatomic molecules must be deduced by a somewhat different method, as the following example for CH_4 shows.

$CH_4(g) \longrightarrow CH_3(g) + H(g)$	$\Delta H = 103$ kcal/mol
$CH_3(g) \longrightarrow CH_2(g) + H(g)$	$\Delta H = 87$ kcal/mol
$CH_2(g) \longrightarrow CH(g) + H(g)$	$\Delta H = 125$ kcal/mol
$CH(g) \longrightarrow C(g) + H(g)$	$\Delta H = 81$ kcal/mol
$CH_4(g) \longrightarrow C(g) + 4H(g)$	$\Delta H = 396$ kcal/mol

The bond dissociation energies for the C—H bond are all different because of the different structures of the parent molecules. To get an average bond energy for the C—H bond, one-fourth of the total energy needed to break all four C—H bonds is taken. Thus, 396/4 kcal/mol = 99 kcal/mol is the bond energy of an average C—H bond. However, because reactions such as $CH_4(g) \longrightarrow C(g) + 4H(g)$ are not especially easy to study, indirect methods are often needed to deduce average bond energies.

An example of the use of experimental heats of actual reactions is given below. The type of experimental data from which average bond energies may be calculated is illustrated for methane.

$$CH_4(g) + 2O_2(g) \longrightarrow CO_2(g) + 2H_2O(g) \qquad \Delta H = -192.2 \text{ kcal}$$
$$CO_2(g) \longrightarrow C(graphite) + O_2(g) \qquad \Delta H = 94.0 \text{ kcal}$$
$$2H_2O(g) \longrightarrow 2H_2(g) + O_2(g) \qquad \Delta H = 114.2 \text{ kcal}$$
$$2H_2(g) \longrightarrow 4H(g) \qquad \Delta H = 206.4 \text{ kcal}$$
$$C(graphite) \longrightarrow C(g) \qquad \Delta H = 170.4 \text{ kcal}$$

$$CH_4(g) \longrightarrow C(g) + 4H(g) \qquad \Delta H = 392.8 \text{ kcal}$$

Therefore, the average C—H bond energy in methane is 392.8/4 = 98.2 kcal.

It is found that the energies calculated or measured for a given bond, say C—H, remain quite constant from molecule to molecule. Thus, although there is a slight variation, the C—H bond in methyl alcohol, H_3C—OH, has about the same bond energy as one of the C—H bonds in propane, H_3C—CH_2—CH_3. The bond energies listed in Table 3.3 can be considered as applicable to all molecules except resonance hybrids, where the bonds may not be uniquely defined. Multiple bonds between the same atoms are stronger than single bonds, but not by integral factors; that is, a double bond is not twice as strong as single bond.

3.17 BOND LENGTHS   The bond length of any bond is the equilibrium distance between the nuclei of the bonded atoms. When the distance between two identical atoms bonded

Table 3.3
Average bond energies, kcal/mol

C—C	83		H—Br	87
C=C	148		H—I	71
C≡C	194		N—N	37
C—H	99		N=N	100
C—N	66		N≡N	225
C=N	147		N—O	53
C≡N	213		N=O	145
C—O	82		N—P	50
C=O	174		N≡P	138
C—P	62		O—O	34
C—S	~60		O=O	119
C=S	114		O—P	120
C—F	102–116		O=S	112
C—Cl	78		P—P	48
C—Br	65		P≡P	117
C—I	57		P—S	55
H—H	103		S—S	63
H—N	93		S=S	84
H—O	110		F—F	37
H—S	88		Cl—Cl	57
H—P	77		Br—Br	45
H—F	135		I—I	36
H—Cl	102			

by a pure single, double, or triple bond has been measured, this distance can be halved to obtain the covalent radius of that atom. The covalent radii (single bonds) have already been noted in Table 2.11.

While there is some variation in bond length from one kind of molecule to another (again resonance hybrids are not considered), the bond length is constant enough to allow the listing of average or normal bond distances. For instance, the O—H bond length is about the same in water (H_2O) and ethyl alcohol (CH_3—CH_2—OH). Table 3.4 lists bond lengths for selected bonds.

Table 3.4
Average covalent bond lengths

bond	length, Å	bond	length, Å
C—C	1.54	H—O	0.96
C—N	1.47	H—N	1.01
C—O	1.42	H—S	1.35
C—F	1.42	C=C	1.33
C—Cl	1.77	C≡C	1.20
C—Br	1.91	C=O	1.21
C—I	2.12	C≡N	1.15
C—H	1.09	C=N	1.28

The higher the bond order, in the C—C, C=C, C≡C series for instance, the shorter the bond length. Therefore, the stronger the bond the shorter the equilibrium internuclear distance although, again, not by integral factors.

3.18 THE FITNESS OF C, H, N, O, P, S

With these ideas about chemical bonding, we are now able to get some insights into the reasons behind the frequency of a relatively few elements in living matter. For the time being, we shall restrict ourselves to those elements that participate in covalent bonding, especially the sextet of elements, C, H, N, O, P, and S, that collectively make up about 99 percent of living matter. For the moment, we shall consider only a few selected characteristics of these six elements. Other aspects of their chemistry, as well as the ions and trace elements found in living things, will be discussed later.

Carbon, Hydrogen, Nitrogen, and Oxygen

The first thing we should realize is that living organisms must have a high degree of stability or durability of structure. This means that the molecules of living organisms must be stable (in that they react slowly except under special biological conditions) but not completely inert. In addition to this internal structural stability, living organisms exhibit the property of complexity. But this is a highly organized complexity, not a complexity in the sense of disorder or chaos. A living organism, even the smallest cell or bacterium, is more complex and highly organized than any nonliving part of nature or any man-made object. Nature has constructed life in all its variety and complexity by the economical use of a few elements.

Aside from the fact that these complex organisms must be held together by some kind of stability factor, their organized complexity must be created, maintained, and repaired by some energy-consuming process. This is where the two elements, S and P, in the second group play an important role. But first we shall consider the most abundant of the six elements: C, H, N, and O.

With the exception of oxygen, these four elements are not the most abundant elements in the earth's crust today. If we leave out questions of what the composition of the earth's surface might have been when life originated on earth (if that is where it originated), we can say that the availability of the elements did not govern the constitution of living things and that some other quality or qualities of these elements especially suited them for participation in life processes and structures. The sum of these qualities has been called *fitness*.

These four elements are among the lightest in the periodic table. We observe that the lighter elements tend to be favored in living matter, almost as though organisms preferentially selected these light elements from the environment. But this alone does not tell us why they are "fit."

So far as covalent bonding goes, these four elements can attain the stable inert gas structure by acquiring, through sharing, one (H), two (O), three (N), and four (C) electrons. Together with their own valence electrons, this process would give them the He or Ne configuration. While other elements, including S and P, can share more than four pairs of electrons, the majority of cases of covalent chemical bonding in living organisms involve no more than four pairs. So the ability to acquire one to four electrons covers the usual range pretty well. However, other groups of atoms could do the same, so there must be something else that confers fitness, especially fitness in terms of the stability of the resulting compounds. This something else is atomic size.

Reference to the table of atomic radii (Table 2.11) shows that the atoms in the first and second periods are the smallest atoms. The size of an atom is, as we have seen before, related to the nuclear charge and to the electronic configuration of the atom, in particular to the number of electron shells. The four atoms under discussion are among the smallest because of the absence of filled (except for $1s^2$) electron shells beneath the valence shell.

Why is smallness important? From our discussion of covalent bonding, it follows that the closer the lobes of the atomic orbitals can get to each other, the better the overlap and the stronger the bond. There is, of course, a limit to this closeness; the nuclei cannot be jammed together too closely or nucleus–nucleus repulsion will begin to predominate. In general, however, the smaller the bonding atoms, the stronger the covalent bond. Refer again to the tables of bond energy (Table 3.3) and bond length (Table 3.4). These four atoms form strong, stable bonds because of their small size. The stable, relatively unreactive bonds confer stability on an organism and help it to retain its organized complexity.

In addition to strong single covalent bonding among these four elements, three of them—O, N, and C—also exhibit a tendency to form multiple bonds as a result of their small size and the existence of p atomic orbitals of the proper energy, which can overlap to form π molecular orbitals. As the p atomic orbitals used to form π bonds are not directed along the line of the nuclei, closeness of approach of the small atoms is especially necessary to π bond formation. In fact, multiple bond formation is pretty well limited to the period containing these three atoms and within this period is restricted to O, N, and C. The major exceptions are S and P. Multiple bonds, in addition to being stronger than single bonds, are also of importance in conferring extra stability to molecules because their presence is essential for resonance in molecules. Pi bond resonance, in turn, causes molecular electronic energy levels to become closer together (see the particle-in-a-box model), so that a molecule with π bond resonance can be electronically excited by promotion of a π electron to a higher energy level with the low-energy quanta available in living organisms. This factor is important in the transmission of energy in living organisms and in the capture of radiant energy in photosynthesis.

The other reason why larger atoms do not participate in the formation of strong covalent bonds is the existence of paired electrons not involved in bonding. These nonbonding electrons include some valence shell electrons as well as those residing in orbitals below the valence shell orbitals. When atoms having such nonbonding electrons try to approach each other to form a covalent bond, these electrons in each atom strongly repel each other and contribute to the weakness of any bond formed by the available bonding electrons in the valence shell. Alternatively, it could be said that the net repulsion of $\sigma\sigma^*$ and $\pi\pi^*$ nonbonding pairs leads to weak bonds. Furthermore, any tendency by larger atoms to form the shorter multiple bonds introduces more electron–electron repulsions. Thus, even neglecting size differences, phosphorus ($1s^2 2s^2 2p^6 3s^2 3p^3$) would be expected to form weaker bonds than nitrogen ($1s^2 2s^2 2p^3$) because of electron–electron repulsion resulting from the nonbonding $2s$ and $2p$ electrons below its valence shell.

Carbon Versus Silicon
The comparative chemistry of carbon and silicon with respect to their role in life affords many examples of how factors of atomic structure and bonding determine fitness. The subject has also provided a platform for some imaginative, if scientifically incorrect, forays into science fiction. Some comparisons of carbon and silicon will be briefly noted here.

Carbon and silicon both belong to the same periodic group and would be expected to have similar chemical properties. Both might be expected to form four covalent bonds and combine with themselves to form chains: —C—C—C—C— or —Si—Si—Si—Si—.

Why is it that silicon is not involved in living material to any extent? The answer can be found by comparing the strength and stability of bonds formed by silicon with that of bonds formed by carbon. The C—C bond is stronger than the Si—Si bond, and silicon does not form multiple bonds because of its larger size. Also, the Si—Si bond is weaker (has a lower bond energy) than bonds formed between silicon and other elements. Compare the bond energies, Si—Si (43 kcal/mol), Si—O (89 kcal/mol), and Si—H (81 kcal/mol). This means that Si—Si bonds will tend to break and form bonds with other atoms, particularly oxygen. Any complex molecule involving Si—Si chains will be unstable in the presence of water or oxygen. No more than six silicon atoms have ever been linked together in a stable chain; there is apparently no limit to the number of carbons that can be joined together.

Carbon–carbon single bonds are about equal to or stronger than single bonds between carbon and other elements. Compare the bond energies: C—C (83 kcal/mol), C—H (99 kcal/mol), C—O (86 kcal/mol). There is little tendency for the C—C bonds in important biological molecules to convert to bonds with other atoms. Finally, the C—H bond is important in biological molecules. The C—H bond has a higher bond energy than the C—O bond, so there is little tendency for the C—H bond to break and the carbon to form

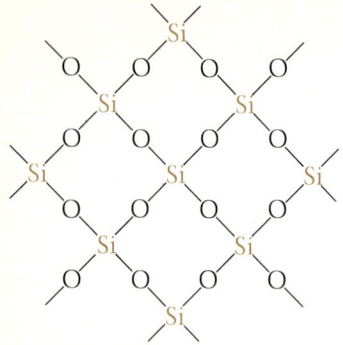

Figure 3.31. Planar view of the network of Si—O—Si bonds in quartz (silica). The ratio of Si to O in this tetrahedral network is given as SiO_2, but there are no individual SiO_2 units. See Figure 13.21 for an illustration of the three-dimensional nature of this network composed of SiO_4 tetrahedra.

a C—O bond by reaction with oxygen at ordinary temperatures. The Si—H bond, on the other hand, reacts with water to form Si—O (silica) bonds and hydrogen.

Maximum formation of multiple bonds by an atom tends to engage all of its valence electrons in bonds. This results in what has been called saturation of valence. For instance, in carbon dioxide, CO_2, all of the valence electrons of carbon are used in bonding:

$$:\ddot{O}::C::\ddot{O}:$$

There are no unshared electron pairs on carbon. This sort of situation causes the CO_2 molecule to be relatively inert to interaction with other molecules or atoms, and CO_2 exists as a comparatively unreactive free gas. The important position of CO_2 in life processes is due primarily to its existence as a gas at ordinary temperatures. Discrete molecules of SiO_2, on the other hand, do not exist, even as a gas. The Si—O bonds are single bonds, and no multiple bonding takes place, thus precluding the independent existence of an SiO_2 molecule. What occurs instead is a network of SiO_4 tetrahedra (Figure 3.31) all joined together to form hard, dense quartz, which does not even melt until the temperature reaches 1600°C. On the basis of these arguments, we can conclude that life based on silicon rather than carbon is highly unlikely.

Sulfur and Phosphorus

The two third-period elements, S and P, lack the small size that makes H, C, O, and N especially fit for their structural role in living organisms. However, the function of sulfur and phosphorus is not quite the same as those of the more popular quartet. Sulfur and phosphorus, but especially the latter, find many of their primary uses in the role of energy transfer and storage in organisms. The very factors that make sulfur and phosphorus less fit for structural use are the same factors that make them especially fit for the other role they do play. Aside from the fact that sulfur and phosphorus can form multiple bonds, their characteristics differ from those of carbon, hydrogen, oxygen, and nitrogen. Their larger size causes their bonds to be weaker and longer than bonds involving only the C, H, O, and N quartet, and the existence of empty orbitals of the same quantum number as the outermost electrons enables them to form more than four covalent bonds. The effect of weaker bonds and of more bonds will be considered briefly.

Bonds to sulfur and phosphorus are weaker and longer because of the larger size of these atoms and because of the underlying nonbonding electrons. These bonds can be more easily broken in chemical processes occurring in the living organism, and part of the energy originally stored in the molecule can be transferred or used to drive other chemical reactions necessary for life. It is fortunate, then, that the bonds are somewhat weak and susceptible to breaking in various reactions. Otherwise, energy could be made available only by disrupting the carbon chain and destroying the organism. Of course, the bond cleavage reactions in the organism are not the same as the reaction used to define the term "bond energy." That is, no gaseous atoms and molecules are involved. But the bond energy is still a useful criterion of the susceptibility of the bond to attack in a variety of chemical reactions.

Phosphorus and sulfur have the electronic configurations

P $1s^22s^22p^63s^23p^33d^0$
S $1s^22s^22p^63s^23p^43d^0$

The empty $3d$ orbitals are shown to indicate their availability to accept electrons. The electrons that can go into the d orbitals may come from (1) unpairing and promotion of paired electrons in the same atom or (2) unshared electron pairs donated by another atom, with coordinate bond formation. According to the first option, phosphorus and sulfur could then exist in two or more electronic states:

P	$1s^2 2s^2 2p^6 3s^2 3p^3 3d^0$	3 unpaired electrons
P*	$(Ne)3s^1 3p^3 3d^1$	5 unpaired electrons
S	$(Ne)3s^2 3p^4 3d^0$	2 unpaired electrons
S*	$(Ne)3s^2 3p^3 3d^1$	4 unpaired electrons
S**	$(Ne)3s^1 3p^3 3d^1 3d^1$	6 unpaired electrons

Alternatively, we might say the existence of the d orbitals allows different kinds of hybridization, without the necessity of specifying unpaired electrons. Phosphorus would then be able to share three or five electron pairs, as in PCl_3 and PCl_5. Sulfur would have the choice of forming two, four, or six covalent bonds, as in SCl_2, SCl_4, and SF_6. This type of "valence shell" expansion for sulfur and phosphorus or other third- and fourth-period elements rarely goes beyond five or six covalent bonds. In IF_7, seven bonds can be formed, presumably because the large size of I prevents crowding of the F atoms.

For biological purposes, the empty $3d$ orbitals in sulfur and phosphorus play a role. These orbitals present an opportunity for the formation of additional coordinate covalent bonds by the acceptance of unshared electron pairs from certain agents such as water. The larger size of sulfur and phosphorus apparently allows the attacking agents to approach the atoms closely to form the coordinate bonds. These bonds may or may not be permanent, but their formation causes the molecule containing the sulfur or phosphorus atom to break up, often with the release of energy needed for vital processes. In the case of sulfur and phosphorus, their larger size and the existence of d orbitals qualifies these two atoms for special functions in living organisms, even though the same properties partly disqualify them for the structural roles played by carbon, hydrogen, nitrogen, and oxygen.

Silicon is larger than sulfur or phosphorus and does not retain the ability to form multiple bonds. It is disqualified for participation in either group of elements. In its compounds, silicon, because of its $3d$ orbitals, shares the susceptibility to attack by agents with unshared electron pairs.

To summarize, the important properties that determine the fitness of the two groups of essential elements are as follows.

Carbon, hydrogen, nitrogen, oxygen:

1. Small size resulting in strong bonds.
2. Few underlying nonbonding electrons.
3. Ability to form multiple bonds (except hydrogen).
4. No energetically attainable empty orbitals in or near the valence shell, resulting in easy saturation of valence and high resistance to attack.

Sulfur, phosphorus:

1. Larger size, resulting in weaker bonds that can be more easily broken in vital reactions.
2. Retention of the ability to form multiple bonds, leading to resonance.

3. Empty $3d$ orbitals, which increase susceptibility to attack and reaction of the molecule in vital processes.

It turns out also that the fitness of these elements is determined as well by their ease of circulation through the biosphere, from land to sea to atmosphere to life, or merely in the cycle from living organism through the physical environment and back to life again. All of these elements, except phosphorus, have gaseous compounds that utilize the atmosphere in their travels through the environment.

There are other elements that have some of the properties of one or more of the six elements just discussed. But during the long chemical evolution of life, only those elements that work best have been selected, while the elements that did not quite measure up were dropped by the wayside so far as large-scale participation in biological systems was concerned.

The properties of these elements render them so fit for their function in life that it is generally believed that life of any kind cannot exist anywhere in the universe where any of these elements is missing. For instance, the Mariner probes of the planet Mars from 1969 to date have detected essentially no nitrogen and only trace amounts of water vapor in the atmosphere of that planet. The absence of nitrogen was sufficient to squash hopes of any kind of life existing on Mars. That is why the search for life on other planets and galaxies is really a search for the presence of these six elements. Also, since these six elements are so important, it is inevitable that we should devote thought in later chapters to their circulation through the biosphere. This circulation involves exchanges of these and other elements between the physical environment and living organisms—between the external and the internal environment of life. It is to the external environment that we now turn.

3.19 SUMMARY

When the total potential energy of two or more atoms decreases as the atoms are brough near each other, chemical bonds form and hold the atoms together.

Two major categories of chemical bonds are ionic and covalent bonds. An ionic bond is the attractive electrostatic force between two oppositely charged ions. In principle, the formation of an ionic bond requires the loss of electrons by one atom and the gain of electrons by another atom to form positive and negative ions. Ionic bonds, therefore, occur most commonly between elements with low ionization energy and elements with high electron affinity.

Covalent bond formation involves the sharing of electron pairs between two atoms. These atoms may be the same or different.

Both ionic bonding and covalent bonding have been rationalized as resulting from the tendency of bonded atoms to attain the particularly stable electronic configuration of the inert gases. Many of the commonly encountered compounds are composed of elements near the ends of the first two eight-membered periods of the periodic table, and these elements will tend to assume the eight-electron outer configuration of argon or neon. This has led to the rule of eight or octet rule, but there are many exceptions to the rule.

Covalent bonding is best explained by the overlapping of atomic or hybrid orbitals on bonding atoms. Overlapping gives molecular orbitals of various energies and symmetries, each of which may hold one electron pair.

Electrons in a lower-energy or bonding molecular orbital encompass two (or more) nuclei rather than being localized around a single atom. This delocalization and the accompanying buildup of electron charge density between the nuclei lead to a decrease in potential energy and constitute the covalent or electron pair bond. Covalent bonding is ultimately electrostatic in nature.

The electrons in a molecule fill the available molecular orbitals to give the molecular electronic configuration of the molecule, in the same way that the electronic configuration of an atom is determined by the progressive occupancy of atomic orbitals.

Many atoms can form stronger and more directional covalent bonds if we imagine the mixing or hybridization of some or all of their valence atomic orbitals. Through the concept of hybridization, we may explain the bonding and geometry of carbon compounds in particular.

Some molecules may be represented by two or more acceptable bond structures with different arrangements of the electrons. These molecules are said to exhibit resonance. Resonance is best regarded as a delocalization of bonding electrons over more than two nuclei in the molecule due to alternative overlapping possibilities of π-type atomic orbitals in the molecule. Electron delocalization leads to increased stability (lower energy) of the molecule and decreased spacing between molecular energy levels.

The six most abundant elements found in living organisms all form covalent bonds with one another. The strength and number of the bonds formed by any element depend on the electronic configurations of the bonded atoms. The smallness of the carbon, hydrogen, oxygen, and nitrogen atoms, and the absence of underlying electron shells, permit close approach of other atoms and maximum overlap of atomic orbitals to form strong single and multiple covalent bonds. The availability of d orbitals in sulfur and phosphorus permits them to form more, but weaker, bonds than the former group of elements.

Questions and Problems

1. Write Lewis structures for the following molecules and ions, noting and explaining any deviations from the "rule of eight."
   a. $SiCl_4$
   b. SCl_4
   c. BF_4^-
   d. Na_2O
   e. SCl_6
   f. NH_2OH
   g. CH_3-CH_2Cl
   h. IF_5
   i. SO_4^{2-}
   j. CsF

2. The sulfur fluorides SF_2, SF_4, and SF_6 exist, but SF_3 and SF_5 do not. Why?

3. The geometry of molecules involving covalent bonding among atoms obeying the rule of eight, that is, those of the first and second periods having four electron pairs about them in their compounds, may be considered from the standpoint of the sharing of corners, sides, and faces of tetrahedra with the nuclei at their centers. The four electron pairs, both shared and unshared, are directed toward the corners of the tetrahedra.
   a. Refer to Figure 3.21 and visualize or sketch the four-sided tetrahedron whose edges are defined by the lines joining the hydrogen atoms at the four corners of the cube. Better yet, obtain some models of tetrahedra.
   b. Satisfy yourself that a single bond between two

carbon atoms, as in $CH_3—CH_3$, $CH_3—CH_2OH$, or $CH_3—CH_2—CH_3$, can be regarded as a sharing of electrons at a corner or apex of two tetrahedra with the carbon atoms at their centers and the other groups at the unshared corners of the tetrahedra.

c. Satisfy yourself that the geometry of doubly bonded carbon atoms, as in $CH_2=CH_2$, can be represented by the sharing of an edge (or two electron pairs at two corners) of adjacent tetrahedra. Note that the geometry of the double bond in $H_2C=O$ can also be visualized in this way, with the two unshared pairs of the oxygen directed at tetrahedron corners in the same plane as the two hydrogens. What about CO_2?

d. Convince yourself that the geometry of groups about triply bonded atoms of first- and second-period elements (again not necessarily carbon) can be pictured by considering the sharing of faces between tetrahedra (the sharing of three electron pairs at three corners). Examples are $CH\equiv CH$, N_2, and HCN. Again, unshared pairs may be directed at corners as well as atoms.

4. Give all suitable electron dot structures for ozone, O_3, and discuss its structure (bond angle, bond lengths, etc.).

5. Draw a Lewis dot structure and an orbital overlap diagram, and predict the structure for hydrogen cyanide, HCN.

6. Why is a triple bond not three times as strong, and a double bond twice as strong, as a single bond between the same pair of atoms?

7. The higher the number of nodes in a molecular orbital, the higher the energy of that molecular orbital. Why should this be the case?

8. In allene, $H_2C=C=CH_2$, the two CH_2 groups lie in planes perpendicular to one another (Figure 3.27).

a. Is electron delocalization possible between the outermost carbon atoms, or are the two double bonds pure, localized double bonds?

b. What would happen to the π_y molecular orbital if the right-hand CH_2 groups were rotated into the same plane (the plane of the page in Figure 3.27) as the left-hand CH_2 group? Do you think that such a rotation might occur spontaneously? Why?

9. Sketch orbital overlap diagrams for the following, indicating where resonance is possible.

a. NO_2 (ONO)

b. NO

c. NO_3^-

10. The C—O bond length in CO_3^{2-} is 1.31 Å. Compare this with the bond lengths for a pure single (C—O) and double (C=O) carbon–oxygen bond, and explain any differences.

11. Verify the four orbital overlap diagrams for the $p_x \pm p_x$ and $p_y \pm p_y$ molecular orbitals of Figure 3.11 by sketching the atomic orbital wave functions and showing their interference and reinforcement.

12. Discuss bond properties such as relative bond energy, bond lengths, and bond orders in the following series.

a. H_2^+, H_2^-, H_2

b. Cl_2, Cl_2^+

c. O_2^+, O_2, O_2^-, O_2^{2-}

d. N_2^{2+}, N_2^+, N_2, N_2^-, N_2^{2-}

Why does the bond in N_2^+ have a higher bond energy than the bond in O_2^+?

13. Contrast the bond orders shown in Table 3.1 with the number of electron pairs between nuclei as predicted by Lewis electron dot diagrams.

14. What do you estimate the bond lengths of the CO bonds in the $HCOO^-$ ion to be? The hydrogen and the two oxygen atoms are each directly bonded to carbon. Explain your estimates.

15. In early periods of the earth's history, life existed under environmental conditions quite different from the present. For instance, the absence of free oxygen molecules meant that organisms utilizing oxygen did not exist. In fact, molecular oxygen would have poisoned some of these anaerobic organisms. Today we know that oxygen-utilizing organisms (aerobes) can extract energy from food much more efficiently than anaerobic organisms. Argue the validity of the following statements.

a. Former environments were particularly fit for life.

b. Our present environment is particularly fit for life and may be the "best of all possible worlds."

c. The fitness of the environment for life at any particular moment is quite as essential as the fitness of organisms achieved by adaptation to the environment.

Some problems involving chemical equations.

A number of problems in later chapters will require the calculation of the amount of reactants used up or the amount of products formed in chemical reactions. The treatment of such problems is discussed in Appendix B.3. Here are a few problems for practice in this type of calculation.

16. Ammonia reacts with oxygen according to the equation

$$4NH_3 + 5O_2 \longrightarrow 6H_2O + 4NO$$

How many grams of NO and H_2O can be formed by the complete reaction of 1.7 g of NH_3?

17. According to the reaction

$$2KMnO_4 + 16HCl \longrightarrow$$
$$2KCl + 2MnCl_2 + 5Cl_2 + 8H_2O$$

how many grams of chlorine can be prepared from 7.30 g of HCl?

18. What amount of aluminum must be used to react completely with 1000 g of iron oxide, Fe_3O_4, in this reaction?

$$8Al + 3Fe_3O_4 \longrightarrow 4Al_2O_3 + 9Fe$$

19. It is desired to prepare 5.00 g of H_2 by the reaction

$$2Al + 6HCl \longrightarrow Al_2Cl_6 + 3H_2$$

How much aluminum must be used?

20. A solution of calcium chloride containing 1.11 g of $CaCl_2$ is mixed with another solution containing 2.55 g of $AgNO_3$. The reaction forms silver chloride:

$$CaCl_2 + 2AgNO_3 \longrightarrow 2AgCl + Ca(NO_3)_2$$

How much silver chloride is formed?

Answers

14. both about 1.3 Å
16. 3.0 g of NO, 2.7 g of H_2O
17. 4.4 g of Cl_2

18. 310 g of Al
19. 44.6 g of Al
20. 2.15 g of AgCl

References

Companion, Audrey L., "Chemical Bonding," McGraw-Hill, New York, 1964.

Margolis, Emil J., "Bonding and Structure," Appleton, New York, 1968.

Pimentel, George C., and Spratley R. D., "Chemical Bonding Clarified Through Quantum Mechanics," Holden-Day, San Francisco, 1969, Chapters 3, 4, 5.

Price, Charles C., "Geometry of Molecules," McGraw-Hill, New York, 1971.

Ryschkewitsch, George E., "Chemical Bonding and the Structure of Molecules," Van Nostrand Reinhold, New York, 1963.

Speakman, J. C., "Molecules," McGraw-Hill, New York, 1966.

Speakman, J. C., "A Valency Primer," Edward Arnold, London, 1968.

TWO

Air

The just completed study of atoms and molecules will in many ways provide the basis for our treatment of the physical and the living realms of the biosphere. Because of the references to the function of the six primary elements in living organisms, it might appear logical to go directly to a more detailed view of the chemistry of life processes. But this would immediately involve us in the complexities of large biological molecules and their chemistry. It is more appropriate to begin with the simpler substances and processes found in the physical environment within which biological structures must exist. The atomic and molecular properties of the substances making up the physical environment are every bit as important in determining fitness for life as are those of the internal environment.

The Greek philosopher Empedocles (ca. 490–435 B.C.) was the first to quarter the world into the four elementary but inert substances Earth, Air, Fire, and Water (all activated by the energetic substances, Love and Strife). We shall adopt these four "elements," not as the elements of matter, but as the major elements of our physical environment. Fire, or energy, provides the link between the inanimate universe and life. Of the three tangible "elements," air is taken up first primarily because the properties of gases, with their smaller and more distinct chemical species, are most easily approached on the microscopic level.

The relative ease of treatment is not the only reason the invisible mantle of air we call the atmosphere has been chosen. The atmosphere interacts chemically and physically with the solid earth and the water to determine the conditions we find fit for life. Within the air itself, we find all of the substances commonly associated with life itself: oxygen, water, and carbon dioxide. The atmosphere is a major way station and transporter in the travels of these compounds through the biosphere.

The Gaseous State

Man, as well as most other living things in the biosphere, moves around near the bottom of an immense ocean of air some 100 miles thick. This relatively thin envelope of gas surrounding the earth is called the *atmosphere*.

4.1 STRUCTURE OF THE ATMOSPHERE

As one moves upward from the earth's surface, the properties of the atmosphere change. The atmosphere has been marked off into a vertical layer structure on the basis of these changes. The most familiar divisions for the atmosphere's vertical structure are based on the variation of temperature with height. The various regions and their temperatures are shown in Figure 4.1. In the *troposphere*, the layer closest to the earth, the temperature decreases, generally at a rate of 3.6°F (2°C) every 1000 ft (305 m). At about 7 miles above the earth (depending on latitude), the temperature decrease ends, and

117

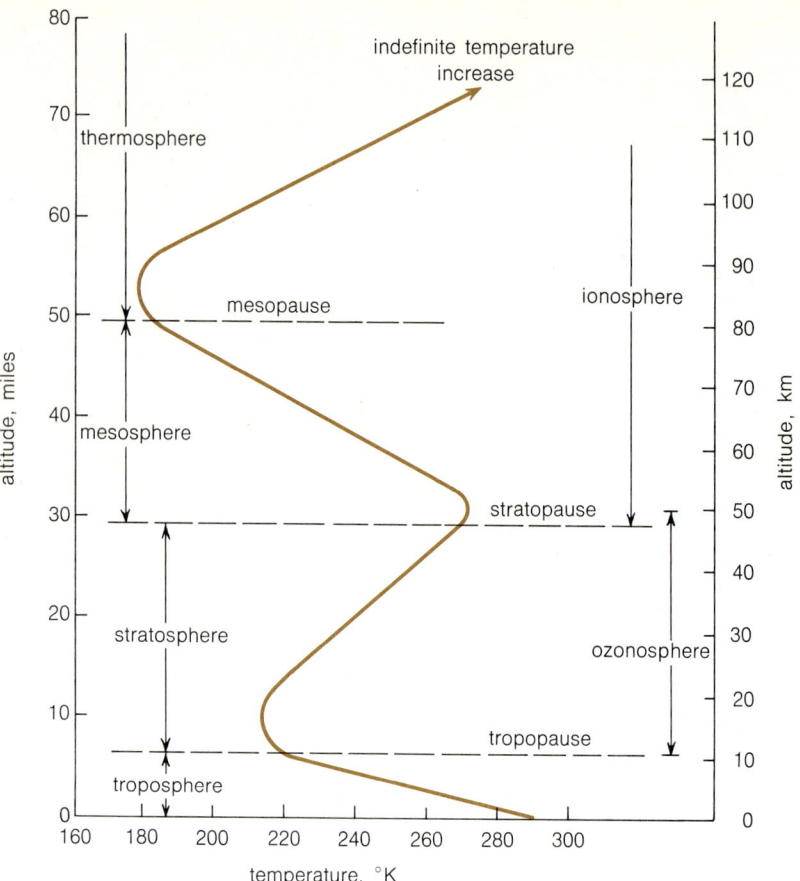

Figure 4.1. Layer structure of the atmosphere.

the temperature remains constant for several miles and then increases slowly with height. This region, where the temperature is either constant or increases with height, is the *stratosphere*. Beyond the stratosphere, at a height of about 30 miles, is the *mesosphere*, where the temperature again decreases with increasing altitude until, once again, at the beginning of the *thermosphere*, the temperature rises. There is no limit to this increase in temperature. The variations in temperature are due to the absorption and reemission of solar radiation by the components of the atmosphere in the various regions. Generally, below the mesopause, the composition of the atmosphere, in terms of the chemical species present, is fairly constant; but above the mesopause, a variety of chemical species occur at various altitudes.

Another designation for atmospheric layers is based on the kind of processes taking place in those regions. These processes are brought about by absorption of radiation. In the ionosphere, many atoms and molecules are extensively ionized. In the higher chemosphere, a great variety of chemical reactions, initiated by radiation, take place. In the ozonosphere, a relatively high concentration (10 ppm) of ozone is formed from oxygen molecules and atoms.

Table 4.1
Nonvariable components of clean
dry air near the earth's surface

component	symbol	content	
		% by volume	ppm by volume
nitrogen	N_2	78.084	
oxygen	O_2	20.946	
argon	Ar	0.934	9340
carbon dioxide	CO_2	0.033	330
neon	Ne	0.001818	18.18
helium	He	0.000524	5.24
methane	CH_4	0.0002	2
krypton	Kr	0.000114	1.14
hydrogen	H_2	0.00005	0.5
nitrous oxide	N_2O	0.00005	0.5
xenon	Xe	0.0000087	0.087

Aside from the formation of ozone in the ozonosphere, we shall be little concerned with the regions of the atmosphere above the troposphere. It is highly unlikely that any life can exist without protection even very far up into the troposphere because of the inability of life to exist in regions of the intense ultraviolet radiation. The troposphere contains about 75 percent of the mass of the entire atmosphere of the earth, including all of the gases, aside from ozone, that have a direct bearing on the fitness of the atmosphere for life on earth.

4.2 COMPOSITION OF THE LOWER ATMOSPHERE

The gaseous chemical elements and molecules comprising the lower atmosphere can be conveniently divided into two groups. In the group listed in Table 4.1 are those whose relative concentration in the mixture we call dry air is fairly constant with respect to location on the surface of the earth and altitude. In the second group, Table 4.2, are those components whose concentration may depend on local conditions, latitude and longitude, or altitude. When we say that the composition is constant, we do not mean that there is the same number of grams or moles of oxygen, for instance, in 1 liter (1 liter = 1000 cubic centimeters, cm^3, = 1000 milliliters, ml) of air at an altitude of 25 miles as at sea level. What is meant is that the ratio of the

Table 4.2
Variable components in air

component	formula	normal content, by volume
water vapor	H_2O	0.1–1%
ozone	O_3	0.02–0.07 ppm
ammonia	NH_3	0.01 ppm
sulfur dioxide	SO_2	0.0002 ppm
carbon monoxide	CO	0.1 ppm
nitrogen dioxide	NO_2	0.001 ppm
hydrogen peroxide	H_2O_2	—
sulfur trioxide	SO_3	—
radon	Rn	—
dust (soot, pollen, sea salt)		

amount of oxygen to the amount of another component is constant. One convenient way of expressing the composition is in terms of percent by volume, which gives the number of liters of one gas contained in 100 liters of the mixture.

EXAMPLE   There are 934 ml of Ar in exactly 100 liters of air at sea level. What is the percent by volume of Ar in the mixture?

SOLUTION   934 ml = 0.934 liters of Ar in 100 liters of air. Therefore, the volume percent of Ar is 0.934 percent.

EXAMPLE   There are 8.0 liters of CO_2 in a mixture of gases having a volume of 30 liters. What is the percent of CO_2 in the mixture?

SOLUTION   If there are 8.0 liters in 30 liters of gas, 100 liters of the gas mixture could contain more CO_2 by a factor $100/30 = 3.3$. Thus, the volume of pure CO_2 in 100 liters of mixture is $3.3(8.0) = 26$ liters in 100 liters of gas, or 26 percent by volume.

It may be easier to use proportions in problems like this (see Appendix A.4):

$$\frac{8.0 \text{ liters of } CO_2}{30 \text{ liters of mixture}} = \frac{x \text{ liters of } CO_2}{100 \text{ liters of mixture}}$$

$$x = \frac{8.0}{30}(100) = 26\%$$

or, in general,

$$\% \text{ volume} = \frac{\text{volume of pure gas}}{\text{total volume of mixture}} \times 100$$

With gases that are present in very small volume percentages, it is more convenient to use parts per million (ppm) to express their concentration in a gas mixture, as is done in the following smog report for a comparatively pleasant day in a large city.

	carbon monoxide ppm	nitric oxide, NO ppm	oxidants (O_3, etc.) ppm
today	18	0.139	0.029
full alert level	70	0.700	0.400

This avoids writing very small decimal numbers. Percent by volume is really parts per hundred (pph). Then, by analogy with the percent by volume, we can see that

$$\text{ppm} = \frac{\text{volume of pure gas}}{\text{total volume of gas mixture}}(1{,}000{,}000)$$

$$= \frac{\text{volume of pure gas}}{\text{total volume}} \times 10^6$$

Thus, $\text{ppm} = \text{pph} \times 10^4 = (\%) \times 10^4$.

EXAMPLE | There are 8.3 ml of Xe in 1200 liters of a gas mixture. What is the volume percent of Xe and what is its concentration in parts per million?

SOLUTION

$$\text{ppm Xe} = \frac{\text{volume of Xe}}{\text{total volume}} \times 10^6$$

$$= \frac{8.3 \text{ ml}}{1200 \text{ liters} (1000 \text{ ml/liter})} \times 10^6$$

$$= \frac{0.0083 \text{ liter}}{1200 \text{ liters}} \times 10^6 = 6.9$$

$$\% \text{ Xe} = \frac{8.3 \text{ ml}}{1200 \text{ liters} (1000 \text{ ml/liter})} \times 100$$

$$= \frac{0.0083 \text{ liter}}{1200 \text{ liters}} \times 100 = 6.9 \times 10^{-4}$$

Note that the units cancel, so percent or parts per million is a dimensionless number.

The same scale is also applicable to weight (mass) concentrations. Weight percent is the same as the weight units of a component in 100 weight units of a mixture. Parts per million is similarly defined with reference to 10^6 weight units of mixture. Thus, if 100 g of a mixture contains 0.002 g of some component, that component is present at 20 ppm by mass.

4.3 COMPOSITION OF THE EARLY ATMOSPHERE

It is of interest to compare the composition of the present-day air to the composition at the time of the origin of the atmosphere. The origin and evolution of the earth's atmosphere and the relation to the origin and history of the earth will be discussed more fully in Chapter 13. At this point, we shall consider briefly the change in composition of the atmosphere through the ages.

No matter how the earth was originally formed about 4.5 billion (4.5×10^9) yr ago, whether from the condensation of very hot gases or from the aggregation of cold cosmic dust, it is generally agreed that the earth was very hot and molten. In this early condition, it is likely that any atmosphere would quickly be lost by escape into space in much the same way that a boiling liquid evaporates (see Section 4.17). As the molten earth cooled and began to solidify, gases (such as CO_2, H_2O, and N_2) that had dissolved in, or chemically combined with, the molten rocks, were given off to form one stage of the earth's primitive atmosphere. No oxygen was present.

As the earth cooled further, the water vapor began to condense, forming the oceans. Still no oxygen. It is in these oceans that life is thought to have originated. Somehow, much later, green plants came into existence. Through photosynthesis, plants could convert carbon dioxide and water to carbohydrates, such as sugar, and oxygen. Over the millions of years, it is thought that green plants have been responsible for the production of all the oxygen gas in our present atmosphere and the maintenance of the oxygen and carbon dioxide concentrations at stable values. This is one of the most striking natural examples of life affecting the physical environment.

The major gases in our present atmosphere, oxygen and nitrogen, and other gases (principally carbon dioxide and water vapor, but also including nitrous oxide and ammonia) containing these two elements in combination with other elements, are linked to oxygen and nitrogen compounds in living and nonliving materials in the oceans and on land. To understand the nature and operation of the mechanisms that facilitate the circulation of these elements through the biosphere, it is necessary to have some knowledge of the properties of gases in general as well as of the specific chemical characteristics of nitrogen and oxygen and their compounds. This chapter is devoted to the general properties of the gaseous state of matter.

4.4 GAS—A STATE OF MATTER

The three common physical states of matter—gas, liquid, solid—of any substance can be differentiated in a number of ways, but one of the most basic is in terms of their internal structure. How are the atoms or molecules making up a substance arranged in space, why are they arranged like this, and what properties does a given arrangement confer on a substance in that state? The structure of a substance in any physical state is determined by the forces between its component particles in much the same way as the structure of a molecule is determined by the forces between its component atoms.

We have not talked about forces between molecules yet, nor about experimental properties of gases. But we shall anticipate some of the results of these discussions by looking briefly at the molecular order in the gaseous, liquid, and solid states of matter. Figure 4.2 gives a rough idea of the arrangement of component particles in the three physical states of matter. The solid state is characterized by closely packed, ordered particles. The surroundings of one particle, say the middle one, look just the same as the surroundings of any other particle, with the exception of those at the edges of the container. This applies both to the immediate and the distant surroundings. There is short-range and long-range order.

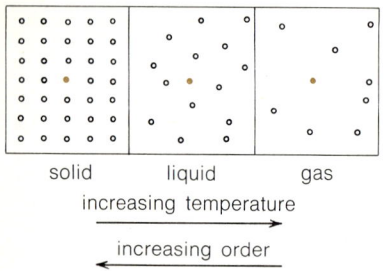

solid   liquid   gas

increasing temperature →

← increasing order

Figure 4.2. Internal structure in the three common states of matter.

As the temperature increases, the forces holding the particles together in their ordered solid arrangement are overcome so that the long-range order disappears, although much of the short-range order remains. A substance is now in the liquid state, and, on the average, the particles are slightly farther from one another.

Finally, with still higher temperatures, all order breaks down. The forces of attraction between particles have been overcome, and the particles are far apart and in random motion. This complete lack of internal order characterizes the gaseous state. We shall see that the measured properties of gases are consistent with the loose arrangement of the randomly moving particles in a gas.

The properties needed to describe gases are mass, volume, pressure, and temperature. All of these properties are related to one another, and the mathematical expression of this relation is called an *equation of state* of the gas. The mass of any substance determines the quantity of matter present. Masses of two different samples of matter may be compared using a balance. The volume of any substance is the amount of space occupied by the substance. A gas, however, does not have a definite volume. A gas always expands to the shape and volume of the container holding the gas. The volume of a gas is the space in which the particles of the gas are free to move about.

For pressure and temperature, operational definitions are in order. Operational definitions describe a quantity or concept by telling how it is measured, but they say nothing about the basic meaning or origin of the quantity.

4.5 PRESSURE

Pressure is defined as the force exerted on a unit of surface. At sea level, the earth's atmosphere exerts a force of about 14.7 lb on a square inch of any surface. This force is equal in magnitude in all directions, so there is no unbalanced force at the earth's surface on an object surrounded by the atmosphere. On a larger scale, of course, there are atmospheric pressure differences caused primarily by uneven heating of the air at different locations on the earth's surface. The resulting pressure gradients cause winds and other air movements responsible for our weather.

The pressure of the atmosphere can be measured with a *barometer* (Figure 4.3). This is a long glass tube closed at one end, which has been filled with mercury and then inverted in a dish of mercury. Some mercury runs out of the tube, but enough remains to form a standing column with its top surface approximately 76 cm (760 mm) above the level of the mercury in the dish. At this height, the pressure at the bottom of the column of mercury in the tube is equal to the pressure of the atmosphere on the surface of the mercury in the dish. The atmospheric pressure holds up or supports a column of mercury approximately 76 cm = 760 mm high. Thus, a barometer rises or falls as the pressure of the atmosphere on the surface of the mercury changes. If the column height is exactly 760 mm, the pressure of the atmosphere is *one atmosphere* (atm).

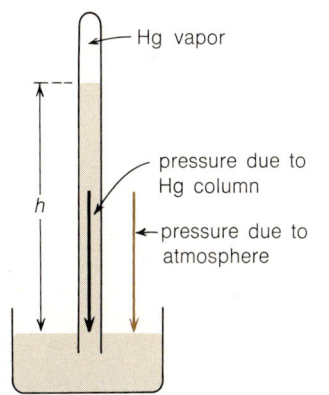

Figure 4.3. Mercury barometer. The pressure due to mercury vapor above the column is negligible.

1 atm = 760 mmHg = 76 cmHg = 30 in. Hg

(Pressure may also be given in torr units. One Torr is 1 mmHg.)

4.6 TEMPERATURE

Our feelings for hotness and coldness result from the relationship between something we call "temperature" and the direction of the flow of heat. We say that body A is at a higher temperature than body B if heat flows from body A to body B, when A and B are put in contact. After a time, body A will have cooled down, and body B will have warmed up to the same temperature, and then heat will no longer flow between them. The two bodies are then said to be in *thermal equilibrium*.

The temperature of a gas is measured with a thermometer that has come to thermal equilibrium with the gas. A thermometer is a device that utilizes a property of one substance that changes with temperature to measure the temperature of other substances. The common capillary tube thermometer uses the change in volume of a liquid, such as mercury or alcohol, with temperature. The scale of the thermometer was originally established by selecting two arbitrary temperatures and dividing the interval between these temperatures into a number of equal parts, or degrees. The freezing point and the boiling point of water at 760 mmHg have been used for this purpose. The three common temperature scales are shown in Figure 4.4.

The Celsius (formerly called centigrade) scale and the Kelvin scale are used in scientific work. *The temperature designated by a capital T that will appear in all of the expressions in this book will refer to the Kelvin scale* (°K). (The degree symbol may be omitted with Kelvin units.) Temperature readings on one scale can be converted to readings on another with the equations

$$°K = °C + 273.15$$
$$°F = \tfrac{9}{5}°C + 32$$

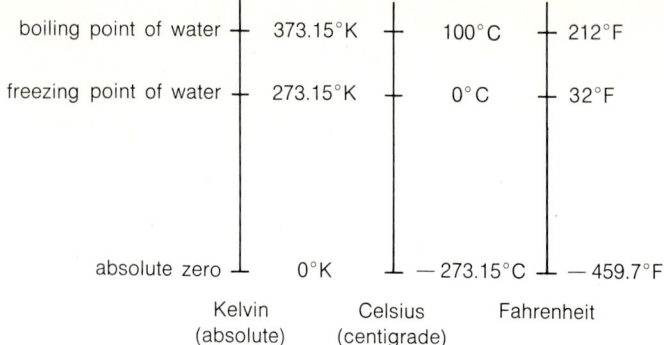

Figure 4.4. Temperature scales.

The Empirical Gas Laws

4.7 *PVT* RELATIONSHIPS
FOR GASES

An empirical law or equation is one that is obtained solely from experimental observations and data. No theory or explanation as to the underlying reason for the relation is involved. The empirical gas laws relating the pressure, temperature, and volume of a given mass of gas are the result of hundreds of years of experience in the measurement of the properties of gases. In 1662, the British physicist and chemist Robert Boyle (1627–1691) formulated the first law relating the pressure and volume of a gas. Boyle's law states that, *at constant temperature, a fixed mass of gas has a volume inversely proportional to the pressure of the gas.* "Inversely proportional" means that as the pressure increases the volume decreases. In symbols,

$$V \propto \frac{1}{P}$$

where $\propto$ means "is proportional to," or "varies as," or

$$V = \frac{K'}{P} \quad \text{and} \quad PV = K' \qquad \text{(at constant } T \text{ and } m\text{)}$$

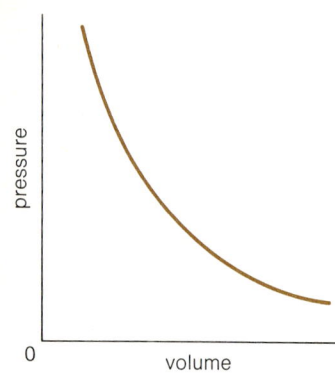

where K' is the proportionality constant. A proportionality constant is a measure of the increase in value of one factor (here V) when the value of another factor (here $1/P$) increases by one unit. A simple example is the relationship between the mass m and the volume V of a substance, $m = kV$, where the proportionality constant is the mass per unit volume of the substance, or its density, usually symbolized by $\rho = k = m/V$. These relations for Boyle's law are shown in graphical form in Figure 4.5.

The second empirical gas law is known as Charles' law, named after the French physicist Jacques Charles (1746–1823). Charles' law relates the volume of a given mass of gas at constant pressure to its temperature. This law states that, *at constant pressure, the volume of a fixed mass of gas is directly proportional to the temperature.*

$$V \propto T$$

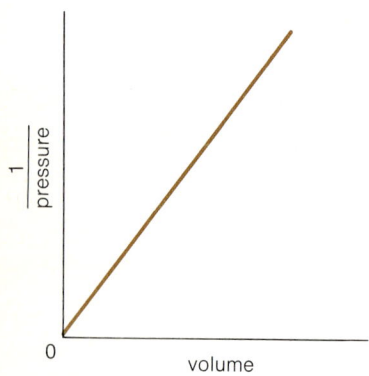

Figure 4.5. Graphical representations of Boyle's law for a fixed amount of gas at constant temperature.

or

$$V = K''T \qquad \text{(constant } P \text{ and } m\text{)}$$

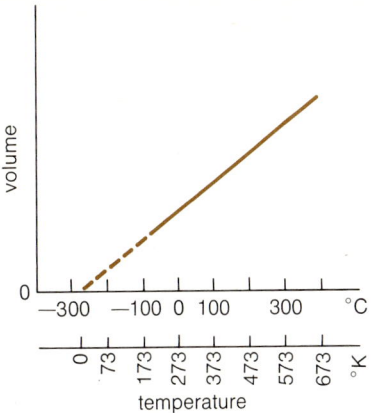

Figure 4.6. Graph of Charles' law for a fixed amount of gas at constant pressure. Extrapolation of the line would show that the volume of a hypothetical gas becomes zero at $-273.15°C$ or $0°K$.

or

$$\frac{V}{T} = K'', \text{ a constant}$$

The graphical relation for Charles' law is shown in Figure 4.6. The hypothetical extension of the V versus T line gives $V = 0$ at some temperature. This temperature comes out to be $-273.15°C$. The absolute zero of temperature, $0°K$, was designated as the temperature at which the volume of a gas would, if it remained a gas at these low temperatures, become zero. The size of the Kelvin degree is the same size as the Celsius degree. The use of temperatures in degrees Kelvin greatly simplifies the form of the gas laws.

The two gas laws can be put in another form if we designate the pressure, volume, and temperature of a given amount of gas in an initial state by P_1, V_1, and T_1. After any change, the new quantities are P_2, V_2, and T_2. Then, since K' and K'' remain constant and equal,

$$P_1 V_1 = K' = P_2 V_2 = K'$$

or

$$P_1 V_1 = P_2 V_2 \qquad (T_1 = T_2)$$

and

$$\frac{V_1}{T_1} = K'' = \frac{V_2}{T_2} = K''$$

or

$$\frac{V_1}{T_1} = \frac{V_2}{T_2} \qquad (P_1 = P_2)$$

4.8 THE COMBINED GAS LAW

What happens to the volume of a given amount of gas when both temperature and pressure are changed? Let us take an amount of gas with initial volume, temperature, and pressure V_1, T_1, and P_1. Consider a two-stage process, in which the pressure of the gas is first changed from P_1 to P_2 but the temperature remains at T_1. How does the volume change in this isothermal process? The new volume, V_x, is obtained from Boyle's law, $P_2 V_x = P_1 V_1$.

$$V_x = \frac{P_1 V_1}{P_2}$$

Now, in the second stage, with the volume V_x at P_1 and T_1, change the temperature to T_2, but leave the pressure at P_2. The final volume, V_2, is found from Charles' law.

$$\frac{V_x}{T_1} = \frac{V_2}{T_2}$$

or

$$V_2 = V_x \frac{T_2}{T_1} = V_1 \frac{P_1}{P_2} \frac{T_2}{T_1}$$

$$= V_1 \frac{P_1 T_2}{P_2 T_1} \qquad \text{(constant } m)$$

This is the combined gas law for a fixed mass of gas. It is usually written

$$\frac{P_1 V_1}{T_1} = \frac{P_2 V_2}{T_2}$$

or, since the term PV/T remains unchanged under these conditions,

$$\frac{PV}{T} = K''', \text{ a constant}$$

A volume change in a fixed amount of gas can then be considered to be the result of two factors, a pressure factor P_1/P_2 and a temperature factor T_2/T_1. If $T_1 = T_2$, the combined gas law becomes Boyle's law. If $P_1 = P_2$, it becomes Charles' law.

Use of the Combined Gas Law   If we have an amount of gas such that its volume is 100 liters when its pressure is 2.00 atm and its temperature is 25°C (298°K), what will be its volume if the pressure is increased to 4.00 atm and the temperature is increased to 35°C (308°K)?

One way to do this problem is to plug numbers into the formula

$$V_2 = V_1 \frac{P_1}{P_2} \frac{T_2}{T_1}$$

to get

$$V_2 = (100 \text{ liters}) \frac{(2.00 \text{ atm})(308°K)}{(4.00 \text{ atm})(298°K)}$$

$$= 51.7 \text{ liters}$$

This is a dangerous procedure, especially if you can't remember the equation. A better approach to the problem is to consider the separate effects of the pressure change (holding T constant) and of the temperature change (holding P constant). Assuming we change only pressure, what should be the ratio of pressures? Will the final volume V_2 be greater or less than V_1?

$$V_2 = V_1 \text{ (pressure ratio)(temperature ratio)}$$

We know that an increase of pressure, as we have here, will decrease the gas volume. So the pressure ratio should be less than unity, that is, 2.00 atm/4.00 atm.

$$V_2 = (100 \text{ liters}) \frac{(2.00 \text{ atm})}{(4.00 \text{ atm})} \text{ (temperature ratio)}$$

Now, what about the contribution from the temperature change? If we ignore the pressure change, an increase in temperature will increase the volume. The temperature ratio should be greater than unity if the volume is to increase. Thus, 305°K/298°K. Finally,

$$V_2 = (100 \text{ liters}) \frac{(2.00 \text{ atm})}{(4.00 \text{ atm})} \frac{(305°K)}{(298°K)}$$

This is the same result as the number-plugging method, except that you don't have to remember whether the pressure factor is P_1/P_2 or P_2/P_1.

4.9 AVOGADRO'S LAW

The relationships just discussed are restricted to a fixed amount of the same gas in two different conditions. Now we shall take into account the amount of the gas. Avogadro's law, formulated by the Italian chemist and physicist Amadeo Avogadro (1776–1856), enables us to relate the temperature, pressure, and volume of a gas to the number of molecules present.

Avogadro's law states that *equal volumes of gases, at the same temperature and pressure, contain equal numbers of molecules or other discrete particles.* This holds true even if gases of different substances are compared. Thus, 10 liters of SO_2 contain the same number of SO_2 molecules as 10 liters of He contain He atoms, if both are at the same temperature and pressure.

This statement is equivalent o saying that equal volumes of gases at the same pressure and temperature contain an equal number of moles n of gaseous substance, since 1 mol is equal to 6.02×10^{23} molecules. Therefore, if $V \propto n$ at constant T and P, and if

$$\frac{PV}{T} = K'''$$

then

$$\frac{PV}{T} \propto n$$

or

$$\frac{PV}{T} = Rn$$

where R is the proportionality constant.

4.10 PERFECT GAS LAW

The equation $PV = nRT$ is the *equation of state for an ideal gas* or the *perfect gas law.* An ideal or a perfect gas can be defined simply as one whose behavior conforms to this equation. Or we can say that an ideal gas follows Boyle's and Charles' laws. Most common gases, at high enough temperatures and at fairly low pressures, act ideally.

The perfect gas law relates P, V, and T for a gas. If the mass or number of moles n of the gas is held constant, the condition or state of the gas can be defined by giving the values of any *two* of the properties P, V, and T. Any two determine the third.

The *molar gas constant* R has the same value for all gases and under all conditions. It is an important universal constant. It is known from experiment that 1 mol of a gas, under temperature and pressure conditions such that it behaves ideally, occupies 22.414 liters at a temperature of 273.15°K (0°C) and a pressure of 1 atm. These particular values of T and P, 273.15°K and 1 atm, are called *standard temperature and pressure, STP.*

If the standard molar volume, 22.414 liters, and STP are substituted in the perfect gas law, the value of R can be found:

$$R = \frac{PV}{nT} = \frac{(1.0000 \text{ atm})(22.414 \text{ liters})}{(1.0000 \text{ mol})(273.15°K)}$$
$$= 0.082057 \text{ liter atm mol}^{-1} °K^{-1}$$

This value of R can be used in the perfect gas equation only if the pressure is expressed in atmospheres, the volume in liters, and the temperature in degrees Kelvin (Kelvins).

The use of the perfect gas equation is illustrated by the following examples.

EXAMPLE | What is the volume of 100 g of CO_2 gas at 2.00 atm and 25°C?

SOLUTION |
$$PV = nRT$$

$$V = \frac{nRT}{P}$$

$$n = \frac{g}{mol.\ wt.} = \frac{100\ g}{44.0\ g/mol} = 2.27\ mol\ of\ CO_2$$

$$V = \frac{(2.27\ mol)(0.082057\ liter\ atm\ mol^{-1}\ °K^{-1})(298°K)}{2.00\ atm}$$

$$= 27.8\ liters$$

The perfect gas equation is often used to determine the molecular weight of gases. Remember that the number of moles of any substance can be obtained by dividing its mass m by its molecular weight M; thus, $n = m/M$. The density ρ of a substance is its mass divided by its volume V; thus, $\rho = m/V$.

EXAMPLE | The density of an unknown gas is 1.429 g/liter, measured at STP. What is its molecular weight M?

SOLUTION |
$$PV = nRT = \frac{m}{M} RT$$

$$M = \frac{m}{V} \frac{RT}{P} = \frac{\rho RT}{P}$$

$$= \frac{1.429\ g}{1.00\ liter} \frac{(0.082057\ liter\ atm\ mol^{-1}\ °K^{-1})(273°K)}{1.00\ atm}$$

$$= 32.0\ g/mol$$

4.11 THEORY OF GASES. MODELS

The perfect gas law, as it has been derived here, was the result of three experimentally deduced, or empirical, laws. These laws say nothing about the structure of a gas, nor do they try to explain why gases behave as they do. Modern science is, however, just as interested in the why and how of material behavior as it is in the description of that behavior. We want to understand how events at the microscopic atomic and molecular level determine the observed behavior of matter. Gases, despite their disordered internal structure, afford a good example of how theories relating microscopic events to macroscopic behavior can be developed and tested.

The theoretical treatment of gases starts off with a model of a gas. A model is a simplified and idealized representation of a real system. The model itself may be a material or mechanical construct, as a model airplane or a ball-and-stick model of a molecule; or it may be some sort of abstraction such as we have in the wave picture of the electron.

The model cannot take everything into account, or its analysis will become unduly complicated. The choice of what factors to ignore in visualizing a model is crucial. Care must be taken in this respect; most of the ecological problems we are having are the result of the use of crude, over-simplified models that neglect too many interrelations within and among systems. These systems may be social, economic, or political as well as the physical systems with which we are concerned at this time.

The model should be easily modified. A model with inflexible properties will have to be abandoned completely if its predictions disagree even slightly with experimental observations.

We can set up a model for a gas if we are guided somewhat by what is experimentally known about gases, such as their compressibility and increase in volume with increasing temperature. We also know that the particles, being atoms or molecules, are very small and have the characteristics commonly expected of a collection of small particles of matter.

Model of an Ideal Gas

The microscopic model of an ideal gas is based on the following assumptions.

1. A gas consists of a very large number of particles, atoms, or molecules, which are regarded as hard spheres.

2. The volume of the particles is negligibly small compared to the total volume (the gas volume) in which they are free to move, or the size of the particles is small in comparison to the average distance between the particles.

3. There are no forces, attractive or repulsive, between the particles.

4. The particles are in constant motion in all directions. Their collisions with the walls of the container are perfectly elastic collisions. There is no net loss of kinetic energy on collision, so the molecules bounce off the wall with the same speed they had when they hit it.

Postulates 2 and 3 are generally considered to define the properties of an ideal gas on the microscopic, or molecular, level.

4.12 KINETIC-MOLECULAR THEORY OF AN IDEAL GAS

The kinetic-molecular theory of gases sets out to determine the pressure that would be exerted by an ideal gas. This pressure is due to the collision of the gas particles with the walls of the container. Pressure, as well as temperature, has no particular meaning when applied to a single molecule, so what must be done is to find the pressure due to collisions with the walls by a large number of molecules. To do this, one must first get some idea of the motion of an average molecule. In effect, kinetic theory applies the laws of physics to the motion of an individual gas particle and then uses the *average* behavior of these particles to predict the macroscopic, or large-scale, properties of pressure, volume, and temperature. Instead of giving the details of the derivation of the pressure equation, we shall discuss briefly the reasoning behind that derivation.[1]

The kinetic-molecular theory of an ideal gas uses the model of an ideal gas to calculate the force exerted by gas particles on one wall of a container. This force is equal to the change in momentum of one gas particle when it bounces off the wall after an elastic collision times the total number of collisions per second by all gas particles that will hit the wall during that time. After taking into account that all the N gas particles in a volume V do not

[1]For a complete derivation, see J. H. Hildebrand, "An Introduction to Molecular Kinetic Theory," Van Nostrand Reinhold, New York, 1963, p. 20; or H. D. Young, "Fundamentals of Mechanics and Heat," McGraw-Hill, New York, 1964, p. 441.

all have the same velocity and are all moving in random directions, the kinetic theory ends up with the equation for the pressure of an ideal gas:

$$P = \frac{1}{3}\frac{N}{V}m(v^2)_{av}$$

or

$$PV = \frac{N}{3}m(v^2)_{av}$$

where N is the total number of gas particles in the volume V, m is the mass of a gas particle, and $(v^2)_{av}$ is the average value of the square of the speed of the gas particle.

This equation represents the end result of the kinetic-molecular theoretical treatment of an ideal gas. In this case, the model is not a physical object, but a mental construct. A more detailed treatment would illustrate how reasoning at the molecular level can, by a kind of statistical or averaging procedure, be used to gain an understanding of such large-scale manifestations of matter as gas pressure.

At this point, we can use our theoretical result to get even more important insights into the behavior of gases. To do this, we combine our result with certain experimentally deduced (empirical) relations. The perfect gas law $PV = nRT$ is the most important of these.

4.13 KINETIC ENERGY AND TEMPERATURE

We now have two equations for PV:

$$PV = nRT \text{ from experiment}$$

and

$$PV = \tfrac{1}{3}Nm(v^2)_{av} \text{ from theory}$$

Equating the two gives

$$nRT = \tfrac{1}{3}Nm(v^2)_{av}$$

or, since the number of moles of gas $n = N/N_A$, the number of molecules in the sample divided by Avogadro's number,

$$RT = \tfrac{1}{3}N_A m(v^2)_{av}$$

But $\tfrac{1}{2}m(v^2)_{av}$ is the average kinetic energy ϵ_k of a single gas molecule, and $\tfrac{1}{2}N_A m(v^2)_{av}$ is the total kinetic energy of a mole of gas E_k, so

$$RT = \tfrac{2}{3}E_k$$

or

$$E_k = \tfrac{3}{2}RT \text{ per mole of gas}$$

and

$$\epsilon_k = \frac{3}{2}\frac{RT}{N_A} = \frac{3}{2}kT \text{ per molecule of gas}$$

where $k = R/N_A$ is called *Boltzmann's constant* and equals 1.380×10^{-16} erg deg^{-1} molecule^{-1} in the most commonly used units. These results show that *the average kinetic energy of a gas molecule is directly proportional to the temperature of the gas,* and that at any given temperature T, lighter molecules will have higher speeds.

4.14 MOLECULAR SPEED AND GAS TEMPERATURE

If we take the square root of the average value of the square of the velocity, $(v^2)_{av}$ appearing in these equations, we get what is called the *root-mean-square* (rms) speed of the gas molecules at a given temperature.

$$\sqrt{(v^2)_{av}} = \text{rms speed} = v_{rms}$$

From these equations, the rms speed is

$$\sqrt{(v^2)_{av}} = v_{rms} = \left(\frac{3RT}{M}\right)^{1/2} = \left(\frac{3kT}{m}\right)^{1/2}$$

where $M = N_A m$ is the molecular weight of the gas, with molecules of mass m.

This important result shows that *the rms speed of a molecule in a gas at temperature $T°K$ is greater the higher the temperature and the lighter the gas molecule.* We shall return to the velocity–temperature relationship in a moment.

4.15 DISTRIBUTION OF MOLECULAR SPEEDS

In a mole of gas there are 6.02×10^{23} molecules, all moving about in a random manner. They collide with the walls of the container and with each other. They stop and go. They travel in all directions with a whole range of speeds. In the kinetic-molecular theory, we realized that it would be impossible to know the exact velocity (direction and speed) of every molecule at any instant. So we worked with an average velocity, knowing that the molecules were really flying around with a whole range of velocities. This is the basis for the statistical approach to any problem, whether it involves molecules or people. We can find the average behavior of a large group of individuals when we are unable to determine the behavior of each separate individual.

A statistician will tell you that there are several different ways to define the term "average." The one with which you are most familiar is technically called the *mean* value. Consider, as a very simple example, 4 gas molecules having speeds 1.00, 2.00, 2.00, and 4.00, respectively. The *mean* speed is

$$v_{mean} = \frac{1.00 + 2.00 + 2.00 + 4.00}{4} = \frac{9.00}{4} = 2.25$$

This mean speed is not the same as the rms speed, which, for these molecules, is given by

$$(v^2)_{av} = \frac{(1.00)^2 + (2.00)^2 + (2.00)^2 + (4.00)^2}{4} = \frac{25.0}{4} = 6.25$$

$$v_{rms} = \sqrt{(v^2)_{av}} = \sqrt{6.25} = 2.50$$

There is one other kind of "average" speed, called the *most probable speed.* In a very simple way, this term tells us the speed of the largest number (or fraction) of the molecules. In our 4-molecule example, 2 would be the most probable speed, since one-half of the molecules have this speed, while only one-fourth have either of the other speeds. A graph might be drawn as in Figure 4.7. The curve connecting the points cannot really be drawn with any

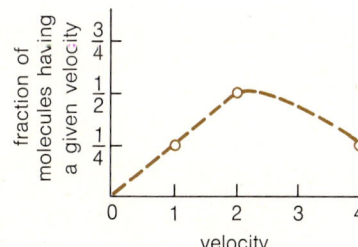

Figure 4.7. Velocity distribution for a small sample (4) of molecules. The most probable speed is 2.

accuracy for this example, but our rough drawing shows a maximum at $v = 2$. The maximum represents the most probable speed. A calculated average more nearly represents the actual values possessed by the members of a group when the number of individuals considered is large or when the individuals have values that are not extremely different.

Table 4.3

Distribution of molecular speeds, 1000 molecules

speed interval, m/sec	number in interval	fraction of total
0–100	5	0.005
100–200	34	0.034
200–300	92	0.092
300–400	138	0.138
400–500	161	0.161
500–600	167	0.167
600–700	144	0.144
700–800	104	0.104
800–900	69	0.069
900–1000	46	0.046
1000–1100	23	0.023
1100–1200	11	0.011
1200–1300	5	0.005
1300–1400	1	0.001

SOURCE: Hugh Young, "Fundamentals of Mechanics and Heat," McGraw-Hill, New York, 1964, p. 451.

When we consider a large number of molecules, we can give the fraction of all molecules that have speed values in a given range. A situation like this is shown in Table 4.3 and in the bar graph of Figure 4.8. As we calculate and plot the fraction of molecules in smaller speed ranges, the distribution curve connecting the tops of the bars becomes smoother and the maximum, repre-

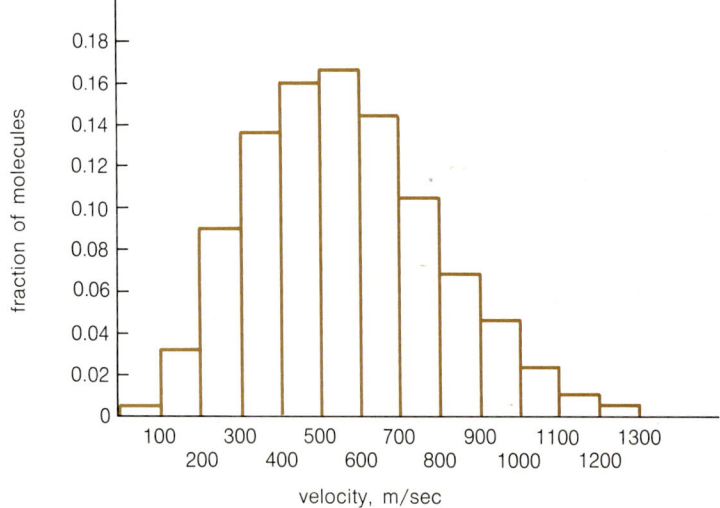

Figure 4.8. Molecular velocity distribution for the sample of Table 4.3. The vertical axis (ordinate) shows the fraction N_v / N_{total} of all molecules having velocities falling within velocity intervals of 100 m/sec. (Data from Hugh Young, "Fundamentals of Mechanics and Heat," McGraw-Hill, New York, 1964, p.451.)

senting the most probable speed, becomes more sharply defined as shown in Figure 4.9.

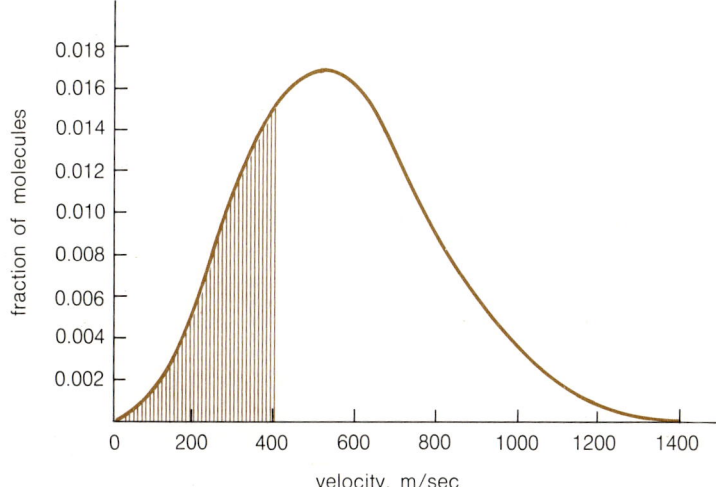

Figure 4.9. Molecular velocity distribution for smaller (10 m/sec) velocity intervals than in Figure 4.8. The fraction of molecules in each interval is one-tenth that in the larger interval of Figure 4.8.

Despite the fact that neither the rms speed nor the mean speed of the gas molecules in a sample at a definite temperature is exactly equal to the most probable speed at the distribution curve maximum, their values do not differ greatly (Figure 4.10), and we shall continue to use our equations for the rms speed.

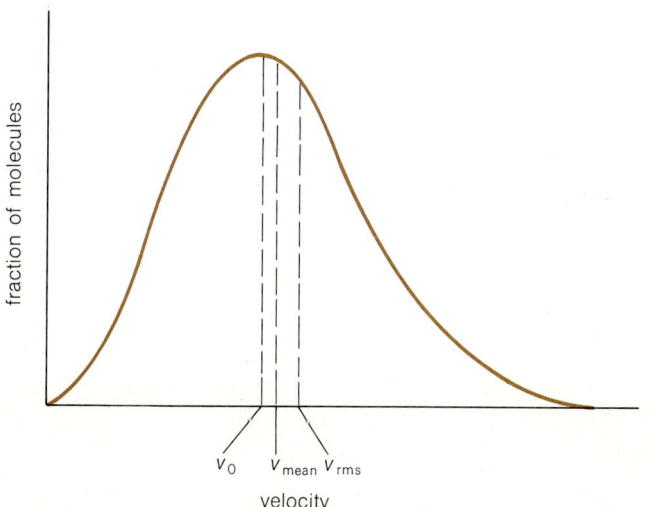

Figure 4.10. Molecular velocity distribution showing the position of the most probable speed v_0, the mean speed v_{mean}, and the rms speed v_{rms}. The speeds are related numerically by $v_{rms} = 1.22v_0$ and $v_{mean} = 1.13v_0$.

4.16 EFFECT OF TEMPERATURE ON SPEED DISTRIBUTION

If molecular speed distribution curves are drawn for a sample of gas at three different temperatures, the results in Figure 4.11 are obtained. (These curves need not be drawn from experimental measurements. A theoretical equation, known as the *Maxwell–Boltzmann distribution equation*, allows one to plot the curves for any temperature.)

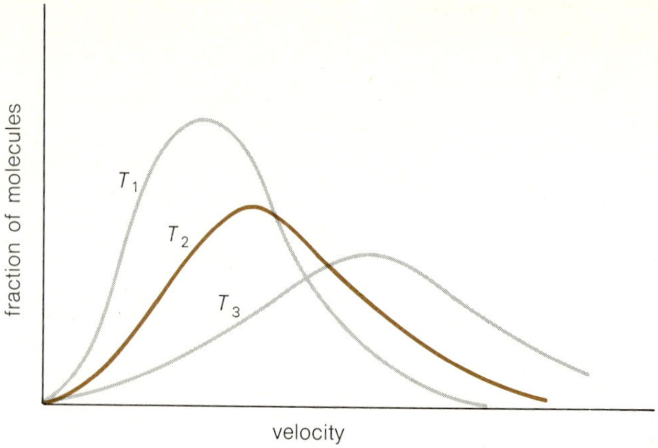

Figure 4.11. Maxwell–Boltzmann molecular velocity distribution curves at different temperatures. At higher temperatures $(T_3 > T_2 > T_1)$, the average velocity and the average kinetic energy of the molecules increase.

In accordance with the kinetic theory, the average speed of the gas molecules increases as the temperature rises. But another important fact appears from these curves. The area under a curve between two speed values is proportional to the fraction of all molecules having speeds in that range. You see that not only does the most probable speed increase with temperature (although the fraction of molecules having the most probable speed decreases), but the number of molecules having higher speeds increases. This fact is of great importance in a number of situations where only those molecules having the highest energies can react because, with an ideal gas, the kinetic energy of a molecule is proportional to the speed, and an energy distribution curve for these molecules would look just like the speed distribution curve and show the same behavior with temperature. These results will be of use in Chapter 17.

4.17 MOLECULAR VELOCITY AND THE EARTH'S ATMOSPHERE

Any body near the surface of the earth experiences a force pulling it toward the center of the earth. This force is the weight (not the mass) of the object, and its magnitude is given by Newton's law of universal gravitation,

$$F = G\frac{m_1 M}{r^2}$$

where

G = the universal gravitational constant = 6.67×10^{-11} m^3/kg sec^2
m_1 = the mass of the object, in kilograms
M = the mass of the earth = 6.0×10^{24} kg
r = distance from the center of the earth to the object, in meters

Any body in the earth's gravitational field has a potential energy

$$U = -\frac{Gm_1 M}{r}$$

relative to the center of the earth.

For a body to escape from the pull of the earth and continue on into space, it must have an upward speed v_e such that the magnitude of its kinetic energy, $\frac{1}{2}m_1 v_e^2$, equals or exceeds its potential energy.

$$\tfrac{1}{2}m_1v_e^2 = \frac{Gm_1M}{r}$$

$$v_e = \left(\frac{2GM}{r}\right)^{1/2}$$

Note that the escape velocity does not depend on the mass of the object.

If we consider an object at the surface of the earth ($r = 6.4 \times 10^6$ m, about 4000 miles), the escape speed comes out to be about 1.1×10^4 m/sec (about 7 miles/sec). That is, it must be given an initial speed of about 7 miles/sec in order to escape from the earth and never return. The initial speed required will decrease the higher the object is above the surface of the earth, but for simplicity we shall consider this figure of 1.1×10^4 m/sec to apply in the troposphere and stratosphere as well.

The main point is that the speed of the molecules in the earth's atmosphere will determine whether a given gas will gradually escape from the earth. The rms speed of a molecule depends on its mass (or molecular weight) and temperature by the equation (Section 4.14)

$$v_{\text{rms}} = \left(\frac{3RT}{M}\right)^{1/2}$$

The velocity will be in units of centimeters per second if M is in grams per mole and R is 8.314×10^7 g cm^2 sec^{-1} mol^{-1} °K^{-1}. Therefore, the lightest gases, such as H_2 and He, will have the highest average speeds at any given temperature. Of course, not all molecules will have speeds in an upward direction, but those that do have the proper direction will stand a better chance of escaping from the earth.

Table 4.4 Velocities of gas molecules at 0°C	gas	molecular weight, g/mol	v_{rms}, m/sec
	hydrogen, H_2	2.02	1838
	helium, He	4.0	1311
	water, H_2O	18.02	615
	neon, Ne	20.2	584
	nitrogen, N_2	28	493
	oxygen, O_2	32	461
	carbon dioxide, CO_2	44	393
	mercury, Hg	200.6	183

The speeds of some gas molecules at 0°C are listed in Table 4.4. The high average speeds of the lightest gases account for their almost complete absence in the earth's atmosphere. Near the top of the atmosphere, the temperature is about 1000°K, and the pressure is quite low. The low pressure means that the distance between molecules is large, so there is little chance that high-speed molecules are deflected downward toward the earth by collisions. If a molecule is moving upward with a speed of about 1.1×10^4 m/sec, it has a good chance of escaping the earth and continuing on into space. At any temperature, there are always going to be some molecules with speeds equal to, or greater than, the escape speed. But there are going to be more light molecules with this speed than heavy molecules. The light ones will, over a period of time, gradually escape from the earth, and the heavy ones will be left behind. Heavy molecules will escape also, but because there are fewer having the required speed, their escape will take much longer.

The foregoing explains two things about the present composition of our atmosphere:

1. Hydrogen and helium escaped rapidly from the earth's atmosphere. This conclusion is based on the assumption that the sun and earth were originally formed from the same material. The relative abundances of hydrogen and helium on earth are now much lower than their abundances in the sun or in the solar system as a whole (see Table 13.1, Chapter 13). It is estimated that oxygen and nitrogen molecules take almost 10^{45} times longer to escape than hydrogen and helium molecules.

2. The earth was formed long ago and was at a high temperature at some stage in its evolution. Otherwise, the heavy inert gases like krypton and xenon would not have had the time and kinetic energy needed for their escape, and we could not account for their present small amounts in the atmosphere. They must have been present in much larger amounts at the formation of the earth. Most of the neon and argon escaped. Water and carbon dioxide did not escape, because they were chemically combined within the rocks.

In addition to these two points, the small surface gravity of our moon results in such a low escape speed (about 1.5 miles/sec compared to the earth's 7 miles/sec) that the moon has lost any atmosphere it might have had.

4.18 REAL GASES

There is no such thing as an ideal gas. An ideal gas is a model, an idealization of a real gas. The details of forces between the gas particles and the finite size of the particles themselves have been ignored in the model of an ideal gas so that the theoretical analysis would not become too complicated at first. However, only real gases exist. Real gases do have forces between their component particles, and the particles do have a finite size. It is fortunate that under certain conditions gases do behave experimentally as we would expect an ideal gas to behave. In general, these conditions are those of fairly low pressure (to about 10 atm) and fairly high temperature. We may apply the preceding empirical and theoretical gas laws under these conditions. But as pressure increases and temperature decreases, we know that gaseous substances must depart from ideal behavior, because these are the very conditions under which gases condense to the liquid and solid state. To achieve the ordered arrangements of particles characteristic of these condensed states, there must be attractive forces between the particles to hold them in these arrangements. The model of a gas must be modified to take these forces into account if the model's range of application is to be extended. And the nature of the forces between the particles must be understood if we are to see why real gases act as they do.

Before taking a microscopic view, however, let us see what the macroscopic behavior of a real gas such as carbon dioxide is like. The experimentally observed behavior of CO_2 at different temperatures is shown in Figure 4.12. Each curve on this graph is an isotherm giving the pressure and corresponding volume of a fixed amount of CO_2 gas at a constant temperature.

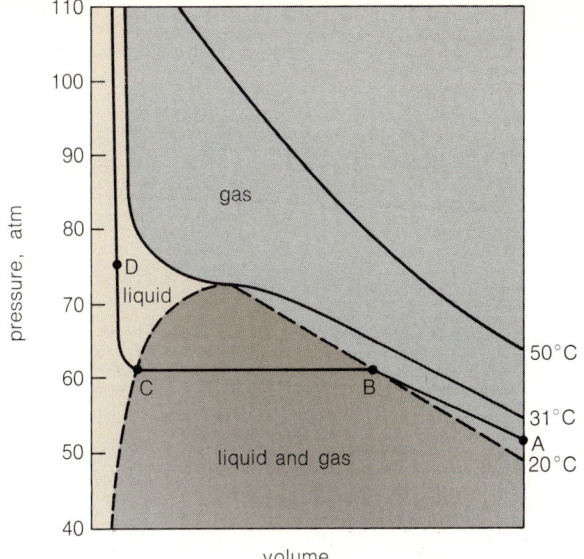

Figure 4.12. Pressure–volume isotherms for CO_2. Ideal behavior is shown at higher temperatures, but below 31°C and 73 atm the gas begins to liquefy as the molecules become closer together and attractive forces become significant. Above 31°C, the kinetic energy of the gas molecules is sufficient to counteract the attractive forces no matter how close together the molecules may be.

At temperatures above about 50°C, the gas behaves ideally. That is, the P versus V curve approaches the hyperbola predicted by Boyle's law, $PV = $ constant. Below 50°C, however, the isotherms show deviations from the ideal behavior at higher pressures, and inflections develop in the isotherms. At 31°C, a flat portion just begins to form and becomes more and more evident as the P–V behavior at lower temperatures is plotted. To see what is happening, consider a gas at point A on the 20°C isotherm. As the gas is compressed, the change in volume is given by the curve ABCD. From A to B, the gas volume gradually decreases with pressure until point B is reached. At B, the CO_2 suddenly begins to condense to the liquid state, and the volume decreases as gas is converted into the more dense liquid. The pressure remains constant until all the gas has condensed at point C. Further increase of pressure results in much less change of volume after point C, because liquids are much less compressible than gases. Any P–V point along the CD portion corresponds to liquid CO_2. Any combination of P and V values on the AB segment corresponds to gaseous CO_2. But along the flat portion from B to C, there are two phases or states of matter present, gaseous and liquid CO_2. The relative portions of gas and liquid change in going from B to C.

All samples of CO_2 gas at temperatures below 31°C show similar types of behavior. But if the gas is above 31°C, no amount of pressure can bring about condensation to the liquid. Above 31°C, CO_2 cannot exist in the liquid state. This maximum temperature above which a gas cannot be liquefied by any pressure increase is known as the *critical temperature* of the gas, and the pressure required for condensation to occur at the critical temperature is the *critical pressure*. Table 4.5 lists some critical parameters for several gases. The most easily condensed gases have the highest critical temperatures.

It is evident from the shapes of the P–V isotherms that a gas deviates greatly from ideal behavior below the critical conditions. What causes condensation? Or put another way, what is less than ideal about the gas? The

Table 4.5
Critical temperatures
and pressures

substance	critical temperature, °K	critical pressure, atm
helium, He	5.2	2.3
hydrogen, H_2	33.2	12.8
nitrogen, N_2	126.0	33.5
carbon monoxide, CO	134.4	34.6
argon, Ar	150.7	48.0
oxygen, O_2	154.3	49.7
methane, CH_4	190.2	45.6
carbon dioxide, CO_2	304.2	72.9
hydrogen chloride, HCl	324.5	81.6
hydrogen sulfide, H_2S	373.5	89.0
ammonia, NH_3	405.5	111.5
chlorine, Cl_2	417.1	76.1
sulfur dioxide, SO_2	430.3	77.7

hint that attractive forces between the gas particles come into play has already been made, so we shall look more closely at the kinds of forces that might exist. Even though these forces are not the strong forces that lead to covalent and ionic bonds, they have their origin in the electronic structure and bonding of atoms and molecules, so we must return to that subject.

4.19 INTERMOLECULAR FORCES

The ionic and covalent bonds between atoms with the proper electronic configurations are the result of comparatively strong attractive forces between individual ions or atoms. In addition to these attractive forces, there are also repulsive forces that come into play when the particles, whether they are bonded or not, are forced together so that their electron clouds overlap and their nuclei repel one another. Both of these strong attractive and repulsive forces are operative over only very short distances. The strong attractive forces leading to bonding are examples of short-range *intramolecular* forces acting between different parts of the same molecule, while the short-range repulsive forces can be either intra- or *intermolecular* forces, the latter acting between or among different molecules.

In talking about the forces between the molecules of a gas, we are obviously interested only in intermolecular forces. Also, because the molecules of gases, ideal or real, are relatively far apart, we shall not be interested in very short-range forces, but in the so-called long-range attractive intermolecular forces that act at somewhat larger distances. There are a variety of such forces between nonbonded atoms and molecules, and they are the ones that lead to deviations from ideal gas behavior and to condensation of gases. They are much weaker than the very short-range chemical or valence forces already discussed. That they are weak forces is shown by the fact that a quite small addition of thermal energy (heat) is sufficient to overcome these forces and convert a liquid to a gas.

The weak, attractive intermolecular forces are sometimes loosely grouped under the heading *van der Waals forces*, in honor of the Dutch physicist J. D. van der Waals (1837–1923; Nobel prize in physics, 1910), who did much work on the behavior of real gases. The weak forces are often further separated into (1) electrostatic forces, (2) inductive forces, and (3) London forces, although all are ultimately electrostatic in nature. All the forces arise from the interactions between the electric charge distributions of adjacent molecules. The charge distribution of greatest interest here is called a dipole.

4.20 DIPOLES AND DIPOLE MOMENTS

An *electric dipole* is a pair of equal and opposite charges, $+q$ and $-q$, separated by a distance d. The dipole moment μ is equal to the product of one charge and the distance between the charges.

$$\mu = qd$$

The dipole moment is a vector and has both magnitude and direction. The positive direction of μ, that is, the direction of the arrow, is chosen from the negative end to the positive end of the dipole (Figure 4.13). (The opposite convention is often used; it makes no difference, as long as we are consistent.)

Dipoles can interact with each other and with ions to give rise to attractive forces between molecules. Before studying this interaction, let us first investigate the origin of dipoles in molecules. To do this requires a return to the concepts of ionic and covalent bonding.

Strong bonds between atoms are not necessarily the pure ionic or the pure covalent bonds discussed in Part One. You will remember that the ionic or electrovalent bond was the result of the electrostatic attraction between two oppositely charged ions. The ions were formed by the complete transfer of electrons from one atom to others. One atom got complete control, so to speak, of one or more electrons from an adjacent atom.

The covalent bond, in one model at least, was pictured as resulting from the equal sharing of a pair of electrons by two atoms. Because of this supposedly 50–50 sharing of two electrons, neither atom appears to have really lost an electron; and it remains neutral, with as many electrons as protons. To take two specific examples of each bond, we can consider a gaseous sodium chloride molecule and a hydrogen molecule in Figure 4.14. It is clear that two ions held together by an ionic bond constitute a true electric dipole. A pure covalent bond, in a homonuclear diatomic molecule, does not give rise to a dipole in the bond.

However, there are all degrees of electron sharing (or transfer) between the two extremes of pure covalent bonding and pure ionic bonding. We have noted that ionic bonding takes place between atoms of widely differing ionization energy, while covalent bonding occurs between atoms having little difference in ionization energy. But what can we say about the bond in hydrogen chloride, HCl?

Experiment shows that the HCl molecule has a dipole moment. But the magnitude of the moment is not as great as would be calculated from $\mu = qd$, with q equal to the electronic charge (i.e., H$^+$ or Cl$^-$ ions with unit charge) and d, the experimentally measured bond distance. The bond in HCl is an example of a polar covalent bond. The two electrons in the shared pair are not shared equally by the H and the Cl atoms. The Cl atom, because of its greater electron affinity, tends to draw the electron pair away from the H atom and closer to itself. But complete transfer of the original H electron to Cl is not attained.

The attraction of the electron pair by the chlorine in HCl produces a separation of the centers of positive and negative charge in the molecule,

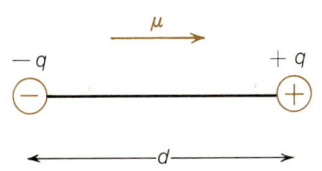

Figure 4.13. Electric dipole. The dipole moment $\mu = qd$.

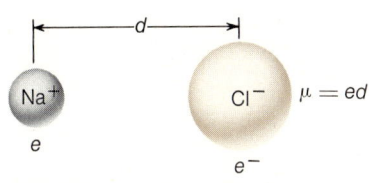

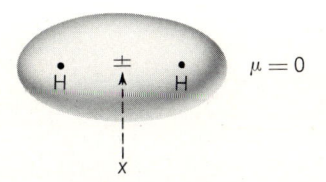

Figure 4.14. Charge distribution and dipole moments in ionic and covalent species. The center of both positive and negative charge is located at *x*.

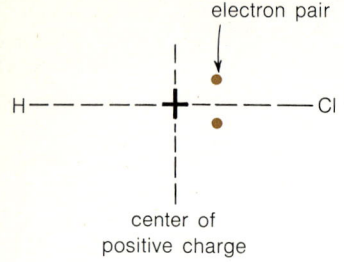

electron pair

H — — — — — + — — — — — Cl

center of
positive charge

Figure 4.15. Charge distribution in a polar covalent bond. The shared pair is displaced from the center of positive charge density by the attraction of chlorine.

resulting in a dipole moment. The situation can be pictured as in Figure 4.15. In the absence of unequal electron attraction, the + and − centers of charge density coincide and $\mu = 0$. In terms of the σ molecular orbital encompassing the two nuclei in HCl, the charge separation can be pictured as resulting from a fattening of the molecular orbital at the Cl end (Figure 4.16).

The net effect of this polarization of the covalent bond is that the H can be regarded as having a small positive charge $\delta+$ on it, smaller than the unit electronic charge it would have if it were a positive hydrogen ion. Similarly, the Cl has a partial negative charge $\delta-$.

$$\overset{\delta+ \quad \delta-}{H-Cl}$$

The same applies in other polar covalent bonds.

Experimentally measured dipole moments for a number of molecules are given in Table 4.6. The units for μ are in debyes, D. One debye is 10^{-18} esu

Table 4.6
Geometry and dipole moment of some gaseous molecules

molecule	geometry	dipole moment, D
diatomic		
HF		1.91
HCl		1.07
HBr	all	0.79
HI	linear	0.38
KF		7.3
KCl		8.0
KBr		9.1
polyatomic		
H_2O	angular	1.84
H_2S	angular	0.92
SO_2	angular	1.62
NO_2	angular	0.39
O_3	angular	0.52
CS_2, CO_2	linear	0
HCN	linear	2.95
SO_3	trigonal planar	0
CH_4, CCl_4	tetrahedral	0
CH_3Cl	tetrahedral	1.87
CH_2Cl_2	tetrahedral	1.54
$CHCl_3$	tetrahedral	1.02

cm; q is measured in electrostatic units (esu), in which the charge of the electron is 4.8×10^{-10} esu; and d is in centimeters, in which a typical bond length is 10^{-8} cm.

4.21 ELECTRONEGATIVITY

The tendency of an atom to attract the bonding electrons in a covalent bond is described by a quantity called the electronegativity of the atom. The greater this tendency, the greater the electronegativity. The electronegativity can be calculated from experimental quantities such as the ionization energy, electron

Figure 4.16. Displacement of the shared pair in a σ molecular orbital makes the orbital fatter at one end.

affinities, or bond energies. Commonly used values of electronegativities are listed in Table 4.7. In general, those elements with the largest electron affinities and ionization energies have the larger electronegativity values and attract the electron pairs in a covalent bond most strongly. The periodic trends in electronegativity should be remembered.

In a diatomic molecule with a single covalent bond, the difference in electronegativities of the bonded atoms is approximately equal to the dipole moment of the molecule. For instance, in HCl, the electronegativity difference is $3.0 - 2.1 = 0.9$ D, compared to the measured moment of 1.07 D.

Table 4.7
Electronegativity values

H	Li	Be	B	C	N	O	F
2.1	1.0	1.5	2.0	2.5	3.0	3.5	4.0
	Na	Mg	Al	Si	P	S	Cl
	0.9	1.2	1.5	1.8	2.1	2.5	3.0
	K	Ca		Ge	As	Se	Br
	0.8	1.2		1.8	2.0	2.4	2.8
	Pb	Sr		Sn	Sb	Te	I
	0.7	1.0		1.8	1.9	2.1	2.5
	Cs	Ba					
	0.7	0.9					

The phrase "percent ionic character" is sometimes used to indicate how much a bond differs from a pure covalent bond with no shifting of bonding electrons. The percent ionic character of the bonds in the hydrogen halides is

HF	40%
HCl	17%
HBr	11%
HI	5%

As a rough approximation, a bond will have more than 50 percent ionic character if the electronegativity difference is greater than 2.1.

Ionic character does not mean that the bond is pure ionic part of the time and pure covalent the rest of the time. What we have here is another resonance phenomenon, although not strictly of the electron delocalization type. The HCl molecule could be written in two resonance forms:

$$H:Cl \longleftrightarrow H^+:Cl^-$$

This ionic–covalent resonance might be expected to stabilize the molecule by increasing the strength of the bond. That this is so is shown in Table 4.8. The hypothetical pure covalent bond energies for HCl, for instance, are calculated by summing one-half the energies of the H—H and Cl—Cl bonds. The difference between the observed energy and the hypothetical energy, $135 - 63 = 72$ kcal/mol, is the resonance energy of HCl. A mole of actual HCl molecules is 72 kcal more stable than a mole of hypothetical HCl molecules with pure covalent bonds.

Table 4.8 Bond energies	bond	hypothetical covalent bond energy, kcal/mol	observed, kcal/mol
	H—F	63	135
	H—Cl	78	102
	H—Br	69	87
	H—I	61	71
	H—O	59	110
	H—N	63	93
	H—C	93	99
	C—O	52	82
	C—F	56	107

4.22 BOND DIPOLE MOMENTS VERSUS MOLECULAR DIPOLE MOMENTS

The dipole moments mentioned so far have been those associated with diatomic molecules. The dipole moment in this kind of molecule must result from the charge displacement in the one bond present. The bond may be single, double, triple, and so on, but, in any case, the dipole moment of the molecule is also a *bond moment*. In the diatomic case, molecular dipole moment and bond dipole moment are the same. But when we come to polyatomic molecules with two or more bonds, each polar covalent bond has its own moment, whose value may be quite constant from one molecule to another. Some representative bond moments are

bond	H—N	H—O	H—S	C—N	C—O	C—Cl	C—Br	C=O
moment, D	1.3	1.5	0.7	1.0	1.2	1.9	1.8	2.7

The dipole moment of the molecule is the resultant of the individual bond moments. The resultant will depend on the geometric structure of the molecule, since the bond moments are vectors and the resultant of two or more vectors depends on their direction. The cases of carbon dioxide, water, and carbon tetrachloride will serve as simple examples.

In CO_2, O=C=O, the two C=O bonds each have a bond moment as shown in Figure 4.17. The two equal C=O bond moments are exactly opposed and cancel to zero because of the linear structure of the molecule. Even though both bonds are polar, the molecule as a whole is nonpolar.

Water is an angular molecule (Figure 4.18). Both O—H bonds have moments, but because of the structure of water, the bond moments do not cancel, and the molecule as a whole has a moment. Thus, H_2O is a polar

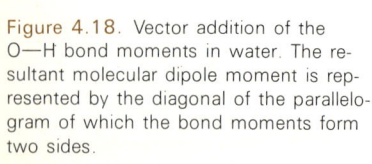

Figure 4.17. The resultant of the C=O bond moments in CO_2 is zero.

Figure 4.18. Vector addition of the O—H bond moments in water. The resultant molecular dipole moment is represented by the diagonal of the parallelogram of which the bond moments form two sides.

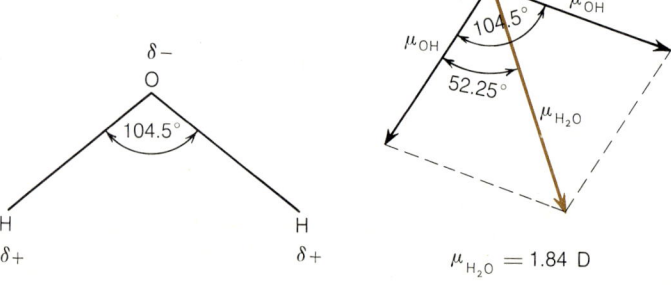

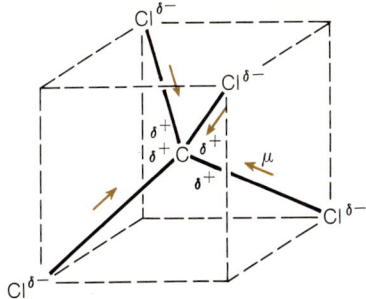

Figure 4.19. The symmetrical CCl_4 molecule has no dipole moment.

molecule, with the negative end of the dipole at the oxygen and the positive end between the two hydrogens.

Carbon tetrachloride, CCl_4, is a symmetrical tetrahedral molecule (Figure 4.19) in which all the C—Cl bond moments cancel. Carbon tetrachloride is nonpolar. Replacement of one of the Cl's by H to form $CHCl_3$, chloroform, disrupts the symmetry, and the bond moments no longer cancel. $CHCl_3$ is a polar molecule.

If one knows the bond moments and the structure of a polyatomic molecule, a good estimate of the molecular dipole may be obtained from vector addition, as shown for the case of water in Figure 4.18. In practice, experimental measurement of molecular dipole moments is used to give valuable information about the geometric structure of the molecule. Table 4.6 lists some simple molecules with their geometry and molecular dipole moment.

4.23 ELECTROSTATIC FORCES

Electrostatic intermolecular forces arise from the interactions among permanent molecular dipoles and ionic charges. If we restrict ourselves to dipoles and ions, there are three kinds of electrostatic interactions: ion–ion, ion–dipole, and dipole–dipole. The first will be considered to belong to the strong ionic-type short-range bonds and will not be discussed here. Besides, there are very few free ionic charges in a gas at ordinary temperatures. The ion–dipole interactions do not apply in a gas made up of identical molecules, but because of their importance in aqueous solutions these interactions will be briefly described in this section.

Dipole–dipole interactions are the most important electrostatic forces in gases made up of polar molecules. The qualitative features of all the interactions can easily be seen by remembering that like charges repel and unlike charges attract and that the closer together the unlike charges, and the farther apart the like charges, the stronger the attractive forces. Two adjacent dipoles, arranged as in Figures 4.20(a) and 4.20(b), are in low potential-energy configurations because their opposite charges are nearer each other. Figures 4.20(c) and 4.20(d) show high-energy (unstable) dipole configurations. If two dipoles are free to arrange themselves, they will tend to line up as in Figure 4.20(a) or 4.20(b). Once they are aligned like this, energy would be required to rearrange them. This energy can be supplied in the form of the

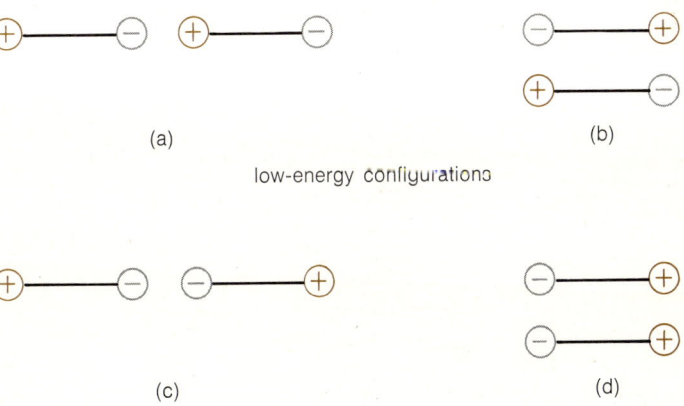

Figure 4.20. Dipole–dipole interactions. Configuration (a) is more stable than (b) under most conditions of dipole dimension and separation.

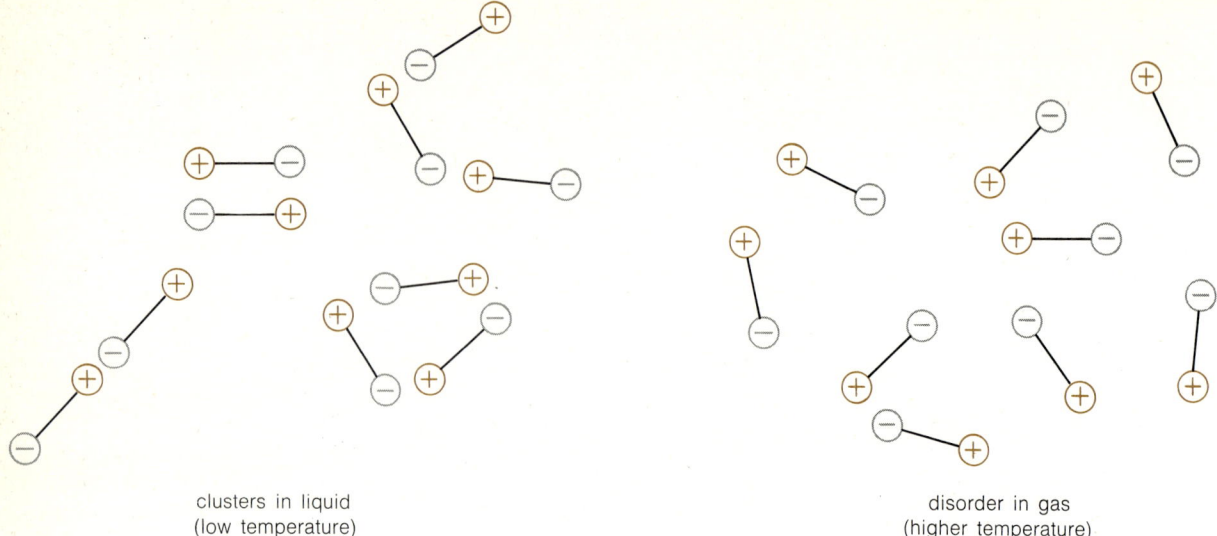

clusters in liquid
(low temperature)

disorder in gas
(higher temperature)

Figure 4.21. Disorientation of dipoles by thermal energy.

kinetic energy of thermal agitation brought about by an increase in temperature. The situation in a liquid and in a gas would be something like that shown in Figure 4.21. Table 4.9 shows that the magnitude of the dipole–dipole electrostatic attractive force is directly proportional to the molecular dipole moment. The attractive force due to dipole–dipole interaction is a major contribution to the deviation from ideal behavior of polar gas molecules such as HCl, H_2O, and NH_3.

Table 4.9
Contributions to intermolecular attraction between like molecules

molecule	dipole moment, D	polarizability, $cm^3 \times 10^{25}$	forces, erg $cm^6 \times 10^{60}$		
			electrostatic	induction	London
H_2	—	7.9	—	—	11.3
Ar	—	16.2	—	—	57
N_2	—	17.6	—	—	62
CH_4	—	26.0	—	—	117
Cl_2	—	46.1	—	—	461
CO	0.12	19.9	0.0034	0.057	67.5
HI	0.38	54	0.35	1.68	382
HBr	0.79	35.8	6.2	4.05	176
HCl	1.07	26.3	18.6	5.4	105
NH_3	1.47	22.1	84	10	93
H_2O	1.84	14.8	190	10	47

Dipoles may also interact with ionic charges. The low-energy, attractive and high-energy, repulsive situations for the interaction of positive and negative ions with a dipole are shown in Figure 4.22.

An electrostatic dipole–dipole interaction known as the hydrogen bond is of particular importance in biological molecules. The hydrogen bond occurs between (and within) molecules where hydrogen is bonded to a highly

(a)   (b)

(c)   (d)

Figure 4.22. Ion–dipole interactions. (a,b) Low-energy (attractive) configurations. (c,d) High-energy (repulsive) configurations.

electronegative atom. Because of the unusually high strength of the hydrogen bond and its importance in water, a detailed discussion of it will be taken up in Chapter 8.

4.24 INDUCED DIPOLES AND POLARIZATION

Electrostatic forces of the dipole–dipole type are not sufficient to explain the intermolecular forces that must exist between symmetrical molecules having no permanent molecular dipole moment. In such nonpolar molecules, the center of negative (electronic) charge coincides with that of the positive (nuclear) charge. Molecules included are those having no polar bonds (Cl_2), symmetrical molecules in which bond moments cancel (CO_2, CH_4), and single atoms. The coincidence of negative and positive charge density may be altered in two ways. One way is through the presence of an electric field due to an adjacent ion or permanent dipole. The other way results from the usual vibrations of the atomic nuclei and the movement of molecular or atomic electrons relative to one another. The first case gives rise to *induced dipoles*. The second case is known as a *transient dipole*. Transient dipoles are discussed in Section 4.25.

To understand the origin and nature of an induced dipole, let's first look at a case that does not really occur in a gas. This is the effect of an ion on a nonpolar molecule. The free nonpolar molecule in the absence of the ion can be drawn as in Figure 4.23, the circle representing the electron charge cloud. In the presence of an electric field from a positive ion, for example, the centers of negative and positive charge in the molecule may be separated by a distance d. The nonpolar molecule is then said to be polarized, and a dipole moment $\mu = \delta d$ has been induced in it. A nearby permanent dipole could also induce a dipole in a polarizable molecule. The polarization in the example shown results from the attraction of the negative electronic charge cloud to the inducing positive charge. The nuclear charges within the molecule are either stationary or repelled. This is an example of *electron polarization*.

The induced molecular moment should not be confused with the permanent dipole moment existing in polar molecules such as HCl or with the permanent bond moments that exist even in many nonpolar molecules. These permanent moments exist in the absence of an external inducing electric field from nearby charges. A molecule is said to be *polarizable*, in the sense of electron polarization, if its electron charge cloud can be distorted and lead to a separation of negative and positive charge within the molecule. The ease of deformation of the electron cloud depends on the freedom of the electrons in the cloud to move in response to the distorting force. In general, the larger

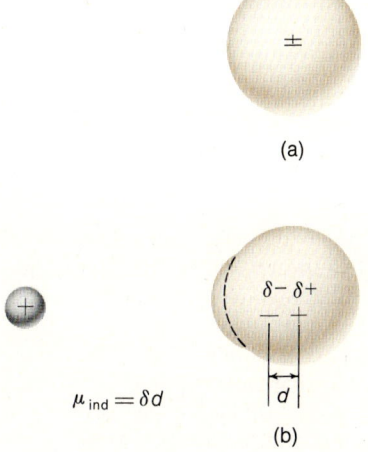

(a)

$\mu_{ind} = \delta d$

(b)

Figure 4.23. Induced dipole. (a) No displacement of electron charge cloud. (b) Polarization by a positive ion with induction of dipole $\mu_{ind} = \delta d$.

the electron cloud associated with a molecule or atom, the greater the polarizability and the greater any induced moment. Thus, Cl_2 is more polarizable than H_2, and Ar is more polarizable than He. The mobility of the electrons in the cloud also determines polarizability. Electrons in π orbitals are more mobile than σ electrons, so unsaturated molecules containing alternating double and triple bonds, in particular, are much more polarizable than saturated systems. As a result, electric effects are easily transmitted from one part of an unsaturated molecule to another. This is of particular importance in the transmission of energy and electric signals along biological molecules. Table 4.9 also lists the polarizabilities of various simple molecules.

The induction of a dipole in a molecule by an adjacent charge or dipole leads to the class of attractive intermolecular forces known as inductive forces. These forces are always quite small, as Table 4.9 indicates. For practical purposes, inductive forces in gases can be neglected.

4.25 TRANSIENT DIPOLES AND LONDON FORCES

As Table 4.9 shows, there is another effect that plays a large part in the molecular interaction between gas molecules and that leads to deviations from ideal behavior. That there is some other effect should be expected, since neither dipole–dipole nor induction interactions can explain the cohesion between atoms of the inert gases or in diatomic nonpolar gases such as H_2 and Cl_2. No permanent dipoles of any kind are found in these molecules, so deviations from ideality ultimately culminating in condensation must be due to some other force. This force, the *London force*, is due to the induction of a dipole in one molecule by a *transient* or instantaneous dipole in another molecule.

In an atom or a nonpolar molecule, the centers of negative electronic and positive nuclear charge coincide *on a time average*. But since the electrons and the nuclei are in constant motion at any particular *instant*, the molecule may find itself with its electronic charge distribution bunched on one side of the center of positive charge so that the overall charge distribution is no longer symmetrical. In a helium atom, the situation might be sketched as in Figure 4.24. The transient dipole in one atom or molecule can induce a dipole in a nearby atom or molecule. The interaction of the induced dipole with the transient dipole then leads to a cohesive or attractive interaction between the two molecules irrespective of their orientation. The magnitude of the induced dipole and, therefore, the magnitude of the interaction will again depend on the polarizability of the atom or molecule, that is, on how easily its electron charge cloud can be deformed.

These transient dipole-induced dipole forces are much more important than one might expect because, unlike the other interactions considered, their effects extend over more than just one pair of molecules; and the transient dipole in one molecule can induce a dipole in any other molecule surrounding it. The surrounding molecules, in turn, can induce dipoles in other molecules, and the effect spreads throughout the substance. Forces involving permanent dipoles are not additive in this manner because of the large number of different possible dipole orientations throughout the substance. As Table 4.9 shows, these London forces are primarily responsible for the cohesion of nonpolar molecules and atoms.

The ideas developed so far can be used to understand why certain gases tend to be less ideal than others. As a rough approximation of nonideality,

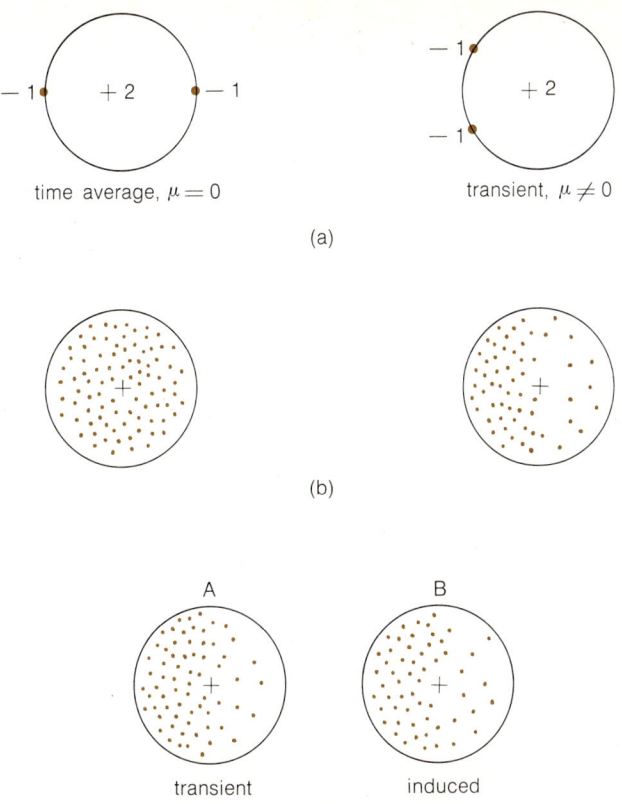

Figure 4.24. Transient dipoles and the origin of London forces. (a) Time average and transient dipoles in a helium atom. (b) Charge cloud picture of (a). (c) Attractive interaction of a transient dipole in helium atom A and an induced dipole of helium atom B.

the critical temperature T_c of a gas can be used. Critical temperatures were listed in Table 4.5. In general, the higher the critical temperature, the more the gas departs from ideal behavior, that is, the stronger the forces between its molecules are.

The different types of intermolecular interactions are listed below.

electrostatic $\begin{cases} \text{ion–ion} \\ \text{ion–dipole} \\ \text{dipole–dipole} \end{cases}$

inductive $\begin{cases} \text{ion–induced dipole} \\ \text{dipole–induced dipole} \end{cases}$ $\left.\begin{array}{c} \\ \\ \\ \end{array}\right\}$ van der Waals forces

London $\begin{cases} \text{transient dipole–induced dipole} \end{cases}$

All but the ion–ion forces are quite weak (the ion–dipole force is about 20 times weaker than an ion–ion interaction, and the others are even weaker). The interacting particles must be comparatively close together ($\sim$4 Å) before any attractive force is evident, although not so close that either very short-range attractive or repulsive forces become active. The weaker forces listed are generally considered to be the van der Waals-type forces.

Intermolecular forces have been described here in the context of the gaseous state. But their importance is by no means limited to gases. They are necessary in any treatment of the mechanism of solution in liquids, changes in states of matter, and the structure of molecules. The detailed structure and function of biological molecules, such as proteins, are dependent on the operation of many of these weak interactions.

4.26 SUMMARY

The earth's atmosphere is a mixture of gases called air. The air in the lower layer of the atmosphere, the troposphere, contains some nonvariable and some variable components. Among the former are nitrogen (78 percent by volume), oxygen (21 percent), and carbon dioxide (0.03 percent or 300 ppm). The most important of the variable constituents is water vapor, whose concentration may range up to 3 percent. The constancy of composition of the nonvariable constituents in dry air indicates that the air is well mixed over the globe and that a balance exists between the addition and removal of the atmosphere's gases.

The internal structure of the gaseous state of matter differs from that of the condensed liquid and solid states by its complete lack of short-range and long-range order among the constituent particles. The rapid random motion of the widely spaced particles of gas is responsible for the observed properties of gases.

At normal temperatures and pressures, the gross properties of gases can be described by the pressure–volume–temperature behavior of gases. Boyle's law, $PV = \text{constant}$, and Charles' law, $V/T = \text{constant}$, summarize the behavior of fixed amounts of many gases. Through the use of Avogadro's law, these two empirical gas laws may be combined to give the equation of state for an ideal gas, $PV = nRT$.

Based on a microscopic model of a gas that postulates that the rapidly moving particles of a gas exert no forces on each other and that the volume of the particles is negligible in comparison to the gas volume, the kinetic theory of ideal gases shows that the pressure and, therefore, the absolute temperature of a gas are proportional to the average kinetic energy of the gas molecules. Additional analysis of molecular velocities shows that the fraction of molecules having speeds or energies above the mean value increases with temperature.

All gases deviate from ideal behavior at sufficiently high pressures and low temperatures. This nonideal behavior culminates in the condensation of the gas and is due to attractive forces among the molecules. The comparatively weak intermolecular forces are all essentially electrostatic in nature and arise from the interaction of permanent or induced unequal charge distributions in adjacent molecules.

The most important charge distribution is the electric dipole arising from the separation of positive and negative charges. A permanent electric dipole and its associated dipole moment may arise when two bonded atoms have different electron-attracting power. The tendency for an atom to draw the electrons in a covalent bond toward it is measured by the electronegativity of the atom. A covalent bond between atoms of different electronegativity is said to be a polar covalent bond or to have ionic character. Polar covalent bonds in a molecule may give rise to polar molecules, that is, molecules having a dipole moment, depending on the geometry of the molecule.

Induced dipoles result from the distortion or polarization of the electron charge clouds in atoms and molecules due to the electric fields of adjacent charge distributions.

Questions and Problems

1. Oxygen is present in air at 20.9 percent by volume. What is its percentage by mass in air? The densities of air and oxygen are 1.29 g/liter and 1.43 g/liter, respectively, at STP.

2. The fundamental units for pressure in the cgs system are $g \, cm/sec^2 = dyn/cm^2$ (dynes per square centimeter). The units of millimeters or centimeters of Hg refer to the hydrostatic pressure exerted on a surface so many millimeters or centimeters below the surface of a column of mercury. The pressure in dynes per square centimeter is given by the formula $P = h\rho g$, where h is the height of the fluid column, ρ is the density of the fluid, and g is the acceleration due to gravity, 980 cm/sec².

   a. What is the pressure, in dynes per square centimeter, at the bottom of a column of mercury ($\rho = 13.66 \, g/cm^3$) that is 76 cm high? The answer will also be the atmospheric pressure that can support a column of mercury this high.

   b. If water ($\rho = 1 \, g/cm^3$) were used as the barometric fluid instead of mercury, what would be the standard atmospheric pressure in cm of H_2O?

   c. Why is it not necessary to specify the diameter or cross-sectional area of the barometer tube in specifying pressure?

   d. What mass of mercury exerts a force on each square centimeter of surface when the pressure is 76 cmHg? What is the mass of atmosphere above 1 cm² of the earth's surface when the atmospheric pressure is 1 atm?

3. What is the volume of 100 g of CO_2 at 2 atm and 25°C? Do this problem in two ways.

4. What is the molecular weight of a gas whose density is 1.429 g/liter at STP? Do this problem in two different ways.

5. At 27°C and 400 mmHg, a gas occupies 150 ml. At what temperature will it occupy 450 ml if the pressure is raised to 800 mmHg?

6. What is the "molecular weight" of air and its density at STP if air is taken to be 76 percent by mass N_2 and 24 percent by mass O_2?

7. A sample of air is analyzed and found to contain 20.0 ppm by volume of carbon monoxide, CO, a toxic air pollutant.

   a. Calculate the density of CO in grams per liter at STP.

   b. How many molecules of CO are there in 1 liter of the polluted air at STP?

   c. What is the concentration, in parts per million

by mass, of CO in this air? Take the density of air to be 1.29 g/liter at STP.

8. In 8 g of a gas there are 3.2 g of sulfur and 4.8 g of oxygen. The density of the gas is 2.47 g/liter at 820 mmHg and 30°C. What is the formula of the gas?

9. The density of a gas at STP is 1.429 g/liter. What is the density at 30°C and 735 mmHg?

10. At what temperature will 3 mol of SO_2 gas exert a pressure of 10.0 atm in a volume of 8.25 liters?

Problems 11–14 consider calculations based on chemical equations. A reading of Appendix B.2 will be helpful.

11. Sulfur dioxide may be prepared by heating iron pyrite in oxygen:

$$3FeS_2 + 8O_2 \longrightarrow Fe_3O_4 + 6SO_2$$

   a. What mass of SO_2 can be prepared from 32 g of O_2 and excess FeS_2 at 780 mmHg and 30°C?

   b. What volume of SO_2, measured at 780 mmHg and 30°C, will be formed from 32 g of O_2?

12. Oxygen gas may be generated in the laboratory by the thermal decomposition of potassium chlorate:

$$2KClO_3 \xrightarrow{\Delta} 2KCl + 3O_2$$

In one experiment, a heated sample of $KClO_3$ decreased in mass by 0.200 g and 153 ml of dry oxygen gas measured at 25°C and 1 atm were collected. From these results, calculate the molar volume of a gas at STP.

13. What volume of chlorine, at STP, will be required to change 15 g of potassium hydroxide into potassium chlorate? Balance the equation first.

$$KOH + Cl_2 \longrightarrow KCl + KClO_3 + H_2O$$

14. Aluminum reacts with sulfuric acid according to the equation

$$2Al + 3H_2SO_4 \longrightarrow Al_2(SO_4)_3 + 3H_2$$

How many grams of aluminum are needed for the production of 5.5 liters of hydrogen at 780 mm Hg and 17°C?

15. Dalton's law of partial pressures states that the total pressure of a mixture of nonreacting gases is equal to the sum of the pressures each gas would exert if it were alone in the volume occupied by the mixture. The pressure of each individual gas is known as its partial pressure. Thus,

$$P_{total} = P_1 + P_2 + P_3 + \cdots$$

where $P_1, P_2, \ldots$ are the partial pressures.

a. Convince yourself that if each gas acts independently in the mixture, its partial pressure is $P_i = N_i m_i (v^2)_{i(av)}/3V$ and $P_{total} = P_1 + P_2 + P_3 + \cdots = \Sigma P_i$.

b. Show that $P_i = n_i RT/V$, where n_i is the number of moles of gas i in a mixture of volume V.

c. Show that the partial pressure P_i of a gas i in a mixture is related to the total pressure by $P_i = n_i P_{total}/n_{total}$, where $n_{total} = n_1 + n_2 + n_3 + \cdots$.

16. Calculate the partial pressures of hydrogen, nitrogen, and oxygen in a mixture with total pressure of 20 atm if there are 20 g of O_2, 14 g of N_2, and 2 g of H_2 present.

17. What will be the volume occupied by a mixture of 10 g of H_2 and 8 g of O_2 at 25°C and 380 mmHg? What is the partial pressure of each gas in the mixture?

18. At 20°C, 1 liter of air can hold a maximum of 0.0173 g of water vapor. What is the partial pressure, in millimeters of Hg, of water vapor in such saturated air?

19. What are the partial pressures of nitrogen and oxygen in air at a pressure of 1 atm?

20. The rate at which a gas can spread (diffuse) throughout a container or leak (effuse) through a tiny hole in a container is, according to Graham's law, inversely proportional to the square root of the molecular weight or the density of the gas.

a. Assuming that the rates of diffusion and effusion are proportional to the molecular velocities, develop an expression relating the rates of diffusion R_1 and R_2 of two different gases to their molecular weights.

b. Show that hydrogen diffuses almost four times as fast as oxygen.

c. The small difference in the rates of effusion of gaseous $^{235}UF_6$ and $^{238}UF_6$ through porous barriers served as the basis of separation of the lighter U-235 isotope from U-238. What is the ratio of effusion times $R_{U\text{-}235}/R_{U\text{-}238}$ for the uranium isotopes?

21. What is the pressure, in atmospheres, in that region of the atmosphere at about 700 km altitude, where the temperature is 1200°K and the density of molecules is 1.00×10^6 molecules/cm³?

22. From the data in Table 4.4, verify that the average molar kinetic energy is very nearly the same for all gases at the same temperature.

23. Use the data in Table 4.4 to sketch a rough curve showing the dependence of molecular velocity on molecular weight at constant temperature.

24. Estimate from the data of Table 4.3 the ratio of the temperature to the molecular weight of the gas molecules.

25. Calculate the rms speeds of
a. H_2 at 1000°K
b. He at 1000°K
c. O_2 at 1000°K

26. A gas is heated from 0°C to 273°C. What is the relative change in the average kinetic energy of its molecules and the rms velocity of its molecules?

27. The rms speed of hydrogen molecules at 0°C is 1838 m/sec. Without using a numerical value for R, find
a. v_{rms} for H_2 at 25°C
b. v_{rms} for N_2 at 25°C

28. The percent ionic character of a single bond in a covalent diatomic molecule may be calculated from the ratio of its measured dipole moment to the dipole moment expected if there were complete charge transfer. In HBr, the bond distance is 1.141 Å and the charge on each atom would be 4.8×10^{-10} esu in a completely ionic structure.
a. What is the percent ionic character of the bond in HBr?
b. Estimate the ionic character of the HBr bond if a 50 percent ionic bond corresponds to an electronegativity difference of 2.1. Compare with the answer to part (a).

29. Estimate the dipole moments of ICl and FBr.

30. Consider the structures of the following molecules and determine which ones are polar: BF_3, ClO_2, NH_3, COS, PH_3, SiH_4, CF_4, OF_2.

Answers

1. 23.1 percent by mass.
2. a. 1.01×10^6 dyn/cm²
   b. 1030 cmHg
   c, 1034 g, 1034 g

3. 27.8 liters
4. 32.0 g/mol
5. 1800°K
6. 28.9 g/mol, 1.29 g/liter

7. a. 1.25 g/liter
    b. 5.4×10^{17} molecules
    c. 19.4 ppm by mass
9. 1.24 g/liter
10. 335°K
11. a. 48 g
    b. 18.2 liters
13. 3.0 liters
14. 4.3 g
16. $P_{O_2} = 5.9$ atm, $P_{N_2} = 4.7$ atm, $P_{H_2} = 9.4$ atm
17. 256 liters
18. 17.5 mmHg
19. $P_{N_2} = 0.78$ atm, $P_{O_2} = 0.21$ atm
20. a. $R_1/R_2 = \sqrt{M_2/M_1}$
    c. 1.0043

21. 1.63×10^{-13} atm
24. $T/M \simeq 15$
25. a. 3.54×10^5 cm/sec
    b. 2.50×10^5 cm/sec
    c. 8.83×10^4 cm/sec
26. $E_{K_2}/E_{K_1} = 2$
    $v_{rms(2)}/v_{rms(1)} = 1.4$
27. a. 1910 m/sec
    b. 511 m/sec
28. a. 12 percent
29. 0.5 and 1.2 from electronegativity differences; measured values are 0.65 and 1.29 D.
30. ClO_2, NH_3, COS, PH_3, and OF_2 are polar.

References

Most introductory chemistry texts have a discussion of the gas laws and their manipulation. See also:

Kieffer, William F., "The Mole Concept in Chemistry," Van Nostrand Reinhold, New York, 1962, Chapters 2, 9.

OTHER REFERENCES

Landsberg, Helmut E., "The Origin of the Atmosphere," *Scientific American*, **189,** 82 (August 1953).

Biogeochemical Cycles—The Nitrogen Cycle

A subtle chain of countless rings
The next unto the farthest brings
Ralph Waldo Emerson—*Nature*

In any functioning ecosystem, there exists a flow of material in the form of chemical elements or compounds and a flow of energy. The study of these flows makes up the field of ecosystem dynamics. The primary source of the energy is the sun. Solar energy powers all living and almost all nonliving processes in the biosphere: the growth and decay of organisms, the weather, the transport of matter from one part of the ecosystem to another. The flow of energy will be discussed in Chapters 16 and 20.

In this chapter, we shall be concerned with the flow of materials in ecosystems, specifically with the flow of the essential elements of life. A little thought should convince you that the natural flow of chemical elements in the ecosystem is cyclic or close to cyclic. Except for negligible amounts of the elements that may enter the biosphere from outer space or from the depths of the earth, or that may be lost by escape from the atmosphere, the total quantity of the elements available to living organisms on earth is constant. The elements—carbon, hydrogen, oxygen, phosphorus, nitrogen, sulfur, and all others essential to life—circulate between the physical and the living environments of organisms in amounts and at rates selected over long evolutionary periods to provide stability to an ecosystem, and to the biosphere as a whole.

There is a cycle for every element. All the cycles are interrelated, so interference with a part of one cycle may affect the operation of all the connected cycles. A study of the transformations of matter and of the energy transfers that power the cycle for each element provides the most sophisticated and basic way to view ecology and to understand man's effect on his environment. Pollution, in particular, can be regarded as the result of human influence on the chemical, biological, and geological processes and energy exchanges involved in natural cycles.

Although our attention will initially be directed to the elements essential for life, it should be remembered that elements not essential to life, and even toxic in sufficient amounts, are also involved in natural cycles. In order to have biological significance in such cycles, a nonessential element (or compound) must be present in higher concentrations in an organism than in the physical environment. There may be several reasons for this biological concentration effect, and these will be pointed out as appropriate.

5.1 BIOGEOCHEMICAL CYCLES

In the biosphere, chemical elements move back and forth between living organisms and the physical environment. There is also a transfer of elements from one organism to another through food chains as well as transport of elements among the regions of the physical environment. The overall cyclic nature of the movement of a hypothetical element X is shown in Figure 5.1.

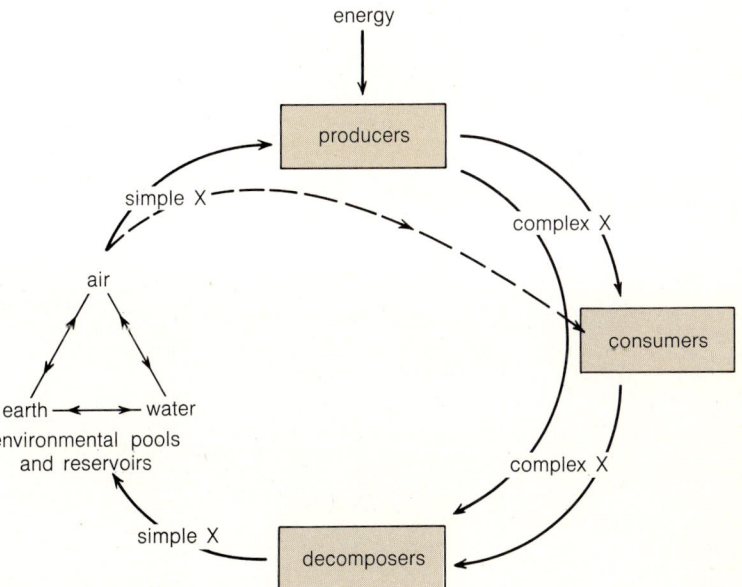

Figure 5.1. Simplified biogeochemical cycle for an element X and its compounds. The passage of X into and out of the physical environment (earth, air, water) and the living environment (producers, consumers, decomposers) is brought about by living organisms. Dashed lines indicate less common pathways.

Producer, and to a lesser extent consumer, organisms absorb X or its simple compounds directly from the physical environment and synthesize more complex X compounds. Consumers use the energy and matter of the complex X compounds of the producer to make their own complex X compounds.

Waste products from living organisms and the bodies of dead organisms are ultimately decomposed by bacteria and fungi. The decay process breaks down the complex protoplasm of the organism to simpler molecules and elements. The simple nutrient molecules pass into parts of the physical environment, where they are stored for various lengths of time until once again taken up by producers to begin the cycle. Such cycles are called *bio-geochemical* cycles, the *bio-* coming from the biotic or living organism stations on the cyclic path, the *geo-* referring to the geological environmental pools (atmosphere, hydrosphere, or lithosphere) of the cycle. Because any element X undergoes chemical reactions as it goes around the cycle again and again, knowledge of the chemistry of *X* and its compounds is important for a full understanding of the cycle.

Living organisms bring about most of the chemical reactions that transform a nutrient element. Thus, not only do elements pass back and forth between the environment and living organisms, but living organisms provide the reaction pathways. In fact, many of the chemical reactions involved in biogeochemical cycles cannot be made to take place with any efficiency unless they occur within living things. Once again, we observe the close dependence of the abiotic (nonliving) and biotic components of the biosphere.

5.2 STABILITY OF BIOGEOCHEMICAL CYCLES

Before considering specific examples, it is well to note that in any actual biogeochemical cycle there is usually more than one path an element or its compound can take from one point to another in the cycle. Some of these paths may involve intermediate stops or cycles within cycles, as diagrammed schematically in Figure 5.2. The alternative pathways are important in maintaining the stable operation of a cycle. For instance, if one reaction path between two points is rendered inoperative for any reason, the flow of nutrient elements can continue by other routes so that the cycle does not break down.

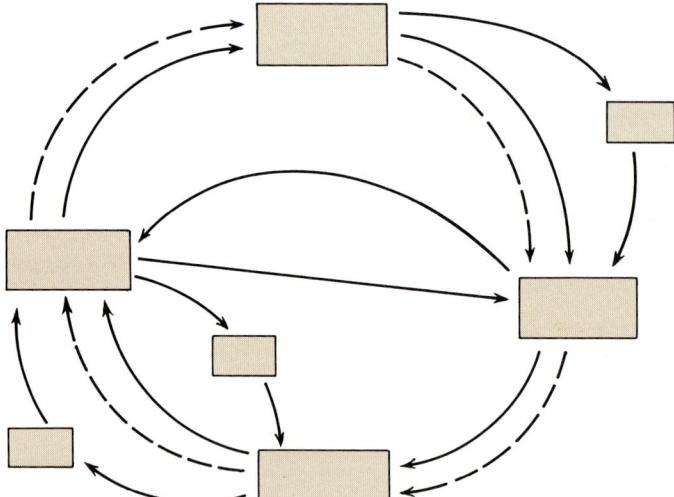

Figure 5.2. Alternative pathways in a hypothetical biogeochemical cycle. The boxes may represent either environmental reservoirs and organisms or the chemical form of an element and its compounds. Less common alternative pathways are shown as dashed lines. Subcycles exist in the center and at the lower right of this cycle.

The cardiovascular system in the human body operates in this fashion. A blockage of a blood vessel (a blood clot) may lead to death or serious impairment if this blood vessel carries all of the blood to a sensitive area. If, however, there are other blood vessels serving the same area, these other vessels will continue the flow of blood. For this reason, physicians prescribe exercise to promote the generation of alternative blood vessels in case a stroke closes down one of them.

Likewise, in a cycle, an increase or decrease in movement of a nutrient along one route is usually compensated by adjustments in the flow rate along other paths. The more pathways a biogeochemical cycle has, the better its ability to regulate itself so that there is no appreciable decrease or buildup of nutrient at any given point. The mechanisms of self-regulation or homeostasis in biogeochemical cycles in ecosystems have their counterparts in the complex metabolic regulation mechanisms in organisms as well as in the mechanisms of many comparatively simple chemical reactions.

It is the intrusion into some of the basic biogeochemical cycles by human activity that causes many of our pollution problems today. Man may influence naturally stable cycles in several ways, and we shall discover numerous examples. In general, however, man may modify balanced cycles at various points in space or time by:

1. Oversimplifying cycles by destroying or by-passing one or more of the pathways among organisms or from organism to environment. For example, by destroying various organisms, pesticides may remove the pathway of part of a cycle.

2. Exceeding the natural ability of a cycle to assimilate substances added to an ecosystem at a high rate or to replace substances removed from an ecosystem faster than they can be replaced. Examples of the former are the effects resulting from excess use of fertilizers and the large-scale introduction of nitrogen oxides and carbon dioxide into the atmosphere by combustion sources. Overharvesting of crops is an example of a too-rapid removal of nutrient elements from the soil ecosystem.

3. Introduction to the environment of substances not found in nature, which do not gain entry into existing biogeochemical cycles or which are incorporated into existing cycles in ways detrimental to organisms. Thus, certain nondegradable wastes such as some plastics accumulate in the environment. Other natural substances such as radioactive strontium-90 from nuclear fission explosions, mercury, and other previously rare materials are finding their way into the environment in increasing amounts, there to enter and accelerate their own cycles. The increased rate of transport of such materials through their cycles often leads to biological concentration of potentially toxic substances in organisms. Strontium-90 is chemically similar to calcium, so the strontium cycle parallels and competes with the calcium cycle. Strontium-90 accumulates like calcium, and replaces calcium, in the bone marrow of man and other vertebrates, where its radioactivity may give rise to cancer and genetic mutations.

5.3 TYPES OF BIOGEOCHEMICAL CYCLES

Biogeochemical cycles are often classified by the geological reservoir containing the greatest amount of a specific nutrient element in the biosphere. For instance, the atmosphere holds the overwhelming proportion of the accessible nitrogen atoms in the biosphere. The nitrogen cycle is an example of a *gaseous biogeochemical cycle*, as is the oxygen cycle. The carbon cycle is also considered a gaseous cycle, because a major part of the cycle occurs in the atmosphere even though the primary reservoir for carbon is the ocean.

The other type of cycle is the *sedimentary cycle*. In this case, the principal reservoir for the element is the solid earth, the rocks and soil, from which the nutrients are released to other parts of the environment and to living organisms by weathering (erosion). Examples of nutrient elements whose cycles are sedimentary are phosphorus and sulfur. Many of the nonessential elements are involved in sedimentary cycles.

The gaseous and sedimentary cycles differ principally in two respects:

1. The sedimentary cycles tend to be simpler and to have fewer alternative routes or reaction paths between portions of the cycle. They are therefore more easily disrupted than gaseous cycles.

2. The nutrient in sedimentary reservoirs tends to be less available to living organisms than the nutrients of gaseous cycles. Sulfur and phosphorus compounds, for instance, are not naturally returned to the cycle as fast as they are removed. They are not efficiently recycled. Since many of these nutrients often control the efficiency of living processes (see limiting factors, Section 11.9), shortages resulting from their failure to recycle may seriously upset organisms. The tendency to develop such shortages because of inefficient recycling makes sedimentary cycles less perfect or less balanced.

5.4 THE NITROGEN CYCLE: A GASEOUS CYCLE

Figure 5.3. The nitrogen cycle. Dashed arrows indicate relatively minor pathways. Several arrows indicate successive processes. The importance of each reservoir and pathway depends on the specific ecosystem.

With this general background on biogeochemical cycles in mind, let us look at a specific gaseous cycle. (Sedimentary cycles will be discussed in Chapter 14.) The nitrogen cycle is the standard example of a gaseous cycle. A diagram of the nitrogen cycle is shown in Figure 5.3. Although somewhat simplified, it encompasses the entire biosphere, not just one ecosystem such as the soil. The major natural reservoir, or storehouse, of nitrogen is the atmosphere. However, this does not mean that nitrogen passes rapidly back and forth between the gas phase and its combined forms in living things. In

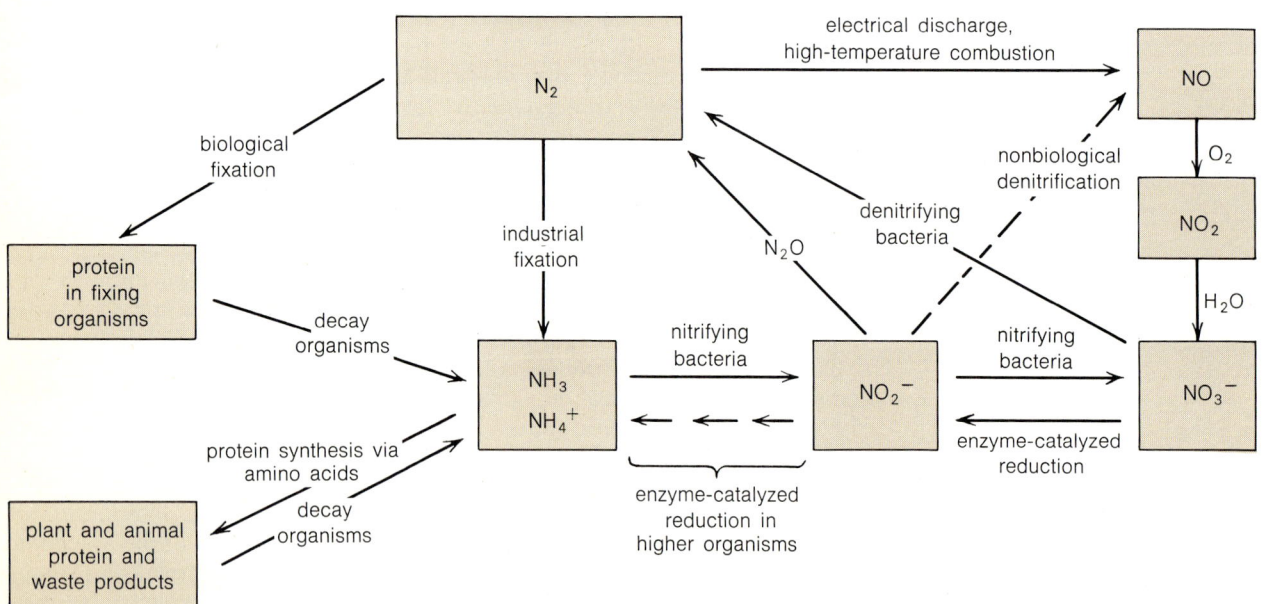

fact, the nitrogen in the atmosphere exhanges rather slowly with organisms. The forms of nitrogen in the soil and in solution are readily available for assimilation by life; and it is these portions of the nitrogen cycle, rather than the atmospheric portions, that are involved in the exchange of nitrogen between living organisms and their environment.

The names and formulas of the major compounds in the nitrogen cycle, as well as some related compounds, are given in Table 5.1. The substances are listed in decreasing magnitude of the oxidation state of the nitrogen atom in each compound.

Table 5.1 Nitrogen compounds in the nitrogen cycle[a]	oxidation state of nitrogen	compound		
		formula	name	
	+5	N_2O_5	dinitrogen pentoxide or nitrogen(V) oxide	oxidized forms
		HNO_3	nitric acid	
		NO_3^-	nitrate ion	
	+4	NO_2	nitrogen dioxide or nitrogen(IV) oxide	
	+3	HNO_2	nitrous acid	
		NO_2^-	nitrite ion	
	+2	NO	nitric oxide or nitrogen(II) oxide	
	+1	N_2O	nitrous oxide or nitrogen(I) oxide	
	0	N_2	nitrogen	
	−3	NH_3	ammonia	
		NH_4^+	ammonium ion	
		$R{-}\overset{\displaystyle H}{\underset{\displaystyle NH_2}{C}}{-}C\overset{\displaystyle O}{\underset{\displaystyle OH}{}}$	α-amino acids	
		$R{-}\overset{O}{\overset{\|}{C}}{-}NH_2$	amides	reduced forms
		$R{-}\overset{\displaystyle}{\underset{\displaystyle H}{N}}{-}\overset{O}{\overset{\|}{C}}{-}R'$	proteins (peptide linkage)	

[a] In the −3 oxidation state compounds, R stands for groups making up the rest of the molecule. These may be large or small organic (carbon-containing) groups or simply hydrogen.

5.5 OXIDATION, REDUCTION, OXIDATION STATE

Oxidation was formerly used to describe a reaction in which oxygen combined with another element or compound. Conversely, reduction meant the removal of combined oxygen. Listed below are some typical oxidation and reduction reactions.

$$\left. \begin{aligned} C + O_2 &\longrightarrow CO_2 \\ 2H_2 + O_2 &\longrightarrow 2H_2O \\ 4Fe + 3O_2 &\longrightarrow 2Fe_2O_3 \\ 2Cu + O_2 &\longrightarrow 2CuO \end{aligned} \right\} \quad \text{oxidations}$$

$$\left. \begin{aligned} Fe_2O_3 + 3H_2 &\longrightarrow 2Fe + 3H_2O \\ CuO + H_2 &\longrightarrow Cu + H_2O \end{aligned} \right\} \quad \text{reductions}$$

Somewhat later, the meanings of oxidation and reduction were broadened to include the removal and addition, respectively, of hydrogen atoms to a substance, as in the oxidation reaction $CH_3OH \longrightarrow CH_2O + H_2$.

Today, we recognize a much more general definition of the chemical processes of oxidation and reduction based on the transfer of electrons from one atom to another. According to the modern definition, *oxidation is the loss of electrons*, and *reduction is the gain of electrons*. The simplest types of reactions illustrating these concepts are those in which actual ions of elements are involved.

$$Na(g) + Cl(g) \longrightarrow NaCl(g)$$

or

$$Cu(s) + 2Ag^+(aq) \longrightarrow Cu^{2+}(aq) + 2Ag(s)$$

In both of these reactions, one element (Na or Cu) gives up electrons to become a positive ion, while another element (Cl or Ag^+) accepts these electrons and assumes a less positive (or more negative) ionic charge. The element losing electrons is said to be oxidized, while the element gaining electrons is reduced. Thus, Na and Cu are oxidized; Cl and Ag^+ are reduced. Neither of these examples involves oxygen or hydrogen. Note one important feature of these reactions: each reaction involves both an oxidation and a reduction. Every loss of electrons by an oxidized substance must be accompanied by a gain of electrons by a reduced species, and the number of electrons lost must equal the number gained. Therefore, any complete reaction of this type is a combination of oxidation and reduction. For short, the name *redox* reaction is commonly applied.

Not all reactions, however, involve ions. Consider the reaction

$$CuO + H_2 \longrightarrow Cu + H_2O$$

Here no free ions are involved, but an oxidation and a reduction still take place. In such cases, the use of the ionic valence to tell whether an element has been oxidized or reduced is abandoned for the concept of *oxidation state*. In compounds containing simple (elemental) ions, the oxidation state is the same as the ionic valence of the element concerned, 2 for Fe in $FeCl_2$, -2 for O in Li_2O, and so on. An element in the free uncombined state, or when combined with itself, such as $Cu(s)$ or H in H_2, has a zero oxidation state.

In compounds with polar covalent bonds, the oxidation state is determined by assuming that all electrons in the bond belong to the more electronegative element. The excess or deficiency of valence shell electrons, with respect to the number of valence electrons the element has as a neutral atom, is the magnitude of the oxidation state. If the bonded atom has more valence electrons than the neutral atom, it has a negative oxidation state. For example, consider the water molecule, H:Ö:H. The oxygen, being more electronegative than hydrogen, is assigned both pairs of bonding electrons (4), which added to the unshared pairs (4) gives a total of 8 valence electrons. But if we look at a neutral oxygen atom, we see that it has 6 valence electrons (we do not consider the inner $1s$ pair), so in H_2O, oxygen has two negative charges in excess and is given a -2 oxidation state. The hydrogens each have no electrons (in water), but as neutral atoms they each have 1 electron. Thus, they have a deficiency of one negative charge each and are assigned a $+1$ oxidation state. The sum of oxidation states for all atoms in any neutral molecule adds up to zero.

As another example, consider hydrogen peroxide, H:Ö:Ö:H. Again, the oxygens are the more electronegative atoms. So the electron pairs in the O:H bonds are assigned to oxygen. The electron pair in the O:O bond is divided equally between the equally electronegative oxygens, 1 electron to each. Each oxygen then has 7 valence electrons around it, while in the neutral atom it has 6. The oxygens each have a -1 oxidation state in peroxides. The hydrogens again have a $+1$ oxidation state, and since H_2O_2 is a neutral molecule, the sum of oxidation states adds up to zero.

Now consider an ion such as the carbonate ion, CO_3^{2-}. A Lewis electronic structure for one of the resonance forms is

$$\left[\begin{matrix} :\ddot{O}:C::\ddot{O} \\ :\ddot{O}: \end{matrix} \right]^{2-}$$

Oxygen is again the more electronegative, so by the above rules each oxygen has 8 valence electrons around it, compared to 6 in the neutral atom. Each oxygen has a -2 oxidation state. The carbon here has no electrons, compared to 4 valence electrons in the neutral atom; it has a $+4$ oxidation state. The sum of all oxidation states in CO_3^{2-} must add up to give the charge on the ion, -2.

It is not necessary to draw the dot structure to determine the oxidation state of an element. A few simple rules, based on the oxidation states for common elements in a variety of compounds, allow us to deduce oxidation states from the formula of a substance.

1. The oxidation state of free elements, whether in atomic or molecular form, is zero. In Ne, H_2, S_8, and so on, the oxidation state of the element is zero.

2. The oxidation state of an element in simple ionic form is equal to the charge on the ion. The oxidation state of Na in Na^+ is $+1$, of Mg in Mg^{2+} is $+2$, of Cl in Cl^- is -1.

3. The oxidation state of oxygen in oxygen compounds is -2. The only exceptions are O_2, O_3 (see rule 1), OF_2 (where it is $+2$), and peroxides such as H_2O_2 and other compounds with —O—O— bonds (where it is -1).

4. The oxidation state of hydrogen is $+1$ except in H_2 and in compounds where H is combined with a less electronegative element. The latter compounds are called *hydrides*, such as NaH, LiH, CaH_2, and the oxidation state of hydrogen is -1.

These rules, together with the general requirement that the sum of the oxidation states of all elements in a substance must add up to the charge on the species, are sufficient to allow the deduction of oxidation states in most common compounds. The operation of these rules is seen in the following examples.

$$N_2O_5$$
$$2(N) + 5(-2) = 0 \qquad 2N = 10$$
$$N = +5$$

$$NH_4{}^+$$
$$N + 4(+1) = +1 \qquad N = -3$$

$$Cr_2O_7{}^{2-}$$
$$2Cr + 7(-2) = -2 \qquad 2Cr = +12$$
$$Cr = +6 \quad \text{or} \quad Cr(VI)$$

Oxidation states are often designated by Roman numerals so as not to give the impression that such an improbable ion as Cr^{+6} exists. The Roman numeral is usually indicated for the less electronegative element in a compound.

In terms of oxidation states, *oxidation is an increase in oxidation state in a positive direction, while reduction is a decrease in oxidation state toward a less positive (or more negative) value,* as shown in Figure 5.4. The concept of oxidation state helps us recognize that oxidation and reduction can include a shift of electrons away from or toward an atom as well as a complete transfer.

In general, the nitrogen cycle consists of a series of oxidation and reduction reactions of nitrogen and certain of its compounds. Almost all of these reactions take place through the agency of living organisms—green plants, algae, bacteria, or fungi—which provide energy to drive many of the reactions or substances to accelerate the reactions, or both. The cycle is an intimate interaction between living organisms and the physical environment, the welfare of one depending on that of the other.

In the nitrogen cycle, elemental or free nitrogen exists almost exclusively in the air. The other forms, aside from proteins and amino acids, reside principally in the soil and the water, where they serve as foods which living organisms convert into body protein. The ecosystems affected by the nitrogen cycle are therefore the terrestrial and marine ecosystems, where the main pollution problems result from an excess of nitrogen in the form of nitrate ion, $NO_3{}^-$. The general chemical properties and reactions of the simpler nitrogen compounds found in the nitrogen cycle will be presented here, followed by a discussion of the major points of the nitrogen cycle.

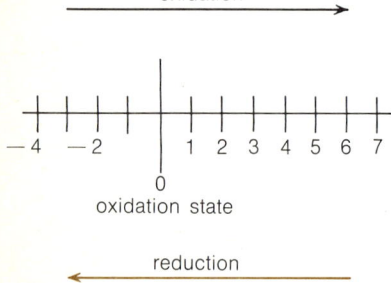

Figure 5.4. Oxidation and reduction processes in relation to oxidation states.

5.6 OCCURRENCE AND PREPARATION OF NITROGEN

Elemental nitrogen, as the diatomic molecule N_2, makes up about 80 percent by volume of the air. The only other concentrated natural source of nitrogen is in the form of the compound sodium nitrate (Chile saltpeter) which is mined in northern Chile. Nitrogen also occurs in plant and animal proteins at an average of about 16 percent, but it is less easily recovered from proteins than it is from the air and from nitrate minerals.

Elemental diatomic nitrogen gas is prepared commercially in large quantities from liquid air by allowing the more volatile nitrogen to evaporate from the mixture. Liquid nitrogen has a lower boiling point than liquid oxygen ($-196°C$ versus $-183°C$), so the vapor (gas) above the boiling liquid air is richer in nitrogen than in oxygen (Section 12.24). Unless further purified, nitrogen obtained from air will be contaminated by the inert gases, which have even lower boiling points than nitrogen.

Small quantities of pure nitrogen gas may be prepared for laboratory use by the thermal decomposition of nitrites such as ammonium nitrite, NH_4NO_2.

$$NH_4NO_2 \xrightarrow{\Delta} N_2 + 2H_2O$$

The Δ symbol indicates that the reaction proceeds at elevated temperatures.

5.7 PROPERTIES OF ELEMENTAL NITROGEN

The nitrogen atom has the electronic configuration $1s^22s^22p^3$ with 3 unpaired p electrons. The atom tends to complete its octet by sharing of electrons in covalent bonds rather than by gaining (or losing) 3 electrons to form ions. In most of its common compounds, therefore, nitrogen is covalently bonded to C, H, O, or other N atoms. However, nitrogen–oxygen bonds are very rare in biological material, and the compounds of interest in living matter, the proteins in particular, consist of nitrogen bonded covalently to hydrogen and carbon. The major exceptions to covalent bonding of nitrogen are the ionic *nitrides* formed with electropositive metals such as lithium and group IIA metals. These nitrides are formed by the direct combination of the elements:

$$3Ca + N_2 \xrightarrow{\Delta} Ca_3N_2$$

The bond energy of the triple bond in the free N_2 molecule is 225 kcal/mol of N_2 molecules:

$$N_2 \longrightarrow 2N \qquad \Delta H = 225 \text{ kcal/mol}$$

The bond energy of a diatomic molecule is the same as the dissociation energy of the molecule. For N_2, this means that 225 kcal of energy are required to break 1 mol of $N{\equiv}N$ bonds and give 2 mol of nitrogen atoms. This value is higher (except for CO) than any other diatomic molecule and makes N_2 an exceptionally chemically unreactive molecule. Because of the relative inertness of N_2, high temperatures are required to initiate reactions of N_2 with other elements and compounds in the laboratory. In living organisms, where nitrogen is essential, such high temperatures are forbidden. To convert free diatomic nitrogen from the atmospheric reserve into more reactive compounds which can be assimilated into organisms through the biogeochemical nitrogen cycle, living organisms use special substances called enzymes (Chapter 17) to speed up the conversion of N_2 to its compounds. Substances that increase the speed of a reaction are known as *catalysts*. The process, whether industrial or biological, of converting N_2 into utilizable (available) nitrogen in compounds is called *fixation*. In the nitrogen cycle, the fixation of nitrogen by organisms forms one of the more vulnerable parts of the cycle.

5.8 COMPOUNDS OF NITROGEN AND OXYGEN

The direct reaction of nitrogen with oxygen is extremely endothermic; that is, a large quantity of heat is absorbed in the reaction

$$N_2(g) + O_2(g) \longrightarrow 2NO(g) \qquad \Delta H = 21.6 \text{ kcal/mol of NO}$$
nitric oxide or
nitrogen(II) oxide

The large input of energy required for this reaction means that it will occur to a significant extent only at the high temperatures of 2000°C or higher, such as are obtained in the electric discharges of arc furnaces, in lightning, and in high-compression internal combustion engines.

As a rough approximation, a reaction requiring a large absorption of heat results in a product that is potentially unstable with respect to its reversion to the reactants. This is because any reverse reaction will be an exothermic reaction. Highly exothermic reactions have a tendency to take place spontaneously. (There are qualifications to this statement. Heat is a particular form of energy, and we have already noted in Section 2.11 that factors other than energy, and heat in particular, are not the whole story to spontaneity of reactions.) In the case of NO,

$$2NO(g) \longrightarrow N_2(g) + O_2(g) \qquad \Delta H = -21.6 \text{ kcal/mol of NO}$$

or $\Delta H = -43.2$ kcal for the reaction as written.

Processes forming other gaseous oxides of nitrogen from the elements are, like the formation of NO, endothermic reactions. Most other elements combine with oxygen exothermically to form oxides. This exothermicity seems to indicate that the bonds between nitrogen and oxygen are unstable and that nitrogen oxides should have a tendency to decompose. This is not actually the case, since nitrogen oxides are fairly stable by themselves. However, as bond energies (Table 3.3) indicate, when nitrogen bonded directly to oxygen is present in molecules containing carbon and hydrogen, the nitrogen preferentially forms bonds with carbon and hydrogen at the expense of its bonds with oxygen.

Nitric oxide can also be prepared by the reaction of *dilute* nitric acid on copper,

$$3Cu + 8H^+ + 2NO_3^- \longrightarrow 3Cu^{2+} + 2NO(g) + 4H_2O$$

(see Appendix D for an explanation of the writing and balancing of ionic redox equations like this), or by the reaction of ammonia and oxygen,

$$4NH_3 + 5O_2 \xrightarrow[\text{Pt}]{500°C} 4NO + 6H_2O$$

at high temperatures using a platinum catalyst (as indicated above and below the arrow, respectively).

Nitric oxide is a colorless gas. But in the presence of oxygen, it readily reacts at ordinary temperatures to give nitrogen dioxide, NO_2, a brown toxic gas responsible for much of the color of photochemical smog (Section 7.4).

$$\underset{\text{colorless}}{2NO} + O_2 \longrightarrow \underset{\text{brown}}{2NO_2}$$

Nitrogen dioxide reacts with water to form nitric acid, HNO_3, and nitric oxide:

$$3NO_2(g) + H_2O \longrightarrow \underset{2HNO_3}{\underbrace{2H^+ + 2NO_3^-}} + NO(g)$$

The set of reactions

(a) $4NH_3 + 5O_2 \xrightarrow[\text{Pt}]{500°C} 4NO + 6H_2O$

(b) $2NO + O_2 \longrightarrow 2NO_2$

(c) $3NO_2 + H_2O \longrightarrow 2H^+ + 2NO_3^- + NO$

comprises the Ostwald process for the manufacture of the industrially important nitric acid, HNO_3.

Nitrogen dioxide, NO_2, may also be produced in the laboratory by the action of *concentrated* nitric acid on copper or silver:

$$Cu + 4H^+ + 2NO_3^- \longrightarrow Cu^{2+} + 2NO_2 + 2H_2O$$

From the standpoint of the nitrogen cycle, the ionic forms of nitrogen oxides—the nitrates, NO_3^- and the nitrites, NO_2^-—are of great importance. Despite the fact that most of the fixed nitrogen in the biosphere is combined in biological or organic forms with carbon and hydrogen, the relatively small amount occurring as nitrates and nitrites is essential for the operation of the cycle.

5.9 NITRIC ACID AND NITRATES

The Ostwald process for the manufacture of nitric acid has already been described. Pure nitric acid is colorless, but it slowly decomposes under the action of heat or sunlight to give brown nitrogen dioxide, NO_2,

$$4HNO_3 \longrightarrow 4NO_2 + O_2 + 2H_2O$$

and this causes the pure acid to become yellow.

In 100 percent nitric acid, molecules of HNO_3 exist, but in water the molecules ionize to give

$$HNO_3 \longrightarrow H^+ + NO_3^-$$

The colorless nitrate ion is a resonance hybrid with a planar structure. The nitrate ion tends to accept electrons from other atomic and molecular species, with the formation of nitrogen or compounds of nitrogen in lower oxidation states. Examples have already been given in the reactions of nitric acid with copper. In these reactions, the nitrate ion is itself reduced.

Nitrate ion also occurs in combination with positive ions in ionic solids. Examples are

$NaNO_3$	$Na^+[NO_3^-]$	sodium nitrate
$Ca(NO_3)_2$	$[NO_3^-]Ca^{2+}[NO_3^-]$	calcium nitrate
NH_4NO_3	$NH_4^+[NO_3^-]$	ammonium nitrate

The solid form of these and most other nitrates is made up of the positive cations and negative nitrate anions. When water is added, the positive and negative ions move apart, and the solution then contains cations and nitrate ions. The rapid ionization of nitrates in water and the resulting characteristic high solubilities of these nitrates have important consequences in the nitrogen cycle.

When metallic nitrates are heated to quite high temperatures, they decompose almost completely.

$$4NaNO_3 \xrightarrow{\Delta} 2Na_2O + 5O_2 + 2N_2$$

Nitrous acid, HNO_2, cannot be obtained in the pure state, and it is known only in aqueous solution where the molecules are partially ionized. The acid is unstable and decomposes slowly to give mixtures of NO, NO_2, and NO_3^-.

The nitrite ion, NO_2^-, is a resonance hybrid. Nitrites, like nitrates, are all quite soluble, with the exception of silver nitrite.

5.10 AMMONIA AND AMMONIUM ION

One of the most important of nitrogen compounds is ammonia, NH_3. Ammonia is produced commercially by the direct combination of nitrogen and hydrogen gas in the Haber process:

$$N_2 + 3H_2 \longrightarrow 2NH_3$$

The difference in energy between product and reactants in this reaction is such that one would expect the direct reaction to take place spontaneously at normal temperatures—that is, the reaction is exothermic and evolves heat—with the production of a good yield of ammonia. Any increase in temperature will give a lower yield of ammonia, because the addition of heat can be shown to shift this reaction toward the elements (Le Chatelier's principle, Section 8.14). However, at 25°C, the combination of N_2 and H_2 is an extremely slow reaction. Increasing the temperature to about 1000°C speeds up the reaction but results in a lower yield of NH_3. As a compromise, the Haber process is run at about 500°C with the aid of an iron oxide or other catalyst. Under these conditions, both the yield and the rate are acceptable. The reaction is also run under a total gas pressure of 150–350 atm, which also increases the yield of NH_3 (Le Chatelier's principle again).

Recently, progress has been made toward a process that may permit the industrial fixation of atmospheric nitrogen as ammonia at room temperature and atmospheric pressure. Certain transition metal compounds, such as those of titanium(II), are able to bind nitrogen molecules. The nitrogen can then be easily reduced to NH_3 and the ammonia released from the compound with the regeneration of the original metal compound. The process parallels that which occurs in the biological fixation of nitrogen by organisms, where the transition metals, molybdenum, cobalt, or manganese, are the operative elements.

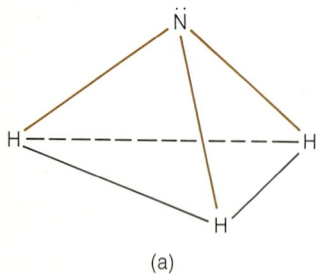

(a)

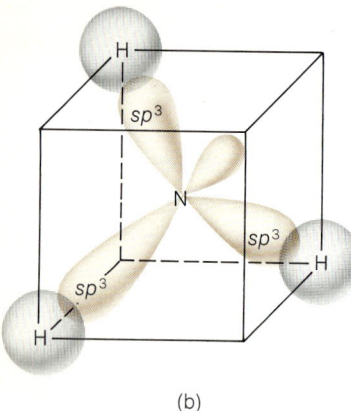

(b)

Figure 5.5. (a) Pyramidal structure of the NH_3 molecule. (b) Tetrahedral hybridization of the nitrogen orbitals in NH_3.

The pyramidal structure (Figure 5.5) of ammonia with H—N—H bond angles of 107° rather than 90° implies that tetrahedral hybridization of the nitrogen 2s and 2p orbitals has occurred to give four essentially equivalent hybrid orbitals directed toward the corners of a tetrahedron. Three of these orbitals are used for bonding to hydrogen. and the unused pair of electrons (formerly the nitrogen's two 2s electrons) is in the unused sp^3 orbital. Or, according to another viewpoint, the four pairs of electrons around nitrogen in the NH_3 molecule try to stay as far from one another as possible and distribute themselves to give the resulting bond angles.

The N—H bonds are highly polar, the negative end of the bond moment being toward the nitrogen. In addition, there is a moment associated with the lone pair of electrons, and this moment reinforces (Figure 5.6) the bond moments to give a highly polar molecule.

Ammonia gas is easily liquefied; its boiling point is a comparatively high −33.4°C. The high boiling point is due to very strong bond dipole–bond dipole interactions of the hydrogen bond variety (Section 8.2), in which the positive hydrogen of an :N—H bond on one NH_3 molecule is attracted to the negative nitrogen end of an N—H bond in another molecule.

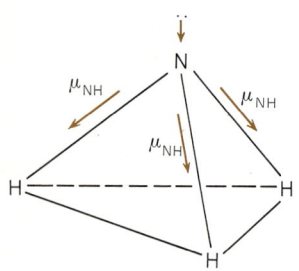

Figure 5.6. Contributions to the dipole moment of NH_3.

$$
\begin{array}{ccc}
& H & & H \\
\delta+ \;\mid\; \delta- & & \delta+ \;\mid\; \delta- \\
H{-}N: & \cdots & H{-}N: \\
\mid & & \mid \\
H & & H
\end{array}
$$

When ammonia gas is dissolved in water, hydrogen bonding between ammonia and water molecules occurs:

$$
\begin{array}{ccc}
& H & \\
\delta+ \;\mid\; \delta- & & \delta+ \;\; \delta- \\
H{-}N: & \cdots & H{-}O: \\
\mid & & \mid \\
H & & H
\end{array}
$$

This interaction accounts for the high solubility of ammonia in water. The resulting solution is often called ammonium hydroxide (NH_4OH) and written as such. Actually, a solution of ammonia gas in water consists mostly of ammonia molecules hydrogen bonded to water, $H_3N \cdots H_2O$, with very few free ammonia molecules or ammonium ions NH_4^+ and hydroxide ions OH^-. The latter ions may be regarded as originating from either of the reactions

$$NH_4OH \rightleftharpoons NH_4^+ + OH^-$$

or

$$NH_3 + H_2O \rightleftharpoons NH_4^+ + OH^-$$

where the double arrows signify that the reactions can go in either direction. If a substance that can furnish hydrogen ions is added to an ammonia solution, ammonium ions are readily formed via the reaction

$$NH_3 + H^+ \longrightarrow NH_4^+$$

5.11 OTHER NITROGEN COMPOUNDS

The only other nitrogen-containing compounds that appear in our diagram of the nitrogen cycle are those occurring in the protoplasm of organisms or in wastes of these organisms. These are the proteins and the much simpler amino acids and amides. Proteins are the end result of the fixation and utilization of nitrogen in living organisms. Some chemistry of these organic substances will be covered in later sections of the book (Sections 15.13–15.15), but only the characteristic chemical combinations of nitrogen given at the bottom of Table 5.1 need be noted now. The important point about these complex substances is that living organisms can synthesize them from simple nitrate and nitrite ions and from ammonia. Other organisms can break them down to these simple molecules and ions.

An Overall View of the Nitrogen Cycle

5.12 SELF-REGULATION IN THE NITROGEN CYCLE

Now let us take another look at the nitrogen cycle and try to point out some of the general features of this cycle. Many of these features are typical of all biogeochemical cycles.

In the first place, the variety and number of possible routes from one nitrogen compound to another (or from one nutrient pool to another) indicates that the nitrogen cycle is fairly "perfect." That is, the nitrogen cycle is self-regulating and can maintain balance when small, short-term natural disturbances affect the flow of nitrogen along one of the pathways. There is little accumulation or depletion of nitrogen in any of the ecosystem pools. In the nitrogen cycle, the only imbalance in the nitrogen budget is believed to be the loss of organic nitrogen to deep, inaccessible, marine sediments. Nitrates formed in these sediments by bacterial decay may eventually be brought to the surface of the earth to form concentrated nitrate deposits.

Instances do develop where the natural homeostatic mechanisms of the nitrogen cycle are incapable of restoring balance. Some of the general ways in which any cycle can be disrupted or disturbed were listed in Section 5.2. In the case of the nitrogen cycle, man has caused local imbalances by drastically altering the amount of nitrate ion in ecosystems. He has done this both by indirectly removing essential nitrates from some ecosystems and by adding too much nitrate to other ecosystems. The problem is essentially one of misplacing nitrates.

On land, proteins and other nitrogen-containing organic matter normally return to the soil. There soil bacteria decompose some of the organic matter, with the release of ammonia and plant-absorbable nitrate ion. Left behind after decay is a complex mixture known as *humus*. Humus maintains the porosity of soil, allowing atmospheric oxygen to penetrate to plant roots and maintaining proper soil drainage. But when crops are continually harvested from a local ecosystem (a corn field, for instance), plant proteins are removed and do not return to the soil at that locality. Without a supply of decomposable organic matter, neither humus nor nitrates are generated in the soil. Man turns to the application of nitrogen fertilizers to increase the plant-utilizable nitrate supply of the soil.

The nitrate ion added directly or indirectly from nitrogen fertilizers contributes nothing to soil porosity. Also, the highly soluble nitrates readily wash out of the soil to enter nearby bodies of water where, along with nitrate from domestic sewage, they may enhance the growth of aquatic plants. When the plants die, decay processes use up the available oxygen in the water. The lowered oxygen content of the water results in adverse effects to marine animals and interferes with the processes whereby natural waters can purify themselves. Although the nitrogen cycle might be considered to have accelerated in order to absorb the nitrates added to the ecosystem, the connected oxygen cycle has fallen far behind. These and other topics dealing with water and soil will be reconsidered in Chapters 12 and 14.

5.13 LIFE IN THE NITROGEN CYCLE

Living organisms are always involved in biogeochemical cycles either as depositories of nutrient elements or as agents bringing about the required chemical transformations. The nitrogen cycle, in particular, has living things deeply incorporated into its operation. All but four of the chemical changes indicated in Figure 5.3 are carried out in nature by organisms, principally bacteria. Without the energy or the enzyme catalysts supplied by these living things, many of the reactions indicated would never take place at the conditions in the soil, water, or organisms.

The chemical properties of nitrogen and its compounds were illustrated earlier by a number of chemical reactions. It should be remembered that most of these reactions were written for conditions pertaining to abiotic conditions in the laboratory or in industry. Thus, the high temperature, the high pressures, and the metallic catalysts used in the Ostwald or Haber processes to synthesize nitrates or ammonia most certainly have no place in the biochemical processes going on inside living organisms. Even if the reactants and the products happen to be identical in the laboratory and in the organism, the reaction must occur by a different mechanism in the two cases, the biochemical process undoubtedly being much more complex and using a greater number of intermediate steps in going from reactant to product. This points up the all-important role of life in the balanced operation of biogeochemical cycles.

Sometimes a process may be carried out by many different organisms. This is the case with the decay of proteins and waste matter, where many nonspecialized organisms can convert proteins and other biological nitrogen compounds to ammonia. In many other cases, however, only one or two rather specialized organisms exist that can bring about the required natural transformation. The nitrifying bacteria that convert ammonia to nitrate are an example; there are only a few specialized organisms that can bring about this oxidation.

Relatively inert elemental nitrogen cannot be used directly by higher plants and animals. A crucial initial step in the utilization of nitrogen is the conversion (fixation) of atmospheric nitrogen gas into compounds that can be assimilated by plant roots for protein synthesis. Only a few species of bacteria found in the soil and water and the blue-green algae found in water can instigate the series of reactions that eventually convert atmospheric nitrogen to available forms. Soluble nitrate ion is the principal form of nitrogen capable of being absorbed by plant roots. Fixing organisms do their job by first incorporating free nitrogen into biological forms, the protein of the multiplying bacteria, for instance. The biological fixing process is not completely understood, but it probably proceeds through an intermediate compound such as ammonia, which can be used to manufacture amino acids and proteins. Upon the death of the fixing organisms, their bodies are acted upon by decay and nitrifying bacteria with the release of nitrate ion. Nitrogen-fixing organisms do not directly convert molecular nitrogen to nitrates.

Nitrogen-fixing organisms are essential to the existence and growth of the higher plants and animals. Biological nitrogen fixation is as important for the utilization of nitrogen in nature as nitrogen fixation by the Ostwald or Haber processes is to the use of nitrogen in industry. In nature, some nitrogen is also fixed by nonbiological means in electric discharges or in photochemical reactions. But it is estimated that the amount of nitrogen fixed in nature by biological means averages about 200 times that fixed by these nonbiological routes.

Important as the nitrogen cycle is for all life, and important as we animals consider ourselves, an inspection of the nitrogen cycle shows that the higher animals are unnecessary; they do not really contribute anything essential to the nitrogen cycle. Only green plants and bacteria must be present.

5.14 ENERGY IN THE NITROGEN CYCLE

As nitrogen circulates through its cycle, it undergoes a series of oxidations and reductions. Energy is required for some of these transformations and is evolved by others. In the majority of cases, the reduction of combined nitrogen requires the input of energy. That is, the conversion of nitrate nitrogen into protein nitrogen by green plants absorbs energy. To understand the endoergic character of nitrogen reduction, the concept of redox reactions involving molecular oxygen should be reviewed.

Every oxidation is accompanied by a reduction. It is possible to separate redox reactions into two separate reactions called *half-reactions*, one for the oxidation, one for the reduction (Section 10.19). Each half-reaction has its own energy change. When the half-reactions and their separate energy terms are added together, the complete redox reaction is obtained. Thus, it is often said that the oxidation of carbohydrates (glucose), $C_6H_{12}O_6 + 6O_2 \longrightarrow 6CO_2 + 6H_2O$, is the major energy-furnishing reaction in living things. It is more correct to say that one portion releases more energy than the other portion. The portion releasing the most energy turns out to be the reduction of oxygen rather than the oxidation of carbon.

In our oxygen-laden atmosphere, molecular oxygen is involved in most of the commonly encountered redox reactions. In the energy-evolving reactions, oxygen acts as an electron (or hydrogen atom) acceptor and is itself reduced to the $-II$ oxidation state in water. Oxygen obtains electrons from a less electronegative element such as carbon or nitrogen. The energy released in the total redox reaction comes primarily from the reduction of oxygen rather than the oxidation of some other element, because the reduction of oxygen is a highly exoergic reaction. The evolution of energy may be attributed to the high electronegativity of oxygen. Conversely, the oxidation of oxygen from $O(-II)$ to $O(0)$ is accompanied by absorption of energy.

Two typical nitrogen reduction reactions in living organisms are

$$2NO_3^- \longrightarrow 2NO_2^- + O_2$$
$$2NO_2^- + 2H_2O \longrightarrow 2NH_4^+ + 3O_2$$

Both require the input of energy, even though the reduction of an electronegative element like nitrogen would, by itself, be energetically favorable. These and other reductions of combined nitrogen where molecular oxygen is evolved require energy, because oxygen is oxidized in the process. An exception is a nitrogen reduction reaction producing molecular nitrogen because of the large amount of energy released in the formation of the N≡N bonds. Thus, the reduction of NO to N_2 and O_2 is exoergic despite the oxidation of oxygen.

The major nonmetallic nutrient elements C, N, and S all occur in living organisms in a reduced state. They all have less oxygen and more hydrogen associated with them in biological molecules than in the oxidized forms in which they exist in the abiotic world as minerals. Phosphorus is the only exception. The reduction of carbon and sulfur accompanying their incorporation into biological material requires, as does nitrogen reduction, some source of energy. The energy held in the reduced molecules will later be released to power living processes.

The existence of these elements in reduced condition in biologically important molecules such as the amino acids suggests that the earth had a reducing atmosphere (perhaps of CH_4, H_2, NH_3) in its early history. That is, the early atmosphere contained substances such as hydrogen that could reduce other substances. Oxidizing substances, in particular oxygen gas, were not present. Under such highly reducing conditions, the reduction of carbon, nitrogen, and sulfur to form biological molecules would tend to be spontaneous.

The energy needed to do the work of nitrogen reduction is supplied, directly or indirectly, by microorganisms associated with a specific reduction transformation in the nitrogen cycle. The enzyme catalysts present in these organisms determine whether the energy can be used for the specific reduction reaction, and this accounts for the specificity of many of the organisms involved. While the needed energy comes initially from the sun, most of it is channeled through energy-rich molecules in these organisms. The biological oxidation of these molecules (mostly carbohydrates) releases energy. Not all of this energy is used for the work of reduction or for building complex molecules. A portion is always lost as heat and is unavailable for doing useful work.

Denitrification, defined as the reduction of nitrate or nitrite ion to gaseous N_2 or N_2O, is an anaerobic (without molecular oxygen) process conducted by bacteria in soil and water. It therefore differs somewhat from the other nitrogen reduction reactions, which involve the formation of molecular oxygen. A typical overall denitrification reaction is

$$\overset{0}{5C_6H_{12}O_6} + \overset{+5}{24NO_3^-} \longrightarrow 30CO_2 + 18H_2O + 24OH^- + \overset{0}{12N_2}$$

This reaction is highly exoergic and supplies energy to the microorganism. Since a reduction of nitrogen takes place, it would appear that this is an exception to the statement that nitrogen reduction requires energy. However, no endoergic oxidation of oxygen is present to override the exoergic reduction of nitrogen. In addition, the formation of N≡N bonds releases energy. Such denitrification processes are essential for the return of nitrogen to the atmosphere.

The oxidation of nitrogen compounds, that of decay and nitrification, furnishes energy to the microorganisms associated with these transformations. These oxidation reactions, though exoergic, may be very slow, and organisms may be needed to speed up the reaction. Some of the energy released is used for the growth and maintenance of the organisms, some is again lost to the surroundings as heat. The redox and energy situations are diagrammed in Figure 5.7.

The energy loss by heat dissipation in both directions in the cycle requires that energy continually be supplied for uninterrupted operation of the cycle.

Figure 5.7. Energy requirements in the nitrogen cycle. Energy requirements refer to reductions of nitrogen in which molecular oxygen is a product. The anaerobic (without molecular oxygen) reductions of NO_3^- and NO_2^- to volatile N_2 or N_2O evolve energy and are shown as dashed lines. The aerobic oxidation of nitrogen involves, by definition, the reduction of molecular oxygen.

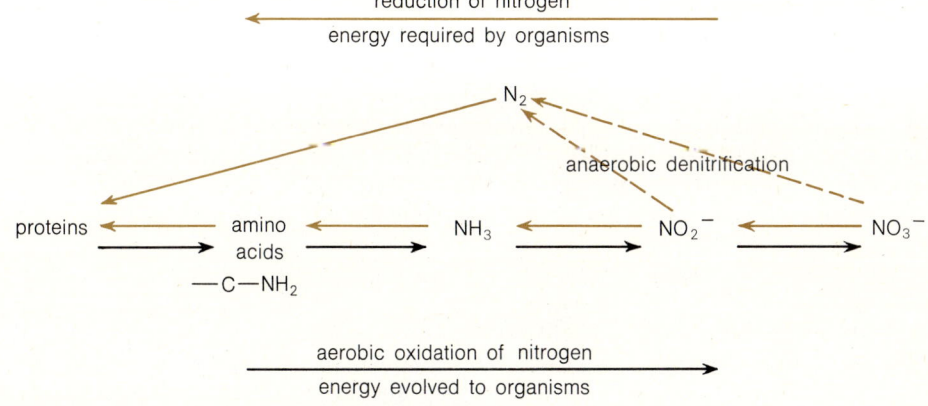

While nitrogen circulates in a cyclic fashion, energy is dissipated as unusable (by the organisms) heat. The flow of energy is one-way; the flow of nutrient elements is cyclic.

5.15 SUMMARY

All chemical elements circulate through the biosphere in cyclic flows from the physical environment to living organisms and back to the environment. The cyclic paths taken by the elements are known as biogeochemical cycles because chemical transformations are involved in the flow of the chemical elements and their compounds through the biological and geological stations of the cycles.

There are two basic types of biogeochemical cycle. In a gaseous cycle, the main environmental reservoir or pool for the element is the atmosphere (or hydrosphere). The carbon, oxygen, and nitrogen cycles are gaseous cycles. In a sedimentary cycle, the major storage reservoir for the element is the earth's crust, from which the element is released by weathering. The phosphorus and sulfur cycles are sedimentary, as are the cycles of most nonessential elements.

The more alternative pathways a biogeochemical cycle provides for any element or compound from one point to another, the better that cycle will be able to adjust to small, short-term disturbances, and the more perfect will be that cycle. Gaseous cycles tend to be self-regulating in this respect, while sedimentary cycles are less so because of their simpler nature and their tendency to lose their element to sediments, where it is inaccessible to organisms and recycling.

Man affects the operation of biogeochemical cycles by oversimplifying their operation and decreasing their stability, by accelerating or retarding various pathways in relation to other pathways in the same or other cycles, and by activating or accelerating formerly insignificant cycles.

The nitrogen cycle is a comparatively complicated gaseous cycle marked by extensive involvement of living organisms in the transport of nitrogen and in the transformations among the many oxidation states of nitrogen.

The oxidation state of an atom is a measure of the extent to which electrons have been lost by that atom in comparison to the neutral atom. An atom that has suffered a loss or a shift of some of its bonding electrons to a more electronegative element is said to be oxidized and to have a positive oxidation state. Reduction is a gain of electrons, and a reduced atom has a lower oxidation state.

In the nitrogen cycle, the oxidation state of nitrogen varies from $+5$ in mineral nitrates to -3 in ammonia and the nitrogen of biological protein molecules. Many of the transformations of nitrogen are mediated by highly specific organisms, among which are the nitrogen-fixing organisms that convert atmospheric nitrogen into forms assimilable by plants. All reductions of combined nitrogen accompanied by the evolution of molecular oxygen require the input of energy. This energy is supplied by living organisms.

Questions and Problems

1. Discuss the factors that determine the stability of biogeochemical cycles. How can natural biogeochemical cycles be disrupted, at least locally?

2. Why are sedimentary-type cycles less perfect or cyclic than gaseous-type cycles?

3. Write electron dot structures for hydroxylamine,

NH_2OH, and for hydrazine, NH_2NH_2. Determine the oxidation states of nitrogen in each compound.

4. Assign oxidation states to each atom in the following compounds: As_2S_5, $S_2O_3^{2-}$, H_3BO_3, $H_2C_2O_4$, $KMnO_4$, K_2MnO_4, $Cr_2O_7^{2-}$, $POCl_3$, CH_4, CH_3CH_3.

5. Carefully read Appendix D (see also half-reactions in Section 10.19). Then balance the following reactions, all of which take place in acid aqueous solution unless otherwise noted. Only the major reactants and products are shown here; you must supply H^+, OH^- or H_2O as needed.
   a. $SO_2 + MnO_4^- \longrightarrow MnO_2 + SO_4^{2-}$
   b. $Ag + HNO_3 \longrightarrow AgNO_3 + NO$
   c. $Zn + NO_3^- \longrightarrow Zn^{2+} + NH_4^+$
   d. $ClO^- + Mn(OH)_2 \longrightarrow Cl^- + MnO_2$ (basic solution)
   e. $ClO_3^- + As_2S_3 \longrightarrow Cl^- + H_2AsO_4^- + S$
   f. $S_2O_3^{2-} + I_2 \longrightarrow S_4O_6^{2-} + I^-$
      (fractional oxidation states are acceptable)
   g. $C_{12}H_{22}O_{11} + H_2SO_4 \longrightarrow CO_2 + SO_2$
   h. $C_6H_{12}O_6 + NO_3^- \longrightarrow CO_2 + N_2O$ (basic solution)
   i. $C_6H_{12}O_6 + Cr_2O_7^{2-} \longrightarrow CO_2 + Cr^{3+}$

6. The photosynthetic "fixation" of carbon carried on by green plants is represented by the overall reaction (unbalanced)

   $$CO_2 + H_2O \xrightarrow{\text{light}} C_6H_{12}O_6 + O_2$$

   Balance this reaction by any method you wish. Assign oxidation states and determine what has been oxidized and reduced.

7. Gold and platinum react with a $3:1$ mole mixture of hydrochloric acid and nitric acid to form nitrogen dioxide and the ions $AuCl_4^-$ and $PtCl_6^{2-}$, respectively. Write balanced equations for each reaction.

8. Hydrogen sulfide reacts with nitric acid to give elemental sulfur and nitric oxide. Write the equation for this reaction.

9. Hydrazine, NH_2NH_2, and some of its derivatives react with various oxidizing agents to yield large amounts of gaseous products at high temperatures. These hydrazine compounds therefore make excellent rocket fuels.
   a. Write the equation for the reaction of liquid hydrazine with liquid hydrogen peroxide. The oxidation state of nitrogen increases by two units.
   b. The descent engine of the lunar module for the U.S. Apollo missions used unsymmetrical dimethylhydrazine, $(CH_3)_2NNH_2$, as a fuel with nitrogen tetroxide, N_2O_4, as the oxidizer. Write the equation for the reaction of these compounds to give gaseous H_2O, CO_2, and N_2.
   c. Monomethylhydrazine, CH_3NHNH_2, is also used as a rocket fuel with N_2O_4 and yields H_2O, CO_2, and N_2. If the thrust of a rocket engine depends only on the mass of gaseous products produced, which is the better fuel, monomethyl- or dimethylhydrazine? Does it make any difference if you express your answers on a per mole or per mass basis?

10. The redox reactions

   $$2H_2O_2 \longrightarrow H_2O + O_2$$

   and

   $$3ClO^- \longrightarrow ClO_3^- + 2Cl^-$$

   involve an element in an intermediate oxidation state that is both oxidized and reduced. Such redox reactions, in which an element reacts with itself, are known as disproportionation reactions. Two other (unbalanced) disproportionation reactions are

   $$NO_2 + H_2O \longrightarrow HNO_3 + NO$$

   and

   $$MnO_4^{2-} + H^+ \longrightarrow MnO_2 + MnO_4^- + H_2O$$

   Devise a method to balance these reactions.

11. The cyanamide process (named after cyanamide, $H_2NC\equiv N$) was the first feasible method for the fixation of atmospheric nitrogen. The process consists of passing nitrogen gas over calcium carbide containing a small amount of calcium oxide,

   $$CaC_2 + N_2 \xrightarrow[\text{CaO}]{1000°C} CaNCN + C$$

   The product, calcium cyanamide, can be applied directly to the soil as a nitrogen fertilizer. It decomposes to give ammonium ions. Calcium cyanamide is an ionic compound. What is the oxidation state of nitrogen in the NCN^{2-} ion?

12. Discuss and illustrate the following statement: "Nitrogen can be cycled because the reduced inorganic compounds of nitrogen can be oxidized by atmospheric oxygen with a yield of useful energy. Under anaerobic conditions, the oxidized compounds of nitrogen can act as oxidizing agents for the burning of organic compounds (and a few inorganic compounds), again with a yield of useful energy."

References

Bormann, F. Herbert, and Likens, Gene E., "The Nutrient Cycles of an Ecosystem," *Scientific American*, **223,** 92 (October 1970).

The entry and outflow of nutrients, including nitrogen, from a forest ecosystem.

Commoner, Barry, "Nature Unbalanced. How Man Interferes with the Nitrogen Cycle," *Scientist and Citizen*, **10,** 9 (January–February 1968).

Commoner, Barry, "Soil and Freshwater: Damaged Global Fabric," *Environment*, **12,** 32 (April 1970).

Delwiche, C. C., "The Nitrogen Cycle," *Scientific American*, **223,** 137 (September 1970).

This entire issue of *Scientific American* is devoted to material cycles and energy flow in the biosphere. It is recommended reading.

Kellogg, W. W., Cadle, R. D., Allen, E. R., Lazrus, A. L., and Martell, E. A., "The Sulfur Cycle," *Science*, **175,** 587 (1972).

This article emphasizes the atmospheric portions of the sulfur cycle.

Odum, Eugene P., "Fundamentals of Ecology," 3rd. ed, Saunders, Philadelphia, 1971, Chapter 4.

Absorption of Energy by Molecules

The thin shell of atmosphere covering the surface of the earth directly provides us with many substances necessary for chemical processes of living organisms. The atmosphere serves to maintain the heat balance of the earth, and its upper layers act as a shield that protects living things from the lethal effects of high-energy solar radiation. This protection comes about because the molecules in the atmosphere absorb electromagnetic radiation. To understand the molecular absorption of energy from the electromagnetic field of radiation, we shall review and extend what we know about the energies of atoms and molecules.

6.1 ENERGIES AND MOTIONS OF MOLECULES

The energy of a molecule may be considered to be made up of contributions from electronic energy, translational energy, rotational energy, and vibrational energy. The electronic energy of a molecule is the sum of the energies of its electrons in their various energy levels. When one or more quanta of energy is added to a molecular system, one or more electrons may be promoted to higher energy levels. A molecule may therefore absorb energy as electronic energy.

The translational energy is the kinetic energy of straight-line motion of a molecule. More properly, we should say this is the energy of translation of the center of mass of the molecule along a straight line. (The center of mass, or gravity, of a system of particles or a body is a point at which the total mass of the system may be thought to be concentrated. Translational motion of the body may be referred to motion of this single point.) The kinetic energy of translation of a molecule of total mass m is $\frac{1}{2}mv^2$, where v is the speed of the molecule. Atoms, as well as molecules, possess electronic and translational energies.

To visualize the other ways in which a molecule (but not an atom) may take up energy, it is helpful to think of the molecule as a mechanical system and enumerate the various ways this system can move in space. In other words, we shall construct a model.

As a simple example, we shall take a diatomic molecule such as HCl and regard it as a system of masses (the atoms) held together by springs (the bonds) as in Figure 6.1. Translational motion of this system has already been noted. But a diatomic molecule can also rotate and vibrate.

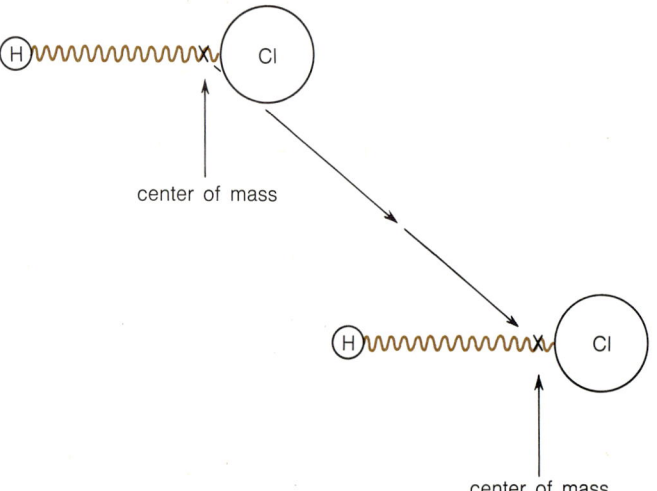

Figure 6.1. Ball-and-spring model of a diatomic molecule, illustrating pure translational motion.

Rotation of the molecule can be pictured apart from translation and vibration, the two other forms of motion of the molecule's nuclei: In pure rotation, we think of the center of mass x of the molecule as being fixed at one point in space. We also think of the bonds, not as flexible springs, but as very stiff springs or as rigid rods. Then a general rotation of the molecule is a pivoting of the fixed, rigid molecule in space about an axis through the center of mass (Figure 6.2).

A general vibration of the molecule can be thought of as a change in the relative configuration of the nuclei in the molecule as a result of a stretching

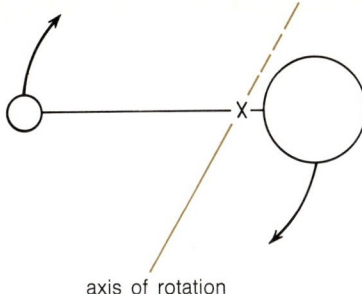

Figure 6.2. Rigid diatomic molecule rotating about an axis through the center of mass. Here the axis of rotation is perpendicular to the plane of the page, and the molecule rotates in the plane of the paper.

Figure 6.3. Stretching and bending vibrations.

and compression of the springlike bonds, or also, in molecules of three or more atoms, a change in angle (bending) between bonds. These types of vibration are shown in Figure 6.3. The usual motion of a molecule in space consists of a simultaneous combination of all three kinds of nuclear motion. This is especially true of relatively unrestrained gas molecules.

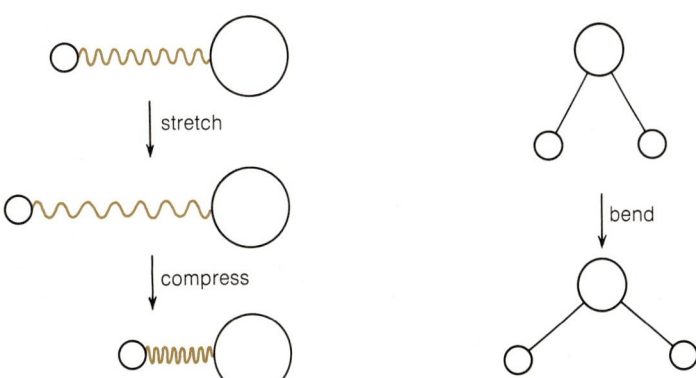

Energy is required to initiate or increase both rotational and vibrational motion. It follows that a molecule may absorb energy to increase its rate of rotation or amplitude of vibration.

Experimentally, molecular energies are investigated by observing the changes in energy of a molecule upon interaction of the molecule with electromagnetic radiation. To interpret the spectral phenomena thus observed, we need to know how the various forms of molecular energies are quantized. What are the allowed energy values, and how far apart are the allowed energy levels?

All of the molecular energies mentioned—electronic, translational, rotational, and vibrational—are quantized. They have discrete, allowed energy levels separated by energy regions inaccessible to the system. You have already seen how atomic and molecular electronic energy levels are quantized. If we, for the time being, accept the fact that the other forms of energy are also quantized, then it is important to know what the relative energy spacings are between the allowed energy levels for the different kinds of energy. To do this in a very rough way, we can return to the ideas of the particle-in-a-box model.

6.2 THE PARTICLE IN A BOX AND THE RELATIVE SPACING OF MOLECULAR ENERGY LEVELS

Two important results emerged from a consideration of the quantized energy levels of a particle in a box. These were that the forbidden region or gap between adjacent allowed energy levels was inversely proportional to (1) the size of the region in which the particle was confined and (2) the mass of the particle.

$$E \propto \frac{1}{l^2 m}$$

On the basis of this result, it is possible to get a rough idea of the spacing between energy levels corresponding to the various forms of molecular energy.

Electronic Energy Levels

In this case the particle in a box is an electron, and the box has the size of an atom or molecule. We shall restrict ourselves here to molecular orbitals that do not encompass more than two or three nuclei, that is, fairly localized electrons. The mass of the particle is very small, and so is the "size" of the box. This means that the spacing between allowed electronic energy levels in an atom and a molecule is quite large. Calculation and experiment show the spacing of electronic levels in molecules to be of the order of 100 kcal/mol (4.34 eV).

Vibrational Energy Levels

With a diatomic molecule as an example, the particles are the atoms at the ends of the bond. Even in a hydrogen molecule, the mass is 1840 times that of an electron. The size of the box is defined by the small displacements from the equilibrium positions of the atoms due to stretching and compression of the bond (Figure 6.4). The box is therefore still quite small, but the mass of the particle is large. The result is that the spacing between allowed vibrational energy levels is smaller than between electronic energy levels. The stiffer the spring (i.e., the stronger the bond), the smaller the box, and the farther apart the vibrational levels. The vibrational spacing is of the order of 5 kcal/mol (0.22 eV).

Rotational Energy Levels

The particle in this instance is the entire molecule, and it has the mass of the entire molecule. The region in which the molecule is free to rotate is roughly equal to the volume of the molecule. The result is that rotational levels are much closer together than even vibrational energy levels. The spacing between rotational levels is of the order of 0.01 kcal/mol (4×10^{-4} eV).

Translational Energy Levels

The particle is again the entire molecule, and the box (for a gas) is the container confining the gas. The size of the "box" is then much, much larger than the size associated with electronic, vibrational, or rotational energies. The gap between translational energy levels is therefore extremely small, almost nonexistent as far as experimental observation of any phenomena that might show transitions from one level to another. The order of spacing is 1×10^{-18} kcal/mol (10^{-20} eV). This energy difference is so small compared to the average kinetic energy of a molecule at room temperature ($\sim 1 \times 10^{-3}$ kcal/mol) that we can ignore any quantization of translational energies. The translational energy levels of a molecule form a continuum of very closely spaced levels.

Figure 6.5 and Table 6.1 gather the results of energy level spacings together and compare the spacings corresponding to the various energies.

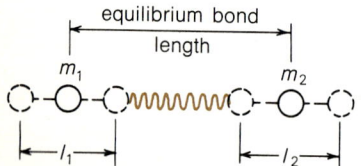

Figure 6.4. Displacement from equilibrium position during a vibration. In HCl, the maximum displacement for $v = 0 \longrightarrow 1$ is about 10 percent of the equilibrium bond length of 1.27Å.

6.3 RADIATION AND MOLECULAR ENERGY LEVELS

In Chapter 2, the emission of radiation by excited atoms was discussed briefly. It was pointed out that energy, in the form of electromagnetic radiation, was emitted by an atom when one or more of its electrons made a transition from one energy state to a lower energy state. The frequency and wavelength of the radiation emitted in the transition process were related to the energy difference by

$$\Delta\epsilon = h\nu = \frac{hc}{\lambda}$$

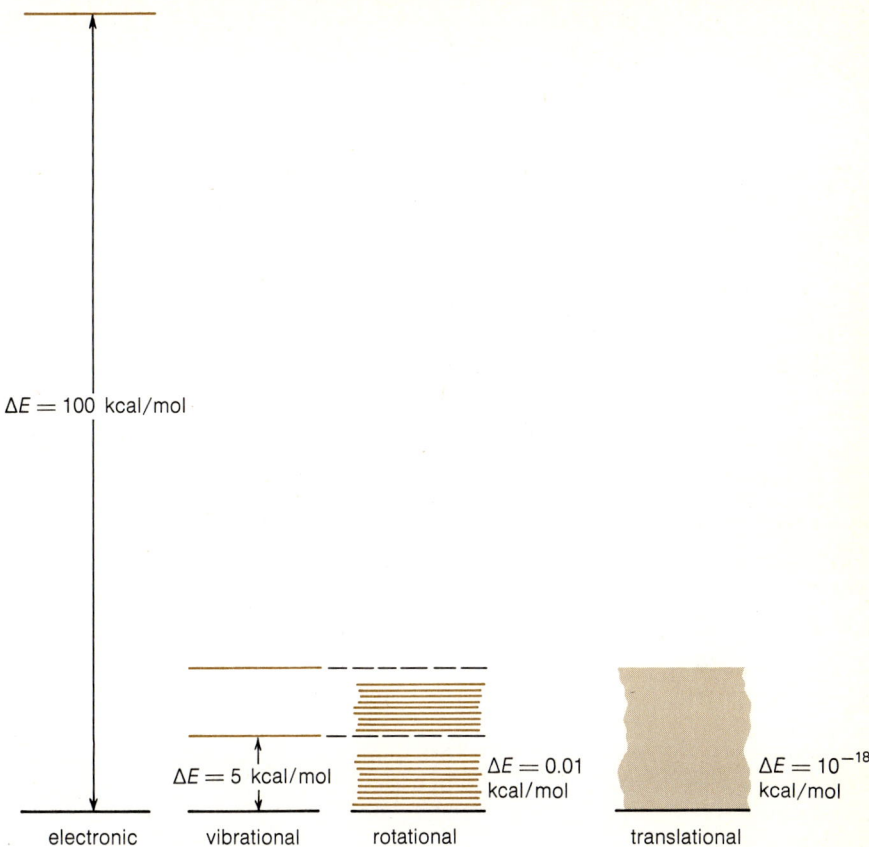

$\Delta E = 100$ kcal/mol

$\Delta E = 5$ kcal/mol

$\Delta E = 0.01$ kcal/mol

$\Delta E = 10^{-18}$ kcal/mol

electronic   vibrational   rotational   translational

Figure 6.5. Relative spacing of molecular energy levels.

This relation applies to any situation where there are transitions between quantized energy levels; it is not restricted to the energy levels of electrons in atoms. We may therefore take the representative values of $\Delta\epsilon$, in ergs per molecule, given in Table 6.1 for the energy spacing of molecular energy levels and calculate the frequency or wavelength of the radiation associated with

Table 6.1
Molecular energy level spacings deduced from the particle-in-a-box model

energy	particle mass, relative	edge size of box	ΔE, relative	ΔE kcal/mol	$\Delta\epsilon$ erg/molecule
electronic	very small, 9×10^{-28} g	very small, $\sim 1 \times 10^{-8}$ cm	very large	100	7×10^{-12}
vibrational	large	very small, 0.1×10^{-8} cm	large	5	3×10^{-13}
rotational	larger	large	small	0.01	7×10^{-16}
translational	largest	very large >1 cm	very, very small	10^{-18}	2×10^{-32}

transitions between these levels. For instance, $\Delta\epsilon$ for vibrational levels is about 10^{-13} erg/molecule, so the wavelength associated with a transition from one vibrational energy level to an adjacent vibrational level is 2×10^{-3} cm. This is in the infrared region, as was shown in Figure 2.7. Changes in electronic energy levels in molecules correspond to radiation in the visible and ultraviolet regions of the spectrum, while rotational energy changes correspond to microwave (radar) radiation. The closer together the energy levels, the longer the wavelength λ and the lower the frequency ν of the radiation that can be absorbed or emitted.

6.4 MOLECULAR ABSORPTION SPECTROSCOPY

So far, only emitted radiation has been mentioned. For an atom or molecule to emit radiation, the atom or molecule must first have had an appropriate amount of energy supplied to it to promote an electron to a higher energy state or to excite the molecule to a more energetic vibrational or rotational energy state. Except in special cases, the atom or molecule returns immediately to a lower energy state, with the emission of radiation characteristic of the energy level spacing.

The wavelength or frequency of this emitted radiation may be determined with an optical or electronic instrument. This, of course, was the procedure that gave some of the first indications of the existence of discrete electronic energy levels in atoms. The analysis of emitted radiation, emission spectroscopy, was mentioned in Section 2.17. Emission spectroscopy is pretty much limited to investigations of atoms, because the high excitation energies needed tend to cause fragmentation of molecules by the breaking of bonds.

Rather than measuring the radiation emitted by an excited molecule, most work in molecular spectroscopy measures the radiation absorbed by a molecule when the molecule makes a transition from a lower to a higher molecular energy level. Changes in the electronic energy of a molecule are brought about by the absorption of visible or ultraviolet radiation. Absorption of infrared and microwave radiation brings about increases in the vibrational and rotational energies, respectively, of most molecules.

The characteristics of the radiation absorbed by a molecule are related to its energy level patterns. The molecular energy levels are, as we shall see, determined by the atomic composition and by the geometric and electronic structure of the molecule. The measurement of absorption of radiation can serve as a window for looking into a molecule.

6.5 MEASUREMENT OF MOLECULAR ABSORPTION

To make use of molecular absorption spectroscopy, not only must the wavelengths or frequencies of the absorbed radiation be determined, but the relative amount of absorption at each frequency should be determined. Even though a molecule may absorb infrared radiation at two or more different wavelengths, for instance, the absorbing power of the substance at the different wavelengths may be quite different. Not only does the amount of absorption at a given wavelength depend on certain internal parameters of the molecule, but it also depends on the number of molecules present in the absorbing mixture.

In a typical absorption experiment, incident radiation from a source appropriate to the chosen region of the electromagnetic spectrum is passed through a sample of a substance in a container that is transparent to the radiation used. The substance may be in the gaseous, liquid, or less commonly, the solid state. After radiation of a given wavelength has passed through the sample, the intensity of the radiation may or may not be diminished. If the measured intensity of the transmitted radiation I is less than that of the incident radiation I_0 at a particular wavelength, absorption of energy from the incident beam must have taken place in the sample. Rather than measuring directly the amount of absorption, the *transmittance T*, where $T = I/I_0$, or the *percent transmittance*, $100 \times T$, may be measured.

The transmittance at a given wavelength is related to the amount of absorbing substance by Beer's law,

$$\log \frac{I_0}{I} = -\log T = \epsilon bc$$

where ϵ is called the absorptivity, b is the thickness of absorbing medium (or the width of the container), and c is the amount (concentration) of absorbing species present. The quantity $-\log T$ is commonly referred to as the *absorbance A*. For a given wavelength of incident radiation, the absorptivity is a constant for a particular absorbing species.

EXAMPLE | Nitrogen dioxide, NO_2, is a common air pollutant. It is strongly absorbent over the blue-green region of the visible spectrum, resulting in transmitted radiation with yellow-red quality (see Table 6.4), which is the reason for the yellow-brown coloration of photochemical smog.

The absorptivity of NO_2 is a maximum at 400 nm, where its value has been found to be 1.13 ppm^{-1} mile^{-1}. If the percent transmittance, at 400 nm, of a 2-mile "sample" of air containing NO_2 is 60, what is the NO_2 content of the air?

SOLUTION

$$100T = 60$$
$$T = 0.60$$
$$-\log T = -\log 0.60 = \log \frac{1}{0.60} = \log 1.66$$
$$= 0.22$$

(See Appendix A.3 for a discussion of logarithms.)

$$0.22 = \epsilon bc = (1.13 \text{ ppm}^{-1} \text{ mile}^{-1})(2 \text{ miles})(c)$$
$$c = \frac{0.22}{(1.13 \text{ ppm}^{-1} \text{ mile}^{-1})(2 \text{ miles})}$$
$$= 0.097 = 0.1 \text{ ppm}$$

Separate instruments are used to determine absorption in the visible–ultraviolet, infrared, and microwave regions, because both the sources of incident radiation and the detectors needed to measure the transmitted radiation in these three common regions are different. In the infrared region, for instance, the source is a ceramic rod heated to a red heat, and the detector is a thermocouple. For visible radiation, the source is a common tungsten-filament incandescent lamp bulb, and the detector is a photomultiplier tube. More details about spectrophotometers can be found in the reference by Barrow given at the end of the chapter.

Since most sources give off radiation having a range of wavelengths in their spectral region, it is necessary to select a small group of wavelengths

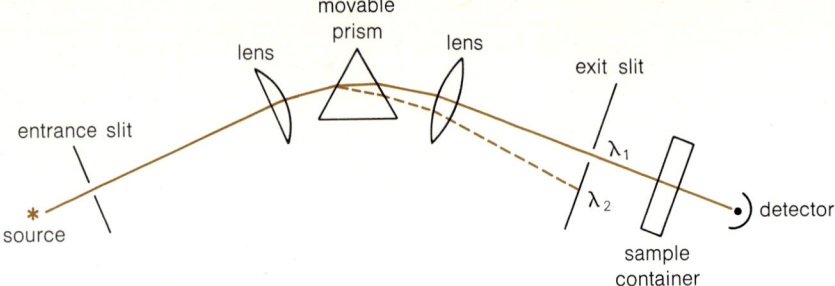

Figure 6.6. Absorption spectrophotometer. Rotation of the prism directs different wavelengths through the sample.

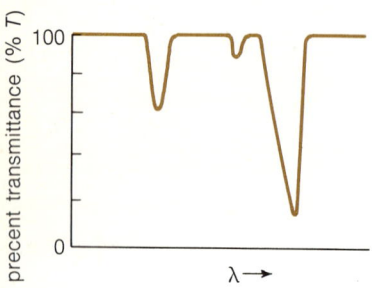

Figure 6.7. Transmittance spectrum of a substance. If absorbance A had been plotted instead, absorption would be indicated by peaks at the same wavelength as the transmittance valleys.

for measurement either before or after they pass through the sample. In the visible–ultraviolet and infrared regions, a relatively monochromatic beam of radiation can be obtained with prisms or gratings to disperse the incident (or transmitted) polychromatic radiation and narrow slits to select a range of wavelengths from the dispersed beam. A schematic diagram of some components in an absorption spectrophotometer is shown in Figure 6.6.

To determine the exact wavelengths at which a single substance or a mixture absorbs or transmits, the usual procedure is to scan the spectrum over a given region to search for the wavelengths of absorption. The wavelength of the incident radiation, for instance, may be slowly and continuously varied so that the detector measures the percent transmittance at a series of wavelengths. A spectrum such as the one in Figure 6.7 may be recorded, showing absorption of energy from the incident beam of radiation at several different wavelengths located by valleys in a plot of percent transmittance versus wavelength.

6.6 ROTATIONAL ABSORPTION SPECTRA OF MOLECULES

The simplest molecular absorption spectra are the result of absorption of radiation in the microwave region of the electromagnetic spectrum. Absorption again corresponds, as we have seen, to transitions between the closely spaced rotational energy levels of the molecule. Unfortunately, the experimental difficulties that arise because of the necessity of excluding other energy changes while measuring rotational energy changes are very hard to overcome, and rotational spectra are not common. Because of this we shall discuss pure rotational spectra only briefly and restrict our attention to diatomic molecules.

The rotation of a molecule can be considered as the rotation of a body made up of two or more particles connected by rigid rods (bonds), so that their relative positions do not change during rotation. For the model of the diatomic molecule shown in Figure 6.8, the rotation occurs about an axis passing through the center of mass. A quantum mechanical treatment shows that the energy levels of such a rigid rotor are given by

$$\epsilon_{\text{rot}} = \frac{h^2}{8\pi^2 I} J(J+1)$$

where $J = 0, 1, 2, \ldots$ is the rotational quantum number and I is called the moment of inertia of the molecule. It is related to the masses, m_1 and m_2, of the atoms and the bond distance r_0 by

$$I = \frac{m_1 m_2}{m_1 + m_2} r_0{}^2$$

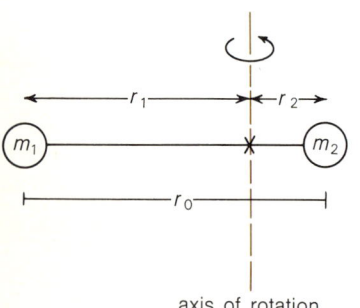

Figure 6.8. Diatomic molecule rotating about an axis through its center of mass. The position of the center of mass x is defined by $m_1 r_1 = m_2 r_2$.

When a molecule absorbs microwave radiation and increases its rotational energy, the rotational quantum number is restricted to an increase of one unit at a time, $\Delta J = 1$, so a particular molecule in rotational energy state ϵ_J may go only to ϵ_{J+1}. Each such transition corresponds to a definite frequency (or wavelength) of absorption,

$$\nu_J = \frac{\Delta\epsilon}{h} = \frac{(\epsilon_{J+1} - \epsilon_J)}{h}$$

The rotational spectrum of a diatomic molecule appears as a series of equally spaced absorption lines on the frequency scale, as shown in Figure 6.9. The frequency separation of the lines is $2h/8\pi^2 I$.

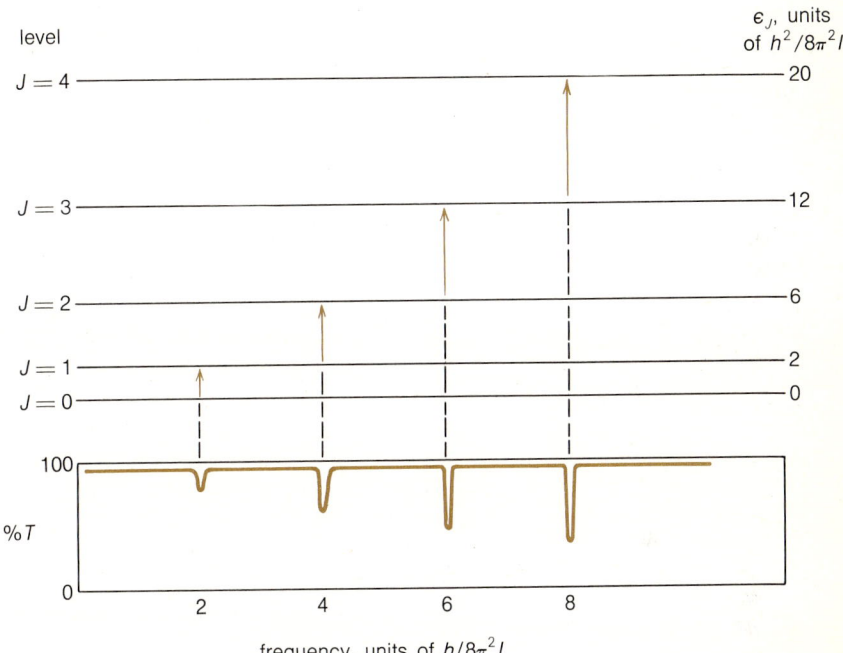

Figure 6.9. Microwave spectrum of a diatomic molecule showing the rotational transitions responsible for the absorption lines. At normal temperatures the molecules may occupy many rotational levels, so several lines appear.

The important thing to note is that the experimental quantity λ or ν of the absorbed radiation is related, through the moment of inertia I, to the bond distance r_0. Even in molecules having more than two atoms, I contains, in addition to the known masses of the atoms, the bond distances. For simple molecules, it is quite easy to calculate r_0 with great precision from the measured frequencies or the uniform spacing of the spectral lines. With more complicated molecules, an idea of the overall shape of the molecule may be obtained from the rotational spectrum because the value of I depends on the bond angles and symmetry of the molecule.

Only those molecules that have a permanent dipole moment may absorb microwave radiation from the electromagnetic field, because they are the only ones that can be caused to rotate by the unbalanced forces of the electric field of the radiation on the charged ends of the dipole. H_2 and Cl_2 do not have a microwave spectrum while HCl, a polar molecule, does.

6.7 VIBRATIONAL ABSORPTION SPECTRA

When a molecule is supplied with radiation of higher-energy quanta in the infrared (heat) region of the spectrum, some of this energy may be absorbed as vibrational energy. Simultaneously with increase in vibrational energy, the molecule will also undergo lower-energy rotational transitions, at least in a gas. We shall consider only pure vibrations in the absence of rotation. Vibration of a molecule includes the stretching of bonds as well as the bending or changing of bond angles. Both of these distortions of the molecular framework are measures of molecular flexibility. The ease of distortion of bond lengths and angles is related to bond strengths, and studies of the absorption of infrared radiation and vibrational spectra can provide valuable information about this molecular parameter. Once again, spectral studies allow one to see into the molecule.

As in the study of rotational motion and energy, it is convenient to begin by looking at a simple classical model, the ball-and-spring system shown in Figure 6.4. We have already likened the bond between atoms to a spring. When the end of the spring is displaced from its equilibrium length by stretching or compression, it exerts a restoring force proportional to the displacement and directed back toward the equilibrium position. The proportionality constant k is called the *force constant* of the spring (or bond), and it is a measure of the stiffness or strength of the spring. If a ball-and-spring system such as this is displaced from its equilibrium position by stretching or compression and then let go, it will vibrate with a frequency ν determined by the force constant of the spring and the connected masses. Again, a quantum mechanical treatment can relate these quantities to the discrete vibrational energy levels of the diatomic system. The result is

$$\epsilon_{\text{vib}} = \frac{h}{2\pi} \sqrt{\frac{k}{\mu}} \left(v + \frac{1}{2} \right)$$

where v is the vibrational quantum number having values $v = 0, 1, 2, \ldots$ and $\mu = m_1 m_2 / (m_1 + m_2)$.

Transitions from one vibrational energy level to an adjacent one result in the absorption or emission of energy quanta in the infrared region of the spectrum. At ordinary temperatures, most molecules are in their ground ($v = 0$) vibrational state and then the $v = 0$ to $v = 1$ transition is the only one responsible for absorption of radiation in the infrared. Experimental measurement of the frequency of infrared absorption by a diatomic molecule permits the determination of the force constant k, which is frequently used as a measure of bond strength.

EXAMPLE

The infrared spectrum of carbon monoxide, CO, consists of one broad line centered at $\lambda = 4.68 \times 10^{-4}$ cm or $\nu = 6.40 \times 10^{13}$ sec^{-1}. This corresponds to the $v = 0$ to $v = 1$ transition in vibrational energy. The energy change is

$$\Delta\epsilon = h\nu = 4.24 \times 10^{-13} \text{ erg}$$

Since

$$\Delta\epsilon_{v=0\to1} = \frac{h}{2\pi} \sqrt{\frac{k}{\mu}}$$

with $\mu = 1.14 \times 10^{-23}$ g, the force constant k is calculated to be 18.4×10^5 dyn/cm.

Table 6.2
Force constants

molecule	k, dyn/cm
H_2	5.2×10^5
D_2	5.3
HF	8.8
HCl	4.8
HBr	3.8
HI	2.9
NO	15.5
F_2	4.5
Cl_2	3.2
Br_2	2.4
I_2	1.7
O_2	11.4
N_2	22.6

Force constants for some bonds in diatomic molecules are given in Table 6.2. Those for homonuclear diatomics were determined by a method known as Raman spectroscopy.

Only those molecules whose molecular dipole moment changes in the course of a vibration (so that there is a net displacement of charge) will interact with the electromagnetic field and be able to absorb energy. Thus, homonuclear diatomic molecules such as H_2 or Cl_2 will not show infrared absorption, since they have no dipole moment at any internuclear separation. Most heteronuclear diatomics have a permanent dipole that changes value with distance between the atoms and will absorb. We shall see that some polyatomic molecules that have no permanent dipole moment in their equilibrium configuration may develop a nonzero moment in the course of a stretching or bending vibration and will show a vibrational spectrum. This does not mean that Cl_2, for instance, has no rotational or vibrational energy levels. It means only that transitions between these levels cannot be brought about by absorption of electromagnetic radiation in the manner discussed.

6.8 INFRARED SPECTRA OF SIMPLE POLYATOMIC MOLECULES

Molecules with more than two atoms can vibrate in ways other than a simple stretching and compression of the one bond. With these molecules, as has been mentioned, there can also be a bending of bond angles in addition to a variety of stretching motions. The overall complicated vibrational motion of a molecule can be thought of as a combination of a number of simpler vibrational motions. These simpler motions are called *degrees of freedom* or *normal modes* of vibration. The number of degrees of freedom depends on the structure of the molecule and on the number of atoms in the molecule. In general, the number of vibrational degrees of freedom is

$3N - 6$ for a nonlinear molecule
$3N - 5$ for a linear molecule

where N is the number of atoms in the molecule.

Each normal mode has its own set of vibrational energy levels with their own spacing. The spacing of levels in bending modes is usually smaller than that in stretching modes. The resistance to bending is smaller and the dis-

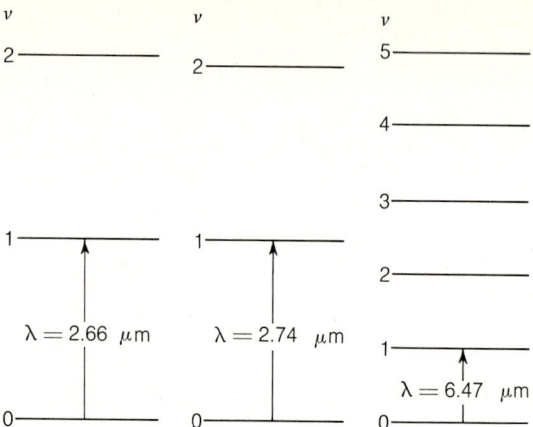

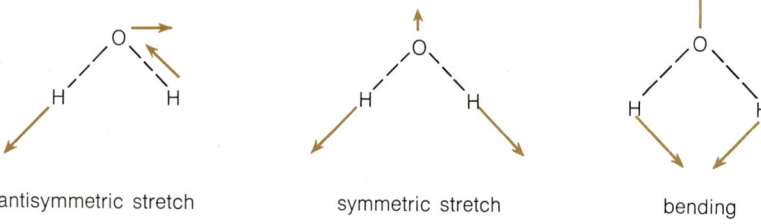

Figure 6.10. Normal modes of vibration of the water molecule. The energy level pattern of each mode is given together with the absorption wavelength (in micrometers) for each mode for the $v = 0$ to $v = 1$ transitions.

antisymmetric stretch   symmetric stretch   bending

placements larger than for stretching, thus effectively increasing the size of the box. The energy level patterns and the normal modes for water are shown in Figure 6.10.

Assuming, as is usually the case, that transitions take place from $v = 0$ to $v = 1$ within each mode pattern, the simple infrared vibration spectrum of H_2O should have three regions of absorption. The experimental spectrum confirms this (Figure 6.14). The linear molecule CO_2 might be expected to have four absorption regions. The normal modes are shown in Figure 6.11. However, the bending mode really is doubly degenerate, consisting of two equally energetic bendings at right angles to one another. Absorption for each of these occurs at the same frequency and only makes the amount of absorption at this frequency twice as great as it would be if the mode were non-degenerate. Furthermore, there is no change in the molecular dipole moment in the symmetric stretch mode, so CO_2 in its simplest infrared spectrum shows only two absorption regions (Figure 6.14).

It has already been mentioned that, simultaneously with vibrational transitions, a number of less energetic rotational transitions occur, since each vibrational energy level has within it a number of closely spaced rotational levels. A molecule in the $v = 0$ vibrational level may be in any one of a number of rotational energy levels. If, as an example, a molecule is in the $v = 0, J = 3$ energy state, a vibrational excitation to $v = 1$ will put the molecule into either a $J = 2$ or a $J = 4$ rotational level of the upper $v = 1$ vibra-

O ----C----O →
←

symmetric stretch
(infrared inactive)

→ ←
←O ----C----O

antisymmetric stretch

O ----C --- O

O ----C----O

} doubly
degenerate
bending

Figure 6.11. The four normal modes of vibration of CO_2.

tional state. Other $\Delta J = \pm 1$ transitions exist for other molecules initially present in states other than $J = 3$. These combined rotation–vibration absorptions result in a series of closely spaced absorption lines rather than just the one line expected from a pure $v = 0$ to $v = 1$, $\Delta J = 0$, vibrational transition. present in states other than $J = 3$. These combined rotation–vibration absorptions result in a series of closely spaced absorption lines rather than just the one line expected from a pure $v = 0$ to $v = 1$, $\Delta J = 0$, vibrational transition. Under low resolution the separate lines do not show up, and the infrared spectrum of gases appears as a broad absorption band, the number of bands being related to the number of normal modes.

6.9 CHARACTERISTIC FREQUENCIES

As the complexity of a molecule increases, the number of vibrational modes and the complexity of the infrared spectrum also increase. Fortunately, the absorption frequency of a particular bond or group of atoms in a large molecule is, to a good approximation, independent of the rest of the molecule, so any molecule containing that bond or group will absorb infrared radiation at approximately the same frequency.

Where absorption frequency does vary from molecule to molecule, the variation can often be related to the specific structure of the molecule or the molecular environment. These constant, or predictably varying, frequencies are known as characteristic or group frequencies, and their use is valuable in the identification of unknown compounds and in the clarification of molecular structure. The wavelengths of some characteristic absorptions of a few bonds and groups are listed in Table 6.3. Through the use of such absorption–structure correlations, it is a simple matter to distinguish the two compounds

$$CH_3CH_2\overset{\overset{\displaystyle O}{\|}}{C}H \quad \text{and} \quad CH_3CH_2\overset{\overset{\displaystyle O}{\|}}{C}-OH$$

by the shift in position of their C=O stretch absorptions and by the appearance of the C—O stretch at 8.0 μm in the latter compound.

6.10 INFRARED ABSORPTION AND THE TEMPERATURE OF THE EARTH

It may at first seem ridiculous to say that the temperature of the earth, that is, the temperature of the lower atmosphere, depends on the vibrations of a few simple molecules. A brief look at why this is so will show how submolecular processes affect global events. The discussion will also show how man may be changing his environment in very slow, very subtle but very important ways.

The temperature of the lower atmosphere depends on the heat budget of the earth. A budget refers to the income and outgo of some quantity. Here the quantity of interest is energy. The temperature at the earth's surface is determined by the amount of solar radiation reaching the surface and by how much of the energy of solar radiation is retained at the surface as thermal (heat) energy. Only about 50 percent of the total solar radiation incident on the top of the atmosphere reaches the earth's surface as direct solar radiation. The remainder of the sun's radiation is absorbed on passage through the atmosphere ($\sim$10 percent) or scattered and reflected back into space by clouds, dust, or other particulate matter ($\sim$40 percent). Thus, 50 percent of the sun's radiation is available as income in the earth's heat budget and consists pri-

Table 6.3
Some characteristic infrared absorptions

bond or group	absorption region, μm^a
—O—H (stretch)	2.70–2.86 in R—OH
—N—H (stretch)	2.86–3.04
$\diagdown$C=O (stretch)	5.75–5.81 in $R-\overset{\overset{\displaystyle O}{\|}}{C}-H$
	5.80–5.87 in $R-\overset{\overset{\displaystyle O}{\|}}{C}-R'$
	and $R-\overset{\overset{\displaystyle O}{\|}}{C}-OH$
C—O (stretch)	8.00 in $R-\overset{\overset{\displaystyle O}{\|}}{C}-OH$
$-\overset{\|}{\underset{\|}{C}}-H$ (stretch)	3.38–3.51 (—CH$_3$, $\diagdown$CH$_2$)
$-\overset{\|}{\underset{\|}{C}}-H$ (bending)	6.74–7.30 (—CH$_2$—, $-\diagdown C$—CH$_3$)

^{a} One micrometer (μm) $= 10^{-6}$ m $= 10^{-4}$ cm. The name *micron* (μ) is also encountered for the micrometer but is supposed to be abandoned. The symbol R corresponds, in general, to the rest of the molecule. More particularly, here R stands for a saturated organic group such as —CH$_3$, CH$_3$—CH$_2$, ...

marily of light in the visible region of the spectrum, with a maximum at about 0.5 μm (Figure 6.12). The visible wavelengths are the ones least susceptible to atmospheric absorption and scattering of the incoming solar radiation.

The comparatively short-wavelength visible light is absorbed (a small portion is reflected) at the earth's surface and is converted directly or indirectly to longer-wavelength infrared (heat) radiation by a variety of processes, including photosynthesis. This infrared radiation represents the energy supplied to the earth's surface by sunlight, and if the temperature of the earth's surface is to remain fairly constant, this energy must be reradiated back into space by the earth. The distribution of the energy radiated into space by the heated earth is shown in Figure 6.13. The maximum occurs at about 12 μm, in the infrared region.

Carbon dioxide and water, and to a lesser extent, ozone and carbon monoxide in the troposphere absorb a portion of the infrared radiation directed into space by the earth. This absorption prevents the escape of some heat into space and traps the energy in the atmosphere with a resultant warming of the atmosphere. Thus, these gases form an insulating blanket over the earth, preventing the escape of a portion of the incoming radiant energy originally absorbed by the earth. During the daytime, the imbalance between income (short λ) and outgo (long λ) leads to a warming of the atmosphere

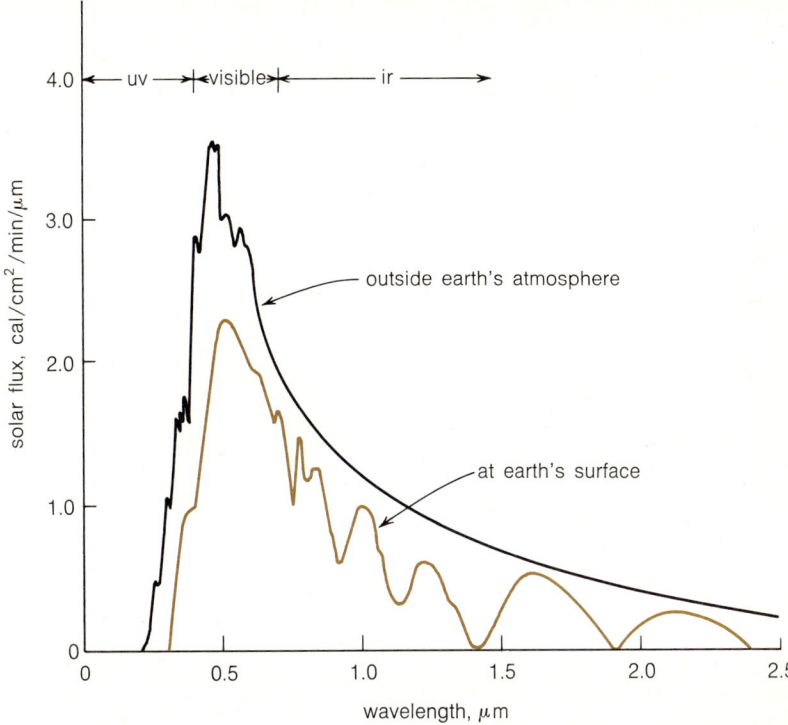

Figure 6.12. Spectral distribution of sunlight. The upper curve approximates the radiation from a body (the sun) at 6000°K (see Figure 2.19). The dips in the lower curve are due to absorption of incoming radiation by atmospheric H_2O, O_3, and O_2. Ozone is responsible for the large attenuation at 0.2–0.3 μm.

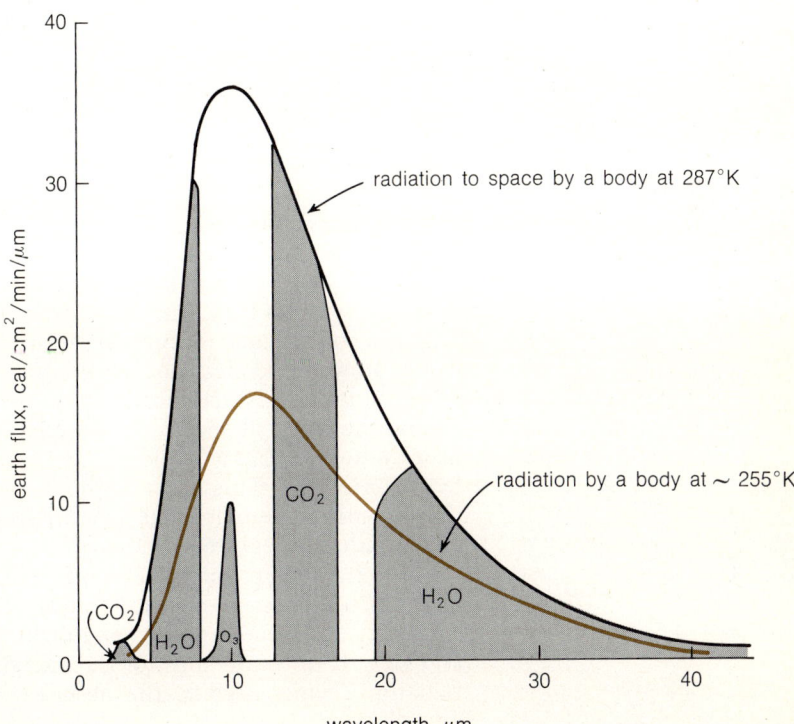

Figure 6.13. Spectral distribution of radiation from the earth to the atmosphere. The earth behaves like a radiating body with a temperature of 255°K with maximum radiation at 12 μm. Regions of absorption of the outgoing radiation by atmospheric H_2O, CO_2, and O_3 are indicated (see also Figure 6.14). Note the long wavelengths of earth radiation compared to solar radiation.

near the surface of the earth. At night, when there is little income, the bottom layer of the atmosphere cools off, but the diurnal (night-to-day) variation depends on the amount of water vapor, in particular, in the layer of air next to the earth. Temperature variations over longer time periods are probably controlled by absorption by CO_2 at higher levels.

All this is due to the absorption of infrared radiation as vibrational energy, principally by H_2O and CO_2. Figure 6.14 shows in more detail the infrared

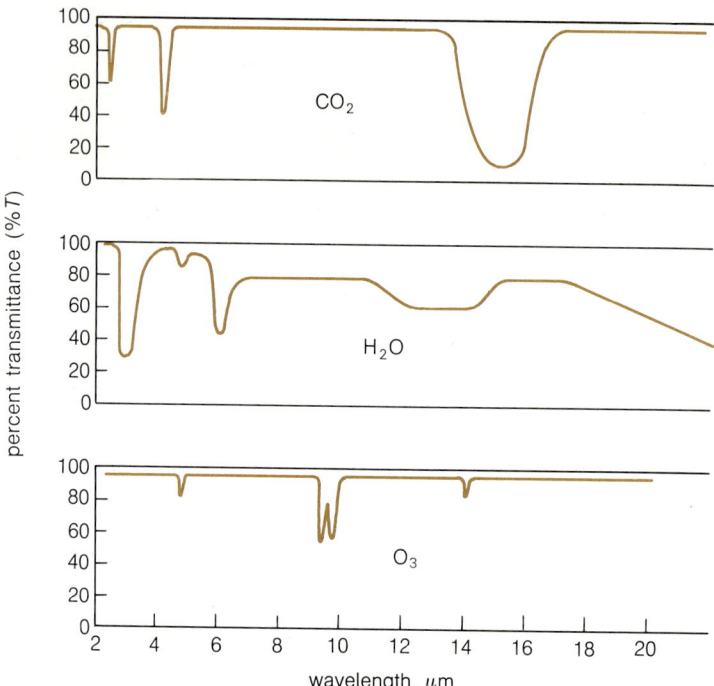

Figure 6.14. Infrared spectra of atmospheric absorbers. The left absorption (~2.5 μm) of CO_2 and the broad absorptions of water to the right of 8 μm are not due to simple normal modes of vibration.

absorption spectra of H_2O, CO_2, and O_3. Carbon monoxide, whose spectrum is not shown, has a single absorption band at 4.68 μm, which may contribute to some heat retention. By comparing the radiation emitted by the earth (Figure 6.13) with these spectra, you can see that the absorption bands of H_2O and CO_2, in particular, occur at positions that lead to absorption and retention in the atmosphere of a good portion of the radiant energy of the earth that would otherwise escape into space in the absence of these molecules. Fortunately, the infrared spectrum of water, which is present in larger amounts near the surface, has a "window" at 12 μm, where the maximum of the earth radiation occurs. This allows some of the radiation to escape into space, so that the atmosphere does not heat up unbearably by trapping the major part of the radiation emitted by the earth.

The trapping of solar energy by the earth is commonly called the *greenhouse effect*, because the same sort of thing happens in a greenhouse or any glass-enclosed space exposed to sunlight. The glass plays the part of the H_2O and CO_2. It transmits most of the energy coming in from the sun as visible radiation. Plants degrade the absorbed visible radiation and reemit it as low-energy infrared radiation, which is trapped in the greenhouse because glass, like H_2O and CO_2, absorbs infrared radiation and prevents its escape from the enclosure. The earth's greenhouse effect is diagrammed in Figure 6.15.

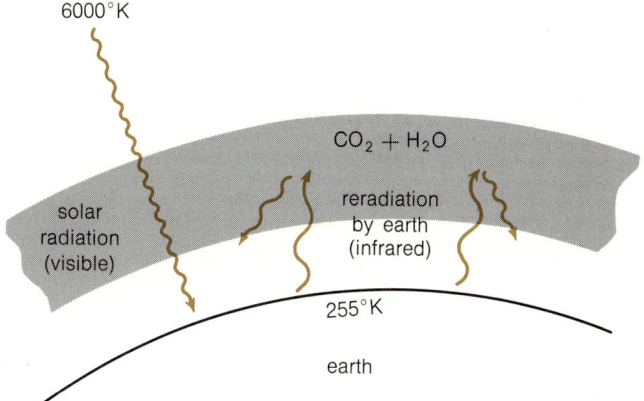

Figure 6 .15. The greenhouse effect. Infrared radiation from the earth to space is absorbed by CO_2 and H_2O molecules in the atmosphere. A portion of the trapped heat may be reradiated back to earth.

As already indicated, water vapor, because of its higher concentration near the ground and its ability to absorb infrared radiation, is primarily responsible for diurnal variations in temperature. At night, in dry areas such as deserts, there is little water vapor in the air to intercept space-bound infrared radiation emitted by the earth that has been warmed up during the daylight hours. The temperature of the earth falls rapidly and by a relatively large amount. In humid areas, on the other hand, the heat of the earth is retained at night, so nighttime and daytime temperatures differ little.

The greenhouse effect is a subject of some apprehension concerning the effect the activities of man may have on the overall heat balance and climate of the earth. As man increases his numbers and as he uses more and more energy to run his industry and technology, he increases the carbon dioxide content of the atmosphere. With the exception of nuclear energy, all of man's large-scale energy sources make use of the combustion of carbon-containing fossil fuels such as coal, oil, and natural gas. The burning of these substances in air produces CO_2 as well as H_2O and CO by reactions such as

$$C + O_2 \longrightarrow CO_2$$
$$2C + O_2 \longrightarrow 2CO \qquad \text{(incomplete combustion)}$$
$$CH_4 + 2O_2 \longrightarrow CO_2 + 2H_2O$$

Major attention has been directed to carbon dioxide because of the rate and the amount of its production and because the natural processes in the carbon cycle do not appear (lately) able to maintain a constant concentration of carbon dioxide in the atmosphere. In fact, measurements have shown that the concentration of CO_2 in the atmosphere has increased by about 11 percent in the period 1880–1940, due primarily to the addition of CO_2 from the burning of fossil fuels. It is estimated that man's activities add about 6 billion tons of carbon dioxide to the atmosphere each year. Other processes add another 2 billion tons.

A major process in the carbon cycle that would tend to remove a large portion (50 percent or more) of the added atmospheric carbon dioxide is absorption by ocean water. The time required for a quantity of carbon dioxide added to the atmosphere to be absorbed by the upper layers of the ocean is believed to be about 5 yr (for the same CO_2 to move from the upper water layers to deeper layers may require 1000 yr). It is unlikely that the solution of CO_2 in the ocean can keep up with its addition to the atmosphere. Another process removing carbon dioxide from the atmosphere is photosynthetic activity by green plants. But man is reducing the coverage of green plants on earth, particularly by clearing of forests, so photosynthetic removal of carbon dioxide may not balance its addition.

Figures for 1880–1940 indicate that, along with the 11 percent rise in CO_2 concentration (addition of about 360 billion tons of CO_2 to the atmosphere by man), the average temperature of the earth has risen by 0.4°C. Other things being equal, a 10 percent increase in CO_2 content will result in a temperature rise of about 0.5°C. Extending this rate of addition of CO_2 to the atmosphere to the year 2000 indicates a temperature increase of 2°C.

There is speculation that even a temperature increase as small as 3.5°C would be sufficient to cause vast changes in the climate of the earth, resulting in the melting of glaciers and polar ice caps with resultant storms and flooding of coastal areas. Things may not be quite this simple, for the processes of climate and weather are the result of the interplay of other factors, as will be considered shortly. However, the possible effects of increased carbon dioxide concentration due to man's activities do show that these activities can have global consequences and that any problems that arise are not just the responsibility of a single city, state, or nation.

It should also be mentioned that, in obtaining energy for his civilization, man produces heat as well as carbon dioxide. The production of heat is not limited to the burning of fossil fuels. Heat is a waste product of any process that converts energy to mechanical work. We are, therefore, adding not only CO_2 to the atmosphere, but heat as well. We are pouring out heat energy that may have originally been locked in chemical bonds and nuclear forces from the "infinite" past. This heat, in addition to the earth's daily solar ration, must somehow be gotten rid of if the heat balance of the earth is not to be disturbed.

In spite of the greenhouse effect of CO_2, the average temperature of the earth has decreased in the period 1940–1967 by about 0.2°C even as the carbon dioxide content increased. Apparently, something must be counteracting any greenhouse effect. This other factor may be air pollution in the form of small particles, both solid and liquid, in the air. These fine particles, called *aerosols*, do not easily settle out by gravitational settling and may spend long periods traveling around the earth before being removed by vertical mixing of the troposphere. They scatter and reflect solar radiation away from the earth, reducing the amount of energy reaching the surface of the earth. These particles can be produced by natural causes such as volcanic eruptions or forest fires or dust storms, but today their major source is human activity. There is evidence that the concentration of small particles is rising.

More recently, there has been concern that the exhaust gases (CO_2 and H_2O) and particulates (ice crystals, sulfates) released by stratospheric flights of supersonic transport airplanes may lead to large-scale climatic changes.

Unlike exhaust products emitted in the troposphere, pollutants released in the stratosphere may remain up there for 1–3 yr because of the absence of appreciable vertical mixing at these heights.

Despite the ups and downs of temperature attributed to CO_2 and to particulates, there are reasons for believing that man's impact on the world climate is insignificant in comparison with the effects of changes in solar activity on our atmosphere. Based on the analysis of the concentrations of oxygen isotopes along a 4700-ft test boring of ice in Greenland (supposedly spanning 100,000 yr of climatic history), a group of scientists have shown that the earth's climate regularly swings between warm and cold periods. These variations are supposed to be the result of a number of periodic variations in the sun's activity. The prediction is that the earth is now entering a cold period, which will last for 40 yr. Around 1985, average temperatures will drop as low as those of the "little ice ages" of the fifteenth and seventeenth centuries. After that, temperatures will rise slowly until 2010, when they will be back to about the level of the 1960's.

This brief account of the factors that may be responsible for climatic change illustrates that there is no simple way to predict exactly what man can or cannot do to his environment, especially in a system so large and complicated as the interaction of the sun, the atmosphere, and the hydrosphere. The most that can be said in cases like this is that man must be prepared to take the consequences if he chooses to ignore the adverse possibilities.

6.11 ELECTRONIC ABSORPTION SPECTRA OF MOLECULES

You already know that the electronic energy transitions brought about by the absorption of radiation in the ultraviolet–visible region involve relatively large energy changes. By analogy with the occurrence of simultaneous rotation–vibration transitions, you might expect that during the transition of one or more molecular electrons from one molecular orbital electronic energy level to a higher one there would be many accompanying rotational and several vibrational transitions. In a spectrophotometer that cannot separate closely spaced lines an electronic transition would appear, not as a narrow absorption line, but as a quite wide band encompassing a range of wavelengths or frequencies of absorption. The same thing causes the widening of vibrational lines in infrared spectroscopy. With a high-resolution ultraviolet–visible instrument, the detailed structure of an electronic absorption band can be seen, and analysis can give much information on moments of inertia, force constants, and dissociation energies of certain bonds, as well as the electronic energy states of molecules.

Since there are a number of electrons in any molecule, but only two may occupy the same molecular orbital, there may be a number of possible electronic transitions, which appear as a number of bands in the spectrum. In general, it is a difficult task to determine which absorption bands correspond to which transitions. In addition, the energy absorbed often leads to dissociation or breakup of the original molecule (photodecomposition) rather than to an actual excited state of the original molecule. This again complicates the spectrum.

Even when a molecule is not dissociated by absorption, the excited state may be much more chemically reactive than the unexcited electronic ground state. Generally, in excitation, 2 electrons in a molecular orbital become unpaired, and 1 electron is excited or promoted to a higher molecular orbital. This has three consequences: (1) there are now 2 unpaired electrons present for pairing with any unpaired electrons on another species; (2) the excited electron is held less tightly to the molecule and can be lost more easily; and (3) the bond involving the original electron pair is weakened by the excitation, and the molecule is more susceptible to dissociation of that bond. Factors such as these lead to the possibility of chemical reactions induced by absorption of light. This subject, photochemistry, will be considered in the following chapter.

As an illustration of the electronic processes that can take place when a molecule absorbs visible–ultraviolet radiation without dissociation, we can look at compounds containing the carbonyl group $>$C=O and consider the electrons in this group. The electron dot structure of the carbonyl group, found in organic compounds called aldehydes and ketones, is

$$
\begin{array}{l}
\text{R} \\
\overset{..}{\underset{..}{\text{C}}} :: \overset{..}{\underset{..}{\text{O}}} \\
\text{R}
\end{array}
$$

One pair of electrons in the double bond and the two pairs in the C—R bonds are involved in σ-type bonds formed with the trigonal (sp^2) carbon hybrid orbitals. The other pair of electrons in the $>$C=O bond are π electrons. The two unshared pairs on the oxygen are the $(2s)^2$ and a $(2p)^2$ of oxygen. They are not involved in bonding and are called nonbonding electrons, and their atomic orbitals are nonbonding (n) orbitals.

It should be remembered that there are also unoccupied molecular orbitals in the molecule. In this compound, the most important of the empty molecular orbitals are antibonding orbitals associated with the bonding σ-type sp^2 (C—O) and π (C—O) molecular orbitals.

The electronic ground state of

$$
\begin{array}{l}
\text{R} \\
\quad \diagdown \\
\qquad \text{C=O} \\
\quad \diagup \\
\text{R}'
\end{array}
$$

can be written as

$$[(1s_C)^2(1s_O)^2(2s_O)^2(\sigma_{CR})^2(\sigma_{CR})^2(\sigma_{CO})^2](\pi_{CO})^2(n_O)^2$$

For excitation at visible–ultraviolet frequencies, the inner electrons enclosed in brackets are not affected, and transitions involving the π_{CO} and lone-pair or nonbonding oxygen electrons are the only possibilities. There are three possibilities for one-electron transitions in the carbonyl group, where an electron is promoted from one molecular orbital to a higher-energy molecular orbital. These transitions are

$n \longrightarrow \pi^*$	One of the nonbonding lone-pair electrons on oxygen is promoted to the antibonding π^* molecular orbital of the C—O bond.
$\pi \longrightarrow \pi^*$	A π electron in the C—O bond is promoted to the π^* antibonding molecular orbital of the same bond.
$n \longrightarrow \sigma^*$	One of the oxygen lone-pair electrons goes to the antibonding molecular orbitals associated with the bonding σ molecular orbitals in the C—R and C—O bonds.

The $n \longrightarrow \pi^*$ absorption is a trademark of substances containing the carbonyl group and can be used to identify these compounds. This kind of transition can occur in other molecules having double bonds and atoms such as N, O, P, and S, which have unshared electron pairs. Some of these compounds play vital roles in biological processes, and $n \longrightarrow \pi^*$ electronic transitions enter into these roles.

Absorption of visible radiation by molecular electronic transitions is responsible for the distinctive color of many substances. This refers to the color of the light transmitted through a comparatively transparent substance such as a solution. The color of the transmitted light is determined by the absorption of certain wavelengths of the incident "white" light. For instance, a water solution of copper(II) ion, Cu^{2+}, appears blue-green because some component of the solution abstracts red light from the white light passing through the solution. The light that emerges has lost its red component and is blue-green.

A general rule is that any colored solution absorbs light of a wavelength that is *complementary* to its transmitted color. A yellow solution absorbs in the blue (indigo) region, a red absorbs in the blue-green region, and a green absorbs in the purple region. Other relations are listed in Table 6.4.

Table 6.4
Absorption and color

absorbed wavelength, nm	color	
	absorbed	transmitted
400–450	violet	yellow-green
450–480	blue	yellow
480–490	green-blue	orange
490–500	blue-green	red
500–560	green	purple
560–580	yellow-green	violet
580–590	yellow	blue
590–640	orange	green-blue
640–700	red	blue-green

In the Cu^{2+} solution, the absorbing species is the hydrated copper(II) ion, $Cu(H_2O)_4^{2+}$. Whether the water molecules are considered to be held to the positive copper(II) ions by electrostatic ion–dipole forces or by covalent forces involving the unshared electrons on the oxygens, the presence of the water of hydration changes the electronic energy levels of the d electrons in the metal ion so that a transition between levels can be induced by red light. Hydrated ions also exist in solid copper(II) sulfate, $CuSO_4 \cdot 5H_2O$. You can show that the hydrated ion is responsible for the color by heating the crystals to drive off the water of hydration. The copper(II) salts then become almost colorless:

$$Cu(H_2O)_4^{2+} \xrightarrow{\Delta} Cu^{2+} + 4H_2O$$
$$\text{blue} \qquad\qquad \text{colorless}$$

The colors of the hydrated ions (aquo complexes) of transition metals are characteristic features of these elements.

An important example of a molecule that dissociates on absorption of radiation is ozone, O_3. Although the vibrational structure of ozone is more important at high altitude in maintaining the heat balance of the atmosphere than is that of carbon dioxide or water, a more important role of ozone in the biological environment is connected with its electronic structure. Measured by its percent composition, ozone is a minor constituent of the atmosphere, but it occurs in relatively high concentration in the upper atmosphere in a belt from about 10 miles to 30 miles in height. Ozone absorbs deadly ultraviolet radiation from the sunlight passing through the atmosphere and prevents a large portion of this high-energy radiation from reaching the surface of the earth. Anyone who has been sunburned by the ultraviolet radiation that does get through will realize that living cells are not resistant to this radiation.

Ozone is formed in the upper atmosphere by the irradiation of oxygen molecules in a photochemically initiated process:

$$O_2 \xrightarrow{h\nu} O + O$$
$$O_2 + O + M \longrightarrow O_3 + M$$

where M is a third body such as an oxygen or nitrogen molecule. The formation of atomic oxygen in the first reaction is brought about by radiation with wavelengths between 176 and 203 nm in the upper atmosphere and at about 210 nm below 20–30 miles.

Ozone absorbs ultraviolet radiation strongly at 200–350 nm, with the maximum absorption at 255 nm. The photochemical reaction associated with this absorption is

$$O_3 \xrightarrow{h\nu \ (200-350 \ nm)} O_2 + O$$

This absorption prevents radiation of less than 290 nm from reaching the earth's surface. Absorption by ozone of most of the solar radiation in the 220–300 nm region results in conversion of this radiation into infrared (heat) radiation. This accounts for the temperature maximum at an altitude of about 30 miles, near the top of the stratosphere.

A further, nonphotochemical, reaction controlling the destruction of ozone in the upper atmosphere is

$$O_3 + O \longrightarrow 2O_2^*$$

where the resulting oxygen molecules are in an excited (*) electronic state.

Ozone also weakly absorbs visible radiation between 450 and 700 nm, with the maximum at 570–610 nm. There are additional weak absorptions in the infrared region at 4.8, 9.6, and 14.1 μm, as we saw earlier. Absorption at 9.6 μm is responsible for a high-altitude greenhouse effect.

The triatomic ozone molecule is nonlinear, with an O—O—O bond angle of 117°. The molecule may be represented by the two resonance structures

Ozone is a highly unstable, highly reactive molecule. Its exothermic reversion to molecular oxygen,

$$2O_3 \longrightarrow 3O_2 \qquad \Delta H = -34.2 \ kcal/mol$$

proceeds with explosive violence. The energy needed for the conversion of oxygen to ozone can be supplied by radiation, as noted before, or by an electric discharge. The distinctive odor of ozone is sometimes detected around electric motors. Even in low concentrations, ozone is a lung irritant. Other properties of ozone will be discussed in connection with air pollution and the concern that the introduction of water vapor and NO into the upper atmosphere by high-flying supersonic aircraft will initiate reactions leading to a decrease in the ozone concentration at these levels. If this does take place, the earth's surface will be exposed to a greater intensity of ultraviolet radiation than at present.

6.13 SUMMARY

The total energy of a molecule is made up of four component energies. These are (1) translational energy, (2) rotational energy, (3) vibrational energy, and (4) electronic energy. Translational energy is the kinetic energy of straight-line motion of the molecule. Rotational energy is associated with the kinetic energy of rotation of the molecule as a whole. Vibrational energy is the kinetic plus potential energy of stretching or bending of bonds within the molecule. The electronic energy of a molecule is due to the kinetic plus potential energy of the electrons in the molecule.

All four types of energy are quantized and can be represented by sets of energy levels. The relative spacing between adjacent energy levels for the four kinds of energies is in the ratio electronic : vibrational : rotational : translational $= 100 : 5 : 10^{-2} : 10^{-18}$. A molecule may absorb energy by interacting with the electromagnetic field of radiation. An increase in molecular electronic energy results when a molecule absorbs ultraviolet or visible radiation. Absorption of infrared radiation increases the vibrational energy of a molecule, while absorption of microwave radiation leads to rotational energy level transitions. The translational energy levels are so closely spaced that their quantization may be ignored.

Molecular absorption spectroscopy investigates the structure of molecules by measuring the frequency or wavelength at which absorption of energy occurs. From the absorption frequency, information may be obtained about the structure and composition of a molecule. The relative amount of absorption at a particular wavelength is related to the concentration of absorbing molecules.

The absorption of radiation by gases in the atmosphere has important consequences for life on earth. Atmospheric carbon dioxide and water vapor contribute to a greenhouse effect by trapping infrared radiation emitted to space by the earth. The addition of large amounts of CO_2 to the atmosphere as a result of human activities has led to concern that global climate may be affected by changes in the heat balance of the earth. However, the heating effect of CO_2 absorption now seems to have been offset, or even reversed, by a cooling effect due to the accumulation of particulate matter in the atmosphere. Ozone in the upper atmosphere strongly absorbs ultraviolet radiation, thus protecting life on the earth's surface from lethal radiation with wavelengths less than 290 nm.

Questions and Problems

1. For light of wavelength 400 nm, find
   a. wavelength in angstroms and centimeters
   b. frequency
   c. photon energy $\Delta\epsilon$ in ergs per quantum (or ergs per molecule)
   d. $\Delta\epsilon$ in electron volts
   e. $\Delta\epsilon$ in kilocalories per mole (of quanta)
   Conversions are 1 erg/molecule $= 1.44 \times 10^{13}$ kcal/mol $= 6.24 \times 10^{11}$ eV.

2. Calculate the wavelengths (or frequencies) of radiation emitted or absorbed for the various types of energy transitions listed in Table 6.1, and identify the general spectral region to which these wavelengths belong.

3. Use $\epsilon = n^2 h^2 / 8 \, ml^2$ from the particle-in-a-box model to calculate $\Delta\epsilon$ for the $n = 1$ to $n = 2$ transition in electronic, vibrational, rotational, and translational energy levels in the HCl molecule. Compare the calculated values with those in Table 6.1. For "box" lengths take $l_{elec} = l_{rot} = $ bond length of HCl $= 1.3$ Å, $l_{vib} = 0.10$ Å, and $l_{trans} = 1$ cm. For particle masses, take $m_{elec} = 9.1 \times 10^{-28}$ g, $m_{vib} = m_{rot} = m_{trans} = m_{HCl} = 6 \times 10^{-23}$ g.

4. A solution of an absorbing species has a transmittance of 0.50 at a certain wavelength.
   a. What is the absorbance of the solution?
   b. If the amount of absorbing species in the solution is doubled, what is the transmittance and absorbance?
   c. If the concentration of the solution in part (a) is halved, what is the transmittance and absorbance?

5. A solution absorbs maximally and minimally at the following wavelengths with the accompanying percent transmittances:

λ	430	460	490	520	600
%T	90	10	70	30	100

   Sketch curves of percent transmittance and of absorbance versus wavelength and compare the two curves.

6. A solution of permanganate ion, MnO_4^-, absorbs strongly between 480 and 580 nm, with maximum absorption at 525 nm.
   a. What color is a permanganate solution?
   b. At 525 nm, the percent transmittance of a given

MnO_4^- solution is 50 percent. What is the absorbance of the solution?
   c. If the solution in part (b) contains 1.30×10^{-4} mol of MnO_4^- per liter of solution, what is the absorptivity (also called the molar absorptivity or molar extinction coefficient) of the solution when in a 1.00-cm container? Give the units of the absorptivity.
   d. The absorbance of a permanganate solution of unknown concentration (in moles per liter) is measured at 525 nm in a 1.00-cm path length cell. If the measured absorbance is 0.40, what is the concentration of the solution?

7. Show that the frequency separation between adjacent rotational absorption lines for a rigid diatomic molecule is constant and equal to $2h/8\pi^2 I$.

8. Compare the force constants listed in Table 6.2 with the bond energies listed in Table 3.3. What general relationship appears to exist?

9. What would happen to the wavelength of radiation associated with the stretching of the singly bonded molecule A—B if
   a. the force constant of the A—B bond is increased?
   b. the masses of the atoms A and B are both increased?

10. The infrared absorption spectrum of HCl is centered at $v = 8.65 \times 10^{13}$ sec^{-1}. In the microwave spectrum, the frequency separation of the absorption lines is 62.7×10^{10} sec^{-1}. Find
    a. the bond length r_0 in HCl
    b. the force constant k in HCl

11. Considering the bond energy of the O_2 molecule, what is the wavelength of radiation having the minimum energy required to dissociate oxygen into two oxygen atoms in their ground states according to the reaction $O_2 + hv \longrightarrow 2O$? Compare your answer with the 176–203 nm range mentioned in Section 6.12, and explain how your answer and the actual values may be reconciled by having one oxygen atom in an excited state of higher energy.

12. Draw an energy level diagram showing the allowed transitions between rotational energy levels in two different vibrational levels of a diatomic molecule. Consider five rotational levels ($J = 0 - 4$ as in Figure 6.9) in each vibrational level $v = 0$ and $v = 1$.

13. An industrialist invents a wonderful process that will make everyone happy and rich. The process

results in the discharge of huge quantities of a gas, but the gas is inert and nontoxic. The gas absorbs infrared radiation very strongly in the 10–14 μm region. Would you approve or disapprove this process for worldwide use?

14. What might happen to the climate if nonpolar diatomic molecules such as oxygen and nitrogen could absorb microwave and infrared radiation?

15. You are given three unlabeled liquid samples. The liquids are an alcohol CH_3—CH_2—CH_2OH; an aldehyde, CH_3CH_2CH; and a ketone,

$$CH_3-\overset{\overset{O}{\|}}{C}-CH_3.$$ How might you tell the liquids apart?

16. Explain how and why the amino acid tyrosine, with the structure

$$HO-\underset{}{\bigcirc}-CH_2-\overset{\overset{H}{|}}{\underset{\underset{NH_2}{|}}{C}}-\overset{\overset{O}{\nearrow}}{C}-OH$$

might be distinguished on the basis of its electronic absorption spectrum from the amino acid, serine, with the structure

$$HO-CH_2-\overset{\overset{H}{|}}{\underset{\underset{CH_2}{|}}{C}}-\overset{\overset{O}{\nearrow}}{C}-OH$$

Answers

1. a. 4000 Å, 4.00×10^{-5} cm
   b. 7.50×10^{14} sec^{-1}
   c. 4.96×10^{-12} erg/quantum
   d. 3.10 eV
   e. 71.5 kcal/mol
3. $\Delta\epsilon_{elec} = 1540$ kcal/mol
   $\Delta\epsilon_{vib} = 4$ kcal/mol
   $\Delta\epsilon_{rot} = 0.02$ kcal/mol
   $\Delta\epsilon_{trans} = 4 \times 10^{-18}$ kcal/mol

4. a. $A = 0.30$
   b. $A = 0.60$, $T = 0.25$
   c. $A = 0.15$, $T = 0.71$
6. b. $A = 0.301$
   c. 2320 liter/mol cm
   d. 1.72×10^{-4} mol/liter
10. a. $r_0 = 1.29$ Å
    b. $k = 4.8 \times 10^5$ dyn/cm
11. 240 nm

References

MOLECULAR SPECTROSCOPY

Barrow, Gordon M., "The Structure of Molecules," W. A. Benjamin, Menlo Park, Calif., 1963.

Clayton, Roderick K., "Light and Living Matter, Vol. 2: The Physical Part," McGraw-Hill, New York, 1970.

Pecsok, Robert L., and Shields, L. D., "Modern Methods of Chemical Analysis," Wiley, New York, 1968, Chapters 8–11.
    Contains material on practical applications and instruments.

Rice, Francis Owen, and Teller, E., "The Structure of Matter," Wiley, New York, 1949, Chapters 10, 11.

Wheatley, P. J., "The Determination of Molecular Structure," Oxford University Press, London, 1959, Chapters II–IV.

THE GREENHOUSE EFFECT AND OTHER ISSUES

Landsberg, Helmut E., "Man-Made Climatic Changes," *Science*, **170**, 1265–1274 (1970).

"Man's Impact on the Global Environment. Assessment and Recommendations for Action." Report of the Study of Critical Environmental Problems, M.I.T. Press, Cambridge, Mass., 1970.

Oort, Abraham H., "The Energy Cycle of the Earth," *Scientific American*, **223**, 54 (September 1970).

Petersen, Eugene K., "The Atmosphere: A Clouded Horizon," *Environment*, **12**, 32 (April 1970). [Condensed from "Carbon Dioxide Affects Global Ecology," *Environmental Science and Technology*, **3**, 1162 (November 1969).]

Plass, Gilbert N., "Carbon Dioxide and Climate," *Scientific American*, **201**, 41 (July 1959).

Singer, S. Fred, Ed., "Global Effects of Environmental Pollution," Springer-Verlag, New York, 1970.

Air Pollution

"O where are you going?" said reader to rider,
"That valley is fatal when furnaces burn,
Yonder's the midden whose odours will madden,
That gap is the grave where the tall return."
W. H. Auden—*"Five Songs, Song V"**

A definition of pollution is not easy, because it depends largely on the qualities of the environment that an individual considers to be of value. Thus, what is a major cause of pollution to one person might be ignored by another. The common definition of a pollutant as something in the environment that adversely affects, directly or indirectly, something of value to man is proper in that it implies that the effects of a true pollutant must be perceived by man—or some men, anyway. But the definition seems somewhat too ego-centric, in implying that only those things that man, taken as a whole, values are worth considering. In the past, mankind has not generally shown any

*Collected Shorter Poems 1927–1957, Random House, New York, 1967. Copyright © W. H. Auden.

198

great concern for life other than his own or that of his domesticated animals. For this reason, or perhaps in spite of it, a more inclusive view of pollution is taken here. *Pollution is any deviation from normal in the natural composition of a part of the environment that adversely affects not just man, but life in general.* Because, in an ecological sense, man and other forms of life are interrelated, the narrower definition of pollution would be acceptable provided that all present and conceivable interconnections are taken into account. So far, man has not shown himself to be particularly adept at recognizing these interrelations. Included in the effects on life are the usual chemical, physical, or biological consequences. For man, however, other less tangible factors (social, cultural, psychological, spiritual, aesthetic, etc.) should be taken into account. Pollutants need not be material substances; noise can be a pollutant, and even electromagnetic waves can be pollutants.

7.1 SOURCES OF POLLUTANTS IN GENERAL

Pollutants are usually added to the environment by man's actions or presence. They are predominantly the waste products of his biology, such as domestic sewage, or of his culture, such as industrial wastes. Most of today's problems are due to man as a pollutant source. In a larger sense, however, some natural substances like pollen or dust in the air, methane gas in mines and near swamps, selenium in plants, and even sunlight adversely affect man and other life, and at times they may be called pollutants.

Many potential pollutants are present as normal constituents of various parts of our environment. Carbon dioxide and ozone are examples. Some of these substances are present in infinitesimal amounts or exist only in isolated locations. Natural processes operate to keep normal constituents of the environment at levels that are fairly constant and harmless. Some biogeochemical cycles serve to prevent possible overconcentrations of potential pollutants by transferring substances from one part of the surroundings to another. Often in this transfer process the substance is put into a form that prevents it from interacting to any great extent with the living world. It has been effectively removed from action, so to speak. Other processes, such as the food chain in the case of DDT, may concentrate pollutants in places or in forms where they do great damage.

Natural processes are of no use in cleansing the environment of many foreign substances. There is no way for nature to decompose many man-made materials and return their elements to the cycles of nature. Such substances will just remain and work whatever harmful effects they can—until they are somehow dispersed or diluted so their action is no longer noticeable. Plastics do not decompose, and they continue to pollute aesthetically our landscape. The early detergents were not affected by sewage treatment, which is based, as is most treatment, on natural processes, and went round and round from sewage plant to river (where they produced suds) to drinking water (with foam on it) to sewage treatment plant (where they gummed up the works).

In summary, man (and sometimes nature) has introduced pollutants into the environment by

1. adding substances at rates exceeding their rate of removal or decomposition by natural processes,

2. introducing substances that are concentrated or converted to more toxic forms by natural processes, and

3. adding substances not removed by natural processes.

It is not surprising that some of these points also appeared in Section 5.2 in connection with man's intrusion on biogeochemical cycles. As will be pointed out at several places in this book, man must either limit the introduction of pollutants or he must invest his resources to augment the natural processes of removal and cleansing.

Some may wish to consider drugs, alcohol, tobacco, and so on as pollutants. They undoubtedly are in certain amounts. But they are personal pollutants administered at the choice of the individual and (supposedly) affecting only that individual, and therefore they are discussed only briefly, if at all, in this book. What we concentrate on are pollutants that affect a number of people, many of whom are not responsible for introducing the pollutant into the surroundings. We are concerned with such large-scale insults to a population of organisms or to an ecosystem rather than with self-willed, self-administered insults to the human organism.

7.2 AIR POLLUTION

Man has used the atmosphere as a sewer almost from the time he learned to make fire, but it was not until man began to congregate in large cities in the fourteenth century that air pollution began to attract attention and complaints. This pollution became critical when coal started to be widely used as a source of heat for homes and power for industry. Most coal contains sulfur. As a result, smoke (particulate matter) and gases such as SO_2 and SO_3 have been the major pollutants in all parts of the industrialized world for over 400 yr. With the advent of liquid and gaseous fuels and the internal combustion engine, new varieties of air pollution have entered the scene.

At one time, the dumping of foreign material into the air was on a small enough scale and consisted of only a few pollutants sufficiently localized that the large but finite volume of the atmosphere and natural mixing processes were able to disperse and dilute the contaminants. Today we are faced with problems caused by a large growth and concentration of population, increased use of energy, and technological advances. All of these have combined to pollute the air we breathe until, in several places, death and severe illness have resulted, and chronic illness and reduced lifespan due to air pollution are ever-present.

Man-made air pollutants cover a wide variety of materials, from wind-blown dust from eroded fields through extremely toxic substances such as mercury fumes, H_2S, Cl_2, and insecticides, to supposedly innocuous gases such as CO_2 and Freon-12 (CCl_2F_2). Some pollutants are localized around industrial facilities and may dissipate rapidly, while other pollutants are poured out in such volumes that large portions of the atmosphere are affected. We shall investigate only a small, but important, part of the air pollution problem. However, it should be borne in mind that any substances other than the normal constituents of clean air that are placed in the atmosphere should be viewed with suspicion. For even when foreign substances are present in the physical environment in amounts near the limits of sensitive detection methods, this does not mean that there is no danger. Many of these substances accumulate, little by little, in a living organism before they reach a concentration at which their toxic effects are noticed. By then it may be too late. Other substances appear to be relatively harmless by themselves, but may exhibit *synergistic effects* with another relatively harmless pollutant. A synergistic effect is one in which the combined toxic effect of the two substances is greater than the sum of the effects of the substances considered individually.

Air pollutants can be divided roughly into two broad classes, *particulate matter* and *gases*. Particulate matter consists of solid particles such as dust or smoke, or fine liquid droplets such as mists or fogs. The sources of air pollutants can be either natural or man-made. For instance, plant life can contribute to the haze over vegetated areas by releasing compounds called *terpenes*, which react with oxygen and sunlight to form visibility-reducing particles. The Smoky Mountains obtain their name from such a natural haze. We are concerned here only with man-made sources, because these are the only sources we can control. Man-made air pollutants are produced primarily by power and heat-generating facilities, transportation, industrial processes, and, in smaller part, by refuse burning. Table 7.1 lists the estimated amounts and sources of the five most common air pollutants from these sources in the United States. Carbon dioxide as well as hundreds of specific pollutants emitted locally by specific activities (mercury smelters and pulp mills, for example) are not included here.

Table 7.1
Estimated emissions of principal air pollutants in the United States, 1968 (millions of metric tons[a] per year)

source	particulates	sulfur oxides	nitrogen oxides	carbon monoxide	hydrocarbons
stationary fuel combustion (electric power plants, space heating)	8.1	22.1	9.1	1.7	0.6
mobile fuel combustion (transportation)	1.1	0.7	7.3	57.9	15.1
combustion of refuse	0.9	0.1	0.5	7.1	1.5
industrial processes	6.8	6.6	0.2	8.8	4.2
solvent evaporation	—	—	—	—	3.9
totals	16.9	29.5	17.1	75.5	25.3

[a] One metric ton = 1000 kg = 2205 lb = 1.102 (short) tons.
SOURCE: "National Emission Reference File," 1970 ed., Compiled by A. Hoffman, National Air Pollution Control Administration, Washington, D. C., 1970.
SEE ALSO: George B. Morgan, Guntis Ozolins, and Elbert C. Tabor, "Air Pollution Surveillance Systems," *Science*, **170**, 291 (1970).

The pollutants listed in the table are, with many others, known as *primary pollutants*. These are substances emitted directly from identifiable sources. The primary pollutants may subsequently react with each other or with normal constituents of the air. Some of the reactions may form harmless substances and serve to remove the primary pollutants from the air. Other reactions may generate more primary pollutant or new harmful substances known as derived or *secondary pollutants*. The biological effects of many of the secondary pollutants may be much greater than that of any of their parent primary pollutants.

Other than elimination of or reduction in quantity of certain specific pollutants generated from localized sources, most of the solutions to the problems of modern air pollution abatement and control are connected with understanding the often complex chemistry of the reactions giving rise to the secondary pollutants and with controlling the emission of the responsible primary pollutants. The chemistry of a polluted air mass depends on numerous factors, including concentration of reactants, solar radiation, conditions of atmospheric circulation, topography, and many others. It has been said that a polluted air mass is a veritable witches' brew of chemical reactions. For this reason, we can give only a brief discussion of the chemistry of two of the more common forms of air pollution.

7.3 POLLUTION BY SULFUR OXIDES

The most common type of air pollution from the fourteenth century to the early twentieth century resulted from the smoke and gases released by the burning of coal. During that time, coal was the major source of heat and industrial power in urban areas.

The lower grades of coal contain sulfur in its reduced form, much of it as iron pyrite, FeS_2. Coal and fuel oils also contain sulfur in organic forms. The sulfur is oxidized to the primary pollutant sulfur dioxide, SO_2, when the fuel is burned:

$$4FeS_2 + 11O_2 \longrightarrow 2Fe_2O_3 + 8SO_2$$

and

$$2H_2S + 3O_2 \longrightarrow 2H_2O + 2SO_2$$

The SO_2 formed is eventually further oxidized to sulfur trioxide, SO_3:

$$2SO_2 + O_2 \longrightarrow 2SO_3$$

Then the SO_3 can combine with water vapor to produce droplets of sulfuric acid, H_2SO_4, with the help of metallic oxide particulates which serve as catalysts.

$$SO_3 + H_2O \xrightarrow{\text{oxides of Mn, Fe}} H_2SO_4$$

The sulfuric acid mist, in turn, reacts with other materials such as metal oxide dusts, MO, and ammonia in the air to form sulfates. Ammonium sulfate, $(NH_4)_2SO_4$, and calcium sulfate, $CaSO_4$, are the major products.

$$H_2SO_4 + MO \longrightarrow MSO_4 + H_2O$$

There are four chemical reactions here leading to the formation of sulfate particulates:

(1) $S + O_2 \longrightarrow SO_2$
(2) $2SO_2 + O_2 \longrightarrow 2SO_3$
(3) $SO_3 + H_2O \longrightarrow H_2SO_4$
(4) $H_2SO_4 + MO \longrightarrow MSO_4 + H_2O$

Reaction Rates and Activation Energies

Things are a bit more complicated than these simple equations indicate. In the first place, not all of these reactions take place with any great speed. When we talk about the speed, rate, or velocity of a reaction, we refer to the amount of reactant that reacts (disappears) per unit time. Alternatively, we could talk about the amount of product that is formed per unit of time. An explosion is an example of a fast reaction. The rusting of iron,

$$4Fe + 3O_2 + 2xH_2O \longrightarrow 2(Fe_2O_3 \cdot xH_2O)$$

is a common slow reaction.

The rates of chemical reactions are governed by a number of factors, among which are the nature and amounts of reactants present, the temperature, and the presence of catalysts. All of these will be discussed in detail in Chapter 17. At this point, we wish to look only at a fundamental quantity called the *activation energy* of a reaction. An understanding of this quantity will help us explain why temperature, catalysts, and later, radiation are important in bringing about chemical reactions.

In order for any chemical reaction to take place, a certain amount of energy must be supplied at least to get the reaction started. This energy, the activation energy E_a, is shown in Figure 7.1 as the energy required to get the

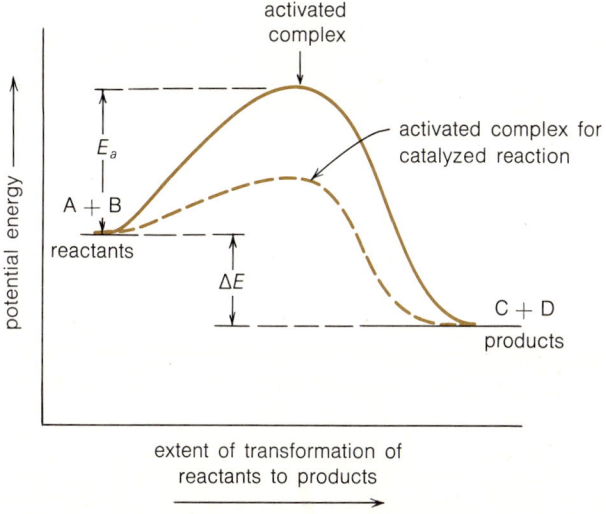

Figure 7.1. Potential energy diagram for the reaction A + B ⟶ C + D. The activation energy E_a is the minimum amount of energy that the reactants must acquire to form the activated complex that transforms to products. The dashed line shows the change in potential energy during the course of a catalyzed reaction where a different activated complex of lower energy is formed. The net energy change ΔE is the same for the uncatalyzed and the catalyzed reaction.

reactants to the top of an energy barrier or hump, from which they can "roll downhill" to become products. The activation energy is considered to be the energy required for the formation of an intermediate grouping of the atoms, or activated complex, which is situated at the top of the energy barrier. Once formed, this complex readily decomposes into the products of the reaction. The activation energy should not be confused with the overall energy change ΔE (or ΔH) for reactions. Activation energies are always positive, but the overall energy change may be positive or negative. Fast reactions generally have small activation energies; slow reactions have large activation energies.

In ordinary chemical reactions, the reactants acquire the necessary activation energy as a result of energy interchanges occurring in the collisions of the moving atoms or molecules. To speed up a reaction, we can do two things (aside from increasing the concentration, and thereby, the number of collisions, of the reactants). We can increase the temperature, which supplies additional thermal energy to the reactant molecules and allows more of them to surmount the activation energy barrier. Reactant molecules also can gain energy by the absorption of radiation, principally in the visible and ultraviolet regions. The other way to increase the velocity of a reaction is to somehow lower the activation energy so that more reactant molecules will have the minimum energy required to form the intermediate activated complex. Substances that lower the activation energy are known as *catalysts*. Catalysts usually act by combining with one or more of the reactants to form a different activated complex of lower energy than the activated complex of the uncatalyzed reaction. The new, more easily formed, complex decomposes to give the products of the reaction plus the original catalytic substance. The catalyst is involved in the reaction but is not permanently changed or used up. In effect, the catalyst provides a different reaction path, with a lower activation energy, from reactants to products. An uncatalyzed reaction might be represented by the equation

$$A + B \longrightarrow A \cdots B \longrightarrow C + D$$

where the high-energy activated complex $A \cdots B$ decomposes into the products, C and D. When a catalyst X is used, a lower energy complex $A \cdots B \cdots X$ is more easily formed,

$$A + B + X \longrightarrow A \cdots B \cdots X \longrightarrow C + D + X$$

The energy relations are shown in Figure 7.1.

Atmospheric temperatures are generally limited to a narrow range of low values so that sufficient thermal energy is unavailable to supply activation energies to atmospheric reactions. Also, with the exception of oxygen, the concentrations of atmospheric reactants are low, and this precludes fast reactions in the absence of photochemical and catalytic effects. Photochemical effects will be discussed shortly. We concentrate now on catalytic effects in the SO_2 type of air pollution. Returning to our SO_2–H_2SO_4–MSO_4 type of air pollution, we can list some of the additional substances in the air that are believed to act as catalysts for some of the four reactions, so that an abundant supply of pollutant products can be generated in a reasonable time. By "reasonable" time is meant a period short enough that sufficient pollutant concentrations are built up to have an adverse effect before natural processes such as rainfall, winds, gravitational settling, and other chemical reactions known as *scavenging processes* can cleanse the air. This time period depends on local conditions, and may range from hours to weeks.

The first reaction, $S + O_2 \longrightarrow SO_2$, does not require a catalyst, because the heat of the combustion process is sufficient to provide the activation energy. The next reaction, the oxidation of SO_2 to SO_3, is inherently the slowest reaction in the S-to-MSO_4 sequence. (Wherever appreciable quantities of the secondary pollutants SO_3, H_2SO_4, and MSO_4 are produced in a given period of time, this oxidation reaction must have been speeded up.) The overall rate of a sequence of reactions depends on its slowest step. The rate of production of sulfuric acid and sulfate salts can be no greater than the rate of oxidation of SO_2 to SO_3.

In the gaseous state, the oxidation of SO_2 to SO_3 takes place very slowly. The reaction is partially promoted by the absorption of light to furnish the required activation energy, but this photochemical reaction alone is not sufficient to form the quantity of SO_3 found in polluted atmospheres. Instead, the oxidation is believed to be catalyzed by ferric oxide, Fe_2O_3, on the metal surfaces of furnaces or other combustion vessels or by iron and manganese sulfates or oxides in particulate matter in the air. This process occurs most readily after some SO_3 has combined with water vapor in the air and minute droplets of sulfuric acid have been formed:

$$SO_3 + H_2O \longrightarrow H_2SO_4$$

Further oxidation of SO_2 to SO_3 takes place in the liquid acid droplets, catalyzed by the iron and manganese compounds. Other substances can aid in the conversion of SO_2 to SO_3 or H_2SO_4. The visibility-reducing sulfuric acid mists thus formed are responsible for the bluish haze in atmospheres polluted with SO_2.

Finally, reaction of the sulfuric acid droplets (generally less than 0.5 μm in diameter and therefore not prone to gravitational settling) with calcium oxide, ammonia, and other substances in the air forms sulfates that eventually settle to the earth.

$$2NH_3 + H_2SO_4 \longrightarrow (NH_4)_2SO_4$$

$$SO_3 + CaO \longrightarrow CaSO_4$$

or

$$H_2SO_4 + CaO \longrightarrow H_2O + CaSO_4$$

It is interesting to note that the two reactions

$$2SO_2 + O_2 \xrightarrow{\text{catalyst}} 2SO_3$$
$$SO_3 + H_2O \longrightarrow H_2SO_4 \qquad \text{(by way of } H_2S_2O_7; \text{ Section 7.8)}$$

are the basis of the industrial manufacture of sulfuric acid by the *contact process*.

The SO_2 type of air pollution is sometimes called "industrial" or "classical" air pollution, because it originates primarily from sulfur-containing coal and oil—fuels that have been superseded in many parts of the world by cleaner and more modern fuels such as natural gas and sulfur-free oils. Nevertheless, in fairly recent times this old-fashioned pollution has been responsible for most of the documented cases of air pollution disasters, as measured by the excess number of deaths above those normally occurring. During a few days in London, in December 1952, the buildup of smoke, fog, and sulfur oxide fumes resulted in the death of 4000 people more than expected statistically. Similar but less deadly episodes have occurred and still occur in Europe and the United States.

7.4 PHOTOCHEMICAL SMOG Since the 1940's, the more affluent parts of the world have experienced a new type of air pollution. This type of pollution, generally called *smog*, is caused primarily by the high-temperature, but incomplete, combustion of fuels containing carbon and hydrogen. The primary source of this type of pollution is the gasoline-powered internal combustion engine. The products of combustion, plus incompletely burned fuel, interact with each other and with secondary pollutants in the presence of sunlight to produce the characteristic pollution found over many highly developed and congested urban areas. This type of pollution was first noticed in Los Angeles, California, in the 1940's and, since then, has appeared throughout the world wherever there are concentrations of people and automobiles. Although the old-fashioned, nineteenth century SO_2 type of air pollution has been reduced in many areas, the newer, twentieth century photochemical smog is on the increase, and its effects cause great concern.

The atmospheric chemical reactions leading to smog are numerous, complex, and incompletely understood. The primary pollutants interact with one another and with sunlight to produce a variety of secondary pollutants, which subsequently interact with each other, with primary pollutants, with sunlight, and with normal constituents of the air to give a still greater array of substances. The possible sequence of reactions depends on atmospheric conditions, concentrations of pollutants, presence of catalysts, reactivity of substances, reaction rates, and a variety of other factors. Investigation of this sequence requires the most sophisticated chemical knowledge and instrumentation.

Nevertheless, even with our limited chemical knowledge, we can see how some of the primary pollutants do react and produce some of the more noticeable secondary pollutants. Just as with the SO_2 type of pollution, the primary pollutants are generally much less offensive to the senses and to life than the secondary pollutants. Indeed, the key to air pollution control lies in intercepting the primary pollutants as they emerge from the source so that they cannot react further. Aside from shutting down the source of the primary pollutant (i.e., forbidding the use of the internal combustion engine), interception of primary pollutants is the only feasible thing to do.

Some Basic Concepts of Photochemistry

As the name "photochemical smog" implies, one of the key processes in the formation of this kind of air pollution involves the absorption of radiation by some substance. In this case, the photochemical reaction sequence is started by the absorption of light by NO_2, nitrogen dioxide, a compound which is important in air pollution and smog formation because of its photochemical dissociation into nitric oxide, NO, and atomic oxygen, O. The atomic oxygen in turn forms the other characteristic constituents of smog, ozone, and oxidation products of hydrocarbons. Before considering the light absorption reaction and the subsequent reactions, we shall briefly treat some of the concepts of photochemistry. Some background on the effect of radiation on the reactivity of molecules has already been given. Here we extend this knowledge a bit.

We have already noted that any chemical reaction pathway has a definite activation energy, the minimum energy that must be supplied to the reactants before they can form a product. For reactions having small activation energies, the thermal energy of molecular collisions may furnish enough energy to the appropriate degrees of freedom of the molecules. These reactions are called *thermal* or *dark* reactions. The reaction potential energy diagram would appear as in Figure 7.1.

Other reactions have larger activation energies. The initial step in the mechanism of the reaction may, for instance, require the rupture of a strong bond or the high-energy excitation of an electron. In cases like this, thermal energy is not sufficient, but the energy transferred to the reactant by the absorption of a quantum of light energy may be large enough to surmount the activation energy barrier and cause reaction. Such photochemical processes require radiation of less than about 800 nm (8000 Å) in the visible and ultraviolet regions. Near the earth's surface, photochemical effects are limited to those initiated by absorption of radiation above about 290 nm, because shorter wavelengths have been absorbed at higher altitudes by atmospheric ozone. Photochemical reactions involve electronic transitions. Light-induced processes are important in the atmosphere, because the comparatively low temperatures preclude thermal activation of significant numbers of reactants.

There are two general rules governing photochemical reactions.

1. Only radiation that is absorbed by a reacting atom or molecule is effective in producing a chemical change.

2. Each atom or molecule taking part in a photochemical reaction absorbs not less than one quantum $h\nu$ of the effective radiation.

The absorption of a photon of visible or ultraviolet radiation leads to an electronically excited atom or molecule. The excited molecule produced in the photon-absorption process, called the *primary process*, may subsequently react in several ways. All the subsequent thermal reactions of the excited molecule or its fragments are called *secondary reactions.* The entire sequence of reactions, primary and later secondary reactions, is called a *photochemical reaction*, and the primary process is said to trigger or *initiate* the later reactions.

An excited molecule may retain its electronic excitation energy for a while, until it loses some of the extra energy by the emission of radiation and returns to a lower energy state. Alternatively, the excited molecule may transfer its excitation energy to other molecules by collision. The other molecules, in turn, become activated and take part in some secondary reaction. The majority of photochemical reactions, however, consist of a primary process in which the absorbing molecule dissociates into two or more fragments. Usually, these fragments are highly reactive and readily able to take part in secondary reactions, which may themselves produce molecules that can undergo further reaction.

The most common reactive fragments produced by primary processes are *free radicals* and atoms. Free radicals are species having unpaired valence electrons and, as such, are highly reactive. Examples of simple free radicals are the methyl radical,

$$\begin{array}{c} H \\ | \\ \cdot C-H \\ | \\ H \end{array} \quad \text{or} \quad \cdot CH_3$$

and the hydroxyl radical,

$$\cdot \ddot{O}-H \quad \text{or} \quad \cdot OH$$

Atomic entities such as $\cdot H$, $\cdot \overline{O} \cdot$, and $\cdot \overline{Cl}|$, although having unpaired electrons, are more properly called atoms rather than free radicals.

Depending on the conditions and the number of substances present, a primary photochemical process may be only the initial step in a bewildering series of secondary reactions. This situation applies in the normal smog-polluted air over any city, and it is no wonder that all of the reactions and substances involved in photochemical smog have not been identified. As in our treatment of SO_2 pollution, we'll discuss only a few of the more common reactions involved.

Chemical Reactions of Smog Formation   The primary photochemical process of greatest importance in smog production is the photodissociation of nitrogen dioxide, NO_2, by light of wavelength less than about 400 nm. The brown NO_2 is initially formed by the slow reaction of the relatively harmless and colorless nitric oxide, NO, with oxygen. The reaction is

$$2NO + O_2 \longrightarrow 2NO_2$$

Nitric oxide is formed in high-temperature combustion processes accompanied by rapid cooling of the exhaust gases. (Not all the NO_2 in the atmosphere can be traced to man-made sources. A good portion must come from as yet unidentified natural sources.) In these reactions, NO is the primary pollutant even though NO_2 is the entity involved in the primary photochemical process.

$$NO_2 + h\nu \longrightarrow NO + O \qquad (1)$$

At photon energies corresponding to radiation of 400 nm and less, the increase in vibrational energy accompanying the electronic transition in the NO_2 molecule is so large that the molecule flies apart.

The atomic oxygen formed in reaction (1) is very reactive and takes part in secondary reactions to give more pollutants. Over 99.5 percent of the atomic oxygen reacts with molecular oxygen to form ozone:

$$O_2 + O + M \longrightarrow O_3 + M \qquad (2)$$

In reaction (2), M is a third body, or a surface, which takes up energy released in the reaction. Once some ozone has been formed, it can rapidly oxidize NO to more NO_2:

$$NO + O_3 \longrightarrow NO_2 + O_2 \qquad (3)$$

Reaction (3) provides an additional source of NO_2 for the formation of ozone by reactions (1) and (2) and regenerates the NO_2 used in reaction (1). The net result is a steady concentration of NO_2 and large quantities of O_3 in the air. The buildup of O_3 is tempered by its removal in reaction (3).

Among the other primary pollutants in the air are saturated and unsaturated (multiple bond) hydrocarbons. They are relatively harmless by themselves but can undergo a wide variety of reactions with the ozone, oxygen atoms, and nitrogen oxides formed in the preceding reactions. Most of the organic hydrocarbons in the air result from the incomplete combustion of fuel. The fuel itself is rarely completely oxidized to CO_2 and H_2O, as the equation for the combustion of 2,2,4-trimethylpentane ("isooctane") would imply:

$$2C_8H_{18} + 25O_2 \longrightarrow 16CO_2 + 18H_2O$$

Instead, a variety of organic compounds plus carbon monoxide are released to the air. Other organic compounds, such as solvents, gasoline, and paints, escape into the atmosphere by evaporation prior to oxidation.

The most reactive organic compounds are those with C=C and C≡C multiple bonds, aldehydes,

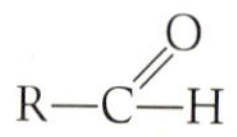

$$R-\overset{\displaystyle O}{\overset{\displaystyle \|}{C}}-H$$

and benzene derivatives. A typical sequence of reactions is shown below, where R again stands for an attached atom or group of atoms; $R = -H, -CH_3,$ $-CH_2CH_3$, and so on.

$$O_3 + RCH=CHR \longrightarrow RC\!\!\overset{\displaystyle O}{\underset{\displaystyle H}{\diagup}} + RO\cdot + H\dot{C}O$$

aldehyde

$$RC\!\!\overset{\displaystyle O}{\underset{\displaystyle H}{\diagup}} + O + h\nu \longrightarrow R\overset{\displaystyle O}{\overset{\displaystyle \|}{C}}\cdot + \cdot OH$$

$$R\overset{\displaystyle O}{\overset{\displaystyle \|}{C}}\cdot + O_2 \longrightarrow R\!-\!\overset{\displaystyle O}{\overset{\displaystyle \|}{C}}\!-\!O\!-\!O\cdot$$

$$R\!-\!\overset{\displaystyle O}{\overset{\displaystyle \|}{C}}\!-\!O\!-\!O\cdot + SO_2 \longrightarrow \text{particulates of indefinite composition}$$

$$R\!-\!\overset{\displaystyle O}{\overset{\displaystyle \|}{C}}\!-\!O\!-\!O\cdot + NO_2 \longrightarrow R\!-\!\overset{\displaystyle O}{\overset{\displaystyle \|}{C}}\!-\!O\!-\!O\!-\!NO_2$$

Therefore, one of the products is

$$CH_3\!-\!\overset{\displaystyle O}{\overset{\displaystyle \|}{C}}\!-\!O\!-\!O\!-\!NO_2$$

peroxyacetylnitrate (PAN), where R = —CH$_3$. The general class of compounds related to PAN is known as peroxyacylnitrates (R = —CH$_3$, —CH$_2$CH$_3$, . . .) and is sometimes collectively known as PAN.

Photochemical smog has characteristic oxidizing properties, as opposed to the reducing character of the SO$_2$ type of air pollution. The oxidant content of photochemical smog includes all oxidizing agents having a greater tendency than molecular oxygen to accept electrons. Ozone is the major component (~90 percent) of photochemical oxidant in smog, with substances like PAN and NO$_2$ making up the remainder. (Nitrogen dioxide is implicated in the oxidation of the pollutant SO$_2$ to SO$_3$, NO$_2$ + SO$_2$ $\longrightarrow$ NO + SO$_3$, in polluted air.)

The variation of pollutant concentration with time of day in a photochemical smog episode is shown in Figure 7.2. Both the amount of human activity injecting primary pollutants (NO, hydrocarbons) into the air and the intensity of solar radiation needed to convert the primary pollutants into secondary pollutants (O$_3$, NO$_2$, aldehydes) vary with the time of day. Photochemical smog may be considered to be the result of a natural process that removes, or scavenges, the primary pollutants from the air. Unfortunately, the products of the natural process are more harmful than the original pollutants. Typical concentrations of pollutants in photochemical smog are listed in Table 7.2.

The role of a major primary pollutant, carbon monoxide, has been ignored in our discussion of photochemical smog. Indeed, until recently CO was believed not to play any part in smog formation. There is now some evidence

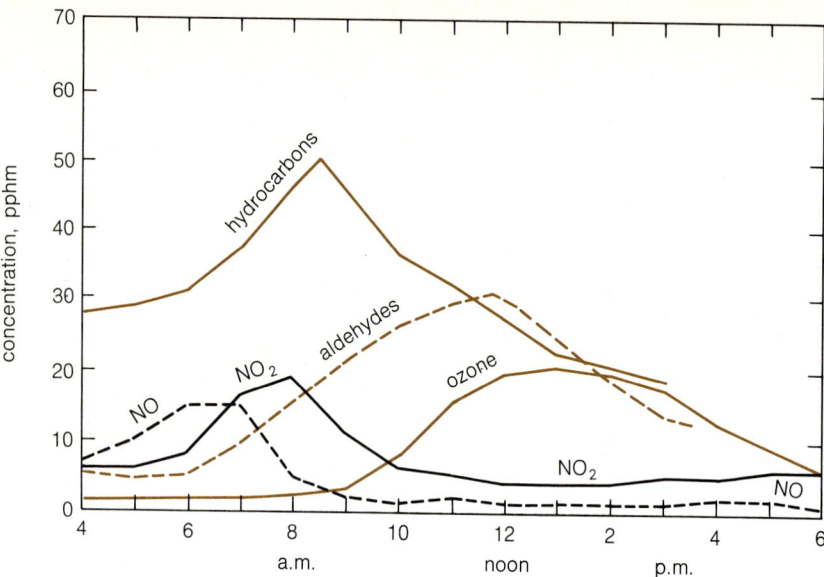

Figure 7.2. Typical concentrations of pollutants during a day of intense smog. Concentrations are given in parts per hundred million (pphm) of air (pphm = ppm × 10^2). The variation of pollutants tends to confirm the mechanism postulated for photochemical smog formation. The initial increase of NO and hydrocarbons is due to morning traffic. This is followed by a decline of NO as it is converted first into NO_2 and then, as solar radiation increases, to ozone. The hydrocarbon concentration falls as more O_3 and NO become available for reaction with unsaturated hydrocarbons to form aldehydes. In the late afternoon photochemical reactions slow, and ozone is removed from the atmosphere by reaction with nitric oxide. (From Philip A. Leighton, "Photochemistry of Air Pollution," Academic, New York, 1961.)

Table 7.2

Typical concentrations in photochemical smog

constituent	concentration, pphm
oxides of nitrogen, NO_x	20
NH_3	2
H_2	50
H_2O	2×10^6
CO	4×10^3
CO_2	4×10^4
O_3	50
CH_4	250
higher saturated hydrocarbons, C_2—	25
C_2H_4 (ethylene)	50
higher unsaturated hydrocarbons, C_3—	25
C_2H_2 (acetylene)	25
C_6H_6 (benzene)	10
aldehydes, RCHO	60
SO_2	20

SOURCE: Richard D. Cadle and Eric R. Allen, "Atmospheric Photochemistry," *Science*, **167,** 246 (1970). Copyright © 1970 by the American Association for the Advancement of Science.

that CO plays a part in the oxidation of NO to NO_2 by way of the reaction sequence

$$CO + \cdot OH \longrightarrow CO_2 + \cdot H$$
$$\cdot H + O_2 + M \longrightarrow HOO \cdot + M$$
$$HOO \cdot + NO \longrightarrow HO \cdot + NO_2$$

The overall reaction for these three steps is, by addition,

$$CO + O_2 + NO \longrightarrow CO_2 + NO_2$$

The symbol M again represents a third body (molecule or surface). The original hydroxyl free radical is believed to be formed in the atmospheric photochemical reaction sequence

$$O_3 + h\nu \ (290–340 \ nm) \longrightarrow O_2 + O$$
$$O + H_2O \longrightarrow 2HO\cdot$$

The presence of CO in polluted air would, according to this scheme, accelerate the formation of NO_2 and, therefore, of O_3 in the air.

Such a series of reactions involving CO (or merely the reaction of CO with $\cdot$OH) could also serve as a removal mechanism, or sink, for the large quantities of CO injected into the atmosphere by human activity and natural sources. As matters stand now, nobody is sure just what happens to all the CO put into the air. Something must be happening to it, because there has been no discernible increase in the average CO concentration in the atmosphere. Some work has suggested that fungi living in soils can act as rapid natural absorbers for large quantities of atmospheric CO. In addition to the possible role of soil microorganisms as a natural sink for CO, it has also been shown that nonbiological chemical reactions in soils may quickly remove huge quantities of SO_2 and NO_2 from the air.

7.5 CLIMATIC AND TOPOGRAPHIC CONDITIONS

For the two kinds of air pollution mentioned so far, it is not enough to have just the primary pollutants and certain other substances present in the air. First the primary pollutants must somehow be contained in an area long enough to react and form the more noxious secondary pollutants. The requisite conditions for this are usually provided by topography and by atmospheric conditions that prevent the dispersal of both primary and secondary pollutants. The possibility of an air pollution episode depends on a breakdown in atmospheric ventilation that would normally dilute and disperse the pollutants. The need for dilution of pollutants is not confined to air pollution. All other kinds of pollution are essentially due to a concentration buildup, or the failure of dilution, of pollutants. Photochemical smog also depends on the amount of sunlight.

A most important atmospheric condition determining the potential for air pollution is the temperature gradient of the atmosphere. We have already noted (Figure 4.1) that the temperature of the atmosphere normally decreases with altitude near the earth's surface. The warming of the air nearer the ground is due to absorption of the earth's infrared radiation by atmospheric CO_2 and H_2O. The rate of change of temperature with altitude is called the environmental or *atmospheric lapse rate*. The actual value of the lapse rate determines whether a parcel of pollutant-carrying air released near the ground will rise to considerable heights and be dispersed and diluted in the atmosphere. First we must consider what happens to a given mass of air as it rises through the atmosphere.

A mass of air (containing pollutants, perhaps) released at the surface of the earth tends to rise so long as its temperature is greater than the surrounding air temperature. If its temperature is less than the temperature of the atmosphere, the air mass will sink. Such behavior is due to the dependence of gas density, and buoyancy, on the temperature of the gas. The density of a gas at a given pressure is inversely proportional to its temperature. However, since atmospheric pressure decreases with altitude, any rising mass of gas will expand. If the rising parcel of gas does not absorb heat from the surrounding atmosphere (adiabatic expansion, Section 16.3), its temperature will drop at a definite rate. For dry air, the temperature of the rising air drops by about 1°C for each 100 m of increase in altitude. This is known as the *adiabatic lapse rate*. It is the comparison between the lapse rate of the atmosphere and the definite adiabatic lapse rate of a rising air mass that determines the ability of the atmosphere to disperse pollutants.

Consider a parcel of dry, polluted air at the earth's surface. The temperature of the parcel is assumed to be equal to or greater than the temperature of the surrounding air at this altitude. As this polluted air mass rises, its behavior will be determined by the prevailing atmospheric lapse rate. If the atmospheric lapse rate is greater than the adiabatic lapse rate of 1°C/100 m, the rising air mass will remain warmer than the (more dense) surroundings, and it will continue to rise. Such behavior leads to vertical mixing and overturning of the atmosphere and to dispersal of pollutants. The atmosphere is said to be unstable.

If the rate of decrease of temperature of the atmosphere is less than 1°C/100 m, the rising air mass will eventually reach an altitude where its temperature is less, and its density greater, than its warmer surroundings. Its rise will slow, and eventually it will begin to sink and return to its starting point. Dispersal of atmospheric pollutants is retarded when the lapse rate is less than 1°C/100 m, and the atmosphere is said to be stable.

An extreme case of a lapse rate less than 1°C/100 m occurs in a *temperature inversion*, where the temperature *increases* with altitude at some level in the lower air layers. The three cases of atmospheric lapse rate versus adiabatic lapse rate are shown in Figure 7.3. During a temperature inversion, there exists an upper layer of air that is warmer than the air layer below. An inversion layer is diagrammed in Figure 7.4. Any rising pollutant-laden gas will most surely be stopped somewhere in the inversion layer, even if the lapse rate below the layer is greater than 1°C/100 m. The bases of such inversion layers occur at relatively low altitudes (0–1600 m). The layer acts as a lid on the rising polluted air, stopping its upward motion. The trapped

Figure 7.3. Conditions for dispersal and trapping of air pollutants. The dashed line represents the constant adiabatic lapse rate of 1°C/100 m for dry air. The solid line is the atmospheric lapse rate for (a) greater than 1°C/100 m, (b) less than 1°C/100 m, and (c) a temperature inversion. Only situation (a) leads to atmospheric instability and effective dispersal of polluted air by vertical mixing. Situation (a) commonly occurs on clear summer days when the earth is being heated. Radiation inversions tend to occur at night as the surface cools and in the winter because of longer nights and decreased solar heating during daylight hours. Other inversions (subsidence inversions) are caused by the sinking of upper air masses and their heating due to compression.

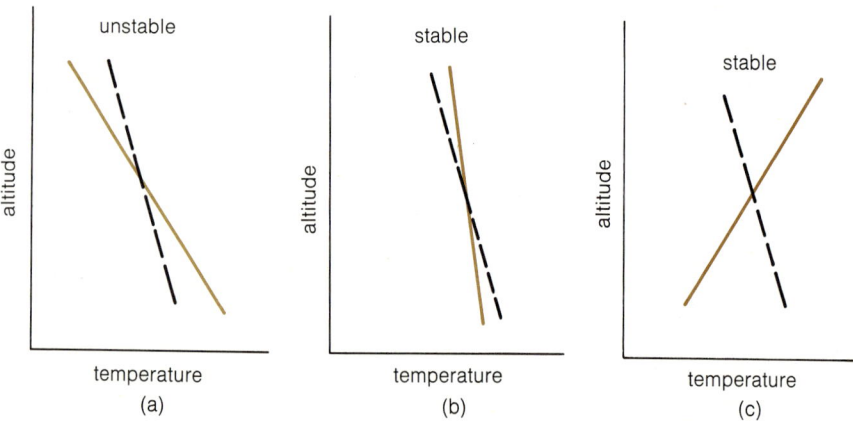

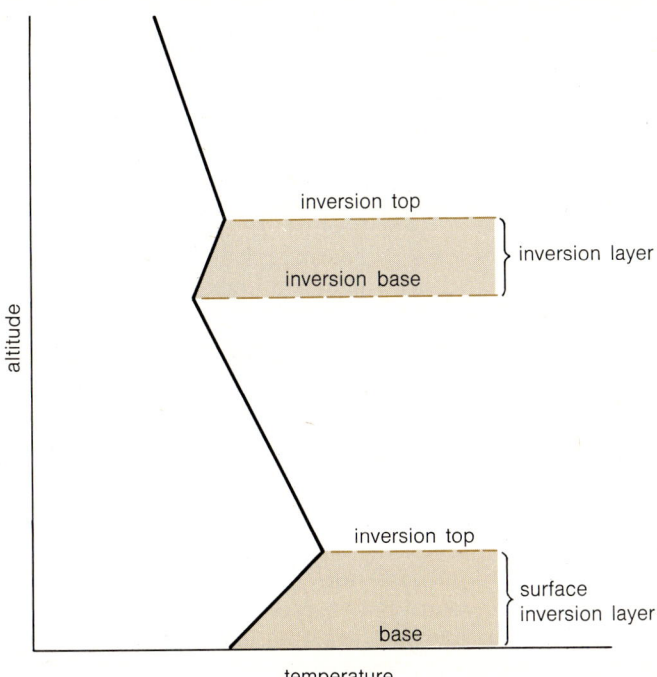

altitude

inversion top

inversion base

inversion layer

inversion top

surface
inversion layer

base

temperature

Figure 7.4. Temperature inversions. Two inversions may exist simultaneously.

pollutants within and under the layer are free to react with each other and with sunlight.

It should be recognized that a temperature inversion is not absolutely necessary for the trapping of pollutants. All that is needed is a lapse rate less than the rate of cooling of a rising, expanding air mass. The presence of an inversion layer only results in having the barrier to the rising polluted air placed at a lower altitude. Topographic features can contribute to the formation of an inversion and to a hindrance of the horizontal convective entry of clean air from the surroundings. Heavier cold air may run down a hill into a valley, displacing the lighter warmer air in the valley upward and leading to an inversion. This is indicated in Figure 7.5 for an area bounded by hills. Again, such distinctive features as hills are not absolutely necessary.

Figure 7.5. Topography and air pollution. When an inversion lies below the hills ringing a valley or basin, both vertical and horizontal dispersion of pollutants is restricted. The inversion seals in the air below it. (Adapted from "Air Pollution and the San Francisco Bay Region," 3rd ed, San Francisco Air Pollution Control District, July 1968.)

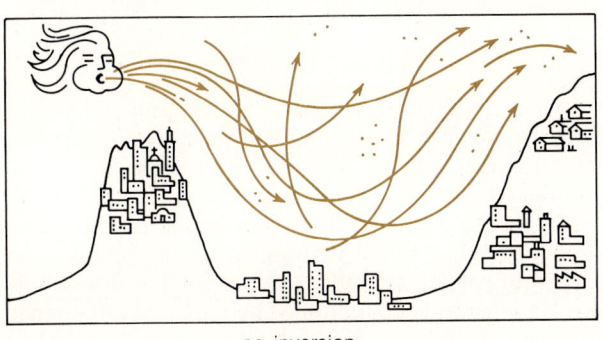

no inversion

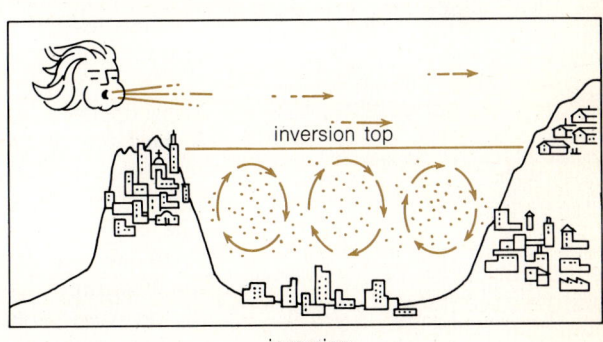

inversion top

inversion

7.6 MAJOR AIR POLLUTANTS AND THEIR SOURCES

There are hundreds of different substances, ranging from asbestos to zinc, that are placed into our air and are classified as air pollutants. They come from just as many sources. Yet it is generally agreed that the major primary air pollutants in terms of the quantities generated are particulates, sulfur dioxide, carbon monoxide, nitric oxide, organic compounds (hydrocarbons), and lead. At any particular place, other pollutants may be more important. Some people would also include carbon dioxide and perhaps water vapor.

The major sources of particulates and sulfur dioxide are industrial operations and fossil-fueled electric power plants. The composition of particulates varies widely and includes sulfuric acid mists, aerosols of organic liquids and solids, metals and metal compounds.

The internal-combustion engine is responsible for most of the carbon monoxide and organic compounds, principally hydrocarbons, placed in the air by man. Both result from the incomplete oxidation of fuel. High-temperature combustion processes in internal combustion engines and in industrial and power-generating plants account for most of the nitrogen oxides, nitric oxide, NO, and nitrogen dioxide, NO_2. The ill-defined mixture is often designated NO_x. Finally, the major source of airborne lead is the automobile engine. The lead comes from the use of organic lead compounds to improve the antiknock properties of gasoline. Each gallon of leaded gasoline contains from 2 to 4 g of lead, as tetraethyl lead, $Pb(C_2H_5)_4$, or tetramethyl lead, $Pb(CH_3)_4$. During the combustion, lead oxide is formed and is reduced to lead metal, which causes pitting of the cylinder walls. Therefore, compounds such as 1,2-dichloroethane,

$$\begin{array}{c} \quad H \quad H \\ \quad | \quad \ | \\ Cl-C-C-Cl \\ \quad | \quad \ | \\ \quad H \quad H \end{array}$$

and 1,2-dibromoethane are added to combine with the lead to form volatile lead halides that leave with the exhaust gases. As a result as much as 75 percent of the lead in the gasoline goes into the atmosphere in the form of solid lead halide salts. These lead compounds make up the bulk of the primary particulate pollutants generated by gasoline sources. The lead aerosol is eventually deposited in water or on land, where it is absorbed into food that humans eat. This represents the primary path for the entry of lead into the body. Amounts taken in by inhalation are small, but only about one-tenth of the ingested lead is absorbed by the body and retained in tissues, compared to at least one-third of the inhaled lead.

7.7 EFFECTS OF AIR POLLUTION

The effects of air pollution on man's environment may be grouped under at least four broad headings: biological, economic, aesthetic, and large-scale ecological or global effects. Included under biological effects are factors involving human health and disease, eye irritation, and damage to vegetation. The costly economic effects of air pollution are associated primarily with corrosion and deterioration of materials and equipment, crop damage, and the cleaning costs resulting from particulate fallout. Aesthetic considerations include discoloration, odor, and visibility reduction. The possibility of change in climate brought about by air pollutants as particulates and CO_2 has already been mentioned in Section 6.10. Many of these effects are interrelated. A biological effect, such as respiratory disease, becomes an economic effect when the hospital bills are added up.

Here we present a brief discussion of each of the adverse effects and the contributing pollutants. Examples of pollutants are generally taken from the previous group of primary pollutants and some of the secondary pollutants generated by them.

Biological Effects No attempt can be made here to cover all of the physiological effects of even the few pollutants selected; the subject is just too large. Effects depend on the concentration of the specific pollutant in the air, length of exposure, other pollutants present, and a whole host of other factors. The effects can be further divided into the faster acting, short-term (acute) effects and the long-term effects caused by prolonged exposure and accumulation of pollutants, effects leading to chronic health problems in man. For further details on the biological and other effects of air pollutants the reference to A. C. Stern (Vol. I) should be consulted.

The principal biological effects of most of the common air pollutants center on their action in the respiratory system of man and animals. Ozone, nitrogen dioxide, and sulfur dioxide and secondary pollutants derived from SO_2 cause respiratory irritation and disease. Generally, the immediate effect of inhalation is swelling of mucous membranes in the respiratory tract, resulting in difficulty in moving air in and out of the lungs or in hindering of absorption of oxygen from air within the lungs. A gas such as sulfur dioxide, which is quite soluble in water, is largely absorbed in the upper (nose and throat) portion of the tract. Little of the gas reaches the lungs unless it is carried further on particulate matter or more rarely as sulfuric acid mist. Ozone and nitrogen dioxide, being less soluble, are able to penetrate to the lungs, where they can cause fluid retention (pulmonary edema) which interferes with the passage of gases from the lung surface into the blood. The result is the same as drowning.

In persons already afflicted with respiratory disease or with heart disease, the resultant respiratory tract irritation or the extra energy expenditure needed for breathing may be fatal. Over a longer period, these and other air pollutants are believed to be contributing factors to the development of chronic respiratory disorders such as bronchitis and emphysema.

Particulates can cause similar respiratory difficulties in the upper airways if the particles are large enough to be stopped by the mucous linings. Smaller particles (0.1–2 μm) are able to penetrate deeply into the lungs. If they are insoluble, they are retained there to cause adverse effects. Particles in combination with pollutant gases can give rise to a synergistic effect, and the effect of the two or more pollutants acting together will be greater than the additive effects of the pollutants taken individually.

Synergism is felt to be the reason for the action of sulfur dioxide when accompanied by particulates. The gas is concentrated on the surface of the particles by a process known as *adsorption*. The gas is probably attracted to the surface of the particle by chemical forces, although adsorption by physical forces (van der Waals type) is also common. As a result, the local concentration of sulfur dioxide is high because the tiny inhaled particles have a large surface area. When the particle with its adsorbed sulfur dioxide lodges in the lung, the gas may leave the particle surface and cause damage to the lung. Synergistic effects such as this are quite common in polluted air, where there are always several pollutants that may interact. There is evidence that CO–NO_x and H_2SO_4 mist–SO_2 are two synergistic combinations. Synergistic effects always should be taken into account when pollutants may interact.

There is some evidence that adsorption of carcinogenic (cancer-producing) pollutants (generally unsaturated aromatic hydrocarbons) on particulates contributes to lung cancer. However, most studies point not to a generalized air pollution, as in urban areas, but to localized events such as cigarette smoking and occupational hazards. An example of an occupational air pollution hazard would be "silo fillers' disease," an acute lung disorder caused by the inhalation of nitrogen dioxide generated by the fermentation of newly stored crops.

If carbon dioxide is excepted, carbon monoxide is by far the major air pollutant in terms of the amount emitted. The toxic effects of carbon monoxide have been known and impressed upon everyone for years. Strangely enough, the presence of carbon monoxide in polluted urban atmospheres has been treated lightly until recently, probably because of the insidious effect of monoxide poisoning and the fact that carbon monoxide, being odorless and colorless and otherwise nonirritating, is not as easily detected by the senses as are ozone, nitrogen dioxide, and sulfur dioxide. The toxic behavior of carbon monoxide on humans results from its great affinity for hemoglobin, the complex iron-containing molecule responsible for the transport of oxygen from the lungs to the tissues through the blood.

The iron of the hemoglobin molecule (Hb) combines with oxygen by covalent bonding according to the equation

$$Hb + O_2 \rightleftharpoons HbO_2$$

The double arrows mean that the reaction can go in either direction, so that the HbO_2 can release its oxygen where it is needed. In the blood, HbO_2 transports oxygen to the tissues, where it is used for cellular respiration. Carbon monoxide has an affinity for hemoglobin that is about 200 times greater than has oxygen, and therefore it displaces oxygen from the hemoglobin–oxygen complex, preventing further combination with oxygen.

$$HbO_2 + CO \rightleftharpoons HbCO + O_2$$

When CO is combined with blood hemoglobin, inhaled oxygen can no longer be carried by the blood to the tissues. Thus, prolonged exposure to carbon monoxide produces a host of distressing symptoms, culminating in death. Nitric oxide has similar effects on hemoglobin. Unlike the other pollutants, carbon monoxide does not accumulate in the body but reaches an equilibrium value determined by the pressure of CO. When the concentration of CO in the air is reduced, the above reaction reverses and the HbO_2 concentration increases.

The nonlethal eye irritation associated with photochemical smog is attributed to the action of the secondary smog constituents, formaldehyde, acrolein ($CH_2{=}CH{-}CHO$), peroxyacetylnitrate (PAN), and related compounds, on the mucous membranes of the eyes.

Little has been said here about the molecular mechanism of the action of toxic air pollutants, because the subject is beyond us at this point. However, carbon monoxide, plus some of the other air pollutants mentioned, may also act to hinder the action of enzymes, complex protein molecules that catalyze metabolic reactions in organisms. By slowing down or stopping one reaction in a chain of reactions, the entire set of necessary reactions will come to a halt, resulting in illness or death for an organism.

Vegetation is more sensitive to most air pollutants than are animals. The role of plants in furnishing food for man and other animals and in keeping the atmospheric oxygen level constant through photosynthesis makes the consequences of their great sensitivity to air pollution very serious. Generally, air pollutants damage the leaf structure of plants, resulting in metabolic imbalance in the plant. The major toxicants for plants are SO_2, O_3, PAN, NO_2, gaseous fluorides, and ethylene, $CH_2=CH_2$, but the oxidants O_3, PAN, and NO_2 are the primary substances causing photochemical smog-type injury to plants. Since plants respond quickly to air pollutants and often develop characteristic symptoms, they have been used for detection and identification of air pollutants.

Finally, increased emission of lead compounds into the air, due principally to the burning of leaded gasolines, is a matter of concern. No other toxic chemical pollutant has accumulated in man to the extent that lead has. Studies have shown that people living in urban areas, where exposure to atmospheric lead is higher, have higher lead concentrations in their blood. Lead is a recognized poison in large enough amounts. Yet there are presently no clinical indications that lead concentrations in urban air are causing harmful effects in humans. The fact that lead is a cumulative poison and that lead levels may not have reached the tolerance limit in humans should not be overlooked. In fact, little is known about the cumulative effects of long-term exposure to the low-level concentrations of many of the numerous pollutants in the air we breathe.

The preceding discussion of the toxic effects of air pollution does not imply that there are not other forms of air pollution that are just as repugnant. Many pollutants not ordinarily considered to be air pollutants spend part of their life—perhaps the most important part as far as their transmission is concerned—in the air. Among these are radioactive fallout, pesticides, and many other substances that, like lead, reach our bodies primarily through our stomachs, not our lungs.

Economic Effects Any action of air pollution resulting in the loss of money or expenditure of money to repair damages is included in economic effects. Thus, economic effects include medical expenses and man-hours of work lost, as well as damage to plants and physical objects. Effects due to corrosion and deterioration are considered briefly and will be confined to the chemical action of pollutants on common materials.

In the chemical attack of pollutants, water vapor in the air is a primary helpmate to the pollutant. Water droplets help to concentrate and dissolve pollutants and provide a medium for the more rapid reactions involving ions rather than neutral atoms and molecules. Solid particulates also may adsorb and concentrate active pollutants.

Some of the more common and expensive effects involve the corrosion of metals. This type of corrosion is produced by gases such as sulfur dioxide, which, with water, can give rise to compounds furnishing hydrogen ions (H^+), notably sulfuric acid. Either by direct reaction or through catalysis by the hydrogen ion, the metal is converted to a compound. This process is especially likely if the metal surface is not protected by a stable surface film or if the film is easily dissolved. Aluminum, for instance, has a tough oxide film, Al_2O_3, which protects it from further attack by air pollutants. (This is unfortunate in view of the number of discarded beverage cans lining our roads.) The tarnishing of silver by hydrogen sulfide in the air is a common effect of air pollution.

Building materials such as marble and mortar are damaged by SO_2–H_2SO_4 air pollutants. Marble is calcium carbonate, $CaCO_3$. The carbonate ion reacts readily with hydrogen ions in the reaction

$$H^+ + CO_3^{2-} \longrightarrow HCO_3^- \xrightarrow{H^+} H_2CO_3 \longrightarrow H_2O + CO_2$$

If the hydrogen ion had been contributed by sulfuric acid, H_2SO_4 ($2H^+ + SO_4^{2-}$ in water), the more soluble calcium sulfate would be the product:

$$2H^+ + SO_4^{2-} + CaCO_3 \longrightarrow CaSO_4 + H_2O + CO_2$$

The volume of the sulfate is greater than that of the carbonate, and the resulting expansion introduces stresses and peeling of surface layers.

The gases NO_2 and CO_2 also react with water to furnish hydrogen ions, which contribute to the dissolution of calcium carbonates.

$$3NO_2 + H_2O \longrightarrow 2H^+ + 2NO_3^- + NO$$

$$CO_2 + H_2O \longrightarrow H^+ + HCO_3^-$$

Other instances of material damage occur in the discoloration of lead-based paints by hydrogen sulfide (through formation of black lead sulfide, PbS) and the deterioration of natural and synthetic textile fibers by sulfur dioxide.

Aesthetic Effects

Aesthetics refers to the appreciation of beauty. With air pollution, the beauty affected is usually natural beauty, and the result is an insult to our senses. The discoloration of the normally blue sky, the blotting out of the horizon by a brown pall, the replacement of natural odors by the stench of air pollutants, and the filth of fallout that settles on surfaces are all aspects of aesthetic deterioration caused by air pollution.

The two significant effects that are further developed here are visibility reduction owing to suspended particulate matter (aerosols) and discoloration attributed to nitrogen dioxide. The amount of visibility reduction depends on a variety of factors, the most important being the size of the suspended particles. High humidity, in combination with particles that can absorb water, will scatter and absorb more light and decrease visibility more than smaller dry particles. Visibility reduction is usually accompanied by a hazy appearance with little contrast between objects.

The brown color of some photochemical smogs is due to a high concentration of nitrogen dioxide, which absorbs visible light broadly between 350–450 nm; the maximum absorption occurs at 400 nm. The transmitted radiation is toward the yellow and gives NO_2 mixtures their characteristic yellow-brown color. The color is usually noted when the NO_2 concentration has reached about 0.25 ppm, at which concentration NO_2 is considered to have reached a level adverse to health.

Large-Scale Effects and
Global Effects

The greenhouse effect of CO_2 is one example of air pollution that may have global ramifications. This is basically a disruption of the carbon cycle in nature. Similar disruptions could possibly occur in the sulfur cycle because of the large quantities of SO_2 emitted into the atmosphere. On a smaller scale, pollutants increase the concentration of airborne condensation nuclei necessary for the formation of fog and some rain. So far only the excess and persistence of fog in urban areas, as contrasted with rural areas, has been attributed to an abundance of particulates, which cause the more persistent type of fog having smaller-sized, less rapidly settling, drops. The possible influence of particulates in reflecting sunlight and cooling the earth has already been mentioned in connection with the greenhouse effect in Section 6.10.

A potential global air pollution problem with very serious consequences for life on earth arises when the effects of supersonic transports (SST's) on the chemistry of the lower stratosphere are considered. These airplanes would fly at 20 km (65,000 ft) in a region of rarefied gases with very little vertical mixing. Combustion products emitted by the engines would remain in the stratosphere for 1–5 yr, compared to several weeks in the troposphere. In addition to the possible effect of ejected water vapor in producing a high-altitude greenhouse effect and also stratospheric clouds, attention has been directed to the effect of combustion products on the ozone concentration in the stratosphere. Originally, it was feared the injection of water vapor into the stratosphere would lead to a reduction of ozone by way of reactions such as

$$H_2O + O \longrightarrow 2HO\cdot$$

and

$$HO\cdot + O_3 \longrightarrow HOO\cdot + O_2$$

Later calculations indicate that nitric oxide (1 ton/hr per plane) may remove ozone 80 times faster than would water vapor by a catalytic reduction cycle:

$$\begin{array}{l} NO + O_3 \longrightarrow NO_2 + O_2 \\ \underline{NO_2 + O \longrightarrow NO + O_2} \\ \quad O_3 + O \longrightarrow 2O_2 \quad \text{(net reaction)} \end{array}$$

The cycle is repeated, so that a small amount of NO can destroy a large quantity of ozone. Based on a fleet of 500 SST's, analysis shows this cycle could be responsible for halving the total ozone content of the stratosphere in a 2-yr period. Such a decrease in the ozone shield would permit lethal ultraviolet radiation below 290 nm in wavelength to reach the earth's surface.

7.8 AIR POLLUTION CONTROL

The average adult male requires about 300 ft³ (8500 liters) of air each day. In the absence of other air pollution control measures, he should continually wear a gas mask if he values his health. And this would protect only his lungs and eyes. Wearing of gas masks is obviously not a welcome solution to air pollution.

Air is a renewable resource, but unlike drinking water it cannot be renewed (purified) at its main point of distribution unless we seriously entertain the idea of enclosing cities in huge air-conditioned domes. Therefore, the only practical way to control air pollutants is to control their emission at the source. The elimination of pollutants at the source can be accomplished in several ways. The most obvious is to prevent the escape of pollutants into the atmosphere. More satisfactory methods would be the substitution or modification of pollution-generating fuels and processes with cleaner counterparts.

In the paragraphs that follow we shall limit our discussion of some air pollution control techniques to the major pollutants already mentioned, that is, SO_2, particulates, NO, CO, and hydrocarbons. These are all primary pollutants. There is no point in trying to control secondary pollutants.

The principal sources of sulfur dioxide and particulates are industrial operations and fossil-fueled electric power plants. Stationary sources like these can use larger and more efficient control procedures than mobile sources. The internal combustion engine in automobiles is the dominant source of pollutants in the mobile source class. It contributes most of the carbon monoxide, nitric oxide, and reactive unsaturated hydrocarbons to our polluted atmosphere.

Particulates

Particulates can be removed from effluents at the source by various simple processes, such as filtration, electrostatic precipitation, collection by means of centrifugal force or gravity, and collection in a liquid (wet scrubbing). Depending on the size of the particulates, one of these methods may be preferable to another. Electrostatic precipitation is one of the most efficient and widely used for trapping particles of less than 2 μm diameter, the ones that travel the farthest in the atmosphere—and into our lungs. In this method, the particle-bearing effluent gas is passed between a pair of electrodes. Usually, one electrode is a wire electrode, and this is surrounded by a large electrically grounded collecting electrode. A high voltage, either positive or negative, applied between the electrodes will cause ionization of the air or gas surrounding the wire electrode. The gas ions will move toward the surrounding collecting electrode and, in the process, transfer their charge to particles with which they collide. Then the charged particles are drawn to the collecting electrode and deposited. Both solid and liquid particles can be removed. The disposal of the collected solid or liquid may bring about a water or solid waste pollution problem if there is no use for the waste material.

Sulfur Dioxide

Several methods are being used, or are under development, for the removal of SO_2. These methods are used by large installations such as utility power plants and are not economical for the smaller industrial or domestic users of fossil-fuel combustion. For the latter, low-sulfur fuels or modified combustion processes must be sought. All of the SO_2 removal processes utilize a chemical reaction of SO_2 with an added reactant. The product of the reaction or reactions can be sold as a by-product if there is a market. Removal and recovery efforts are hampered by the low concentrations of SO_2 in effluent gases. Some of the more promising processes are described in equation form.

1. *Limestone process.* Limestone ($CaCO_3$) is heated in the furnace to form calcium oxide:

$$CaCO_3 \xrightarrow{\Delta} CaO + CO_2$$

The calcium oxide reacts with SO_2 to form solid sulfites and sulfates:

$$CaO + SO_2 \longrightarrow CaSO_3$$

$$2CaO + 2SO_2 + O_2 \longrightarrow 2CaSO_4$$

The solids can be removed from the effluent gas with an electrostatic precipitator. The major problem remaining is the disposal of the solid calcium sulfate.

In another process, molten alkali metal (Na, K) carbonates react with the SO_2 in stack gases to form alkali metal sulfites and sulfates. Reaction of these salts with carbon monoxide and hydrogen produces hydrogen sulfide and regenerates the carbonates. The H_2S may be processed to give salable sulfur or sulfuric acid and the carbonates can be recycled.

2. *Catalytic conversion to SO_3 and H_2SO_4*

$$2SO_2 + O_2 \xrightarrow{V_2O_5} 2SO_3$$

The SO_3 is absorbed in 98 percent H_2SO_4 (oleum) to form fuming sulfuric acid, $H_2S_2O_7$, (pyrosulfuric acid), which reacts with water to give commercially valuable 100 percent H_2SO_4.

$$SO_3 + H_2SO_4 \text{ (98\%)} \longrightarrow H_2S_2O_7(l)$$

$$H_2S_2O_7 + H_2O \longrightarrow 2H_2SO_4(l)(100\%)$$

3. *Metal oxide absorption (alkalized alumina) process.* Sulfur dioxide is absorbed on beads of an alumina (Al_2O_3)-sodium oxide (Na_2O) mixture:

$$4SO_2 + Na_2O + Al_2O_3 \longrightarrow Na_2SO_3 + Al_2(SO_3)_3$$

The sulfites thus formed are treated with an H_2–CO mixture, which regenerates the alumina and sodium oxide and produces H_2S and CO_2.

$$Na_2SO_3 + Al_2(SO_3)_3 + 4H_2 + 8CO \longrightarrow$$
$$Na_2O + Al_2O_3 + 4H_2S + 8CO_2$$

The H_2S is oxidized to sulfur by a series of reactions:

$$2H_2S + O_2 \longrightarrow 2H_2O + 2S$$

$$2H_2S + 3O_2 \longrightarrow 2H_2O + 2SO_2$$

$$2H_2S + SO_2 \longrightarrow 2H_2O + 3S$$

4. *Absorption.* Sulfur dioxide may be absorbed in a solid or liquid and later regenerated as pure SO_2. If there is a market for the by-products (H_2SO_4, S, SO_2) of any of these cleansing processes, they will be much more attractive. If there is no market, then the by-products will constitute another waste disposal problem. The magnitude of the marketing and disposal problem can be seen from the fact that a large (800 MWe) power plant would produce 750 tons of 80 percent sulfuric acid a day (catalytic process) or 180 tons of sulfur per day (alkalized alumina process).

Power plants produce most of the SO_2 effluents, and of this, coal-fired plants contribute almost all of the SO_2 as well as particulates. By the year 2000, electric power consumption is expected to increase by about sevenfold. Although nuclear powered plants may then generate more than half of the electricity required in the United States, fossil-fuel (mostly coal) powered plants will generate about three times the power that they are called upon to produce today. It is obvious that pollution control measures for these SO_2 and particulate emitting sources will be badly needed. A reduction in our use of power for so-called luxury purposes (home air conditioners, power lawn mowers, electric pencil sharpeners) will probably do little to alleviate the problem, because they comprise a very small fraction of total electric power consumption. Because of the pollution tendencies of power generating plants, including nuclear power, there is growing resistance in the United States to the construction of these plants near populated areas where the power is most needed. Considering the need for increased power-generating facilities in the future, some sort of compromise will be required.

Pollutants from Internal Combustion Engines

The internal combustion engine is the largest contributor to air pollution in urban areas; the percentage contribution varies with locality. The exhaust from these engines accounts for about 65 percent of the pollutants emitted. These emissions consist of nitric oxide, carbon monoxide, various hydrocarbons, and lead salts. Assuming, for the present, that we are stuck with the gasoline-powered internal combustion engine, pollution control measures can begin by modifying the engine, modifying the fuel, or preventing the pollutants from being emitted. All three of these avenues of attack are being followed.

Modification of the gasoline fuel will shortly result in the removal of all lead antiknock compounds from gasoline. To retain some of the advantageous combustion properties conferred on gasoline by lead (i.e., the higher octane ratings), the molecular structure of the hydrocarbons in gasoline will have to be modified by additional refining procedures, which may add a few cents a gallon to the price of gasoline. In addition to fuel modification, engine operating characteristics are being altered so that there will be no need for leaded gasolines. Modified low-compression engines also will help to decrease the amount of nitric oxide generated, because lower compression of the air–gasoline mixture means reduced combustion temperatures. The absence of lead is important for two reasons. The most obvious is the complete elimination of lead particulate emissions. The other is that lead interferes with the operation of control devices for the reduction of all the other exhaust emissions.

The lead problem is a good illustration of the kind of vicious circle encountered when we attempt to control one pollutant. The removal of lead from gasolines will probably necessitate the addition of more aromatic (benzene derivatives) hydrocarbons to gasoline to maintain the octane rating. The addition of aromatic hydrocarbons leads to higher concentrations of carcinogenic hydrocarbons, such as 1,2-benzopyrene (Figure 15.2), in auto emissions. Reduced combustion temperatures in low-compression engines also permit the escape of more of the unused aromatic and unsaturated hydrocarbons. These hydrocarbons are not only carcinogenic, but have a greater tendency to form photochemical smog. Yet if lead is not removed, catalytic converters for the control of hydrocarbon emissions will not work. How do we break into the cycle?

Efforts to reduce the amounts of carbon monoxide and hydrocarbons in exhaust gases center on methods to complete the oxidation of these wastes so that they are more fully converted into carbon dioxide and water. Two devices show promise. One, the exhaust-manifold thermal reactor, injects air into the hot exhaust gases as they leave the engine and retains the hot gas mixture long enough for complete oxidation of the carbon monoxide and hydrocarbons to take place. The other device, the catalytic converter, also injects additional air into the exhaust gases but continues oxidation at a lower temperature through the use of a catalyst such as platinum metal or various metal oxides. There are several problems connected with the choice and operation of a suitable catalyst, but the biggest stumbling block is the poisoning of the catalysts by lead. This difficulty will disappear when nonleaded gasolines return to general use.

Nitric oxide emissions can be reduced by decreasing the amount of air in the air–gasoline vapor combustion mixture and by using lower combustion temperatures. One way to effect NO reduction is to recirculate a portion of the exhaust gas back into the combustion chamber. Catalytic reduction of NO to N_2 and O_2 is also a possibility. Such a reduction would amount to a reversal of the reaction whereby NO is formed during the high-temperature combustion process,

$$N_2 + O_2 \rightleftharpoons 2NO$$

High temperature favors the formation of NO, low temperature favors the formation of N_2 and O_2. Unfortunately, the rate of decomposition of NO to N_2 and O_2 is very slow at any temperature in the absence of a catalyst. A possible catalyzed reduction reaction would be

$$2NO + 2CO \xrightarrow{\text{Cu}} N_2 + 2CO_2$$

This NO reduction would have to occur before any oxidation processes used to lower monoxide and hydrocarbon concentrations. Overcoming the difficulties in stopping NO emissions is imperative, because NO is the key substance in the formation of photochemical smog.

As with lead, the decrease of nitrogen oxides (NO_x) is not as straightforward as it seems. Obvious methods, such as reducing the amount of air or lowering the combustion temperature, will lead to incomplete combustion and an increase in carbon monoxide and hydrocarbon emissions. Furthermore, there is evidence that nitrogen oxides are inhibitors of free radical reactions. They act to stop free radical reactions and terminate photochemical processes by removing the radicals $RO\cdot$. On the other hand, NO_x also initiates photochemical reactions leading to smog. There is probably a certain concentration of NO_x where initiation is a maximum and inhibition is a minimum. It is important that NO_x removal devices eliminate 80–90 percent of the NO_x rather than about 50 percent, because intermediate removal may place the NO_x concentration in the atmosphere at this optimum concentration for smog formation.

The examples of lead and NO_x removal illustrate a common problem that arises in pollution abatement and control. Often, the removal of a pollutant from the environment or its replacement by another substance (the replacement of phosphates in detergents by other compounds is an example) leads to serious problems. Very often, a trade-off or compromise is imposed, if not by the chemistry of the situation, then by social or economic factors. As another example, air pollution control devices are expected to decrease the efficiency of operation of internal combustion engines by about 17 percent, resulting in greater fuel consumption.

Despite the efficacy of any of the exhaust emission control devices mentioned here, or of any others that may come along, many observers believe that residents of urban areas will never again breathe smogfree air. No control device can eliminate all the pollutant emissions from the internal combustion engine. Considering the explosive growth in the number of automobiles on the road and our driving habits (if you live a mile or two from work or school, do you drive, walk, or use public transportation?), it seems unlikely that smog control measures can keep up with the growth in sources of emission. The only possibilities are restrictions on the use of automobiles or the development of different power plants (gas turbine, steam engines) or fuels (electricity, liquid gas, etc.). For many, restrictions on the use of automobiles conflict with what we call our way of life.

7.9 MORE ON CATALYSTS AND ADSORPTION

The terms "catalyst" and "adsorption" have been used several times in this chapter. We said that a catalyst is a substance that accelerates a reaction by lowering the activation energy for the reaction. The catalyst emerges unchanged after the reaction. Adsorption is used to describe the concentration increase of a gas (or a liquid, or a dissolved substance) in the boundary region between two states of matter. In the reactions previously described, the adsorption process was always of some gaseous substance on the surface of a solid. The phenomena of catalysis and adsorption are closely connected in the catalysis of a gaseous reaction by a solid surface as, for example, in the catalytic converter for the decrease of carbon monoxide and hydrocarbon emissions.

A solid surface catalyzes a gaseous reaction by adsorbing reactant molecules. The adsorption of a reactant molecule onto the surface may proceed through the formation of a chemical bond between a surface atom and a reactant molecule. The solid surface and the reactant molecule together form an intermediate species that decomposes to give the reaction products. In the presence of the solid, the activation energy needed to form this intermediate is much less than that needed to form a decomposable intermediate between reactant molecules in a homogeneous gas reaction. The solid surface has lowered the activation energy of the reaction.

There are two important ways that the presence of the solid surface increases the ease of formation of the intermediate and the speed of the reaction. The adsorption process may strain and weaken some of the bonds in the reactant molecules, and this may aid the required decomposition of the intermediate. Second, the catalyst will aid the reaction by bringing reactant molecules close together on its surface, where the chance of intermediate formation will be much greater than if the molecules were floating around in the gas.

Because the adsorption of a reactant on a solid catalyst usually involves the formation of a chemical bond, the electronic structures of the surface atom and of the reactant molecule are powerful influences on the degree of adsorption, catalysis, and the course of the reaction. Catalysts are often quite specific; for the same reactants, one catalyst may give an entirely different set of products than another catalyst. The best all-around catalysts for the greatest number of reactions are the transition metals Sc–Ni, Y–Pd, La–Pt, and their oxides. These are all expensive materials. Most of these metals have partially filled d orbitals, which are connected with their catalytic activity. Catalytic poisons are substances that are strongly and preferentially adsorbed on a surface or otherwise block reactant molecules from a surface.

As an example of the possible mechanism of a surface-catalyzed reaction, consider the oxidation of carbon monoxide,

$$2CO + O_2 \longrightarrow 2CO_2$$

Let M stand for an atom of the solid surface. The surface first may adsorb oxygen,

$$-\overset{|}{\underset{|}{M}}-\overset{|}{\underset{|}{M}}- \; + \; O_2 \longrightarrow -\overset{\overset{\textstyle O}{|}}{\underset{|}{M}}-\overset{\overset{\textstyle O}{|}}{\underset{|}{M}}-$$

Then the resulting oxide surface may react with carbon monoxide

$$-\overset{|}{\underset{|}{M}}-O + CO \longrightarrow -\overset{|}{\underset{|}{M}}-O \cdot CO \longrightarrow -\overset{|}{\underset{|}{M}} + CO_2$$

intermediate

The surface-catalyzed oxidation of CO is much faster than the homogeneous gas reaction at the same temperature and reactant concentrations.

There are now catalytic heaters that use a platinum catalyst (coated on asbestos fiber to increase surface area) for the oxidation of hexane, C_6H_{14}. The catalyzed reaction takes place on the surface, and no flame is produced because the combustion occurs at a temperature lower than that required to sustain a flame. No CO is formed, because the catalytic oxidation goes completely to CO_2 and H_2O. The heat produced by the catalytic oxidation is identical to that produced by the usual (assuming complete oxidation) thermally activated combustion.

7.10 OTHER PROBLEMS IN POLLUTION CONTROL

As industry and automobiles and other sources of air pollution proliferate and pump more and more effluent into a finite volume of air, the effects of pollution become more widespread. Air pollution respects no political boundaries; airplane pilots report that, at times, a pall of smog extends a thousand miles east from California. Sweden has experienced damaging acid rains caused by sulfur oxides injected into the atmosphere in England and in Germany's industrial Ruhr Valley. So the fight against air pollution takes on international dimensions. However, the first actions to solve the problems of air pollution must begin on the local level through education of individual voters.

In one respect, the return of clean air to our country is more a matter of economics than it is of technology. At the present level of air pollution, we have the technical resources and knowledge at hand to cleanse our air. But all of the air pollution control devices and measures mentioned here cost money, and our governments and the individual citizen must decide if clean air is worth the cleaning expense. Who will pay for clean air? It is generally conceded that the additional costs of pollution control processes, new or modified fuels, and new power sources will have to be passed on to the consumer. Only if pollution control laws are enacted and enforced will everyone be treated equally. Otherwise, the company that does not practice pollution control will be able to offer its goods for sale at a price lower than its more conscientious competitor. It is up to the citizen to decide first if he is willing to pay the price and then to urge the passage and enforcement of pollution control legislation.

Thus, another aspect of air pollution control is our attitudes and values. When we decide to spend money for pollution control or for the elimination or replacement of sources of pollution, we have made a value judgment—a choice of clean air as opposed to lower prices or convenience. Looked at in this way, the pursuit of clean air is a reflection of personal values or priorities. And polluted air is a personal insult—an insult perpetrated by industries, by governments, and more often by your fellow citizens (including yourself) because it is a violation of a most basic right—the right to breathe clean air.

7.11 SUMMARY

Pollution is defined as a deviation from the natural composition of a part of the environment resulting in adverse effects on life. Pollution is usually brought about by the addition to the environment of waste products of human activity. When the waste products are not efficiently assimilated, decomposed, or otherwise removed by the natural biological and physical processes of the biosphere, adverse effects may result as the pollutants accumulate or are converted to more toxic substances.

Air pollutants may be liquid and solid particulate matter or gases. Primary air pollutants are defined as those substances emitted directly from an identifiable source. Secondary pollutants are substances derived from primary pollutants by chemical reactions.

The two major types of air pollution are due to sulfur oxides derived from combustion of fossil fuels and to photochemical smog resulting from nitrogen oxides and hydrocarbons emitted by internal combustion engines. The older sulfur dioxide type of air pollution is brought about by the conversion of the primary pollutant sulfur dioxide to sulfur trioxide and then to sulfuric acid and sulfate particulates. Atmospheric reactions in the conversion sequence are often speeded up by the presence of catalysts, which provide reaction paths with lower activation energies.

The key reaction in the formation of photochemical smog is the light-induced decomposition of nitrogen dioxide,

$$NO_2 + h\nu \longrightarrow NO + O$$

followed by the formation of ozone and the reaction of ozone or atomic oxygen with hydrocarbons to give a wide variety of products.

Air pollution episodes are brought about when meteorologic conditions prevent vertical mixing in the atmosphere and dispersal of the pollutants. Such conditions occur when the rate of cooling of a rising mass of polluted air is greater than the atmospheric lapse rate.

The six major primary air pollutants are carbon monoxide, sulfur dioxide, hydrocarbons, nitric oxide, particulates, and lead. The effects of air pollutants may be biological, economic, aesthetic, or global. Pollutants may exhibit synergism, when the combined effect of two or more harmful substances is greater than the total effect of the substances considered one at a time.

The control and abatement of air pollution must be based on the reduction of the emission of primary pollutants at their source or the conversion to cleaner fuels or combustion processes. Control technology exists for many air pollutants, particularly those from stationary sources, but the cost will be high.

Questions and Problems

1. Sketch a potential energy diagram for a reaction having a negative activation energy. Explain why it is unlikely that chemical reactions will exhibit such behavior.

2. For toxicological purposes, it is often more convenient to know the weight of a pollutant per unit volume of a gas than its concentration in parts per million. The number of micrograms (μg) per cubic meter (m^3) of air is widely used. If conditions of 25°C and 760 mm Hg are taken, show that conversion from ppm to $\mu g/m^3$ is given by

$$\mu g/m^3 = \frac{10^3 \, (\text{ppm})(\text{molecular weight of pollutant})}{24.5}$$

3. If all the sulfur oxides emitted in 1968 (Table 7.1) were sulfur dioxide and all were converted into calcium sulfate by the limestone process, how many metric tons of calcium sulfate would have had to be disposed of in 1968?

4. Photochemical oxidants are determined as those substances that liberate iodine from a potassium iodide solution according to the reaction

$$O_3 + 2I^- + H_2O \longrightarrow I_2 + O_2 + 2OH^-$$

The iodine released is spectrophotometrically measured. All oxidants are calculated as ozone. If 7500 liters (STP) of air are passed through the iodide solution with the liberation of 6.7×10^{-5} mole of I_2, what is the oxidant concentration (as O_3) in parts per million and in micrograms per cubic meter in the air?

5. The SO_2 content in air may be determined by passing the air through water to produce a solution of sulfurous acid, H_2SO_3. The subsequent reaction of this solution with potassium permanganate, $KMnO_4$, establishes the amount of SO_2 absorbed from the air.

   a. Write the balanced equation for the direction reaction of SO_2 with water to produce H_2SO_3.

   b. Write the balanced equation for the reaction of H_2SO_3 (or SO_3^{2-}) with $KMnO_4$ (or MnO_4^-) to form $MnSO_4$ and H_2SO_4, among other products (see Appendix D).

   c. If 1000 liters of air are passed through water and the resulting solution requires 2.5×10^{-5} mol of $KMnO_4$ for complete reaction (part b) what is the parts per million by volume of SO_2 in the air? Assume, for simplicity, STP conditions.

6. The average concentration of particulate lead in a city with highly polluted air is 5 $\mu g/m^3$. If an adult respires 8500 liters of air daily and if 50 percent of the particles below 1 μm in size are retained in the lungs, calculate how much lead is absorbed in the lungs each day. Of the particulate lead in urban air, 75 percent is less than 1 μm in size.

7. If the atmospheric lapse rate is 2°C/1000 ft, as given in Chapter 4, is the atmosphere typically stable or unstable with respect to mixing of dry air?

8. The atmospheric lapse rate on a given day is 0.60°C/100 m. A parcel of polluted dry air at ground level is 2°C warmer than its surroundings and rises. At what altitude will the upward motion of the polluted air cease?

Answers

3. 63×10^6 metric tons
4. 0.20 ppm
5. c. 1.4 ppm

6. 16 μg
8. 500 m

References

"Cleaning Our Environment—The Chemical Basis for Action," American Chemical Society, Washington, D. C., 1969.
   Section 1 (pp. 21–92) is a comprehensive discussion of all aspects of air pollution, with particular emphasis on control technology. Written for the nonscientist, but with many references to more advanced work.

Faith, W. L., "Air Pollution Control," Wiley, New York, 1959.

Haensel, Vladimir, and Burwell, Robert L., Jr., "Catalysis," *Scientific American*, **225**, 46 (December 1971).
   An introduction to the mechanism of catalysis with many examples, including the Haber process and catalysis in the production of high-octane gasoline.

Junge, C. E., "Air Chemistry and Radioactivity," Academic, New York, 1963.

    An advanced treatment of the reactions of atmospheric components.

Lowry, William P., "The Climate of Cities," *Scientific American*, **217,** 15 (August 1967).

O'Sullivan, Dermot A., "Air Pollution," *Chemical and Engineering News*, 38 (June 8, 1970).

    A good introductory article.

Stern, Arthur C., Ed., "Air Pollution," 2nd ed, Academic, New York, 1968. Vol. I: "Air Pollution and Its Effects"; Vol. II: "Analysis, Monitoring, and Surveying"; Vol. III: "Sources of Air Pollution and Their Control."

    Volumes I and III will be the references of choice for those delving deeper into air pollution.

Stoker, Stephen H., and Seager, Spencer L., "Environmental Chemistry. Air and Water Pollution," Scott, Foresman, Glenview, Ill., 1972.

THREE

Water

Of all substances on earth, the simple triatomic molecule H_2O is one of the most important for all life. This statement applies to the external surroundings as well as to the internal intercellular makeup of living organisms.

The water of the abiotic (nonliving) portion of the biosphere is known as the *hydrosphere*. The hydrosphere includes all water at the surface of the earth in both liquid and solid form: salty seas, fresh-water lakes, groundwaters and rivers, and polar ice caps. Altogether, about 71 percent of the earth's surface is covered by water. The oceans themselves account for about 98 percent of the mass of the hydrosphere. There is much to suggest that the hydrosphere was the birthplace of life on earth. But wherever life originated, it developed in the seas before moving onto the land. Yet today, the waters of the earth produce less life of all kinds than the smaller land area—a factor to be considered by those who look to the oceans for man's ever-increasing food needs.

In its interaction with the atmosphere, the hydrosphere plays a key role in determining climate and weather. Water vapor transferred from the hydrosphere to the atmosphere is the primary connecting link in the processes that move water and energy from one locality to another. Bodies of water can absorb large quantities of heat energy without significant temperature change, thereby stabilizing the temperature of neighboring land areas as well as the temperature of aquatic organisms. Large bodies of water also dissolve, dilute, and dissipate many chemical substances, again stabilizing the external environment of aquatic organisms. However, there is a limit to the amount of heat and chemicals that man can expect the hydrosphere to accept without harmful effects to life.

Within living organisms, water makes up 80–90 percent of the weight of the cell, which is the fundamental unit of life. Thus, one could consider a living organism to be a system designed around water. Almost all the chemistry of biological systems occurs in aqueous solution, with water serving various functions: as an inert medium for vital chemical reactions, as a participant in reactions, or as a transporter of dissolved substances or heat.

We might infer that this simple compound of hydrogen and oxygen is important to life solely on the basis of its abundance in the biosphere and in living organisms. But even more significant is the remarkable fitness of water for the conditions of life within and without the living organism. No other substance even approaches water in the appropriateness of its properties to living organisms and their interactions with their surroundings. Water is not only fit but also unique to all conceivable forms of life in the universe. It shows, more than any other substance, how the fitness of the environment is determined by the properties of its constituents.

All of the unique properties of water are due to its structure, which is in turn the result of its bonding, which itself is predetermined by the electronic structures of the hydrogen and oxygen atoms. In this part of the book, we shall see how these factors determine the behavior of water in its three physical states. We shall compare the properties of water to those of other liquids and, in doing so, define some of the general characteristics of the liquid state. Finally, we shall consider some aspects of water, both clean and polluted, in our environment.

Physical Properties of Water

The same law that shapes the earth-star shapes the snow-star. As surely as the petals of a flower are fixed, each of these countless snow-stars comes whirling to earth, pronouncing thus, with emphasis, the number six. Order, ...

And they all sing, melting as they sing of the mysteries of the number six—six, six, six. He takes up the waters of the sea in his hand, leaving the salt; He disperses it in mist through the skies; He recollects and sprinkles it like grain in six-rayed snowy stars over the earth, there to lie till He dissolves its bonds again.

Henry David Thoreau—
Journal, January 5, 1856

Substances are characterized and distinguished from one another by their properties. The many properties of matter can be divided into the two broad categories of *physical* and *chemical properties.* Physical properties are those such as density, odor, color, hardness, melting point, and a host of others that occur without any change of the substance from one distinct chemical compound to another, that is, with no change in composition. Density, or mass per unit volume, is obviously a physical property, since no change in composition is involved in its measurement. Likewise, the temperature at which a substance melts is a physical property so long as the molecules remain intact during the melting process. A number of the physical properties are often more narrowly classified as *thermal properties,* those associated with the addition of heat to the substance. Melting and boiling points are among the thermal properties of substances.

Chemical properties are characteristics exhibited by a substance when it undergoes a change in composition. In general, chemical properties are associated with the making and breaking of covalent and ionic bonds. The temperature at which a compound decomposes into its elements or simpler compounds is a chemical property, as is the reactivity of a substance with other substances under specified conditions. Thus, a chemical property distinguishing gold from copper is the inertness of gold to nitric acid. Often, a clear delineation between chemical and physical properties is not easy. How much of a substance will dissolve in a specified volume of water, its solubility, may or may not be a physical property, depending on whether or not chemical bonds are broken or formed in the solution process. In this chapter, we shall be concerned with some physical properties of water that contribute to the role water plays in life.

8.1 THE STRUCTURE OF THE WATER MOLECULE

The structure and bonding in the water molecule were considered briefly in Section 3.12. The nonlinear molecule has an H—O—H bond angle of 104.5°. The O—H bond distance is 0.96 Å. The departure of the H—O—H bond angle from the 90° expected from pure $H(1s)$—$O(2p)$ overlap is attributed to (1) partial hybridization of the oxygen $2s$ and $2p$ orbitals to give a near sp^3 or tetrahedral hybridization or (2) repulsion among the four electron pairs, both the two shared and the two unshared pairs, around the oxygen atom, so that the four electron pairs are as far apart from one another as possible. With four electron pairs, this arrangement should also approach the tetrahedral configuration. The water molecule has the configuration shown in Figure 8.1. The charge clouds around the two unshared electron pairs in the oxygen octet are directed at opposite corners of one face of the cube.

The consequences of the structure and electronic charge distribution in the water molecule are far-reaching. In the first place, even if the unshared pairs are ingnored, we have seen that the water molecule has a dipole moment as a result of its bond moments and its angular structure (Figure 4.18). The dipole moment of H_2O is further increased by the extended charge cloud of the lone pair electrons, in a manner similar to that noted in the NH_3 molecule (Figure 5.6). The moments due to the lone pair clouds reinforce the H—O bond moments.

Another consequence of the charge distribution about the oxygen atom is that the charge clouds of the two lone pairs present a negative charge in their specific directions of extension. As a result, the partial negative charge associated with the oxygen atom attracts, at most, two positive entities along the tetrahedral directions associated with the lone pair clouds. This directional feature is pertinent when it comes to relating the structures of liquid water and ice to several of their properties.

8.2 THE HYDROGEN BOND

Hydrogen bonds have been mentioned very briefly before—in the discussion of intermolecular forces (Section 4.23) and in connection with ammonia (Section 5.10). It is now time to consider the hydrogen bond in more detail, because its occurrence in water determines, more than anything else, the unique fitness and unusual, even anomalous, properties of this common

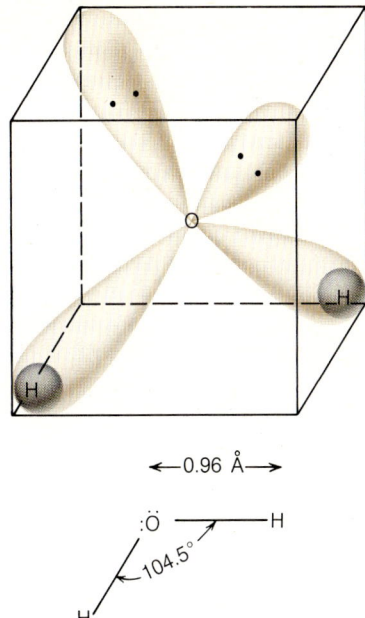

Figure 8.1. Structure of the water molecule. The charge clouds of the unshared electron pairs of oxygen have decided directional character.

←—0.96 Å—→

substance. In fact, were it not for the hydrogen bond, water would not exist as a liquid at earth temperature, and life would be unknown on this planet.

In a hydrogen bond, a single hydrogen atom forms a bridge between two electronegative atoms. The bridged atoms may be on the same or on different molecules. Water and other molecules having a hydrogen atom bonded to a small, highly electronegative atom with unshared electron pairs, such as F, O, and N, fulfill the requirements for participating in the formation of hydrogen bonds between molecules. Examples are

$$\begin{array}{cc} \overset{\delta+}{H}\!-\!\overset{\delta-}{\ddot{O}}\!:\text{---}\overset{\delta+}{H}\!-\!\overset{\delta-}{\ddot{O}}: \\ \underset{\delta+}{|} \qquad \underset{\delta+}{|} \\ H_{\delta+} \qquad H_{\delta+} \end{array}$$

$$\overset{\delta+}{H}\!-\!\overset{\delta-}{F}\text{---}\overset{\delta+}{H}\!-\!\overset{\delta-}{F}$$

$$R\!-\!C\!\!\begin{array}{c} \overset{\delta-}{O}\text{---}\overset{\delta+}{H}\!-\!\overset{\delta-}{O} \\ \diagdown \quad \diagup \\ \underset{\delta-}{O}\!-\!\overset{\delta+}{H}\text{---}\underset{\delta-}{O} \end{array}\!\!C\!-\!R$$

The symbol δ indicates a partial charge on atoms due to bond polarity. The hydrogen bond is denoted by a dashed line to indicate that it is not the same as a regular covalent bond.

The hydrogen bond need not bridge identical molecules. This was shown in Section 5.10 by the interaction of ammonia with water. The hydrogen bonding that is important in determining the structure of proteins (Section 15.14) illustrates bonding between nonidentical groups:

$$C\!=\!\overset{\delta-}{O}\text{---}\overset{\delta+}{H}\!-\!N$$

$$C\!-\!\overset{\delta+}{O}\!-\!H\text{---}\overset{\delta-}{O}\!=\!C$$

$$N\text{---}\overset{\delta-}{H}\!-\!\overset{\delta+}{O}\!-\!C$$

$$-N\text{---}\overset{\delta-}{H}\!-\!\overset{\delta+}{N}$$

In some instances, hydrogen bonds may form between carbon and oxygen or nitrogen, but only when carbon is attached to a highly electronegative atom, as in

$$H\!-\!C\!\equiv\!N\text{---}H\!-\!C\!\equiv\!N$$

Also, Cl and S may sometimes be at one end of a hydrogen bond if the other end is occupied by an electronegative O, N, or F atom.

We can show that the hydrogen bond is not the same as an ordinary electron pair covalent bond by comparing the bond energies for hydrogen bonds and normal covalent bonds. As we noted, the bond energy is the average energy required to break a particular bond and separate the bound atoms into gaseous fragments. For example, the bond energy of the covalent O—H bonds in water is one-half the heat of the reaction of

$$H_2O(g) \longrightarrow 2H(g) + O(g)$$

The heat of this reaction is 220 kcal/mol of water molecules. Thus, the bond energy of the O—H bond is about 110 kcal/mol of O—H bonds.

Most covalent bond energies are quite high—in the range of 50–100 kcal/mol of bonds. On the other hand, experimental values for the energies of hydrogen bonds are lower, about 2–8 kcal/mol of O—H bonds. Hydrogen bonds are therefore weaker by a factor of 10 or so than conventional bonds, either covalent or ionic. They are, however, stronger than van der Waals forces (~0.6 kcal/mol) by a factor of about 10. (Table 8.1 lists the energies of various hydrogen bonds.) The hydrogen bond is essentially electrostatic in nature—it is a dipole–dipole type of interaction. The size of the electronegative atom affects the electrostatic interaction and the hydrogen bond energy. The smaller this atom, the closer one hydrogen bonded molecule can approach the small positive hydrogen of the other molecule, and the more intense the negative charge density of the electronegative atom. As a result, the strongest hydrogen bonds are formed in HF.

Table 8.1
Energies and lengths of hydrogen bonds

bond		energy, kcal/mol	approximate bond length, Å
O—H---O		3–7	2.7
F—H---F	in $(HF)_6$	6–7	2.26
O—H---N		4–7	—
N—H---O		3–4	—
N—H---N	(in NH_3)	1.3	3.38
N—H---N	(in other cases)	3–5	—
N—H---F		5	—
C—H---N	in $(HCN)_2$	3.2	—
C—H---O		2.6	—

Experimental studies of bond lengths provide a further indication that the hydrogen bond is different from a conventional bond. Using water as our example again, the distance between covalently bonded oxygen and hydrogen atoms in the water molecule in ice is 0.99 Å. The distance between the hydrogen atom of one water molecule and the hydrogen bonded oxygen atom of an adjacent water molecule is 1.77 Å, so that the hydrogen atom is unsymmetrically situated and each water molecule retains its identity.

The discussion so far seems to imply that a hydrogen bond may involve only two molecules. But there are other possibilities. In the first place, hydrogen bonds can connect many molecules. Such an arrangement occurs in water and ice and is shown in Figure 8.2. Furthermore, hydrogen bonds can form intramolecularly, that is, a hydrogen atom bridges two atoms in the same molecule. Intramolecular hydrogen bonding occurs in protein molecules.

8.3 STRUCTURE AND PROPERTIES OF WATER AND ICE

The unique physical properties of water that will concern us are all related to the structure of liquid and solid water in bulk. We must therefore move from a consideration of the bonding within a single water molecule or among

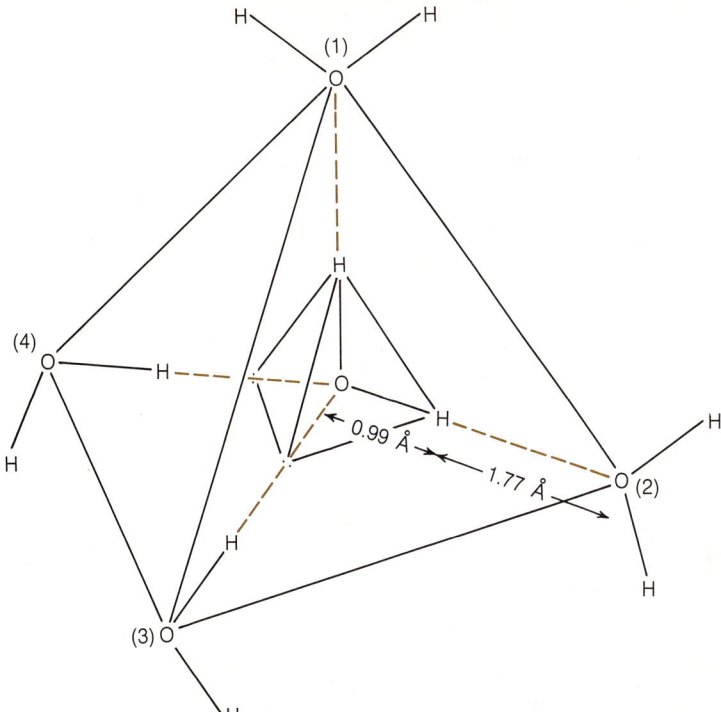

Figure 8.2. Tetrahedral arrangement of water molecules in ice. Hydrogen bonds are shown as dashed lines and are directed tetrahedrally in accordance with the direction of extension of the lone electron pair clouds. (Adapted from J. T. Edsall and J. Wyman, "Biophysical Chemistry," Academic, New York, 1958.)

a few molecules to the arrangement of a large number of water molecules in ice and water.

The structure of ice, as determined by X-ray diffraction experiments, is shown in Figures 8.2 and 8.3. Both figures show the positioning of the oxygen atom of one water molecule at the center of a tetrahedron. A central oxygen is surrounded tetrahedrally by four other oxygen atoms at a distance of 2.76 Å; any one oxygen atom takes part in four hydrogen bonds. When the type of arrangement in Figure 8.2 is extended over many molecules, the structure

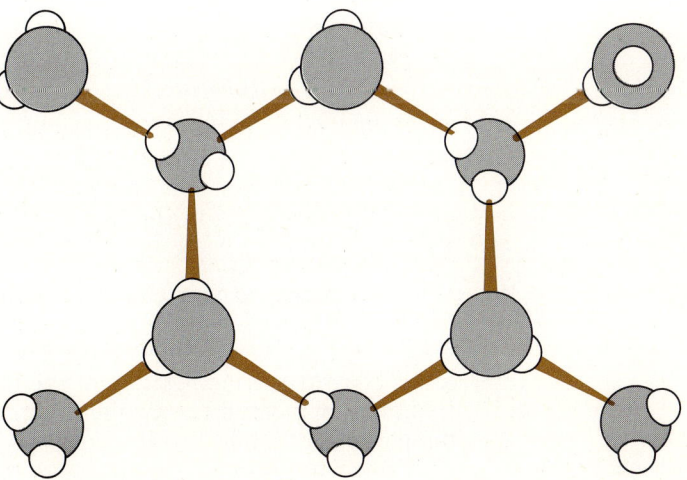

Figure 8.3. Structure of ice. This extended view shows the layer structure of ice and the hexagonal channels that account for the low density of ice relative to water. The hexagonal arrangement of water molecules is reflected in the six-sided shape of snow flakes.

in Figure 8.3 results. Such a regular array or *lattice* of molecules or atoms is characteristic of most substances in the solid state; such solids are called *crystalline*. Figure 8.3 also reveals the porous or open nature of the ice structure, which contributes to the low density of ice in comparison with liquid water. In effect, we might imagine that each water molecule in a piece of ice is attached by hydrogen bonds to all its closest neighbors and thus becomes part of one huge molecule.

When ice melts, only about 15 percent of the hydrogen bonds are broken. As the temperature of the water rises, more and more hydrogen bonds are broken because of their low bond energy. In water vapor or steam, almost all the hydrogen bonds are broken, and gaseous water consists mostly of free single molecules. Even at the boiling point, however, there is evidence that liquid water retains a significant number of hydrogen bonds between water molecules. In liquid water at any temperature between the melting (or freezing) and boiling point, there exist hydrogen bonded portions of liquid. The colder the water, the more extensive the arrays of water molecules connected by hydrogen bonds in an open icelike lattice structure. Interspersed among the three-dimensional rafts of hydrogen bonded molecules are free water molecules not involved in hydrogen bonding, although they may still be subject to other electrostatic interactions with other molecules. Within the hydrogen bonded portions of the liquid, the hydrogen bonds are continually breaking and reforming.

The porous three-dimensional structure of the ice crystal, the gradual breakdown of structure by melting, and the extra energy required to break the comparatively strong hydrogen bonds in liquid water are factors that are responsible for the many anomalous physical properties of water. Let's define and examine these properties of water and their relation to the fitness of the environment. These properties are

density of ice and water

heat capacity
heat of vaporization
heat of fusion $\left.\right\}$ thermal properties
boiling point
freezing point

surface tension
viscosity
solvent properties

All of these properties are affected in one way or another by hydrogen bonding. Solvent properties are also affected by the polar nature of the water molecule, exclusive of hydrogen bonding, and can involve chemical changes, depending on what is dissolving in the water.

8.4 DENSITY OF ICE AND WATER

When ice melts, its open crystalline structure begins to collapse as hydrogen bonds are ruptured. The resulting water at 0°C has a more compact structure than ice at the same temperature. Thus, water at 0°C is more dense than ice at 0°C. As the temperature of the liquid water is increased by the addition of more thermal energy, the breaking of hydrogen bonds continues, and the portions of liquid retaining the open, icelike structure begin to decrease in size while the amount of more tightly packed nonhydrogen bonded water increases. We therefore expect the density of water to increase even more as the temperature goes above 0°C, and this is what occurs until a maximum density of 0.999973 g/cm³ or 1.00000 g/ml (for pure water) is reached at 3.98°C. Thereafter, the density of water begins to decrease. The dependence of density on temperature is presented in Figure 8.4. The decrease in density above 3.98°C probably arises from the usual thermal expansion induced by increased molecular agitation, a phenomenon that occurs in any substance. Up to 3.98°C, this normal expansion must be overshadowed by the breakdown of the open, icelike structure.

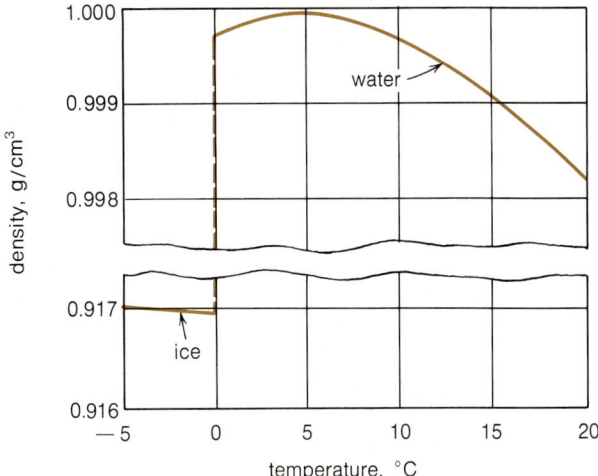

Figure 8.4. Density of distilled water. Density maxima also occur with fresh water and dilute sea water. The presence of dissolved ionic compounds (salts) causes the temperature of maximum density (∼ − 3.5°C) to be below the freezing point (∼ 1.9°C), so that the density of normal sea water increases continually with decreasing temperature.

Most solids are denser than their liquids. Aside from ice, there are only a few solids bearing a similar density relation to their liquid form. Bismuth metal is one of them. None of these solids is in any way related to ice or water, nor could they possibly take the place of water in nature. Among related substances, only ammonia, NH_3, and hydrogen fluoride, HF, form strong hydrogen bonds. Their molecules can associate with each other by hydrogen bonding, but only in chains or rings and not in the open three-dimensional networks that give ice and water their unique properties.

Because the density of ice is less than that of water at the freezing point, 0°C, and because the density of water has a maximum at about 4°C, freshwater lakes in temperate climates do not freeze from the bottom up. In the autumn, as the temperature of the surface waters of lakes and ponds cools toward 4°C, the water sinks to the bottom and displaces the warmer bottom water to the surface. Eventually, the top and bottom waters are at the same density (at 4°C), and this circulation stops. As the surface water cools from 4°C to the freezing point, its density decreases relative to the underlying water, so it remains at the top and freezes. The top layer of ice insulates the underlying water from heat loss to the atmosphere, thereby slowing down the further freezing of the body of water.

Imagine what would happen if ice were denser than water. Ice formed at the surface would sink to the bottom; more ice would form at the top and sink, until eventually the entire body of water would be frozen solid at the end of a cold winter—or at least it would have a thick layer of ice at the bottom. Not only would this kill fish and other aquatic organisms by encasing them in ice, but the lakes, ponds, and rivers would rarely thaw by the end of summer. This late thawing would result from the comparatively low thermal conductivity of water, which would mean that heat would be poorly transferred from the top water layer to the ice beneath.

The sinking of the denser 4°C water and the resultant displacement of warmer bottom layers to the top produces convection currents that help to bring bottom water rich in nutrients (from decay at the bottom) to the top and move oxygen-rich surface water to the deoxygenated bottom. This overturn occurs in the fall.

The expansion of water on freezing has significant implications in the weathering of rocks on the earth's surface. The shattering of rock by frozen water aids erosion, one of the primary agents for soil formation and the circulation of the nutrient elements of the sedimentary cycles.

8.5 HEAT CAPACITY OF WATER

We begin our discussion of the thermal characteristics of water and other substances with a consideration of the amount of heat required to increase the temperature of a specified amount of the substance by 1°C. This quantity of heat or thermal energy is called the *heat capacity*. If the specified quantity of material is 1 g, the heat capacity is commonly called the *specific heat*. If 1 mol (1 g-mol. wt.) is taken, we have the molar heat capacity. The most common unit of heat quantity is the *calorie* (cal), the amount of heat required to raise the temperature of 1 g of water from 14.5°C to 15.5°C. The calorie is sometimes called the "gram calorie" or the "small calorie" to distinguish it from the *kilocalorie* (1 kcal = 1000 cal), which is also called the "kilogram calorie" or the "large calorie." The calorie used by weight-conscious people is really a kilocalorie.

The specific heat of water is therefore 1 cal/g °C. Strictly speaking, this value applies only in the temperature interval 14.5–15.5°C, but for most purposes we can take the specific heat of liquid water as constant over the range of 0–100°C. Similarly, we can assume that the specific heat or the molar heat capacity of most pure substances is constant over a reasonable temperature range, provided that there are no changes of physical state (melting, vaporization, etc.) within the temperature range. Table 8.2 lists the specific heats of some representative gases, liquids, and solids. Water has a high specific heat relative to most other substances, so a large amount of heat is needed to cause an appreciable increase in the temperature of water. Ammonia is the only liquid that has a higher specific heat.

The heat capacity of a substance is related to the ease with which energy is absorbed by the molecules of that material. We have seen that molecules can absorb energy as translational, rotational, vibrational, and electronic energy. The more ways the molecules of a substance can absorb energy while undergoing a temperature change of 1°C, the higher is the heat capacity of the substance. Liquid water, in addition to having the usual intramolecular degrees of freedom, has extra vibrations associated with the weak intermolecular hydrogen bonds. Because of the weakness of these bonds and the resultant close spacing of the vibrational energy levels, energy is easily absorbed as significant populations of water molecules are promoted to excited vibra-

Table 8.2
Specific heat of substances
(at constant volume), cal/g °C

substance	
aluminum (s) (20°C)	0.21
ammonia (l)	1.23
copper (s)	0.093
ethyl alcohol (l)	0.58
glass (s)	0.12
hydrogen, $H_2(g)$	2.4
ice	0.48
mercury. (l)	0.033
oxygen (g)	0.16
steel (s)	0.11
water (l)	1.00
water vapor (g) (300°K)	0.471

tional states at relatively low temperatures. At the extreme, absorption of such vibrational energy results in the breaking of the hydrogen bonds.

Temperature is one of the most critical factors affecting life on earth. Living organisms must be protected from wild fluctuations of temperature, particularly their internal temperature. We tend to forget—since man seems to be so adaptable to temperature extremes in his physical environment, at least for short times—that most other organisms are much less adaptable. Many organisms can survive only within a narrow range of environmental temperatures. Temperature changes in the physical environment not only directly affect the biological process in these organisms, but they also bring about changes in the environment that adversely affect the organism. The high heat capacity of water tends to minimize temperature fluctuations. This minimization is especially evident within bodies of water, so fish and other aquatic animals rarely have to tolerate large temperature changes. The presence of large bodies of water also moderates the climate of surrounding land areas. The large heat capacity of water contributes to the transfer of energy from one locality to another by way of ocean currents and through the evaporation and subsequent condensation of water.

Water, a predominant substance in organic tissues, absorbs heat and helps to regulate the temperature of living things. Such temperature regulation is critical, because the chemical processes of life evolve substantial quantities of energy, much of it given off as heat. Without the thermostating action of cellular water in absorbing some of this heat with minimum temperature change, the internal temperature of the organism would rise. One result of a radical internal temperature change might be the breakdown (denaturation) of the highly organized proteins in living systems. We shall return to the consequences of environmental temperature increase when thermal pollution of natural waters in considered in Section 12.13.

8.6 HEAT OF VAPORIZATION

The vaporization of any liquid is an endothermic process. The quantity of heat absorbed when a given amount of a liquid is converted from the liquid state to the vapor state at a given temperature is the heat of vaporization, ΔH_{vap}. The heat of vaporization is usually given in calories (or kilocalories) per gram or per mole of liquid vaporized. For water, for instance, the heat of vaporization is the energy required for the following transformation at constant temperature and pressure:

$$H_2O(l,\ T,\ P) \longrightarrow H_2O(g,\ T,\ P) \qquad \Delta H_{vap}$$

Vaporization means that molecules in the liquid state are escaping from the liquid and going into the vapor or gaseous state. This process takes place at any temperature where a liquid can exist, so the temperature T in this equation is not restricted to the temperature we commonly call the boiling point, namely 100°C for water under a pressure of 1 atm.

At 100°C and 1 atm pressure, we have the thermochemical equation

$$H_2O(l,\ 100°C,\ 1\ atm) \rightleftharpoons H_2O(g,\ 100°C,\ 1\ atm)$$

$$\Delta H_{vap} = 540\ cal/g$$

$$= 9.72\ kcal/mol$$

The double arrows signify that this transformation can occur in either direction. The reverse of this equation corresponds to condensation, in which case the heat of condensation ΔH_{cond} is -540 cal/g, indicating that heat is evolved. The magnitudes of ΔH_{vap} and ΔH_{cond} of water are unusually high compared to the values for other liquids. Some values for heats of vaporization are presented in Table 8.3.

We can understand the anomalously high value for the heat of vaporization of water by looking at the process of vaporization on the molecular level. In the liquid state, the molecules are close enough together for a variety of intermolecular attractive forces to operate. For a molecule to escape through the surface of a liquid into the vapor phase, where the molecules are relatively far apart and attractive forces are not very effective, energy must be supplied to do the work of pulling the molecules apart. The thermal energy required at constant temperature to convert 1 mol of liquid to 1 mol of vapor is determined partially by the strength of the attractive forces among the molecules in the liquid state. The stronger the intermolecular forces, the greater the heat of vaporization will be and vice versa. Table 8.3 shows that water and other strongly hydrogen bonded liquids have high heats of vaporization, with water having the highest value of all.

The heat absorbed (or evolved) in transitions of state is often called a *latent* heat, because it is stored in the molecules until the reverse change of state occurs, when the "latent" heat is recovered. Ordinarily, the addition of heat to a substance increases its temperature, because the average kinetic energy of the atoms or molecules is increased. However, when there is a change of state as, for instance, at the boiling point or melting point, it is found that there is no temperature change until all of the more condensed state is gone. In vaporization of water, this means that at 1 atm pressure there will be no increase of temperature above 100°C for a water–steam system, even though heat is continually added, until all the water has been converted into steam. After this, the temperature of the steam will increase as heat is added. Another example of change of state without temperature increase is the melting of ice. If temperature is plotted versus time as heat is gradually added to ice at a constant rate and then to water, we obtain a heating plot

Table 8.3
Heats of vaporization of liquids

substance	formula	ΔH_{vap}, cal/g
water, 0°C		595.9
water, 20°C	$H{\diagup}^{O-H}$	584.9
water, 100°C, normal b.p.[a]		540.0
methyl alcohol, 64.7°C, normal b.p.	$\begin{matrix} H \\ \vert \\ H-C-OH \\ \vert \\ H \end{matrix}$	262.8
ethyl alcohol, 78.3°C, normal b.p.	$\begin{matrix} H \\ \vert \\ H_3C-C-OH \\ \vert \\ H \end{matrix}$	204
chloroform, 61°C, normal b.p.	$\begin{matrix} Cl \\ \vert \\ H-C-Cl \\ \vert \\ Cl \end{matrix}$	59
carbon tetrachloride, 76.7°C, normal b.p.	$\begin{matrix} Cl \quad Cl \\ \diagdown C \diagup \\ \diagup \quad \diagdown \\ Cl \quad Cl \end{matrix}$	46
hexane, 68.7°C, normal b.p.	$\begin{matrix} H \ H \ H \ H \\ \vert \ \vert \ \vert \ \vert \\ H_3C-C-C-C-C-CH_3 \\ \vert \ \vert \ \vert \ \vert \\ H \ H \ H \ H \end{matrix}$	79
ammonia (l), −33.6°C, normal b.p.	NH_3	327.4
neon (l), −246°C	Ne	21

[a] The normal boiling point is the temperature at which the liquid boils when subjected to a pressure of 1 atm.

similar to Figure 8.5. The horizontal portions indicate the melting (or freezing) point and boiling point of water under the experimental condition of 1 atm pressure. Experimental plots like this are widely used to determine the melting and boiling points of substances.

At temperatures other than the transition temperatures, added thermal energy appears as increased kinetic energy (translational, rotational, some vibrational) of the molecules. Because temperature is a measure of the average molecular kinetic energy, the temperature increases. At the transition temperatures, however, a portion of the added thermal energy must be used to overcome intermolecular attractive forces. This energy appears as potential energy (some vibrational, some electronic). Another portion of the added energy goes to increase the kinetic energy of molecules in the liquid. Ordinarily, this would result in an increase of temperature, but those molecules having higher kinetic energies are just those that escape from the liquid into the vapor. They carry their excess kinetic energy with them; the average kinetic energy of the molecules remaining in the liquid does not increase, and the temperature remains constant.

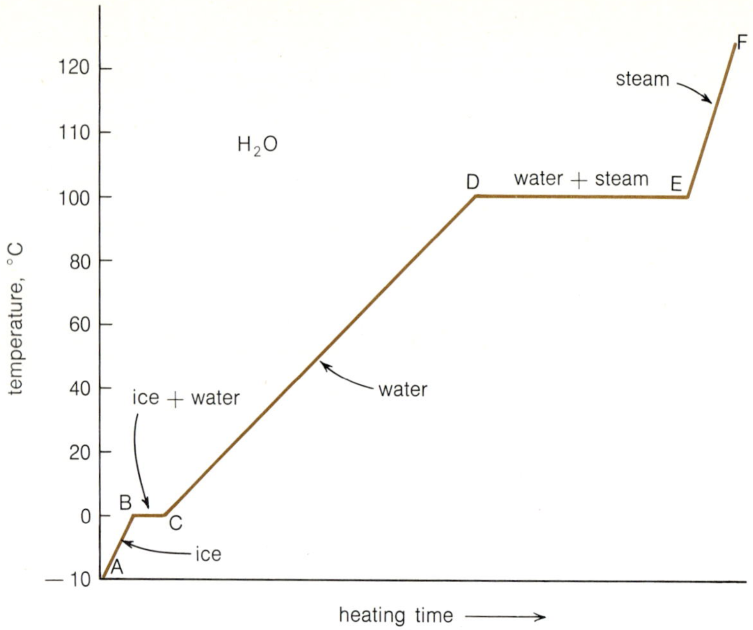

Figure 8.5 Heating curve for water. Heat is added at a constant rate. The rate of climb (slope) of the rising portions AB, CD, and EF is inversely proportional to the heat capacities of ice (0.50 cal/g °C), water (1.0 cal/g °C), and water vapor (0.44 cal/g °C), respectively. The lengths of the level segments BC and DE are proportional to the heats of fusion and vaporization, respectively, and to the mass of the sample.

It is easy to see why water, with its numerous intermolecular hydrogen bonds in the liquid state, exhibits a high heat of vaporization when it is converted to the essentially free collection of individual molecules in the vapor. The high heat of vaporization of water is another factor that makes water useful in temperature regulation and consequent stabilization of the physical environment of organisms. Even more than the high heat capacity of water, it is the heat of vaporization and its reverse, the heat of condensation, that stabilize the temperature of organisms and of their environment. The vaporization of 1 g of water absorbs about 585 times as much heat without a resultant temperature rise as the same mass of water undergoing a temperature rise of 1°. Evaporation of sweat from perspiring animals and transpiration (the vaporization of excess water) by plants are the principal methods used to cool these organisms. The heat necessary for this vaporization is removed from the organism.

The vaporization and condensation of water are also of prime importance in stabilizing the temperature of the larger physical environment and in the storage and transport of large amounts of heat energy. The evaporation of water from bodies of water can absorb large amounts of heat from the atmosphere. The heat is stored in the water vapor and transported to other locations, where it can be released to the surroundings on condensation of the vapor to liquid water.

8.7 HEAT OF FUSION When solids melt, heat is absorbed—another endothermic process, like vaporization. Conversely, when liquids freeze, thermal energy is evolved—an exothermic process, like condensation. In general, we can write, assuming constant temperature and pressure,

$$\text{solid} \longrightarrow \text{liquid} \qquad \Delta H_{\text{fus}}$$

The heat absorbed in melting a specified amount of solid substances is the *heat of fusion*. As with the heat of vaporization, the common units used are calories per gram or calories (or kilocalories) per mole. Values for the heats of fusion of some solids are listed in Table 8.4. As compared to most materials, other than ammonia and certain ionic solids, the heat of fusion of water is high.

Table 8.4
Heats of fusion

substance	melting point, °C	ΔH_{fus}, cal/g
ice	0	79.7
NH_3	−75	108.1
methyl alcohol	−97	16
carbon tetrachloride	−23	4
mercury	−39	3

The heat of fusion can be partly related, in a rough way, to the energy required to overcome intermolecular forces in the solid. Since only 15 percent of the hydrogen bonds are broken when ice melts, the heat of fusion of water can be expected to be quite a bit less than its heat of vaporization.

Ice melts to water with the absorption of 80 cal of heat per gram. There is no temperature change during melting until all the ice has been converted to water. Going in the other direction, water freezes, without temperature change, to ice with the evolution of 80 cal/g. The thermal regulation of the environment by ice or water results from this isothermal absorption or evolution of heat. The amount of heat that must be given up to the cooler atmosphere by a freezing body of water both moderates the atmospheric temperature and prevents the rapid freezing of the water. Similarly, the heat absorbed from warm surroundings by ice tends to cool the surroundings and moderate the temperature rise of the surroundings.

8.8 HEAT OF SUBLIMATION

The direct conversion of a solid to a vapor or gas is called *sublimation*. Depending on the conditions of temperature and pressure, many substances are capable of sublimation.

The equation for sublimation at constant temperature and pressure is

$$\text{solid} \longrightarrow \text{gas} \qquad \Delta H_{sub}$$

The heat absorbed is the heat of sublimation. This direct process can be thought of as the sum of two consecutive processes—melting followed by vaporization.

$$
\begin{array}{ll}
\text{solid } (T, P) \longrightarrow \text{liquid } (T, P) & \Delta H_{fus} \\
\text{liquid } (T, P) \longrightarrow \text{gas } (T, P) & \Delta H_{vap} \\
\hline
\text{solid } (T, P) \longrightarrow \text{gas } (T, P) & \Delta H_{sub} = \Delta H_{fus} + \Delta H_{vap}
\end{array}
$$

At 0°C, the value of ΔH_{sub} for water is 12.2 kcal/mol. If all of the hydrogen bonds in water are broken in the sublimation of water and other types of attraction between water molecules are neglected, the energy of 1 mol of hydrogen bonds comes out to be 6.1 kcal/mol, since there are two hydrogen bonds per water molecule. Water does not sublime at ordinary pressures. Carbon dioxide (dry ice) is a common solid that changes directly to a vapor on heating at normal pressures.

8.9 VAPOR PRESSURE OF LIQUIDS

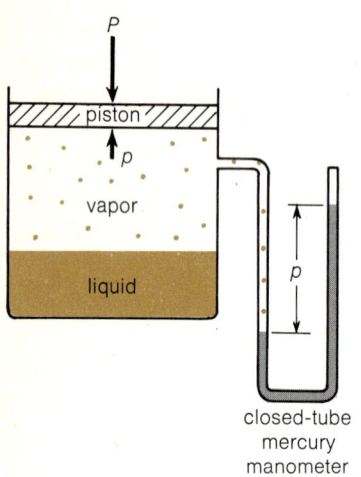

Figure 8.6. Vaporization of a liquid in a closed system. If the external pressure P and the vapor pressure p are equal, the liquid and the vapor are in equilibrium. If P is greater than p, condensation occurs; and if P is less than p, vaporization takes place until equilibrium is established. The vapor pressure p is given directly by the difference in levels of mercury (or other liquid) in the closed-tube manometer.

Molecules are continually leaving the surface of a pure liquid and entering the vapor (gaseous) state. In an open system, that is, one where the escaping molecules are free to get far away from the liquid surface, the liquid will eventually evaporate completely. If the liquid in such an open system is in an insulated container, so that heat cannot enter from the surroundings, the temperature of the liquid will fall as evaporation takes place. The temperature drops because the more energetic molecules are the ones leaving the liquid, and the average kinetic energy of the molecules left behind decreases. (This situation should not be confused with the constant-temperature vaporization of Section 8.6, where heat was continually added to the liquid.)

We now compare this open system, in which the evaporation of the liquid proceeds indefinitely, with the case of a liquid in a closed container (Figure 8.6) with no other substance present. Vapor can no longer escape to the surroundings but is confined to a limited space above the liquid. At first, liquid vaporizes, the number of molecules of vapor increases, and the pressure of the vapor in the confined space increases. But after a time, the number of molecules leaving the liquid surface per unit time equals the number returning to the liquid from the vapor, the average number of molecules in the vapor space above the liquid becomes constant, and the pressure due to the molecules in the vapor also becomes constant so long as liquid and vapor are both present.

The equality of the rates of vaporization and condensation in a closed system constitutes a condition of *dynamic equilibrium*, in which the amount of neither state increases at the expense of the other. Dynamic equilibrium occurs when opposing forces or processes continue to operate but cancel among themselves. This kind of equilibrium is distinct from *static equilibrium*, in which the opposing processes each completely stop rather than cancel each other.

The constant pressure at equilibrium is called the *vapor pressure* of the liquid. The value of the vapor pressure of a pure liquid in equilibrium with its vapor depends only on the temperature. Once the temperature is fixed, so is the equilibrium vapor pressure and vice versa. The vapor pressure does not depend on the relative amounts of liquid and vapor so long as both are present. Neither does it depend on the surface area of the liquid.

The vapor pressure of a liquid is a measure of the *escaping tendency* of the molecules in the liquid. For a given liquid, the tendency of the molecules to escape from the liquid into the vapor increases as the temperature of the liquid increases, since a greater fraction of molecules have the minimum energy required to escape (Figure 4.13). The escaping tendency of the molecules in two different liquids at the same temperature may be compared by comparing their vapor pressures at that temperature. In general, liquids with strong intermolecular forces have low vapor pressures and are less *volatile*. Table 8.5 lists the equilibrium vapor pressures at different temperatures for a number of liquids.

Reversible Reactions and Equilibrium

The general reaction liquid $\rightleftharpoons$ gas can go in either direction, as we noted. It is an example of a reversible reaction—a physical reaction in this case. There are many other processes, both chemical and physical, that may be considered to be reversible so long as reactants and products are not separated from each other. The entire subject of equilibrium is concerned with this type of reac-

Table 8.5
Vapor pressures of liquids (mmHg)

temperature, °C	H_2O	CCl_4	acetone $CH_3{-}\overset{\overset{O}{\|}}{C}{-}CH_3$	ethyl ether $(C_2H_5)_2O$	ethyl alcohol C_2H_5OH	n-octane $CH_3(CH_2)_6CH_3$	chloroform $CHCl_3$
0	4.58	33		185	12	3	
10	9.21	56	116	292	24	6	100.5
20	17.54	91	185	442	44	10	159.6
30	31.82	143	283	647	79	18	246.0
40	55.32	216	421	921	135	31	366.4
50	92.51	317	613	1277	222	49	526.0
60	149.38	451	866		353	78	739.6
70	233.7	622	1200		542	118	1019
80	355.1	843			813	175	
90	525.8	1122			1187	253	
100	760.0	1463				354	
boiling point, °C	100	76.8	56.10	34.6	78.3	125.7	61.0

tion, which we shall write with double arrows to indicate that the products of a process can transform back to reactants:

$$\text{reactants} \rightleftharpoons \text{products}$$
$$H_2O \rightleftharpoons H^+ + OH^-$$

When the rate of the forward reaction (reactants $\longrightarrow$ products) is equal to the rate of the backward reaction (products $\longrightarrow$ reactants), a condition of dynamic equilibrium is attained. At the equilibrium point, the amounts of reactant and product (both of which must still be present) remain constant— neither increases at the expense of the other. There is no net change in the system.

8.10 LIQUID–VAPOR EQUILIBRIUM: THE VAPOR PRESSURE CURVE

The vapor pressure (or vaporization) curve is a convenient way of displaying the variation of equilibrium vapor pressure with temperature. Figure 8.7 shows the vapor pressure curve for water from 0°C to the critical point (374°C, 218 atm), where liquid water can no longer exist in equilibrium with water vapor. As before, we assume that only water vapor is in contact with the liquid, so that the pressure on the liquid is entirely due to the vapor. Any pair of pressure and temperature values lying on the curve represents a condition of equilibrium between liquid water and water vapor. Points such as A and D in the area above the curve correspond to the existence of liquid alone or to a nonequilibrium condition of vapor condensing to liquid. Points such as C or E below the curve lie in the region of pressure and temperature values where only vapor may exist or, if there should happen to be any liquid, the liquid will completely vaporize. To visualize the situation at these points, imagine liquid water and water vapor enclosed in a cylinder fitted with a weightless, frictionless piston on which some external pressure P can be applied (Figure 8.6). Starting with the liquid–vapor equilibrium situation at

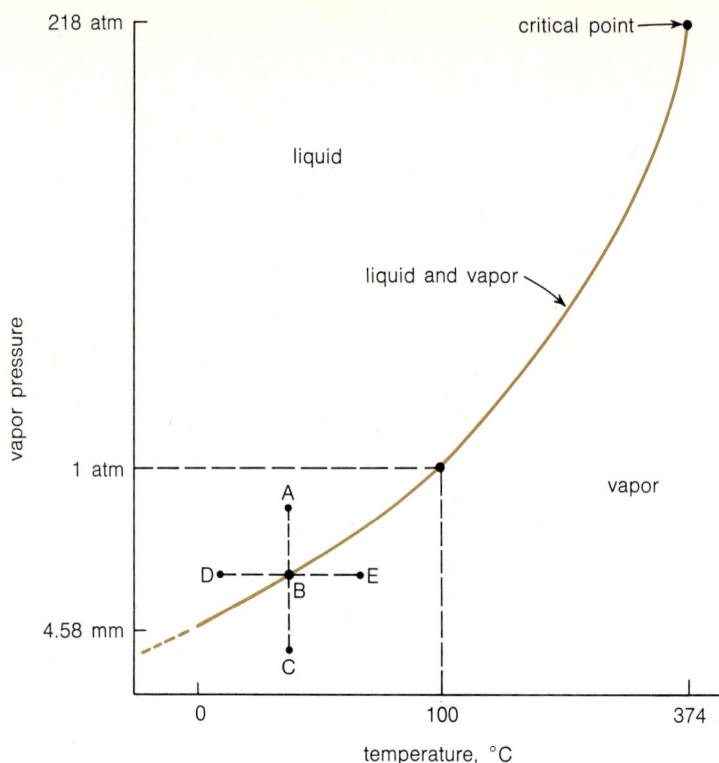

Figure 8.7. Vapor pressure of water (not drawn to scale). Water and water vapor are in equilibrium only at points on the vapor pressure curve.

point B on the curve, where $P = p$, the external pressure P is increased isothermally to P_A. If it is kept at the higher value P_A, all the vapor will condense. If, on the other hand, starting at point B, the pressure P is decreased below the equilibrium value P_B to P_C and held there, vaporization of all the liquid occurs.

The boiling point of a liquid is the temperature at which its equilibrium vapor pressure p is equal to the externally applied pressure P. At the boiling point, the vapor pressure is sufficient to counteract the external pressure, so visible bubbles of vapor form within the liquid. Therefore, the vapor pressure curve also gives the boiling points of a liquid at a series of external pressures if the vertical axis is thought of as giving the external pressure P rather than the vapor pressure p.

Figure 8.8 is a boiling point curve. It shows the temperature at which the vapor pressure of a liquid (water) is equal to any external pressure P. The lower the external pressure, the lower the boiling point of the liquid. Since liquid and vapor are in equilibrium at the boiling point, the boiling temperature will remain constant as long as liquid remains. The *normal boiling point* of a liquid is the temperature at which the vapor pressure of the liquid is 1 atm or 76 cmHg.

If water is boiled in an open container, the external pressure is atmospheric or barometric pressure. At sea level, this is approximately 1 atm, so water would boil near its normal boiling point of 100°C. But the pressure of the atmosphere decreases by about one-half for each $3\frac{1}{2}$ miles (5.64 km)

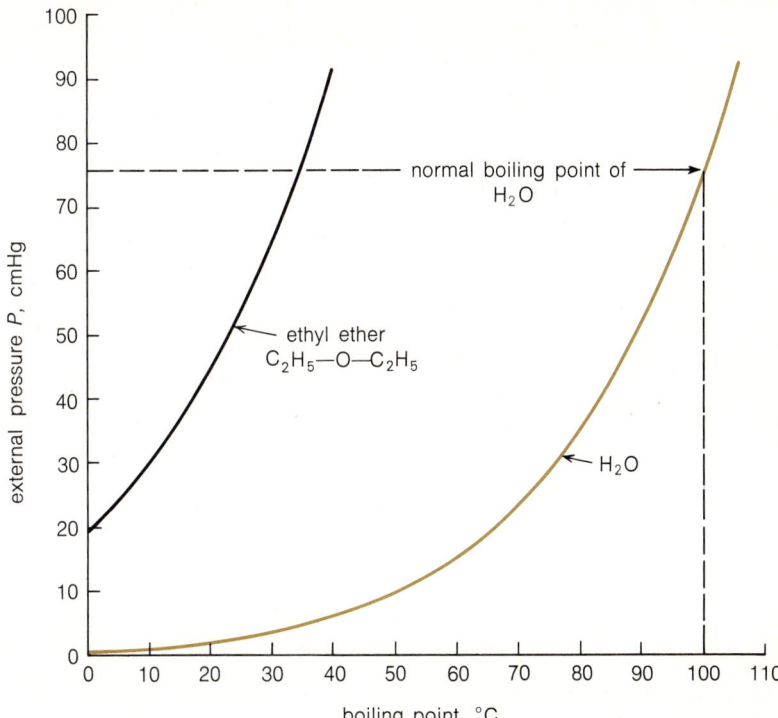

Figure 8.8. Boiling point curve for water. Points on the line give the temperature at which the vapor pressure is equal to the external pressure. For comparison, the boiling point (or vaporization) curve of the more volatile ethyl ether (normal b.p. 34.6°C) is also shown.

in altitude, and boiling water at this altitude could get no warmer than about 82°C. Since the rate of a chemical reaction usually decreases with decreasing temperature, the time required to cook food properly at this altitude will be longer than the cooking time at sea level. The situation can be remedied with a pressure cooker. The pressure of the trapped steam is greater than the atmospheric pressure, so the boiling point is increased and the cooking time decreased.

8.11 SOLID–LIQUID–VAPOR EQUILIBRIA FOR PURE SUBSTANCES

In addition to liquid–vapor equilibrium, we could discuss solid–vapor and solid–liquid equilibrium in terms of *PT* plots that show conditions necessary for equilibrium between two phases. It is common to consolidate the *PT* conditions for all three kinds of equilibrium into one diagram, called a *phase diagram*. The phase diagram for a pure substance summarizes all the *PT* data needed to determine relationships among the solid, liquid, and gaseous states of the substance. As such, it contains, in one way or another, much of the information about latent heats of transitions between physical states, boiling point, melting point, effect of pressure on transition points, and so on, that are helpful in characterizing pure substances.

A *phase* is defined as any part of a system that is uniform in chemical composition and physical properties and that is separated from other parts of the system by definite bounding surfaces. Ice, liquid water, and water vapor are three phases of water. When speaking about a pure substance the phases are often the same as the three physical states of matter, and we may interchange the words "phase" and "state" in our previous discussions. In other cases, however, such as when we have a mixture of sand and sugar, oil and water, or different solid forms of the same substance (such as graphite and carbon), then we have two solid, two liquid, and two solid phases, respectively. From now on we shall use the word "phase" for "state."

Figure 8.9 is the phase diagram of water. Curve OA is the vaporization curve already discussed in detail (Figure 8.8). As before, values of water vapor pressure and temperature lying on OA represent conditions of equilibrium between liquid water and its vapor. Curve OC is the sublimation curve for the solid–vapor equilibrium. This curve is obtained by plotting the vapor pressure of ice in equilibrium with water vapor versus temperature. Again, each point on this curve represents a condition of equilibrium between ice and water vapor.

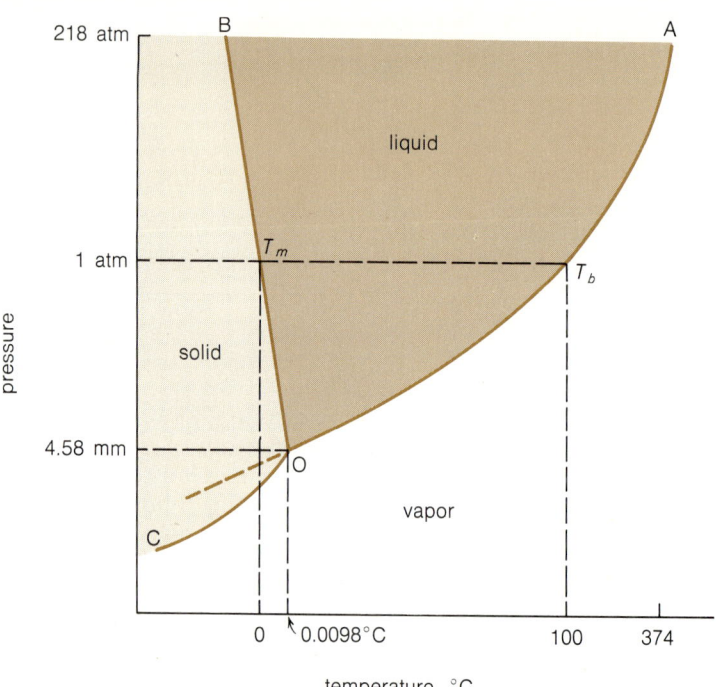

Figure 8.9. Phase diagram for water. For lines OA and OC, the pressure axis is the vapor pressure p. For OB, the pressure axis gives the external pressure P. The tilt of line OB to the left of vertical is exaggerated to show the decrease of melting point with pressure. The continuation of the line OA past the triple point O represents the supercooling of water, the cooling of the liquid below its freezing point without solidification taking place. The dashed line gives the vapor pressure of the supercooled liquid in this nonequilibrium or metastable state.

Curve OB is called the melting point (fusion) or *freezing point curve*. It gives the conditions under which water and ice can exist in equilibrium with each other according to the equilibrium

$$H_2O(s,\ T,\ P) \rightleftharpoons H_2O(l,\ T,\ P)$$

Since no vapor is involved, the pressure P here is the external pressure applied to the ice–water system, not an equilibrium vapor pressure. Curve OB gives the melting point of ice (or the freezing point of water) at various external pressures and reveals the influence of pressure on the melting point. Water is somewhat unusual (along with bismuth and antimony) in having a melting point that decreases with increasing pressure. For ice, the melting point decreases about 0.01°C for each 1 atm increase in pressure. The more common behavior for the melting point and pressure is depicted in the phase diagram

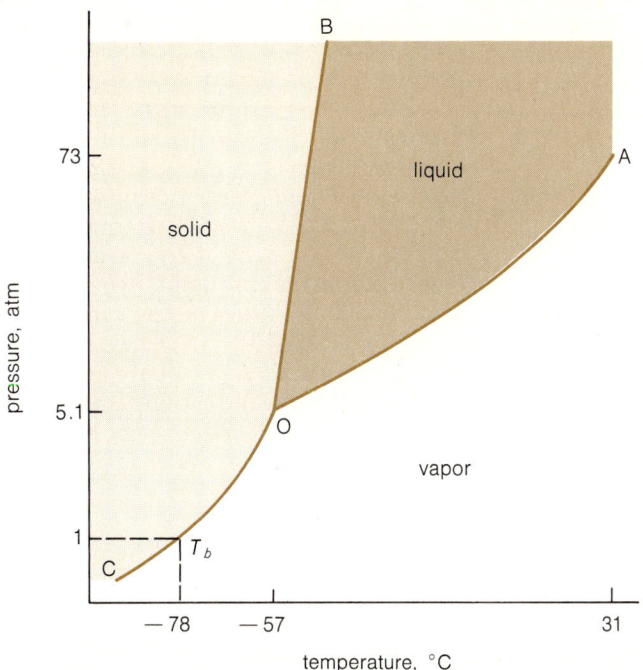

Figure 8.10. Phase diagram for carbon dioxide.

(Figure 8.10) for carbon dioxide, where the melting point curve OB leans to the right of the vertical axis.

The *normal melting point* of a solid is the temperature at which solid and liquid phases are in equilibrium (have the same vapor pressure) at an external atmospheric pressure P of 1 atm. Experimentally, it is the temperature at which liquid first appears when the solid is slowly heated under 1 atm pressure. As Figure 8.5 indicated, the temperature remains constant until all the solid has melted. The melting point is a key parameter for the determination of the purity of a substance, because the presence of even a small amount of impurity changes the melting point. The normal melting point T_m of ice is 0°C. Table 8.4 shows that ice has an anomalously high melting point compared with other substances, again the result of the extra thermal energy required to break a portion of the hydrogen bonds in ice.

The areas between the curves in the phase diagrams of Figure 8.9 and Figure 8.10 correspond to regions where conditions permit the existence of one phase only. The interpretation of pressure and temperature changes on systems represented by PT values in these regions or on the curves of these phase diagrams follows the same reasoning as applied to the vaporization curve of water, Figure 8.7.

The point where all three curves meet in the phase diagrams is known as the *triple point*. For water, this occurs at 0.0098°C and 4.58 mmHg vapor pressure and at no other point. These are the only conditions where all three phases of water may be in equilibrium with one another. It is not the same as the normal melting point, where only the solid and liquid phases need be in equilibrium and where the external pressure is 1 atm.

This completes the investigation of the major thermal properties of water.

It is apparent that the hydrogen bond is responsible for all the unusual thermal properties of water. But the fitness, the uniqueness, of water for life on this planet is not dictated by any one of these properties. Rather, it is the sum of all these thermal properties, plus a few more nonthermal characteristics, that leave water unequaled in fulfilling the requirements of living organisms. You can understand the necessity of water in life processes if you assume that some liquid phase is required for the transport of dissolved nutrient compounds in both the external physical environment and the internal cellular environment of organisms. If you were to decide only on the basis of melting and boiling points, what other substance exists naturally as a liquid in sufficient quantities on the earth or occurs so commonly in three distinct forms? Around 540 B.C., Thales, the first Greek philosopher, thought about such questions. Based, no doubt, on his observation of the abundance of water and the daily transformations among the various phases of water, he reached the conclusion that water might be altered into all kinds of solid, liquid, and gaseous forms, making water the single world substance underlying all things on earth.

8.12 HOMOGENEOUS AND HETEROGENEOUS EQUILIBRIA

The phase changes of water and other substances are examples of *heterogeneous equilibria*, meaning that the reversible reactions involve two or more phases, or states, of matter. For pure water, the processes or reactions first considered were physical transitions. Yet chemical reactions also can be heterogeneous. Examples of physical and chemical heterogeneous equilibria are

$$H_2O(s, 0°C, 1 \text{ atm}) \rightleftharpoons H_2O(l, 0°C, 1 \text{ atm}) \quad \text{physical}$$
$$Ag^+(aq) + Cl^-(aq) \rightleftharpoons AgCl(s) \quad \text{chemical}$$
$$C(\text{diamond}) \rightleftharpoons C(\text{graphite}) \quad \text{chemical}$$

When a reaction proceeds within one phase, or when an equilibrium involves only one phase, then the reaction or the equilibrium is *homogeneous*. Examples are

$$2NO_2(g) \rightleftharpoons N_2O_4(g) \quad \text{chemical}$$
$$NH_3(aq) + H_2O(l) \rightleftharpoons NH_4^+(aq) + OH^-(aq) \quad \text{chemical}$$

8.13 DYNAMIC EQUILIBRIUM AND Le CHATELIER'S PRINCIPLE

Dynamic equilibrium, as distinguished from static equilibrium, has already been defined as a balanced condition resulting from the cancelling of opposing ongoing processes. Almost all of the equilibrium situations we see around us are dynamic equilibria. In simple chemical and physical systems, the rates of forward and reverse reactions are equal at equilibrium. However, the conditions of dynamic equilibrium are common in such diverse fields as human behavior, economics, psychology, sociology, and, of course, ecosystems. The predator–prey relationship mentioned in Chapter 1 is a good example of a population balance maintained by the equality in the birth rate of prey and their rate of death by predators. Other equilibrium–maintaining or *homeostatic mechanisms* will be brought up as we go along.

In any discipline, a major concern is the study of equilibrium in systems, both large and small. The determination of the factors defining equilibrium in a given system, the ways the systems may become imbalanced, and the methods available for restoring the system to equilibrium are important in all sciences, including ecology. As you might guess, ecosystems are considerably more complicated than the systems we shall explore right now. At the moment, our concern is with simple chemical and physical equilibria, but as we go along you should be conscious of the similarities among all instances of dynamic equilibrium. The broad principles are the same; only the mechanism and number of interconnected forces or processes involved differ.

Now we focus on the changes taking place in an equilibrium system when the system is thrown off balance. A qualitative general description of what happens is provided by Le Chatelier's principle (1888), which can be stated: *A change in any of the factors that determine the equilibrium conditions of a system will cause the system to change in such a direction as to reduce, counteract, or overcome the effect of the change.*

The phase equilibria of water have already shown that temperature and pressure are two of the basic variables needed to specify a condition of equilibrium. In chemical equilibria, a knowledge of a third factor, the amounts of reacting substances present in the system, is also necessary. With simple systems, we illustrate the influence of changes in pressure, temperature, and amount of substance on closed systems in a state of dynamic equilibrium.

EXAMPLE

The phase equilibrium

$$H_2O(l, T, P) \rightleftharpoons H_2O(g, T, P) \qquad \Delta H_{vap}$$

This is an endothermic reaction; ΔH_{vap} is positive.

EFFECT OF PRESSURE CHANGE. If the pressure P of the water vapor is increased, or, what is the same thing, the volume of the system is decreased, condensation occurs. (Since water vapor is assumed to be the only gas present, its vapor pressure is the total pressure. See Section 11.7.) Either of these effects can be brought about by pushing down the piston in Figure 8.6. To counteract either of these equivalent departures from the P–T values at equilibrium, some water vapor will condense, decreasing the number of moles of water vapor until the vapor pressure

$$P_{H_2O} = \frac{n_{H_2O(g)}RT}{V_{H_2O(g)}}$$

returns to its original equilibrium value. Another, simpler way of viewing this is to say that the equilibrium shifts in the direction of smaller volume, to liquid water, or to the left as the equation is written. In either case, the original equilibrium conditions of pressure and temperature tend to be reestablished.

If the pressure is decreased by pulling up the piston, liquid water vaporizes to increase the number of moles of water vapor in the larger volume and return the vapor pressure to the original value. The equilibrium shifts in the direction of greater volume, to water vapor, or to the right as the equation is written.

EFFECT OF TEMPERATURE CHANGE. The effect of a temperature change on dynamic equilibria depends on the sign of the heat change accompanying the reaction. Of course, in a reversible reaction, the ΔH of the forward reaction is equal in magnitude but opposite in sign to the ΔH of the reverse reaction. For the vaporization of water, the forward reaction is endothermic. It is often helpful in applying Le Chatelier's principle to write reactions as though the thermal energy were one of the reactants:

$$H_2O(l) + \Delta H_{vap} \rightleftharpoons H_2O(g)$$

A temperature increase corresponds to the addition of heat. To reduce the effect of the added heat, the equilibrium shifts in the direction that will absorb this extra heat.

In the vaporization of water, a temperature increase shifts the equilibrium to the water vapor side (the right), and more liquid vaporizes. Conversely, a decrease in temperature favors the condensation or exothermic reaction, because the equilibrium shifts (to the left) to make up for the heat removal associated with a drop in temperature.

EXAMPLE | The phase equilibrium

$$H_2O(s, 0°C) \rightleftharpoons H_2O(l, 0°C) \qquad \Delta H_{fus}$$

EFFECT OF PRESSURE. A pressure increase causes the equilibrium to shift in the direction of smaller volume. Due to the expansion of ice, a given mass of liquid water has a smaller volume than the same mass of ice, so some ice will melt. Equivalently, under high pressures, ice melts at a lower temperature than usual (or water freezes at a lower temperature). The lowering of freezing point with pressure is not large, being only 0.0075°C per 1 atm increase in pressure. The melting point of ice under a pressure of 1000 atm is −7.4°C. In ice skating, the pressure of the skate blade on the ice causes melting of a thin surface layer of ice, and the blade glides over the resulting film of liquid water.

EFFECT OF TEMPERATURE. Temperature increase favors the endothermic or melting reaction, thus causing ice to melt. The application of Le Chatelier's principle to the phase equilibria of water provides the same qualitative results as does an interpretation of the phase diagram of water (Figure 8.10), but the large collection of experimental data represented by the phase diagram is not needed.

Le Chatelier's principle applies only if all of the substances involved in the equilibrium system are present and in contact with one another. Otherwise, there is no equilibrium system to talk about.

EXAMPLE | The homogeneous equilibrium

$$N_2(g) + 3H_2(g) \rightleftharpoons 2NH_3(g) \qquad \Delta H = -22.08 \text{ kcal}$$

The direct combination of nitrogen and hydrogen in the industrial Haber process was discussed briefly in Section 5.10. There we said that low temperature and high total pressure conditions were best for a high yield of ammonia.

EFFECT OF PRESSURE. An increase in total pressure, leading to a decrease in volume, will, according to Le Chatelier's principle, shift the equilibrium to the side having the smallest volume, the right side here. The side having the least number of moles of *gaseous* molecules is favored by a decrease in volume. For this reason the yield of ammonia in this reaction is increased by a decrease in volume, because the ammonia occupies less volume than the reactants.

EFFECT OF TEMPERATURE CHANGE. The synthesis of ammonia is an exothermic reaction. Therefore, an increase in temperature of the system tends to drive the reaction to the nitrogen and hydrogen side, thereby decreasing the yield of ammonia (Figure 8.11). The same conclusion is reached by considering

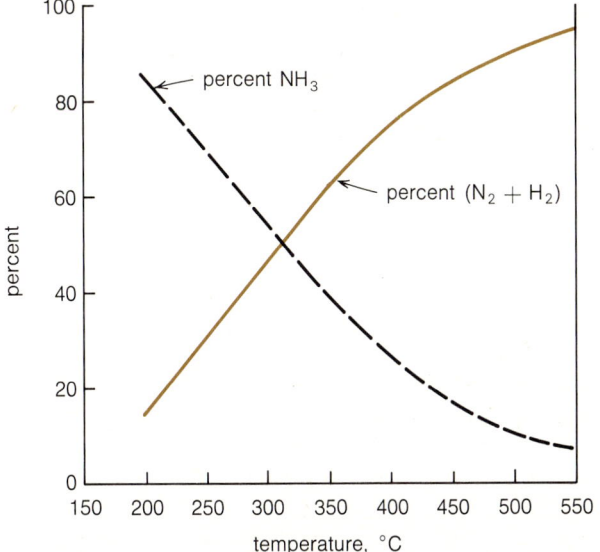

Figure 8.11. Effect of temperature on the composition of the $N_2 + 3H_2 \rightleftharpoons 2NH_3$ equilibrium mixture (total pressure 100 atm). More ammonia is formed at lower temperatures, but the reaction is very slow.

heat as a product of the forward reaction. To obtain as much ammonia as possible from given starting amounts of nitrogen and hydrogen, the temperature should be kept as low as possible. But the union of N_2 and H_2 is much too slow at a low temperature (about 25°C), even though a lot of ammonia would eventually be formed. The use of a catalyst to speed the union at low temperatures is not completely satisfactory either. So some ammonia yield is sacrificed by raising the reaction temperature and using a catalyst—both of which increase the rate of synthesis. Catalysts do not affect the equilibrium amounts of the reactants or products in a reversible reaction. They affect only the rate of attainment of the equilibrium condition. *The addition of a catalyst to an equilibrium mixture of some system will not cause the equilibrium to shift in any direction.*

EFFECT OF AMOUNT OF REACTANTS. Change in the amount of a reactant will cause an equilibrium system to shift in a direction counteracting the change. If the amount (concentration) of a substance on the left side of the equation of a reversible reaction at equilibrium is increased, the forward reaction will be favored so as to use up the added reactant and produce more product; the equilibrium will shift to the right side of the equation of the reaction. For the ammonia synthesis, this means that the addition, at constant volume, of more nitrogen or hydrogen or both to a gaseous equilibrium system of N_2,

H_2, and NH_3 will bring about further combination of N_2 and H_2 to produce more ammonia. If additional ammonia had been introduced at constant volume, the reversible reaction would have shifted to the left, with subsequent decomposition and removal of some of the added ammonia.

Conversely, the removal, at constant volume, of any substance will cause the equilibrium system to shift to produce more of this substance. So, in the ammonia synthesis it is common to remove the product continually as it is formed, either by solution in water or by liquefaction if the total pressure is high enough. The system continually shifts to the right or the ammonia side to make up for the loss in ammonia.

EXAMPLE | The homogeneous equilibrium

$$N_2(g) + O_2(g) \rightleftharpoons 2NO(g) \qquad \Delta H = 43.2 \text{ kcal}$$

This reaction was introduced in our discussion of air pollution.

EFFECT OF PRESSURE. A decrease in volume caused by a change in total pressure of the equilibrium mixture of N_2, O_2, and NO will have no effect on the amount of the three gases at equilibrium. Because both sides of the equation for the reversible reaction have the same number of molecules of gas, there is no way the composition of the mixture can change to yield a smaller volume.

EFFECT OF TEMPERATURE. The synthesis of nitric oxide is a highly endothermic reaction. According to Le Chatelier's principle, if an equilibrium mixture of N_2, O_2, and NO exists at some temperature, an increase in temperature will cause the formation of more NO. Even at very high temperatures (2000–3000°K), however, there is little NO present at equilibrium; that is, N_2 and O_2 do not react very completely.

It is evident that at ordinary temperatures (25°C) nitrogen and oxygen in the air will react to produce very little NO, even if the rate of their reaction were appreciable. However, in the cylinder of an internal combustion engine, N_2 and O_2 are subjected to high temperatures, and they do combine to give considerably more NO than at lower temperatures. If we assume that high-temperature equilibrium is established in the combustion chamber, the high-temperature equilibrium amount of NO is vented to the much cooler atmosphere as exhaust. At lower temperatures, the equilibrium lies even farther on the N_2–O_2 side, and we might expect the NO concentration to decrease to its low-temperature equilibrium value. But the rapid attainment of the smaller low-temperature equilibrium value of NO is prevented by the slow rate of the decomposition of NO to N_2 and O_2 (even though the decomposition is highly exothermic). And the rate of the decomposition reaction falls rapidly as the temperature of the exhaust gas decreases. Therefore, the amount of NO released to the atmosphere more nearly approximates the amount produced at high temperature. High-temperature combustion processes, accompanied by rapid cooling of the exhaust gases, are, as we noted in Section 7.8, the chief source of the primary air pollutant, nitric oxide.

Question: What is the effect of adding N_2 or O_2 to this equilibrium system at constant volume? The effect of adding NO?

EXAMPLE | The heterogeneous equilibrium

$$O_2(g) + H_2O(l) \rightleftharpoons (O_2 \text{ dissolved in } H_2O)(l)$$

This equation represents the solution of oxygen gas in water, an important process in nature.

EFFECT OF PRESSURE. If water vapor is ignored and the only gas present is oxygen, an increase in pressure leading to a decrease in the volume of the oxygen gas in contact with the water will result in a shift of the gas–liquid equilibrium system to the solution side. More oxygen will dissolve in the water. The effervescence of carbonated beverages when opened is an example of the shift of equilibrium position to the gas side when volume is increased (or the pressure of the CO_2 in equilibrium with solution is decreased).

The mass of any gas dissolved in a given volume of any liquid is proportional to the pressure of that particular gas above the solution. This is a statement of a general relation known as Henry's law (Section 11.7).

EFFECT OF TEMPERATURE. An increase in temperature will result in the escape of dissolved oxygen from the liquid solution phase to the gaseous oxygen phase, so that water will hold less oxygen at higher temperatures. The reason for this is that most gases dissolve in water with the release of heat, ranging from 2 to 20 kcal/mol of gas:

$$\text{gas} + \text{water} \rightleftharpoons \text{solution} + \text{heat}$$

According to Le Chatelier's principle, an increase in temperature (an addition of heat) will disturb this equilibrium, and the reversible reaction will shift to the gaseous oxygen side to counteract or reduce the temperature increase. The escape of carbon dioxide from carbonated beverages when they warm up is a well-known phenomenon, as is the formation of air bubbles in water when the water is warmed.

EFFECT OF AMOUNT OF REACTANT. The addition of more oxygen, at constant volume, results in the solution of more oxygen in the water. This is equivalent to an increase in pressure.

EXAMPLE | The heterogeneous equilibrium

$$2C(s) + O_2(g) \rightleftharpoons 2CO(g)$$

EFFECT OF PRESSURE. A decrease in volume, resulting from an increase in pressure, of a closed system containing carbon, oxygen, and carbon monoxide in equilibrium will cause the decomposition of some of the carbon monoxide and the formation of more oxygen. This behavior illustrates that a reversible reaction at equilibrium is driven in the direction having the smallest number of *gaseous* molecules when volume is decreased. The solids or liquids present are incompressible in comparison to any gases, so only the relative numbers of gas molecules on either side determine the equilibrium shift.

EFFECT OF TEMPERATURE. The oxidation of carbon is a well-known exothermic reaction. The heat of reaction ΔH is negative, and the heat can be considered as a product of the reaction $2C(s) + O_2(g) \rightleftharpoons 2CO(g) + \Delta H$. An increase in temperature (addition of heat) will favor the decomposition of CO.

EFFECT OF AMOUNT OF REACTANTS. First let us consider the gases in this equilibrium. The addition of oxygen or the removal of carbon monoxide, while keeping the volume of the system constant, will cause the generation of more carbon monoxide. But what happens if solid carbon is added or removed from the system? Nothing. A change in amount of reactant or product at constant volume is possible only for gases (and dissolved substances) because of the low compressibility of solids and liquids. The addition of a pure solid or liquid to an equilibrium system involving that solid or liquid necessarily increases the volume of that phase by a proportionate degree.

Actually, the amounts of gaseous reactants and products referred to in previous examples have been, not masses, but concentrations defined as mass (or moles) per unit volume of the gas phase. When more gas is added or removed at constant volume of the gaseous phase, the concentration of the gaseous reactant or product changes, and it is this change in concentration that affects the equilibrium. However, when a pure solid or pure liquid is added to a system, the volume of the solid or liquid phase increases, but the concentration of the pure solid or liquid remains constant. Similarly, the removal of some, but not all, of a pure solid or liquid phase does not affect the concentration, and there is no effect on the equilibrium.

With these examples in mind, we can now summarize the effects of changes of volume or total pressure, temperature, and amount (concentration) of reacting substance on simple physical and chemical equilibria.

Volume or Total Pressure   Decrease in volume at constant temperature (or increase in total applied pressure) of a closed system at equilibrium favors the reaction leading to a decrease in volume. The effect is most pronounced when gases are involved. Then the equilibrium is displaced to the side of the equation having the smaller number of gas molecules.

Temperature   Increase in temperature of a system at equilibrium disturbs the equilibrium so that, for a reversible reaction, the endothermic reaction and its products are favored.

Amount of Reactant   If the volume of an equilibrium system is kept constant, addition of a reacting substance favors the reaction that consumes this substance. This applies only to gases and to reactants in solution.

These examples have shown how Le Chatelier's principle can be applied to various equilibria. In no way do these examples explain or justify the validity of the principle. To do this in a rigorous way, we would need to apply some theoretical arguments based on chemical energetics and the relationship of reactant and product concentrations at equilibrium, or less rigorously, on the rates of simple chemical reactions (the law of mass action). These approaches will be taken up in Chapters 16 and 17.

8.14 SUMMARY   Water is one of the most abundant substances on the surface of the earth and the most abundant substance in living organisms. The ability of the water molecule to form comparatively strong hydrogen bonds between molecules confers a number of exceptional physical properties on liquid water and ice. Among these properties are the low density of ice, the density maximum of water, and the high heat capacity, heat of vaporization, and heat of fusion of water. The special properties resulting from the presence of hydrogen bonds enable water to play a unique role in the biosphere.

A liquid and its vapor are in a state of dynamic equilibrium when the opposing processes of vaporization and condensation cancel one another so that no net change occurs in the liquid–vapor system. In such a state, the pressure of the vapor above the liquid is constant and depends only on the temperature of the system. The vapor pressure of a liquid is also related to intermolecular forces such as hydrogen bonds.

The conditions of pressure and temperature for equilibrium between liquid and vapor, as well as other kinds of heterogeneous equilibria, are conveniently summarized in the phase diagram of a substance or a system. Such diagrams also show how the boiling and melting points of a substance are affected by external pressure.

When a system in a state of dynamic equilibrium is disturbed or displaced from equilibrium, the system changes in a way to nullify or oppose the effect of the disturbance and to reestablish equilibrium. This is Le Chatelier's principle.

Questions and Problems

1. How much heat is absorbed when 10 g of ice at $-5°C$ are converted to steam at $100°C$?

2. Each year, 423,000 km^3 of water are evaporated from the seas and the land.
   a. Taking the heat of vaporization of water to be 580 cal/g, how much absorbed energy does this represent?
   b. What percentage of the total solar energy intercepted each year by the earth and its atmosphere, namely 1.31×10^{24} cal/yr, does the answer to part (a) represent? (Of the total energy intercepted, only about 50 percent is actually available at the earth's surface.)

3. The normal boiling points of the hydrides of elements in the same periodic group as oxygen are $-61°C$ for H_2S, $-42°C$ for H_2Se, and $-2°C$ for H_2Te. What might be the expected boiling point of H_2O? Explain the discrepancy.

4. a. What is the pressure in a 1-liter container containing both $H_2O(l)$ and $H_2O(g)$ at $100°C$?
   b. What is the pressure in a 1-liter vessel containing only 0.5 g of H_2O at $100°C$?

5. A 10-liter closed container initially has nothing except 10 g of ice in it. The temperature is raised to $70°C$. What is then the pressure of water vapor within the container, and how much liquid water is present in the container?

6. At approximately what temperature does water boil under an external pressure of 900 mmHg, 760 mm Hg? 500 mmHg?

7. Heat is added at a constant rate of 50 cal/sec to a 1-mol sample of benzene. The normal boiling point of benzene is $80°C$, and its ΔH_{vap} is 7.35 kcal/mol. The freezing point is $5.5°C$, and $\Delta H_{fus} = 2.36$ kcal/mol. The average heat capacities in calories per mole per degree for solid, liquid, and gaseous benzene are 24, 32, and 21, respectively. Draw the heating curve for benzene from $0°C$ to $100°C$.

8. a. Under what conditions can water be made to sublime?
   b. Under what conditions can solid CO_2 be melted to liquid CO_2?

9. What are the temperatures of the following systems?
   a. $H_2O(l)$ and $H_2O(g)$ in equilibrium at 1 atm pressure
   b. $H_2O(l)$ at 1 atm external pressure
   c. $H_2O(l)$ and $H_2O(s)$ in equilibrium under 1 atm external pressure

d. $H_2O(l)$, $H_2O(s)$, and $H_2O(g)$ in equilibrium with one another

e. ice at 1 atm pressure

10. The equilibrium $2NO_2(g) \rightleftharpoons N_2O_4(g)$ is exothermic, with $\Delta H = -14.67$ kcal. Nitrogen dioxide is brown; nitrogen tetroxide is colorless. Would you expect the color of photochemical smog to be deeper on a warm day or on a cool day?

11. For each of the following reactions, describe the shift in equilibrium resulting from (1) an increase in temperature and (2) a decrease in pressure (or an increase in volume) of the reaction mixture. All substances are gases unless otherwise noted.

a. $4NH_3 + 5O_2 \rightleftharpoons 4NO + 6H_2O + $ heat

b. $PCl_5 \rightleftharpoons PCl_3 + Cl_2$ $\quad \Delta H = 32.8$ kcal

c. $C(s) + O_2 \rightleftharpoons CO_2$ $\quad \Delta H = -97$ kcal

d. $C(s) + CO_2 \rightleftharpoons 2CO$ $\quad \Delta H = 41.4$ kcal

e. $CO + H_2O \rightleftharpoons CO_2 + H_2 + $ heat

12. For the reaction

$$C(s) + CO_2(g) \rightleftharpoons 2CO(g)$$

what is the effect of

a. adding CO at constant volume, that is, increasing the concentration of CO?

b. adding C(s)?

c. introducing a catalyst to the reaction mixture?

Answers

1. 7220 cal
2. a. 2.46×10^{23} cal
   b. ~19 percent

4. b. 0.85 atm
5. 234 mmHg, 8 g H_2O

References

PHYSICAL PROPERTIES OF WATER

Buswell, Arthur M., and Rodebush, Worth H., "Water," *Scientific American*, **194,** 76 (April 1956).

Edsall, John T., and Wyman, Jeffries, "Biophysical Chemistry," Vol. I, Academic, New York, 1958, Chapter 2.

Henderson, Lawrence J., "The Fitness of the Environment," Macmillan, New York, 1913. (Reprinted by Beacon Press, Boston, 1958.)

The chapter on water has been widely acclaimed, even though nothing was known of hydrogen bonds when this book was written.

EQUILIBRIUM

Campbell, J. A., "Why Do Chemical Reactions Occur?," Prentice-Hall, Englewood Cliffs, N. J., 1965.

Solutions and Solubility

The physical properties of water that we discussed in Chapter 8 are obviously major factors in the special fitness of water. But if we stop there, we have hardly gotten our feet wet as far as the fitness of water for life is concerned. We have yet to explore the ability of water to dissolve an unusually large number of substances. Water as a solvent is indispensable and almost unique in its role within and around the cells of organisms.

As a solvent, water plays two major roles within organisms. It acts as the transport medium for molecules and ions containing nutrients and regulatory substances that circulate among cells, through cell walls, and within cells. Furthermore, water participates in most of the biochemical reactions occurring in organisms. This participation is possible because water can ionize substances, and the resulting ions are required by cells for their energy-producing processes. The fluids in living cells contain many ions dissolved in water, and the chemistry of living systems is largely developed around water's solvent properties.

In the external environment of organisms, water again serves as a transport agent for many dissolved substances. Some of these substances are nutrients needed by relatively immobile marine organisms that absorb the nutrients from the waters constantly bathing them. Dissolved materials are also carried from the air to the sea by rains and from the earth to the sea by runoff water. When water carries dissolved materials from the earth to the sea, it decomposes rocks, deposits sediments, and so forth. These processes are important steps in the operation of biogeochemical cycles of the sedimentary type.

Man is not unaware of the capability of water to dissolve and disperse his biological and technological wastes. He uses this property, together with the natural processes that eventually return all substances to the seas, to remove these wastes from his living area. In doing so, he is accelerating certain steps in biogeochemical cycles and, in general, slowly making the earth's oceans his communal garbage dump.

Many of these aspects of water in nature will be brought up again in subsequent chapters of the text. First we must understand what is meant by the term "solution" and why water is so good at forming solutions with so many substances.

9.1 SOLUTIONS

When something dissolves in water or any other substance (not necessarily a liquid), a solution is formed. A *solution* is defined as a homogeneous mixture of two or more components. The component or substance whose physical state is preserved in the solution is known as the *solvent*. The other substance is called the *solute*. There can be more than one solute in a solution. If both the solvent and solutes have the same physical state as the solution, the component that is present in excess is generally called the solvent, but the division into solvent and solute(s) is not always distinct.

As in Chapter 8, the term "homogeneous" means that the solution consists of only one phase: solid, liquid, or gas. In general, this means that the solute(s) is dispersed uniformly throughout the solvent on an atomic, ionic, or molecular scale, so that there are no distinct boundaries between phases. Figure 9.1

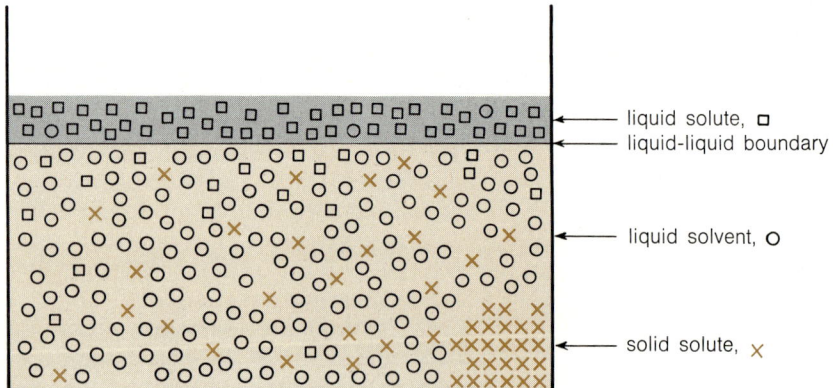

Figure 9.1. Representation of a solid (**X**) and a liquid (□) dissolving in a liquid solvent (O). Some of the solvent also has dissolved in the liquid solute.

liquid solute, □
liquid-liquid boundary

liquid solvent, O

solid solute, X

illustrates schematically a solid and a liquid solute undergoing solution in a liquid solvent.

The opposite of a homogeneous solution is a heterogeneous mixture consisting of distinct phases with observable boundaries between the phases. Examples of heterogeneous mixtures are sand and water (a solid and a liquid phase), sugar and salt (two solid phases with distinct boundaries between the grains of sugar and salt), and water and oil (two liquid phases that, even if shaken vigorously, still possess easily observable boundaries). The latter two examples show that individual phases, even of the same physical state, have uniform and distinct chemical and physical properties.

The term "mixture" indicates that the composition of the solution (i.e., the relative proportions of the components) may vary. A mixture does not need to have a definite chemical composition as do chemical compounds. Some examples of solutions are listed here.

1. A solid in a liquid solvent: sugar dissolved in water.
2. A gas in a liquid solvent: oxygen gas dissolved in water; ammonia in water.
3. A liquid in a liquid: ethyl alcohol in an equal volume of water. (In this particular example, there is no clear distinction between solvent and solute.)
4. A gas in a gas: air.
5. A solid in a solid: alloys such as brass (copper and zinc).

There are other kinds of solutions (gas or liquid in solid solvent), but they do not concern us here. Our attention will be confined primarily to liquid solutions and, among these, to aqueous solutions, those in which water is the solvent.

9.2 MECHANISM OF SOLUTION

When a solute dissolves in a solvent, its component particles end up randomly dispersed throughout the system. To accomplish this dispersion, three processes must occur. First, the solute particles must be separated from each other. Second, some solvent particles must be separated from each other to provide room for the dispersed solute particles. These two processes involve the disruption of solute–solute and solvent–solvent attractive forces, respectively. Such forces are known collectively as *cohesive* forces. Third, *adhesive* forces between solute and solvent particles come into play if dissimilar particles show an attraction for each other. The energy released by the attraction of solvent and solute provides some of the energy required to separate the particles of the solute and of the solvent.

With respect to energy, a solute will tend to dissolve appreciably in a solvent if the energy evolved in forming the adhesive interactions is equal to or greater than the energy needed (absorbed) for the disruption of the cohesive forces in the solute and the solvent. That is, considering only energy factors, for a solution to result, the attraction between solute and solvent particles must be at least equal to the sum of the solute–solute and solvent–solvent attractions. On the other hand, if there are very strong cohesive forces in either solute or solvent, the intermingling of solute and solvent may be negligible and no solution will result. For instance, a nonpolar gas such as methane (CH_4) is only slightly soluble in water, because CH_4 and H_2O molecules do not interact, whereas methyl alcohol (CH_3OH) is very soluble, because it can hydrogen bond with the water molecules. To look at the mechanisms of solution formation a little more closely, we present two broad categories based on the relative strengths of cohesive and adhesive forces.

Strong adhesive forces can arise from two phenomena that are not often easily distinguishable, namely, chemical reaction and solvation of the solute.

Chemical reaction Evidence of a strong attraction between solute and solvent exists when a definite chemical reaction occurs upon mixing the components. For instance, solid sodium metal is soluble in water as shown in the reaction:

$$2Na(s) + 2H_2O(l) \longrightarrow 2Na^+(aq) + 2OH^-(aq) + H_2(g)$$

The sodium hydroxide exists in the excess water as sodium and hydroxide ions. In such cases, the substances in the resulting solution are different from the substances originally mixed.

Another example, in which the chemical reaction is less noticeable, occurs with the solution of the covalent molecule of hydrogen chloride gas (HCl) in water:

$$HCl(g) + H_2O(l) \longrightarrow H_3O^+(aq) + Cl^-(aq)$$

Here $H_3O^+(aq)$, called the hydronium ion, is formed by the attraction of the polar water molecule for the positive hydrogen ion derived from HCl. It is a hydrated proton, $H^+(H_2O)$.

Solvation Solute–solvent interaction leading to solution may occur without chemical reaction. This type of interaction is called *solvation* of solute particles by the solvent. When water is the solvent, this process is called *hydration*. Hydration is of particular importance with respect to charged solute particles. An example will illustrate the process.

Common salt, sodium chloride, is a solid composed of sodium ions and chloride ions. When NaCl is placed in water, the ions attract polar water molecules and separate. The negative chloride ions attract the positive (hydrogen) end of several water dipoles, while the positive sodium ions attract the negative (oxygen) end of several water molecules.

$$Cl^- + yH_2O \longrightarrow Cl^-(H_2O)_y \quad \text{or} \quad Cl^-(aq)$$

$$Na^+ + xH_2O \longrightarrow Na^+(H_2O)_x \quad \text{or} \quad Na^+(aq)$$

The water molecules are bound to the ions by ion–dipole interactions, as shown in Figure 9.2. The number of water molecules (x or y) in the primary hydration layer around an ion depends on the size and charge of the ion; but there are rarely more than six water molecules. Hydration of the solute ions releases energy that usually compensates for the energy needed to overcome cohesive forces.

Although aqueous solutions of HCl and NaCl both contain hydrated ions, the ions result from different solution processes. The difference is due to the types of bonding in HCl gas and in solid NaCl. Solid NaCl consists of sodium and chloride ions held together in the crystal by simple ionic bonds. The ions are not formed by dissolving NaCl in water; they are merely separated. But no ions are initially present in the polar covalent HCl molecule, and ions

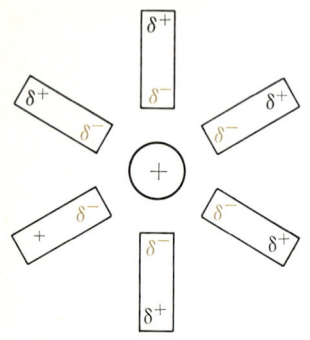

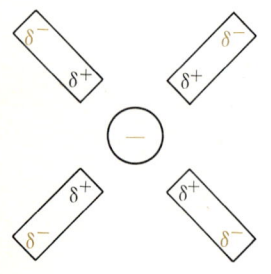

Figure 9.2. Hydration of ions in solution. The primary hydration layer, consisting of the water molecules immediately surrounding the ion, is shown.

are formed and hydrated only when water is added. Thus, different species exist in HCl gas and in aqueous HCl.

For ease in writing equations, it is customary not to show water of hydration attached to ions in aqueous solution. All ions are hydrated in water, and it is sufficient to show this by writing $H^+(aq)$ or $Na^+(aq)$ or $Cl^-(aq)$. The abbreviation (aq) is also used to show that a molecular species is present in aqueous solution. Since most of the ionic reactions of interest take place in water, it is even more usual to write simply H^+, Na^+, Cl^-, and so on and take the presence of water for granted. Thus, the solution of HCl and NaCl is simplified to

$$HCl \rightleftharpoons H^+ + Cl^-$$

and

$$NaCl \rightleftharpoons Na^+ + Cl^-$$

In addition to its ability to solvate ions, water has two other features that contribute to its capacity to dissolve ionic and polar covalent compounds. Both of these features, dielectric constant and hydrogen bonding, are related to the polar nature of the water molecule.

Dielectric constant. Coulomb's law for the electrostatic force of attraction F between oppositely charged ions, $+q$ and $-q$, that are a distance r apart, is

$$F = -\frac{q^2}{Kr^2}$$

The term K is called the dielectric constant of the medium between the ions. The dielectric constant of a vacuum is unity, and it is almost this for air. The dielectric constant of water is 80, which is unusually large for liquids, as Table 9.1 shows. This means that the attractive force between two ions

Table 9.1
Dielectric constants of liquids (20°C)

water, H_2O	80
methyl alcohol, CH_3OH	33
ethyl alcohol, C_2H_5OH	24
carbon tetrachloride, CCl_4	2.2
chloroform, $CHCl_3$	4.8
benzene, C_6H_6	2.3
ammonia, NH_3	17
hydrogen cyanide, HCN (16°C)	123
hydrogen fluoride, HF (0°C)	84

in an aqueous medium is decreased some 80-fold compared to the force in the ionic crystal. When an ionic solid is placed in water, the entry of water of hydration between ions reduces their cohesive force and allows them to be separated from each other more easily. The high dielectric constant of water and other highly polar liquids is the result of polar molecules lining up against the electric field between the ions and diminishing part of the field strength, as shown in Figure 9.3.

Hydrogen bonding. Even in the absence of chemical reaction, water may dissolve polar covalent compounds to varying extents because of the possibility

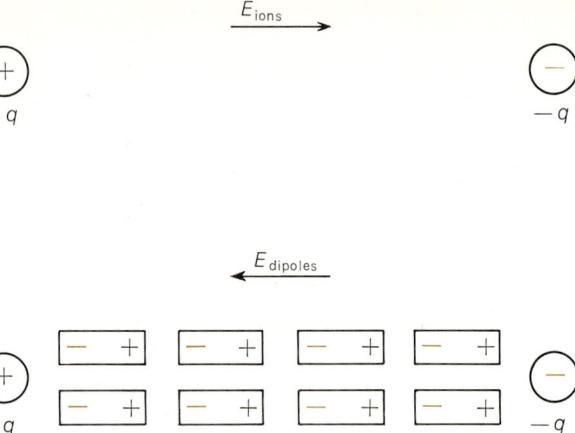

Figure 9.3. Effect of water dipoles on the electrostatic force between two ions. The electric field E_{ions} between the two ions in a vacuum is directed to the right. The alignment of water dipoles in the electric field induces an electric field $E_{dipoles}$ directed to the left. The resultant field between the ions in the presence of the polarized water dipoles at 20°C is $E_R = E_{ions} - E_{dipoles} = \frac{1}{80} E_{ions}$. Because the attractive force between the two ions is $F = Eq$, this force is decreased by a factor of 80 due to the water molecules.

of dipole–dipole interactions between the polar solute and polar solvent molecules. Generally, however, these adhesive forces are not as strong as the cohesive forces due to hydrogen bonding in water itself. Exceptions occur with polar covalent solutes that are able to hydrogen bond with water molecules. Such molecules contain O—H or N—H bonds or oxygen or nitrogen atoms carrying negative charges. Examples are

$$
\begin{array}{cc}
 & \text{O—H} \\
 & | \\
 & \text{H} \\
 & \vdots \\
\text{H} & \text{R—O---H—O} \\
| & | \quad\quad | \\
\text{H—N---H—O} & \text{H} \quad\;\; \text{H} \\
| \quad\quad | & \vdots \\
\text{H} \quad\;\; \text{H} & \text{O—H} \\
 & | \\
 & \text{H} \\
\text{ammonia} & \text{alcohols, ROH}
\end{array}
$$

$$
\begin{array}{cc}
 & \text{CH}_3 \\
 & | \\
 & \text{CH}_3\text{—C}=\text{O---H—O} \\
\text{CH}_3\text{—O---H—O} & \quad | \quad\quad | \\
\quad\quad\; | \quad\;\; | & \quad \text{H} \quad\quad \text{H} \\
\quad\;\; \text{CH}_3 \;\; \text{H} & \quad | \\
 & \text{H—O} \\
\text{dimethyl ether} & \text{acetone}
\end{array}
$$

Cohesive and Adhesive Forces Equal

If the difference between cohesive and adhesive forces is slight, a solution can still be obtained. Strong solute–solvent forces are not necessary for solution, provided that the cohesive forces are as weak as the adhesive forces. In this situation, solute particles disperse throughout the solvent and form a solution. Dispersion occurs because physical and chemical systems naturally tend to assume a more disorderly or more probable state, even if there is no overall energy change. Thus, nonpolar molecules of one substance readily dissolve nonpolar molecules of another substance, such as when nonpolar dry cleaning fluids dissolve grease. Under such circumstances, where any inter-molecular forces are weak, adhesive forces need only be greater than or equal to cohesive forces. The tendency toward disorder is measured by a quantity called *entropy*, which is discussed in detail in Chapter 16.

9.3 CONCENTRATION OF SOLUTIONS

To have any meaningful discussion of solutions, we must specify the relative amounts of solute and solvent. In a qualitative sense, we can say that a solution containing a small amount of dissolved solute in a comparatively large amount of solvent is a *dilute* solution. Conversely, if the amount of solute is large compared to the amount of solvent, then the solution is *concentrated*.

Eventually, we shall need to know the exact amount of solute dissolved in a specified amount of solvent or solution. This quantitative specification is the *concentration* of the solution. The concentration of a solution can be given in several ways, and we shall define four of the more widely used expressions of concentration. Other concentration terms, which are used in certain areas of study, will be defined as needed. You have already encountered two of these less general terms: parts per million and percent by volume. The most commonly used concentration terms are molarity (M), molality (m), mole fraction (X), and percent by weight (% or wt %).

Molarity (*M*)

The molarity of a solute is defined as the number of moles of solute in 1 liter (1000 ml) of solution. In equation form, an operational definition of molarity is

$$M = \frac{\text{moles of solute}}{\text{liters of solution}}$$

A $1M$ solution of a substance will have 1 mol of that substance dissolved in enough solvent to make 1 liter of solution, or 0.5 mol of solute dissolved in enough solvent to make 500 ml of solution, and so on.

Molality (*m*)

The number of moles of solute dissolved in 1000 g of *solvent* specifies the molality of a solution. In equation form,

$$m = \frac{\text{moles of solute}}{\text{kilograms of solvent}}$$

$$= 1000 \frac{\text{moles of solute}}{\text{grams of solvent}}$$

If 0.5 mol of solute has been dissolved in 1000 g of solvent, a $0.5m$ solution of solute results.

Mole Fraction (*X*)

The mole fraction of a component (solute or solvent) in a solution is the ratio of the number of moles of that component to the total number of moles of all components present in the solution. If n_1 is moles of solvent and n_2, n_3, ... stand for moles of solutes 2 and 3, then the mole fraction of solvent is

$$X_1 = \frac{n_1}{n_1 + n_2 + n_3 + \cdots}$$

and the mole fraction of solute 2 is

$$X_2 = \frac{n_2}{n_1 + n_2 + n_3 + \cdots}$$

The mole fractions of all components in a solution always add to one:

$$X_1 + X_2 + X_3 + \cdots = 1$$

If a solution consists of 1 mol of a solute and 4 mol of solvent, the mole fraction of solute is $1/(1 + 4) = 0.2$, and the mole fraction of solvent is $4/(1 + 4) = 0.8$.

Percent Composition (%) Percent by weight (as opposed to percent by volume) is usually meant when the percent of a solute is specified. Percent by weight is the number of parts, by weight, of solute in 100 parts, by weight, of *solution*. Stated differently, the percent solute is 100 times the fraction of the total solution weight that is solute. In equation form, the weight percent of a solute 2 is

$$\% = 100 \times \frac{w_2}{w_1 + w_2 + w_3 + \cdots}$$

where w_1 is the weight of solvent in the solution and w_2, w_3, . . . are the weights of solutes. For instance, if 100 g of solution contains 25 g of solute, the solution is 25 percent solute and 75 percent solvent.

A few examples can best show how these definitions can be applied to various situations.

EXAMPLE A solution is made by dissolving 18.0 g of glucose ($C_6H_{12}O_6$) in enough water to make 200 ml of solution. What is the molarity, the percent composition, the molality, and the mole fraction of this solution?

SOLUTION *Molarity* (M). The molecular weight of glucose is 180 g/mol; 18.0 g then represents 18.0 g/180 g/mol or 0.100 mol of glucose in 200 ml of solution. By proportion, the number of moles of glucose that would be present in five times this much solution, or in 1000 ml = 1 liter is the molarity M and is given by

$$\frac{0.100 \text{ mol}}{200 \text{ ml}} = \frac{M \text{ mol}}{1000 \text{ ml/liter}}$$

$$M = 0.100 \text{ mol} \left(\frac{1000 \text{ ml/liter}}{200 \text{ ml}} \right)$$

$$= 0.500 \text{ mol/liter} = 0.500M \ C_6H_{12}O_6$$

Another way to find the molarity is to use the relation

$$M = \frac{\text{moles of solute}}{\text{liters of solution}}$$

Then

$$M = \frac{18.0 \text{ g}/180 \text{ g/mol}}{0.200 \text{ liter}}$$

$$= \frac{0.100 \text{ mol}}{0.200 \text{ liter}} = 0.500 \text{ mol/liter}$$

$$= 0.500M$$

Percent (%). The percent composition cannot be found unless the density of the solution is known. If the density is 1.032 g/ml, then the 200 ml of solution weighs 200 ml × 1.032 g/ml = 206 g. The weight fraction of glucose in the solution is

$$\frac{18.0 \text{ g}}{206 \text{ g}} = 0.0875$$

The percent by weight of glucose in the solution is 0.0875 × 100 = 8.75 percent $C_6H_{12}O_6$.

The percent could also have been found by proportion. If there are 18.0 g of glucose in 206 g of solution, then

$$\frac{\%}{100} = \frac{18.0 \text{ g}}{206 \text{ g}}$$

$$\% = 18.0 \left(\frac{100}{206}\right) = 8.75\%$$

Molality (m). Knowing the weight of the solution from the percent calculation, we can find the weight of solvent. It is 206 g − 18.0 g = 188 g = 0.188 kg of H_2O. By proportion, the number of moles of glucose in 1000 g of H_2O is the molality m and is given by

$$\frac{m}{1000} = \frac{0.100 \text{ mol}}{188 \text{ g}}$$

$$m = 0.100 \text{ mol} \left(\frac{1000 \text{ g}}{188 \text{ g}}\right)$$

$$= 0.532m \ C_6H_{12}O_6$$

Or, using the equation

$$m = \frac{\text{moles of solute}}{\text{kilograms of solvent}}$$

the molality is

$$m = \frac{0.100 \text{ mol}}{0.188 \text{ kg}} = 0.532m \ C_6H_{12}O_6$$

Mole fraction (X). The number of moles of glucose is 0.100 mol. The number of moles of water is

$$\frac{188 \text{ g}}{18.0 \text{ g/mol}} = 10.4 \text{ mol}$$

(It is just a coincidence that the molecular weight of water is the same as the weight of glucose.)

The total number of moles in the solution is

10.4 mol of H_2O + 0.100 mol of glucose = 10.5 mol

The mole fraction of glucose, X_g, is

$$X_g = \frac{0.100 \text{ mol}}{10.5 \text{ mol}} \simeq 0.0100$$

The mole fraction of water, X_w, is

$$X_w = \frac{10.4 \text{ mol}}{10.5 \text{ mol}} \simeq 0.990$$

As a check, the mole fractions of all components should add to one: $0.99 + 0.01 = 1.00$.

EXAMPLE | A solution is made by dissolving 2.92 g of NaCl in 100 g of water. Since the weight of solvent (H_2O) is given, find the molality, percent, and mole fraction.

SOLUTION | *Molality.* The molecular weight (or formula weight) of NaCl is 58.4 g/mol, so the number of moles of NaCl present is 2.92 g/58.4 g/mol = 0.0500 mol. The molality can be calculated by proportion or through the use of an equation. Using proportion,

$$\frac{m}{1000 \text{ g}} = \frac{0.0500 \text{ mol}}{100 \text{ g}}$$

$$m = 0.0500 \text{ mol} \left(\frac{1000 \text{ g}}{100 \text{ g}} \right)$$

$$= 0.500m \text{ NaCl}$$

Or, using the equation

$$m = \frac{\text{moles of solute}}{\text{kilograms of solvent}}$$

$$= \frac{0.0500 \text{ mol}}{0.1000 \text{ kg}} = 0.500m \text{ NaCl}$$

Percent. The weight fraction of NaCl is

$$\frac{2.92 \text{ g}}{2.92 \text{ g} + 100 \text{ g}} = 0.0284$$

The percent of NaCl is

$$0.0284 \times 100 = 2.84\% \text{ NaCl}$$

Mole fraction. The number of moles of NaCl present in the solution is 0.0500 mol. The total number of moles is 0.0500 mol + 5.55 mol = 5.60 mol. The mole fraction of NaCl is

$$X_{\text{NaCl}} = \frac{0.0500 \text{ mol}}{5.60 \text{ mol}} = 0.00893$$

Molarity. To find the molarity, we must know the final volume of the solution or relevant data to calculate the solution volume. For dilute aqueous solutions, the molarity and the molality are approximately equal.

EXAMPLE | A solution has a solute concentration of 0.50M. The solute has a molecular weight of 40 g/mol. How many moles and how many grams of solute are there in 100 ml of this solution?

SOLUTION | Again, we can use either proportion or the relationship $M = $ moles/volume. By proportion, since there is 0.50 mol of solute in 1.0 liter (1000 ml) of solution, the number of moles n of solute in 100 ml (0.10 liter) of solution is given by

$$\frac{n}{0.10 \text{ liter}} = \frac{0.50 \text{ mol}}{1.0 \text{ liter}}$$

$$n = 0.50 \text{ mol}\left(\frac{0.10 \text{ liter}}{1.0 \text{ liter}}\right)$$

$$= 0.050 \text{ mol in } 0.10 \text{ liter}$$

Either milliliters or liters may be used so long as consistent units are maintained throughout the problem.

If the equation is used, the result is

$$\text{moles of solute } n = M \times \text{volume in liters}$$

$$= 0.50 \text{ mol/liter} \times 0.10 \text{ liter}$$

$$= 0.050 \text{ mol}$$

The grams of solute in 100 ml are

$$0.050 \text{ mol} \times 40 \text{ g/mol} = 2.0 \text{ g}$$

In addition to concentration problems involving only one solution, we shall often encounter problems concerning the dilution of a solution by the addition of more solvent, and also the mixing of two nonreacting solutions. A few examples are given here.

Dilution | The addition of more solvent to a solution does not affect the number of moles of solute originally present. The number of moles of solute initially present, n_i, will be the same as the number of moles after dilution, n_f. Because this number of moles equals the initial molarity M_i multiplied by the initial volume V_i of the solution in liters, we can write

$$n_i = n_f$$
$$M_i V_i = M_f V_f$$

where the subscripts i and f refer to the initial and final solutions.

EXAMPLE | What is the molarity of a solution made by dilution of 0.50 liter of 0.20M solution to a final volume of 1.5 liters?

SOLUTION |
$$M_i V_i = M_f V_f$$

$$M_f = M_i \times \frac{V_i}{V_f}$$

$$= 0.20M\left(\frac{0.50 \text{ liter}}{1.5 \text{ liters}}\right) = 0.067M$$

Mixing of Solutions | When two or more solutions having a solute in common are mixed and no chemical reaction occurs, the number of moles of that solute in the mixture is the sum of the moles of solute in the solutions before mixing, and the new molarity is $(n_1 + n_2)$/total volume.

EXAMPLE | What is the molarity of the solution resulting from the addition of 200 ml of 0.15*M* NaCl to 100 ml of 0.50*M* NaCl?

SOLUTION | Total number of moles of NaCl = $(0.15M)(0.2 \text{ liter}) + (0.50M)(0.10 \text{ liter}) = 0.080 \text{ mol.}$

$$\text{total volume} = 300 \text{ ml} = 0.30 \text{ liter}$$

$$\text{final molarity } M = \frac{\text{total moles}}{\text{total volume } (l)}$$

$$= \frac{0.080}{0.30} = 0.27M$$

9.4 ELECTROLYTES AND NONELECTROLYTES

We have already noted that ions result when some substances are dissolved in water. Some of these are substances in which ions preexisted, for example, ionic compounds such as sodium chloride; others are covalently bonded molecules such as HCl. The extent to which these substances yield ions when dissolved in a solvent, and the resulting electrical behavior of the solutions, provides a convenient basis for classifying solutes, especially in aqueous solutions.

The term *electrolyte* is applied to those substances whose solutions (not necessarily in water) contain positive and negative ions. There are two subclasses of electrolytes, strong electrolytes and weak electrolytes. Substances whose solutions contain no ions are called *nonelectrolytes*. These terms describe the electrical behavior of the solutions as well as the substances themselves.

The electrical behavior of solutions, particularly their resistance to the passage of an electric current, provides much information about the characteristics and number of species in solution, the bond structure of the solute, and the nature of the solvent. Unlike metals, in which current is carried by electrons, solutions—especially aqueous solutions—conduct electricity by means of the motion of positive and negative ions flowing in opposite directions under an applied voltage. The two types of conduction are called, respectively, metallic and electrolytic conduction.

An experimental set-up for determining the conductivity of a solution is shown in Figure 9.4. Two chemically inert electrodes are immersed in the solution and are connected to a source of alternating current. Joined in series with the electrodes and the current source is a current-measuring device, either a meter or an ordinary light bulb. If there are ions in the solution, negative ions move toward the positive electrode, where they transfer electrons to the electrode, and positive ions move toward the negative electrode, where they pick up electrons. This transfer of charge constitutes the current flowing through the solution. (The polarity of the electrodes periodically changes when alternating current is used, but this circumstance does not change this description. An alternating current is used in conductivity measurements to avoid chemical changes at the electrodes. See Section 13.20 for more details.)

When there are ions in the solution, these ions carry current between the electrodes. The more ions that are present, the greater the amount of charge flow per unit time, the higher the current, and the brighter the light bulb. The current that flows is a measure of the conductivity of the solution. No current flow occurs with a nonelectrolyte.

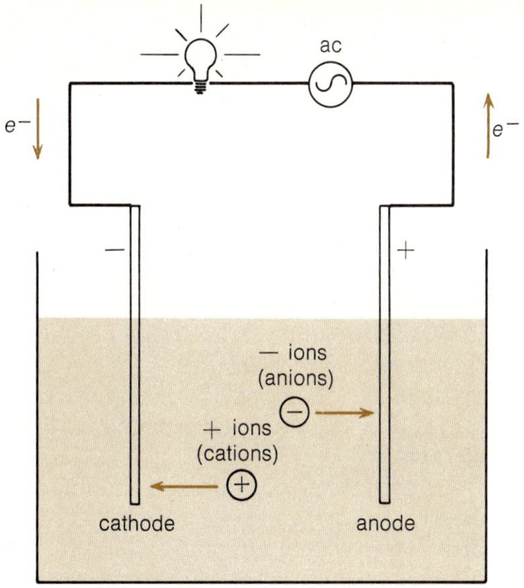

Figure 9.4. Apparatus for detection of electric conductivity of a solution. The greater the conductivity, the brighter the light. The electrode polarity shown here corresponds to an instantaneous polarity when alternating current is used. Electrons move through the external (metallic) circuit in the direction shown, with the indicated polarity. In the solution, positive ions called cations move to the negative electrode, called the cathode, where they are reduced; negative ions called anions move to the positive electrode, called the anode, where they are oxidized.

The existence of ions in a particular solution depends on the nature of both the solute and solvent. The observed conductivity depends on factors such as ion concentration, size, and charge. The latter two factors influence the speed at which the ions move. Soluble substances with predominantly covalently bonded atoms generally have discrete molecules as the units of crystal structure. These substances will maintain their molecular integrity when dissolved and produce a mixture of electrically neutral molecules. Because no ions are present to serve as charge carriers, the solution is a nonconductor, and the solute is a nonelectrolyte. Examples of nonelectrolytes are sugar ($C_{12}H_{22}O_{11}$), acetone (CH_3COCH_3), and alcohols (ROH).

However, if some bonds in the solute molecule are polar covalent, it is possible that solution in a polar solvent (e.g., water) will break the bonds and result in ionic fragments. The formation of ions from dissolved covalent compounds is known as *ionization*. Then the solution may conduct electricity, and the solute is an electrolyte. If all or almost all solute molecules dissolved in the solvent ionize, then the solute is a *strong electrolyte* and the light bulb in Figure 9.4 will glow brightly, because the solution is a good conductor. When only a few of the dissolved solute molecules ionize, the solute is a *weak electrolyte* and the resulting solution is a poor conductor. For instance, HCl is a strong electrolyte in water. The compound $HgCl_2$ and organic acids (RCOOH) such as acetic acid are weak electrolytes because only a small portion of the dissolved molecules ionize. In most covalent compounds, bond dissociation in water usually occurs at the most nearly ionic bond as determined from electronegativity differences.

$$Na\text{---}OH \longrightarrow Na^+ + OH^-$$

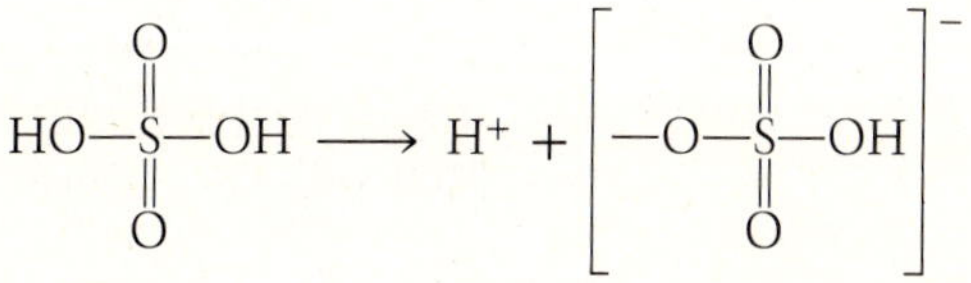

$$RC-OH \longrightarrow \left[R-\overset{\textstyle}{\underset{\textstyle O}{C}}-O- \right]^{-} + H^{+}$$

All ions here are assumed to have water of hydration attached to them.

Polar covalent substances will ionize in water if the energy gained by hydration of the ions is larger than that needed to break the bond. Some other examples besides HCl are HBr and HNO_3, in which the small hydrogen ion has a high hydration energy in comparison to the bond strength.

In ionic compounds such as NaCl, ions already exist in the solid state. In water, the oppositely charged ions separate or dissociate because the electrostatic force of attraction between them is reduced. It is strictly correct to say that ionic compounds dissociate in water while polar covalent compounds ionize, but the terms "dissociation" and "ionization" are usually used interchangeably.

Compounds, regardless of solubility, in which all the *dissolved* molecules dissociate (or ionize) are generally called strong electrolytes. For example, $BaSO_4$ is a strong electrolyte; but it is only sparingly soluble in water, and its most concentrated solution may contain only as many ions as a solution of a weak electrolyte. As a result, $BaSO_4$ solutions are poor conductors.

Although we are concerned only with aqueous solutions, we should mention that a substance that is a strong electrolyte in water may be a nonelectrolyte in some other solvent. For instance, HCl ionizes completely in water, but it remains in the molecular form in a nonpolar solvent such as benzene (C_6H_6) or carbon tetrachloride. Also, in solvents with low dielectric constants as well as in concentrated aqueous solutions, solvated ions tend to form ion pairs held together by weak electrostatic attraction. The hydration layers surrounding each ion in the ion pair remain fairly distinguishable and do not merge appreciably. The two hydrated ions move near each other but do not become strongly bonded to each other. The net charge of the ion pair need not be zero. Examples of ion pairs are $Ca^{2+} \cdot CO_3^{2-}$ [or $CaCO_3(aq)$], $Na^+ \cdot CO_3^{2-}$, and $Mg^{2+} \cdot HCO_3^-$. Ion pair formation is favored by small, highly charged ions.

Unless otherwise noted, the rest of this chapter deals only with solutions of strong electrolytes, that is, with substances that ionize or dissociate completely when dissolved in water. Many of the compounds that behave this way are called *salts*. Salts will be defined as ionic compounds composed of positive metal ions (or NH_4^+, but not H^+) and negative nonmetal ions (but not OH^-).

9.5 SOLUBILITY EQUILIBRIA

Just as it is not sufficient to rely on qualitative terms such as "dilute" and "concentrated," it is not enough to say that a substance is soluble, slightly soluble, or insoluble in a solvent. A quantitative measure of solubility is needed. To obtain such a measure, we can look at what happens when we add so much of a solute to a solvent that no more of the solute will dissolve. For instance, sugar is soluble in water but it cannot be added to water continually without some eventually settling to the bottom of the container as a solid phase that will not dissolve no matter how much the solution is stirred. A state of heterogeneous equilibrium has been reached. There are two phases present: solid solute and liquid solution. Two reactions are going on, and there is a state of dynamic equilibrium. What are these opposing reactions?

When a solution will not dissolve any more solute, and when pure solute still is in contact with the solution, the two reactions that are going on at equal rates are the dissolving of solute and the precipitation or departure of the dissolved solute from the solution. The equilibrium process can be written as

$$\text{solvent} + \text{solute} \rightleftharpoons \text{solution}$$

or simply

$$\text{solute} \rightleftharpoons \text{solution}$$

Specific examples are

$$C_{12}H_{22}O_{11}(s) \rightleftharpoons C_{12}H_{22}O_{11}(aq)$$
$$NaCl(s) \rightleftharpoons Na^+ + Cl^-$$
$$CaCl_2(s) \rightleftharpoons Ca^{2+} + 2Cl^-$$

In these equations, the ions on the right side of the arrows are hydrated ions in solution. At equilibrium no more solute will dissolve, because solute particles (either ions or molecules) are leaving the surface of the solid to go into the solution as fast as solute particles in solution are returning to the excess solid phase and condensing on its surface. The opposing processes are shown schematically in Figure 9.5.

A state of heterogeneous equilibrium can also occur when a liquid or a gas is dissolved in water. For a liquid, the equilibrium is

$$\text{pure liquid} \rightleftharpoons \text{liquid in solution}$$

As an example, consider the two-phase system of oil and water. At equilibrium (Figure 9.6), the solution may be described by the equations

$$\text{oil in water} \rightleftharpoons \text{oil (in oil)}$$
$$\text{water in oil} \rightleftharpoons \text{water (in water)}$$

A similar equilibrium will be set up for a gas–liquid system such as

$$O_2 \text{ (in air)} \rightleftharpoons O_2 \text{ (in water)}$$

When no more solute will dissolve in a solution at a given temperature, the solution is said to be *saturated* with that solute. If more solute will dissolve,

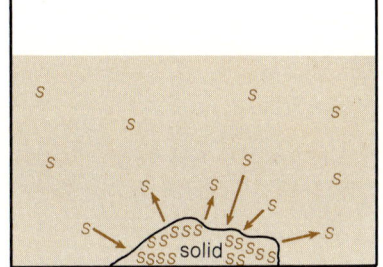

Figure 9.5. Dynamic equilibrium in solution. Solid solute and saturated solution are in equilibrium, because the rate of passage of solid solute into solution equals the rate of deposition of dissolved solute particles s on the solid.

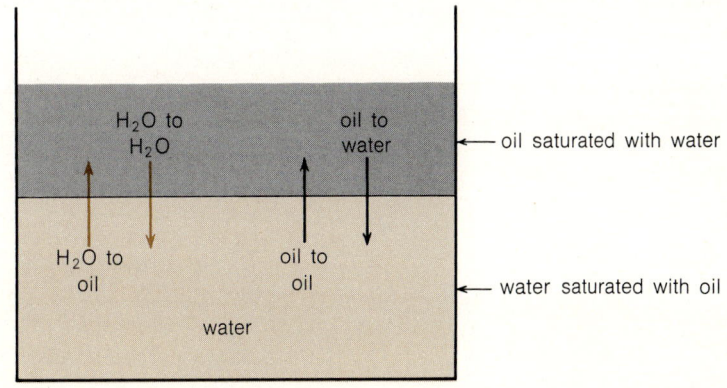

Figure 9.6. Two immiscible liquids in equilibrium with one another. Liquids such as oil and water, which are only sparingly soluble in each other, are said to be immiscible. The oil layer is a saturated solution of water in oil; the water layer is saturated with oil.

the solution is called *unsaturated*. When a saturated solution is in contact with excess solid solute, the solute and the solution are in equilibrium. Do not confuse dilute with unsaturated. A saturated solution of a sparingly soluble substance will also be dilute. Conversely, it is possible to have an unsaturated solution of a highly soluble substance that is concentrated.

9.6 THE SOLUBILITY PRODUCT

Consider an ionic solid in equilibrium with its solution at some temperature. As an example, let us choose silver chloride (AgCl), which is highly insoluble. Whatever AgCl does dissolve goes into solution as (hydrated) silver and chloride ions. Hence, excess solid is in contact with the solution.

$$AgCl(s) \rightleftharpoons Ag^+ + Cl^-$$

When equilibrium has been established between the solid and the solution, the concentrations of silver and chloride ions in the solution no longer change. It has been determined experimentally that the product of their concentrations is a constant, which is called the *solubility product constant*, or simply *solubility product*, and is denoted by K_{sp}:

$$K_{sp} = [Ag^+][Cl^-]$$

where the brackets represent concentrations in moles per liter and in this equation represent equilibrium concentrations.

For a solute that gives more than two ions, such as calcium phosphate, $Ca_3(PO_4)_2$, the solubility product constant is equal to the product of the concentrations of the ions, each raised to a power equal to its coefficient in the balanced equation for the dissociation of the solid.

$$Ca_3(PO_4)_2(s) \rightleftharpoons 3Ca^{2+} + 2PO_4^{3-}$$

$$K_{sp} = [Ca^{2+}]^3[PO_4^{3-}]^2$$

In general, for a compound M_mA_a,

$$M_mA_a \rightleftharpoons mM^{a+} + aA^{m-}$$

$$K_{sp} = [M]^m[A]^a$$

The value of the solubility product for any salt must be determined experimentally.

EXAMPLE | A saturated solution of calcium fluoride (CaF_2) at 25°C has a concentration of F^- ion equal to $4.6 \times 10^{-4}M$. The dissociation equation

$$CaF_2 \rightleftharpoons Ca^{2+} + 2F^-$$

indicates that there should be only half this much Ca^{2+}, so $[Ca^{2+}) = 2.3 \times 10^{-4}M$.

$$
\begin{aligned}
K_{sp} &= [Ca^{2+}][F^-]^2 \\
&= (2.3 \times 10^{-4})(4.6 \times 10^{-4})^2 \\
&= (2.3 \times 10^{-4})(21.2 \times 10^{-8}) \\
&= 49 \times 10^{-12} \\
&= 4.9 \times 10^{-11}
\end{aligned}
$$

The solubility product of a salt changes with temperature; the direction of change depends on the heat of solution. Table 9.2 lists the K_{sp} values for a number of salts at 25°C.

If, instead of specifying the concentration of one ion in a saturated solution, we give the salt's solubility in grams per liter or moles per liter, K_{sp} still can be calculated.

EXAMPLE

A 0.014-g sample of silver chromate, Ag_2CrO_4, dissolves in 0.50 liter of solution. What is the K_{sp} of Ag_2CrO_4?

SOLUTION

The molecular weight (or gram-formula weight) of silver chromate is 332 g/mol; hence, 0.014 g is 0.014 g/(332 g/mol) = 4.2×10^{-5} mol dissolved in 0.50 liter or 8.4×10^{-5} mol dissolved in 1 liter.

According to the dissociation equation

$$Ag_2CrO_4 \rightleftharpoons 2Ag^+ + CrO_4^{2-}$$

the number of moles of CrO_4^{2-} is equal to the number of moles of Ag_2CrO_4 that dissolve. And for every mole of Ag_2CrO_4 dissolved, there are 2 mol of Ag^+ produced. Therefore,

$$[CrO_4^{2-}] = 8.4 \times 10^{-5} M$$
$$[Ag^+] = 2[CrO_4^{2-}] = 1.7 \times 10^{-4} M$$

and

$$K_{sp} = [Ag^+]^2[CrO_4^{2-}]$$
$$= (1.7 \times 10^{-4})^2(8.4 \times 10^{-5})$$
$$= 2.4 \times 10^{-12}$$

The solubility of a salt in moles per liter is often represented by the letter s. Thus, for our Ag_2CrO_4 solution, $[Ag^+] = 2s$,

$$s = [CrO_4^{2-}] = \frac{[Ag^+]}{2}$$

and in a saturated solution of Ag_2CrO_4, $K_{sp} = 4s^3$.

9.7 COMMON ION EFFECT

The previous examples apply to solutions containing ions derived from only one slightly soluble salt. When the concentration of one or more of the ions derived from a slightly soluble salt is augmented by the presence in solution of a strong electrolyte that furnishes an ion in common with the salt, the solubility of the salt decreases. To understand why this is so, let's first examine this common ion effect in a qualitative way. Using AgCl again as an example, we have

$$AgCl(s) \rightleftharpoons Ag^+ + Cl^-$$

If a saturated solution of AgCl has excess chloride ions contributed from some dissolved sodium chloride in the solution, Le Chatelier's principle tells us that the equilibrium represented by this equation is shifted to the left. In other words, the dissociation is repressed, and the amount of AgCl in solution is decreased. The AgCl concentration in moles per liter is now given

Table 9.2
Solubility products at 25°C

Compound	formula	K_{sp}
aluminum hydroxide	$Al(OH)_3$	5×10^{-33}
aluminum phosphate	$AlPO_4$	5.8×10^{-19}
barium carbonate	$BaCO_3$	5.1×10^{-9}
barium sulfate	$BaSO_4$	1.0×10^{-10}
calcium carbonate	$CaCO_3$	4.7×10^{-9}
calcium chromate	$CaCrO_4$	7.1×10^{-4}
calcium fluoride	CaF_2	4.9×10^{-11}
calcium hydroxide	$Ca(OH)_2$	5.5×10^{-6}
calcium oxalate	CaC_2O_4	2.1×10^{-9}
calcium phosphate	$Ca_3(PO_4)_2$	2.0×10^{-29}
calcium sulfate	$CaSO_4$	2.4×10^{-5}
chromium(III) hydroxide	$Cr(OH)_3$	7.0×10^{-31}
copper(I) sulfide	Cu_2S	1.0×10^{-48}
copper(II) carbonate	$CuCO_3$	2.5×10^{-10}
copper(II) hydroxide	$Cu(OH)_2$	2.2×10^{-20}
copper(II) sulfide	CuS	8×10^{-36}
iron(II) carbonate	$FeCO_3$	3.5×10^{-11}
iron(II) hydroxide	$Fe(OH)_2$	8.0×10^{-16}
iron(II) sulfide	FeS	5×10^{-18}
iron(III) hydroxide	$Fe(OH)_3$	6×10^{-38}
iron(III) phosphate	$FePO_4$	1.3×10^{-22}
iron(III) sulfide	Fe_2S_3	1×10^{-88}
lead bromide	$PbBr_2$	4.6×10^{-6}
lead carbonate	$PbCO_3$	1.5×10^{-13}
lead chloride	$PbCl_2$	1.6×10^{-5}
lead chromate	$PbCrO_4$	2.0×10^{-16}
lead hydroxide	$Pb(OH)_2$	4.0×10^{-15}
lead iodide	PbI_2	7.1×10^{-9}
lead sulfate	$PbSO_4$	1.7×10^{-8}
magnesium carbonate	$MgCO_3$	1×10^{-5}
magnesium hydroxide	$Mg(OH)_2$	1.1×10^{-11}
magnesium phosphate	$Mg_3(PO_4)_2$	$\sim 1 \times 10^{-27}$
manganese(II) carbonate	$MnCO_3$	8.8×10^{-11}
manganese(II) hydroxide	$Mn(OH)_2$	1.6×10^{-13}
mercury(I) chloride	Hg_2Cl_2	1.3×10^{-18}
mercury(II) chloride	$HgCl_2$	6.1×10^{-15}
mercury(II) sulfide	HgS (black)	3.0×10^{-52}
silver acetate	$AgC_2H_3O_2$	2.3×10^{-3}
silver bromide	$AgBr$	5.2×10^{-13}
silver chloride	$AgCl$	1.6×10^{-10}
silver chromate	Ag_2CrO_4	2.4×10^{-12}
silver iodide	AgI	8.3×10^{-17}
strontium carbonate	$SrCO_3$	7×10^{-10}
tin(II) hydroxide	$Sn(OH)_2$	1.6×10^{-27}
tin(IV) hydroxide	$Sn(OH)_4$	1.0×10^{-5}
zinc hydroxide	$Zn(OH)_2$	7×10^{-18}
zinc sulfide	ZnS	$\sim 1 \times 10^{-22}$

only by the silver ion concentration in the solution. The chloride ion concentration is greater than the solubility of AgCl, because a portion of the chloride is not derived from dissolved AgCl.

EXAMPLE | The solubility of AgCl in pure water is

$$s = [Ag^+] = [Cl^-]$$
$$= \sqrt{K_{sp}} = (1.6 \times 10^{-10})^{1/2} = 1.3 \times 10^{-5}M$$

In $0.010M$ KCl solution, the solubility of AgCl is still given by $s = [Ag^+]$. But $[Cl^-]$ now equals $0.010 + 1.3 \times 10^{-5} \simeq 0.010M$. Since the solution is still saturated with AgCl, the solubility product must still apply:

$$K_{sp} = [Ag^+][Cl^-]$$
$$\text{solubility of AgCl, } s = [Ag^+] = \frac{K_{sp}}{[Cl^-]}$$
$$s = \frac{1.6 \times 10^{-10}}{0.010} = \frac{1.6 \times 10^{-10}}{1.0 \times 10^{-2}} = 1.6 \times 10^{-8}M$$

The solubility of AgCl is 1300 times less in $0.010M$ KCl than in water.

9.8 PRECIPITATION OF SLIGHTLY SOLUBLE SALTS

The situations considered so far have been those where equilibrium between the dissolved salt and the solid salt prevailed. In such saturated solutions, the solubility product K_{sp} equals the product of the ionic concentrations in the saturated solution. A different situation occurs when solutions that separately contain the ions of a slightly soluble compound are mixed. This is a common occurrence in chemical processes in the laboratory and in nature. For example, suppose that solutions of $AgNO_3$ and NaCl are mixed. Depending on the concentrations of the ions in the resulting mixture, some of the Ag^+ and Cl^- ions may come together and precipitate from the solution as solid AgCl. In particular, we must determine whether the product of the concentrations of the ions (each to the appropriate power) of a slightly soluble salt in the resulting solution is greater or less than the numerical value of K_{sp}. The product of the ion concentrations in a solution is called the *ion product* and has the same form as K_{sp}, that is, $[M]^m[A]^a$, except that the concentrations are not necessarily the equilibrium concentrations found in saturated solutions.

We first consider the case where the ion product is greater than K_{sp}. This is generally a short-lived nonequilibrium situation, because the solution *and* the solid cannot be at equilibrium unless the ion product and K_{sp} are equal. The two phases reach equilibrium by the precipitation of the solid, which reduces the concentration of ions in solution until the ion product equals K_{sp} for the solid. The downward adjustment of the ion product and the resulting precipitation of solid often occur very rapidly, so that one could almost say that the ion product cannot exceed K_{sp} for a slightly soluble salt. Nevertheless, it is helpful to think about an instantaneous ion product that can exist in solution for a period of time before readjustment by precipitation takes place. During this time, which is usually but not always very brief, the solution is said to be *supersaturated* and holds more solute in solution than does a saturated solution at the same temperature.

If the ion product in solution is less than K_{sp}, the solution is *unsaturated* with respect to the ions of the salt. Equilibrium can be attained if enough solid is added to the system so that some (not all) of the solid dissolves and brings the concentration of ions in solution up to the point where the ion product equals K_{sp} and a saturated solution is produced. The only way we can be sure the solution is saturated is if some solid is present in equilibrium with the solution.

The relations between the ion product and K_{sp} for a slightly soluble salt and its solution may be summarized as follows:

1. If the ion product is less than K_{sp}, the solution is unsaturated. No solid phase is present, no precipitation occurs, and if solid phase is added, some or all of it will dissolve.

2. If the ion product equals K_{sp}, the solution is just saturated; no precipitate will form. Any solid precipitate present is in equilibrium with the solution and will not dissolve.

3. If the ion product is greater than K_{sp}, precipitation will take place until equality is reached and the solution is saturated. If no precipitation occurs for some time, the solution is in the unstable state of supersaturation.

A few numerical examples will illustrate the use of these rules in predicting precipitation.

EXAMPLE | Will precipitation of calcium sulfate, $CaSO_4$, take place when 100 ml of $4.0 \times 10^{-3} M$ calcium chloride, $CaCl_2$, are mixed with 100 ml of $2.0 \times 10^{-3} M$ sodium sulfate, Na_2SO_4?

SOLUTION | The ions in the two mixed solutions and the possible reaction are

$$Ca^{2+} + 2Cl^- + 2Na^+ + SO_4^{2-} \rightleftharpoons CaSO_4(s) + 2Na^+ + 2Cl^-$$

The equilibrium position for this ionic reaction lies well over to the right, because the formation of the insoluble $CaSO_4$ removes calcium and sulfate ions from the reaction. The sodium and chloride ions do not combine because of the high solubility of NaCl.

From Table 9.1, the solubility product of $CaSO_4$ is $K_{sp} = 2.4 \times 10^{-5}$. The instantaneous (inst) concentrations (before any precipitation would occur) of Ca^{2+} and SO_4^{2-} in the final mixture, with doubled volume, are

$$[Ca^{2+}]_{inst} = \frac{\text{mol of } Ca^{2+} \text{ in 100 ml}}{\text{total volume in liters}} = \frac{MV}{0.200 \text{ liter}}$$

$$= \frac{(4.0 \times 10^{-3})(0.100)}{0.200}$$

$$= 2.0 \times 10^{-3} M$$

$$[SO_4^{2-}]_{inst} = \frac{(2.0 \times 10^{-3})(0.100)}{0.200}$$

$$= 1.0 \times 10^{-3} M$$

The ion product is

$$[Ca^{2+}]_{inst} [SO_4^{2-}]_{inst} = (2.0 \times 10^{-3})(1.0 \times 10^{-3})$$
$$= 2.0 \times 10^{-6}$$

The ion product is less than K_{sp}, so the resulting mixture is unsaturated with respect to $CaSO_4$. No precipitate forms.

Question: Will precipitation of $BaSO_4$ occur when 100 ml of $1.0 \times 10^{-3}M$ barium chloride, $BaCl_2$, are mixed with 50 ml of $3.0 \times 10^{-2}M$ sodium sulfate, Na_2SO_4?

Fractional Precipitation

If a solution containing Ba^{2+} is added to a solution containing Cl^- and SO_4^{2-}, the only precipitate that forms is $BaSO_4$ because of the great difference in solubility between $BaSO_4$ and $BaCl_2$, the latter being very soluble. However, if the two salts that might precipitate are both quite insoluble, it is not always so obvious which one will precipitate first. The following example illustrates this.

EXAMPLE

A solution contains $0.010M$ Cl^- and $0.0010M$ CrO_4^{2-}. A solution of $AgNO_3$ is added to this solution.

a. Which will precipitate first, silver chromate, Ag_2CrO_4, or silver chloride, AgCl?

b. What will be the concentration of the first negative ion to precipitate, at the moment the other negative ion starts to precipitate?

SOLUTION

a. For Ag_2CrO_4,

$$K_{sp} = [Ag^+]^2[CrO_4^{2-}] = 2.4 \times 10^{-12}$$

The $[Ag^+]$ needed to precipitate silver chromate is

$$[Ag^+] = \left(\frac{2.4 \times 10^{-12}}{0.0010}\right)^{1/2} = 4.9 \times 10^{-5}M$$

For AgCl,

$$K_{sp} = [Ag^+][Cl^-] = 1.6 \times 10^{-10}$$

The Ag^+ concentration required to precipitate silver chloride is

$$[Ag^+] = \frac{1.6 \times 10^{-10}}{0.010} = 1.6 \times 10^{-8}M$$

Therefore, AgCl will precipitate first.

b. As more Ag^+ is added, more AgCl will precipitate. As this happens, $[Cl^-]$ will decrease but $[Ag^+]$ will increase. The silver ion concentration will increase until, eventually, $[Ag^+]^2[CrO_4^{2-}]$ just exceeds K_{sp} for silver chromate, which will start to precipitate along with the AgCl. The solution will then be saturated with both silver chromate and chloride, so *the solubility product expressions for both salts must be satisfied.* The concentration of silver ion needed to just saturate the solution with silver chromate was $[Ag^+] = 4.9 \times 10^{-5}M$ as determined by the K_{sp} of silver chromate. (The chromate ion concentration is $[CrO_4^{2-}] = 0.0010M$ *just as* silver chromate begins to form.) Therefore, from the K_{sp} of silver chloride,

$$[Cl^-] = \frac{K_{sp}}{Ag^+} = \frac{1.6 \times 10^{-10}}{4.9 \times 10^{-5}} = 3.3 \times 10^{-6}M$$

From this it can be seen that chloride ion can be well removed from a mixture of chloride ion and chromate ion by the selective precipitation of silver chloride.

It is possible for a solution to be supersaturated if care is taken not to disturb the solution. A common method of preparing a supersaturated solution is to make a saturated solution of the salt at an elevated temperature. This solution is separated from any solid that might remain and is slowly cooled to a lower temperature. If the salt is one whose solubility increases with temperature, the solution will contain more than the equilibrium amount of dissolved salt at the lower temperature. The solution will remain clear and supersaturated until disturbed by the addition of a crystal of solid, a bit of dust, or by scratching the inside of the container—all of which introduce nuclei for the formation of precipitate crystals. Salts such as sodium thiosulfate (hypo) and sodium acetate exhibit supersaturation very well.

The oceans appear to be supersaturated with calcium carbonate, $CaCO_3$, at least near the surface. Supersaturated $CaCO_3$ solutions are quite stable for long periods. This supersaturation may be more apparent than real, however, because the concept of the solubility product in terms of molar concentrations breaks down in concentrated ionic solutions, and sea water is concentrated in this sense. It turns out that molar concentrations of ions do not truly represent the effective ionic concentrations in solution. In solutions with high ion concentrations, an effective concentration known as the "activity" of the ions replaces the molarity. One result of high ionic concentration is the *salt effect:* the presence of other electrolytes not having an ion in common with a slightly soluble salt increases the solubility of the salt. The existence of ion pairs containing carbonate ions paired with Na^+, Mg^{2+}, and Ca^{2+} ions in sea water also has a bearing on the saturation characteristics of the solution.

9.9 FACTORS AFFECTING SOLUBILITY EQUILIBRIA AND THE DISSOLUTION OF PRECIPITATES

The equation for the dissolution of a precipitate of a slightly soluble salt MA is

$$MA(s) \rightleftharpoons M^+ + A^-$$

Application of Le Chatelier's principle to this ionic equation tells us that the removal of either of the ionic products from the scene of action so they cannot participate in the reverse (precipitation) reaction will cause this equilibrium to shift to the right, and more solid MA will dissolve to make up for the loss of ions.

The removal of any one of the products causes any reversible reaction to shift to the right, or product, side to make up for the loss of product, but with ionic products such as we have here we can enumerate several specific ways to remove the ionic products:

1. Formation of a weak electrolyte or nonelectrolyte.
2. Formation of an even more insoluble precipitate.
3. Changing the form of one of the ionic products (by oxidizing or reducing it).

As an example of the last, if insoluble iron(III) hydroxide, $Fe(OH)_3$, dissolves according to

$$Fe(OH)_3 \rightleftharpoons Fe^{3+} + 3OH^-$$

a solution containing a reactant that would convert Fe^{3+} to Fe^{2+} would help dissolve iron(III) hydroxide.

Only the first of these methods will be considered in more detail. The weak or nonelectrolyte may take several forms. The only requirement is that it tie up one of the ions involved in the solubility equilibria more tightly than the solid salt does. An example of this method of shifting the solubility equilibrium is the stepwise process

$$CaCO_3(s) \rightleftharpoons Ca^{2+} + CO_3^{2-}$$
$$H^+ + CO_3^{2-} \rightleftharpoons HCO_3^-$$
$$H^+ + HCO_3^- \rightleftharpoons H_2CO_3$$
$$H_2CO_3 \longrightarrow H_2O + CO_2(g)$$

The solubility of an insoluble carbonate such as $CaCO_3$ is increased in a solution containing many hydrogen ions (an acidic solution), because the reaction of CO_3^{2-} and H^+ eventually results in the formation of the weak electrolyte carbonic acid, H_2CO_3. The subsequent decomposition of H_2CO_3 to the nonelectrolyte water and gaseous CO_2 increases the solubility even more.

Similarly, the slightly soluble salt silver acetate ($AgC_2H_3O_2$ or AgAc for short) is soluble in a solution of hydrogen ions because the competing reaction

$$Ac^- + H^+ \rightleftharpoons HAc$$

forms the weak electrolyte hydrogen acetate (acetic acid).

The positive ion (the cation) of a slightly soluble salt may be removed from circulation through the formation of a charged or uncharged soluble complex. Such complexes consist of a positively charged central metal ion surrounded by negative ions or polar molecules called *ligands*. The ligands bond to the metal ions through coordinate bonds (Section 3.13 and Figure 3.20). Some simple examples of complex ion formation are

$$Ag^+ + 2NH_3 \rightleftharpoons Ag(NH_3)_2^+$$
$$Cu^{2+} + 4NH_3 \rightleftharpoons Cu(NH_3)_4^{2+}$$
$$Ag^+ + 2Cl^- \rightleftharpoons AgCl_2^-$$

The formation of the complex *ammine* and *chloro* ions here involves a displacement of their waters of hydration:

$$Ag(H_2O)_2^+ + 2NH_3 \rightleftharpoons Ag(NH_3)_2^+ + 2H_2O$$

Depending on the stability of these complexes in solution, that is, on the extent of their dissociation into the component ions on the left side of the equation, they may remove some of the metal ions on the right side of the solution equation. The dissociation of a complex ion is denoted by a reaction such as

$$Ag(NH_3)_2^+ \rightleftharpoons Ag^+ + 2NH_3$$

and, in analogy with the solubility product, a constant known as the instability constant K may be written:

$$K = \frac{[Ag^+][NH_3]^2}{[Ag(NH_3)_2^+]} = 6.3 \times 10^{-8}$$

Unlike K_{sp}, the instability constant includes a concentration term in the denominator, because the complex ion is not a solid but exists in solution.

In some instances, the addition of an ion common to the slightly soluble salt can result in the formation of stable complex ions. This effectively removes the common ion and can result in solution of the salt. The equilibrium

$$AgCl(s) + Cl^- \rightleftharpoons AgCl_2^-$$

shows why AgCl dissolves in concentrated HCl or NaCl solution. The effect of complex ion formation on the solubility of a salt is illustrated by the following example.

EXAMPLE | What is the solubility of AgCl in a 0.50M NH$_3$ solution?

SOLUTION | Two equilibrium reactions are involved here:

$$AgCl(s) \rightleftharpoons Ag^+ + Cl^-$$
$$K_{sp} = [Ag^+][Cl^-] = 1.6 \times 10^{-10}$$
$$Ag(NH_3)_2^+ \rightleftharpoons Ag^+ + 2NH_3$$
$$K = \frac{[Ag^+][NH_3]^2}{[Ag(NH_3)_2^+]} = 6.3 \times 10^{-8}$$

The solubility s of the AgCl will be equal to the concentration of Cl$^-$ in solution when equilibrium is established. Also,

$$s = [Cl^-] = [Ag^+]_{total} = [Ag^+] + [Ag(NH_3)_2^+] \simeq [Ag(NH_3)_2^+]$$

since almost all of the Ag$^+$ from the dissolved AgCl becomes complexed.

The assumption is made that, at equilibrium, [NH$_3$] = 0.50M. In other words, the amount of NH$_3$ ligand incorporated in the complex is negligible compared with the initial 0.50M NH$_3$. From $K = 6.3 \times 10^{-8}$,

$$[Ag(NH_3)_2^+] = \frac{[Ag^+](0.50)^2}{6.3 \times 10^{-8}}$$

so that

$$s = [Cl^-] = \frac{(0.50)^2[Ag^+]}{6.3 \times 10^{-8}}$$

and from K_{sp},

$$K_{sp} = 1.6 \times 10^{-10} = [Ag^+] \frac{(0.50)^2[Ag^+]}{6.3 \times 10^{-8}}$$
$$[Ag^+]^2 = 40 \times 10^{-18}$$
$$[Ag^+] = 6.3 \times 10^{-9} M$$

so that

$$s = [Cl^-] = \frac{(0.50)^2(6.3 \times 10^{-9})}{6.3 \times 10^{-8}} = 2.5 \times 10^{-2} M$$

Therefore, the solubility of AgCl is some

$$\frac{2.5 \times 10^{-2}}{(1.6 \times 10^{-10})^{1/2}} = \frac{2.5 \times 10^{-2}}{1.3 \times 10^{-5}} \simeq 2000 \text{ times greater}$$

in $0.50M$ NH_3 solution than in pure water.

9.10 ENERGETIC FACTORS INFLUENCING THE SOLUBILITY OF SALTS

Cohesive forces within the ionic crystal of a solute decrease its solubility. Adhesive interactions between the solute ions and the solvent increase solubility. If the solute is an ionic crystal, one of the more important factors related to cohesion is the *lattice energy*. This is defined as the energy that is released when the individual gaseous ions are brought together to form the solid crystal:

$$M^+(g) + A^-(g) \longrightarrow MA(s)$$

If energy is liberated in forming the solid, then energy is required for the reverse process of breaking up the solid crystal. The larger the lattice energy, the greater the cohesive forces in the solid.

After the breakup of the solid crystal into gaseous ions, the ions become hydrated. Thus, we have first

$$MA(s) + \text{lattice energy} \longrightarrow M^+(g) + A^-(g)$$

followed by

$$M^+(g) + A^-(g) \xrightarrow{\text{H}_2\text{O}} M^+(aq) + A^-(aq) + \text{hydration energy}$$

The second reaction releases energy when water molecules are attracted to the ions. This is an adhesive interaction and the larger the energy of hydration of the ions, the more soluble the ionic solid is likely to be. These two energy factors are not the whole story of solubility. The heat of solution is the difference between the lattice energy and the hydration energy, but the heat of solution does not allow us to calculate the absolute solubility of a substance. Nevertheless, some relationships between solubility and structure can be found by looking at these two energy terms.

In general, higher lattice energies and decreased solubility will result with

1. increasing size of the negative ion
2. increasing ionic charge on either ion
3. eighteen-electron (noninert gas) configuration rather than eight-electron configuration of the positive ion.

All of these factors lead to increased polarization of the ions, principally the negative ion, with increased contribution of the London–van der Waals forces to the lattice energy (or increased covalence of the bond between the ions). Thus, solubility decreases in the series AgF, AgCl, AgBr, AgI (rule 1); $Mg(OH)_2$ is less soluble than LiOH (rule 2); and AgCl is less soluble than NaCl (rule 3). Rule (2) also partially explains why salts such as MgO, CuS, $BaSO_4$, $SrCO_3$, and $AlPO_4$, containing highly charged ions, are often insoluble.

There are a number of exceptions to these rules; and in any case, the rules are useful for estimating the relative solubilities of salts in a series only "if all other effects are equal." The qualitative solubility characteristics of a number of common compounds are summarized in Appendix E.

9.11 HARD WATER

Water containing calcium, magnesium, or iron(II) (Fe^{2+}) ions in solution is called *hard water* and may precipitate insoluble salts of these metals under certain conditions. Calcium ions derived from the passage of natural waters through and over limestone ($CaCO_3$) are the most common metallic ions in hard water. If hydrogen carbonate (HCO_3^-) ions are associated with the metallic ions in solution, the hardness is temporary, or carbonate, hardness. Other ions, such as chloride (Cl^-) and sulfate (SO_4^{2-}), contribute permanent, or noncarbonate, hardness.

Hardness in water is objectionable for several reasons—all associated with the precipitation of the insoluble metallic salts when the water is heated or when soaps are added. Both temporary and permanent hardness lead to the formation of boiler scale in pipes, tanks, or boilers when water is boiled. With temporary hardness the Ca^{2+}, for instance, is held in solution as the soluble hydrogen carbonate salt. The hydrogen carbonate ion is in equilibrium with carbonate ion and carbonic acid,

$$2HCO_3^- \rightleftharpoons H_2CO_3 + CO_3^{2-}$$

Furthermore, H_2CO_3 is unstable and is easily decomposed by heat:

$$H_2CO_3 \xrightarrow{\Delta} H_2O + CO_2$$

Boiling a solution containing HCO_3^- ions shifts the HCO_3^-–CO_3^{2-} equilibrium to the carbonate side because of the removal of a product, H_2CO_3, by thermal decomposition and subsequent escape of CO_2 from the hot water, where its solubility is less. With increased CO_3^{2-} concentration the precipitation of insoluble calcium carbonate is favored:

$$Ca^{2+} + CO_3^{2-} \longrightarrow CaCO_3(s)$$

Even in water exhibiting only permanent hardness (due to Ca^{2+} and SO_4^{2-}), scale may result because of the decrease in solubility of $CaSO_4$ with temperature, since the heat of solution of $CaSO_4$ is negative.

Insoluble calcium and other metal salts are precipitated by the addition of soaps to hard water. Soaps are sodium or potassium salts of organic compounds called fatty acids and have the general formula

$$\left[CH_3(CH_2)_{8-16}C \begin{matrix} \diagup O \\ \diagdown O \end{matrix} \right]^- Na^+ \quad \text{or} \quad \left[RC \begin{matrix} \diagup O \\ \diagdown O \end{matrix} \right]^- Na^+$$

The cleansing properties of soap come about because the nonpolar $—CH_3$ end of the negative soap ions cluster about the water-insoluble oil, dirt, and grease particles (see Figure 12.8) and emulsify or break up these particles. The outer portion of the cluster of negative soap ions with the emulsified dirt particle at the center consists of the polar $\left(—C \begin{matrix} \diagup O^- \\ \diagdown O \end{matrix} \right)$ end of the soap ions. This end is soluble in water, and the whole cluster or micelle, as it is called, can be floated away with the water. If soaps are added to hard water, a precipitate results because the calcium, magnesium, and iron salts, such as $[RCOO]_2^-Ca^{2+}$, are insoluble in water and form a scum that adheres to clothes, or the bath tub, and clogs equipment. To use soap in hard water, it is first necessary to add enough soap to precipitate all the metal ions. Then more soap must be added to do the cleaning. This wastes soap in addition to producing the objectionable curdy precipitate. The cleansing action of detergents, through wetting and emulsification of oily dirt particles, is identical to that of soaps, but detergents do not form insoluble compounds with the heavier metal ions in hard water.

Commercially, the hardness of water is removed, or the water softened, by removal of the calcium and other metal ions either as insoluble precipitates or by tying up the metallic ions in complexes. Temporary hard water can be softened by boiling. As before, this precipitates calcium carbonate, and removal of the precipitate leaves behind soft water. This method is not practical on a large scale. Permanent hardness cannot be removed by boiling. Temporary hardness also can be removed by the addition of calcium hydroxide, $Ca(OH)_2$ (slaked lime). The hydroxide ion produced on solution of the calcium hydroxide reacts with the HCO_3^-,

$$HCO_3^- + OH^- \longrightarrow H_2O + CO_3^{2-}$$

and the carbonate ion precipitates the Ca^{2+} as $CaCO_3$. The addition of excess calcium hydroxide also brings about the removal of Mg^{2+} as insoluble $Mg(OH)_2$. Ammonia also functions in a similar way:

$$HCO_3^- + NH_3 \longrightarrow NH_4^+ + CO_3^{2-}$$

Precipitation methods are also used for softening permanent hard water. Among the substances used for this are sodium carbonate, Na_2CO_3 (soda ash or washing soda), sodium tetraborate, $Na_2B_4O_7$ (borax), and trisodium phosphate, Na_3PO_4, which precipitate calcium as the insoluble carbonate, tetraborate, or phosphate, respectively. All these compounds also undergo a reaction (see hydrolysis, Section 10.13) with water,

$$PO_4^{3-} + H_2O \rightleftharpoons HPO_4^{2-} + OH^-$$

The OH^- helps to convert any HCO_3^- present to CO_3^{2-}, and the calcium carbonate is precipitated along with the phosphate.

The removal of calcium ion by complexing is accomplished by phosphorus compounds called polyphosphates containing the metaphosphate group PO_3^-. The general formula of this type of compound is $(NaPO_3)_n$, sodium polymetaphosphate, with $Na_6P_6O_{18}$ or $(NaPO_3)_6$, sodium hexametaphosphate as a specific example. These compounds do not precipitate insoluble calcium salts from hard water. Instead, they form stable complex ions with calcium ion,

$$Na_6P_6O_{18} + 2Ca^{2+} \longrightarrow Ca_2P_6O_{18}{}^{2-} + 6Na^+$$

The complex may have the structure

Complexes such as this, where a group attaches to the central metal ion at two or more positions, are called *chelates*. (Chelates are of especial importance in the complexing or chelation of heavy metals in the environment and in the body. Two important chelates are the iron-containing heme groups that occur in hemoglobin and the magnesium chelate in chlorophyll.)

Ion Exchange A most effective water softening method, which might be considered as a complexing process, is *ion exchange*. Basically, ion exchange involves the substitution or exchange of a hardness-producing cation such as Ca^{2+} for another cation such as Na^+ that does not contribute to hardness. Insoluble, high molecular weight, polymeric substances called ion exchange resins are commonly used for this purpose. If the resin R contains the exchangeable ion Na^+, the reversible exchange process is

$$2R^-Na^+ + Ca^{2+} \rightleftharpoons [R^-]_2Ca^{2+} + 2Na^+$$

The first ion exchange materials were naturally occurring minerals called zeolites consisting of SiO_4 and AlO_4 tetrahedra forming an open or porous framework within which exchangeable cations such as Na^+ could be accommodated. A typical formula is $NaAlSi_2O_6 \cdot H_2O$. Later, synthetic zeolites (trade name Permutit) were developed. Most ion exchange materials today are synthetic organic polystyrene polymers with active groups such as

$$-\overset{\overset{\displaystyle O}{\|}}{\underset{\underset{\displaystyle O}{\|}}{S}}-O^- \quad \text{or} \quad -\overset{\overset{\displaystyle O}{\|}}{C}-O^-$$

If we let the rest of the molecule be represented, as before, by R, except for the active group, the sodium salt of the resin is

$$[R-SO_3^-]Na^+ \quad \text{or} \quad [R-COO^-]Na^+$$

Hydrogen salts (acids) are also in use:

$$[R-SO_3^-]H^+$$

In use, the hard water is passed slowly over the resin and the exchange takes place:

$$2[R-SO_3^-]Na^+ + Ca^{2+} \rightleftharpoons [R-SO_3^-]_2Ca^{2+} + 2Na^+$$

or

$$2[R-SO_3^-]H^+ + Ca^{2+} \rightleftharpoons [R-SO_3^-]_2Ca^{2+} + 2H^+$$

In general, highly charged ions tend to displace less highly charged ions. When the resin has been spent, that is, converted to the calcium salt, it can be regenerated by allowing it to stand in a concentrated Na^+ or H^+ solution so that the reversible equilibria above are shifted to the left.

The ion exchange resins illustrated have been cation (positive ion) exchangers. Anion (negative ion) exchange resins are also available. The active group is positively charged, and the exchangeable ions are negative. A typical active group is

$$[R-CH_2N^+(CH_3)_3]X^-$$

where X^- is the exchangeable anion, such as OH^-, which may be exchanged for another anion.

The passage of water containing ions through a mixture of a cation exchange resin with exchangeable H^+ and an anion exchange resin with exchangeable OH^- results in deionized water. Let R' stand for for the resin including the active group, M^+ for any cation, and X^- for any anion. Then

$$R'H^+ + R'OH^- + M^+ + X^- \rightleftharpoons R'M^+ + R'X^- + H_2O$$

The displaced hydrogen and hydroxide ions have combined to form water, and few ions are left in the treated water.

Aside from the tendency to remove highly charged cations, the cation exchange resins mentioned above are not specific for any cation. Other cation exchange resins have been formulated that contain metal chelating groups that will retain only specific metal ions. They find use in the laboratory and in the treatment of industrial sewage containing toxic metal wastes.

Another way to soften hard water is to boil the water and condense the steam. This distillation procedure leaves the nonvolatile salts behind. Distillation is economical for the production of drinking water from highly saline waters, but the method does not lend itself to the economical softening of water for large-scale use.

9.12 DISTRIBUTION OF A SOLUTE BETWEEN IMMISCIBLE SOLVENTS

A solute soluble in each of a pair of immiscible liquids will divide itself between the two liquid solvents when added to the two-phase system. Eventually, a dynamic equilibrium will be reached where the rate of passage of solute molecules across the interface from solvent A to solvent B equals the rate of passage of solute from solvent B to A. At this equilibrium point, the ratio of the concentration (as grams per milliliter or as moles per liter) of solute in solvent A to that in solvent B is a constant K_D, called the distribution or partition coefficient.

$$K_D = \frac{[\text{solute}]_A}{[\text{solute}]_B} = \frac{M_A}{M_B}$$

The value of the distribution coefficient depends only on the temperature and on the nature of the solvents and the solute. It does not depend on the quantities of the two immiscible solvents, on their area of contact, or on the total amount of solute present. In fact, if so much solute is present that both solvents are saturated, the distribution coefficient is given by the ratio of the solubilities of the solute in each solvent.

$$K_D = \frac{s_A}{s_B}$$

This simple form for K_D holds only if the solute has the same molecular weight in both solvents. If there is more than one solute present, the distribution law is obeyed for each solute with its own value of K_D.

Distribution coefficients for a few solute–solvent pair systems at 25°C are listed below.

solute: I_2 solvent system: CCl_4 (A)
H_2O (B)

$$K_D = \frac{[I_2]_A}{[I_2]_B} = 85$$

solute: I_2 solvents: CS_2 (A)
H_2O (B)

$K_D = 600$

solute: Br_2 solvents: CCl_4 (A)
H_2O (B)

$K_D = 23$

A nonpolar solute distributes itself so that it will have a greater concentration in the less polar solvent. This is the basis of extraction of nonpolar solutes from aqueous solution by shaking the solution with a nonpolar solvent that is immiscible with water.

The distribution of solutes between immiscible solvents has a number of important consequences in living organisms. In particular, molecules that are nonpolar or nearly so will concentrate in the lipids (fats, oils, sterols) of the body rather than in the less fatty parts of the body. For instance, the anesthetic ether concentrates in the nervous tissues and brain rather than in the less fatty tissues and blood.

More important from an ecological standpoint are the concentration of the relatively nonpolar chlorinated hydrocarbon pesticides such as DDT and related compounds in animal fats and plant waxes, where their breakdown by biological catalysts is absent or very slow. Repeated exposure to these water-insoluble, fat-soluble substances leads to a concentration buildup in the fatty tissues and organs (liver, brain), with possible physiological damage. Moreover, these substances are passed from prey to predator with further concentration at each step in a food chain. The possible action, effects, and biological concentration of such compounds are discussed in detail in Chapter 17.

9.13 SUMMARY

Solutions are homogeneous mixtures of two or more components. The component present in the largest amount is usually called the solvent, and the other components are solutes. The solution of one substance in another is favored by large adhesive forces between solvent and solute particles. Water is an unusually good solvent for many solutes, because the polar nature of its molecule leads to large attractive forces between solute particles and water molecules. These forces are associated with the hydration of ions and with hydrogen bonding. The high dielectric constant of water also favors the dissociation of ionic solutes.

The quantity of solute in a solution, the concentration of the solution, is commonly given by the molarity of the solution. The molarity is the number of moles of solute per liter of solution. Other concentration units such as molality, mole fraction, and percent by weight are often convenient.

Precipitation of a salt occurs when the product of the concentrations of its ions, the ion product, in solution exceeds its solubility product. Dissolution of a precipitate takes place if the ion product is less than the solubility product. Therefore, the solubility of a slightly soluble salt can be increased by the removal of the products of the dissolution equation $MA(s) \rightleftharpoons M^+ + A^-$. This equilibrium can be shifted to the right side by the formation of weak electrolytes or nonelectrolytes.

Hard water is due to Mg^{2+}, Ca^{2+}, and Fe^{2+} ions. These cations are accompanied by HCO_3^- (temporary, or carbonate, hardness) or Cl^- and SO_4^{2-}

(permanent, or noncarbonate, hardness). Water softening is the removal of the metallic cations by various methods: precipitation, complex formation, or ion exchange.

Questions and Problems

1. What would be the major cause of the solubility of
   a. zinc metal in hydrochloric acid?
   b. sodium hydroxide in water?
   c. ethyl alcohol, C_2H_5OH, in water?
   d. chloroform, $CHCl_3$, in water?
   e. $CHCl_3$ in CCl_4?
2. Which would you expect to be more soluble in water, $CHCl_3$ or CCl_4? Why?
3. "Like dissolves like" is a rule of thumb for the solubility of one substance in another. Discuss the applicability of this rule and give examples.
4. In solutions of constant H^+ concentration (other things being equal), what would be the expected trends in solubility in the following groups of compounds?
   a. $Ca_3(PO_4)_2$, $Ca(HPO_4)_2$, $CaHPO_4$
   b. $Fe(OH)_3$, $Fe(OH)_2$
   c. $FePO_4$, $Fe_3(PO_4)_2$
5. When 7.35 g of $CaCl_2 \cdot 2H_2O$ are dissolved in water to make 200 ml of solution, the density of the resulting solution is 1.023 g/ml. Calculate
   a. molarity
   b. percent of $CaCl_2 \cdot 2H_2O$
   c. percent of $CaCl_2$
   d. molality
   e. the molar concentration of Ca^{2+} and of Cl^- in the solution
6. a. Show that the concentration of water in pure water is about 55.5M.
   b. Show that the concentration of water in a concentrated solute solution is no longer constant and equal to 55.5M.
7. The concentrated hydrochloric acid found in the laboratory is 39 percent by weight HCl. The density is 1.192 g/ml. What is the molarity and the mole fraction of HCl in this solution?
8. a. What volume of 0.250M NaOH can be made from 5.00 g of NaOH?
   b. How many grams of NaOH are there in 25.00 ml of this solution?
9. The molar concentration of cadmium ion in a

sample of water is found to be $5.0 \times 10^{-6}M$. What is the concentration of Cd^{2+} as parts per million by mass?
10. What volume of water must be added to 250 ml of 1.25M NaCl to make it 0.50M?
11. To 50 ml of 0.50M H_2SO_4, 100 ml of 0.25M H_2SO_4 are added. What is the molarity of the final solution if its volume is 150 ml?
12. A solution is made by mixing 125 ml of 0.10M KCl and 175 ml of 0.050M $CaCl_2$ to give 300 ml of solution. Find the molar concentrations of potassium, sodium, and chloride ion in the solution.
13. Which of the following electrolytes would ionize and which would dissociate in water? CsCl, HCl, KF.
14. Which is less soluble in water, silver chloride or silver chromate?
15. At 20°C, the solubility of lead fluoride is 0.064 g in 100 ml of solution. What is the solubility product?
16. What is the solubility of iron(II) hydroxide in distilled water?
17. a. What is the solubility of calcium hydroxide in water?
    b. What is the solubility in 0.05M NaOH?
    c. What is the solubility in 0.10M $CaCl_2$?
    d. Will the solubility of $Ca(OH)_2$ in 0.05M NaCl be greater or less than the solubility in water?
    e. Will the solubility in 50 percent ethanol (ethyl alcohol) be greater or less than in water?
18. Of the two inexpensive chemicals, NaOH and Na_2CO_3, which one will best remove magnesium ions from hard water? Calcium ions?
19. The hydroxide ion concentration of a water is $5 \times 10^{-11}M$. What is the maximum amount of iron(III) ion, in mg/liter, that can exist in this water?
20. Fluoridated water supplies contain a maximum of about 1 mg/liter F^-. What is the maximum hardness, if measured as mg/liter of Ca^{2+}, that can be tolerated in the water if a fluoride concentration of 1 mg/liter is to be attained by addition of NaF?

21. Will a precipitate of $PbSO_4$ be formed if 20.0 ml of $2.0 \times 10^{-4}M$ $Pb(NO_3)_2$ is mixed with 80.0 ml of $1.0 \times 10^{-4}M$ Na_2SO_4?

22. A solution contains $0.01M$ Ca^{2+} and $0.05M$ Ba^{2+}. Sulfate ion is added to the solution. What salt precipitates first? What is the cation concentration of that salt when the second salt starts to precipitate?

23. Would you expect silver chloride to dissolve in
    a. nitric acid, HNO_3?
    b. concentrated hydrochloric acid, HCl?
    c. concentrated sodium chloride solution?
    d. ammonia solution, $NH_3(aq)$?
    e. ammonium nitrate solution, NH_4NO_3?
    f. acetic acid?
    Explain your answers.

24. The solubility of $AgCl$ in $0.004M$ Cl^- solution is a minimum. With increasing chloride ion concentration, the solubility of $AgCl$ rises. Explain qualitatively the reason for the minimum and the cause of the increased solubility.

25. 100 ml of $0.05M$ $AgNO_3$ are to be mixed with 100 ml of $0.05M$ KBr. What should the ammonia concentration of the resulting solution be to prevent the formation of a $AgBr$ precipitate?

26. The following reactions occur in the excess lime–soda ash water softening process:
    a. removal of carbonate (HCO_3^-) hardness by addition of $Ca(OH)_2$ and precipitation of $CaCO_3$:

    $$Ca^{2+} + HCO_3^- + \cdots \longrightarrow$$

    $$Mg^{2+} + HCO_3^- + \cdots \longrightarrow Mg^{2+} + CO_3^{2-}$$

    b. addition of excess $Ca(OH)_2$ to remove Mg^{2+} as $Mg(OH)_2$ precipitate and carbonate as $CaCO_3$:

    $$Mg^{2+} + CO_3^{2-} + \cdots \longrightarrow$$

    $$Mg^{2+} + SO_4^{2-} + \cdots \longrightarrow Ca^{2+} + SO_4^{2-} + \cdots$$

    c. Addition of Na_2CO_3 to remove noncarbonate calcium hardness remaining from (b):

    $$Ca^{2+} + SO_4^{2-} + \cdots \longrightarrow$$

    Complete and balance all the reactions above.

27. Temporary hardness due to $Ca(HCO_3)_2$ can be removed by adding $Ca(OH)_2$ solution. Does the addition of Ca^{2+} substitute more permanent hardness for temporary hardness? Explain by considering what happens to the calcium ions.

28. 100 ml of water containing 0.025 g of iodine are to be extracted with 50 ml of carbon tetrachloride. Using concentrations in grams per milliliter, calculate the weight of iodine remaining in the water layer after
    a. one extraction with 50 ml of CCl_4
    b. two successive extractions with 25 ml portions of CCl_4
    Note which method is the most efficient for a given total volume of extracting solvent.

29. Olive oil is a liquid fat. The distribution coefficient for DDT between olive oil and water is reported to be 923. The solubility of DDT in water is 1.2 ppb or 1.2 μg/liter of water. Calculate the solubility of DDT in olive oil in molarity and in parts per million. The molecular weight of DDT is 354.5 g/mol.

Answers

5. a. $0.250M$
   b. 3.59 percent
   c. 2.71 percent
   d. $0.254m$
   e. $[Ca^{2+}] = 0.250M$; $[Cl^-] = 0.500M$
7. $13M$, $X = 0.25$
8. a. 500 ml
   b. 0.250 g
9. 0.56 ppm
10. 375 ml
11. $0.33M$
12. $[K^+] = 0.042M$; $[Na^+] = 0.029M$; $[Cl^-] = 0.10M$

15. $K_{sp} = 7.0 \times 10^{-8}$
16. $s = 5.8 \times 10^{-6}M$
17. a. $s = [Ca^{2+}] = 1.2 \times 10^{-2}M$
    b. $2.2 \times 10^{-3}M$
    c. $3.7 \times 10^{-3}M$
19. ~0.03 mg/liter
20. 700 mg/liter
22. $[Ba^{2+}] = 4.2 \times 10^{-8}M$
25. $[NH_3] = 8.7M$
28. a. 0.000575 g
    b. 0.000049 g of I_2 remain in H_2O
29. $3.2 \times 10^{-6}M$, 1.1 ppm

References

Bard, Allen J., "Chemical Equilibrium," Harper & Row, New York, 1966, Chapters 1, 2, 4, 5.

Butler, J. N., "Solubility and pH Calculations," Addison-Wesley, Reading, Mass., 1964, Chapter 2.

Fischer, Robert B., and Peters, D. G., "Chemical Equilibrium," Saunders, Philadelphia, 1970, Chapter 2.

Proton and Electron Transfer

The positive hydrogen ion or proton plays a vital role in natural processes. Two major constituents of the biosphere, water and carbon dioxide, are largely responsible for both the widespread presence and the effects of hydrogen ions. Hydrogen ions are involved in many natural systems, including the regulation of soil fertility, the catalysis of biological reactions, the regulation of the composition of sea water, the chemistry of the blood, the weathering of rock, and the formation of soil.

The regulation of hydrogen ion concentration is a significant factor in nature, because organisms may be greatly affected by small changes in its concentration in both their external and internal environments. For instance, variations in the hydrogen ion concentration in natural waters can affect marine organisms directly or through chemical changes in their environment. Hydrogen ions are highly reactive with the cellular proteins that act as catalysts in living organisms (enzymes). Changes from the optimum hydrogen ion concentration in cellular fluids affect the structure and metabolic activity of these proteins.

It is difficult and perhaps unjustified to pick any one of the roles of hydrogen ions in nature as the single reason for going into such a wide-ranging chemical subject. So, as we have done with other chemical concepts with a similarly wide applicability, we discuss hydrogen ions and their solutions in a single chapter.

The largest class of reactions of hydrogen ions involves the transfer of the proton from one substance (an acid) to another substance (a base). Another transfer process, that of electrons from one substance to another, is also of great significance in natural systems. Such electron transfer, or oxidation–reduction reactions, are involved in most of the energy-yielding reactions in organisms, as we saw in Chapter 5. Many of the redox reactions in living systems involve the removal of hydrogen atoms (proton plus electron) from one compound and their transfer to another compound, so that there are certain similarities between acid–base and oxidation–reduction reactions. The second part of this chapter extends the treatment of redox processes begun in Chapter 5 to a more quantitative level.

Acids and Bases

Definitions of Acid and Base

10.1 THE PROTON OR BRØNSTED–LOWRY CONCEPT

An early (Arrhenius) concept of acids and bases defined an acid as a hydrogen-containing substance that produced hydrogen ions (or more correctly, hydrated hydrogen ions or hydronium ions, H_3O^+) when dissolved in water, whereas bases yielded hydroxide ions, OH^-, in water. Thus, HCl is an acid because it ionizes in water,

$$HCl \rightleftharpoons H^+ + Cl^-$$

and sodium hydroxide is a base in water because it dissociates,

$$NaOH \rightleftharpoons Na^+ + OH^-$$

These definitions of acid and base were eventually superseded by much more general definitions that included additional substances and lifted the restriction to aqueous solutions. According to the proton concept, or the Brønsted–Lowry system, of acids and bases, an acid is defined as a *proton donor* and a base as a *proton acceptor*. Because a proton is just a hydrogen ion, an acid is still some molecule or ion that can give up a hydrogen ion although it does not have to do this in water. With the understanding that water is not necessary, the following could be examples of acids:

$$HCl \rightleftharpoons H^+ + Cl^-$$
$$HNO_3 \rightleftharpoons H^+ + NO_3^-$$
$$HCO_3^- \rightleftharpoons H^+ + CO_3^{2-}$$
$$H_2O \rightleftharpoons H^+ + OH^-$$
$$NH_4^+ \rightleftharpoons H^+ + NH_3$$
$$Al(H_2O)_6^{3+} \rightleftharpoons H^+ + [Al(OH)(H_2O)_5]^{2+}$$
$$H_3O^+ \rightleftharpoons H^+ + H_2O$$

A general reaction may be written to show that the loss of a proton from an acid generates a base, called the *conjugate base of the acid*:

$$acid_1 \rightleftharpoons H^+ + base_1$$

Whether substances will function as acids depends on the presence of a basic substance that will take up the proton. Some possible bases, on the left side of the equations, are

$$CO_3^{2-} + H^+ \rightleftharpoons HCO_3^-$$
$$HCO_3^- + H^+ \rightleftharpoons H_2CO_3$$
$$OH^- + H^+ \rightleftharpoons H_2O$$
$$NH_3 + H^+ \rightleftharpoons NH_4^+$$
$$H_2O + H^+ \rightleftharpoons H_3O^+$$

In general,

$$base_2 + H^+ \rightleftharpoons acid_2$$

where $acid_2$ is the conjugate acid of the base reactant, $base_2$.

Inspection of these two lists shows that certain substances on the left side of the equations (neglecting the conjugate acids and bases on the right for the moment) can act as acids or bases. H_2O and HCO_3^- are examples of such *amphiprotic* substances. (The name "amphoteric" is usually applied to oxide and hydroxide compounds, such as $[Al(OH)(H_2O)_5]^{2+}$, that can act as an acid or base. See Section 10.5.)

For any substance to function as an acid, it must be paired with a base other than its conjugate base. The proton must be accepted by something, just as the electrons given up by the oxidized substance in a redox reaction must be accepted by some substance undergoing reduction. Therefore, every time an inherently acidic substance acts as an acid, some inherently basic substance must act as a base. The reactions in the preceding lists are merely half-reactions, two of which add up to a complete reaction of the general form

$$acid_1 + base_2 \rightleftharpoons base_1 + acid_2$$

Some examples are

$$H_3O^+ + OH^- \rightleftharpoons H_2O(l) + H_2O(l) \tag{1}$$
$$H_2O(l) + NH_3(g) \rightleftharpoons OH^- + NH_4^+ \tag{2}$$
$$H_2O(l) + CO_3^{2-} \rightleftharpoons OH^- + HCO_3^- \tag{3}$$
$$HCl(g) + H_2O(l) \rightleftharpoons Cl^- + H_3O^+ \tag{4}$$

$$\underset{\text{acetic acid}}{CH_3\overset{O}{C}\!\!-\!\!OH} + H_2O(l) \rightleftharpoons \underset{\text{acetate ion}}{CH_3\overset{O}{C}\!\!-\!\!O^-} + H_3O^+ \tag{5}$$

$$HCl(g) + NH_3(g) \rightleftharpoons (Cl^- + NH_4^+) \longrightarrow NH_4Cl(s) \tag{6}$$

All but equation (6) involve water. Equation (1) is the common reaction of an acid with a hydroxide base. In equations (2) and (3), water functions as an acid and donates a proton to the base. In equations (4) and (5), water acts as a base and accepts a proton to form the hydrated hydrogen ion, H_3O^+. (The general formula for hydrated H^+ is $H(H_2O)_n^+$. The species with $n = 4$ is probably present in greatest amount, but it is common practice to write simply H_3O^+.) Equation (6) for the reaction of HCl and NH_3 in the absence of water shows that hydroxide ions need not be involved if we use the Brønsted–Lowry definitions of acid and base. Equations (2) and (3) show that a base in water does indeed produce OH^- ions, as the Arrhenius definition said. However, these ions do not necessarily originate from the base itself. The only time the OH^- ions come directly from the base is when they already exist in the base and need only be separated from the associated cation, such as when a metallic hydroxide is dissolved in water. Equations (4) and (5) show that, in aqueous solutions, both definitions of an acid coincide if we realize that a hydrogen ion in water must assume the hydrated form H_3O^+.

10.2 THE LEWIS CONCEPT

One other view of acids and bases, the Lewis definition, is often useful. According to this definition, an acid is any chemical species capable of accepting an *electron pair* from another species. The species donating the electron pair to the Lewis acid is a Lewis base. The Lewis definition is much more general than the protonic definition and includes as acids or bases all the substances that are so classified in the protonic definition and many others besides. The number of reactions classified as acid–base reactions is also greatly enlarged under the Lewis definition, since essentially any process involving the formation of a coordinate bond is an acid–base reaction. An example of the reaction of a Lewis acid with a Lewis base is the reaction of SO_2 with metallic oxides or alkali metal carbonates.

$$CaO + SO_2 \longrightarrow CaSO_3$$

Acid–base reactions of this type are used to remove SO_2 from stack gases in air pollution control (Section 7.8).

The price of the greater generality of the Lewis concept of acids and bases is greater complexity. Since we shall have no need for greater generality than the proton concept provides, we shall not pursue the Lewis concept further.

Strengths of Acids and Bases

10.3 IONIZATION CONSTANTS

Acids were originally defined, and can still be characterized, by the following properties: They have a sour taste; they turn the color of a dye, litmus, from blue to red; they react with electropositive metals to give hydrogen gas. Bases, on the other hand, are known by the slippery feel of their aqueous solutions, a bitter taste, and the turning of litmus from red to blue. And, of course, acids react with bases and vice versa. As we might guess, these properties of acids and bases *in water* are due to the presence of hydrogen ion and hydroxide ion, respectively. The intensity of the various properties is a function of the ease with which a proton is donated to water (acting as a base) by the acid or the tendency of the base to accept a proton from water (acting as an acid).

The strength of an acid is measured by its tendency to give up a proton, and the strength of a base is measured by its tendency to accept a proton.

If we confine our attention to aqueous solutions of acids and bases, the extent to which the following reversible reactions go to completion in water will give an idea of the strength of the general acid HA and the general base B.

$$\text{acid}_1 + \text{base}_2 \rightleftharpoons \text{base}_1 + \text{acid}_2$$

$$HA + H_2O \rightleftharpoons A^- + H_3O^+$$

$$H_2O + B \rightleftharpoons OH^- + HB^+$$

Strong acids and bases have ionization equilibria lying far over on the right side of these equations.

Expressions called *equilibrium constants* can be written for both of these ionization reactions. Thus, we have for the acid,

$$K = \frac{[H_3O^+][A^-]}{[HA][H_2O]}$$

where the quantities in brackets are the molar concentrations of the various species when equilibrium has been established. In dilute solutions, the concentration of water, $[H_2O]$, is a constant and can be included with K as

$$K[H_2O] = \frac{[H_3O^+][A^-]}{[HA]} = K_a$$

K_a is the *acid ionization constant* of HA.

We can make a further simplification by realizing that every H_3O^+ ion results from a proton (or H^+) released by the acid HA; so $[H_3O^+]$ can be written simply as $[H^+]$ if we remember that simple H^+ ions are not really present in solution:

$$K_a = \frac{[H^+][A^-]}{[HA]}$$

This is the same form of K_a that would have resulted from the reaction

$$HA \rightleftharpoons H^+ + A^-$$

where the water has been omitted. The equations for the reaction of all proton-donating acids with water can be written this way, if it is kept in mind that water enters into the reaction. Some examples follow.

$$HF \rightleftharpoons H^+ + F^- \qquad K_a = \frac{[H^+][F^-]}{[HF]}$$

$$H_2CO_3 \rightleftharpoons H^+ + HCO_3^- \qquad K_a = \frac{[H^+][HCO_3^-]}{[H_2CO_3]}$$

$$HCO_3^- \rightleftharpoons H^+ + CO_3^{2-} \qquad K_a = \frac{[H^+][CO_3^{2-}]}{[HCO_3^-]}$$

$$\underset{\text{(HAc)}}{CH_3COOH} \rightleftharpoons H^+ + \underset{\text{(Ac}^-)}{CH_3COO^-} \qquad K_a = \frac{[H^+][CH_3COO^-]}{[CH_3COOH]}$$

A base, NH_3 for instance, accepts protons from water according to

$$NH_3 + H_2O \rightleftharpoons NH_4^+ + OH^-$$

The base ionization constant for NH_3 is, with simplifications,

$$K_b = \frac{[NH_4^+][OH^-]}{[NH_3]}$$

(You may see K_b for ammonia written as $K_b = [NH_4^+][OH^-]/[NH_4OH]$, because of the older belief that all the NH_3 was hydrated as $NH_3 \cdot H_2O$ or NH_4OH, which ionized as

$$NH_4OH \rightleftharpoons NH_4^+ + OH^-$$

We shall not use this ionization equation or the form of K_b resulting from it.)

The simplest bases are those such as NaOH that already include the hydroxide ion. Aside from the dissociation of the Na^+ and OH^- ions in the solid upon solution, no other reaction with water is necessary to form hydroxide ions. An equilibrium constant for the reaction of a hydroxide like this could be written

$$MOH \rightleftharpoons M^+ + OH^-$$

$$K_b = \frac{[M^+][OH^-]}{[MOH]}$$

However, most of the simple hydroxides are those of the active alkali and alkaline earth (groups IA and IIA) metals. These hydroxides are all strong electrolytes, which dissociate completely in solution so that no MOH molecules exist in solution. With $[MOH] = 0$, K_b makes no sense, and base ionization constants are not used with these strong electrolytes. The same reasoning applies to acidic strong electrolytes such as HCl and HI, which essentially ionize completely in dilute solutions.

At a given temperature, the ionization constants for acids and bases in dilute aqueous solution have a constant numerical value so long as all the chemical species in the equilibrium constant expression exist together in the solution. If the concentration of any one of the species in solution is varied, the concentration of other species will change so that the value of K remains the same. This is an example of Le Chatelier's principle. For instance, consider a solution of acetic acid, HAc for short.

$$HAc \rightleftharpoons H^+ + Ac^-$$

$$K_a = \frac{[H^+][Ac^-]}{[HAc]}$$

If $[H^+]$ is decreased by any means, more HAc will ionize to produce more H^+ and Ac^-. The increase in $[Ac^-]$ and the decrease in [HAc] will counteract the decrease in $[H^+]$, and K_a will remain constant at its equilibrium value. In a qualitative sense, the removal of the product H^+ causes the equilibrium to shift to the right.

The numerical values of the ionization constants for acids and bases in water can be determined experimentally by various methods, all of which amount to a measurement of the equilibrium concentrations of the chemical species in the ionization constant expressions. Some values are listed in Table 10.1. The numerical magnitude of K_a is a measure of the tendency of an acid to donate protons to water. The value of K_b is a measure of the tendency of a base to accept protons from water. The larger the value of K_a or K_b, the stronger is the acid or the base, respectively. Aqueous solutions of strong acids have higher H^+ concentrations than solutions of weak acids of the same total acid concentration (i.e., $[HA] + [H^+]$).

EXAMPLE | The equilibrium concentrations of H^+, Ac^- and HAc molecules in a $0.010M$ solution of acetic acid are found to be

$$[H^+] = 4.2 \times 10^{-4}M$$

$$[Ac^-] = 4.2 \times 10^{-4}M$$

$$[HAc] = 9.6 \times 10^{-3}M$$

What is the ionization constant of acetic acid?

SOLUTION | The experimentally determined equilibrium concentrations are inserted directly into the expression for K_a:

$$K_a = \frac{[H^+][Ac^-]}{[HAc]} = \frac{(4.2 \times 10^{-4}M)^2}{9.6 \times 10^{-3}M}$$

$$= 1.8 \times 10^{-5}$$

Table 10.1
Ionization constants (25°C) for weak acids and weak bases in aqueous solution

acids		K_a
acetic	CH_3COOH (HAc)	1.8×10^{-5}
boric	H_3BO_3	$K_1{}^a = 6.0 \times 10^{-10}$
carbonic	H_2CO_3	$K_1 = 4.2 \times 10^{-7}$
	HCO_3^-	$K_2 = 4.8 \times 10^{-11}$
hydrocyanic	HCN	4.0×10^{-10}
hydrofluoric	HF	6.9×10^{-4}
hydrogen sulfide	H_2S	$K_1 = 1.0 \times 10^{-7}$
	HS^-	$K_2 = 1 \times 10^{-14}$
hypochlorous	HClO	3.2×10^{-8}
chlorous	$HClO_2$	1.1×10^{-2}
nitrous	HNO_2	4.5×10^{-4}
phosphoric	H_3PO_4	$K_1 = 7.5 \times 10^{-3}$
	$H_2PO_4^-$	$K_2 = 6.2 \times 10^{-8}$
	$HPO_4{}^{2-}$	$K_3 = 1 \times 10^{-12}$
sulfuric	H_2SO_4	$K_1 \simeq 10$ (strong)
	HSO_4^-	$K_2 = 1.2 \times 10^{-2}$
sulfurous	H_2SO_3	$K_1 = 1.3 \times 10^{-2}$
	HSO_3^-	$K_2 = 5.6 \times 10^{-8}$
ammonium ion	NH_4^+	5.5×10^{-10}
water	H_2O	$1.8 \times 10^{-16} = \dfrac{1.0 \times 10^{-14}}{55.6}$

Table 10.1 (*Continued*)	some weak organic acids		K_a
formic	HCOOH		2.1×10^{-4}
oxalic	COOH		$K_1 = 3.8 \times 10^{-2}$
	COOH		$K_2 = 5.0 \times 10^{-5}$

lactic

$$\begin{array}{c} CH_3 \\ | \\ HC-OH \\ | \\ COOH \end{array} \qquad 1.4 \times 10^{-4}$$

citric

$$\begin{array}{ll} H_2C-COOH & K_1 = 8 \times 10^{-4} \\ HO-C-COOH & K_2 = 2 \times 10^{-5} \\ H_2C-COOH & K_3 = 4 \times 10^{-7} \end{array}$$

tartaric

$$\begin{array}{ll} COOH & K_1 = 1.1 \times 10^{-3} \\ HC-OH & K_2 = 6.9 \times 10^{-5} \\ HC-OH & \\ COOH & \end{array}$$

bases		K_b
ammonia	NH_3	1.8×10^{-5}
methyl amine	CH_3NH_2	4.4×10^{-4}

^{a} Ionization constants K_1, K_2, . . . for acids with more than one ionizable hydrogen correspond to the successive ionization steps of the acid (see Section 10.11).

10.4 RELATIVE STRENGTHS OF ACIDS AND BASES

Strong acids and strong bases are strong electrolytes that ionize or dissociate almost completely into ions in water solution. Although we earlier wrote the equation for the ionization of HCl in water with double arrows to indicate the possibility of equilibrium, the value of K_a for HCl (about 10^3) in water reveals that conversion to hydrogen (hydronium) and chloride ions is effectively 100 percent complete in dilute solution. We could abandon the double arrows and write

$$HCl \longrightarrow H^+ + Cl^-$$

Conversely, acetic acid and carbonic acid are weak electrolytes and, therefore, weak acids that do not ionize completely in water. Acetic acid, for instance, ionizes only to the extent of about 1 percent in $0.1M$ solution (of every 100 HAc molecules placed in the solution, only one ionizes). The degree of ionization of a weak electrolyte such as HAc increases rapidly as the solution becomes more dilute (Figure 10.1). The small extent of ionization of a not too dilute acetic acid solution, where its concentration is greater than $10^{-2}M$, means that acetate ion is a strong base because it has a greater tendency to combine with H^+ than water does.

$$HAc \rightleftharpoons H^+ + Ac^-$$

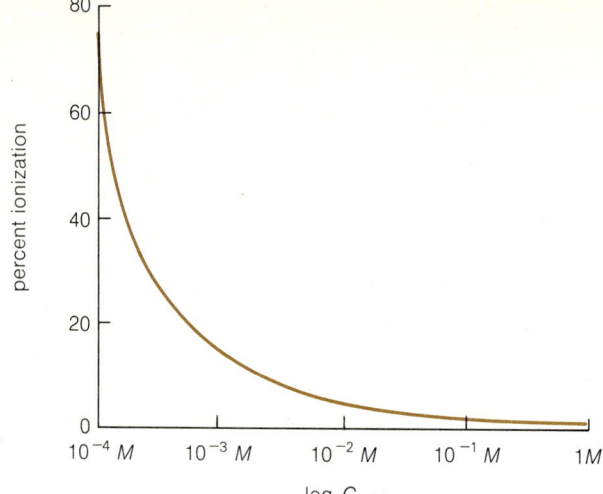

Figure 10.1. Effect of concentration on percent ionization of weak electrolyte. The concentration of acetic acid covers the range $10^{-4}M$ to $10^0 = 1M$. In infinitely dilute solutions, all weak electrolytes are completely ionized.

Equations written with a larger arrow in one direction than the other remind us that strong acids and bases have weak conjugate bases and acids, respectively.

Table 10.2
Relative strengths of acids and bases

	conjugate acid		conjugate base
strongest acid			**weakest base**
	H_2SO_4	$\rightleftharpoons$	$H^+ + HSO_4^-$
	HCl	$\rightleftharpoons$	$H^+ + Cl^-$
	HNO_3	$\rightleftharpoons$	$H^+ + NO_3^-$
	H_3O^+	$\rightleftharpoons$	$H^+ + H_2O$
	HSO_4^-	$\rightleftharpoons$	$H^+ + SO_4^{2-}$
	CH_3COOH	$\rightleftharpoons$	$H^+ + CH_3COO^-$
	H_2CO_3	$\rightleftharpoons$	$H^+ + HCO_3^-$
	NH_4^+	$\rightleftharpoons$	$H^+ + NH_3$
	HCO_3^-	$\rightleftharpoons$	$H^+ + CO_3^{2-}$
	H_2O	$\rightleftharpoons$	$H^+ + OH^-$
weakest acid			**strongest base**

The relative strengths of a series of acids and their conjugate bases are shown in Table 10.2. Acids on the left side will react with bases located diagonally below them to produce a weaker acid and a weaker base.

$$HCl + H_2O \longrightarrow H_3O^+ + Cl^-$$

strong acid   weak base   weaker acid   weaker base
                           than HCl   than H_2O

Another reaction proceeding almost 100 percent to completion is

$$H_3O^+ + CH_3COO^- \rightleftharpoons CH_3COOH + H_2O$$

moderately strong acid	moderately strong base	weaker acid than H_3O^+	weaker base than CH_3COO^-

Because this reaction proceeds in the direction written, the reverse reaction, the ionization of acetic acid in water, takes place only to a very small extent.

Table 10.2 could have been constructed by investigating whether appreciable reaction took place between acid–base pairs. Or the table could have been built up simply by ordering all the acids on the left according to decreasing acid dissociation constants.

10.5 CHEMICAL CONSTITUTION AND ACID STRENGTH

An indication of the relative proton-donating tendency of similar compounds can be obtained from a consideration of bond type and strength. We consider here only the so-called oxy acids, whose structures contain the hydroxyl group —O—H bound to another atom by a bond ranging from ionic to covalent. Some examples are NaOH, $Al(OH)_3$, H_2CO_3, and H_2SO_4, whose structures are

$$Na—O—H$$

$$Al \begin{matrix} O—H \\ O—H \\ O—H \end{matrix}$$

$$O{=}C \begin{matrix} O—H \\ O—H \end{matrix}$$

$$\begin{matrix} O \\ \| \\ S \\ \| \\ O \end{matrix} \begin{matrix} O—H \\ O—H \end{matrix}$$

All of these have the general formula $O_xE(OH)_y$, where E is the central atom. If the E—O bond of the E—O—H group breaks when the compound is placed in water, hydroxide ions are released and the substance is a base. If the O—H bond of the same group ruptures, H^+ (or H_3O^+) ions result in solution and the compound is an acid. Whether the O—H or the E—O bond ruptures depends on the electronegativity of the central element E.

If E is highly electronegative, bonding electrons in the O—H bond will be drawn away from the hydrogen toward the E—O bond. This weakens the O—H bond, leaving the proton hanging in midair, so to speak, where it can be easily picked off by H_2O to form H_3O^+:

$$E—O{+}H + H_2O \longrightarrow E—O^- + H_3O^+$$

Such an acidic reaction will be given by "hydroxides" of the electronegative nonmetals on the right side of the periodic table. The oxy compounds of nonmetals such as C, S, and N are therefore acids in water. The elements tend to remain in solution as soluble oxy anions (CO_3^{2-}, SO_4^{2-}, NO_3^-, etc.) of the acids. The presence of additional electronegative oxygens attached to E weakens the O—H bond even further.

As the central element E decreases in electronegativity, electrons are no longer withdrawn from the O—H bond. For the more electropositive elements (the alkali and alkaline earth metals), the electrons in the E—O bond are drawn toward oxygen, thus weakening the E—O bond and causing its rupture to give hydroxide ions in water:

$$E{+}O—H + H_2O \longrightarrow E(H_2O)^+ + OH^-$$

The oxy compounds of the more active metals are bases. After reaction with water, the hydrated cations tend to remain in solution.

Amphoterism Central elements E with intermediate electronegativity ($\sim$1.7) tend to form oxy compounds that can act as acids or bases; that is, they are amphoteric. Their oxy compounds (hydroxides) are usually highly insoluble in water, but they will dissolve in either strongly acidic or strongly basic solution. Aluminum hydroxide is an example:

$$Al(OH)_3(s) + 3H^+ \longrightarrow Al^{3+} + 3H_2O$$

$$Al(OH)_3(s) + OH^- \longrightarrow Al(OH)_4^- \text{ or } AlO_2^- + 2H_2O$$

(These reactions can be written as

$$Al(OH)_3(H_2O)_3(s) + 3H_3O^+ \longrightarrow Al(H_2O)_6^{3+} + 3H_2O$$

and

$$Al(OH)_3(H_2O)_3(s) + OH^- \longrightarrow Al(OH)_4(H_2O)_2^- + H_2O$$

to show the actual hydrated species involved. The reaction with H^+ or H_3O^+ proceeds stepwise through $Al(OH)_2^+$ or $Al(H_2O)_4(OH)_2^+$ and $Al(OH)^{2+}$ or $Al(H_2O)_5(OH)^{2+}$ to the final triply charged product.) In effect, an amphoteric hydroxide dissociates in two ways in water:

$$H^+ + AlO_2^- + H_2O \rightleftharpoons \begin{Bmatrix} Al(OH)_3(s) \\ \text{or} \\ HAlO_2 \cdot H_2O \end{Bmatrix} \rightleftharpoons Al^{3+} + 3(OH)^-$$

If a strong base is added to the mixture, the OH^- from the base will react with H^+ to form water. The equilibrium will be shifted to form AlO_2^-, and $Al(OH)_3$ will dissolve. If H^+ from an acid is added, it will react with OH^- in the mixture, again forming water and shifting the equilibrium to the Al^{3+} side. The $Al(OH)_3$ will again dissolve. Zinc hydroxide is also amphoteric.

In general, if an E—O—H group exists, an acid is formed in water if the E—O bond is predominantly covalent; if the E—O bond is ionic, a base is formed; and if the E—O bond is intermediate, a water-insoluble amphoteric hydroxide results.

Acidic and Basic Oxides Similar reasoning can be used to explain the acidic nature of covalent nonmetal oxides (SO_2, SO_3, CO_2) in water and the basic nature of ionic metallic oxides if it is assumed that a hydroxide intermediate is formed upon solution in water.

$$CO_2 + H_2O \rightleftharpoons H_2CO_3 \text{ or } CO(OH)_2 \rightleftharpoons H^+ + HCO_3^-$$
$$Na_2O + H_2O \longrightarrow (2Na{+}O—H) \longrightarrow 2Na^+ + 2OH^-$$

Amphoteric oxides such as Al_2O_3 and ZnO are insoluble in water but soluble in strong acids or bases according to equations such as

$$Al_2O_3 + 6H^+ \longrightarrow 2Al^{3+} + 3H_2O$$

and

$$Al_2O_3 + 2OH^- + 3H_2O \longrightarrow 2Al(OH)_4{}^-$$

Acidity of Hydrated Cations

The behavior of hydrated cations in solution may be predicted by arguments based on the strength of bonds. If we consider a hydrated ion to have a structure

$$E \leftarrow O \underset{H}{\overset{H}{\diagup}}$$

the relative strengths of the E←O and of the O—H bonds will depend on the polarizing power of the atom E. The polarizing power is greatest for an atom of high ionic charge-to-ionic size ratio (this ratio is often called the ionic potential) and is proportional to the electronegativity of the ion. If E is a large metal ion of low charge (Na^+, K^+, Ca^{2+}), the E←O bond is weak, and the ion remains free in solution as a loosely hydrated ion. Hypothetical highly charged, small, positive ions such as S^{4+}, S^{6+}, and C^{4+} have a high charge-to-size ratio and a high polarizing ability. They pull electrons from the O—H bond so that the hydrogens can be easily removed and leave an oxy anion ($SO_3{}^{2-}$, $SO_4{}^{2-}$, $CO_3{}^{2-}$) in solution. As before, cations of intermediate electronegativity (and ionic potential) tend to produce insoluble amphoteric hydroxides by reaction with water. Among these ions are Al^{3+}, Pb^{2+}, Sn^{2+}, Zn^{2+}, and Cr^{3+}. Their hydrated ions give acidic reactions in water:

$$Al(H_2O)_6{}^{3+} + H_2O \rightleftharpoons Al(H_2O)_5OH^{2+} + H_3O^+$$

In a slightly basic medium such as sea water (pH = 7.5–8.4), this reaction can be carried two more steps to $Al(H_2O)_3(OH)_3$, which, as already pointed out, is really insoluble $Al(OH)_3$ with waters of hydration. The formation of such insoluble hydroxides in sea water and their sedimentation has been roughly correlated with the abundance of dissolved species in sea water (Section 11.3).

10.6 COMMON ACID–BASE REACTIONS

Ionization. The most common acid–base reaction is ionization in water, where water acts as an acid or a base. Ionization reactions have already been discussed in detail.

Neutralization. The reaction of equivalent amounts of an acid and a base, neither of which is water, is called *neutralization*. The products of this reaction are water and a solution of an ionic compound, a salt, in water. Under certain circumstances, the resulting solution is neutral: that is, the concentration of hydrogen ions (hydronium ions) is equal to the concentration of hydroxide ions. While, strictly speaking, only such a result should be termed neutralization, in practice the term is used more broadly. The discussion of hydrolysis that follows shortly will show why the reaction of equivalent amounts of acid and base does not always produce a neutral solution. We begin by considering the neutralization reactions of different combinations of strong and weak acids and bases.

1. Strong acid + strong base neutralization

$$HCl + NaOH \longrightarrow H_2O + NaCl$$

This is the molecular equation. In solution, both HCl and NaOH are completely ionized so that (ignoring the hydration of the H^+ of HCl to H_3O^+) we have

$$H^+ + Cl^- + Na^+ + OH^- \longrightarrow H_2O + Na^+ + Cl^-$$

The sodium and chloride ions take no part in the reaction, and they can be omitted to give a net ionic equation

$$H^+ + OH^- \longrightarrow H_2O$$

(or

$$H_3O^+ + OH^- \longrightarrow 2H_2O)$$

This net equation applies to any strong acid–strong base reaction in aqueous solution. The result of the reaction of equivalent amounts of HCl and NaOH is a solution of sodium chloride, a salt of a strong acid and a strong base.

2. Weak acid + strong base neutralization

$$HAc + NaOH \rightleftharpoons H_2O + Na^+ + Ac^-$$

In water, NaOH is completely dissociated but HAc (acetic acid), being a weak acid, exists mostly in the molecular form. Again, Na^+ does not participate in the reaction, so the net equation can be written

$$HAc + OH^- \rightleftharpoons H_2O + Ac^-$$

With sodium hydroxide, the result is a solution of sodium acetate in water. Acetate ion is a fairly strong base, and its tendency to react with water to reform HAc has a slight reversal effect on this neutralization. This reverse reaction is an example of hydrolysis and results in a slightly basic solution when equimolar amounts of NaOH and HAc are mixed.

3. Weak acid + weak base neutralization

$$HAc + NH_3 \rightleftharpoons NH_4^+ + Ac^-$$

The production of water is obscured here, but it may be brought in by writing the separate ionization equations and adding them:

$$HAc + H_2O \rightleftharpoons H_3O^+ + Ac^-$$
$$\underline{NH_3 + H_2O \rightleftharpoons OH^- + NH_4^+}$$
$$HAc + NH_3 + 2H_2O \rightleftharpoons NH_4^+ + Ac^- + H_3O^+ + OH^-$$

The hydronium ion and hydroxide ion products react essentially to completion to give water,

$$H_3O^+ + OH^- \longrightarrow 2H_2O$$

so that $2H_2O$ appears on both sides of the equation and need not appear in the net equation. The ionic products are a weak acid (NH_4^+) and a weak base (Ac^-), both of which can partially react with water to reverse the neutralization reaction:

$$NH_4^+ + H_2O \rightleftharpoons NH_3 + H_3O^+$$
$$Ac^- + H_2O \rightleftharpoons HAc + OH^-$$

These are hydrolysis reactions and are discussed below.

Hydrolysis. Hydrolysis is generally described as the reaction of a substance with water. Water ionizes very slightly according to the equation

$$H_2O \rightleftharpoons H^+ + OH^-$$

(or

$$2H_2O \rightleftharpoons H_3O^+ + OH^-)$$

The hydrogen ions may react with anions that are conjugate bases of weak acids (acetate, carbonate, etc.) to reform the weak acid in what is effectively a reversal of the neutralization reaction:

$$CO_3^{2-} + H_2O \rightleftharpoons HCO_3^- + OH^-$$

Similarly, a cation that is the conjugate acid of a weak base may react with the hydroxide ions of water to form the weak base:

$$NH_4^+ + H_2O \rightleftharpoons NH_3 + H_3O^+$$

(or

$$NH_4^+ \rightleftharpoons NH_3 + H^+)$$

Such hydrolysis reactions are really nothing more than the reactions of ionic acids or bases with water. Anions of strong acids (Cl^-) and cations of strong bases (Na^+, K^+) do not undergo hydrolysis.

Reaction of acids with solutions of metallic carbonates and hydrogen carbonates

$$H^+ + CO_3^{2-} \rightleftharpoons HCO_3^-$$
$$H^+ + HCO_3^- \rightleftharpoons H_2O + CO_2$$

Only the first step occurs with weak acids such as H_2CO_3 or with very dilute acids. Stronger acids carry the reaction through the second step to H_2O and CO_2. Natural waters dissolve carbon dioxide from the air and from decay processes in the soil to produce H_2CO_3 solutions, which dissolve limestone according to the equation

$$H_2CO_3 + CaCO_3 \rightleftharpoons Ca^{2+} + 2HCO_3^-$$

This equilibrium explains the appearance of formations in limestone caves (Figure 10.2).

Dissolution of insoluble salts in acids

$$AgAc(s) + H^+ \longrightarrow Ag^+ + HAc$$

Reactions of acidic and basic oxides
1. Strong acids + metallic (basic) oxides

$$2H^+ + CaO \longrightarrow Ca^{2+} + H_2O$$
$$6H^+ + Fe_2O_3 \longrightarrow 2Fe^{3+} + 3H_2O$$

Figure 10.2. Limestone stalactites and stalagmites. When CO_2-charged solution, saturated with $CaCO_3$, trickles through cracks into a cave, the CO_2 pressure is released and the solution loses CO_2. Water also begins to evaporate. The loss of CO_2 (or H_2CO_3) and water both cause an increase in CO_3^{2-}, and $CaCO_3$ begins to precipitate as stalactites hanging from the ceiling and as stalagmites rising from the floor of the cave. (Courtesy of National Park Service. Photo by Fred Mang Jr.)

2. Water + basic (metallic) oxides

Basic or ionic oxides of the more active metals can react with the weak acid water to produce a solution containing hydroxide ions. A typical net reaction is

$$H_2O + CaO \longrightarrow Ca^{2+} + 2OH^-$$

These oxides contain O^{2-} ions, which react with the hydrogen ions derived from the slight ionization of water:

$$H_2O \rightleftharpoons H^+ + OH^-$$
$$Ca^{2+} + O^{2-} + H^+ + OH^- \longrightarrow Ca^{2+} + 2OH^-$$

3. Water + acidic (nonmetal) oxides

Acidic or covalent oxides may be considered to react with the weak base water by the acceptance of an electron pair from the hydroxide ion from water:

$$\overset{\displaystyle O}{\underset{\displaystyle O}{\overset{\|}{\underset{\|}{C}}}}{\leftarrow}OH^- + H^+ \longrightarrow HCO_3^- + H^+ \rightleftharpoons H_2CO_3$$

4. Acidic oxides + basic oxides

Acidic and basic oxides may react in the absence of water:

$$\underset{O^{2-},\text{ Lewis base}}{CaO} \quad + \quad \underset{\text{Lewis acid}}{SO_3} \quad \longrightarrow \quad CaSO_4$$

(see Sections 7.2 and 10.2).

Quantitative Aspects of Acid–Base Equilibria

In discussing solubility of substances and solutions, we saw that it was not enough to say that a compound was soluble or insoluble or that a solution was concentrated or dilute. We needed some quantitative measure of solubility (K_{sp}) and of concentration (molarity). The situation is the same here. We would like some quantitative measure of the hydrogen ion or hydroxide ion in solutions of acids or bases. The use of acid and base ionization constants provides an avenue into this subject. Let us look first at the ionization of water itself.

10.7 THE ION PRODUCT OF WATER Water is an amphiprotic substance and can act as an acid or a base. In the equation

$$H_2O + H_2O \rightleftharpoons H_3O^+ + OH^-$$

one water molecule acts as an acid, the other as the base. In line with our agreement to simplify notation by writing $[H^+]$ for $[H_3O^+]$, by taking for granted the role played by water, we write

$$H_2O \rightleftharpoons H^+ + OH^-$$

for the self-ionization of water.

The ionization takes place to a very small but finite degree, as shown by the very small electric conductivity of pure water. Water is really a very weak electrolyte rather than a nonelectrolyte. The ionization constant for water is

$$K = \frac{[H^+][OH^-]}{[H_2O]}$$

As before, $[H_2O]$ is a constant in pure water and dilute solutions, and we combine it with K to obtain

$$K_w = K[H_2O] = [H^+][OH^-]$$

K_w is called the ion product of water.

The numerical value of K_w can be determined from conductance or other measurements. Its value at 25°C is 1.0×10^{-14}:

$$K_w = [H^+][OH^-] = 1.0 \times 10^{-14}$$

This value of K_w applies at all times in pure water and dilute solutions of electrolytes. In pure water, the concentrations of H^+ and OH^- must be equal, because for every H^+ produced by ionization of a water molecule, one OH^- is formed. Therefore, in pure water,

$$[H^+] = [OH^-]$$
$$[H^+]^2 = 1.0 \times 10^{-14}$$
$$[H^+] = 1.0 \times 10^{-7} \text{mol/liter } (M)$$
$$[OH^-] = 1.0 \times 10^{-7} M$$

A *neutral* solution is one in which the concentrations of H^+ and OH^- are equal, each being $1.0 \times 10^{-7} M$ at 25°C. Neutrality is not limited to pure water. Solutions of other substances in water also may be neutral.

A solution in which $[H^+]$ is greater than $[OH^-]$ is called *acidic*; a solution in which $[OH^-]$ is greater than $[H^+]$ is *basic* or alkaline. In every case, however, the product $[H^+][OH^-] = 1.0 \times 10^{-14}$ applies at 25°C. If the concentration of hydrogen ion exceeds $1.0 \times 10^{-7} M$, the hydroxide ion concentration must be less than $1.0 \times 10^{-7} M$. At higher temperatures, the degree of ionization of water increases, as does K_w and the concentration of H^+ and OH^- in a neutral solution.

10.8 pH To avoid the necessity of working with a wide range of small exponential numbers, we often use a logarithmic scale to denote low hydrogen ion concentrations in the range of about $[H^+] = 1-10^{-14} M$. This scale is called the *pH scale*. For our purposes, pH is defined as the negative exponent of 10, which gives the hydrogen ion concentration

$$[H^+] = 10^{-pH}$$

or

$$pH = -\log_{10}[H^+]$$

(See Appendix A.3 for a discussion of logarithms.) In pure water, or any neutral solution, where $[H^+] = 1.0 \times 10^{-7}$, the pH is 7.00. Acidic solutions have pH values less than 7, and basic solutions have pH values above 7 at 25°C. Table 10.3 shows $[H^+]$ and $[OH^-]$ and pH values for a series of solutions ranging from pH = 0.0 to pH = 14.0. The pH scale extends below zero (i.e., $[H^+] = 10 M = 10^1$, pH = -1.0) and above 14 ($[H^+] = 1 \times 10^{-15}$, $[OH^-] = 10 M = 10^1$, pH = 15.0), but such strongly acidic or basic situations are rare in natural systems. A further definition, $pOH = -\log_{10}[OH^-]$ is sometimes used. Thus, pH + pOH must equal 14.0.

Table 10.3 The pH scale	$[H^+]$	$[OH^-] = \dfrac{1 \times 10^{-14}}{[H^+]}$	pH	
	$1 \times 10^0 = 1M$	$1 \times 10^{-14}M$	0.0	↑
	$1 \times 10^{-1} = 0.1M$	$1 \times 10^{-13}M$	1.0	increasing
	$1 \times 10^{-2} = 0.01M$	$1 \times 10^{-12}M$	2.0	acidity
	$1 \times 10^{-3} = 0.001M$	$1 \times 10^{-11}M$	3.0	
	⋮			↑
	$1 \times 10^{-7}M$	$1 \times 10^{-7}M$	7.0	neutral
	⋮			↓
	$1 \times 10^{-12}M$	$1 \times 10^{-2}M$	12.0	increasing
	$1 \times 10^{-13}M$	$1 \times 10^{-1}M$	13.0	basicity
	$1 \times 10^{-14}M$	1×10^0M	14.0	↓

10.9 SOLUTIONS OF STRONG ACIDS AND STRONG BASES

We can find the hydrogen ion concentration of solutions of only strong acids or of strong bases without recourse to ionization equilibria because of the assumed 100 percent ionization of these strong electrolytes.

EXAMPLE | What is the pH of a solution of 0.010M HCl?

SOLUTION |

$$HCl \longrightarrow H^+ + Cl^-$$
$$[H^+] = [H^+]_{\text{from HCl}} + [H^+]_{\text{from water}}$$

The amount of hydrogen ion resulting from the ionization of water is negligible in comparison with that produced by the complete ionization of HCl:

$$[H^+] = [H^+]_{\text{from HCl}} = 0.010M$$
$$pH = -\log(0.010) = -\log 1.0 \times 10^{-2} = 2.00$$

In this example, the approximation that all of the hydrogen ion in solution can be considered as coming only from the acid is justified in view of the weakness of H_2O as an acid and the concentration of the strong acid. The result is the repression of the ionization of water,

$$H_2O \rightleftharpoons H^+ + OH^-$$

by the acid. The hydrogen ion formed from the ionization of water is equal to the hydroxide ion concentration in solution. This is only $1.0 \times 10^{-12}M$ and is obviously negligible compared to 0.010M. We shall continue to use this approximation for the examples in this book. It is only in very dilute solutions that the contribution of water ionization to $[H^+]$ or $[OH^-]$ need be considered. For instance, the pH of $1.0 \times 10^{-7}M$ HCl solution is 6.79, not 7.00. The value 6.79 can be obtained by setting

$$[H^+][OH^-] = (1.0 \times 10^{-7} + [OH^-])([OH^-])$$

and solving for $[OH^-]$ using the quadratic equation.

10.10 SOLUTIONS OF WEAK ACIDS AND WEAK BASES

For the time being, we shall limit consideration to weak acids and bases that donate or accept only one proton. These are called *monoprotic* (or monoacidic and monobasic) acids and bases. Weak acids and bases do not ionize completely upon solution, so there will be an equilibrium between ionization products and the un-ionized acid or base.

EXAMPLE | What is the hydrogen ion concentration and pH of 0.50M acetic acid?

SOLUTION | The ionization equation is

$$HAc \rightleftharpoons H^+ + Ac^-$$

The acid ionization constant K_a is

$$1.8 \times 10^{-5} = \frac{[H^+][Ac^-]}{[HAc]}$$

If we neglect the hydrogen ions from the slight ionization of water, every H^+ produced by the ionization of HAc is accompanied by one Ac^-. Therefore, at equilibrium,

$$[H^+] = [Ac^-] \quad \text{(ionization of } H_2O \text{ neglected)}$$

The concentration of the acid solution, 0.50M, refers to the moles of un-ionized HAc molecules initially used to make up a liter of solution. It is the *total* acetate concentration, $0.50 = [HAc] + [Ac^-]$. At equilibrium, the concentration of HAc molecules is less than 0.50M because some of the molecules have ionized. For each molecule (or mole) of HAc that ionizes, one molecule (or mole) of H^+ and of Ac^- are produced. At equilibrium, the actual concentration of un-ionized HAc is

$$[HAc] = 0.50 - [H^+]$$
$$= 0.50 - [Ac^-]$$

It is often helpful to summarize these conditions under the ionization equation:

	HAc	$\rightleftharpoons$	H^+	+	Ac^-
initially	0.50		0		0
equilibrium	$0.50 - [H^+]$		$[H^+]$		$[Ac^-] = [H^+]$

We have then

$$1.8 \times 10^{-5} = \frac{[H^+]^2}{0.50 - [H^+]}$$

Solution of this equation for $[H^+]$ requires the quadratic formula. However, with a sufficiently weak acid ($K_a = 10^{-3}$ or less) in a not too dilute solution, the concentration of hydrogen ion will be small compared to 0.50M and can be neglected, so $[HAc] \simeq 0.50M$.

With this second approximation,

$$1.8 \times 10^{-5} = \frac{[H^+]^2}{0.50}$$

and

$$[H^+] \simeq \sqrt{9.0 \times 10^{-6}} = 3.0 \times 10^{-3}$$
$$pH = -\log(3.0 \times 10^{-3})$$
$$= -\log 3.0 - \log 10^{-3}$$
$$= -0.48 + 3 = 2.52$$

The approximation holds because

$$0.50 - 3.0 \times 10^{-3} \simeq 0.50M$$

Provided that the two approximations made in the example above hold for a solution of a general weak monoprotic acid HA, we can write

$$[H^+] \simeq \sqrt{K_a C_a}$$

where C_a is the concentration of the acid.

A corresponding equation for the hydroxide ion concentration in a solution of a weak monoprotic base such as NH_3 of concentration C_b is

$$[OH^-] \simeq \sqrt{K_b C_b}$$

10.11 POLYPROTIC WEAK ACIDS AND BASES

Acids and bases capable of donating and accepting 2 or more protons are *polyprotic*. The protons are lost or gained in consecutive steps, each of which has its own ionization constant. The stepwise ionization of carbonic acid illustrates this:

$$H_2CO_3 \rightleftharpoons H^+ + HCO_3^- \qquad K_{a_1} = 4.2 \times 10^{-7}$$
$$HCO_3^- \rightleftharpoons H^+ + CO_3^{2-} \qquad K_{a_2} = 5.0 \times 10^{-11}$$

It is not permissible to assume complete dissociation of H_2CO_3 according to the equation

$$H_2CO_3 \rightleftharpoons 2H^+ + CO_3^{2-}$$

The acid ionization constants K_{a_1}, K_{a_2}, ... become successively smaller by several factors of 10 because of the difficulty of removing a proton from the less positive species produced in the first step. Similarly, the successive base ionization constants become smaller, as illustrated by the polyprotic base, carbonate ion.

$$CO_3^{2-} + H_2O \rightleftharpoons OH^- + HCO_3^- \qquad K_{b_1} = 1.8 \times 10^{-4}$$
$$HCO_3^- + H_2O \rightleftharpoons OH^- + H_2CO_3 \qquad K_{b_2} = 2.2 \times 10^{-8}$$

The calculation of the H^+ (or OH^-) concentration in a solution of a weak polyprotic acid (or base) can become complicated if all equilibria are taken into account. However, if the first ionization constant is far greater than the second, it is permissible to assume that the first ionization is the only one contributing to the H^+ (or OH^-) concentration.

EXAMPLE

At 25°C and at the partial pressure (see Section 11.7) of CO_2 in normal air, the concentration of CO_2 dissolved in water in equilibrium with the atmosphere is $1.0 \times 10^{-5}M$. What is the pH of CO_2-saturated water?

SOLUTION

All of the CO_2 dissolved in the water is considered to form carbonic acid according to the equations

$$CO_2(g) + H_2O \rightleftharpoons \{CO_2(aq)\} \longrightarrow H_2CO_3$$

Ignoring the second ionization constant of H_2CO_3, we write

$$H_2CO_3 \rightleftharpoons H^+ + HCO_3^- \qquad K_1 = 4.2 \times 10^{-7}$$
$$= \frac{[H^+][HCO_3^-]}{[H_2CO_3]}$$

where $H_2CO_3 = 1.0 \times 10^{-5}M$. (Actually, only about 0.25 percent of the dissolved CO_2 forms H_2CO_3, but the value of K_1 used here takes this into account and allows the calculation to be made on the assumption of complete conversion to H_2CO_3.) Assuming negligible H^+ from the ionization of water, we have

$$[H^+] = [HCO_3^-]$$

$$4.2 \times 10^{-7} = \frac{[H^+]^2}{1.0 \times 10^{-5}}$$

Again, in the denominator we have used the approximation that the initial amount of total dissolved CO_2 (as H_2CO_3) is large in comparison with the amount ionizing to give HCO_3^-:

$$[H^+]^2 = 4.2 \times 10^{-12}$$
$$[H^+] = 2.0 \times 10^{-6}$$
$$pH = -\log(2.0 \times 10^{-6})$$
$$= -(0.30 - 6.00) = 5.70$$

This example explains why natural waters exposed to the air are slightly acidic in the absence of other acidic or basic solutes. This is why the effects of H^+ are so common in geological and fresh-water systems. Sea water is slightly alkaline (average $pH \simeq 8.1$) because of hydrolysis (Section 10.13) of dissolved carbonates and hydrogen carbonates.

10.12 A WORD ABOUT APPROXIMATIONS

In preceding examples, and in those to follow, we use a number of approximations concerning the relative magnitude of solute concentrations. These approximations are not always valid. Fortunately, it is often possible to tell from the results based on these approximations whether the problem should be recalculated without some of the simplifying assumptions. When there are a number of species in solution and their concentrations are not negligible, other algebraic equations must be included in the calculation. These are, in addition to all equilibrium constant expressions pertaining to processes in solution, material balance and charge balance or electroneutrality equations. *Material balance* means merely that there can be no more and no less of a substance in solution in its various forms than was put into the solution originally. *Charge balance* means that the sum of the negative ionic charges in a solution must equal the sum of the positive ionic charges in the solution. The charge balance condition says simply that no solution may have a net positive or negative charge.

The example of the carbonic acid solution of Section 10.11 can be used to give an idea of mass and charge balance equations. In this problem, the equilibrium conditions are given by K_w and the two ionization constants K_1 and K_2 in the general case. The mass balance equation is

$$C_{acid} = [H_2CO_3]_{total} = 1.0 \times 10^{-5}M$$
$$= [H_2CO_3] + [HCO_3^-] + [CO_3^{2-}]$$

The charge balance (electroneutrality) equation is

$$[H^+] = [OH^-] + [HCO_3^-] + 2[CO_3^{2-}]$$

The smallness of K_{a_2} leads to the approximation that $[CO_3^{2-}]$ is negligible in comparison with $[HCO_3^-]$. Also, $[OH^-]$ is negligibly small because the OH^- comes only from the ionization of water and this ionization is repressed by the presence of H^+ in the acidic solution. With these approximations, the charge balance becomes $[H^+] \simeq [HCO_3^-]$ and then, if K_1 is small and C_a fairly large (just barely applicable in this example), the mass balance becomes $1.0 \times 10^{-5}M \simeq [H_2CO_3] + [H^+] \simeq [H_2CO_3]$. You can see that although we have not explicitly used mass and charge balance equations in our examples, we have made use of the ideas. When the approximations used in our examples do not apply, the simple calculations cannot handle all situations and the computations get more involved.

10.13 HYDROLYSIS

We have described hydrolysis broadly as the reaction of a substance with water. In common types of acid–base reactions, the consequences of hydrolysis are an excess of either H^+ or OH^- ions in the resulting "neutral" solution. In this section, we are interested in the acidic or basic nature of solutions of salts that result from the neutralization of an acid by a base. Examples of such salts are

$NaCl$, KNO_3, NaI:	salts of a strong acid and a strong base
$NaAc$, K_2CO_3, $Ca_3(PO_4)_2$:	salts of a weak acid and a strong base
NH_4Cl, $FeCl_3$, $Al(NO_3)_3$:	salts of a strong acid and a weak base
NH_4Ac, NH_4CN:	salts of a weak acid and a weak base

Solutions of these salts, all of which are strong electrolytes, will be neutral, acidic, or basic depending on the acidic and basic strength of the ions in solution. Solutions of salts of strong acids and strong bases are neutral (pH = 7); salts of strong acids and weak bases give acidic solutions; salts of weak acids and strong bases afford basic solutions; and weak acid–weak base salts have solutions with pH's near 7. In the following, we examine in detail the behavior of a solution of a salt of a weak acid and a strong base.

Sodium acetate (NaAc) is a typical salt of a weak acid and a strong base. Since the salt ionizes completely in solution,

$$NaAc \longrightarrow Na^+ + Ac^-$$

we must consider each ion of the salt in turn. There is no tendency for Na^+ to react with water (or with the OH^- from water) to form NaOH, because NaOH is a strong electrolyte. Thus, there is no equilibrium involving Na^+.

The acetate ion, on the other hand, is a fairly strong base. It reacts with water according to the equation

$$Ac^- + H_2O \rightleftharpoons HAc + OH^-$$

An equilibrium constant, the hydrolysis constant (really the base dissociation constant of Ac^-), can be written for this reaction:

$$K_h = \frac{[HAc][OH^-]}{[Ac^-]}$$

If both numerator and denominator are multiplied by $[H^+]$, the resulting expression becomes

$$\frac{[HAc]}{[Ac^-][H^+]} \times \frac{[H^+][OH^-]}{1} = \frac{K_w}{K_a}$$

or

$$\frac{K_w}{K_a} = \frac{[HAc][OH^-]}{[Ac^-]}$$

If the salt NaAc is the only solute present and if the contribution of water ionization to $[OH^-]$ is neglected, the hydrolysis reaction shows

$$[HAc] = [OH^-]$$

so that

$$\frac{K_w}{K_a} = \frac{[OH^-]^2}{[Ac^-]}$$

And if the initial concentration of the salt is large enough so that the acetate ions converted to HAc are negligible in comparison with the salt concentration, then

$$[Ac^-] \simeq C_{salt}$$

so that

$$[OH^-] = \sqrt{\frac{K_w C_{salt}}{K_a}}$$

or

$$[H^+] = \sqrt{\frac{K_w K_a}{C_{salt}}}$$

Since $K_w = 10^{-14}$ and $K_a \simeq 10^{-5}$ for HAc, the pH of a $0.1M$ NaAc solution will be about 9, a basic solution.

An alternative way of viewing the hydrolysis of acetate ion is to consider the reaction of Ac^- with H^+ derived from water:

$$
\begin{array}{ccccc}
NaAc & \longrightarrow & Ac^- & + & Na^+ \\
 & & + & & + \\
H_2O & \rightleftharpoons & H^+ & + & OH^- \\
 & & \Updownarrow & & \downarrow \\
 & & HAc & & \text{no reaction}
\end{array}
$$

The removal of H^+ by combination with Ac^- results in more OH^- than H^+ in the solution.

The reaction of the cation of a salt of a strong acid and a weak base with water results in the generation of H^+ ions in solution. With ammonium chloride (NH_4Cl) as an example, the reaction of NH_4^+ can be written

$$NH_4^+ + H_2O \rightleftharpoons NH_3 + H_3O^+$$

or

$$NH_4Cl \longrightarrow \quad NH_4{}^+ \quad + \quad Cl^-$$
$$+ \qquad\qquad +$$
$$H_2O \rightleftharpoons \quad OH^- \qquad H^+$$
$$\Updownarrow \qquad\qquad \downarrow$$
$$NH_3 + H_2O \quad \text{no reaction}$$

The hydrolysis constant (the acid dissociation constant of $NH_4{}^+$) is

$$\frac{[NH_3][H^+]}{[NH_4{}^+]} \quad \text{(writing } H^+ \text{ for } H_3O^+\text{)}$$

Multiplying by

$$\frac{[OH^-]}{[OH^-]}$$

and setting

$$[NH_3] \simeq [H^+] \quad \text{and} \quad [NH_4{}^+] \simeq [C_{salt}]$$

gives

$$[H^+] = \sqrt{\frac{K_w C_{salt}}{K_b}}$$

where K_b is the base ionization constant of NH_3 in this particular example.

EXAMPLE   Sodium carbonate, Na_2CO_3, is the most common substitute for phosphates (sodium tripolyphosphate) in detergents. One of the problems with Na_2CO_3 and some other phosphate replacements is the caustic nature of their solutions and the resultant danger of alkali burns of the esophagus or eyes in children who play with the detergents or their solutions.

One detergent contains 65 percent by weight of Na_2CO_3. The amount of this detergent used in a wash load produces a solution that is about $0.01M$ in Na_2CO_3. What is the pH of the solution?

SOLUTION   Carbonate ion hydrolyzes according to the equation

$$CO_3{}^{2-} + H_2O \rightleftharpoons HCO_3{}^- + OH^-$$

for which the hydrolysis constant is

$$\frac{[HCO_3{}^-][OH^-]}{[CO_3{}^{2-}]} = \frac{K_w}{K_a}$$

where K_a is the ionization constant for the acid $HCO_3{}^-$ and is the same as the second ionization constant, K_{a_2}, of H_2CO_3, 4.8×10^{-11}. Setting $[HCO_3{}^-] = [OH^-]$ and $[CO_3{}^{2-}] \simeq C_{salt} = 0.01M$,

$$[OH^-] = \sqrt{\frac{K_w C_{salt}}{K_{a_2}}}$$

or

$$[H^+] = \sqrt{\frac{K_w K_{a_2}}{C_{salt}}} = \left(\frac{1.0 \times 10^{-14} \times 4.8 \times 10^{-11}}{1 \times 10^{-2}}\right)^{1/2}$$

$$= (48 \times 10^{-24})^{1/2} = 7 \times 10^{-12}$$

$$pH = 12 - 0.8 = 11.2 \quad \text{(quite strongly basic)}$$

In this problem, the subsequent hydrolysis of HCO_3^- produced from the hydrolysis of CO_3^{2-} is ignored because it is so much smaller than the hydrolysis of CO_3^{2-}.

As the above example shows, the calculation of ion concentrations in solutions of polyprotic bases such as CO_3^{2-} and PO_4^{3-} can often be simplified by ignoring the base ionization (or hydrolysis) steps beyond the first. However, when the hydrolyzing species is amphiprotic, another reaction must be considered. Ions such as HCO_3^-, HPO_4^{2-}, and $H_2PO_4^-$ are examples of species that can function as acids or bases. The solutions resulting from the hydrolysis of such amphiprotic species may be acidic, basic, or even neutral depending on the relative strengths of the substance as an acid and as a base. For instance, hydrogen carbonate ion can hydrolyze in two ways, as an acid,

$$HCO_3^- + H_2O \rightleftharpoons CO_3^{2-} + H_3O^+ \quad K = K_{a_2} = 4.8 \times 10^{-11}$$
$$(\text{or } HCO_3^- \rightleftharpoons CO_3^{2-} + H^+)$$

and as a base,

$$HCO_3^- + H_2O \rightleftharpoons H_2CO_3 + OH^- \quad K_b = \frac{K_w}{K_{a_1}} = 2.4 \times 10^{-8}$$

where K_{a_2} and K_{a_1} are the second and first acid ionization constants of H_2CO_3, respectively.

Both of these reactions take place, but not to identical extents. A solution of $NaHCO_3$ is basic (pH $\simeq$ 8), because HCO_3^- is a stronger base than acid.

10.14 EFFECT OF HYDROLYSIS ON SOLUBILITY

A slightly soluble salt containing an anion that is the conjugate base of a weak acid will have its solubility increased by the hydrolysis of that anion. In such cases, the solubility depends on the hydrolysis constant as well as the solubility product. As a simple example, consider the solution equilibria for slightly soluble silver acetate, AgAc,

$$AgAc(s) \rightleftharpoons Ag^+ + Ac^- \quad K_{sp}$$

Acetate ion hydrolyzes:

$$Ac^- + H_2O \rightleftharpoons HAc + OH^- \quad K_h$$

The true solubility of AgAc is determined by both of these equilibria, not by K_{sp} alone. That is, the true solubility s of AgAc is

$$s = [Ag^+] = [Ac^-] + [HAc]$$

rather than $s_0 = [Ag^+] = [Ac^-]$ as determined only from the solution equilibrium. For AgAc and for other salts of weak acids of relatively large K_a (greater than about 10^{-6}), the effect of hydrolysis on solubility is negligible in solutions of low $[H^+]$. On the other hand, the solubility of salts of polyprotic weak acids is strongly affected by hydrolysis, because the small second or third acid ionization constants of the weak acids means that hydrolysis will be extensive. Common examples of affected salts are sulfides (ZnS, FeS), carbonates ($CaCO_3$), and phosphates [$Ca_3(PO_4)_2$]. The following equilibria for carbonates and phosphates show that hydrolysis increases the solubility of such salts, because one of the products of the dissolution equilibrium is removed by the hydrolysis reaction:

$$CaCO_3(s) \rightleftharpoons Ca^{2+} + CO_3^{2-}$$
$$CO_3^{2-} + H_2O \rightleftharpoons HCO_3^- + OH^-$$
$$Ca_3(PO_4)_2(s) \rightleftharpoons 3Ca^{2+} + 2PO_4^{3-}$$
$$PO_4^{3-} + H_2O \rightleftharpoons HPO_4^{2-} + OH^-$$

Further hydrolysis equilibria involving H_2CO_3 or $H_2PO_4^-$ and H_3PO_4 may enter if the hydrogen ion concentration in the solution is higher than in pure water. The effect of increasing hydrogen ion concentration on solubility of these salts is clear from these equilibria. On the basis of Le Chatelier's principle, added hydrogen ions either react with OH^- to remove one of the products of the hydrolysis reaction, or, if hydrolysis is considered to be a reaction of the basic anion with H^+, added hydrogen ions increase the concentration of one of the reactants in the hydrolysis equation. Conversely, the salts should be less soluble in basic environments. Solubility–pH considerations such as these are important in geological, agricultural, and other problems, and they will be discussed further in Chapters 13 and 14.

10.15 NEUTRALIZATION, TITRATION, AND INDICATORS

The controlled addition of accurately measured volumes of a solution of known concentration to a second solution with which the added solution reacts is a procedure known as *titration*. If the equation for the reaction is known and if the point can be determined at which just enough of the first solute has been added to react with all of the solute in the second solution, then the concentration of reacting solute in that solution can be found. Titration is a common procedure in the chemical analysis of many substances.

The point of complete reaction between the added solution and the titrated solution is the *equivalence point* of the reaction. The equivalence point in the titration of 100 ml of 0.1M HCl with 0.1M NaOH will occur when 100 ml of the NaOH had been added. If 0.2M NaOH had been used instead, the equivalence point would be at 50 ml of added 0.2M NaOH. Both of these are consistent with the requirement that the H^+ of the HCl and the OH^- of the NaOH react on a 1-to-1 mole basis.

$$H^+ + OH^- \longrightarrow H_2O$$

The result of complete reaction in either case is a solution of sodium chloride in water. Because Na^+ and Cl^- do not hydrolyze, this solution at the equivalence point should be neutral, with a pH of 7.

The titration of a solution of weak acid, such as HAc, with NaOH also has an equivalence point when equal numbers of moles of HAc and NaOH have reacted:

$$HAc + OH^- \rightleftharpoons H_2O + Ac^-$$

Now, however, the resulting solution is composed of dissolved sodium acetate alone, and we have seen that the pH of this solution is not 7 because of hydrolysis of the acetate ion. Even though we say that the original HAc has been completely neutralized by the base, the resulting solution is not neutral; it is basic. Similarly, the equivalence point in the titration of a weak base and a strong acid is characterized by an acidic solution.

Figure 10.3 shows the course of the pH change as monoprotic bases of various strengths are added to various monoprotic acids in titration experiments. In each case, the equivalence point is considered to lie at the middle of the sharply rising portion of the curve. The equivalence point can be detected in many titrations by the use of colored weak acids, known as indicators, that change color at a pH within the vertical portions of the titration curves. Table 10.4 lists a few common indicators, with their colors

Table 10.4 Acid–base indicators	indicator	pH values of the color change range	color in acid solution	color in basic solution
	methyl orange	3.1–4.4	red	yellow
	methyl red	4.2–6.1	red	yellow
	phenol red	6.8–8.1	yellow	red
	phenolphthalein	8.0–9.8	colorless	red

and the pH range of their color change. The pH at which the indicator changes color is known as the *end point* of the titration. The indicator end point and the equivalence point are brought as close together as possible by a proper choice of indicator. Thus, according to Figure 10.3(b) and Table 10.4, phenolphthalein would give a better indication of the equivalence point in a strong base–weak acid titration than methyl orange.

10.16 BUFFER SOLUTIONS

Solutions of a weak acid and a salt containing the anion of the acid, or solutions of a weak base and its salt, are called *buffer solutions*. Buffer solutions tend to resist large changes in their pH when H^+ or OH^- ions are added or when the solution is diluted. Many natural systems require an environment of comparatively constant pH and utilize natural buffer systems to maintain this pH. We shall discuss some natural buffers after examining the mechanism of buffer action.

As a simple example of a buffer solution, we take a solution containing acetic acid and sodium acetate. The sodium acetate is completely dissociated. Not so for the weak electrolyte, acetic acid, where the ionization equilibrium

$$HAc \rightleftharpoons H^+ + Ac^-$$

applies.

The hydrogen ion concentration of the weak acid–salt solution is determined by the acid dissociation constant of the weak acid:

$$K_a = \frac{[H^+][Ac^-]}{[HAc]}$$

It is important to realize here that $[H^+] \neq [Ac^-]$. Most of the Ac^- concentration is furnished by the strong electrolyte NaAc. In fact, if the concentration of the salt in solution is C_s, then $[Ac^-] \simeq C_s$ since the presence of Ac^- from the salt represses the ionization of HAc (common ion effect). Also, because

Figure 10.3. Acid–base titration curves. Equivalence points (EP) are indicated, as are the pH's of solutions of maximum buffer capacity.

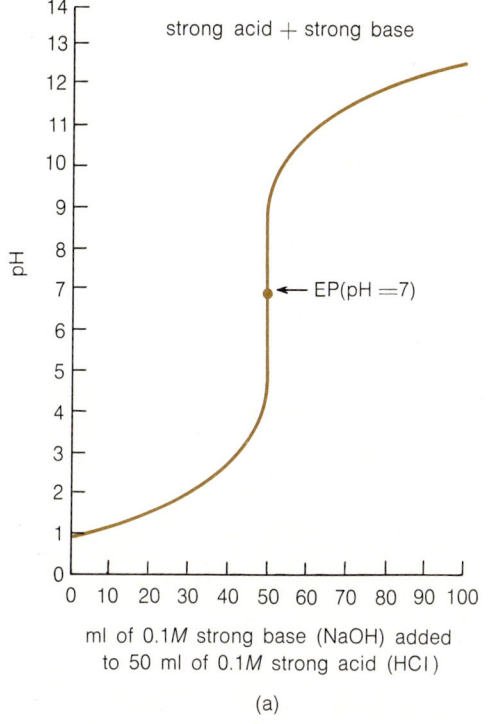

ml of 0.1M strong base (NaOH) added to 50 ml of 0.1M strong acid (HCl)

(a)

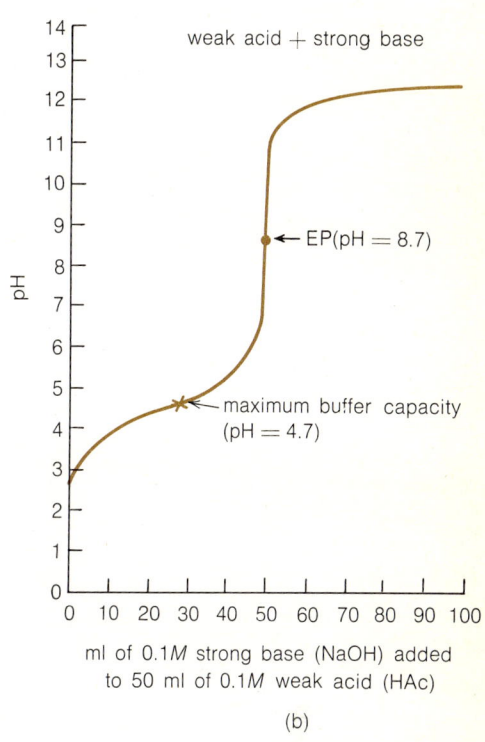

ml of 0.1M strong base (NaOH) added to 50 ml of 0.1M weak acid (HAc)

(b)

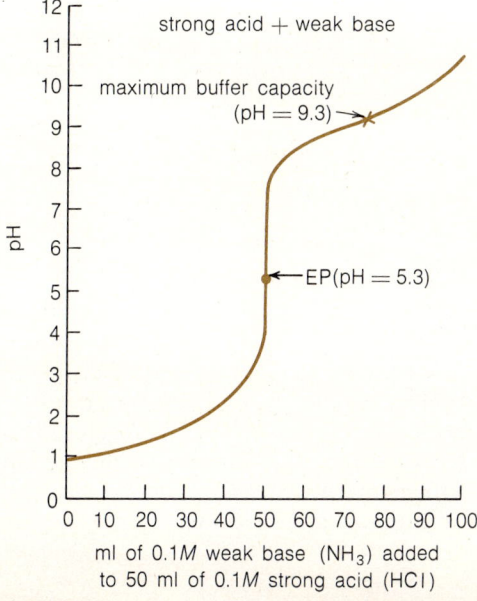

ml of 0.1M weak base (NH$_3$) added to 50 ml of 0.1M strong acid (HCl)

(c)

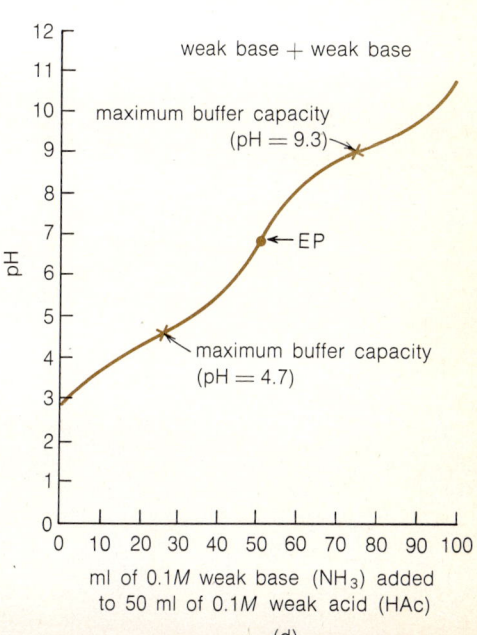

ml of 0.1M weak base (NH$_3$) added to 50 ml of 0.1M weak acid (HAc)

(d)

the ionization of HAc is repressed, we can set $[HAc] \simeq C_a$, where C_a is the initial concentration of HAc molecules in the solution. With these approximations, the hydrogen ion concentration of the solution is

$$[H^+] = K_a \frac{C_a}{C_s}$$

Now let's see why this solution can resist changes in its pH that might be brought about by the addition of H^+ or OH^- from some external source. Most added hydrogen ions will be removed by the reaction

$$H^+ + Ac^- \rightleftharpoons HAc$$

in which the acetate ions from the salt act as a base, and most added hydroxide ions will be neutralized by the reaction

$$OH^- + HAc \rightleftharpoons H_2O + Ac^-$$

We can consider a buffer solution to have a reserve basicity from the Ac^- to neutralize added H^+ and a reserve acidity in the form of HAc to neutralize added OH^-. The action of HAc as a weak electrolyte is responsible for both of these mechanisms.

EXAMPLE | As a numerical example, consider an acetic acid–sodium acetate buffer solution, where the concentration of acid and salt are equal, $C_a = C_s$.

$$[H^+] = K_a$$
$$= 1.8 \times 10^{-5} \qquad \text{for HAc}$$
$$pH = 4.74$$

Now, to be specific, let $C_a = C_s = 0.100M$, and let there be 100 ml of this buffer mixture to which are added an amount of base equivalent to 1 ml of $0.100M$ NaOH. An equivalent amount of HAc is converted to Ac^-.

$$[HAc] = \frac{(0.100M)(0.100 \text{ liter}) - (0.100M)(0.001 \text{ liter})}{0.101 \text{ liter}}$$
$$= 0.099M = C_a$$
$$[Ac^-] = \frac{(0.100M)(0.100 \text{ liter}) + (0.100M)(0.001 \text{ liter})}{0.101 \text{ liter}}$$
$$= 0.101M = C_s$$
$$[H^+] = (1.8 \times 10^{-5})\frac{0.099M}{0.101M} = 1.76 \times 10^{-5}$$
$$pH = 4.75$$

The pH changes by only 0.01 unit. By comparison, the addition of 1 ml of $0.100M$ NaOH to 100 ml of pure water would change the pH from 7 to 11, a change of 4 units.

The greatest resistance to change in pH, or the greatest buffer capacity for a buffer solution, occurs when the concentrations of the weak acid (or the weak base) and its salt are equal. The hydrogen ion concentration at that point is just K_a (or $[OH^-] = K_b$). The points of maximum buffer capacity are marked with crosses on the titration curves of Figure 10.3. Note that these are the points of minimum rate of change in pH.

10.17 BUFFER SYSTEMS IN NATURE

Most organisms require a comparatively constant environment with respect to quantities such as temperature, osmotic pressure, and pH if they are to function properly. This constancy applies both to the internal environment within and around the cells and to the external environment of the organism, although the organism will be more sensitive to changes in its internal body fluids. Nature has provided all organisms with a variety of buffer systems for the stabilization of the pH of their internal fluids, and we briefly examine the factors contributing to the pH regulation in human blood. In Chapter 11, we shall discuss the buffer systems in the ocean.

The Blood Buffer System

The metabolic processes or organisms continually form acidic substances that, in the absence of buffer systems and other regulatory and hydrogen ion removal mechanisms, would soon lower the pH of body fluids and disrupt the activity of the organism. Chief among the acidic products is carbon dioxide, but organic acids (lactic, etc.) and other inorganic acids (phosphoric, sulfuric) are also liberated in the tissues. One of the mechanisms for pH regulation in the human body is the action of buffer systems in body fluids, principally in the blood.

The blood of humans contains several weak acid–salt (or weak acid–conjugate anion base) pairs that contribute to its overall buffer capacity and the maintenance of an acid–base balance. The most important of these are the $H_2CO_3 \rightleftharpoons HCO_3^-$ and the protein hemoglobin $HHB \rightleftharpoons Hb^-$ systems. Proteins carry a variety of weakly acidic groups and can function as both weak acids or weak bases. The H_2CO_3–HCO_3^- system assumes most of the buffering load in blood plasma, while the hemoglobin system is the more important buffer in the red blood cells. Other buffer pairs in body fluids are $H_2PO_4^-$–HPO_4^{2-}, the oxyhemoglobins, $HHbO_2$–HbO_2^-, and other protein buffers.

Carbon dioxide is delivered to the blood by the tissues at a (partial) pressure of about 50–70 mmHg. Most of the CO_2 diffuses through the plasma into the red blood cells, where it reacts quickly with water to form carbonic acid. This reaction is catalyzed (in both directions) by an enzyme, carbonic anhydrase. Hydrogen ions derived from H_2CO_3 ionization are buffered by hemoglobin. The reactions involved are

$$CO_2 + H_2O \underset{\text{catalyzed}}{\overset{\text{enzyme}}{\rightleftharpoons}} H_2CO_3 \rightleftharpoons HCO_3^- + H^+$$

$$H^+ + Hb^- \rightleftharpoons HHb$$

In the lungs, the pressure of CO_2 has decreased to about 40 mmHg, and these reactions shift to the left and ultimately form CO_2, which is exhaled. Buffering of hydrogen ions added directly to the blood at the cells is handled by HCO_3^-.

$$H^+ + HCO_3^- \rightleftharpoons H_2CO_3 \longrightarrow H_2O + CO_2 \text{ (exhaled)}$$

The rapid removal of CO_2 from the body is possible only because it is a gas. Thus, the buffering properties resulting from the weakly acidic nature of H_2CO_3 and HCO_3^- are augmented by the gaseous nature of CO_2.

The acid–base balance of the blood as partially determined by these and other buffer components is intimately involved in the mechanism of oxygen and carbon dioxide transport in blood. All of these are interrelated in the physical chemical processes making up the detailed chemistry of respiration.

In respiration, oxygen combined with hemoglobin is carried from the lungs to the cells. The degree and rate of this combination is determined mainly by the dissociation of hemoglobin as a weak acid. This dissociation equilibrium is determined by the pH of blood and this, in turn, varies with the partial pressure of CO_2 in the blood. At the heart of the matter are subtle changes in the structure of the protein hemoglobin molecule (Section 15.14) that affect its oxygen-carrying and ionization properties. Justice cannot be done to the chemical control of respiration by this brief discussion. We shall say only that the interlocked and competing equilibria create an automatic mechanism tending to control respiratory rates and oxygen transport to meet the needs of the body. The homeostatic mechanisms of respiration are miniature analogs of large-scale homeostasis activities in the biosphere as a whole.

Oxidation and Reduction

10.18 THE ACTIVITY SERIES OF METALS

One of the characteristics of acids is their ability to react with active metals to form hydrogen gas:

$$M + 2H^+ \rightleftharpoons M^{2+} + H_2$$

This is not an acid–base reaction in the sense of the proton or Brønsted–Lowry definition, but it is in terms of other, more general concepts. Since the oxidation states of the metal M and hydrogen have changed, we should recognize this as a redox reaction in which the metal is oxidized and the hydrogen reduced. In this example, 2 electrons from the metal atom have been completely transferred, one to each hydrogen ion. If we restrict ourselves to the proton concept, we can say that acid–base and redox reactions are analogous insofar as one involves a transfer of protons and the other a transfer of electrons. Some other points of similarity will appear soon. We should remember, however, that a redox reaction need not involve a complete transfer of electrons. For instance, we saw in Chapter 5 that in the reaction

$$2H_2 + O_2 \longrightarrow 2H_2O$$

the hydrogen electrons were merely shifted toward the more electronegative oxygen. For purposes of assigning oxidation states, however, we considered the bonding electrons to belong to oxygen.

The tendency for hydrogen ion to be displaced from a solution by a metal depends, among other things, on the nature of the metal and on the concentration of hydrogen ions in the solution. If several different metals are placed in solutions with $[H^+] = 1M$, the result will be that some metals will generate hydrogen and some metals will not react at all. If we order the metals with the most reactive metals first and the nonreactive ones last, with hydrogen separating the two groups, we obtain a list that is variously called the activity, reactivity, displacement, or the electromotive series and is shown in Table 10.5.

The position of hydrogen in Table 10.5 applies to reaction in $1M$ acid solution. In less concentrated solution, some of the metals above hydrogen will not react. In water, the small hydrogen ion concentration is not sufficient for the reaction of lead or cadmium. In hot water, magnesium and aluminum would displace hydrogen were it not for protective oxide coatings (tarnish)

Table 10.5
Activity series of the metals

potassium	K	
sodium	Na	
calcium	Ca	
magnesium	Mg	
aluminum	Al	
zinc	Zn	increasing
chromium	Cr	activity
iron	Fe	
cadmium	Cd	
nickel	Ni	
tin	Sn	
lead	Pb	
hydrogen	H_2	
copper	Cu	
mercury	Hg	
silver	Ag	
gold	Au	

on the metals. The most reactive metals at the top of the list displace hydrogen even from cold water. (Actually, the reactive metals of groups IA and IIA do not react with the hydrogen ions in water; they are oxidized by water itself, e.g.,

$$Ca + 2H_2O \longrightarrow Ca^{2+} + 2OH^- + H_2$$

The result is a basic solution.)

Now let us look more closely at a specific redox reaction, the reaction of an acid with zinc:

$$Zn + 2H^+ \longrightarrow Zn^{2+} + H_2$$

Zinc is oxidized to Zn^{2+}, and its oxidation state increases from 0 to $+2$. Hydrogen is reduced from an oxidation state of $+1$ to a lower state of 0.

Redox reactions can be broken down into two half-reactions, one involving a loss of electrons (oxidation), the other a gain of electrons (reduction). In the present example, the two half-reactions are

$$\begin{array}{ll} Zn \longrightarrow Zn^{2+} + 2e^- & \text{(oxidation)} \\ 2e^- + 2H^+ \longrightarrow H_2 & \text{(reduction)} \\ \hline Zn + 2H^+ \longrightarrow Zn^{2+} + H_2 & \text{total redox reaction} \end{array}$$

The electrons lost from Zn are taken up by H^+ in much the same way that protons lost by an acid are taken up by a base. (We do not usually think of breaking acid–base reactions into half-reactions, but we do it every time we write an acid ionization equation as

$$HA \rightleftharpoons H^+ + A^-$$

This is an acid half-reaction.)

Half-reactions also can be written for redox reactions involving polar covalent species. For the reaction

$$H_2S + Cr_2O_7{}^{2-} \longrightarrow S + Cr^{3+} \qquad \text{[unbalanced]}$$

the balanced half-reactions are

$$H_2S \longrightarrow S + 2H^+ + 2e^- \qquad \text{[oxidation of } S^{2-}]$$

and

$$Cr_2O_7{}^{2-} + 14H^+ + 6e^- \longrightarrow 2Cr^{3+} + 7H_2O \qquad \text{[reduction of Cr(VI)]}$$

To obtain the total redox reaction, the first half-reaction must be multiplied by 3 before adding to the second half-reaction, so that the total number of electrons lost is equal to the number gained. The complete balanced equation is $3H_2S + Cr_2O_7{}^{2-} + 8H^+ \longrightarrow 3S + 2Cr^{3+} + 7H_2O$. Rules for writing half-reactions and combining them to get the total balanced redox reaction are discussed briefly in Appendix D.

10.19 STRENGTHS OF OXIDIZING AND REDUCING AGENTS

In a redox reaction, the oxidized substance or element contributes electrons that reduce some other species. Thus, the oxidized species is the *reducing agent*, whereas the reduced species is the *oxidizing agent*. For instance, in the Zn–H^+ reaction, zinc metal is the reducing agent and H^+ is the oxidizing agent.

A redox process involves a competition for electrons between oxidizing and reducing agents, much as an acid–base reaction involves a relative competition for hydrogen ions between an acid and a base. A substance is a strong oxidizing agent when it has a great affinity for electrons; the stronger reducing agents are those substances that tend to give up electrons that they already possess. Cesium and lithium metals are good reducing agents, whereas fluorine gas is a very strong oxidizing agent.

Another analogy with acid–base reactions is that the loss of electrons by a strong reducing agent produces a weak conjugate oxidizing agent; a strong oxidizing agent forms a weak reducing agent when it gains electrons. For instance, the strong reducing agent potassium gives up an electron easily:

$$K \longrightarrow K^+ + e^-$$

The K^+ cation has little tendency to accept an electron to reform the K atom; it is a very weak oxidizing agent.

Whether a substance will function as a reducing agent or as an oxidizing agent depends on the other substance with which it is reacting (again note the parallel with acid–base reactions). The activity series in Table 10.5 indicates that all the metals above hydrogen function as reducing agents with $1M$ H^+. The H^+ acts as the oxidizing agent. But hydrogen ion does not act as an oxidizing agent with metals below it in this series. Instead, if hydrogen gas is reacted with a solution of copper(II) ions under the proper conditions, the reaction that takes place is

$$H_2 + Cu^{2+} \longrightarrow 2H^+ + Cu$$

Hydrogen gas is a better reducing agent than copper. The strongest reducing agents are at the top of the activity series.

The reaction of a metal with hydrogen ion only locates the metal in the group above or below hydrogen. To find the position of a metal within each group, additional reactions must be tried to determine the relative reducing power of the metal. Because metallic copper will displace silver ion from solution by the reaction

$$Cu + 2Ag^+ \longrightarrow Cu^{2+} + 2Ag$$

copper has a stronger tendency to lose electrons to form Cu^{2+} than Ag has to form Ag^+. There is no detectable reaction when silver metal is added to a Cu^{2+} solution, so copper goes above silver in the activity series. In a similar manner, mercury plates out on copper placed in a solution of Hg_2^{2+} ions, but silver does not react with the same solution:

$$Cu + Hg_2^{2+} \longrightarrow 2Hg + Cu^{2+}$$
$$Ag + Hg_2^{2+} \longrightarrow \text{no reaction}$$

Mercury must lie between copper and silver in the activity series.

The same procedure can be extended to redox reactions involving non-metals and their anions, as well as more complex compounds, so the activity series is no longer limited to metals. A portion of such a series, written to show the half-reaction for the reduction of the oxidized form of each pair or couple, is given in Table 10.6. This is called an *electrode reduction potential series* for reasons that will soon become apparent. The strongest reducing agents are at the upper right-hand side of this series; the strongest oxidizing agents are at the lower left. Written in this way, the oxidized form of any

Table 10.6

Standard reduction potential series (acid or neutral aqueous solution)[a]

oxidized form		reduced form	E^0, volts
$K^+ + e^-$	$\rightleftharpoons$	K	-2.93
$Mg^{2+} + 2e^-$	$\rightleftharpoons$	Mg	-2.37
$Zn^{2+} + 2e^-$	$\rightleftharpoons$	Zn	-0.76
$Fe^{2+} + 2e^-$	$\rightleftharpoons$	Fe	-0.44
$2H^+ + 2e^-$	$\rightleftharpoons$	H_2	0.00
$SO_4{}^{2-} + 4H^+ + 2e^-$	$\rightleftharpoons$	$H_2SO_3 + H_2O$	0.17
$Cu^{2+} + 2e^-$	$\rightleftharpoons$	Cu	0.34
$I_2 + 2e^-$	$\rightleftharpoons$	$2I^-$	0.54
$Fe^{3+} + e^-$	$\rightleftharpoons$	Fe^{2+}	0.77
$Hg_2{}^{2+} + 2e^-$	$\rightleftharpoons$	$2Hg$	0.79
$Ag^+ + e^-$	$\rightleftharpoons$	Ag	0.80
$NO_3{}^- + 4H^+ + 3e^-$	$\rightleftharpoons$	$NO + 2H_2O$	0.96
$Br_2 + 2e^-$	$\rightleftharpoons$	$2Br^-$	1.09
$O_2 + 4H^+ + 4e^-$	$\rightleftharpoons$	$2H_2O$	1.23
$Cl_2 + 2e^-$	$\rightleftharpoons$	$2Cl^-$	1.36

[a] Half-reactions are often tabulated as oxidations, with the reduced form on the left side of a couple. Then the reduced form of any couple will reduce the oxidized form of any couple lying *below* it.

couple will oxidize the reduced form of any couple lying above it in the reduction potential series. If the oxidized and reduced forms in different couples are joined by a line, a "Z" is formed to show the products of the spontaneous redox reaction. As an example, let's see what will happen if silver metal is placed into dilute nitric acid. We already know that silver will not displace H^+ from an acid solution, but we can now see that nothing should happen, because the spontaneous reaction is the reduction of Ag^+ by H_2, not the reverse. Since

$$2H^+ + 2e^- \rightleftharpoons H_2$$

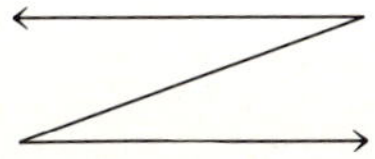

$$Ag^+ + e^- \rightleftharpoons Ag$$

or

$$2Ag^+ + H_2 \longrightarrow 2Ag + 2H^+$$

However, the series tells us that silver metal can be oxidized by the nitrate ion of the nitric acid, because

$$Ag^+ + e^- \rightleftharpoons Ag$$

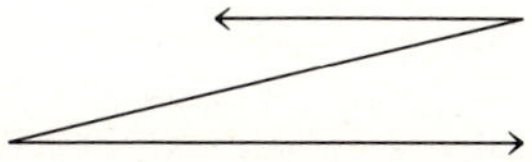

$$NO_3{}^- + 4H^+ + 3e^- \rightleftharpoons NO + 2H_2O$$

or

$$3Ag + NO_3{}^- + 4H^+ \longrightarrow 3Ag^+ + NO + 2H_2O$$

10.20 ELECTRODE POTENTIALS

An oxidation–reduction or redox reaction can be carried out by a method other than directly mixing the reactants together. The reaction

$$Zn + 2H^+ \longrightarrow Zn^{2+} + H_2$$

serves as an example. The reaction proceeds by way of the half-reactions

$$Zn \longrightarrow Zn^{2+} + 2e^-$$
$$2H^+ + 2e^- \longrightarrow H_2$$

Instead of adding zinc metal directly to a hydrogen ion solution, we could construct an electrochemical or galvanic cell consisting of an electrode of zinc metal immersed in a solution of zinc ions and an inert electrode of platinum metal bathed in hydrogen gas and immersed in a solution of hydrogen ions. The two electrodes and their surrounding solutions are contained in separate vessels connected by a "salt bridge" containing a KCl solution; the two electrodes are connected with a wire (Figure 10.4). The salt bridge permits

Figure 10.4. An electrochemical cell. The electrodes are metal strips through which electrons enter and leave the solution. In a current-producing cell such as this one, the anode is often called the negative terminal and the cathode the positive terminal.

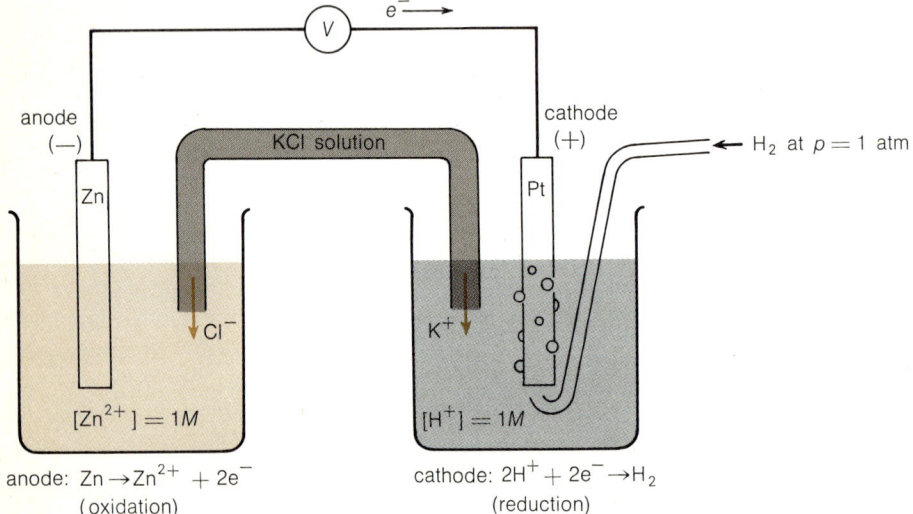

anode: $Zn \rightarrow Zn^{2+} + 2e^-$
(oxidation)

cathode: $2H^+ + 2e^- \rightarrow H_2$
(reduction)

the passage of K^+ or Cl^- ions into or out of the two solutions, so electroneutrality is conserved as H^+ and Zn^{2+} ions are discharged and formed, respectively.

The only way that electrons can be transferred from the zinc to the hydrogen ion is through the wire. In other words, when the reaction $Zn + 2H^+ \longrightarrow Zn^{2+} + H_2$ occurs, there will be a flow of electrons from the zinc electrode to the hydrogen electrode. This flow constitutes an electric current in the wire. The driving force for the flow of electrons from zinc to the hydrogen electrode is an indication of the relative reducing power of zinc with respect to hydrogen gas and is measured by the potential difference, in volts, between the two electrodes. The zinc electrode, at which oxidation occurs, is called the *anode* of the electrochemical cell; the hydrogen electrode, at which reduction occurs, is the *cathode*. For any cell (including electrolysis cells, discussed in Section 13.20), oxidation takes place at the anode; reduction at the cathode.

If the zinc and hydrogen ions in their respective compartments or half-cells are at unit molarity (more strictly, at unit activity) and if the H_2 gas is at 1 atm pressure, the measured electromotive force (emf) or potential of this cell is 0.76 volt. This value represents the sum of the *standard electrode potentials* E^0 of the individual electrodes. In terms of the half-reactions,

$$
\begin{array}{ll}
\text{Zn} \longrightarrow \text{Zn}^{2+} + 2e^- & E_{\text{Zn}}{}^0 \\
\underline{2\text{H}^+ + 2e^- \longrightarrow \text{H}_2} & \underline{E_{\text{H}_2}{}^0} \\
\text{Zn} + 2\text{H}^+ \longrightarrow \text{Zn}^{2+} + \text{H}_2 & E_t{}^0 = E_{\text{Zn}}{}^0 + E_{\text{H}_2}{}^0
\end{array}
$$

(Standard potentials refer to conditions when all substances are at unit activity. For solutes, the approximation that activity equals molar concentration will be sufficient for our purposes.)

By convention, $E_{\text{H}_2}{}^0$, the standard single electrode potential of the H_2 electrode when $[\text{H}^+] = 1M$ and $P_{\text{H}_2} = 1$ atm, is zero. Therefore, the standard single electrode potential $E_{\text{Zn}}{}^0$ of the Zn–Zn^{2+} couple is 0.76 volt when $[\text{Zn}^{2+}] = 1M$ and refers to the oxidation half-reaction $\text{Zn} \longrightarrow \text{Zn}^{2+} + 2e^-$. The value for the opposite reduction reaction $\text{Zn}^{2+} + 2e^- \longrightarrow \text{Zn}$ is -0.76 volt and is referred to, and tabulated, as the standard reduction potential of the Zn^{2+}–Zn couple. The standard single electrode (reduction) potentials of other couples may be determined in the same way with respect to the hydrogen electrode. Selected values are tabulated in Table 10.6.

Now, if we set up a cell composed of a zinc electrode bathed in a solution of zinc ions at $1M$ concentration and of a copper electrode immersed in a solution of copper ions at $1M$, the electrode reactions and their potentials are

$$
\begin{array}{ll}
\text{Zn} \longrightarrow \text{Zn}^{2+} + 2e^- & E_{\text{Zn}}{}^0 = 0.76 \text{ V} \\
\text{Cu}^{2+} + 2e^- \longrightarrow \text{Cu} & E_{\text{Cu}}{}^0 = 0.34 \text{ V}
\end{array}
$$

The total cell reaction and emf are

$$
\text{Zn} + \text{Cu}^{2+} \longrightarrow \text{Zn}^{2+} + \text{Cu} \qquad E_t{}^0 = 0.76 + 0.34 = 1.10 \text{ V}
$$

The measured emf of the cell should be 1.10 volts. The positive value of the emf indicates that the cell reaction proceeds spontaneously in the direction written. *This is general; a positive emf means a spontaneous reaction.* The emf of the reverse nonspontaneous reaction $\text{Zn}^{2+} + \text{Cu} \longrightarrow \text{Zn} + \text{Cu}^{2+}$ would be -1.10 volts.

For one further example, the standard reduction potential of the Zn^{2+}–Zn couple is -0.76 volt; the standard reduction potential of the Ag^+–Ag couple is 0.80 volt. These potentials refer to the half-reactions

$$
\begin{array}{ll}
\text{Zn}^{2+}(M = 1) + 2e^- \longrightarrow \text{Zn} & E_{\text{Zn}}{}^0 = -0.76 \text{ V} \\
\text{Ag}^+(M = 1) + e^- \longrightarrow \text{Ag} & E_{\text{Ag}}{}^0 = 0.80 \text{ V}
\end{array}
$$

both written as reductions. In the spontaneous reaction, Zn is oxidized to Zn^{2+} by Ag^+ so that, to obtain the total cell reaction from the half-reactions, the zinc couple must be written as an oxidation, the sign of its potential reversed, and the Ag^+–Ag reaction doubled to balance electron loss and gain:

$$
\begin{array}{ll}
\text{Zn} \longrightarrow \text{Zn}^{2+} + 2e^- & E_{\text{Zn}}{}^0 = 0.76 \text{ V} \\
\underline{2(\text{Ag}^+ + e^- \longrightarrow \text{Ag})} & \underline{E_{\text{Ag}}{}^0 = 0.80 \text{ V}} \\
\text{Zn} + 2\text{Ag}^+ \longrightarrow \text{Zn}^{2+} + 2\text{Ag} & E_t{}^0 = 0.76 + 0.80 = 1.56 \text{ V}
\end{array}
$$

Note that the E^0 value is not changed when a multiple of a half-reaction is taken. The higher emf for this reaction, compared to that of the Zn–Cu^{2+} reaction, shows that Ag^+ is a better oxidizing agent than Cu^{2+}. Again, the superscript zero on E_t^0 means the concentrations of silver and zinc ions are unity (standard conditions).

10.21 EFFECT OF CONCENTRATION ON CELL EMF

The standard potentials discussed above refer to reactions where the concentrations of ionic reactants and products are unity. If other concentrations are employed, the resulting single electrode potentials and total cell emfs are no longer the standard values. A treatment based on thermodynamics yields the *Nernst equation*, which indicates what the nonstandard values will be. Consider the general reaction

$$aA + bB \longrightarrow cC + dD$$

which could be broken up into half-reactions

$$aA \longrightarrow cC + ne^-$$
$$bB + ne^- \longrightarrow dD$$

The Nernst equation says that the total reaction emf at 25°C when A, B, C, D are not at unit concentration is

$$E_t = E_t^0 - \frac{0.0591}{n} \log \frac{[C]^c[D]^d}{[A]^a[B]^b} \tag{1}$$

where the quantities in brackets are the actual concentrations involved and n is the number of electrons transferred in each half-reaction. (Since the equation is balanced, the number of electrons in each half-reaction must be the same.)

To illustrate the use of the Nernst equation, we shall take the zinc–silver ion reaction with the concentrations of zinc ion and silver ion as $0.05M$ and $0.1M$, respectively. Using the standard potential values, we found before that for this reaction $E_t^0 = 1.56$ volts. The lower emf of the cell with the ion concentrations chosen is

$$E_t = E_t^0 - \frac{0.0591}{2} \log \frac{[Zn^{2+}]}{[Ag^+]^2}$$
$$= 1.562 - \frac{0.0591}{2} \log \frac{0.05}{(0.1)^2}$$
$$= 1.56 - \frac{0.0591}{2} \log 5$$
$$= 1.56 - 0.02$$
$$= 1.54 \text{ V}$$

The pure solids Zn and Ag are not included in the concentration ratio, because their concentrations are constant (and at their standard concentrations of unity.) This same equation can be used to determine quantitatively how the potential of a single electrode will change when concentrations are varied from the standard values. For instance, the half-reaction

$$O_2 + 4H^+ + 4e^- \longrightarrow 2H_2O(l)$$

controls the oxidizing environment of sea water. Its standard reduction potential is 1.23 volts, meaning that, relative to the standard hydrogen electrode, this half-reaction proceeds as a reduction when $P_{O_2} = 1$ atm and $[H^+] = 1M$. At other than standard conditions, this reduction half-reaction would have a potential

$$E = 1.23 - \frac{0.0591}{4} \log \frac{1}{P_{O_2}[H^+]^4} \tag{2}$$

$$= 1.23 + 0.015 \log P_{O_2}[H^+]^4$$
$$= 1.23 + 0.015 \log P_{O_2} - 0.06 \text{ pH}$$

The use of the pressure of oxygen as a concentration term reflects the increase in solubility of a gas with increasing pressure of the gas (Section 8.13, and Section 11.7). At oxygen pressures other than 1 atm and at pH's other than pH = 0, the potential of this important half-reaction could be found by placing the new values of P_{O_2} and pH into this equation.

To obtain a qualitative understanding of the effect of concentration, we can use Le Chatelier's principle to see that a decrease in P_{O_2} from 1 atm to the normal atmospheric partial pressure of O_2 (0.2 atm) would tend to favor the oxidation direction, $H_2O \longrightarrow O_2$, and this would mean a decreased tendency for the reduction half-reaction to occur. The reduction potential would decrease from 1.23 volts to some lower value. Similarly, increasing the pH is the same as decreasing the concentration of hydrogen ion reactant, again causing the equilibrium to shift to the oxidized form of the couple and leading to a decrease in the reduction potential. Because the redox potential of many reactions is dependent on the pH of the reaction medium, the pH of a medium can be determined quickly and accurately by measuring the emf of a pH-dependent reaction in the medium. This is the principle behind a pH meter. Also, because the potential of a reaction is a function of the concentrations of reactants and products, the measured potential indicates the amounts of substances present at any time in a reaction mixture.

Cell emf at Equilibrium   The total cell emf E_t measures the tendency for a cell reaction to occur. When a cell reaction has attained equilibrium, there is no further tendency for the net reaction to proceed and E_t becomes zero. (This conclusion is reached from experience and is justified by thermodynamic reasoning, Sections 16.10 and 16.11.) Thus, at equilibrium, the Nernst equation, equation (1), becomes

$$E_t{}^0 = \frac{0.0591}{n} \log \frac{[C]^c[D]^d}{[A]^a[B]^b}$$

where the terms in brackets are now the *equilibrium* concentrations.

10.22 REDOX IN NATURAL ENVIRONMENTS   The interplay of pH and redox potential is a matter of some importance, because their joint effect determines both the direction and extent of a number of common redox reactions in the laboratory and in nature. The ability of the environment to oxidize a chemical species may depend not only on the presence of an appropriate oxidizing agent, but on the pH of the reaction medium. Considerations of pH and redox potential are common in aquatic, soil, and geological chemistry, to name just a few broad areas. Two examples are given here, one from the chemistry of natural waters, the other from soil chemistry.

Redox Conditions in Water

In the natural waters of the seas, lakes, and streams, the half-reaction

$$O_2 + 4H^+ + 4e^- \longrightarrow 2H_2O \qquad E^0 = 1.23 \text{ V}$$

controls the oxidizing nature of the aquatic environment. This, in turn, controls the chemical species present in the water as well as the kinds of living organisms that may exist. The standard reduction potential given for this couple refers to $[H^+] = 1M$ (pH = 0) and $P_{O_2} = 1$ atm. At usual conditions of water in equilibrium with the atmosphere, the potential of this couple is less than 1.23 volts. The actual value can be found from equation (2) in Section 10.21 with $P_{O_2} = 0.2$ atm and $[H^+] = 10^{-7}M$. The calculated value is $E = 0.81$ volt. At lower oxygen pressure and at high pH, the potential for this half-reaction will be even less, although the variation with P_{O_2} is quite small. The higher the positive value of E, the greater the tendency for other chemical species in solution to become oxidized. Conversely, the lower the value of E, the more likely that reduced species can be formed and continue to exist in aqueous solution.

Theoretically, at equilibrium, a stronger oxidizing agent than oxygen cannot exist in water because it would oxidize water to oxygen. Such an oxidizing agent is chlorine gas:

$$Cl_2 + 2e^- \longrightarrow 2Cl^- \qquad E^0 = 1.36 \text{ V}$$

Chlorine should oxidize water, because it has a reduction potential greater than 0.81 volt (pH = 7) or even 1.23 volts (pH = 0). The combination of half-reactions is

$$
\begin{array}{lll}
2(Cl_2 + 2e^- \longrightarrow 2Cl^-) & & E^0 = 1.36 \text{ V} \\
2H_2O \longrightarrow O_2 + 4H^+ + 4e^- & & E^0 = -1.23 \text{ V (pH = 0)}; \\
& & = -0.81 \text{ V (pH = 7)} \\
\hline
2H_2O + 2Cl_2 \longrightarrow 4Cl^- + O_2 + 4H^+ & & E_t^0 = 0.13 \text{ V (pH = 0)} \\
& & = 0.55 \text{ V (pH = 7)}
\end{array}
$$

The total potential is positive and indicates that the reaction of Cl_2 with water should go in the direction written. In real systems, chlorine gas, $Cl_2(aq)$, can exist in water for periods of time only because of the extreme slowness of this reaction. Low pH shifts the equilibrium to the left and favors $Cl_2(aq)$. There are other reactions of Cl_2 with water that also need to be considered in deciding whether Cl_2 can be present in water. A more important reaction of Cl_2 in water is

$$Cl_2 + H_2O \longrightarrow HClO + H^+ + Cl^-$$

which is favored by high pH.

The oxidizing nature of natural waters is indicated by giving the reduction potential E of the O_2–H_2O couple, as measured against the standard hydrogen electrode:

$$H_2(P_{H_2} = 1 \text{ atm}) \rightleftharpoons 2H^+(1M) + 2e^- \qquad E^0 = 0.00 \text{ V}$$

As the oxygen content of the water decreases and P_{O_2} becomes very small, the reduction potential of the O_2–H_2O couple decreases. However, the maximum pH range of natural waters (4–9) sets a lower limit to which the reduction potential can decrease. This minimum value is about -0.4 volt and is determined by the half-reaction

$$2H^+ + 2e^- \rightleftharpoons H_2$$

which has a reduction potential of -0.414 volt at pH $= 7$ and $P_{H_2} = 1$ atm. In water, no reducing agent strong enough to reduce the hydrogen ion of water to hydrogen gas can exist. Because of a number of factors, the measured reduction potential of natural waters at pH $= 7$ is generally lower than both the end values of 0.81 volt and -0.41 volt. The actual range is between $+0.3$ volt for aerated water and -0.6 volt for oxygen-depleted (anoxic) bottom waters in contact with organic matter. As mentioned, high positive values of E mean an environment of high oxidizing power. The reduction potential E is a measure of the ability of the environment to accept electrons from a reducing agent, just as the pH of an environment is a measure of the tendency to take up protons from an acid. Sea water is a mildly oxidizing solution in which many elements exist in their higher oxidation states.

EXAMPLE | If the reduction potential of a natural water is 0.30 volt, what will be the predominant form of dissolved iron, Fe^{2+} or Fe^{3+}, if only the equilibrium $Fe^{3+} + e^- \longrightarrow Fe^{2+}$ with $E^0 = 0.77$ volt is considered?

SOLUTION | The pertinent half-reactions are

$$
\begin{array}{ll}
O_2 + 4H^+ + 4e^- \longrightarrow 2H_2O & E^0 = 1.23 \text{ V} \\
4Fe^{2+} \longrightarrow 4Fe^{3+} + 4e^- & E^0 = -0.77 \text{ V} \\
\hline
O_2 + 4H^+ + 4Fe^{2+} \longrightarrow 4Fe^{3+} + 2H_2O & E_t^0 = 1.23 - 0.77
\end{array}
$$

$$E_t = E_t^0 - \frac{0.0591}{4} \log \frac{[Fe^{3+}]^4}{[Fe^{2+}]^4 P_{O_2} [H^+]^4}$$

$$= 1.23 - \frac{0.0591}{4} \log \frac{1}{P_{O_2}[H^+]^4} - 0.77 - \frac{0.00591}{4} \frac{[Fe^{3+}]^4}{[Fe^{2+}]^4}$$

We are looking for the concentrations at equilibrium, so $E_t = 0$. The first two terms on the right side of the last equation represent the value of the O_2–H_2O single electrode reduction potential in the water. This value is 0.30 volt, so the equation reduces to

$$E_t = 0 = 0.30 - 0.77 - 0.0591 \log \frac{[Fe^{3+}]}{[Fe^{2+}]}$$

$$\log \frac{[Fe^{3+}]}{[Fe^{2+}]} = \frac{-0.47}{0.0591} \simeq -8$$

$$\frac{[Fe^{3+}]}{[Fe^{2+}]} \simeq 10^{-8}$$

Thus, Fe^{3+} must be in low concentration relative to Fe^{2+} in this water. Note that this result applies only in acidic water. At about pH $= 3$, $Fe(OH)_3$ begins to precipitate, and equilibria such as

$$3H_2O + Fe^{2+} \rightleftharpoons Fe(OH)_3 + 3H^+ + e^-$$

and

$$Fe(OH)_2 + OH^- \rightleftharpoons Fe(OH)_3 + e^-$$

must be considered as the pH increases. In weakly basic ocean waters, Fe(III), in one form or another, is the predominant iron species because of such pH-dependent half-reactions. These reactions more than counterbalance the tendency of high pH (low H^+) to lower the reduction potential of the O_2–H_2O couple.

A general formula giving the relative concentrations of species in a couple can be obtained from the reasoning used in the above example. This equation is

reduction potential of medium = E^0_{red} of couple

$$-\frac{0.0591}{n} \log \frac{[\text{reduced form}]}{[\text{oxidized form}]}$$

The same general equation could also have been obtained, and the example done more directly, by recognizing that at equilibrium $E_t = 0$ so that

$$E_t = E_{red}(O_2\text{–}H_2O) + E_{ox}(\text{couple}) = 0$$

or

$$E_{red}(O_2\text{–}H_2O) = E_{red}(\text{couple})$$

Sulfate ion is the predominant form of sulfur in well-aerated water. In the absence of oxygen, in anoxic waters where the reduction potential has become negative and the environment reducing rather than oxidizing, SO_4^{2-} may be reduced to S(II) as HS^- or H_2S gas by the half-reaction

$$SO_4^{2-} + 10H^+ + 8e^- \rightleftharpoons H_2S + 4H_2O \qquad E^0 = -0.30 \text{ V}$$

This half-reaction already shows, by Le Chatelier's principle, that H_2S should be favored by acidic environments. If we combine the O_2–H_2O couple with this half-reaction, we get

$$2(O_2 + 4H^+ + 4e^- \rightleftharpoons 2H_2O) \qquad\qquad E = +0.3 \text{ to } -0.6 \text{ V}$$
$$\underline{H_2S + 4H_2O \rightleftharpoons SO_4^{2-} + 10H^+ + 8e^- \qquad\quad E = +0.3 \text{ V}}$$
$$2O_2 + H_2S \rightleftharpoons SO_4^{2-} + 2H^+ \qquad\qquad E_t = +0.6 \text{ to } -0.3 \text{ V}$$

This equation again shows the tendency for H_2S formation under acidic conditions. It also shows that H_2S will exist at low oxygen concentrations when the reduction potential of the O_2–H_2O couple, or the potential of the environment, is below about -0.3 volt. This does not mean that H^+ reduces SO_4^{2-} to H_2S; rather, under these highly reducing conditions, there are reducing agents available to reduce SO_4^{2-} to H_2S. Chief among these reducing agents is organic matter that has accumulated on the bottoms of stagnant lakes and deep marine basins, where the lack of oxygen has prevented the usual oxidative decay. If the organic matter is represented by CH_2O, a typical reaction is

$$SO_4^{2-} + 2CH_2O + 2H^+ \rightleftharpoons H_2S + 2CO_2 + 2H_2O$$

Such reactions are catalyzed by the enzymes of anaerobic bacteria existing in these reducing environments. The bacteria use the energy evolved by the reactions for their own metabolic processes. Under even more highly reducing conditions in fresh waters, CO_2 may be reduced to methane, CH_4.

Redox Conditions in Soils: The Availability of Manganese to Plants

Growing plants require manganese in very small amounts. Soluble manganese in the form of Mn^{2+} is absorbed from the soil by the plant roots. The amount of soluble manganese available to the roots depends on both pH and redox environment in the soil. The half-reaction for the reduction of Mn(IV), as insoluble MnO_2, to soluble Mn^{2+} in an acid medium is

$$MnO_2 + 4H^+ + 2e^- \rightleftharpoons Mn^{2+} + 2H_2O \qquad E^0 = 1.23 \text{ V}$$

or, on combination with the O_2–H_2O couple,

$$2MnO_2 + 4H^+ \rightleftharpoons 2Mn^{2+} + O_2 + 2H_2O$$

Either the half-reaction or the total reaction shows that manganese availability will be increased by an acidic soil environment or a low oxygen condition such as would be found in a poorly aerated (waterlogged or compacted) soil.

10.23 SUMMARY

According to the proton concept, an acid is defined as a proton (hydrogen ion) donor and a base as a proton acceptor. A substance can function as an acid only when it reacts with another substance acting as a base. Water is an example of a compound that may act as an acid or a base under different circumstances. Most commonly, the acidic and basic behavior of substances in aqueous solution is referred to their reaction with water, the water functioning as a base or acid, respectively.

Factors such as the electronegativity, size, oxidation state, and charge of atoms or ions in a compound affect the bond type and strength of bonds to hydrogen and determine whether a substance will act as an acid, a base, or both in water solution. The strength of an acid is measured by its tendency to donate protons; the strength of a base, by its tendency to accept protons. The ionization constants of acids and bases are quantitative measures of their strengths relative to water. Strong acids and bases are extensively ionized in aqueous solution whereas weak acids and bases ionize in water to only a small extent.

Pure water is itself slightly ionized into H^+ and OH^- ions. In any aqueous solution, $[H^+][OH^-] = 1.0 \times 10^{-14}$ at 25°C, and in pure water and in any neutral solution, $[H^+] = [OH^-] = 1.0 \times 10^{-7} M$. The hydrogen ion concentration in solutions is often given as the pH of the solution, defined by $pH = -\log_{10}[H^+]$. Neutral solutions have pH = 7. In acidic solutions $[H^+]$ is greater than $[OH^-]$, and the pH is less than 7. Basic solutions have pH's greater than 7.

Salts derived from the neutralization of acids and bases of unequal strength may give acidic or basic solutions as a result of the hydrolysis of the ions. Salts of strong acids and weak bases give acidic solutions. Salts of weak acids and strong bases give basic solutions.

Buffer solutions resist changes in pH caused by the addition of small amounts of acids or bases. Most buffer solutions contain a weak acid and a salt of the weak acid. The H_2CO_3–HCO_3^- system is active in the buffering of blood.

Redox reactions may be thought of as consisting of two half-reactions corresponding to reduction and oxidation. The reduction half-reaction involves only a gain of electrons; the oxidation half-reaction involves only a loss of electrons. The total redox reaction is the sum of the half-reactions, each with the same gain or loss of electrons.

Substances differ in their tendency to give up or accept electrons. Those, like active metals, that easily lose electrons are good reducing agents. Others, such as nonmetals like Cl_2, are good oxidizing agents. The relative oxidizing and reducing powers of metals and their ions may be determined by constructing an activity or displacement series. A quantitative measure of oxidizing and reducing strength and the spontaneous direction of redox reactions is provided by single electrode potentials. Single electrode potentials are usually tabulated as standard reduction potentials for a half-reaction. The standard potential for a half-reaction is its potential measured relative to a hydrogen electrode in an electrochemical cell at 25°C when all concentrations and pressures are unity. Good reducing agents have high negative reduction potentials; good oxidizing agents have high positive reduction potentials.

The total potential for a redox reaction is the sum of the single electrode potentials of the half-reactions occurring at the electrodes. If the emf of a redox reaction is positive, the reaction has a tendency to go spontaneously as written.

The emf and the direction of a redox reaction depend on the concentration of reactants and products. The effect of concentration can be determined qualitatively by application of Le Chatelier's principle and quantitatively by the Nernst equation.

Questions and Problems

1. Write an equation for the reaction of a strong acid and a weak base. Identify the salt formed.
2. Will acid strength increase or decrease in the following series?
   a. HIO_3, $HBrO_3$, $HClO_3$
   b. $ClOH$, $OClOH$ ($HClO_2$), O_2ClOH ($HClO_3$), O_3ClOH ($HClO_4$)
3. Arrange the following substances in the order of increasing base strength: H_3PO_4, $Al(OH)_3$, $NaOH$, H_2SO_4, $HClO_4$, $Mg(OH)_2$, H_2SiO3.
4. Write equations for the reaction of the following oxides with water: K_2O, CO_2, BaO, SO_3, ZnO.
5. Table 10.2 was interpreted as follows: acids on the left side will react with bases located diagonally below them to form weaker acids and bases. On the basis of equilibrium arguments, why should this be so?
6. Show that the hydroxide ion concentration in a solution of a weak base such as NH_3 of concentration C_b is given by $[OH^-] = \sqrt{K_b C_b}$.
7. Calculate how many water molecules and how many hydrogen ions are in 1 liter of pure water at 25°C. Take the density of water as 0.997 g/ml.
8. The pH of a 0.12M solution of a weak monoprotic acid is 4.80.
   a. What is $[H^+]$?
   b. What is K_a for the acid?

9. Find the pH of a 0.001M KOH solution.
10. Calculate the hydrogen ion concentration in a saturated solution of $Ca(OH)_2$.
11. The percent ionization of a species in solution is defined as
$$\frac{\text{amount ionized}}{\text{amount initially un-ionized}} \times 100$$
   Calculate and compare $[H^+]$ for 0.1M and 1M acetic acid solutions. Compare the percent ionization of HAc in the two solutions. Explain the change in percent ionization with concentration.
12. What is the sulfate and hydrogen ion concentration in a 0.5M H_2SO_4 solution? Assume complete ionization to HSO_4^- but only partial ionization to SO_4^{2-} (This problem requires the use of the quadratic formula.)
13. a. At 25°C and 1 atm pressure, the solubility of H_2S gas in water is 0.10M. What is the pH of a saturated solution of H_2S in water?
    b. Use the second ionization constant of H_2S and the hydrogen ion concentration in part (a) to find $[S^{2-}]$ in the saturated solution.
14. a. Compared to the value of $[S^{2-}]$ in a saturated solution of H_2S in water, qualitatively what change will occur in $[S^{2-}]$ if a 0.1M HCl solution is saturated with H_2S? A 0.1M NaOH solution?

b. Justify your qualitative statements by calculation of $[S^{2-}]$ in the two solutions in part (a).

15. Show that the acid ionization constant K_a of a weak acid is related to the base ionization constant K_b of its conjugate base by $K_a = K_w/K_b$. Use H_2CO_3 as your example.

16. Use the second ionization constant of H_2CO_3 to find $[CO_3^{2-}]$ in a saturated solution in equilibrium with the atmosphere. Note your assumptions.

17. Considering the equilibria

$$CO_2 + H_2O \rightleftharpoons H_2CO_3 \rightleftharpoons H^+ + HCO_3^-$$

and

$$HCO_3^- \rightleftharpoons H^+ + CO_3^{2-}$$

what species, CO_2 (or H_2CO_3), HCO_3^-, or CO_3^{2-}, will predominate in highly basic solutions? In highly acidic solutions? In near neutral solutions of carbonates in water?

18. What is the pH of a $0.1M$ NH_4Cl solution?

19. a. Write two equations for the hydrolysis of hydrogen carbonate ion in water.
    b. Write an equation for the reaction of one hydrogen carbonate ion (as an acid) reacting with another hydrogen carbonate ion (as a base).
    c. On the basis of the equilibrium in (b), explain the dissolution of $CaCO_3$ in a saturated solution when CO_2 is added and the precipitation of $CaCO_3$ when a solution of $Ca(HCO_3)_2$ is heated.

20. How much $0.15M$ HCl must be added to 100 ml of $0.20M$ ammonia solution to reach the equivalence point? What is the hydrogen ion concentration and pH at the equivalence point?

21. Explain why phenolphthalein is a good indicator for the titration of a weak acid with a strong base, but methyl orange is a preferred indicator for the titration of a weak base by a strong acid.

22. Verify the titration curve for the titration of 50 ml of $0.10M$ HCl with $0.10M$ NaOH by calculating hydrogen ion concentrations at 0 ml, 25 ml, 50 ml, 50.1 ml, and 100 ml of NaOH added.

23. Varying volumes of the strong acid $0.1M$ HCl are added to 100 ml of the weak base $0.1M$ NH_4OH in a titration procedure.
    a. Without doing any calculations, what acid–base system would you use and what concepts would you consider to determine the pH of the solution resulting from the addition of the following total amounts of acid to the base:
       1. no HCl

2. 50 ml of $0.1M$ HCl
3. 100 ml of $0.1M$ HCl
4. 120 ml of $0.1M$ HCl
    b. Estimate the pH of the final solutions in (1) and (3) above.

24. Calculate the hydrogen ion concentration in the solutions obtained by the addition of the following volumes of $0.10M$ NaOH to 100-ml samples of $0.10M$ HAc:
    a. no NaOH
    b. 50 ml of NaOH
    c. 100 ml of NaOH
    d. 120 ml of NaOH

25. Calculate the hydrogen ion concentration of mixtures made by adding
    a. 100 ml of $0.10M$ HAc to 100 ml of $0.05M$ NaOH
    b. 100 ml of $0.15M$ NaOH to 100 ml of $0.10M$ HAc
    c. 100 ml of $0.10M$ HAc to 50 ml of $0.20M$ NaOH.

26. A $0.10M$ solution of a weak acid HX has a pH of 2.38. What is the pH of an equimolar solution of the sodium salt of the conjugate base of the acid and the acid?

27. Two acids of approximately $10^{-2}M$ are titrated separately with a strong base. The pH's at the end points are

HA: pH $= 9.5$
HB: pH $= 8.5$

    a. Which acid is the stronger?
    b. Which of the bases, A^- or B^-, is the stronger base?
    c. Estimate K_a for each acid.

28. a. In the titration of a polyprotic base such as CO_3^{2-} with HCl, there are said to be two equivalence points. To what completed reactions do these equivalence points correspond?
    b. If 100 ml of $0.10M$ Na_2CO_3 are titrated with $0.10M$ HCl, find
       1. the approximate pH at the start of the titration
       2. the volume of HCl added and the approximate pH at the first equivalence point
       3. the HCl volume added and the pH at the second equivalence point

29. Write equations for the action of a $H_2PO_4^- - HPO_4^{2-}$ buffer mixture toward added hydrogen and hydroxide ions. If the acid ionization constant for $H_2PO_4^-$ is 6.2×10^{-8}, what is the pH of a mixture containing equal amounts (moles) of NaH_2PO_4 and Na_2HPO_4?

30. If the $H_2PO_4^- - HPO_4^{2-}$ acid–conjugate base pair were the dominant buffer mixture in blood plasma, what would be the optimum pH of plasma?

31. The $H_2CO_3 - HCO_3^-$ pair is the dominant buffer system in blood plasma. If the normal pH of blood is 7.4, what is the ratio of H_2CO_3 to HCO_3^-?

32. Show that the hydrogen ion concentration of a solution of a salt of a weak base and a weak acid such as ammonium acetate is given by

$$[H^+] = \sqrt{\frac{K_w K_a}{K_b}}$$

33. What is the pH of a 0.100M solution of H_2PO_4? Note that K_{a_1} is quite large.

34. Explain the existence of two points of maximum buffer capacity in Figure 10.3(d) for solutions resulting from the titration of NH_3 with HAc.

35. Show that the hydrogen ion concentration of a buffer solution containing a weak base and a salt of the weak base is given by

$$[H^+] = \frac{K_w}{K_b} \frac{C_{salt}}{C_{base}}$$

36. a. What is the hydrogen ion concentration in a solution made by mixing 50 ml of 0.20M NH_3 and 50 ml of 0.10M NH_4NO_3?
    b. What is the hydrogen ion concentration when 10 ml of 0.10M HCl are added to the solution of part (a)?

37. Which elements are the best oxidizing agents? Why?

38. In the reaction $ClO_3^- + SO_3^{2-} \longrightarrow Cl^- + SO_4^{2-}$,
    a. What element (with its oxidation state) is reduced?
    b. What element is oxidized?
    c. What substance is the reducing agent?
    d. What substance is the oxidizing agent?
    e. Balance the reaction.

39. For the following sets of reactants, tell where a redox reaction will take place. Where reaction occurs, complete and balance the reactions.

$$\begin{array}{c} Cl_2 + I^- \longrightarrow \\ Fe^{3+} + I^- \longrightarrow \\ Fe^{2+} + H^+ \longrightarrow \\ Br_2 + Cl^- \longrightarrow \\ Ag^+ + Br^- \longrightarrow \end{array}$$

40. Five metals, A, B, C, D, and E, exhibit the following properties:

Only A, C, and D react with hydrochloric acid. When D is added to solutions of the ions of the other metals, B, E, and A are displaced from solution.

Metal E displaces B from solution.

Arrange the five metals in an activity series with hydrogen ion.

41. Should it be possible to extract bromine from sea water containing Br^- by treatment of the water with chlorine gas?

42. What will be the products of the reaction, if any, of copper metal with dilute sulfuric acid?

43. Write balanced equations for the reaction of
    a. copper with nitric acid
    b. sulfurous acid with nitric acid

44. The total redox reaction in a cell of a lead storage battery is

$$Pb + PbO_2 + 4H^+ + 2SO_4^{2-} \rightleftharpoons 2PbSO_4 + 2H_2O$$

The voltage delivered by one cell is about 2 volts when the concentration of sulfuric acid electrolyte is 3.7M. Explain what happens to the voltage when the water produced in the reaction dilutes the sulfuric acid.

45. The emf E_t of an electrochemical cell is zero when the cell reaction has come to equilibrium and all concentrations are the equilibrium values. In this case,

$$E_t^0 = \frac{0.0591}{n} \log K$$

Show that a negative E_t^0 corresponds to a reaction (when all concentrations equal unity) that does not proceed to an appreciable extent.

46. Calculate the solubility product constant for AgCl from the half-reactions

$$Ag^+ + e^- \longrightarrow Ag \qquad E^0 = 0.80 \text{ V}$$

and

$$AgCl(s) + e^- \longrightarrow Ag^+ + Cl^- \qquad E^0 = 0.22 \text{ V}$$

47. Fuel cells convert chemical energy directly into electric energy, as any electrochemical cell does. In a fuel cell, however, the reducing and oxidizing agents are fed continuously (as gases) to the cell as electricity is produced. The gaseous reaction

$$2H_2 + O_2 \longrightarrow 2H_2O$$

proceeds by way of the half-reactions

$$2H_2 + 4OH^- \longrightarrow 4H_2O + 4e^-$$
$$O_2 + 2H_2O + 4e^- \longrightarrow 4OH^-$$

in an aqueous sodium hydroxide solution. How would the voltage delivered by this cell be affected by (a) increasing the pressure of hydrogen or oxygen? (b) by failure to remove H_2O as it is formed? (c) by decreasing the concentration of the NaOH solution?

48. Consider the reaction

$$2Ag^+ + Cd \longrightarrow 2Ag + Cd^{2+}$$

The standard reduction potentials for the Ag^+–Ag and Cd^{2+}–Cd couples are 0.80 volt and -0.40 volt, respectively.
   a. What is the standard potential $E_t{}^0$ for this reaction?
   b. For the electrochemical cell in which this reaction takes place, which electrode is the negative electrode?
   c. Will the total emf of the reaction be more posi-

tive or more negative if the concentration of Cd^{2+} is 0.10M rather than $1M$?
   d. If $[Cd^{2+}] = 0.10M$ and $[Ag^+] = 0.01M$, what will E_t become?

49. Verify that the half-reaction

$$2H^+ + 2e^- \longrightarrow H_2$$

has a reduction potential of -0.414 volt at pH $= 7$ and $P_{H_2} = 1$ atm.

50. In lake water at pH $= 5$ and a reduction potential of 0.30 volt, what concentration of Cu^{2+} ion can exist in equilibrium with metallic copper?

51. What reduction potential must an environment have in order for the concentrations of Fe^{2+} and Fe^{3+} to be equal?

52. Iron(II) ion released by weathering is stable at low pH values. When it reaches the higher pH of the seas, it is easier to oxidize to Fe^{3+} and undergoes hydrolysis and precipitation. Explain why this is so.

53. What happens to the oxidizing ability of natural waters when (a) dissolved oxygen is depleted and (b) the waters become alkaline?

Answers

7. 3.34×10^{21} water molecules
   6.02×10^{16} hydrogen ions
8. $[H^+] = 1.6 \times 10^{-5}M$
   $K_a = 2.1 \times 10^{-9}$
9. pH $= 11$
10. $[H^+] = 4.5 \times 10^{-13}M$
11. 0.42 percent in $1M$ HAc; 1.3 percent in 0.1M HAc
12. $SO_4{}^{2-} = 0.012M$
13. a. pH $= 4$
    b. $[S^{2-}] = 1 \times 10^{-14}M$
16. $[CO_3{}^{2-}] = K_{a_2}$
18. pH $= 5.1$
20. 133 ml; $[H^+] = 6.9 \times 10^{-6}M$; pH $= 5.16$
22. pH $= 1.48$ at 25 ml; 10 at 50.1 ml
24. a. $1.3 \times 10^{-3}M$
    b. $1.8 \times 10^{-5}M$
    c. $1.9 \times 10^{-9}M$
    d. $1.1 \times 10^{-12}M$
25. b. $4.0 \times 10^{-13}M$
    c. $1.5 \times 10^{-9}M$

26. pH $= 3.75$
28. b. 1. 11.7
       2. 100 ml; 8.3
       3. 200 ml; 3.9
29. 7.2
30. 7.2
31. 0.10
33. 1.62
36. a. $[H^+] = 2.8 \times 10^{-10}M$
    b. $[H^+] = 3.7 \times 10^{-10}M$
40. C, D, A, H^+, E, B
48. a. $E^0 = 1.20$ volts
    b. Cd
    c. more positive
    d. 1.10 volts
50. $[Cu^{2+}] \simeq 0.05M$
51. 0.77 volts

References

ACIDS AND BASES

Bard, Allen J., "Chemical Equilibrium," Harper & Row, New York, 1966, Chapter 3.

Butler, J. N., "Solubility and pH Calculations," Addison–Wesley, Reading, Mass., 1964, Chapters 1, 3, 4.

Drago, Russell S., and Matwioff, N. A., "Acids and Bases," Raytheon Education Co., Boston, 1968.

Fischer, Robert B., and Peters, Dennis G., "Chemical Equilibrium," Saunders, Philadelphia, Chapter 3.

Robbins, Omer, Jr., "Ionic Reactions and Equilibria," Macmillan, New York, Chapters 1, 4, 12.

OXIDATION AND REDUCTION

Bard, *op. cit.*, Chapter 6.

Fischer, *op. cit.*, Chapter 5.

Krauskopf, Konrad B., "Introduction to Geochemistry," McGraw-Hill, 1967, Chapter 9.

This chapter discusses redox conditions in natural environments. Geochemists refer to oxidation potentials but assign a negative sign to spontaneous reactions. The result is that their oxidation potentials are equal to our reduction potentials. Chapter 2 presents acid–base and carbonate equilibria.

The Sea

Chemically, the face of our planet, the biosphere, is being sharply changed by man, consciously, and even more so, unconsciously. The aerial envelope of the land as well as all its natural waters are changed both physically and chemically by man. In the twentieth century, as a result of the growth of human civilization, the seas and the parts of the oceans closest to shore become changed more and more markedly. Man must now take more and more measures to preserve for future generations the wealth of the seas which so far have belonged to nobody.

W. I. Vernadsky (1943)

The study of the hydrosphere is commonly divided into two parts, oceanography and limnology. Oceanography has as its concern all aspects of the salty seas. This includes the physical, chemical, geological, geographical, and biological phenomena of the seas. Because all these areas interact, oceanography is really a combination of all the basic sciences. It is difficult to separate chemical oceanography or marine chemistry from marine biology or physical oceanography or any of the other subdivisions one may encounter.

Limnology is the study of all aspects of natural fresh waters: the inland surface waters of our lakes, streams, and rivers. These waters form a tiny, but essential, part of the hydrosphere. Although the oceans represent approximately 98 percent of the mass of the earth's water and are an obvious part of the environment of life, the role of fresh terrestrial waters in shaping our environment should not be dismissed because of their comparatively small mass and area. Not only are these fresh waters of economic value for industry and for transportation, but they are also the primary source of drinking water. Furthermore, fresh-water runoff from the land replenishes the seas and carries nutrients to the sea, along with a growing variety of pollutants. The next chapter will examine some of the chemistry of fresh waters, particularly their pollution and restoration.

The present chapter concentrates on selected aspects of the chemistry of sea water and examines some of the ecological consequences of this chemistry. Such consequences are not necessarily limited to the marine ecosystem, because the sea is a complex dynamic system of interacting chemical, physical, biological, and geological processes with connecting links to climate, atmospheric composition, and several of the major biogeochemical cycles. In these respects, the sea plays a key role in the regulation of the entire biosphere. And what is done on land affects the sea because of the general drift of all materials down to the seas, which act as the catch basins of the earth.

Composition of Sea Water

It is appropriate to begin a discussion of the hydrosphere with the chemical composition of sea water, because the composition is determined largely by living processes in the biosphere. At the same time, the composition of the sea is a condition for life in the biosphere. The sea shares this same reciprocal relationship with the atmosphere and the soil.

Sea water is an aqueous solution of dissolved salts, gases, and organic molecules. In addition to dissolved matter, there are suspended solids of both organic and inorganic origin in sea water. Our main concern now is with the dissolved salts. These salts are usually divided into two categories. One group, the major constituents, includes 11 ionic species that make up over 99.9 percent of the dissolved solids in sea water. The other group, the minor constituents, is composed of all other dissolved elements in sea water. (The use of the term "salts" for the individual ions derived from salts is technically incorrect, but we shall follow the custom of oceanographers in this chapter.)

11.1 MAJOR CONSTITUENTS AND SALINITY

The 11 major constituents of sea water are those with concentrations above 0.001 g/kg (1 ppm) of sea water. The major constituents are tabulated in Table 11.1. For comparison, the concentrations of the species in average river water is also tabulated. (Although silica, SiO_2, is not a major constituent of sea water, it is also listed for comparison because of its importance in river water.) Sea water is predominantly a solution of sodium chloride, whereas river water is mostly a much more dilute solution of calcium hydrogen carbonate and sulfate, comparatively rich in silica. In the sea, the concentrations of Ca^{2+}, HCO_3^-, and silicon are lowered relative to other chemical species by incorporation into organisms and the sediments they subsequently form.

A distinguishing feature of the major constituents of sea water is their constancy of concentration *relative* to each other in all samples of sea water. Although the absolute concentration of any of the major constituents may vary from place to place or from time to time, the *ratios* of the concentrations of these constituents are nearly constant from one sample of sea water to

Table 11.1

Major constituents of sea and river water

constituent	sea water[a]		river water	
	g/kg of water	molarity	g/kg of water	molarity
chloride, Cl^-	19.35	0.55	0.0078	0.00022
sodium, Na^+	10.76	0.47	0.0062	0.00027
sulfate, SO_4^{2-}	2.71	0.028	0.011	0.00012
magnesium, Mg^{2+}	1.29	0.054	0.0041	0.00034
calcium, Ca^{2+}	0.41	0.010	0.015	0.00038
potassium, K^+	0.39	0.010	0.0024	0.000059
hydrogen carbonate, HCO_3^-	0.14	0.0023	0.059	0.00096
bromide, Br^-	0.067	0.00083	—	—
strontium, Sr^{2+}	0.008	0.0015	—	—
boron, as H_3BO_3	0.004	0.00043	—	—
fluoride, F^-	0.001	0.00007	—	—
silicon, as SiO_2	—	—	0.0130	

[a] For water of 35‰ salinity.

another. Geological evidence indicates that sea water has maintained a nearly constant composition for at least the past 100 million years. The relative constancy of composition of the major constituents is the result of the normal physical circulation and mixing processes going on in the interconnected oceans and seas of the earth. As a consequence of the uniformity of composition, the determination of the concentration of any one major constituent in sea water permits the estimation of the concentration of any other major constituent in that sample of water.

In discussing the concentration of the major ionic constituents of sea water, it is common to make use of two arbitrarily defined terms, *salinity* and *chlorinity*. Salinity is a measure of the total salt concentration in sea water. If some small technicalities that arise in the exact definition of the term are neglected, salinity is approximately equal to the total weight, in grams, of dissolved salts in 1 kg of sea water (Figure 11.1). The most obvious way to determine the salinity of a sample of sea water would be to evaporate 1 kg of water to dryness and weigh the residue. However, because some of the salts would decompose during this heating, an alternative method of determining salinity has been devised which makes use of chlorinity. The chlorinity is equal to the total amount, in grams, of chloride, bromide, and iodide in 1 kg of sea water. The chlorinity is determined by the precipitation of silver chloride with silver nitrate solution. Insoluble silver bromide and silver iodide are also precipitated in this test and are not separated from the chloride. The ionic reactions are

1 kg of sea water

35 g of salts

Figure 11.1. A kilogram of sea water and the approximately 35 g of dissolved salts contained in it.

$$\left.\begin{array}{c} Cl^- \\ Ag^+ + Br^- \\ I^- \end{array}\right\} \longrightarrow \left\{\begin{array}{l} AgCl(s) \\ AgBr(s) \\ AgI(s) \end{array}\right.$$

The mixed precipitate is assumed to be all AgCl. Since the chlorinity value obtained also includes Br^- and I^-, it is somewhat greater than the actual Cl^- concentration and is really a measure of the total halide ion (Cl^-, Br^-, I^-) concentration.

Both salinity and chlorinity are reported in units of grams per kilogram of water, which is equivalent to parts per thousand and is given the symbol ‰, read per mil. The relation between salinity, S‰, and chlorinity, Cl‰, is

$$S‰ = 1.805 \ Cl‰ + 0.03$$

Through this empirical relationship the hard-to-measure salinity can be found from a relatively easy determination of chlorinity with silver nitrate or by electric conductivity measurements. A salinity value of 35‰ and a chlorinity value of about 19‰ are average values for sea water. The concentration of salts in sea water is usually given by referring the substances to water of a standard chlorinity or salinity of 19‰ or 35‰, respectively. By comparison, the salinity values for hard fresh water and soft water are about 0.5‰ and 0.06‰, respectively.

The equation relating the salinity and the chlorinity is a consequence of the constancy of the relative proportions of the major ions in sea water. Even though the salinity of a series of sea water samples may vary, the total dissolved salt content bears much the same relation to the halide ion content from one sample to another. The only restriction is that the water samples be taken in the open sea, away from land areas where dilution by land drainage or other conditions could upset the uniformity of proportions of the major constituents. Table 11.2 illustrates this constancy of relative composition with respect to Cl‰ for sea water from a number of locations.

11.2 MINOR CONSTITUENTS

Sea water probably contains all of the naturally occurring elements, even if some of them have not yet been detected in water. As it is, over 70 minor constituent elements in addition to those contained in the major constituents have been identified in sea water. Unlike the major constituents, the concentrations of the minor constituents can vary widely and independently of each other with location, depth, and time, so their concentration ratios are nowhere near constant.

The low concentration of the minor constituents in sea water is the result of their low concentration in the earth's crust or interior, from which they must all originate, or their participation in biological and nonbiological marine processes, which remove them from the dissolved state. Some of the minor constituents enter the sea as solids, which settle rapidly.

The same biological and nonbiological processes contributing to the low concentration of minor constituents also lead to their variable concentration. These processes remove constituents from solution for lengths of time that prevent their full participation in the mixing of ocean waters. Biological incorporation of an element into marine plants and animals is the major cause of variable concentration. The movement and distribution of the minor constituents may be considered to depend on a biological component resulting from the movement of living and dead organisms. The biological component overpowers the physical circulation component that determines the distribution of the major constituents.

Even if an element is not involved in a biogeochemical cycle, its concentration may vary because of its removal from solution by sedimentation processes such as precipitation or coprecipitation (adsorption or ion exchange on existing precipitates that subsequently settle out). Several of the nonbiological processes may be indirectly influenced by biological activity. For instance, copper can precipitate as the sulfide in deeper waters devoid of

Table 11.2

Major constituent concentration-to-chlorinity ratios for various oceans and seas

ocean or sea	$\frac{Na}{\text{‰ Cl}}$	$\frac{Mg}{\text{‰ Cl}}$	$\frac{K}{\text{‰ Cl}}$	$\frac{Ca}{\text{‰ Cl}}$	$\frac{Sr}{\text{‰ Cl}}$	$\frac{SO_4}{\text{‰ Cl}}$	$\frac{Br}{\text{‰ Cl}}$
N. Atlantic	—	—	0.02026	—	—	—	0.00337–0.00341
Atlantic	0.5544–0.5567	0.0667	0.01953–0.0263	0.02122–0.02126	0.000420	0.1393	0.00325–0.0038
N. Pacific	0.5553	0.06632–0.06695	0.02096	0.02154	—	0.1396–0.1397	0.00348
W. Pacific	0.5497–0.5561	0.06627–0.0676	0.02125	0.02058–0.02128	0.000413–0.000420	0.1399	0.0033
Indian	—	—	—	0.02099	0.000445	0.1399	0.0038
Mediterranean	0.5310–0.5528	0.06785	0.02008	—	—	0.1396	0.0034–0.0038
Baltic	0.5536	0.06693	—	0.02156	—	0.1414	0.00316–0.00344
Black	0.55184	—	0.0210	—	—	—	—
Irish	0.5573	—	—	—	—	0.1397	0.0033
Puget Sound	0.5495–0.5562	—	0.0191	—	—	—	—
Siberian	0.5484	—	0.0211	—	—	—	—
Antarctic	—	—	—	0.02120	0.000467	—	0.00347
Tokyo Bay	—	0.0676	—	0.02130	—	0.1394	—
Barents	—	0.06742	—	0.02085	—	—	—
Arctic	—	—	—	—	0.000424	—	—
Red	—	—	—	—	—	0.1395	0.0043
Japan	—	—	—	—	—	—	0.00327–0.00347
Bering	—	—	—	—	—	—	0.00341
Adriatic	—	—	—	—	—	—	0.00341

SOURCE: R. A. Horne, "Marine Chemistry," Wiley, New York, 1969, p 152.

dissolved oxygen but rich in hydrogen sulfide. Such conditions can be produced by an overload of decomposing organic matter near the bottom. Some major constituents (especially HCO_3^- and SO_4^{2-}) are absorbed by organisms. But such biological activity does not noticeably affect their local concentrations because of the large amounts of each present.

Of the minor constituents, nitrogen (as NO_3^-) and phosphorus (as PO_4^{3-}) will receive major attention, because they are required for the building of all protoplasm. Their low concentration often limits biological activity in water. Other minor constituents, notably the trace elements Fe, Mn, Cu, Co, I, and others in the parts per billion range, are essential to specific organisms and absorption by marine life may deplete the water of them. As mentioned in Chapter 5, it is not necessary that an element be essential to life in order to be absorbed by an organism and be regenerated when that organism dies and decomposes, that is, to have a biogeochemical cycle of its own. Nonessential, and toxic, elements such as the heavy metals Hg and Pb are also absorbed by organisms. In fact, the heavy metals constitute the largest group of toxic elements absorbed by organisms.

The concentration of each of the minor constituents in sea water is very small. When these elements are incorporated into marine organisms, they are concentrated many times over in the organism. Sometimes this concentration amounts to over 2 million times the concentration of the element in the water. High biological concentration is shown especially by the transition metals such as Ti, Cu, and V, which can form specific chelate molecules in certain organisms. Concentration factors increase if an absorbed element or compound is not broken down in the organism or excreted in metabolic processes. Concentration of an element begun by direct absorption from the water by producer organisms can be intensified in consumer organisms that eat the producers. In the marine ecosystem these concentration effects are particularly strong, because complicated marine food chains (webs) have many steps and concentration occurs at each step (Section 20.18).

11.3 RESIDENCE TIME AND RELATIVE REACTIVITIES

The low overall concentrations and the variation of concentration from place to place of the dissolved minor constituents of sea water can be correlated with their higher degree of chemical reactivity compared to major elements in the marine environment. In comparing the relative reactivities of elements in the sea, use is made of the concept of *residence time*. This concept is based on the oceans of the world being in a steady state wherein the rate of addition of an element to the seas is just balanced by the rate of its removal from solution by deposition in insoluble sediments. Using this assumption, a residence time may be defined equal to the total amount of the element in solution in all the seas divided by the amount introduced from the land by rivers in a year. (Nearly all of the chloride and about 53 percent of the sodium in river water is cycled from the sea as salt crystals via the atmosphere and rain. Since such cyclic salts are not generated by the land, their contribution must be subtracted from the total amount carried by river water.) The residence time calculated in this manner is equal to the average time, in years, that an element remains in solution in the sea before it is permanently removed from solution by some sedimentation process. Residence times are given in Table 11.3 for selected major and minor elements.

	element	abundance, mg/liter	principal species	residence time, yr
long residence times or indeterminate residence times (greater than 10^6 yr)	Na	10,500	Na^+	2.6×10^8
	Mg	1,350	Mg^{2+}, $MgSO_4{}^a$	4.5×10^7
	Ca	400	Ca^{2+}, $CaSO_4$	8.0×10^6
	K	380	K^+	1.1×10^7
	Sr	8	Sr^{2+}, $SrSO_4$	1.9×10^7
	C	28	$HCO_3{}^-$, H_2CO_3, $CO_3{}^{2-}$, organic compounds	—
	H	108,000	H_2O	—
	O	857,000	H_2O, $O_2(g)$, $SO_4{}^{2-}$	—
	N	0.5	$NO_3{}^-$, $NO_2{}^-$, $NH_4{}^+$, $N_2(g)$, organic compounds	—
	P	0.07	$HPO_4{}^{2-}$, $H_2PO_4{}^-$, $PO_4{}^{3-}$, H_3PO_4	—
	S	885	$SO_4{}^{2-}$	—
	Cl	19,000	Cl^-	—
intermediate residence times 10^3–10^6 yr	Zn	0.01	Zn^{2+}, $ZnSO_4$	1.8×10^5
	Cu	0.003	Cu^{2+}, $CuSO_4$	5.0×10^4
	Ni	0.002	Ni^{2+}, $NiSO_4$	1.8×10^4
	Mo	0.01	$MoO_4{}^{2-}$	5.0×10^5
	Si	3	$Si(OH)_4$, $Si(OH)_3O^-$	8.0×10^3
short residence times (less than 10^3 yr)	Al	0.01	—	1.0×10^2
	Fe	0.01	$Fe(OH)_3(s)$	1.4×10^2
	Cr	0.00005	—	3.5×10^2

^{a} $MgSO_4$, $CaSO_4$, etc. are dissolved ion pairs.
SOURCE: Data selected from Edward D. Goldberg, in "The Sea," Vol. 2, M. N. Hill, Ed., Wiley, New York, 1963, Chapter 1, p 4.

In general, the major constituents of sea water have long residence times, some comparable to the estimated age of the oceans themselves, 10^9 yr. Some major constituents such as chlorine, sulfur, and carbon may be considered to have indefinite residence times, because they were present in large amounts at the origin of the oceans or, more likely, because they are supplied to the oceans by undersea and volcanic emissions not related to runoff from the land. Many of the minor elements have short residence times of less than 10^3 yr. Such elements are removed from the dissolved state rapidly by a variety of sedimentation processes. Elements with long residence times tend to be unreactive in the marine environment and are not readily incorporated into sediments by geochemical or biological processes.

For the most part, sedimentation processes leading to short residence times are the result of chemical reactions not directly involving biological activity. Living organisms are involved in sedimentation processes if they absorb an element and if that element remains locked in the organism when it dies and sinks to the bottom. Calcium and silicon are two elements removed by biological sedimentation processes (although calcium has a long residence time because $CaCO_3$ redissolves in deep water). Elements such as phosphorus and nitrogen uniformly absorbed by organisms are usually readily returned to solution by decomposition of the dead organism. The concentration of these elements tends to be higher nearer the sea floor, where decay occurs, than at the surface, but this has no effect on the residence time.

The quantity of different elements in sea water is not proportional to their concentration in river water. An element's concentration in sea water is less in proportion to the ease with which the element is removed from solution by chemical or biological processes in the sea. Unsuccessful attempts have been made to correlate reactivity in removal processes, and therefore residence times, with the ionic charge-to-radius ratio, with the acidity of hydrated ions (Section 10.5), and with the solubilities of compounds of the elements. Most of the elements (except Ca) exist in sea water at concentrations less than that required for precipitation of their most insoluble salts, so solubility considerations do not account for the low but steady concentrations. Even in the case of calcium, where the upper layers of the sea appear to be supersaturated with respect to $CaCO_3$, addition of Ca^{2+} does not result in precipitation. The major removal processes for dissolved material entering the seas are considered to be adsorption of ions on suspended matter or sediments (Section 14.9) plus biological removal processes.

Mixing Time of Oceans

The variability of concentration of the minor constituents is explained by a comparison of their residence times with the mixing time of the oceans. The mixing time of an ocean is of the order of 1000 yr. This is the average time that a water molecule or a solute originally at the surface of the ocean spends on its circulatory trip to the deep water and back to the surface again. This mixing tends to smooth out differences in concentrations of solutes within oceans and to some extent from ocean to ocean. Elements with residence times in excess of 10^3–10^4 yr will remain in solution long enough to participate in this oceanic mixing. Their concentrations will not differ greatly from place to place—nor will their concentrations relative to one another differ greatly. All of the major constituents just discussed have long residence times in this respect, and their relative amounts in sea water are quite constant.

Elements having short residence times in comparison with the mixing time will not be able to mix uniformly before they are removed from solution as sediments. Therefore, their concentrations will vary from ocean to ocean and even from place to place within a single ocean. Many of the minor constituents have short residence times and resulting variable concentration because of the high reactivity of their ions with respect to sedimentation in the marine environment. Certain artificial radioactive isotopes, such as those of cesium-137 and strontium-90, which have been introduced into the sea as wastes from tests of nuclear explosives or from nuclear reactors, have long residence times. Although they are not removed from solution by sedimentation, they can be absorbed and concentrated by living organisms during the short life of the organism, leading to concentration of these toxic isotopes by the organism and by any consumer that eats the original organism. The persistence and long residence times of such hazardous isotopes makes them especially detrimental to life. This is true of strontium-90, whose chemical similarity to calcium causes it to accumulate in bone material where it can contribute to leukemia.

The shortest residence time is not less than 100 yr, indicating that the rate of formation of insoluble sediments in the sea is very slow. The slowness of many reactions is a characteristic of marine chemistry and is partially due to the extremely low concentration of many of the reactants in sea water. The chemistry of the sea has been described as largely the chemistry of obscure reactions at extreme dilution in a concentrated salt solution. Most of the important reactions in the sea take place at phase boundaries between the ocean and the atmosphere, the ocean and solid phases (suspended matter or sediments), and the ocean and living matter. Undoubtedly, at phase boundaries the reactants are concentrated by adsorption or biological processes leading to increased reaction rates. Thus, solid phases present in the water drive or control many of the important reactions in the sea. Because so many of these reactions are obscure and very slow, marine chemistry does not lend itself to the simple distilled water chemistry treatment of preceding chapters. As a result, we shall not attempt to deal with marine chemistry in a very quantitative fashion.

11.4 COMPARISON OF SEA WATER AND BODY FLUIDS

The body fluids, such as the blood and spinal fluids of both marine and land animals, bear a striking similarity to the chemical composition of sea water. The similarity is particularly evident with respect to the major constituents in the fluids of marine invertebrates. Table 11.4 lists the absolute and relative concentrations of the major ions in sea water, land animals, fresh-water animals, and salt-water animals. The salt-water animals are generally divided into the invertebrates (lobster, mollusk, sea cucumber), the bony fish, and another group composed of sharks and rays. The listing for fish represents both salt- and fresh-water fish. The relative concentrations (with Na^+ as 100) indicate that organisms still maintain the approximate ionic balance of sea water in their vital fluids. For years, this similarity has been taken as evidence that all life began and evolved in the sea. Those organisms that eventually left the sea took with them an internal environment resembling the sea, an environment that has become less concentrated in salts than sea water through the years of evolution but that retains about the same relative composition of major constituents.

Although the ionic balance of sea water and body fluids may be similar, the salinity or the total amount of dissolved salts of sea water and of the body fluids of some organisms differ considerably. Table 11.4 also gives the sum of the absolute concentrations of ions in millimoles per liter for each organism. This value is proportional to the salinity of the body fluid. The values for land animals are in the range 200–300; for bony fish, both fresh- and salt-water, about 400; and for marine (salt-water) invertebrates, about 1000. The value for sea water itself is about 1000. If we assume that the concentration of other ions is negligible and that no other complicating effects exist, the salinity of sea water approximately matches that of marine invertebrates but is quite a bit higher than that of other organisms, both aquatic and terrestrial. The difference in salinity of the external environment (the sea) and the internal environment (the body fluid) of an organism has significant ecological consequences connected with the tendency of water to flow through the cell

Table 11.4
Ionic concentrations in sea water and body fluids[a]

	Na^+	K^+	Ca^{2+}	Mg^{2+}	Cl^-	SO_4^{2-}	total
sea water	470	10	10	50	550	40	1130
	(100)	(2)	(2)	(11)	(120)	(9)	
fresh water	0.30	0.06	0.40	0.20	0.20	0.10	1.26
	(100)	(20)	(130)	(70)	(70)	(30)	
vertebrates							
man	145	5.1	2.5	1.2	103	2.5	259
(blood serum)	(100)	(3.5)	(1.7)	(0.83)	(71)	(1.7)	
frog (plasma)	103	2.5	2.0	1.2	74	1.9	185
	(100)	(2.4)	(1.9)	(1.2)	(72)	(1.8)	
fish (fresh and salt water)	200	6	2	4	160	—	372
	(100)	(3)	(1)	(2)	(80)	—	
invertebrates							
insect	100	15	10	20	40	—	185
	(100)	(15)	(10)	(20)	(40)	—	
crustacean (lobster, crab)	500	10	15	15	500	15	1055
	(100)	(2)	(3)	(3)	(100)	(3)	
salt-water mollusc	500	10	10	50	550	30	1150
	(100)	(2)	(2)	(10)	(110)	(6)	
fresh-water mollusc	15	0.3	10	0.3	10	0.7	36
	(100)	(0.2)	(7)	(0.2)	(7)	(0.5)	

[a]Concentrations are expressed in millimoles per liter of water or body fluid and as relative concentrations (parentheses) referred to $Na^+ = 100$. Values have been rounded off in most cases.
SOURCE: Absolute concentration values selected from C. Ladd Prosser and Frank A. Brown, Jr., *Comparative Animal Physiology*, 2nd ed., Saunders, Philadelphia, 1961, p 60.

membranes of the organism to equalize the salt concentration on each side of the cell. This phenomenon is known as osmosis.

11.5 OSMOSIS *Osmosis* is a property of solutions that is related to the diffusion of solvent molecules (in this case, water) through a barrier that does not allow the ready passage of solute particles. Barriers permeable to one substance but not to another are termed *semipermeable*. All membranes surrounding cells in living organisms exhibit selective permeability to water, both that in body fluids and that in the external environment. Water will cross the membrane in both directions, but the net flow is from the more dilute (in solute) solution to the more concentrated (in solute) solution, and thus the volume of the solution having more solute will increase. This net direction of flow conforms with the natural tendency for molecules to diffuse from a region where their concentration is high to a region of lower concentration. For example, consider that pure solvent is separated from a solution by a semipermeable membrane

Figure 11.2. Osmosis and osmotic pressure. (a) The flow of water molecules through the membrane into the solution is greater than the reverse flow from solution to pure water. The hydrostatic pressure due to the column h of expanded solution increases the rate of flow of water from solution to pure water until, at equilibrium, the opposite flows are equal. (b) The osmotic pressure of the solution is equal to the hydrostatic pressure Π that must be applied to the solution to equalize the rate of flow to and from solution and produce a net flow of zero.

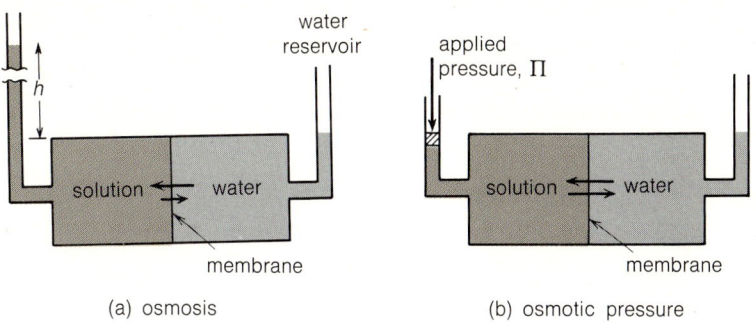

(a) osmosis   (b) osmotic pressure

as in Figure 11.2(a). Solvent will pass through the membrane and enter the solution compartment, driving the solution level up the left-hand tube until the hydrostatic pressure of the column of dilute solution on the left is sufficient to counter the diffusion pressure of solvent molecules passing into the solution through the membrane. The hydrostatic pressure at equilibrium when solvent molecules are crossing the membrane in both directions at equal rates is the *osmotic pressure* Π *of the solution.*

To measure the osmotic pressure of a solution, it is not necessary to wait until the flow stops and equilibrium is established. Hydrostatic pressure can be applied with a piston on the solution side. The pressure necessary to prevent or reverse the flow of solvent into the solution compartment is the osmotic pressure of the solution, as shown in Figure 11.2(b). Of course, a solution has no osmotic pressure in the absence of a semipermeable membrane.

The osmotic pressure of a solution depends on the number of solute particles, whether ions or molecules, per unit volume. We can calculate the numerical value of the osmotic pressure Π, in atmospheres, of a solution from

$$\Pi = MRT$$

where M is the total molar concentration of solute particles, T is the temperature in degrees kelvin, and R is the ideal gas constant, 0.08205 liter atm/deg mol.

EXAMPLE   The osmotic pressure of a 0.20M sucrose solution at 20°C is

$$\Pi = MRT$$
$$= \left(0.20 \, \frac{mol}{liter}\right)\left(0.08205 \, \frac{liter \ atm}{mol \ °K}\right)(293°K)$$
$$= 4.8 \ atm$$

EXAMPLE   The calculated osmotic pressure of a 0.010M sodium chloride solution at 20°C takes into account that there are 0.020M of particles in solution (0.010M Na^+ and 0.010M Cl^-), so

$$\Pi = MRT$$
$$= (0.020)(0.08205)(293)$$
$$= 0.48 \ atm$$

At 20°C, the measured osmotic pressure of 35‰S sea water is 24.52 atm. The osmotic pressure of a 1M solution of an un-ionized compound is calculated to be 24.05 atm. Sea water, therefore, behaves osmotically as though it contained about 1 mol of particles per liter.

11.6 REGULATION OF OSMOTIC PRESSURE

Marine organisms whose body fluids have approximately the same osmotic pressure as the external water have no particular problems with diffusion of water into or out of their bodies or cells. However, organisms having body fluids differing in osmotic pressure from the external medium, whether sea water or fresh water, must have methods to prevent this flow of water. The differences of osmotic pressure between internal and external media are, of course, due primarily to differing ionic concentrations.

If the salinity and therefore the osmotic pressure of body fluid are lower than that of the external medium, water will tend to leave the organism and flow to the external medium where its concentration is lower. Such an imbalance is observed in bony fish in both salt and fresh water. These fish are presumed to have originated in fresh water, and those now living in salt water still retain the low osmotic pressure body fluids that originally better adapted them to a fresh-water environment. Such organisms must have some way to prevent outward flow of water (by coverings impermeable to water) or make up for loss of water from their cells or the cells will eventually collapse. The bony fish of the sea swallow water to make up for that lost to the sea. Along with the swallowed sea water come excess (to the fish) salts, especially chlorides, which are excreted through the gills. With these fish, the problem is to maintain a low internal osmotic pressure, and this is done by conserving or replacing water and excluding or rejecting salts. With fresh-water fish and other fresh-water organisms, the opposite situation applies. Their body fluids have a much higher osmotic pressure than the external medium. In the absence of regulatory mechanisms, this situation would lead to entry of water into the cells, followed by swelling and bursting of the cells. In general, maintenance of the normal high osmotic pressure of body fluids here requires the disposal of the water that does enter. Fresh-water fish drink little, their skin is relatively impervious to water, and what inward flow of water does take place is disposed of as large quantities of dilute (with respect to salts) urine. The urine still contains more salts than the fresh-water medium, and this salt loss is made up from the food and by absorption from the water.

Where an osmotic pressure imbalance is normal, the maintenance of this situation requires the expenditure of energy by the organism. The energy is used for the work of excreting salts to the even saltier sea, of excreting water as urine, and so forth. Aquatic organisms have different capabilities for furnishing energy for osmoregulation, especially as the salinity of the external medium changes. Some of them have a great degree of tolerance to changes in salinity. Included among these are organisms living in coastal regions and near river mouths, where the salinity varies with tides and land runoff. Salmon and eel are completely tolerant to salinity variations and are able to pass to and from fresh and salt water without damage.

Other organisms, both animal and plant, are sensitive to small changes in salinity and may die if they are transported to regions of higher or lower salinity or if the salinity of their external medium changes by natural or artificial (pollution) means. Salinity and osmotic pressure of the sea thus impose a considerable ecological effect on the distribution, range, and well-being of marine organisms.

Before leaving the subject of salinity of sea water, we can mention a few more ecological consequences of salinity, in addition to osmotic pressure. Indirectly related to salinity is the presence of various ions in the water. Because of its very low salinity, fresh water may be devoid of some of the ions necessary for life and reproduction of various aquatic organisms. So, aside from their troubles in maintaining osmotic equilibrium, certain organisms may be even more troubled by the lack of essential ions in fresh water (or even in inland salt lakes having an osmotic pressure similar to the seas).

The greater the salinity of sea water, the greater is its density. Thus, salinity could have a bearing on the buoyancy of some marine organisms. More important by far is the effect of salinity on ocean currents. The salinity of sea water is one of the major factors, along with temperature and wind, in controlling the circulation and mixing of sea waters. Any process that increases the salinity of surface sea water, such as freezing (leaving the salts behind in solution) or evaporation, leads to convection currents caused by the sinking of the now denser surface water.

11.7 DISSOLVED GASES IN SEA WATER

In addition to ions and a very small amount of dissolved organic material, sea water contains dissolved gases. The principal gases are oxygen, nitrogen, and carbon dioxide. Of these, oxygen and carbon dioxide are of equal importance for the support of marine life. The seas provide the primary environmental pool or reservoir for carbon dioxide in the biosphere. Dissolved molecular nitrogen is of no significance biologically, because it appears that both nitrogen-fixing and nitrogen-liberating bacteria, though present in sea water, contribute little to the nitrogen cycle in marine ecosystems.

The solubility of a gas in a liquid depends on several factors. Among the most important are temperature, pressure, and the nature of the gas and the solvent, the latter being particularly important if there is the likelihood of a chemical reaction between gas and solvent. In Section 8.13, we noted that the temperature dependence of the solubility of a gas in a liquid depends on the heat of solution. We pointed out that an increase in temperature decreased the solubility of most gases in water. The greater the heat of solution, the greater is the influence of temperature; and gases that react chemically with water, as carbon dioxide does, show a much greater decrease in solubility in water with temperature increase than do inert nitrogen and oxygen.

We also pointed out in Section 8.13, on the basis of Le Chatelier's principle, that an increase in gas pressure causes a greater mass of gas to dissolve in a given volume of liquid. That brief treatment assumed that only the one gas was present in the space above the liquid, so the total pressure was equal to the pressure of that one gas. In most situations, two or more gases are present, and the total pressure is the sum of the partial pressures of the gases present.

Partial Pressure

When two or more nonreacting gases are present in a mixture of gases, the total pressure P_t is the sum of the partial pressures of the individual gases.

$$P_t = P_1 + P_2 + P_3 + \cdots$$

This is an expression of *Dalton's law of partial pressures*. The partial pressure of each gas is equal to the pressure that each gas would exert if it were present alone in the volume V_t occupied by the mixture. In terms of the ideal gas law, the partial pressure of a gas is

$$P_1 = \frac{n_1 RT}{V_t}$$

where n_1 is the number of moles of gas 1.

The partial pressure of a gas in a mixture is related to the volume fraction V_1/V_t of the gas in the mixture by

$$P_1 = \frac{V_1}{V_t} P_t$$

If the total atmospheric pressure is 760 mmHg and the percents by volume of N_2, O_2, and CO_2 in dry air are 78.08, 20.05, and 0.033 respectively, the partial pressures of the three gases in the atmosphere are $P_{N_2} = 594$ mmHg, $P_{O_2} = 159$ mmHg, and $P_{CO_2} = 0.25$ mmHg.

Henry's Law   The relationship of the solubility of a gas to its pressure is given by *Henry's law*, which states that at a constant temperature, the solubility of a gas in a liquid is directly proportional to its partial pressure in the gas phase. The gas and the solution are assumed to have attained solubility equilibrium. In equation form, Henry's law is

$$C = kP$$

where C is the amount of gas dissolved when its equilibrium partial pressure in the gas above the solution is P. The value of the proportionality, or Henry's law, constant k depends on the units used for C (mole fraction, grams per unit volume, moles per liter of solution, etc.), and for P (atm, mmHg, etc.) the particular solute–solvent system, and the temperature. The value of k is determined experimentally; and if the gas does not react chemically with the liquid, k should remain constant over a range of pressures, especially at low pressures.

EXAMPLE   A liter of water (not sea water) at 25°C dissolves 0.0283 liter of oxygen (measured at STP) when the pressure of oxygen in equilibrium with the solution is 1 atm. From the solubility of oxygen given, we can find the proportionality constant in Henry's law with $P_{O_2} = 1$ atm, because

$$k = \frac{C}{P_{O_2}} = \frac{0.0283}{1} = 0.0283 \text{ liter/liter atm}$$

The solubility of oxygen in liters (STP) of gas in 1 liter of water at 25°C when the pressure of oxygen is that prevailing in a normal dry atmosphere with $P_{O_2} = 159$ mmHg is

$$C = kP_{O_2} = (0.0283)\left(\frac{159}{760}\right) = 0.00592 \text{ liter/liter } H_2O$$

or 5.92 ml/liter.

At equal partial pressures nitrogen is about half as soluble as oxygen, while carbon dioxide is about 27 times more soluble than oxygen. The high solubility of carbon dioxide in water is the result of a chemical reaction with the water, as we saw in Chapter 10.

The effect of decreasing the partial pressure of a gas is especially apparent with bottled carbonated beverages. When a soda pop bottle is opened, the partial pressure of the carbon dioxide in equilibrium with the solution drops, and dissolved carbon dioxide escapes from the solution. If the beverage happens to be near the freezing point, the sudden escape of carbon dioxide may cause the beverage to freeze, because the escape of gas is an endothermic process that absorbs heat. If the escape of gas is rapid, this heat energy can come only from the solution (adiabatic process), thereby cooling it below the freezing point.

The solubility of gases in water is decreased by the presence of dissolved electrolytes. For this reason, sea water will not dissolve as much oxygen or carbon dioxide as pure water under the same conditions of temperature and pressure. In contrast to the solubility dependence on the external partial pressure, the solubilities of gases in water are only slightly dependent on the hydrostatic pressure of the water at different depths.

The concentration of dissolved oxygen in sea water varies from zero to about 8.5 ml of O_2 (measured at STP) per liter of sea water. The concentration of oxygen and other biologically active constituents of sea water varies from place to place in the sea. We now examine some of the factors governing concentration distribution.

11.8 VERTICAL DISTRIBUTION OF DISSOLVED NUTRIENTS

We have noted that the concentration of dissolved material in sea water can vary in space and time. Although the relative composition of the major constituents (salts) in sea water is fairly constant, the salinity of water at a given place or time in an ocean may decrease because of dilution of the water by land runoff, rain, or melting ice. Water at another place or time in the same ocean may have a high salinity because of evaporation of surface water or removal of surface water by freezing, which leaves the salts behind in solution. Further variations in salinity can result from the action of ocean currents. High density, high salinity water formed at the surface by evaporation or freezing can sink and flow along the bottom and come to the surface thousands of miles away.

Right now, however, we are more interested in the variation of concentration of certain minor constituents, specifically those that serve as food for marine organisms. Among marine organisms, the microscopic, photosynthetic plants (phytoplankton) are of foremost importance, because they are the first link in the food chain for all other organisms in the seas. Without them there would be no life in the seas.

The major nutrients for phytoplankton are oxygen, carbon dioxide, nitrogen in the form of nitrate ion (NO_3^-), and phosphorus as PO_4^{3-}. An obvious factor determining the amount of dissolved oxygen and carbon dioxide is temperature. In the absence of other factors, cold water will contain more dissolved gas. For this reason, surface water at the poles will pick up more oxygen from the atmosphere. The higher density of this water, because of low temperature or surface freezing or both, will cause it to sink and recharge the bottom waters with life-giving oxygen. (Sea water exhibits no temperature of maximum density. See Figure 8.4.)

Biological activity is undoubtedly the most important single factor influencing the distribution of these nutrients. Dissolved oxygen is consumed in the respiration of nearly all marine animals and plants. It is also used up in the biological oxidation of organic matter from the decomposition of dead organisms or from the excretions of living organisms, and other minor processes. Oxygen is generated in the upper layers of the sea by photosynthetic activity of marine plants. It also enters the surface waters by absorption from the atmosphere. The net result of the processes using oxygen and those producing it is a general decrease in dissolved oxygen concentration in going from surface to bottom waters. In the surface layers, the production of oxygen by photosynthesis and absorption from the air is greater than the removal by marine organisms. As depth increases, oxygen gained from photosynthesis and atmospheric absorption falls off, while consumption by organisms and decay continues and even increases. In the absence of deep water currents to transport oxygen from other locations or the sinking of cold oxygen-rich waters, the oxygen content at the bottom may fall to zero. Fortunately, such anoxic conditions are quite rare and occur only in isolated locations such as the Black Sea and the deep fjords of Norway. In the more common case, the oxygen content of the water drops sharply, goes through a minimum at 100–1000 m, and then begins to increase as the oxygen-rich polar waters flowing along the bottom are encountered.

The situation with carbon dioxide is somewhat the reverse. Carbon dioxide is produced by respiration of organisms and by bacterial decomposition of organic matter at lower levels. It is used up in photosynthetic activity only in the surface layers. One might expect that the concentration of dissolved carbon dioxide would increase with depth because of decay and decreasing temperatures. Because the chemistry of carbon dioxide in water is subject to so many factors, such as hydrogen ion concentration and salinity, the expected variation with depth can be used only as a rough guide.

Phosphates and nitrates are two critical nutrient ions needed by marine plants. The concentration in sea water of these two ions is not large (Figure 11.3). They are incorporated into the bodies of growing microscopic marine plants (phytoplankton) in the upper layers, where photosynthesis occurs. Animals may eat some of the plants. When the plants or animals die, their remains sink to deeper water. On the way down and at the bottom, biological decomposition of the phosphorus and nitrogen-containing biological molecules takes place, and the organic nitrogen and phosphorus are regenerated and released as inorganic nitrate and phosphate ion. As a result, most of the phosphorus and nitrogen in sea water exists in deep water (below 1000 m) as PO_4^{3-} or HPO_4^{2-} and NO_3^-. These ions tend to accumulate near the bottom, until vertical mixing of the water mass brings them to the surface where absorption by organisms can begin again. The general downward movement of these nutrients from upper to deeper layers causes the concentration of NO_3^- and PO_4^{3-} to increase with depth, but usually with a slight maximum at about 1000 m (Figure 11.3). If there should be a rapid growth of plants at the surface, without subsequent vertical mixing, all of the nitrate and phosphate could conceivably be removed from the upper layers, thus rendering the sea relatively sterile at this level. There is the same tendency for all other biologically essential nutrient species derived from decay to accumulate in the bottom layers until they are brought to the surface by vertical mixing processes. Nutrients such as these also undergo a seasonal and daily variation in concentration.

Depth and light are two other related factors influencing concentration distribution. Photosynthesis requires sunlight. Sunlight sufficient for photo-

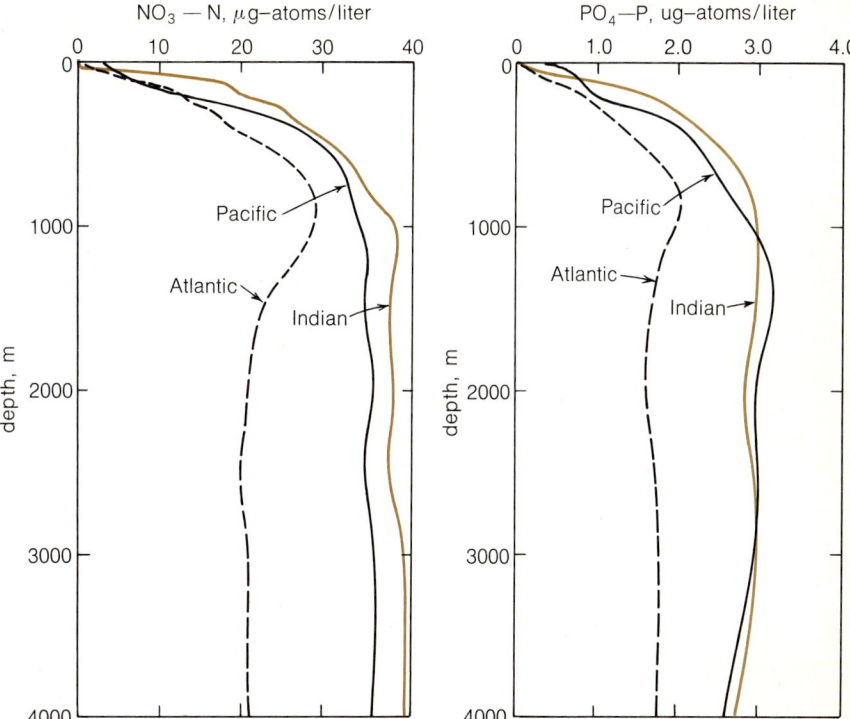

Figure 11.3. Vertical distribution of nitrate and phosphate in the open sea. Concentrations are given in microgram-atoms of N and P per liter (molarity of N and P $\times 10^6$). Maximums occur because of mixing processes and decay at intermediate depths. [From H. V. Sverdrup, Martin W. Johnson, and Richard H. Fleming, "The Oceans: Their Physics, Chemistry, and General Biology," Prentice-Hall, Englewood Cliffs, N.J., 1942 (copyright ⓒ renewed 1970), pp 241, 242 (Figures 48 and 50).]

synthesis rarely penetrates below about 100 m. The surface layer of sea water in which there is sufficient light for photosynthetic growth of marine plants is called the *euphotic zone*. The production of oxygen and the consumption of carbon dioxide and other nutrients by photosynthetic plants in the euphotic zone decrease with depth as the sun's light is absorbed by water. Because photosynthetic activity of autotrophic plants that can obtain their carbon from carbon dioxide predominates over all other biological activity in the surface layers of the sea, the environmental factors influencing photosynthesis generally determine the amounts of the important nutrients in the euphotic zone. Other factors related to the ones just mentioned are the seasonal and diurnal variations of light and temperature.

11.9 LIMITING FACTORS FOR MARINE LIFE

One of the fundamental concepts of ecology is that of *limiting factors*. This concept is really a combination of two more specific ideas, the *law of the minimum* and the *law of tolerance*.

The law of the minimum was first formulated in 1840 by the foremost organic chemist of the day, Justus von Liebig (1803–1873), to account for the growth of agricultural crops. In its present form, the law states that the growth of an organism is limited by the amount of whichever essential nutrient is present in smallest quantity relative to the needs of the organism. For example, a marine organism requires some minimum amount of carbon dioxide, nitrate ion, phosphate ion, several metal ions, and other nutrients for its growth. If all the required nutrients except one, phosphate for example, are present in amounts above the minimum required by that organism, the growth and reproduction of the organism is determined by the phosphate, the substance present in minimal quantity. The limiting nutrient is the one that runs out first. It makes no difference if all other nutrients are plentiful. The absence or below-minimum concentration of only one will prevent the growth of the organism. If this one nutrient is present in amounts just above the minimum needed for growth, the organism will grow until the amount of that nutrient drops below the minimum. The factors that limit the growth of an organism often include other environmental factors besides chemical materials. A minimum amount of light, for instance, is needed for photosynthesis even if all nutrients are available.

From a purely chemical standpoint, the law of the minimum as applied to nutrients is a reflection of the stoichiometric or weight relationships of chemical reactions. In the reaction $2H_2 + O_2 \longrightarrow 2H_2O$, the reactants combine in a mole (or atom) ratio of 2 hydrogen to 1 oxygen (or in a weight ratio of 1 to 8). The quantity of water formed is determined by that reactant present in least amount relative to the required ratio. If the mole ratio of hydrogen to oxygen is less than 2, hydrogen is present in least amount and is the limiting factor for the production of water by this reaction. Water is formed until the hydrogen is used up and an excess of oxygen remains.

The atomic composition of phytoplankton may be represented by the formula $(CH_2O)_{106}(NH_3)_{16}H_3PO_4$, and the synthesis of phytoplankton protoplasm by the equation

$$106CO_2 + 16NO_3^- + HPO_4^{2-} + 122H_2O$$
$$+ \text{ trace elements } + \text{ energy } (h\nu) \longrightarrow$$
$$(CH_2O)_{106}(NH_3)_{16}H_3PO_4 + 138O_2$$

This equation shows that carbon, nitrogen, and oxygen react in the proportions $106:16:1$ by atoms or $41:7.2:1$ by weight. Experiment has shown that the decrease in C, N, and P concentration in sea water during phytoplankton growth is in the atomic ratio of $105:15:1$. Likewise, the decomposition of phytoplankton increases C, N, and P in the same atomic ratio of $105:15:1$. These results confirm that the variation in concentration of C, N, and P in sea water is almost entirely the result of the synthesis and decomposition of organic matter. Analysis of average sea water shows that C (as CO_3^{2-}), N, and P are available in the atom ratios of $1000:15:1$, so that carbon is in great excess and never becomes limiting. Nitrogen and phosphorus are present in the ratio $15:1$. This ratio of $15:1$ is the same as the experimentally determined ratio of the simultaneous uptake of N and P in photosynthesis. Because both N and P lie on the threshold of the minimum amounts required for growth, either one is a candidate for the limiting factor, depending on local conditions.

The law of tolerance recognizes that too much of a good thing might be just as bad as too little. A law of the maximum may apply to certain environmental factors; there is a range of tolerance within which an organism prospers, but at either end of this range, the organism experiences serious difficulties. Our previous discussion of salinity and osmotic regulation indicates that there is a narrow range of salinity values that some organisms can tolerate. Some organisms can tolerate a wide range of temperatures, others can tolerate only a narrow range.

By combining the ideas expressed by these two laws, we can arrive at the concept of limiting factors by stating that any one of several factors contributes to the absence or failure of an organism if that factor is outside the organism's limits of tolerance. Any factor that slows down the growth or distribution of an organism is a limiting factor. In a more general sense, any factor that upsets the potential balance in an ecosystem is a limiting factor. The interaction among environmental factors is often complex, and it is not always easy to pin down a single limiting factor, although there must always exist one limiting factor for a given organism at any one time and place. An oversupply of one nutrient in an ecosystem can lead to a deficiency of another nutrient. In general, however, the deficient nutrient is usually called the limiting factor in a situation such as this.

Limiting factors can be physical and chemical in nature (temperature, light, nutrient concentration) or biological (predator–prey relations, population density, competition). We consider only some of the physical-chemical factors here. What the probable limiting factors will be in a given case depends on the given ecosystem. In a terrestrial ecosystem (a forest, for example), water and temperature are major limiting factors. In the sea, water is abundant so it is not going to be a limiting factor. Marine plants are greatly affected by the light available, but in the euphotic zone, mineral nutrients such as nitrate, phosphate, and even trace elements such as silicon and iron are usually limiting factors. Some of the limiting factors for marine life are now examined in more detail.

Nutrients As Limiting Factors

As the previous discussion showed, nitrogen and phosphorus are the most likely limiting factors in the surface waters of the open sea because of their generally low concentration in these waters and because the ratio of their concentrations approaches that required for their simultaneous absorption in the synthesis of phytoplankton. Of the two, phosphorus is more often the limiting nutrient in the open sea because of its lower concentration. The low concentration of phosphorus is believed to be a result of the low solubility of calcium phosphate $Ca_3(PO_4)_2$ in the deeper waters of the seas.

Although the amounts of N and P are greater, and may not be limiting in deep water, their low concentration in most places in the upper layers of the open sea puts severe limits on the growth of phytoplankton. Only in a few places in the world's seas is there sufficient mixing of the surface waters with nutrient-rich deeper water so that phytoplankton growth does not strip the upper layers of nutrients. In all other areas, plant growth tends to exhaust the upper layers of nitrate or phosphate. The depletion of the upper layers in turn limits all other marine organisms, because the phytoplankton are the primary food producers of the marine ecosystem.

Nutrient-rich deeper waters are brought to the surface to replenish the surface layers by processes involving currents in the atmosphere and the ocean and the rotation of the earth. The simplest vertical mixing or overturn occurs in colder climates when the colder, denser, oxygen-rich surface waters sink to a lower level and displace underlying nutrient-rich but oxygen-poor waters upward. Another important mixing process is called *upwelling*. Upwelling occurs when warmer, lighter coastal water is blown away from the coasts by winds. Colder, denser, nutrient-rich water from below rises to replace the surface water. Upwelling is responsible for the high productivity of the seas off the coast of Peru. The sea life produced in these waters supports a large bird population, which nests on islands where guano is deposited. The guano is rich in nitrate and phosphate and has been mined extensively for fertilizer. The return of phosphate to the land by birds is one of the few ways phosphorus, which has no natural volatile compounds, can escape from its cycle within the sea and return to land. Areas of natural upwelling comprise only 0.1 percent of the area of the oceans, but they supply 50 percent of the fish catch. By contrast, the biological desert that is 90 percent of the ocean gives only 1 percent of the fish catch.

Surface layers of tropical waters in the open seas are generally nutrient-poor and cannot support much life. Little vertical mixing can take place because the warm, less dense, surface waters cannot sink into and displace the underlying denser water upward. As a result, the surface is low in nutrients and life. Tropical waters have the blue color of biological deserts, while productive or fertile waters have a green-brown coloration caused by yellowish metabolic products of phytoplankton in the waters. More than 90 percent of the basic organic (plant) material that feeds marine animals is made by plant photosynthesis in the euphotic zone. Phytoplankton perform about 70 percent of all the photosynthesis carried out on earth. Phytoplankton are small and have a short life span of days or hours. In a given period of time, the total production of living organic matter by marine plants in the most fertile but restricted parts of the ocean is higher than that of fertile land areas, but at any one instant the amount of marine plants is very small. (Ecologists would say that the productivity is high but the standing crop biomass is very low.) The low productivity of sea water is a direct result of the low concentration of nutrients in the surface layers and of the fact that the primary food production by the plants is limited to a thin surface layer rarely deeper than 100 m. The plants are present in such low concentration that man would have to spend considerable energy to gather them for food.

Because of the small quantity of producer organisms present at any one time, the animals that live on the producers compose only a small part of the living matter in the seas. These animals live almost entirely in the upper 1500 m of the oceans and are usually concentrated near the coasts, where upwelling or land drainage provides nutrients for growth of phytoplankton. The living substance of marine animals is limited by the amount of energy they can obtain by eating the primary food supply of the plants. As will be discussed in detail in Section 20.18, only 10–15 percent of the organic substance of a producer is transferred to a consumer that eats the producer. And only 10–15 percent of the substance of the primary consumer is passed on to a secondary consumer in a food chain. Therefore, the total weight of fish and other consumers that eat the phytoplankton must be only about 10 percent of the total weight of the phytoplankton available as food, and this total weight is small. Because of the restrictions due to the low nutrient concentrations and to the low transfer of organic substance at each step in a food chain, it appears that man will not be able to obtain any more than a small fraction of his food needs by hunting sea life. The often-voiced hopes of mankind of feeding his ever-increasing numbers with food from the seas are misplaced.

These pessimistic remarks about obtaining more food from the sea apply to our present practice of hunting and gathering sea food from its natural habitat, such as is done in present methods of fishing. Much more food can be produced if special nutrient-rich areas are established for the controlled growth of marine organisms. For the most part, sea farming would consist of artificial fertilization of sea water in restricted environments such as lagoons and ponds. As an example, cold nutrient-rich water from the depths could be pumped to the surface of the oceans and used to fill ponds in which marine plants and animals are grown. Before entering the ponds, the cool water might be used for air conditioning or as coolant for electric power plants. Controlled dumping of sewage might be used to convert sterile parts of the sea to the production of food. But the large-scale fertilization of the seas in an attempt to raise more and more food could very well lead to disastrous consequences because of the interrelations among the sea, the atmosphere, and climate.

Dissolved oxygen is limiting to marine animals in certain cases, especially in the depths of the oceans where it is consumed by bacterial decomposition processes. The limiting effect of oxygen and its relation to other environmental factors in aquatic ecosystems is discussed in Chapter 12.

Light As a Limiting Factor

Sunlight is a limiting factor for photosynthetic growth of phytoplankton and is an indirect limiting factor for the consumers that eat these plants. The limiting nature of light is connected with its diminishing intensity with depth and turbidity attributable to suspended matter. Profuse growths of plants in the surface can cut off light to underlying layers and indirectly limit their own growth. This latter effect would constitute a biological limiting factor.

Temperature As a Limiting Factor

Temperature influences the amount of dissolved gas in water, the rate of growth and reproduction of marine organisms, and the rate of sinking of plants suspended in the water. Temperature even influences the size of animals in the sea. The temperature changes at any one place in the ocean are small compared to temperature variations on land. Marine organisms vary in their tolerance to temperature changes, and temperature will be limiting to those with narrow ranges of tolerance. The above limiting factors will be encountered again in Chapter 12 when we consider the effects of water pollution.

11.10 pH REGULATION IN THE SEA

The surface waters of the seas are slightly alkaline, with a pH of 8.1 ± 0.2. Local pH values in the sea can vary from slightly acid (~ 6.9) to values somewhat greater than 8.3. Prevailing opinion is that the pH of the oceans has been essentially constant over geological time, as it must have been to provide a constant environment for the evolution and development of life in the seas. Sea water contains an abundant concentration of hydrogen carbonate and carbonate ions. At the pH of the sea, HCO_3^- is the dominant species, comprising about 90 percent of the total dissolved carbon (Figure 11.4). It is natural to assume that these anions of weak acids, together with carbonic acid derived from the solution of atmospheric carbon dioxide in sea water, make up the components of the buffer system of the ocean. The pertinent equilibria are

$$CO_2(g) \rightleftharpoons CO_2(aq)$$
$$\left. \begin{array}{c} CO_2(aq) + H_2O \rightleftharpoons H_2CO_3 \\ H_2CO_3 \rightleftharpoons H^+ + HCO_3^- \end{array} \right\} \quad CO_2(aq) + H_2O \rightleftharpoons$$
$$HCO_3^- \rightleftharpoons H^+ + CO_3^{2-} \qquad\qquad\qquad H^+ + HCO_3^-$$

In addition, the solubility equilibrium for calcium carbonate is sometimes included:

$$CaCO_3(s) \rightleftharpoons Ca^{2+} + CO_3^{2-}$$

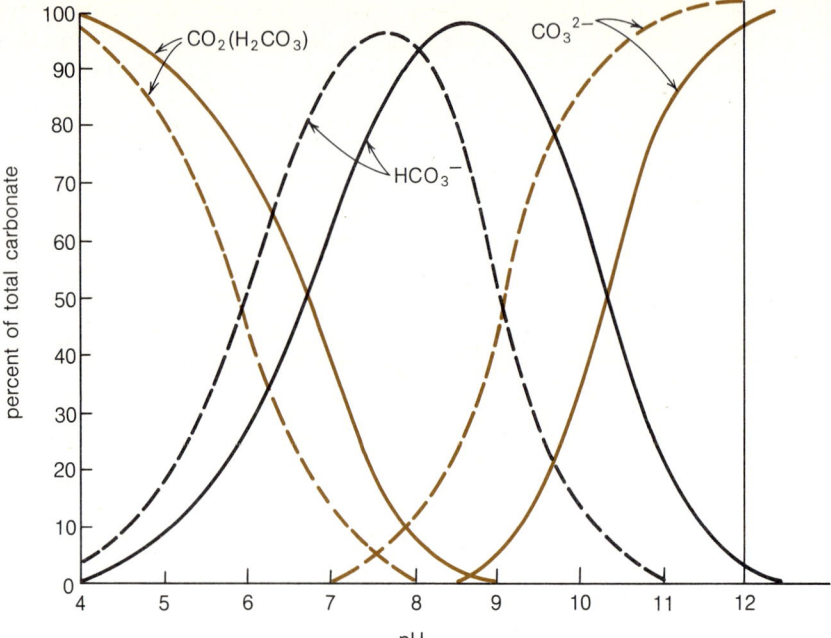

Figure 11.4. Relative abundances of carbonate species in distilled water (solid lines) and sea water (dashed lines) as a function of pH. The curves for each species are shifted to the left in sea water (35‰S) because of a decrease in ionization constants with electrolyte concentrations.

The addition of an acid to sea water is countered by the reactions

$$H^+ + CO_3^{2-} \rightleftharpoons HCO_3^-$$

and

$$H^+ + HCO_3^- \rightleftharpoons H_2CO_3 \longrightarrow H_2O + CO_2$$

Excess H^+ also will be removed, if solid $CaCO_3$ is present, by the reaction

$$CaCO_3 + H^+ \longrightarrow Ca^{2+} + HCO_3^-$$

The addition of a base (OH^-) will be buffered by

$$OH^- + HCO_3^- \rightleftharpoons CO_3^{2-} + H_2O$$

and

$$OH^- + H_2CO_3 \rightleftharpoons HCO_3^- + H_2O$$

The increase in CO_3^{2-} may also cause the precipitation of $CaCO_3$ if the water is close to saturation.

The carbonate buffer system is controlled at one end by the partial pressure of CO_2 in the atmosphere and at the other end by solid $CaCO_3$, where present, in the bottom sediments. If CO_2 is added to the atmosphere, as is being done in ever-increasing amounts by the combustion of fossil fuels, a portion variously estimated at one-half to two-thirds will be absorbed by the ocean or incorporated into vegetation. Because the atmosphere and the upper layers of the ocean are in equilibrium with respect to CO_2, any increase in the equilibrium partial pressure of CO_2 in the atmosphere will require the solution of considerable amounts of CO_2 in the ocean. To double the partial pressure of CO_2 in the atmosphere would be easy if no CO_2 were absorbed by the sea. However, since CO_2 is absorbed by the large volume of the sea, the maintenance of twice the partial pressure of CO_2 *in equilibrium* with the sea would require the addition to the atmosphere of many more times the original CO_2 content of the atmosphere. If the oceanic carbonate system is in equilibrium with atmospheric carbon dioxide, an increase in the CO_2 content of the atmosphere will, according to Le Chatelier's principle, increase the hydrogen ion content of surface sea water. However, this same increase in H^+ will decrease the CO_3^{2-} concentration (ignoring dissolution of solid carbonates) by reversing the reaction

$$HCO_3^- \rightleftharpoons H^+ + CO_3^{2-}$$

thereby counteracting some of the original H^+ increase. The ocean should therefore be a major reservoir and sink of carbon dioxide produced by consumption of fossil fuels. The problem is that only the upper layers of the seas can come to equilibrium with added atmospheric CO_2 in relatively short times of the order of years. Complete equilibrium with and absorption by the entire ocean depends on the mixing of the upper and lower layers and contact with the carbonate sediments of the bottom, a process requiring on the order of 1000 yr. There will be a considerable time lag in absorption of increased atmospheric CO_2 because of the slow rates of attainment of complete equilibrium. Because of this slowness, the climatic effects of CO_2 may become apparent despite the existence of the ocean as a large sink for the added CO_2.

The carbonate buffer system of the oceans reacts comparatively rapidly to sudden additions of acids or bases. This is true in the upper layers of the ocean and if the additions of acid, for instance, from undersea volcanoes, are not continuous and large scale. The carbonate buffer system is now believed by some to lack the capacity to buffer the large additions of acids, in particular, those that must have occurred over long geological time periods. The long-term buffering of the ocean may be controlled by sediments on the ocean bottoms. Because of the long mixing times and the small area of contact between the sediments and water, this type of buffering action is very slow. Nevertheless, the buffer capacity of the sediment system, its ability to neutralize large quantities of acid, is very much larger than the carbonate buffer system. The sediments in question are compounds of aluminum, silicon, and oxygen called clays (see Section 14.8), which have the ability to adsorb cations such as Na^+ and K^+ on their surfaces. These cations are readily exchanged for hydrogen ions in sea water. The transition from one kind of clay mineral to another may also absorb hydrogen and other ions. The interaction of the deep sea water with clay sediments constitutes a much larger but slower acting buffer system, which may, in reality, be the controlling buffer of the oceans.

11.11 POLLUTION OF THE SEAS

The sea provides a relatively constant chemical and physical environment for life. Marine organisms do not have to put up with the large variations of temperature and amounts of nutrients that bother terrestrial life. This happy state of affairs is due partly to the physical properties of water itself and partly to chemical and biological processes that maintain a constant chemical environment despite the natural and continual downhill movement of minerals and nutrients from the land to the seas.

Mankind is contributing, consciously and unconsciously, to a disruption of some of the natural balancing mechanisms of the marine ecosystem by polluting the seas. He does this directly by dumping pollutants into the sea and indirectly by polluting the air and the soil with substances that eventually find their way to the sea. It is not the purpose here to outline the general causes, effects, and remedies of water pollution. This topic is taken up in Chapter 12. Rather, we give a brief rundown of some of the more publicized instances of ocean pollution.

When man consciously dumps pollutants in the ocean, he usually has at the back of his mind the belief that the seas are so large, so deep, and so well mixed that the waste eventually will be dissolved, diluted, dispersed, and decomposed—or at the very least, hidden from view. This belief assumes a knowledge of currents, marine chemistry, and other oceanographic subjects that would do a professional oceanographer proud. Man's belief in the ability of the seas to absorb all the wastes of the air, the land, and all life on earth is not justified.

The "out-of-sight, out of mind" philosophy is nowhere more evident than in the use of the seas as a dumping ground for some of the most toxic and dangerous materials known to man. In 1968, 48.2 million tons of waste were dumped at sea from 20 cities on the Pacific, Atlantic, and Gulf coasts of the United States. Of this tonnage, 80 percent was dredging spoils from harbor- and river-deepening operations, and some of these spoils consisted of large amounts of municipal and industrial wastes. Industrial wastes (chemicals, acids, solvents, oils, cleaners, etc.) comprised 10 percent of the dumped wastes, followed by sewage sludge at 9 percent. Construction and demolition debris, refuse and garbage, and outdated military explosives and chemical warfare agents accounted for the remaining 1 percent. Disposal of wastes at sea is expected to increase considerably as time goes on. In every instance of ocean dumping, little is known of long-term effects on the marine environment or on marine organisms. Only time will tell if we can get away with such treatment of the seas.

The pouring of domestic and industrial sewage into the seas or into our lakes and rivers, which empty into the sea, is an even more widespread and seemingly approved practice (pathogenic bacteria cannot live very long in sea water, but this was not known when the practice began). As far as the seas themselves are concerned, the diversion of the nutrients contained in domestic sewage to the sea contributes a major change in the biogeochemical cycle of some of the nutrient elements, phosphorus being one of them. Before the population density of man increased, untreated human wastes were returned directly to the land. This practice became a bit odorous when people became too numerous (and disease producing because of pathogenic organisms on food), so nearby streams or other bodies of water were then used to carry wastes away. At one time this disposal method worked, but now, with increased population and industry along the coasts and other waterways, it fails. There is simply too much waste for the natural dispersal and decomposition mechanisms of the sea to handle. And many of these wastes are synthetic materials that do not decompose.

In the sea many of these nutrients, derived from the land through a cycle involving the eating of crops, accumulate and are very slowly, if at all, returned to the land. The result on the land is an overall loss of nutrients in the soil and a decrease in fertility. In the sea, as the process goes on, an alteration

Figure 11.5. Ocean pollution caused by an oil well erupting 6 miles offshore in the Santa Barbara Channel. Before the leaks could be stopped, some 200,000 gallons of oil escaped. (Courtesy of Federal Water Pollution Control Administration, U. S. Department of the Interior.)

in the chemical balance may occur, with unknown but possibly serious consequences for marine life.

Some coastal cities are considering the compaction of garbage into dense blocks that will be dumped into the deep sea. But studies have shown that the rate of microbial decomposition of the organic waste is some 10–100 times slower in the deep sea than at comparable temperatures on land and probably even slower when compacted. This raises the question of what might happen if the blocks of garbage began to come apart on the bottom. (Most organic matter is degraded on the way down to the bottom, partially explaining the maximum in nutrient concentration at about 1000 m.) Even if decomposition does occur at the bottom, it might use up the oxygen there and harm or eliminate aerobic bottom-dwellers. Again we see that there are no easy answers.

Two examples of the possible effect of added nutrients in sewage or in runoff from fertilized lands are the destruction of the kelp forests off the coast of southern California and the appearance of the deadly red tides that periodically occur in warm coastal waters. In the first case, amino acids found in sewage effluent partly fed sea urchins, predators of the valuable kelp plant, so that the urchins stayed around and overgrazed the kelp. In the latter case, excess nutrients are believed to cause explosive growth of a microscopic sea animal that secretes a nerve toxin that paralyzes the breathing apparatus of fish.

Industrial sewage containing toxic heavy metal ions also enters the seas. As has been pointed out, many of these ions are concentrated by organisms, often with pathological effects to the organisms or to predators. Some of the metals represented here are copper, mercury, lead, nickel, and the rare earths. Many ions of heavy metals are removed from solution by adsorption on natural particulate matter so that, in some measure, the oceans cleanse themselves of these metals. Nevertheless, natural means are probably not capable of coping with the greater amounts added by industry.

Other pollutants are not placed in the seas or in waters leading to the seas but are carried to the oceans through the air. Among these are lead and persistent pesticides. Off the coast of southern California, the surface concentration of lead is almost six times that in deep water. The high surface concentration is just the reverse of the concentration gradient for other plant nutrient metals. The high surface concentration is due to lead derived from the use of leaded gasoline. The lead concentration in the sea has increased about tenfold in the 40 years since lead was introduced into gasolines. DDT and other persistent pesticides are also carried through the atmosphere and by surface runoff to the sea, where they are absorbed by marine organisms. Radioactive fallout from nuclear explosions is another well-known air pollutant that finds its way to the sea.

Pollution of the seas by oil is becoming a major problem. Most of this pollution happens accidently, as it did when the tanker *Torrey Canyon* broke up in the English Channel in 1967, spilling about 100,000 tons of crude oil, or when an offshore oil well in the Santa Barbara Channel in California spilled 1,000 tons of oil over several days in 1969 (Figure 11.5). Other events of oil pollution are intentional acts perpetrated by ships that empty or wash their oil tanks at sea. No method has yet been devised to clean up spilled oil covering the water. The use of detergents to emulsify and disperse oil causes more problems for surface plants and animals than it cures.

Other examples of ocean pollution could be given, but these should be sufficient to show that the seas have not been treated with respect by man despite his repeated pronouncements that he looks to the seas for part of his food needs, minerals, and even for living room. The sea has always affected man in many ways. We now know that man can influence the seas, and one of the ways he does this is to regard it as a gigantic waste basket. When nature uses it as its ultimate waste basket, natural processes keep it from overflowing. Man should realize that he does not have the same privileges that nature has. He should be especially careful of his treatment of coastal waters, particularly at river mouths and in tidal flat wetlands, which are the most fertile areas of the sea for marine life and birds.

The sea has been called the last wilderness on earth. Like most wilderness areas today, greater and greater demands are being placed on it. But time is short. Several noted marine scientists have estimated that at current pollution rates it may be only 25–50 yr before there will be no life in the world's oceans. But with the growing recognition that the sea is a complex of interacting systems of biological, chemical, and physical components, man may still have time to learn to use the oceans wisely.

11.12 SUMMARY   The ocean is a complex ecosystem, which interacts with other regions of the biosphere. In these interactions, the ocean regulates many of the environmental conditions, such as the climate and the composition of the atmosphere. The ocean is able to maintain a certain homeostatic effect on the biosphere partly because its own chemical composition is relatively constant.

Sea water is essentially a concentrated ($\sim 0.5M$) sodium chloride solution containing almost all the other elements. Those elements and their compounds present in high concentrations display a constancy of relative concentration in all seas because of their participation in the worldwide circulation and mixing of the waters. The concentration of any or all of the major constituents is related to the salinity of sea water. The salinity is a measure of the concentration of dissolved salts in grams per kilogram (‰). Average sea water has a salinity of 35‰. The minor constituent elements in sea water often exhibit variable concentrations resulting from their removal from solution by sedimentation processes, mostly involving biological activity.

The residence time of an element in the ocean is the average time that an element entering the seas remains in solution before it is removed by sedimentation. The concept of residence time assumes that the oceans are in a steady state wherein the rate of addition of an element is equal to its rate of removal by sedimentation. Elements exhibiting high reactivity with respect to sedimentation in the marine ecosystem have low residence times. If the residence time is less than the mixing time of the oceans ($\sim 10^3$ yr) the element will exhibit variable concentrations in the seas of the world. Major constituents have long residence times (greater than 10^3 yr); many minor constituents removed from solution for long periods will exhibit short residence times.

Osmosis occurs when water diffuses through a semipermeable membrane separating two solutions of differing concentration. Water flows from the less concentrated (in solute) to the more concentrated solution. The pressure that must be applied to a solution to just prevent the passage of water into the solution from pure water is the osmotic pressure of the solution. The osmotic pressure of sea water is determined by its salinity. When the salinity or osmotic pressure of the body fluids of marine organisms differs from that of the external medium, the ability of marine organisms to maintain this difference determines their tolerance to salinity and their habitat.

The solubility of a gas in sea water depends on the partial pressure of the gas, the temperature, the salinity of the water, and the nature of the gas. Henry's law relates solubility to partial pressure.

Dissolved nutrients such as nitrate, phosphate, oxygen, and carbon dioxide are required for marine life, particularly for the floating phytoplankton that are the first step in marine food chains. When the concentration of an essential nutrient approaches or falls below the minimum required for the growth and reproduction of an organism, that nutrient becomes the limiting factor for that organism. The productivity of the open sea is limited primarily by the low concentrations of phosphorus or nitrogen in the thin euphotic zone. For this reason, most of the ocean is a biological desert.

The carbonate buffer system of the seas effectively regulates the pH of the waters and is a major factor in the maintenance of the chemical integrity of the oceans. The buffer system also is believed to regulate the carbon dioxide content of the atmosphere.

Pollution of the seas threatens to upset the chemical and biological processes regulating the oceanic environment. In view of the part the seas play in biogeochemical cycles (which also involve atmospheric and terrestrial components), such breakdowns could have serious consequences for mankind.

Questions and Problems

1. What is the concentration of sodium, in grams per kilogram, in sea water with a salinity of 33‰? Use Table 10.1.

2. A sample of sea water has a concentration of $0.470M$ Na^+ and $0.054M$ Mg^{2+}. If another sample of sea water contains $0.500M$ Na^+, what is the Mg^{2+} concentration?

3. A 25.00-ml sample of sea water is titrated with $0.250M$ $AgNO_3$ solution, using K_2CrO_4 as an indicator, until the presence of excess Ag^+ is shown by the reddish color of Ag_2CrO_4 at the end point. The density of the sea water is 1.028 g/cm^3. Find the chlorinity and salinity of the sea water if 53.50 ml of $AgNO_3$ solution were used to complete the precipitation of silver halides.

4. Derive a relationship between salinity and chlorinity for a solution of NaCl, and no other salt, in water.

5. A recipe for artificial sea water is NaCl, 26.518 g/kg; $MgCl_2$, 2.447 g/kg; $MgSO_4$, 3.305 g/kg; $CaCl_2$, 1.141 g/kg; KCl, 0.725 g/kg; $NaHCO_3$, 0.202 g/kg; NaBr, 0.083 g/kg. All of these salts are dissolved in water to make a total weight of 1000 g. The density of the solution is 1.0276 g/cm^3.
   a. What are the chlorinity and salinity of the solution?
   b. What is the osmotic pressure of the solution at 20°C, assuming 100 percent dissociation of all salts?

6. The total volume of all oceans is 1372×10^6 km^3, and the average density of the water is 1.03×10^{12} kg/km^3. The earth's rivers add water to the oceans at a rate of 30×10^{15} kg of water per year. Using the values in Table 10.1, calculate the residence time of

a. K^+ (noncyclic)
b. Na^+ (53 percent cyclic)

7. Show that the partial pressure P_i of a gas in a gaseous mixture is equal to $P_i = X_i P_t$, where X_i is the mole fraction of the gas in the mixture and P_t is the total pressure of the mixture.

8. If the partial volume V_i of a gas in a mixture is the volume that it would occupy if it were confined alone at the temperature T and total pressure P_t of the mixture, show that its partial pressure is $P_i = V_i P_t / V_t$.

9. If the partial pressures of N_2, O_2, and CO_2 in 1 liter of dry air at 0°C and 1 atm pressure are 594 mm, 159 mm, and 0.25 mm, respectively, what will be their partial pressures if the mixture is compressed to 0.5 liter at 0°C?

10. A gaseous mixture consisting of 0.010 mol of helium, 0.050 mol of hydrogen, and 0.02 mol of oxygen is enclosed in a 1-liter container. The hydrogen and oxygen are made to react to form water, after which the temperature of the mixture is 25°C. What is the final pressure of the mixture in the container after reaction? The vapor pressure of water at 25°C is 23.5 mm.

11. It was once proposed to extract oxygen from air by repeatedly dissolving air in water, heating the water, and expelling, collecting, and redissolving the air in cool water. Show how successive expulsion and dissolution of air might enrich the oxygen content by comparing the volume percentages of nitrogen, oxygen, and carbon dioxide in dry air at 25°C with the composition of air dissolved in fresh water in equilibrium with the atmosphere at 1 atm pressure. The Henry's law constants for the gases in liters (STP) of gas dissolved by a liter of water

at 25°C when the gas pressure is 1 atm are $k_{N_2} = 0.0143$ liter/liter atm, $k_{O_2} = 0.0283$ liter/liter atm, and $k_{CO_2} = 0.759$ liter/liter atm.

12. The concentration of phosphate, nitrate, and other nutrients in surface sea water is inversely proportional to the dissolved oxygen content of the water (except in highly anoxic water). What could be the cause of this?

13. On the assumption that carbonate sediments are involved in a carbonate buffer system, what would be the effect on carbonate solubility if the partial pressure of atmospheric CO_2 increases? What will be the overall effect on the pH of the solution? Explain your answers.

14. The dissolved carbon dioxide content of deep ocean water corresponds to higher partial pressures of carbon dioxide than warmer surface water. Surface water appears to be supersaturated with calcium carbonate, but the deep water is not saturated. The heat of solution of calcium carbonate is about -2.9 kcal/mol. What are some of the factors that might contribute to the increased solubility of $CaCO_3$ in deep water?

15. What is the pH at 25°C of water saturated with CO_2 at a pressure of 1 atm? The Henry's law constant for CO_2 is 0.759 liter (STP) per liter of water at 25°C.

16. The capacity of a natural water to neutralize acids is an important consideration, because a decrease in pH may disturb life processes in the water. Acids (H^+) are neutralized by anions of weak acids in the water. The most important of these anions are HCO_3^-, CO_3^{2-}, and OH^-, although other bases such as HBO_3^-, NH_3, and PO_4^{3-} may be present. The neutralizing power of a water is given by its *alkalinity*, which is the number of millimoles (moles $\times 10^{-3}$) of hydrogen ion required to react with all bases in a liter of water.

For the important bases, the reactions are

$$H^+ + OH^- \rightleftharpoons H_2O$$
$$H^+ + CO_3^{2-} \rightleftharpoons HCO_3^-$$
$$H^+ + HCO_3^- \rightleftharpoons H_2CO_3 \longrightarrow$$
$$CO_2 \text{ (1 atm)} + H_2O$$

The total alkalinity, in millimoles per liter, is given by

$$\text{alkalinity} \times 10^{-3} = [HCO_3^-]$$
$$+ 2[CO_3^{2-}] + [OH^-] - [H^+]$$

Because $[H^+] = [OH^-]$ in pure water, pure water has an alkalinity of zero. The alkalinity of a water sample may be determined by titration with a strong acid. Two successive end points may be observed, one with phenolphthalein at pH = 8.3 and a later one with methyl orange at pH = 4.5. For a water containing only HCO_3^- and CO_3^{2-} and titrated with strong acid,

a. Explain the two equivalence points in the titration.

b. What concentrations may be obtained from the volume of acid required to reach the phenolphthalein end point? From the total volume of acid needed to reach the methyl orange end point?

c. (optional) If 100 ml of a natural water require 2.0 ml of 0.0200M HCl to reach the phenolphthalein end point and 41.0 ml (total) for the methyl orange end point, what are the concentrations of HCO_3^- and CO_3^{2-} in the water, and what is its alkalinity?

17. Sea water is slightly basic (pH = 8.1). The average content of dissolved iron in river water is about 1 ppm, but in sea water it is only about 0.008 ppm or 8 ppb. What happens to iron- $(Fe^{3+}-)$ bearing river water when it enters the sea?

Answers

1. 10.14 g/kg
2. 0.057M
3. Cl = 18.4‰
5. 19.05‰Cl, 34.42‰
6. a. K^+; 8×10^6 yr
   b. Na^+; 170×10^6 yr
9. N_2, 1188 mm; O_2, 318 mm; CO_2, 0.50 mm
10. 0.52 atm

11. N_2: 65 percent in water, 78 percent in air
    O_2: 34 percent in water, 21 percent in air
    CO_2: 1.4 percent in water; 0.03 percent in air
15. pH = 3.9 for 0.034M H_2CO_3
16. c. $[CO_3^{2-}] = 0.40$ mmol/liter
    $[HCO_3^-] = 7.4$ mmol/liter
    alkalinity = 8.2 mmol

References

Chave, Keith E., "Chemical Reactions and the Composition of Sea Water," *Journal of Chemical Education*, **48,** 148–151 (1971).

Coker, Robert E., "This Great and Wide Sea," Harper & Row, New York, 1954, Chapters 6, 7.

Coker, Robert E., "Streams, Lakes, Ponds," Harper & Row, New York, 1954, Part I, especially Chapter 6 ("The Basic Nutrients in Water").
> Both of the books by Coker are excellent introductions to the ecology of aquatic systems.

Harvey, H. W., "The Chemistry and Fertility of Sea Waters," Cambridge University Press, London, 1960.

Hedgepeth, Joel, "The Oceans: World Sump," *Environment*, **12,** 40 (April 3, 1970).

Hood, Donald W., Ed., "Impingement of Man on the Oceans," Wiley, New York, 1971.

MacIntyre, Ferren, "Why the Sea Is Salt," *Scientific American*, **223,** 104–115 (November 1970).
> An excellent article illustrating the complexity of ocean chemistry and the interrelationships to life, the atmosphere, and the lithosphere. Good discussion of the buffer systems in the ocean. Many of the topics discussed apply also to Chapters 5, 7, 10, 12, and 14.

Martin, Dean F., "Marine Chemistry," Vol. 2, Marcel Dekker, New York, 1970.

Marx, Wesley, "The Frail Ocean," Ballantine Books, New York, 1967.

Mason, Brian, "Principles of Geochemistry," 3rd ed., Wiley, New York, 1966, Chapter 7.

Pinchot, Gifford B., "Marine Farming," *Scientific American*, **223,** 15–21 (December 1970).

Redfield, Alfred C., "The Biological Control of Chemical Factors in the Environment," *American Scientist*, **46,** 205–221 (September 1958).
> A classic paper presenting evidence to show that the nitrate content of sea water and the oxygen content of the atmosphere are determined by the organic activity in the biogeochemical cycles of phosphorus, nitrogen, and oxygen in the oceans. The overall controlling factor is the solubility of phosphates in sea water.

Scientific American, **221** (September 1969).
> This entire issue is devoted to the ocean.

Sverdrup, H. V., Johnson, Martin W., and Fleming, Richard H., "The Oceans. Their Physics, Chemistry, and General Biology," Prentice-Hall, Englewood Cliffs, N. J., 1942. Chapters III, VI–VIII, XVI, XVII.
> The standard, and most comprehensive, book on this broad subject.

Turekian, Karl K., "Oceans," Prentice-Hall, Englewood Cliffs, N. J., 1968, Chapter 6.

Water Quality and Water Resources

What chemistry!
That the winds are really not infectious,
That this is no cheat, this transparent green-wash
    of the sea, which is so amorous after me,
That it is safe to allow it to lick my naked body all
    over with its tongues,
That it will not endanger me with the fevers that have
    deposited themselves in it,
That all is clean, forever and forever.
That the cool drink from the well tastes so good, ...

Walt Whitman—*This Compost*

The Earth has been called the water planet because of its abundant supply of liquid water and because of the importance of water in all its forms in the physical and biological processes of the biosphere. The water of the primeval seas was the birthplace of all life on earth. The water present in these ancient oceans after life formed is the same water we have today—and the only water we shall ever have. Fortunately, water is an inexhaustible and an eminently recyclable natural resource. Water is the most abundant substance on earth, and there is enough to meet all conceivable needs of man and nature if used wisely.

Nevertheless, we all know that water shortages are becoming more and more common. How can this be? The answer to this paradox lies in the problems of water distribution or management and water quality. Water is not distributed uniformly over the face of the earth. Some regions have abundant water supplies from rainfall, groundwater, or natural bodies of water. Other areas are arid deserts. It would seem that problems of water distribution for human activities would not arise if these activities were confined to the regions of abundant water supply. Unfortunately, for a variety of reasons, this is not the case. We find large population centers or intensive agriculture or industry in areas of obviously minimal natural water supply. Sometimes this is the result of poor planning in the location and growth of these activities. But whatever the reason, usable water must be transported from distant watersheds and stored somewhere along the line. The dams, canals, and aqueducts that make this possible are engineering marvels, but they do have direct and indirect ecological consequences, some of which are chemical in nature.

The other problem, that of water quality, plagues communities that are literally surrounded by water. We know that sea water usually cannot be used for drinking, industry, or agriculture. But why do communities situated on large fresh-water lakes or rivers suffer water shortages? Most of the time, the answer is pollution of the natural water supply.

In this chapter, we look at the sources of water on earth, at the sources and effects of water pollution, and at some of the common methods used to treat waste water before it is released to natural bodies of water. Finally, we discuss some methods for purifying natural waters before they are used by man.

12.1 THE HYDROLOGIC CYCLE

Nature has its own water distribution and transportation system. The cyclic path of water through the biosphere is called the *hydrologic or water cycle*. It describes the path taken by water in moving from the seas to the atmosphere and to the land and, eventually, back to the seas. The hydrologic cycle differs somewhat from the biogeochemical cycles in that only one chemical compound is involved, and the changes undergone by this substance are principally the physical processes of evaporation from land and water surfaces, transpiration from plants, precipitation as rain or snow, and condensation. A pictorial representation of the hydrologic cycle is shown in Figure 12.1. Evaporation of large amounts of water from water surfaces, mainly the oceans, is the key process in the cycle. This water eventually returns to the earth from the atmosphere as rain, snow, or dew. Approximately three times as much of this precipitated water falls on the oceans as on land surfaces. Of that portion falling on land, part flows over the surface to rivers, lakes, and oceans; part returns almost immediately to the atmosphere by evaporation from land and water surfaces or by transpiration from vegetation; and part percolates into the ground. The water that enters the ground follows different paths. A portion is stored by capillary action in the upper layers of the soil, there to provide moisture for vegetative growth. Another portion seeps deep into the ground until it meets impervious rock layers or the groundwater table. The groundwater table is the upper surface of the water saturating the earth. Most of the groundwater is soon discharged at the earth's surface as springs, seeps, or into natural bodies of water, but some is stored underground for long periods.

The sun provides the energy for the water cycle. Within the cycle, evaporation and precipitation are the driving forces, and the energy absorbed and

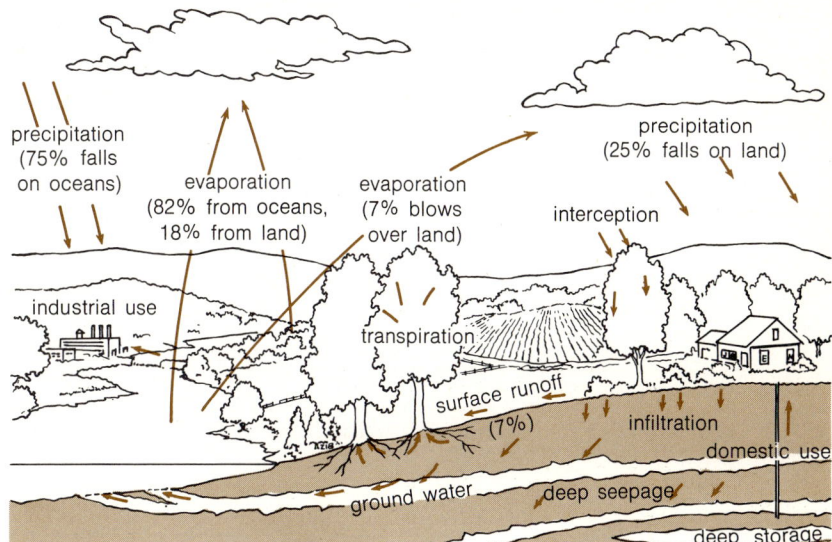

Figure 12.1. The hydrologic or water cycle. The figures are the percentages of the total mass of water circulated among the various reservoirs each year. The 25 percent falling on the land represents 99×10^3 km³/yr. The total amount of water in the oceans is 1350×10^6 km³ compared to 0.2×10^6 km³ in lakes and rivers and 8.4×10^6 km³ in soil and groundwater.

released by these processes causes the cycle to be a major factor in the earth's weather and climate. Water is transported around the earth as a vapor as part of the general atmospheric circulation. Within the hydrosphere, large quantities of water are moved by ocean currents and by rivers.

As a result of human activity, small amounts of water can be diverted from one pathway to another. Water can be diverted from surface runoff to irrigation (and transpiration) or to recharge water tables that may have been lowered by extensive pumping. Man can impede or accelerate certain parts of the cycle by, for example, storing water behind dams or by inducing precipitation by cloud seeding. However, man has not yet had any appreciable effect on the overall hydrologic cycle.

Water Quality

The reasons for man's diversion of water are many, but three of the most important are his living needs, his industrial needs, and his agricultural needs. When the water has been used for these purposes and is released by man to its natural cycle it has undergone a change, usually a deterioration in quality. This is another effect, aside from path diversion, that man has on the hydrologic cycle—he changes the quality of the water by using natural waters as a carrier and as a receptacle for wastes.

12.2 SEWAGE AND WASTES   Wastes can be roughly grouped into three large categories: domestic sewage, industrial wastes, and agricultural wastes. The term *sanitary sewage* includes both domestic sewage and industrial wastes when the latter are permitted to enter public sewerage systems. Domestic sewage is the waste water from normal living activities; waste from kitchens, toilets, and laundries. Domestic sewage is almost all (99.9 percent) water, but we ordinarily associate it primarily with human excrement and associated pathogenic organisms. Domestic sewage also includes food wastes, soaps and detergents, paper, and numerous other substances and objects.

Industrial waste, part of which may enter the municipal sewerage system, is even more complex in composition. Included are all manner of soluble and insoluble organic, biological, inorganic, and mineral substances. Many of these substances are so poisonous, infectious, corrosive, inflammable, or otherwise dangerous and objectionable that they should not be allowed to become part of sanitary sewage. Other industrial effluents are considered to be so innocuous that they can be discharged directly into natural bodies of water. Heated water, even if pure, is also an industrial pollutant.

Important constituents of agricultural wastes contributing to water pollution are fertilizers carried from the land as soluble salts in surface runoff, in ground seepage or adsorbed on eroded soil particles; pesticides; sediments from eroded soil; and animal wastes carried from feedlots by surface or groundwater. Agricultural wastes rarely find their way into public sewerage systems and, although some of it is treated at the site, most enters the environment untreated.

12.3 PROBLEMS FROM DISPOSAL OF UNTREATED WASTES

It is clear to us that something must be done about domestic sewage with its disease-producing organisms and its odors. But prior to the 1840's, this type of sewage was either thrown in the streets or spread on the land. With the growth of population centers these methods became untenable, and the removal of human wastes by the water-carriage system through sewers (open and closed) was introduced. This method, however, only carried the wastes away from the cities before dumping them, still untreated, on the land or into natural waters, and problems of water pollution continued. The dilution, dispersion, and self-purification abilities of natural bodies of water were exceeded as more and more wastes were dumped into them. And since water for domestic use must usually be obtained from the same natural sources that receive untreated sewage, part of the waters contaminated with disease-producing organisms and all the other characteristics of domestic waste found their way back to humans, along with predictable public health problems.

In addition to pollution of natural waters by pathogens, domestic sewage also deposits considerable amounts of oxidizable organic material into natural water courses. When the sewage load is small, organisms in the water can break down such compounds by oxidizing them with atmospheric oxygen dissolved in the water in reactions catalyzed by enzymes in their own bodies. In effect, the organisms are using the sewage as food. If too much sewage is added to the waters, this biological oxidation can use up all the dissolved oxygen in the water. The result will be that aquatic organisms, including fish, that require oxygen will die and be replaced by organisms needing little or no free oxygen for their life processes.

As industrial wastes began to be added to public sewers and to natural waters, the recovery processes of nature were taxed even more. Some of the industrial wastes were oxidizable by natural organisms in the water and contributed to deoxygenation of the water. Other ingredients were not removed at all by natural processes, and some of the wastes were toxic to the purification organisms and to other life. Natural processes had difficulty in coping with ever-increasing and ever more complex waste discharges as society developed. It became necessary to treat sewage, and installations designed for that purpose were put into operation about 1890.

Sanitary Sewage Parameters

In preparation for a discussion of some of the physiological and ecological effects of water-borne wastes, we shall give an account of common sanitary sewage treatment procedures used in municipal waste water disposal. Thus, you will have some idea of the successes and failings of currently used methods. Before doing this, though, we should first become familiar with the terms that describe the strength of sanitary sewage, the efficiency of its treatment, and the degree of deterioration of polluted natural waters.

12.4 SUSPENDED SOLIDS

The simplest criterion of water contamination is the amount of solid material carried by the water. Although the solid material could include both dissolved and undissolved solids, the latter is usually taken as a measure of sewage strength. Those undissolved solids that can be removed by filtration through ordinary filter paper are called *suspended solids* and usually include the coarser particles that would settle by gravity in calm water as well as those that remain in suspension. The lower size limit is about 1 μm (10^{-6} m). Below this size, the particles are called *colloidal* and will pass through the filter. The colloidal size range is from about 1 μm to 1 nm (10^{-9} m).

The suspended solid content of water is given as the weight of filterable solids in 1 liter of water. In a medium-strength municipal sewage, the suspended solid content is about 300 mg/liter (500 mg/liter are colloidal or dissolved). About two-thirds of the suspended solids are organic in nature, and the remainder is mineral (inorganic). Of the organic matter, approximately half is carbohydrate, 40 percent nitrogen-containing proteins, and 10 percent fats.

12.5 BIOCHEMICAL OXYGEN DEMAND (BOD)

The biological oxidation of the organic matter, in all forms, purifies the water and depletes molecularly dissolved oxygen in natural waters. The oxygen requirements for the degradation process form the basis for another very important criterion of water contamination, namely, the oxygen demand. We have already seen that almost all living organisms are dependent on oxygen, either free or combined, for their maintenance, growth, and reproduction. Those organisms that require free molecular oxygen are called *aerobic;* those that can get along without free oxygen are *anaerobic*. Some anaerobes are poisoned by free oxygen and require its complete absence.

Biological sewage treatment and the desired natural self-purification of streams, rivers, and lakes depend on the degradation of organic matter by aerobic bacteria in conjunction with other aerobic organisms. In the ideal case, organic matter is completely oxidized by these organisms to CO_2, H_2O, and to relatively innocuous materials such as nitrates, phosphates, and sulfates. However, if the water does not contain sufficient dissolved oxygen for the aerobic decay organisms, anaerobic organisms, which are always present, take over and produce objectionable products. The products of anaerobic decomposition processes of fermentation and putrefaction are disagreeable or toxic gases such as methane, CH_4, or hydrogen sulfide, H_2S. The sulfide generated by anaerobic decomposition reacts with metal ions in natural waters to form black precipitates, thus resulting in black water. Anaerobic decomposition is also much slower than the aerobic process. For these reasons, a satisfactory concentration of dissolved oxygen must be maintained in sewage treatment so that aerobic biological oxidation can proceed. It is also necessary to know how much the oxygen content of natural waters would be depleted by the oxidizable organic matter in the water because, as has been pointed out, the depletion of dissolved oxygen will lead to adverse consequences.

The potential oxygen requirement of sewage and of sewage polluted waters is measured by the *biochemical oxygen demand* (BOD). As the name implies, this is related to the amount of oxygen needed by aerobic organisms for the breakdown of organic matter in the water. The BOD is the most significant parameter for the determination of sewage strength and the efficiency of pollution control processes because it is a measure of the decomposable organic material in the water, a quantity that is itself not directly measurable. The BOD of a water sample is given by the weight of dissolved oxygen consumed by a liter sample when incubated in the dark for 5 days at 20°C. Although the bacteria in the water normally take at least 20 days to complete the oxidation of organic matter, 70–80 percent of the decomposable matter is oxidized in 5 days, and the shorter period has been chosen as a standard. In practice, a sample of water is diluted with water containing sufficient oxygen and nutrients to enable the various organisms to thrive for the test period. Sometimes a seed of bacteria is added to ensure the presence of sufficient organisms of the type present in the undiluted sample.

The numerical value of the BOD, usually in milligrams of oxygen consumed per liter of sample, is determined by measuring the dissolved oxygen content of the sample before and after the 5-day incubation period. The BOD is the difference between these dissolved oxygen values when appropriate corrections are made for sample dilution and any organic material that might have been added in the diluting water. The BOD is often reported in parts per million; that is, in mg O_2/1000 g, which is about equivalent to milligrams per liter for fresh water *only*.

The BOD test does not include oxygen consumed by nitrifying bacteria in reactions such as

$$2NH_3 + 3O_2 \longrightarrow 2NO_2^- + 2H^+ + 2H_2O$$

or

$$2NO_2^- + O_2 \longrightarrow 2NO_3^-$$

because these bacteria are usually present in such small numbers that they do not consume appreciable oxygen for 8–10 days under BOD test conditions. BOD values are reported as the 5-day BOD, sometimes written as BOD_5, or as the total (or ultimate) BOD. The latter is the BOD for the total oxidation of decomposable material (excluding oxidation of nitrogen compounds). Only the 5-day values are used here.

BOD values can range from 0 to 20,000 mg/liter. The BOD of average domestic raw sewage is 150–250 mg/liter. Wastes from food-processing plants such as canneries can go as high as 6000 mg/liter, while untreated effluent from wood pulp and paper mills may reach 15,000 mg/liter.

If the BOD of a natural body of water is above 5 mg/liter, the water is seriously polluted. The purification, or recovery from pollution, of natural waters depends on a number of factors. Fast moving, turbulent streams absorb atmospheric oxygen faster than sluggish streams or ponds. The capacity of turbulent streams to renew their oxygen supply as dissolved oxygen is withdrawn for bacterial decomposition is much greater than the aeration capacity of deeper or calmer waters. So, even if two bodies of water have the same BOD, the capacity of one for self-purification may be much larger than the other's. The BOD merely gives a measure of the pollution imposed on water at any given time when adequate oxygen is available. Care must be taken in the interpretation of the standardized BOD measurements, because the actual capacity of water to recover from pollution depends on the chemical makeup of the wastes, the oxygen balance situation (oxygen withdrawal versus oxygen supply from the atmosphere or photosynthetic plants), temperature, populations of microorganisms present in the natural waters, and many other factors.

12.6 CHEMICAL OXYGEN DEMAND (COD)

Another indicator of oxygen demand, the *chemical oxygen demand* (COD) is sometimes used. The COD is the total quantity of oxygen needed for the oxidation to CO_2 and H_2O of the organic matter in water (again excluding nitrogen oxidation). A strong chemical oxidizing agent such as dichromate ion in acid solution is generally used; a typical reaction, one for a sugar, is

$$C_6H_{12}O_6 + 4Cr_2O_7^{2-} + 32H^+ \longrightarrow 6CO_2 + 22H_2O + 8Cr^{3+}$$

Many organic compounds are not biologically degradable (biodegradable), and these substances are not measured in the BOD test but are oxidized in the COD test. But some organic compounds, such as benzene, are not affected in either test, so the COD still does not give a complete indication of organic material. The difference between COD and BOD can give a measure of the nonbiodegradable substances. The more rapid COD test often can be used to obtain BOD values if the tests have been accurately correlated.

None of these measurements have any bearing on the number of pathogenic bacteria or viruses in sewage or polluted water. Bacterial contamination is measured by the coliform bacteria count, a bacteriological test indicating the amount of human fecal matter recently entering the water at a nearby point. Coliform bacteria themselves are not pathogenic, but they serve as an important indicator organism because their origin in the human intestinal tract is the same as the pathogenic organisms of typhoid, cholera, dysentery, and a whole host of other microorganisms with unsavory reputations. Furthermore, because coliform bacteria do not survive long in natural waters, their presence indicates recent pollution.

The BOD test and other tests for determining the oxygen-demanding strength of domestic sewage cannot be used with confidence to measure the strength of industrial waste waters. Industrial wastes have an almost infinite variety of ingredients, many of which are not biodegradable but which have a high COD. For example, these tests will not reveal the presence of toxic heavy metals, or organic and inorganic complexes in industrial sewage (many of which are, but should not be, mixed with domestic sewage).

12.7 DISSOLVED OXYGEN (DO)

The actual amount of dissolved oxygen in a water sample must be found for a BOD determination. The *dissolved oxygen* (DO) in water is a crucial parameter in its own right, because it gives the capacity of a sample of water to assimilate an imposed pollution load, with or without further supply of oxygen to the water. The DO indicates which species of organisms will survive and prosper in an aquatic community. As a general rule, water containing less than 5 mg of oxygen per liter will not satisfactorily support a healthy community of organisms in warm water. The solubility of oxygen in fresh water under its normal atmospheric partial pressure is about 8.5 mg/liter at 25°C.

The ecological effects of a low DO are discussed in Section 12.11. It should be apparent that unpolluted water with a BOD of zero could have a very low dissolved oxygen content and be unsuitable for life other than anaerobic bacteria. It is, in fact, the low solubility of oxygen that limits the ability of natural waters to purify themselves and makes it necessary to provide treatment for wastes before they are allowed to enter natural waters.

The dissolved oxygen in water may be determined chemically by a test (the Winkler method) based on the oxidation of manganese(II) ion, Mn^{2+}, in alkaline solution by the dissolved oxygen according to the reaction

$$Mn^{2+} + 2OH^- + \tfrac{1}{2}O_2 \longrightarrow MnO_2 + H_2O$$

The resulting solution is then acidified and allowed to react with iodide ion, I^-, which is oxidized to elemental iodine, I_2, by the previously formed MnO_2:

$$MnO_2 + 2I^- + 4H^+ \longrightarrow Mn^{2+} + I_2 + 2H_2O$$

Thus, the amount of I_2 formed is related to the quantity of oxygen that has reacted by these two equations. To determine the amount of iodine, the solution is titrated with sodium thiosulfate, $Na_2S_2O_3$, solution until all the iodine has reacted according to the ionic equation

$$2S_2O_3^{2-} + I_2 \longrightarrow S_4O_6^{2-} + 2I^-$$

The end point of this reaction is shown by the disappearance of a color associated with the free iodine. If 200-ml samples of water are used and the sodium thiosulfate solution concentration is $0.025M$, the milliliters of thiosulfate solution required is equal to the quantity of dissolved oxygen in milligrams per liter. A much faster electrochemical method for determining dissolved oxygen is also available.

Sanitary Sewage Treatment

Treatment of sanitary sewage has two major objectives: (1) to remove oxygen-demanding organic waste from sewage and convert it to an unobjectionable form and (2) to remove infectious organisms. The relative importance of these two objectives varies from one sewage treatment plant to another. Probably very few waste disposal plants succeed completely in either objective.

12.8 PRIMARY TREATMENT AND SLUDGE DIGESTION

In a typical sewage disposal plant, the incoming raw sewage is first screened to remove larger debris; then the flow rate of the sewage is decreased to allow coarse particles such as sand and grit to settle out. Finally, the liquid effluent from these two processes is allowed to flow very slowly through large settling or sedimentation tanks for a retention time of anywhere from 30 min to 10 hr depending on factors such as the strength of the sewage. This sedimentation process removes about 60 percent of the suspended solids in the sewage. Suspended solids are principally small organic particles of carbohydrates, fats, and proteins. These substances, if allowed to enter natural waters, are the ones that use up oxygen in the natural biological oxidation processes. Suspended solids do not include the larger objects that were removed earlier. The screening and sedimentation procedure is called *primary treatment*. The liquid effluent from the sedimentation tank is sometimes released to natural watercourses as is, but usually the water is treated with chlorine gas, Cl_2, or calcium hypochlorite, $Ca(OCl)_2 \longrightarrow Ca^{2+} + 2OCl^-$. Both Cl_2 and OCl^- are potent oxidizing agents and disinfectants. Unfortunately, chlorination of the effluent does not kill all viruses, and those remaining will be discharged into waters that may be used again. Other disinfectants such as ozone are being studied.

The settled solids, called primary sludge, although mostly (93–95 percent) water, represent a considerable solid waste diposal problem. The solid content is further reduced by about 70 percent by subjecting the sludge to anaerobic bacterial decomposition. This biological decay, called *sludge digestion*, produces methane gas, CH_4, which is often collected and used as fuel to generate electricity for the sewage plant or to supply heat to the anaerobic sludge digesters, which work more rapidly at higher temperatures. The remaining water is more easily separated from digested sludge than from primary sludge.

The anaerobic decomposition of sugar, for instance, follows the overall equation

$$C_6H_{12}O_6 \longrightarrow 3CO_2 + 3CH_4$$

Similarly, more complex and insoluble carbohydrates, proteins, and fats in sludge are all first converted to organic acids (RCOOH) by bacteria and their enzymes. Then these acids are fermented by the anaerobic bacteria to form carbon dioxide and methane:

$$CH_3C\overset{\displaystyle O}{\underset{\displaystyle OH}{\diagdown}} \xrightarrow[\text{enzymes}]{\text{bacterial}} CO_2 + CH_4$$

The CO_2 can act, as NO_3^- did for denitrifying bacteria, as a source of oxygen (or as a hydrogen-atom acceptor) with resultant reduction of CO_2 to give more methane:

$$8H \text{ (from the organic compounds)} + CO_2 \longrightarrow CH_4 + 2H_2O$$

After digestion the separated liquid is mixed with effluent from the primary settling tanks, and the digested and dewatered sludge is sometimes dried and used as a fertilizer or soil conditioner. Usually, it is disposed of as landfill or dumped in the sea.

A septic tank is essentially a combined primary sedimentation–primary sludge digestion tank. The settled sludge undergoes anaerobic digestion, and the effluent liquid is distributed to underground disposal fields and allowed to percolate through the soil (Figure 12.2). Here aerobic soil bacteria may complete the degradation of organic matter. Even though some disease organisms are removed by sedimentation and digestion, a large number can enter the soil if the disposal field is not properly designed; or if soil conditions are not right, then contamination of groundwater can result. This can be hazardous if domestic water supply is obtained from wells. It is estimated (1970) that one-third of the population of the United States is serviced by septic tanks, while the rest use municipal sewer systems. The number using the more hazardous cesspools or privies is relatively negligible.

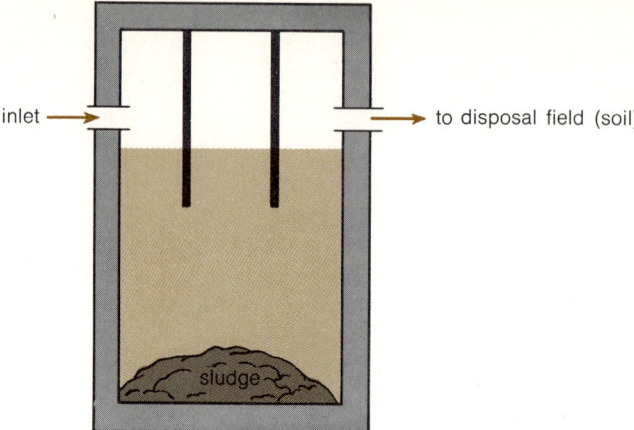

inlet →   → to disposal field (soil)

sludge

Figure 12.2. Septic tank.

12.9 SECONDARY TREATMENT   Primary treatment is considered to provide only minimal treatment, because it lowers the BOD of the raw sewage by only 30–35 percent and removes only 60 percent of the suspended solids. However, in all the communities with sewer systems in the United States in 1969, 10 percent of the waste water was completely untreated and 30 percent got only primary treatment. The other 60 percent received *secondary treatment* of the liquid effluent from the primary treatment. Secondary treatment is based on a continuous aerobic biological oxidation, that is, oxidative decomposition by microorganisms, principally bacteria. There are three common methods of secondary treatment, differing mainly in the method of aeration.

In the *trickling filter* method of secondary treatment, liquid effluent from the primary settling tanks trickles downward through a 5–10 ft deep bed of loosely packed rocks or plastic shapes about 3 in. in diameter (Figure 12.3). Adhering to the large surface area of this ballast is a large mass of microbial organisms (plus worms, fungus, larvae, protozoa) that biologically degrade the soluble organic, and oxygen-demanding, compounds in the effluent. The mass of microorganisms forms a slime on the filter, and this slime periodically sloughs off and is settled out in secondary sedimentation tanks called humus tanks. This secondary or humus sludge undergoes anaerobic decomposition in the same manner as primary sludge, and the liquid effluent from the sedimentation and digestion tanks is often further purified by filtration and chlorination before being released into streams or lakes.

In the trickling filter and in the other secondary treatment processes to be described, the degradation of organic matter by bacteria is catalyzed by enzymes secreted by the bacterial cells. On the average, the major atomic makeup of organic matter in the protoplasm of living things can be denoted as $(CH_2O)_{106}(NH_3)_{16}H_3PO_4$, so the overall equation for aerobic decay is

$$(CH_2O)_{106}(NH_3)_{16}H_3PO_4 + 106O_2 \longrightarrow$$
$$106CO_2 + 16NH_3 + 106H_2O + H_3PO_4 + energy$$

In the presence of nitrifying bacteria, the ammonia will, after a period of time, be further oxidized to nitrate:

$$NH_3 + 2O_2 \xrightarrow[\text{bacteria}]{\text{nitrifying}} NO_3^- + H^+ + H_2O$$

Figure 12.3. A trickling filter. Primary effluent is continuously applied to the bacteria-coated bed of stones by the slowly rotating arms. In the background is an oxidation pond.

Ammonia is the predominant form of nitrogen in secondary sewage effluents, but the nitrifying bacteria in natural waters can make NO_3^-, which is important as a plant nutrient, as is phosphate, PO_4^{3-}. Sulfur-containing substances, such as proteins, may be oxidized to sulfate, SO_4^{2-}. The oxidation reaction produces energy, which is used by the decay microorganisms for assimilation or incorporation of nutrients for respiration, growth, and reproduction of new organisms.

The *activated sludge* process of secondary treatment is slightly more popular than the trickling filter. In this process, the waste water from primary treatment is mixed and aerated with compressed air (or pure oxygen) in long channels or tanks (Figure 12.4). The sludge formed by the flocculent mass of aerobic bacteria settles out in sedimentation tanks, and the liquid effluent is discharged possibly, but not always, after chlorination. Most of the sludge, containing active bacteria, is returned to the aeration tank to innoculate incoming sewage, because any given volume of waste water can produce only about 5 percent of the sludge required for its treatment. The remainder of the sludge is treated as before.

Another secondary process, the *oxidation pond*, utilizes a symbiotic algae–bacteria relationship in shallow (less than 5 ft deep) outdoor settling

Figure 12.4. A portion of an activated sludge aeration tank. The tanks are about 15 ft deep. Air under pressure is forced into the bottom of the tank by a number of units, one of which is shown raised for cleaning.

ponds (Figure 12.3). The photosynthetic growth of the algae provides oxygen for the bacteria, and the bacteria break down the complex organic compounds in the sewage so the algae can assimilate them.

Secondary treatment methods can remove 90–95 percent of the BOD and up to 95 percent of the suspended solids in raw sanitary sewage. The activated sludge process can remove 90 percent of the bacteria in raw waste, and if the secondary effluent is chlorinated, 99 percent of the bacteria can be removed. As noted earlier, chlorination is not very effective against viruses, and their removal is still an unsolved problem.

Secondary sewage treatment is essentially a part of an existing natural cycle, whether it be of nitrogen, phosphorus, carbon, oxygen, or any other essential element. Secondary treatment uses the same biological mechanisms and organisms for decomposition as nature does. In secondary treatment, technology merely augments that part of the natural cycle connecting complex compounds with their simpler inorganic constituents. Augmentation by man is necessary to prevent stress caused by the uncommonly large demands that would be placed on the natural cycles of receiving waters if sewage were released untreated. In the absence of aeration of oxygen-demanding wastes in sewage plants, the oxygen supply of natural waters would soon be depleted, most of the essential organisms in the natural cycles would die, and the cycles would collapse. But, as we shall see in Section 12.12, our very success in helping nature handle the large load placed on the decomposition step of a cycle places a heavy burden on another step of the cycle, the assimilation or growth step, in which organisms absorb the inorganic nutrients generated in sewage plants. We have speeded up one part of a cycle without making any provision for a controlled increase in another part. Nature speeds up the other parts of the cycle, but the result is not a happy one so far as man or nature is concerned.

12.10 TERTIARY AND ADVANCED TREATMENT

The main limitation of the common secondary treatment processes is their inability to remove certain pollutants: inorganic compounds such as the plant nutrients NO_3^- and PO_4^{3-}, toxic heavy metals, synthetic organic compounds that are not biodegradable, certain radioactive materials, and heat. To remove these and other materials, the effluent from secondary treatment may be further processed in *tertiary treatment*. A more inclusive term, *advanced treatment*, applies to treatment that can do a better job than the usual primary and secondary treatment. Advanced process may even replace certain steps or sequences of primary and secondary treatment.

The separation of solids and liquids in primary treatment is essentially a physical process relying on gravity. Secondary treatment is a biological process. Tertiary treatment, in all its many forms, is essentially chemical in nature. Among the chemical processes used are precipitation of insoluble compounds, adsorption, complexing and chelation of metals, chemical coagulation, chemical oxidation (as opposed to biological oxidation), ion exchange, and several electrical methods.

With suitable tertiary treatment, a few sewage treatment plants in the United States take in domestic sewage and produce an effluent that is safe to drink. This effluent has been used to make lakes for recreational use, including swimming (Figure 12.5). One such plant is at Lake Tahoe, California. Here the secondary effluent is treated with CaO to remove solids and phosphates. Then NH_3 is removed with forced air followed by successive

Figure 12.5. Water activities in a lake formed from effluents from a tertiary treatment plant. (Department of Water Resources, State of California.)

filtration through coarse coal, medium sand, fine garnet, and activated charcoal. Finally, the tertiary effluent is chlorinated.

Tertiary treatment methods, as well as modifications to primary and secondary processes, are particularly important for the treatment of industrial wastes, which often contain chemical compounds not affected by ordinary primary and secondary treatment. Because some of the toxic chemicals in industrial waste impair the biological processes of secondary treatment, these substances must be removed before the industrial waste is combined with municipal sewage for further treatment. Alternatively, the waste may be subjected to complete treatment at the industrial site before discharge to receiving waters. Industry can alleviate its waste treatment problems by treatment and reuse of water within the plant or by developing new or modified manufacturing procedures to reduce the amount or change the character of the wastes.

Most tertiary treatment processes are tailored to the removal of a specific contaminant, and they illustrate a wealth of basic chemical theory and practice. Many of the methods used in tertiary treatment are also employed in the treatment of water supplies (Section 12.16) and in desalination of saline waters (Sections 12.26 and 12.27). Also, as we discuss some of the effects of water pollution, we shall note tertiary methods used to remove selected contaminants. Although a considerable amount of technological and scientific knowledge, effort, and expense has been applied to the protection of our natural waters, much remains to be done to cope with the magnitude of the waste water disposal problems facing a highly developed society. Apart from tertiary and advanced treatment methods, present-day sewage treatment procedures have not changed appreciably in 60–70 yr.

With these methods, we are today faced with at least two major problems. One is how to cope with the ever-increasing volume of domestic sewage. The other problem concerns the growing variety of industrial and agricultural wastes. The industrial growth of the past quarter century has generated more than 12,000 toxic chemicals, with 500 more new compounds, some of which are toxic, being developed each year. These substances and the wastes from their manufacture are showing up in the nation's water at an increasing rate. Adequate treatment methods may or may not exist for these substances. Even with the best efforts at treatment (or all too often, with the best efforts to avoid the expense of treatment), these compounds find their way into the public water and food supplies and, from there, into newspaper headlines.

Related to both these problems is the inescapable fact that the planet has a finite and connected volume of natural water serving as the ultimate sink for all the untreated wastes of our civilization. Fresh water, in particular, is our principal source of usable water. We cannot afford continually to contaminate this supply, even without the expected increase in demand for fresh water (from 410 billion gal/day in 1970 to 1300 billion gal/day in 2020).

Some Effects of Water Pollution

The effects of water pollution are often grouped under such headings as chemical (oxygen demand, toxic effects), physical (temperature changes, sediments, surface films), and biological (disease, disturbance of species structure of aquatic community, unchecked plant growth). Aesthetic, recreational, and economic factors are included in most of these. We cannot attempt a review of all effects, and our discussion is limited to effects resulting from oxygen depletion, eutrophication, thermal pollution, and toxic pollutants.

12.11 OXYGEN DEPLETION EFFECTS

The major cause of dissolved oxygen depletion in natural water is the aerobic biological decomposition of degradable organic materials. Some inorganic (mineral) substances, if present, also exhibit a high oxygen demand. Among these are sulfite ion, SO_3^{2-}, sulfide ion, S^{2-}, and iron(II) ion, Fe^{2+}, which take part in oxygen-consuming reactions such as

$$2SO_3^{2-} + O_2 \longrightarrow 2SO_4^{2-}$$
$$S^{2-} + 2O_2 \longrightarrow SO_4^{2-}$$
$$4Fe^{2+} + O_2 + 2H_2O \longrightarrow 4Fe^{3+} + 4OH^-$$

We have already discussed aerobic decomposition and its depletion of dissolved oxygen. We should reemphasize that biological decomposition continually takes place in even the purest natural waters. It is only when the supply of dissolved oxygen cannot keep up with the demand that the dissolved oxygen content reaches adverse levels. The BOD of secondary sewage treatment effluent is usually low enough to avoid exceeding the renewal capacity of natural receiving waters. When untreated sanitary sewage is discharged directly or along with storm (surface runoff) sewage in cities having combined sewers, then the fate of the receiving waters is in doubt. Another major source of BOD in natural waters is organic waste in the runoff from feedlots for cattle and other farm animals.

A dissolved oxygen concentration of about 5 mg/liter is generally considered necessary for the support of a healthy aquatic community. By "healthy" is meant a community of aquatic animals made up of a large diversity of species, each species having a moderate population. One of the first effects of a decrease in dissolved oxygen is the disappearance of those organisms having little tolerance for the lower oxygen levels. For instance, trout require at least 7 mg/liter DO. Bass can get along with 3 mg/liter, but carp and catfish can get along in polluted or sluggish waters having a DO of only 1 mg/liter. If the dissolved oxygen falls below these levels, these fish begin to have difficulties in respiration, reproduction, defense from predators, or resistance to injury or disease. The fish may be able to exist for extended periods at or below these levels, but they will be under stress. The less tolerant organisms will migrate out of the region of depressed oxygen concentration if they can, or they will die. In either case, the area is left to the more tolerant organisms. This process results in an extraordinary growth in population of the more tolerant species because of freedom from competition or predation by the eliminated species. A short period of hot weather or low water in a stream or pond may be sufficient to reduce the oxygen supply to levels that will eliminate nontolerant species from the water.

Even in natural waters where the dissolved oxygen content of the surface or free-flowing water signifies ideal conditions (5–7 mg O_2/liter), the water at the bottom may be severely depleted of oxygen because mixing currents are absent or because of an excess of oxygen-demanding organic debris on the bottom. It is at the bottom that many organisms such as insect larvae exist and serve as food for fish. A low DO at the bottom may mean the disappearance of the food needed by bottom-feeding fish.

When the DO decreases, the organisms that are tolerant of low oxygen concentration multiply and become the dominant species in the community. These organisms serve as biological indicators of the degree of oxygen depletion in the water. When the oxygen levels become quite low (below 1 mg/liter), the dominant species are generally lower forms of life such as protozoa, larvae, worms, and rough fish such as carp.

Below 1–2 mg/liter DO, anaerobic bacterial decomposition of organic matter begins to take precedence over aerobic decay. After the free oxygen has been sufficiently depleted, the NO_3^- produced in aerobic decomposition can serve as a source of oxygen in a bacterial denitrification reaction,

$$(CH_2O)_{106}(NH_3)_{16}H_3PO_4 + 84.8HNO_3 \longrightarrow$$
$$106CO_2 + 42.4N_2 + 148.4H_2O + 16NH_3 + H_3PO_4$$

where the NH_3 may be further oxidized to N_2 by nitrate:

$$5NH_3 + 3HNO_3 \longrightarrow 4N_2 + 9H_2O$$

Following nitrate depletion, sulfur bacteria can utilize SO_4^{2-} in the water for the reaction

$$(CH_2O)_{106}(NH_3)_{16}H_3PO_4 + 53SO_4^{2-} \longrightarrow$$
$$106CO_2 + 53S^{2-} + 16NH_3 + 106H_2O + H_3PO_4$$

In water with a pH below 8, sulfide, S^{2-}, forms hydrogen sulfide gas. In basic solution (sea water), insoluble sulfides of metals may discolor the water or coat the bottom. In fresh water, an additional anaerobic decomposition process, the formation of methane (marsh gas) may occur.

12.12 EUTROPHICATION

Eutrophication refers to the processes occurring in an aquatic ecosystem as a greater supply of nutrients or foods becomes available to the ecosystem from outside sources. As the ecosystem becomes well fed in this way, the growth of organisms and the amount of organisms present increase until limited by the food supply or other factors. The eutrophication of aquatic ecosystems is part of the natural process of ecological succession occurring in all ecosystems. Ecological succession will be discussed in more detail in Section 20.19. Briefly, succession refers to the changes in the structure and function of an ecosystem as time goes on and as the ecosystem ages to its final mature or climax state.

As an ecosystem develops, many changes take place, due primarily to the influence of the organisms themselves in changing the chemical, physical, and biological characteristics of the environment. These changes permit other organisms to enter and gain a foothold in the particular ecosystem. As time goes on, some organisms are eliminated from the community and others enter, but, overall, the number of different species represented in the community increases. There is an increase in the amount of both living organic matter (biomass) and dead organic matter (detritus) and an associated movement from a simple autotrophic system (one populated by producers living on simple inorganic nutrients) to a more complex and diversified heterotrophic system of consumers feeding on producers.

With respect to aquatic ecosystems, especially the more self-contained lakes, those in the early stages (oligotrophic lakes) contain few inorganic or mineral nutrients derived from normal surface runoff from the land. Their waters are clear and cold and support little life. As nutrients are added, more primary producers such as plankton and algae can be supported; and they, in turn, can support more and different consumer organisms. The production of organic matter eventually exceeds its rate of decomposition by bacterial respiration, and the excess accumulates at the bottom. The lake has reached the eutrophic or well-fed stage. In time, the sediments will begin to fill the lake and convert it into a marsh and eventually a meadow. During the aging or eutrophication process, the dissolved oxygen content, particularly at the lower levels, is reduced. The natural eutrophication of a lake, fed only with the normal supply of nutrients from surrounding land areas, normally takes thousands of years. The use of lakes as receiving waters for the wastes of man is resulting in artificial or *cultural eutrophication*, and this speeds up the natural successional process. Because of the entry of sewage and agricultural fertilizers into one of our biggest sewers, Lake Erie, it is estimated that the lake has gone through 50,000 yr of the natural eutrophication process in only 50 yr. The overfertilization of Lake Erie and other lakes results in rapid growth or blooms of aquatic plants. When these organisms die, the aerobic decomposition of their bodies strips dissolved oxygen even from the surface waters. (Figure 12.6). The light-blocking rafts of algae on the surface also can prevent oxygen-restoring photosynthesis by plants in lower layers. Anaerobic conditions become increasingly common at the bottom of the lake. All the usual

Figure 12.6 Dead fish in oxygen depleted waters of a eutrophied lake. (Courtesy of U. S. Department of the Interior, Bureau of Sport Fisheries & Wildlife. Photo by Nelius B. Nelson.)

and deplorable consequences of oxygen depletion result from accelerated eutrophication.

The added nutrients that cause eutrophication and resultant oxygen depletion of natural waters can vary from one aquatic ecosystem to another. It depends on which particular element is limiting in a given situation. Carbon (as CO_2), metals ions, sulfur, phosphorus, nitrogen, or any other essential element may accelerate growth. Because of the limiting nature of phosphorus in the growth of both fresh- and salt-water plants, excessive additions of phosphorus are usually blamed for accelerating eutrophication. Nitrogen is often ranked behind phosphorus as a culprit. Although opinion is not unanimous that one or the other of these two elements is the limiting factor, we shall see where they come from and what can be done to prevent their excessive introduction into natural waters.

If we regard phosphate ion, $PO_4{}^{3-}$, and nitrate ion, $NO_3{}^-$, (and maybe ammonium ion, $NH_4{}^+$) as the assimilable forms of the two elements, we see that they are normal products of both the aerobic and anaerobic degradation of the organic matter in sanitary sewage. However, over half of the phosphorus in municipal sewage comes not from human waste but from phosphate salts used in laundry detergents. Another source of phosphates entering natural waters is agricultural fertilizers. Animal wastes also contribute to phosphates in agricultural runoff water.

Nitrogen occurs in sanitary sewage primarily as ammonia unless sufficient nitrifying bacteria, oxygen, and time are available to oxidize NH_3 to nitrite and nitrate. These conditions are not usually met in secondary treatment processes. A major source of nitrogen is runoff of chemical fertilizers from farms into surface waters. Fertilizers usually contain nitrogen as ammonia, ammonium ion, or nitrate ion. All these forms are soluble, and they are easily washed from the soil (especially $NO_3{}^-$), unlike phosphorus, which tends to be held in the soil (see Section 14.14). Another agricultural source of nitrogen is animal wastes from feedlots and barnyards. Even when this waste does not run directly off into natural waters, nitrogen can be carried through the air as ammonia and give nitrogen overdoses to lakes a mile or more away. Additional nitrogen also can enter water from air polluted with oxides of nitrogen from combustion sources.

Removal of Phosphates from Secondary Effluents

Even though the activated sludge process can remove 90–95 percent of the phosphorus from raw sewage, most of the phosphate returns to solution when the sludge settles. Until we learn how to keep the phosphate in the settled sludge, tertiary chemical methods offer the best solutions for removing phosphates from sewage. A number of methods are listed below.

1. Precipitation of insoluble phosphates with
   a. aluminum sulfate, $Al_2(SO_4)_3 \cdot 18H_2O$

$$Al^{3+} + PO_4{}^{3-} \longrightarrow AlPO_4(s)$$

   b. sodium aluminate, $NaAlO_2$

$$H_2O + H^+ AlO_2{}^- \longrightarrow Al^{3+} + 3OH^-$$
$$Al^{3+} + PO_4{}^{3-} \longrightarrow AlPO_4(s)$$

   c. with calcium oxide (lime)

$$CaO + H_2O \longrightarrow Ca(OH)_2(s)$$
$$Ca(OH)_2(s) \longrightarrow Ca^{2+} + 2OH^-$$
$$3Ca^{2+} + 2PO_4{}^{3-} \longrightarrow Ca_3(PO_4)_2(s)$$

2. Removal with an anion exchange resin.

3. Removal of phosphate compounds from detergents.

To appreciate the problems involved in the removal of phosphates from detergent cleansing mixtures, we shall have to digress somewhat to see what detergents are, how they work, and what phosphates have to do with them.

Surface Activity and Phosphate Detergents

The word *detergent* describes a variety of substances used for cleansing purposes. Common soap as well as many more modern materials are detergents. All detergents are large molecules that have a polar, water-soluble, or *hydrophilic* (water-loving) end and a nonpolar, water-insoluble, or *hydrophobic* (water-hating) end which is soluble in fats, oils, greases. Substances whose molecules have both a hydrophilic and a hydrophobic portion are variously called by the more general names of surfactants, surface-active, or capillary-active compounds. These names arise because these substances are adsorbed at the surface of an aqueous solution of the surfactant so that their concentration in the surface layer of the solution, the air–liquid interface, is greater than their concentration in the lower layers of the solution. Such substances lower the surface tension of the solution.

The phenomenon of surface tension arises because a molecule in the surface of a liquid experiences a greater attractive (or cohesive) force from the more numerous liquid molecules below it than it does from the vapor-phase molecules above the surface. This imbalance is simply a result of the greater density of liquid molecules beneath the surface. Because of this inward pull, the surface of a liquid tends to contract to the smallest possible surface area (a sphere), and the liquid resists any increase in its surface area. In some ways, the liquid acts as though it had a skin over its surface, and to increase the surface area requires that work be done to stretch this skin. The surface tension of a liquid is proportional to the magnitude of the cohesive forces within the liquid. Thus, it should be no surprise that water, with its hydrogen bonds, has a high surface tension compared to other liquids. The surface tension of a liquid also depends on the other medium in contact with the surface (air, vapor, a solid, or another liquid). The term "surface tension" commonly refers to the liquid in contact with vapor-saturated air, while the term "interfacial tension" is used in other situations, particularly that between immiscible liquids. When a liquid wets the walls of a capillary tube, it rises in the tube and its rise is proportional to its surface tension. Liquids that do not wet the tube, such as mercury in a glass tube, show a fall in level; but, the decrease in height is still proportional to the surface tension.

The wetting of a solid by a liquid or the spreading of one immiscible liquid on another is governed by the interfacial tension between the two phases. If this tension is low, then little energy is required to form a new surface between the phases and wetting or spreading occurs readily. Low interfacial tensions occur if the molecules in the two phases have an attraction for each other. For instance, water will wet glass but will not wet a waxed or oiled surface, because the nonpolar wax or oil molecules have no attraction

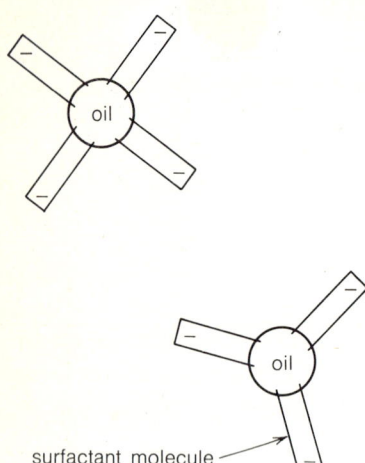

surfactant molecule

Figure 12.7. Surfactant molecules attached to oil droplets. The nonpolar ends are attached to the droplet. The outward-directed charged ends give the droplets like charges, which prevent their coalescence.

for the polar water molecules. Instead, the water forms beads on the surface. Similarly, a nonpolar hydrocarbon will form a lens on a water surface unless it has a polar or ionic group in the molecule that is attracted to water. In this case, the organic molecule will orient itself with its polar end in the water and spread over the surface. The polar or ionic group lowers the interfacial tension by increasing the attraction between water and the rest of the hydrocarbon molecule. This is exactly what soap molecules do in washing. They increase the attraction between water and the oils, greases, and fats that hold dirt in by being adsorbed at the oil–water interface with their polar or ionic ends in the water and their long hydrocarbon tails in the oil. They act as wetting agents to allow the water to wet and spread over the oil-encapsulated dirt.

Once the oily material has been wetted by water with the help of a surface-active substance, the oil can be broken up into small droplets. Again, the formation of numerous small oil droplets suspended in aqueous solution is brought about by the interfacial tension-lowering properties of a surfactant molecule such as soap, because the breakup of a large oil droplet into a number of small droplets—an emulsion—results in an increase in the total area of the oil–water interfacial surface. The surfactant decreases the energy needed to form the emulsion. Once the emulsion has been formed, the surfactant molecules prevent the oil from coalescing by orienting themselves around each droplet with their nonpolar ends in the droplet and their charged ends pointing outward into the water phase, as shown in Figure 12.7 for surfactant molecules with a negative charge on one end. This action stabilizes the oil-in-water emulsion and keeps the droplets dispersed because the like-charged droplets repel each other. The combined wetting–emulsifying–dispersing action by a detergent molecule or ion breaks up the oily substances acting as binders for dirt and allows the dirt and its binder to be floated away.

In Chapter 9, we mentioned that soaps have the failing of forming insoluble precipitates with the calcium and magnesium ions in hard water. They also tend to revert, in acid or neutral solution, to their parent fatty acids, which have no detergent action. For these reasons synthetic detergents (syndets), which are not affected by hard or acidic water, were introduced in the 1940's and have become increasingly popular. Although soap is technically a detergent, the name as commonly used now refers to the synthetic detergents only.

There are at least three general kinds of syndets (anionic, cationic, and nonionic), but the anionic ones are the most popular. The molecules of two anionic syndets and one soap are shown below.

An alkyl benzene sulfonate (ABS) syndet

$$CH_3{-}CH{-}CH_2{-}CH{-}CH_2{-}CH{-}CH_2{-}CH{-}CH_3$$

with CH_3 groups and a benzene ring bearing $O{=}S{=}O$, O^-, $[Na^+]$

A linear alkylsulfonate (LAS) syndet

$$CH_3-(CH_2)_9-CH-CH_3$$

A soap

$$CH_3-(CH_2)_{8-16} \quad C \underset{O_-}{\overset{O}{\diagup}} \quad \left[Na^+\right]$$

Branched-chain syndets of the ABS type are not biodegradable, or only very slowly so, and pass through the secondary stages of sewage treatment to appear in groundwaters, streams, and even tap water, where their presence is revealed by foaming. The foam also interferes with the operation of sewage treatment facilities. The use of ABS syndets was discontinued, under government pressure, in 1965, and they were replaced by LAS-type syndets, which are almost completely biodegradable.

Phosphorus compounds are mixed with syndet formulations for household use for a variety of reasons, with water softening being the main reason. The most widely used phosphorus compound in solid detergent mixtures is sodium tripolyphosphate (STP), $Na_5P_3O_{10}$. In solution, the tripolyphosphate hydrolyzes to produce a basic solution:

$$P_3O_{10}{}^{5-} + H_2O \rightleftharpoons HP_3O_{10}{}^{4-} + OH^-$$

(Liquid detergents usually contain sodium or potassium pyrophosphates such as $K_4P_2O_7$, which apparently hydrolyze to $PO_4{}^{3-}$ more slowly than tripolyphosphates.) Some conversion of tripolyphosphate to (ortho)phosphate may occur,

$$P_3O_{10}{}^{5-} + 2H_2O \rightleftharpoons 3PO_4{}^{3-} + 4H^+$$

but such reactions are slow and make no contribution to detergent action. As a water softener, STP functions at high tripolyphosphate concentrations by forming soluble complexes with Ca^{2+} and Mg^{2+} and with ions of iron and manganese, which may stain fabrics. In the presence of Ca^{2+}, some syndets form insoluble calcium salts, but the main reason for removing these metal ions is to prevent the formation of insoluble metal carbonates from water with temporary (hydrogen carbonate) hardness as temperature is raised or as the basicity of the wash water is increased.

In addition to its softening action, STP also (1) neutralizes acids that may be associated with solid surfaces; (2) helps to suspend and disperse water-insoluble substances (i.e., dirt) and prevent their redeposition on fabrics during rinsing, and (3) saponifies fats or fatty acids and converts them into soaps and prevents the precipitation of the resulting soaps with Ca^{2+} and Mg^{2+}.

Phosphate Substitutes

Many household syndet mixtures contain over 50 percent by weight poly-phosphate (12–13 percent phosphorus), and in order to reduce the amount of phosphorus entering natural water through domestic sewage there has been action to remove sodium tripolyphosphate from detergents. A number of substances have been used as replacements for polyphosphates. Most, however, cannot match all the accomplishments of the phosphates and serve mainly as replacements for the water-softening capabilities of the phosphorus compound. Two widely used phosphate substitutes are sodium carbonate, Na_2CO_3, and sodium metasilicate, Na_2SiO_3. Both are salts of very weak acids, and their solutions are considerably more basic (pH $\simeq$ 11) than solutions of sodium tripolyphosphate (pH $\simeq$ 9–10), so there is a hazard of caustic burns. (Sodium tripolyphosphate is a salt of the stronger tripolyphosphoric acid, $H_5P_3O_{10}$, not of the weaker phosphoric acid, H_3PO_4.) Both carbonates and silicates soften water by precipitating the hardness-causing ions as insoluble salts.

The ecological consequences of most phosphate substitutes are not well known. The case of the sodium salt of nitrilotriacetic acid (NTA), $N(CH_2-COONa)_3$, is an example. In 1970, great hopes were held for the use of NTA, which, like STP, forms soluble complexes with hardness-producing ions, and many manufacturers began to use it in their detergents. Soon, however, NTA was found to have several potential disadvantages, and it was voluntarily removed from the market pending further tests. One disadvantage is that NTA complexes heavy metal ions such as Cd^{2+} and Hg^{2+}, which are found in galvanized pipes and lake bottoms. Although NTA is normally degraded in aerobic sewage treatment and loses its chelating properties, it does not degrade under anaerobic conditions found in septic tanks. Thus, the complexed heavy metals can be carried down to the groundwater table and brought back up as drinking water from wells and springs. The NTA–heavy metal combination was shown to have teratogenic effects with test animals. Teratogenicity refers to the induction of fetal malformations and birth defects by substances ingested by the pregnant animal (see Section 19.15). The potential teratogenicity of NTA–metal apparently arises from the replacement of essential Zn^{2+} (also complexed by NTA) in the fetus by Cd^{2+} or Hg^{2+} (see Section 12.14). Other potential hazards from NTA come from its aerobic breakdown product, NO_3^-, which may appear in drinking water. Excess NO_3^- in the diet can lead to a pathological condition (Section 12.14), and NO_3^- in aquatic ecosystems may accelerate eutrophication if nitrogen is the limiting nutrient. A further complication is that bacterial action may convert NTA into cancer-causing nitrosamine compounds (Section 19.10). So the search continues for an acceptable phosphate replacement.

The Phosphate Controversy

The phosphorus-caused eutrophication controversy is just that. Some scientists claim that substances other than PO_4^{3-} are limiting factors in many cases. For instance, carbon dioxide from decomposing organic waste has been suggested as the controlling nutrient causing excess growth of photosynthetic algae in certain aquatic ecosystems. And it is generally agreed that nitrogen is a limiting nutrient in some lakes and in many, if not all, estuaries and coastal waters.

Another line of reasoning points out that, even if phosphorus is the culprit in the cultural eutrophication of lakes, removing polyphosphates from detergent formulations will not immediately solve any problems. In the first place, it is said, phosphates from human wastes and agricultural fertilizer runoff contribute more than enough phosphorus to satisfy the requirements of algae growth. This requirement is only about 0.05 mg/liter of PO_4^{3-}. The argument is that removal of phosphorus compounds from detergents will not bring the phosphorus level in lakes below this critical level above which algal blooms occur. Furthermore, no quick results can be expected, because the excess phosphorus already committed to lakes still remains there, being continually recycled except for the amount locked in insoluble sediments. Thus, the waters are vast reservoirs of phosphorus capable of supporting algal growth long after surface input is shut off. The natural phosphorus cycle has no provision for the escape of phosphorus from water as a gas and very little provision for the transport of phosphorus from the hydrosphere via organisms. Phosphorus, of all the major elements for life, tends to accumulate in the waters.

Despite this discouraging prospect, it does seem important to stop the flow of a large part of the earth's phosphorus resources into the sea. Whether or not phosphorus is the limiting nutrient in a specific system, attention has been centered on its removal because it is the most readily controllable element. We do not know how to prevent the entry of nitrogen (blue-green algae can fix N_2 and transform it to NO_3^-) and carbon into waters from the atmosphere. Phosphorus may not be the element present in shortest supply relative to the algae's need—and therefore not limiting. But it can be made the limiting element if its supply is choked off to a point low enough to bring about decreased algal growth. The ultimate solution for phosphorus, at least, may be restricting the production and use of phosphate detergents and its complete removal by tertiary sewage treatment processes. Tertiary treatment will be necessary, because the usual bacterial (secondary) treatment of domestic sewage (with or without detergents) removes, in the sludge, 80 percent of the carbon in sewage compared to only 24 percent of the nitrogen and 29 percent of the phosphorus. The resulting secondary effluent has a C:N:P atom ratio of 20:19:1, low in carbon relative to its atomic ratio in protoplasm (which the algae can obtain from the air) but high in N and P. The effluent is therefore particularly ideal for the growth of photosynthetic algae. (For further discussion of this point see the reference by Stumm and Morgan.)

Removal of Nitrogen   Most of the nitrogen in effluent from domestic sewage treatment plants is in the form of NH_3 or NH_4^+. These forms can be removed biologically by first converting NH_3 to NO_3^- with nitrifying bacteria followed by bacterial denitrification of the NO_3^- to nitrogen gas. Methanol (CH_3OH) or some other oxidizable organic compound must be present to provide energy for the nitrogen reduction process. The denitrification step alone can be used if nitrate ion is the predominant form of nitrogen in the effluent. This is the case where fertilizer runoff from farm lands must be treated.

Ammonia or ammonium ion can be removed chemically from secondary effluent by forcing air through alkaline effluent. At higher pH values, $[H^+]$ is low, and the equilibrium shifts

$$NH_4^+ \rightleftharpoons NH_3 + H^+$$

toward the NH_3 side. The gas is then carried off by the flow of air. Nitrogen as ammonia or as nitrate can also be removed from water by growing algae in large shallow ponds. The algae assimilate the nitrogen and are later removed by filtration.

Other methods in use or under study for the removal of nitrates, phosphates, and other mineral ions from water may also be employed for the desalination of sea water or the demineralization of brackish groundwater. Their description is taken up in Section 12.20 and subsequent sections.

12.13 THERMAL POLLUTION

In Chapter 11, we briefly noted the role of temperature as a limiting factor for marine life. The influence of temperature changes on aquatic ecosystems is receiving increased attention because of the projected discharge of large volumes of heated water from power plants into rivers, inland lakes, and the oceans (Figure 12.8). The discharge of waste heat into natural waters can bring about thermal pollution in a number of ways.

Figure 12.8. Infrared image of the heated effluent from an electric power plant. Discharged coolant water enters the river through a canal (bottom). The warm effluent (~30°C) appears lighter than the normal river water (~20°C). (Courtesy Environmental Sciences Branch, HRB-Singer, Inc.)

But let us first see why thermal pollution is expected to become more evident in the years to come. Electric power plants are the biggest contributors to heat discharge into natural waters. In these plants, thermal energy is released by the burning of fossil fuels or by nuclear fission reactions. This energy heats water and produces steam, which drives turbines for the generation of electricity. Heat is being converted into the mechanical energy of the turbine rotors, and the mechanical energy is transformed into electric energy by generators. The second law of thermodynamics (Sections 16.6 and 20.16) states that the conversion of heat energy into mechanical energy, or work, cannot take place with 100 percent efficiency. Some of the thermal energy is inevitably left unchanged and is wasted. The exact portion of the heat that goes unutilized depends on the specific operating characteristics of the generating plant, particularly the temperatures at which heat is absorbed (the boiler temperature) and the temperatures at which the waste heat is discharged (the receiving waters). The greater the difference between these two temperatures, the greater the efficiency of conversion of heat into mechanical and electrical energy.

When steam has passed through the turbines, it must be cooled and condensed so it can be returned to the boiler for revaporization. Water drawn into the condenser provides this cooling and absorbs the heat of condensation of the steam. The heated cooling water is then discharged to receiving waters outside the plant unless some provision has been made for cooling and recirculating it. The need for large volumes of cooling water is the reason power plants are almost always located close to a body of water.

The expected doubling of electric power consumption each decade will be met by construction of more and larger generating plants, many of them nuclear fueled. Only about 40 percent of the thermal energy obtained by combustion of fossil fuels in a power plant is converted into electricity. The rest is dumped into the environment as waste heat in the cooling water. Nuclear-fueled plants are even less efficient than fossil fuel plants, because their uranium fuel cannot be raised to the high temperatures obtained by burning coal, oil, or gas. As compared to most fossil fuel plants, they waste about 50 percent more energy as heat (or require 50 percent more cooling water) per unit of electrical energy produced. The ability of streams and lakes to absorb the enormous amounts of waste heat without a temperature rise detrimental to aquatic life is in serious doubt.

The temperature increase of receiving waters depends on several factors including current, wind speed, and air temperature. Generally, the condenser cooling water leaving a plant is 5.5–17°C (10–30°F) above that of the intake water. This means that the water outfall in summer months may reach a temperature of 40–50°C (100–115°F). The temperature of large areas of the receiving waters, especially at the surface, may be increased proportionately, particularly if the cooling water is not drawn from the cooler lower levels of lakes or seas. In addition to decreasing the solubility of oxygen in the water, an immediate result of increased water temperature is an increase in velocity of the chemical reactions occurring in the aquatic ecosystem. As a rough approximation, the rate of a chemical reaction doubles for each 10°C rise in temperature (Section 17.7). The chemical reactions of most interest here are the oxygen-consuming metabolic processes of living organisms. The quantity of dissolved oxygen withdrawn and its rate of withdrawal are both increased as respiration and reproduction by all forms of aerobic life accelerate. A temperature increase therefore increases the biochemical oxygen demand of the water and reduces the capacity of the water to assimilate organic wastes.

In addition to their tolerance limits for dissolved oxygen, all aquatic organisms have definite temperature tolerances. Some have wide degrees of tolerance, some have narrow limits. Whatever the limits, there is an optimum temperature, usually closer to the maximum limit, where the organism functions best. The optimum temperature and the upper and lower limits of tolerance may vary with each stage in the life cycle of the organism. Also, most aquatic animals are cold-blooded. There is no regulation to constant internal temperature as is found in warm-blooded animals. All changes in the temperature of the water are rapidly reflected in a change in internal body temperature.

There are other effects of increased temperature on the respiration of fish. At higher temperatures, the hemoglobin of a fish's blood has less affinity for oxygen, and this decreases the efficiency of oxygen transport to tissues. In some fish, carbon dioxide acts as a stimulant to the respiratory system, increasing the rate of respiration and increasing oxygen withdrawal from the water. High carbon dioxide concentrations in water can result from enhanced respiration and decay at elevated temperatures.

The outright heat death of organisms by water temperatures at the upper limit of tolerance is relatively rare. The cause of heat death may be connected with the coagulation of body proteins (similar to the cooking of an egg) or the destruction of essential enzyme catalysts (which are also proteins). The highest temperatures that North American fish can tolerate range from approximately 20.5 to 30.6°C (77 to 97°F), the exact temperature varying with the species of fish. In general, however, waters warmer than 30.4°C (93°F) are uninhabitable for all fish in the United States. The upper tolerance limits allow for the acclimatization of fish to the higher temperatures. Fish can adjust to higher temperatures if the temperature rise is moderate and very gradual.

Although the optimum temperature for a single organism may be some definite value, the most favorable temperature for a complete aquatic community is an average of the optimum temperatures of the many organisms. The optimum temperature of the community may be above or below the tolerance limits of certain species, which is why some organisms are excluded from the community. When the average temperature is increased, the community structure will be altered by the death or migration of some organisms and by the entry of new species into the community. A temperature increase can result in the disappearance of certain predator species with resulting unchecked growth of other organisms. Or heat may kill or drive off the principal food supply of predator species.

Oceans are better able to absorb the discharge of cooling water, so thermal effects on marine organisms and communities are not as pronounced as those in fresh-water communities. Nevertheless, the use of saline sea water for cooling can result in other adverse consequences to marine life. The corrosive action of salt water carries toxic heavy metal ions from condenser cooling tubes into the sea, where organisms can assimilate and concentrate the metals. The high fertility of coastal waters also results in biological growths in condenser tubing, and these are removed by flushing with a variety of cleaning agents, including detergents, acids, chlorine, and other biocides, which are discharged into the sea.

The addition of controlled quantities of heat to natural waters can have beneficial effects. When this is the case, it is called thermal addition rather than thermal pollution. The use of warmed coolant water drawn from nutrient-rich depths for sea farming in lagoons is a good example of thermal addition. Certain fish grow better in warm water. Shellfish such as oysters grow better in warm water, probably because increased temperatures influence the rate at which calcium carbonate can be precipitated and incorporated into the shell. Some fish prefer warmer water, and power plant coolant outfalls are often good fishing spots.

Cooling ponds or cooling towers can be employed to control thermal pollution through the cooling of effluent water before its release. A cooling pond is a shallow artificial lake through which the warm water is passed slowly until its temperature has decreased sufficiently for reuse or for discharge to receiving waters. The disadvantage of such ponds is their size. For the large power plants that will be built, a pond of about 2 to 3 square miles in area would be required. Cooling towers are of two types. In the evaporative cooling tower (Figure 12.9), the heated water is allowed to fall over baffles in the lower part of the tower, where it is exposed to air moving up through the open chimneylike tower. Heat of vaporization carries away excess thermal energy,

Figure 12.9. Evaporative cooling towers. These are hyperbolic natural draft towers in which the draft of air upward through the concrete shell results from the great height of the tower and the difference in density of warm moist air within the tower and cooler outside air entering the base. In another type of evaporative tower, air is drawn up through the tower by fans at the top. The towers shown here are each large enough at their bases to hold a football field, and they are 437 ft high. They are located at TVA's Paradise steam plant in Kentucky which, with an electrical generating capacity of 2558 MW, is the largest fossil fuel-powered plant in operation anywhere in the world today. (Courtesy Tennessee Valley Authority.)

and the cooled water is either discharged or recycled. In the closed-circuit or dry cooling tower, the heated water is confined to pipes and is cooled by radiative heat exchange, with air drawn up through the tower by large fans. This is the same principle as is used in fan-cooled automobile radiators. No water is lost by evaporation. As is always the case, such measures cost money, and the cost is paid by the energy-consuming public.

12.14 TOXIC SUBSTANCES

A wide range of toxic pollutants find their way into natural waters. Municipal sewage treatment methods often have no effect on these substances, and much industrial sewage is untreated or only partially treated before discharge to receiving waters or to municipal sewers. In many cases, specific chemical processes must be used by an industry to remove a given substance from its waste water. A number of these substances are toxic to aquatic organisms and to humans, so they must be kept out of domestic water supplies.

The ions of heavy metals are among the most serious of the toxic pollutants finding their way into natural waters. Among these metals are cadmium, lead, mercury, nickel, antimony, and arsenic. Most of these metals are cumulative poisons capable of being assimilated, stored, and concentrated by organisms that are exposed to low concentrations of the metals or their compounds for long periods or repeatedly. Eventually, the buildup of the metal in tissues will be sufficient to cause noticeable physiological effects. Here we use some of the origins and effects of mercury and cadmium in water as illustrations of pollution by toxic substances. We also add a brief discussion of nitrate poisoning.

Mercury

Mercury metal and its compounds have long been known to be highly toxic to living organisms. Mercury and other heavy metals and their compounds deactivate protein molecules that serve as enzyme catalysts by combining with —SH and —OH groups in the molecule. Among humans, one of the many distressing symptoms of mercury poisoning is mental instability, often resulting in paranoia and other emotional disturbances. Such symptoms were formerly encountered among employees in the felt hat industry, where mercury(II) nitrate, $Hg(NO_3)_2$, was used to treat felts. The situation gave birth to the phrase "mad as a hatter."

As a water pollutant, mercury and its compounds originate from a number of industrial and agricultural sources. The biggest industrial contributors to mercury water pollution are chlorine gas–sodium hydroxide plants. In such plants, mercury metal serves as the electrode (cathode) material in the electrical decomposition (electrolysis) of sodium chloride solution. Sodium metal is then liberated at the cathode (instead of H_2), forms an amalgam with the mercury, and is later brought in contact with water to yield H_2 and NaOH. The mercury is recycled but 100 percent efficiency is not attained, and the waste waters from the process contain metallic mercury and mercury(II) chloride, $HgCl_2$, which are often discharged into receiving waters. Mercury compounds are also widely used as catalysts in the manufacture of plastics such as urethanes and polyvinylchlorides. Other industrial uses of mercury are for electric equipment and in antifouling paints.

Organic mercury compounds have been widely used as fungicides to prevent fungus attack of seeds and as slimicides to prevent fouling of paper mill machinery. Some of the more popular compounds are ethylmercuric chloride, C_2H_5HgCl; phenylmercuric acetate (PMA), $C_6H_5HgOCCH_3$, and the fungicide Panogen,

$$CH_3Hg-\underset{\underset{\displaystyle H}{|}}{N}-\underset{\underset{\displaystyle ||}{O}}{\overset{\overset{\displaystyle H}{|}}{C}}-\underset{\underset{\displaystyle H}{|}}{N}-C\equiv N$$

In 1969, 6 million pounds of mercury were consumed by industry and agriculture in the United States. Estimates are that each year one-third of this amount is not recovered and is lost to the air, water, and soil.

Mercury can enter organisms in its elemental form, as a liquid or a vapor; in its inorganic combined forms (Hg_2Cl_2, $HgCl_2$, etc.); or as its organic compounds. The organic compounds appear to have the greater toxic effect on the organism because of their ease of diffusion through biological membranes and their resultant tendency to attack the central nervous system, and their tendency to accumulate in the body because of their long retention times as compared to inorganic mercury compounds. Elemental mercury and many of its inorganic compounds are quite insoluble and had been expected to sink harmlessly to the bottom of receiving waters, but it now appears that anaerobic bacteria in the bottom sediments can convert these forms to highly volatile and toxic methyl derivatives such as monomethylmercury(II) chloride, CH_3HgCl, and dimethylmercury(II), $(CH_3)_2Hg$. The "methyl mercury" released to the water can be absorbed directly by aquatic organisms (by both producers and consumers) or passed up through the food chain to fish, birds, and humans with accompanying biological concentration. Another possible way for insoluble mercury to dissolve is through the formation of soluble complexes with organic molecules derived from sewage or dead organisms.

In the United States, federal regulations state that there must be no trace of mercury fungicides on foodstuffs and that fish containing more than 0.5 mg/kg (0.5 ppm) of mercury may not be sold or given away. There have been many instances of confiscated fish and of the closing of inland waters to commercial and sport fishing.

The above mercury-in-water comments apply principally to inland waters. The mercury in the oceans is apparently from natural sources such as the degassing of the earth's crust, although a comparatively small amount of mercury is contributed through the atmosphere from human activities such as fossil fuel combustion, cement making, and ore roasting. The concentration of methyl mercury in ocean fish has not risen appreciably in the last century, although our more sensitive modern methods of detection often make it seem otherwise because we find mercury present where we had no previous knowledge of it.

Mercury should be removed from industrial wastes at the plant site by a suitable chemical treatment, perhaps by precipitation of an insoluble compound such as mercury(II) sulfide, HgS. Even then, the treated effluent should be recycled rather than discharged to natural waters or to municipal sewers.

Cadmium

Another highly toxic metal whose presence in water is causing concern is cadmium. Cadmium can originate in waste waters from mining operations or in water supply systems from the corrosion of zinc-galvanized pipes and mains. Zinc is normally accompanied by the chemically similar cadmium and also by lead. If soft water is transported through galvanized pipes, some or all of these metals will dissolve:

$$Cd + 2H^+ \longrightarrow Cd^{2+} + H_2$$

The acidity of water results from the absorption of carbon dioxide:

$$CO_2 + H_2O \rightleftharpoons H^+ + HCO_3^-$$

Soft water, because of the low concentration of hydrogen carbonate and carbonate ions, does not repress the H_2CO_3 dissociation.

Zinc plays a vital role in the breakdown of fats in the body. The metal is a necessary constituent of one of the enzyme catalysts in the fat metabolism process. Cadmium replaces zinc, but it renders the enzyme inactive by altering the molecular structure of the enzyme in a subtle way. As a result, fats accumulate in the circulatory system, resulting in high blood pressure and heart disease.

Ingestion of cadmium and lead also interferes with a kidney function regulating mineral levels in the blood and bones. The result is a sizable loss of minerals from bones. The condition resulting from the abnormalities in mineral (phosphate and calcium, mostly) metabolism is similar to the disease called rickets. The withdrawal of calcium and other minerals from skeletal structures results in a disease characterized by the easy fracture of bones. The Japanese first traced the cause of the disease to pollution of river water by cadmium and lead from mining operations and the ingestion of these metals in drinking water or in crops irrigated with the contaminated water. The painful nature of the disease is suggested by its Japanese name, "itai itai," which translates as "ouch ouch."

Nitrates

As a final example of a toxic water pollutant, we consider nitrate ion, NO_3^-, and nitrite ion, NO_2^-. Nitrite ion can oxidize Fe(II) in the hemoglobin molecule to Fe(III). The product is called methemoglobin and does not combine with oxygen, so it is unable to transport oxygen to body tissues. The resulting disease is called methemoglobinemia or nitrogen cyanosis and has symptoms of labored breathing, blue skin coloring, and suffocation. Nitrite can pass into the blood stream from the stomach. Nitrite is not as common in foods and water as nitrate, but bacteria found in the stomachs of infants and ruminant animals can reduce nitrates to nitrites and cause the disease. Nitrates can be ingested in water with a high nitrate concentration resulting from leaching of overfertilized fields or oxidation of organic sewage. Certain plants such as spinach can absorb high NO_3^- concentrations from overfertilized fields. When fed to babies under a year old, such foods may cause methemoglobinemia.

12.15 POLLUTION CONTROL IN AN EFFLUENT SOCIETY

Not too many years ago, few people paid much attention to the importance of sewage and industrial waste water treatment and disposal. Similarly, few people were conscious of the slow deterioration of their streams and lakes. Sewage treatment plants went about their business quietly, and their good works were taken for granted. Nobody noticed what was happening until they discovered that there were no more fish left in their favorite fishing hole or found that they were swimming in garbage. This brief rundown of waste water treatment and effects of pollution gives you some idea of how important it is that domestic and industrial and even agricultural waste waters be adequately treated and safely disposed of. Sanitary engineers and others concerned with waste water disposal in government and industry have the difficult job of seeing to it that our natural water resources are protected from the effects of water pollution. Public health officials have the responsibility for the protection of the public from many effects of water pollution. Sometimes there are even people who worry about the fish. We have the responsibility of seeing to it that all of these people have the backing and the money to meet their responsibilities effectively. As a starting point, you might want to arrange a visit to your local sewage treatment plant to see first hand what is happening there.

Technology is probably capable of solving most of our present water pollution problems given the tens of billions of dollars required for construction of the proper facilities. However, if our population, industry, and consumption (and waste) continue to grow, it seems likely that radically new and more efficient methods of treatment will be needed together with a fundamental change in our present attitudes, which have helped to give us the name "the effluent society."

Water Resources

Conservation has been called applied ecology and is today defined more in terms of the rational use of the environment to provide a high quality of life for mankind than in terms of preserving the environment in its natural state. The problems of pollution are but one aspect of ecology and conservation. Another aspect is the management and wise use of our natural resources to ensure a continuing supply.

The close relationship between pollution and wise use of natural resources is well illustrated by our use and misuse of water. Our utilization of this inexhaustible but finite water resource, particularly fresh water, is increasingly hindered because of the deterioration in quality of natural waters. It is almost literally true that some waters in the United States are used two or three times—the water drunk by one person having already been used by another person upstream. Recycling and reuse of natural resources is a subject dear to the hearts of modern conservationists; and, in a sense, much water is being recycled. Unfortunately, after its first use it is usually returned to the stream in poorer quality than it was when withdrawn. The downstream consumer is then forced to greater efforts to renovate the quality of the water.

To solve the problems of increased use of water, we have been required to turn to water purification techniques that, in conjunction with waste water treatment, close the fresh-water cycle so far as human consumption is concerned. Even this closure does not always satisfy the need for usable water, and less common sources of water must be tapped. We now consider common methods for purifying fresh water for domestic use and discuss how water can be obtained from new sources.

12.16 MUNICIPAL WATER TREATMENT

Few cities have natural water supplies for domestic and industrial use that require no treatment whatsoever. Because of the increased waste load imposed on ground and surface waters that constitute the raw water supplies of most communities, water must be treated to assure disease prevention and palatability.

The operations for purifying urban water supplies are essentially the same as those for treating waste waters. Both purification of raw natural waters and sewage treatment deal, in a sense, with contaminated water, the difference being only in the quality of the incoming water and the quality of the outgoing water effluent. The more common processes of water purification are listed here.

1. Chlorination for the purpose of disinfection and oxidation of substances such as phenols that impart an objectionable taste or odor to water.

2. Aeration to remove undesirable tastes and odors owing to volatile substances (such as H_2S) and to remove undesirable metal ions (such as Fe^{2+} and Mn^{2+}) by oxidation to their less soluble forms:

$$4Fe^{2+} + O_2 + 4H^+ \longrightarrow 4Fe^{3+} + 2H_2O$$
$$Fe^{3+} + 3OH^- \longrightarrow Fe(OH)_3(s)$$

3. Adsorption on activated carbon for removal of taste and odor-imparting organic compounds.

4. Sedimentation of settlable solids.

5. Chemical coagulation of nonsettlable suspended and colloidal solids.

6. Filtration through sand bed filters to remove natural suspended solids or precipitates from previous treatment steps.

7. Addition of softening or corrosion-reducing chemicals.

All of these processes need not be used in a water treatment plant. The number and sequence of the treatment processes depend on the quality of the raw water and on the desired quality of the effluent. Any given operation can be employed at several stages of treatment. All of the operations listed have been explained with respect to their role in waste water treatment or are reasonably self-explanatory except for chemical coagulation. Coagulation is common in water treatment and involves some basic chemistry, so we shall discuss it in more detail.

12.17 CHEMICAL COAGULATION OF COLLOIDAL PARTICLES

Chemical coagulation removes a variety of solids and other nonsettleable matter of colloidal dimensions that are responsible for the turbidity (cloudiness) and color of water. The treatment often incidentally causes the removal of taste- and odor-producing substances and of phosphates.

The lower size limit of colloidal matter (10^{-9} m) is about the size of certain complex molecules of rubber or proteins, whereas the upper limit (10^{-6} m) approaches the particle size that may be distinguished in an ordinary light microscope. Although colloidal particles are not visible in such a microscope, they may be detected very readily when a suspension or a dispersion of colloidal particles is viewed at right angles to a light beam passing through it. The beam of light is clearly visible because of the preferential scattering of blue light by the particles. This phenomenon is called the *Tyndall effect*. The effect is easily observed in a sunbeam passing through a smoky room.

The characteristic properties of colloidal particles or of colloidal dispersions are the result of the particles' small size and the large surface area of a given mass of colloidal particles. For instance, the surface area of a cube with sides 1 cm long is only 6 cm² ($= 6 \times 10^{-4}$ m²). If the cube material is reduced to colloidal-sized cubes of side length 1×10^{-8} m, the surface area increases to 600 m². Since many chemical, physical, and biological processes take place at or on surfaces, the study of the colloidal state is an important part of surface science.

There are eight kinds of colloidal dispersions of one finely divided substance in another. Names and examples of some are listed in Table 12.1. At this point, we are concerned only with systems of solids dispersed in water.

Table 12.1
Colloidal systems

dispersed phase	in	dispersion medium	name	common example
solid	in	liquid	sol	AgCl in water (hydrophobic) gelatin in water (hydrophilic)
solid	in	gas	aerosol, solid aerosol	smoke
solid	in	solid	solid sol	ruby glass (gold in glass)
liquid	in	liquid	emulsion	milk (fat in water)
liquid	in	gas	liquid aerosol	fog
liquid	in	solid	solid emulsion	—
gas	in	liquid	foam	whipped cream
gas	in	solid	solid foam	—

These sols may be divided into two types, hydrophobic and hydrophilic. In hydrophobic sols, the dispersed solid has little attraction for the dispersion medium. Hydrophilic sols resemble gelatin; the dispersed phase has a strong affinity for water. The colloidal dispersions or sols encountered in water treatment (except in sludge) usually behave like hydrophobic sols.

There are two reasons why solid hydrophobic sols do not settle as do dispersions of larger visible particles. Because of their small size, they have a high kinetic energy in all directions. This energy counteracts the small gravitational force acting on the particles just as it does with gas molecules. Second, the surface of hydrophobic sols usually carries an electric charge as a result of the preferential adsorption of ions by the surface. The charge on all the sol particles is of the same sign, and repulsion between like charges keeps the particles from sticking together during their many collisions. If there were no charge to prevent sticking, enough particles would soon come together or coalesce by successive collisions to form larger settleable particles. The process of coalescence to form larger particles is called coagulation or flocculation.

One way to bring about the coagulation of colloidal particles is to neutralize the charge on the particles so that they can coalesce and settle. If the particles are negatively charged, as are most of the sols encountered in water and sewage treatment, the particles must acquire positive ions to neutralize the charge; the higher the positive charge on the neutralizing ions, the more effective their coagulating power. The relative coagulating power of trivalent to bivalent to univalent ions is about 500 to 7 to 1. For this reason, aluminum sulfate, $Al_2(SO_4)_3$, incorrectly called alum, which yields Al^{3+} ions on solution,

$$Al_2(SO_4)_3 \rightleftharpoons 2Al^{3+} + 3SO_4^{2-}$$

is widely used as a chemical coagulant. Some of the aluminum ions neutralize the negative surface charge on the colloidal particles and begin the coagulation process. Iron(III) chloride or sulfate is also used, Fe^{3+} serving the same function as Al^{3+} in inducing coagulation.

Most of the aluminum ions react with hydroxide ions from lime to form insoluble aluminum hydroxide or hydrated aluminum oxide,

$$Al^{3+} + 3OH^- \longrightarrow Al(OH)_3(s) \text{ or } \{Al_2O_3 \cdot xH_2O\}$$

The aluminum hydroxide itself is originally formed as a colloidal sol with a positive charge owing to excess aluminum ions on the surface. The positively charged hydroxide sol is usually much in excess of the amount required to neutralize the original negative sol, so the excess $Al(OH)_3$ sol must be co-agulated too. After a time, the negatively charged SO_4^{2-} ions added with the alum neutralize the excess positive charge on the hydroxide and the hydroxide settles out. Negatively charged sols of silica or of organic materials called polyelectrolytes are often employed to aid the coagulation of the hydroxide. The aluminum hydroxide settles out as a gelatinous, sticky precipitate that also pulls down suspended noncolloidal materials, and this clarifies the water. Some bacteria are also removed by coagulation. The use of Al^{3+} or Fe^{3+} for coagulation also helps to remove dissolved phosphates because of the insolubility of aluminum and iron(III) phosphates. Both ions are used in tertiary treatment for the removal of phosphates in waste water.

12.18 THE WATER SUPPLY AND DEMAND PROBLEM

For the most part, the natural waters to which we have referred have been the fresh waters of our inland ground and surface waters. These fresh waters supply all the water for human consumption and are also the receiving waters for disposal of sewage effluent except in coastal cities.

The total amount of fresh water on earth is relatively constant, and its source is precipitation in the form of rain or snow. Our present fresh-water supply is that portion of precipitation that appears as ground and surface water runoff. Of the approximately 4300 billion gal (U.S.)/day average rainfall in the continental United States, only one-fourth appears as potentially usable runoff—the rest evaporates. This average runoff is therefore the upper limit of our conventional direct fresh-water supply.

By 1980, the withdrawal of water from these fresh-water supplies in the United States will be 600 billion gal/day; by 2000 it will be 900 billion gal/day. These figures amount to 56 percent and 84 percent of the theoretical supply limit, respectively. Even though about 70 percent of the water withdrawn is directly returned to surface waters, its quality has usually deteriorated significantly, and renovation of the water is not a simple or inexpensive task.

Our increasing demand for water is due to population growth, increasing industrial and irrigation uses of water, and the desire or necessity of large parts of the population to live in regions of unpredictable water supply or of poor water quality. Our estimates of supply and reasons for greater water demand refer to the United States, but the problems of water availability, quality, distribution, and cost are of worldwide significance. And though, at first glance, the solutions to these problems appear to be in the realms of science and technology, the solutions also involve important social, economic, and political considerations.

Solutions to the problems of water availability and supply could be found by reducing per capita consumption or reducing population generally or regionally. Methods for increasing the efficiency of use of existing conventional fresh-water supplies could be reducing evaporative or seepage losses in storage and transport, developing industrial processes and crops using less water, and recycling and reuse of water. Finally, the total amount of available water could be increased by turning to less conventional sources. We examine two of these sources, clouds and the ocean. Both of these are, of course, not uncommon sources so far as nature's water cycle is concerned. They are, in fact, the primary and secondary sources, respectively, of fresh ground and surface water. For man's needs, though, rain clouds are once removed from his usual raw water supply, and the oceans are twice removed. By seeding the clouds and engaging in induced rain making, man attempts to control where and when rain may fall. It is possible that more moisture may be caused to leave the atmosphere as rain or snow by these efforts, but the principal effect is merely a temporal and spatial shift in the hydrologic cycle. If saline sea water can be economically converted to fresh water, however, the evaporation and precipitation steps in the natural hydrologic cycles can be by-passed—or at least brought down to earth where needed.

12.19 RAIN MAKING

A moist air mass is saturated with water vapor, or has 100 percent relative humidity, when the partial pressure of water vapor in the air equals the vapor pressure of water at the temperature of the air. (Relative humidity is the ratio of the partial pressure of water vapor in the air to the vapor pressure of water at the temperature of the air.) If an air mass is not saturated with water vapor, the partial pressure of the water vapor is less than the vapor pressure of water at the air temperature. The air can be saturated with water vapor if it is cooled to a temperature where the vapor pressure of water becomes equal to the partial pressure of the water vapor in the air. At this temperature, called the *dew point*, water vapor just begins to condense to liquid water and a cloud is formed. In cloud formation, the temperature decrease is brought about by the cooling of a rising air mass. If the temperature is less than the dew point, water vapor continues to condense into water droplets that might coalesce and fall as rain. In most clouds, however, the droplets do not coalesce, and they remain too small to fall. If the temperature of the cloud falls below the freezing point of liquid water but the water does not freeze, the liquid water is in a supercooled condition (see Figure 8.9). Supercooled water is in an unstable equilibrium state similar to the dissolved solute in a supersaturated solution. The true stable form of water at the low temperature is the solid state. In the supercooled condition, the vapor pressure of the liquid water is greater than the vapor pressure of the stable ice form at the same tempera-ture. If ice crystals are present in the supercooled cloud, they grow rapidly in size as water vapor evaporates from the supercooled water droplets and deposits on the ice crystals because the vapor pressure of a condensed phase, liquid or solid, is a measure of the escaping tendency of its molecules into the vapor. The supercooled liquid water preferentially evaporates, and the molecules are transferred to the solid ice phase. Eventually, the ice crystals grow large enough to fall, thereby creating rain or snow.

For precipitation to occur according to this mechanism, ice crystals must be present. In some instances, ice crystals form spontaneously if the tempera-ture is low enough (-20 to $-40°C$). The temperature can be lowered by seeding supercooled clouds with solid CO_2 (dry ice), resulting in precipitation. In general, however, the formation of ice crystals in a cloud requires the presence of nuclei on which the ice may condense. Normally, there are particles in the air to serve as freezing nuclei. Particles of clay are believed to be the most common and the most effective of the naturally occurring nuclei. In addition, salt crystals from the ocean and a number of other naturally occurring particulates are also able to act as nuclei. When such nuclei are absent or, for some reason, do not cause precipitation at the time and place desired, cloud seeding with artificial nuclei has been comparatively successful. One of the most common nucleating substances is silver iodide. The silver iodide crystal is similar in structure to the ice crystal, and the similarity is important in serving as an effective base for the condensation and growth of ice. Silver iodide is easily vaporized on the ground and carried into the air by the wind and updrafts (Figure 12.10). The ice crystal description of rain formation explains rainfall from supercooled clouds. In the tropics, supercooling may not exist. In the warm tropical clouds, liquid water may condense on a few nuclei to form large rain drops at high altitudes. At lower elevations, smaller droplets can form by collision with droplets torn from the falling large drops (Figure 12.11).

Rain making by cloud seeding has not fulfilled expectations, but research is continuing. The formation of rain is still an imperfectly understood process requiring much more study. The powerful forces in the atmosphere and in weather systems may not easily yield to a kilogram or two of dry ice or a

Figure 12.10. Silver iodide cloud-seeding generator. Bottled propane is used as fuel, and the silver iodide is introduced in an acetone solution. The acetone burns off, leaving the silver iodide in the air to serve as nuclei for snow crystals. (Courtesy U. S. Department of the Interior, Bureau of Reclamation. Photo by W. L. Rusho.)

few grams of silver iodide. When rain or snow has resulted from cloud seeding, it has often been too much or in the wrong place or both. There are legal questions concerning responsibilities for flooding and the rights of those whose rain may be stolen before the clouds get to their lands. Rain making by cloud seeding has some distance to go before it can be regarded as a suitable source of fresh water.

12.20 SALINE WATER AND DESALINATION PROCESSES

If we realize that 71 percent of the earth's surface is covered by salt water to an average depth of $2\frac{1}{2}$ miles and that this water constitutes 98 percent of all the water on earth, it is no wonder that the conversion of saline water to fresh water has attracted great interest as the population and development of arid regions has increased.

The term *saline* water applies to any water containing over 1000 ppm

| Table 12.2 | | |
Classification of saline waters	concentration of dissolved solids (salts), ppm	classification
	1,000–3,000	slightly saline
	3,000–10,000	moderately saline
	10,000–35,000	very saline
	more than 35,000	brine

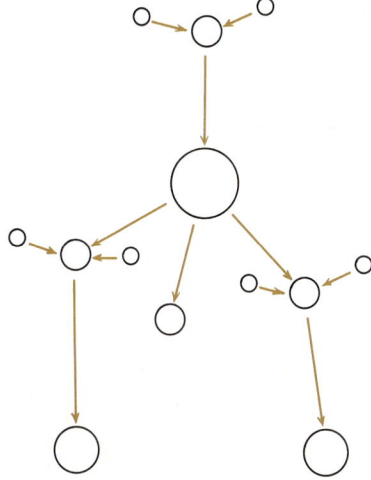

Figure 12.11. Coalescence and breakup of falling water droplets. Collision of small droplets formed at high altitude results in large drops. Coalescence of droplets from the breakup of descending large drops gives intermediate-sized drops at low altitudes.

of salts. This is equivalent to a salinity of 1‰ or 1 g of salts per kilogram of water. Brackish water contains more than 2000 ppm (2‰S) of salts and is usually found in groundwaters or waters of estuaries where rivers enter the sea. Another classification avoiding the imprecise name "brackish" is that used by the U.S. Geological Survey and listed in Table 12.2. Saline water is generally not suitable for human consumption, industry, or agriculture. The salinity of drinking (potable) water should be less than 500 ppm according to the U.S. Public Health Service. The maximum salinity permissible for human consumption is 1000 ppm, although many public water supplies provide water in excess of 2000 ppm as a matter of necessity. The nature of the dissolved salts has a lot to do with the fitness of the water. A high proportion of sodium or magnesium sulfates may cause even waters of salinity below 500 ppm to be unfit for drinking. On the other hand, if the salinity is due primarily to sodium chloride, water of high salinity may not be unfit for drinking. Agricultural irrigation water should not contain more than 2000 ppm of salts even though water containing several thousand parts per million dissolved solids is sometimes used when some of the salts can later be leached out of the soil by the use of large amounts of much less saline water applied to well-drained lands (see Section 14.16). Saline water is generally unsuitable for industrial use because of its corrosiveness and because of its hardness and scaling properties. Except for cooling purposes, industrial water usually has to be of equal quality to drinking water and in many cases must be much purer.

The separation of water and dissolved salts is variously called salt- (or saline-) water purification or conversion, or desalting, desalination, or desalinization. Hundreds of processes have been suggested, but few are satisfactory from the standpoints of energy and economic requirements and rate of conversion of saline to usable water. Desalination processes are often separated into two groups. In one group are those that remove the water from the dissolved salts and leave the salts behind. The other group of processes removes the salts from the water and leaves the water, or at least a more dilute solution, behind. A number of the methods in use are listed in Table 12.3). Other methods are under investigation, so this list is not to be considered complete. Most of the processes listed under the distillation and crystallization headings represent different engineering approaches to bring about the fundamental physical change desired. Before examining some of these processes, let us look first at some of the properties of solutions that will give us some insight into the separation of solvent and solute. These properties, classified under the name *colligative properties*, are all related to the vapor pressures of the solutions.

Table 12.3
Desalination processes

I. Processes removing water
   A. Distillation
      1. Multiple effect
      2. Multistage flash
      3. Multieffect–multistage
      4. Vapor compression
      5. Solar evaporation (below boiling point)
   B. Crystallization (freezing)
      1. Vacuum freezing–vapor compression (direct freezing)
      2. Secondary refrigeration (indirect freezing)
      3. Hydrate formation
   C. Reverse osmosis

II. Processes removing salts
   A. Electrodialysis
   B. Ion exchange
   C. Transport depletion (nonselective membrane electrodialysis)

12.21 VAPOR PRESSURES OF SOLUTIONS

The vapor in equilibrium with a solution contains molecules of all the components of the solution that have a tendency to leave the solution and go into the vapor phase. In such a case, the total pressure of the vapor in equilibrium with a solution at a given temperature is equal to the sum of the partial pressures of the components in the vapor. This is just Dalton's law of partial pressures, $P_t = P_A + P_B + \cdots$. For example, the vapor pressure of a solution of benzene (C_6H_6) and toluene ($C_6H_5CH_3$) at 20°C is 40 mm Hg. Of the 40 mm, 25 mm is due to benzene molecules in the vapor and 15 mm is contributed by toluene molecules. If the solutions are ideal, then the partial pressure of any component in the *vapor* is related to its concentration in the *solution* and its vapor pressure as a pure substance by *Raoult's law*,

$$P_A = P_A^0 X_A$$

where P_A is the partial pressure of component A in the *vapor*, P_A^0 is the vapor pressure of the pure component at the temperature of the solution, and X_A is the mole fraction of component A in the *solution*.

Ideality in liquid solutions assumes that all interactions between components of the solution are equal. Ideal solutions are just as rare as ideal gases, but the concept is useful just the same. A truly ideal solution would obey Raoult's law over the entire concentration range from pure A to pure B, if we restrict ourselves to solutions of two components. That is,

$$P_t = P_A^0 X_A + P_B^0 X_B$$

throughout the entire concentration range from pure A to pure B. Ideal behavior is rare, but for many solutions the component in excess obeys Raoult's law and behaves ideally in very dilute solutions as the mole fraction of the solute approaches zero. Ideal behavior is illustrated by the solid lines in Figure 12.12.

Raoult's law shows that the vapor pressure of a component in a solution

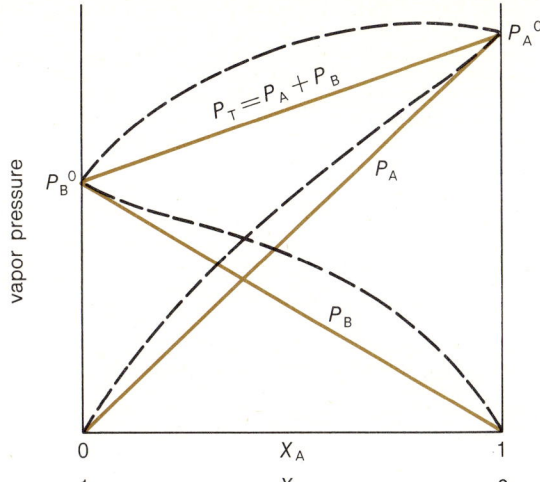

Figure 12.12. Vapor pressure curves for an ideal solution (solid lines) and for a solution exhibiting positive deviation from ideality (dashed lines). Positive deviations lead to a maximum total vapor pressure and a minimum boiling point at a definite solution composition.

is less than its value as a pure substance. For component A, for instance, $P_A = P_A^0 X_A$ and P_A is less than P_A^0 because X_A is always less than unity except in pure A. The same applies to any other component in the solution if it has a finite vapor pressure P^0 in the pure state.

If only one of the components of a solution is volatile (has a finite vapor pressure in the pure state), then its vapor pressure is also the total vapor pressure P_t of the solution,

$$P_t = P_A^0 X_A + P_B^0 X_B = P_A^0 X_A = P_A$$

if $P_B^0 = 0$. If we consider B to be solute, then B is a nonvolatile solute. Salts are typical nonvolatile solutes, so saline water is a solution of a nonvolatile solute in water.

Whether a solute is volatile or not, its presence in a solution lowers the partial vapor pressure of the solvent by decreasing the number of solvent molecules in a unit of surface area of the solution and reducing the opportunity, relative to that in pure solvent, of solvent molecules escaping into the vapor. The relative lowering of the escaping tendency of solvent molecules from the solution is proportional to the mole fraction of solute molecules.

The decrease in escaping tendency of component A (solvent) is proportional to the decrease $P_A^0 - P_A$ in vapor pressure due to the solute. If the solute is nonvolatile, the decrease in vapor pressure is also the decrease in total vapor pressure of the solution as a whole resulting from the presence of solute. From Raoult's law for the solvent A,

$$P_A = P_A^0 X_A$$
$$\Delta P_A = P_A^0 - P_A$$
$$= P_A^0 - P_A^0 X_A$$
$$= P_A^0 (1 - X_A)$$

or

$$\Delta P_A = P_A^0 X_B$$

where

$$X_B = \frac{n_B}{n_A + n_B}$$

But for a dilute solution, where n_B is small compared to n_A,

$$X_B = \frac{n_B}{n_A + n_B} \simeq \frac{n_B}{n_A} = \frac{w_B M_A}{w_A M_B}$$

where w is the weight in grams of substance in the solution and M is the molecular weight.

Also, since the molality of the solution,

$$m_B = \frac{1000 w_B}{M_B w_A}$$

$$X_B = \frac{m_B M_A}{1000}$$

Therefore,

$$\Delta P_A = P_A{}^0 \frac{m_B M_A}{1000}$$

and ΔP_A is proportional to the molality of the solution, the proportionality constant $P_A{}^0 M_A / 1000$ depending on the solvent.

12.22 COLLIGATIVE PROPERTIES OF SOLUTIONS

The lowering of the vapor pressure of a liquid solvent by dissolved nonvolatile solutes is responsible for a number of properties of solutions. The vapor pressure depression results in an increase in boiling point of the solution as compared to the pure solvent, a decrease in freezing point of the solution, and gives rise to the osmotic pressure exhibited by solutions. All of these four phenomena—vapor pressure lowering, boiling point elevation, freezing point depression, and osmotic pressure—are called colligative properties because their value for a particular solvent depends only on the number of solute particles per unit volume in the solution and not on the nature of those particles.

The boiling point elevation and the freezing point depression can be understood from Figure 12.13, where the vapor pressure–temperature relations of a pure solvent and a dilute solution are given. The difference in normal boiling points of solution and solvent, $T_b - T_b{}^0 = \Delta T_b$, is proportional to the vapor pressure lowering at the normal boiling point of the pure solvent $T_b{}^0$ if the two vapor pressure curves remain parallel near the boiling point, as they do. Therefore, since we have already seen that ΔP is proportional to the molality m of the solution, ΔT_b must also be proportional to m:

$$\Delta T_b \propto \Delta P \propto m$$

so

$$\Delta T_b \propto m$$
$$\Delta T_b = K_b m$$

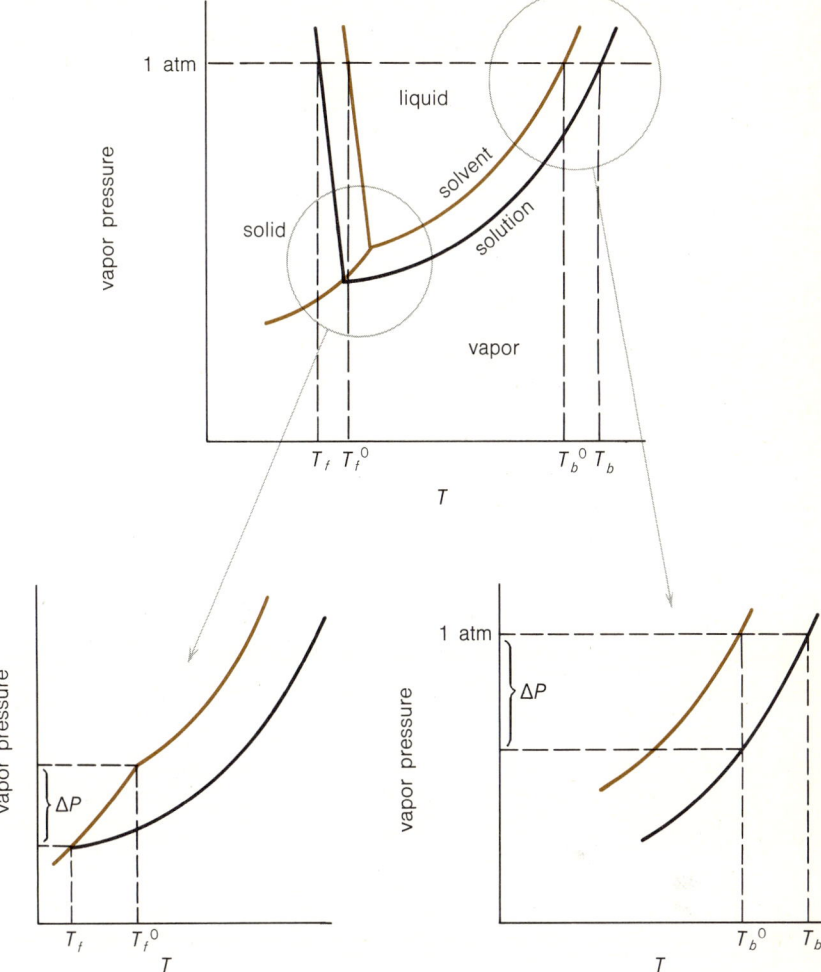

Figure 12.13. Vapor pressure curves at the freezing point and the boiling point for water, and an aqueous solution of a nonvolatile solute illustrating freezing point depression and boiling point elevation. Triple points are used in place of normal freezing points to indicate T_f in the exploded view.

where the proportionality constant K_b is called the molal boiling point elevation constant. Its value depends only on the solvent.

Qualitatively, the increased boiling point of a solution of nonvolatile solutes can be justified by the following argument. At the normal boiling point of the pure solvent, the vapor pressure of the *solution* is less than 1 atm. In order for the solution to boil, its vapor pressure must be raised to 1 atm. The vapor pressure is increased by adding thermal energy to increase the rate of escape of solvent molecules. The increase in temperature required to do this is ΔT_b. A similar analysis for the freezing point depression can be made if it is assumed that the triple points are changed in the same way as the normal freezing points. Then $\Delta T_f = T_f^0 - T_f$ is the lowering of the freezing point and

$$\Delta T_f = K_f m$$

where K_f is the molal freezing point depression constant and depends on the solvent alone. This treatment assumes that the only solid freezing out of the solution is pure solid solvent. Some values of K_b and of K_f for several solvents are listed in Table 12.4. These constants should not be interpreted as meaning that a 1 m solution of sugar in water will boil at 100.52°C at 1 atm pressure. A 1 m solution is much too concentrated to follow the $\Delta T_b = K_b m$ relationship, which applies only to dilute solutions.

Table 12.4
Molal boiling point and freezing point constants

solvent	normal b.p. (°C)	K_b, °C/molal	normal f.p. (°C)	K_f, °C/molal
water	100	0.52	0	1.86
chloroform	61.3	3.63		
carbon tetrachloride	61	5.0		
ethyl alcohol	78.4	1.22		
benzene	80.1	2.53	5.4	5.10
camphor			173	40.0

Vapor pressure depressions are very small and difficult to measure for dilute solutions, but boiling point elevations and freezing point depressions are quite easy to measure. A few examples will serve as illustrations of the kind of information obtainable from boiling point and freezing point measurements.

EXAMPLE | When 1.55 g of an unknown compound is dissolved in 75.0 g of chloroform, the boiling point of the solution is 0.860°C higher than that of pure chloroform. What is the apparent molecular weight of the unknown compound?

SOLUTION |
$$\Delta T_b = K_b m$$
$$\Delta T_b = 0.860°C \qquad K_b = 3.63°C/molal$$
$$m = \frac{0.860}{3.63} = 0.247 m$$

There are 1.55 (1000/75.0) = 20.6 g of solute in 1000 g of chloroform; this 20.6 g represents 0.247 mol of the unknown compound. One mole of compound corresponds to

$$20.6\left(\frac{1}{0.247}\right) = 83.6 \text{ g}$$

The molecular weight is 83.6 g/mol.

The same result could have been obtained using the equation $\Delta T_b = K_b m$ with the substitution of

$$m = \frac{1000 w_B}{M_B w_A}$$

In practice, it is usually preferable to determine molecular weights by freezing point depression because of the possibility of decomposition of compounds at the higher boiling temperatures.

EXAMPLE | If 7.30 g of NaCl is dissolved in 250 g of water, what is the freezing point of the resulting solution?

SOLUTION

$$\Delta T_f = K_f m \qquad K_f = 1.86°C/molal$$

The molality of NaCl in the solution is the number of moles of NaCl in 1000 g of water. There are $7.30(1000/250) = 29.2$ g NaCl in 1000 g of water, or

$$\frac{29.2 \text{ g}}{58.4 \text{ g/mol}} = 0.500 \text{ mol NaCl}$$

The solution is $0.500m$ in NaCl. But NaCl is a strong electrolyte and dissociates almost completely to give two ions for each NaCl formula unit:

$$NaCl \longrightarrow Na^+ + Cl^-$$

The colligative properties depend on the *number* of particles in solution. In a $0.500m$ NaCl solution, there is 1.00 mol of particles in 1000 g of solvent (0.500 of Na^+ and 0.500 of Cl^-), so the above solution is $1.00m$ in terms of particle concentration. The freezing point depression is then

$$\Delta T_f = (1.86)(1.00) = 1.86°C$$

rather than the $0.93°C$ that would be obtained if NaCl did not dissociate at all. Since the freezing point of the pure solvent is $0°C$, the freezing point of the solution is $-1.86°C$.

EXAMPLE

A $0.0030m$ solution of acetic acid, HAc, has a freezing point of $-0.0061°C$. What fraction of HAc has dissociated?

SOLUTION

If the acid were completely dissociated according to the equation $HAc \longrightarrow H^+ + Ac^-$, the freezing point of the $0.0030m$ solution would have been $-2(0.0030)(1.86) = -0.0011°C$. If none of the HAc molecules dissociated, ΔT_f would have been $0.0056°C$. If x is the number of moles of HAc that do dissociate, then the concentrations of all species at equilibrium are

$$HAc \rightleftharpoons H^+ + Ac^-$$
$$m - x \qquad x \qquad x$$

The total number of moles of all species is

$$m - x + x + x = m + x$$

The value of x can be found from the actual freezing point:

$$T_f = 0.0061°C = (1.86)(0.0030 + x)$$
$$x = \frac{0.0005}{1.86} = 0.00027 \text{ mol}$$

Therefore, of the $0.0030m$ HAc, only

$$\frac{0.00027}{0.0030} \times 100 = \frac{0.027}{0.0030} = 9.0 \text{ percent}$$

has dissociated.

The other colligative property, osmotic pressure, has been discussed in Section 11.5. It also can be related directly to the depression of vapor pressure by a solute. Measurement of the osmotic pressure of solutions of high molecular weight polymers is widely used to determine the molecular weight of such substances.

After this excursion into colligative properties, we are ready to return to saline water conversion. Generally, we consider the desalting of sea water, but desalination is also applied to saline ground and surface waters as well.

12.23 ENERGY REQUIREMENTS IN DESALINATION

Energy must be expended to convert salt water into fresh water. The necessity for this energy can be visualized by considering the vapor pressures of sea water and of fresh water. The numerical values of many of the properties of 35‰S sea water are similar to those of a 0.5m NaCl solution. The vapor pressure of sea water, (or of 0.5m NaCl), P is obviously lower than that of pure water, P^0, but not much lower. If a beaker of sea water and a beaker of pure water are placed together in a closed container (Figure 12.14), the greater escaping tendency (higher vapor pressure) of the pure water results in the spontaneous passage of water from the pure-water beaker to the sea-water beaker and an increase in volume of the sea water.

To reverse this spontaneous process and cause water to leave the saline solution and condense in the pure water, energy must be supplied to work against the pressure difference. The desalination process may be thought of as occurring in three steps:

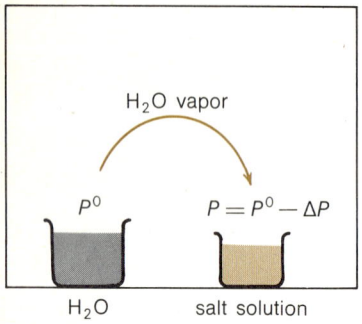

Figure 12.14. Pure water has a higher vapor pressure than a salt solution. Water will transfer spontaneously to the salt solution.

$$H_2O(l, P) \longrightarrow H_2O(v, P) \tag{1}$$
$$H_2O(v, P) \longrightarrow H_2O(v, P^0) \tag{2}$$
$$H_2O(v, P^0) \longrightarrow H_2O(l, P^0) \tag{3}$$

If the energy of vaporization in step 1 is regained in the condensation of step 3, the only energy required is that for the compression of water vapor from the pressure P to the higher pressure P^0 in step 2. This energy corresponds to that needed to separate water from the ions in solution. The minimum energy for desalination of 35‰S sea water turns out to be about 2.7 kW hr or 2300 kcal/1000 gal(U. S.) of fresh water.

12.24 DESALINATION BY DISTILLATION

Distillation is a two-stage process in which a liquid is first vaporized by boiling, followed by cooling and condensation of the vapors. If the liquid is a solution of a nonvolatile solute, the vapor consists only of the volatile solvent, and the condensed vapor (the *distillate*) is pure solvent. The nonvolatile solid solute remains behind. Because the salts in sea water are nonvolatile, this simple process has been long used for producing fresh from saline water.

If the solution contains two or more volatile and miscible components, the vapor will be a mixture. The vapor will have a greater concentration of the solution component having the lower boiling point (the higher vapor pressure). At any solution composition, the partial pressures of the two (or more) volatile components in the vapor is given by Raoult's law:

$$P_A(\text{vapor}) = P_A^0 X_A \text{ (liquid solution)}$$

and

$$P_B (\text{vapor}) = P_B^0 X_B \text{ (liquid solution)}$$

where the P^0's are those at the boiling temperature of the mixture boiling under a total pressure $P_t = P_A(v) + P_B(v)$.

The composition of the *vapor* is given by Dalton's law of partial pressures.

$$X_A \text{ (vapor)} = \frac{P_A(v)}{P_t}$$

$$X_B \text{ (vapor)} = \frac{P_B(v)}{P_t}.$$

At any given external pressure, the boiling point elevation of a salt-water solution corresponds to the extra energy needed to establish equilibrium between water vapor and solution, considering that it takes no net energy to produce pure water from pure water.

In the distillation of large quantities of saline water for human, industrial, or agricultural use, the major problems are economic. Even though the vapor pressure difference between fresh and salt water is very small, this difference is related to the minimum energy required for desalting the water. When hundreds of thousands of gallons of fresh water are distilled, the cost of supplying the energy can become substantial. And the actual cost of desalting is usually approximately four times that of the theoretical minimum. Various processes, all based on the simple evaporation–condensation distillation principle, have been designed to improve the efficiency of the distillation process in terms of cost and rate of water conversion. Plant designs primarily attempt to prevent the loss of heat energy at various steps and to improve heat transfer. Distillation of sea water is merely a miniature earthbound version of the huge hydrologic cycle.

Multiple Effect Distillation

In most distillation processes, the incoming preheated salt water is boiled by contact with metal heat exchange tubes containing steam. If the steam is at 100°C, the salt water must be at a lower temperature for heat to pass into it. Since the normal boiling point of sea water is about 100.5°C, the boiling temperature must be lowered below 100°C by reducing the pressure in the evaporator. The efficiency of energy utilization in distillation is increased by using the heat recovered from the condensation of water vapor from the first boiling step for the boiling of salt water in a succeeding evaporator (Figure 12.15).

Usually, the residue from the first boiling is boiled again using the vapors from the first boiling as the heat supply. Because the residue is more concentrated than the original sea water and its temperature has also decreased, the successive boilings must be carried out at continually lowered pressures. Each of the boilings or evaporations is called an *effect*.

Multiple Stage Flash Distillation

In the popular multistage flash distillation (Figure 12.16), previously heated sea water is caused to boil rapidly by introducing it into a chamber where the pressure is below the equilibrium vapor pressure at the sea-water temperature. Unevaporated sea water at a lowered temperature passes on to another evaporator where it is rapidly boiled or "flashed" at a still lower pressure. Incoming sea water is preheated in each of the evaporators, and the sea-water coils serve as condensation surfaces.

Vapor Compression

In the vapor compression method, externally supplied steam does not heat the incoming sea water. Instead, sea water is boiled by heat evolved from the compression of vapors in a later stage of the process. The condensation of

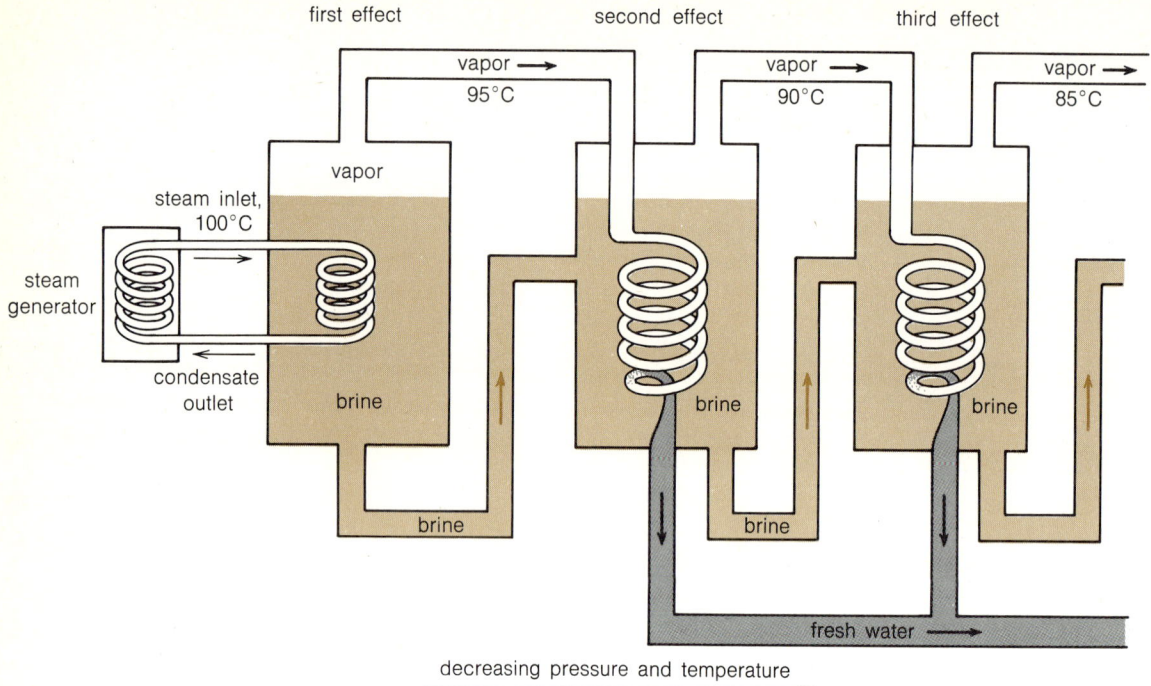

Figure 12.15. Multiple effect distillation. Vapor from the first evaporator (effect) condenses in the second; the heat of condensation is used to boil brine at a lower temperature and pressure in the second evaporator. This process is repeated through several effects until most of the heat supplied in the first effect is recovered.

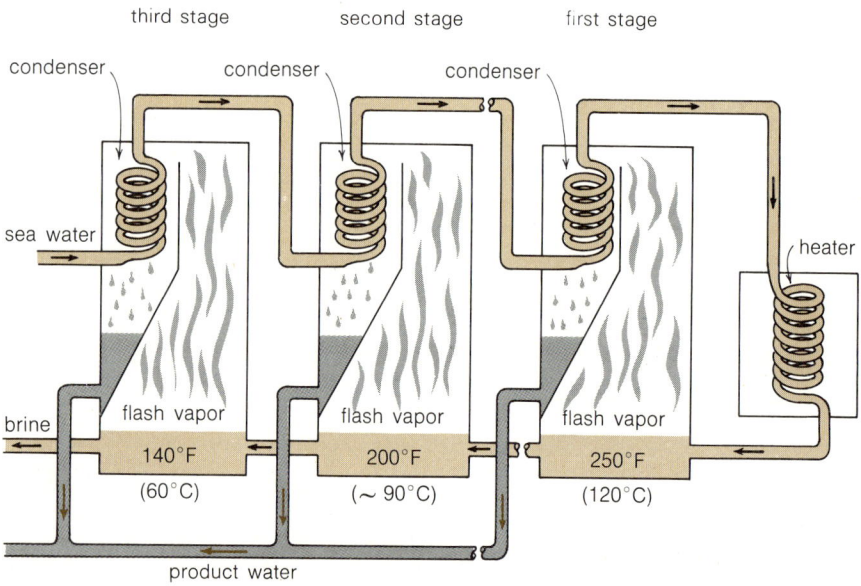

Figure 12.16. Multistage flash distillation.

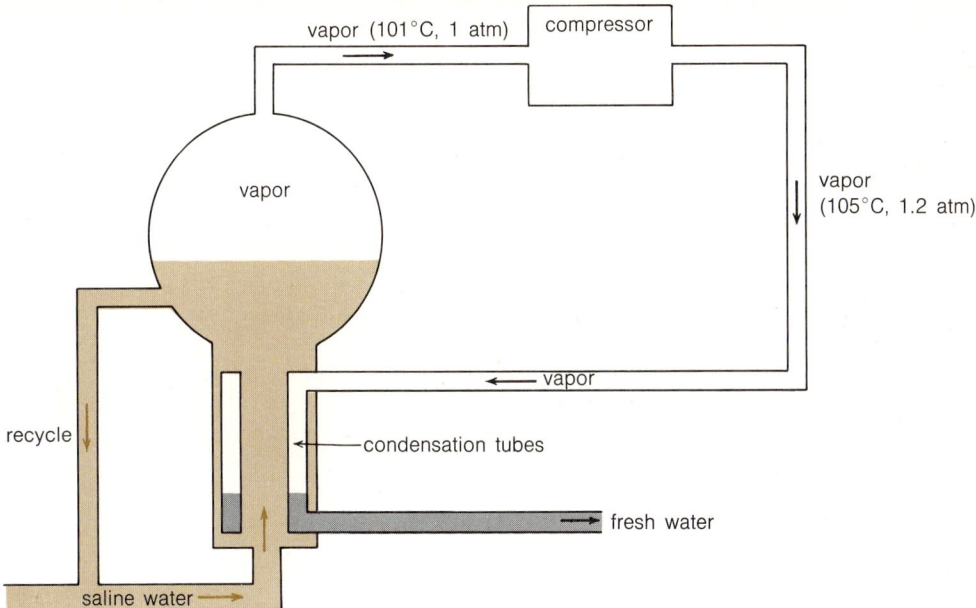

Figure 12.17. Vapor compression distillation. Only one effect is shown in this simplified diagram.

the heated, compressed vapor when it gives up heat to the incoming saline water produces fresh water (Figure 12.17). The only thermal energy required is that needed to get the process going. After that the only energy required is the mechanical or electrical energy to power the compressor.

Solar Distillation   Evaporation with a solar still is illustrated in Figure 12.18. The salt water does not boil, but vaporizes slowly and condenses on the cooler glass or plastic surface. Although the energy requirement is nil, the rate of conversion of salt to fresh water is small unless large surface areas are provided for the evaporation process.

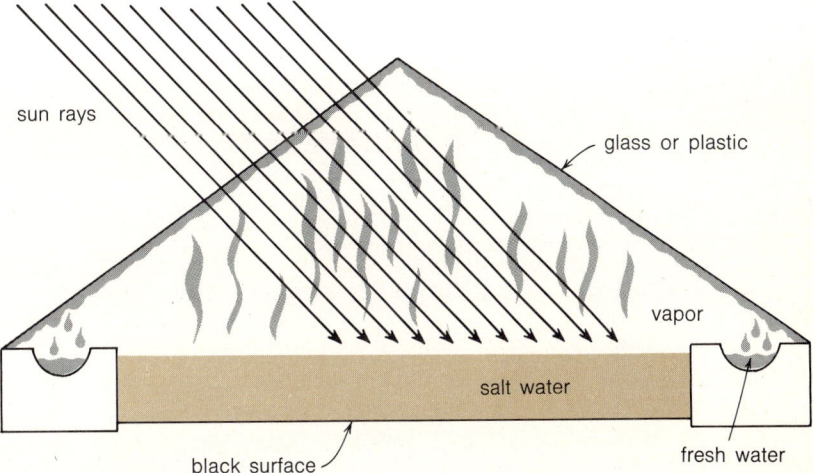

Figure 12.18. A simple solar still.

Distillation processes are responsible for about 95 percent of the saline-water conversion in the world today. Thermal distillation processes will become even more common and less costly as they are coupled with fossil fuel and nuclear electric power generating plants to utilize the waste heat (see Sections 12.13 and 20.16) from these plants.

12.25 DESALINATION BY CRYSTALLIZATION (FREEZING)

When a salt-water solution is slowly cooled to its freezing point, the solid that freezes out is pure ice and not a solid mixture of ice and salt. Desalination by crystallization is based on the separation of the ice crystals from the unfrozen brine mixture. The freezing process has certain advantages because the latent heat of fusion of water is much less than its heat of vaporization, and the lower temperatures reduce corrosion and scaling problems. As with distillation, the magnitude of the energy required to convert salt water to fresh water by freezing is proportional to the freezing point depression of the solution, that is, to the extra energy required to reduce the temperature of the solution from 0°C to the actual freezing point.

Engineering features of freezing processes are designed to make use of the heat given up in the freezing of the solution for the subsequent melting of the ice crystals produced. As in distillation, these measures attempt to prevent the waste of heat and to improve the economics of the process. We shall not go into these design problems here but shall discuss only the basic freezing processes.

In the vacuum freezing–vapor compression or direct freezing process, water acts as its own refrigerant. Precooled sea water is flash evaporated at a low pressure. The rapid evaporation removes heat (the heat of vaporization) from the sea water, and with proper design enough heat can be extracted to freeze the sea water.

In the secondary refrigeration method (indirect freezing), a volatile liquid such as butane, C_4H_{10}, is evaporated in contact with cool sea water. The heat needed for butane vaporization is taken from the water and causes freezing of the solution. Butane and water from the melted ice are immiscible and readily separated.

Desalination by hydrate formation depends on the formation of a solid crystalline compound of water and the refrigerating agent. These hydrate compounds are members of a general type of compound called *clathrates* in which a molecule of one component is enclosed within a cage composed of molecules of the other component. Water molecules form such a cage enclosing other molecules, among which are certain hydrocarbons such as propane, C_3H_8, and the inert gases. If propane is used as a secondary refrigerant in the indirect freezing process, a clathrate solid will form about 6°C above the normal freezing point of ice. The clathrate (Figure 12.19) is produced by water molecules forming a cage holding one propane molecule. Salt ions are excluded. The solid clathrate can be separated from the brine and melted, and the pure water and immiscible propane separated.

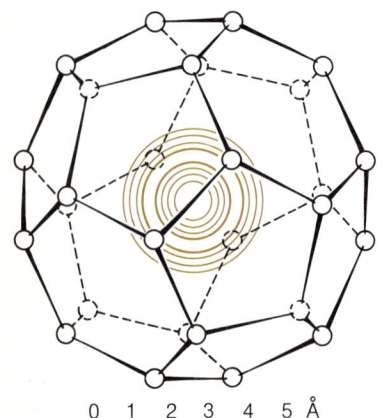

0 1 2 3 4 5 Å

Figure 12.19. A clathrate (cage) compound. The small circles represent water molecules; the large central body is a hydrocarbon molecule.

12.26 REVERSE OSMOSIS

Osmosis was introduced in Section 11.5, where it was shown that water tends to diffuse through a semipermeable membrane separating a salt solution and pure water. The spontaneous direction of flow is from the pure water to the salt solution so that the solution tends to become more dilute. The osmotic pressure of the salt solution was the pressure required on the solution side

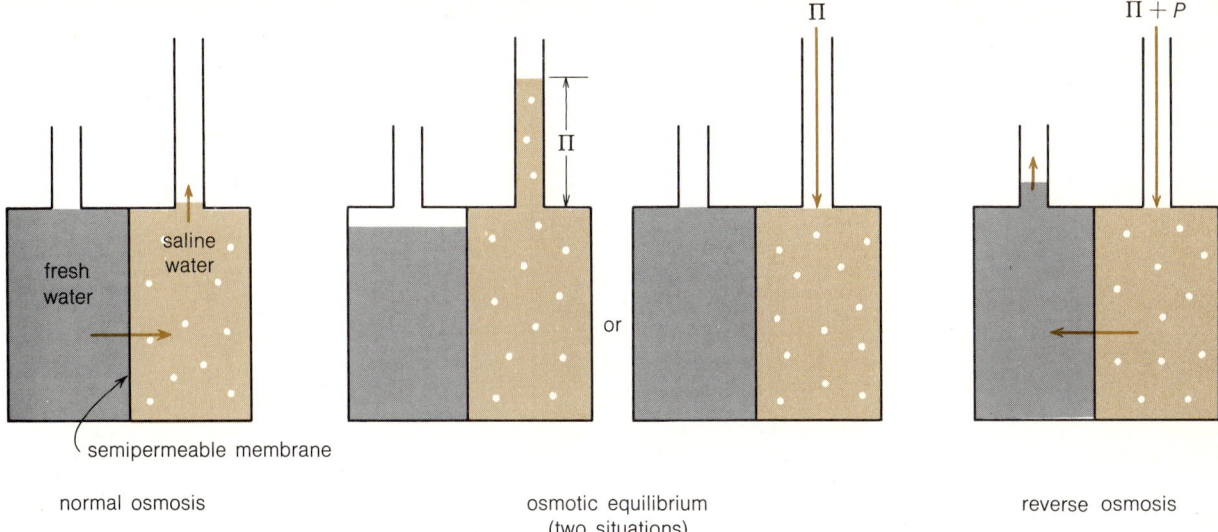

normal osmosis

osmotic equilibrium
(two situations)

reverse osmosis

Figure 12.20. Reverse osmosis.

to just prevent the flow of water into the solution. By exerting a hydrostatic pressure greater than the osmotic pressure of the solution on the solution, the direction of water flow may be reversed and water will flow out of the solution, leaving the salts behind. All of these situations are sketched in Figure 12.20.

Most present-day membranes are made of cellulose acetate. Semipermeable membranes do not necessarily function by a simple sieve or filter-type mechanism, but their exact mechanism of action is uncertain.

Since there is no phase change involved in saline-water conversion by reverse osmosis, the only energy consumed is that required by the pumps supplying the hydrostatic pressure. The energetics of the process are favorable compared to distillation and freezing. Nevertheless, reverse osmosis is limited by the need for strong membrane material and membrane support to withstand the high pressures. Opposed to this requirement is the need for high water permeability and surface area so that an adequate rate of conversion is realized. For these reasons, reverse osmosis is employed for desalination of slightly and moderately saline (brackish) water but is not yet sufficiently developed for economic application to sea water. Reverse osmosis is also under study for removing phosphates and nitrates from secondary effluents in waste treatment plants and from agricultural drainage waters.

12.27 ELECTRODIALYSIS

Electrodialysis is another membrane process but with an electric field rather than hydrostatic pressure as the driving force. Instead of membranes impermeable to all ions, electrodialysis employs ion-selective membranes made of ion exchange material. One kind of membrane is permeable only to negative ions (anion permeable), while another kind is permeable only to positive ions (cation permeable). Both kinds of membrane must be able to conduct electricity, and the rate of passage of water across the membranes must be less than the rate of migration of ions. Figure 12.21 illustrates a simple electrodialysis cell. Cations migrate through the cation-permeable membrane to the

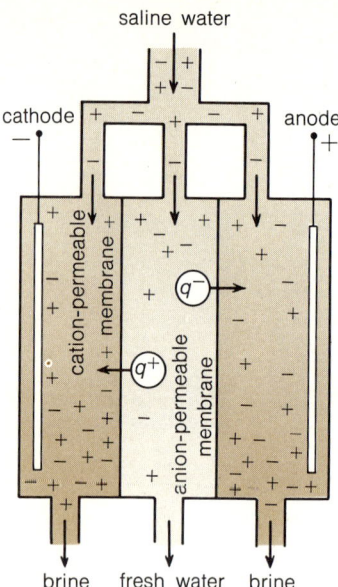

Figure 12.21. Electrodialysis.

negative electrode, anions go to the positive electrode and deplete the solution in the central compartment of salts. In practice, stacks of alternating cation- and anion-permeable membranes are employed so that the solution in alternate compartments is depleted of ions.

Electrodialysis is widely used for brackish water. However, the electric power requirements mount with the salinity of the water; and the economic factor, combined with anion-permeable membrane deterioration and fouling problems caused by organic material precludes desalination, at present, of sea water in high-volume plants. Electrodialysis can be, and is, used to remove dissolved inorganic ions (i.e., PO_4^{3-}) from secondary sewage effluent.

12.28 TRANSPORT DEPLETION AND ION EXCHANGE

Transport depletion is a variation of electrodialysis. The anion-permeable membrane is replaced by a nonselective membrane, as shown in Figure 12.22.

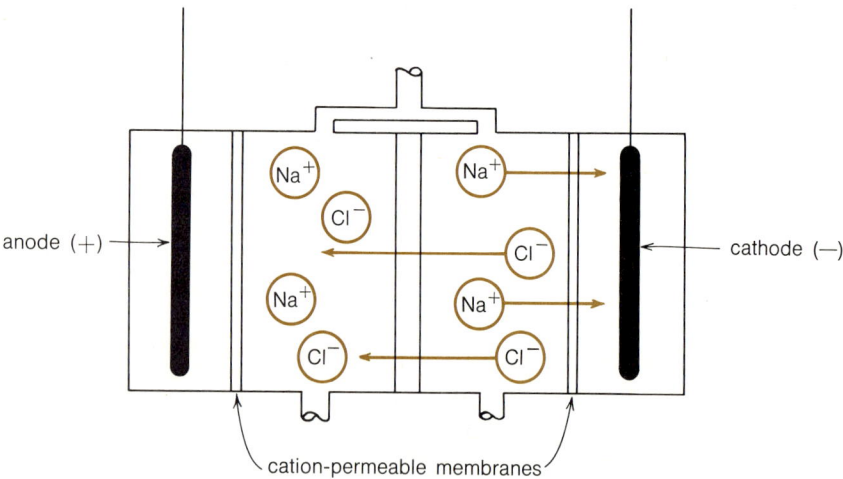

Figure 12.22. Transport depletion.

The right-hand compartment is depleted of ions as the cations move through the cation-permeable membrane to the negative electrode. The anions move toward the positive electrode, and the left-hand compartment is enriched with salts. Aside from replacement of the anion-permeable membrane, transport depletion suffers from the same high power requirements with highly saline sea water as does electrodialysis.

Descriptions of ion exchange, ion exchange resins, and demineralization of water were given in Section 9.11 in connection with water softening. Ion exchange is not economically feasible for desalting large volumes of sea water because of the high salt concentration and the need for frequent regeneration of the exchange resins. It is used for small, emergency sea-water desalination units and for the purification of slightly saline waters.

12.29 COST OF DESALTED WATER

In 1970, more than 850 desalination plants were in operation or under construction around the world. Most were distillation plants. One of the largest, a flash distillation plant, had a capacity of 7.5 million gal/day (mgd). Desalination plants in the planning stage have a fresh-water output of up to 150 mgd. These large plants will be combined electric power generation–desalting plants and may be able to produce fresh water for 25¢ to 15¢ per 1000 gal, although the cost of water from existing plants is more like 85¢/1000 gal. For comparison, tertiary treatment (complete recycling) of secondary sewage effluent in large 100-mgd plants can produce drinking water for 34¢/1000 gal, and general irrigation water for 13–17¢/1000 gal. When labor, distribution, treatment of raw water from conventional sources, and other costs associated with the supply of municipal water are included, this 25–15¢ cost should be competitive with fresh water from conventional sources. Because treatment and distribution are the major items in the cost of delivered municipal water, desalination can be expected to become economically more popular for coastal cities as their nearby sources of fresh water become more polluted. It appears, however, that the value of irrigation water to the farmer is in the range 3–10¢/1000 gal (delivered), and even large desalting plants cannot meet that figure. For the next 20 yr, at least, the cost of water from present state-of-the-art desalination processes will be at least ten times greater than the value of water to agriculture.

12.30 SUMMARY

Water is an inexhaustible natural resource that is continually recycled in the natural hydrologic cycle. Water shortages do arise, however, because of inadequacies in water distribution and because of degradation of water quality resulting from pollution.

The major sources of pollution are domestic, industrial, and agricultural wastes. Such wastes, if untreated, can irreparably impair the quality of natural receiving waters by introducing pathogenic organisms, toxic substances, excess plant nutrients, and organic substances whose decomposition results in oxygen depletion.

The strength of sanitary sewage and of the effluent from treatment plants is measured by such parameters as the suspended solid content, biochemical oxygen demand (BOD), and the chemical oxygen demand (COD). The dissolved oxygen (DO) content of water must be known for the determination of the BOD, and it is also a key factor in deciding the adequacy of natural waters for the life of fish and other aquatic organisms.

Sanitary sewage treatment usually consists of primary and secondary treatment. In primary treatment, solids are allowed to settle and are removed from the water. Secondary treatment consists of biological treatment of the primary effluent utilizing aerobic bacteria to decompose oxygen-demanding organic wastes. Tertiary or advanced treatment can be applied to the secondary effluent to effect further removal of solids, organic substances, dissolved salts and plant nutrients, and toxic substances.

Some major effects of water pollution are oxygen depletion caused by high BOD wastes, eutrophication resulting from addition of limiting plant nutrients to natural waters, thermal pollution due to disposal of heated water into water courses, and poisoning by toxic substances in untreated industrial and agricultural wastes. Eutrophication is attributed largely to phosphate nutrients in secondary effluent. Most of the phosphates are contributed by sodium tripolyphosphate, a major constituent in household detergents.

To satisfy the increasing demand for water for domestic, industrial, and agricultural purposes, natural, but polluted, fresh waters must be purified, and new or less common sources of water must be utilized.

Municipal treatment of natural waters for domestic purposes makes use of a variety of disinfection, sedimentation, filtration, and softening processes. Chemical coagulation is a common process for the clarification of water by the removal of nonsettleable solids of colloidal (10^{-6}–10^{-9} m) dimensions. In this process, the negative charge on colloidal sol particles is neutralized with positive Al^{3+} or Fe^{3+} ions so the suspended particles can stick together and precipitate.

Rain making and large-scale desalination are two comparatively new approaches used to augment the supply of fresh water. In rain making, clouds are seeded with condensation nuclei on which water may condense or freeze.

Desalination of saline water is accomplished either by removing dissolved salts from the water or by extracting fresh water from the salt solution. The energy expended in all desalination processes depends on the vapor pressure of a salt solution relative to that of pure water, and through the vapor pressure, on the boiling point, freezing point, or osmotic pressure of the solution. Because the vapor pressure of an aqueous solution of a nonvolatile solute is always less than the vapor pressure of pure water, the boiling point and osmotic pressure of a salt solution are higher, and the freezing point is lower than those of pure water. Boiling points, osmotic pressures, and freezing points of solutions are known as colligative properties.

The most popular desalination methods are variations on the simple distillation of salt solutions wherein the water vapor is condensed at a lower temperature. Other methods utilize the direct or indirect freezing of salt solutions and the formation of hydrate (clathrate) compounds. Membrane processes such as reverse osmosis and electrodialysis are appropriate for the removal of dissolved salts from moderately saline water. Except in arid regions, desalination of saline water is currently more expensive than the complete recycling of waste water by tertiary treatment.

Questions and Problems

1. What is the equilibrium concentration of oxygen gas in fresh water at 20°C when exposed to air at a pressure of 760 mm? The Henry's law constant for oxygen in water at 20°C is 0.0314 liter (STP)/liter atm. Express your answer in milligrams of oxygen per liter.

2. A waste water sample is known to contain 300 mg/liter of glucose, $C_6H_{12}O_6$, as the only oxidizable waste. The chemical oxygen demand of this water is the milligrams of oxygen necessary to oxidize the glucose in 1 liter to carbon dioxide and water.
   a. What is the theoretical chemical oxygen demand of this glucose solution?
   b. In the laboratory determination of COD, dichromate ($Cr_2O_7^{2-}$) is used as the oxidizing agent. Write the balanced reaction for the oxidation of glucose to CO_2 and H_2O by dichromate ion in acid solution. The dichromate is reduced to Cr^{3+}.
   c. In practice, the COD is determined by adding excess dichromate to the water sample to ensure as complete an oxidation as possible. The amount of dichromate reduced is found from a redox titration with standard Fe^{2+} solution. If we assume standard dichromate can be added to just oxidize the matter present and that the volume of $0.0420M$ $K_2Cr_2O_7$ solution required to do this is 10.0 ml for a 50.0-ml waste water sample, what would be the COD of the water? The quantity of dichromate can be related to the oxygen equivalent by the equations for the oxidation of glucose by oxygen and by dichromate.
   d. How would the BOD_5 of the waste water sample in part (c) compare with the COD if (1) glucose were the only pollutant? and (2) the water contains, in addition to glucose, a quantity of nitrogenous matter and some cellulose waste?

3. Verify that the milliliters of $0.025M$ sodium thiosulfate used in the Winkler determination of dissolved oxygen in a 200-ml sample of water represents the DO in milligrams per liter of water.

4. Two 10-ml samples of a waste water are diluted to 300 ml with aerated water containing inorganic nutrients and seeded with bacteria. One sample is analyzed immediately for dissolved oxygen, but the second sample is incubated for 5 days at 20°C before its dissolved oxygen is determined. The results are $(DO)_0^{20°C} = 7.90$ mg/liter, $(DO)_5^{20°C} = 1.00$ mg/liter. Taking sample dilution into account, what is BOD_5 of the waste water? Ignore any BOD associated with the dilution water.

5. Daytime pH values as high as 10 have been observed in shallow waters where algae are growing profusely. At night, the pH of the water decreases. Explain these observations.

6. If the atomic composition of photosynthetic phytoplankton (algae) is taken as $(CH_2O)_{106}(NH_3)_{16}$ H_3PO_4, then inorganic nitrogen and phosphorus nutrients (mainly as nitrate and phosphate) are removed from waters during the growth of these organisms in the atomic ratio $N:P \simeq 16:1$. When the organisms die and decompose, nitrogen and phosphorus are returned to the water in the same ratio, so unpolluted natural waters should maintain the same concentration ratio as the living organic matter. However, allowing for variations, we shall take the average atomic ratio in algae and in water as $N:P = 10:1$. If an analysis of a sea water shows an $N:P$ atomic ratio of $3:1$, would banning the use of phosphate-containing detergents in the drainage area prevent or reverse cultural eutrophication?

7. The composition of photosynthetic aquatic organisms is commonly given in terms of the ratios of the number of atoms of carbon, nitrogen, and phosphorus in their tissue as $C:N:P = 106:16:1$. Analysis of organic matter and of water usually affords the weights, not the number of atoms, of these elements. What would be the corresponding weight ratios among these three elements?

8. The reaction

$$(CH_2O)_{106}(NH_3)_{16}H_3PO_4 + 138O_2 \xrightarrow[\text{growth}]{\text{decay}}$$

$$106CO_2 + 122H_2O + 16HNO_3 + H_3PO_4$$

represents aerobic bacterial decomposition and growth of photosynthetic marine organisms (algae).
   a. If phosphorus is limiting, how many grams of algae will be produced by the photosynthetic assimilation of 1 mg of phosphorus from the water?
   b. When the algae mass formed in (a) sinks to the bottom and decomposes, approximately what BOD will it exert?
   c. How much phosphorus and nitrogen would be released to the water by the complete consumption of the oxygen in 1 liter of water at 25°C?

9. The permissible upper limit for cadmium in drinking water as set by the U. S. Public Health Service is 0.01 mg/liter (0.01 ppm = 10 ppb). If a waste water known to contain cadmium is saturated at a constant pH = 7 with hydrogen sulfide at 25°C and 1 atm, would the undiluted effluent meet the Cd standards for drinking water? If the pH = 0, would the same process produce an effluent meeting the Cd standards? K_{sp} for CdS is 7×10^{-27}.

10. Silver ion is a bactericide. Silver iodide is used in cloud seeding. The U. S. Public Health Service sets

a 0.05 mg/liter upper limit for silver in drinking water. Discuss some of the considerations that might be in order in view of these facts. Include the possible effects on grazing or ruminant animals.

11. Give two reasons why the biological magnification or concentration of environmental contaminants such as heavy metals and persistent pesticides is more likely to involve aquatic organisms than terrestrial organisms.

12. The partial pressure of water vapor in a sample of air at 20°C is 10 mm.
   a. What is the relative humidity?
   b. What is the approximate dew point for this air?
   c. What is the absolute humidity, that is, the mass of water vapor present in 1 liter of the moist air, at the relative humidity in part (a)?

13. When 0.5550 g of a nonvolatile solute of molecular weight 110.1 g/mol is dissolved in 100.0 g of a solvent whose freezing point is 45.000°C and whose molecular weight is 94.10 g/mol, the resulting solution freezes at 44.618°C. What is the molal freezing point constant for the solvent?

14. A $0.01m$ solution of Hg_2Cl_2 boils at 100.0154°C. What is the molality of ions in the solution? Can you deduce what ions are most likely to be present in the solution?

15. What would be the freezing point of a $0.01m$ aqueous solution of a weak acid HX with an ionization constant at 0°C of 1.0×10^{-4}?

16. Verify by calculation that the normal boiling point of sea water, approximated by $0.5m$ NaCl, is about 100.5°C.

17. What is the vapor pressure of 35‰S sea water at 25°C? Assume that sea water is $0.50m$ NaCl and that the NaCl is completely dissociated.

18. Ethanol and methanol form near-ideal solutions with one another. If the mole fractions of ethanol and methanol in a solution are 0.40 and 0.60, respectively, what is the vapor pressure of the solution and what is the composition of the vapor in equilibrium with the liquid at 20°C? The vapor pressure of pure ethanol at 20°C is 44 mm; of pure methanol, 89 mm.

19. What pressure, in atmospheres, is required to just force 25°C sea water through a semipermeable membrane in a reverse osmosis desalination process? Take the molar concentrations of the major ions in sea water as Na^+, $0.47M$; K^+, $0.01M$: Ca^{2+}, $0.01M$; Mg^{2+}, $0.054M$; Cl^-, $0.55M$; SO_4^{2-}, $0.030M$.

20. Give possible explanations for the following:
   a. Groundwaters in the vicinity of feedlots are alkaline.
   b. Highly saline water flowing into reservoirs (as into Lake Powell on the Colorado River) hastens the sedimentation of silt, thereby shortening the lifetime of the reservoir. (Silt is negatively charged colloidal SiO_2.)

Answers

1. 9.37 mg O_2/liter
2. a. 320 mg/liter
   c. 404 mg/liter
4. $BOD_5 = 207$ mg/liter
7. C:N:P = 41:7.2:1
8. a. 0.12 g   c. 0.059 mg P/liter
   b. 140 mg of O_2 d. 0.43 mg N/liter
9. a. Yes, $[Cd^{2+}] = 6.1 \times 10^{-15}$ mg/liter
   b. No, $[Cd^{2+}] = 0.61$ mg/liter

12. a. 57 percent
    b. 11.4°C (Table 8.5)
    c. 0.0098 g/liter
13. $K_f = 7.60$°C/molal
14. $m = 0.03m$
15. $\Delta T_f = 0.021$°C
17. 23.34 mm
18. $P_t = 71$ mm, $X_{methanol} = 0.75$
19. $\Sigma c = 1.12M$, $\Pi = 27.4$ atm

References

GENERAL

"Cleaning Our Environment. The Chemical Basis for Action," The American Chemical Society, Washington, D. C., 1969, pp 93–162.

Swenson, H. A., and Baldwin, H. L., "A Primer on Water Quality," Document No. 1965 0-789-120, U. S. Department of the Interior, Geological Survey, U. S. Government Printing Office, Washington, D. C., 1965.

Warren, Charles E., "Biology and Water Pollution Control," Saunders, Philadelphia, 1971.
   Heavy on the ecology, light on the chemistry, but interesting reading.

WASTE WATER TREATMENT

Clark, John W., Viessman, Warren, Jr., and Hammer, Mark J., "Water Supply and Pollution Control," 2nd ed, International Textbook Co., Scranton, Pa., 1971.

Fair, Gordon M., Geyer, J. C., and Okun, D. A., "Water and Wastewater Engineering," Vol. 2, "Water Purification and Wastewater Treatment and Disposal," Wiley, New York, 1968.

Kardos, Louis T., "A New Prospect," *Environment*, **12**, 10–21, 27 (March 1970).
> Disposal of disinfected secondary effluent on land and its use for irrigation.

"A Primer on Waste Water Treatment," Document No. 1969 0-335-309 (CWA-12), U. S. Department of the Interior, Federal Water Pollution Control Administration, U. S. Government Printing Office, Washington, D. C., 1969.

Sawyer, Clair N., and McCarty, Perry L., "Chemistry for Sanitary Engineers," 2nd ed., McGraw-Hill, New York, 1967.
> All the fundamental chemistry needed for an understanding of waste water and municipal water treatment.

EFFECTS OF WATER POLLUTION
Oxygen depletion

Coker, Robert E., "Streams, Lakes, Ponds," Harper & Row, New York, 1968, Chapter 10.

Eutrophication and phosphates

Grundy, Richard D., "Strategies for Control of Man-Made Eutrophication," *Environmental Science and Technology*, **5**, 1184–1190 (1971).

Hammond, Allen L., "Phosphate Replacements: Problems with the Washday Miracle," *Science*, **172**, 361–363 (1971).
> The role of phosphates in detergents and its substituents. Discussion of the controversy about the limiting nature of phosphates.

Mitchell, Dee, "Eutrophication of Lake Water Microcosms: Phosphate versus Nonphosphate Detergents," *Science*, **174**, 827–829 (1971).
> Evidence is presented to show that cultural eutrophication also results from the use of nonphosphate detergents.

Ryther, John H., and Dunstan, William M., "Nitrogen, Phosphorus, and Eutrophication in the Coastal Marine Environment," *Science*, **171**, 1008–1013 (1971).
> Investigation of a system where nitrogen is the limiting nutrient.

Stumm, Werner, and Morgan, James J., "Aquatic Chemistry. An Introduction Emphasizing Chemical Equilibria in Natural Waters," Wiley, New York, 1970, pp 428–439.
> This is a highly quantitative book. Most of it is rough going for students of general chemistry, but the few pages referred to here give a readable discussion of eutrophication based on the stoichiometric requirements of photosynthesis by algae and the degradation of organic matter by bacterial respiration.

Toxic substances (mercury)

Dunlap, Lloyd, "Mercury: Anatomy of a Pollution Problem," *Chemical and Engineering News*, July 5, 1971, pp 22–34.
> A good article for background on the mercury problem.

Environmental Research, **4**, (March 1971).
> An issue devoted to the hazards of mercury in the environment.

Goldwater, Leonard J., "Mercury in the Environment," *Scientific American*, **224**, 15–21 (May 1971).

Hammond, Allen L., "Mercury in the Environment: Natural and Human Factors," *Science*, **171**, 788–789 (1971).

Novick, Sheldon, "A New Pollution Problem," *Environment*, **11**, 2–9 (May 1969).
> Plus two other articles in the same issue.

Stokinger, H. E., "Sanity in Research and Evaluation of Environmental Health," *Science*, **174**, 662–665 (1971).
> This article scolds the ''ruinous concept of 'zero tolerance' for pollutants'' and suggests that too many substances (including mercury) are having unnecessarily severe standards placed on them, or being banned outright, on the basis of questionable evidence. A provocative article, well worth reading in connection with any discussion of environmental pollutants.

Thermal pollution

Clark, John R., "Thermal Pollution and Aquatic Life," *Scientific American*, **220**, 19–27 (March 1969).

Woodson, Riley D., "Cooling Towers," *Scientific American*, **224**, 70–78 (May 1971).

DESALINATION

"The A-B-Seas of Desalting," Document No. 1968 0-306-916, U. S. Department of the Interior, Office of Saline Water, U. S. Government Printing Office, Washington, D. C., 1968.

Spiegler, K. S., Ed., "Principles of Desalination," Academic, New York, 1966.

Spiegler, K. S., "Salt-Water Purification," Wiley, New York, 1962.

OTHER

Mason, B. J., "The Growth of Snow Crystals," *Scientific American*, **204**, 120–131 (January 1961).
> Nucleation of ice crystals by dust and silver iodide.

FOUR

The Earth and the Soil

With this section we come to the last tangible member of the trio composing our physical environment of air, water, and earth. The processes in the solid earth and its interactions with the atmosphere and hydrosphere involve more basic chemistry, perhaps, than the processes within the air and water. The chemistry of the earth is more comprehensive, but also more complicated, first because of the greater number of interacting elements and compounds and, second, because of the variety of interfaces at which reaction can occur. For although we say solid earth (or, at least, think of it as solid), reactions take place between many different solid, liquid, and gaseous phases on and within the earth, so that a whole series of heterogeneous chemical and physical equilibria face us.

In the next two chapters, we shall follow our ecological theme and deal with selected aspects of chemical science that affect the environment. The ecological connections may, at times, seem a bit tenuous, because most living organisms are related only indirectly to the solid earth—except as a thing to stand on—primarily through plants growing on the earth. But modern man, in addition to his dependence on the soil for food resources, relies on the earth for mineral resources.

In Chapter 13, we shall consider the earth as a source of minerals for our civilization and in Chapter 14, we shall examine the very thin but very significant layer of soil on the solid earth.

The Earth and Its Minerals

In this chapter, we shall be concerned with the elemental composition of the earth, and with the abundance, distribution, and combined form of the elements. All of these were determined by the conditions existing at the time of the formation of the earth and during its early history. The same conditions also have a bearing on the possible origin of life on earth, as we shall see in Section 15.2.

13.1 FORMATION AND EARLY HISTORY OF THE EARTH

According to one of the currently popular theories of the origin of our solar system, the sun, the planets, and all other bodies in the solar system had a common beginning in a vast cloud of cold gas and cosmic dust particles. The sun itself was probably formed by the accumulation and compression of gas and solid particles near the center of this cloud. The planets were formed by the gradual aggregation of the solid dust particles and larger fragments from the turbulent cloud surrounding the rotating sun. According to this picture, known as the cold accretion hypothesis, the primitive earth was initially cold and solid because the dust cloud was assumed to be at a relatively low temperature. The hypothesis is supported by the relative abundance of some of the elements on earth. For instance, the earth has very small amounts of both the light volatile elements such as helium, neon, and free hydrogen and the heavy inert gases, xenon and krypton, compared to the amounts of these gases in the sun and in the rest of the universe. If the earth had been formed all at once from hot, incandescent gases torn from the sun (another theory), then the gravitational attraction of the earth would have been too small to hold the lighter elements; but the heavier Xe and Kr should have been present in the primitive hot earth in the same abundance as in the sun and should not have escaped from the earth to the extent they apparently have. In the cold accretion hypothesis, the original small, cool particles could not have held much Xe or Kr to begin with because of the very small gravitational force on the particles.

Additional support for the cold accretion hypothesis is that the only light elements retained in any large degree by the original particles were such elements as carbon, nitrogen, oxygen, and some hydrogen. Free volatile compounds of these elements, such as CO_2, H_2O, and NH_3, would have been lost from the original, small, cold particles; however, in part, these elements were present as nonvolatile hydrates, carbonates, or adsorption complexes. In this form, they were retained by the particles. At higher temperatures (300–400°C), such substances would have decomposed to their volatile constituent compounds which, like the lighter inert gases, would have escaped from the small particles. As the earth formed, just such a temperature increase occurred, but simultaneously with the temperature increase there was an increase in size and mass of the earth, which led to greater gravitational attraction. In the larger gravitational field, even the volatile compounds of C, N, and O would have been retained.

The formation of the cool earth as an individual body within the solar system took place about 5 billion (5×10^9) yr ago, as determined by isotopic analysis (Section 20.5) of meteorite material. The sun itself is about 5–6 billion yr old. In a geological sense, the closeness in age of the sun and earth is reasonable if both were formed from the same cosmic cloud. Following the accretion of particles, the primitive earth began to heat up. Energy for heating was supplied by the conversion of gravitational energy of the compacting particles into heat and by the trapping of the energy of spontaneous decay of radioactive elements such as ^{40}K and ^{235}U. After a time, perhaps 600 million to a billion yr, the temperature in the interior of the primitive earth must have risen to near the melting point of iron. With the attainment of a molten state in the interior, and possibly extending to the surface at various times and places, a rearrangement of materials took place by the gravitational flow of denser components toward the center of the earth. As we shall soon see, this process was partially responsible for the internal layered structure (Figure 13.1) of the earth, which consists of, broadly speaking, a dense core (density 10–15 g/cm^3), a thick homogeneous mantle (density 4–6 g/cm^3), and a thin, heterogeneous crust of still lighter materials (average density 2.8 g/cm^3).

The heating and cooling processes leading to solidification of the mantle

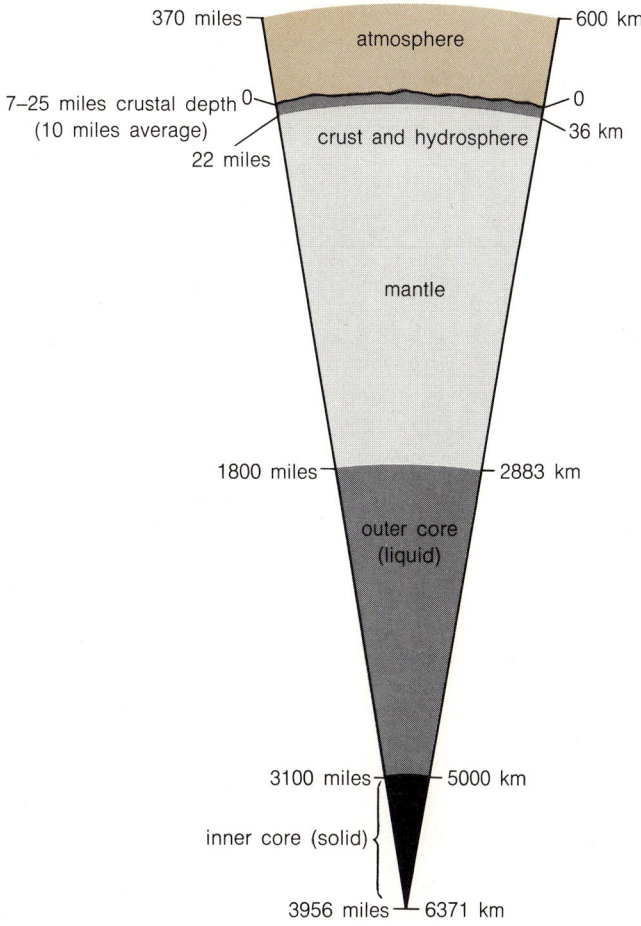

370 miles — 600 km

atmosphere

7–25 miles crustal depth 0 — 0

(10 miles average) — 36 km

22 miles

crust and hydrosphere

mantle

1800 miles — 2883 km

outer core
(liquid)

3100 miles — 5000 km

inner core (solid)

3956 miles — 6371 km

Figure 13.1. Internal structure of the earth. The crust is thinner (about 7 miles average) under ocean basins, and up to about 40 miles thick under mountain ranges.

and a fairly stable crust probably took from 1 to 1.5 billion yr. Then, if the age of the earth is approximately 5 billion yr, the age of the oldest rocks in the earth's crust is about 3.5 billion yr. (The age of rocks refers to their most recent solidification.) Thus, for about 3.5 billion yr, the surface of the earth has been subject to weathering, erosion, and sedimentation that modify the crust, cover older rocks with a veneer of sedimentary rocks, and form soil. These processes will be examined in Chapter 14. Right now, we are more interested in the elemental composition of the earth in general and of the crust in particular.

13.2 ABUNDANCE OF THE ELEMENTS

The relative abundance of selected elements in various bodies in the universe is listed in Table 13.1. These data were obtained by a variety of methods and assumptions. For instance, the elemental abundances in the sun were determined by spectroscopic methods with suitable corrections to relate the composition of the sun's observable surface to its interior. The entries under "meteorites" are weighted averages of the abundances of the elements in meteorites ranging from the "irons," with as high as 98 percent iron, to the

Table 13.1
Relative abundances of elements
in the earth and the universe
(atoms per 10,000 atoms of Si)

element	continental crust	universe	meteorites	whole earth	solar atmosphere
rock-forming elements					
Si	10,000	10,000	10,000	10,000	10,000
Al	3,050	950	880	940	500
Fe	1,010	6,000	8,900	13,500	1,200
Mg	950	9,100	8,700	8,900	7,900
Ca	1,030	490	890	330	450
Na	1,050	440	450	460	630
K	540	30	70	40	16
Mn	20	70	70	40	25
Ti	120	20	50	20	15
Ni	1	270	460	1,000	260
P	35	100	70	100	69
volatile elements					
H	1,400	4.0×10^8		84	3.2×10^8
O	29,500	215,000	35,000	35,000	290,000
N	2	66,000		0.2	30,000
C	17	35,000		70	166,000
S	8	3,750	1,140	1,000	6,300
F	33	16		3	
Cl	4	90		32	
inert gases					
He		3.1×10^7		3.5×10^{-7}	5×10^7
Ne		86,000		12×10^{-7}	
Ar		1,500		$5,900 \times 10^{-7}$	
Kr		0.51		0.6×10^{-7}	
Xe		0.04		0.05×10^{-7}	

SOURCES:

Continental crust: Konrad Krauskopf, "Introduction to Geochemistry," McGraw-Hill, New York, 1967, p 603.

Universe: H. E. Suess and H. C. Urey, "Abundances of the Elements," *Reviews of Modern Physics*, **28,** 53 (1956); quoted by Krauskopf, *op. cit.*, p. 603.

Meteorites: V. M. Goldschmidt, *Skrifter Norske Videnskaps-Akad. Oslo. I. Mat.-Naturv. Kl.* **4,** 99–101 (1937); quoted by Brian Mason, "Principles of Geochemistry," 2nd ed., Wiley, New York, 1958, p 19.

Whole earth: Brian Mason, *op. cit.*, pp 51, 202.

Solar atmosphere: Lawrence H. Aller, "The Abundance of the Elements," Wiley, New York, 1961; quoted by Brian Mason, "Principles of Geochemistry," 3rd ed., Wiley, New York, 1966, p 16.

"stones," some of which resemble magnesium and iron silicate rocks similar to those in the mantle. The composition of the whole earth (essentially the mantle and core) is, in turn, based on analysis of these same meteorites, since it is believed that either the earth was formed from meteoritic material or the meteorites are pieces of a vanished planet similar to the earth. Abundances in the universe are combinations of values obtained from spectra of the sun and other stars and from analyses, direct and indirect, of meteorites and the earth.

So far as the nonvolatile rock-forming elements are concerned, the composition of the whole earth does not differ much from that of the universe. In percent by weight, 90 percent of the earth is composed of only four elements: Fe, O, Si, and Mg, and 99 percent of the earth is made up of these and only 11 more elements: Ni, S, Ca, Al, Na, K, Cr, Co, P, Mn, and Ti. These elements are not evenly distributed within the earth. Table 13.1 shows that the crust has a much larger share of the lighter nonvolatile elements (Al, Ca, Na, and K) than the rest of the earth. In addition, the crust, which makes up only 1 percent of the earth's mass, contains most of the oxygen. The earth's core is predominantly iron, with about 10 percent dissolved nickel.

13.3 PRIMARY DIFFERENTIATION OF THE ELEMENTS

The elements have been unequally distributed or, as geochemists say, differentiated within the earth. Part of this differentiation may have been controlled by gravity, but the chemical characteristics of the elements were of greater significance in determining whether an element would have its greatest abundance in the core, mantle, crust, or atmosphere. An understanding of the differentiation of the elements begins with the assumption that the overall composition of the earth resembles that of meteorites. These visitors from interplanetary space are mostly O, Fe, Si, and Mg, with smaller amounts of S and other elements. In the meteorites, and presumably in the earth too, the proportions of these elements are such that oxygen is greatly in excess of sulfur. Also, Fe, Si, Mg, and other more electropositive elements are in excess of O and S, with Fe more abundant than Mg and Si.

In a wholly or partially molten earth, electronegative oxygen would react completely with the Mg and Si, leaving the excess iron uncombined to form a heavy liquid iron phase, which would concentrate at the core. The combination of oxygen and sulfur with magnesium, silicon, and some of the iron would form another liquid phase of silicates of iron and magnesium. The silicates are substances that may be thought of as being composed of SiO_4^{4-} units. Their varied structures and compositions will be examined shortly. A separate sulfide phase is often added, but we shall assume that this is dispersed throughout the iron metal and silicate phases. Presumably, these liquid phases—iron, silicate, and perhaps sulfide—all would be immiscible with one another, and the lighter silicate phase would float on the denser liquid iron core. For completeness, a gaseous phase, the atmosphere, is sometimes added.

The other metals were distributed between the iron core and the silicate phase on the basis of their chemical affinity for iron and for silicates. The redox reaction

$$M + Fe \text{ silicate (or oxide)} \rightleftharpoons M \text{ silicate (or oxide)} + Fe$$

can be considered to govern the form and distribution of a metal M. Metals more readily oxidized (better reducing agents) than iron displaced iron from the silicate phase and were concentrated in the mantle and crust as silicates or oxides. More electronegative metals than iron were reduced by iron and, as metals, displaced from the silicate phase to the iron core, where they alloyed with the iron.

Elements congregating in the silicate phase were generally those that formed ionic bonds. That is, they were the elements of either high or low electronegativity that tended to attain the 8-electron inert gas configuration by electron transfer. Elements of intermediate electronegativity tended to go into the iron (and sulfide) phases. These elements were, in general, those near the middle of the periodic table; they tended to form covalent (and metallic) bonds. The grouping of the elements according to the phase in which they concentrate is called Goldschmidt's geochemical classification (Table 13.2).

Table 13.2
Geochemical classification of the elements

iron phase	sulfide phase	silicate phase	atmosphere
Fe Co Ni	Cu Ag (Au)^a	Li Na K Rb Cs	H N (C) (O)
Ru Rh Pd	Zn Cd Hg	Be Mg Ca Sr Ba	(F) (Cl) (Br) (I)
Re Os Ir Pt Au	Ga In Tl	B Al Sc Y rare earth metals	inert gases
Mo Ge Sn C P	(Ge) (Sn) Pb	(C) Si Ti Zr Hf Th	
(Pb) (As) (W)	As Sb Bi	(P) V Nb Ta	
	S Se Te	O Cr W U	
	(Fe) (Mo) (Re)	(Fe) Mn	
		F Cl Br I	
		(H) (Tl) (Ga) (Ge)	
		(N)	

^{a}Parentheses around a symbol indicate that the element belongs primarily in another group, but has some characteristics that relate it to this group.

Such classifications are all approximate, because there were many factors operating in a molten earth. What is important is that a primary differentiation of elements on the basis of chemical affinity for the liquid phases took place after the gravitational separation into an iron-rich core, silicate-rich mantle and crust, and a gaseous phase. The distribution of elements between the liquid iron and silicate phases was not controlled by gravity or volatility. Otherwise, a heavy but electropositive element such as uranium would have been concentrated, not in the crust of the earth, but in the core.

Following this primary differentiation in the molten phases, the earth began to cool and the molten rock (magma) slowly solidified. The crystallization process led to another, or secondary, differentiation based on various factors in the solid–liquid phase equilibria of the freezing process. One of the important factors was the size of the ions incorporated into the crystalline minerals formed. To see how this additional fractionation of elements came about, we need some information about the structure of solids.

Crystalline Solids

The vast majority of solid substances are characterized by an ordered, geometrical arrangement of their constituent particles, be they atoms, ions, or molecules. The ordered three-dimensional arrangement is in contrast to the increased internal disorder in liquids and gases, and confers on solids their characteristic features of rigidity and strength, definite volume, and distinctive shape of some pieces of the solid. These pieces are, of course, the crystals of the solid. Because practically all solids are crystalline, it is common to equate the solid state with the crystalline state, although there are some materials generally thought of as solids, like glass, that have very little of the long-range internal order of crystalline solids and, therefore, have no definite crystal shape. Such noncrystalline "solids" are said to be *amorphous*. In internal structure, they more nearly resemble a supercooled liquid than a true solid.

It is difficult to deduce information about the internal structure of a crystal from its external appearance, although the study of the external features of some crystals provides support for the idea of order in solids and gives information about some of the microscopic details of arrangement. The overall shape of a crystal depends strongly on the conditions under which it grows, even though the angles between the surfaces or faces of the differently shaped crystals usually remain the same. In addition, most naturally grown crystalline solids usually consist of a large number of imperfectly formed single crystals that have grown together into a polycrystalline mass. It is more useful to have some way of looking directly at the internal arrangement of the component particles.

13.4 X-RAY DIFFRACTION

The most powerful tool for the study of the internal structure of crystals is X-ray diffraction, which has been used to determine the structural parameters of crystals and molecules from simple monatomic solids up to extremely complex proteins. The simple description of X-ray diffraction given here is intended only to show how one of the crystal parameters is measured and how the results justify the concept of order in solids.

We assume a crystal with a high degree of order, its identical atoms or ions lying in rows an equal distance d from each other as shown in Figure 13.2(a). The rows shown are the end views of planes or layers of the units (atoms, ions) in the crystal, as shown in Figure 13.2(b). Other planes with different spacings and with different densities of units in the layers can be drawn; end views of some of these other planes are shown in Figure 13.2(c).

X-Rays can be generated by directing high-speed electrons against a metal target. These electrons knock electrons out of the metal atoms from the inner $K(n = 1), L(n = 2), \ldots$ shells of the metal atoms in the target. Outer electrons then fall into the energy levels vacated by the ejected inner electrons and, as they do so, they emit high-energy X-ray radiation of several characteristic frequencies or wavelengths. By using suitable filters, one can sort out a single wavelength and then direct this monochromatic beam onto the surface of the crystal under investigation. Within limits, any desired wavelength of X-rays can be obtained by using different target and filter materials. If a copper target and a nickel filter are used, X-rays with a wavelength of 1.542 Å are obtained. The wavelengths of X-rays are about equal to the distances between atoms or ions in a crystal, and therefore noticeable diffraction effects are produced.

To get an X-ray diffraction pattern from a crystal, a narrow beam of X-rays is directed onto the crystal as shown in Figure 13.3. The layers of crystal particles are thought of as reflecting planes that act as mirrors for the incident X-ray beam. In the figure, the layers have been drawn in cross section with horizontal lines representing end views of the parallel layers, all containing the same kinds of atoms and all having the same surface density of atoms. A parallel beam of X-rays is incident on the set of planes from the left. Two

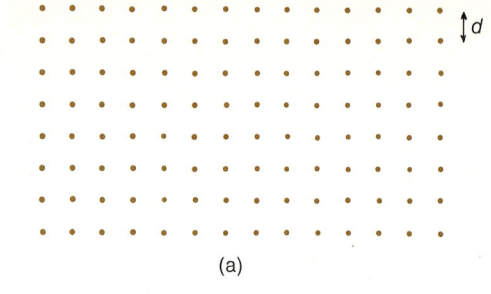

(a)

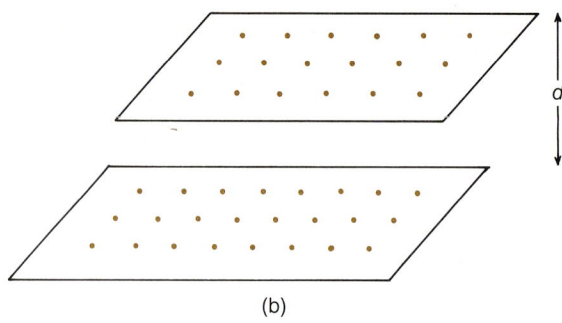

(b)

Figure 13.2. Lattice points and planes in crystals. (a) Interplanar spacing, *d*. (b) Planes of lattice units. (c) End view of different families of planes in the same crystal. Even without the aid of the lines in the figure, the existence of various planes with different spacings can be recognized by holding the figure near eye level and sighting down the rows of points. Several rows with different linear point densities will be seen by rotating the figure. The same kind of effect can be noted with tree trunks in an orchard or grape stakes in a fallow vineyard by sighting down the rows in different directions.

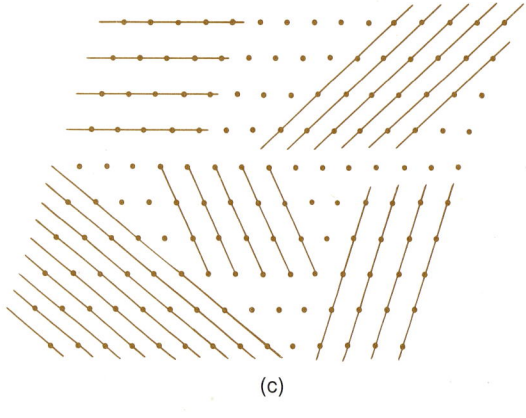

(c)

Figure 13.3. Derivation of Bragg's law, $n\lambda = 2d \sin \theta$. The path length of the ray DEF is $2l$ longer than that of ABC. But $l = d \sin \theta$, so the path difference is $2l = 2d \sin \theta$. The condition for reinforcement is that the difference in path be an integral number *n* of wavelengths λ, or $n\lambda = 2d \sin \theta$.

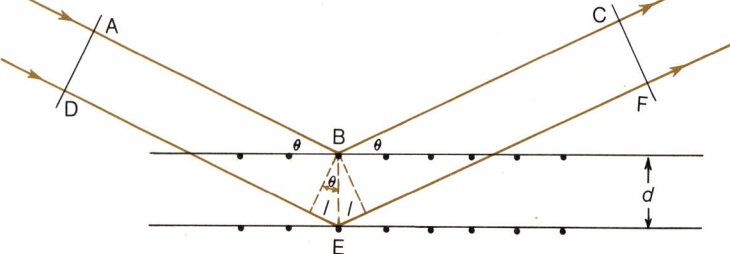

rays of the beam have been separated from each other for purposes of illustration. Some of the radiation from the X-ray beam will be reflected from the uppermost plane as shown by the path ABC in Figure 13.3. Some of the rays in the same beam (or part of the radiation) will penetrate and be reflected by an underlying plane as illustrated by the segment DEF. These rays will be in phase at a given point after reflection if their difference in path is an integral number n of wavelengths λ. The two rays will then reinforce each other, and constructive interference (i.e., high intensity of the reflected beam) will result. The condition for reinforcement is

$$n\lambda = 2d \sin \theta$$

For a given set of layers, the interplanar spacing d is fixed. As the glancing angle θ is increased, there will be a number of angles for maximum reinforcement such that $2d \sin \theta$ equals, successively, λ, 2λ, 3λ, . . . , for $n = 1$, 2, 3, ... If λ and the order n of a reflection maximum is known and θ is measured for a given n, then the spacing d between planes can be calculated. In its simplest form, this is what is done to determine this important parameter in simple crystals. The consistent results of X-ray diffraction studies of crystals support our assumption of the existence of definite order in crystals. X-ray diffraction can also be used for the identification of substances having known crystal structures.

The actual mechanism leading to X-ray diffraction effects is more complex than was used in the foregoing discussion. The X-ray beam interacts with the electrons of the crystal particles, inducing vibrating dipoles that emit or scatter X-radiation in all directions. The X-rays scattered by different particles interfere, and the degree of interference depends on the distance between the particles. Each particle (or plane) in effect acts as a separate source or slit in a Young's diffraction-type experiment (Section 2.15). Moreover, the intensity of the scattered radiation depends on the electronic configuration of the atoms or ions, and this fact can be of use in the X-ray analysis of crystals containing two or more different atoms and in identifying the type of atom causing the scattering.

13.5 BONDING IN SOLIDS   The bonds between the units in a crystalline solid may be ionic, covalent, weak van der Waals and London type, or metallic, a type of bonding we have not discussed yet. In an ionic solid such as NaCl, the sodium and chloride ions are held together in an ordered crystalline arrangement by electrostatic forces between the ions. In diamond, however, the bonds between carbon atoms are strong covalent bonds; thus, diamond and the crystals of many other nonmetallic elements are covalent solids. In molecular solids, the units are distinct covalent molecules held together in the crystal structure by weaker intermolecular forces. In ice, for instance, the water molecules are joined by hydrogen bonds. In solid carbon dioxide, the nonpolar CO_2 molecules are weakly held in position by London forces. These forces also operate in other solids composed of nonpolar molecules (O_2, N_2, I_2, etc.) and of inert gas atoms. Stronger dipole–dipole interactions operate in molecular crystals composed of polar molecules (CO, NH_3, etc.). Table 13.3 summarizes some features of the various types of solids.

Table 13.3
Types of solids

type of solid	structural units	bonds between units	properties	examples
ionic	atomic or molecular ions	nondirectional electro-static (coulombic)	brittle, hard, high melting point	NaCl, KNO_3
molecular	polar molecules	weakly directional van der Waals, or direc-tional H bonds	soft, low melting point, stable	H_2O, $CHCl_3$
	nonpolar molecules (or atoms)	nondirectional, weak to strong London		I_2, CO_2, Ne
covalent	nonmetal atoms	highly directional covalent	very hard, very high melting point	C (diamond) SiO_2 (quartz) SiC (carborundum)
metallic	metal atoms	nondirectional electro-static; metallic; some covalent character with transition metals	elasticity, high electrical and thermal conduc-tivity	Na, Cu, Au

Metallic Bonding   Most of the elements are classified as metals (Figure 13.4). The atoms of metals are characterized—even defined—by their electropositive nature. Their elec-tropositive character is the result of the low ionization energy of the atoms, and the low ionization energies are a reflection of the small numbers of electrons, usually 1 to 3, in the valence shell orbitals of these elements. These valence electrons are easily removed because they are in high-energy orbitals well screened from the nucleus by inner-shell electrons.

Chemically, the electropositive nature of metals is expressed in the high reducing power of the elemental atom and by the tendency of metal atoms to assume a positive oxidation state in nature. Most people do not think of metals in these chemical terms. They think, instead, of the characteristic physical properties of more massive pieces of solid metal. Among these prop-erties are high electrical and thermal conductivity, metallic luster, and me-chanical characteristics such as malleability, ductility, and high mechanical strength. These properties must be explainable on the basis of the bonding interactions within the metal crystal. Because of the large number of neigh-boring atoms surrounding any one atom in metals (up to 14), it is evident that an atom cannot be bound to its neighbors by ordinary localized covalent bonds. And ionic bonding would not be expected between identical metal atoms.

Many of the properties of metallic solids can be explained qualitatively by delocalized covalent bonding. The large outer electron orbitals of a metal atom in the crystal overlap with those of its neighbors to form large molecular orbitals encompassing all the atoms in the crystal, as in Figure 13.5. Bonding electrons are not localized between atoms, but are pictured as being free to move throughout the entire crystal as a sort of electron gas surrounding the cation cores. The freedom of movement of these electrons explains the ease with which the charge and the thermal energy of electrons can be carried

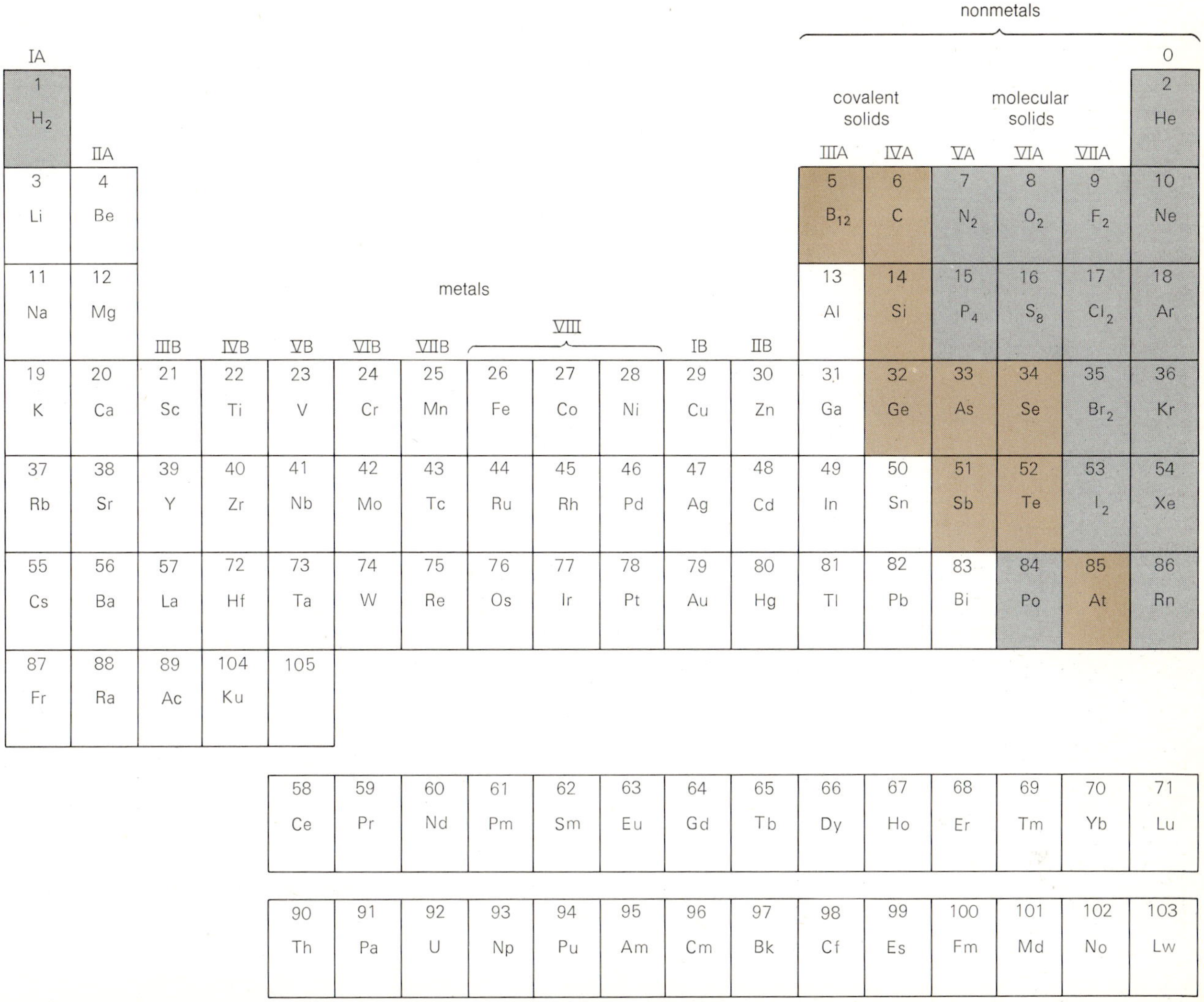

Figure 13.4. Solid forms of the elements. The elements Ge, As, Se, Sb, Te, and At have properties intermediate between metals and nonmetals.

from one part of the metal to another. The nondirectional character of the delocalized electrons in the molecular orbital explains the mechanical properties; when the position of an atom in the crystal is changed by deformation, the atom immediately forms bonds with its new neighbors. The new bonds are just as strong as the old ones, and the crystal remains intact. This is in contrast to the brittle nature of ionic crystals.

The electrical conductivity of metals and the way metals differ from nonconductors or insulators can be explained more fully by the energy band model of metals. In the valence shell of each metal atom, there are several unoccupied atomic orbitals in addition to those filled by the small number of valence electrons. Overlap of these empty orbitals, as well as of the valence electron orbitals, takes place when the atoms are packed together in the crystal. The amount of overlap depends on the extension of the orbitals in a particular metal

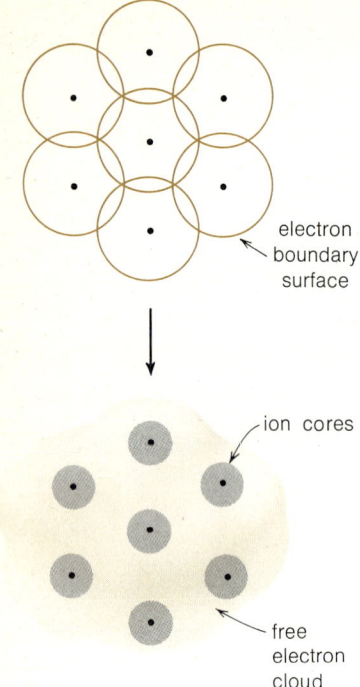

Figure 13.5. Electron delocalization in metals.

atom and on the closeness of packing of the atoms. The overlap of unoccupied atomic orbitals generates molecular orbitals just as does the overlap of occupied valence electron orbitals. The situation can be pictured with metal atoms having a valence shell composed of s and p orbitals. This would include any metal in the first groups of the periodic table and aluminum. In general, the combination, by overlapping, of a total of N atomic orbitals on N atoms leads to N very closely spaced molecular orbitals. The degree of interaction and the spread of energies of the molecular orbitals is determined by the extent of overlapping. Where there is no overlapping, no molecular orbitals are formed, and the atomic orbitals stay as they are. Where there is little overlapping, the resulting molecular orbitals differ little in energy from each other and from the component atomic orbitals. All of these features are illustrated when N metal atoms interact at their equilibrium distances in a crystal.

In Figure 13.6, N atoms with 2s and 2p orbitals in the valence shell have been brought together to form a crystalline solid. The number of valence electrons is immaterial at present. The atoms are not close enough for the underlying 1s atomic orbital in each atom to interact, so the N 1s orbitals retain their original energy. The more diffuse 2s atomic orbitals on each atom do overlap, as do the degenerate 2p atomic orbitals on each atom. The interaction of the N 2s atomic orbitals forms an energy band of N closely spaced molecular orbital energy levels, while the 3N 2p atomic orbitals of the atom overlap to give an energy band of 3N molecular orbital levels. The closely spaced sets of N molecular orbitals derived from the 2s atomic orbitals, and of the 3N molecular orbitals from the 2p atomic orbitals, are called *energy bands*. Each molecular orbital within any band can hold 2 electrons. Thus, the capacity of the lower band is 2N electrons and that of the upper band is 2 × 3N = 6N electrons.

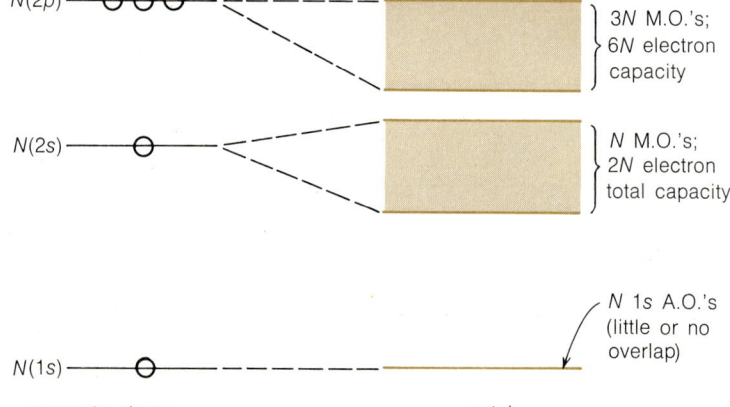

Figure 13.6. Energy band formation in metals. The decrease in energy of the 2s and 2p bands is due to electron delocalization.

An alternative representation of the energy bands is shown in Figure 13.7. Here each atom is thought of as being at the bottom of a one-dimensional box. The walls of the box prevent the interaction of the inner atomic orbitals of the atom, but the higher-energy atomic orbitals can overlap and spread over several boxes.

Electrical conductivity in a metal, or any other electronic conductor, arises from the acceleration of electrons by an externally applied electric field. Electron acceleration can occur only if the energy of the electron can be increased, and this requires that the electron be promoted into a higher energy level. If there is no empty level of higher energy easily accessible to the electron, the material will not conduct electricity.

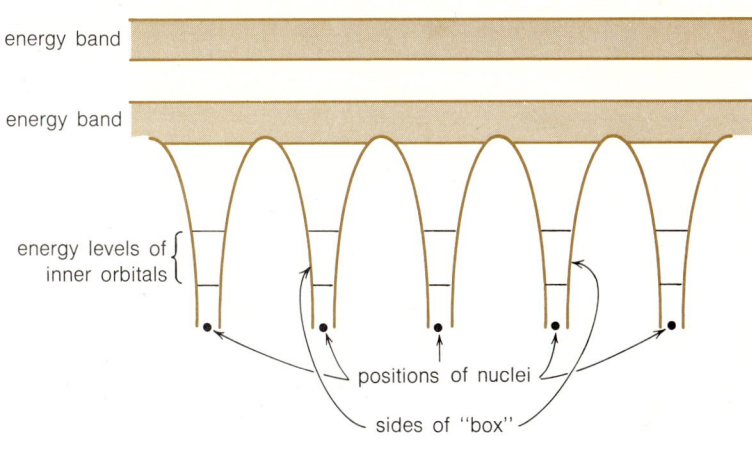

energy band

energy band

energy levels of { inner orbitals

positions of nuclei

sides of "box"

Figure 13.7. Energy bands in a metal. The individual nuclei create deep potential energy wells. Overlap of electron orbitals on adjacent atoms may occur for the higher-energy outermost electrons.

Now let us pick the element lithium and see how its electrons fill up the allowed energy bands. Lithium has the configuration $1s^2 2s^1$. The 1s orbitals in the lithium atoms interact little when the atoms are at their lattice sites in a lithium crystal. They remain pretty much as N atomic orbitals, each holding the two 1s electrons per atom. The N 2s electrons go into the first energy band, only half filling it. This leaves $N/2$ empty energy levels just above the topmost electron. These levels are available to accept an accelerated electron; therefore, lithium is a conductor. The case of partially filled valence electron levels is shown in Figure 13.8(a), where the occupied inner-shell electron levels have been omitted.

Going on to Be ($1s^2 2s^2$) or any of the other alkaline earth elements, we see that there are now $2N$ valence electrons to be placed in the valence electron band. This should fill up the band, and for the top electron to be promoted it would have to have enough energy to jump the gap to the lowest molecular orbital level in the upper empty band. If this gap is large, only a very intense electric field could supply enough energy to promote electrons to the upper or conduction band. The material would be a nonconductor or an insulator. This does not happen with these metals, because the lower valence band and the upper conduction band overlap. Promotion of electrons from the lower band into overlapping levels of the upper band is easy, and the alkaline earth elements are conductors. The overlapping band and large gap situations are shown in Figure 13.8(b) and 13.8(c).

A semiconductor has the same band pattern as a nonconductor, but the forbidden energy gap between the lower valence electron band and the upper empty conduction band in a semiconductor is very much smaller, so excitation of a valence electron to an upper level is relatively easy. Excitation may be supplied by thermal energy, the resulting conductivity increasing with temperature (the conductivity of a metal decreases with temperature). Conduction is due to both motion of promoted electrons in the upper band and the promotion of electrons left behind in the lower band into the energy levels or *holes* vacated by the promoted electrons. The band structure of a semiconductor is shown in Figure 13.8(d). Silicon, germanium, selenium, and a number of metallic oxides are semiconductors.

The conductivity of a semiconductor may be altered by the presence in the crystal of impurity atoms, which provide extra energy levels in the forbidden

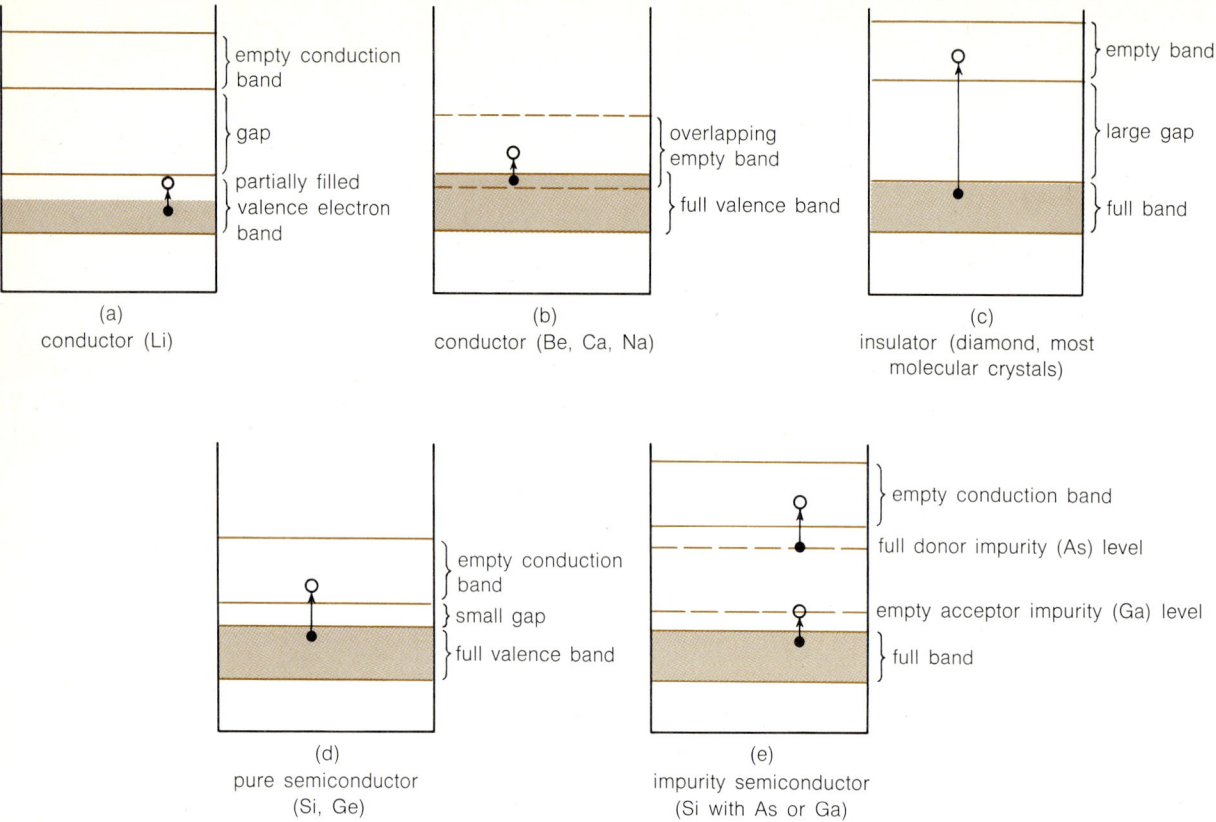

Figure 13.8. Energy band models for electric conductors, insulators, and semiconductors. Arrows show the direction of movement of electrons from an occupied level (solid circles) to an empty level (open circles).

region—much like inserting an extra rung in a ladder. If an impurity atom such as P, As, or Sb is introduced into a Si or Ge semiconductor, it establishes an energy level close to the bottom of the upper conduction band. These *donor* levels contain electrons (the fifth valence electrons) that can be easily excited into the conduction level. Conversely, if impurity atoms having only 3 valence electrons (Al, Ga) are substituted for some Si or Ge atoms, they produce an *empty* energy level just above the lower filled band. Electrons from the filled band can be easily excited to this empty acceptor level, and conductivity ensues by the movement of valence electrons into the holes left in the lower band. Both cases are shown in Figure 13.8(e).

Of all the kinds of interactions discussed above, only the covalent (and hydrogen bonds) are strongly directional in nature, and only they are subject to saturation of valence, being able to act only among a definite number of atoms. The other types of bonding are essentially nondirectional and, in solids, can act between an indefinite number of adjacent particles. Thus, the attractive electrostatic force between a positive ion and a nearby negative ion is not diminished by a number of anions at the same distance. In fact, the more nearby negative ions with which a positive ion can interact, the lower the energy and the more stable the group of ions. The potential energy of interaction U between a cation and an anion depends on the distance between them, r, and the magnitude of the charge on each ion, q^+ and q^-, according to the relation

$$U = \frac{q^+ q^-}{r}$$

Before discussing the structure of solids made of dissimilar ions, let's first see how identical ions or atoms, regarded as hard spheres, can pack together to fill space as completely as possible.

13.6 PACKING OF SPHERES   Crystallinity is the result of the tendency of the atoms, ions, or molecules in a substance to pack together as closely as possible and to be arranged in a highly ordered manner. This tendency is particularly evident when the forces between these units are nondirectional and when the only limit to the number of units that may surround and contact another unit is determined solely by the size of the units. These requirements are met in a number of metallic crystals where each spherical cation is surrounded by 12 identical nearest neighbor spheres, each of which touch the central sphere. The number of nearest neighbors of a unit in a crystal is called its *coordination number* (CN). In substances other than metals, however, the nearest neighbors need not be identical to the surrounded unit, nor need they touch it.

We shall now see in more detail how identical spheres can pack together as closely as possible. We start by considering a layer of identical hard spheres as shown in Figure 13.9. This planar arrangement with six spheres surrounding

Figure 13.9. Closest packing of spheres in three layers.

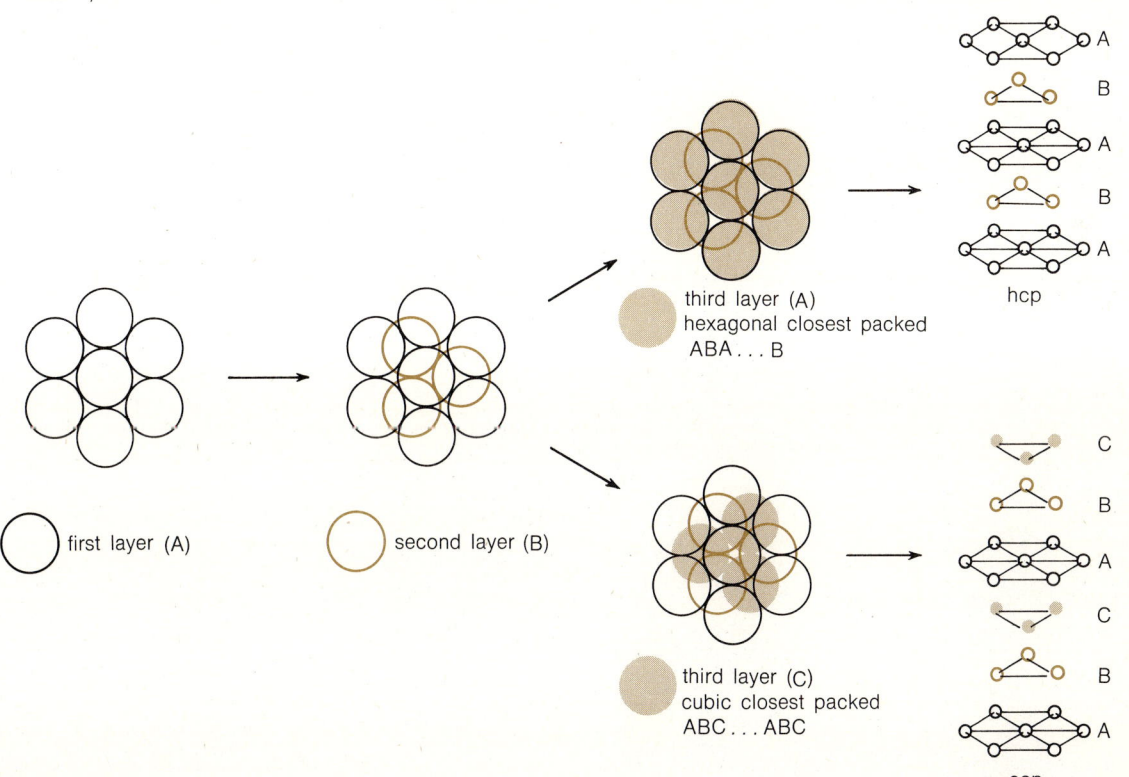

first layer (A)

second layer (B)

third layer (A)
hexagonal closest packed
ABA...B

hcp

third layer (C)
cubic closest packed
ABC...ABC

ccp

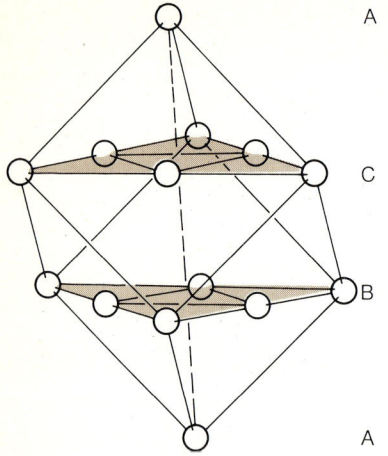

Figure 13.10. Face-centered cubic perspective of the cubic closest packed arrangement. The interior planes are at right angles to the body diagonal shown as a dashed line.

the center one is obviously the most efficient use of space in two dimensions. Now an additional layer of spheres is stacked on top of this first layer. There is only one way to place this second layer (we shall call it the B layer) on the first (the A layer); spheres in the B layer must go into alternate depressions in the A layer.

When an additional layer of spheres is stacked on top of the second layer, we find that there are two possible arrangements. On the one hand, the third layer can be placed so that its spheres are directly above those in the bottom (A) layer. This ABAB . . . sequence can be continued indefinitely and is called the *hexagonal closest packed* (hcp) *arrangement*. On the other hand, the third layer can be positioned with its spheres in the pockets of the second (B) layer that are not over the spheres in the A layer (i.e., over holes in the A layer). This is called *cubic closest packing* (ccp) and has the sequence ABC.

In both of these closely packed structures, the coordination number of any sphere is 12. In both, the maximum utilization of space is realized; that is, the ratio of the volume of the spheres to the total volume of the space in which they are packed is 0.74, so there is only 26 percent empty space in these structures.

If the cubic closest packed arrangement is extended to the fourth layer, forming the sequence ABCA . . . , and examined from a different angle as shown in Figure 13.10, the arrangement looks more symmetrical. In this view, the layers with the highest surface density are perpendicular to the body diagonals of the cube. The new perspective evokes the name *face-centered cubic structure* (fcc), where there are units at the center of each cube face as well as at the corners. It is sometimes more convenient to regard the ABCABC . . . arrangement as a face-centered cubic structure rather than as a cubic closest packed structure.

There is one more closest packed arrangement that we can visualize. This structure, the *body-centered cubic* (bcc), is not as closely packed as the hcp or the ccp, and uses only 68 percent of the space available. As seen in Figure 13.11, the number of nearest neighbors (at the cube corners) is only eight, but there are six more (at the centers of adjacent cubes), which are only

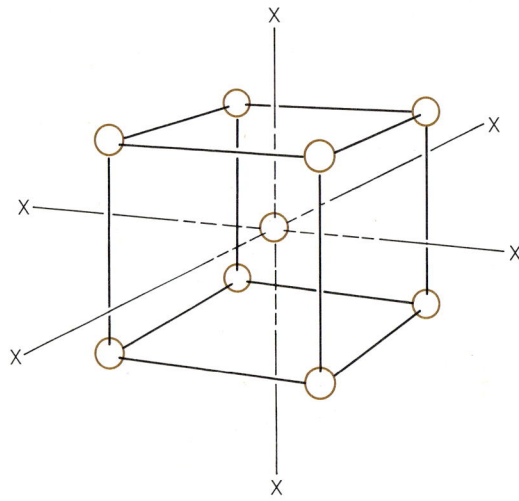

Figure 13.11. Body-centered cubic unit cell. One atom is in the center of the cube.

slightly farther away. In this case, the coordination number is sometimes regarded as 14. A number of metals do form crystals with their atoms in this relationship to one another.

13.7 UNIT CELLS
The unit cell of a crystal is the simplest structural unit representing the repeating internal order of the crystal. The unit cell is, so to speak, the brick used to construct the entire crystal. For a given crystal, all of the unit cells are identical in size and shape, although there may be latitude in the choice of the unit cell, just as there may be in the choice of bricks to construct a building. We have already generated three kinds of unit cell: the hexagonal, the face-centered cubic, and the body-centered cubic. There are 14 unit cells (Figure 13.12) belonging to seven more general crystal systems, which represent the general shapes of the unit cells in terms of the edge lengths and the angles between the edges.

The points shown in the unit cells of Figure 13.12 can represent different atoms or ions or can denote the centers of more complicated molecules or other composite entities, or merely geometric points. It is only necessary that the unit cell represent the smallest repeating units that can be put together to construct the entire crystalline solid, whatever its size. In very complicated structures (silicate minerals, proteins), it may be necessary to include hundreds of atoms in the unit cell before the smallest, most symmetrical, repeating unit is found. When the unit cells are stacked together, they generate a geometric framework of points called the *space lattice*, as shown in Figure 13.13. Each point of intersection in the lattice has exactly the same surroundings as any other point. The actual arrangement of the atoms is given by the *crystal structure*, which may or may not correspond to a simple unit cell.

13.8 HOLES IN CLOSEST PACKED STRUCTURES
Despite the fact that closest packing of identical spheres uses up to 76 percent of the space in a unit cell, there are still empty regions or holes into which smaller atoms or ions can fit. The structures of several ionic compounds having ions of different size and charge can be generated by imagining the smaller ions (usually the cations) as residing in these holes. To illustrate this point, we shall examine the face-centered cubic structure and its holes.

Consideration of the face-centered cubic structure in Figures 13.14(a) and 3.14(b) shows that there is a hole at the very center of the cube and that the coordination number of the center of this hole is 6; that is, there are six nearest neighbors surrounding the center of this empty site. Because the centers of these six closely packed spheres locate the vertices of an octahedron, this hole is called an octahedral hole. If you imagine a larger portion of the crystal lattice, you will see that the centers of 12 other octahedral holes are located at the midpoints of the unit cell edges. These are represented by X's in the figure. Each of the closely packed spheres is also surrounded octahedrally by six holes.

In addition to octahedral holes, there are smaller tetrahedral holes, each with a coordination number of 4. A tetrahedral hole in the face-centered unit cell is outlined in Figure 13.14(c). There is one tetrahedral hole in the center of each octant of this unit cell. There are eight such holes around each closely packed sphere, so that each sphere is in the center of a cube with holes at each corner (similar to a body-centered unit cell if there were atoms in the holes).

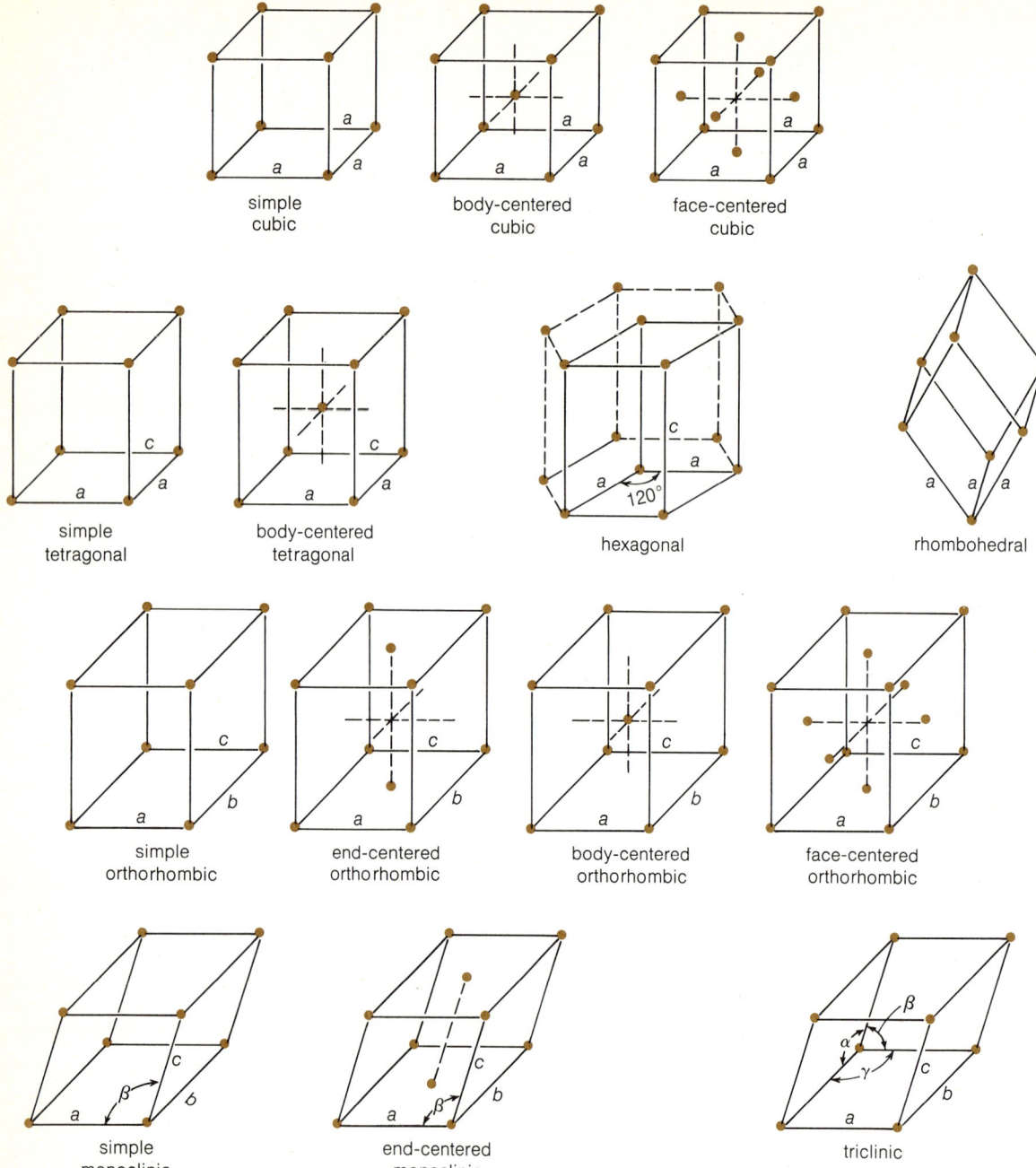

simple
cubic

body-centered
cubic

face-centered
cubic

simple
tetragonal

body-centered
tetragonal

hexagonal

rhombohedral

simple
orthorhombic

end-centered
orthorhombic

body-centered
orthorhombic

face-centered
orthorhombic

simple
monoclinic

end-centered
monoclinic

triclinic

Figure 13.12. The 14 unit cells. Each unit cell belongs to one of seven different crystal systems: cubic, tetragonal, hexagonal, rhombohedral, orthorhombic, monoclinic, or triclinic.

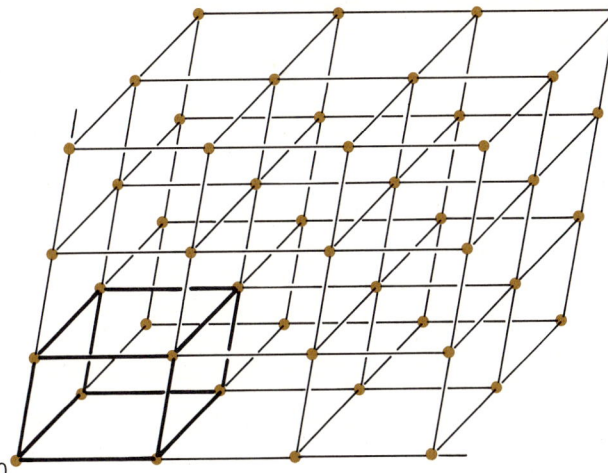

Figure 13.13. A three-dimensional space lattice of points generated by the stacking of unit cells.

0

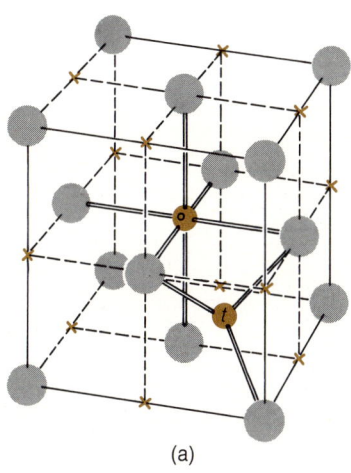

(a)

Figure 13.14. Octahedral and tetrahedral holes in a face-centered cubic unit cell. The centers of octahedral holes are denoted by O and x's. Centers of tetrahedral holes are designated by t. Parts (b) and (c) outline one octahedral hole and one tetrahedral hole.

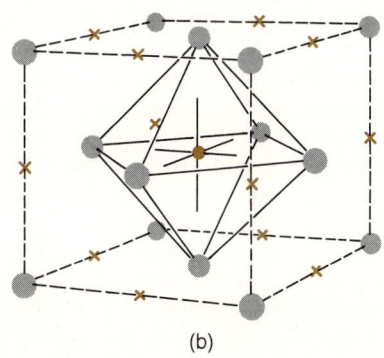

(b)

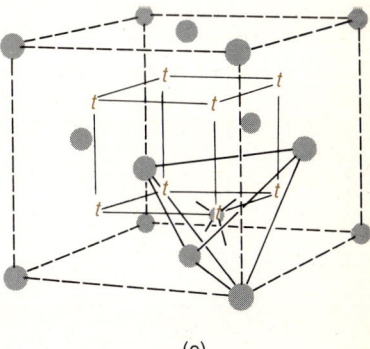

(c)

13.9 IONIC CRYSTALS By considering how some ions might order themselves into a closely packed arrangement and how other ions might fit into the holes in these structures, we can arrive at the unit cells of a number of common ionic solids. In dealing with spherical ions of different size, it is not possible to achieve close packing of all the ions. In general, the anions are larger than the cations. The larger anions arrange themselves into a closely packed structure, and the smaller cations position themselves in the holes. The cubic closest packed (or face-centered cubic) structure is the most common one for anions in ionic compounds.

Radius Ratios Before giving examples of the crystal structures of ionic compounds, we must consider some other complications connected with crystals having ions of different sizes. There is a geometric restriction on the number of ions that can be grouped around a central ion. This restriction governing the coordination number of the central ion is best given in terms of the ratio of the radius of the central cation to that of the anion. Consider a central ion in an octahedral hole surrounded by six other ions. Figure 13.15 shows a top

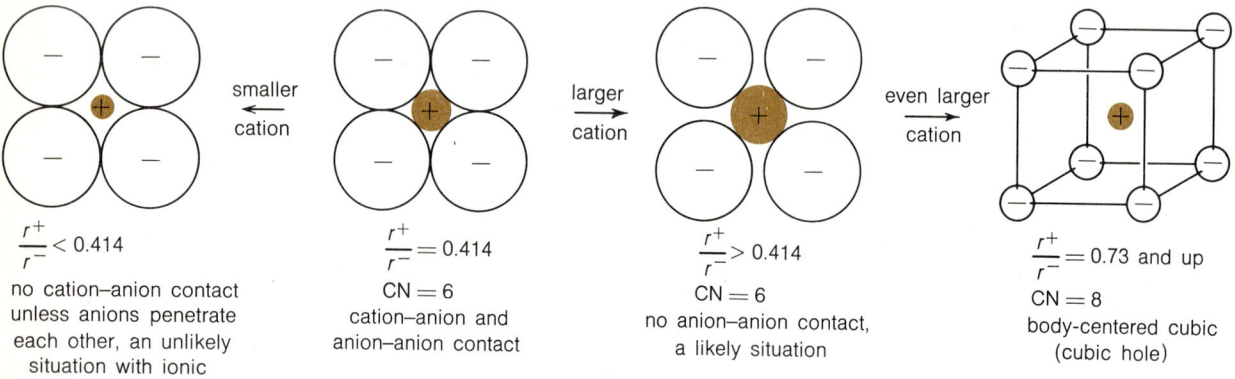

$$\frac{r^+}{r^-} < 0.414$$

no cation–anion contact unless anions penetrate each other, an unlikely situation with ionic bonding

$$\frac{r^+}{r^-} = 0.414$$

CN = 6
cation–anion and anion–anion contact

$$\frac{r^+}{r^-} > 0.414$$

CN = 6
no anion–anion contact, a likely situation

$$\frac{r^+}{r^-} = 0.73 \text{ and up}$$

CN = 8
body-centered cubic (cubic hole)

smaller cation ← larger cation → even larger cation

Figure 13.15. Ionic radius ratios and coordination numbers. Octahedral coordination (CN = 6) is favored for radius ratios in the range 0.41–0.73. Above 0.73, cubic coordination (CN = 8) is favored; below 0.41, tetrahedral coordination (CN = 4) of the cation is favored.

view of this octahedral configuration. The relative sizes of the cation and surrounding anions are such that the four (six, in three dimensions) anions can just touch the cation and their neighbor anions. The radius ratio required in this arrangement is calculated to be $r^+/r^- = 0.414$. If the cation size were increased so that the ratio became greater than 0.414, the anions could still touch the central ion, but they could no longer touch each other. This situation would be energetically favorable, because the repulsive anion–anion interactions would decrease, while the attractive cation–anion forces would still be at full strength. An octahedral structure with r^+/r^- equal to, or greater than, 0.414 would be stable. If, however, the cation had an even greater size relative to the anion, so that the ratio r^+/r^- became 0.73, then eight anions could arrange themselves around the cation and just touch the cation. Such an arrangement would be energetically more favorable than the octahedral coordination of six anions because of the increased number of attractive cation–anion interactions. This structure would be cubic, with the cation at the center of the cube.

On the other hand, if the central cation were of a smaller size, so that

r^+/r^- became less than 0.414, either the anions would have to penetrate each other (their electron clouds would overlap) in order to crowd in and maintain contact with the central cation *or* the cation would break contact and rattle around in the larger octahedral hole. Neither of these alternatives is likely with ions, for they result in either greatly increased anion–anion repulsion or greatly decreased cation–anion attractions. Instead, the anions rearrange to a tetrahedral configuration with fewer cation–anion contacts but with larger anion–anion distances to cut down the destabilizing repulsive forces. The limiting radius ratio for a cation in a tetrahedral hole is 0.22.

Within the radius ratio range 0.22–0.41, the tetrahedral configuration is the most stable; within the range 0.41–0.73 the octahedral is most favorable (Figure 13.16). There may be overlap in these values, so ions with a radius

coordination	radius ratio limits	spatial arrangement
tetrahedral CN = 4	0.22–0.41	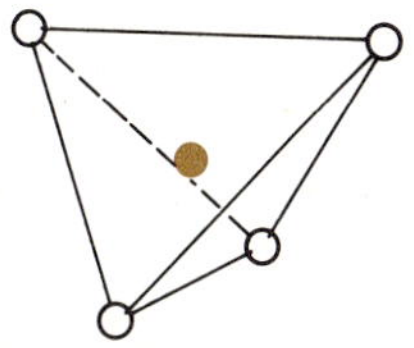
octahedral CN = 6	0.41–0.73	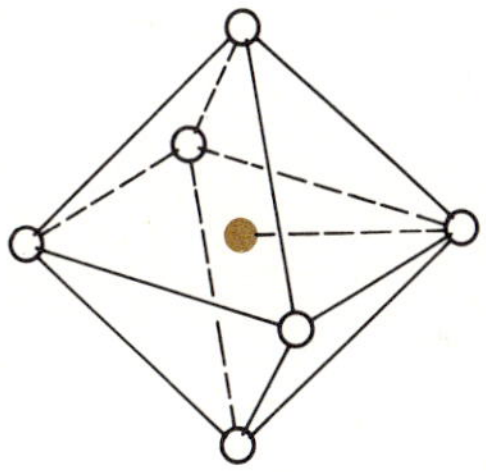
cubic CN = 8	0.73 and up	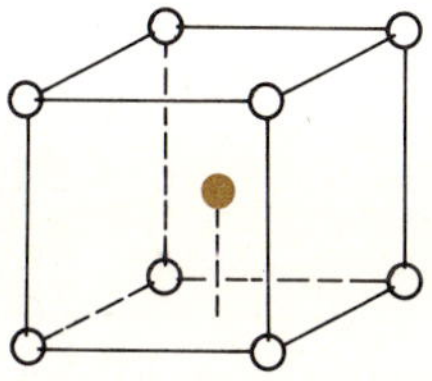

Figure 13.16. Radius ratio limits for cation coordination situations commonly found in ionic lattices.

ratio of, say 0.50, may be in the tetrahedral configuration even though the radius ratio value says that the octahedral configuration is the most stable. For reasons already given, it is not likely that an ion combination with a radius ratio of 0.50 will go into the cubic configuration with the cation surrounded by eight anions.

The radius ratio value indicates the optimum coordination number for the cation in an ionic solid. Usually, if the coordination number of the central ion is less than that predicted by the radius ratio, the bonds between the central atom and the surrounding atoms are predominantly covalent in nature.

Table 13.4
Ionic radii and coordination numbers

ion	radius[a] r, Å	r/r$_{O2-}$	predicted CN with O^{2-}	observed CN with O^{2-}
		cations		
Cs^+	1.67	1.19	12	12
Rb^+	1.47	1.05	12	8–12
Ba^{2+}	1.34	0.96	8	8–12
K^+	1.33	0.95	8	8–12
Ag^+	1.26	0.90	8	8,10
Sr^{2+}	1.12	0.80	8	8
Th^{4+}	1.02	0.73	6,8	6,8
Ca^{2+}	0.99	0.71	6	6,8
Cd^{2+}	0.97	0.69	6	6,8
U^{4+}	0.97	0.69	6	6,8
Na^+	0.97	0.69	6	6,8
Cu^+	0.96	0.69	6	6,8
Mn^{2+}	0.80	0.57	6	6
Fe^{2+}	0.74	0.53	6	6
V^{3+}	0.74	0.53	6	6
Zn^{2+}	0.74	0.53	6	4,6
Co^{2+}	0.72	0.51	6	6
Mo^{4+}	0.70	0.50	6	6
Ni^{2+}	0.69	0.49	6	6
Li^+	0.68	0.49	6	6
Ti^{4+}	0.68	0.49	6	6
Mg^{2+}	0.66	0.47	6	6
Fe^{3+}	0.64	0.46	6	6
Cr^{3+}	0.63	0.45	6	6
W^{6+}	0.62	0.44	6	4,6
Mo^{6+}	0.62	0.44	6	4,6
Al^{3+}	0.51	0.36	4	4,6
Si^{4+}	0.42	0.30	4	4
P^{5+}	0.35	0.25	4	4
Be^{2+}	0.35	0.25	4	4
S^{6+}	0.30	0.21	3	4
B^{3+}	0.23	0.16	3	3,4
		anions		
I^-	2.20			
Br^-	1.96			
S^{2-}	1.85			
Cl^-	1.81			
O^{2-}	1.40			
F^-	1.36			

[a] Cation radii are for six-coordinated ions.

Covalent character deforms the ions, brings them closer together, and results in deviations from the radius ratio rules. If the coordination number is equal to or greater than the rules specify, the bonds are ionic. An example is ZnO, which crystallizes with a cation coordination number of 4, rather than 6, as the radius ratio predicts. The Zn—O bonding has considerable covalent character (the electronegativity difference between Zn and O is about 1.8). Covalent character and coordination numbers of less than 4 will be favored by small, highly charged (polarizing) cations and by large polarizable anions. Coordination numbers above 4 indicate predominantly ionic bonding, while at CN = 4, both ionic and covalent bonding are about equally likely.

Because a structure may have a cation coordination number lower than predicted by the radius ratio, the use of these ratios in predicting the structure of crystals is not foolproof. Basically, what we would like to be able to do is to tell what holes in a cubic closest packed structure of anions (generally) should be assigned to the cations. A couple of examples will illustrate the use of the radius ratio concept. For reference, some ionic radii are given in Table 13.4. Other entries in the table are for later use.

In NaCl, the radius of Na^+ is 0.97 Å and that of Cl^- is 1.81 Å; the radius ratio is 0.54. The chloride ions are in the fcc (or ccp) arrangement. The sodium ions will go into all the octahedral holes surrounded by six chloride ions if the chloride ions are moved out of their closest packed positions to enlarge the octahedral holes. The tetrahedral holes are much too small for the sodium ions. The unit cell of sodium chloride is shown in Figure 13.17(a). There is one octahedral hole for each anion in the fcc structure, so the number of Na^+ and Cl^- are equal and give the NaCl stoichiometry required for electric neutrality. The unit cell of NaCl, and of most other ionic salts of the MX type, is this interpenetration of two face-centered cubic arrangements of anions and cations.

In CsCl, the radius ratio is 0.92. This value is within the range required for cubic coordination of Cs^+ by eight Cl^-. The unit cell of cesium chloride is the body-centered cubic one shown in Figure 13.17(b).

We shall not pursue the generation and justification of crystal structures further, but you should now be convinced of the availability of holes in structures and realize that ions may reside in some or all these holes, subject to size considerations. When oppositely charged ions are involved, cations must be present to neutralize the ionic charge of the anions.

Silicate Minerals

We turn now to the composition and structure of the silicate minerals found in the earth's crust and mantle. Both the chemical composition and crystal structure of the silicates can become exceedingly complex. They are important to us because of the significant role they play in determining the suitability of soils for growing plants. The Si—O—Si bond is the major link in building the skeletons of the molecules of the mineral (inorganic) world, just as the C—C bond is the primary linkage in the organic world. In Chapter 14, we shall see how the silicate minerals exhibit their fitness in soil fertility and structure. In this chapter, we shall explore the structure of silicates.

First let us clarify what we mean by a mineral. The term *mineral* refers

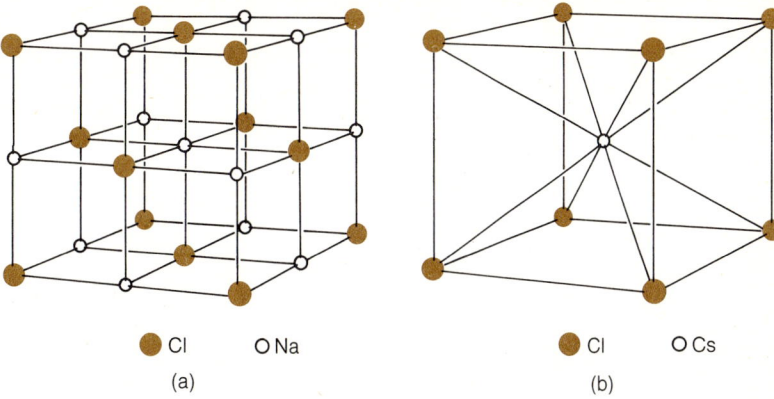

Figure 13.17. Unit cells. (a) Face-centered cubic unit cell of NaCl. (b) Body-centered cubic unit cell of CsCl. Lattice units are shown separated for clarity, but in the crystal the units are close together and there is little empty space. Note that each ion at a corner is shared by eight unit cells, each ion on an edge by four unit cells, and each ion on a face by two unit cells. Thus, in one unit cell of NaCl there are Na⁺ (center) $+ \frac{12}{4} Na^+$ (edges) $+ \frac{8}{8} Cl^-$ (corners) $+ \frac{6}{2} Cl^-$ (faces) or $Na_4Cl_4 = NaCl$, which is the required 1 to 1 stoichiometry.

● Cl   ○ Na

(a)

● Cl   ○ Cs

(b)

to a naturally occurring solid having either a definite chemical composition or at least a limited range of compositions. Each mineral also has a specific crystal structure. According to the definition, oil and coal are not minerals, although they are often called mineral resources in the wider sense that they are nonliving naturally occurring substances of use to man. Rocks are usually coherent mixtures of the crystals of different minerals. Most of the earth's crust consists of silicate minerals made up of silicon, oxygen, and various metallic elements. Minerals are usually known by their common names rather than their chemical names. Thus, the common name for the mineral with chemical composition $CaCO_3$ is not calcium carbonate but may be either calcite or aragonite, depending on the crystal structure. In calcite, the $CaCO_3$ has crystallized in a hexagonal structure, but in aragonite it has crystallized in the orthorhombic structure.

Calcite and aragonite illustrate the phenomenon of *polymorphism*, the existence of the same substance in more than one crystal form. Such a substance is said to be polymorphic, and forms with the same composition but different structures are polymorphs. Diamond (cubic system) and graphite (hexagonal) are polymorphic forms of carbon. The particular crystal structure assumed by a substance depends on the conditions at the time of its crystallization. The dense diamond is formed under high-pressure conditions within the earth. Graphite, with a less dense open structure, is the energetically stable form of carbon at all temperatures and normal pressures. If it were not for the extreme slowness of the C(diamond) $\longrightarrow$ C(graphite) transition (due to a high activation energy), diamond would crumble to black, powdery graphite.

13.10 STRUCTURE OF SILICATES

The fundamental structural unit of the numerous and varied group of silicate compounds is the SiO_4 tetrahedron in which silicon is surrounded by four oxygen atoms. In the simplest silicate compounds, the SiO_4 unit occurs as the anion, SiO_4^{4-}. The tetrahedrally coordinated silicon is bonded to oxygens by strong bonds. The electronegativity difference between Si and O is 1.7, which indicates that the Si—O bond is less than 50 percent ionic. It is common, however, to talk about silicon and oxygen in silicates as though they were the ions Si^{4+} and O^{2-}, and this practice will be followed here although it is probably better to think of Si and O as being in the +4 and −2 oxidation states. The Si^{4+} ion is in a tetrahedral hole surrounded by four

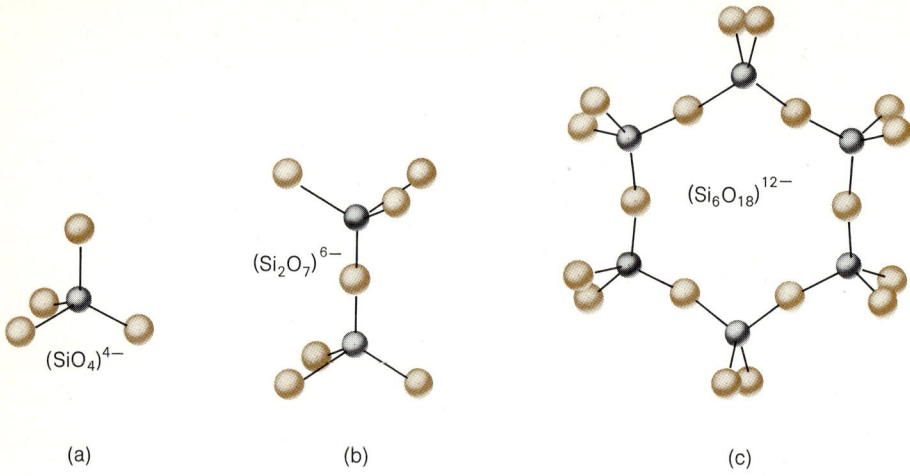

Figure 13.18. Silicate anion units with definite numbers of silicate tetrahedra.

closely packed oxide ions, O^{2-}, which are in mutual contact. The radius ratio $r_{Si^{4+}}/r_{O^{2-}}$ is 0.30.

The bewildering complexity of the naturally occurring silicate compounds can be explained on the basis of (1) the various ways in which the SiO_4^{4-} tetrahedra can be linked to each other by sharing oxygens at some of their corners and (2) the substitution of like-sized cations of different elements for one another in various holes in the oxygen lattice of these structures.

Table 13.5
Structure of silicate minerals

classification	structural arrangement	anion unit	examples
nesosilicates	independent tetrahedra	$(SiO_4)^{4-}$	forsterite, Mg_2SiO_4
sorosilicates	two tetrahedra sharing one oxygen	$(Si_2O_7)^{6-}$	akermanite, $Ca_2MgSi_2O_7$
cyclosilicates	closed rings of tetrahedra, each sharing two oxygens	$(Si_3O_9)^{6-}$, $(Si_4O_{12})^{8-}$, $(Si_6O_{18})^{12-}$	beryl, $Al_2Be_3Si_6O_{18}$
inosilicates	continuous single chains of tetrahedra, each sharing two oxygens	$(SiO_3)^{2-}$	pyroxenes: enstatite, $MgSiO_3$, diopside, $CaMg(SiO_3)_2$
	continuous double chains of tetrahedra sharing alternately two and three oxygens	$(Si_4O_{11})^{6-}$	amphiboles: anthophyllite, $Mg_7(Si_4O_{11})_2(OH)_2$
phyllosilicates	continuous sheets of tetrahedra, each sharing three oxygens	$(Si_4O_{10})^{4-}$	talc, $Mg_3Si_4O_{10}(OH)_2$ kaolinite, $Al_4Si_4O_{10}(OH)_8$
tektosilicates	continuous framework of tetrahedra, each sharing all four oxygens	(SiO_2)	quartz, SiO_2; albite, $NaAlSi_3O_8$

SOURCE: Adapted from Brian Mason, "Principles of Geochemistry," 3rd ed, Wiley, New York, 1966, p 83, Table 4.3.

Independent SiO_4^{4-} Anion Units [Figure 13.18(a)]

Naturally occurring silicate minerals made of discrete silicate anions and charge-neutralizing cations in octahedral holes are called *nesosilicates* by geologists and were formerly called *orthosilicates* by chemists (Table 13.5). Their general formula is M_2SiO_4, where M represents a divalent cation such as Mg^{2+}, Fe^{2+}, Mn^{2+}, and sometimes Ca^{2+}. Common minerals, with their names, are Mg_2SiO_4, forsterite; Fe_2SiO_4, fayalite; $(Mg, Fe)_2SiO_4$, olivine; and the garnets $M_3^{2+}M_2^{3+}(SiO_4)_3$, in which trivalent cations reside in cubic holes. The formula for olivine should be interpreted to mean that the relative amounts of magnesium and iron(II) ion may vary from one sample of olivine mineral to another, with Mg^{2+} usually present in greater amount, but there must always be two such cations for each SiO_4^{4-} anion.

The nesosilicate structures can be thought of as a close packed network of oxide ions containing silicon ions in some of the tetrahedral holes and divalent metal cations in some of the octahedral holes. If there are two tetrahedral holes and one octahedral hole for each O^{2-} anion, then only one-eighth of the available tetrahedral holes will be occupied by Si^{4+} and only one-half of the octahedral holes will contain divalent cations in order to maintain charge neutralization.

Two Linked Tetrahedra [Figure 13.18(b)]

Two tetrahedra may link by the sharing of one oxygen between them to give discrete anions with the formula $(Si_2O_7)^{6-}$. Examples of such compounds are $Ca_2MgSi_2O_7$ and $Sc_2Si_2O_7$. Minerals in this group are known as *sorosilicates*.

Ring Structures of Linked Tetrahedra

Rings of three to six silicate tetrahedra are known. The general anion formula is $(SiO_3)_n^{2n-}$. In these *cyclosilicates*, such as benitoite, $BaTiSi_3O_9$, and beryl, $Al_2Be_3Si_6O_{18}$, the silicon ions define the plane of the ring. The $(Si_6O_{18})^{12-}$ anion ring of beryl is shown in Figure 13.18(c). Metal ions between the rings bind them together.

Indefinite Anion Chain Structures (Figure 13.19)

Silicate tetrahedra may form anion chains of indefinite length by the sharing of two oxygen atoms between tetrahedra. The chains may be single or double and are characteristic of minerals in the *inosilicate* group. Compounds with single chains, with the general anion formula $(SiO_3)_n^{2n-}$, are called *metasilicates* by chemists, and the minerals are classified as *pyroxenes*. The double-chain minerals are classified as *amphiboles* and have the general anion formula $(Si_4O_{11})_n^{6n-}$. In both the single- and double-chain substances, divalent metal ions in octahedral holes bind adjacent single or double chains together. These metal ion–silicate bonds are not as strong as the Si—O bonds within the chains, so these minerals cleave easily parallel to the chains. As a result, many of the substances are fibrous.

Amphiboles have, in addition to octahedral cation sites, cubic (CN = 8) and 10- or 12-fold coordinated cation sites. Anion sites, usually occupied by OH^-, also are available at the center of the hexagonal rings.

Sheet or Layer Structures (Figure 13.20)

The extension of the double-chain structure of the amphiboles in two dimensions by the sharing of three oxygens of each tetrahedron with adjacent tetrahedra results in the laminar structures of the *phyllosilicates* with the general anion formula $(Si_4O_{10})_n^{4n-}$. Individual flat sheets are bonded together by electrostatic forces between metal ions in octahedral sites and the oxygens of the tetrahedra. Again, these interplanar forces are not strong and the layers

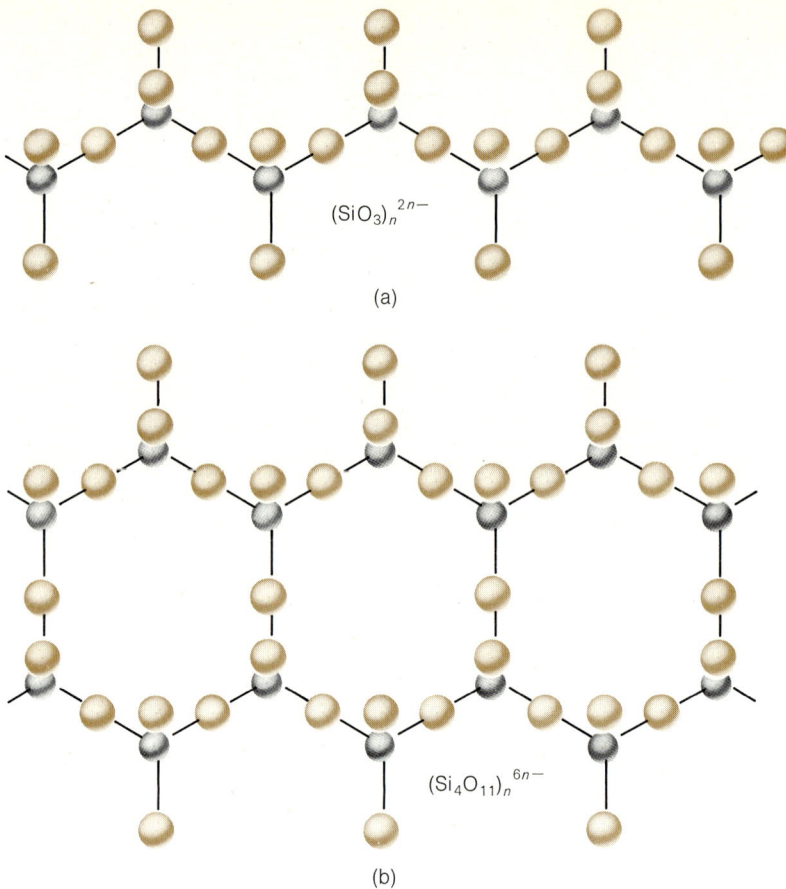

$(SiO_3)_n^{2n-}$

(a)

$(Si_4O_{11})_n^{6n-}$

(b)

Figure 13.19. Single and double silicate chains as found in pyroxene (a) and amphiboles (b).

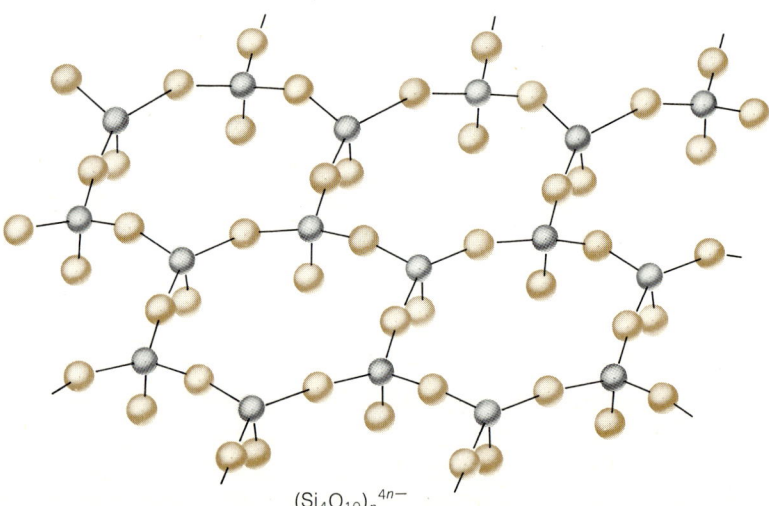

Figure 13.20. Sheet structure of phyllosilicates.

$(Si_4O_{10})_n^{4n-}$

can be easily separated. Asbestos, a name given to the fibrous forms of many minerals, is predominantly chrysotile, a fibrous form of the phyllosilicate serpentine, $Mg_6Si_4O_{10}(OH)_8$.

Other sheet silicates characterized by easy cleavage are talc, $Mg_3Si_4O_{10}(OH)_8$; mica, $KAl_3Si_3O_{10}(OH)_2$; and clays such as kaolinite, $Al_4Si_4O_{10}(OH)_8$. In all of these, OH^- ions are held in the hexagonal holes of the sheets, as they were in the amphiboles. The Mg^{2+} and Al^{3+} are held in octahedral coordination between two silicate sheets so that a sublayer of SiO_4 tetrahedra is connected by shared oxide ions to a sublayer of MgO_6 or AlO_6 octahedra. In mica, in addition, one-fourth of the silicon in the tetrahedral sublayer is replaced by aluminum. [This feature might be better shown by giving mica as $KAl_2(Si_3AlO_{10})(OH)_2$.] The K^+ in mica are sandwiched into 12-fold coordination sites between sets of the double layers of AlO_6 octahedra and SiO_4 tetrahedra. We shall return to the structure of aluminosilicates in Section 14.8.

Framework Silicates (Figure 13.21)

If silicate tetrahedra share all their four oxygens with one another, a three-dimensional structure composed entirely of strong Si—O—Si bonds will be formed. This is the structure of silica, SiO_2, of which quartz is an important form. As with the layer minerals, aluminosilicates are commonly formed by the replacement of silicon with aluminum in tetrahedral sites. To make up

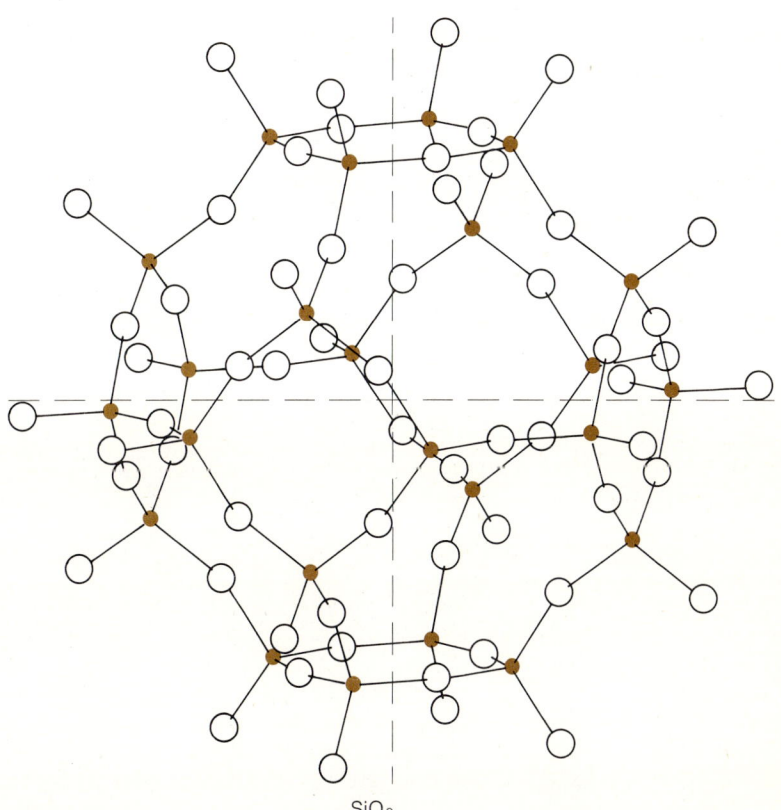

SiO_2

Figure 13.21. Framework structure of silicate tetrahedra in silica. (From L. G. Berry and Brian Mason, "Mineralogy: Concepts, Descriptions, Determinations," Freeman, San Francisco, 1959, p. 466, Figure 15.4.) Copyright © 1959.

for the smaller positive charge on aluminum, a positive charge from a cation must be added to larger holes in the lattice for each silicon replaced by aluminum. Such a process forms the *feldspars*, the major constituents of igneous rocks in the earth's crust. Examples are orthoclase, $KAlSi_3O_8$; albite, $NaAlSi_3O_8$; and anorthite, $CaAl_2Si_2O_8$. The zeolites, used for ion exchange, are also silicate network minerals found in cavities and veins in certain igneous rocks.

13.11 ISOMORPHISM AND ATOMIC SUBSTITUTION

Different compounds are said to be *isomorphous*, or to exhibit isomorphism, if they have the same, or closely related, crystal structures. Isomorphism is widespread among silicate minerals, and examples of isomorphs have already been given to illustrate some of the previous silicate structures. For example, those silicate minerals constructed from independent SiO_4 tetrahedra and divalent cations form the isomorphous olivine group of minerals. Olivines crystallize in the orthorhombic system. The mica group is another purely isomorphous series in the monoclinic system. Another group, the pyroxene group, consists of two isomorphous series. Members of each of these series are listed in Table 13.6.

Table 13.6
Isomorphous series

group	formula
olivine (orthorhombic)	Mg_2SiO_4
	Fe_2SiO_4
	$(Mg, Fe)_2SiO_4$
	Mn_2SiO_4
	$CaMgSiO_4$
	$CaMnSiO_4$
mica (monoclinic)	$KAl_2(AlSi_3O_{10})(OH)_2$
	$NaAl_2(AlSi_3O_{10})(OH)_2$
	$KMg_3(AlSi_3O_{10})(OH)_2$
	$K(Mg, Fe)_3(AlSi_3O_{10})(OH)_2$
	$KLi_2Al(Si_4O_{10})(OH)_2$
pyroxene	$MgSiO_3$ ⎫ orthorhombic
	$(Mg, Fe)SiO_3$ ⎭
	$MgSiO_3$ ⎫
	$(Mg, Fe)SiO_3$ ⎪
	$CaMgSi_2O_6$ ⎬ monoclinic
	$NaAlSi_2O_6$ ⎪
	$LiAlSi_2O_6$ ⎭

Isomorphism results because anions and cations of the same relative sizes and, therefore, the same radius ratio tend to crystallize in the same crystal system, especially if the same total number of ions is involved. Chemical similarities between isomorphous compounds are incidental to the requirement of similarity in ionic size. For example, although NaCl and PbS are chemically dissimilar, they are isomorphous because of similar cation/anion radius ratios. For our purposes, however, we need be concerned only with absolute sizes, because we are discussing silicate structures where the sizes of the tetrahedral and octahedral holes among the oxygens are fairly constant.

Substitution in Octahedral Holes

In the olivine series, the metal ions reside in octahedral sites in the lattice. If we wish to see if Fe^{2+} and Mg^{2+} can substitute for one another in this series, we need only look to see if their ionic sizes, 0.74 Å and 0.66 Å, are similar. In the same series, we see that other divalent substitutions, such as Mn^{2+} (0.80 Å) and Ca^{2+} (0.99 Å) are less desirable because of their larger size relative to Mg^{2+} and the resultant distortion of the crystal lattice, so there is little substitution of Mg^{2+} or Fe^{2+} by Ca^{2+} or Mn^{2+}. Substitution is not common for ions with radii differing by more than 15 percent, although minerals formed at higher temperatures can extend this tolerance limit to 20–40 percent.

Cations having different charges may substitute for each other, although little substitution occurs when the charge difference is greater than one. When ions have approximately the same radii, and might be expected to replace each other about equally, the ion of higher charge is favored for inclusion into a lattice with a suitable site. For example, although Li^+, Fe^{2+}, and Mg^{2+} are about the same size, Li^+ is rejected in favor of the divalent ions in olivine, $(Mg, Fe)_2SiO_4$. Furthermore, Mg^{2+} is present in greater amount in this mineral, because it is the smaller of the two divalent ions available. Small size and higher cation charge both result in stronger ionic bonds between the substituent cation and the oxygens of the lattice.

Another reason that replacement of iron or magnesium by lithium in olivine crystal structures is not favored is because of the requirement of electric neutrality. If a univalent ion replaces a divalent ion, an excess negative charge results. This charge must be neutralized by replacing a divalent cation with a trivalent cation elsewhere in the lattice. In the olivine example, another iron or magnesium ion would have to be replaced by a trivalent cation such as Al^{3+} to give a neutral composition of $AlLiSiO_4$.

If the limiting radius ratio for mutual contact of six O^{2-} and an octahedrally coordinated cation is taken, the radius of an octahedral site surrounded by closest packed (touching) O^{2-} is $r^+ = 0.414 r_{O^{2-}} = 0.414$ (1.40 Å) = 0.58 Å. The examples given show that ions with radii above this minimum value do fit into octahedral holes in silicates, and it would appear that the silicates have an expanded or expandable lattice. In fact, the octahedral hole in silicates has a radius of 0.7–0.8 Å.

Substitution in Tetrahedral Holes

In addition to substitution or interchange of cations in octahedral sites in silicates, an important replacement also takes place in the tetrahedral site occupied by silicon. Table 13.4 shows that the radius of the abundant Al^{3+} ion allows it to enter both octahedral and tetrahedral coordination sites in silicate lattices. The aluminum ion, with a radius of 0.51 Å, is the smallest ion that can fit into an octahedral hole, where it interacts ionically with the surrounding oxygens. It is, at the same time, the largest ion that can exchange for silicon in the tetrahedral sites (although this violates the 15 percent rule) where it acts covalently as a nonmetal. [Compare this with the amphoteric nature of $Al(OH)_3$.] The minimum radius of a tetrahedral site surrounded by closest packed oxide ions is (0.22)(1.40 Å) = 0.31 Å; so, here again, the surrounding oxygens have moved away from their closest packed positions.

Aluminum is the only common substitute for silicon in silicates, and it appears frequently in the aluminosilicates derived from the double-chain amphiboles, the sheet structure micas, and the framework feldspar silicates. As before, when a trivalent aluminum replaces a quadrivalent silicon, the resulting charge imbalance must be made up by another substitution elsewhere in the crystal lattice. For instance, electrical neutrality is maintained when Al is substituted for Si in albite, $NaAlSi_3O_8$, by the coupled substitution of Ca^{2+} for Na^+ to give anorthite, $CaAl_2Si_2O_8$. Alternatively, one could say that Al^{3+} replaces Si^{4+} to balance the substitution of Na^+ by Ca^{2+}. The general formula for the series of compositions between albite and anorthite may be written

$$(Na, Ca)(Si, Al)AlSi_2O_8$$

to show that a number of compositions are possible, depending on the extent of substitution. Another example is $CaMgSi_2O_6$, where both Mg^{2+} and Si^{4+} can be partially replaced by Al^{3+} to give $CaAl_2SiO_6$.

Other Factors in Isomorphous Replacement

Although ionic size is the most important factor, the nature of the bonds between the substituent cation and the coordinating ions must be considered. Isomorphous replacement will occur only if the new substituent ions form the same kind of bonds with the lattice as the ions they are replacing. As an example, Ag^+ ($r = 1.26$ Å) does not substitute for K^+ ($r = 1.33$ Å) in potassium silicate minerals, because silver forms covalent rather than ionic bonds with oxygen. Similarly, Cd^{2+} ($r = 0.97$ Å) does not replace Ca^{2+} in minerals. The importance of bond type is also illustrated in the geochemical classification of elements (Table 13.2).

Some structures allow more substitution than others. Those structures with more open lattices can accommodate a larger variety of substituents or allow extensive exchange of one ion for a larger one. The sheet silicates have crystal structures with larger cation sites than the denser and more closely packed nesosilicates with their independent silicate tetrahedra. The sheet silicates therefore allow a greater variety and extent of substitution without undue lattice distortion.

The great opportunity for isomorphous substitution in the silicates accounts for the early difficulties in the interpretation of their widely varying chemical composition. The confusion resulting from the change in composition in the same crystal was not dispelled until X-ray studies showed how various ions could fit into the same unit cell.

Solid–Liquid Phase Equilibria

We know from crystal structure which different ions can substitute for one another in a crystal lattice, but we must look elsewhere to find out how this substitution came about. Since it is highly unlikely that different ions exchange in solid substances, we must assume that the compositions of the igneous rocks in the earth's mantle and crust were determined in the molten state and locked in when the melt solidified. To appreciate how the solidification of this molten silicate rock can lead to an understanding of the distribution of the elements in the resulting igneous rock, it is helpful to know something about the heterogeneous (solid–liquid) equilibria of the freezing process.

Molten rock is hardly a simple chemical system. Instead of the one or two substances needed to specify the composition of the equilibrium systems discussed in earlier chapters, there are a minimum of some 10 to 15 elements that determine the composition of a simple molten silicate rock system. Since we are interested only in some general principles of solid–liquid equilibrium systems, however, we can restrict ourselves to simple two-component (binary) systems for illustrative purposes.

13.12 EUTECTIC MIXTURES

In Section 12.22, we discussed the depression of the freezing point of a solvent by a dissolved solute. Let us now take a liquid solution composed of two components, A and B. If the solution is composed mostly of A, then B can be called the solute and A the solvent. If A and B are miscible in the liquid state, we can also have solutions where B is present in excess, in which case B is the solvent. We shall assume that A and B are miscible in all proportions in the liquid state and, as in our earlier treatment of solutions, that when a solution of A and B freezes, the solid freezing out is pure solvent (A or B).

If the freezing points of solutions containing A in excess are plotted against composition, a curve similar to the one on the left side of Figure 13.22 is

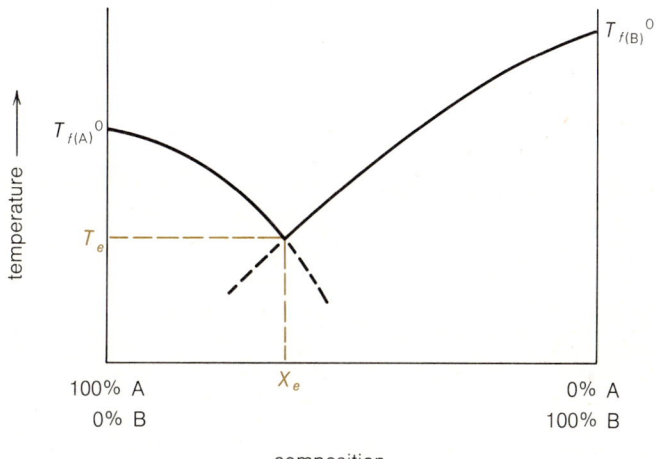

Figure 13.22. Eutectic behavior. Freezing point curves for the solvents in solutions of two immiscible solids meet at the eutectic point (T_e, X_e).

obtained. Points on this curve correspond to the temperature at which pure solid A begins to separate from the solution whose composition is given on the horizontal axis. Then, if solutions of mostly B as solvent are taken and the temperatures at which pure solid B freezes out are plotted, the freezing point curve on the right side of Figure 13.22 is obtained. Under the conditions specified, the separate freezing point curves of A and B will meet at some point. At this temperature, called the *eutectic point*, both solid A and solid B will freeze out as a heterogeneous or two-phase mixture of solid A crystals and solid B crystals. That is, A and B are immiscible in the solid state. The eutectic temperature is the lowest temperature at which any liquid mixture of A and B will freeze. Below the eutectic point, no liquid phase can exist.

The temperature–composition diagram of Figure 13.22 is a phase equilibrium diagram and it gives the conditions under which different phases may be in equilibrium with each other. In Figure 13.23, the diagram for a eutectic mixture is labeled with the phases present in each area. For any set of temperature–composition values above the freezing point curves AC and CB, only liquid mixtures of the two components may exist. Below the horizontal line

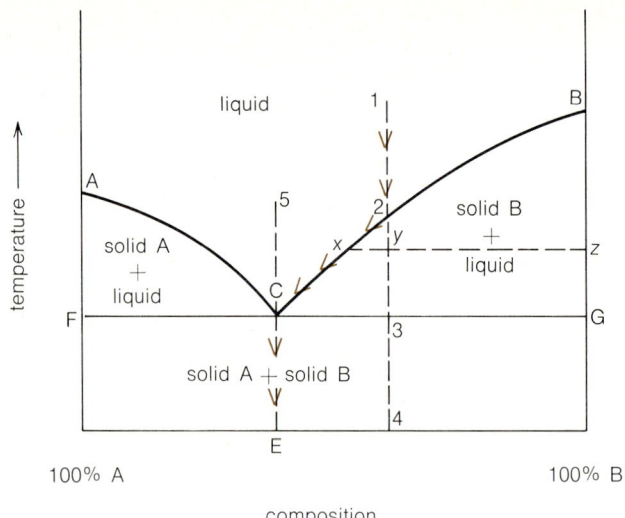

Figure 13.23. Temperature–composition phase diagram for a simple two-component eutectic mixture.

FCG, only solid mixtures of A and B may exist. In the areas ACF and BCG, solid A or solid B, respectively, can be in equilibrium with a liquid mixture. For example, if the temperature and overall composition of a solid B–liquid mixture is located at point y, solid B is in equilibrium with a liquid mixture whose composition corresponds to that of point x on the freezing point curve BC. By equilibrium we mean that the amounts of liquid mixture and solid B remain constant and neither phase increases at the expense of the other. In the equilibrium system at point y, the proportion of liquid phase to solid (B) phase is given by the ratio yz/xy of the lengths of the segments.

The phase diagram of Figure 13.23 can be used to examine the behavior of a liquid system upon cooling. If a liquid of composition and temperature given by the point 1 is cooled, solid B begins to freeze out when the temperature has decreased to that corresponding to point 2. As the liquid continues to cool (from 2 to 3), its composition and corresponding freezing temperature move from point 2 along the curve to point C, the eutectic point. During this time, more solid B will separate from the liquid and the liquid becomes richer in component A. The total composition of solid + liquid, of course, remains constant. At the eutectic point, which represents the composition shown at E and the temperature at F, solid A freezes out along with the rest of solid B. Solid A and B then freeze out in the same proportion as A and B exist in the remaining liquid. For overall compositions and temperatures below the horizontal line, no liquid remains. The reverse sequence of changes would take place if a solid mixture with composition given by point 4 were slowly heated, the first liquid appearing at the temperature and composition corresponding to point C. The liquid would become richer in component B as more solid B melted.

If the composition of the liquid corresponded to that at point 5, no solid would separate until the eutectic temperature was reached. At that point, solid A and solid B would separate out simultaneously, and the liquid would solidify completely at this temperature. Only after all liquid had frozen would the temperature of the mixture decrease further. The cooling behavior of a liquid mixture with the eutectic composition is similar to that of a pure compound, as the next paragraph will show; however, the eutectic composition varies with external pressure and so cannot represent a pure substance.

The freezing point–composition diagram can be obtained by interpreting the change in slope of the cooling curves (temperature versus time) for solutions of various compositions. The way this is done is explained by Figure 13.24, where the phase diagram of a bismuth–cadmium mixture is shown

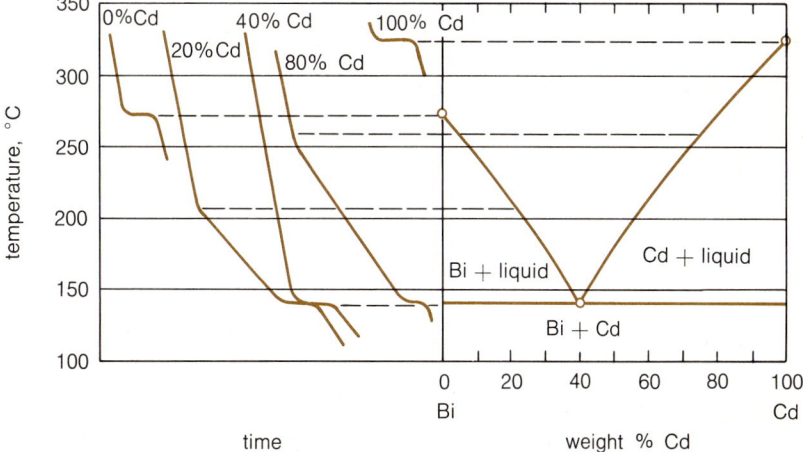

Figure 13.24. Cooling curves and the eutectic phase diagram for the Bi–Cd system. (Adapted from Farrington Daniels and Robert A. Alberty, "Physical Chemistry," 3rd ed, Wiley, New York, 1966, p. 157, Figure 5.15.)

together with cooling curves for liquid mixtures of various compositions. The cooling curves of pure liquid Bi or pure liquid Cd show a uniform decline until the pure component begins to freeze out. Then the temperature remains constant, and the cooling curve horizontal, until all pure Bi or Cd has solidified. If a liquid mixture of Bi and Cd is cooled, the cooling curve of the mixture will show a uniform decline until either pure Bi or Cd begins to freeze out. At this point, the rate of fall of temperature will decrease, and the slope of the cooling curve will decrease. The temperature at which this occurs is the freezing point of Bi or Cd in the mixture, and represents a point on the freezing point diagram. Further cooling of the mixture will eventually show a horizontal portion at the temperature where both components are freezing out. This horizontal portion occurs at the eutectic temperature. The length of the horizontal portion (the eutectic halt) depends on the composition of the mixture, being longest for a mixture having the eutectic composition. If a liquid mixture having exactly the eutectic composition is cooled, there will be no change in slope of the cooling curve until the horizontal portion at the eutectic temperature is reached. Here both Bi and Cd begin to solidify at the same time.

The example of bismuth–cadmium mixtures shows that a solid mixture having a eutectic composition has a melting point below the melting points of either pure component. The low melting point of eutectic mixtures has considerable practical importance in the formulation of mixtures of metals (alloys) for casting and soldering purposes. One alloy, Wood's metal, is a eutectic mixture of 50 percent bismuth (m.p. 271°C), 27 percent lead (m.p. 327.5°C), 13 percent tin (m.p. 232°C), and 10 percent cadmium (m.p. 321°C).

The melting point of the solid eutectic mixture is only 66°C. Wood's metal is used to make easily fusible valve plugs in automatic sprinkler systems. Practical jokers sometimes fashion spoons out of Wood's metal and then wait breathlessly for a victim to come along to stir a cup of hot coffee.

13.13 SOLID SOLUTIONS

Solids that crystallize in different crystal systems tend to show eutectic behavior because the dissimilar crystalline solids are immiscible in the solid state. The solid substance crystallizing from the melt consists of a mixture of separate crystals of the components. It is a true two-phase mixture. Other substances, on the other hand, may be completely miscible in the solid state, so that the solid mixture is a true homogeneous or one-phase mixture in which the components are distributed throughout each other on an atomic or molecular level. Such solid mixtures are called *solid solutions* because they are analogous to liquid solutions. Substances forming solid solutions tend to have similar crystal structures, that is, they are isomorphous. But not all isomorphous substances form solid solutions with each other. Also, there are examples of solid solutions of nonisomorphous substances.

Among isomorphous substances, solid solutions are more liable to form if the similar crystal structures also have similar unit cell volumes or dimensions. Again, in analogy with miscible liquids, solids will tend to be miscible if the A–A, B–B, and A–B interactions are of about the same magnitude, that is, if the solution is ideal.

When a solid solution of two components can form, it is possible for the freezing point of one component to be raised by the presence of the other component. Solid solution formation leads to phase diagrams of the form shown in Figure 13.25.

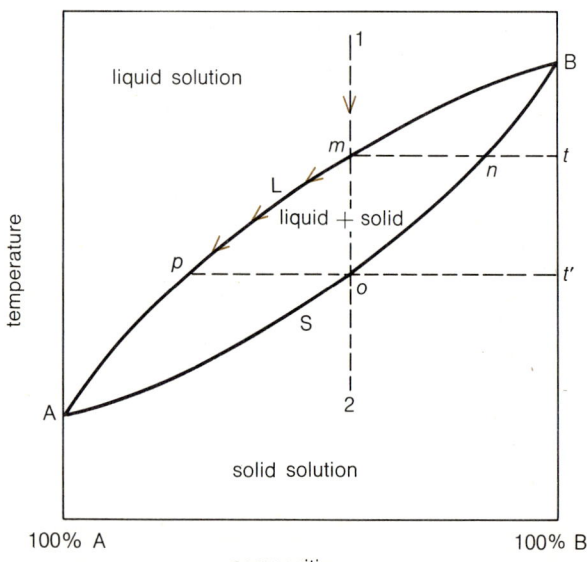

Figure 13.25. Solid–liquid phase diagram. The components are completely miscible in both the liquid and solid phase. A continuous series of solid solutions is formed.

If a liquid having the composition at point 1 is cooled, the solid that crystallizes at point *m* is neither pure component A or B nor does it have the composition of the liquid. Instead, the first solid to separate has the composition corresponding to point *n*. The composition of the liquid in equilibrium with this first-crystallized solid corresponds to that of point *m*. As crystals of composition *n* are formed, the liquid becomes richer in the lower-melting component A, since the crystals contain relatively more B and the freezing point of the liquid progressively decreases. The total composition of the solid–liquid system remains constant along the line 1–2, and the compositions of the liquid and solid in equilibrium at any point on this line between points *m* and *o* are given by the intersections of a horizontal line through this point with the equilibrium curves. The liquid becomes richer in component A as crystals freeze out along the curve *no*. The freezing point of the liquid decreases along the line *mp*. As the temperature falls, the amount of solid increases, until at point *p* the last trace of liquid is just about to disappear and solidification is complete. The entire solid mass has the uniform composition of the original liquid, although the composition of the solid phase changes continuously from *n* to *o* until complete solidification occurs.

This description assumes that the crystals remain in contact with the liquid until solidification is complete in such a way that complete equilibrium between the crystals and the liquid melt is maintained. The solid continually reacts with the liquid as cooling progresses so that, at a given temperature, all the crystals have the same composition. If, however, the crystals freezing out are removed from the vicinity of the cooling liquid, those crystals cannot adjust their concentrations to the required equilibrium values as further cooling takes place. The end result is a mass of crystals of nonuniform composition, ranging from B–rich (the first to solidify) to A–rich (the last to crystallize). Thus, solid solutions may have a range of compositions determined by (1) the composition of the original liquid melt (equilibrium cooling) or (2) conditions leading to nonequilibrium cooling from a liquid of given composition. This is not the case for solids freezing out of a eutectic mixture.

A eutectic system corresponds to complete immiscibility of solid components. A solid solution system corresponds to complete miscibility in the solid phase. There are all degrees of behavior between these extremes, but we need be concerned here only with the formation of solid solutions, because almost all rock-forming minerals are involved in the formation of solid solutions with isomorphous mineral compounds. For instance, fayalite, Fe_2SiO_4, (m.p. 1205°C) and forsterite, Mg_2SiO_4, (m.p. 1890°C) enter into solid solution to produce the mineral olivine, $(Mg, Fe)_2SiO_4$, in which the ratio of Mg to Fe can vary from one specimen to another. Forsterite and fayalite would correspond to the pure components A and B, respectively, in Figure 13.25, and the solid crystallizing from molten solutions of the two minerals would be the solid solution, olivine.

The feldspars, albite, $NaAlSi_3O_8$, and anorthite, $CaAl_2Si_2O_8$, are two other isomorphous minerals that form solid solutions upon the cooling of the molten mixtures within which they occur. This leads to other feldspars of intermediate composition. The wide range of the compositions of the more abundant elements in specific minerals is caused by isomorphous substitution in the crystal lattices formed in solid solutions. The examples used here to illustrate solid–liquid equilibria are simple two-component systems. Now let us see what may happen when the complex silicate system of molten rock cools and solidifies to form the earth's mantle and crust.

Crystallization of Magma

We now return to the planet Earth in its molten stage. Excess iron has concentrated in the core. On top of the core floats a complex mixture of oxygen, silicon, aluminum, iron, magnesium, calcium, sodium, potassium, and a host of less concentrated trace elements, all under high pressure. They compose the magma that will solidify to give the igneous rock of the mantle and crust. Another important ingredient of this molten rock is water. The crystallization of this complex mixture is, as might be expected, a complicated process about whose details we can only guess. Nevertheless, through analysis of natural rock specimens and investigations of artificial silicate melts in the laboratory, the general features have been put together. Here, in an oversimplified version of the cooling process, we see only the barest outline.

13.14 CRYSTALLIZATION SERIES

The complex, multicomponent magma will not freeze at a single temperature. Instead, solids will form over a wide range of temperatures. One mineral after another will separate out, and after solidification is complete, rocks containing many different minerals will result. The solidified minerals will have different melting points, densities, crystal structures, and chemical compositions. All of these differences will contribute to some differentiation of the elements present. For instance, the compounds with high melting points will solidify first. It turns out that these compounds have a high density, so they will settle to the bottom of the cooling silicate mass, where they will form the main part of the mantle. Such substances have a composition different from that of the remaining melt, and the remaining liquid will change in composition as solids freeze out, becoming less concentrated in the elements composing the precipitating solid. At the end of the cooling period, the remaining relatively light liquid will solidify and form the upper part of the mantle and parts of the crust.

The sequence of crystallization in cooling magma has been fairly well worked out. Two crystallization series have been identified. One involves a continuous solid solution (isomorphous) series of the high-melting Ca-feldspar, $CaAl_2Si_2O_8$, and the lower-melting Na-feldspar, $NaAlSi_3O_8$. The other includes a series of nonisomorphous silicates of iron and magnesium that successively partially react with the remaining liquid to form the next mineral phase listed. The sequence of solidification for each is shown in Figure 13.26 for the relatively low pressures (a few thousand atmospheres) expected in the top few kilometers of the earth's crust.

Fundamentally, the sequence of crystallization is determined by the stability or lattice energy of the possible solids. This stability is related to the strength of bonding in the crystal. The solid with the strongest bonding at a given temperature will have the highest melting point and will freeze out first as the magma cools.

13.15 SECONDARY DIFFERENTIATION OF THE ELEMENTS

In Section 13.3, we discussed primary geochemical differentiation, a process in which the distribution of the elements between the earth's iron core and silicate mantle was controlled by the oxidizing property of an element relative to iron. Secondary differentiation of the elements is concerned with the further distribution of the elements within the minerals comprising the mantle and crust. Secondary differentiation is determined by the role of the various elements in solid–liquid equilibria as the magma solidifies.

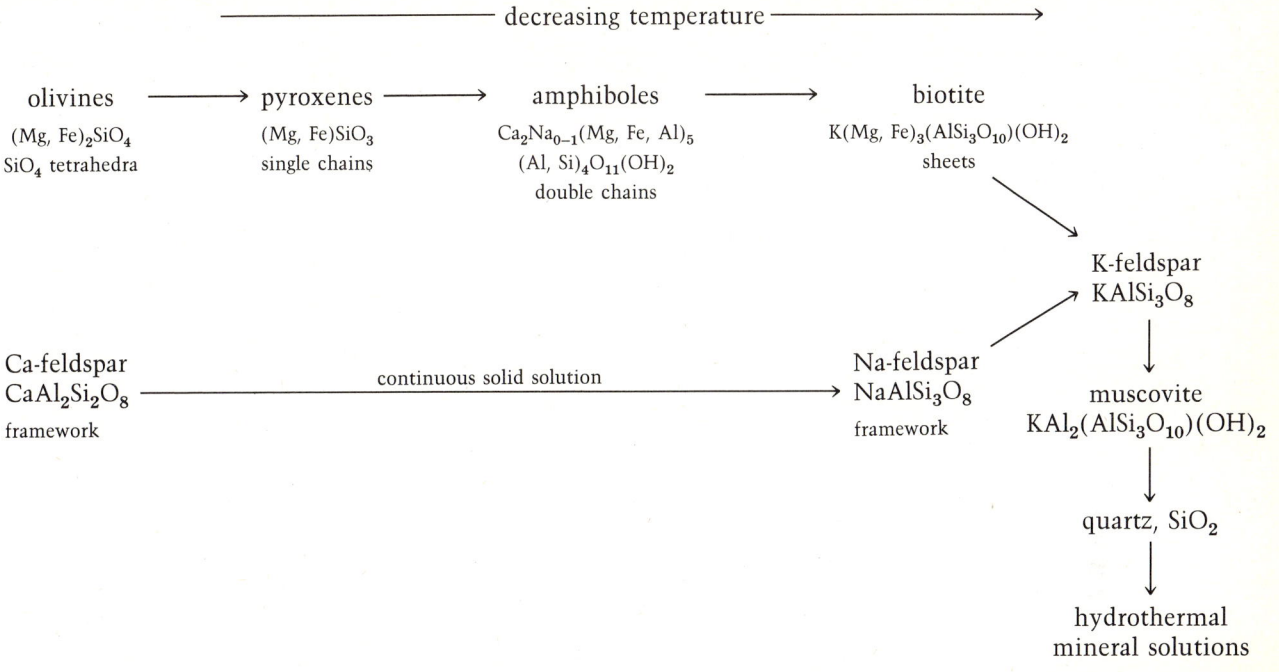

$\longleftarrow$ decreasing temperature $\longrightarrow$

olivines $\longrightarrow$ pyroxenes $\longrightarrow$ amphiboles $\longrightarrow$ biotite

$(Mg, Fe)_2SiO_4$ $(Mg, Fe)SiO_3$ $Ca_2Na_{0-1}(Mg, Fe, Al)_5$ $K(Mg, Fe)_3(AlSi_3O_{10})(OH)_2$

SiO_4 tetrahedra   single chains $(Al, Si)_4O_{11}(OH)_2$ sheets

double chains

K-feldspar
$KAlSi_3O_8$

Ca-feldspar   continuous solid solution   Na-feldspar
$CaAl_2Si_2O_8$ $\longrightarrow$ $NaAlSi_3O_8$

framework   framework   muscovite
$KAl_2(AlSi_3O_{10})(OH)_2$

quartz, SiO_2

hydrothermal
mineral solutions

early-forming minerals with high   late-forming minerals with lower
melting points and crystal densities   melting points and crystal densities
(may concentrate in mantle)   (may concentrate in crust)

Figure 13.26. Sequence of mineral crystallization from a cooling magma. The feldspar crystallization series is a continuous solid solution series in which the early-formed crystals change continuously in composition by reaction with the melt (assuming equilibrium conditions). In the olivine–biotite sequence, an early-formed solid phase, which is itself a member of a shorter continuous solid solution series, reacts with the melt at a fairly definite temperature to form a new mineral with a different crystal structure. The exact crystallization sequence depends on the original composition of the magma and on the conditions of cooling. This general crystallization sequence is known to geologists as Bowen's reaction series.

For instance, we can see from the crystallization sequence in Figure 13.26 that some of the major elements (Mg, Fe, Ca) will tend to concentrate in the early-formed minerals but that others (Al, Si, K, Na) will appear in greater abundance in the lighter late-forming minerals. This leads to the description of the earth's mantle as being composed primarily of dense, early-formed ferromagnesian minerals of the olivine type, and the crust as being composed of light ferromagnesian silicates (lower crust) and the even lighter silicates of the alkali and alkali earth metals (upper crust).

Bond strength governs the entry of substituent cations into a crystalline lattice, just as it governs the sequence of crystallization itself. The bonding strength of cations with a silicate lattice can be estimated with the following rules, which have already been mentioned in connection with isomorphous substitution. The rules assume that there is no significant change in bond type.

1. Two cations with similar radius (within 15 percent) and the same charge can enter a given crystal lattice with equal ease.

2. Of two cations with approximately the same radii and with the same charge, the ion with the smaller radius will form stronger bonds and will preferentially enter the lattice.

3. Of two cations with the same radii but differing by one unit in charge, the ion with the higher charge will enter the lattice more readily because of stronger bond formation.

4. Where the cation interacts with the lattice through ionic bonding, the element with the lower electronegativity will be favored because the larger electronegativity difference with oxygen leads to a stronger ionic bond.

The operation of some of these rules can be illustrated with a few of the major and minor elements. Of the olivine minerals, Mg_2SiO_4 should have stronger cation–silicate lattice bonds than Fe_2SiO_4 because of the smaller size and lower electronegativity of Mg^{2+}. This causes the crystal to be more stable and more dense, to have a higher melting point, and to separate from the melt first. Magnesium should therefore be more concentrated than iron in the early-formed olivine group, as well as in the early members of all the other ferromagnesian minerals. In the feldspars, the calcium ion enters the lattice in preference to the sodium ion because of the higher charge on Ca^{2+}, even though the radii of the two ions are almost equal.

The crystal structures of the ferromagnesian silicates and the feldspars, and the order in which they crystallize, determine the differentiation of the less abundant elements. As the major elements form their own minerals, isomorphous substitution for the major elements by the minor elements in the crystals may take place subject to the rules governing competition among various ions. When a mineral separates from magma containing a variety of other ions, the crystal lattice sorts out or screens these ions. It allows into its lattice only those ions having the requisite ionic size, provided the charge and bond type qualifications are met. This is the process responsible for the concentration of minor elements in some minerals but not in others.

The minor cations Rb^+, Ba^{2+}, and Pb^{2+} are matched in ionic size only by K^+ among the major mineral-forming elements. They could substitute for potassium in biotite and K-feldspar, and their concentration should be expected to increase in the same manner as potassium in the crystallization sequence. Thus, potassium and these three metals are all concentrated in late-forming rocks in the crust. The larger Rb^+ concentrates in the even later-formed potassium minerals because their less dense structure permits easier entry by this ion.

Manganese (Mn^{2+}), Sc^{3+}, Ti^{3+}, and V^{3+} all have radii close to that of Fe^{2+} and Ca^{2+}. However, the large electronegativity difference between calcium and the silicate lattice and the resultant binding strength of calcium prevents exchange of the electropositive Ca^{2+} and these less electronegative elements. These minor metals are found as isomorphous substituents only in iron minerals. Magnesium is substituted for by Cr^{3+}, Ni^{2+}, and Co^{2+} in ferromagnesian silicates. The concentration of all of these metals and of iron decreases in the later-forming minerals within the crystallization sequence.

There are many elements that do not readily substitute for a major element in a silicate lattice. Two groups of such elements can be distinguished. In one group are those cations having such great differences in ionic size or charge between themselves and the major metals that they are not accepted into a silicate lattice until other competitive ions have been placed. These elements are the ones that occur mixed with the very last-formed solids, quartz and feldspar. Among these elements are B^{3+} (0.23 Å), Be^{2+} (0.35 Å), W^{6+} (0.62 Å), Sn^{4+} (0.69 Å), Th^{4+} (1.02 Å), U^{4+} (0.97 Å), Cs^+ (1.67 Å), Li^+ (0.68 Å), Rb^+ (1.47 Å), and the rare earths. They are concentrated in the lowest-melting fraction of the silicate melt (pegmatite), whose main constituents are water, Na, K, Al, and SiO_2.

The other group of metals contains those forming predominantly covalent bonds with oxygen. For this reason they cannot substitute for the major, ionically bonded elements in the crystal lattice. These important elements, among them Cu^+, Ag^+, Cd^{2+}, Hg^{2+}, and Zn^{2+}, often turn up as sulfide compounds in minerals of their own.

In addition to these left-over metals, the last portions of the cooling magma will contain large quantities of water along with other stable but volatile molecules such as CO_2, HCl, HF, N_2, and some sulfur compounds. There will be soluble salts of various metals as well.

13.16 ORE DEPOSITS   During the final solidification stages of the residual fluid, volatile components are made available to the as-yet unformed oceans and to the early atmosphere. Minor elements left behind in the melt will form concentrated deposits of their compounds. Many of these deposits are important sources of the elements we use in modern technology. Such deposits are called *economic mineral deposits* if the element can be profitably extracted from it. The terms *ore* or *ore deposit* imply a source of metals, but they are often used more broadly to describe sources of nonmetallic elements such as sulfur. We use "ore" in the metallic sense. Whether an element can be profitably extracted depends on the concentration of the deposit, the chemical and physical nature of the deposit, and also on the market demand for the element itself.

The mechanism of ore deposition of minor elements is a matter of speculation and controversy. Those elements differing greatly in size and charge from the major elements eventually find their way into a silicate lattice and crystallize out of the final silicate melt as oxides or as complex silicate minerals such as beryl, $Be_3Al_2Si_6O_{18}$. Some may form their own minerals. The elements tending to form covalent bonds and sulfides often appear as narrow vein deposits. The hydrothermal hypothesis suggests that these ore deposits are (1) precipitated from a water-rich gas containing volatile metal compounds (principally the chlorides) that later react with hydrogen sulfide or (2) precipitate as sulfides from a hot aqueous liquid. Evidence indicates that many sulfide deposits originate from magmatic material that is transported appreciable distances from the source before cooling and deposition of ore material takes place. Other evidence shows that most deposits are crystallized from aqueous solution. Therefore, both of the above mechanisms, in succession, may be involved. In some cases, it is believed that the sulfide is transported in aqueous solution; but the extreme insolubility of metallic sulfides, even at high temperatures, makes this hard to explain unless the metallic sulfides are held in solution by complexes or other means.

Not all ore deposits are derived directly from magmatic cooling. In many cases, important ore deposits result from physical disintegration and chemical breakdown of igneous rocks. In some instances, insoluble compounds either already present in the rock or produced by chemical change of the minerals there are left behind as the more soluble portions are leached away. The insoluble residues form *residual* or *placer* deposits. Chemical weathering caused by the action of acidic carbon dioxide solution may also form soluble compounds that are carried some distance before being deposited as precipitates both above and below the surface of the ground. The precipitates form *chemical sedimentary deposits*.

Sedimentary and residual deposits are important for the more abundant metals, which appear as major constituents with their own minerals. An important example of a residual ore deposit is bauxite rock, which is rich in hydrated aluminum oxides, $Al_2O_3 \cdot 2H_2O$. Bauxite is formed by the chemical weathering of aluminum-rich igneous rocks at the earth's surface. Chemical changes induced by rainfall convert the other common elements (Na, K, Ca, Mg, but not Fe) to soluble compounds, and they are leached from the upper layers. Remaining behind is a clay mineral such as kaolinite, $Al_4Si_4O_{10}(OH)_8$, rich in silica and aluminum hydroxide. If iron is present, its compounds are also left behind. If the percolating water is near neutral (pH = 5–9), the silica component of the kaolinite also dissolves, and only the aluminum (and iron) oxide remains. If the iron content is low, the residual deposit is called bauxite. Bauxite is found principally in tropical climates where rainfall is or was abundant.

A chemical sedimentary deposit of hematite, Fe_2O_3, forms the important Lake Superior region ore deposits. These deposits originally contained substantial amounts of silica, which ordinarily would have rendered the deposit unprofitable. Through geologic time, however, chemical weathering due to water percolating through the upper layers leached out the silica and other soluble impurities. This process left an enriched ore of about 55 percent iron in the upper leached layers, which have now been exhausted by mining. New methods have been employed to concentrate the vast quantities of underlying, unleached taconite ore (15–40 percent iron) so that iron can be profitably extracted.

Some of the chemical processes of weathering will be discussed in the next chapter. At present, it is enough to see that intricate and imperfectly understood chemical reactions are responsible for the concentration of elements within the earth and for the formation of even more concentrated deposits from which elements can be economically extracted. After a short trip to the moon, we shall examine some of the problems connected with our utilization of natural resources, particularly metals.

13.17 SOME LUNAR GEOCHEMISTRY

On July 21, 1969, with the landing of the U. S. Apollo 11 spacecraft, man set foot on the moon for the first time. As a result of this and subsequent moon landings, samples of the lunar surface have been returned to earth and subjected to minute chemical analyses with the hope of resolving the questions of the moon's origin and history. Much has since been learned from the ages and chemical compositions of a variety of moon rocks.

One of the igneous moon rocks was found to be 4.15 billion yr old, but most of the samples were 3.2–3.8 billion yr old. As stated earlier in the chapter, the oldest earth rock is 3.5 billion yr old. The age of rocks, again, refers to the date of their solidification from a melt. The moon itself is believed to have formed about 4.6 billion yr ago, and to have evolved in five distinct stages. Initially, a cold accretion of solid matter took place. Then heating occurred to melt the *outer* surface of the newly formed moon to a depth of only a few hundred kilometers (the moon has no dense, metallic core). In the third stage, the top part of the molten surface solidified into the original crust. Then, over a period of millions of years, the crust was bombarded by meteorites, which formed large craters and basins ringed by mountains. Finally, magma (lava) from beneath the crust flowed into the basins and solidified to form the lunar seas or maria. According to this sequence of events, there were two periods of solidification—one for the original crust and one for the maria—separated by a billion or so years. The 4.15 billion-yr-old rock is considered to be part of the original crust, while the younger rocks represent the solidified volcanic lava. The age difference of the two types of rock is the period between their formation as postulated in this sequence.

During both the solidification of the original crust and the lunar seas, processes of geochemical differentiation occurred as a result of partial melting and fractional crystallization. The 4.15 billion-yr-old rock came from the lunar highlands. It is rich in calcium plagioclase feldspar minerals which, in the case of the lunar rock, is almost pure anorthite, $CaAl_2Si_2O_8$. Apparently, plagioclase was one of the first-formed minerals (along with olivine) when the lunar magma cooled. Its density is less than the underlying lunar magma, so it came to the surface to make up the original crust.

The composition of the rocks from the lunar maria is considerably different from the rocks of the lunar highlands and from similar rocks (basalts) on earth. The dense, solidified lavas filling the maria are rich in pyroxenes and ilmenite, $FeTiO_3$. Compared to terrestrial igneous rocks, these rocks are about 10 times more abundant in high-melting elements such as Ti, Cr, and the rare earths; about 100 times less abundant in the low-melting or volatile elements such as Na, K, Cl, Ni, Au, Ag, and Pt; and about equal in Fe, Si, Ca, Mg, and Al. The composition of such lunar rocks is again the result of complex chemical fractionation in the cooling lava that filled the lunar basins. The difference in chemical composition between earth and moon rocks indicates that the initial composition of their magma was different and that it is unlikely the two bodies were ever one. That is, one of the theories of the moon's origin—that the moon was torn from the earth—is rendered unlikely. Two other theories, that the moon was a wandering planet captured by the earth and that the moon was formed along with the earth, are neither proved nor disproved by the results so far.

Natural Resources

13.18 RENEWABLE AND NONRENEWABLE RESOURCES

From man's point of view, resources are the things he needs to survive and progress. The desired level of existence, the standard of living, and the complexity of civilization and state of technology often determine what will be a resource and what will not. A resource need not be tangible. Thus, some men may regard silence as an essential resource. The designation of a resource also depends on point of view. Some may regard open space and wilderness as a resource, while others may feel they can survive only with the resources available in a dynamic but crowded city.

In this section, we shall assume agreement upon what constitutes a resource—or at least a *natural resource*, namely the naturally occurring supplies of materials on which a culture draws for its food, shelter, fabrics, and energy. Natural resources are commonly divided into two groups, *renewable* and *nonrenewable*. Renewable resources are those having the inherent capacity to replenish and maintain themselves provided they are managed wisely. These are usually living resources such as plants and animals and the food, wood, and fiber products derived from them. The ability of such resources to renew themselves depends largely on the rate at which they are used. With increasing population and increasing per capita consumption of resources because of advanced technology, the rate of consumption of many renewable resources is approaching their finite rate of regeneration. Whether many resources, such as trees and food crops, remain adequate to our needs will depend on the kind of resource management applied as well as on the total consumption. Decreased total consumption could come about through a decrease in per capita consumption even as the population grows (with foods, in the extreme case, this is called starvation) or through limiting population growth.

Our major concern here is with nonrenewable resources. These are the ones that cannot be replaced, and whose total amounts are, as far as we know, fixed. Some of these resources do regenerate themselves but at a rate so slow as to be insignificant when compared with their rate of consumption. The fossil fuels are prime examples. The world's initial supply of recoverable crude oil was about 2,000 billion barrels, all formed in the 600-million-yr period that life has been abundant on earth. Yet oil is being withdrawn from the earth at a current rate of about 15 billion barrels/yr. Very roughly, it would take 2.3 million yr to renew the annual consumption of oil (assuming that only 50 percent of all the oil on earth is recoverable).

Nonrenewable resources can be further divided into *reusable* and *depleting* (or nonreusable) resources. The first group is by far the smaller, and its foremost member is water.

Depleting resources are, in general, all of those nonliving resources taken from the earth, including the oceans and the air. Once used, many of these resources are permanently consumed and lost forever. Again, oil and other fossil fuels are common examples. They are changed into chemical forms from which recovery is impossible. Other depleting resources, such as the metals, even if not appreciably changed in chemical form, are so dispersed and dissipated throughout the environment during and after their use that recovery is either impossible or presently economically unrewarding. In this sense, they are "consumed."

The naturally occurring nonliving resources are usually referred to as *mineral resources*. They include metals and their ores, and nonmetals and their compounds. The latter category includes such resources as sulfur, phosphates, sodium chloride, sand, asbestos, fossil fuels, water, and various gases.

The atmosphere supplies us with the gases, O_2, N_2, and the inert gases (except helium, which is "mined" from the earth). Atmospheric gases are reusable resources. The oceans now serve as a commercial source of sodium and chlorine (derived from sodium chloride), bromine, magnesium, and, of course, water. The dispersed and very often dilute resources in the atmosphere and the seas are present in such large total amounts as to be an essentially inexhaustible supply. In contrast, the earth's crust contains many mineral resources in relatively concentrated deposits thanks to the complex geological processes mentioned in Section 13.16; however, the total amounts of these concentrated crustal resources may be quite limited. Because of the variety of mineral resources in the earth and the availability of ore deposits, the earth's crust supplies most of our mineral resources. Our discussion emphasizes the crust as a source of metallic elements.

13.19 EXTRACTION OF METALS

World consumption rates of the elements (Table 13.7) shows there is little relation between the amount of an element that is used and its average abundance. Some of our most widely used metals are quite rare, whereas some abundant metals have little demand.

Table 13.7

Annual world consumption of the elements (tons)

C	10^9–10^{10}
Na, Fe	10^8–10^9
N, O, S, K, Ca	10^7–10^8
H, F, Mg, Al, P, Cl, Cr, Mn, Cu, Zn, Ba, Pb	10^6–10^7
B, Ti, Ni, Zr, Sn	10^5–10^6
Ar, Co, As, Mo, Sb, W, U	10^4–10^5
Li, V, Se, Sr, Nb, Ag, Cd, I, rare earths, Au, Hg, Bi	10^3–10^4
He, Be, Te, Ta	10^2–10^3

SOURCE: U. S. Bureau of Mines.

The availability of metals depends on the ease of their extraction from the minerals of which they are constituents. Although iron and aluminum have large abundances, the major portions are contained in hard, high-melting silicate minerals. Extraction of metal in any quantity from silicates is very difficult and expensive. Extraction from their oxides is much easier and more economical. Some less abundant metals also occur widely as substituents in silicate minerals of the major elements, but their extraction is likewise difficult because of both the intractability of the silicates and the low metal concentration. Fortunately, many of the less abundant metals also occur in easily worked sulfide or oxide deposits. Gold and platinum, though relatively rare, commonly occur in the uncombined or native form; but this, of course, does not mean that their extraction is easy.

Many sources have a certain percentage of the metal below which extraction is unprofitable, but the percentage is not always the deciding factor. For instance, magnesium is more easily obtained from the sea, where it is present at 0.13 percent, than from olivine, where it is present at 30 percent. In addition to factors already mentioned in Section 13.16, the line between unprofitability and profitability depends on the available technological processes for extraction and the availability of other sources of the metal, or a substitute metal.

Because of their electropositive nature, most metals occur in their ores in the positive oxidation state. The extraction or winning of the metal from these ores is a process of reduction. The chemical process, in essence, merely involves the addition of electrons to the metal; in practice, however, this is not always a simple or cheap procedure on a massive scale. In a commercial operation, the ore must be excavated from the earth, concentrated by removal of unwanted material, reduced to the metal, and the metal further refined or treated. We shall briefly consider some of the common reduction processes for selected metals and their common ores.

Metals release energy when they are oxidized. Reduction of the metal to the zero oxidation state therefore requires the input of energy, which is generally higher for the more reactive metals such as Ca, Mg, Be, Li, and Al. The noble metals (Au, Pt, etc.) are the most easily reduced. Most common ores are composed of metallic oxides or sulfides, but they are sometimes converted to chlorides or fluorides before reduction. The electronegativity difference between the metal and the anion (O^{2-}, S^{2-}, Cl^-, or F^-) in these compounds is a measure of the ionic character of the bond in the crystal, the heat of formation of the substance, and the resultant difficulty of reduction of the combined metal. Thus, metals with higher electronegativities, such as the noble metals, which tend to form covalent compounds, are most easily reduced. Because of the ease of extraction of the noble metals from their compounds, they were among the first to be discovered and used, even though their abundance is small.

Among the less active metals whose principal ores are sulfides, some can be obtained directly by heating the ore in a stream of oxygen. This process is called *roasting*, or more properly, *roast-reduction*. An example is

$$HgS + O_2 \longrightarrow Hg + SO_2$$
cinnabar

Copper is also produced by the reaction of the sulfide, Cu_2S, with oxygen, but preliminary ore concentration and purification steps are necessary to remove iron(II) sulfide impurities. After this is done, oxygen is blown through molten copper(I) sulfide:

$$Cu_2S + O_2 \longrightarrow 2Cu + SO_2$$

In the case of certain sulfide ores, a two-step process is required: conversion of the ore to the oxide by roasting and reduction of the oxide with carbon:

$$2ZnS + 3O_2 \longrightarrow 2ZnO + 2SO_2(g)$$
$$ZnO + C \longrightarrow Zn(g) + CO(g)$$

At the temperatures employed, zinc is liberated as the vapor and is condensed. Lead may be produced from galena, PbS, by similar reactions.

Among the metals that occur naturally as oxides, iron is the most important. The ore, whether Fe_2O_3 or Fe_3O_4, is progressively reduced to FeO and then to Fe by carbon monoxide in a blast furnace:

$$FeO + CO \longrightarrow Fe + CO_2$$

The major part of the carbon monoxide reducing agent is formed in the reaction

$$2C + O_2 \longrightarrow 2CO$$

Some of the FeO is also reduced directly by the coke from which the carbon monoxide is produced:

$$FeO + C \longrightarrow Fe + CO$$

These reactions concern only the basic reduction process for the production of the crude or pig iron. A variety of other reactions and processes are employed to remove silicon, sulfur, phosphorus, and other impurities and to regulate the amount of carbon in the iron. Steel is iron containing a small controlled amount of alloying agents, chiefly carbon, that confer important mechanical properties on the steel.

Carbon reduction of oxides of some of the more reactive (electronegativity 1.5–1.7) metals such as titanium and uranium is not feasible for several reasons. One of these is the tendency for such metals to combine with carbon to form carbides such as TiC and UC$_2$. Other quite reactive metals have such high melting points that the carbon reduction method is not feasible. Among these metals are tungsten (W, m.p. 3410°C) and molybdenum (Mo, m.p. 2610°C). Titanium and uranium are commonly produced by converting their oxides to halide salts and reducing these with an even more reactive metal:

$$TiO_2 + 2Cl_2 + 2C \xrightarrow{800°C} TiCl_4 + 2CO$$
$$TiCl_4 + 2Mg \longrightarrow Ti + 2MgCl_2$$
$$UF_4 + 2Ca \longrightarrow U + 2CaF_2$$

The higher-melting tungsten and molybdenum are produced by the reduction of their oxides with hydrogen gas:

$$WO_3 + 3H_2 \longrightarrow W + 3H_2O$$
$$MoO_3 + 3H_2 \longrightarrow Mo + 3H_2O$$

The ores of the highly reactive and low-melting alkali and alkaline earth metals, and aluminum, cannot be reduced by carbon, hydrogen, or most of the other chemical reducing agents. Such metals are commonly produced by electrolytic reduction of a molten halide salt. Electrolytic reduction is also an important method of refining the less active metals (those that will not displace hydrogen from water). Because of the utility of this method, we discuss it in some detail.

13.20 ELECTROLYSIS

In Section 10.20, we saw how electric energy could be produced by a spontaneous oxidation–reduction reaction proceeding in an electrochemical cell. Chemical energy was converted directly into electric energy, represented by the potential difference between the electrodes of the cell. The reverse of this—the use of electric energy to bring about a nonspontaneous chemical reaction—is called *electrolysis*.

We also saw, in Section 9.4, that a potential difference applied to two electrodes dipping into an electrolyte resulted in an electric current being carried by the ions of the electrolyte. Electrons are the charge carriers in the wires between the electrodes in the external circuit. The circuit will not be complete and current will not flow unless there is some way to remove electrons from the *cathode* (the electrode connected to the negative pole of the battery or power source), and some way to give up electrons to the *anode* of the electrolysis cell. The removal of electrons is accomplished by a reduction half-reaction at the cathode of the electrolysis cell. At the anode, electrons are put back into the external circuit by an oxidation half-reaction. Thus, electrons flow into the electrolyte at the cathode and flow out at the anode.

Consider the electrolysis of molten sodium chloride. Figure 13.27 shows an electrolytic cell containing molten NaCl. Two inert electrodes dip into the melt and are connected to one another through a battery or other direct current power supply. Electrons pass through the external circuit from anode

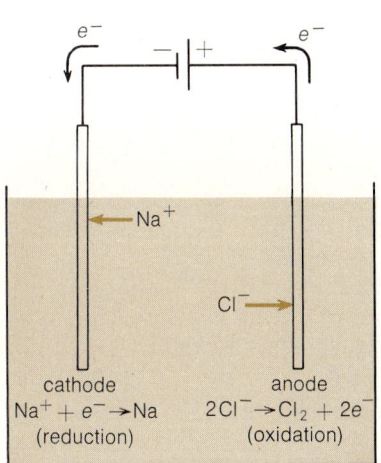

Figure 13.27. Electrolysis of molten NaCl.

to cathode. As with electrochemical cells, reduction occurs at the cathode and oxidation at the anode.

Within the molten NaCl, positive sodium cations move to the cathode, where they pick up electrons and are reduced:

$$Na^+ + e^- \longrightarrow Na \qquad \text{(reduction at the cathode)}$$

Negative chloride anions move to the anode, where they are oxidized to chlorine gas:

$$2Cl^- \longrightarrow Cl_2 + 2e^- \qquad \text{(oxidation at the anode)}$$

The electrons given up in the oxidation are taken up by the anode. The complete cell reaction is the sum of the half-reactions occurring at each electrode after equalization of electron gain and loss·

$$2Na^+ + 2Cl^- \xrightarrow{\text{electrolysis}} 2Na + Cl_2$$

Sodium is produced commercially by the electrolysis of molten sodium chloride. Electrolysis of molten chlorides is the best method for the commercial production of other active metals such as lithium, calcium, strontium, barium, and magnesium. Magnesium and chlorine are obtained from sea water by the following reaction sequence:

$$CaCO_3 \text{ (from oyster shells)} \longrightarrow CaO + CO_2$$
$$CaO + H_2O \longrightarrow Ca^{2+} + 2OH^-$$
$$Mg^{2+} \text{ (in sea water)} + 2OH^- \longrightarrow Mg(OH)_2(s)$$
$$Mg(OH)_2 + H^+ + Cl^- \longrightarrow Mg^{2+} + 2Cl^- + 2H_2O$$
$$Mg^{2+} + 2Cl^- \xrightarrow{\text{electrolysis}} Mg + Cl_2$$
$$\text{(molten)}$$

Aluminum is produced by electrolysis of purified alumina, Al_2O_3, obtained from bauxite. In the Hall–Héroult process, the Al_2O_3 (m.p. 2045°C) is dissolved in molten cryolite, Na_3AlF_6, and electrolyzed at a temperature of only 1000°C.

Electrolysis of Aqueous Solutions

The electrolysis of molten salts is particularly simple because there is usually only one chemical species to be reduced and only one to be oxidized. The electrolysis of aqueous solutions presents complications because there may be several species capable of being oxidized or reduced. Electrolysis of aqueous solutions is little used for the extraction of metals from their ores, but it is a common method for the refining of impure metals.

A solution of sodium chloride illustrates some of the features of the electrolysis of aqueous solutions. The electrodes are inert material, such as platinum, and do not enter into any electrode reactions. In an NaCl solution, the ionic and molecular species present are Na^+, Cl^-, H^+, OH^-, and H_2O. Of these, Cl^-, OH^-, and H_2O might undergo oxidation at the anode in the half-reactions

$$2Cl^- \longrightarrow Cl_2 + 2e^-$$
$$4OH^- \longrightarrow O_2 + 2H_2O + 4e^-$$
$$2H_2O \longrightarrow O_2 + 4H^+ + 4e^-$$

At the cathode the following reductions could conceivably take place:

$$Na^+ + e^- \longrightarrow Na$$
$$2H_2O + 2e^- \longrightarrow H_2 + 2OH^-$$
$$2H^+ + 2e^- \longrightarrow H_2$$

The particular reactions that may occur at either electrode depend on a number of interrelated factors, among which are the standard reduction potentials of the species present, the concentrations of the species, and the relative rates of the reactions at the electrode surfaces. From our previous considerations of oxidation and reduction, we would expect the strongest reducing agent present to be preferentially oxidized at the anode. This species can be readily determined from the table of redox potentials and is the substance having the more positive oxidation potential (or the more negative reduction potential). Actually, this expectation is not always fulfilled, because it does not take all factors into account. In the case of the electrolysis of a concentrated NaCl solution, experiment shows that the chloride ion is oxidized at the anode even though the oxidation of H_2O to O_2 has a more positive (standard) oxidation potential; that is, $E = -1.36$ volts for $2Cl^- \longrightarrow Cl_2$ and $E = -1.23$ volts for $H_2O \longrightarrow O_2$. Because we have been interested in oxidations, we have referred to the potential for the oxidation half-reaction rather than to reduction potentials and have used the fact that reactions, whether oxidation or reduction, with higher positive potentials are more likely to occur. The use of standard potentials is, of course, a gross approximation in these examples.

Similarly, we would expect the substance preferentially reduced at the cathode to be the strongest oxidizing agent present. Its reduction half-reaction has the most positive reduction potential. The standard reduction potential for sodium ion is the highly negative -2.71 volts, while the reduction potentials for water and hydrogen ions in neutral solutions are both only -0.41 volt. The higher concentration of water molecules makes the reduction of water faster at the cathode than the reduction of H^+. The two electrode reactions for concentrated NaCl solution are then

$$2Cl^- \longrightarrow Cl_2 + 2e^- \qquad \text{(anode)}$$
$$2H_2O + 2e^- \longrightarrow H_2 + 2OH^- \qquad \text{(cathode)}$$

The net ionic reaction is the sum of these half-reactions:

$$2Cl^- + 2H_2O \longrightarrow H_2 + Cl_2 + 2OH^-$$

For every two Cl^- discharged, two OH^- are produced to maintain electrical neutrality of the solution. As electrolysis proceeds, NaCl is replaced by NaOH.

In a dilute NaCl solution, or in one having a less easily oxidized anion than Cl^- (such as $SO_4{}^{2-}$ or F^-), the anode reaction would be the oxidation of water to oxygen. The two electrode reactions would be

$$2H_2O \longrightarrow O_2 + 4H^+ + 4e^- \qquad \text{(anode)}$$
$$\underline{4H_2O + 4e^- \longrightarrow 2H_2 + 4OH^- \qquad \text{(cathode)}}$$
$$6H_2O \longrightarrow O_2 + 2H_2 + 4H_2O$$

or

$$2H_2O \longrightarrow O_2 + 2H_2 \qquad \text{(cell reaction)}$$

The hydrogen and hydroxide ion products combine to form water. The ions of the salt are not involved in either electrode reaction and serve only as charge carriers between the electrodes.

Finally, if the solution is acidified with H_2SO_4 (not HCl), the hydrogen ion concentration becomes high enough that the reduction of H^+ will occur at the cathode:

$$2H_2O \longrightarrow O_2 + 4H^+ + 4e^- \quad \text{(anode)}$$
$$\underline{4H^+ + 4e^- \longrightarrow 2H_2} \quad \text{(cathode)}$$
$$2H_2O \longrightarrow O_2 + 2H_2 \quad \text{(cell reaction)}$$

EXAMPLE | A solution contains dissolved $ZnSO_4$ and $CuSO_4$ with about equal concentrations of each cation. If the solution is electrolyzed with inert electrodes, which metal will be preferentially reduced at the cathode?

SOLUTION | The standard reduction potentials for Zn and Cu are

$$Zn^{2+} + 2e^- \longrightarrow Zn \quad E^0 = -0.76 \text{ V}$$
$$Cu^{2+} + 2e^- \longrightarrow Cu \quad E^0 = +0.34 \text{ V}$$

Copper(II) ion is the more easily reduced cation and is the strongest oxidizing agent present. Copper metal is plated out on the cathode before zinc. This type of process is the basis for the electroseparation of these metals in solution.

Electrolysis of solutions with an electrode that enters into one of the electrode reactions is the basis for the electrorefining of metals. Copper is commonly refined electrolytically. A bar of impure copper obtained from ore reduction is made the anode, and a sheet of pure copper is used as the cathode. A copper(II) sulfate solution is the electrolyte. Upon electrolysis the electrode reactions are

$$Cu \text{ (impure)} \longrightarrow Cu^{2+} + 2e^- \quad \text{(anode)}$$
$$Cu^{2+} + 2e^- \longrightarrow Cu \text{ (pure)} \quad \text{(cathode)}$$

Copper electrolytically dissolves at the anode and is deposited on the pure cathode. Now, assume that the impurities in the copper anode are metals such as iron, which is more active than copper, and silver or gold which are less active. The standard reduction potentials of the metals involved are

$$Cu^{2+} + 2e^- \longrightarrow Cu \quad E^0 = 0.34 \text{ V}$$
$$Ag^+ + e^- \longrightarrow Ag \quad E^0 = 0.80 \text{ V}$$
$$Fe^{2+} + 2e^- \longrightarrow Fe \quad E^0 = -0.44 \text{ V}$$

By proper adjustment of the applied voltage, only copper and the more active iron will be oxidized and go into solution at the impure anode. As the anode "dissolves," unoxidized silver will drop to the bottom of the cell, where it may be recovered as a valuable by-product. At the cathode, only the copper(II) ion will be reduced to the metal, while the more difficultly reduceable iron(II) ion will remain in solution. The net result is the dissolution of impure copper at the anode and the deposition of pure copper on the cathode.

13.21 FARADAY'S LAWS OF ELECTROLYSIS

The amount of electricity passing through an electrolytic cell and the amount of material liberated at the electrodes are related by Faraday's laws of electrolysis.

1. The mass of an element or other substance deposited, dissolved, or liberated at an electrode is proportional to the quantity of electricity passed through the cell.

2. The mass of an element or other substance deposited, dissolved, or liberated by a given quantity of electricity is proportional to the chemical equivalent weight of the substance.

For an element, the equivalent weight as used in Faraday's laws is the atomic weight divided by the valence or change in oxidation state of the element in the cell reaction. That is, the (gram-) equivalent weight of any element is the weight liberated or dissolved when Avogadro's number of electrons are transferred in its cell reaction.

The quantity of electricity (or charge) is given in units of coulombs. The number of coulombs passed per second is the current I, in amperes. The charge Q, in coulombs, passing through the cell in t sec is the product of the current in amperes and the time in seconds, $Q = It$. The quantity of charge associated with Avogadro's number N of electrons is the *faraday*. One faraday is therefore equal to $Ne = (6.0225 \times 10^{23}$ electrons$)$ $(1.6019 \times 10^{-19}$ coulomb/electron$)$ or 96,489 coulombs.

An amount of charge equal to 96,489 coulombs will result in the deposition or dissolution of one gram-equivalent weight of an element at the electrode of an electrolytic cell. Thus, 1 faraday will bring about the deposition of 107.87 g of silver, or 65.37/2 g of zinc, or 26.98/3 g of aluminum, or 1.008 g of hydrogen, or 8.00 g of oxygen from the appropriate cell containing solutions of these elements in their common oxidation states.

If Ne coulombs result in the deposition of 1 equivalent weight of an element, then $Q = It$ coulombs will, by proportion, result in the deposition of It/Ne equivalents. Since the number of equivalents is the mass of the element divided by its equivalent weight, the mass deposited is

$$m = (\text{equiv. wt.})\frac{It}{Ne} = (\text{equiv. wt.})\frac{It}{\mathfrak{F}}$$

where $\mathfrak{F} = 96,489$ coulombs $= 1$ faraday.

We can use Faraday's law to get an idea of how much energy is needed to produce aluminum by the Hall-Héroult process. The charge required to produce 1 lb (453.6 g) of aluminum is

$$Q = It = \frac{453.6 \text{ g}}{8.994 \text{ g/equiv. wt.}} \times 96,489 \text{ coulombs/equiv. wt}$$

(coulomb $=$ amp sec)

The energy required to transport this amount of charge against a potential difference of V volts is QV. The potential difference is the voltage applied to the electrolytic cell and is about 6.0 volts in the Hall–Héroult process. The energy needed to deposit 1 lb of aluminum is

$$QV = \frac{453.6 \text{ g}}{8.994 \text{ g/equiv. wt.}} \times (96,489 \text{ coulombs/equiv. wt.})(6.0 \text{ V})$$

energy $= QV = 2.91 \times 10^7$ joules or watt seconds (W sec)

$$= (2.91 \times 10^7 \text{ W sec}) \left(10^{-3}\frac{\text{kilowatt}}{\text{watt}}\right)\left(\frac{1}{3600 \text{ sec/hr}}\right)$$

$= 8.1$ kilowatt hours (kW hr) of electric energy per pound of aluminum

This figure is the *minimum* energy requirement. The actual energy consumption is more like 10–12 kW hr/lb. A typical Hall–Héroult cell produces about 600 lb of aluminum a day and uses roughly 6000 kW hr of energy per day. Some appreciation of the energy requirements for the production of aluminum may be gained by comparing the 6000 kW hr *daily* energy consumption of *one* cell with *annual* per capita electricity consumption of about 8000 kW hr in the United States in 1970. The high energy cost of aluminum, along with its nondegradability, make it a somewhat less than ideal container material from an ecological point of view.

As an example of the energy needs for metal reduction, aluminum is a bit extreme because of its low equivalent weight. The high energy requirement for aluminum is the reason that most aluminum plants are located in regions having plentiful supplies of low-cost hydroelectric power. Of all the electric power generating processes, the power provided by falling water is by far the cleanest in terms of air and water pollution, but power dam sites are limited and there is increasing pressure to curtail the construction of dams at available sites because of the ecological, social, and aesthetic consequences.

For any element, the quantity of energy needed and the cost of extraction increase as recourse is made to lower-grade or more intractable ores. The only factors that have reversed this general trend have been increased mechanization and automation and greater efficiency in the use of available energy resources. There is, however, a limit to all of these, and we can expect the cost of metal production to rise. It is inevitable that man will come to rely on ores of lower and lower grade as time goes on. Total consumption of metals will increase rapidly because of population growth and greater utilization of metals (and other resources) resulting from increasing technology and consumer economies. Per capita consumption of many metals will increase even as population grows, as long as the metals are available.

Man's ability to extract metals from the earth serves as one example of how mankind as a whole is becoming a mighty geological force and is transforming the biosphere into a new geological state called the noösphere, the biosphere dominated by the mind of man (see the references by Vernadsky at the end of the chapter). According to this view, man creates new biogeochemical processes that did not exist before. For instance, he changes the biogeochemical history of chemical elements by bringing into existence metals such as aluminum and calcium, which have never before been known in their native state. Of course, man's effects do not stop here; he also modifies other parts of the biosphere.

13.22 CONSERVATION OF RESOURCES

Throughout man's history, the most available and highly concentrated natural resources have always been depleted before alternative sources were tapped. The situation in a highly industrialized and technologically advanced nation like the United States illustrates this. High-grade ores of copper, iron, and other metals have been used up. This situation has been countered in two ways. First, lower-grade domestic ores are being processed with the help of technological know-how and increased energy input. Second, high-grade ores are imported into the United States. However, this cannot continue indefinitely, for as the exporting nations develop, they will undoubtedly want the ores for themselves; and a struggle for the earth's remaining high-grade ores may ensue. Whatever the outcome, huge demands will be placed on earth resources of all kinds, and we should proceed to make more efficient use of them.

IRON AND IRON-ALLOY METALS

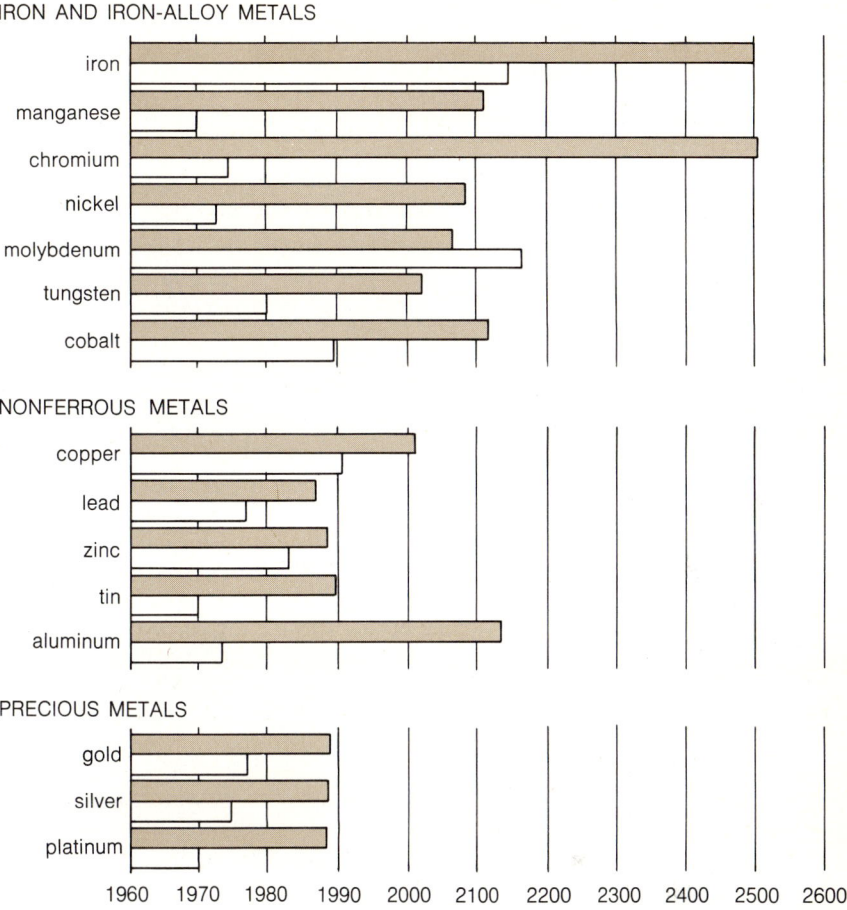

Estimated lifetimes of some metallic ore reserves are indicated in Figure 13.28. These estimates take into account expected population growth, rising per capita consumption, discovery of new ore deposits of lower but workable grade, and the creation of advanced extraction technology. Other sources may give different lifetime estimates because there are obviously many imponderables involved. For instance, the reserves of iron ore are considered by some to be enough to last several centuries because of the successful winning of iron from low-grade taconite-type ores.

Whatever the expected lifetime of ore reserves, it is finite, and the day will come when depletion of ores becomes serious. To meet that day, or to forestall it, several possibilities exist.

1. Alternative sources not considered in the estimated lifetimes. Ordinary rock contains almost all the elements needed in a technologically advanced society. In 100 tons of an average igneous rock such as granite, there are approximately 8 tons of aluminum, 5 tons of iron, 180 lb of manganese, 40 lb of nickel, 20 lb of copper, and 4 lb of lead. Ordinary clay could substitute for bauxite as a source of aluminum. Carbon, sulfur, and phosphorus could be extracted from other sedimentary rocks. All this would take enormous

quantities of energy; but, assuming an almost unlimited energy supply, the elements could be extracted from these materials.

The oceans will be mined much more extensively than at present. Currently, only sodium, chlorine, bromine, and magnesium are extracted in commercially significant quantities from the sea, as already noted. The sea is a potential source of many other elements, among them K, Ca, Sr, I, F, S, B, P, H, and O. Nodules having high percentages of MnO_2 cover large areas of the ocean floors, especially in the Pacific. If the technological and ecological problems involved in dredging these accretions from the sea can be solved, they could yield large quantities of manganese metal.

2. Use of alternative materials. Substitution of one metal for another is a common practice today. Substitution may be advantageous for reasons of supply, economy, or quality. Whatever the reasons, the consumption of one metal is lessened and the depletion of its reserves is postponed. Aluminum is now used widely as a replacement for the less available, more expensive, and heavier copper. As a coinage metal, silver is being replaced by copper and nickel. In some situations, metals are being replaced entirely by plastics.

3. Salvage and recycling. Salvage and reuse of metals and other resources has been widely practiced in the United States only during wartime, when the rate of mining and extraction of minerals could not keep up with consumption. In more relaxed times, it is the fashion to throw things away. There has been little incentive to reclaim metals, except in special circumstances, because of the expense of collection and processing. Technological innovations have so far made the extraction of metals from low-grade ores cheaper than reclamation of metals from discarded items. This will no doubt change as ore deposits dwindle.

The potentially reclaimable quantities of metal differ from metal to metal, but steel can serve as an example. Of the approximately 600 kg annual per capita steel production in the United States, only 410 kg reaches the consumer. The difference, 190 kg, is returned to the furnaces as recycled scrap from fabricating operations. Of the 410 kg, 80 kg remains in use for up to 25 yr average and is unavailable for salvage. This leaves about 330 kg potentially available for recycling, but only 40 percent of this is ever recovered and returned to the steel mill as scrap. The other 60 percent is lost, mostly buried in dumps or scattered about the landscape. In 1966, scrap iron and steel accounted for about 45 percent of total annual production, while corresponding figures for copper and zinc were 42 percent and 25 percent, respectively. Most recycled scrap is obtained from sources affording easy collection of relatively uncontaminated metal. Currently, most solid waste reclaimers pass up mixed sources such as municipal wastes. Even where collection and processing may be done with little effort, economic incentive may be absent. For instance, only 6 percent of the junked automobiles in the United States are processed for scrap, because it is cheaper to extract iron from low-grade ores.

Household and industrial solid waste collected by disposal agencies in the United States amounted to nearly 350 million tons in 1969. The cost of collection and disposal was about $4.5 billion. About 10 percent of the collected waste was incinerated, but most of the waste was disposed of in open land dumps. Disposal of this solid waste is one of our most pressing environmental problems. Aside from the high cost of collection and disposal, we are rapidly running out of disposal sites for this ever-increasing amount of rubbish and garbage. In any event, these dumps lead to a variety of problems such as air and water pollution, creation of ugly landscapes, and health hazards. Although proper burial of wastes in landfill operations often creates valuable land, it can also preempt areas, especially along shorelines, that might be better off if left for use by wildlife and for recreation. On the other hand, many wastes such as rubber, metals, and minerals of potential value to future generations might be profitably stored in specific disposal sites, which could become a source to be mined later. Schemes such as the compacting of wastes into blocks to be sunk in the ocean are only temporary solutions and, in any case, have unknown ecological consequences.

If processed instead of being discarded, the 1969 household refuse collected in the United States could yield 10 million tons of iron and steel, 1 million tons of nonferrous metals, and 15 million tons of glass and other recoverable nonmetallic materials. For these reasons, such rubbish has been called "urban ore." In addition, the garbage, plastics, and other organic constituents can be converted into fuel gas and oil. For instance, it has been shown that 1 ton of old rubber tires can be converted into 140 gal of oil and 1500 ft^3 of high-grade gas. Similar conversion processes are available for garbage and animal wastes. Such treatment of municipal refuse is presently too costly for private industry, but cities could make an acceptable profit from the sale of recovered products. This would solve many of the solid waste disposal problems and would help conserve natural resources.

13.23 SUMMARY

At some point in its early history, the earth was molten, at least in its interior. Within the molten material, a primary differentiation or fractionation occurred that distributed the elements between the iron and silicate phases—and thus between the core and the mantle of the earth—on the basis of their oxidizing properties relative to iron. A further or secondary differentiation of the element also took place among the various minerals constituting the mantle and crust, but this distribution depended on the crystal structure of the minerals.

Solids are composed of units (atoms, ions, molecules) held in relatively fixed positions. In crystalline solids, the fixed units form an ordered three-dimensional arrangement or crystal structure. The smallest, most symmetrical, portion of the crystal structure that can be used to build up the crystal is known as the unit cell. X-Ray diffraction is the major tool for the determination of the internal structure of crystals.

Bonding between the structural units in solids may be ionic, covalent, van der Waals, or metallic; and each type of bonding confers different properties on the solid. Metallic bonding, for instance, is an extreme case of electron delocalization resulting in the characteristic electrical conductivity of metals. Ionic crystals can often be regarded as close packed arrays of the larger anions, with smaller cations fitting into appropriate interstices in the anion lattice. Occupancy of the cation sites in an ionic lattice is determined largely by the cation-to-anion radius ratio.

Silicate minerals make up most of the rocks in the earth's mantle and crust. The silicates may be classified structurally on the basis of the manner of linkage of SiO_4^{4-} tetrahedral anion units. The chemical composition of the silicates, even of a given crystal structure, shows wide variation because of the ability of different cations to occupy a variety of holes in the oxide ion lattice and to substitute for one another. Substitution or exchange of cations in the same crystal structure leads to isomorphism. The ability of one cation to substitute for another in a crystal lattice is controlled mainly by similarity in cation size.

The composition of a mineral is fixed when that mineral freezes out of the cooling magma. The composition is determined, however, by the nature of the components of the magma, by their concentration in the magma, and by the conditions of cooling. If the components of a melt are immiscible in the solid state, they exhibit eutectic behavior, and the solid phases freezing out of the melt will be limited to the pure components or to eutectic mixtures of fairly definite compositions. If the components are miscible in the solid phase, solid solutions result. The composition of the solid solution freezing out depends on the original composition of the melt under equilibrium conditions, and it may have a whole range of compositions under non-equilibrium cooling conditions. Most silicate minerals are solid solutions of varying composition.

During the crystallization of a complex magma, minerals crystallize in the order of their thermal lattice stability. Other elements may be retained in the lattices of the minerals of the major elements if their ions can fit into the mineral lattices. Such a process leads to the secondary differentiation of the elements among early-formed and late-formed silicate minerals.

The course of magmatic crystallization leads to some ore deposits, and processes of weathering and sedimentation lead to others. Ore deposits are examples of depleting nonrenewable natural resources. Other natural resources may be classified as renewable and as reusable but nonrenewable resources.

The extraction of most metals requires the reduction of the metal. One metallurgical reduction process used for the more active metals is electrolysis of fused salts. Electrolysis of aqueous solutions is widely used for metal plating and refining. Faraday's laws of electrolysis permit the calculation of the amount of chemical reaction produced by a given amount of electricity in electrolysis.

All of our depleting resources have finite lifetimes. The lifetimes can be extended by conservation measures such as the practice of salvage and re-cycling.

Questions and Problems

1. From the ionic radii given in Table 13.4,
   a. What general statement can you make about the relative sizes of the cations of a multivalent element? What is the reason for this?
   b. What statement can be made concerning the ionic radii of cations having the same electronic configuration (i.e., Na^+, Mg^{2+}, ... , S^{6+})? Explain.
   c. Describe and explain the trends in ionic radii for cations in the same periodic group. Do the same for anions.

2. In addition to strictly metallic bonding, transition metals such as chromium and iron are considered to have some covalent bonding arising from the overlap of atomic d orbitals. How might the presence of such localized bonding affect mechanical properties such as hardness and brittleness as well as the melting points of such metals, in comparison to the properties of nontransition metallic solids?

3. Show that the limiting ionic radius ratio r^+/r^- for a coordination number of 4 is 0.414 (sin $45° = 0.707$).

4. Zinc and magnesium ions have about the same radii, 0.74 and 0.66 Å, respectively. Zinc exhibits coordination numbers of 4 and 6, but Mg^{2+} exists only with CN = 6. Account for the difference.

5. In the fcc unit cell, there are four octahedral holes and eight tetrahedral holes for each four closely packed ions (usually anions) in the unit cell. Dis-

cuss the probable disposition of ions (coordination number, which ions are in how many of what kind of holes) in the unit cells of

a. RbI
b. CsF
c. $SrCl_2$
d. ZnS

6. a. Construct a temperature–composition phase diagram for the A–B solid–liquid system, given the following information:

1. Pure A freezes at 600°C.
2. Pure B melts at 800°C.
3. The lowest freezing point of any mixture of A and B occurs at 400°C at a composition of 40 mol percent A.

Label all areas with the phases present.

b. What is the composition of the liquid phase when the overall composition of the A–B mixture at 500°C is 20 mol percent A?

c. Sketch cooling curves for liquid A–B mixtures having the following compositions:

1. 100 mol percent A
2. 40 mol percent A
3. 80 mol percent A

Indicate what happens at certain temperatures on these curves.

7. Two phase diagrams are given below, one for a eutectic mixture and one for a solid solution.

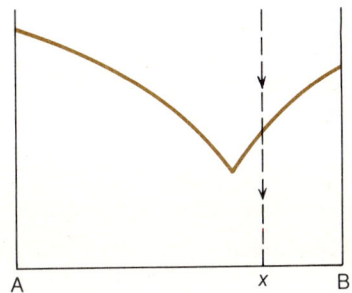

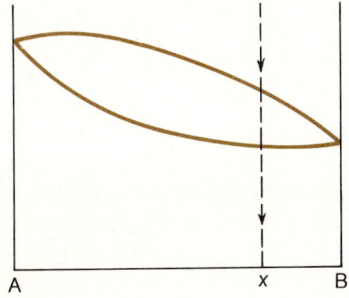

a. Explain why the solidification of a liquid with composition x in each case will give solids of at most two definite compositions from eutectic behavior but may give a wide range of solid compositions if the mixture forms solid solutions (assume nonequilibrium removal of solids).

b. Explain why a change in the composition of the melt due to removal of solids from the melt has little or no effect on the eutectic solid compositions but does affect the composition of the solid solutions.

8. The electric power consumption by air conditioners on very hot days has contributed to power failures in recent years. A possible way to avoid this power drain during periods of peak power demand in the daytime is to have an air conditioning system that freezes a substance at night when power demand is low. The next day the frozen substance would be used for cooling, with the only power consumption being that needed to circulate the air. Thus, the system would act as a refrigerator at night and store "coolness" for use the next day. A substance being considered for use in the low-temperature reservoir is a eutectic mixture of sodium sulfate, sodium chloride, and ammonium chloride in water which freezes at 13°C (55°F). What properties of a eutectic mixture would make it especially suitable for such a purpose?

9. Lunar soil contains ilmenite, a mineral with the composition $FeO \cdot TiO_2$. It has been suggested that moon explorers might obtain water and oxygen from lunar soil by reducing it with hydrogen (brought from the earth) according to the reactions, beginning with soil heated to 1300°C in a solar furnace,

$$FeTiO_3 + H_2 \longrightarrow H_2O + Fe + TiO_2$$

$$2H_2O \xrightarrow{\text{electrolysis}} 2H_2 + O_2$$

How much water and oxygen could be obtained from 50 kg of soil if the soil is taken to be 5 percent ilmenite and only the FeO is considered to react?

10. Discuss reasons for the lack of isomorphous substitution of

a. Cu^+ for Na^+
b. Cd^{2+} for Na^+
c. Li^+ for Na^+
d. Li^+ for Mg^{2+}

11. A coulometer is a device for the accurate measurement of the quantity of electricity. In a copper

coulometer, an acidified solution of cupric sulfate is electrolyzed. If a constant current flowing for a period of 1 hr deposits 0.1000 g of copper on the cathode, what was the value of the current?

12. Describe the probable electrode reactions in the electrolysis of the following solutions. Inert platinum electrodes are used unless otherwise stated.
   a. $1M$ $CuSO_4$
   b. $1M$ $AgNO_3$ with silver electrodes
   c. $0.1M$ $CdBr_2$
   d. $1M$ HCl
   e. $1M$ KOH
   f. $0.1M$ NaF

13. The total cell reaction for an *electrochemical* (not electrolysis) cell is

$$Zn + 2Ag^+ \longrightarrow Zn^{2+} + 2Ag$$

What weight of the zinc electrode is used up when the cell delivers 0.500 amp for a duration of 4 hr? What weight of silver is deposited? What will happen to the voltage of this cell as it discharges? How could the cell be recharged to its original voltage?

14. What is the minimum applied voltage necessary to begin plating out the metal in
   a. $0.01M$ $AgNO_3$
   b. $1M$ $AgNO_3$
   c. $1M$ $CuSO_4$
   Assume that oxygen at 1 atm (instead of 0.2 atm) is evolved from a platinum anode, all solutions are neutral, and the cathode is made of the same metal as the solution cation.

15. Define the terms "anode" and "cathode" as applied to electrochemical and electrolytic cells. Do the same for "plus" and "minus" terminals in each kind of cell.

Answers

9. 420 g of H_2O, 370 g of O_2
11. 1.2 amp
13. 2.44 g of Zn, 8.05 g of Ag

14. a. 0.13 volt
    b. 0.016 volt
    c. 0.48 volt

References

GEOCHEMISTRY (COMPOSITION OF THE EARTH, SILICATES, GEOCHEMICAL DIFFERENTIATION)

Ahrens, Louis H., "Distribution of the Elements in Our Planet," McGraw-Hill, New York, 1967.

Ernst, W. G., "Earth Materials," Prentice-Hall, Englewood Cliffs, N. J., 1969.

Krauskopf, Konrad B., "Introduction to Geochemistry," McGraw-Hill, New York, 1967, Chapters 5, 20, 21.

Mason, Brian, "Principles of Geochemistry, 3rd ed., Wiley, New York, 1966. Chapters 2, 3, 4, 11.

THE SOLID STATE (CRYSTAL STRUCTURE AND BONDING)

Bragg, Sir Lawrence, "X-Ray Crystallography," *Scientific American*, **219,** 195–205 (September 1967).

Holden, Alan, "The Nature of Solids," Columbia University Press, New York, 1965.

Mott, Sir Nevill, "The Solid State," *Scientific American*, **217,** 80–89 (September 1967).

Smith, Cyril Stanley, "Materials," *Scientific American*, **217,** 69–79 (September 1967).

LUNAR GEOCHEMISTRY

Brown, G. M., "Geochemistry of the Moon," *Endeavour*, XXX, 147–151 (September 1971).

Husain, Liaquat, Schaeffer, Oliver A., and Sutter, John F., "Age of a Lunar Anorthosite," *Science*, **175,** 428–430 (1972).

Mason, Brian, "The Lunar Rocks," *Scientific American*, **225,** 49–58 (October 1971).

Wood, John A., "The Lunar Soil," *Scientific American*, **223,** 14–23 (August 1970).

MINERAL RESOURCES

Brown, Harrison, "Human Materials Production As a Process in the Biosphere," *Scientific American*, **223,** 194–208 (September 1970).

Brown, Harrison, Bonner, James, and Weir, John, "The Next Hundred Years," Viking, New York, 1963.
   Some sections are out of date, perhaps only temporarily, but the chapters on mineral resources are not.

Flawn, Peter T., "Environmental Geology," Harper & Row, New York, 1970, Chapters 4, 5, 6.
  Considers the conservation and management of earth resources, land, air, and water as well as minerals.
National Academy of Sciences, "Resources and Man," Freeman, San Francisco, 1969, Chapters 2, 6.
Skinner, Brian J., "Earth Resources," Prentice-Hall, Englewood Cliffs, N. J., 1969.

MISCELLANEOUS

Vernadsky, W. I., "Problems of Biogeochemistry, II," *Transactions of the Connecticut Academy of Arts and Sciences*, **35,** 483–517 (1944).
  A discussion of the differences between living matter and inert natural bodies in the biosphere.
Vernadsky, W. I., "The Biosphere and the Noösphere," *American Scientist*, **33,** 1–12 (1945).
  A condensation of the previous reference.

Soil

The fields of Nature long prepared and fallow,
  the silent, cyclic chemistry,
The slow and steady ages plodding, the unoccupied
  surface ripening, the rich ores forming beneath;
Walt Whitman—*Song of the Redwood Tree*

Soil has been defined as that thin layer of the earth's crust in which biological activity (i.e., life) takes place. Man is most directly concerned with the life represented by rooted plants. These plants capture the sun's energy in photosynthesis and use the energy to manufacture food from simple inorganic substances such as water, carbon dioxide, and minerals. Plants are the producers in the biosphere, and all animals, including man, are absolutely dependent on them for food in one way or another.

Plants growing in the soil occupy the first trophic (food) level in a food chain. The next level is occupied by primary consumers, which eat the plants. They, in turn, are eaten by a variety of carnivores. The final or top trophic level in this simplified food chain is sometimes considered to be occupied by decomposer organisms. These microorganisms break down and transform the remains and wastes of the organisms on lower levels to simpler substances that can be utilized by the producers. Alternatively, the decomposers can be thought of as a separate chain linking the producers and the consumers, as indicated in Figure 14.1.

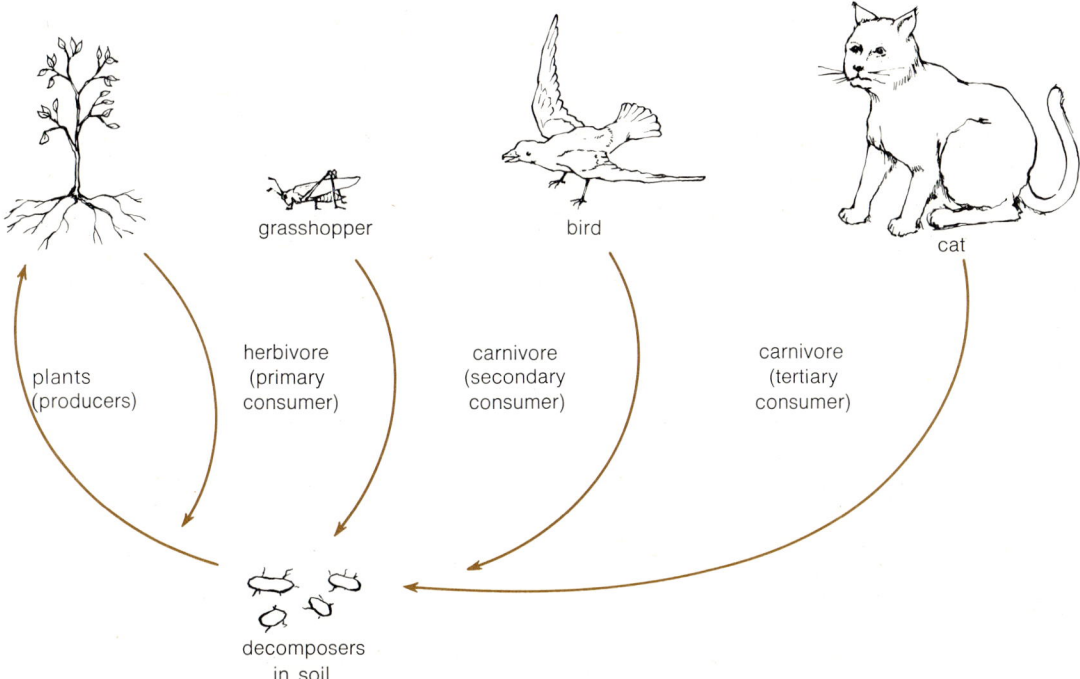

grasshopper   bird   cat

plants
(producers)

herbivore
(primary
consumer)

carnivore
(secondary
consumer)

carnivore
(tertiary
consumer)

decomposers
in soil

Figure 14.1. A simple food chain, indicating the role of decomposer organisms in the soil in the recycling of nutrients.

Plants transform simple inorganic substances into complex organic compounds. The decomposer organisms reverse this process and power a nutrient recycling operation by returning to the soil in readily assimilated forms the nutrients removed by the plants. Plants obtain all their essential nutrients, except for CO_2, from the soil through their roots. Decomposers also perform their tasks in the soil. The soil is therefore an important reaction site for two of the crucial pathways in the biogeochemical cycles of carbon, nitrogen, sulfur, phosphorus, and other elements. If conditions in the soil do not permit the functioning of plant roots or of decomposers, these cycles will break down.

Because the soil is so intimately related to our most important renewable resource, food, the ecological significance of soil and soil chemistry is apparent. Indeed, the soil is considered to be such an important part of the biosphere that it is sometimes called the *pedosphere*, that portion of the solid earth or lithosphere in which life exists. By whatever name it is called, the soil is a dynamic ecosystem, sometimes thought of as a living organism with a function, rather than as an inert layer.

There are at least two ways to approach the study of soils. Soil can be regarded as a separate physical part of our natural environment and the science of *pedology* is the study of its physical, chemical, and biological properties. Another approach to the study of soils places more emphasis on the importance of soil in the growth of plants and the production of food. It is difficult and, in our ecological context, unwise to separate the two viewpoints. We shall begin with the pedological approach, but our discussion of soil structure and fertility will bring in the plant–soil relationship. Although few details of plant physiology are included here, some commentary will be given in Chapter 18 on the green plant.

In this chapter, some aspects of soil chemistry involved in the formation, fertility, and physical structure of soils will be examined. As with geo-chemistry, the chemistry of soil science is a large subject so, of necessity, only selected topics can be included here. Most of the basic chemistry needed in soil science has been developed in preceding chapters, and the present chapter will serve as an illustration of how certain chemical principles can be used to explain some important features of soils.

Soil Formation

Soil is formed from surface rocks by a long and slow geological process. The process is cyclic, because the rocks are changed to unconsolidated particles which may be eventually cemented together by other chemical and physical mechanisms to yield new rocks. The first half of this cycle—the change from consolidated rock to unconsolidated particles—is the initial step in soil forma-tion. As with most natural cycles, physical, chemical, and biological forces all work together, and we shall examine each in turn. The pace of the geo-logical soil-producing process is far too slow to be of any use to man, and this means that soil is a depleting natural resource when subjected to erosion or loss of fertility.

14.1 PHYSICAL WEATHERING

Soil formation begins with the *weathering* of surface rocks. In its most general sense, weathering refers to the changes in consolidation and composition of rocks by exposure to conditions at the earth's surface. Physical weathering first disintegrates the massive rocks into smaller fragments by mechanical stresses. Some of the more important processes are described here.

1. When rocks originally formed under high pressure within the earth or by sedimentation on ocean bottoms are brought to the surface by uplift or exposed by erosion, the decrease in pressure may allow expansion and the formation of cracks.

2. The growth of foreign crystals within cracks in the rock may exert very high pressures and lead to more fragmentation. The expansion of water on freezing is the most common mechanism here. An important agent in dry climates is salt crystal growth resulting from evaporation of saline water in rocks. The formation of calcium sulfate crystals in cracks of calcium carbonate-containing rocks exposed to moist air contaminated with SO_2 (Section 7.7) might be regarded as physical weathering which has contributed to the deterioration of monuments in industrial areas.

3. Uneven thermal expansion of the various minerals in the rock as the surface temperature changes may set up stresses leading to cracks. This process is thought to be significant on the moon, where the temperature variations are quite large, but on earth, temperature changes probably only affect rocks weakened by other forces.

4. Pressure caused by plant roots is also a factor in physical weathering.

14.2 CHEMICAL WEATHERING   After physical weathering has fragmented the rock and increased the surface area, chemical processes at the surface become important. Chemical weathering brings about a change in the composition of the rocks and their constituent minerals, principally through the reaction of the rock with water and dissolved substances. Chemical weathering causes decomposition of some of the minerals of the parent rock, with the release of various products. Some of these products will become reactants for the synthesis of new minerals, while others may be removed in solution by percolating water (leaching). Leached materials may be deposited in deeper layers of the soil or in the underlying bedrock, or they may be completely removed to distant locations. In any case, a residue will be left behind by the leaching action. Chemical weathering is therefore a modification of the parent rock involving both the decomposition of the original minerals and synthesis of new mineral compounds.

The major chemical reactions of rock weathering depend in large part on the physical and chemical nature of the parent rock as the reactant material. Climate, topography, presence of vegetation, and the time available for reaction are other factors contributing in turn to the type, direction, and speed of reaction in weathering. These factors influence the amount of water available for reaction with the rock, the rate of water supply by precipitation and drainage, the temperature of reaction, and the kinds and amounts of reactive solutes carried by the water. The interplay of all of these factors determines the kinds of soils that will eventually be formed from a parent rock material. Even where the parent rocks are the same, different soils may be formed.

The chemical processes most frequently involved in rock weathering are hydrolysis and oxidation, but there are others as well. Most weathering processes involve a combination of two or more reactions. The presence of large amounts of water is important for both hydrolysis and oxidation. Water is the carrier of reactant species to the rock surface and also helps to drive these reversible reactions to completion by transporting products away from the reaction sites. The examples given will concentrate on the weathering of igneous silicate rocks rather than on sedimentary rocks (those formed from the cementation of grains of former rocks or the precipitation from solution of minerals) or metamorphic rocks (those formed by the change of previously existing rocks of any origin under extreme pressure and temperature).

Hydrolysis   Hydrolysis, the reaction of rock minerals with water, or more accurately, with the ions derived from water, is the most important of all chemical weathering reactions. The reaction of most interest is that of the abundant aluminosilicate minerals (feldspars, micas) with hydrogen ions. The resulting decomposition reaction gives products from which the important clay minerals are synthesized. Clays are one of the chief components of soils. The overall reaction of potassium-feldspar, $KAlSi_3O_8$ (orthoclase), with water can be written as

$$2KAlSi_3O_8 + 11H_2O \longrightarrow Al_2Si_2O_5(OH)_4 + 4H_2SiO_4(aq) + 2K^+ + 2OH^-$$

kaolinite—a   silicic acid
clay mineral

[In reactions we shall write the formula of kaolinite as $Al_2Si_2O_5(OH)_4$ instead of as $Al_4Si_4O_{10}(OH)_8$.]

This reaction shows the increased alkalinity resulting from the hydrolysis of silicates in general. Alternatively, to show that the hydrogen ion is a significant reactant, the overall reaction can be written as

$$2KAlSi_3O_8 + 2H^+ + 9H_2O \longrightarrow Al_2Si_2O_5(OH)_4 + 4H_4SiO_4(aq) + 2K^+$$

Other feldspar minerals, as well as micas, undergo similar reactions to yield kaolinite or other clay minerals as well as soluble silicic acid and cations (K^+, Na^+, or Ca^{2+}).

The actual formation of clays by hydrolysis is more complex than is indicated by these reactions. The aluminosilicates are believed to decompose with the release of metal cations and the formation of soluble hydrous oxides of Si and Al from the SiO_4^{4-} and AlO_4^{5-} tetrahedra. The silicon and aluminum oxides then combine to give a clay by a synthesis reaction whose mechanism is obscure. As we shall see shortly, the aluminum atom in clays exists in octahedral coordination with O^{2-} and OH^-. The change of aluminum from tetrahedral coordination in the aluminosilicate parent mineral to octahedral coordination in clay is the distinguishing feature of this weathering process.

The rate of weathering by hydrolysis is increased in an acid environment and at higher temperature. Additional hydrogen ions may be supplied by organic acids formed from decomposition of vegetation or by carbon dioxide dissolved in the water. In the latter case, the hydrolysis of K-feldspar would be described by the overall reaction

$$2KAlSi_3O_8 + 2H_2CO_3 + 9H_2O \longrightarrow$$
$$Al_2Si_2O_5(OH)_4 + 4H_4SiO_4 + 2K^+ + 2HCO_3^-$$

The clays formed in all such reactions are quite stable and are often considered to be the end products of weathering, but under conditions of high temperature and heavy rainfall such as are found in the tropics, clays weather further with the leaching of the silica component, leaving behind hydrated aluminum oxides in bauxite rocks.

The hydrolysis of olivine illustrates the weathering of a simpler silicate mineral:

$$Mg_2SiO_4 + 4H^+ \longrightarrow 2Mg^{2+} + H_4SiO_4(aq)$$
$$\text{silicic acid}$$

Silicic acid is such a weak acid that it is often written as $Si(OH)_4$ or as soluble, hydrated silica, $SiO_2 \cdot 2H_2O$. In the above equation, silicic acid might be considered simply as SiO_2 dissolved in water.

The ions released in these weathering reactions may be absorbed by plants (K^+, Ca^{2+}), incorporated into other minerals, or washed into the sea (Na^+, Ca^{2+}, HCO_3^-).

Hydrolysis is often preceded by a hydration reaction at the surface of the rock. In hydration, complete water molecules are adsorbed by a mineral in the rock. This may lead to expansion or swelling of the surface layers, thus facilitating the entry of hydrogen ions. Swelling and flaking due to hydration is often considered to be a process in physical weathering. Weathering by hydrolysis can also be viewed as a cation exchange reaction, with H^+ exchanging for Ca^{2+} or Na^+ (Sections 14.9 and 14.12).

Oxidation Oxidation is secondary in importance to hydrolysis, but is distinctive for rock minerals containing iron(II). If Fe(II) is oxidized to Fe(III) while it is still part of the crystal lattice of the mineral, some other cation must leave the lattice to maintain electrical neutrality. This leaves empty cation positions in the lattice, which weaken the structure and make it more subject to disintegration and decomposition. The reaction of Fe(II) in the lattice and the departure of cations from the lattice is not easily accomplished. It is more likely that Fe(II) is dissolved out of the mineral by water and then oxidized by dissolved oxygen:

$$2Fe^{2+} + \tfrac{1}{2}O_2 + 2H^+ \longrightarrow 2Fe^{3+} + H_2O$$

This is followed by hydrolysis of the Fe(III):

$$Fe^{3+} + 3H_2O \longrightarrow Fe(OH)_3 + 3H^+$$

The $Fe(OH)_3$ may also be considered as a hydrated iron(III) oxide, $Fe_2O_3 \cdot nH_2O$. The overall oxidation of iron(II) to iron(III) oxide may be written

$$2Fe^{2+} + \tfrac{1}{2}O_2 + 2H_2O \longrightarrow Fe_2O_3 + 4H^+$$

This reaction is aided by the hydrolysis of the Fe^{3+} and by the low solubility of the Fe_2O_3 [or $Fe(OH)_3$]. Although high acid concentration should favor the initial Fe^{2+}–Fe^{3+} oxidation, the overall reaction is shown to be favored by low acid concentration.

Oxidation reactions such as these are responsible for the common red colors of iron-containing rocks. They are also responsible for the troublesome occurrence of acid mine drainage, which pollutes water and soils. When the common iron(II) sulfide minerals pyrite and marcasite (polymorphous forms of FeS_2), are exposed during mining operations, mostly for coal, both the iron and sulfur components of the minerals are oxidized in a series of reactions. Both minerals consist of Fe^{2+} and $(-S{=}S-)^{2-}$ units. The sulfur is in the -1 oxidation state. Initial oxidation of the S_2^{2-} to sulfate releases sulfuric acid:

$$FeS_2 + \tfrac{7}{2}O_2 + H_2O \longrightarrow Fe^{2+} + 2SO_4^{2-} + 2H^+ \tag{1}$$

Then Fe^{2+} is oxidized to Fe^{3+},

$$2Fe^{2+} + \tfrac{1}{2}O_2 + 2H^{2+} \longrightarrow 2Fe^{3+} + H_2O \tag{2}$$

and the Fe^{3+} hydrolyzes to form more acid,

$$Fe^{3+} + 3H_2O \rightleftharpoons Fe(OH)_3(s) + 3H^+ \tag{3}$$

The low solubility of $Fe(OH)_3$ is responsible for the high production of acid by the oxidative weathering of pyrites.

A further reaction of FeS_2 with Fe^{3+} exists:

$$FeS_2 + 14Fe^{3+} + 8H_2O \longrightarrow 15Fe^{2+} + 2SO_4^{2-} + 16H^+ \tag{4}$$

This produces more acid and additional Fe^{2+}, which can be recycled through the reaction sequence.

Sulfuric acid may react further with the more reactive marcasite to form H_2S and sulfur:

$$FeS_2 + 2H^+ + SO_4^{2-} \longrightarrow Fe^{2+} + H_2S + S + SO_4^{2-}$$

Finally, if the sulfuric acid encounters clays (section 14.8) that contain Al and perhaps K, then aluminum sulfate, $Al_2(SO_4)_3$, and potassium aluminum sulfate (alum), $KAl(SO_4)_2 \cdot 12H_2O$, can result. Aluminum ion may hydrolyze when the pH rises to form a gelatinous aluminum hydroxide that appears as a white precipitate in still waters or as a coating on stream banks.

The sulfuric acid affects the pH of natural streams, damaging aquatic life and rendering the water unsuitable for domestic and industrial use. The insoluble, gelatinous $Fe(OH)_3$ coats stream and lake beds, interfering with bottom-feeding aquatic organisms. The $Fe(OH)_3$ precipitate formed in reaction (3) serves as a reservoir of Fe(III) for both reactions (3) and (4). Thus, if the production of Fe(III) and of acid from FeS_2 through the sequence of reactions (1), (2), and (3) is eventually halted, acid can still be formed in reactions (3) and (4) with Fe(III) derived from the precipitate.

Acid mine drainage may continue for decades from active or abandoned mines. In the United States, it is estimated that coal mines annually contribute the equivalent of 8 million tons of sulfuric acid to streams. Of this, only half is neutralized by the natural alkalinity of the waters; the rest degrades thousands of miles of streams.

Several methods have been proposed to combat acid mine drainage. Because the oxidation reactions will not take place in the absence of oxygen, underground mines, both active and abandoned, might be filled with a non-oxidizing gas such as nitrogen or natural gas. Or acidic water leaving the mine might be treated with an ion exchange resin. The oxidation of Fe^{2+} to Fe^{3+} is catalyzed by iron-metabolizing bacteria, and ways of interfering with the action of these organisms have been proposed. At present, the most common method of treating mine drainage is neutralization with lime (CaO) or limestone ($CaCO_3$), but this produces a sludge of $CaSO_4$, which presents a disposal problem.

Carbonation Chemical weathering by reaction of minerals with carbonic acid is important in rock containing carbonates of calcium (limestone) or of calcium and magnesium (dolomite). The reaction

$$CaCO_3 + H_2CO_3(aq) \longrightarrow Ca^{2+} + 2HCO_3^-$$

produces the hardness-causing hydrogen carbonate ions.

Stability of Minerals Under Chemical Weathering The minerals in igneous rocks formed by solidification of molten magma have varying susceptibility to chemical weathering. In general, the minerals that were first to crystallize from the molten magma (Figure 13.26) react most readily with water and its solutes at the earth's surface. The least reactive igneous rocks are those composed of the late-crystallizing minerals. Thus, the calcium feldspars and the olivines weather rapidly, while minerals such as quartz and the potassium feldspars are most resistant to weathering. Broadly, we can consider those minerals formed under conditions furthest removed from prevailing conditions at the earth's surface to be the most unstable to both physical and chemical weathering processes. The minor elements incorporated in the crystals of the minerals will be released and appear in soils in proportion to the susceptibility of the mineral to weathering. Most of the trace elements needed by plants (Zn, Mo, Cu, Co, Mn) tend to concentrate in the more easily weathered minerals.

The factors that determine the presence of the minor elements as substituents in minerals in igneous rocks, namely ionic size, ionic charge, and bond type, were discussed in Section 13.15. More will be said shortly about the role of the minor elements in soils, but we can already begin to see how the purely chemical mechanisms involved in the formation and composition of igneous rocks can affect life.

14.3 SEDIMENTS AND THE FATE OF RELEASED IONS

The elements in the minerals of rocks decomposed by chemical weathering are released to the surroundings as ions, and they may or may not take part in other reactions to form sedimentary rocks. Thus, released aluminum and silicon cations form hydrous oxides or hydroxides which, in turn, synthesize clay minerals. Released iron eventually forms a hydrous oxide or hydroxide. Sodium ions, on the other hand, remain soluble as the hydrated cation and are washed into the sea. The general fate of the released ions can be predicted from their polarizing power or their electronegativities (Section 10.5). Elements with low electronegativities tend to remain in solution; elements with high electronegativities form soluble oxy anions; elements with intermediate electronegativities form insoluble hydroxides or related compounds (hydrolyzates). Thus, hydrous oxides of Al and Si react to form clays and Fe^{2+}, after its oxidation to Fe^{3+}, forms hydrated Fe_2O_3.

Subsequent to the formation of soluble ions, further reactions may occur to form insoluble compounds or otherwise immobilize the ions. Potassium ions are strongly adsorbed on clay particles and removed from solution. Calcium is incorporated into marine organisms and eventually appears as a carbonate in the form of the sedimentary rock, limestone. Other ions may leave solution as the solution evaporates, as in inland seas. Some ions will be incorporated into the crystal lattices of synthesized minerals such as clays. Many of these removal processes occur after the ions have been transported from the land to the sea and therefore determine the residence time of the ions in the sea.

Mechanical deposition or chemical precipitation of some compounds leads to formation of certain sedimentary rocks, some of which are listed in Table 14.1 along with their associated trace elements. Sedimentary rocks overlay about 80 percent of the igneous and metamorphic rocks on the continents. The most abundant sedimentary rocks are the shales, which comprise 80–85 percent of the continental sediments. Shales are composed of clay minerals. Thus, most soils will continue their development on shale rocks, and the most important mineral components of soils will be the clays.

The sedimentary deposits also provide important ores such as bauxite and iron oxides to supplement the hydrothermal ore deposits. Chemical weathering, then, in addition to being an essential initial step in the formation of agricultural soil, concentrates mineral resources for our use.

14.4 BIOLOGICAL ACTIVITY IN SOIL FORMATION

The unconsolidated rock fragments and minerals resulting from physical and chemical weathering are called the *regolith*. The regolith consists of varying proportions of slightly weathered fragments of original rock, residues of intense weathering (SiO_2), synthesized materials (clays), and other decomposition products (Fe_2O_3, etc.). The regolith is not soil. As the parent material of soil, it is only one of the four essential ingredients necessary for the development and continued existence of a true soil. The formation of a soil from the mineral regolith requires the presence of living organisms and organic compounds.

Table 14.1
Deposition of sedimentary rocks with their associated trace elements

process of sedimentation	major constituents	type of product	main rock types	associated trace constituents
	Si	→ resistates (SiO₂)	sandstones	Zr, Ti, Sn, rare earths, Th, Au, Pt
	Al Si K	→ hydrolyzates	shales (clays) and bituminous shales bauxites	V, U, As, Sb, Mo, Cu, Ni, Co, Cd, Ag, Au, Pt, B, Se Be, Ga, Nb, Ti
	Fe Mn	→ oxidates	iron ores manganese ores	V, P, As, Sb, Mo, Se Li, K, Ba, B, Ti, W, Co, Ni, Cu, Zn, Pb
	Ca Mg Fe	→ carbonates	limestones, dolomites	Ba, Sr, Pb, Mn
	K Na Ca Mg	→ evaporates	salt deposits	B, I
	Na Mg	→ sea		B, Al, I, Br, F, Rb, Li

SOURCE: Firman E. Bear, Ed., "Chemistry of the Soil," © 1964 by Litton Educational Publishing, Inc. Reprinted by permission of Van Nostrand. Reinhold Company.

Plants soon gain a foothold in the regolith and begin to manufacture carbon-containing organic compounds from the CO_2 in the air, the water percolating through the developing soil, and chemicals released by weathering. The first plants are usually simple forms, such as lichens, but they will be followed by larger plants. At death, the remains of the plants are fed upon, disintegrated, and decomposed by a myriad of living organisms in the young soil. These organisms may range from larger animals such as rodents through insects and earthworms to fungi and bacteria. Bacterial decomposition of the remains of plants and of other organisms releases acids such as H_2CO_3, HNO_3, H_2SO_4, and various organic acids to the soil. Bacteria aid in this decomposition and also act as catalysts for reactions such as the oxidation of iron and manganese. Other acidic substances are exuded from plant roots and other organisms. All these substances derived from biological activity contribute to the continued weathering of rocks and formation of soil.

14.5 CONSTRUCTIVE AND DESTRUCTIVE WEATHERING

In a first stage of soil formation, chemical weathering causes the decomposition of the original, or primary, rock minerals and the synthesis of new, or secondary, minerals. The main chemical agents are water and carbon dioxide, and the most important products are the clay minerals. Often included in this *constructive* stage of soil formation is the entry of organic matter into the soil as living organisms and its decomposition by bacteria found in every soil.

A second and *destructive* stage in the development of soils is often defined. This stage results from continued weathering over a long period of time or from increased weathering due to high rainfall or temperature. In the destructive phase, most of the primary minerals have disappeared and the weathering process now acts on the secondary or synthesized minerals, of which the clays are the most important. Eventually, the forces of weathering in the destructive stage of soil formation will dissolve or wash away the clays. Soil is therefore a stage in the weathering of rocks at the surface of the earth. Soil has been called rock on its way to the ocean.

The Soil Profile   As the agents of weathering do their work on the upper portion of a soil, a layered structure develops. The layers, or *horizons*, are characteristic of soils and, taken together, constitute the soil *profile*. Most soils contain three horizons labeled A, B, and C (Figure 14.2). The uppermost A horizon, or the topsoil, is characterized by maximum weathering and leaching of minerals. The A horizon also contains most of the living and nonliving organic matter of the soil. The B horizon, or the subsoil, is a zone of intermediate weathering. In the B horizon are deposited both soluble minerals and some small solid particles washed down from the A horizon. The A and B horizons are considered to be the important portions of a true soil and together average 3–4 ft in depth in a temperate climate. Beneath the B horizon is the C horizon, which is composed largely of fragments of unweathered parent rock. So far as the formation of clay minerals is concerned, the A horizon represents a soil undergoing destructive weathering. The B horizon, however, is in the constructive stage. Some soils have a comparatively thin uppermost O horizon composed solely of organic material representing the remains of plants and animals.

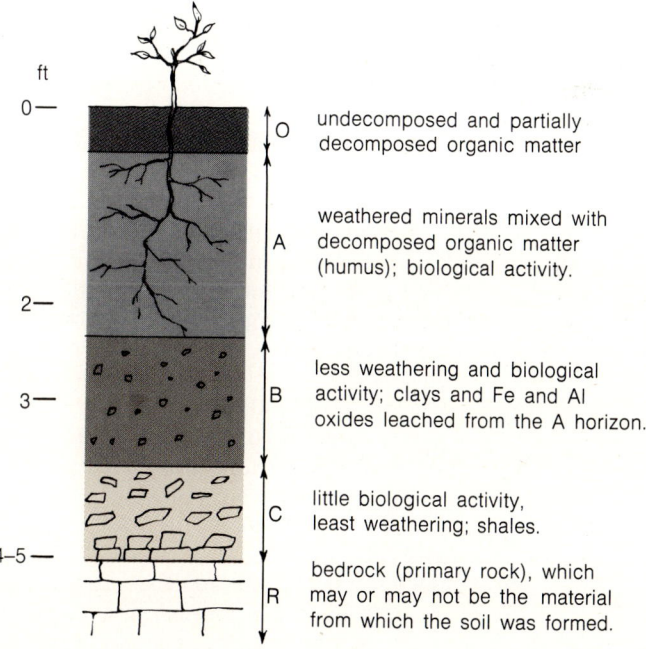

Figure 14.2. A soil profile. Depths are approximate and vary with the type of soil. The top layers are often called the surface (top) soil; the B layer is called subsoil.

ft

0 —

O   undecomposed and partially decomposed organic matter

A   weathered minerals mixed with decomposed organic matter (humus); biological activity.

2 —

3 —

B   less weathering and biological activity; clays and Fe and Al oxides leached from the A horizon.

C   little biological activity, least weathering; shales.

4–5 —

R   bedrock (primary rock), which may or may not be the material from which the soil was formed.

Soil Components

In addition to mineral matter, soil is composed of organic matter, water, and air. The proportions of the four major soil components vary from place to place and from time to time, but the following breakdown is representative.

	volume percent	
mineral matter	35–45%	} 50%
organic matter	15–5%	
water (soil solution)	15–35%	} 50%
air (O_2, N_2, CO_2)	35–15%	

14.6 ORGANIC MATTER (HUMUS)

Not all of the decomposition of organic matter in soils proceeds to the end products of carbon dioxide, water, and inorganic forms of nutrient elements such as NO_3^-, PO_4^{3-}, and SO_4^{2-}. There are a variety of intermediate decay products, among which are some that are quite resistant to further breakdown. These slowly decomposed substances, derived from the more resistant parts of the organisms in the soil, make up the organic mixture known as *humus*. Humus is of great importance in maintaining the physical structure of soils. Along with the clay minerals, it plays the major role in determining the chemical properties of soils.

14.7 SOIL SOLUTION AND SOIL AIR

The spaces between the mineral and organic soil particles are occupied by variable proportions of water and air. The water, together with its dissolved substances, is called the *soil solution*. The soil solution is the direct source of most of the nutrients required by plants. The nutrients are absorbed through the miles of roots that can be associated with a single plant. Water itself is taken in through tiny projections, the root hairs, on the roots. A mineral nutrient must assume a soluble form in order to be available to plants.

The oxygen of the soil air is necessary for the respiration of aerobic organisms, including plants, growing in the soil. Soil air may be several hundred times more concentrated in carbon dioxide than the atmosphere as a result of decomposition of organic matter and other metabolic processes of soil organisms. The carbonic acid produced contributes to rock weathering and to the release of additional nutrients from soil particles. Most of the carbon dioxide adsorbed by plants in photosynthesis enters the plant from the atmosphere, not from the soil solution.

Of the four major components of a soil, only the soil solution and air are essential to the growth of plants. This is indicated by the flourishing of aquatic plants and by the practice of hydroponics or solution culture, in which the roots of plants are immersed in an aerated solution of nutrients in the ionic forms directly assimilated by plant roots. This solution, plus the nitrogen, oxygen, and carbon dioxide of the air, and perhaps a little physical support, are all that a healthy plant requires.

Closely related to solution culture is sand culture. Here the roots are allowed to grow in an inert sand or gravel. The sand supplies no nutrients, but these are provided in an added nutrient solution. The sand provides mechanical support for the plant and aeration of the roots as solution is periodically added and allowed to drain off slowly.

Neither solution nor sand culture makes any use of solid mineral matter or of any kind of organic matter that is supposedly essential to a true soil.

What, then, is the purpose of these components? Aside from the mechanical support they give the plants, the solid mineral and organic matter in a soil determine the fertility and structure of a soil. Soil fertility refers to the ability of a soil to serve as a reservoir for the chemical nutrients required for plant growth and to its ability to make these nutrients easily available to plants in an assimilable form.

Soil structure refers to the way individual small soil particles are grouped or attached to each other to give larger aggregates. The nature of these aggregates determines, among other things, the water retention and aeration properties of a soil. Before taking up the subjects of soil fertility and structure in detail, we must examine the properties of one of the mineral components of soils, the clays, which play a dominant role in the physical and chemical foundations of fertility and structure.

14.8 THE CLAY MINERALS   Most of the important chemical and physical processes in soils occur at the surface of soil particles. Because of their large surface area per unit weight, the most important particles are those in the colloidal size range (10^{-6}–10^{-9} m). In a soil the clay minerals and humus exhibit the greatest surface activity in this regard.

Clay minerals are hydrated (or hydrous) aluminum silicates. As seen in Section 13.10, they are members of the layer or sheet-type silicates. The clays are made up of alternating layers of (1) SiO_4 tetrahedra linked by the sharing of three oxygens and (2) octahedral sheets of Al^{3+} in octahedral coordination with close packed O^{2-} and OH^- anions. The octahedra are linked together by the sharing of O^{2-} and OH^- between adjacent octahedra. The tetrahedral and octahedral layers are linked together by the sharing of some oxygens, as shown in the representations of the structure of three of the more important clay minerals in Figures 14.3–14.5.

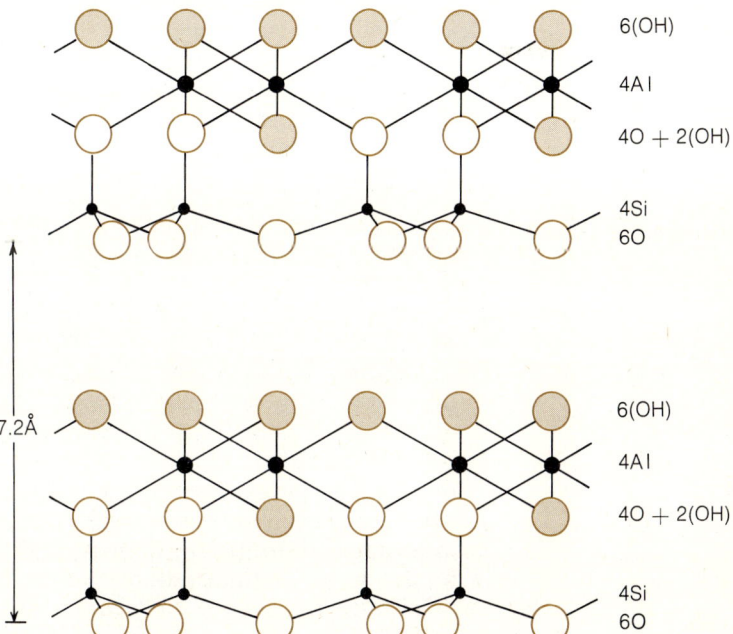

6(OH)

4Al

4O + 2(OH)

4Si
6O

7.2Å

6(OH)

4Al

4O + 2(OH)

4Si
6O

Figure 14.3. Kaolinite, $Al_4Si_4O_{10}(OH)_8$, a two-layer clay mineral.

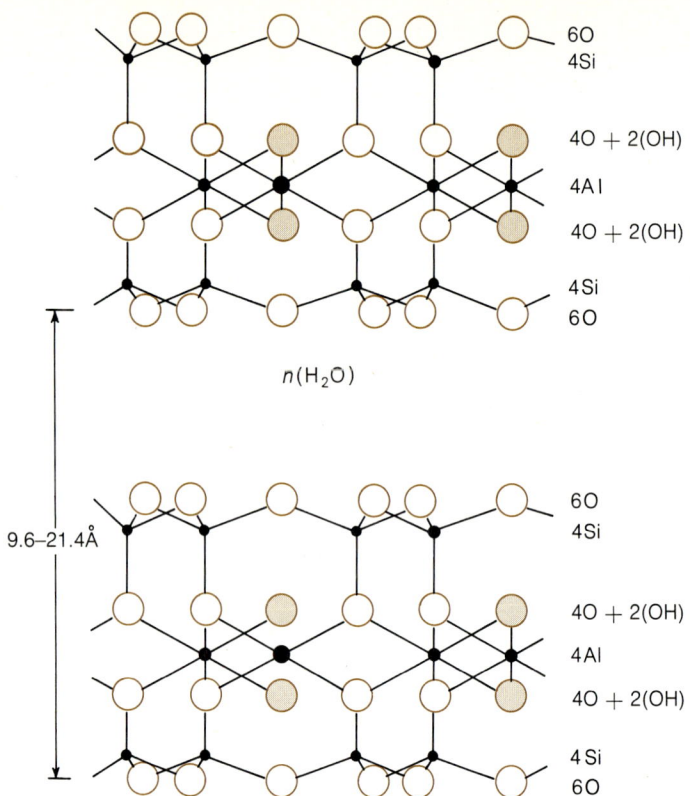

6O
4Si

4O + 2(OH)

4Al

4O + 2(OH)

4Si
6O

$n(H_2O)$

9.6–21.4Å

6O
4Si

4O + 2(OH)

4Al

4O + 2(OH)

4Si
6O

Figure 14.4. Montmorillonite, $Al_4(Si_4O_{10})_2(OH)_4 \cdot n(H_2O)$, a three-layer clay mineral.

Kaolinite, Figure 14.3, is a two-layer structure made up of one tetrahedral layer and one octahedral layer. The formula is $Al_4Si_4O_{10}(OH)_8$, and the figure indicates the placement of atoms (or ions). The two-layer units or double sheets are stacked on one another with the O^{2-} of one double sheet facing the OH^- of the other. The attraction between the O^{2-} and OH^- in adjacent double sheets is fairly strong as a result of hydrogen bonding, $O^{2-}\text{---}H\text{—}O^-$.
$$\delta+$$

In montmorillonite, $Al_4(Si_4O_{10})_2(OH)_4 \cdot nH_2O$, an octahedral aluminum layer is sandwiched between two tetrahedral silicon layers. Stacking of this three-layer structural unit, shown in Figure 14.4, presents the O^{2-} of one unit to the O^{2-} of an adjacent three-layer unit. The absence of hydrogen bonding between the oxygen anions in adjacent units means that the units can be easily separated. Replacement of silicon by aluminum in the tetrahedral layers of montmorillonite results in excess negative charge in these surface layers. Because of their size, potassium ions can fit into openings surrounded by oxygens on the surface of adjacent layers. The electrostatic attraction of K^+ and O^{2-} on either side holds the layers together. The result is a member of the *illite* group of clay minerals with varying amounts of potassium. The structure of the illite clays is represented in Figure 14.5 by muscovite, in which maximum substitution of Si by Al has resulted in maximum K content. The formation and sedimentation of the illite clays from kaolinite and montmorillonite washed into the ocean is believed to be the main mechanism for

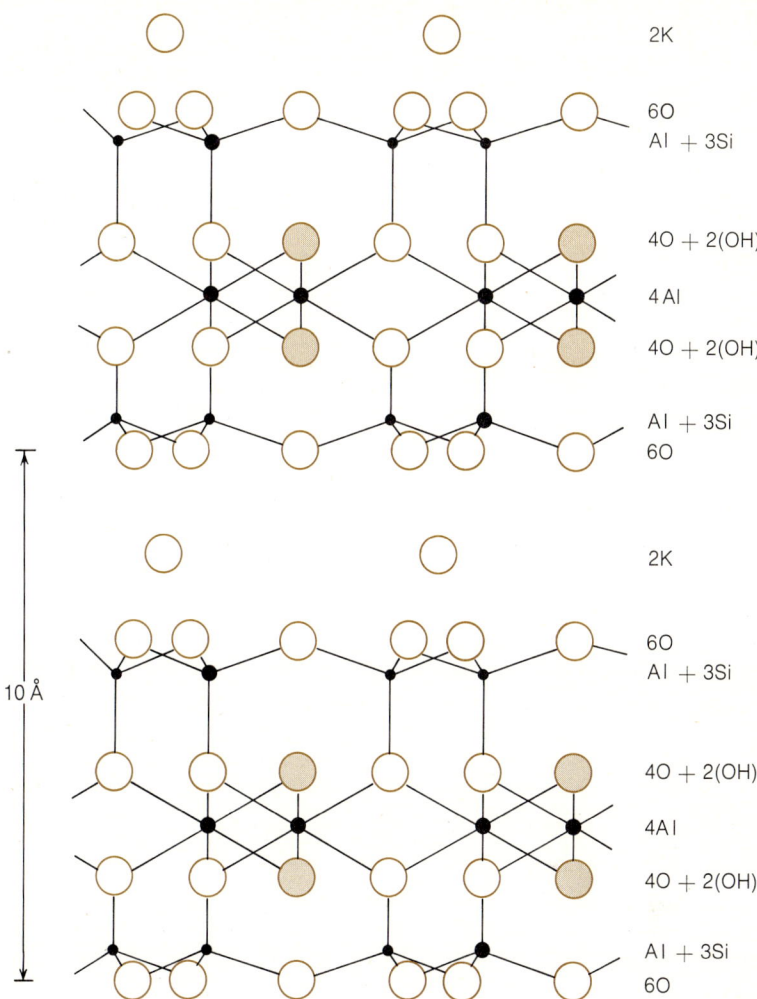

Figure 14.5. An illite clay. The composition illustrated corresponds to that of muscovite, $K_2Al_2(Si_6Al_2)O_{20}(OH)_4$. Muscovite, while not considered to be a clay, has the same structure as illite.

the removal of K^+ added to the seas by surface runoff. Magnesium ion concentration in the sea is also regulated by its incorporation into another clay mineral, chlorite.

14.9 CATION EXCHANGE PROPERTIES OF CLAYS

The clay minerals, along with humus, play a significant role in soil chemistry because of their ability to adsorb cations on their surfaces. These cations, many of them plant nutrients, are exchangeable with other cations in the soil solution. The resultant cation exchange equilibria are deeply involved in almost every phase of soil fertility.

Clays attract, adsorb, and hold cations because their surfaces exhibit an excess negative charge. The origin of this negative charge varies somewhat with the type of clay mineral. Kaolinite crystals assume a pH-dependent negative charge on their external surface because of ionization of hydrogen ions from the OH groups on one side of the two-layer units:

HO surface $\rightleftharpoons$ O$^-$ surface + H$^+$

The three-layer montmorillonite obtains most of its negative surface charge by complete or partial isomorphous substitution of Mg^{2+}, Fe^{3+}, or Zn^{2+} for Al^{3+} in the octahedral layers of the lattice. In addition to replacement of octahedral Al^{3+} by other ions, some of the tetrahedral Si^{4+} may be replaced by Al^{3+}. Except for the replacement of Al^{3+} by Fe^{3+}, these substitutions all result in a negative charge, located mostly in the center layer of the three-layer unit. This charge is pH-independent.

The net negative charge of the montmorillonite crystals causes cations to be adsorbed on the surface of the crystals. The available surface area for cation adsorption is particularly large because of the weak attraction between the three-layer units. The three-layer units in montmorillonite are easily forced apart by water molecules (the nH_2O in the formula), which are adsorbed on the surface by hydrogen bonds. The entry of water causes swelling and opens a pathway for cations to the negatively charged internal surfaces of the colloidal clay. Montmorillonite has a much greater negatively charged surface available for the adsorption of cations than kaolinite. The nonswelling kaolinite is pretty much limited to cation adsorption at the ionized —OH sites on its external surface. The comparison is shown in Figure 14.6. Other clays have swelling and negative surface charge characteristics intermediate to kaolinite and montmorillonite.

As already stated, the sites of negative charge on the surface of colloidal clay particles are the sites for adsorption of cations. In a soil, these cations come from the soil solution in equilibrium with the clay particles. In a natural soil, cations slowly enter the soil solution through weathering of the mineral solids of the soil. The adsorbed cations can be easily and rapidly exchanged for other cations in the soil solution. In effect, the adsorption of cations at the pH-dependent sites (—OH) on clays has already involved an exchange of H^+.

$$H^+|O^-\text{ surface} + M^+ \rightleftharpoons M^+|O^-\text{ surface} + H^+$$

We shall denote the surface of a clay (or humus) particle by a vertical line with the adsorbed cations on one side. Among the cations adsorbed on clays are Ca^{2+}, Al^{3+}, H^+, K^+, Na^+, Mg^{2+}, and NH_4^+. These cations can all enter into exchange equilibria with one another. As an example,

$$\left.\begin{array}{l}H^+\\H^+\end{array}\right| + Ca^{2+} \rightleftharpoons Ca^{2+}| + 2H^+$$

Exchange equilibria such as this determine important aspects of soil fertility and acidity. Cation exchange reactions are the most important of the chemical reactions in soils. As the cation concentration in the soil solution changes, the exchange equilibria adjust in accordance with Le Chatelier's principle. Thus, if calcium ions, important plant nutrients, are adsorbed on clay, a decrease in pH of the soil solution will bring about desorption of Ca^{2+} and its entry into the soil solution where it can be absorbed by plant roots. The colloidal clay particles, therefore, act as a reservoir of exchangeable nutrient cations that can be made available to the plants as needed, primarily through exchange for hydrogen ions in the soil solution. In a natural soil, exchangeable hydrogen ions are provided by water percolating through the soil and by acids formed by the microbial decomposition of organic matter. We also see that the acidity of a soil has a direct bearing on the availability of cations of nutrient elements to plants, that is, to the fertility of the soil.

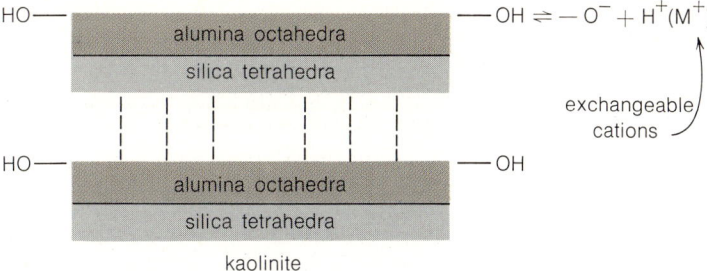

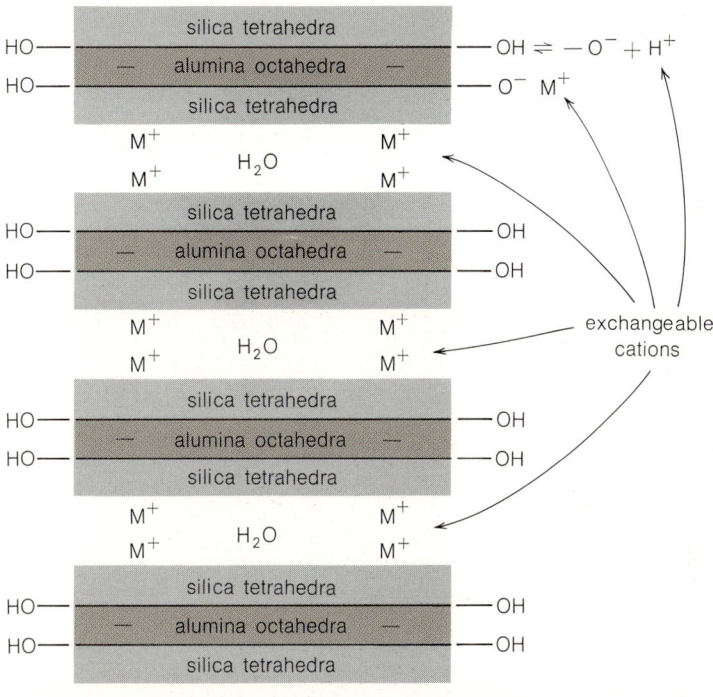

Figure 14.6. Surface charge and cation exchange sites on clays.

Soil Fertility

Soil fertility, the ability of the soil to supply plants with their essential nutrient elements, directly determines what crops can be grown on a soil and the nutritional value of these crops for man and animals. The ecology of plants and their food value in turn have an effect on the agricultural and industrial activity of a region, on the health of the population, and even on the prevailing political philosophy of an area (see the Kellogg reference at the end of the chapter).

We have seen that the fertility of a soil is determined by chemical weathering, the original rock materials, and the degree and intensity of weathering. The intensity of weathering depends heavily on climatic factors such as temperature and amount of rainfall. But climate has its major influence on plant growth in an indirect manner so that, contrary to popular opinion, it is not the only factor, and very often not even a direct factor, in the ecology of plants. In most cases, the fertility of a soil directly determines what plants will grow in it, because different types of plants vary considerably in the kinds and amounts of nutrients they require from the soil. For instance, alfalfa requires rather large amounts of calcium, while cotton does not.

The fertility of a soil is also affected by the kind of vegetation growing in the soil, by the amounts of organic matter in the soil, and by the physical structure of the soil. All of these factors will be mentioned again as we consider certain aspects of soil fertility and productivity. First let's see what essential plant nutrients the soil must supply.

14.10 PLANT NUTRIENTS

All plants require at least 17 nutrient elements for their growth. Nine of these elements, the macronutrients, are needed in relatively large quantities, while the rest, the micronutrients, are needed only in very small, or trace, amounts. These elements are listed in Table 14.2 together with the form in which they

Table 14.2
Plant nutrients[a]

element	assimilable form	source
	macronutrient elements	
carbon	CO_2, CO_3^{2-}, HCO_3^-	air, organic matter, mineral solids
hydrogen	H^+, OH^-	water–soil solution
oxygen	O_2	air
nitrogen	NH_4^+, NO_3^-, NO_2^-	organic matter
phosphorus	HPO_4^{2-}, $H_2PO_4^-$	organic matter, mineral solids
potassium	K^+	mineral solids
calcium	Ca^{2+}	mineral solids
magnesium	Mg^{2+}	mineral solids
sulfur	SO_3^{2-}, SO_4^{2-}	organic matter, mineral solids
	micronutrient (trace) elements	
iron	Fe^{2+}, Fe^{3+}	mineral solids
manganese	Mn^{2+}, Mn^{4+}	mineral solids
boron	BO_3^{3-}, $B_4O_7^{2-}$	mineral solids
molybdenum	MoO_4^{2-}	mineral solids
copper	Cu^+, Cu^{2+}	mineral solids
zinc	Zn^{2+}	mineral solids
chlorine	Cl^-	mineral solids
cobalt	Co^{2+}	mineral solids

[a] In addition to the listed elements, others may also be essential to some or all plants. Among these are vanadium (as VO_4^{3-}), iodine (I^-), and strontium (Sr^{2+}).

can be assimilated by plants. Also listed are the major sources of these elements in the soil. The form of an element in a source is not necessarily the form that is directly usable by a plant. For instance, phosphorus in organic combination or in highly insoluble metallic phosphates such as $Ca_3(PO_4)_2$ is not available to plants until converted to water-soluble HPO_4^{2-} or $H_2PO_4^-$. Similarly, calcium, magnesium, and potassium, although they may be present in the solid mineral form in large amounts, do not generally become available to plants except through the soil solution. Plants, however, cannot tolerate a soil solution with a high ionic concentration. Fortunately, exchangeable cations adsorbed and stored on clay and humus particles are made available to the soil solution as needed for absorption by the roots, so that the plants are not limited to the nutrients in true solution.

Clay and humus serve as intermediates between unweathered minerals and plants. Decomposition or dissolution of minerals releases cations that are adsorbed on the colloidal particles of clay and humus. These adsorbed nutrient cations are eventually exchanged for hydrogen ions provided by the plant roots. In some cases, the exchangeable cation may be transferred directly from the clay particle to the root and need not go into the soil solution.

Soil fertility therefore depends not on the total amount of a nutrient element present in a soil, but on its amount in available forms. Deficiencies in nutrient elements may be made up by adding available forms directly to the soil in fertilizers or by adjusting soil conditions so that elements present in unavailable forms are readily converted to available forms.

14.11 DEFICIENCY AND AVAILABILITY OF NUTRIENTS

A soil may be deficient in one or more essential elements for a number of reasons. An element may be absent or in short supply because it was not present in the original rocks, as is often the case with micronutrient elements. A nutrient element may be present in an unavailable form because it is locked in the lattice of an unweathered mineral or is in an otherwise insoluble form. The most important contributors to infertility of soils, however, are leaching and crop removal. In leaching, the removal of soluble nutrient elements by water percolating through the soil decreases the available adsorbed cation nutrient elements. The loss by leaching is particularly serious when the water is acidic, because the hydrogen ions readily replace the adsorbed cations on the clay and humus. This really amounts to accelerated chemical weathering and explains why extensively weathered soils tend to infertility because of the loss of major and trace elements. In a highly weathered soil, the nutrient reserves are low or nonexistent. As we shall see, the addition of fertilizers is a reversal of the weathering process. In general, a soil undergoing weathering gradually increases to a maximum in fertility and then declines as weathering progresses.

Continued growth and removal of crops from the land also results in soil deficiencies of important nutrient elements. Under undisturbed conditions, where vegetation is not removed, the remains of plants or the remains and wastes of animals that have fed on the plants return to the soil. Then the nutrients that were extracted from the soil by the plants are returned to the soil. When crops are harvested and removed from the land, this nutrient cycle is broken; and, depending on the mineral reserves and availability of nutrients, the productivity of the soil for that particular crop at least can decline rapidly. It is mostly for this reason that both mineral and organic fertilizers are added to crop lands. The continual removal of organic matter in the form of crops also causes a deterioration of the physical structure of the soil, which affects soil fertility. Modern soil management practices, therefore, are as concerned with the maintenance of soil structure by the return of part of the vegetation to the soil or by addition of manure as with the replenishment of plant nutrients with fertilizers. As we saw in Section 12.12, one of the results of our breaking the soil–plant nutrient cycle in this way is the eutrophication of our streams and lakes (see also Section 14.14).

14.12 SOIL ACIDITY

The acid–base characteristics of a soil are sometimes called the *soil reaction.* Soils can be acidic, neutral, or alkaline, but, since the majority of soils are acidic, we shall use the term "acidity" rather than the more general term "reaction."

Soils have a pH somewhere within the range of 4–8.5, but the majority of plants grow best in a soil with a pH between 6 and 7. The pH of a soil is usually measured by mixing the dry soil with water (frequently one part soil to one part water). After stirring and standing for a time, the pH of the mixture is measured with a pH meter.

Most of the acidity of a soil results from the presence of exchangeable hydrogen and aluminum ions adsorbed on colloidal clay and humus particles. In clays, exchangeable hydrogen ions may reside either in the —OH groups at the crystal edges or at interlayer positions where they may have been exchanged for previously adsorbed cations such as Ca^{2+}, Mg^{2+}, or K^+. Aluminum always exists in clays as a lattice constituent. In this form, however, aluminum is not easily exchangeable. Under acid conditions, some of the aluminum may dissolve from the lattice and be adsorbed in an exchangeable form that we shall consider to be a hydrated aluminum ion, $Al(H_2O)_6^{3+}$, or simply Al^{3+}. In the soil solution, the aluminum ions in equilibrium with the adsorbed aluminum ions can undergo hydrolysis and contribute to acidity:

$$Al(H_2O)_6^{3+} + H_2O \rightleftharpoons H_3O^+ + Al(H_2O)_5OH^{2+}$$

or

$$Al^{3+} + H_2O \rightleftharpoons H^+ + Al(OH)^{2+}$$

Hydrogen ions may also be derived from acidic groups (—COOH, —OH, etc.) of the organic humus particles in soils. In any case, the clay and humus particles are directly or indirectly proton donors to the soil solution. In the case of hydrogen ions adsorbed on clay particles, an equilibrium exists between adsorbed ions and those in the soil solution. The adsorbed hydrogen (and aluminum) ions constitute a potential or *reserve acidity* much like the undissociated hydrogens of a weak acid in a buffer solution. The ions in the soil solution comprise the so-called active acidity.

A number of sources may provide the exchangeable hydrogen ions adsorbed on clay particles. The slightly ionized water is, of course, one source. More important are the acids produced by the decomposition of organic matter by soil organisms and exuded from plant roots (Section 14.4). Less common sources of hydrogen ion in soils are mineral oxidation, microbial oxidation of fertilizers containing reduced forms of nitrogen or sulfur, and nitrogen and sulfur oxides in the air carried to the soil in rainfall.

Hydrogen and aluminum ions are held more tightly to the colloidal clay particles than other cations, but they are exchangeable for these other cations subject to cation exchange equilibrium considerations already discussed in Section 14.9. Thus, according to the equation

$$\left. \begin{matrix} H^+ \\ H^+ \end{matrix} \right| + Ca^{2+} \rightleftharpoons Ca^{2+} \Big| + 2H^+$$

if the concentration of calcium ions (or of Mg^{2+}, K^+, Na^+, or other exchangeable cations) in the soil solution is high, adsorbed hydrogen ions will be replaced by the other cations. Although the exchange process results in the

2 CaO → 2 Ca⁺ + O₂

entry of hydrogen ions into the soil solution, the overall result is a more alkaline soil if we assume that the liberated hydrogen ions are carried away by drainage or are neutralized by an anion. Thus, the acidity of a soil is determined mainly by the nature of the cations adsorbed on the clay (and humus) particles and accordingly is more a measure of the reserve acidity of the dry soil than a measure of the acidity of the soil solution.

Adsorbed metallic cations (and NH_4^+) contribute to the alkalinity of soils by means of hydrolysis. Again, using calcium as a representative cation,

$$Ca^{2+}| + 2H_2O \rightleftharpoons \left.\begin{matrix} H^+ \\ H^+ \end{matrix}\right| + Ca^{2+} + 2OH^-$$

In this reaction, the hydrogen ions are derived from water and, as before, removal of the soluble products by leaching will lead to the development of an acid soil. Where stronger acids than water are present in the percolating water, the exchange of metallic cations for hydrogen is accelerated. If carbonic acid from the decomposition of organic matter is present, the reaction becomes

$$Ca^{2+}| + 2H_2CO_3 \rightleftharpoons \left.\begin{matrix} H^+ \\ H^+ \end{matrix}\right| + Ca^{2+} + 2HCO_3^-$$

All of the above reactions can be represented by the general reaction

$$M^{m+}| + mH^+ \rightleftharpoons \left.\begin{matrix} H^+ \\ H^+ \end{matrix}\right| + M^{m+}$$

where M^{m+} is a representative cation. An exchange equilibrium of this sort represents the phenomenon of soil dynamics and illustrates:

1. How colloidal particles serve as reservoirs for nutrient cations which can be exchanged for H^+ and made available to plants.

2. The reversion of a soil to an acidic nature where there is sufficient water to remove exchanged cations and their salts from the soil. Only in regions of limited rainfall or of poor drainage do soils tend to be alkaline.

3. Why the addition of calcium and magnesium compounds to soils (liming) decreases soil acidity.

Effect of Soil Acidity on Plant Growth

The acidity of a soil determines to a large extent the availability of nutrient elements to plants, because the solubility of available forms of the elements varies with change in soil pH. The effect of pH will differ for various nutrient elements, depending on the form of the element most easily assimilated by plants, and with the presence of other cations. The availability of aluminum (as Al^{3+}) and iron (as Fe^{3+}) will be high at a pH of 4 because of the solubility of the oxides in an acidic environment. The dissolved concentration of these and other trace elements (except Mo and B) may become so high at low pH that they become toxic to plants.

The availability of phosphorus is affected by pH in two ways. The $H_2PO_4^-$ ion is believed to be the form of phosphorus most readily assimilated by plants. In acid soil solutions (pH $\simeq$ 4), $H_2PO_4^-$ would be expected to be present in high concentrations. Under the same acidic conditions, however, high concentrations of iron and aluminum can contribute to low phosphorus availability through precipitation of highly insoluble iron and aluminum phosphates. At high pH, phosphorus availability is reduced by formation of insoluble $Ca_3(PO_4)_2$ and other calcium phosphate compounds (apatites, Section 14.14). Maximum phosphorus availability occurs at soil pH's between 6 and 7. The relationship between soil pH and the availability of a number of plant nutrients in mineral soils is shown in Figure 14.7.

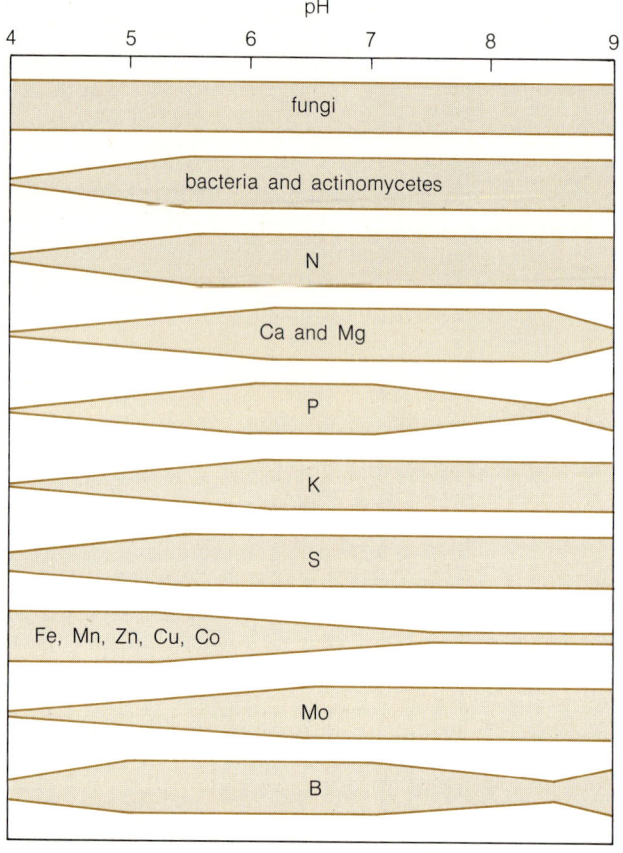

Figure 14.7. Nutrient availability and soil pH. The width of the bonds is proportional to the availability of the nutrient at a given pH. Low pH also decreases the activity of some soil microorganisms (top bands). (From Harry O. Buckman and Nyle C. Brady, "The Nature and Properties of Soils," 7th ed, Macmillan, New York, 1969, p. 396.) © Copyright, The Macmillan Company, 1969.

Even though the availability of an exchangeable cation such as calcium from a mineral such as limestone, $CaCO_3$, would be high in an acidic environment, a low soil pH is an indication of both extensive replacement of adsorbed cations by hydrogen and of the absence of reserves of exchangeable cations in soil minerals. With respect to exchangeable nutrient cations and also with respect to other nutrients whose minerals are soluble at low pH, a highly acid soil is a sign of an infertile soil. This is especially true in humid regions, where leaching will gradually remove one or more of the base-forming nutrient cations such as Ca^{2+} and Mg^{2+} from the soil. All processes leading to infertility are accompanied by the replacement of nutrient cations with hydrogen and the gradual acidification of the soil.

14.13 MICRONUTRIENTS AND SOIL FERTILITY

The concept of limiting factors introduced in Section 11.9 was originally developed to explain the nutritional needs of growing plants. Agricultural chemists at that time (1840–1860) were concerned mostly with the macronutrient elements used in large quantities by plants. Today we know that such macronutrients are not necessarily the limiting factors to plant growth. As methods of detection became more sensitive, the need for minute quantities of certain elements, the trace or micronutrient elements, in plant growth was established. Very often one of these elements is the one in shortest supply relative to the plant's need for it despite the very, very small amounts needed by the plant.

The micronutrient elements are used in plants as parts of enzymes that catalyze specific metabolic processes in growing plants. Each enzyme goes about its business in a remarkably efficient manner, so that only a few such molecules are required to carry the plant through that part of its life cycle regulated by any particular enzyme-catalyzed reaction. A single enzyme molecule often contains only one micronutrient element atom out of hundreds of other atoms. This accounts for the tiny amounts of these elements required for plant growth and for the absence of growth when the essential element is not available.

As with all nutrient elements, there are tolerance limits. If an essential element is missing in the soil or is present in an unavailable form, plant growth suffers even if all other nutrients are available in adequate amounts. If, on the other hand, too high a concentration becomes available, toxic effects on the plants, or on the animals eating the plants, may arise. Such effects are not limited to the essential nutrients. Plants, like animals, may concentrate nonessential elements. The livestock disease known as the "blind staggers" results from selenium accumulations in certain plants growing in selenium-rich soils.

Assuming that the essential micronutrients are present in the soil in some form, available or not, the problem is that of making the *optimum* amount available to a particular plant. Soil acidity, reaction with other substances in the soil, and oxidation state are among the factors regulating availability of micronutrients. We shall briefly examine a few micronutrients in terms of their availability in different soil environments.

Iron is a part of a plant enzyme that participates in the production of chlorophyll, the green pigment essential to photosynthetic activity. A representative humid-region soil may contain a total concentration of about 25,000 ppm (2.5 percent) Fe in its surface layers. Varying amounts of this total will be available to plants, depending on chemical conditions in the soil. If the iron is fixed in the soil in an insoluble or otherwise unavailable form, plants may suffer from an iron-deficiency disease called chlorosis, characterized by yellow foliage.

In addition to its presence in the unweathered, ferromagnesian silicate minerals, iron commonly occurs in a weathered soil as the oxide, hydrous oxide, or hydroxide. The iron may exist in either the reduced Fe(II) or oxidized Fe(III) state, depending on soil aeration and moisture conditions. Both forms are assimilable by plants, but the oxidized form is much less soluble than the reduced form at common soil pH. The oxides of each become more soluble at lower pH values. $Fe(OH)_3$ precipitates at pH's above 3, while $Fe(OH)_2$ precipitates at pH = 6 or higher. Therefore, under highly acid conditions, the available iron concentration may be so high as to be toxic. This will be especially true in poorly aerated, water logged, or compacted soils. On the other hand, alkaline soil conditions may result in iron deficiency. Iron deficiency is brought about by some commonly used agricultural practices. Addition of phosphate fertilizers, reduction of soil acidity by addition of calcium and magnesium compounds, and accumulation in the soil of excess Zn, Mn, or Cu contribute to iron deficiency. Thus, chlorosis may be common in citrus orchards because of high concentrations of soil copper built up over years of use of copper fertilizers and copper fungicides. The copper ions interfere with the absorption of iron by plants.

Iron can be maintained in an available form in the presence of precipitating agents such as phosphate or hydroxide ions by incorporating the iron into an organic chelate compound. The metal chelate is soluble, and the iron is readily available to the plant. At the same time, the chelated iron is protected from fixation by reaction with many other soil substances. Common synthetic chelating agents for Fe(III) are ethylenediaminetetraacetic acid (EDTA, Sequestrene, Versene) and diethylenetriaminepentaacetic acid (DTPA). The structure of EDTA and its iron complex is shown in Figure 14.8. Incidentally, nitrilotriacetic acid (NTA), mentioned in Section 12.12 as a substitute for phosphates in detergents, is used as a chelating agent in the

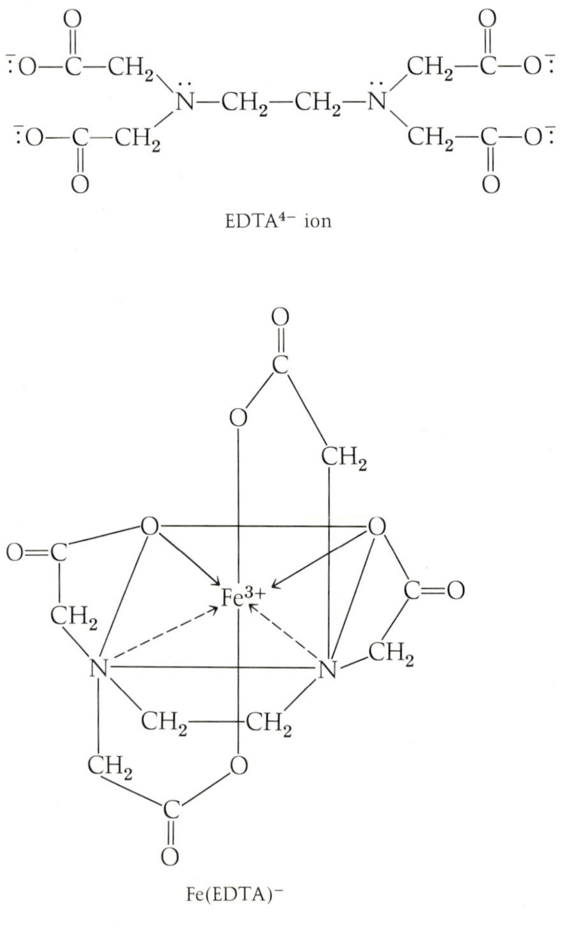

EDTA^{4-} ion

Figure 14.8. Structure of ethylenediaminetetraacetic acid (EDTA) ion and its Fe(III) chelate.

Fe(EDTA)$^-$

application of zinc to soils. Natural chelates are believed to be abundant in soils, the chelating agents being derived from the organic matter in the soil.

Boron is another plant micronutrient. It occurs in a representative humid region surface soil at about 50 ppm. The tolerance limits of most plants to available boron is very small; levels exceeding 1 ppm of available boron in the soil solution are toxic to most plants. Fortunately, most of the boron in soils is in the unavailable form of highly insoluble primary minerals. The available form is primarily the tetraborate $B_4O_7{}^{2-}$ anion in calcium, magnesium, and sodium borates. Boron deficiencies result from crop removal and from leaching of the soluble forms from acid soils. Deficiencies also occur at high soil pH values, because boron is fixed by clays under these conditions.

Borates, in the form of sodium tetraborate decahydrate (borax), $Na_2B_4O_7 \cdot 10H_2O$, are widely used as water softeners, alone or in detergent mixtures. The presence of borates in sewage effluents may be a critical factor in the use of effluents for irrigation because of the toxicity of boron to plants.

The oxidation state of an element also affects its availability, generally by way of solubility. The combined effect of oxidation state and pH on the availability of Mn(II) was mentioned in Section 10.22.

14.14 MACRONUTRIENTS AND SOIL FERTILITY

The three most common macronutrient elements applied to soils in commercial fertilizers are nitrogen, phosphorus, and potassium. They are called the fertilizer elements and are often applied together in commercial fertilizer mixtures. The composition of these mixtures as given by the fertilizer grade or guarantee is the percentage of *total* nitrogen (as N), *available* phosphorus (as P_2O_5) and *water-soluble* potassium (as the hypothetical compound K_2O) in the mixture. Thus a 4–8–8 fertilizer contains 4 percent total N, 8 percent available P_2O_5, and 8 percent soluble K_2O. Only a few of the more ecological aspects of commercial fertilizers furnishing these nutrients will be considered here.

Nitrogen Fertilizers

The total amount of nitrogen in a representative mineral soil is small (less than 0.5 percent). Most of the nitrogen is in the organic form, and it is converted slowly to the available forms by decay organisms and nitrifying bacteria. The principal available form of soil nitrogen is the nitrate ion, followed by the ammonium ion. At any given time, only a very small percentage of the total soil nitrogen content is available to plants in these forms.

Nitrogen losses from the soil result from crop removal, leaching of soluble forms, erosion, and bacterial denitrification of $NO_3{}^-$ and $NO_2{}^-$ to gaseous forms. In addition to these total losses, the positive ammonium ion may be rendered unavailable to plants by fixation on (or in) clay and organic matter particles.

The primary source of soil nitrogen is the atmosphere. Molecular nitrogen is converted to available forms by nitrogen-fixing soil bacteria and by electric and photochemical processes in the atmosphere. Where removal of nitrogen from soils exceeds the amount added by primary sources, the nitrogen content of the soil must be replenished by artificial fertilization. Most commercial nitrogen fertilizers are synthetic products produced from ammonia gas, which is made from atmospheric nitrogen by the Haber process. Man has, to a large extent, taken over the role of the nitrogen-fixing bacteria.

Some common inorganic nitrogen fertilizers are listed below.

anhydrous liquid ammonia	NH_3
ammonium nitrate	NH_4NO_3
urea	$H_2N-\underset{\underset{O}{\|\|}}{C}-NH_2$
nitrogen solutions	NH_4NO_3 or urea in dilute ammonium hydroxide
ammonium sulfate	$(NH_4)_2SO_4$
calcium nitrate	$Ca(NO_3)_2$

All inorganic nitrogen fertilizers are soluble, and the nitrogen is totally available to plants.

Ammonia and urea furnish ammonium ions by hydrolysis:

$$NH_3 + H_2O \rightleftharpoons NH_4^+ + OH^-$$
$$CO(NH_2)_2 + 2H_2O \rightleftharpoons 2NH_4^+ + CO_3^{2-}$$

The ammonium ions may be absorbed directly by plants, but in most cases they are converted to nitrates in a two-step process catalyzed by bacterial enzymes:

$$2NH_4^+ + 3O_2 \rightleftharpoons 2NO_2^- + 2H_2O + 4H^+$$
$$2NO_2^- + O_2 \rightleftharpoons 2NO_3^-$$

Specialized groups of soil bacteria operate in each of these steps, one group being able only to convert NH_4^+ to NO_2^-, while the other group is restricted to the second step. The overall reaction is

$$NH_4^+ + 2O_2 \rightleftharpoons 2H^+ + NO_3^- + H_2O$$

This reaction shows that nitrogen fertilizers yielding ammonium ions will lower soil pH values except in soils having high carbonate content. The increased acidity could increase the concentration of some plant nutrients, conceivably to toxic levels.

There are other ecological consequences of overfertilization by nitrogen fertilizers.

1. Overapplication of nitrogen fertilizers can result in high runoff of nitrates to natural waters because all nitrates are soluble and because, like most negative ions, they are not held on soil particles. Nitrogen fertilizers can therefore be a major contributor to eutrophication of lakes and streams.

2. High mineral nitrogen concentrations in soils due to overapplication of fertilizers have a direct effect on soil microorganisms. Heavy applications of ammonium nitrogen fertilizers to highly alkaline soils can result in accumulation of *nitrite* in the soil. Apparently, high ammonium concentrations poison the bacteria (*Nitrobacter*) responsible for the oxidation of NO_2^- to NO_3^-. Not only are high nitrite concentrations directly toxic to some plants, but certain plants can absorb and store quantities of nitrite. As pointed out in Section 12.14, ingestion of nitrites can lead to methemoglobinemia in animals. Also, in cases of high *nitrate* nitrogen fertilization, where high nitrate absorption can take place in many plants, airborne bacteria or bacteria found in the intestines of infants and ruminant animals can convert nitrate to nitrite with the same adverse results as direct nitrite ingestion.

3. Application of large amounts of nitrate nitrogen fertilizers to soils can depress the activity of nitrogen-fixing microorganisms and may eventually lead to their disappearance from the soil. Fortunately, it is easy to reestablish nitrogen-fixing organisms in the soil by innoculation of the soil or legume seeds with the bacteria. For the added bacteria to survive, however, organic materials such as manure must be present to provide food for the bacteria; and other decay bacteria must be present to break down the organic matter for the nitrogen-fixing bacteria. Thus, in a natural soil, as opposed to an artificially fertilized one, organic matter and living organisms are intimately related. Also, heavy application of nitrates, which serve as food for denitrifying soil bacteria, will encourage the loss of nitrogen from the soil.

Phosphorus Fertilizers   As with nitrogen, the total amount of phosphorus in a soil is quite low, the average being about 0.06 percent in the United States. The total phosphorus content is made up of varying proportions of inorganic phosphate minerals and organic phosphorus compounds. Of the total amount of phosphorus, only about 1 percent is available phosphorus, mostly as the soluble $H_2PO_4^-$ ion. The rest of the phosphorus is tied up chemically in forms unavailable to plants. Complicating the problem of phosphorus fertilization is the tenacious fixation of added soluble phosphorus in the soil due to reaction with soil components. Fixation is one of the major sources of loss of available soil phosphorus, the other being crop removal.

Phosphorus, because of its low total amount in soil and its low availability to plants, is involved in more cases of soil infertility than any other element. That is, phosphorus is a major limiting factor on land as well as in water. Before going into the use of phosphorus compounds as fertilizers, we look at the phosphorus cycle in nature.

The Phosphorus Cycle   The biogeochemical cycle of phosphorus (Figure 14.9) is a typical sedimentary cycle. By far the largest reservoir of phosphorus in nature is the phosphate rocks of the lithosphere. These rocks, which may be igneous, metamorphic, or sedimentary, all contain phosphorus as the important mineral *apatite*. Apatite is a calcium phosphate with varying compositions, usually somewhere between fluorapatite, $[Ca_3(PO_4)_2]_3CaF_2$, and hydroxyapatite, $[Ca_3(PO_4)_2]_3Ca(OH)_2$. The composition of apatite can be summarized as $Ca_5(PO_4)_3(OH, F)$. The conversion of the phosphorus in this highly insoluble form to more or less soluble inorganic forms (PO_4^{3-}, HPO_4^{2-}, $H_2PO_4^-$, or H_3PO_4) by weathering and natural erosion is a slow process. But man has accelerated this step greatly by mining mineral phosphates and converting them into soluble phosphate fertilizers. A common phosphate fertilizer, known as superphosphate, is a mixture of soluble calcium dihydrogen phosphate and calcium sulfate. Superphosphate is made from apatite (rock phosphate) by treatment with dilute sulfuric acid. If, for simplicity, we denote the raw rock phosphate as $Ca_3(PO_4)_2$, the reaction is

$$Ca_3(PO_4)_2 + 2H_2SO_4 \longrightarrow Ca(H_2PO_4)_2 + 2CaSO_4$$

Once the unavailable phosphorus in the rock is converted into more soluble forms, it enters into the cyclic part of the phosphorus cycle and circulates between inorganic and organic (biological) forms. This portion of the cycle is shaded in Figure 14.9. Only a small proportion of the available phosphorus

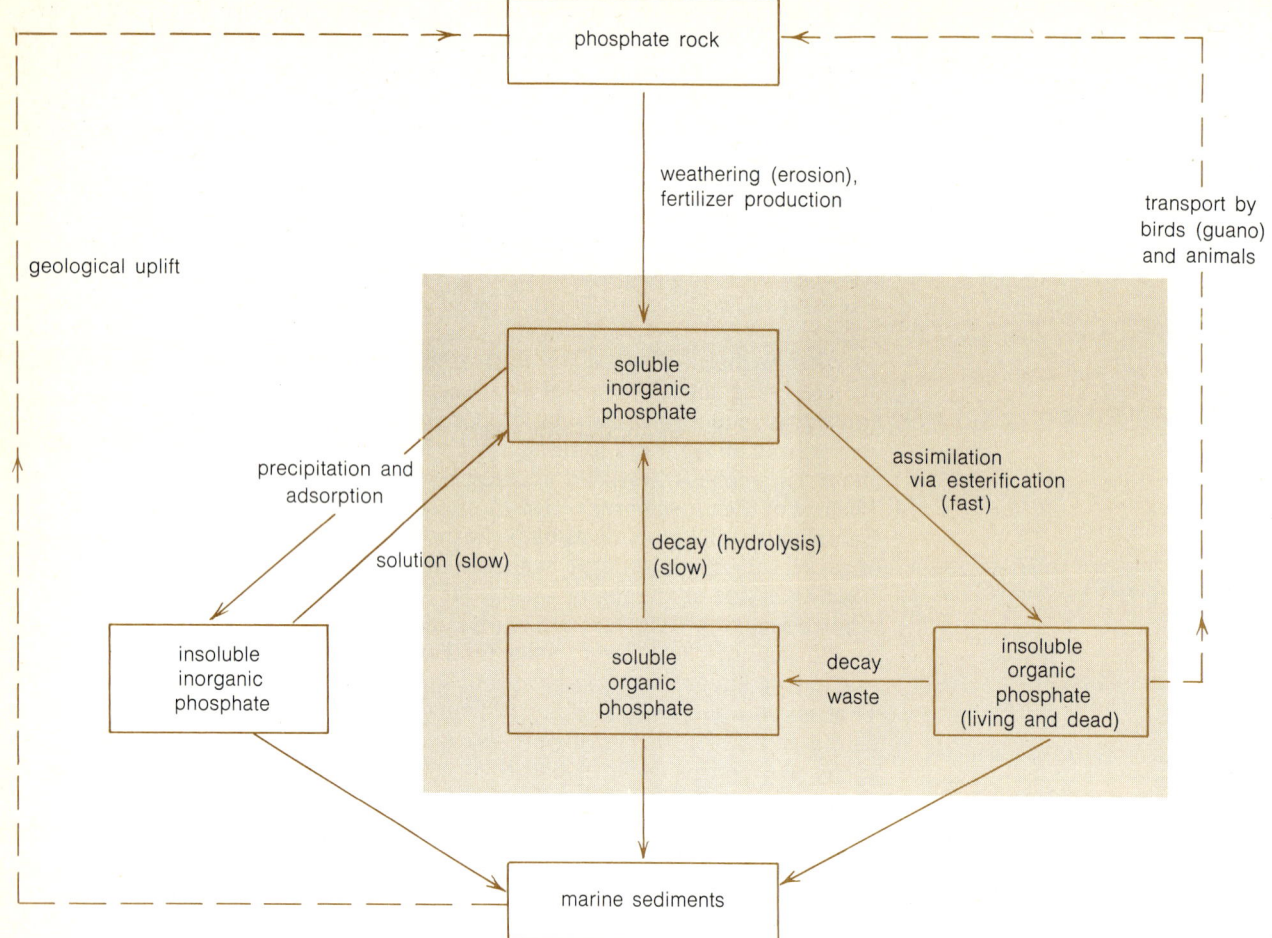

Figure 14.9. The phosphorus cycle. Dashed arrows indicate minor phosphorus transportation pathways.

continues in this cyclic process, because various mechanisms in lakes, oceans, and the soil tend to remove large amounts of phosphorus from the total cycle for long periods of time. The major removal occurs through deposition of phosphorus compounds in the sea. It is estimated that every year 3.5 million tons of phosphorus are washed into the sea and trapped in unavailable marine sediments. The return of phosphorus from sea to land to complete the overall cycle depends on slow or infrequent processes. Birds or animals eating sea organisms may transfer phosphorus from the sea to the land in their wastes or remains, as in the guano deposits in Peru. Large-scale, but infrequent transfers of phosphates from marine apatite sediments come about through geological uplift.

These biological and geological sea-to-land transfers do not match the sedimentary land-to-sea processes, and this makes the phosphorus cycle less perfect, with an overall one-way flow from the land to the sea. The one-way flow of phosphorus is further enhanced because phosphorus, unlike the other macronutrients of the biosphere, has no naturally occurring volatile forms (with the possible exception of phosphine, PH_3, formed by bacterial action), so that the atmosphere cannot serve as a route from the sea to the land. Even sulfur, although involved in a sedimentary cycle, has naturally occurring compounds (SO_2 and sulfate particulates) that travel through the atmosphere. Its earthbound nature makes the phosphorus cycle simpler but less stable.

It should be pointed out that, because of the highly sedimentary nature of the phosphorus cycle, phosphorus in natural waters is being locked into sediments of both inorganic (such as insoluble $FePO_4$) or organic nature (remains of algae), from which its return to the water is slow (3–15 days). The result is that the concentration of available phosphorus in the water is low. Because other nutrients such as nitrates and sulfates are often in plentiful supply in water, if not naturally then from atmospheric pollution, phosphorus will often be the limiting factor for aquatic plant growth. Any phosphorus added in sewage will be rapidly (5 min) assimilated by phytoplankton with resultant eutrophication.

Phosphorus, unlike other major nonmetallic elements in the biosphere, undergoes no decrease in oxidation state when it is incorporated into living matter. Both in its common mineral form and in organic combination, phosphorus is in the oxidized +5 state. Instead of a change in oxidation state, phosphorus undergoes a reaction called *esterification* when it enters living matter. Whereas nitrogen reduction is limited to certain organisms, phosphate esterification is a universal property of all life. In phosphate esterification, phosphorus is linked through oxygen to a carbon (or nitrogen) atom:

$$\underset{\text{organic acid}}{\overset{\overset{\textstyle O}{\|}}{RC}-OH} + \underset{\underset{\text{phosphoric}}{\text{acid}}}{\overset{\overset{\textstyle OH}{|}}{HO-\underset{\underset{\textstyle O}{\|}}{P}-OH}} \longrightarrow \overset{\overset{\textstyle O}{\|}}{RC}-O-\underset{\underset{\textstyle O}{\|}}{\overset{\overset{\textstyle OH}{|}}{P}}-OH + H_2O$$

Phosphate esterification requires energy. In living organisms, this energy is supplied by a separate oxidation reaction, and the added energy is stored in the phosphate ester for later use by the organism. The energy is released in a hydrolysis reaction that is essentially the reverse of esterification. These aspects of phosphorus utilization in organisms will be treated in Section 16.17.

Phosphorus Fixation in Soils Most of the phosphorus added to soils in phosphate fertilizers is held tightly in the soil in forms unavailable to plants. These forms are principally highly insoluble compounds formed by reaction of iron, aluminum, or calcium ions with the dihydrogen phosphate, $H_2PO_4^-$, commonly found in the more soluble fertilizers:

$$Al^{3+} + H_2PO_4^- + H_2O \rightleftharpoons$$
$$2H^+ + Al(OH)_2H_2PO_4 \qquad K_{sp} = 2.8 \times 10^{-29}$$

In an acid soil, Al^{3+}, Fe^{3+}, or Fe^{2+} ions will tend to be present in the soil solution or on the surface of hydrous oxides of these metals. In alkaline soils, exchangeable Ca^{2+} or $CaCO_3$ will be present, and the following reactions can lead to insoluble calcium phosphate:

$$Ca(H_2PO_4)_2 + 2Ca^{2+} \rightleftharpoons Ca_3(PO_4)_2 + 4H^+$$
$$Ca(H_2PO_4)_2 + CaCO_3 \rightleftharpoons Ca_3(PO_4)_2 + 2CO_2 + 2H_2O$$
$$Ca(H_2PO_4)_2 + CaCO_3 + 3H_2O \rightleftharpoons 2CaHPO_4 \cdot 2H_2O + CO_2$$

Because of soil fixation reactions such as these, leaching or runoff of phosphate fertilizers from the land cannot be a major source of phosphate eutrophication in natural waters unless the phosphate is carried away on eroded soil particles themselves. Where phosphates do cause eutrophication, their principal source must be domestic sewage.

Potassium Fertilizers

Potassium, the third fertilizer element, is present in great total amount in soil ($\sim$2 percent), but a large proportion of it is in unavailable forms in original minerals or trapped in illite clays. Because potassium salts are soluble, available potassium is readily lost by leaching; but most of the potassium is removed with crops. Potassium chloride, KCl, is a common potassium fertilizer.

Sulfur

Sulfur is not so commonly applied to soil in commercial fertilizers as nitrogen, phosphorus, or potassium. Replacement of soil sulfur removed in crops has generally been incidental, because common phosphorus and nitrogen fertilizers [superphosphate and $(NH_4)_2SO_4$] also contain sulfur. The recent trend to fertilizers with higher concentrations of P and N but less S is, however, resulting in more instances of sulfur deficiency.

A considerable amount of sulfur needed by crops comes from the atmosphere, principally as SO_2 from the combustion of coal and other sulfur-rich fuels. Atmospheric sulfur can be absorbed directly by growing plants, or taken into the soil by direct absorption or dissolved in rain and snow. Where soil sulfur is low, most of the sulfur requirements of plants can be satisfied by direct absorption from a sulfur-rich atmosphere. With the progress that has been made in decreasing SO_2-type air pollution by removal of SO_2 from effluent gases and switching to cleaner fuels, there is some concern that the decrease in the supply of atmospheric sulfur to plants and soils will bring about more instances of sulfur deficiency, particularly in areas away from industrial centers. Nevertheless, a more prevalent problem in the United States is the adverse effect of high atmospheric SO_2 concentrations on sensitive plants.

Fertilizers As Nonrenewable Resources

With the exception of nitrogen fertilizers derived from manufactured ammonia, all fertilizers are made from nonrenewable mineral resources. Phosphorus fertilizers, for instance, are made from phosphate rock and, in the United States, the per capita consumption of phosphate rock is over 200 lb (91 kg) per year. Whether apatite reserves will remain adequate as world population grows and the demand for more fertilizer to grow more food increases remains to be seen. In the meantime, our land-based concentrated phosphorus deposits are being shifted to the sea and lakes or dispersed in unavailable forms in the soil. Efforts should be made to recover phosphorus wasted in sewage effluents and to recycle phosphorus in other ways.

Soil Structure and Soil Conservation

14.15 SOIL STRUCTURE

Loss of soil productivity is associated with a deterioration in soil fertility or structure or both. Both characteristics of a soil may be affected by either the addition or the subtraction of something from a productive soil. Soil structure relates to the aggregation of the individual colloidal soil particles—clay and humus—into larger particles or granules. The pore spaces between the granules in a productive soil should be such that adequate drainage of water can occur through the soil. If the soil structure is characterized by particle aggregates that are very small, the pores will be too small, and this will lead to a compacted soil with excessive water retention and poor aeration. On the other hand, if the aggregates are too large, the larger pores will permit water to sink rapidly into lower soil layers where it would be beyond the reach of roots. The ideal pore size is one that allows water to be drawn upward from lower levels by capillary action.

Proper soil aggregation and porosity depend on the binding together of the individual soil particles by organic compounds produced by bacterial and fungal action on organic matter in the soil. Larger organisms, particularly earthworms, contribute to soil structure by mixing the soil, adding organic matter, and aiding drainage and aeration.

In a natural soil, soil fertility cannot exist without the proper structure. Both of these properties are dependent on the existence of organic matter in the soil and on the presence of soil organisms that feed on the organic matter and thereby recycle nutrients and build soil structure. The preservation of both fertility and structure is necessarily linked with the preservation of the population of living organisms in a soil.

14.16 SOIL CONSERVATION

With this background, we can now examine a few aspects of soil conservation such as erosion, presence of toxic substances, and crop removal.

Erosion

Soil erosion, the disintegration and transportation of rock and soil by wind or water, is a normal geological process integral to the formation of soil. Normal erosion also slowly removes the worn out, nutrient-leached upper layers of the top soil so that plant roots can reach underlying mineral-rich layers. But, when erosion accelerates to the point where the top soil is lost faster than new topsoil can be formed from the subsoil, then both soil fertility and soil structure suffer. Nutrients are removed, but more important is the loss of soil structure in that hard, compacted lower horizons are exposed. Prevention of soil erosion is what most of us have been taught to regard as soil conservation.

Accelerated soil erosion is prevented by maintaining soil structure and fertility so that stable soil aggregates will form. With proper soil structure, water will be absorbed by the soil instead of carrying topsoil away in surface runoff. The larger, more stable, soil aggregates will be more resistant to wind erosion because of their size and because of their higher moisture retention. Any practice that decreases the soil organic matter will contribute to decreased erosion resistance. The Dust Bowl of the Great Plains of the United States received its name during a drought period in the 1930's after excessive plowing exposed to the wind dry, structureless soil depleted of organic matter by continued crop removal.

Soil Pollution and Toxic Substances

A soil is polluted when substances in the surface soil produce changes in the chemical, physical, or biological properties of the soil and adversely affect the quality and quantity of crops. Among the soil pollutants are many natural and synthetic substances that can be toxic to soil organisms, plants, or herbivores. The adverse effects of some substances on soil fertility, plants, and animals feeding on the plants has been mentioned in previous sections of this chapter. Here only a few aspects of the effect of toxic substances on soil organisms will be treated.

The soil comprises an individual ecosystem populated by an enormous number of animals and minute plants, exclusive of plant roots. Impressive numbers of different species of organisms are present in any sample of productive soil; and, in a natural soil, these organisms are responsible for continued soil formation and for soil structure and fertility. The addition of organic fertilizers and the return or retention of part of the crop to the land are measures used to provide food for these beneficial organisms. Other substances, particularly insecticides, herbicides, and fungicides, are finding their way into soils in ever larger quantities and may act as soil pollutants by interfering with the growth of various soil organisms. Although such substances have been shown to reduce both the numbers of organisms and the number of species present, it appears that the affected soil organisms can reestablish themselves in the soil and little permanent damage is done. Recovery is more rapid if the toxic substance is transient and decomposes through natural processes before too many organisms have been affected. The persistence of certain toxic compounds, such as the chlorinated hydrocarbon insecticides (i.e., DDT), create a more serious problem. Furthermore, these substances or their toxic residues can accumulate in soil animals (earthworms) and be transferred in potentially toxic amounts to birds and other predators farther up the food chain.

Abnormally high concentrations of salts can be classified as soil pollutants because of their toxic effects or their detrimental effect on soil structure. Soils containing excessive amounts of soluble salts derived mainly from Ca^{2+}, Mg^{2+}, K^+, Na^+, Cl^-, SO_4^{2-}, and HCO_3^- are called *saline* soils. Soils in which large amounts of adsorbed but exchangeable sodium ions predominate are termed *alkaline* or *sodic*.

The pH of saline soils is usually no more than 8.5, so the injurious effects are due to direct toxic effects or, more commonly, to the high osmotic pressure of the soil solution. High osmotic pressure reduces the ability of the plants to absorb water from the soil solution. Sodic soils have even higher pH values, often approaching 10. The adsorbed sodium ion is more easily replaced by hydrolysis than other adsorbed cations:

$$Na^+| + H_2O \rightleftharpoons H^+| + Na^+ + OH^-$$

In addition, sodic soils usually contain sodium carbonate, and hydrolysis of the carbonate ion also contributes to the high pH:

$$CO_3^{2-} + H_2O \rightleftharpoons HCO_3^- + OH^-$$

Not only are the high concentrations of sodium, carbonate, hydrogen carbonate, and hydroxide ions directly toxic to plants and soil organisms, but the presence of adsorbed sodium results in breakdown of the soil aggregates necessary for proper soil drainage and aeration.

Saline and sodic soils result from the accumulation in the soil of salts formed from mineral weathering. The salts are delivered to the soil either by normal runoff from higher elevations or in irrigation waters. Such soils are generally located in regions of low rainfall and poor soil drainage. Under these conditions, salts are not leached out of the soil and accumulate in the lower levels of the soil or in groundwater. Heavy use of even nonsaline irrigation water in arid region soils will intensify the salt concentration in surface soils by dissolving the salts in the lower layers, transporting the salts to the surface, and depositing the salts as the water evaporates. Heavy irrigation may also raise the level of saline groundwater to the root level.

Saline soils can be reclaimed and maintained by construction of an extensive drainage system followed by leaching of the salts with large quantities of nonsaline irrigation water. On the other hand, sodic soils cannot be reclaimed by leaching with nonsaline water alone, but exchangeable sodium in sodic soils can be removed by replacement with Ca^{2+} from calcium salts or H^+ from sulfuric acid. Sodium carbonate also reacts with added Ca^{2+} and H^+ to give a soluble sodium product that can be leached out of the soil:

$$2Na^+| + Ca^{2+} \rightleftharpoons Ca^{2+}| + 2Na^+$$
$$Na_2CO_3 + Ca^{2+} \rightleftharpoons CaCO_3 + 2Na^+$$
$$Na_2CO_3 + H_2SO_4 \rightleftharpoons CO_2 + H_2O + Na_2SO_4$$

It has been shown that many plants can be grown with saline and sea water, provided that a very sandy soil with rapid drainage is used to prevent the accumulation of sodium and magnesium chlorides (see the reference by Boyko at the end of the chapter). In addition, sodium is not adsorbed on sand.

Crop Removal   All plants mine the soil beneath them for mineral nutrients and return these same nutrients to the soil when they die. In any soil–plant system, there is a distribution of total available nutrient supply between the soil and the plants. In some systems, such as in a forest, most of the total nutrient supply is contained in the plants, and the soil is relatively poor in nutrients. In such cases, the harvesting and removal of the vegetation may seriously reduce the fertility of the soil, and in all cases, continued harvesting of any crop will eventually strip the soil of one or more of the essential nutrients unless fertilizers are used to replenish the supply.

The soil–plant relationship in tropical climates is especially important because of the common belief that food for the world's growing population can easily be grown on cleared jungle land. This belief arises from the mistaken view that climate rather than soil fertility is directly responsible for the growth of plants. Despite what may be inferred from the luxurious jungle growth, the high temperature and heavy rainfall of the tropics make the land highly unsuitable for intensive agriculture.

Some soils of the tropics are among the most highly weathered soils known. From what we know of weathering and soil fertility, we might guess that these soils are relatively infertile. These soils (latosols) get that way because the decay of dropped foliage is very rapid due to high temperatures. The nutrients released from the fast-disappearing organic matter are then quickly leached from the upper horizons of the soil by heavy rainfall. Although one might expect a highly acid solution due to the products of decay, the solution percolating through the upper soil is near neutral (pH = 5–9) because of rapid hydrolysis of calcium, magnesium, and other base-forming cations adsorbed on humus. Under these conditions, silica derived from the breakdown of clay is more soluble than the accompanying aluminum and iron oxides, and it is preferentially leached from the top layers. The upper soil layer becomes low in silica and rich in iron and aluminum (Figure 14.10).

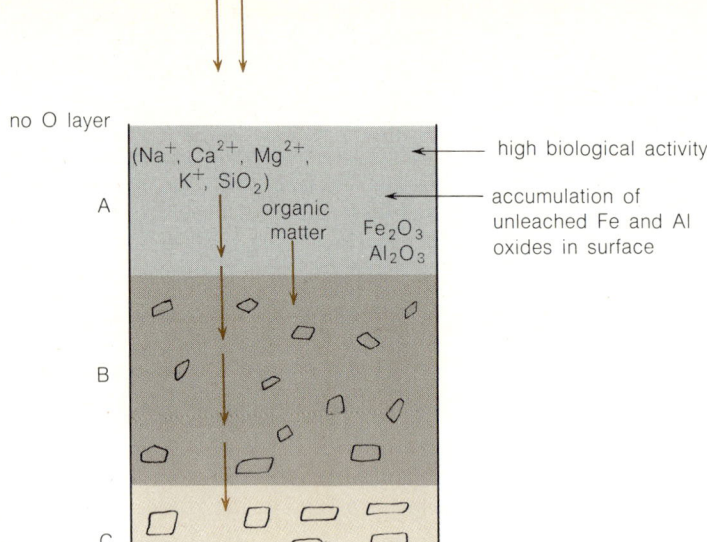

warm, heavy rainfall

no O layer

A — $(Na^+, Ca^{2+}, Mg^{2+}, K^+, SiO_2)$ — organic matter — Fe_2O_3 Al_2O_3

high biological activity

accumulation of unleached Fe and Al oxides in surface

B

C

Figure 14.10. Soil development leading to a lateritic surface soil in the tropics. The B and C horizons may be thin or missing. The soil is very porous. In addition to the metal ions and silica, most of the organic matter is leached out of the A horizon.

When iron predominates, the soil is called a laterite; when aluminum predominates, we have bauxite.

Except for a thin organic layer on the surface, tropical soils are very low in nutrients because of the rapid leaching, aided by the characteristic high porosity of these soils. For their natural productivity these soils rely on a uniform recycling of nutrients from vegetation to soil and back again. Clearing the jungle of its natural growth and then cultivating crops would rapidly exhaust the soil. Massive amounts of fertilizer would be needed because of the rapid leaching and the enormous phosphorus-fixing power of the iron and aluminum oxides. Furthermore, clearing of the natural jungle growth exposes the surface to the sun. The resulting loss of moisture can cause the iron-rich soil to harden into a bricklike crust, which is highly prone to erosion. Sophisticated soil management practices are necessary if such tropical soils are to be cultivated.

The use of herbicide defoliants by the United States in Viet Nam has apparently caused large areas of the tropical soil to become infertile. Under natural conditions, foliage is dropped slowly and uniformly throughout the year. Despite rapid decay and leaching, there is always a small but continuing supply of nutrients to the soil. Defoliation causes all the leaves to drop at once and, aside from the fact that the soil is exposed, all of the nutrients in the soil–plant system are immediately subject to rapid leaching from the soil, leaving no continuing supply to maintain soil fertility. Rapid defoliation of this kind has been called "nutrient dumping."

14.17 SUMMARY

The formation of soil begins with the disintegration of surface rocks by physical weathering. This is followed by processes of chemical weathering, which decompose the primary minerals and, in most cases, result in the synthesis of secondary minerals. The most important components of soils, the clay minerals, are formed by the hydrolysis of feldspars. Biological activity in the soil speeds the progress of both physical and chemical weathering.

The more easily weathered primary minerals are the early-formed minerals in the magma crystallization sequence. The minor elements incorporated in these minerals tend to be relatively abundant in soils compared to the minor elements incorporated in the more resistant, late-forming minerals.

A true soil has four major components: mineral matter, organic matter, the soil solution, and the soil air. Of the mineral component, the clay minerals are the most significant constituent because they function as reservoirs of plant nutrients. This role results from the cation exchange properties of the clays.

Soil fertility and soil structure determine the suitability of a soil for growing plants. Soil fertility refers to the ability of a soil to supply plants with their essential nutrient elements in a form that the plants can assimilate. Soil fertility is closely related to soil acidity, because the pH of the soil often determines whether a plant nutrient will be present in an available (soluble) or unavailable form.

The elements essential to the growth of plants may be divided into the micronutrient or trace elements and the macronutrient elements. Members of both groups are removed from the soil by leaching, crop removal, and erosion. Continued soil fertility depends on their replacement by natural recycling processes or by application of commercial fertilizers. Natural processes depend on the return of organic matter to the soil and its decomposition by microorganisms in the soil. Of the elements commonly added to the soil in commercial fertilizers, phosphorus is most often involved in cases of soil infertility because it is strongly fixed in insoluble, and therefore unavailable, forms by reaction with other soil components.

The aggregation of colloidal soil particles of clay and humus into larger particles determines soil structure and the resultant drainage and aeration properties of a soil. Loss of soil organic matter leads to deterioration of soil structure and subsequent loss of fertility and susceptibility to erosion. Structure and fertility of soils are also affected by soil pollutants.

Questions and Problems

1. In comparison to calcium and magnesium oxides, hydroxides, and carbonates, the sulfates and chlorides are not good liming agents for soils and may even decrease the pH of the soil solution. Why?
2. The capacity of a liming material to neutralize soil acidity is expressed by its *neutralizing value* or *power*. Pure calcium carbonate is assigned a neutralizing value of 100 percent.
   a. What would be the neutralizing value of pure calcium oxide. That is, how much more or less hydrogen ion will CaO neutralize than an equal weight of $CaCO_3$?

   b. What is the neutralizing value of a commercial ground limestone containing 80 percent $CaCO_3$ and 14 percent $MgCO_3$?
3. The cation exchange capacity (C.E.C.) of a soil is a measure of the ability of the soil to adsorb exchangeable cations. The C.E.C. may be defined as the number of millimoles (mmol) of unipositive cations on 100 g of soil. Montmorillonite clay has a C.E.C. of 80–150 mmol/100 g; kaolinite about 3–15 mmol/100 g; and humus up to 250 mmol/100 g. Soils may range from a C.E.C. of almost zero (fine sands) to about 60 for soils high in organic matter.

One method for determining the total C.E.C. of a soil is as follows:

(1) Replace all natural soil cations with NH_4^+ by treating the soil with $1M$ NH_4Ac.

(2) Wash excess NH_4Ac from soil, without displacing adsorbed NH_4^+, with ethanol.

(3) Replace adsorbed NH_4^+ with successive portions of acidified 10 percent NaCl.

(4) Analyze the NaCl wash liquid for displaced NH_4^+ by treating the wash liquid with NaOH, distilling, and titrating the distillate with standardized HCl.

a. Write equations for the exchange equilibrium reactions in (1) and (3).

b. Write equations for the treatment in (4).

c. If the distillate from the wash liquid from 20 g of soil requires 40.0 ml of $0.10M$ HCl to reach a methyl red end point, what is the C.E.C. of the soil?

d. Why do you think alcohol instead of water is used in (2) to remove excess NH_4Ac solution from the soil?

4. Why are acid soils common in humid regions but rare in arid regions?

5. Give possible explanations for the following:

a. Sodium ions are held on clay particles less strongly, and are desorbed more easily, than Ca^{2+}, H^+, or Mg^{2+} ions.

b. Pond bottoms of calcium-rich clays may be permeable to fresh water but may be made impervious to fresh water by filling the pond with sea water or brine for several days.

6. Sulfur dioxide-type air pollution can contribute to a small but continuous loss of soil fertility, and an increase in soil acidity. How might this happen in humid regions?

7. Explain the following:

a. Addition of nitrogen to soils as $(NH_4)_2SO_4$, NH_4NO_3, $CO(NH_2)_2$, or NH_3 increases soil acidity, but addition as $NaNO_3$ or $Ca(NO_3)_2$ does not increase acidity, and may even decrease it slightly.

b. Nitrogen added as NO_3^- is lost from the soil more rapidly than nitrogen applied as NH_4^+.

8. A fertilizer mixture contains 100 lb of ammonium monohydrogen phosphate, $(NH_4)_2HPO_4$, and 100 lb of KCl. What is the approximate fertilizer guarantee of the mixture?

9. The solubility products of $Fe(OH)_2$ and $Fe(OH)_3$ are 10^{-17} and 10^{-38}, respectively.

a. Why should $Fe(OH)_3$ be less soluble?

b. If the concentrations of Fe^{2+} and Fe^{3+} are each $10^{-5}M$, at what pH will each hydroxide just begin to precipitate?

10. Domestic sewage wastes (solids and liquids) draining through soil may cause the appearance of Fe^{2+} in groundwaters. Why?

11. In a study in which secondary sewage effluent was sprayed on forest soil, the amount of phosphorus in the soil water at 1-ft depth was decreased by 98 percent of the phosphorus content in the effluent in the first year, 86 percent in the second year, and somewhat less in the third year. On the other hand, when effluent was sprayed on agricultural crops harvested yearly, the decrease in phosphorus remained about the same each year. Can you explain these results?

12. An Eskimo in the Arctic decides to take advantage of the long hours of summer sunshine to grow a crop. Knowing that the soil is poor in available nitrogen, phosphorus, and sulfur, he spreads caribou manure, but this does no good. What is wrong? How might the soil be made productive?

13. The San Joaquin Valley of California includes millions of acres of the most fertile, most highly cultivated agricultural land in the United States. Because of heavy irrigation and resultant leaching, salts are carried down to the groundwater. To prevent deterioration of groundwater by salts (and pesticides), shallow drainage systems are used to intercept irrigation water before it reaches groundwater. A 300-mile Master Drain has been proposed to collect waste water from all the interceptor drains and discharge it into San Francisco Bay at a point where several rivers enter the Bay. What effects might this procedure have on San Francisco Bay and the delta area where the rivers enter the Bay? How might adverse effects be prevented? [Frank M. Stead, "Desalting California," *Environment*, **11**, 2 (June 1969).]

14. The chelating agent EDTA is a weak tetraprotic acid. The $EDTA^{4-}$ anion is the actual complexing agent. Why should $Fe(EDTA)$ chelates not be applied to soil if the soil pH is very low or very high?

15. a. At what pH would $Fe(OH)_3$ begin to precipitate from a solution containing a total initial concentration of $0.050M$ Fe^{3+} and $1.0M$ EDTA as the $EDTA^{4-}$ ion? The equilibrium constant for the formation of the chelate is

$$Fe^{3+} + EDTA^{4-} \longrightarrow Fe(EDTA)^-$$
$$K = 1.3 \times 10^{25}$$

Assume that $Fe(EDTA)^-$ formation is virtually complete. The solubility product constant for $Fe(OH)_3$ is 4×10^{-38}.

b. At what pH would $Fe(OH)_3$ precipitate from a $0.050M$ Fe^{3+} solution if no EDTA were present?

Answers

2. a. 179 percent
   b. 96.6 percent
3. c. 20 mmol/100 g
8. 20–55–65 or 4–11–13

9. $Fe(OH)_2$ at pH = 6; $Fe(OH)_3$ at pH = 3
15. a. pH = 10.3
    b. pH = 2.0

References

Albrecht, William A., "Soil Fertility and the Human Species," *Chemical and Engineering News*, **21**, 221–227 (1943).

Bear, Firman, Ed., "Chemistry of the Soil," 2nd ed, Van Nostrand Reinhold, New York, 1964.
   Chapter 8, "Trace Elements in Soils," ties in especially well with the material in Chapter 13 of this text.

Bloom, Arthur L., "The Surface of the Earth," Prentice-Hall, Englewood Cliffs, N. J., 1969, Chapter 2 ("Rock Weathering").

Bormann, F. Herbert, and Likens, Gene E., "The Nutrient Cycles of an Ecosystem," *Scientific American*, **223**, 92–101 (October 1970).
   Loss of soil fertility from a forest ecosystem as a result of logging, air pollution, and other factors.

Boyko, Hugo, "Salt-Water Agriculture," *Scientific American*, **216**, 89–96 (March 1967).

Brady, Nyle C., Ed., "Agriculture and the Quality of Our Environment," American Association for the Advancement of Science, Washington, D. C., 1967.
   A collection of articles on the effect of agriculture on our environment and on the effects of air, water, and soil pollution on agriculture.

Buckman, Harry O., and Brady, Nyle C., "The Nature and Properties of Soils," 7th ed., Macmillan, New York, 1969.
   A general treatment of soils as applied to agriculture.

Edwards, Clive A., "Soil Pollutants and Soil Animals," *Scientific American*, **220**, 88–99 (April 1969).

Keller, W. D., "The Principles of Chemical Weathering," Lucas Brothers, Columbia, Mo., 1955.

Keller, W. D., "Geochemical Weathering of Rocks; Source of Raw Materials for Good Living," *Journal of Geological Education*, **14**, 17–22 (1966).

Kellogg, Charles E., "Soil," *Scientific American*, **183**, 30–39 (July 1950).

McLaren, A. D., "Biochemistry and Soil Science," *Science*, **141**, 1141 (1963).

McNeil, Mary, "Lateritic Soils," *Scientific American*, **211**, 96–102 (November 1964).

"Soil. The Yearbook of Agriculture 1957," U. S. Department of Agriculture, Washington, D. C., U. S. Government Printing Office, 1957.
   A well-written selection of articles covering all aspects of agricultural soil science.

Yaalon, Dan H., "Weathering Reactions," *Journal of Chemical Education*, **36**, 73 (1959).

FIVE

Life

Of the ancient elements—earth, air, fire, and water—the preceding chapters have slighted only fire. This shortcoming will be remedied in following chapters, when the fires, the energies that drive life, civilization, and the entire biosphere are considered. Our general aim in Part Five, however, is to discuss that awesome combination of the elements, whether ancient or modern, that we call life.

In Part Five we shall see that, in addition to our physical or planetary environment, there is another part of our surroundings that comprises all the chemistry, and more, of our nonliving or abiotic environment. This is our biological environment, and it includes, in addition to all life on earth, the most intimate environment of all, our own bodies. The members of this biological community are all tied together by a vast and intricate web of relationships. The members of this community of living systems are related to the physical environment in many ways, some of which we have already discussed.

As we have pointed out, a number of these ecological relationships are chemical in nature. To gain an even better understanding of the connections between the external and internal aspects of living things, we need to embark on a more detailed and organized discussion of the processes of life. In doing so, our assumption will be that at the molecular level the ongoing processes of life obey the same chemical and physical principles as the inanimate world of rocks, oceans, and air; that life is merely a near-perfect coordination of electronic, atomic, and molecular interactions taking place in, and caused by, the highly organized structure of living organisms.

The material of the next five chapters will draw on a number of fields of study—biochemistry, molecular biology, biophysics, biophysical chemistry, physiological chemistry, medicine, and others. Hopefully, we shall gain some appreciation of what has been called the internal ecology of our bodies as well as some understanding of the present excitement of research into the basic processes of life, that most relevant of all subjects.

15

The Molecules of Life

The question, "What is life?", seems at first to be a trivial question with a self-evident answer. We tend to think of life as a very commonplace phenomenon in our world, so common that we are insensitive to its definition or characteristics. Only now, as we begin to journey into outer space, has it occurred to us that life might be a unique event in our solar system and perhaps in the universe. Even on earth, life as we know it is confined to a thin layer on the surface of the earth. The community of living things in this layer, the biosphere, extends a few yards into the ground and a few miles below the surface of the oceans or into the atmosphere. This thin layer, very splotchy in parts, contains an amazing variety and number of what are called living organisms.

**15.1 CHARACTERISTICS
OF LIFE**

Despite the variation existing in this huge collection of living organisms, animals ranging from protozoa to whales, plants ranging from algae to giant sequoias, and all the microscopic bacteria, there are certain basic characteristics usually considered to be present in all of them. So, rather than attempt a true definition of life, if any exists, we shall be content to look first at some of its characteristics.

First, we have already called these living objects organisms. Organisms are highly structured, highly ordered, highly integrated, but very complex systems. We can think of some inanimate objects, such as crystals, that also display the high order of living organisms, but a little thought will show that these objects are not highly structured or very complex.

Organisms are self-contained entities set off from their surroundings by some kind of boundary or limiting surface through which they can interact with their environment. The high degree of organization coupled with the interaction with the surroundings is one of the major characteristics of living systems. In fact, this interaction is absolutely necessary if the organism is to maintain the high degree of organization necessary for life. We shall consider this interaction as consisting of the absorption of energy by the organism from its surroundings. The absorbed energy is used by the organism to maintain its present state of organization as well as to grow to even more complex and highly organized states. The flow of energy and materials into and out of the system for the maintenance of order and of growth is called *metabolism* and involves all of the chemical reactions that take place within the living organism.

Another major characteristic of life is its ability to reproduce its own kind, to make replicas of itself for generation after generation. The new biological system, or organism, is always produced from one of its own kind and is identical, except for minor variations, to its parent. If any one characteristic could be said to define life, it would probably be this property of *self-replication*.

More limited characteristics follow from structural organization, interaction with surroundings, metabolism, and reproduction. These qualities, such as growth and reaction to an external stimulus, are often listed as characteristics simultaneously possessed by life, with the realization that nonliving matter may also possess some, but not all, of these properties. For instance, a crystal can grow, certain systems (oil droplets) can grow and reproduce by division, some machines may be able to reproduce themselves at least once. And, at certain times and under certain conditions, some nonliving systems show all of the so-called characteristics of life.

It is because of this inability to list a set of properties possessed by life alone that no clear definition of life exists to permit us to draw a sharp dividing line in the present hazy area between the living and the nonliving. A better approach might be to abandon the search for the place, if there is any, where present nonlife ends and life begins and to try to describe the living state in terms of its behavior over long periods of time. Then we see that the apparent self-duplication of living organisms also contains a component allowing for change, for adaptation through evolution, that reproducing nonliving systems do not have. Those changes, or *genetic mutations*, proving useful to the existence and the stability of the organism are themselves reproduced. Thus, a living system is any self-replicating, metabolizing system capable of mutation.

15.2 THE ORIGIN OF LIFE

Even if a perfect definition of life were available, the question of the origin of life would remain. This question has occupied man throughout history, and a brief discussion of the current status of the problem will provide an entry into the molecular world of living things.

As late as the middle of the nineteenth century, there was still a widespread belief in the *spontaneous generation* of life. According to this idea, living things—insects and other animals, plants, bacteria—could arise spontaneously, fully formed, from inanimate, nonliving matter, as well as from live parent organisms. Maggots originated in rotten meat, flies in manure, frogs in mud, mice from wheat and sweaty underwear. Sometimes the concept of a vital or supernatural force was invoked to explain what made this sort of thing happen. Although it was generally accepted, after about 1670, that the larger forms of life did not arise by spontaneous generation, it was not until the 1860's that Louis Pasteur (French biochemist, 1822–1895) showed that microorganisms also could not originate spontaneously. Today, not only has this particular concept of the spontaneous generation of life been abandoned, but so have the former ideas that the operations of living things are presided over by some mysterious life force. Rather, it is believed that the workings of living organisms are controlled by the same laws of chemistry and physics that apply to inanimate matter. True, some of these laws are very poorly understood, as is life itself, but in principle the "how" of life, its mechanisms, and even its origin may be explained by the laws governing electrons, atoms, and molecules.

Abiotic Origin of Life

Studies of the origin of life on earth provide striking illustrations of how life may have come about as a result of ordinary chemical reactions of the simple molecules composing the earth's early atmosphere. As we shall see later in this chapter, living organisms are composed of several kinds of large molecules. Among these molecules are the proteins and the nucleic acids. The nucleic acids contain the hereditary information passed on from one generation of the organism to the next. They contain the instructions telling the cells of the organism what protein molecules to manufacture and when. Protein catalysts called enzymes regulate all the chemical activities of the organism, including the synthesis of additional nucleic acids. If a living organism is to develop, it is safe to assume that one or the other of these substances, at least, must be present. Current experimental evidence is that these large molecules could have been formed by nonbiological means from simple molecules early in the history of the earth.

Large protein molecules are built up of many small units (amino acids). Ignoring for the time being how the individual units combine to form the protein molecules, let us consider how the units could have originated in the first place. We must go back hundreds of millions of years in the history of the earth to see how such simple molecules as the carbon-, hydrogen-, oxygen-, and nitrogen-containing amino acids might have been formed.

The early atmosphere of the earth was quite unlike the present atmosphere. At that time, 3.0–3.5 billion yr ago, the primeval atmosphere was probably composed of a reducing mixture of water vapor, hydrogen, nitrogen, methane (CH_4), and ammonia. Whatever the exact composition of the early atmosphere may have been, free oxygen was not present, and no ozone existed at higher levels to shield the earth from high-energy ultraviolet radiation.

No life of any kind existed, nor did even simple organic molecules except

CH_4. Yet simple organic molecules such as the amino acids were formed in the primitive earth environment, and eventually these molecules must have combined to form proteins and other large molecules. Finally, the large molecules organized themselves into arrangements having attributes of life—the first primitive living cells from which all other life on the planet evolved. The stage from the formation of the earth to the emergence of the first self-replicating, mutable chemical system is the period of *chemical evolution*. This stage lasted about 1.5 billion yr.

The products of some of the nearly infinite variety of chemical reactions during this stage probably were not stable under the prevailing conditions and decomposed. Other products, even if unstable, somehow escaped breakdown, perhaps by dissolving in the oceans, and served as the precursors of more complex substances. Most of these substances existed in the seas, and the waters formed a sort of organic broth or dilute primordial soup within which the final steps of chemical evolution could be played out.

These ideas are no longer pure speculation. Laboratory studies have shown that organic molecules essential to the structure and function of living organisms can be made abiotically under conditions thought to have prevailed during the early lifetime of the earth. On the primitive earth, the energy necessary for the synthesis of complex organic molecules from simple ones was most likely supplied by ultraviolet radiation or through electric discharges from lightning. Heat or ionizing radiation from radioactive elements also may have been locally effective. Since 1953, a number of laboratory investigations have shown that amino acids and other biologically important molecules are formed when mixtures of CH_4, H_2O, H_2, NH_3, and other simple molecules are exposed to ultraviolet radiation, electric discharges, or heat. Other studies have shown that these simple organic molecules can further form larger proteinlike molecules under conditions believed to have prevailed on the early earth. Under certain conditions, these more complex molecules have been observed to organize themselves into cell-like objects possessing some of the attributes of life, such as growth and reproduction. It is conceivable that continuing chemical evolution in the substances and reactions of these objects could have resulted in precellular living entities possessing self-replication and mutability. Following this stage, biological evolution would take over from chemical evolution, as indicated in Figure 15.1. Somewhere during the course

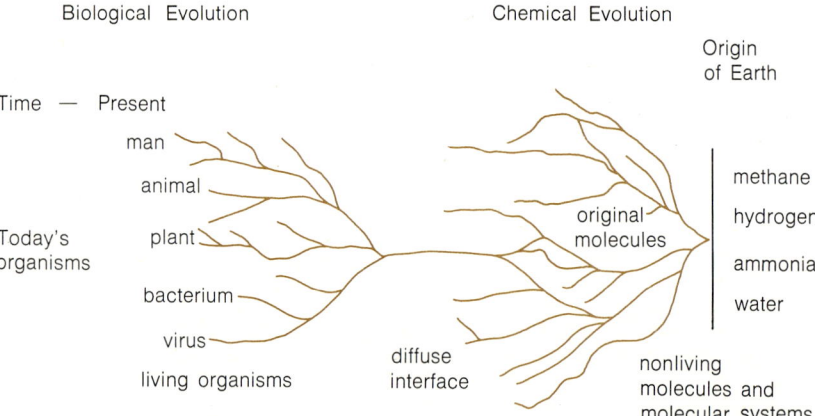

Figure 15.1 Stages in the origin of life from nonliving matter. (From Melvin Calvin, "Chemical Evolution," Oxford University Press, London, 1969. With permission of the author.)

of biological evolution, photosynthetic organisms would develop, release oxygen to the atmosphere, and pave the way for the entry of the more efficient aerobic organisms. As the oxygenated atmosphere developed, the accompanying increase of atmospheric ozone absorbed the short-wavelength ultraviolet radiation. This absorption put a stop to chemical evolution and required the biological evolution of more complex organisms capable of using longer wavelength solar radiation. Thus do life and the physical environment control the development of each other.

To a great extent, the elemental composition of living systems must have been determined by the availability of elements on the earth before life appeared. That is, in the beginning, the environment determined the makeup of life almost completely. Once life had appeared, however, natural selection could begin to operate, and the living systems could adapt to slightly different environments (or to the one environment as the properties of that environment *slowly* changed).

Thus, the present composition of living systems is the ultimate result of their past history, extending back to their abiotic origin. The nature of the basic elements and materials placed chemical and physical restrictions on the ways these systems could vary. The basic makeup—the chemical simplicity and economy of life—are the result of the nature of elementary nonliving matter and the way in which this material has been brought together in the origin of this planet and of life itself. The diversity and complexity of life are the result of slow mutation and natural selection as organisms undergo adjustment to their environment.

Inherent in the chemical origin of life is the idea that the essential molecules and perhaps even the structure of the first living organisms were formed without the intervention of any other life or vital force. We shall not pursue the fascinating subject of the chemical origin of life much further here. Details can be found in the references at the end of this chapter. But as we go along, we shall see that the structure, and therefore the function, of living things is inherent in the chemical and physical properties of the molecules from which they are made. It is almost as if life must have been the natural outcome given the substances and conditions available on the early earth.

The abiotic origin of life need not have been confined to the earth or even to our solar system. A number of molecules that play important roles in chemical evolution have already been detected by microwave spectroscopy (radioastronomy) in the vast gas and dust clouds of outer space. Among these are HCN and formaldehyde, $H_2C{=}O$, both known to play key intermediate roles in the synthesis of amino acids. The existence of about a dozen of these and other molecules (NH_3, H_2O, H_2, CO, CH_3OH, and $H_2N{-}CH{=}O$; even a complex chlorophyll-like molecule, $MgC_{46}H_3ON_6$, has been reported) in interstellar space indicate that the requisite materials for the construction of life are present throughout the universe, awaiting only the proper conditions to begin the course of chemical evolution. That the beginnings of chemical evolution may have already occurred in space is suggested by the detection of 16 amino acids inside a meteorite that fell in Australia in 1969. Five of the acids were members of the group of 20 found in all life on earth, but their molecular structures showed that they could not have been of biological origin. The other 11 amino acids are rarely found in living systems on earth, suggesting that they were formed in outer space rather than on earth. The ideas and the evidence of the chemical origin of life would seem to be a resurrection of the theory of spontaneous generation, but there are obvious differences between the two concepts.

Organic Chemistry

Living systems are invariably made up largely of very complex chemical building blocks. These large molecules are then organized into complex structures called cells, the fundamental units of living organisms. Before tackling the cell, however, we should learn something about the molecules that make up the cell. Most of the chemicals of life are classified as *organic* compounds. This name is a carry-over from the days when organic substances were thought to be formed only by organized living systems. Inorganic substances were thought to be formed and to exist apart from the processes of life. Today, numerous compounds that are made by living organisms can also be made in the laboratory from mineral substances without the intervention of a vital force (aside from that of the chemist who does the mixing). Thus, the correspondence between "organic" and "living" no longer applies.

Organic chemistry today is considered to be the chemistry of the compounds of carbon whether they are formed biologically or not. Inorganic chemistry is a broad classification of the chemistry of compounds of elements other than carbon. Even then there is considerable overlapping, because strictly inorganic chemicals, such as metals, form compounds with carbon. Whether these substances are organic or inorganic depends on the point of view of the person investigating their properties, their processes of formation, and their uses. In any event, the division of chemistry into inorganic, organic, and physical and analytical chemistry has broken down as chemists have seen how artificial and arbitrary it is to compartmentalize knowledge in this way.

There are approximately 4 million different carbon-containing compounds, most of which involve combinations of carbon and hydrogen together with other elements. The great number of carbon and hydrogen compounds is accompanied by an almost limitless variety of chemical and physical properties. The carbon atom and the properties of the C—C and C—H bonds are primarily responsible for this state of affairs. Carbon is unique in being able to combine with itself to form chains of strong C—C bonds. Then again, the only single bond stronger than the C—H bond is the rare (in nature) C—F bond. The substitution of hydrogen by other elements, the inviolable four-covalence of carbon, the ability of carbon to form strong multiple bonds with itself and some other elements, and the possibility of different structural and geometric arrangements of the same collection of atoms all contribute to the multiplicity of organic compounds.

Hydrocarbons

Our brief study of organic chemistry begins with an examination of the structures of the covalent compounds of carbon and hydrogen. These *hydrocarbons* are the basis of all other organic compounds. The two major groups of interest are the aliphatic and the aromatic hydrocarbons.

Aliphatic hydrocarbons are open-chain (acyclic) compounds, which are subdivided into three groups on the basis of bond multiplicity.

1. Alkanes. Aliphatic hydrocarbons formally derived from methane, CH_4. All carbon–carbon bonds are single bonds.

2. Alkenes. Aliphatic hydrocarbons containing one or more C=C linkages. Formally derived from ethene, $H_2C=CH_2$.

3. Alkynes. Aliphatic hydrocarbons with one or more C≡C bonds. Formally derived from acetylene, HC≡CH.

There are also carbon-atom ring structure (alicyclic) hydrocarbons with properties similar to the open-chain compounds.

Aromatic hydrocarbons are conjugated cyclic (ring) hydrocarbons based on benzene, C_6H_6, as the parent compound. Conjugated hydrocarbons contain two or more multiple bonds separated by, or alternating with, single bonds.

15.3 ALKANE HYDROCARBONS

The alkanes all have the general formula C_nH_{2n+2}. If the parent member of the alkane group is methane, CH_4, then the other members may be considered to be derived by the successive addition of CH_2 units. This results in a *homologous series*, one in which each compound differs from the preceding one in the same way, in this series one CH_2 unit. As seen from the structures in Table 15.1, this series might also be thought of as being formed by the successive replacement of an end hydrogen atom by a —CH_3 or methyl group. Alkanes are called *saturated* hydrocarbons because all the C—C bonds are single bonds.

Table 15.1

Homologous series of straight-chain (normal, *n*) alkanes

name	structural formula	condensed structural formula	molecular formula, C_nH_{2n+2}
methane	$H-\overset{\displaystyle H}{\underset{\displaystyle H}{C}}-H$	CH_4	CH_4
ethane	$H-\overset{\displaystyle H}{\underset{\displaystyle H}{C}}-\overset{\displaystyle H}{\underset{\displaystyle H}{C}}-H$	CH_3CH_3	C_2H_6
propane	$H-\overset{\displaystyle H}{\underset{\displaystyle H}{C}}-\overset{\displaystyle H}{\underset{\displaystyle H}{C}}-\overset{\displaystyle H}{\underset{\displaystyle H}{C}}-H$	$CH_3CH_2CH_3$	C_3H_8
n-butane	$H-\overset{\displaystyle H}{\underset{\displaystyle H}{C}}-\overset{\displaystyle H}{\underset{\displaystyle H}{C}}-\overset{\displaystyle H}{\underset{\displaystyle H}{C}}-\overset{\displaystyle H}{\underset{\displaystyle H}{C}}-H$	$CH_3CH_2CH_2CH_3$	C_4H_{10}
n-pentane			C_5H_{12}
n-hexane			C_6H_{14}
n-heptane			C_7H_{16}
n-octane			C_8H_{18}
n-nonane			C_9H_{20}
n-decane			$C_{10}H_{22}$

Neither the name "straight-chain" nor the two-dimensional structural formulas given for the alkanes in Table 15.1 should be interpreted as giving the actual three-dimensional configurations of the molecules. In the first place, all the bond angles in a saturated hydrocarbon are 109°28′, which is characteristic of the single sp^3 hybrid bond. Also, the entire molecule can assume any number of twisted and kinked conformations. The representation in the table shows only that the carbon atoms are linked end to end.

Isomers Isomers are molecules having the same molecular formula (composition) but differing in at least one chemical or physical property. The most common type of isomerism is structural isomerism, which results from the different order in which atoms are bonded to one another. Among the alkanes, structural isomerization becomes possible when butane is reached. The two possible isomers of butane, both with the formula C_4H_{10}, are

$$CH_3CH_2CH_2CH_3 \qquad \text{n-butane (butane)}$$
$$CH_3CHCH_3 \qquad \text{isobutane (methylpropane)}$$
$$| $$
$$CH_3$$

The names given are the common and systematic (in parentheses) names. The isomers of pentane are

$$CH_3CH_2CH_2CH_2CH_3 \qquad \text{n-pentane (pentane)}$$
$$CH_3CHCH_2CH_3 \qquad \text{isopentane (2-methylbutane)}$$
$$|$$
$$CH_3$$

$$\begin{array}{c} CH_3 \\ | \\ CH_3{-}C{-}CH_3 \\ | \\ CH_3 \end{array} \qquad \text{neopentane (2,2-dimethylpropane)}$$

Hexane has five structural isomers, one of which is

$$\begin{array}{c} CH_3 \quad CH_3 \\ | \qquad | \\ CH_3CH{-}CHCH_3 \end{array} \qquad \text{(2,3-dimethylbutane)}$$

The systematic names illustrate how branched-chain alkane hydrocarbons are named by choosing the longest unbranched chain as the parent hydrocarbon and numbering the carbons of this long chain so that the groups attached to the chain have the lowest set of numbers. If the attached groups have only C and H atoms, they are alkyl groups with names ending in -yl.

methyl	$-CH_3$
ethyl	$-CH_2CH_3$
n-propyl	$-CH_2CH_2CH_3$
isopropyl	$-CH\big\langle{}^{CH_3}_{CH_3}$

A more complex example is

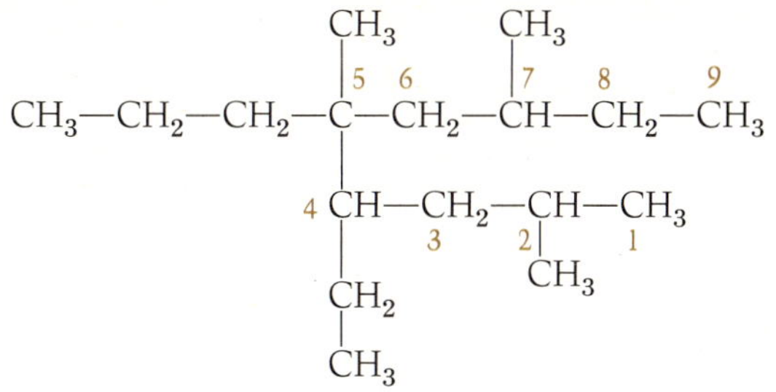

$$CH_3 \quad\quad CH_3$$

$$CH_3-CH_2-CH_2-\overset{\underset{|}{5}\quad 6}{C}-CH_2-\overset{7}{CH}-\overset{8}{CH_2}-\overset{9}{CH_3}$$

$$4\ \overset{}{CH}-CH_2-\overset{3}{CH}-\overset{1}{CH_3}$$

$$CH_2$$

$$CH_3$$

4-ethyl-2,5,7-trimethyl-5-*n*-propylnonane

Complete rules for the systematic naming of organic compounds can be found in any organic chemistry text.

Properties of Alkane Hydrocarbons

Chemically, the saturated hydrocarbons are quite unreactive at ordinary temperatures. For this reason, they are often given the name *paraffin* (little affinity) hydrocarbons. Their most characteristic and valuable reaction is combustion with oxygen. If the oxidation is complete, the products are carbon dioxide and water. The alkanes are the main components of petroleum and natural gas fuels and supply most of the energy for today's civilization.

The alkanes are all nonpolar molecules and are completely insoluble in water for all practical purposes. Physically, the alkanes from C_1 to C_4 are gases at normal conditions, C_5–C_{17} are liquids, and C_{18-on} are solids. Boiling points decrease with increase in chain branching among isomers.

Alkanes do not appear as such in living systems. However, most other organic compounds can be considered to be *formally* derived from alkanes by the replacement of one or more hydrogen atoms by more reactive, or functional, groups. These groups will be discussed shortly.

15.4 UNSATURATED ALIPHATIC HYDROCARBONS

Alkene Hydrocarbons

The alkenes have one or more carbon–carbon double bonds. The parent compound is ethylene, $CH_2=CH_2$. The ending *-ene* is characteristic of all members of the alkene series, just as the *-ane* ending designates corresponding members of the alkane series. Some of the simpler linear alkenes are listed in Table 15.2. Alkenes are also called *olefins*.

The double bond in alkenes may be thought of as resulting from the removal of two hydrogen atoms from adjacent carbons:

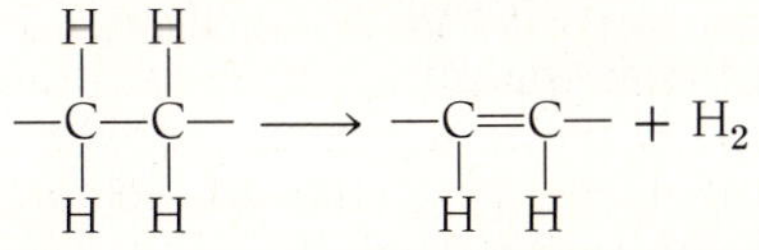

$$
\begin{array}{c}
\text{H} \quad \text{H} \\
| \quad\ | \\
-\text{C}-\text{C}- \\
| \quad\ | \\
\text{H} \quad \text{H}
\end{array}
\longrightarrow
\begin{array}{c}
\\
-\text{C}=\text{C}- \\
| \quad\ | \\
\text{H} \quad \text{H}
\end{array}
+ \text{H}_2
$$

The alkenes are unsaturated hydrocarbons. The term "unsaturated" indicates the presence of carbon–carbon double or triple bonds in the molecule. On the basis of the above reaction, unsaturated compounds were originally described as not containing the maximum number of hydrogen atoms.

Structural isomers of the alkenes can arise by varying the placement of the double bond and by varying the placement of alkyl groups. This is shown in Table 15.2 for three butene isomers. Other alkenes may have two, three, or more double bonds; they are called dienes, trienes, . . ., or *polyenes*, as

Table 15.2
Some simple alkene
hydrocarbons

name		structural formula	condensed structural formula	molecular formula, C_nH_{2n}
common	systematic			
ethylene	ethene	H H | | C=C | | H H	$CH_2=CH_2$	C_2H_4
propylene	propene	H H H | | | C=C—C—H | | H H	$CH_2=CHCH_3$	C_3H_6
α-butylene	1-butene		$CH_2=CHCH_2CH_3$	C_4H_8
β-butylene	2-butene		$CH_3CH=CHCH_3{}^a$	C_4H_8
isobutylene	2-methyl-1-propene		$CH_2=\overset{\overset{\displaystyle CH_3}{\textstyle\|}}{\underset{2}{C}}-\underset{3}{CH_3}$ 1	C_4H_8

^{a}Only structural isomers are considered. Another type of isomerism, called geometric isomerism, occurs with 2-butene. In geometric isomers the sequence of atoms are identical, but they differ in their arrangement in space. The two geometric isomers of 2-butene are

$$
\begin{array}{ccc}
\overset{\displaystyle H}{\underset{\displaystyle H_3C}{}}\!\!\diagup\!\!C=C\!\!\diagdown\!\!\overset{\displaystyle H}{\underset{\displaystyle CH_3}{}} & \text{and} & \overset{\displaystyle H}{\underset{\displaystyle H_3C}{}}\!\!\diagup\!\!C=C\!\!\diagdown\!\!\overset{\displaystyle CH_3}{\underset{\displaystyle H}{}}
\end{array}
$$

cis-2-butene trans-2-butene

Two distinct isomers exist because conversion of one to the other would require the rotation of one H—C—CH$_3$ group by 180° about the double bond. Such rotation is restricted by an energy barrier arising from the necessity of breaking the π-type p atomic orbital overlap during the rotation. In other words, the π part of the double bond must be broken (see Figures 3.26 and 15.11).

a group. Conjugation (alternation) of the double bonds can lead to π electron delocalization and increased stability of the molecule.

Properties of Alkenes

The physical properties of alkenes are similar to those of the alkanes, but the chemical reactivity of the alkenes is considerably greater than the alkanes because of the unsaturated nature of the double bond. The carbon atoms joined by the double bond are hybridized in sp^2 (trigonal) fashion. The lobes of one of the sp^2 orbitals of each atom overlap to form a σ bond, while the unhybridized p atomic orbitals on each carbon overlap to form the π bond that constitutes the remainder of the double bond. Although the entire double (σ + π) bond is stronger than a single (σ) C—C bond, the π bond portion of the C=C bond is considerably weaker than the σ portion. The π electron pair can often form a stronger σ-type bond with an electron-seeking reagent such as H$^+$. The hydrocarbon acts as a Lewis base; that is, it is an electron pair donor (Section 10.2). The result is the conversion of the double (σ + π) bond into two single (σ) bonds, a C—C and a C—H bond. The other carbon atom completes its octet by reaction with an electron pair donor, such as Cl$^-$, originally associated with the added hydrogen. This type of addition reaction is the major characteristic of the double bond.

Alkyne Hydrocarbons   Aliphatic hydrocarbons containing carbon–carbon triple bonds are called alkynes. The two triply bonded carbon atoms are hybridized in *sp* (digonal) fashion, and the triple bond consists of a σ and two π portions. As with the alkenes, the multiple bond allows the addition of other reagents and is responsible for the reactivity of the alkynes. The industrially important acetylene (ethyne), $CH{\equiv}CH$, is the first member of the alkyne series (C_nH_{2n-2}).

15.5 AROMATIC HYDROCARBONS   The parent substance of the large class of aromatic hydrocarbons and their derivatives is benzene, C_6H_6. Benzene has a ring structure that may be represented as

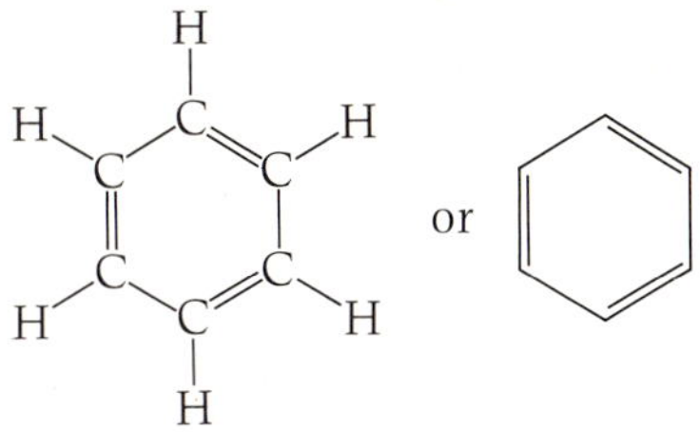

but that should not be taken literally to indicate alternating single and double bonds. Some examples of aromatic hydrocarbons are given in Figure 15.2.

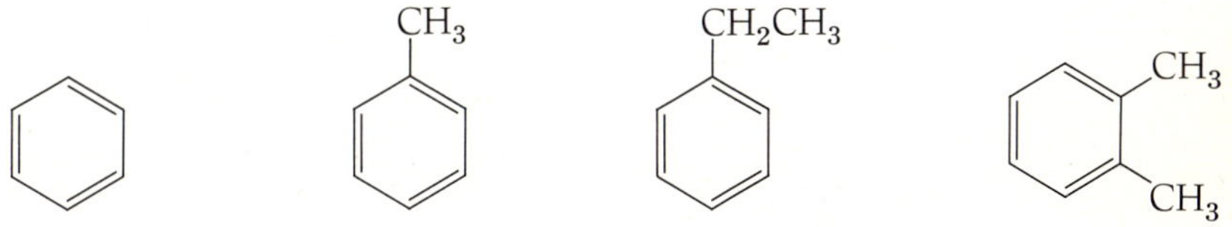

benzene, C_6H_6

toluene
(methylbenzene)
$C_6H_5CH_3$

ethylbenzene
$C_6H_5CH_2CH_3$

ortho-xylene
(1,2-dimethylbenzene)
$C_6H_4(CH_3)_2$

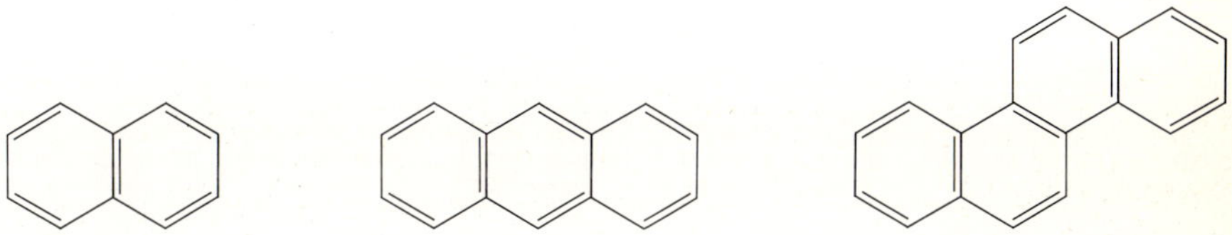

naphthalene

anthracene

1,2-benzophenanthrene
(1,2-benzopyrene)

Figure 15.2. Some aromatic hydrocarbons. Systematic names are in parentheses.

   The chemical properties of aromatic hydrocarbons are determined by the electronic structure of the benzene molecule. Despite the apparently unsaturated structure of benzene, addition reactions that result in the disappearance of double bonds are not characteristic of benzene. Rather, the characteristic reaction is the substitution of a hydrogen atom on the ring by another atom or group, as in the reaction

$$C_6H_6 + Cl_2 \longrightarrow C_6H_5Cl + HCl$$

The reason for this is found in the high stability of the conjugated π bond system of the benzene ring. Benzene is an example of a resonance hybrid, and the structure drawn above is only one contributing structure.

Electron Delocalization in Benzene

The concept of resonance or electron delocalization was introduced in Section 3.15 with the carbonate ion. Now let us investigate electron delocalization in benzene. Benzene has six trigonally hybridized carbon atoms at the apices of a planar hexagon. The geometry and the σ bonding are shown in Figure 15.3. Each carbon atom has three covalent σ-type bonds, two to adjacent carbons and one to a hydrogen atom. Now, remember that trigonal or sp^2 hybridization leaves one p atomic orbital alone. We shall call this the p_y orbital.

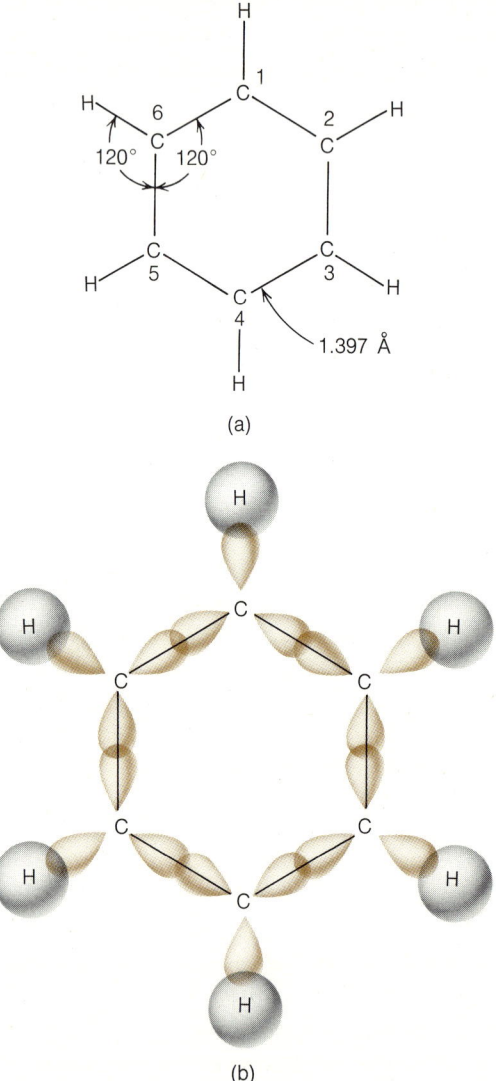

(a)

(b)

Figure 15.3. Geometry and σ bond overlap in benzene, C_6H_6.

Each carbon atom has its one unhybridized p_y orbital perpendicular to the plane containing the six carbons and the six hydrogens. The lobes of each of these p_y orbitals extend above and below the plane of the carbon hexagon. Each of these p_y orbitals contains 1 unpaired electron. With the assumption that the signs of all p_y orbitals are as shown in Figure 15.4, there are several ways in which they can be overlapped in pairs. Structure I shows the formation of π molecular orbitals between atoms 2 and 3, 4 and 5, and 6 and 1. Other p_y orbital overlapping schemes are shown for structures II, III, IV, and V.

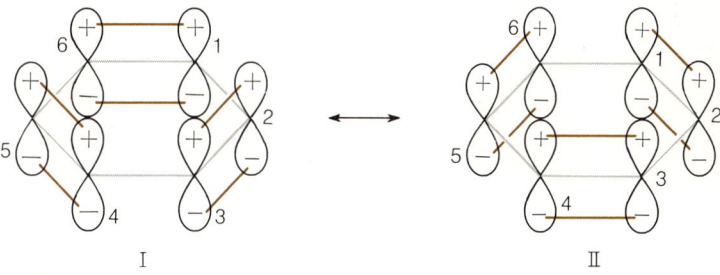

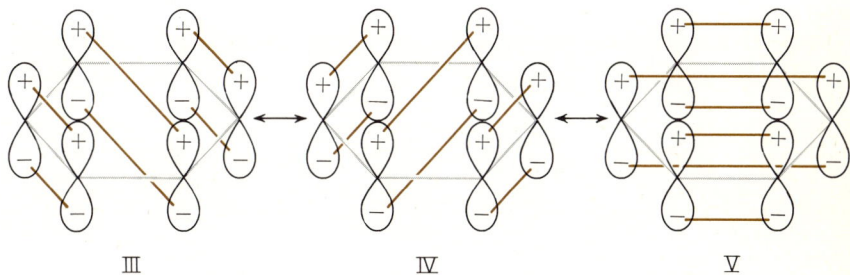

Figure 15.4. Pairwise overlapping of p_y atomic orbitals in benzene. The y axis is perpendicular to the plane of the ring.

Structures III–V are relatively unimportant because of the large distances between carbon atoms that are not adjacent. This distance means poor overlap, weak bonds, and a high-energy or relatively unstable structure. Structures I and II are the important ones, and they are equal in energy. Thus, it might appear that the benzene molecule could exist in one form having double bonds between atoms 6 and 1, 2 and 3, and 4 and 5, or in another form having double bonds between atoms 1 and 2, 3 and 4, and 5 and 6. The bond structures would be

As pointed out in Section 3.15, neither structure exists as such. In benzene, as in CO_3^{2-}, each carbon p_y orbital overlaps with both adjacent p_y orbitals to form one π molecular orbital encompassing all six carbon nuclei. This is the lowest energy or π molecular orbital for the benzene molecule. It holds 2 paired electrons—the other 4 p_y electrons go into two other bonding molecular orbitals of higher energy, which result from other combinations of the six p_y atomic orbitals. These three π-type molecular orbitals of the ground state are shown in Figure 15.5. In addition to these three bonding molecular orbitals, there are three other antibonding molecular orbitals (six atomic orbitals combine to give six molecular orbitals). The antibonding orbitals hold no electrons in the lowest energy state of the benzene molecule and do not concern us here.

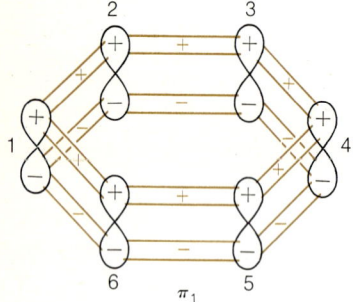

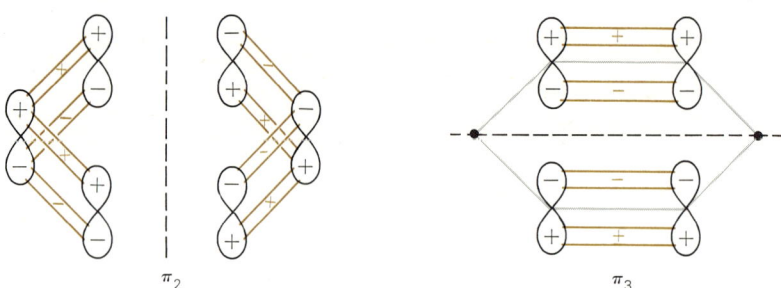

Figure 15.5. The three bonding molecular orbitals of benzene. Dashed lines indicate nodal planes. The lowest energy molecular orbital has no nodes between nuclei; π_2 and π_3 each have one nodal plane and are degenerate. The highest energy molecular orbital (not shown here) has six nodal planes.

The benzene molecule is chemically inert compared to a similar molecule, cyclohexatriene, which can be pictured but which does not exist:

cyclohexatriene   benzene

In cyclohexatriene, the bond orders would be 1 and 2 alternately. In benzene, the bond order of each C—C bond can be said to be about $1\frac{1}{2}$—somewhere between a pure single and a pure double bond. All C—C bonds in benzene are equivalent.

15.6 FUNCTIONAL GROUPS

Most organic compounds can be thought of as being derived from hydrocarbons by the replacement of one or more hydrogen atoms by another atom or group of atoms. The replacing groups are usually more chemically reactive than the hydrocarbon portion of the molecule and determine the characteristic chemistry of the derived compound. For this reason, these groups are called functional groups. Table 15.3 lists a number of the more important functional groups, the generic names of the compounds they form, and some specific examples. In this table, R stands for the aliphatic remainder or side chain of a molecule. In certain cases, it may also stand for hydrogen. Only compounds containing one functional group (other than alkyl) are listed. Many compounds contain more than one function group, and the groups may be the same or different.

Table 15.3
Some general classes of organic compounds and their functional groups

class	functional group	general compound formula	examples[a]
alkene	$\diagdown$C=C$\diagup$ double bond	$\underset{R'}{\overset{R}{\diagdown}}$C=C$\underset{R'''}{\overset{R''}{\diagup}}$	$CH_2{=}CHCH_3$, propylene (propene)
alcohol	—OH, hydroxyl	ROH	CH_3OH, methyl alcohol (methanol) C_2H_5OH, ethyl alcohol (ethanol)
aldehyde	$\overset{O}{\overset{\|}{-}}$CH, carbonyl	$\overset{O}{\overset{\|}{R}}CH$	$H\overset{O}{\overset{\|}{C}}H$, formaldehyde (methanal) $CH_3\overset{O}{\overset{\|}{C}}H$, acetaldehyde (ethanal) $C_2H_5\overset{O}{\overset{\|}{C}}H$, propionaldehyde (propanal)
ketone	$-\overset{O}{\overset{\|}{C}}-$ carbonyl	$R\overset{O}{\overset{\|}{C}}R'$	$CH_3\overset{O}{\overset{\|}{C}}CH_3$, acetone or dimethyl ketone (2-propanone) $CH_3\overset{O}{\overset{\|}{C}}C_2H_5$, methyl ethyl ketone (butanone)
carboxylic acid	$-\overset{O}{\overset{\|}{C}}{-}OH \rightleftharpoons -\overset{O}{\overset{\|}{C}}{-}O^-$	$R\overset{O}{\overset{\|}{C}}OH$	HCOOH, formic acid CH_3COOH, acetic acid (ethanoic acid) C_2H_5COOH, propionic acid (propanoic acid)
ester	$-\overset{O}{\overset{\|}{C}}{-}O-$ ester linkage	$R\overset{O}{\overset{\|}{C}}{-}O{-}R'$	$CH_3\overset{O}{\overset{\|}{C}}{-}O{-}C_2H_5$, ethyl acetate
ether	—O— ether linkage	R—O—R' or RCH_2—O—CH_2R'	CH_3—O—C_2H_5, methyl ethyl ether C_2H_5—O—C_2H_5, ethyl ether or diethyl ether (ethoxyethane)
amide	$-\overset{O}{\overset{\|}{C}}{-}NH_2$, amide	$R\overset{O}{\overset{\|}{C}}{-}NH_2$	$CH_3\overset{O}{\overset{\|}{C}}{-}NH_2$, acetamide

Table 15.3
Some general classes of organic compounds and their functional groups (*Continued*)

class	functional group	general compound formula	examples[a]
amine	$-NH_2 \rightleftharpoons -NH_3^+$, amino	RNH_2, primary amine	$C_2H_5NH_2$, ethylamine $C_6H_5NH_2$, aniline
		$\begin{matrix} R \\ \diagdown \\ \quad NH, \text{ secondary amine} \\ \diagup \\ R' \end{matrix}$	
		$\begin{matrix} R \\ \diagdown \\ R'-N, \text{ tertiary amine} \\ \diagup \\ R'' \end{matrix}$	

phenol
  (an aromatic
  alcohol)
	Ar—OH
	"Ar" = aromatic residue
		Ar—OH	C_6H_5OH, phenol, carbolic acid
			  (hydroxybenzene)
			
			OH—⬡—OH, hydroquinone
			  (1,4-dihydroxybenzene)

mercaptan	—SH sulfhydryl or thiol	R—SH	C_2H_5SH, ethyl mercaptan (ethanethiol)
disulfide	—S—S— disulfide bridge	R—S—S—R′	CH_3SSCH_3, dimethyl disulfide

[a] Names of examples include the common name and, in parentheses, the systematic name.

Reactions of Organic Compounds

Organic reactions differ from the reactions of inorganic compounds in two respects; organic reactions are generally slower at ordinary temperatures and, unless catalyzed by highly specific enzymes, organic reactions can exhibit a variety of reaction paths and products. There are a great many organic reactions, but they may be gathered into a smaller number of general reaction classes. We have already mentioned two general reaction types, addition and substitution. Another major reaction class is *elimination*. These classes are illustrated by simple examples.

1. Substitution. Replacement of one group or atom attached to a carbon atom by another group or atom:

$$Cl_2 + CH_3CH_3 \xrightarrow{h\nu \text{ or } \Delta} CH_3CH_2Cl + HCl$$

2. Addition. An increase in the number of groups attached to multiply bonded carbon atoms in a reactant:

$$CH_3-CH=CH_2 + HCl \longrightarrow CH_3-\underset{\underset{Cl}{|}}{CH}-CH_3$$

3. Elimination. Removal of a group attached to a carbon with an increase in the degree of unsaturation:

$$CH_3{-}CH_2{-}OH \xrightarrow{H_2SO_4, \Delta} CH_2{=}CH_2 + H_2O$$

In the next few chapters, we shall talk about some of the chemical reactions going on inside living organisms. The classification of these reactions as substitution, addition, and elimination, although technically correct, is not always useful. Instead, we shall classify organic reactions in living systems according to the type of enzymes involved. While there are some 700 different enzymes known, they fall into six major classes based on the type of reaction they catalyze. Therefore, if we are interested in biological reactions, these six classes will include all the reaction types we are likely to encounter.

The six enzyme classes and the general reaction types are listed below. Following this list are selected examples of reactions included in these broad reaction classifications.

1. Oxidoreductases catalyze oxidation–reduction reactions.

2. Transferases catalyze reactions involving the transfer of a functional group from one molecule to another molecule.

3. Hydrolases catalyze hydrolysis reactions.

4. Lyases catalyze reactions involving the addition of groups to C$=$C double bonds or the removal (elimination) of groups to produce a double bond.

5. Isomerases catalyze isomerization (rearrangement) reactions.

6. Ligases or synthetases catalyze the cleavage of a molecule or the joining (condensation) of two molecules.

There are other classification schemes for enzymes that would have served our purposes just as well, but this is now the most preferred, more modern classification. In the following discussion of specific reactions in each class, only the overall reaction is given. The illustrative reactions are not necessarily biological, enzyme-catalyzed, reactions; they are cited only as simple illustrations of various reaction types. Note that some of the reactions could have been placed under different classifications.

15.7 OXIDATION–REDUCTION REACTIONS

Redox reactions involving organic compounds are best viewed from the simpler and older definitions of oxidation and reduction: oxidation as the addition of oxygen or the removal of hydrogen, and reduction as the removal of oxygen or the addition of hydrogen. Electron transfer is not immediately evident in organic reactions because of the common absence of ions and because oxidation states of carbon are not commonly used in organic chemistry. Most biological redox reactions involve the transfer of hydrogen. The general reaction can be written

$$AH_2 + B \rightleftharpoons A + BH_2$$

Here B, the oxidizing agent, is the hydrogen *acceptor*. This reaction can be thought of as the transfer of two hydrogen atoms with their 2 electrons from A to B so that, in effect, A has lost 2 electrons and has been oxidized. Specific examples of organic oxidations are the oxidations of alcohols to aldehydes and of aldehydes to acids.

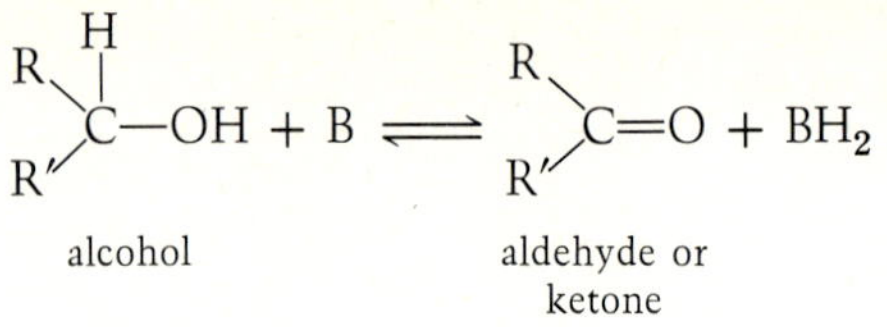

$$R\diagdown\underset{R'}{\overset{H}{C}}-OH + B \rightleftharpoons \underset{R'}{\overset{R}{C}}=O + BH_2$$

alcohol　　　　　　aldehyde or
　　　　　　　　　　ketone

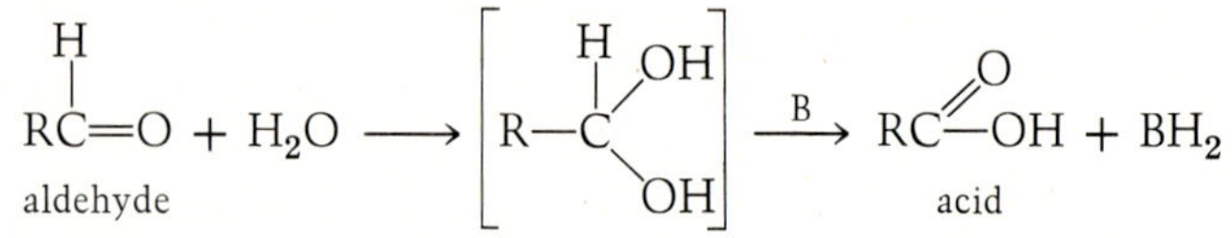

$$\underset{H}{RC}=O + H_2O \longrightarrow \left[R-\underset{OH}{\overset{H \diagup OH}{C}} \right] \xrightarrow{B} RC\diagup_{OH}^{O} + BH_2$$

aldehyde　　　　　　　　　　　　　　　acid

Redox reactions such as this are also called *dehydrogenation reactions*. Besides being redox reactions, they are also elimination-type reactions. The hydrogen acceptor B may be oxygen itself, in which case BH_2 is water.

The following complete oxidation sequence from an alkane to carbon dioxide and water illustrates that oxidation numbers can, if desired, be applied to carbon in organic compounds.

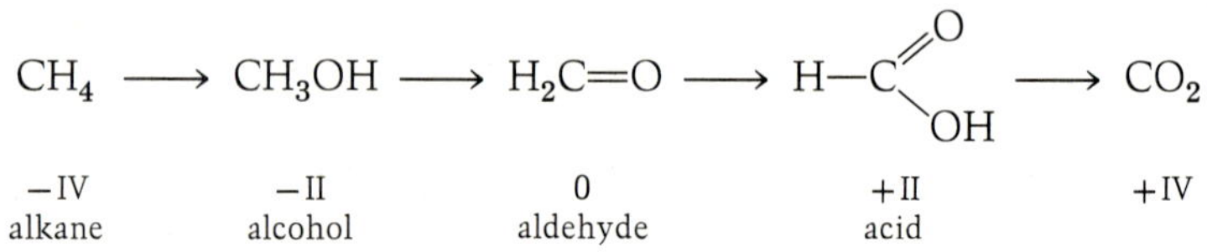

$$CH_4 \longrightarrow CH_3OH \longrightarrow H_2C=O \longrightarrow H-C\diagup_{OH}^{\diagup O} \longrightarrow CO_2$$

　$-IV$　　　　$-II$　　　　　0　　　　　　$+II$　　　　　$+IV$
　alkane　　　alcohol　　　aldehyde　　　acid

Exercise. Verify the assignment of oxidation states in the above sequence by using electron dot structures of the molecules.

15.8 GROUP TRANSFER REACTIONS

Group transfer reactions are reactions of the general type

$$RB + C \rightleftharpoons RC + B$$

where C is the group transferred. Since C may be originally attached to some other group, say R′, the general reaction could also be written as

$$RB + CR' \rightleftharpoons RC + BR'$$

Some commonly transferred groups are

phosphoryl　　　$-\underset{OH}{\overset{\overset{O}{\|}}{P}}-OH$

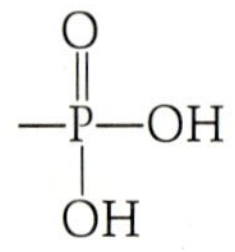

amino　　　$-NH_2$

acyl　　　$-\overset{\overset{O}{\|}}{C}-R$

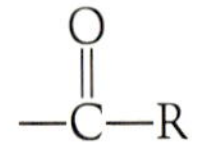

glycosyl　　　$-CH-O\diagdown_R$

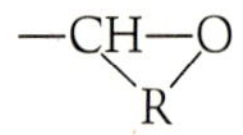

Illustrations are

$$RO \overset{O}{\underset{OH}{\overset{\|}{-}P}} -OH + HOR' \rightleftharpoons ROH + HO-\overset{O}{\underset{OH}{\overset{\|}{P}}} -OR'$$

phosphate ester   alcohol   alcohol   phosphate ester

$$R \overset{}{\underset{H}{-}C}=O + NH_2CH_2R' \underset{\text{transamination}}{\rightleftharpoons} RCH_2NH_2 + R'\overset{}{\underset{H}{C}}=O$$

aldehyde   amine   amine   aldehyde

$$R-S-\overset{O}{\overset{\|}{C}}-CH_3 + HSR' \rightleftharpoons RSH + CH_3-\overset{O}{\overset{\|}{C}}-SR'$$

thiol ester   mercaptan   mercaptan   thiol ester

$$R'-CH-O + R'''H \rightleftharpoons R'H + R'''-CH-O$$
$$\quad\quad\ \backslash_{R''}\qquad\qquad\qquad\qquad\qquad \backslash_{R''}$$

glycoside   glycoside

Reactions in which the hydroxyl group is transferred from water constitute a separate class of enzyme reactions and will be considered next. Most of the reactions in metabolism are group transfer reactions.

15.9 HYDROLYSIS REACTIONS

Water is the only biological solvent, and hydrolysis reactions involve the decomposition of a molecule by reaction with water. The term hydrolysis as applied to this reaction has a somewhat broader meaning than when it was used in Sections 10.6 and 10.13. Hydrolysis reactions may be further subdivided by the type of bond attacked in the hydrolyzed molecule.

1. Peptide bond in proteins:

$$RC \overset{O}{\overset{\|}{-}}NH-R' + H_2O \rightleftharpoons RC \overset{O}{\overset{\|}{-}}OH + R'NH_2$$

amide or protein   acid   amine

2. Glycoside bond in saccharides (see Section 15.18):

$$RC-O-CR' + H_2O \rightleftharpoons RCOH + R'COH$$
$$\overset{|}{H}\quad\ \ \overset{|}{H}\qquad\qquad\qquad \overset{|}{H}\qquad\ \ \overset{|}{H}$$

sugar-O-sugar   sugar   sugar
(disaccharide)   (monosaccharides)

3. Ester linkage:

$$RC \overset{O}{\overset{\|}{-}}OR' + H_2O \rightleftharpoons RC \overset{O}{\overset{\|}{-}}OH + R'OH$$

ester   acid   alcohol

$$R-O \!\!\not{\,}\!\! P-OH + H_2O \rightleftharpoons ROH + HO-P-OH$$

$$\underset{\text{phosphate ester}}{} \qquad \underset{\text{alcohol}}{} \quad \underset{\text{phosphoric acid}}{}$$

$$RO-P-OR' + H_2O \rightleftharpoons RO-P-OH + R'OH$$

$$\underset{\text{phosphate diester}}{} \qquad \underset{\text{phosphate ester}}{} \quad \underset{\text{alcohol}}{}$$

The same enzymes that catalyze the three hydrolysis reactions of part 3 also catalyze the reverse *esterification* reactions, since all of these reactions are reversible.

15.10 ADDITION TO DOUBLE BONDS OR ELIMINATION TO FORM DOUBLE BONDS

1. Decarboxylation (elimination of CO_2):

$$\underset{\text{acid}}{RCOOH} \rightleftharpoons RH + CO_2$$

$$\underset{\text{pyruvic acid}}{CH_3-C-C-OH} \rightleftharpoons \underset{\text{acetaldehyde}}{CH_3CH} + O=C=O$$

$$\underset{\text{amino acid}}{R-CH-C-OH} \rightleftharpoons \underset{\text{amine}}{R-CH_2} + O=C=O$$

$$\underset{\text{pyruvic acid}}{H_3C-C-C-OH} + H_2O + B \rightleftharpoons \underset{\text{acetic acid}}{H_3C-C-OH} + BH_2 + CO_2$$

$$\underset{\substack{\text{oxidative} \\ \text{decarboxylation}}}{}$$

2. Dehydration (elimination of water):

$$\underset{\text{alcohol}}{R_2C-C-OH} \rightleftharpoons \underset{\text{alkene}}{R_2C=C} + H_2O$$

15.11 ISOMERIZATION REACTIONS

Certain enzymes catalyze the interconversion of isomers. Such reactions may involve structural isomers, but we shall use as our example a more common situation dealing with optical isomers. Optical isomers, like geometric isomers, arise because of different arrangements of atoms in space. Consider the isomerization of the amino acid alanine.

$$
\begin{array}{c}
\text{COOH} \\
| \\
\text{H—C*—NH}_2 \\
| \\
\text{CH}_3
\end{array}
\quad\rightleftharpoons\quad
\begin{array}{c}
\text{COOH} \\
| \\
\text{H}_2\text{N—C*—H} \\
| \\
\text{CH}_3
\end{array}
$$

D-alanine   L-alanine

The two forms of alanine are called optical isomers because they rotate the plane of polarized light in opposite directions. Optical isomerism usually occurs when a carbon atom (the starred atom in alanine) in a molecule is bonded to four different atoms or groups of atoms. Optical isomers are mirror images of one another, but they are not superimposable. (The dashed vertical line between the two alanines represents the mirror or plane of the reflection.) Your right and left hands exhibit the same type of symmetry relationship. The different structures of the two alanines are made clearer if three-dimensional representations are compared as in Figure 15.6. Almost all the

Figure 15.6. Optical isomers of alanine. The two isomers are mirror images of one another. A common convention, illustrated here, is to let a solid line indicate a bond in the plane of the paper, a dashed or dotted line indicate a bond behind the plane of the paper, and a solid triangular bond indicate a bond directed in front of the plane of the paper toward the viewer. The D and L notation says nothing about the direction of rotation of the plane of polarized light by either isomer (see Section 15.17).

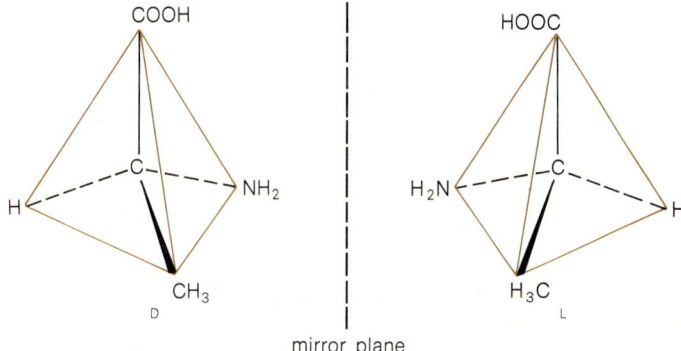

mirror plane

amino acids synthesized by living organisms have the L configuration (on earth, at least), and all naturally occurring amino acids in proteins are L. Laboratory syntheses of amino acids produce an optically inactive mixture of D and L forms. Somewhere in the course of the chemical evolution of life, the L form of amino acid was chosen, to the best of our knowledge by chance, and essentially all biochemical reactions and biological structures of amino acids have since been restricted to this configuration.

15.12 CONDENSATION REACTIONS

At this point, we consider very briefly only one type of condensation reaction, the formation of certain simple polymers. In the next section, we shall see that the natural polymers, proteins, carbohydrates, and nucleic acids, can all be regarded as polymers formed in condensation reactions.

A polymer is a large molecule built up by repetition of simple chemical units derived from some small *monomer* units. In a condensation polymer, the smaller monomer units link together with the elimination of water or some other small molecule. The monomer units must have at least two functional groups. For instance, an alcohol with two hydroxyl groups can condense repeatedly,

$$\text{HO—R}\boxed{\text{OH + H}}\text{O—R—OH} \longrightarrow$$
$$\text{HO—R—(O—R)—OH} + H_2O$$

$$\text{HO—R—O—R}\boxed{\text{OH + H}}\text{O—R—OH} \longrightarrow$$
$$\text{HO—R—(O—R)—(O—R)—OH} + H_2O$$

to give, after n steps,

$$\text{HO—R(—O—R—)}_n\text{OH}$$

where (—O—R—) is the repeating unit.

Amino acids, with the general formula

$$\text{H}_2\text{N—CH—C}\overset{\displaystyle O}{\underset{\displaystyle OH}{}}$$

$$\underset{R}{}$$

can similarly condense, with the elimination of water:

$$\text{H}_2\text{N—CH—C} \quad + \quad \text{N—CH—C} \longrightarrow$$

$$\text{H}_2\text{N—CH—C—NH—CH—C} + H_2O$$

The condensation reactions here might also be viewed as a dehydration reaction, the reverse of reaction 1 in Section 15.9.

The Molecules of Life

On the molecular level, the organization and complexity of living systems is indicated by the structural complexity of the large organic polymers or macromolecules always found in these systems. These biological molecules have molecular weights in the range 10^3–10^9. Yet we find also that the simplicity and economy of nature is reflected in these components of life. Living systems use six of the chemical elements to build a relatively small number of simple molecules which, in turn, serve as the building blocks of biological polymers. The total number of giant molecules is very large, but they fall into four major classes: proteins, carbohydrates or polysaccharides, nucleic acids, and lipids. The remainder of this chapter describes the chemical constitution and molecular configuration of the molecules in each class. The structure of each molecule determines its function in living systems.

Proteins

Proteins are the most important and the most abundant of the polymers found in living organisms. All proteins are built up of α-amino acid molecules, which have the general structure

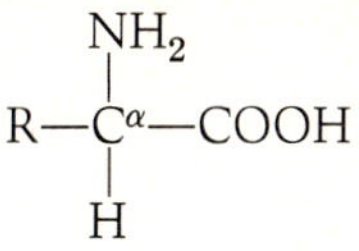

$$R-\overset{\displaystyle \overset{NH_2}{|}}{\underset{\displaystyle \underset{H}{|}}{C^\alpha}}-COOH$$

15.13 AMINO ACIDS The α-amino acids have the amino group attached to the same carbon atom (the α-carbon) as the carboxyl (—COOH) group. It is possible to produce amino acids in the laboratory in which the amino group is farther away from the carboxyl group, but the only amino acids found in natural protein molecules are the α-amino acids. Twenty different α-amino acids, in various combinations, are commonly found in proteins. Except for proline, the naturally occurring amino acids differ only in the nature of the R group or amino acid side chain attached to the common $H_2N-CH-COOH$ portion. The 20 α-amino acids are listed in Figure 15.7 together with their common abbreviations.

Figure 15.7. Amino acids.

I. Nonpolar R groups	**I. Nonpolar R groups (continued)**
alanine (ala)	phenylalanine (phe)
valine (val)	tryptophan (try)
leucine (leu)	methionine (met)
isoleucine (ileu)	glycine (gly)
proline (pro)	**II. Polar R groups**
	serine (ser)

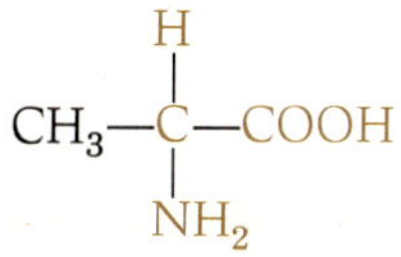

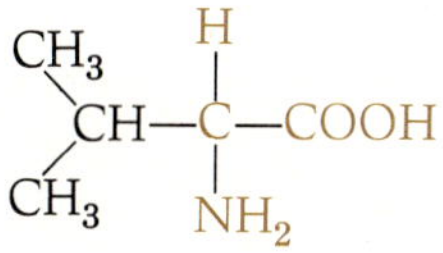

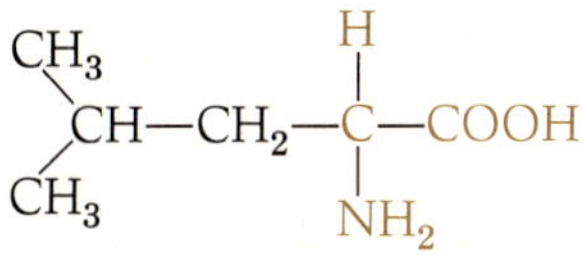

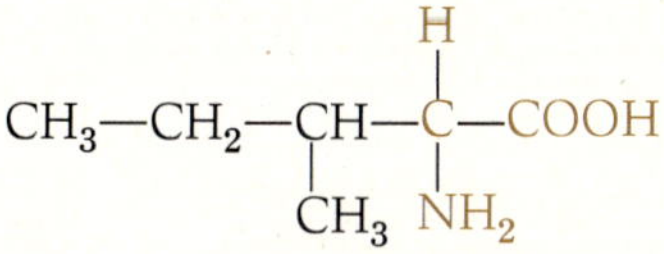

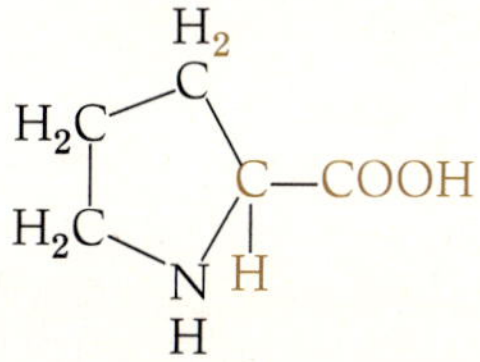

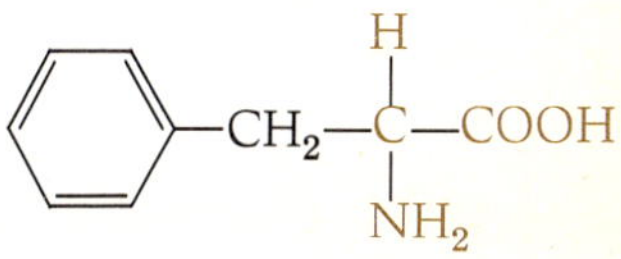

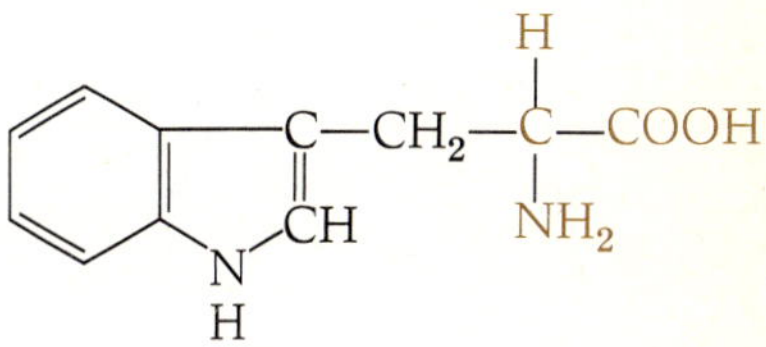

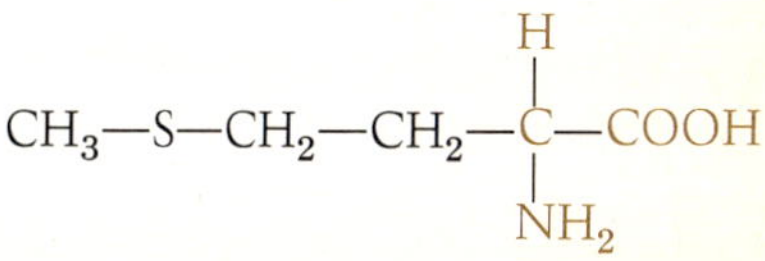

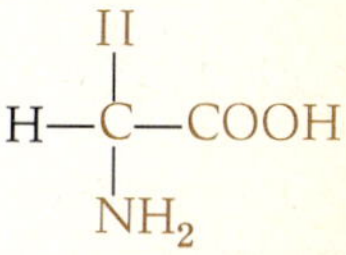

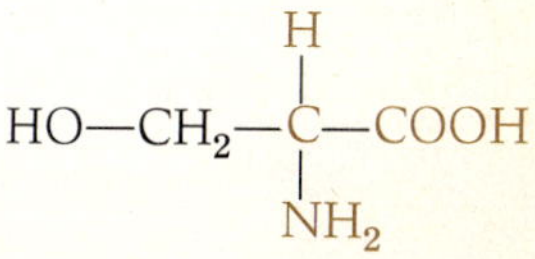

Figure 15.7. (*Continued*)

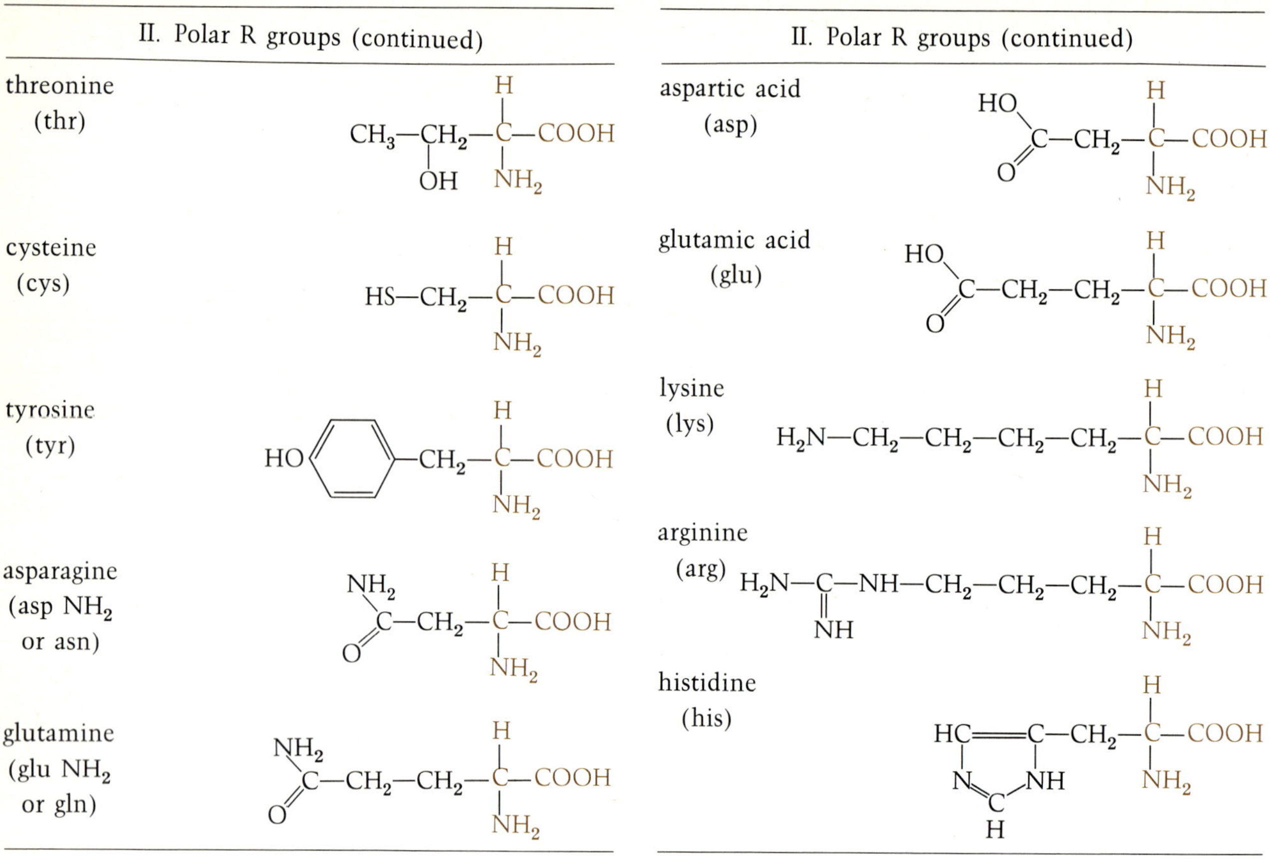

II. Polar R groups (continued)	II. Polar R groups (continued)

threonine
(thr)

$$CH_3{-}CH_2{-}\underset{\underset{OH}{|}}{\overset{\overset{H}{|}}{C}}{-}COOH \quad (NH_2)$$

aspartic acid
(asp)

$$\underset{O}{\overset{HO}{\diagdown}}C{-}CH_2{-}\underset{\underset{NH_2}{|}}{\overset{\overset{H}{|}}{C}}{-}COOH$$

cysteine
(cys)

$$HS{-}CH_2{-}\underset{\underset{NH_2}{|}}{\overset{\overset{H}{|}}{C}}{-}COOH$$

glutamic acid
(glu)

$$\underset{O}{\overset{HO}{\diagdown}}C{-}CH_2{-}CH_2{-}\underset{\underset{NH_2}{|}}{\overset{\overset{H}{|}}{C}}{-}COOH$$

tyrosine
(tyr)

$$HO{-}\bigcirc{-}CH_2{-}\underset{\underset{NH_2}{|}}{\overset{\overset{H}{|}}{C}}{-}COOH$$

lysine
(lys)

$$H_2N{-}CH_2{-}CH_2{-}CH_2{-}CH_2{-}\underset{\underset{NH_2}{|}}{\overset{\overset{H}{|}}{C}}{-}COOH$$

asparagine
(asp NH_2
or asn)

$$\underset{O}{\overset{NH_2}{\diagup}}C{-}CH_2{-}\underset{\underset{NH_2}{|}}{\overset{\overset{H}{|}}{C}}{-}COOH$$

arginine
(arg)

$$H_2N{-}\underset{\underset{NH}{\|}}{C}{-}NH{-}CH_2{-}CH_2{-}CH_2{-}\underset{\underset{NH_2}{|}}{\overset{\overset{H}{|}}{C}}{-}COOH$$

glutamine
(glu NH_2
or gln)

$$\underset{O}{\overset{NH_2}{\diagup}}C{-}CH_2{-}CH_2{-}\underset{\underset{NH_2}{|}}{\overset{\overset{H}{|}}{C}}{-}COOH$$

histidine
(his)

$$\underset{\underset{H}{\overset{|}{C}}}{\underset{N\diagdown\diagup NH}{HC{=\!=\!=}C}}{-}CH_2{-}\underset{\underset{NH_2}{|}}{\overset{\overset{H}{|}}{C}}{-}COOH$$

Acid–Base Behavior of Amino Acids

Figure 15.7 shows the un-ionized forms of amino acids. In aqueous solution, however, the amino acids always exist as ions because of the ionization of the basic —NH_2 group or the acidic —COOH group or both. Whether either or both of these groups exist in appreciable concentrations as —NH_3^+ or —COO^-, that is, the nature and extent of ionization, depends on the pH of the medium. If the pH is low, relatively more of the amino groups will accept protons to give —NH_3^+; if the pH is high, the carboxyl groups will tend to become —COO^- by loss of a proton, and the amino acid will assume a net negative ionic charge. At some intermediate pH, the extent of ionization of the amino and the carboxyl groups in an amino acid may be equal, in which case the acid will possess no net ionic charge and will not move in an electric field. This pH is called the *isoelectric point*. These points can be summarized by the equilibria

$$\underset{pH \simeq 1}{^+NH_3{-}\overset{\overset{R}{|}}{CH}{-}COOH} \rightleftharpoons \underset{pH \simeq 6}{^+NH_3{-}\overset{\overset{R}{|}}{CH}{-}COO^-} \rightleftharpoons \underset{pH \simeq 11}{NH_2{-}\overset{\overset{R}{|}}{CH}{-}COO^-}$$

At physiological pH's within the body (6–7), amino acids exist as dipolar ions.

Most amino acids have isoelectric points between 5.1 and 6.5. Those that do not, have additional acidic groups on their side chains (asp, glu) and have a net negative charge at pH = 6–7 or have additional basic groups (lys, arg, his) that ionize at pH = 6–7 to give a net positive charge. Thus, at pH = 6, the dicarboxylic amino acid asparagine has a net negative charge due to ionization of two carboxyl groups:

$$
\begin{array}{c}
COO^- \\
| \\
CH_2 \\
| \\
{}^+NH_3—CH—COO^-
\end{array}
$$

Condensation of Amino Acids   Amino acids condense with one another through the elimination of the elements of water between the carboxyl group of one acid and the α-amino group of another acid, as was illustrated in Section 15.12. The

$$
\begin{array}{c}
—NH—C— \\
\parallel \\
O
\end{array}
$$

linkage between condensed amino acid units is called a *peptide bond*. The condensation product, called a *peptide*, may contain two, three, or several amino acids, but it always has an amino group at one end and a carboxyl group at the other so that additional amino acids can be tacked on, head to tail. Continued condensation produces a *polypeptide* composed of many more amino acid units. The molecular backbone is

$$
—C^\alpha—C—N—C^\alpha—C—N—C^\alpha—C—N—
$$

or

$$
(—C^\alpha—C—N—)_n
$$

When the number of units n in the polypeptide chain reaches about 50, or a molecular weight of about 6000, a *protein* is generally considered to be formed, although, as we shall soon see, there is more to protein structure than a mere string of amino acid units.

Although a protein may consist of more than one polypeptide chain, we shall consider only single-chain proteins for the time being. Figure 15.8 represents a single-chain protein, the enzyme ribonuclease. Also shown are four disulfide bridges, connecting cystine residues in different parts of the 124-unit chain. The molecular weight is 13,700.

15.14 THE STRUCTURE OF PROTEINS   Protein molecules have several levels of organization, and each will be discussed separately.

Primary Structure   The order or sequence of the amino acid residues in a protein molecule is its *primary structure*. This order is exactly the same in any one kind of protein molecule, and each different protein molecule in any organism has a unique primary structure. The slightest variation from this order, for example by the substitution of one amino acid residue for another, may render the molecule useless for its biological role in the organism. Thus, the protein molecule insulin in humans always has the same sequence of 52 amino acids. The insulin molecule in other species, sheep for example, has almost the same

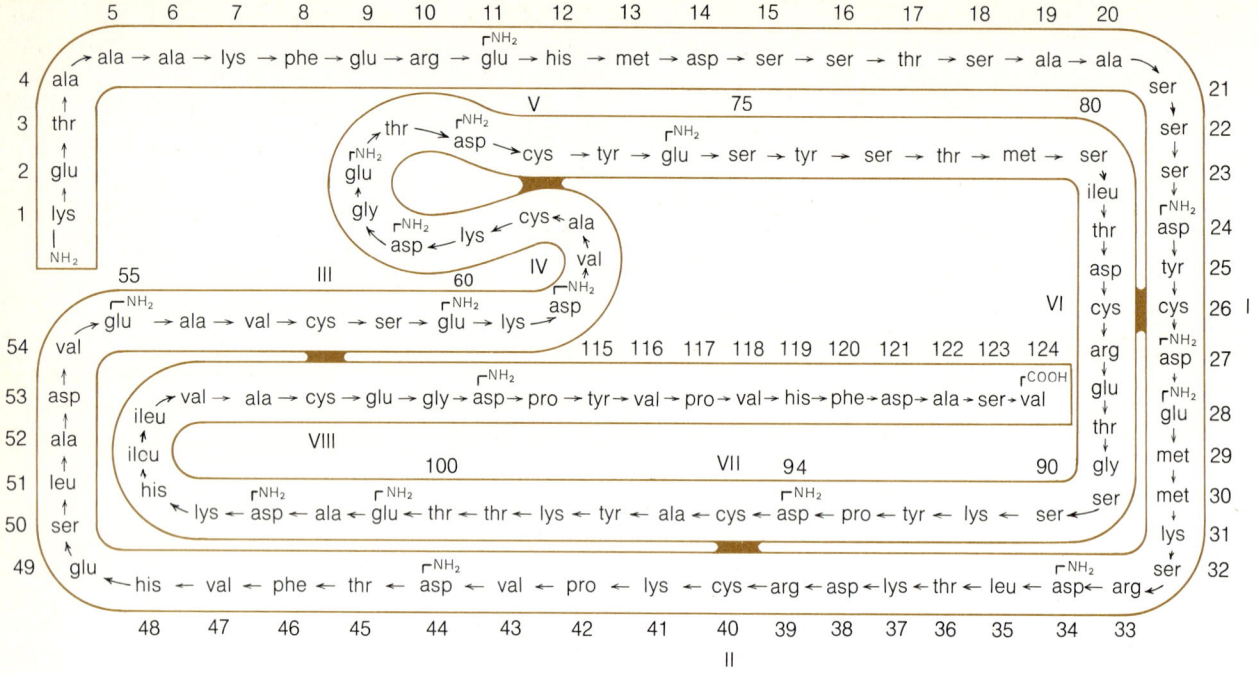

Figure 15.8. The primary structure of ribonuclease. The dark areas are covalent disulfide bridges between cysteine residues. [From D. G. Smyth, W. H. Stein, and S. Moore, *Journal of Biological Chemistry,* **238,** 227 (1963).]

arrangement of the 52 amino acids. Although there is only a difference in the positioning of four amino acid residues in the insulin molecules of the different species, sheep insulin will not function in the human body.

Each organism is endowed with the ability to construct its own protein molecules with the proper sequence of units. Mutations are expressed when an organism lacks the inherited information to synthesize a protein with the necessary primary structure. The chemistry of the organism may malfunction even if only one amino acid residue is out of place.

The primary structure of proteins is important because the specific sequence of amino acids determines the other structural characteristics of the molecule. And ultimately, the complete structure of the protein determines how well, if at all, the molecule can carry out its assigned task in a highly integrated living system.

Secondary Structure

The backbone or set of —C—C—N— segments of a polypeptide chain does not form a straight line because of the angles between the bonds. Neither does the entire molecular backbone lie in one plane. In the most important form of a protein molecule, the chain forms a spiral as though it were wound about the outside surface of a cylinder. Analogies are the threads of a screw or the coils in an extensible telephone cord. This configuration, called the *α-helix*, is shown in Figure 15.9. At normal physiological pH, the helix is held together, at least in the absence of a highly polar solvent, by hydrogen bonds between the —C=O of one amino acid and the —N—H group of another acid four amino acid units farther along the peptide chain (i.e., there are three units included in one hydrogen bonded loop). The twisted configuration is dictated by the fact that the six atoms in the peptide group

Figure 15.9. The α-helix in proteins. Two right-handed helixes are shown. Most proteins exist in this form. A left-handed helix is shown for comparison. The dashed lines are hydrogen bonds: —N—H---O═C—

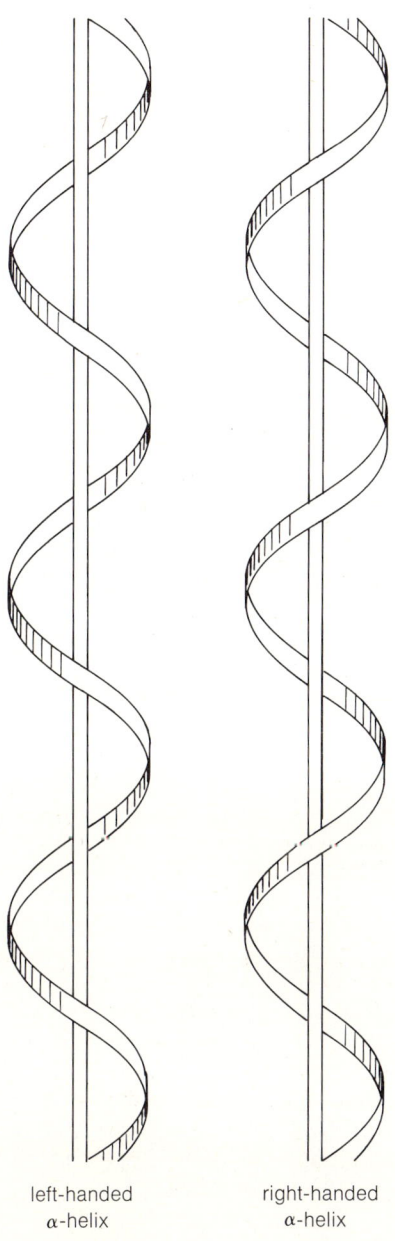

left-handed
α-helix

right-handed
α-helix

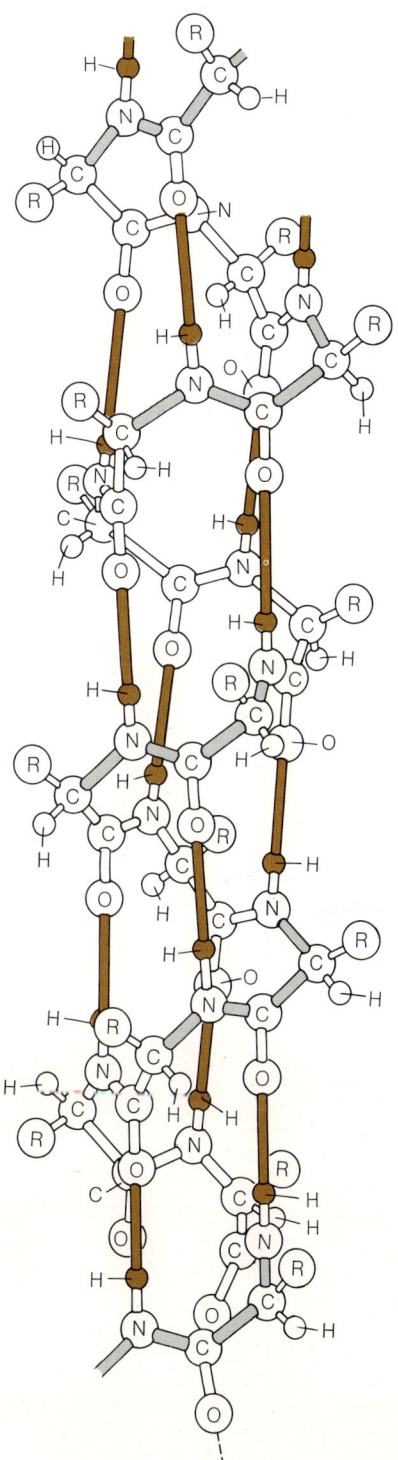

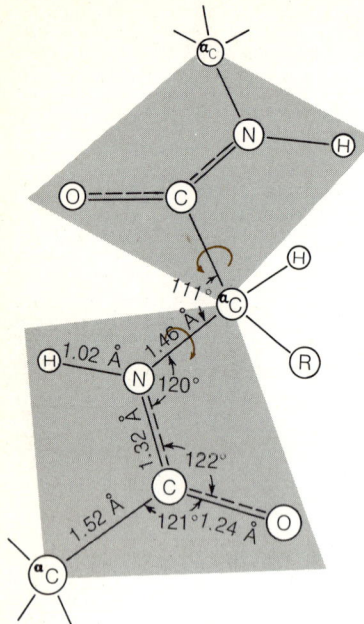

Figure 15.10. Bond angles and distances in the peptide group. Two groups are shown. The six atoms in each group lie in a plane, each plane sharing an α-carbon atom with an adjacent plane. The two planes here can be moved relative to each other only by rotation about the ${}^{\alpha}$C—C and the ${}^{\alpha}$C—N bonds. A polypeptide resembles a chain of flat plates (the peptide groups) connected by α-carbon atom swivel points.

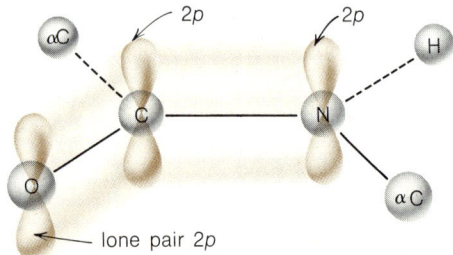

must all lie in the same plane as a consequence of partial double bond character in the C—N bond. The experimentally determined dimensions in the peptide group are shown in Figure 15.10. The bond lengths of the C=O and the C—N bonds are, respectively, longer and shorter than the lengths to be expected if the C=O bond were a pure double bond (1.23 Å) and the C—N bond were a pure single bond (1.49 Å). The discrepancies can be explained by the resonance bond structures

so that the C—N bond has some double bond character. Resonance (electron delocalization) stabilization is greatest when all six atoms of the peptide group lie in the same plane. Only then can the π-type p atomic orbitals on O, C, and N overlap maximally to form a π molecular orbital spread over these three atoms (Figure 15.11). The planar configuration is the low-energy one, and any departure from coplanarity of the six atoms requires the input of energy. (The same energy barrier to rotation about double bonds is responsible for geometric isomerism, as noted in Table 15.2.) Because the C—N bond does have double

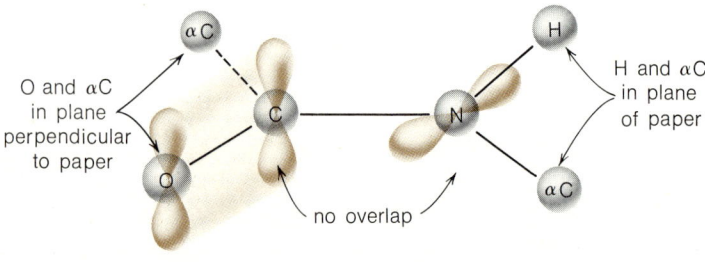

planar (low energy)

Figure 15.11. The planar configuration (top) of all six atoms in the peptide group has maximum electron delocalization of the $2p$ electrons of O, C, and N. Rotation of the H and α-carbon atoms out of the plane destroys the overlap between C and N.

O and αC in plane perpendicular to paper

H and αC in plane of paper

no overlap

nonplanar (high energy)

bond character, and because of resonance, the C and O atoms on one side must remain in the same plane with the H and C on the other side, so the only place a peptide chain may be twisted is at the single (σ) bonds to and from the α-carbon atoms.

Many conformations of the polypeptide chain can be generated by rotation about the σ bonds to the α-carbons. The α-helix configuration corresponds to the one allowing formation of hydrogen bonds to every N—H group in the chain and allowing the N—H axis on one amino acid residue to point directly at the C=O axis on the opposite residue. The α-helix has 3.6 amino acid residues per turn and rises 5.4 Å per turn. The R side chains stick out of the helix, away from the axis, as shown in Figure 15.9. Ball-and-stick models of proteins give the impression that there is a lot of empty space in the helix, but this is not the case. Scale models show that the interior of the helix is completely filled.

Most proteins do not have a helical configuration throughout their entire length, since there are a number of factors that may prevent the formation of an α-helix at certain points in the molecule or that may disrupt the intrachain hydrogen bonds stabilizing the helix. Where the helical structure breaks down, the straight protein chain may turn or take the form of a random coil.

Our main interest in later chapters will be with the globular (soluble in aqueous media) proteins that act as enzyme catalysts in living systems. The globular proteins all have a partial α-helix form. Besides being catalysts, proteins play many other roles in life. The proteins of skin, tendons, or cell walls play a structural role. Others, like insulin, are metabolic regulators known as hormones. Some transport oxygen in the blood or muscles and still others are involved in muscle contraction. This list does not exhaust the many important functions of proteins in life.

Opposed to the globular proteins are the insoluble fibrous proteins. An example of this class is α-keratin, the protein of hair, wool, and nails. A strand of keratin consists of three keratin molecules, each of which has the α-helix form, wound about each other. Numerous disulfide bonds between the separate chains add strength. The breaking of the disulfide bridges and their reforming after hair is set is the basis for many hair styling treatments. The α-keratin molecules in a fiber attain a new configuration if the fiber is subjected to moist heat and stretching. The new molecular configuration, the β-configuration (Figure 15.12) consists of extended, nonhelical protein molecules joined to adjacent molecules by hydrogen bonds between C=O of one molecule and N—H of the other. There are no intrachain hydrogen bonds. The extension of the α-keratin pulls the coils apart and breaks the intrachain hydrogen bonds. The β-configuration is the one found in silk. This structure is far less common in nature than the α-configuration.

Tertiary Structure   Natural globular proteins are not long, stretched out filamentlike molecules. Evidence shows that most globular proteins have compact, almost spherical shapes. To achieve this compact structure, the polypeptide chain folds back on itself in a complex, but, for a given protein, highly characteristic manner. Figure 15.13 shows the folded structure of the protein myoglobin. This three-dimensional structure and its stabilization are considered the tertiary structure of the molecule. As we shall see in this chapter and in more detail in Chapter 17, the shape of the convoluted molecule determines its biological function

—NH₂ terminal

—COOH terminal

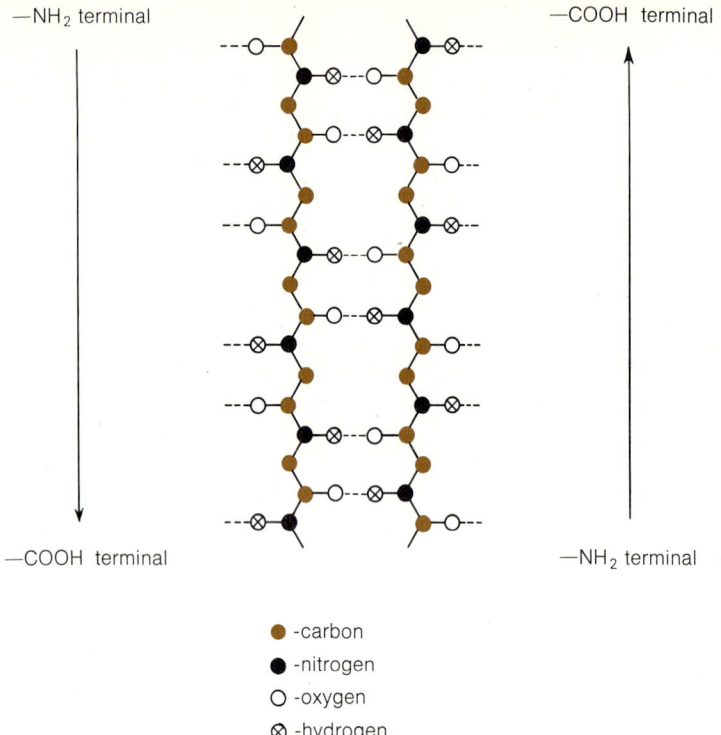

—COOH terminal

—NH₂ terminal

Figure 15.12. Extended protein molecules hydrogen bonded to one another, as found in the β-configuration in silk. This is the antiparallel arrangement, in which adjacent chains are oriented in opposite senses with respect to the direction from the —COOH to the —NH₂ terminals. Side chains are not shown.

● -carbon
● -nitrogen
○ -oxygen
⊗ -hydrogen

and activity. There are several kinds of interactions believed to give rise to the tertiary structure of a protein. In general, these interactions are due to forces between the side chains (R) of the amino acid residues in the polypeptide chain. The interacting side chains may be linearly far removed from one another along the chain but brought close together by folding. The interactions may be covalent, polar, ionic, or apolar. Examples of each type are summarized in Figure 15.14. The apolar or hydrophobic bond is believed to be the most common of the noncovalent interactions. The occurrence of this type of interaction has been attributed to an increase in stability resulting from the displacement of highly organized water molecules normally oriented about nonpolar side chains. For present purposes, we can use a variation of the "like dissolves like" theme and consider that the nonpolar portions of nearby side chains coalesce much like oil drops. The first group of eight amino acids listed in Figure 15.7 have nonpolar side chains. The forces operating in such hydrophobic bonds are weak van der Waals forces.

The precise three-dimensional conformation assumed by a protein molecule is the one of lowest energy consistent with the environment of the molecule. This point is supported by the observation that protein molecules in aqueous media fold in a manner to bury the hydrophobic or nonpolar side chains within the interior folds, where they are protected from exposure to water. The hydrophilic or polar groups are located mainly on the external surfaces of the folded protein, in contact with the solvent.

We shall say more about the tertiary structure of proteins in Chapter 17 when we discuss the catalytic activity of enzymatic proteins. It will be seen that there are active sites or groups of amino acid side chain residues on the

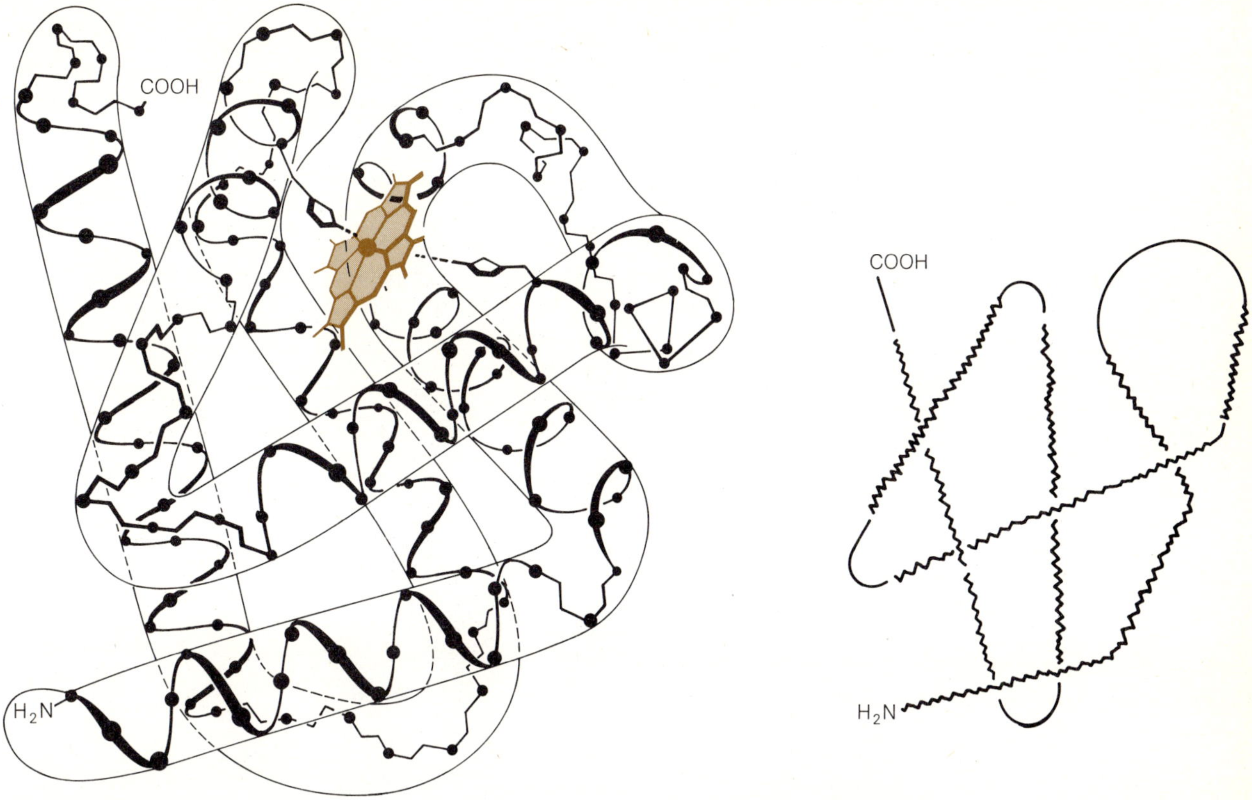

Figure 15.13. The structure of the myoglobin molecule. There are eight straight stretches of α-helix connected by nonhelical curves. The top view shows how the iron-containing heme group is held between the two nearest portions of the chain. The lower illustration indicates the portions of α-helix (wavy lines) and nonhelix. (Top illustration from Richard E. Dickerson, X-Ray Analysis and Protein Structure, in "The Proteins," Vol. II, 2nd ed, Hans Neurath, Ed., Academic, New York, 1964.)

folded surfaces of an enzyme that possess the required catalytic activity and that must be accessible to a reactant approaching the enzyme molecule. The removal or replacement of a critically situated amino acid in the polypeptide chain may destroy the biological usefulness of the enzyme, either by altering a crucial portion of the active site or, more likely, by altering one of the interactions necessary for the precise tertiary structure of the molecule. The

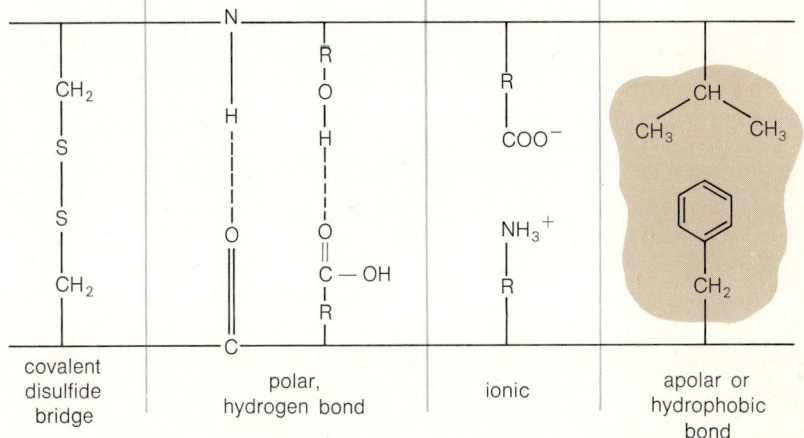

Figure 15.14. Some side-chain interactions responsible for tertiary structure in proteins.

resultant change in three-dimensional conformation may destroy the active site or move it to an inaccessible location. In either case, the enzyme molecule can no longer perform its biological task. The amino acid sequences in portions of the chain not involved in active sites or in the determination of tertiary structure can often be varied without affecting the biological function of the molecule, provided that, in general, amino acids with side chains of similar polarity are substituted.

Quaternary Structure

The highest level of protein structure is observed only in proteins composed of two or more noncovalently bonded polypeptides. Most proteins with a molecular weight over 50,000 are of this type. The relationship of one chain

Figure 15.15. Hemoglobin, a four-chain protein with quaternary structure. (From Richard E. Dickerson and Irving Geis, "The Structure and Action of Proteins," Harper & Row, New York, 1969. Copyright © 1969 by Richard E. Dickerson and Irving Geis. By permission of the authors and publisher.)

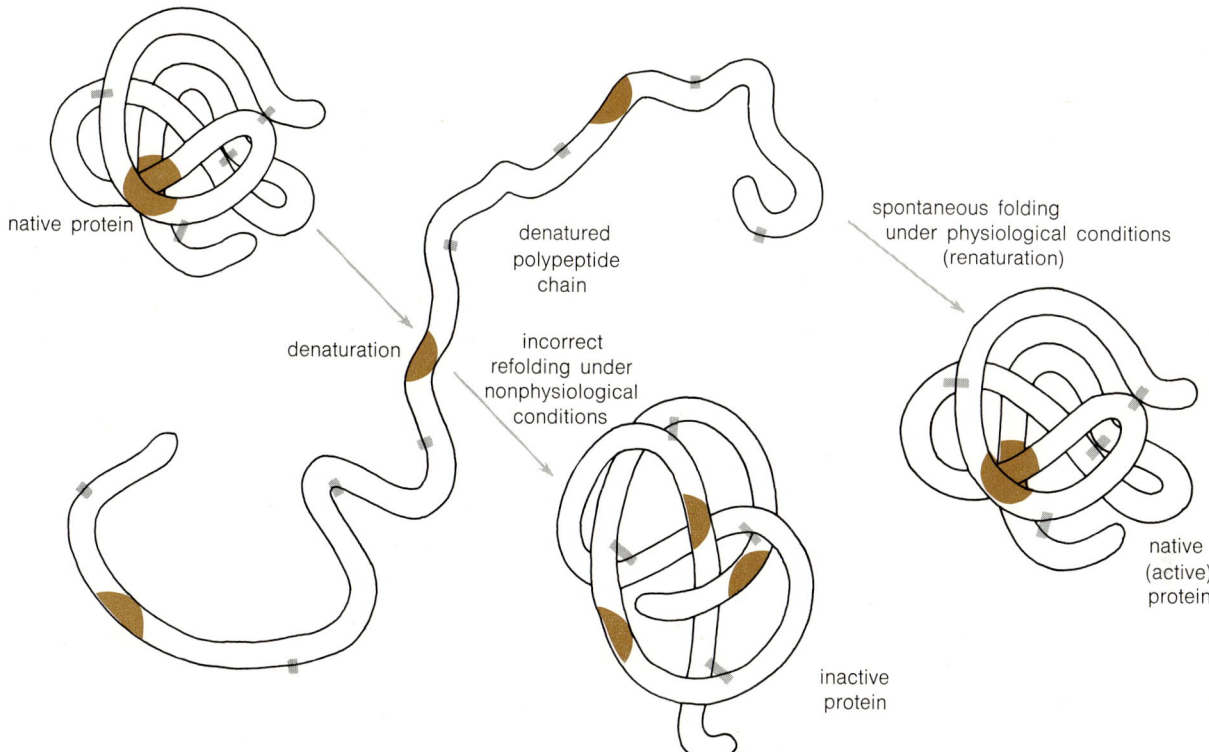

native protein

denaturation

denatured
polypeptide
chain

incorrect
refolding under
nonphysiological
conditions

spontaneous folding
under physiological conditions
(renaturation)

native
(active)
protein

inactive
protein

Figure 15.16. Loss of tertiary structure of a native protein by denaturation results in loss of biological activity. Activity is regained if renaturation occurs under the proper conditions. Refolding may also take place under other conditions, but biological activity may not be regained. The small rectangles represent sulfhydryl groups or disulfide bridges. The shaded areas are portions of the active site of the native protein.

to another, or the way they fit into the folds of each other, constitutes the quaternary structure of the entire complex molecule. Hemoglobin (mol wt 64,500), the oxygen-transporting protein in blood, has four polypeptide chains, none of which is covalently bonded to any other. This molecule is shown in Figure 15.15. Each chain encloses an iron-containing heme group. The interactions holding the individual chains together in such proteins are of the same kind as are responsible for tertiary structure, with the exclusion of the covalent disulfide bonds. When one of the heme groups absorbs an oxygen molecule, the quaternary structure of the entire molecule changes as two of the subunit chains move closer together, without much accompanying change in the tertiary structure of the subunits. The change in quaternary structure somehow causes the other heme groups to take up successive oxygen molecules more and more readily.

15.15 DENATURATION OF PROTEINS

The biological activity of natural or native proteins, especially the highly specific enzymes, depends on their three-dimensional conformation. If the protein molecule is caused to unfold, it will lose its activity. The tertiary structure of a protein is determined primarily by weak interactions of the hydrogen bonding and hydrophilic type. Certain rather slight changes in the normal environment can easily disrupt these stabilizing forces and allow the molecule to unfold. The inactivated protein is said to be *denatured* (Figure 15.16).

Native proteins can be denatured by heat, change in pH of the medium, or by a variety of chemicals. When heated above about 50°C, most proteins become denatured. The coagulation of egg white by cooking is a common example. Chemicals such as urea and alcohols denature proteins by breaking the weak bonds and by providing an environment in which the unfolded configuration is more stable than the folded one. In this way, alcohol denatures bacterial protein and acts as a disinfectant. Heavy metal ions such as Pb, Hg, Ag, and As denature proteins by reacting with the sulfhydryl (—SH) or disulfide bridges to form covalent metal compounds.

Changes in the pH of the environment can lead to protein insolubility by altering the charge balance on the molecular surface. Each protein molecule has acidic and basic groups in the side chains (R) of the amino acid units of the polypeptide chain. In an acid medium, the acidic groups such as —COOH will be uncharged, but the basic groups such as —NH$_2$ will have a positive charge. In a basic medium, the situation will be reversed, the acidic groups being negatively charged. Thus, like their individual amino acid units (Section 15.13), most protein molecules have a net positive charge at low pH and a net negative charge at high pH. In either situation, the like-charged molecules repel each other and do not come together into insoluble aggregates. However, at the isoelectric point, there is a maximum concentration of molecules on which the *net* charge is zero, because the number of positive charges on basic sites is equal to the number of negative charges on acidic groups. Then the protein molecules clump together and precipitate. While this is not an example of denaturation by unfolding of the molecule, the protein is obviously not of use in the insoluble form. This is one reason why regulation of pH in body fluids is important. The optimum pH should not coincide with the isoelectric points of any of the soluble proteins in the blood or other fluid. The curdling of milk is caused by the lowering of the pH to the isoelectric point of the principal milk protein, casein, by acids formed by bacteria. Changes in pH can also affect the helical structure of proteins and through this, the tertiary structure. If —COOH groups on some of the amino acid side chains ionize and become negatively charged in basic solution, the electrostatic repulsion of the negative groups may prevent the close order of the helix.

The denaturation of essential proteins—the enzymes, hormones, antibodies, transport and structural proteins—by poisons and environmental pollutants is one of the common ways in which the physical environment can adversely affect living organisms. We shall return to this subject in Section 17.15, with special reference to enzymes.

Renaturation   A denatured protein molecule can often regain all or part of its biological activity spontaneously if the original biological conditions of pH and temperature are reestablished. The molecule refolds itself, of its own accord, to the original native configuration. This refolding is called *renaturation* (Figure 15.16). Reversible denaturation indicates that the protein molecule contains all the information it needs to restore itself. This information is contained in the amino acid sequence and the interactions of the side chains of the amino acids. The molecule apparently flops around, trying various bonding possibilities and conformations until it eventually finds the folded structure of lowest energy, the active native form. According to this view, it should be possible, in principle, to uncook an egg. Even when energetically favorable original conditions are restored, however, the rate of renaturation may be so slow as to be essentially irreversible. The rate of renaturation of some proteins can often be accelerated catalytically.

15.16 THE STUDY OF PROTEINS

The primary structure of a protein is found by first identifying the amino acids at each end of the molecule. Then the protein is partially hydrolyzed into a number of small peptide fragments (oligopeptides) whose amino acid sequence can be determined by chemical analysis. Hydrolysis of the protein by different procedures results in fragments whose amino acid sequences overlap. Identification of overlapping segments of the fragments permits the piecing together of the fragments to give the primary structure of the original protein. Amino acid identification and sequence determination procedures are now done by automated instruments.

Secondary, tertiary, and quaternary structure determinations depend more on physical procedures than on chemical methods. The most powerful of these physical methods is X-ray crystallography. The analysis of the X-ray pattern of crystalline proteins is many times more difficult than the analysis of a simple NaCl crystal. The tedious computations would not have been possible without the help of high-speed electronic computers. Several Nobel prizes have been awarded for work leading to the determination of the structures of proteins and other biological molecules.

Carbohydrates

Carbohydrates (carbon hydrates) have the general formula $C_x(H_2O)_y$, hence the name, which should not be taken literally as denoting their structure. Carbohydrates, like proteins, play a variety of roles in living systems, but the two most important are as energy sources and as structural components.

15.17 MONOSACCHARIDES

The simplest carbohydrates are known as *monosaccharides* or simple sugars. They are the building blocks of polymeric carbohydrates or *polysaccharides*, just as amino acids are the units of proteins. Monosaccharides may have from three to seven carbon atoms. Glyceraldehyde is a three-carbon monosaccharide, called a triose, with the structure

$$
\begin{array}{c}
\overset{\displaystyle O}{\underset{\displaystyle |}{\overset{\displaystyle \|}{C}H}} \\
H-\overset{\displaystyle |}{\underset{\displaystyle |}{C}}{}^{*}-OH \\
CH_2OH
\end{array}
$$

D-glyceraldehyde

Glyceraldehyde and all other natural sugars are optically active because of the presence of carbon atoms (starred) joined to four different groups. Chemical evolution has resulted in the incorporation of the D configuration in living systems, rather than the L isomer, as was the case with amino acids. (The optical isomers of glyceraldehyde are the reference standards for D and L configurations of other molecules. Thus, D-alanine has the same configuration in space as the glyceraldehyde isomer arbitrarily designated as D if the —COOH of the acid corresponds to the —CHO of the aldehyde, the —NH₂ group of the acid to the —OH of the aldehyde, and the —CH$_3$ (or —R) group of the acid to the —CH$_2$OH group of the aldehyde.)

The five-carbon and six-carbon monosaccharides (pentoses and hexoses, respectively) contain biologically important members. The hexose D-glucose (dextrose) is the most common in polysaccharides. The structure of glucose may be written in straight-chain form, but in nature the molecule turns back on itself, like a dog chasing its tail, and the chain form occurs in equilibrium with the ring form shown in Figure 15.17. In Figure 15.17, the ring is con-

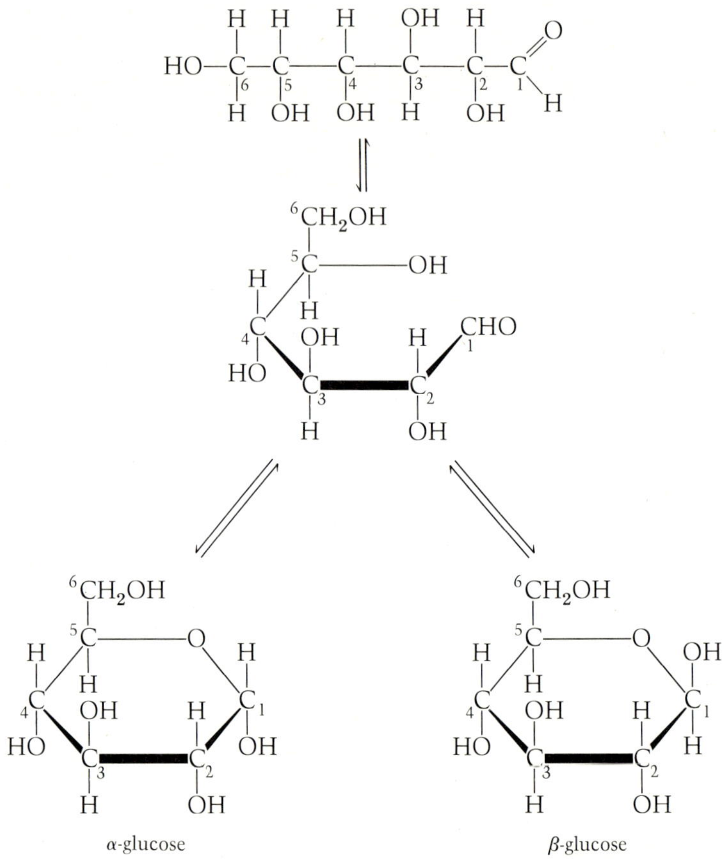

Figure 15.17. Ring forms of glucose. Both of the ring forms are in equilibrium with the linear form. The α- and β-isomers differ only in the position of the hydroxyl group on carbon atom 1. Under natural conditions, the ring forms predominate almost exclusively.

α-glucose   β-glucose

sidered to lie in a plane perpendicular to the page. The heavier line between atoms 2 and 3 is the side of the ring closest to the reader. The —CH$_2$OH, —H, and —OH groups lie above or below the plane of the ring. In actuality, the ring is not planar but has a folded or puckered (chair) form.

There are many possible isomers of the glucose ring structure. In the D series, the —CH$_2$OH group lies above the ring; in the L series (not found in nature), it lies below the ring. Different arrangements of the —OH groups at carbon atoms 2, 3, and 4 give other isomers in the D series. If the —OH on carbon atom 1 is on the same side of the ring (above the ring) as carbon atom 6, the β-configuration results; if on the opposite side (below the ring), the α-configuration results.

Fructose, another hexose, has the structure

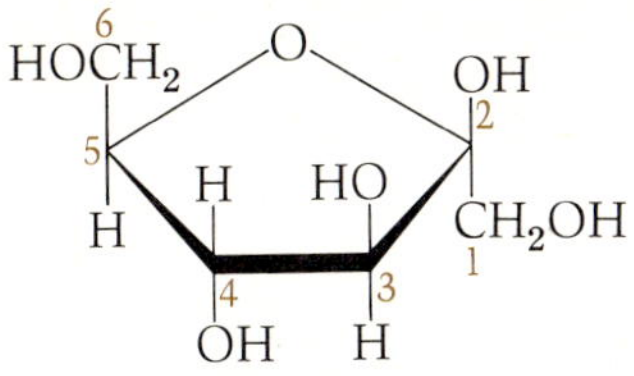

β-D-fructose

The C's in the ring have been omitted.

Condensation of two monosaccharides will give a disaccharide. With glucose, the condensation proceeds through the elimination of water between the —OH group of carbon atom 1 of glucose and an —OH group of another sugar molecule. Common cane sugar, sucrose, is a disaccharide of glucose and fructose:

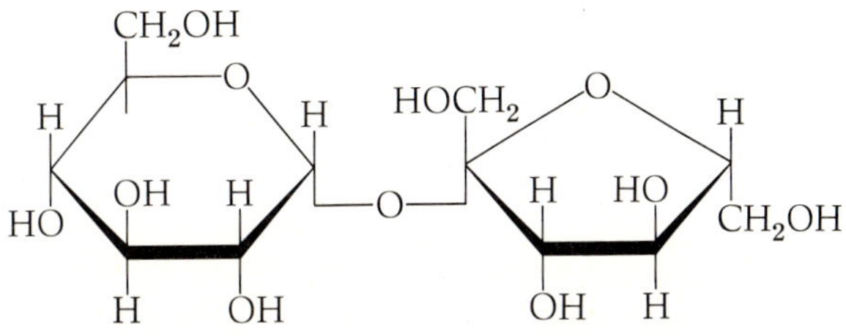

D-glucose   D-fructose

sucrose

There are two other monosaccharides that we shall encounter. Both are five-carbon sugars or pentoses. They are important constituents of the nucleic acid group of large biological molecules. These two sugars are ribose and deoxyribose, the latter lacking one of ribose's oxygen atoms:

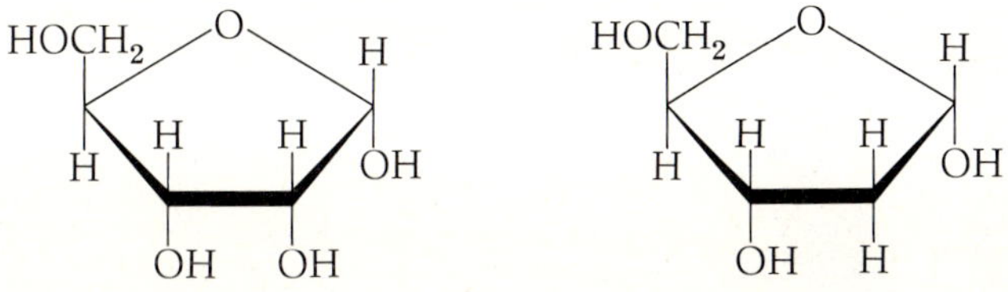

ribose   deoxyribose

15.18 POLYSACCHARIDES   The C—O—C linkage formed between two joined simple sugar units is called a *glycosidic* linkage. It plays the same role in polysaccharides that the peptide bond does in proteins. Continued condensation of monosaccharide units gives polysaccharides. The most common polysaccharides are amylose and cellulose, both made up of glucose units. Amylose is a major constituent of starch and glycogen, the principal food- and energy-storage compounds of the plant and animal worlds, respectively. Cellulose is the major structural component of plants.

Amylose is a polymeric molecule made up of α-glucose units bonded at the 1 and 4 positions. Cellulose is composed of β-glucose units also bonded

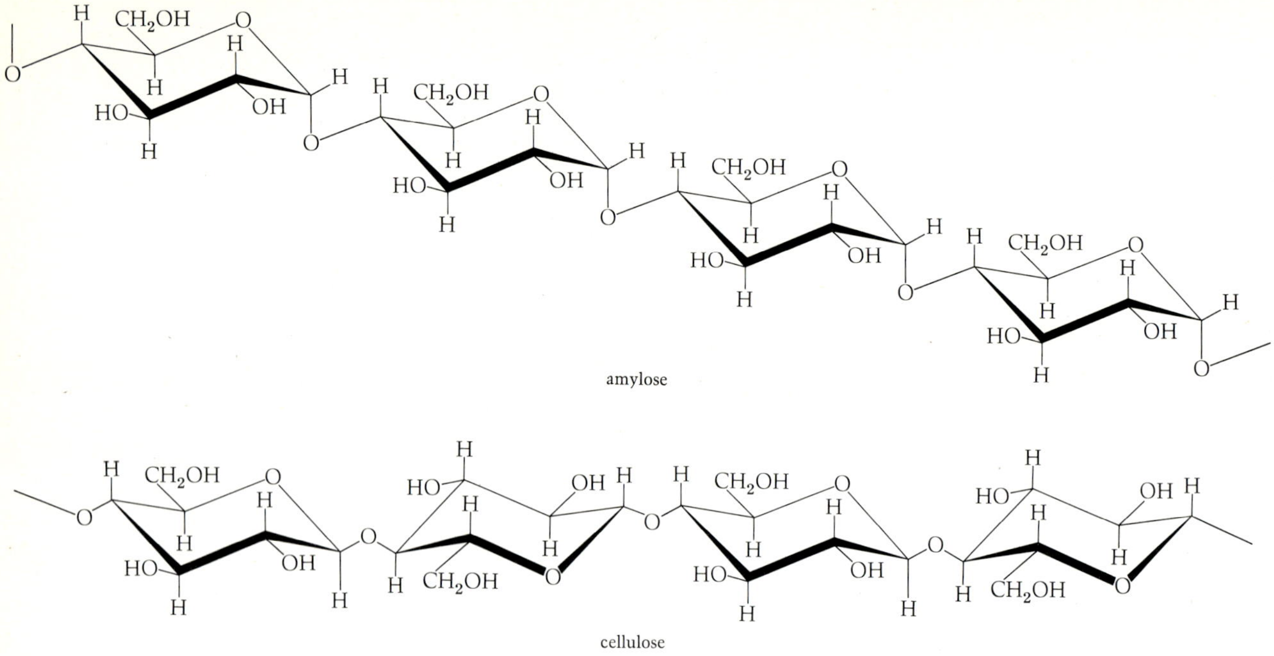

amylose

cellulose

Figure 15.18. Portions of the amylose and cellulose molecules. In cellulose, each glucose unit is rotated by 180° from its adjacent units.

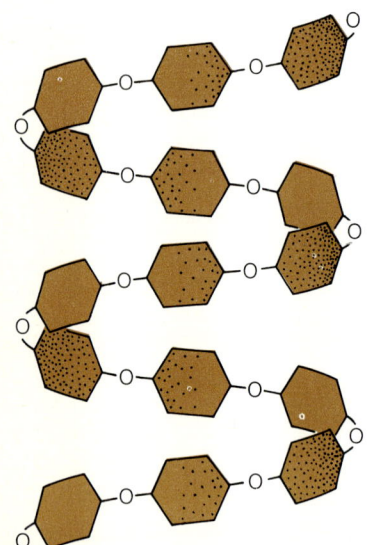

Figure 15.19. The helical configuration of the amylose molecule leads to little adhesion between amylose chains.

at the 1 and 4 positions. Segments of both molecules are shown in Figure 15.18, where the puckered forms of the glucose units are used. The 1,4-α-chains of amylose assume a helical conformation, with six glucose rings per turn of the helix (Figure 15.19). The hydroxyl groups of the glucose units are distributed about the surface of the helix and tend to interact with the solvent (water) rather than with hydroxyl groups of adjacent chains. Because of this interaction, amylose is nonfibrous and slightly soluble.

Cellulose, on the other hand, is nonhelical. The difference between amylose and cellulose is a result of the α-glycosidic bonding in amylose and the β-glycosidic bonding in cellulose. The hydroxyl groups in a cellulose chain are on one side or another of a flat surface, as it were, rather than distributed thinly over the surface of a cylinder. Parallel cellulose molecules therefore have a greater area of contact available between the —OH groups of the molecules (the contact area between two flat plates is much greater than that between two cylinders). The hydrogen bonds between the —OH groups on adjacent cellulose molecules cross-link the chains (Figure 15.20) and pack them into crystalline bundles which, in turn, eventually twist into insoluble fibers. The contrast in physical properties between amylose and cellulose is due entirely to a seemingly trivial difference in bonding between the glucose units.

The rigidity of wood is due to the hydrogen bonding between cellulose molecules. If a piece of wood is dipped in liquid ammonia or exposed to gaseous ammonia under pressure, the small NH_3 molecule penetrates the crystal lattice of cellulose and breaks the hydrogen bond cross links. Ammonia is a stronger base than the ROH groups $(NH_3^+ + —ROH \longrightarrow NH_4^+ + —RO^-)$ and removes the hydrogen atoms. When the bonds are broken, the cellulose chains can slide past one another and the wood can be bent into any shape.

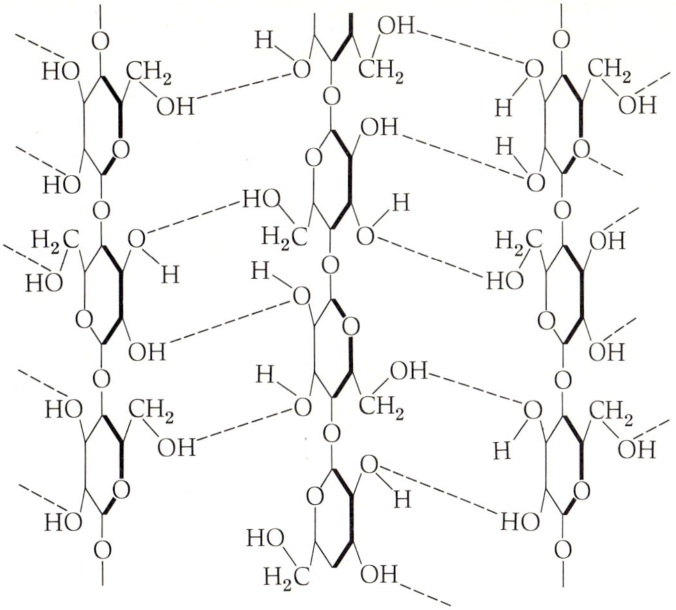

Figure 15.20. Hydrogen bonding between parallel cellulose molecules. (Adapted from H. R. Mauersberger, Ed., ''Matthews' Textile Fibers,'' 6th ed, Wiley, New York, 1954, p 73, Figure 10.)

When the ammonia evaporates, the hydrogen bonds reform in new positions and the wood retains its new shape, just like hair setting.

Polysaccharides can be hydrolyzed into their simple monosaccharide sugar units in the same way that proteins may be broken down by hydrolysis in acidic solution to their constituent amino acids. In living organisms, the digestion of polysaccharides is catalyzed by enzymes. Humans and most other carnivores are not supplied with digestive enzymes to hydrolyze the β-links in cellulose in a reasonable time. However, soil organisms and many herbivores such as cattle and horses as well as termites can digest cellulose. The digestive enzyme in herbivores and termites is supplied by symbiotic microorganisms living in their digestive systems.

Nucleic Acids

Along with proteins, the class of polymeric substances known as nucleic acids are the very stuff of life. The nucleic acids in a living system contain in their chemical composition and three-dimensional structure all the information that system needs to reproduce its own kind. One kind of nucleic acid macromolecule is the principal component of the genes, the conveyors of hereditary information from cell to cell of a living organism and from one generation of an organism to the next. The genes, in turn, are the components of the chromosomes. Another nucleic acid directs the synthesis of proteins and of new nucleic acids in the organism.

We shall go into some of the details of molecular genetics in Chapter 19, where the basic relationship of the higher structure of nucleic acids to their genetic function will be explored. Here we discuss only the chemical composition and primary structure of nucleic acids.

15.19 STRUCTURE OF NUCLEIC ACIDS

The structure of the nucleic acids resembles the polymeric proteins and polysaccharides in that all three types of macromolecules are built up of repeating units joined by bonds resulting from the dehydration condensation of the units. The unit of primary structure in a nucleic acid is somewhat more complex than the amino acid or monosaccharide units of the other two polymers. The complex building blocks of nucleic acids are called *nucleotides,*

Figure 15.21. Components of nucleotides.

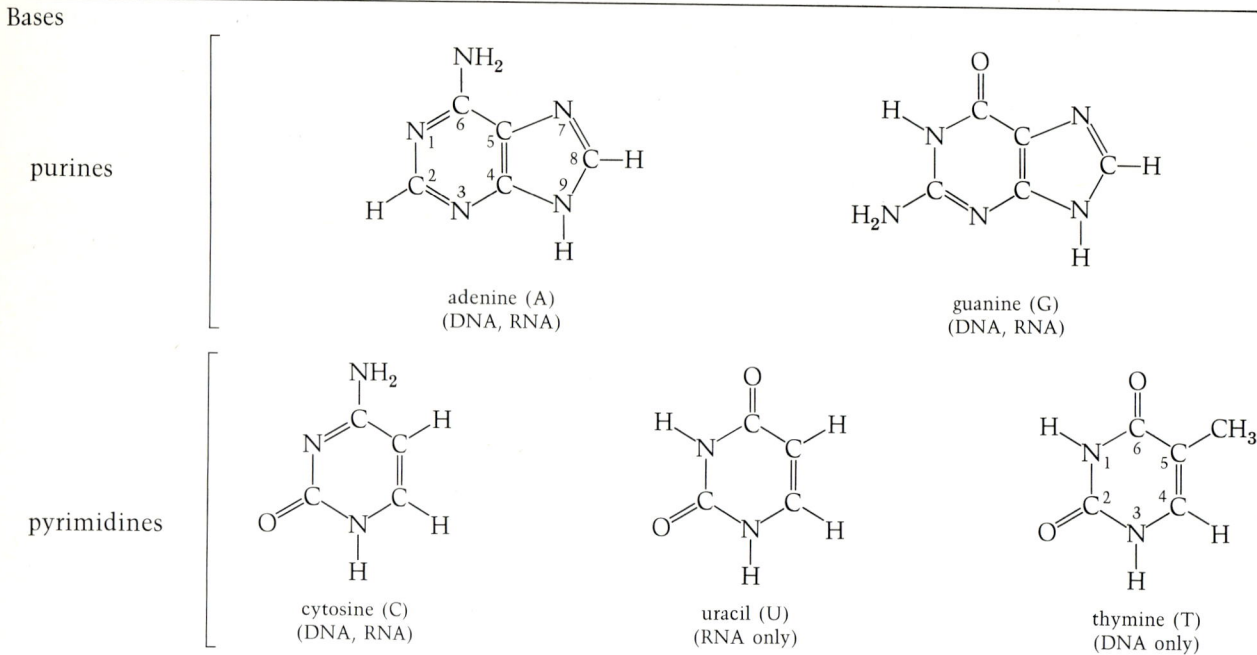

Bases

purines

adenine (A)
(DNA, RNA)

guanine (G)
(DNA, RNA)

pyrimidines

cytosine (C)
(DNA, RNA)

uracil (U)
(RNA only)

thymine (T)
(DNA only)

Pentoses

deoxyribose (DNA only)

ribose (RNA only)

Phosphate

phosphoric acid
(DNA, RNA)

and another name for the nucleic acids is *polynucleotides*. Each nucleotide is made up of three component molecules linked together, again by linkages resulting from dehydration condensation. The subunits of a nucleotide are (1) a complex organic base, (2) a simple pentose monosaccharide, and (3) a phosphate group. The bases that occur in natural nucleic acids are members of groups of nitrogen-containing ring compounds called *purines* and *pyrimidines*. The pentose sugars are ribose and deoxyribose. A base, a pentose, and phosphoric acid condense to form a base–sugar–phosphate nucleotide unit. The nucleotide components are summarized in Figure 15.21. Figure 15.22

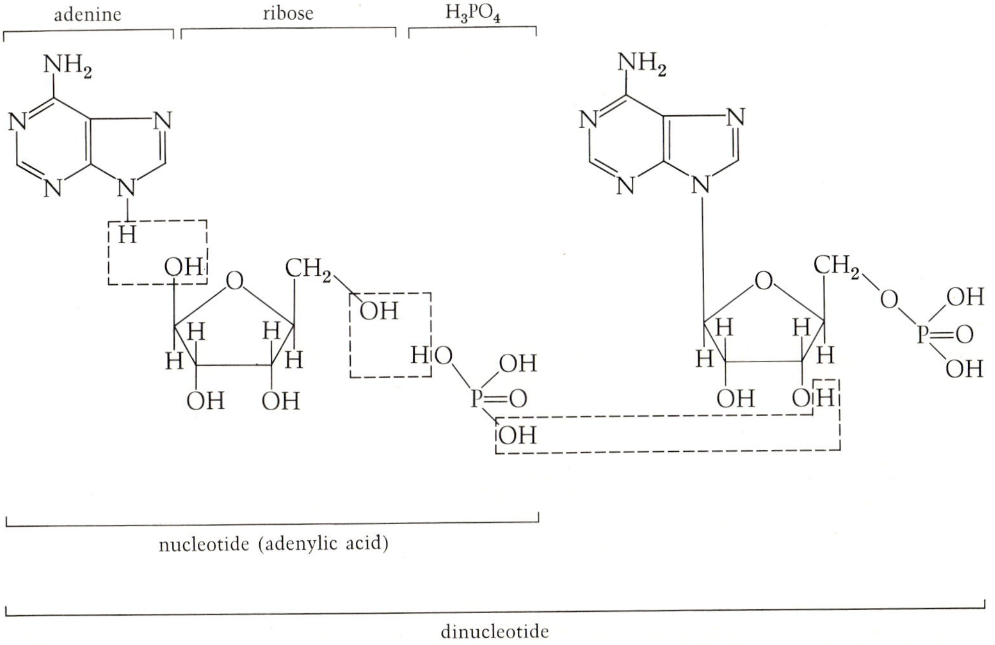

Figure 15.22. Nucleotide and polynucleotide formation by condensation of nucleotide constituents and nucleotide units.

illustrates how a nucleotide unit may be considered to result from condensation of three of these subunits and how two or more nucleotides may combine by condensation to begin a linear polynucleotide chain.

There are two kinds of nucleic acids. One, deoxyribonucleic acid (DNA), is made up of nucleotide units composed of deoxyribose, phosphate, and any one of the bases adenine (A), guanine (G), cytosine (C), or thymine (T). The other nucleic acid, ribonucleic acid (RNA), contains ribose, phosphate, and any one of the bases adenine (A), guanine (G), cytosine (C), or uracil (U). The differences between DNA and RNA are in the sugar subunits and in the fact that uracil is found only in RNA and thymine only in DNA. Figure 15.23 shows part of a DNA chain. Actual DNA molecules may contain up to 10^9 nucleotide units. RNA molecules are considerably smaller.

Like the proteins, the nucleic acids permit an enormous number of differences in sequence and identity of the base side chains. We shall see in Section 19.7 that it is the base sequence in a strand of DNA that contains the genetic information telling the organism whether it is to manufacture

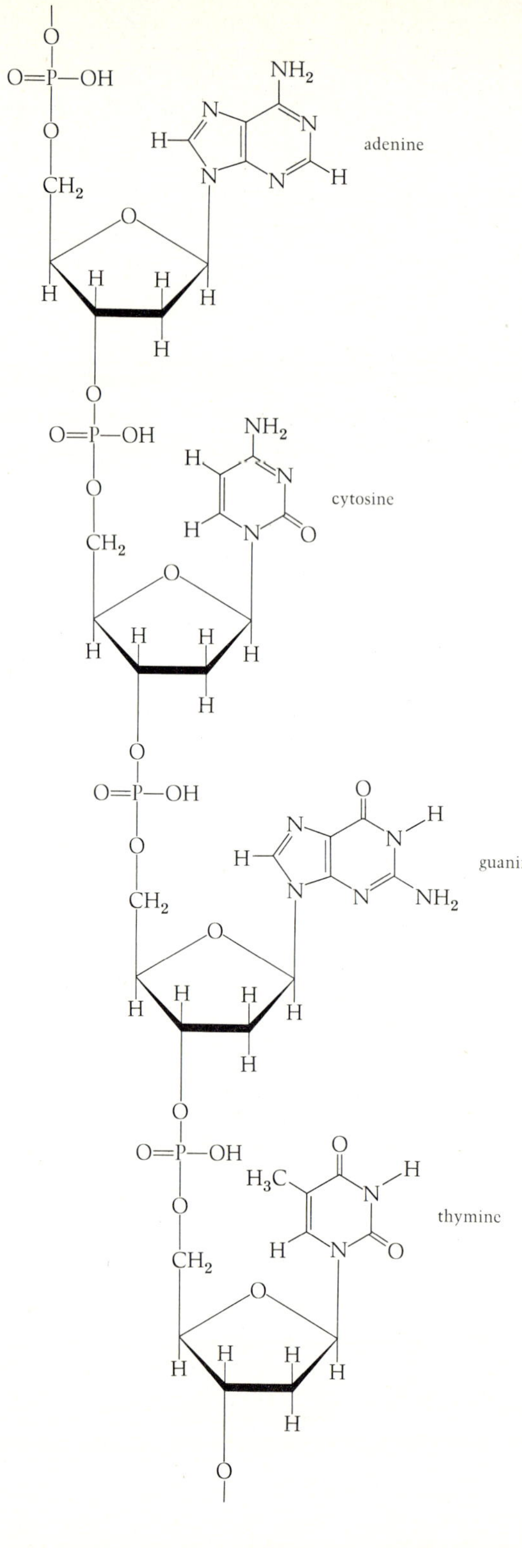

Figure 15.23. A segment of a DNA polynucleotide chain.

a protein enzyme ultimately leading to one of any number of genetic characteristics such as brown or black hair.

Lipids

Lipids are a diverse class of substances soluble in nonpolar organic solvents but insoluble in water. Lipids, in general, are sometimes called fats, but the term is incorrect because fats are strictly esters of the trihydroxy alcohol glycerol and long-chain fatty acids.

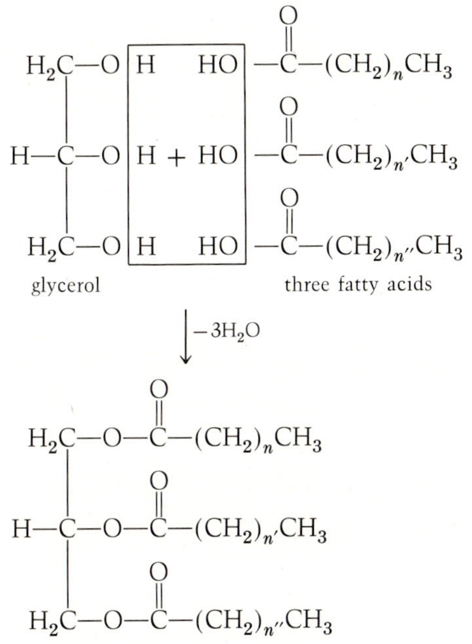

$$
\begin{array}{c}
H_2C\!-\!O\!\left[H \qquad HO\right]\!-\!\overset{\displaystyle O}{\overset{\|}{C}}\!-\!(CH_2)_nCH_3 \\[2mm]
H\!-\!\overset{|}{C}\!-\!O\!\left[H + HO\right]\!-\!\overset{\displaystyle O}{\overset{\|}{C}}\!-\!(CH_2)_{n'}CH_3 \\[2mm]
H_2\overset{|}{C}\!-\!O\!\left[H \qquad HO\right]\!-\!\overset{\displaystyle O}{\overset{\|}{C}}\!-\!(CH_2)_{n''}CH_3
\end{array}
$$

glycerol   three fatty acids

$$\downarrow -3H_2O$$

$$
\begin{array}{c}
H_2C\!-\!O\!-\!\overset{\displaystyle O}{\overset{\|}{C}}\!-\!(CH_2)_nCH_3 \\[2mm]
H\!-\!\overset{|}{C}\!-\!O\!-\!\overset{\displaystyle O}{\overset{\|}{C}}\!-\!(CH_2)_{n'}CH_3 \\[2mm]
H_2\overset{|}{C}\!-\!O\!-\!\overset{\displaystyle O}{\overset{\|}{C}}\!-\!(CH_2)_{n''}CH_3
\end{array}
$$

fat (a triglyceride)

Typical fatty acids are palmitic ($n = 14$) and stearic ($n = 16$). The long-chain aliphatic group may also be unsaturated.

    Phospholipids are fats in which one of the fatty acid groups has been replaced by an esterified phosphate group.

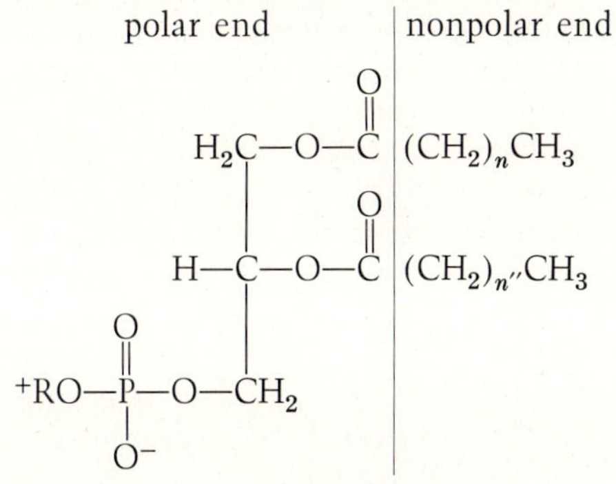

polar end | nonpolar end

$$
\begin{array}{c}
H_2C\!-\!O\!-\!\overset{\displaystyle O}{\overset{\|}{C}}\;\Big|\,(CH_2)_nCH_3 \\[2mm]
H\!-\!\overset{|}{C}\!-\!O\!-\!\overset{\displaystyle O}{\overset{\|}{C}}\;\Big|\,(CH_2)_{n''}CH_3 \\[2mm]
\overset{\displaystyle O}{} \\
{}^+RO\!-\!\overset{\displaystyle \|}{\underset{\displaystyle O^-}{P}}\!-\!O\!-\!CH_2
\end{array}
$$

phospholipid

The phosphate group is free to rotate away from the rest of the molecule. The group R is usually a small, polar entity such as $-CH_2-CH_2-NH_3^+$.

The steroids, such as cholesterol, are another physiologically important group of lipids.

The lipids are not true polymers like the proteins, polysaccharides, and nucleic acids, because their molecules do not retain two of the original functional groups ($-OH$ and $-COOH$) required to carry on further condensation. The biologically important feature of the lipids, and the phospholipids in particular, is their dual hydrophobic and hydrophilic nature. The hydrophobic fatty acid ends of phospholipid molecules are repelled by water and tend to associate with themselves. The charged polar ends are attracted to water. Because of these properties, lipids are believed to arrange themselves into bimolecular layers that play central roles in the structure of biological membranes, as shown in Figure 15.24.

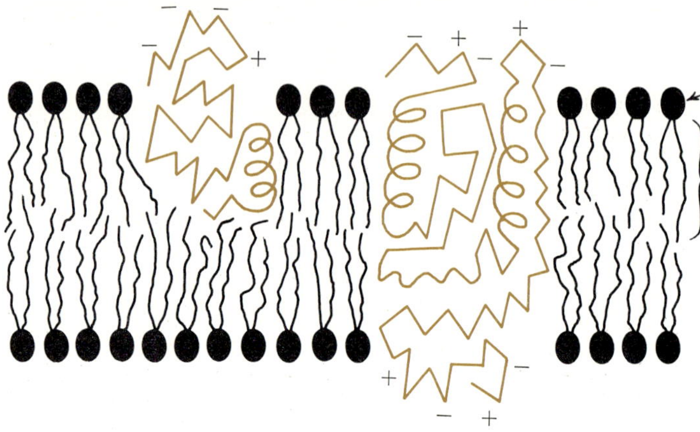

Figure 15.24. The dual hydrophilic–hydrophobic structure of phospholipids and their arrangement in a mosaic model of a biological membrane. In the newer mosaic model, globular proteins are embedded in a discontinuous fluid phospholipid bilayer. The polar groups of the proteins protrude from the membrane into the aqueous environment. Nonpolar residues of the proteins are embedded in the bilayer. Some proteins may span the entire membrane.

polar (charged) end of lipid molecule

nonpolar fatty acid residue

15.21 SUMMARY

One of the characteristics of living organisms is their high degree of structural and functional organization. Structural organization is directly reflected at the molecular level by the complex structures of the large molecules that make up the bulk of living systems. The detailed structures of these macromolecules determine their biological function.

The four major classes of the large molecules of life are the proteins, nucleic acids, carbohydrates, and the smaller lipids. All are compounds of carbon and fall within the realm of organic chemistry, the chemistry of carbon compounds whether of biological origin or not. The carbon atom has unique properties, which lead to its central role in living systems. Among these properties are its ability to form strong covalent bonds with itself and with hydrogen, among other elements. The quadrivalence of carbon, the directional character of its covalent bonds, and the ability to form carbon–carbon chains all contribute to the multitude and structural versatilty of the relatively stable (in terms of their speed of reaction) molecules of life.

Organic compounds can be conveniently organized and studied in classes whose members share common structural or compositional features. Thus, the large number of carbon–hydrogen or hydrocarbon compounds can be subdivided into the saturated alkanes, the unsaturated alkenes and alkynes, and the conjugated aromatic hydrocarbons. Many derivatives of each of these classes can result from the replacement of hydrogen atoms by more reactive functional groups. The chemical and physical properties of organic compounds depend on (1) the number and arrangement (structural isomerism) of carbon atoms in the molecule; (2) the nature and position of atoms and functional groups attached to the carbon atom chain; (3) the nature of the bonds (multiplicity, polarity) in the molecule; and (4) the spatial arrangement of the atoms in the molecule (optical and geometric isomerism).

The reactions of organic substances can be classified in various ways. We have chosen to classify the general reactions into six broad types that are catalyzed by enzymes. The large molecules of life are all formed from smaller organic units by a dehydration condensation reaction.

Proteins are formed from the condensation of amino acids, R—CHNH$_2$—COOH, to produce the repeating peptide unit,

$$-CHR-\overset{\overset{O}{\|}}{C}-NH-.$$ The primary structure of a protein is the sequence of amino acid units along the chain. This sequence determines all the other properties of the protein molecule. The secondary structure of a protein molecule refers, in most cases, to the helical conformation of the chain that results from hydrogen bonding between —NH and —C=O groups on the chain. The folding and twisting of the protein chain gives rise to tertiary structure. Multichain proteins exhibit quaternary structure. The forces giving rise to secondary through quaternary structure are predominantly weak interactions such as hydrogen bonds and van der Waals forces, and it is these forces that are responsible for the cohesion, structure, and function of biological polymers. When these forces are disrupted, denaturation occurs, the higher structure of the protein dissolves, and loss of biological activity results.

Carbohydrates or polysaccharides are produced by the condensation of monosaccharides or simple sugars such as glucose. Nucleic acids or polynucleotides are composed of condensed nucleotide units. A nucleotide unit, in turn, is composed of a simple pentose sugar, a complex organic base, and a phosphate group, all of which are joined by dehydration condensation. There are two kinds of nucleic acids of interest to us: DNA and RNA. They differ in the monosaccharide unit and in their organic base composition.

Lipids comprise a wide variety of water-insoluble, fatty substances whose distinctive properties result from the dual hydrophobic–hydrophilic nature of their molecules. One class of lipids, the phospholipids, is an important structural component of membranes.

Questions and Problems

1. Write the condensed structural formulas and give the systematic names for all the structural isomers of hexane.

2. What is the relation between two molecules whose structural formulas are written as

   a. Cl—CH(H)(H)—CH(H)(H)—Cl and Cl—CH(H)(Cl)—CH(H)(H)—H

   b. CH$_3$—CH(CH$_3$)—CH$_3$ and CH$_3$—CH$_3$—CH$_3$

   c. H—CH(H)(H)—CH(H)(Cl)—Cl and Cl—CH(H)(H)—CH(H)(Cl)—H

3. Give systematic names for the following compounds.

a. $CH_3CH=CHCH_2CH_2CH_3$

b. $CH_3\overset{\displaystyle CH_3}{\underset{\displaystyle CH_3}{C}}\text{----}\overset{\displaystyle CH_3}{C}HCH=CH_2$

c. $CH_3CH=CHCH=CHCH_3$

4. Give the structure of
   a. 2,2-dimethyl-3-hexene
   b. 5-methyl-3-heptyne

5. What is the name and structure of the alkyne hydrocarbon that follows acetylene in the homologous series?

6. To what class of hydrocarbons do the following belong?
   a. C_4H_8
   b. C_6H_{14}
   c. C_7H_8
   d. C_4H_6

7. To what compound class do the following belong?
   a. CH_3COOCH_3
   b. 4,5-dimethyl-3-heptanol
   c. C_4H_9OH
   d. $C_6H_5CH_2COOH$
   e. isopropyl propionate
   f. $CH_3NHC_2H_5$
   g. CH_3CHNH_2COOH
   h. $C_3H_5(OOCC_{15}H_{31})_3$
   i. $(CH_3)_2CHCONH_2$

8. Complete the following with the expected reaction:
   a. $CH_3CH_3 + Br \longrightarrow$
   b. $CH_2=CH_2 + Br_2 \longrightarrow$
   c. $CH\equiv CH + Br_2 \longrightarrow$
   d. $C_6H_6 + Br_2 \longrightarrow$

9. What are the oxidation states of each carbon atom in CH_3CH_3 and CH_3COOH?

10. a. Draw the orbital overlap diagram for $ClHC=CHCl$, 1,2-dichloroethene, and discuss the energy barrier to the rotation of one $Cl\text{---}\underset{|}{C}\text{---}H$ group about the double bond.

   b. Draw the orbital overlap diagram of the substituted allene,

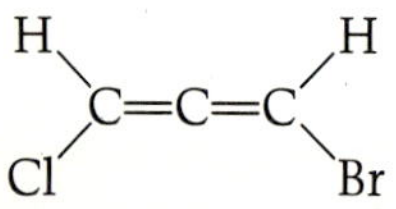

$$\underset{\displaystyle Cl}{\overset{\displaystyle H}{\diagdown}}C=C=C\underset{\displaystyle Br}{\overset{\displaystyle H}{\diagup}}$$

   Note that this molecule has nonsuperimposable mirror images and optical isomers even though there is no carbon atom with four different attached groups.

11. Complete or write out the following reactions. Name all reactants and products and indicate what type or class of reaction is involved.
   a. $CH_2=CH_2 + H_2O \longrightarrow CH_3CH_2OH$
   b. $HC\equiv CH + ? \longrightarrow CH_3\overset{\displaystyle O}{\overset{\displaystyle \|}{C}}\text{---}OH$
   c. $HC\equiv CH + ? \longrightarrow CH_3CHCl_2$
   d. preparation of ethyl ether from ethyl alcohol
   e. $2CH_3OH + O_2 \longrightarrow 2HCHO + 2H_2O$
   f. acetic acid + ethyl alcohol $\longrightarrow$
   g. $? + ? \longrightarrow CH_3\overset{\displaystyle O}{\overset{\displaystyle \|}{C}}NH_2 + H_2O$
   h. $R\overset{\displaystyle O}{\overset{\displaystyle \|}{C}}\text{---}NH\text{---}\underset{\displaystyle R}{\overset{}{C}}H\text{---}\overset{\displaystyle O}{\overset{\displaystyle \|}{C}}\text{---}NH\text{---}R \xrightarrow{\text{complete hydrolysis}}$ three products
   i. reaction of a primary amine with water
   j. $CH_3Cl + NH_3 \longrightarrow CH_3NH_2 + ?$
   k. $CH_3CHO + H_2 \longrightarrow$
   l. benzene + $CH_3Cl \longrightarrow$
   m. $(n + 1) HOOC(CH_2)_4COOH +$
      $(n + 1) H_2N(CH_2)_6NH_2 \xrightarrow{\Delta}$ nylon + $2nH_2O$

12. a. Representing the ionization of an amino acid by the two equations

$$^+NH_3RCHCOOH \longrightarrow {}^+NH_3RCHCOO^- + H^+$$

and

$$^+NH_3RCHCOO^- \longrightarrow NH_2RCHCOO^- + H^+$$

   with the ionization constants $K_1 = K_{COOH}$ and $K_2 = K_{NH_{3^+}}$, respectively, show that the hydrogen ion concentration at the isoelectric point of such a monoamino, monocarboxylic amino acid is given by

$$[H^+]_I = \sqrt{K_1K_2}$$

   b. The ionization constants for alanine and many other amino acids are in the range $K_1 = 10^{-2}\text{--}10^{-3}$ and $K_2 = 10^{-9}\text{--}10^{-10}$. Specifically, for alanine, $K_1 = 4.47 \times 10^{-3}$ and $K_2 = 2.04 \times 10^{-10}$. What is the isoelectric pH of alanine?

13. A mixture of the amino acids glycine, lysine, alanine, arginine, and glutamic acid is placed in an electric field between two electrodes. If the pH of the mixture is 6, which acid migrates toward the positive electrode (anode), which toward the nega-

tive electrode (cathode), and which do not move in the field?

14. In what direction, if any, would the following peptides migrate in an electric field when in a solution of pH $\simeq$ 6?
    a. lys-gly-ala-gly
    b. his-gly-glu-ala
    c. gly-glu-ala-glu

15. a. Write an equation for the hydrolysis of the dipeptide

$$H_2N-CH_2-\overset{\overset{\displaystyle O}{\|}}{C}-NH-\overset{\overset{\displaystyle CH_3}{|}}{CH}-COOH$$

    and name the products.

16. Suppose that partial hydrolysis of a peptide yields the following fragments: asp-glu, glu-his-leu-cys, aspNH$_2$-val-asp, his-leu-cys-gly, and val-asp-glu. What is the primary structure of the peptide?

17. Complete hydrolysis of a peptide yields alanine, valine, and cysteine in the proportions 1:2:2. Partial hydrolysis gives the fragments ala-val, val-cys-val, and cys-val-cys. What is the amino acid sequence in the peptide?

Answers

1. five isomers
12. b. 6.02
13. to anode: glu
    to cathode: arg, lys
    no movement: gly, ala

14. a. to cathode
    b. no movement
    c. to anode
16. aspNH$_2$-val-asp-glu-his-leu-cys-gly
17. ala-val-cys-val-cys

References

LIFE AND THE ORIGIN OF LIFE

Bernal, J. D., "The Origin of Life," World, Cleveland, 1967.

Handler, Philip, Ed., "Biology and the Future of Man," Oxford, New York, 1970, Chapter 5.
    This book is also recommended reading for other topics covered in this text.

Keosian, J., "The Origin of Life," 2nd ed., Van Nostrand Reinhold, New York, 1968.
    A short review.

Oparin, A. I., "The Chemical Origin of Life," Charles C. Thomas, Springfield, Ill., 1964.
    A short book by the founder of modern studies on the origin of life.

Wald, George, "The Origin of Life," *Scientific American*, **191,** 44–53 (August 1954).

ORGANIC CHEMISTRY

White, Emil H., "Chemical Background for the Biological Sciences," 2nd ed., Prentice-Hall, Englewood Cliffs, N. J., 1964.

MOLECULES OF LIFE

Barry, J. M., and Barry, E. M., "Introduction to the Strucutre of Biological Molecules," Prentice-Hall, Englewood Cliffs, N. J., 1969.

Bennett, Thomas Peter, and Frieden, Earl, "Modern Topics in Biochemistry," Macmillan, New York, 1966, Chapters 1–3, 7–9.

Dickerson, Richard E., and Geis, Irving, "The Structure and Action of Proteins," Harper & Row, New York, 1969.

Doty, Paul, "Proteins," *Scientific American*, **197,** 173–184 (September 1957).

Kendrew, John C., "The Three-Dimensional Structure of a Protein Molecule," *Scientific American*, **205,** 96–110 (December 1961).
    X-Ray crystallography of proteins.

Merrifield, R. B., "The Automatic Synthesis of Proteins," *Scientific American*, **218,** 56–74 (March 1968).

Pauling, Linus, Corey, R. C., and Hayward, R., "The Structure of Protein Molecules," *Scientific American*, **191,** 51–59 (July 1954).

Perutz, M. F., "The Hemoglobin Molecule," *Scientific American*, **211,** 64–76 (November 1964).

Rhinesmith, H. A., and Cioffi, L. A., "Macromolecules of Living Systems," Van Nostrand Reinhold, New York, 1968.

Smith, C. U. M., "Molecular Biology," M.I.T. Press, Cambridge, Mass., 1968.

Stein, William H., and Moore, Stanford, "The Chemical Structure of the Proteins," *Scientific American*, **204,** 81–92 (February 1961).

MISCELLANEOUS

Lightly written books useful as an introduction to cells, biochemistry, molecular biology, and other topics covered in Chapters 15–19.

Asimov, Isaac, "Life and Energy," Bantam, New York, 1965.

Butler, J. A. V., "Inside the Living Cell," Science Editions, New York, 1959.

Gerard, R. W., "Unresting Cells," Harper & Row, New York, 1961.

Jevons, F. R., "The Biochemical Approach to Life," 2nd ed., Basic Books, New York, 1968.

Energy and Life

You cannot step twice into the same rivers; for freshwaters are ever flowing in upon you.
Heraclitus

Life is a pure flame, and we live by an invisible sun within us.
Sir Thomas Browne—
Hydriotaphia

Every living organism takes in matter and energy from its surroundings. This is a necessity if the organism is to maintain and repair its ordered structure, grow, move, and reproduce. The ultimate source of the energy required by all living systems is the sun. Photosynthetic producer organisms use sunlight directly to manufacture complex molecules from simple molecules. In the synthesized carbohydrates, proteins, and fats, solar energy is stored and transferred to other organisms (the consumers or heterotrophs) as chemical energy to be extracted for the living needs of the organism.

All the chemical reactions occurring in an organism in connection with the production and use of energy are considered under the general heading of *metabolism*. In a more general sense, the term metabolism refers to all the chemical reactions and associated energy transformations taking place within the cells of living organisms.

16.1 CELLS The cell is the fundamental unit of structure and function in almost all living things. If we ignore the few apparent exceptions, it is proper to say that the cell is the basic unit of life much as an atom is the basic unit of matter. Although cells may differ considerably in size, shape, and details of function, from the single cells of unicellular organisms to the specialized cells of some complex multicellular organisms, there are certain features all have in common. Here we first consider some of the common structural features of the cells in more advanced organisms.

Figure 16.1 is a simple representation of a highly generalized cell, showing only those structural units whose function will be mentioned in the next few chapters. This cell has an outer plasma membrane that isolates the contents of the cell from the external environment. The interior of the cell is divided into two compartments. The inner body, the *nucleus*, is set off from the larger outer space, the *cytoplasm*, by a double nuclear membrane (except for bacteria and blue-green algae). Within the cytoplasm, there are a number of other bodies, or *organelles*. Figure 16.1 shows only three kinds of organelles, the *mitochondria, ribosomes,* and portions of an extensive membrane system (the endoplasmic reticulum) whose surface is often covered with ribosomes. In plant cells, other highly structured bodies, the *chloroplasts,* are found in the cytoplasm.

Our interest in the cell centers around its activity as a miniature but very intricate chemical factory where complex energy-laden molecules are de-

Figure 16.1. Schematic drawing of an animal cell.

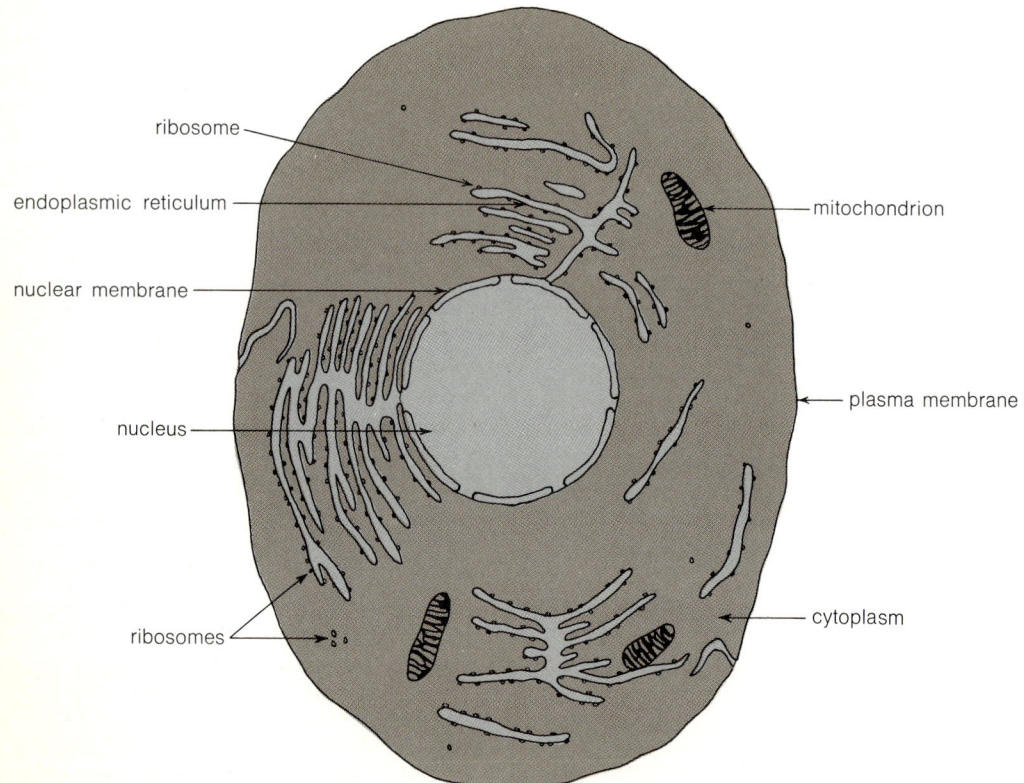

graded, and the energy released in the oxidative degradation process is used to synthesize other biological molecules and to perform various kinds of work. In this cellular chemical factory, the nucleus plays the part of the control room or of the board chairman. Within the nucleus are the master instructions in the form of DNA molecules, which tell the rest of the cell what to do and when to do it. The nucleus sends RNA molecule messengers out of the nuclear inner sanctum through pores to the granular ribosomes, which are also largely composed of RNA. At the ribosomes, additional protein is synthesized in accordance with the instructions delivered by the RNA messenger. Some of the proteins made at the ribosomes function as enzyme catalysts for the specialized chemical reactions going on in other compartments or laboratories of the chemical factory. The biosynthesis of protein is aided by the transportation and storage capacities of the folded membrane channels for the reactants and products of biosynthesis. The membranes of the reticulum also may contain enzymes required for protein synthesis at the ribosome.

The mitochondria are the power-generating compartments of the cell. Here fats and carbohydrates are oxidized, and the chemical energy released during oxidation is used to synthesize high-energy (ATP) molecules, which serve as the common fuel for all other energy-requiring processes of the cell. Figure 16.1 shows only a few mitochondria in the cell, but a single cell may contain from 50 to 5000 mitochondria.

The various membrane systems of the cell are not merely passive semipermeable skins used to package different areas of the cell. In most cases, they are the seat of active chemical reactions involving transport across the membrane or biosynthesis near the membrane. The outer cell membrane, for instance, must be the site of energy-consuming processes needed to move substances against a concentration gradient, that is, from a region of lower concentration to a region of higher concentration. The K^+ ion concentration in the cytoplasm is generally higher than that in the fluid outside the cell, while the situation is the reverse for Na^+. Since the cell membrane is freely permeable to both ions, some energy-requiring pumping mechanism must bring in K^+ and force out Na^+.

Energetics

16.2 THE FIRST LAW OF THERMODYNAMICS

The subject of thermodynamics, or more generally, energetics, is concerned with the way energy is transformed and transferred to or from one specific portion of the universe. The portion of the universe on which we focus our attention is called a *system*. A system is separated from its surroundings by real or imaginary boundaries which may allow the exchange of energy and matter with the surroundings. Most systems considered in elementary energetics are *closed* systems, which are defined as systems that cannot exchange matter with their surroundings. Living systems, as we shall see later, do exchange matter with their surroundings.

It will be helpful to select a specific system to illustrate the interactions between system and surroundings. A gas enclosed in a cylinder with a movable piston constitutes a rather uninteresting but illustratively useful system. If the gaseous system is in thermal contact with surroundings at a higher temperature, heat (q) will flow into the system. The thermal (kinetic molecular) energy of the gas molecules will increase. The thermal energy of the surroundings will decrease in the same amount. As a result of the initial increase in thermal energy, the system may be able to do work (w) on the surroundings. In the simple system considered here, the heated gas could expand and do mechanical work by moving the piston against the external pressure of the surroundings or by increasing the height of a massive piston in the earth's gravitational field. In other situations, the system might perform electric work by moving a charge in an electric field.

Heat is a manifestation of random molecular motion, while work corresponds to ordered motion on the macroscopic level. As we shall see later, it is highly improbable that the heat added to a real system will be converted completely into work and transferred out of the system. Instead, some of the original thermal energy added to the system will be retained within the system as *internal energy* (E) of the system. Internal energy is an all-inclusive term signifying all forms of kinetic and potential energy connected with the internal motions and composition of a system. On a microscopic level, internal energies correspond to the kinetic energies of translation, vibration, and rotation of molecules and the potential energies of charges in the electric force fields of other charges.

If E_1 is the internal energy of the system before the inflow of heat from the surroundings and E_2 is the final internal energy of the system after heat q has been added and work w has been done, then the net increase in the system's internal energy is equal to the energy represented by the difference between the heat added (gained) and the work done (lost):

$$E_2 - E_1 = \Delta E = q - w$$

The symbol ΔE without a subscript always means the change in energy of the system, ΔE_{sys}.

This equation is a formulation of the *first law of thermodynamics* or, what is the same thing, the *law of conservation of energy*, because it asserts that the energy added as heat must show up within the system as other forms of energy or as work done by the system. Writing the equation in the form

$$q = \Delta E + w$$

shows more clearly that the sum of the energy retained in the system and that expended as work must exactly equal the energy originally put into the system as heat.

The form of the first law used here requires adherence to the following sign conventions:

1. The sign of the numerical value of q is positive when heat is absorbed by the system; the sign is negative when heat is evolved by the system.

2. The work w is positive when work is done by the system on the surroundings; w is negative when work is done on the system by the surroundings.

With these conventions, an increase in internal energy of a system will have a positive sign.

EXAMPLE | The net energy change for the system plus surroundings is
$$\Delta E_{total} = \Delta E_{sys} + \Delta E_{surr}$$

By the first law, the quantity $q - w$ equals the net energy *loss* by the surroundings, $-\Delta E_{surr}$, so that

$$\Delta E_{sys} = -\Delta E_{surr}$$

or

$$\Delta E_{sys} + \Delta E_{surr} = 0 = \Delta E_{total}$$

Because the system plus surroundings may, if we wish, encompass the entire universe,

$$\Delta E_{universe} = 0$$

This is the standard formulation of the law of conservation of energy, namely, the total energy of the universe remains constant.

The subscripts 1 and 2 in $E_2 - E_1 = \Delta E$ correspond to the initial and final states of a system that has undergone a change. The state of a system is defined by specifying the values of a sufficient number of its measurable properties such as temperature, pressure, volume, and quantity of matter such that, for any set of these variables, all other remaining properties are fixed in value. This implies that the system is in a condition of thermal and mechanical equilibrium, so that the same temperature and pressure values apply uniformly throughout the system. This is not always the case in real systems, but we shall consider it to be so unless nonequilibrium conditions are specified.

Functions of State   The change in internal energy of a system has a very important property, namely, in going from one state to another, ΔE depends only on the initial and final states and not on the way the system was taken from one state to the other. This follows from the fact that each state of the system as defined by giving its T, P, V, and so on has a definite internal energy. As an example, consider a mole of gas having a pressure P_1, a volume V_1, and a temperature T_1. If its temperature is changed to T_2, the new pressure and volume of the gas will be P_2 and V_2. The change in internal energy of the gaseous system will be

$$E(P_2, V_2, T_2) - E(P_1, V_1, T_1) = \Delta E$$

There are many ways in which the final state of the gaseous system might be reached, as shown in Figure 16.2. To understand the smaller graphs in Figure 16.2, we must know that the work done on the surroundings by an expanding gas is proportional to the area under the curve in the PV diagram representing the course of expansion of the gas. Three fairly simple paths from P_1, V_1, T_1 to P_2, V_2, T_2 are shown. Each corresponds to a different amount of work done by the expanding system as well as to a different amount of heat absorbed from the surroundings. Despite the variation in w and q for the different paths, the difference $q - w$ for all paths from the same initial to the same final state is the same for all paths and is equal to ΔE. If this were not so, it would be possible to go from the initial state to the final state by one path that absorbs an amount of internal energy less than would be evolved in the return from the final to the initial state by another path. The net result would be the creation of energy from nothing. The continued cycling of the system would create energy and permit perpetual motion, something that is contrary to all experience.

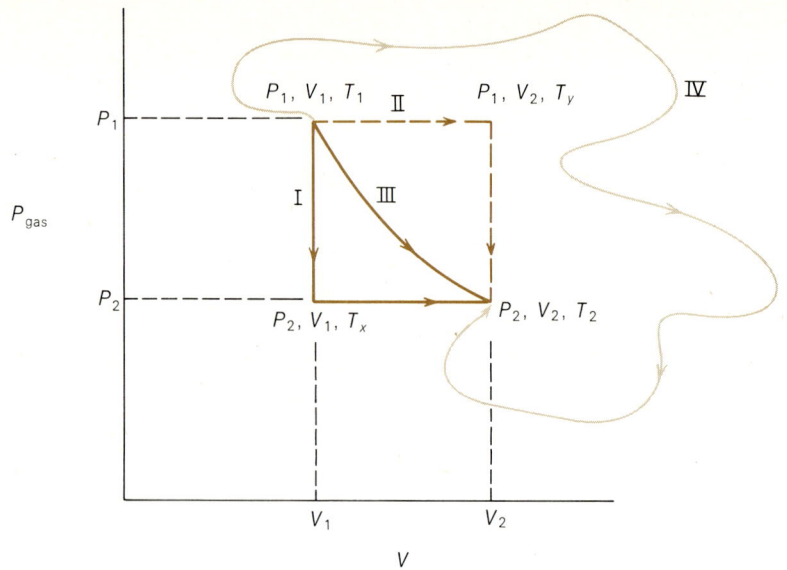

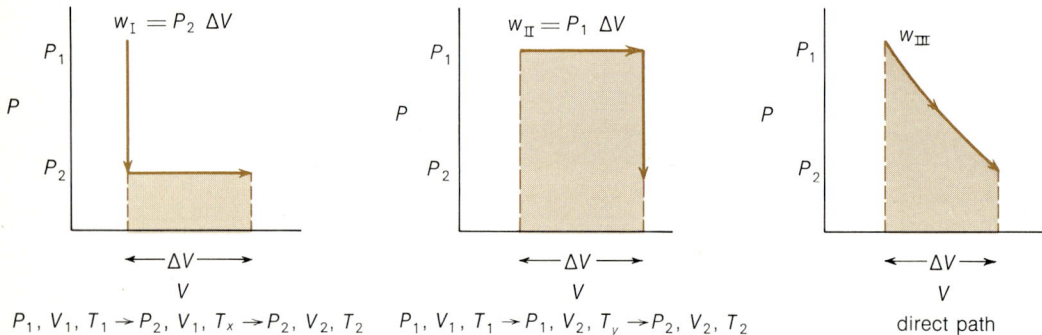

Figure 16.2. Various paths between states of a gas. The area of the shaded portions under the curves is proportional to the work done by the expanding gas.

Quantities such as the internal energy E that are independent of path from one state of a system to another are called *thermodynamic functions* or *functions of state*, because they depend only on state of the system. Heat q and work w are not path independent and are not functions of state except under special circumstances.

16.3 APPLICATIONS OF THE FIRST LAW TO GASES
Work Done by Expansion of a Gas at Constant Pressure

The mechanical work done when a gas expands against a constant external pressure, P_{ext}, is

$$w = P_{ext}(V_2 - V_1) = P_{ext} \, \Delta V$$

As an example, if a gas enclosed in a cylinder (Figure 16.3) with a frictionless, weightless piston expands from 1.00 liter to 2.00 liters against a constant external pressure (outside the cylinder) of 10 atm, the work done by the gas is

$$w = P_{ext} \, \Delta V = (10.0 \text{ atm})(1.00 \text{ liter}) = 10.0 \text{ liter atm}$$

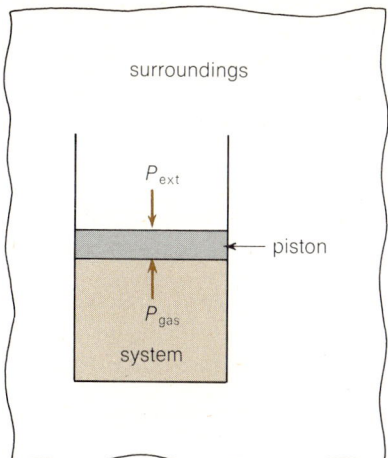

Figure 16.3. A simple system, a gas in a cylinder fitted with a frictionless piston. The gas is the system; the cylinder, the piston, and everything else are the surroundings.

In thermal units, since 1 liter atm equals 24.2 cal, $w = 242$ cal. Note that the pressure used is the resisting pressure of the surroundings, not the pressure of the gas inside the cylinder.

Only in the special case in which the pressure of the gas is a constant and only infinitesimally more than the external pressure can the pressure of the gas itself be used to calculate the work of expansion. This is called a *reversible process*, because the expansion process can always be reversed and changed to a compression by increasing the external pressure by an infinitesimal amount. This implies that (1) the gas is always in equilibrium so that the same value of P_{gas} applies throughout the gas volume and (2) the reversible expansion process takes an infinite amount of time. The work done in reversible and irreversible processes is

$$w_{rev} = P_{gas} \Delta V$$
$$w_{irrev} = P_{ext} \Delta V$$

Since P_{ext} is always less than P_{gas} by a finite amount in any actual (irreversible) expansion of a gas occurring at a finite rate, the hypothetical reversible process always does more work but very, very slowly. All natural processes are irreversible in the thermodynamic sense and spontaneous; that is, they can be thought of as taking place without any outside help.

EXAMPLE | How much work is done by a gas with an initial pressure of 1 atm expanding into a vacuum, that is, a so-called free expansion against a constant pressure $P_{ext} = 0$?

SOLUTION | The work done is a function of the external pressure alone, which is zero,

$$w = P_{ext} \Delta V = 0$$

No work is done by the expanding gas. The free expansion of a gas is an irreversible process.

Adiabatic Expansion of a Gas | An adiabatic system is a thermally insulated one that allows no transfer of heat to or from the system (between the system and its surroundings). In an adiabatic process, $q = 0$ and

$$\Delta E = -w = -P_{ext} \Delta V$$

If a gas is permitted to expand adiabatically, the energy required for the work of expansion must be drawn from the internal energy of the gas, since no energy in the form of heat may enter the system from the surroundings. The internal energy of the gas decreases on expansion. Because the internal energy of an ideal gas, at least, depends only on the temperature (by the kinetic theory of ideal gases, $E = \frac{3}{2}RT$ or $\Delta E = \frac{3}{2}R \Delta T$, Section 4.13), the temperature of the gas must fall during an adiabatic expansion. Conversely, an adiabatic compression causes an increase of temperature. Adiabatic conditions can be approached even for uninsulated systems if the rate of heat flow into or out of the system is too slow to equalize temperatures rapidly.

EXAMPLE   An ideal monatomic gas at 10.0 atm and 300°K expands adiabatically against a constant external pressure of 1.00 atm until $P_{2(gas)} = P_{ext}$. What is the final temperature of the gas?

SOLUTION

$$q = 0$$
$$\Delta E = -w$$
$$\Delta E = n\frac{3}{2}R(T_2 - T_1)$$
$$w = P_{ext}(V_2 - V_1)$$

so that

$$n\frac{3}{2}R(T_2 - T_1) = -P_{ext}(V_2 - V_1)$$

Expressions for the initial and final volumes are given by the perfect gas law,

$$V_1 = \frac{nRT_1}{P_1} = \frac{nR(300°K)}{10 \text{ atm}}$$

$$V_2 = \frac{nRT_2}{P_2} = \frac{nRT_2}{1 \text{ atm}}$$

Then

$$n\frac{3}{2}R(T_2 - 300°K) = -\left(nRT_2 - \frac{300nR}{10}\right)$$

Cancelling R and n and solving for T_2 gives

$$T_2 = 192°K$$

Processes at Constant Pressure (Enthalpy)   Most chemical reactions, and especially those in living organisms, take place at conditions of constant atmospheric pressure. If only mechanical work, $w = P \Delta V$, against the constant external pressure is done, the heat absorbed at constant pressure, q_P, is given by the first law as

$$q_P = \Delta E + P \Delta V$$

or

$$q_P = E_2 - E_1 + PV_2 - PV_1$$
$$= (E_2 + PV_2) - (E_1 + PV_1)$$
$$= H_2 - H_1$$
$$= \Delta H$$

Here a new quantity, $H = E + PV$, called the *heat content* or *enthalpy* of the system, has been introduced. The enthalpy change is equal to the heat absorbed in a constant pressure process where only $P \Delta V$ work is done. Like internal energy, enthalpy is a function of state, so ΔH is independent of the path between two states of a system.

16.4 THERMOCHEMISTRY AND HEAT OF REACTION

In previous chapters, we have often used heats of reaction in our discussions. Those reactions that gave off heat were exothermic and had a negative ΔH. Endothermic reactions absorbed energy when reactants formed products and had a positive ΔH because the products possessed more energy than the reactants. These heats of reactions (ΔH's) apply to reactions run under conditions of constant pressure and are what we now call enthalpy changes. Examples of heat changes for a number of physical and chemical changes have been specifically pointed out. Among these have been heats of vaporization, fusion, sublimation, and solution as well as the heat quantities corresponding to bond energies.

In some instances, the heats of several reactions have been added together to give the heat of a composite reaction, which can be used, for example, to compute bond energies (Section 3.16). The justification for this procedure is based on the fact that ΔH is independent of path. Since enthalpy is a function of state, it makes no difference how the reactants in a chemical or physical reaction are transformed to products. For the reaction

$$\text{reactants} \longrightarrow \text{products}$$
$$\text{(state 1)} \qquad \text{(state 2)}$$
$$\Delta H = H_{\text{prod}} - H_{\text{reac}}$$

ΔH will be the same whether reactants are transformed directly to products or the reactants go to products by a number of intermediate steps. To illustrate this point, which is sometimes called Hess' law of constant heat summation, consider the combustion of glucose at 1 atm and 25°C:

$$C_6H_{12}O_6(s,\ 1\ \text{atm},\ 298°K) + 6O_2(g,\ 1\ \text{atm},\ 298°K) \longrightarrow$$
$$6CO_2(g,\ 1\ \text{atm},\ 298°K) + 6H_2O\ (l,\ 1\ \text{atm},\ 298°K)$$

$$\Delta H_{\text{comb}} = -673.0\ \text{kcal/mol}$$

Suppose the heat of formation ΔH_f^0 of glucose is desired. This heat is defined as the enthalpy change for the reaction in which 1 mol of glucose is formed from its elements in their most stable form at 1 atm pressure and 25°C, their standard states.

$$6C(\text{graphite}) + 6H_2(g) + 3O_2(g) \longrightarrow C_6H_{12}O_6(s) \qquad \Delta H_f^0(\text{glucose})$$

Here the zero superscript on ΔH_f^0 refers to the standard states at 1 atm and 25°C.

Similar formation reactions for $CO_2(g)$ and $H_2O(l)$ can be written:

$$C(\text{graphite}) + O_2(g) \longrightarrow CO_2(g) \qquad \Delta H_f^0(CO_2) = -94.1\ \text{kcal/mol}$$
$$H_2(g) + \tfrac{1}{2}O_2(g) \longrightarrow H_2O(l) \qquad \Delta H_f^0(H_2O) = -68.3\ \text{kcal/mol}$$

By convention, the heat of formation of an element (from itself) is zero.

The experimental measurement of ΔH_f^0((glucose) by direct combination of the elements is not possible. However, by using the property of independence of path of ΔH, we can calculate ΔH_f^0(glucose) by devising a suitable but imaginary round-about route to glucose. This indirect path may be constructed by adding together (1) the reverse of the combustion equation for glucose, (2) six times the formation equation of CO_2, and (3) six times the formation equation for H_2O as shown below. Conditions of 1 atm pressure and 25°C are assumed.

$$6CO_2(g) + 6H_2O(l) \longrightarrow C_6H_{12}O_6(s) + 6O_2(g)$$
$$\Delta H = 673.0 \text{ kcal}$$

$$6[C(\text{graphite}) + O_2(g) \longrightarrow CO_2(g)]$$
$$6\,\Delta H_f^0(CO_2) = 6(-94.1) \text{ kcal}$$

$$6[H_2(g) + \tfrac{1}{2}O_2(g) \longrightarrow H_2O(l)]$$
$$6\,\Delta H_f^0(H_2O) = 6(-68.3) \text{ kcal}$$

$$\overline{6C(\text{graphite}) + 3O_2(g) + 6H_2(g) \longrightarrow C_6H_{12}O_6(s) \qquad \Delta H_f^0(\text{glucose})}$$

where

$$\Delta H_f^0(\text{glucose}) = 673.0 + 6(-94.1) + 6(-68.3)$$
$$= 673.0 - 974.4$$
$$= -301.4 \text{ kcal/mol}$$

In practice, heats of combustion are determined by burning a substance in pure oxygen inside a closed vessel called a *bomb calorimeter*. The measured heat evolved is actually $q_V = \Delta E_{comb}$, but this is converted to $q_P = \Delta H_{comb}$ by

$$\Delta H = \Delta E + P\,\Delta V$$
$$= \Delta E + (\Delta n)RT$$
$$[\text{since } PV = nRT \text{ and } P\,\Delta V = \Delta n(RT) \text{ at constant } T]$$

where Δn is the change in the number of moles of *gas* involved in the reaction. For the combustion of glucose, $\Delta n = n_{CO_2} - n_{O_2} = 6 - 6 = 0$ so that $\Delta H_{comb} = \Delta E_{comb}$.

The reaction for the formation of glucose from its elements is an imaginary reaction. The fact that we can add up a series of reactions, some real, some not, to get an imaginary reaction should help us remember that enthalpy changes are determined solely by the initial and final states of a system. The set of reactions that were combined to compute the enthalpy changes have nothing to do with the actual pathway of the total reaction. The pathway or mechanism is the actual sequence in which the molecules come together to form the products in the naturally occurring reaction. Just because a reaction might be written A $\longrightarrow$ M does not mean that A transforms directly to M in one reaction step. There may be a hundred intermediate steps between A and M in the actual transformation. The application of Hess' law tells us nothing about these intermediate steps.

16.5 SPONTANEITY AND DIRECTION OF PROCESSES

At one time, it was believed that processes took place by themselves, or spontaneously, only in the direction corresponding to a decrease in "energy." Thus, water always ran downhill because its potential energy decreased. Or two or more atoms combined to form a stable molecule because the "energy" of the molecule was less than the combined energy of the individual atoms. Based on such reasoning and first law calculations, it was reasonable to conclude that processes involving a decrease in ΔH (or ΔE), that is, exothermic processes, would be spontaneous, whereas the reverse endothermic process would not take place without some energy supply from the surroundings.

Further experience has shown the sign of ΔH is not a good criterion for the prediction of the direction of a process, particularly for nonmechanical and thermal processes where the internal atomic and molecular structure of the system undergoes rearrangement. There are many instances of a system spontaneously changing from an initial state to a final state with a zero or a positive change in enthalpy or internal energy, as illustrated by the following two examples.

EXAMPLE | What is the change in internal energy for the expansion of an ideal gas against a resisting pressure of $P_{ext} = 0$, that is, into a vacuum?

SOLUTION | In this case, $w = P_{ext} \Delta V = 0$ no matter what the final volume of the gas. Whether the expansion is carried out isothermally ($\Delta T = 0$) or adiabatically ($q = 0$), the change in internal energy comes out to be $\Delta E = 0$. Yet this kind of process is obviously spontaneous.

EXAMPLE | What is the enthalpy change for the melting of ice at 10°C?

SOLUTION | The melting of ice at 10°C is a spontaneous process, but the enthalpy change is positive:

$$H_2O(s, 10°C) \longrightarrow H_2O(l, 10°C) \qquad \Delta H = 1.5 \text{ kcal/mol}$$

Other examples could be given, but these two are sufficient to show that energy changes alone do not control the direction of a reaction.

Spontaneous Processes Are Irreversible

An important characteristic of spontaneous processes is their unidirectional property. Such processes are never found to proceed spontaneously in the opposite direction even though all the conditions of temperature, pressure, and so forth appropriate to the original state of the system are restored. Thus, a gas originally confined to one-half of a cylinder and then allowed to expand into a vacuum in the other half is never known to gather itself back spontaneously into the original half of the cylinder, leaving the other side empty. And water at 10°C has never been observed to freeze spontaneously at that temperature, any more than water has ever been known to run uphill of its own accord. All these processes are inherently irreversible, meaning that the system cannot be restored to the original state without an appreciable interaction with the surroundings which results in a change in the surroundings. The only way the expanded gas in the cylinder can be stuffed back into the original half of the cylinder, or the water refrozen, or water made to run uphill is through the expenditure and loss of energy by the surroundings to operate a compressor, a refrigerator, or a pump, respectively. By contrast, reversible processes can be reversed without any resultant changes in the system or the surroundings.

Another way to distinguish reversible and irreversible processes is in terms of the magnitude of the changes needed to reverse a process once it has begun. Reversible processes need only infinitesimal changes to reverse the events taking place. If a gas is expanded against an external pressure only infinitesimally less than the pressure of the gas, the expansion may be stopped and reversed by only infinitesimally increasing the external pressure until it is greater than the gas pressure. However, to reverse the free expansion of a gas at a finite pressure into a vacuum requires the imposition of a finite, external pressure. The melting of ice at 0°C may be reversed by making the temperature infinitesimally less than 0°C, but the melting of ice at 10°C cannot be arrested or reversed until the temperature has been decreased by a finite 10°C or more. Therefore, the expansion of a gas against an external pressure only slightly less than the gas pressure at any instant is a reversible change. But if the external pressure is appreciably less than the instantaneous gas pressure, the expansion is irreversible. The melting of ice at 0°C is reversible; at 10°C, the melting process is irreversible.

Spontaneous processes may be performed in a reversible or irreversible manner, although all naturally occurring processes are both spontaneous and irreversible to some extent. The concept of irreversibility is implicit in the term "spontaneous" as it is used here. The redox reaction of zinc with Cu^{2+} ion solution is irreversible if the zinc is added directly to the Cu^{2+} solution. But if an electrochemical cell composed of zinc and copper electrodes dipping into solutions of their ions is set up, the redox reaction

$$Zn + Cu^{2+} \rightleftharpoons Cu + Zn^{2+}$$

may be run reversibly by imposing an opposing voltage between the electrodes only infinitesimally greater or smaller than the electromotive force of the cell. By slightly changing the applied opposing voltage, the cell reaction may be reversed at any time.

Most of us have some intuitive feeling for many changes that we feel to be irreversible. If this were not so, we would not be amused by some of the things we see when a motion picture is run backwards. Some motions seem reasonable whether the film is run forward or backward. But other actions, such as a diver returning from the water to the diving board or a pie reassembling itself on the face of a victim and returning to the hand of the thrower, interest us because we are not used to seeing such things happen in real life. The likelihood of really seeing such events is vanishingly small. They are highly *improbable*, although, it turns out, *not impossible*. This is one clue to the nature of spontaneous processes; they are changes to a more probable state of a system. Another clue may be discerned if we think about how improbable it is to see order suddenly restored from a completely disordered or chaotic situation. The normal course of events is from order to disorder.

16.6 DISORDER AND PROBABILITY

It is not difficult to see that spontaneous processes correspond to an increase in the disorder of a system if we choose processes where the loss of order is apparent. The expansion of a gas into a larger volume leads to a more disordered state because the molecules are now allowed to roam around in a bigger space. They are not sorted according to position the way they were before expansion. Similarly, in the melting of ice or the vaporization of water, a loss of order or of molecular organization takes place. The formation of a solution by the mixing of two gases or of a solute and solvent are examples of processes leading to increased disorder because, after mixing, the molecules are no longer sorted into separate volumes containing only one kind of molecule. All molecules are mixed up with one another in a state of greater disorder or randomness.

In general, any system where the molecules are sorted according to energy, velocity, position, or kind of molecule is a more ordered situation than one where the molecules have evened out the energy or positional differences. Spontaneous reactions correspond to the passage from the microscopically ordered to the disordered state.

As we shall see, it is not probable that order will appear spontaneously from disorder. This is part of the reason that heat always flows from a warmer body to a cooler body. When the bodies are at different temperatures, their molecules are grouped into two average energy classes. But after heat flow has established thermal equilibrium, the molecules are no longer sorted according to velocity. Similarly, mechanical work involves a component of order at the molecular level (the molecules in a piston, for instance, have a net component of motion in the direction of motion of the piston), while heat corresponds to thermal energy associated with completely random molecular motions. Because spontaneous processes proceed in the direction of greatest disorder, work may be transformed spontaneously into heat, but the complete transformation of heat into work is not possible. The only way that order can be obtained from disorder is by the expenditure of energy from the surroundings.

Having obtained a qualitative idea of disorder, we can now look at the relation of disorder and probability on a slightly more quantitative level. We shall see that the systems or states with the greatest microscopic disorder are also the most probable states.

Probability of Distributions

To begin, let us refresh our ideas of how objects may be arranged or distributed in various ways and see how certain distributions occur more often and are therefore more probable than other arrangements. If four coins are tossed, the possible arrangements of heads and tails are as shown in Table 16.1. For purposes of deducing the different ways of distributing the four coins among the heads and tails categories, we have assumed that the coins are distinguishable, perhaps by being labeled 1, 2, 3, and 4. However, the 16 different

Table 16.1
Distribution of heads and tails

T T T T	one way to get 4 T
H T T T T H T T T T H T T T T H	four ways to get 1 H, 3 T
H H T T H T T H T T H H H T H T T H H T T H T H	six ways to get 2 H, 2 T
T H H H H T H H H H T H H H H T	four ways to get 3 H, 1 T
H H H H	one way to get 4 H

arrangements of coins yield only five different combinations of heads and tails as determined by the number of heads and tails in each combination. The distribution of two heads and two tails can occur more times, or is more probable, than any of the other four distributions. The relative probability of the 2 H, 2 T distribution is $\frac{6}{16}$; that of either the 3 H, 1 T or 3 T, 1 H distribution is $\frac{4}{16}$ each; and the relative probability of either the 4 H or 4 T distribution is only $\frac{1}{16}$ each. The sum of the relative probabilities for all distributions equals unity. It is clear that the arrangement of 2 H and 2 T is the most probable result for the flipping of four coins, because this particular distribution of heads and tails can occur in more ways. The arrangement of 2 H and 2 T is also less ordered than other distributions. With a much larger number of coins, or a great number of tosses of the same coin, the distribution giving equal numbers of heads and tails would still be the most probable. In fact, it becomes more and more probable as either the number of coins or the number of tosses increases. The number of possible ways of achieving

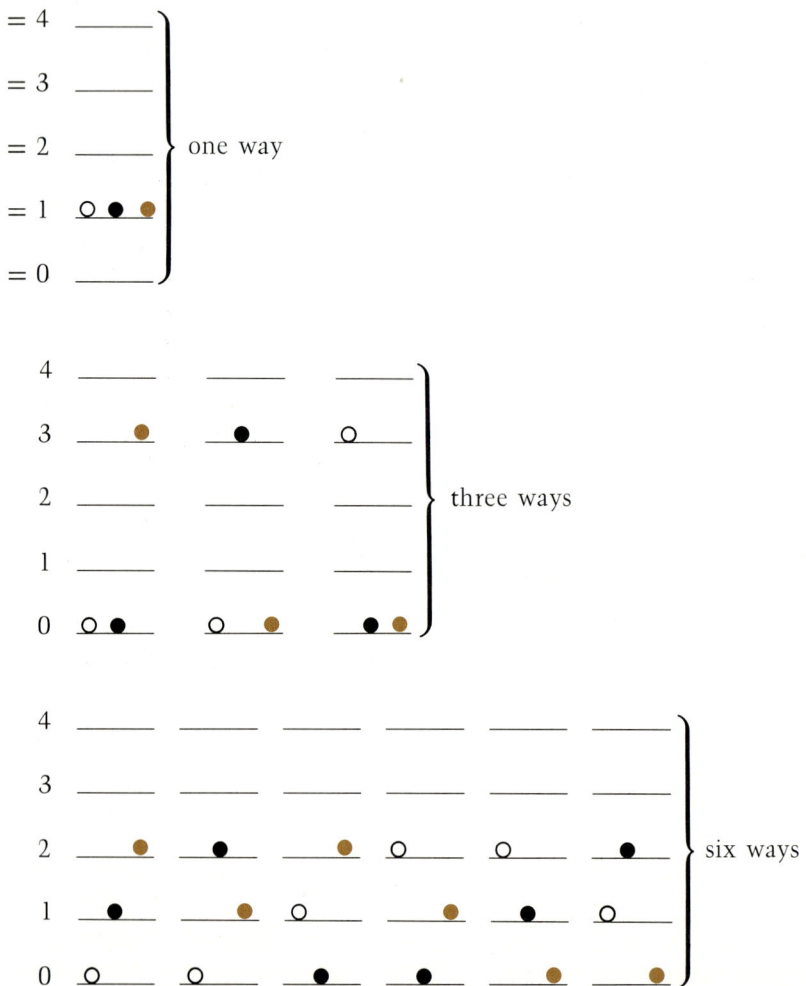

Figure 16.4. Distribution of three distinguishable molecules with a total energy of three units over molecular energy levels.

a distribution having N_H coins with heads and N_T coins with tails with a total of N coins is given by

$$W = \frac{N!}{N_H! N_T!}$$

where ! is read (not exclaimed) "factorial" and means a series of multiplications as in the example $4! = 4 \times 3 \times 2 \times 1$. This formula was obtained from a more general formula that gives the number of ways that N distinguishable objects can be distributed among a number of piles, groups, or levels so that there are N_1 objects in the first group, N_2 in the second, and so on. With the coins, there are only two groups, $N_1 = N_H$ and $N_2 = N_T$, since there are only two factorial terms in the denominator.

If, instead of coins, molecules are considered, the disorder and probability of a system will depend on how N molecules can be distributed over a series of allowed energy levels representing the quantized electronic, vibrational, rotational, and translational energies of the molecules. Consider three identical but distinguishable molecules A, B, and C with allowed energy levels $\epsilon_0 = 0$, $\epsilon_1 = 1, \epsilon_2 = 2, \epsilon_3 = 3, \epsilon_4 = 4, \ldots$ energy units. The total energy of the system of three molecules is constant and equal to three energy units. Figure 16.4 shows how the three molecules can distribute themselves among the energy levels. The distribution with $N_0 = 1$, $N_1 = 1$, and $N_2 = 1$ is seen to be the most probable. This probability can also be calculated from

$$W = \frac{N!}{N_0! N_1! N_2! N_3! \cdots}$$

with $N = 3, N_0 = 1, N_1 = 1, N_2 = 1, N_3 = N_4 = \cdots = 0$, with $0! = 1$. Then

$$W = \frac{3 \times 2 \times 1}{1 \times 1 \times 1} = 6$$

For later use, suppose that the same three molecules have the same total energy but that there are more available energy levels in the range $\epsilon = 0$ to $\epsilon = 3$, as shown in Figure 16.5. In addition to the previous ten ways of distributing the molecules among the levels ϵ_0, ϵ_1, ϵ_2, and ϵ_3, there will be other possibilities when molecules will be in the levels $\epsilon_{0.5}$, $\epsilon_{1.5}$, $\epsilon_{2.5}$ as well. For instance, the configuration $N_0 = 0, N_{0.5} = 1, N_1 = 1, N_{1.5} = 1$ can occur in six ways. Other energy level occupancies involving the additional half-integral levels could be deduced, and they would add to the total number of ways that three units of energy could be shared by the three molecules so the sum of all the W's would be greater than the ten ways in Figure 16.4.

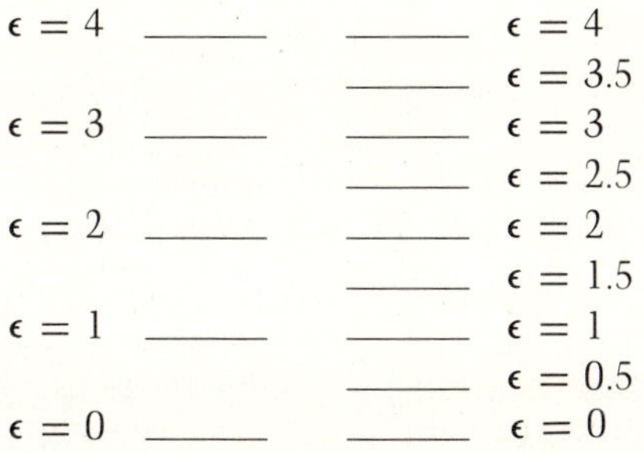

Figure 16.5. An increase in the number of energy levels within a given energy range will allow a greater number of distributions of molecules over the levels.

$\epsilon = 4$ _____   _____ $\epsilon = 4$

_____ $\epsilon = 3.5$

$\epsilon = 3$ _____   _____ $\epsilon = 3$

_____ $\epsilon = 2.5$

$\epsilon = 2$ _____   _____ $\epsilon = 2$

_____ $\epsilon = 1.5$

$\epsilon = 1$ _____   _____ $\epsilon = 1$

_____ $\epsilon = 0.5$

$\epsilon = 0$ _____   _____ $\epsilon = 0$

Finally, consider what happens when both the number of molecules and their total energy increases. If there are 20 particles with a total energy of 20 units, all of the molecules could be in the $\epsilon_1 = 1$ energy level. The probability of this distribution is $W = 1$. Another possible distribution is $N_0 = 10$, $N_1 = 4$, $N_2 = 3$, $N_3 = 2$, $N_4 = 1$, for which the value of W is greater than 10^8, showing that this distribution of molecules among energy states is much more probable than the former.

In any system composed of a large number of molecules, the distributions that can occur in the greatest number of ways (with the largest W) are those in which the lower energy states are most densely populated. Among all these high-probability distributions, there is one having a maximum probability W consistent with the number of molecules and their total energy. All natural systems (if left alone) tend to approach this internal condition of maximum probability, and they do so by spontaneous irreversible processes. The state of maximum probability is also the state with the most microscopic disorder. Since W is a measure of the probability of the molecular state of a system, it is also a measure of the disorder of that system.

An isolated system is one in which no matter or energy can be exchanged with the surroundings. Then $\Delta H = \Delta E = 0$, and a spontaneous process may occur if the system goes from a state of low microscopic probability (order) to a state of higher probability (disorder). In other words, the system tends to molecular distributions of high W. If the number of energy levels within a given energy range can change (as in Figure 16.5), the system will tend spontaneously to the state having the greater number of available energy levels. This state will be the one with the greater number of possible distributions of the molecules over the energy levels.

16.7 ENTROPY

In thermodynamics, the term *entropy* (S) is used as a measure of the degree of disorder or randomness of a system. Entropy is related to disorder, or probability, by the formula

$$S = 2.303k \log P$$

where k is Boltzmann's constant and P is the total number of all possible distributions of molecules over the available energy levels of the system. The value of P is equal to the sum of the individual W's, ΣW, for distributions having fixed N_1, N_2, N_3, and so on. For the system in Figure 16.4, $P = 1 + 3 + 6 = 10$.

We are not particularly interested here in absolute values of the entropy, but only in entropy changes. The change in entropy when a system goes from one state to another is

$$\Delta S = 2.303k \log \frac{P_2}{P_1}$$

If the probability or the randomness of the final state P_2 is greater than that of the initial state P_1, then ΔS is positive. A positive ΔS therefore corresponds to a spontaneous reaction if $\Delta H = \Delta E = 0$.

16.8 THE SECOND LAW OF THERMODYNAMICS

The second law of thermodynamics can be stated in many ways. We choose to express it in the following form.

The total entropy change, $\Delta S_{total} = \Delta S_{sys} + \Delta S_{surr}$, for any process must be either zero or positive.

If the process is reversible, $\Delta S_{total} = 0$; if it is spontaneous (irreversible), ΔS_{total} is positive. For an isolated system that cannot interact with its surroundings, $\Delta S_{surr} = 0$ and ΔS_{sys} must be either zero or positive.

A common approach to the second law starts with an operational definition of an entropy change,

$$\Delta S_{sys} = \frac{q_{rev}}{T}$$

where q_{rev} is the heat absorbed by the system (or lost by the surroundings) at temperature T when the system goes reversibly from state 1 to state 2. Even if an irreversible process takes place, it is always possible to devise a reversible path between states, calculate q_{rev}, and find ΔS_{sys}. Because ΔS is independent of path, ΔS_{sys} calculated by the reversible path is equal to ΔS_{sys} for the irreversible process. (In an irreversible process, however, the heat absorbed by the surroundings is $-q_{irrev}$.) By the first law,

$$q_{rev} - w_{rev} = q_{irrev} - w_{irrev}$$

Since w_{rev} is always greater than w_{irrev}, q_{rev} must always be greater than q_{irrev}. Then

$$\Delta S_{total} = \Delta S_{sys} + \Delta S_{surr}$$

$$= \frac{q_{rev}}{T} + \frac{-q_{irrev}}{T}$$

and ΔS_{total} is greater than zero for a spontaneous change.

While the total entropy change ΔS_{total} must increase or remain constant in any process, the second law permits a decrease in the entropy of the *system* in a spontaneous process so long as the system is not isolated. Failure to realize this has led to statements that claim living processes violate the second law. Living organisms are highly ordered and therefore highly improbable systems. The fact that they maintain and increase their order through growth, with a decrease in entropy of the living system (including all life on earth), merely means there is an even greater increase in entropy in the surroundings. This increase in entropy of the surroundings originates in the flow of low-entropy radiation (by high-energy photons) from the sun to the earth and the radiation of an equal amount of energy by high-entropy radiation (low-energy photons) to space by the earth.

16.9 FREE ENERGY AND SPONTANEITY

Two factors contribute to the spontaneity of any physical or chemical process. For processes at constant pressure, these factors are ΔH and ΔS. (From now on, it will be understood that ΔS without a subscript refers to the system.) It then follows for such processes that a decrease in ΔH is a sole criterion for spontaneity if $\Delta S = 0$. If $\Delta H = 0$, then an increase in ΔS is a prerequisite for spontaneous change. In the first case, the system tends to the state of lowest possible enthalpy; in the second case, the system changes until the state of highest possible entropy is attained.

Mechanical systems undergoing processes involving little or no change in internal molecular order, such as a ball rolling down a hill, are governed by enthalpy, since this represents the potential energy of the system. Where processes are accompanied by significant internal rearrangements, the entropy change may govern the direction of the process. Examples are the expansion of gases, the dissolving of solids, and the combination of atoms into molecules.

It is not uncommon to find processes in which the enthalpy and entropy effects work at cross purposes. An example is the vaporization of water. At any temperature, the vaporization of water is an endothermic process with a positive ΔH. In the absence of any other effects, water vapor should condense spontaneously at any temperature, because liquid water has the lower energy. On the other hand, the entropy effect favors the vaporization of liquid water and, again in the absence of other effects, water should vaporize spontaneously because of the greater freedom, randomness, disorder, or entropy of the molecules in the gas phase. At 1 atm pressure, the opposing enthalpy and entropy effects exactly cancel each other at a temperature of 100°C. The system is then in equilibrium, because no overall change in the system takes place. At lower temperatures the enthalpy effect is dominant, and water vapor condenses. At higher temperatures the entropy effect predominates, and liquid water vaporizes. It is always true that the entropy effect becomes more important at higher temperatures.

A new thermodynamic function of state called the *Gibbs free energy*, or just free energy for short, has been devised to relate the competing enthalpy and entropy factors quantitatively. The free energy G is defined as

$$G = H - TS$$

The change in free energy for a system undergoing a process at constant temperature and pressure is

$$\Delta G = \Delta H - T\,\Delta S$$

Since totally spontaneous processes are accompanied by negative ΔH and positive ΔS, it is safe to conclude that spontaneity is accompanied by a decrease in free energy. Conversely, a strictly nonspontaneous process would have a positive ΔH and a negative ΔS so that ΔG would be positive.

16.10 EQUILIBRIUM
When a system undergoes a spontaneous change, its free energy G decreases until it reaches a minimum. At the minimum, the opposing tendencies of enthalpy and entropy change are just balanced. That is, $\Delta H = T\,\Delta S$ and $\Delta G = 0$. This is the state of equilibrium toward which all spontaneous reactions proceed. Figure 16.6 illustrates the free energy for a chemical system at various stages of the reaction. Starting with pure reactants, a reaction will proceed spontaneously, the free energy of the system decreasing down the left-hand curve until the minimum is reached. At that point, the reaction is in a state of dynamic equilibrium, and the concentrations of reactants and products remain constant and are those appropriate to the equilibrium constant of the reaction. The other side of the diagram shows that pure products can also react spontaneously to form the same equilibrium mixture as did the pure reactants. Thus, the reaction

reactants $\rightleftharpoons$ products

can proceed spontaneously in either direction, depending on the starting concentrations in the reaction mixture. This is why double arrows are used.

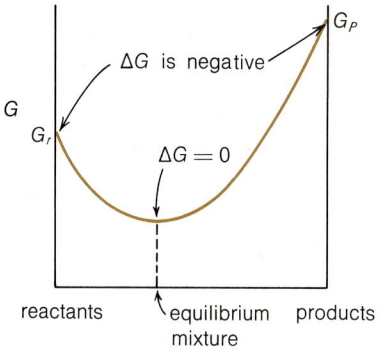

Figure 16.6. Variation of free energy in the course of a chemical reaction.

A reaction such as this is called a reversible reaction because the reaction may be made to proceed in either direction by a small change in conditions when the reaction is at the equilibrium point at the bottom of the curve. Here $\Delta G = 0$ for any slight departure of the system from the minimum. [A horizontal portion can be thought to exist at the minimum point in the curve. More accurately, the rate of change of free energy with extent of reaction or the slope of the curve (Figure 17.1) is zero at the minimum.] The reaction is not reversible on either side of the minimum point without a *large* change in conditions.

The free energy change ΔG is a single criterion for the tendency of a reaction to take place (at constant temperature and pressure). If the statement is made that all natural processes go in the direction of lower "energy" (exoergic process), then the energy referred to should be the free energy, not enthalpy or internal energy, because ΔH or ΔE can increase or decrease in a spontaneous reaction just as ΔS can. To summarize,

ΔG is negative for a spontaneous change.

$\Delta G = 0$ for a system in equilibrium.

ΔG is positive for a process that would not take place except with the supply of energy from the surroundings.

Either ΔG or ΔS_{total} can be used as a criterion for spontaneity, but ΔG is preferred because heat exchanges with the surroundings need not be considered.

EXAMPLE   Calculate ΔG for the following reactions at 1 atm pressure, given ΔH_{fus} and $\Delta S_{\text{H}_2\text{O}}$ for each reaction.

	ΔH_{fus}, cal/mol	$\Delta S_{\text{H}_2\text{O}}$, cal/°K mol
(1) $H_2O(s, 10°C) \longrightarrow H_2O(l, 10°C)$	1530	5.60
(2) $H_2O(s, 0°C) \longrightarrow H_2O(l, 0°C)$	1440	5.30
(3) $H_2O(s, -10°C) \longrightarrow H_2O(l, -10°C)$	1340	4.90

SOLUTION   For reaction (1),

$$\Delta G = 1530 \text{ cal/mol} - (283°K)(5.60 \text{ cal/°K mol})$$
$$= 1530 - 1590$$
$$= -60 \text{ cal/mol}$$

For reaction (2),

$$\Delta G = 1440 \text{ cal/mol} - (273°K)(5.30 \text{ cal/°K mol})$$
$$= 1440 - 1440$$
$$= 0$$

For reaction (3),

$$\Delta G = 1340 \text{ cal/mol} - (263°K)(4.90 \text{ cal/°K mol})$$
$$= 1340 - 1290$$
$$= 50 \text{ cal/mol}$$

The values of ΔG bear out our expectations. Ice at 10°C melts spontaneously. Ice and water are in equilibrium at 0°C, and ice does not melt spontaneously at $-10°C$.

The Jumping-Bean Model   The combined role of enthalpy and entropy in regulating the direction and extent of a process is well shown by the jumping-bean model due to Blum (1968). Mexican jumping beans are imagined to hop over a barrier occasionally from one compartment to another compartment. Several possible arrangements of the two compartments are shown in Figure 16.7. The beans or molecules have been distributed among "energy" levels for purposes of later discussion. Variations in the relative level of the compartment bottoms and compartment sizes permit a number of different configurations. The difference in level of the compartment floors corresponds to the difference in enthalpy ΔH (at equilibrium) of the initial and final states. Differences in compartment size are a partial measure of the difference in entropy ΔS between the initial and

Figure 16.7. Two-compartment situations for the jumping-bean model.

final states, the final state being the one where the beans have distributed

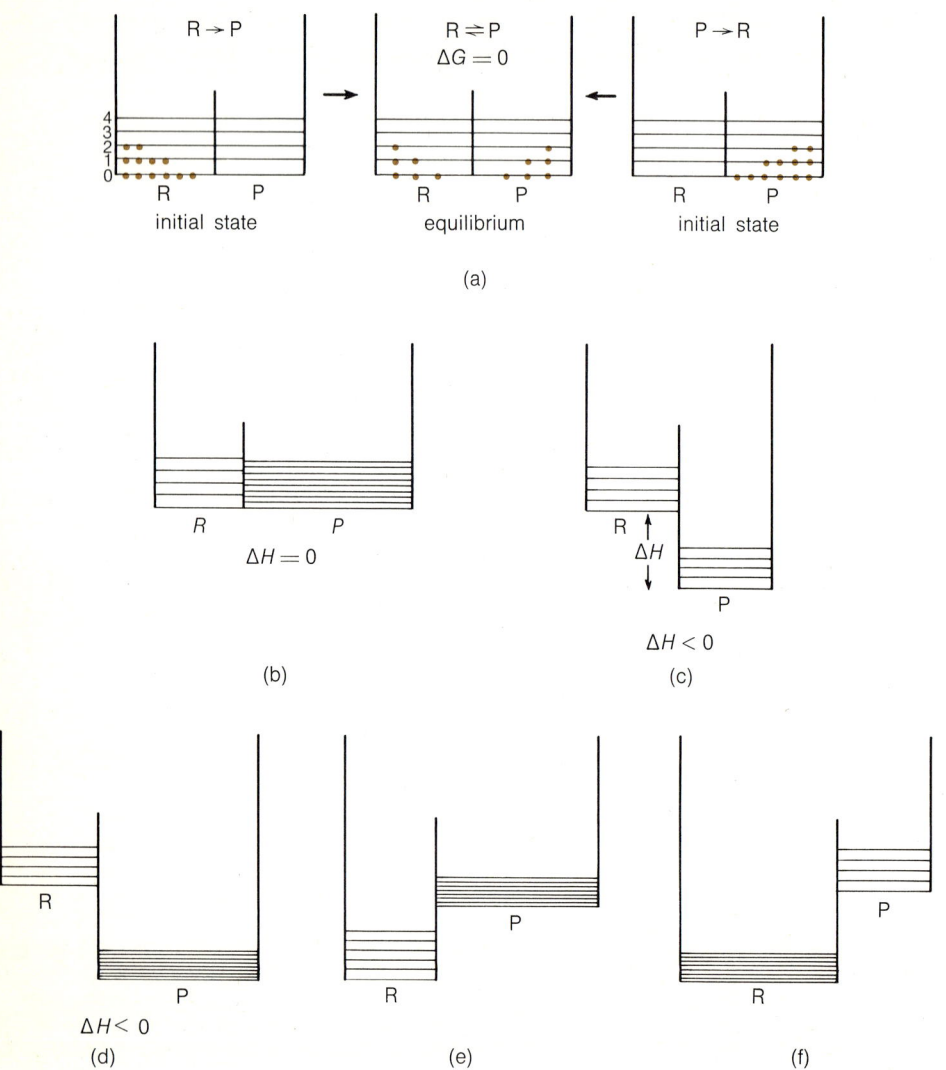

themselves between both compartments. This will be the equilibrium state, and as many beans will jump from R to P per unit time as jump from P to R. The ratio of beans in the two compartments remains constant with time in the final equilibrium state.

Situation (a): Here $\Delta H = 0$. The beans move out of the initial (reactant, R) state by jumping over the barrier until there are equal numbers in each compartment at equilibrium. Then $\Delta H = 0 = \Delta S$ and $\Delta G = 0$. The entropy effect causes the beans to distribute themselves, because there are more ways of distributing the 12 particles between two compartments than there are of distributing the particles in one compartment. You can show that the maximum number of distributions occurs when there are equal numbers of particles in each compartment. We need not consider the levels in each compartment for this case, but there are also more ways of distributing six particles over two sets of levels than there are for 12 particles over one set of levels, keeping the energy (H) of the system constant. The distributions of particles shown in the figure reflect the distribution of maximum probability with the lowest levels the most populated (Section 16.6)

Situation (b): Even though $\Delta H = 0$, the increased size of the P compartment brings the energy levels closer together (Section 2.22) so that there are more available levels on this side. The closer the levels, the greater the possible number of distributions over these levels. Therefore, the entropy effect is increased here and more particles will be found in compartment P at equilibrium. The equilibrium constant P/R will be greater than one. A bean jumping into compartment P as equilibrium is established will be less likely to jump back to R.

Situation (c): Here compartment P lies at a lower level than compartment R, so ΔH is negative, implying spontaneity for the jump from R to P. All the beans do not eventually end up in P, however. Some of the more lively beans will have enough energy to make the more energetic jump from P to R, but this will happen less often than the R-to-P jump. At equilibrium, with more beans in P than in R (the equilibrium constant is greater than unity), the rate of movement in both directions will be equal. Since both compartments have the same size and spacing of levels, entropy does not contribute appreciably to the spontaneous direction of the process, but entropy does prevent all the beans from ending up on the lower energy side.

Situation (d): Here both ΔH and ΔS will be favorable for spontaneous movement from R to P. At equilibrium, an even greater proportion of the beans will be in P than was the case in situation (c).

Two other situations, (e) and (f), are shown in Figure 16.7 for consideration in Problem 11.

In all these examples, ΔH is comparable to the difference in level between the bottoms of the compartments. The height of the barrier between compartments is analogous to the activation energy of the reaction ΔH_a. This energy barrier has no effect on ΔG or on the final equilibrium position. The activation energy affects only the rate at which equilibrium is established from either direction. Unlike ΔH, the activation energy differs in magnitude, but not in sign for the forward (R $\longrightarrow$ P) and reverse (P $\longrightarrow$ R) processes. Activation energy will be discussed in more detail in the next chapter. The free energy change for a process tells us nothing about the velocity of a reaction. Many cases are known where ΔG is large and negative (i.e., $H_2 + O_2 \longrightarrow H_2O$ and the oxidation of glucose) yet the reaction, though spontaneous, is negligibly slow at room temperature. Most complex biological molecules, such as carbohydrates and proteins, are thermodynamically unstable, but they do not decompose or oxidize rapidly unless an enzyme catalyst is present.

16.11 FREE ENERGY AND THE EQUILIBRIUM CONSTANT

The more negative the free energy change for a reaction, the further the reaction will go toward completion before equilibrium is established. The extent of a reaction is measured by its equilibrium constant K. For a hypothetical reaction,

$$a A + b B \longrightarrow c C + d D$$

the free energy change and the concentrations of products and reactants at any time, not just at equilibrium, are related by

$$\Delta G = \Delta G^0 + 2.303 R T \log \frac{[C]^c [D]^d}{[A]^a [B]^b}$$

Here ΔG^0 is the standard free energy change for the reaction in which reactants in their standard states at 298°K and 1 atm pressure or $1M$ concentration go to products in their standard states:

$$a A(1M) + b B(1M) \longrightarrow c C(1M) + d D(1M) \qquad \Delta G^0$$

If the concentrations of reactants and products are the equilibrium concentrations, then $\Delta G = 0$ and

$$\Delta G^0 = -2.303 R T \log K$$

where K is the equilibrium constant for the reaction. This equation shows that a reaction having a negative standard free energy change will proceed spontaneously to an equilibrium position in which K is greater than unity. The equilibrium will lie on the product side. Conversely, in a reaction with a positive ΔG^0, little product will be formed at equilibrium and K will be less than unity.

The standard free energy change ΔG^0 for a reaction may be calculated from known values of standard free energies of formation, just as was done for heats of reaction in Section 16.4. Another common way to find ΔG^0 and, through it, the equilibrium constant for a reaction, is to determine the standard emf (E^0) of the reaction in an electrochemical cell. The E^0 and ΔG^0 values are related by

$$\Delta G^0 = -n \mathcal{F} E^0 \qquad \text{(also } \Delta G = -n \mathcal{F} E)$$

and

$$\log K = \frac{n \mathcal{F} E^0}{2.303 R T} = \frac{n}{0.059} E^0 \qquad \text{at 25°C}$$

The equation

$$\Delta G = \Delta G^0 + 2.303 R T \log \frac{[C]^c [D]^d}{[A]^a [B]^b}$$

also indicates how a change in the concentration of reactants or products may cause a reaction at equilibrium to be displaced from its equilibrium position and become spontaneous in one direction or the other. If the concentration of reactants is increased (or the concentration of products decreased by removal) from their equilibrium values, the last term in this equation will become smaller in absolute magnitude (exclusive of sign) than ΔG^0. If ΔG^0 were negative, ΔG would become negative and the reaction would proceed in the direction of C and D until equilibrium was again established and $\Delta G = 0$ once more. The value of the equilibrium constant would not change. For instance (using an arbitrary $\Delta G^0 = -30$), at equilibrium $\Delta G = 0 = -30 + (+30)$. On changing the concentration of the equilibrium mixture, $\Delta G = -30 + (-20) = -10$ and reaction occurs. If ΔG^0 were positive, then the log term would be negative at equilibrium [$\Delta G = 0 = +30 + (-30)$]. Addition of A or removal of C or D would make the log term more negative so that ΔG would become negative, $\Delta G = +30 + (-40) = -10$, and reaction would again take place to form more C and D until equilibrium ($\Delta G = 0$) once again was established. All this is, of course, just a quantitative way of justifying what Le Chatelier's principle says will happen when equilibrium is displaced by the addition or removal of reactants or products. The smaller ΔG (the closer to $\Delta G = 0$), the more nearly reversible the reaction will be.

16.12 FREE ENERGY AND USEFUL WORK

The free energy change for a reaction tells us which way the reaction will go and how far. It also tells us how much useful work the reaction system can do on the surroundings if the reaction is carried out reversibly. Useful work is defined as all kinds of work (electric, chemical, mechanical) other than the work done by the system in expanding gases against the atmosphere, that is, $w_{useful} = w_{total} - P\Delta V$. The appropriate relationship is[1]

$$-\Delta G = w_{rev} - P\Delta V$$

Because the work obtainable from a reversible, infinitely slow, process is the maximum work possible, $w_{useful} = w_{rev} - P\Delta V$ applies only to a reversible process. All known reactions are irreversible to some extent, and since w_{irrev} is always less than w_{rev}, the useful work available from a spontaneous reaction in going from its initial state to its equilibrium state is always smaller than the hypothetical $w_{rev} - P\Delta V$.

[1] The derivation of this relationship is

$$\Delta G = \Delta H - T\Delta S \quad \text{at constant } T \text{ and } P$$
$$\Delta H = \Delta E + P\Delta V \quad \text{at constant } T \text{ and } P$$

Then

$$\Delta G = \Delta E + P\Delta V - T\Delta S \quad \text{at constant } T \text{ and } P$$
$$\Delta E = q_{rev} - w_{rev} \quad \text{for a reversible process}$$

Since

$$\Delta S = \frac{q_{rev}}{T}$$

$$\Delta E = T\Delta S - w_{rev}$$

Substituting for ΔE in ΔG,

$$\Delta G = T\Delta S - w_{rev} + P\Delta V - T\Delta S$$
$$-\Delta G = w_{rev} - P\Delta V \quad \text{for a reversible process at constant temperature and pressure}$$

If ΔG is negative, w_{useful} is positive, meaning that work is done by the system. If ΔG is positive, w_{useful} is negative, and work must be done on the system. Therefore, only spontaneous reactions can do work. In particular, a system at equilibrium, with $\Delta G = 0$, can do no work. This is our first clue that the chemical systems in living organisms are not at equilibrium, for they would not be able to supply the energy needed for the work of motion, building complex molecules, transporting solutes and charges against concentration and potential gradients, and so on.

In an irreversible or spontaneous reaction occurring at a finite rate, ΔG is the same as if the reaction has been carried out reversibly. In this case, however, $w_{useful} = w_{irrev} - P\,\Delta V$ and is less than $-\Delta G$. For a spontaneous isothermal reaction, that portion of ΔG that does not appear as useful work is evolved as heat and is wasted.[2]

In a reaction at constant temperature and pressure, the enthalpy change $(\Delta H = \Delta G + T\,\Delta S)$ corresponds to the total energy change. In any reaction, reversible or spontaneous, under these conditions, the portion $T\,\Delta S$ of the total energy evolved is always unavailable for doing work. This portion is dissipated to the surroundings as heat. The other portion ΔG is classed as "available" energy, capable of doing useful work. If the process is irreversible, not all of the decrease in free energy appears as work. The difference between ΔG and the actual useful work done, $-(\Delta G - w_{actual})$, appears as heat dissipated to the surroundings along with $T\,\Delta S$. This extra heat passing to the surroundings causes ΔS_{surr} to be greater than $-\Delta S_{sys}$, so that ΔS_{total} is greater than zero. No useful work is done by this heat in an isothermal process. For a totally irreversible system ($w_{useful} = 0$), the whole of ΔH is lost as heat and no useful work is done.

In a sense, energy that escapes from the system as $P\,\Delta V$ work and as heat is no longer available within the system to do useful work. Useful work will be done by a system only if there is some means of transferring, utilizing, storing, or otherwise drawing off by nonthermal means, all or part of its free energy decrease as work against an opposing force other than the atmosphere. This implies a slowing up of the approach to equilibrium of an exoergic

[2] $\Delta G = \Delta H - T\,\Delta S \qquad (T, P \text{ constant})$
$\qquad = \Delta E + P\,\Delta V - T\,\Delta S$
$\qquad = q - w_{total} + P\,\Delta V - T\,\Delta S$
$\qquad = q - w_{useful} - P\,\Delta V + P\,\Delta V - T\,\Delta S$
$\qquad = (q - w_{useful}) - T\,\Delta S$

or

$\qquad q = \Delta G + T\,\Delta S + w_{useful}$

for the heat evolved in any process at constant temperature and pressure.
For a completely reversible process,

$\qquad q = T\,\Delta S$

and

$\qquad -\Delta G = w_{useful} = w_{rev}$

For a completely irreversible process where no useful work is done by the system,

$\qquad q = \Delta G + T\Delta S$

and the extra heat evolved is represented by $\Delta G = -w_{useful}$. In other processes that are not completely irreversible, a smaller portion of ΔG will be converted to heat.

reaction and corresponds to a more reversible reaction. Two examples will illustrate these points.

EXAMPLE

$$Zn + Cu^{2+}(1M) \longrightarrow Zn^{2+}(1M) + Cu$$
$$\Delta G^0 = -51.0 \text{ kcal}$$
$$T \Delta S^0 = 2.0 \text{ kcal at } 25°C$$
$$\Delta H^0 = -53.0 \text{ kcal}$$

When zinc is added directly to a Cu^{2+} solution, the reaction is completely irreversible and no useful work is done. The total energy drop $\Delta H = -53.0$ kcal is dissipated as heat. When an electrochemical cell is set up and the cell is discharged reversibly and slowly by opposing the cell emf with another emf only infinitesimally smaller, the useful work hypothetically derived from the cell reaction is 51.0 kcal.

EXAMPLE

$$C_6H_{12}O_6(s) + 6O_2(g) \longrightarrow 6H_2O(l) + 6CO_2(g)$$
$$\Delta G^0 = -688 \text{ kcal/mol}$$
$$\Delta H^0 = -673 \text{ kcal/mol}$$
$$T \Delta S^0 = 15 \text{ kcal/mol } (25°C)$$

When glucose is burned in the open air, no useful work is done. Heat energy equivalent to

$$-688 + 15 = -673 \text{ kcal}$$

is lost to the surroundings. This is the measured heat of combustion. Under completely reversible conditions, 688 kcal of useful work could be done and 15 kcal of heat would be *absorbed* from the surroundings.

From these two examples, it is apparent that the quantity of work possible and the amount of heat lost as "unavailable" (for doing work) energy depend on how a reaction is carried out.

Glucose is a major energy-providing molecule for living organisms. Glucose is not oxidized directly by a one-step combustion to CO_2 and H_2O in organisms. If this were the case, very little energy would be available to do the work of the organism. Almost all of the energy would be evolved as heat and would overcome the temperature-regulating mechanisms (which require energy, as work, for their operation anyway), and the living system would suffer a heat death. Instead, the biological oxidation of glucose proceeds by a number of intermediate steps, some exoergic, some endoergic. The exoergic steps release free energy in small amounts capable of being easily utilized as useful chemical work within the complex living system. Each step is comparatively reversible because of its small ΔG. This means that a small change in concentration, temperature, pressure, or small application of energy can change the value of ΔG by an appreciable percentage. As a result, the complete oxidation of glucose to CO_2 and H_2O is considerably more reversible than the free combustion of glucose. The metabolic degradation of glucose supplies more useful energy or work to the organism and less unusable and potentially destructive heat. About 40 percent of the maximum possible work, that is, 40 percent of $\Delta G = -688$ kcal/mol, is transformed into useful work in the stepwise biological oxidation of glucose. Over a period of time, of course, all of the energy transferred as work is eventually degraded to heat in other living processes, so that the total heat produced by 1 mol of glucose ingested as fuel or food is equal to the amount evolved in the free combustion of glucose. The point is that in living systems the heat is not evolved in one sudden burst as it is in combustion. It is to the energy transformations in living systems that we now turn.

Energy Transformations in Living Systems (Bioenergetics)

Considered as an isolated system, a living organism, with its high degree of order, is a highly improbable event in the universe. Furthermore, the maintenance or increase of order in an *isolated* living system would violate the second law of thermodynamics. From another viewpoint, the energy required to perform the work of living organisms must be derived from spontaneous chemical reactions going on in the living system. Starting out with a given quantity of reactants, these reactions should supply usable energy until they reach equilibrium. At that time, the living machine should run down and die.

16.13 OPEN SYSTEMS AND THE STEADY STATE

Both of the problems mentioned are imaginary, because living organisms, or their cells, are not isolated systems. We have already seen (Section 16.8) that living organisms are not isolated from their surroundings with respect to energy and do not violate the second law. Organisms also exchange matter with their surroundings. Living systems are therefore examples of *open systems* as opposed to isolated systems. The entry of matter into living systems provides both energy and structural material for growth and regeneration. We are concerned primarily with the energy (as food, for example) intake as we see why organisms do not readily run down and die.

The chemical energy contained in the bonds of the food molecules is gradually released to the organism by oxidation of the food. The overall oxidation reaction can be indicated, in general, as

$$AH_2 + B \underbrace{\longrightarrow \longrightarrow \longrightarrow \longrightarrow}_{\text{intermediate steps}} A + BH_2 + \text{work} + \text{heat}$$

where B may, or may not, be oxygen. The overall reaction is exoergic, although not all of the intermediate steps are exoergic. The energy-providing reactions of living systems are spontaneous reactions that cannot be permitted to reach a state of equilibrium with the environment. Living systems must be continually supplied with matter and energy so that they can maintain this nonequilibrium state. Over a period of time (but not too short or too long), living thermodynamic systems are at least in a *steady state*, meaning that the rate of entry of matter and energy into the system from the surroundings is balanced by the rate of removal of matter and energy from the system. Under these conditions, there is a continual flow of matter and energy through the system.

A wind-up clock may be used to compare an isolated system tending toward equilibrium with a nonequilibrium steady state system. If the initial state of the clock is the fully wound state, it will tend to the rundown state of lowest (free) energy if left unattended. Normally, a clock is wound before it dissipates all the energy stored in the spring. A living system is more like a self-winding watch that is wound a little bit very often to maintain a constant tension in the spring. In a living system, tension in the "spring" is maintained by a continuous import of energy from the surroundings. The entry of energy holds the living system at a certain distance from the true equilibrium condition that it would attain if left alone. Certain homeostatic or self-regulating processes in the organism see to it that a constant energy flow (constant tension or distance from true equilibrium) is maintained. For all life, solar radiation is the ultimate force that winds the mainspring.

Life is therefore a continual transformation of free energy into useful work. The living system absorbs energy in one way or another from its surroundings and transforms this energy, on the molecular level, into various kinds of work. If the source of external energy stops or if there is a breakdown in the mechanisms used for transforming the energy into work, death eventually results. As a clock may stop or malfunction if a bearing or gear in its power train breaks down, so also will a living organism experience trouble if only one of the many chemical reactions in its energy transformation chain ceases to operate properly. In a larger sense, the same things can be said about the energy and matter transfers by way of biogeochemical cycles in the steady state biosphere.

16.14 SOURCES OF ENERGY IN LIVING SYSTEMS

Heterotrophic organisms obtain their energy by the oxidation of complex, energy-rich carbon compounds such as carbohydrates and fats. In these compounds, carbon is in one of its lower oxidation states. The lower the oxidation state, or the greater the H/C ratio, the greater the energy store of the compound. As before, the complex carbon compound may be represented as AH_2 and the ultimate oxidizing agent (or electron or hydrogen acceptor) as B, so that

$$AH_2 + B \longrightarrow A + BH_2$$

In the aerobic heterotrophs, oxygen is the ultimate oxidizing agent and

$$AH_2 + \tfrac{1}{2}O_2 \longrightarrow A + H_2O$$

If $AH_2 = [CH_2O]$, where $[CH_2O]$ represents a unit of a carbohydrate molecule, complete oxidation is

$$[CH_2O] + O_2 \longrightarrow CO_2 + H_2O \qquad \begin{aligned}\Delta G &\simeq -120 \text{ kcal/mol} \\ \Delta H &\simeq -112 \text{ kcal/mol}\end{aligned}$$

Anaerobic heterotrophs may use inorganic oxidizing agents other than oxygen. Some of the anaerobes, called *facultative* anaerobes, may use oxygen and the other oxidizing agent interchangeably. Strict, or *obligate*, anaerobes cannot use oxygen at all and can live only in the absence of oxygen. They represent early members of the evolutionary sequence of life. Denitrifying bacteria are facultative anaerobes; they can oxidize carbohydrates with nitrate ion or with oxygen. Sulfate-reducing bacteria and the CO_2-reducing, methane-forming bacteria are strict anaerobes.

Autotrophic organisms build complex carbon compounds by the reduction of CO_2, which serves as the source of carbon. This requires an outside source of energy. The largest and most important group of autotrophs is the photosynthetic plant, which obtains its energy from solar radiation. The overall equation for photosynthesis is

$$CO_2 + H_2O \xrightarrow{h\nu} [CH_2O] + O_2$$

Other autotrophs can use the energy of other oxidation reactions to power this reduction of CO_2, by the process called *chemosynthesis*. For instance, some bacteria use the energy derived from the oxidation of reduced sulfur (as S, H_2S, or thiosulfates $S_2O_3{}^{2-}$) for the CO_2 reduction:

$$2H_2S + O_2 \longrightarrow 2S + 2H_2O$$
$$2S + 2H_2O + 3O_2 \longrightarrow 2SO_4{}^{2-} + 4H^+$$
$$S_2O_3{}^{2-} + H_2O + 2O_2 \longrightarrow 2SO_4{}^{2-} + 2H^+$$

Nitrifying bacteria use O_2 to oxidize NH_3 to $NO_2{}^-$ or $NO_2{}^-$ to $NO_3{}^-$ to obtain energy, while iron bacteria can get energy from the reaction

$$4Fe^{2+} + 4H^+ + O_2 \longrightarrow 4Fe^{3+} + 2H_2O$$

The use of an exoergic reaction to drive an endoergic reaction is an example of coupling, a feature to be discussed in detail in Section 16.17. For the remainder of this chapter, only heterotrophic aerobes will be considered. In Chapter 18, photosynthesis will be discussed.

16.15 METABOLISM   The large carbon-containing food molecules taken in by the heterotrophic or consumer organisms are broken down into their smaller component mono-saccharides, amino acids, and fatty acids by hydrolysis (digestion). (See Figure 16.8.) The smaller molecules can easily pass through the membranes into the cells. There, the many hydrolysis products are partially oxidized to the few different molecules shown in Figure 16.8 II. These few molecules are further oxidized to carbon dioxide by dehydrogenation in a continuous oxidation cycle known as the Krebs cycle. The hydrogen atoms removed during the cyclic oxidation are passed along a series of compounds known as the respiratory chain, ultimately combining with atmospheric oxygen to form water. During the cyclic oxidation and the passage of hydrogens along the respiratory chain, some of the free energy of the exoergic reactions in the sequence is used to generate a particular energy-rich molecule (ATP), which serves as an intermediate in the conversion of the energy to various kinds of chemical, mechanical, osmotic, and electric work in the organism. Unlike a power plant or heat engine, which converts all the free energy of fuel into heat and then converts the heat into work, a living cell avoids the wasteful intermediate heat step. The free energy of food is converted directly to and stored as chemical free energy until needed for work.

The aerobic degradation of foodstuffs to CO_2, H_2O, and energy is called *respiration*. *Catabolism* is the name applied to this degradative phase of metabolism. Some of the free energy supplied by the oxidative breakdown is utilized as the chemical work necessary in the biosynthesis of proteins and other large molecules. The constructive phase of metabolism is called *anabolism*. The sequence of chemical reactions in cells or organisms connecting initial reactants with products is called a *metabolic pathway*. We now examine some aspects of the metabolic pathway for the degradation of glucose to water and carbon dioxide. We shall see that the path from reactant to product proceeds through a large number of steps in which the product of one reaction

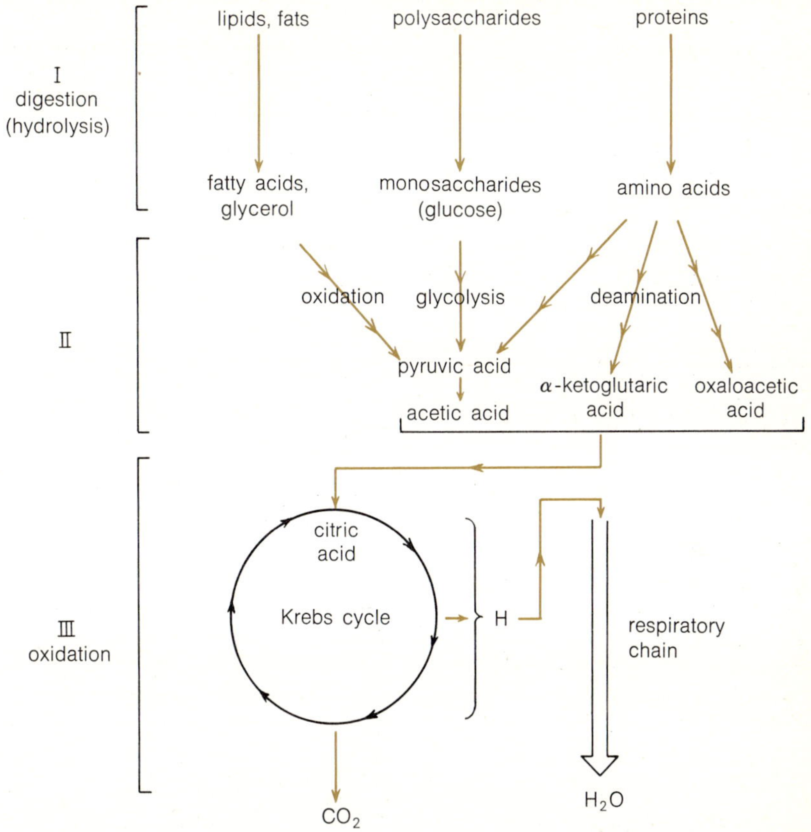

Figure 16.8. Stages in degradative metabolism.

is the reactant for the following reaction. These intermediate substances are called *metabolites*. Some of the metabolites, in addition to undergoing further degradative oxidation, are also starting points for biosynthetic metabolic pathways for the formation of amino acids, proteins, nucleic acids, and so on.

16.16 BIOLOGICAL OXIDATION AND REDUCTION

The metabolic pathway for the degradation of glucose will be clearer if we first review some concepts of oxidation–reduction and thermochemistry. Most of the energy-yielding reactions in biological systems are oxidations. As already noted, oxidation in biological systems may be viewed as resulting, simply, from the loss of hydrogen atoms (dehydrogenation) or, more generally, from the loss of electrons. A case in which the hydrogen atom transfer viewpoint is adequate is shown in the general equation $AH_2 + B \rightleftharpoons A + BH_2$; where B, the hydrogen acceptor, is the oxidizing agent. An example requiring the use of the electron transfer concept is

$$C-Fe^{2+} + A-Fe^{3+} \rightleftharpoons C-Fe^{3+} + A-Fe^{2+}$$

where C and A are groups attached to the irons. When a hydrogen atom is transferred, its electron goes along with it so there is no fundamental difference between these two cases.

Metabolic pathways are usually not written as a series of chemical equations like this. It is more convenient to represent these and other reactions by arrows, for the forward reaction,

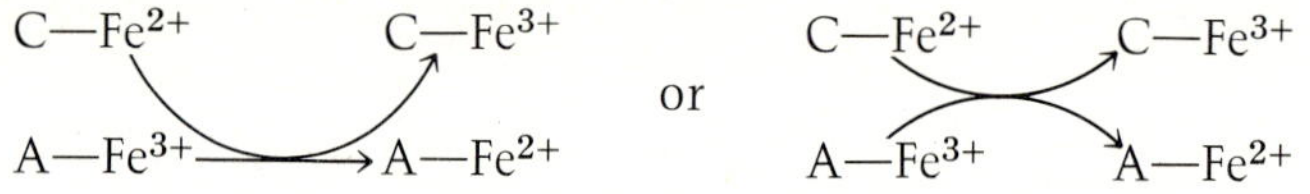

or

and

Figure 16.9. Nicotinamide adenine dinucleotide, NAD. Flavin adenine dinucleotide, FAD $\rightleftharpoons$ FADH$_2$, is another hydrogen acceptor found in living systems. Both NAD and FAD are associated with a protein portion. The complete unit is an enzyme, while the NAD and FAD are known as coenzymes (Section 17.12).

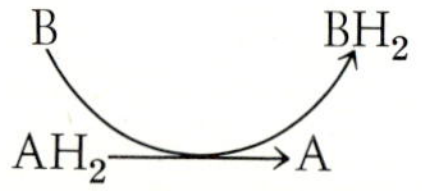

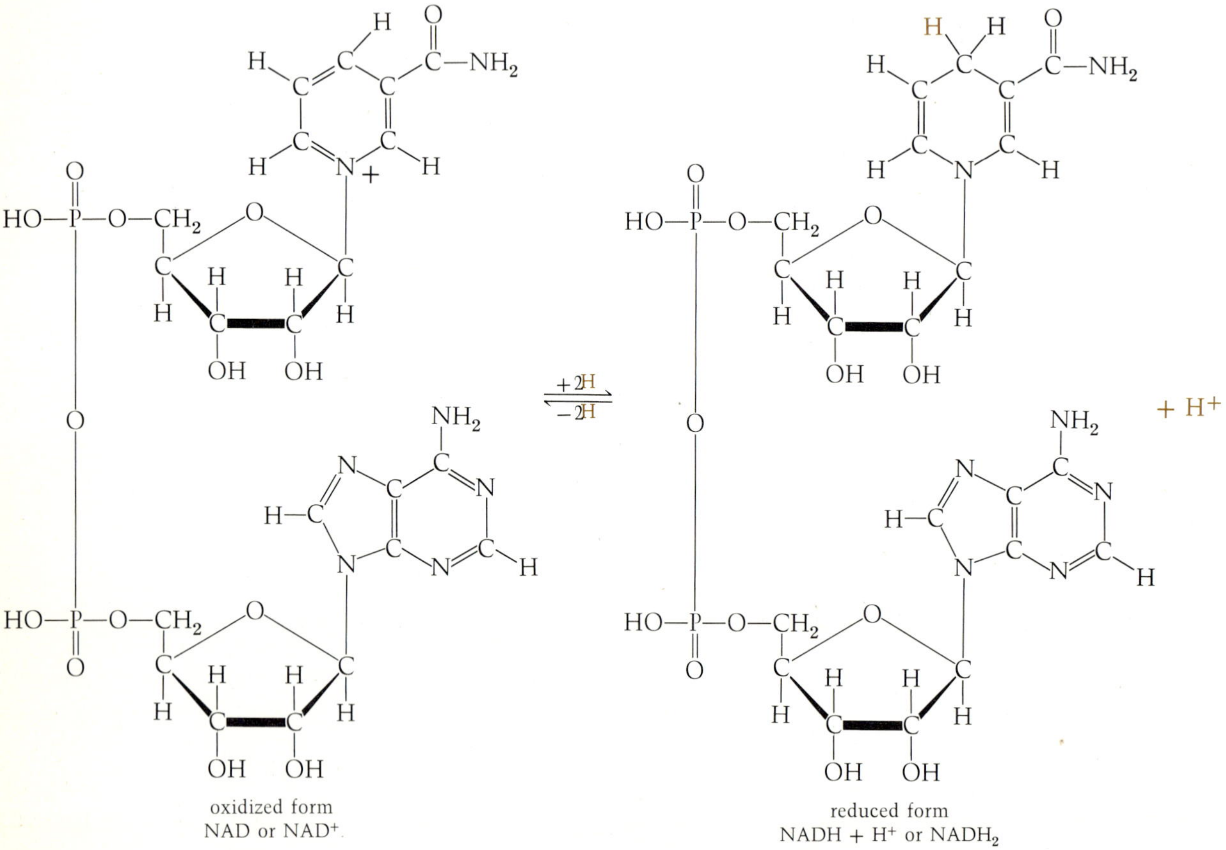

This sort of representation is not limited to redox reactions.

The most important electron acceptors we shall encounter are the cytochrome pigments. These are proteins with an iron atom that may exist in the Fe(II) or Fe(III) oxidation state.

oxidized form
NAD or NAD$^+$.

reduced form
NADH + H$^+$ or NADH$_2$

The hydrogen transferring agents are dinucleotides. One of them is illustrated in the oxidized and reduced form in Figure 16.9. An example of a biological oxidation using NAD (nicotinamide adenine dinucleotide) as the oxidizing agent is the oxidation of malic to oxaloacetic acid:

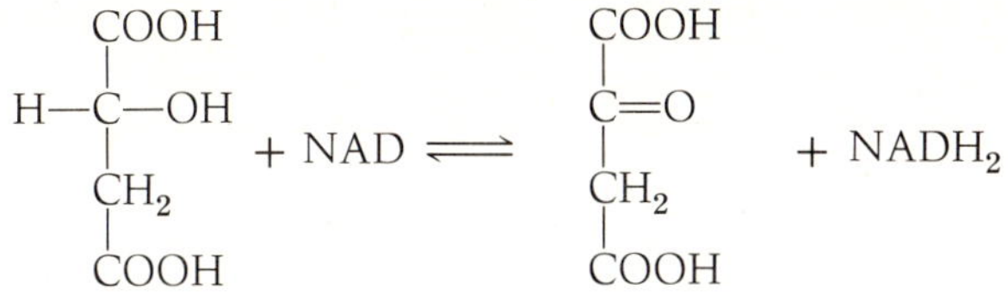

$$
\begin{array}{cc}
\begin{array}{c}
\text{COOH} \\
| \\
\text{H---C---OH} \\
| \\
\text{CH}_2 \\
| \\
\text{COOH}
\end{array} + \text{NAD} \rightleftharpoons &
\begin{array}{c}
\text{COOH} \\
| \\
\text{C}=\text{O} \\
| \\
\text{CH}_2 \\
| \\
\text{COOH}
\end{array} + \text{NADH}_2
\end{array}
$$

malic acid   oxaloacetic acid

or

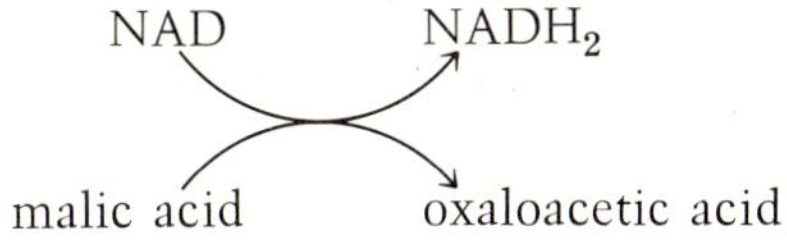

malic acid   oxaloacetic acid

16.17 COUPLED GROUP TRANSFER REACTIONS

A metabolic pathway consists of a series of consecutive reactions, a product of one reaction becoming a reactant for the next reaction:

$$ A \longrightarrow B \longrightarrow C \longrightarrow D \longrightarrow \longrightarrow \longrightarrow M $$

The overall change in free energy for the sequence A to M must be negative. Nevertheless, individual reactions within the pathway may have positive ΔG's. For such endoergic reactions to occur, they must be connected or coupled to an exoergic reaction by a common substance or intermediate. Consider two general examples.

EXAMPLE

$$
\begin{aligned}
A &\rightleftharpoons B + X & \Delta G_1{}^0 &= +3 \text{ kcal} & (1) \\
B &\rightleftharpoons C + Y & \Delta G_2{}^0 &= -7 \text{ kcal} & (2)
\end{aligned}
$$

B is the common intermediate in these two reactions. Reaction (1) is endoergic and will not occur spontaneously. How can C ever be formed if reaction (1) will not form the needed reactant B? First we can add the two equations and their free energies (since G is a function of state) to get

$$
\begin{aligned}
A \longrightarrow C + X + Y \qquad \Delta G_3{}^0 &= \Delta G_1 + \Delta G_2 & (3) \\
&= (3 - 7) \text{ kcal} \\
&= -4 \text{ kcal}
\end{aligned}
$$

The total reaction is exoergic, with $\Delta G = -4$ kcal, so that A should react spontaneously to form C. The fact that ΔG_3 is negative might be enough to justify the production of C, but we can look at the problem from some other angles.

On the basis of Le Chatelier's principle, we can say that the exoergic reaction (2) continually removes reactant B from reaction (1). This causes more B to be formed from A in reaction (1). Furthermore, if the reactions are coupled via substance B, it is permissible to combine equilibrium constants K_1 and K_2,

$$ K_1 = \frac{[B][X]}{[A]} $$

K_1 is less than 1, say about 0.01.

$$K_2 = \frac{[C][Y]}{[B]}$$

K_2 is greater than 1, about 10^5.

$$K_3 = K_1 \times K_2 = \frac{[C][X][Y]}{[A]} = 10^3$$

The combined constant K_3 shows that the concentration of products at equilibrium is considerably greater than that of reactants. That is, reaction (3) proceeds appreciably to the right.

EXAMPLE

$$\begin{array}{ll} A \rightleftharpoons B + X & \Delta G^0 = -7 \text{ kcal} \\ B \rightleftharpoons C + Y & \Delta G^0 = +3 \text{ kcal} \end{array}$$

This example differs from the previous example in that the first reaction is exoergic, while the second is not. Just the same, the sum of the ΔG's is negative, so the coupled reaction A $\longrightarrow$ C + X + Y can take place spontaneously to give C. In this case, the first reaction continually adds reactant B to the second equation and, according to Le Chatelier's principle, "pushes" the second reaction to the right. This coupling between exoergic and endoergic reactions can happen only if they share a common component. Otherwise, there is no energy linkage between them. Through coupled reactions, free energy is transferred from one chemical compound to another in metabolic reactions without the dissipation of the free energy change as heat.

Adenosine Triphosphate (ATP)

One molecule is invariably involved in all energetically coupled biological reactions. This molecule is the mononucleotide *adenosine triphosphate* (ATP) shown in Figure 16.10 in its completely ionized form. ATP is known as a

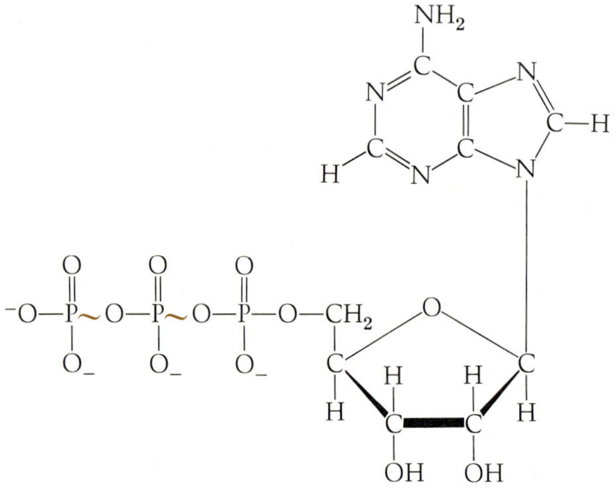

Figure 16.10. Adenosine triphosphate, ATP. The completely ionized form is shown. The un-ionized form has H's on all the negatively charged oxygens. At pH = 7, three of the phosphate hydroxyl groups are completely ionized, but one of the terminal hydroxyls is only 80 percent ionized. The bonds denoted by wavy lines are the bonds broken in the hydrolysis of ATP and ADP.

adenosine triphosphate, ATP

high-energy or energy-rich compound because of its relatively large negative free energy of hydrolysis. The hydrolysis reaction for ATP in the presence of Mg^{2+} is, in abbreviated form,

$$A\!-\!R\!-\!O\!-\!\overset{\displaystyle O}{\underset{\displaystyle O^-}{\overset{\|}{P}}}\!-\!O\!\sim\!\overset{\displaystyle O}{\underset{\displaystyle O^-}{\overset{\|}{P}}}\!-\!O\!\sim\!\overset{\displaystyle O}{\underset{\displaystyle O^-}{\overset{\|}{P}}}\!-\!O^- + H_2O \xrightarrow{Mg^{2+}}$$

(ATP)$^{4-}$

$$A\!-\!R\!-\!O\!-\!\overset{\displaystyle O}{\underset{\displaystyle O^-}{\overset{\|}{P}}}\!-\!O\!\sim\!\overset{\displaystyle O}{\underset{\displaystyle O^-}{\overset{\|}{P}}}\!-\!O^- + HPO_4^{2-} + H^+ \qquad \Delta G^{0'} = -7.3 \text{ kcal}$$

(ADP)$^{3-}$

adenosine diphosphate

or

$$ATP + H_2O \underset{\phantom{Mg^{2+}}}{\overset{Mg^{2+}}{\rightleftharpoons}} ADP + P_i + H^+$$

where P_i = inorganic phosphate.

The relative instability of ATP with respect to hydrolysis is the result of negative charge repulsion in ATP^{4-} and resonance stabilization of the hydrolysis products, ADP and P_i. At pH = 7, most ATP molecules are completely ionized and have four closely spaced negative charges. The electrostatic repulsion among these charges is lessened when ATP hydrolyzes to ADP. In addition, there is a charge repulsion between the negatively charged product ions, which hinders their recombination. The hydrolysis products also have more resonance forms between the two of them than does ATP alone. The increased delocalization of the electrons in ADP and P_i favors hydrolysis. ADP is also a high-energy molecule and undergoes hydrolysis by cleavage of the remaining —O∼P bond:

$$ADP + H_2O \rightleftharpoons AMP + P_i \qquad \Delta G^{0'} = -7.3 \text{ kcal}$$

adenosine   adenosine
diphosphate   monophosphate

However, AMP is not a high-energy compound:

$$AMP + H_2O \rightleftharpoons A + P_i \qquad \Delta G^{0'} = -3.4 \text{ kcal}$$

The primes on $\Delta G^{0'}$ indicate that $[H^+] = 10^{-7}M$, rather than that $[H^+] = 1M$ as in the standard thermodynamic state. A pH of $7([H^+] = 10^{-7}M)$ is the biochemical standard state for hydrogen ion concentration.

The hydrolysis of ATP does not actually occur in living cells. Instead, ATP participates directly or indirectly in group transfer reactions. A simple example will illustrate this point. The degradative metabolic pathway of glucose begins with the transfer of a phosphoryl group,

$$-\overset{\displaystyle O^-}{\underset{\displaystyle O^-}{P}}\!\!\!=\!\!O \qquad \text{or} \qquad -\overset{\displaystyle OH}{\underset{\displaystyle OH}{P}}\!\!\!=\!\!O$$

to a glucose molecule. This group is usually symbolized as ⓅP and is often (incorrectly) called a phosphate group. The direct transfer of this group from inorganic phosphate is endoergic,

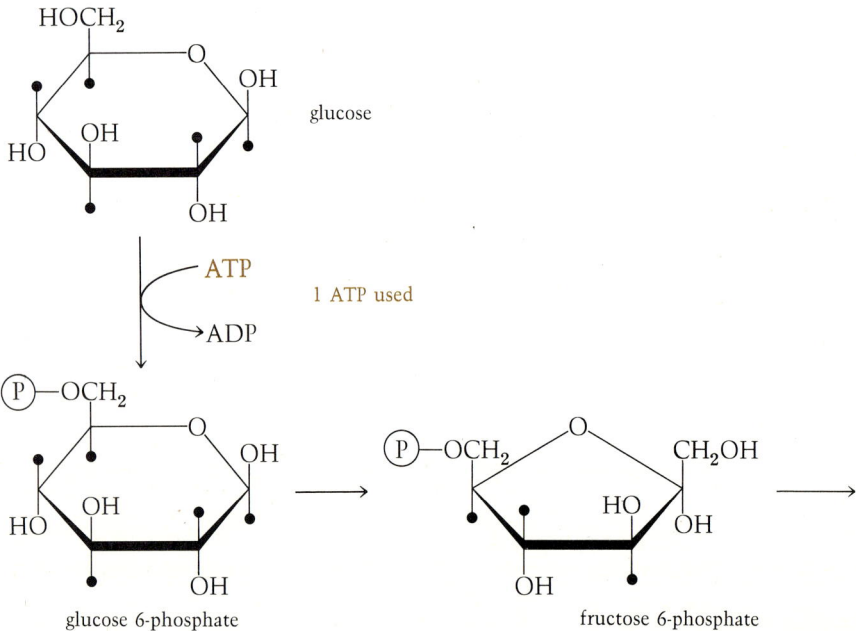

$$\text{glucose} + HPO_4^{2-} (P_i) \longrightarrow \text{glucose 6-phosphate} + H_2O$$

with a $\Delta G^{0\prime}$ of about 4 kcal. If this reaction is considered to be coupled to the hydrolysis of ATP, the overall reaction is exoergic:

$$
\begin{array}{ll}
\text{glucose} + P_i \rightleftharpoons \text{glucose 6-phosphate} + H_2O & \Delta G^{0\prime} \simeq 4 \text{ kcal} \\
ATP + H_2O \rightleftharpoons ADP + P_i & \Delta G^{0\prime} \simeq -7 \text{ kcal} \\
\hline
\text{glucose} + ATP \rightleftharpoons \text{glucose 6-phosphate} + ADP & \Delta G^{0\prime} \simeq -3 \text{ kcal}
\end{array}
$$

or

$$\text{glucose} \xrightarrow{\quad ATP \quad\quad ADP \quad} \text{glucose 6-phosphate}$$

This reaction has been broken into two component reactions only for purposes of free energy calculation. In the biological reaction, no inorganic phosphate P_i is involved, nor are there two separate reactions. The ATP transfers the PO_3^{2-} group directly to glucose. However, there are other instances of coupled reactions in which each reaction does really occur as a separate process. An example in the metabolic pathway for the breakdown of glucose (Figure 16.11)

Figure 16.11. Glycolysis.

fructose 6-phosphate

ATP

1 ATP used

ADP

(P)—OCH_2 ... H_2CO—(P)

fructose 1,6-diphosphate

HO OH

OH

$\begin{matrix} H \\ C \end{matrix}$=O

2 H—C—OH

H_2C—O—(P)

glyceraldehyde 3-phosphate

H_2C—OH

C=O

H_2C—O—(P)

2 dihydroxyacetone phosphate

NAD

NADH$_2$

O
‖
C—O~(P)

2 H—C—OH

H_2C—O—(P)

1,3-diphosphoglycerate

ADP

2 ATP's gained

ATP

COOH

2 H—C—OH

H_2C—O—(P)

3-phosphoglyceric acid

COOH

2 C—O~(P)

CH_2

2-phosphoglyceric acid

2 ATP's gained

ADP   ATP

COOH

2 C=O   pyruvic acid

CH_3

Figure 16.11. *(Continued)*

is the coupling of the two consecutive reactions,

$$\text{glyceraldehyde 3-phosphate} + \text{NAD} + P_i \rightleftharpoons$$
$$\text{1,3-diphosphoglycerate} + \text{NADH}_2 \qquad \Delta G^{0\prime} = 1.5 \text{ kcal}$$

and

$$\text{1,3-diphosphoglycerate} + \text{ADP} \rightleftharpoons$$
$$\text{3-phosphoglyceric acid} + \text{ATP} \qquad \Delta G^{0\prime} = -4.5 \text{ kcal}$$

The second reaction is coupled to the endoergic first reaction through the common intermediate, 1,3-diphosphoglycerate, a high-energy compound. The exoergic reaction uses up this intermediate and pulls the endoergic reaction along. The sum of the two reactions is exoergic:

$$\text{glyceraldehyde 3-phosphate} + \text{NAD} + P_i + \text{ADP} \rightleftharpoons$$
$$\text{3-phosphoglyceric acid} + \text{NADH}_2 + \text{ATP} \qquad \Delta G^{0\prime} = -3.0 \text{ kcal}$$

1,3-Diphosphoglycerate is an example of a phosphate compound whose $\Delta G^{0\prime}$ of hydrolysis (-11.8 kcal) is larger than that of ATP. The reaction of ADP with such compounds regenerates ATP. In general,

$$\text{R—O}\sim\textcircled{P} + \text{ADP} \rightleftharpoons \text{R—O}^- + \text{ATP}$$

or

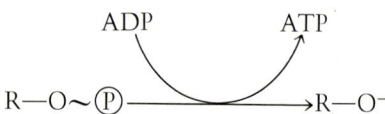

Thus, the free energy of hydrolysis of ATP is intermediate in value to that of other biologically important molecules. It is this intermediate position that permits the ADP–ATP system to both donate and accept "high-energy" $-\text{PO}_3^{2-}$ groups and to act as the carrier of energy from higher energy compounds to lower energy compounds.

16.18 METABOLISM OF GLUCOSE

The oxidation of glucose to CO_2 and H_2O takes place in three main stages. In the first stage, called *glycolysis*, glucose is split into two molecules of pyruvic acid. During this process, two pairs of hydrogen atoms are transferred to NAD and a net of two ATP molecules are produced per glucose molecule. The metabolic pathway is shown in Figure 16.11. Not much of the free energy of the glucose molecule is released in this initial partial oxidation to two molecules of pyruvic acid, only about 37 kcal of the 686 kcal total for complete oxidation to CO_2 and H_2O. Of the 37 kcal, the two molecules of ATP produced represent a store of 2(7.3) kcal. The difference (37 − 15) kcal is lost as heat.

The Krebs Cycle

The second stage of oxidative metabolism is represented by the cyclic oxidation of acetic acid, CH_3COOH, in the form of a complex high-energy derivative acetyl coenzyme A, abbreviated acetyl Co-A or $CH_3C\sim S-CoA$. This is formed from pyruvic acid and NAD by the reaction

$$CH_3\overset{O}{\overset{\|}{C}}COOH + NAD + CoA{-}SH \longrightarrow$$

coenzyme A

$$CH_3\overset{O}{\overset{\|}{C}}{\sim}S{-}CoA + NADH_2 + CO_2$$

acetyl coenzyme A

The structure of coenzyme A is shown in Figure 16.12. Acetyl coenzyme A enters the cyclic oxidation process variously called the Krebs cycle, the tricarboxylic acid (TCA) cycle, or the citric acid cycle, where decarboxylation (removal of CO_2 from —COOH groups) and oxidation (dehydrogenation) occur. The Krebs cycle is outlined in Figure 16.13.

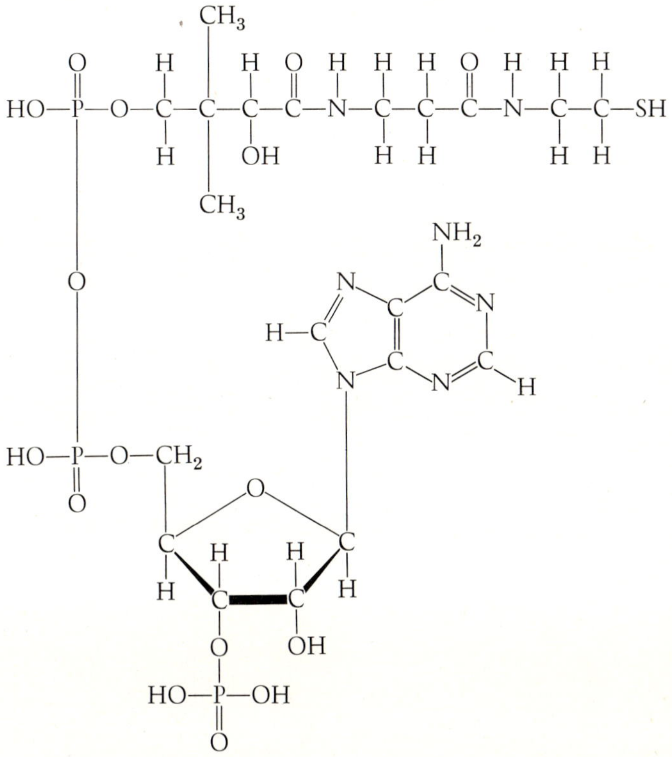

Figure 16.12. Coenzyme A.

Acetyl coenzyme A is a high-energy compound like ATP and can transfer its acetyl group, $CH_3\overset{O}{\overset{\|}{C}}{-}$, to oxaloacetic acid to produce citric acid and reform the original coenzyme A. The series of hydrogen atom and CO_2 removals eventually regenerates oxaloacetic acid and the cycle begins again.

A total of five pairs of hydrogen atoms and two molecules of CO_2 are produced from one pyruvic acid molecule in the Krebs cycle. The overall reaction, starting with acetic acid, is

Figure 16.13. The Krebs or tricarboxylic acid cycle. A series of decarboxylations and oxidations produces five pairs of hydrogen atoms and one ATP molecule for every pyruvic acid molecule entering the cycle via coenzyme A. Since two pyruvic acid molecules result from each glucose molecule, the yield is ten pairs of hydrogen atoms and two ATP molecules per glucose molecule.

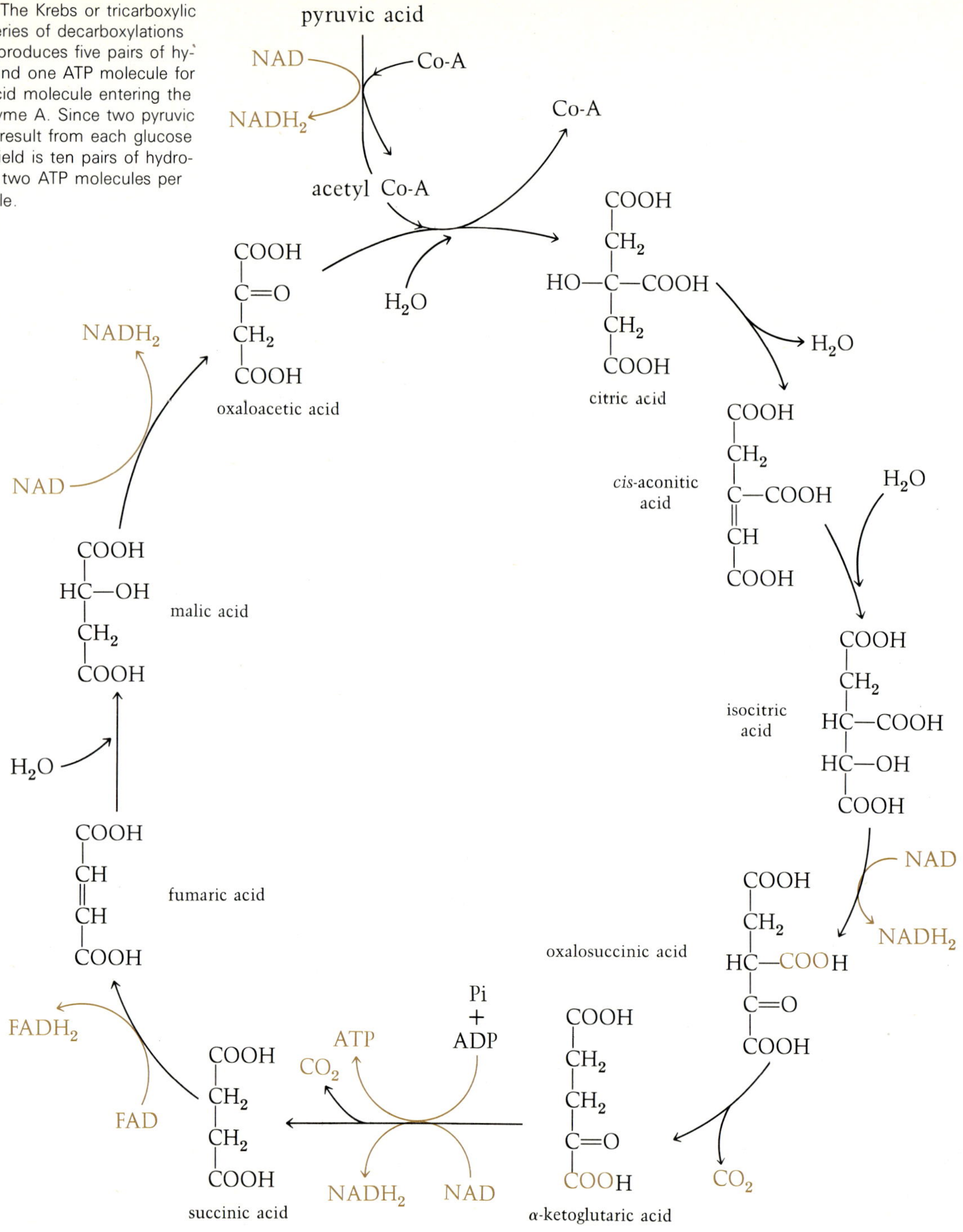

$$CH_3COOH + 2H_2O \rightleftharpoons 2CO_2 + 8H$$

with the other two hydrogens coming from the pyruvic acid $\longrightarrow$ acetyl Co-A reaction. One ATP molecule is formed in the cycle for each pyruvic acid molecule, or two ATP's per glucose molecule.

Entry to the Krebs cycle is not restricted to glucose. Fatty acids and some amino acids are also converted to acetyl coenzyme A. Other amino acids are transformed to other compounds such as oxaloacetic acid or α-ketoglutaric acid, which participate in the cyclic oxidation process.

The Respiratory Chain   All of the hydrogen atoms taken up by the oxidizing agents, such as NAD, in both glycolysis and the Krebs cycle are now fed into the *respiratory* or *electron transport* chain, the third and final stage. Oxidizing agents remove the hydrogens and their electrons from $NADH_2$. The electrons are passed from one oxidizing agent to another until, at the end of the chain, they are handed over to molecular atmospheric oxygen, the strongest oxidizing agent. The reduced oxygen, plus the hydrogen ions, form water. During the passage of the electrons to successively stronger oxidizing agents, the free energy of the electron decreases in small amounts. The successive decreases in free energy are coupled at certain points to the production of ATP molecules, which serve as the immediate energy source for all other endoergic processes in the cell. The passage of hydrogen atoms and electrons down the respiratory chain is shown in Figure 16.14.

A total of 12 electron pairs traverse the respiratory chain for every molecule of *glucose* oxidized. The total free energy decrease in the respiratory chain portion of the complete degradative pathway is then $12(-52 \text{ kcal}) = -624$ kcal/mol of glucose. Since the free energy of combustion of glucose is -686 kcal, almost all of the free energy available in the oxidation of glucose is made available during the respiratory chain portion of the metabolic pathway. The coupled formation of ATP from ADP uses part of this energy. Each pair of electrons going down the chain produces an ATP molecule as it passes each coupling point. Ten pairs pass three coupling points, two pairs pass two points, so that $10(3) + 2(2)$ or 34 ATP's are produced *in the respiratory chain*. For each *glucose* molecule, two more ATP's are produced in the Krebs cycle and two during glycolysis. The total of 38 ATP's represent about 38(7.3 kcal) or 277 kcal of free energy available to the cell for useful work. This is about 40 percent of the 686 kcal potentially available per mole of glucose. The remaining 60 percent is lost as heat. This efficiency of energy conversion is made possible by the many small exoergic steps as the electron cascades down the respiratory chain from one oxidizing agent to another. Larger energy drops could not be coupled efficiently to ATP production.

The hydrogen transfer agents (NAD and FAD) and electron transfer agents (the cytochromes) can be listed in order of their reduction potentials (standard at pH $= 7$):

$$NADH + H^+ + e^- \rightleftharpoons NADH_2 \qquad E^{0\prime} = -0.32 \text{ V}$$
$$FAD + 2H^+ + 2e^- \rightleftharpoons FADH_2 \qquad E^{0\prime} = -0.05 \text{ V}$$
$$\text{Cyt b (Fe}^{3+}) + e^- \rightleftharpoons \text{Cyt b (Fe}^{2+}) \qquad E^{0\prime} = 0.04 \text{ V}$$
$$\text{Cyt c (Fe}^{3+}) + e^- \rightleftharpoons \text{Cyt c (Fe}^{2+}) \qquad E^{0\prime} = 0.25 \text{ V}$$
$$\text{Cyt a (Fe}^{3+}) + e^- \rightleftharpoons \text{Cyt a (Fe}^{2+}) \qquad E^{0\prime} = 0.29 \text{ V}$$
$$2H^+ + \tfrac{1}{2}O_2 + 2e^- \rightleftharpoons H_2O \qquad E^{0\prime} = 0.82 \text{ V}$$

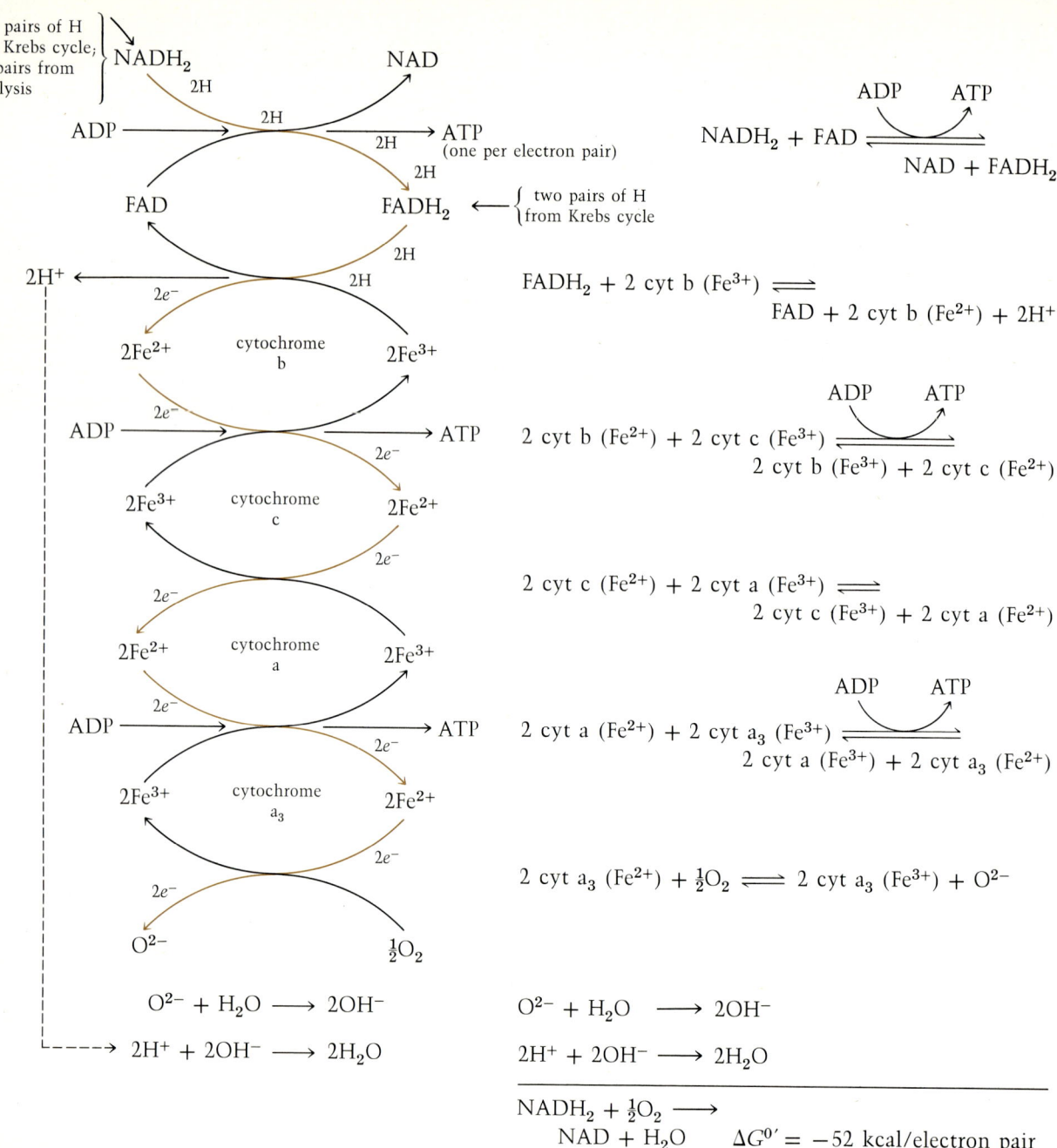

$$FADH_2 + 2 \text{ cyt b (Fe}^{3+}) \rightleftharpoons$$
$$FAD + 2 \text{ cyt b (Fe}^{2+}) + 2H^+$$

$$2 \text{ cyt b (Fe}^{2+}) + 2 \text{ cyt c (Fe}^{3+}) \xrightleftharpoons[\text{ADP}]{\text{ATP}}$$
$$2 \text{ cyt b (Fe}^{3+}) + 2 \text{ cyt c (Fe}^{2+})$$

$$2 \text{ cyt c (Fe}^{2+}) + 2 \text{ cyt a (Fe}^{3+}) \rightleftharpoons$$
$$2 \text{ cyt c (Fe}^{3+}) + 2 \text{ cyt a (Fe}^{2+})$$

$$2 \text{ cyt a (Fe}^{2+}) + 2 \text{ cyt a}_3 \text{ (Fe}^{3+}) \xrightleftharpoons[\text{ADP}]{\text{ATP}}$$
$$2 \text{ cyt a (Fe}^{3+}) + 2 \text{ cyt a}_3 \text{ (Fe}^{2+})$$

$$2 \text{ cyt a}_3 \text{ (Fe}^{2+}) + \tfrac{1}{2}O_2 \rightleftharpoons 2 \text{ cyt a}_3 \text{ (Fe}^{3+}) + O^{2-}$$

$$O^{2-} + H_2O \longrightarrow 2OH^-$$
$$2H^+ + 2OH^- \longrightarrow 2H_2O$$

$$NADH_2 + \tfrac{1}{2}O_2 \longrightarrow$$
$$NAD + H_2O \qquad \Delta G^{0\prime} = -52 \text{ kcal/electron pair}$$

Figure 16.14. The hydrogen and electron transfers of the respiratory chain. Production of ATP is coupled to electron transfer at these points. The hydrogen carrier, coenzyme Q, has been omitted between FAD and cytochrome b.

The strongest oxidizing agents are, as usual, those having the more positive reduction potentials. The free energy change for a complete redox reaction involving any two half-reactions is obtained from

$$\Delta G^{0\prime} = -n\mathcal{F}E^{0\prime}{}_{total}$$

For the complete sequence $E^{0\prime}{}_{total} = 0.32 + 0.82 = 1.14$ volts and

$$\Delta G^{0\prime} = -(2)(23.06 \text{ kcal/V})(1.14 \text{ V})$$
$$= 51.8 \text{ kcal}$$

The difference between the reduction potentials in the series is proportional to the ΔG of the steps in the respiratory chain.

The electron carriers in the respiratory chain, the cytochromes, are high molecular weight (13,000 to 370,000) proteins (enzymes). Each molecule contains one heme group (Figure 16.15) with its iron atom. This group is similar to that found in hemoglobin, the oxygen carrier of the blood. The protein portion of the molecule differs among the various heme proteins. Unlike the iron in hemoglobin, which retains the Fe^{2+} state whether combined with an oxygen molecule or not, the iron atom in the cytochromes undergoes cyclic oxidation and reduction. There are a number of cytochromes. The last two cytochromes, a and a_3, are sometimes lumped together and called *cytochrome oxidase*. Cytochrome a_3 also contains a copper atom.

The action of the cytochromes is effectively inhibited by some common respiratory poisons. Some, like cyanide, inhibit oxidation of Fe^{2+} by holding the iron in the reduced state. Others, such as narcotics and barbiturates, block the reduction of Fe^{3+}. Cyanide and H_2S at low concentrations and CO at high concentrations inhibit cytochrome oxidase (cyt a + cyt a_3). All of these poisons prevent the use of oxygen by the cells and also stop the energy-generating machinery of the organism. Carbon monoxide also interferes with the transport of oxygen to the cells by hemoglobin, so if one effect of CO does not

Figure 16.15. A metal-containing porphin is called a porphyrin. The planar molecule binds the metal at four points. Two additional coordination positions to the metal are available above and below the plane of the ring to give an octahedral (CN = 6) complex. In hemoglobin and myoglobin, one of the extra positions of the heme prosthetic group is used to attach a protein molecule. The sixth position is either empty or binds water. The water is easily displaced by π-bonding substituents, the most important being molecular oxygen.

In the cytochromes, the iron is six coordinated, with four bonds to the porphin ring and two permanent covalent bonds to attached proteins.

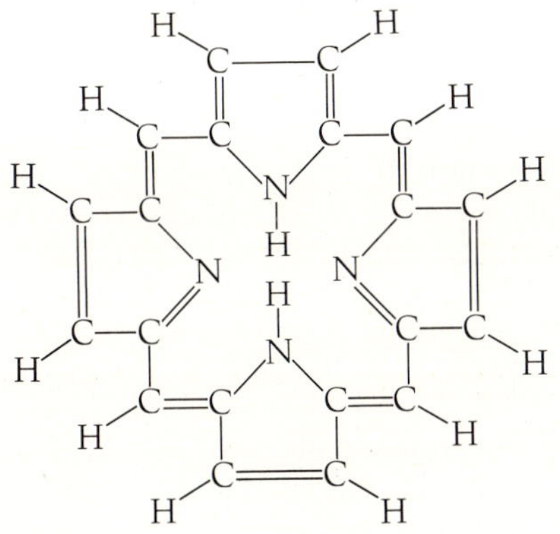

porphin

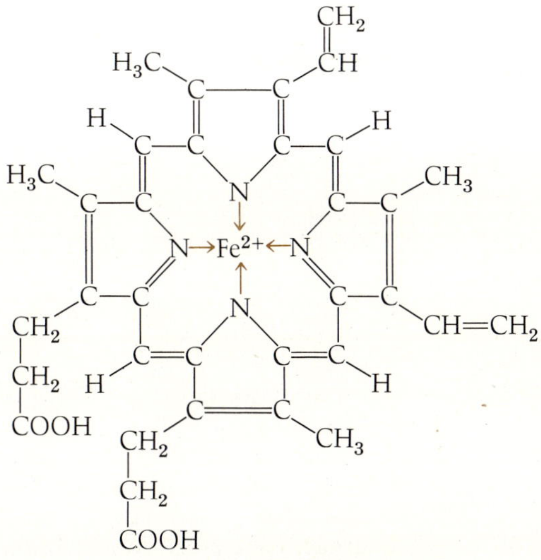

a porphyrin, heme

put the system out of commission, the other may. Another chemical, 2,4-dinitrophenol, uncouples the mechanism whereby ATP is made from ADP in the respiratory chain. The total free energy drop along the chain is then dissipated as heat. It has been noted that this compound might provide the basis for a reducing drug for overweight people, since the energy for the body would have to be obtained from the body's excess fats and carbohydrates. Unfortunately, the compound cannot be used safely for this purpose because it has undesirable side effects.

Other Factors in Metabolism

The metabolic pathways illustrated here are not meant for detailed analysis. It is enough that they give us an idea of the complicated but quite elegant and economical way in which one kind of living cell goes about obtaining energy. By an integrated series of reactions, a cell unlocks the energy of food molecules and distributes it to a number of energy-rich molecules that serve as the energy currency of the organism. All of these chemical reactions, and many more, go on continually in a living cell. The reactions of glycolysis take place in the cytoplasm, but those of the Krebs cycle and respiratory chain are all localized in the mitochondria. In particular, the hydrogen and electron transfer agents of the respiratory chain are all firmly attached in close proximity to each other on the inner membrane of this organelle. This close, structured arrangement apparently is necessary for the transfer of hydrogens and electrons from one molecule to an adjacent one. This is another example of the need for highly ordered structures in living systems. Because most of the energy of food is converted to the prime energy-storing and transporting molecules of life (ATP) in the mitochondrion, this cellular organelle is rightly called the powerhouse of the cell.

Other substances besides the normal food molecules find their way into organisms. Some of these do not find entry into any metabolic pathway and are eliminated unchanged. Some are partly degraded to end products that may or may not be harmful to the organism. For example, a metabolic product of the artificial sweetener cyclamate (a sodium salt of cyclohexylsulfamic acid) has been found to be formed in 20–30 percent of all those ingesting the sweetener. The product, cyclohexylamine, damages the DNA molecules of the cell and may cause genetic damage, so cyclamate has been banned. Other instances of this kind will be mentioned later.

The malfunctioning of most metabolic pathways, both degradative and biosynthetic, is caused by errors in the regulation of one or more of the interlocked reactions. Each chemical reaction in a metabolic pathway is catalyzed by a specific enzyme. The overall rate and direction of the entire array of reactions is controlled by the interplay of checks and balances on the rates of individual reactions. It is to the subject of reaction rates and the function of enzymes that we turn next.

16.19 SUMMARY

Living systems require a continual flow of energy through them in order to maintain, increase, and repair their ordered structures. Metabolism refers to the totality of the chemical reactions and energy transformations occurring in the cells of living organisms. The energy transformations can be interpreted on the basis of the laws of thermodynamics. Thermodynamics, or more generally, energetics, considers the exchanges of energy in its various forms between systems and their surroundings. The first law of thermodynamics is a statement of the law of conservation of energy: the total energy of the universe (a system and all its surroundings) is constant. Applied only to a system, the first law states that any addition of thermal energy (heat) to a system must be compensated by an increase in the internal energy of the system or by work done by the system on the surroundings, or by a combination of both. In symbols, $q = \Delta E + w$. The change in internal energy ΔE of a system is, like some other thermodynamic quantities, dependent only on the initial and final states of the system. Such quantities are called functions of state.

The second law of thermodynamics states that the total entropy change of a system and its surroundings, $\Delta S_{total} = \Delta S_{sys} + \Delta S_{surr}$, must be either zero or positive. The entropy of a system is a measure of the internal, microscopic disorder of the system. All natural processes are spontaneous (irreversible) and proceed in a direction of positive ΔS_{total}. In an isolated system, a positive ΔS_{sys} corresponds to a spontaneous process.

The tendency for all systems to undergo spontaneous processes leading to minimum energy or maximum entropy or both is described by the change in free energy ΔG of the system. Spontaneous processes occur with a decrease in free energy until an equilibrium is reached, at which point the free energy of the system is a minimum. The free energy change is related to ΔS and ΔH (the energy change for a process at constant temperature and pressure) by $\Delta G = \Delta H - T \Delta S$. The free energy change is also related quantitatively to the equilibrium constant of a reversible reaction.

The useful work that can be done by a system is a maximum for a reversible process, meaning an infinitely slow transition through a series of equilibrium states. An irreversible, spontaneous process provides less useful work and evolves more useless heat than the reversible process. A system in equilibrium can do no useful work.

Degradative metabolic processes in living systems extract maximum useful work from a normally highly spontaneous oxidation process by releasing the chemical energy in controlled increments in a series of comparatively reversible (small ΔG) steps. Some of the energy evolved is used to drive endoergic steps in the metabolic pathway; some is stored in high-energy compounds such as ATP for later use in biosynthesis and other energy-requiring processes.

Through the controlled release of energy and the import of matter and energy, living systems maintain themselves in a steady state, held away from the equilibrium position.

Questions and Problems

1. For the phase transition

$$H_2O(l, 1 \text{ atm}, 100°C) \longrightarrow H_2O(g, 1 \text{ atm}, 100°C)$$

calculate ΔE, w, ΔH, ΔS, ΔS_{total}, and ΔG. Assume that $H_2O(g)$ behaves as an ideal gas.

2. Show that the heat absorbed or evolved in a constant volume process (q_V) with no other work done is equal to ΔE.

3. The atmospheric pressure at 1 km altitude is 0.90 atm and at sea level it is 1 atm. A parcel of dry air undergoes an adiabatic (irreversible) expansion on rising from sea level to 1 km. What will be its change in temperature, ΔT, after arriving at 1 km altitude? Note that air is a diatomic gas with $E = \frac{5}{2}RT$ ($\frac{1}{2}RT$ each from three translational degrees of freedom and two rotational degrees of freedom; the vibrational degrees do not absorb energy

at the temperatures considered here). Take $P_{ext} = 1$ atm. Compare your answer with the adiabatic lapse rate of 9.86°C/km $\simeq$10°C/km in Section 7.5.

4. An oceanographer collects a sample of sea water at a depth of 8000 m in a thermally insulated sampling device. After the sample is raised to the surface, its temperature is measured and found to be 1.075°C. The oceanographer, however, notes its temperature as 2.000°C. What reason might there be for this?

5. Show that, in general, the heat of any reaction may be calculated from the molar heats of formation of the reactants and products by the equation $\Delta H_{rx} = \Sigma n\, \Delta H_{f(prod)} - \Sigma n\, \Delta H_{f(reac)}$. That is, for a general reaction

$$aA + bB \longrightarrow cC + dD$$

$$\Delta H = c\, \Delta H_{fc} + d\, \Delta H_{fD}$$
$$- a\, \Delta H_{fA} - b\, \Delta H_{fB}$$

6. Calculate the heat of formation of PCl_5, given the following heats of reaction:

$$2P(s) + 3Cl_2(g) \longrightarrow 2PCl_3(g)$$
$$\Delta H = -151.8 \text{ kcal}$$

$$PCl_5(g) \longrightarrow PCl_3(g) + Cl_2(g)$$
$$\Delta H = 32.8 \text{ kcal}$$

7. The Bombardier beetle drives off attackers by ejecting a hot, repellant spray. The spray is generated by the mixing of an aqueous solution of 10 percent hydroquinone (a reducing agent) and 25 percent hydrogen peroxide (the oxidizing agent) with enzymes in a glandular compartment. The overall reaction is

$$\text{hydroquinone} \left(\text{HO} - \bigcirc - \text{OH} \right)(aq)$$

$$+ H_2O_2(aq) \longrightarrow \text{quinone}$$

$$\left(O = \bigcirc = O \right)(aq) + 2H_2O(l)$$

with $\Delta H = -48.5$ kcal/mol. Any additional H_2O_2 above that required for this reaction is oxidized enzymatically by the reaction

$$H_2O_2(aq) \longrightarrow H_2O(l) + \tfrac{1}{2}O_2(g)$$

with $\Delta H = -22.6$ kcal/mol. Pressure resulting from the oxygen expels an irritating solution of quinone.

a. If the quantity of reacting solution is 1 mg and it initially contains 0.91 μmol (i.e., 0.91 × 10^{-6} mol) of hydroquinone and 7.40 μmol of H_2O_2, how much heat is evolved per milligram of reacting solution?

b. If the 1 mg of ejected solution approximates water in its heat capacity and heat of vaporization, how much solution is vaporized after heating from 25°C to 100°C in each ejection? (Shaving cream containers that dispense self-heated cream use this same principle and employ H_2O_2 as the oxidizing agent.)

8. A combustion reaction in a calorimeter generates temperatures much higher than 25°C. However, if sufficient time is allowed for the combustion products to cool down to 25°C, the measured heat of combustion corresponds to reactants (25°C) $\longrightarrow$ products (25°C). Use either a general or specific example to show that this is true on the basis of Hess' law of constant heat summation. You may use a specific reaction as an illustration.

9. a. Justify the statement that the heat of a (gaseous) reaction may be calculated by adding the bond energies for all bonds broken and subtracting the bond energies of all bonds formed in the reaction, that is, $\Delta H_{rx} = \Sigma(\text{bonds broken}) - \Sigma(\text{bonds formed})$. To do this you may use simple examples such as

$$2H_2(g) + O_2(g) \longrightarrow 2H_2O(g)$$

and

$$CH_4(g) + Cl_2(g) \longrightarrow CH_3Cl(g) + HCl(g)$$

and the bond energies listed in Table 3.3. Justify the statement by combining equations such as $4H \longrightarrow 2H_2$, not merely by calculating ΔH_{rx} from the formula given.

b. Check your calculation of ΔH_{rx} from bond energies by comparing with ΔH_{rx} calculated from heats of formation. Standard heats of formation for the pertinent substances are $H_2O(g)$, -57.8 kcal/mol; $CH_4(g)$, -17.9 kcal/mol; $CH_3Cl(g)$, -19.6 kcal/mol; $HCl(g)$, -22.1 kcal/mol.

10. If, in Figure 16.7(a), there were originally four distinguishable particles (1, 2, 3, 4), calculate the number of ways the particles can be placed in one compartment only and compare with the number of ways they could be distributed between two compartments, two to a compartment. Ignore the levels shown in the compartments of the figure. Which

arrangement is more probable and has the larger entropy?

11. Describe what would happen in situations (e) and (f) of Figure 16.7 as the system goes from the initial state (all beans in R) to the equilibrium state. Compare the relative number of particles in R and P at equilibrium in the two cases.

12. The energy necessary for desalination of water can be reduced to that required for the isothermal compression of water vapor over the salt solution to water vapor at the vapor pressure of pure water (Section 12.23).

$$H_2O(g, P = 23.34 \text{ mm}, 25°C) \longrightarrow$$
$$H_2O(g, P^0 = 23.76 \text{ mm}, 25°C)$$

a. Convince yourself that the free energy change for a reversible isothermal compression at constant volume is given by $\Delta G = V \Delta P$. (Hint: use the relations $G = H - TS$, $H = E + PV$, and $q_{rev} = T \Delta S$ with only $P \Delta V$ work possible.)

b. Calculate the free energy expenditure for the above water vapor compression process. Express your answer in kilowatt hours per 1000 U.S. gallons. (One U.S. gallon = 4405 g of H_2O; 1 liter atm = 2.815 kW hr.)

13. The reaction

$$2Cu^+ \rightleftharpoons Cu^{2+} + Cu$$

is in equilibrium at 25°C when the concentrations are $[Cu^{2+}] = 0.0100M$ and $[Cu^+] = 7.75 \times 10^{-5}M$.

a. What is the standard free energy change ΔG^0 for the reaction?

b. What is the standard emf, E^0_{total}, for the reaction?

14. The potassium ion concentration in blood plasma is about $0.0050M$, and the concentration in muscle cell fluid is about $0.15M$. The plasma and intracellular fluid are separated by the cell membrane, which we assume is permeable only to K^+.

a. What is the free energy change at 25°C for the transfer of 1 mol of K^+ from the blood plasma into the cellular fluid? $\Delta G = 0$.

b. Where does this energy come from when such active transport of K^+ from a region of lower concentration to a region of higher concentration is required?

c. If there is an electrical potential difference across the cell membrane that opposes the potential difference resulting from the unequal concen-

trations of K^+, what is the magnitude of this membrane potential?

15. A reaction A $\rightleftharpoons$ B has a standard free energy change, $\Delta G^0 = -0.41$ kcal.

a. Will the reaction proceed spontaneously when $[A] = 1M$ and $[B] = 1M$?

b. What situation pertains when $[B] = 2M$ and $[A] = 1M$? What is ΔG?

c. When A and B are mixed and have initial concentrations $[A] = 1M$ and $[B] = 3M$, in what direction will the reaction proceed?

16. Draw resonance structures for HPO_4^{2-}.

17. For the oxidation of glucose in solution,

$$C_6H_{12}O_6(aq) + 6O_2(g) \longrightarrow$$
$$6CO_2(g) + 6H_2O(l)$$

a. Calculate ΔG^0_{298}, given the standard free energies of formation:

$$\Delta G_f^0(\text{glucose}) = -217.0 \text{ kcal/mol}$$
$$\Delta G_f^0(CO_2) = -94.3 \text{ kcal/mol}$$
$$\Delta G_f^0(H_2O) = -56.7 \text{ kcal/mol}$$

b. What is the free energy change for the reaction when the glucose concentration is $0.01M$ and the partial pressures of O_2 and CO_2 are 0.2 atm and 0.05 atm, respectively?

18. If the "standard" free energy of hydrolysis of ATP, $\Delta G^{0'}$ is -7.3 kcal at pH $= 7$ and 25°C,

a. Calculate the equilibrium constant for the hydrolysis reaction.

b. Calculate the free energy of hydrolysis, $\Delta G'$, under cellular conditions, where $[ATP] = 0.005M$, $[ADP] = 0.0005M$, and $[P_i] = 0.005M$. Retain pH $= 7$ and $T = 298°K$.

19. Inspect each of the steps in glycolysis and in the Krebs cycle and describe the type of reaction taking place (oxidation, group transfer, rearrangement, etc.). Some steps may involve two reaction classes.

20. What will be the sign of the entropy change for the denaturation of a protein? Explain.

21. A biological reaction A + B $\rightleftharpoons$ C + D has a $\Delta G^{0'}$ of 0.10 kcal/mol for the production of C and D at 25°C and pH $= 7$. Can the spontaneous formation of C be enhanced by coupling this reaction with

a. a reaction in which C is converted spontaneously to another substance E?

b. an exoergic reaction that produces C from some other substance F?

22. The hydrolysis of acetyl coenzyme A is exoergic,

$$\text{acetyl CoA} + H_2O \rightleftharpoons CH_3COO^- + H^+ + CoA$$

with $\Delta G^0 = -3.7$ kcal/mol.

a. What is $\Delta G^{0'}$ for this reaction at 25°C and pH = 7?

b. What is ΔG for this reaction at 25°C and pH = 7 when all substances are present at $0.01M$?

Answers

1. $\Delta E = 8.98$ kcal/mol
   $w = 746$ cal/mol
   $\Delta H = 9.72$ kcal/mol
   $\Delta S = 26$ cal/°K mol
   $\Delta S_{total} = 0$
   $\Delta G = 0$
3. 9°C/km
6. -108.7 kcal/mol
7. a. -0.19 cal/mg
   b. 0.21 mg
9. for $2H_2O$: -115 kcal
   for CH_3Cl: -23 kcal

10. one compartment: $W = 1$
    two compartments: $W = 6$
12. b. 2.98 kW hr/1000 gal
13. a. -8.49 kcal
    b. 0.368 volt
14. a. 2.0 kcal/mol
    c. 87 mV
17. a. $\Delta G^0 = -689.0$ kcal
    b. $\Delta G = -691.2$ kcal
18. a. $K' = 2.3 \times 10^5$
    b. $\Delta G' = -11.8$ kcal
22. b. -16 kcal/mol

References

THERMODYNAMICS WITH SOME METABOLISM

Blum, Harold F., "Time's Arrow and Evolution," 3rd ed., Princeton University Press, Princeton, N. J., 1968, Chapters II, VII.

Hargreaves, G., "Elementary Chemical Thermodynamics," Butterworths, London, 1961.

Klotz, I., "Energy Changes in Biochemical Reactions," Academic, New York, 1967.

Linford, J. H., "An Introduction to Energetics with Applications to Biology," Butterworths, London, 1966.

Morris, J. Gareth, "A Biologist's Physical Chemistry," Addison-Wesley, Reading, Mass., 1968, Chapters 7, 8, 9.

Nash, L. K., "Elements of Chemical Thermodynamics," Addison-Wesley, Reading, Mass., 1962.

Pimentel, George C., and Spratley, Richard D., "Understanding Chemical Thermodynamics," Holden-Day, San Francisco, 1969.

METABOLISM WITH SOME THERMODYNAMICS

Cloud, Preston, and Gibor, Aharon, "The Oxygen Cycle," Scientific American, **223,** 110–123 (September 1970).

Lehninger, Albert L., "Bioenergetics," W. A. Benjamin, Menlo Park, Calif., 1965.

Lehninger, Albert L., "How Cells Transform Energy," Scientific American, **205,** 62–73 (September 1961).

Ramsay, J. A., "The Experimental Basis of Modern Biology," Cambridge University Press, London, 1965, Chapters 9–20.

Watson, James D., "Molecular Biology of the Gene," 2nd ed, W. A. Benjamin, Menlo Park, Calif., 1970, Chapters 2, 5.

Reaction Kinetics and Enzymes

. . . There is an ecology of the world within our bodies. In this unseen world minute causes produce mighty effects; the effect, moreover, is often seemingly unrelated to the cause, . . . "A change at one point, in one molecule even, may reverberate throughout the entire system. . ."

Rachel Carson—*Silent Spring*

The preceding chapter talked a lot about initial and final states in reactions the free energy difference between these two states, and the direction and extent of reaction. But this is not the whole story of getting from reactant to product. If we liken a chemical reaction to an automobile trip from one place to another, then such matters involve only the starting point and ultimate destination, whose latitude and longitude would be state functions, as it were. But other matters would concern us in planning the trip. We would undoubtedly want to consider what route to take and how long we would spend on the road.

Chemists ask the same questions about reactions. The route of a chemical reaction concerns the path or paths taken by reactant molecules when they undergo the changes that produce product molecules. We have already seen the very complicated path involved in the biological oxidation of a glucose molecule to carbon dioxide and water. Such a sequence of reactions describes each step in the reaction path from reactants to products.

Then too, every chemical reaction has a definite speed by which reactants transform to products. This reaction rate depends on a number of factors. Among the more important are the intrinsic reactivity of the reactants, the concentration of substances present, the temperature, and the presence of catalysts. The rate of change of reactants into products is intimately related to the reaction mechanism. It is important to realize that the equilibrium thermodynamics of the last chapter tells us nothing about the velocity or the mechanism of a reaction. The free energy change merely tells us whether a reaction *could* happen, not whether it *will* happen. We have already noted this dichotomy in the case of the energetically favorable but very slow transformation of diamond to graphite.

Reaction Rates

17.1 THE RATE EQUATION

The rate of a chemical reaction is a function of the concentrations of some or all of the reacting substances. The mathematical relationship between the rate of a reaction and the concentration of reacting substances is called the rate law or the *rate equation*. Every rate equation has the functional form

$$\text{rate} = k[\text{M}]^m[\text{N}]^n \cdots$$

The proportionality constant k is the *rate constant*. The quantities in brackets are the concentrations of any substances affecting the rate of the reaction. They are not equilibrium concentrations, and they change with time as the reaction proceeds, as does the reaction rate. Such rate-affecting substances are not limited to those we normally call the reactants on the left side of a reaction equation; the rate may also depend on the concentrations of reaction products, or even on the concentrations of substances that do not appear in the net reaction equation. The exponents $m, n, \ldots$ may be positive or negative whole numbers or fractions. *In general, the rate equation—the concentration terms and the exponents—has no relation to the balanced equation for the reaction.* This being the case, the rate equation for a particular reaction can be determined only by experimental measurement of the effect of concentration changes on the reaction rate. The concentration dependence of the rate cannot be predicted, except in special cases, from the reaction equation.

17.2 REACTION RATE

So far the word "rate" has been used in a very general sense, while our attention has been directed to the functional form of the right side of the rate equation. The left side of the rate equation has not been completely specified.

The rate of a reaction is commonly measured and expressed as the change in concentration of a reactant or product per unit time. For homogeneous gas phase reactions, the concentrations can be expressed as molarity or as partial pressures. In other homogeneous reactions, concentrations are usually expressed as molarity. In heterogeneous reactions, the surface area of a condensed phase corresponds to its concentration, so that an increase in the degree of subdivision of the liquid or solid phase may drastically increase the rate, as happens in the explosion of coal dust. Unless otherwise specified, we shall limit ourselves to homogeneous reaction systems.

The rate can be expressed as the rate of disappearance of a reactant or as the rate of formation of a product. To illustrate, consider the reaction

$$2A + B_2 \longrightarrow 2AB$$

(Remember, this equation tells us nothing about the right side of the rate equation.) The average rate can be expressed in three ways:

1. The average rate can be expressed as the change in concentration of reactant A per unit time. That is, if the concentration of A at time t_1 is $[A]_1$ and at a later time t_2 is $[A]_2$, the average rate is

$$\frac{[A]_2 - [A]_1}{t_2 - t_1} = -\frac{\Delta[A]}{\Delta t}$$

The negative sign is used to satisfy the convention that rates always be positive numbers. It is needed here because the concentration of A decreases with time, so that $[A]_2 - [A]_1 = \Delta[A]$ is negative.

2. The average rate can be expressed as the rate of disappearance of B_2,

$$\text{average rate} = -\frac{\Delta[B_2]}{\Delta t}$$

3. The average rate can be expressed as the rate of formation of AB,

$$\text{average rate} = \frac{\Delta[AB]}{\Delta t}$$

Note here the absence of a minus sign, since $[AB]_2 - [AB]_1$ is positive while AB is being produced.

The average rates are related to one another by

$$-\frac{\Delta[A]}{\Delta t} = -2\frac{\Delta[B_2]}{\Delta t} = \frac{\Delta[AB]}{\Delta t}$$

because, according to the overall equation of the reaction, A is disappearing and AB is being formed at twice the rate that B_2 is being used up. For this reason, the substances used to determine the rate should be specified.

The term "average rate" has been used in the above example because we have assumed a comparatively lengthy time interval $t_2 - t_1$. As normally understood, however, the rate of a reaction refers to the rate at some particular instant, not the average rate over a time interval. The difference between the instantaneous rate and the average rate is illustrated in Figure 17.1, where the concentration of a reactant is plotted against time. The instantaneous rate at any time, t_i, is equal to the negative of the slope (rise/run) of the tangent to the curve at the point t_i. Mathematically, the slope of the tangent is symbolized by $d[R]/dt$. The instantaneous rate and the average rate are not necessarily equal. Nevertheless, the previously stated equations relating rates to concentrations hold for both average and instantaneous rates.

It is apparent that the rate of the chemical reaction of Figure 17.1 decreases as the reaction proceeds. However, the rate constant k for any given reaction remains the same over the course of the reaction at a given temperature.

Figure 17.1. Average rates and instanta-neous rates. The average rate in the time interval t_0 to t_1 is $-\Delta[R]_1/\Delta t = -\Delta[R]_1/t_1$. The instantaneous rate at time t_1 is given by the slope of the line tangent to the curve at t_1. With the origin of axes used here, the instantaneous rate at t_1 is $[R']/t'$. Similarly, the average rate for the interval t_0 to t_2 is $\Delta[R]_2/t_2$, whereas the instantaneous rate at t_2 is $[R'']/t''$. In general, average and instantaneous rates are not equal. Compare the automobile driver who covers 1 mile in 1 min at a constant speed of 60 mph with another driver who drives 30 mph for $\frac{1}{2}$ min and 90 mph for another $\frac{1}{2}$ min. He also goes 1 mile in 1 min at an *average* speed of 60 mph, but at no time (except while accelerating to 90 mph) is his instantane-ous speed 60 mph.

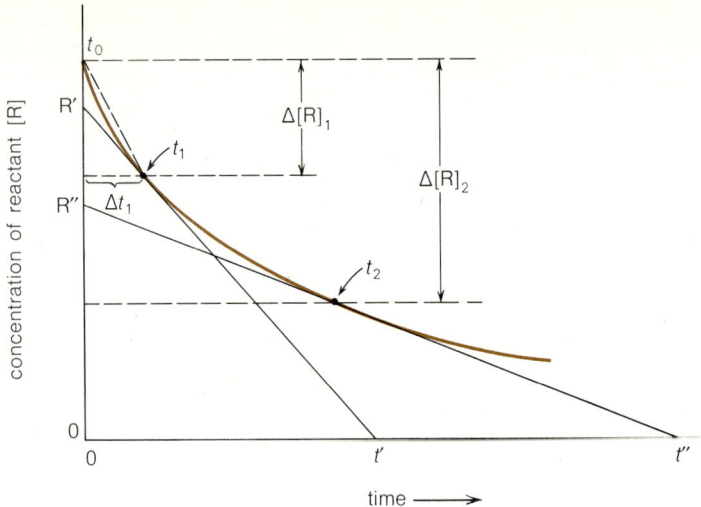

17.3 DETERMINATION OF THE RATE EQUATION

The rate equation for a reaction is established by the experimental determi-nation of the reaction rate at different concentrations of reacting substances. These experiments provide information for the deduction of the values of the exponents $m, n, \ldots$ in the general rate equation, rate $= k[M]^m[N]^n \cdots$. The overall *order* of a reaction is the sum of the exponents, $m + n + \cdots$. The order with respect to an individual reactant is also specified. Thus, m is the reaction order with respect to M, n is the order with respect to N, and so on. For simple reactions, the orders may be determined by measuring how the reaction rate varies with reactant concentration. The reaction order, like the rate equation, can be determined only from experimental data. The exponents $m, n, \ldots$ are not equal to the stoichiometric coefficients of reactant species that appear in the reaction equation except in special cases.

A reaction $A + B + C \longrightarrow P$ proceeds according to the rate equation

$$\text{rate} = \frac{d[P]}{dt} = k[A][B]^2$$

Table 17.1
Rate data for A + B + C ⟶ P[a]

experiment	$[A]_0$	$[B]_0$	$[C]_0$	initial rate $= \left(\dfrac{dP}{dt}\right)_0$ mol/liter min
1	0.10	0.10	55	0.020
2	0.20	0.10	55	0.040
3	0.30	0.10	55	0.060
4	0.10	0.20	55	0.080
5	0.10	0.30	55	0.180
6	0.10	0.10	54	0.020

[a] The reactant concentrations are the initial con-centrations and the rate is the initial rate of reac-tion. This is the rate at $t = 0$ immediately after the reactants are mixed together.

as determined from the experimental data in Table 17.1. This reaction is first order in A, second order in B, zero order in C, and third order overall. Zero order means that the concentration of C has no effect on the rate, at least in the range of concentrations used to determine the rate equation. For this particular reaction, doubling the concentration of A while keeping [B] constant would double the reaction rate. On the other hand, doubling the concentration of B while keeping [A] constant would quadruple the rate, and tripling the concentration of B would increase the rate by nine times. Table 17.1 illustrates this behavior.

EXAMPLE | From experiments 1, 2, and 3 in Table 17.1, we see that the reaction order for A is 1. From experiments 4, 5, and 6, we see that the order with respect to B is 2. Thus, rate $= k[A][B]^2$. The rate constant k can now be calculated from any of the experiments. Using run 4, $k =$ rate$/[A][B]^2 = 0.080/(0.10)$ $(0.20)^2 = 20$ liters^2/mol^2 min. With k known, the rate can be calculated for any concentration of A and B. If $[A]_0 = 0.40$ and $[B]_0 = 0.40$, then the initial rate is 1.3 mol/liter min.

EXAMPLE | The concentration of reactant A at various times during the reaction A $\longrightarrow$ B + C is given in Table 17.2. Determine the order of the reaction and calculate an approximate rate constant.

Time, sec	[A], M	average rate $=$ $-\Delta[A]/\Delta t$, mol/liter sec	$[A]_{av}$, M	$\dfrac{\text{average rate}}{[A]_{av}}$, per sec	$\dfrac{\text{average rate}}{[A]^2}$, liter/mol sec
0	0.10				
		0.0025	0.075	0.033	0.44
20	0.050				
		0.00063	0.038	0.017	0.44
60	0.025				
		0.00016	0.019	0.0084	0.44
140	0.013				

Table 17.2
Rate data for A $\longrightarrow$ B + C

SOLUTION | The average rates for the time intervals have been calculated and are listed in the table. The rate equation of the reaction, using average rates and concentrations, is

$$\text{average rate} = k[A]_{av}{}^a$$

where a is the reaction order. Because k is a constant at a given temperature, the quantity (average rate$/[A]_{av}{}^a$) should be constant for the correct value of a (assumed here to be 0, 1, or 2). The average rate obviously changes, so a cannot be zero. The quantity (average rate$/[A]_{av}$) also is not constant, so $a \neq 1$. With $a = 2$, (average rate$/[A]_{av}{}^2$) is constant and equal to 0.44 liter/ mol sec. The approximate value of k is also 0.44.

Of course, instantaneous values of the rate could have been obtained by graphing [A] versus t and measuring slopes. A more accurate value of $k = 0.50$ would have been obtained from the instantaneous rates.

The basis of a common method (the isolation method) for determining the order of a reaction with respect to one reactant is to have the other reactants present in such large excess that their concentration changes do not affect the reaction rate as the reaction proceeds. Thus, in the reaction $A + B + C \longrightarrow P$ in Table 17.1, the initial concentration of C was held constant, or nearly so, so that it does not appear to affect the initial rate. Not only is C held constant, but it is in large excess, so that during the reaction its concentration remains virtually constant and is absorbed into the rate constant. (In the example of Table 17.1, the maximum concentration decrease in C would amount to only $0.30M$ if the reaction were first order in C.) The data of Table 17.1 could illustrate a reaction between solutions of A and B, where C would represent the solvent which also takes part in the reaction. If the concentration of C were comparable to that of A and B, changes in [C] might be found to influence the rate, and the reaction would no longer be zero order with respect to C. Thus, in runs 1 and 6 of Table 17.1, if $[C]_0$ were $0.10M$ and $0.20M$, respectively, the new initial rate of run 6 would be double that of run 6, indicating that the reaction is also first order in C and that the actual rate equation is rate $= k'[A][B]^2[C]$. An actual example of a pseudo-first-order reaction is the hydrolysis (inversion) of sucrose,

$$C_{12}H_{22}O_{11} + H_2O \longrightarrow \underset{\text{glucose}}{C_6H_{12}O_6} + \underset{\text{fructose}}{C_6H_{12}O_6}$$

which is found to be a first-order reaction in aqueous solution,

$$\frac{d[C_6H_{12}O_6]}{dt} = k[C_{12}H_{22}O_{11}]$$

In nonaqueous solution, where water is not present in excess, the reaction is second order, with a rate equation

$$\frac{d[C_6H_{12}O_6]}{dt} = k'[C_{12}H_{22}O_{11}][H_2O]$$

When water is present in excess, $[H_2O]$ is constant and $k'[H_2O] = k$.

In practice, the order and rate constant characterize a particular reaction, not the actual rates as determined from the slopes of tangents to the concentration versus time curves. A variety of techniques exist for finding the orders and rate constants of reactions, but we shall not go into this subject here.

17.4 REACTION MECHANISMS

Most chemical reactions proceed from reactants to products through a series of intermediate reaction steps. The sum of the reaction steps must be equal to the balanced equation for the overall reaction. A reaction taking place by a series of steps is known as a multistep or complex reaction, and the individual reaction steps are elementary reactions. In an elementary (one-step) reaction, the reactants transform directly into products. The sequence of elementary reactions in a complex reaction constitutes the mechanism of the complex reaction.

Only for an elementary reaction is the rate equation directly related to the balanced reaction equation, for only then do the concentration terms and their exponents correspond to the reactants and the stoichiometric co-efficients in the reaction equation. The overall stoichiometric equation for a complex reaction tells us nothing about the mechanism of the reaction; it only tells us the identity of the initial reactants and final products and their mole proportions.

The rate equation of a reaction is determined by the mechanism of the reaction. For an elementary reaction, the rate equation exponents are equal to the coefficients in the balanced equation. But if the rate equation exponents and the stoichiometric coefficients in a reaction equation are equal, this is not proof that the reaction is elementary. A discrepancy between the stoichiometric equation and the experimental rate equation is proof that the reaction is a multistep process. As illustrations of these statements, consider the following gaseous reactions and their experimental rate equations:

(1) $Br_2 \xrightarrow{k_1} 2Br$ $\qquad\qquad$ rate $= k_1[Br_2]$

(2) $H_2 + 2I \xrightarrow{k_2} 2HI$ $\qquad\qquad$ rate $= k_2[H_2][I]^2$

(3) $2NO + Br_2 \xrightarrow{k_3} 2NOBr_2$ $\qquad$ rate $= k_3[NO]^2[Br_2]$

(4) $4HBr + O_2 \xrightarrow{k_4} 2H_2O + 2Br_2$ $\qquad$ rate $= k_4[HBr][O_2]$

Reactions 1 and 2 are elementary reactions; their rate equations can be written directly from their reaction equations. Reaction 3 is a complex reaction, and it is a coincidence that the net equation and the rate equation are correlated. Reaction 4 is obviously a multistep process. Except when devising reaction mechanisms, the reaction equations we write and encounter most often are the overall equations of complex reactions, so we are not justified in writing rate equations on the basis of the stoichiometric equations.

For a particular reaction, we must assume that all mechanisms are possible but not equally probable. The overall reaction will occur faster by some mechanisms than by others, and the measured rate will usually be determined by the more rapid mechanism, since it is this mechanism that will be primarily responsible for the conversion of reactants to products. When this is the case, there may be, within the fastest mechanism, an elementary reaction step that is much slower than the other steps. When such a step exists, the reaction rate and the rate equation of the overall reaction will depend primarily on that slow step, which is called the *rate-determining step*. The rate equation for the overall reaction will have one term. If, on the other hand, there is no single slow step, the rate equation for a reaction may be quite complicated. The effect of a single rate-determining step is most apparent in mechanisms and processes (like metabolism) involving consecutive reaction steps.

17.5 SOME COMPLEX REACTIONS

The mechanism of a complex reaction is rarely determined directly from the experimental rate equation. Instead, a mechanism is guessed at, and a rate equation is deduced from the set of elementary reactions. If the predicted and the observed rate equations are identical, then the mechanism is at least consistent with the experimental rate data, although not necessarily the only possible mechanism or even the true mechanism. Some examples are given to illustrate this procedure.

A. The reaction between nitric oxide and hydrogen,

$$2NO + 2H_2 \longrightarrow N_2 + 2H_2O,$$

has the experimental rate equation

$$-\frac{d[NO]}{dt} = k[NO]^2[H_2]$$

The overall order is three, not four. The rate equation is inconsistent with the stoichiometric equation—a sure sign that this is a complex reaction. One possible mechanism of this reaction might consist of two elementary steps,

(1) $2NO + H_2 \longrightarrow N_2 + H_2O_2$ slow
(2) $H_2O_2 + H_2 \longrightarrow 2H_2O$ fast

Hydrogen peroxide is an intermediate product in this proposed mechanism. If the first step is assumed to be very slow compared to the second, the first step limits the overall reaction rate because it determines the rate of supply of H_2O_2 to the faster second step. The rate equation for the first step is the same as the experimentally determined rate equation.

Another possible mechanism for this same reaction might be

$$(1) \qquad 2NO \underset{k_{-1}}{\overset{k_1}{\rightleftharpoons}} N_2O_2$$

$$(2)\ N_2O_2 + H_2 \xrightarrow{k_2} N_2O + H_2O \qquad \text{slow}$$

$$(3)\ N_2O + H_2 \xrightarrow{k_3} N_2 + H_2O$$

with equation (2) being the slowest, and therefore rate-determining, step. Then

$$-\frac{d[NO]}{dt} = k_2[N_2O_2][H_2]$$

But the concentration of N_2O_2 should be proportional to its rate of production in reaction (1), through the equilibrium constant K,

$$[N_2O_2] = K[NO]^2$$

so that

$$-\frac{d[NO]}{dt} = Kk_2[NO]^2[H_2]$$

This agrees with the experimental rate equation with $k = Kk_2$. Other mechanisms could be devised to agree with the rate equation, so it is apparent that the experimental rate equation does not pinpoint the exact mechanism. In the absence of other experiments, the most the experimental rate equation can do is to aid in the elimination of some postulated mechanisms.

B. A much more complicated example is the formation of hydrogen bromide by the reaction $H_2 + Br_2 \longrightarrow 2HBr$. The experimental rate equation is

$$\frac{d[HBr]}{dt} = \frac{k[H_2][Br_2]^{1/2}}{1 + (k'[HBr]/[Br_2])}$$

One is hard put to assign an order to this reaction. The proposed mechanism for this reaction is a free radical (atom) chain reaction:

(1) $\qquad Br_2 \xrightarrow{k_1} 2Br\cdot$

(2) $\quad Br\cdot + H_2 \xrightarrow{k_2} HBr + H\cdot \qquad$ slow

(3) $\quad H\cdot + Br_2 \xrightarrow{k_3} HBr + Br\cdot$

(4) $\quad H\cdot + HBr \xrightarrow{k_4} H_2 + Br\cdot$

(5) $\qquad 2Br\cdot \xrightarrow{k_5} Br_2$

Reaction (1) initiates the series of reactions. Steps (2) and (3) propagate the series by regenerating the reactive bromine atoms. Step (4) is also a propagation step, because it regenerates a reactive atom. It also uses up product (HBr), however, so this step inhibits the formation of HBr. Step (5) terminates the chain of reactions.

The rate law based on this mechanism is

$$\frac{d[HBr]}{dt} = \frac{2k_2(k_1/k_5)^{1/2}[H_2][Br_2]^{1/2}}{1 + (k_4/k_3)([HBr]/[Br_2])}$$

The deduced rate law and the experimental rate law have the same form because

$$2k_2\left(\frac{k_1}{k_5}\right)^{1/2} = k = \text{a constant}$$

and

$$\frac{k_4}{k_3} = k' = \text{another constant}$$

Although reaction (2) is a slow step, it does not completely determine the form of the derived rate equation. The [HBr] term in the denominator indicates its inhibition of the overall reaction. When the concentration of a substance appears in the denominator of a rate equation, that substance inhibits the reaction. There are many examples of inhibitors or chain terminators in everyday life. Tetraethyl lead, used in gasoline, functions as a chain terminator by combining with reactive free radicals generated in the combustion process. Some spoilage retardants added to foods inhibit the free radical oxidation process in foods.

17.6 REVERSIBLE REACTIONS AND THE EQUILIBRIUM CONSTANT

Reversible or opposing reactions are classed as complex reactions in chemical kinetics. Thus, the general reversible reaction

$$a\mathrm{A} \underset{k_r}{\overset{k_f}{\rightleftharpoons}} b\mathrm{B}$$

is a complex reaction. If the forward and reverse steps are elementary reactions, rate equations may be written directly from the reaction equation. The rate of the forward reaction is

$$-\frac{d[A]}{dt} = k_f[A]^a$$

The rate of the reverse reaction is

$$\frac{d[A]}{dt} = k_r[B]^b$$

The overall rate of production of A is

$$\frac{d[A]}{dt} = k_r[B]^b - k_f[A]^a$$

The equilibrium constant for the reversible reaction is obtained if (1) the rate of the forward reaction is equated to the rate of the reverse reaction, since this is the criterion for dynamic equilibrium *or* (2) if the overall rate of production (or disappearance) of A (or B) is set equal to zero, since this is the result of dynamic equilibrium. Figure 17.2 shows how the velocities

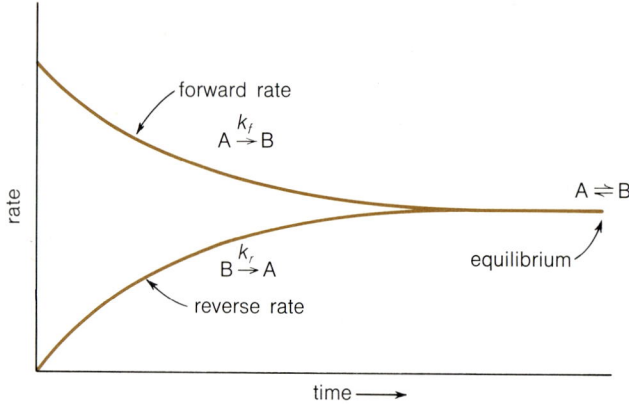

Figure 17.2. The rates of the forward and reverse reactions in a reversible reaction approach each other as reactants are used up and become equal when equilibrium is established.

of the forward and reverse reactions approach each other with time, starting out with pure reactant. Either way, the ratio of the rate constants at equilibrium is obtained.

$$\frac{k_f}{k_r} = \frac{[B]_{eq}^b}{[A]_{eq}^a}$$

Because k_f and k_r are constants at a given temperature and because the concentrations are now those at equilibrium, the equilibrium constant K is identical to k_f/k_r:

$$K = \frac{k_f}{k_r} = \frac{[B]_{eq}^b}{[A]_{eq}^a}$$

In this generalized example, both the forward and reverse reactions were assumed to be elementary, one-step processes, and the rate equations for the opposing reactions could be written directly from the stoichiometric reaction equation. This cannot be done with complex reversible reactions. Just the same, it is found that equating the *experimentally* determined rate equations for the forward and reverse overall reactions in any reversible reaction, no matter what the mechanism, does yield the same equilibrium constant that would be written from the balanced equation of the overall reaction. Furthermore, the rate equations deduced from a postulated mechanism consistent with the experimental rate equation also yield the equilibrium constant.

17.7 EFFECT OF TEMPERATURE ON REACTION RATE

An increase of temperature almost always increases the rate of a reaction. This means that the rate constant k is temperature dependent. The relation between k and the reaction temperature is given by an empirical equation called the Arrhenius equation,

$$k = Ae^{-E_a/RT}$$

where E_a is known as the energy (or heat) of activation, A is a constant for a particular reaction and is almost temperature independent, and e is a universal constant equal to 2.71828, the base of natural logarithms (Appendix A.3). In logarithmic form, this equation is

$$\log k = \log A - \frac{E_a}{2.303RT}$$

so that a graph of $\log k$ versus $1/T$ would give a straight line with a slope of $-E_a/2.303R$ as shown in Figure 17.3. With $R = 1.987$ cal/mol deg, E_a is in calories per mole.

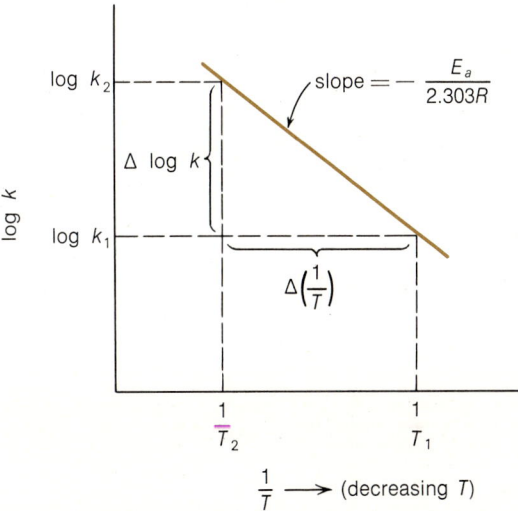

Figure 17.3. The measured slope of the straight line is

$$\text{slope} = \frac{\Delta \log k}{\Delta(1/T)}$$

$$= \frac{\log k_2 - \log k_1}{(1/T_2) - (1/T_1)}$$

Then

$$E_a = -2.303R \times \text{slope}$$

The activation energy E_a determined this way corresponds to the heat or enthalpy of activation, not the free energy of activation.

Instead of graphing a series of k values for different temperatures, it is often sufficient to take two k values, k_1 and k_2 at T_1 and T_2, and use the equation

$$\log k_2 - \log k_1 = \log \frac{k_2}{k_1} = -\frac{E_a}{2.303R}\left(\frac{1}{T_2} - \frac{1}{T_1}\right) = \frac{E_a}{2.303R}\left(\frac{T_2 - T_1}{T_1 T_2}\right)$$

to calculate E_a. In Section 12.13, it was mentioned that, *as a very rough approximation*, the speed of a chemical reaction in solution doubles with a

10°C rise in temperature. For this to be true for a reaction at room temperature, the energy of activation E_a must be about 12.5 kcal/mol, and this is
a typical value for the activation energy of many rate-determining steps in
metabolic processes.

Theories of Reaction Rates

So far in this chapter the effects of concentration and temperature have been
described, but we have not tried to explain them. To explain the observed
effects, we need to take a brief look at two related theories of the mechanism
of elementary reactions.

17.8 THE COLLISION THEORY

A chemical reaction may involve the making or breaking of bonds or both.
The collision theory of reaction rates says that the conditions necessary for
electronic and atomic rearrangements within or between reacting molecules
result from collisions between reacting molecules. A collision accomplishes
two things. First, it brings the molecules close together so that a reshuffling
of atoms can take place. That is, bonds can easily be formed between atoms
in adjacent molecules. Second, the transfer or pooling of the kinetic energy
of colliding molecules provides energy needed for the breaking of bonds.

According to the collision theory, the rate of a chemical reaction is
determined by two factors. One is clearly the total number of collisions per
unit time between reacting molecules. The other factor concerns the effectiveness of the collisions. Let us worry first only about the total number of
collisions.

If we consider an elementary reaction between two molecules
$(A + B \longrightarrow P$ or $A + A \longrightarrow P)$, the number of collisions of A with B or
of A with A is proportional to the number or concentration of reacting
molecules. The concentration dependence of the rate of reaction is

rate $\propto$ number of collisions

and

number of collisions $\propto$ concentration of molecules

Using k, the rate constant, as the proportionality constant, simple considerations show the rate of reaction to be given by equations such as

rate $= k[A][B]$

or

rate $= k[A][A] = k[A]^2$

or, in general, for an elementary process involving the simultaneous collision
of a molecules of A, b molecules of B, and so forth,

rate $= k[A]^a[B]^b \cdots$

The kinetic theory of gases can be used to deduce an expression for the
total number of collisions between reactant molecules in a gas at a given
temperature. The frequency of collisions often comes out to be as much as
10^{12}–10^{15} times as high as any measured reaction rate. Evidently, not every

collision leads to formation of product molecules. One reason for this discrepancy might be the requirement of a particular geometric orientation of the colliding molecules, but this steric influence by itself cannot explain the large difference between collision frequency and observed reaction rate.

The activation energy E_a in the Arrhenius equation is an important clue to the cause of the lower reaction rates. Even before kinetic theory was applied to rate theory, it was believed that reactant molecules had to become activated by the absorption of an extra amount of energy above their average energy before they could react. From the kinetic theory of ideal gases (Section 4.15), we know that the molecules do not all have the same energy. The largest fraction of the molecules have some average energy, but there are always some that have energies considerably larger than the average, as illustrated in Figure 17.4, where the fraction of molecules having various energies is plotted

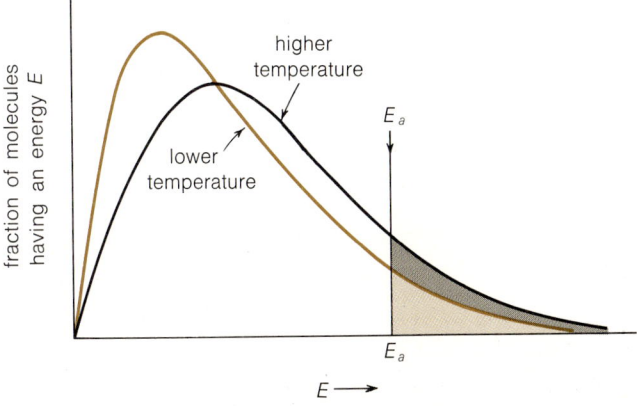

Figure 17.4. As the temperature increases, a greater number of gas molecules have energies greater than E_a. The area under the curves beyond E_a is equal to the fraction of molecules with energy greater than E_a. The height of the curve above the energy axis at E_a is also a measure of the fraction of molecules possessing that minimum energy needed for reaction.

versus energy for two different temperatures. As pointed out in Section 4.16, these curves have the same form as the speed distribution plots of Figures 4.10 and 4.11. At a given temperature T, the fraction of the total number of gas molecules having an energy in excess of the activation energy E_a is given by the area beyond E_a under the curve. This fraction is given to a good approximation by the relation

$$\text{fraction with energy greater than } E_a = e^{-E_a/RT}$$

Figure 17.4 shows that only a small fraction of the total number of molecules has the requisite activation energy to produce a reaction. The figure also shows how the number of molecules having energies equal to or greater than E_a increases with temperature. This explains why the reaction rate increases with temperature.

If we realize that the rate constant k represents the reaction rate when all reactants are at unit concentration then, at these concentrations,

$$k = (\text{total number of collisions})(\text{fraction of collisions with the minimum energy } E_a)$$

$$= Ze^{-E_a/RT}$$

where Z is the total number of collisions, or the collision frequency. This relation has the same form as the Arrhenius equation. Usually, the factor A

of the Arrhenius equation is equated to pZ, where p is a factor introduced to account for any deviation of the experimental k from the k calculated by collision theory. The term p is a typical fudge factor, which, at least in part, takes into consideration the steric factors affecting the effectiveness of collisions.

17.9 THE TRANSITION STATE THEORY

The transition state theory builds on the collision theory by considering that the colliding molecules come together to form an *activated complex* or *transition state* of high potential energy. The difference in potential energy of the activated complex and the reactant molecules is equal to the activation energy. This relationship is shown in Figure 17.5. The complex is formed by

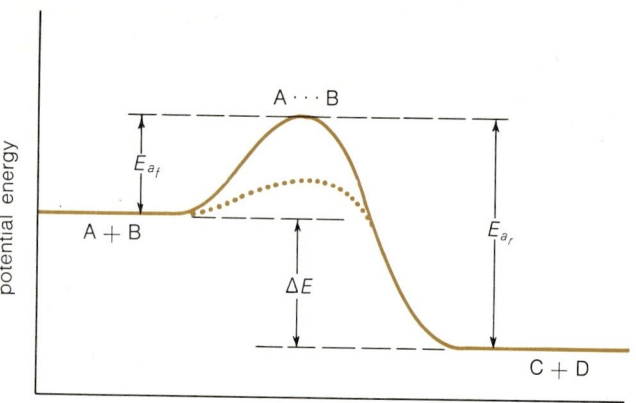

Figure 17.5. Potential energy diagram for the chemical reaction $A + B \longrightarrow C + D$. The potential energy of the system is plotted against the reaction coordinate, which is a function of the distances between reactant and product species. The reactants A and B must have an extra energy E_{a_f} between them before they can form the activated complex $A \cdots B$, which then decomposes to the products C and D. The overall change in energy for the reaction $A + B \longrightarrow C + D$ is given by ΔE. The activation energy for the reverse reaction, $C + D \longrightarrow A + B$, is E_{a_r}. The dotted potential energy curve represents a catalyzed reaction with a lower activation energy but with the same overall energy change and equilibrium constant as the uncatalyzed reaction.

reactant molecules possessing the necessary activation energy. For example, an elementary reaction of NO with O_3 may be written

$$NO + O_3 \rightleftharpoons \underset{O}{\overset{O \diagdown O \cdots O}{\diagup} N} \rightleftharpoons NO_2 + O_2$$

The activated complex is shown with dashed bonds to indicate that these bonds may be in the process of forming and breaking. The activated complex is in equilibrium with NO and O_3, and it has equal probability of reforming reactants or of forming the products, NO_2 and O_2. It is the frequency with which the complex decomposes to products that determines the overall rate of the reaction. According to the transition state theory, the rate of reaction is equal to the concentration of activated complexes at the top of the energy hump times the rate of their crossing the barrier, that is, decomposing in the product direction. The theory is also called the absolute rate theory because, in principle, it shows how to calculate both the concentration and frequency of decomposition of the activated complex from fundamental physical properties such as bond distances and molecular energy states of the reactants and of the most stable activated complex. However, our only interest now is in the nature and energy of the activated complex.

Referring again to Figure 17.5, we can see that there are two activation energies, one for the forward reaction and one for the reverse reaction. The difference $(E_{a_r} - E_{a_f})$ between them is equal to the thermodynamic change in internal energy ΔE for the overall reaction $A + B \longrightarrow C + D$. Instead of internal energy E, we can just as well talk about enthalpy changes ΔH and heats of activation or free energy changes ΔG and free energies of activation. (Strictly, the rate of reaction depends on the free energy of activation, $G_a = H_a - TS_a$. The temperature dependence of the rate is related only to H_a, and this is the quantity found from the plot of $\log k$ versus $1/T$.) The general form of Figure 17.5 applies to all three energy terms. Figure 17.5 can then be considered to show an exoergic (negative ΔG) forward reaction. The height of the energy barrier, G_a, for the forward reaction determines the rate of the reaction. If G_a (or E_a or H_a) is large, the reaction will be slow despite a large negative ΔG for the spontaneous forward reaction. This explains why thermodynamically spontaneous reactions are often very slow. It also shows why the reverse, endoergic, reaction will be slower still, since the activation energy must be at least equal to the positive free energy change and greater than the activation energy of the forward reaction. The difference in rates (until equilibrium is reached) accounts for the direction of the spontaneous reaction.

Molecularity of Reactions

The *molecularity* of a reaction is equal to the number of molecules coming together to form the activated complex in an elementary process. The molecularity is determined by the mechanism of the reaction and, except for elementary reactions, is not equal to the order of the reaction. The molecularity of a reaction is sometimes identified with the number of molecules that collide simultaneously in an elementary process, but this can be different from the number of molecules making up the activated complex. For instance, in a unimolecular gas reaction, a single reactant molecule is energized by collision with another molecule, but the two molecules do not stick together in the activated complex. Whichever definition of molecularity is used, the term applies only to elementary reactions. The molecularity of a reaction is a very small integer, usually 1 or 2. Very rarely it is 3. The simultaneous collision of three bodies is improbable, either on the basis of experience or on the basis of the loss of entropy represented by the formation of the activated complex.

17.10 CATALYSIS

The collision theory amply explains the effect of concentration and temperature on reaction rates. The transition state theory is useful for explaining the effect of catalysts on rate. Catalysis was introduced briefly in Section 7.3, where it was pointed out that catalysts increase reaction rates by lowering the activation energy of the reaction, although there are some cases where catalysts may speed up a reaction by increasing the number of collisions. When a catalyst lowers the activation energy, it does so by combining with a reactant molecule to form an activated complex with an energy lower than the complex in the uncatalyzed reaction (Figures 17.5 and 7.1). A hypothetical uncatalyzed reaction might be written

$$A + B \longrightarrow A\text{---}B \longrightarrow C + D$$

where the activated complex is A---B. If a catalyst X affects this reaction, it may do so by forming a different, lower energy complex:

$$A + B + X \longrightarrow A\text{---}B\text{---}X \longrightarrow C + D + X$$

Whatever the exact mechanism, the catalyst is both a reactant and a product and does not appear in the overall stoichiometric equation for the reaction. However, the concentration of a catalyst may well appear in the numerator of the rate equation, since the catalyst increases the rate of a complex reaction by lowering the activation energy of the slowest rate-determining step and is somehow involved in that step.

A catalyst does not affect the equilibrium position of a reversible reaction, because it has no effect on the free energy change of the reaction. The initial (reactant) and the final (product) states are identical for the uncatalyzed and the catalyzed reactions (Figure 17.5). Therefore, ΔG and K are the same. Only the rate of approach (in either direction) to equilibrium is increased by the presence of a catalyst.

Inhibitors decrease the rate of a reaction, and they may do so in various ways. It is not possible to force a reaction to proceed by a mechanism having a higher activation energy than the uncatalyzed reaction. In going from reactants to products, a reaction always picks that mechanism whose rate-determining step (if any) has the lowest activation energy. In the systems we shall discuss in the next section, inhibitors act by destroying a catalyst already present in the system, not by raising the activation energy of an uncatalyzed reaction. In chain reactions, inhibitors act as chain terminators by deactivating the free radicals needed for chain propagation steps.

Enzymes

17.11 BIOLOGICAL FUNCTION OF ENZYMES

Enzymes are biological catalysts. All enzymes are composed in whole or in part of protein molecules ranging in molecular weight from around 10,000 to 1 million. These large catalytic molecules play a central role in the functioning of every living system for a number of reasons. At the very heart of life, enzymes are the tools whereby living processes obey the instructions carried by the genetic apparatus of every organism. On a molecular level, enzymes determine, directly or indirectly, the stability of a living system and control the coordination of the chemical reactions of life. The presence or absence of certain enzymes or changes in the concentration, availability, or activity of enzymes determines cellular differentiation and the development of organisms. The presence and concentration of enzymes is in turn dictated by genetic factors. That is, in a multicellular organism, the genes call forth enzymes that control the protein syntheses that will determine what kind of a differentiated, or specialized cell, a newly formed cell will become, as well as its rate of growth and location in relation to the other different specialized cells in the organism.

On a more specific level, enzymes are required in every cell because almost all of the organic reactions of cellular metabolism are extremely sluggish in the absence of a catalyst. This sluggishness is due partly to the fact that organisms are composed of complex carbon compounds that are comparatively unreactive, although rich in energy. The slowness of reaction is fortunate, because fast reactions would rapidly approach an equilibrium state from which no work could be obtained. Slow reactions are essential if a steady state is to be maintained. But when a rapid release of energy is needed, some reactions must be accelerated. In a reaction flask in the laboratory, some of these reactions might be speeded up by raising the temperature or increasing the concentration of a reactant such as hydrogen ion. But an organism can tolerate only a narrow range of temperatures and pH, otherwise many protein molecules would be destroyed. Therefore, the measures used to speed reactions in the laboratory cannot be used in living systems and, instead, reactions can be accelerated in organisms only by a controlled lowering of the activation energy with enzyme catalysts.

Enzymes possess two advantages over the inorganic catalysts used in the laboratory and in industry for the same reactions. First, enzymes are much more efficient; they lower the activation energy more than inorganic catalysts. Second, enzymes are highly specific in their action; most of them catalyze only a single chemical reaction. Very often, this specificity is limited to one particular reactant molecule or even to one optical isomer of a given compound. Because of their high specificity, enzyme-catalyzed reactions give only one set of products from given reactants. Unlike the organic chemist in his laboratory, the organism is not burdened with the separation of a variety of reaction products and intermediates resulting from different reaction paths (side reactions). The specificity of enzymes is the result of the detailed three-dimensional structure of their molecules.

The extreme importance of enzymes in making the chemical reactions of life take place, coupled with the dependence of catalytic activity on subtle details of molecular conformation, makes enzymes prime leverage points for environmental influences on living systems, as we shall see in the following sections.

17.12 THE STRUCTURE OF ENZYMES: COFACTORS

Because enzymes are proteins, their stabilities and three-dimensional structures are governed by the factors discussed in Chapter 15.

Certain enzymes are catalytically active only if a smaller nonprotein group, or *cofactor*, is associated with the large protein portion. The cofactor may be either an organic group or a metallic ion such as K^+, Ca^{2+}, or Cu^{2+}. An organic cofactor is usually called a *coenzyme* if it is not held tightly to the protein portion of the enzyme and is easily detached. Tightly bound organic cofactors are usually known as *prosthetic groups*.

As we already know, an enzyme functions by forming an intermediate complex with a reactant molecule. Metal ion cofactors may serve as connecting links between the enzyme molecule and the reactant molecule, called the *substrate*, or between the protein and organic coenzyme parts of an enzyme. In other cases, the metal ion may take a direct part in the catalytic activity, as do the iron ions in the heme prosthetic groups of the cytochromes.

The organic cofactors, the coenzymes, are usually involved in group transfer reactions. Thus, the coenzymes NAD and FAD are associated with the enzyme catalyzing the transfer of a hydrogen atom from one compound to another. Coenzyme A, while technically not a coenzyme because it combines with the substrate rather than the enzyme, is involved in the transfer of an acyl group.

Many coenzymes are related to vitamins. Vitamins are organic compounds needed in small amounts for normal metabolism. Many organisms, especially the more highly developed heterotrophs, are unable to synthesize these compounds and must obtain them, plus required metallic cofactors, in their diet if nutritional diseases are to be avoided. Although the same compound may be needed by several species, it is generally termed a vitamin only for a species that cannot synthesize it, or that cannot synthesize it in adequate amounts. The vitamin nicotinamide or niacin, a B vitamin, is a constituent of coenzyme A. Because of the high catalytic activity of a single molecule, a comparatively small number of enzyme molecules are required, and the normal dietary input of vitamins is likewise small. Similar considerations apply to metals that serve as cofactors. The catalytic activity of enzymes is given by the *turnover number*, the number of substrate molecules converted to product per minute by a single enzyme molecule. Turnover numbers range from 100 to tens of millions. Continual replacement of vitamins is necessary, because enzyme molecules eventually become degraded as they go through their catalytic cycles.

17.13 THE MECHANISM OF ENZYME ACTION

A simple mechanism for the action of an enzyme that agrees with the observed rate equation is the following: an enzyme E and a substrate S form a complex ES, which in turn decomposes to products P in a near-irreversible rate-determining step. This mechanism, the Michaelis–Menten mechanism, is

$$E + S \underset{k_{-1}}{\overset{k_1}{\rightleftharpoons}} ES$$

$$ES \xrightarrow{k_2} E + P$$

In particular, this mechanism can be used to show that the rate of the overall reaction is proportional to the substrate concentration when [S] is small but that the rate is independent of [S] (zero order) at high substrate concentrations. That is,

$$\text{rate} = \frac{dP}{dt} \propto [E]_t [S] \qquad \text{for [S] small}$$

$$\text{rate} = \frac{dP}{dt} \propto [E]_t \qquad \text{for [S] large}$$

where $[E]_t$ is the total enzyme concentration $= [E]_{\text{free}} + [ES]$. The experimental results are shown in Figure 17.6 for two enzyme concentrations. A limiting rate is reached at high substrate concentrations, because all the enzyme is combined in the complex, while at low substrate concentrations, not all the enzyme is being utilized.

This sort of kinetic reasoning helps to establish the validity of the enzyme–substrate complex mechanism. The existence of the transitory complex has even been detected in some systems by sensitive experimental methods. However, an explanation of the high catalytic efficiency and specificity of enzyme catalysts will be found at the molecular level.

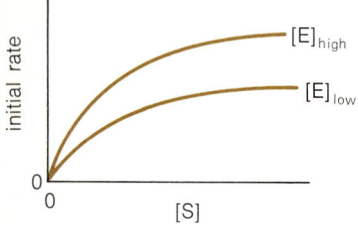

Figure 17.6. Variation of initial rates with substrate concentration at two values of total enzyme concentration. The initial rates of enzyme-catalyzed reactions increase with substrate concentration until a maximum rate is reached when all the available enzyme is tied up in the enzyme–substrate complex.

17.14 THE ENZYME–SUBSTRATE COMPLEX

The origin of the catalytic activity of enzymes lies in the closeness of approach of the large, intricately folded enzyme molecule and the smaller substrate molecule. The enzyme–substrate complex is formed by an interlocking of the two molecules in a manner analogous to the fit of two adjacent pieces in a jigsaw puzzle or, to use a common analogy, the fit of a key in a lock. Both the catalytic activity and the high specificity of the enzyme molecule are consequences of this close union.

Enzyme molecules may be single protein molecules with tertiary structure, or they may be composed of more than one protein molecule and exhibit quaternary structure. In either case, the enzyme molecule folds itself into a globular shape having a surface with an irregular topography. X-Ray studies have verified the proposal that the substrate molecule establishes a close union with the enzyme surface by fitting into a specific cleft or depression in the surface. The substrate molecule is anchored to the enzyme surface by various interactions between the two molecules. These binding forces are usually weak attractions of the hydrogen bond, van der Waals, hydrophobic, or polar variety. These interactions are effective only at short distances, so it is necessary that there be both an overall accurate spatial fit of the two molecules to establish close approach and a number of coincident interacting groups in both molecules. Figure 17.7 is a schematic picture of an irregular but unique groove in an enzyme molecule cradling a substrate molecule within it. Groups along the substrate molecule interact with complementary groups arrayed along the depression in the enzyme surface. The formation of the complex requires

Figure 17.7. A polysaccharide substrate molecule (dark) fits into a depression in an underlying enzyme molecule (lighter). The substrate is held to the enzyme by hydrogen bonds (eight of which are shown in color), and by a number of hydrophobic interactions (the vicinity of one hydrophobic interaction is shown). The four shaded circles represent the catalytically active groups in the enzyme molecule that act on a specific region of the substrate molecule and bring about its rupture at the position denoted by the wavy line.

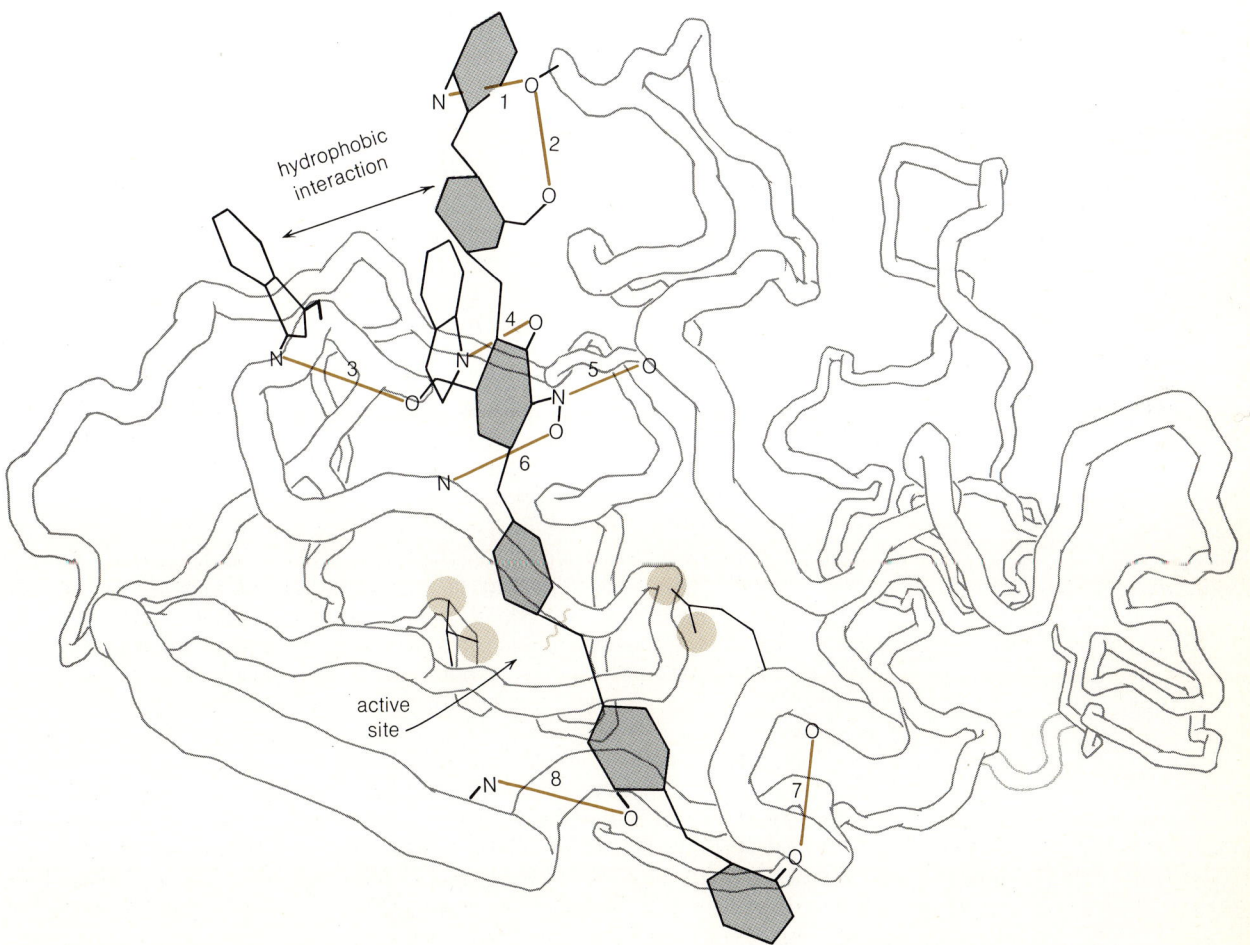

near-exact matching of shape and interacting groups. The small region of the larger enzyme surface along which the substrate is bound is known as the *active site* of the enzyme. At some point, reactive groups in the enzyme molecule are in close contact with the substrate molecule. While these reactive groups need not contribute to the attachment of the substrate, they do react with a bond or groups in the substrate and promote the making or breaking of a substrate bond in that vicinity. This is the reaction catalyzed by the enzyme. The product or products of the reaction do not satisfy the requirements of close fit and so escape from the enzyme surface. The region of the enzyme surface where the reactive groups exert the catalytic influence is also sometimes called the active site. As seen in Section 15.14, the catalytically active groups of the enzyme molecule may be spaced far apart in a stretched-out enzyme molecule but will be concentrated in a given region of the enzyme surface when the enzyme is folded into its distinctive tertiary structure.

The interaction of the enzyme as a whole or of a few reactive groups somehow changes the chemical properties of the substrate molecule and enhances the reactivity at some point in the molecule. Several mechanisms could be postulated. The enzyme might withdraw electronic charge from some point in the substrate, or the binding of substrate to enzyme might crowd certain groups of the substrate, causing distortion of the substrate molecule. This would result in strain, destabilization, and high reactivity (activation) of a group or bond in that region.

Whether the active site is considered to be the larger portion of an enzyme surface into which the substrate molecule nestles and binds, or the smaller region adjacent to the catalytically reactive groups of the enzyme, anything affecting the tertiary structure of the active enzyme may act as an inhibitor by decreasing or destroying the enzyme's catalytic power. On the other hand, inactive enzymes may be activated by some substances. These activators (coenzymes among them) apparently also change the conformation of the enzyme molecule. The above ideas of enzyme activity and specificity are contained in two theories of enzyme–substrate interaction. The older theory, the *lock-and-key* theory, regards the enzyme surface as a rigid template (the lock) into which a substrate with a complementary surface fits. This idea is illustrated in Figure 17.8. The newer *induced fit* theory considers the enzyme molecule to be flexible to the extent that it can change the conformation of its active site on the approach of a substrate molecule, bringing its binding and catalytic sites into a closer alignment with the complementary sites of the substrate. This alignment may consist of the orientation (or directional polarization) of electron orbitals of reacting groups of the substrate and the catalytic groups of the enzyme, a process that has been called "orbital steering." Alteration of enzyme structure could also lead to localized distortion and activation of the substrate. Such conformational changes have been detected by optical and X-ray methods.

The role played by molecular fit, complementarity of surfaces and binding groups, active sites, and change in conformation in the mechanisms by which different biological molecules recognize and interact with each other is not limited to enzyme catalysts. Similar considerations apply in studies of how protein antibody molecules in organisms interact with selected foreign substances (antigens), how the protein synthesis system is switched on or off by certain molecules, and how the genetic apparatus of the cell replicates itself.

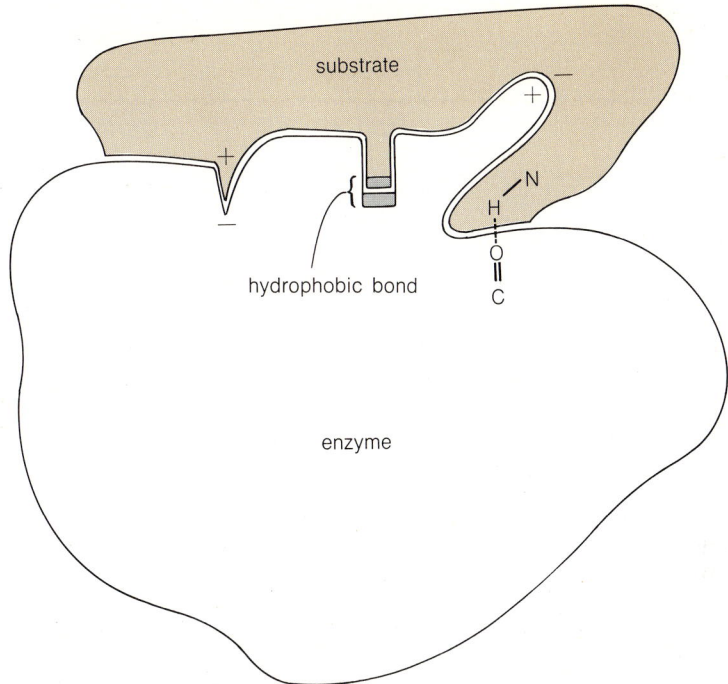

Figure 17.8. Lock-and-key model of the enzyme–substrate complex.

Despite the many existing theories of enzyme function, all of them combined cannot yet explain the enormous catalytic power of enzymes. Enzymes can speed up a chemical reaction by factors of 10^9–10^{20} over an uncatalyzed reaction or even a reaction catalyzed by typical inorganic catalysts. Exactly how they do this is still beyond comprehension.

17.15 ENZYME INHIBITION

A variety of substances can depress or completely destroy the catalytic activity of enzymes. The most drastic enzyme inactivation mechanism is complete denaturation of the protein. But even more subtle ways of inhibition may come about. We first look at some general features of enzyme inhibition and then turn to a discussion of some specific examples of substances that function as inhibitors. Some of these are potent poisons for living systems; others are life-saving drugs.

Types of Enzyme Inhibition

Inhibition processes can be broadly classed as reversible and irreversible. In *reversible inhibition*, some inhibitor molecule binds reversibly to the enzyme molecule or to the enzyme–substrate complex. Reversible inhibition occurs by two different mechanisms related to the binding site of the inhibitor molecule. In the first kind of reversible behavior, *competitive inhibition*, the inhibitor molecule binds at the active site of the enzyme, where it blocks the approach and binding of the substrate molecule. This implies some structural similarity of the inhibitor and the substrate molecules, since they compete for the same active site on the enzyme. Competitive inhibition can be reversed by increasing the concentration of substrate. Figure 17.9 illustrates competitive inhibition schematically.

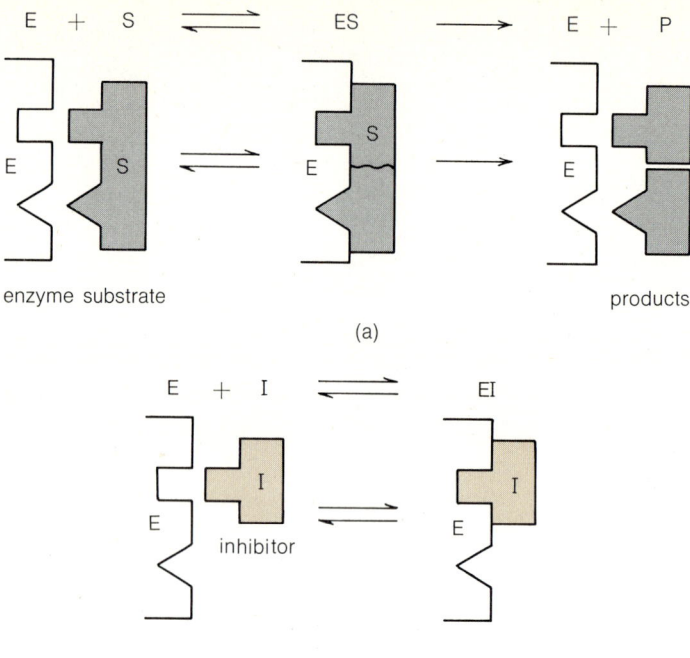

$$E \;+\; S \;\rightleftharpoons\; ES \;\longrightarrow\; E \;+\; P$$

enzyme substrate   products

(a)

$$E \;+\; I \;\rightleftharpoons\; EI$$

inhibitor

(b)

Figure 17.9. Competitive inhibition. (a) No inhibition. (b) Inhibition. An inhibitor I, with some structural similarity to the substrate S, blocks the active site. The inhibitor shown here might also be one of the products of the enzyme-catalyzed reaction.

Noncompetitive inhibition, the other kind of reversible inhibition, occurs when the inhibitor molecule binds to the enzyme at some location other than the active site. The inhibitor may bind to the enzyme alone or to the enzyme–substrate complex. No competition of the inhibitor and substrate for the same site is involved, but the presence of the inhibitor affects the active site of the enzyme so that it cannot combine with the substrate. This may come about because of a change in the structure of the catalytically active site caused by a reversible change in tertiary structure induced by the binding of the inhibitor. The inhibitor may either reduce the binding of the substrate to the enzyme or prevent the decomposition of the enzyme–substrate complex. These situations are illustrated in Figure 17.10.

Irreversible inhibition results when an inhibitor reacts with an enzyme to form a highly stable derivative according to the equation

$$E + I \longrightarrow EI$$

The inhibitor may bind irreversibly at the active site or at another location where its presence modifies the structure of the active site.

17.16 EXAMPLES OF ENZYME INHIBITORS
Certain Antibacterial Drugs

A classic example of a competitive inhibitor is sulfanilamide, the basis for several sulfa drugs. Its structure is shown in Figure 17.11 along with that of *para*-aminobenzoic acid (PABA), which is a part of folic acid, an essential coenzyme (vitamin) for a metabolic reaction in many kinds of bacteria. A bacterial enzyme catalyzes the incorporation of PABA during the formation of folic acid by combining with and transferring PABA into the folic acid product. Sulfanilamide competes with the structurally similar PABA at the active site of the enzyme and prevents the incorporation of PABA into the

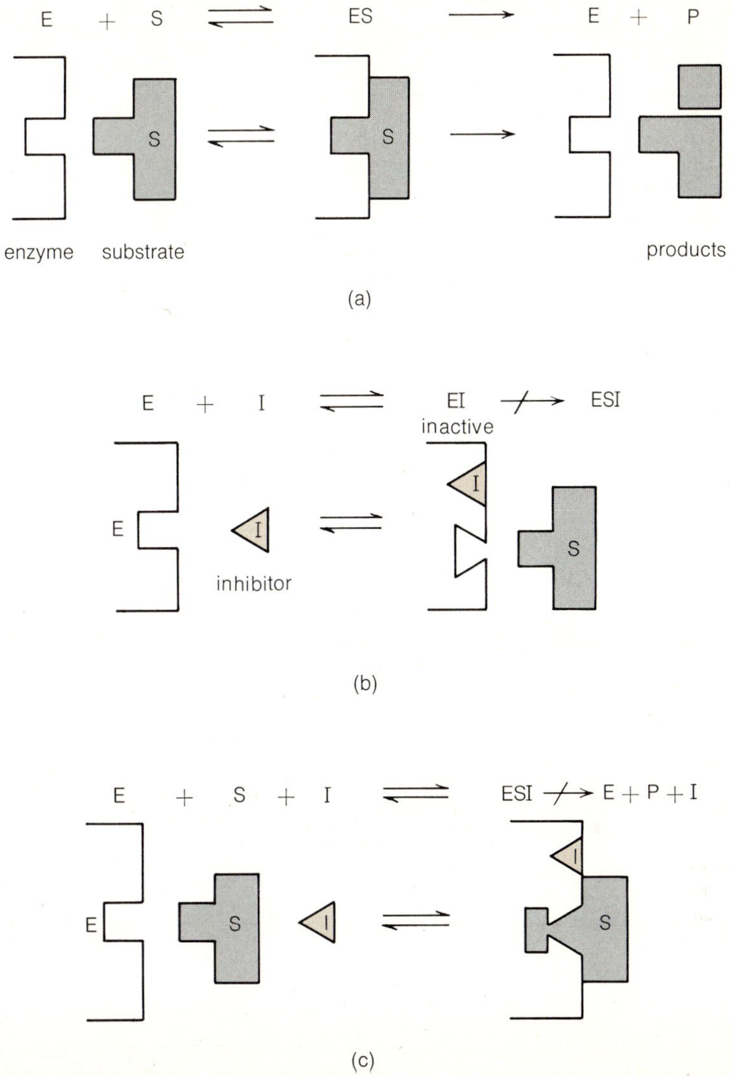

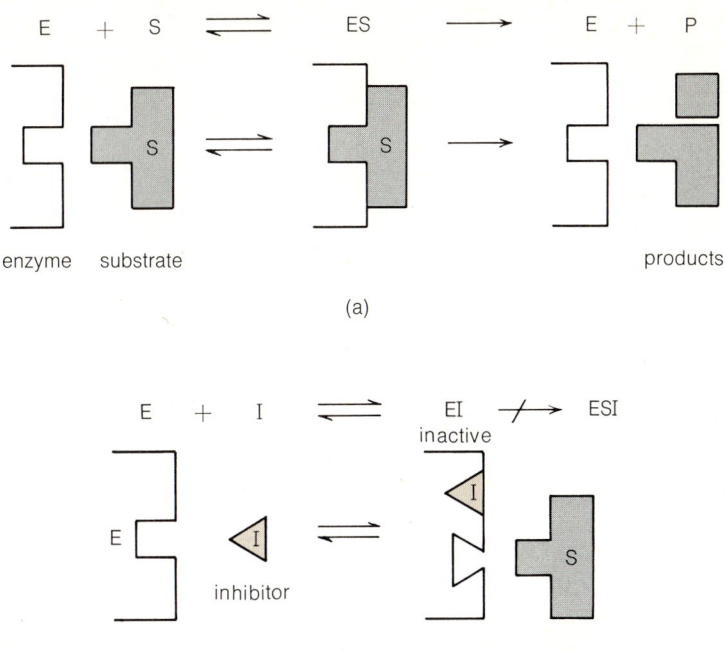

Figure 17.10. Noncompetitive inhibition. (a) No inhibition. An inhibitor I binds at a location distant from the active site and either destroys the active site (b) or prevents the enzyme–substrate complex from decomposing (c).

coenzyme. The bacteria may die from the nutrient deficiency disease. Humans, unlike bacteria, do not synthesize folic acid but obtain the required quantities ready made in their diet. Thus, sulfa drugs do not interfere with human metabolism.

Sulfanilamides are examples of *antimetabolites*, chemicals that are related

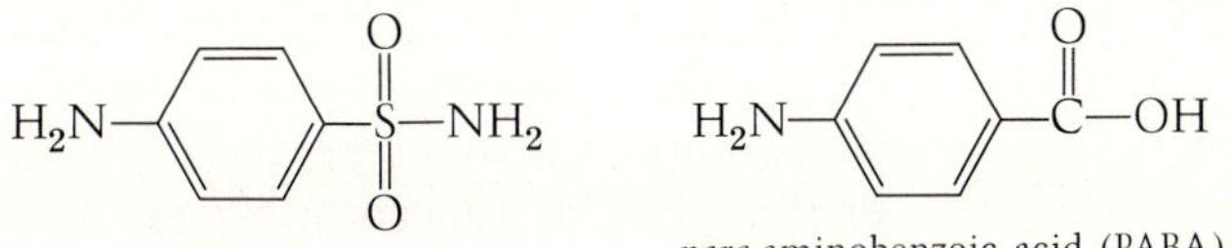

Figure 17.11. Sulfanilamide, a sulfa drug, is similar to *para*-aminobenzoic acid and competes for the active site of an enzyme. Other sulfa drugs are obtained by replacing a hydrogen on the —SO_2—NH_2 group with other groups.

sulfanilamide
(*para*-aminobenzenesulfonamide)

para-aminobenzoic acid (PABA)

in structure to normal metabolic products but that inhibit important metabolic processes.

Other antibacterial drugs are synthesized by living microorganisms for their own protection from other organisms and can often be used by man for medicinal purposes. These drugs are called *antibiotics*. The mechanism of action of antibiotics is varied, but one, penicillin, probably acts as an enzyme inhibitor (not necessarily a competitive inhibitor) at a metabolic step in the synthesis of bacterial cell walls. Inhibition of the synthesis of cell walls kills newly formed bacteria because the osmotic pressure of the cell interior is normally higher than that of the exterior fluid. Without a strong cell wall, the remaining cell membrane cannot withstand the pressures generated by the entry of water, and the cell bursts.

Metals The toxic effects of metals, particularly the heavy metals such as lead, mercury, and arsenic, on living systems have been referred to repeatedly in previous chapters. The adverse environmental effects of these and other metals become more publicized every day. Pollution by toxic metals is considered by many to be a much more serious and much more insidious environmental health problem than that due to all other pollutants. This concern arises from the mounting quantities of metals flowing into the environment from industrial operations, fossil fuel combustion, and inadequate waste disposal. Helping in our perception of the problem are better medical diagnosis and more sensitive analytical methods for the detection of trace amounts of the metals in organisms and in the environment. Many of the toxic effects of these metals are undoubtedly due to interference in enzyme-catalyzed metabolic reactions. One of the common ways that heavy metals inactivate enzymes is by combining with sulfhydryl (—SH) groups on cysteine amino acid residues of proteins:

$$\text{protein—SH} + \text{M}^+ \rightleftharpoons \text{protein—S—M} + \text{H}^+$$
$$\text{M}^+ = \text{Ag}^+, \text{Hg}_2{}^{2+}$$

$$\text{protein}\begin{matrix}\diagup \text{SH} \\ \diagdown \text{SH}\end{matrix} + \text{M}^{2+} \rightleftharpoons \text{protein}\begin{matrix}\diagup \text{S} \\ \diagdown \text{S}\end{matrix}\text{M} + 2\text{H}^+$$

$$\text{M}^{2+} = \text{Cd}^{2+}, \text{R—As}\diagup\diagdown, \text{Pb}^{2+}, \text{Hg}^{2+}$$

The —SH groups may be located at the active site, or they may be at some distance from the active site where they nevertheless help determine the three-dimensional structure of the enzyme. The latter constitutes noncompetitive inhibition. Other mechanisms of deactivation by heavy metal ions may include complete denaturation by rupture of —S—S— cross links or removal of charge on the protein. Additionally, a metal ion may displace an essential metal ion cofactor in an enzyme, as Cd^{2+} does with Zn^{2+} (Section 12.14). These metallic cofactors play an important part in the proper conformation of an active enzyme. A wide variety of toxic metals contribute, or show potential of contributing, to health impairment. Among these are beryllium, nickel, and vanadium, in addition to those already mentioned. We now look at some additional details of lead poisoning.

Lead acts as an inhibitor of several enzymes required in the biosynthesis of the iron-containing heme group. This prosthetic group, as we know, is a critical part of hemoglobin, the oxygen transporter, and the electron-carrying cytochromes. Lead inhibits these enzymes by combining reversibly with —SH groups. The pertinent metabolic pathway has six steps, each enzyme catalyzed. The enzymes in two of these steps are known to be inhibited by lead, and two other steps may be inhibited by lead. Studies have shown that an enzyme in one of the former steps is inhibited at all concentrations of lead, including the concentrations known to be carried by everyone. Yet there are no observable effects on the total biosynthesis at these concentrations. Apparently, this particular step is not the slowest step in the six-membered metabolic pathway. If it were, the production of heme would decrease at any lead concentration. In effect, there is enough of this particular enzyme to absorb low lead concentrations without noticeably affecting the entire pathway. However, at higher lead concentrations, the enzyme of the slowest step is inhibited and anemia results. It is entirely possible that we are all being slowly poisoned by lead or by any number of other environmental poisons. So far there has been no noticeable effect because the toxicant has not been stored and concentrated to its threshold level in the tissues or because the rate of excretion has approached the intake rate.

The symptoms of lead poisoning are varied, and not all of the effects may be due to inhibition of heme production. For instance, lead is attracted to the bones, where it may interfere with calcium metabolism. Lead also damages the central nervous system.

It is interesting to note that lead poisoning may have been a factor in the decline and fall of the Roman Empire. The widespread use of lead vessels and lead-glazed pots for storing and preparing acidic grape juice and wine, among other things, is supposed to have led to a gradual poisoning of the upper classes. Today, lead poisoning is common among children living in surroundings where they eat chips of lead-based paint flaking off dilapidated buildings. For most of us, the principal source of environmental lead has been leaded gasoline.

Metal poisoning can be treated effectively in many instances through the administration of chelating agents. Chelating agents are molecules such as the heme group that are capable of donating more than two pairs of electrons to a metal ion to form coordinate bonds between the complexing molecule (the chelate ligand) and the metal ion. Chelating agents generally bind to the central metal ion through the lone-pair electrons of nitrogen, oxygen, or sulfur atoms. The effectiveness of a chelating agent such as ethylenediaminetetraacetic acid (EDTA) as an antidote for metal poisoning depends on its ability to bind the metal atom more strongly than the metal is bound to the enzyme under physiological conditions. The chelate of Pb^{2+} with EDTA, or $Pb(EDTA)^{2-}$, has the same structure as $Fe(EDTA)^-$ shown in Figure 14.8. If a highly stable chelate is formed, the metal is readily excreted from the body. Chelates have been used as antidotes for lead, copper, mercury, iron, nickel, and radioactive plutonium, strontium, and cesium poisoning.

Chelating compounds and other complexing agents may themselves act as inhibitors of metal-containing enzymes. Thus, EDTA may inhibit enzymes with cofactors of Mg^{2+}, Fe^{2+}, and other divalent metal ions either by forming a complex with the metalloenzyme itself or by completely removing the metal cofactor from the enzyme. Complexing agents such as CN^-, S^{2-}, and CO also form inactive complexes with metal-containing enzymes. At higher complexing agent concentrations, the metal may be completely removed from the enzyme as a stable complex. At low concentrations, such chelating and complexing agents act as noncompetitive inhibitors. Inhibition is irreversible at high concentrations.

Acetylcholinesterase Inhibitors: Nerve Poisons

The most potent poisons of living organisms act by inhibiting enzymes. Of these poisons, the most powerful are those that affect the central nervous system directly. This part of an organism easily suffers irreversible and permanent damage because nerve cells, unlike other body cells, are not continually replaced. Even those poisons such as carbon monoxide that do not act directly on the nerves ultimately kill because of indirect effects on the nervous system; that is, CO deprives the brain of oxygen.

The enzyme acetylcholinesterase is intimately involved in the transmission of nerve impulses and the contraction of muscles. Compounds that inhibit the action of this enzyme are widely used as insecticides and are the basis of the nerve gases of chemical warfare. Before discussing the effects of acetylcholinesterase inhibition, let us first see how nerves work. When sensory cells associated with the senses of sight, touch, or taste receive a signal from the external or internal environment of an organism, an electric signal or impulse is transmitted along a long fiberlike portion of nerve, the axon, to the central nervous system (the brain and spinal cord in vertebrates). From there, another electric impulse is transmitted along an axon to another cell such as a muscle cell. The junction between the end of the axon and the muscle cell is called a neuromuscular junction or synapse. At the junction, the electric signal in the axon is converted to a chemical transmission to the muscle cell. The signal transmitted chemically to the muscle cell causes contraction of the muscle. Figure 17.12 illustrates the path of the impulse.

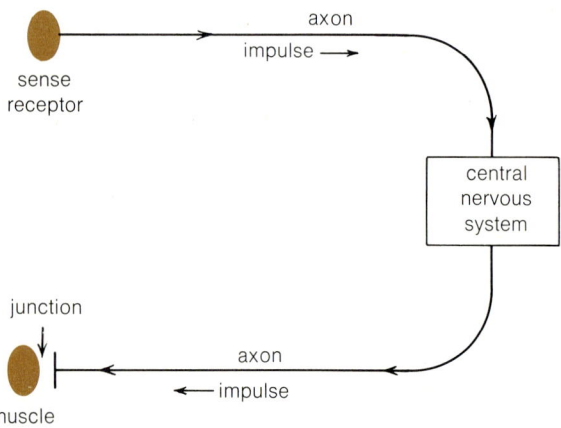

Figure 17.12. An electrical nerve impulse arriving at the neuromuscular junction triggers the release of acetylcholine. The acetylcholine chemically transmits the signal across the junction to the muscle cell.

The arrival of the electrical nerve impulse at the junction causes the release of a transmitter substance, acetylcholine, which stimulates the muscle cell on the opposite side of the junction. The muscle cell remains stimulated (contracted) and unable to become stimulated again unless the acetylcholine is quickly removed from the junction region. The enzyme acetylcholinesterase acts to remove acetylcholine by hydrolyzing it into inactive fragments of choline and acetic acid.

Acetylcholinesterase has a serine amino acid residue, $-CH_2-OH$, at its active site. Inhibitors act by forming a stable, covalently bonded ester with the hydroxyl group of the serine residue. The reaction, and therefore the inhibition, are irreversible. Some of the common acetylcholinesterase inhibitors are listed here.

1. Organophosphates

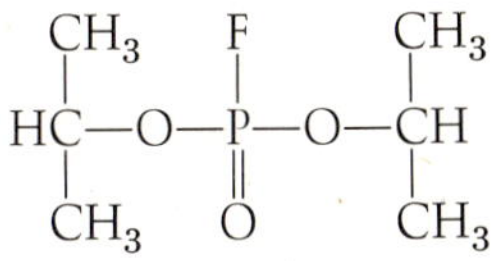

diisopropyl phosphofluoridate (nerve gas)

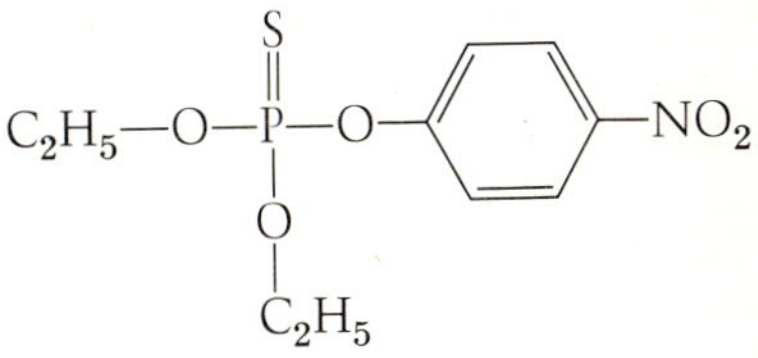

parathion (insecticide)

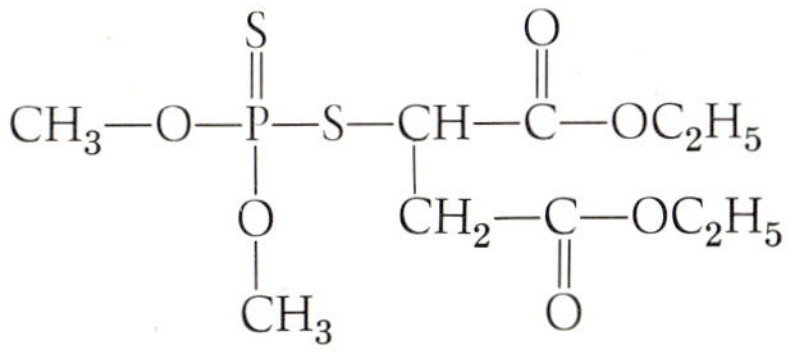

malathion (insecticide)

2. Carbamates (esters of carbamic acid, $H_2N-\overset{O}{\overset{\|}{C}}-OH$)

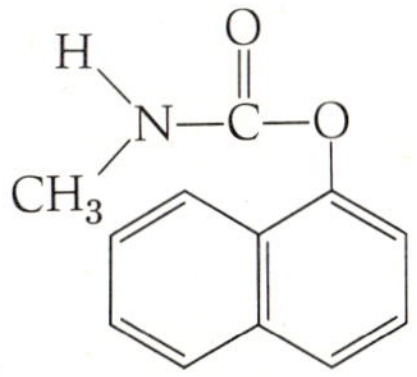

carbaryl (Sevin) (insecticide)

The reactions of these inhibitors involve the transfer of a group from the inhibitor to the enzyme. If the enzyme is represented simply by E—OH, a typical reaction would be

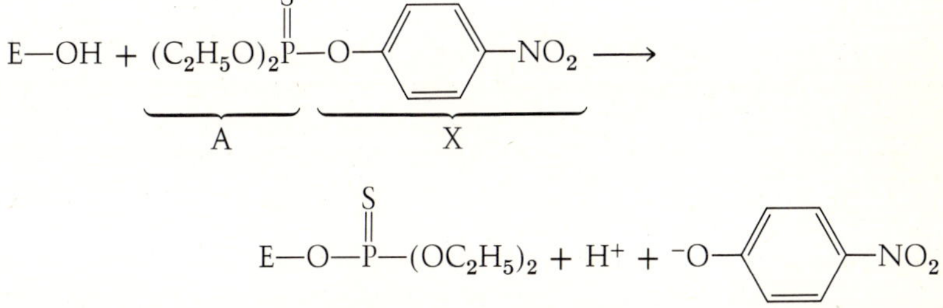

In general, each inhibitor transfers a group A $[(RO)_2PS-,\ -\overset{O}{\overset{\|}{C}}NHCH_3]$ to

the enzyme, and another group X leaves, so the inhibition reaction can be symbolized by

$$EOH + AX \rightleftharpoons EOH—AX$$
$$EOH—AX \longrightarrow EOA + X^- + H^+$$

The inhibition of acetylcholinesterase prevents the degradation of acetylcholine generated by the first nerve impulse. Additional impulses arriving at the junction cause a buildup of acetylcholine, leading first to overstimulation of the nerves and muscles, then to paralysis and death by asphyxiation due to respiratory failure. The latter is the mechanism of death by acetylcholinesterase inhibitors in mammals, but the exact cause of death in insects is still unknown.

The mode of action, on the molecular level, of the organophosphate and carbamate insecticides is well established, at least with vertebrates. By contrast, the mechanism of the much older, much larger, and widely used class of insecticides, the chlorinated hydrocarbons, is not known with any certainty. The best-known chlorinated hydrocarbon is DDT (*p,p'*-dichlorodiphenyl-trichloroethane),

Other common chlorinated hydrocarbon insecticides are chlordane, lindane, heptachlor, toxaphene, aldrin, and dieldrin. All of them are believed to act on the nervous system, perhaps by affecting the permeability of the axon membrane to sodium and potassium ions. The propagation of the electrical nerve impulse along the axon is associated with changes in the permeability of this membrane with respect to these two ions. Possibly, the chlorinated hydrocarbons may inhibit the enzymes acetylcholinesterase or adenosine triphosphatase (ATPase), both of which are believed to be involved in axon permeability. The ATPases are a group of enzymes that catalyze the cleavage of ATP to ADP and inorganic phosphate. This is the energy-producing reaction coupled to so many of the energy-requiring processes in an organism. Among these processes are the ion pumping mechanisms used to transport sodium and potassium ions through the axon membrane against high concentration gradients.

17.17 OTHER EFFECTS OF PESTICIDES

The story of the ecological impact of widespread use of insecticides or pesticides in general is a perfect example of the interplay of chemical and biological factors at all levels, from the molecular to the ecosystem. The molecular mechanism, if known, of action of a pesticide is only the beginning of a sequence of events that have had disastrous ramifications for many living organisms aside from the target insects. As far as the general public is concerned, the first warning of the environmental contamination resulting from pesticide use came from Rachel Carson's *Silent Spring*. Published in 1962, *Silent Spring* was considered by some to be merely an emotional book filled with scientific half-truths. Dr. Carson was ostracized by many of her fellow scientists. Nevertheless, this kind of approach was just what was needed, because the only other warnings were hidden from the eyes of almost everyone in abstruse scientific papers. Since that time, there has been a flood of information and concern about the effects of pesticides, chiefly the chlorinated hydrocarbon pesticides such as DDT. Most chemical pesticides are broad-spectrum agents that kill a wide variety of useful insects as well as the intended target pest. This disrupts the entire local ecosystem by removing or starving the predators needed to keep pest populations in check. More important, the chlorinated hydrocarbons, although generally less toxic than the organophosphorus and carbamate insecticides, are highly persistent. This means that they are not quickly degraded chemically or biologically to nontoxic products by weathering or by enzyme degradation in microorganisms in the soil or water. Such pesticides are called "hard" pesticides. The persistence of DDT and other hard pesticides is aided by other factors such as their extremely low solubility in water and low vapor pressure. The low solubility is a result of the nonpolar nature of the chlorinated hydrocarbons. It was pointed out in Section 9.12 that these substances enter the fatty tissues (particularly the liver) of organisms, where they are stored intact until such time as these tissues are drawn upon as a source of energy. Then the pesticide or one of its degradation products is released and may poison the organism. The organism that dies in this way is not likely to be the original target, but an animal that has accumulated pesticide from its food source, as for example, a bird that eats insects. This is an example of delayed expression of the poison.

The need for chemical pesticides, their method and frequency of use, and their ecological consequences depend in large part on man's food needs, the need to control insect-transmitted diseases (primarily malaria), and on his structuring of the agricultural ecosystem. These subjects will be considered in the next chapter. There are, however, two important consequences of pesticide usage that can be briefly considered here. These are the development of resistance to insecticides and the effect of insecticides and related chemicals on the reproduction of birds. Both are related to enzymes.

Resistance

A discouraging side effect of the widespread and frequent use of synthetic insecticides is the ever-increasing number of insect species showing resistance to one or more of these insecticides. Resistance is defined as the ability of insects to tolerate doses of an insecticide that would have previously been lethal to most members of that species. Put another way, resistance means more insects in a given population survive exposure to the insecticide. There are now over 200 insect species resistant to previously effective insecticides.

Resistance may come about in several ways. The insect may show behavioral resistance, meaning the development of the ability to avoid lethal exposure to the poison. This may come about because the insecticide irritates the insect and causes it to move away from the treated area. More serious is resistance resulting from a truly biological adaptation. Insect strains with certain physical features that protect it from the poison, a thick "skin" for instance, may evolve. Or insect strains possessing the biochemical ability to detoxify the poison may be selected. Such biochemical resistance may come about because the insect possesses an enzyme that rapidly breaks the insecticide down to nontoxic metabolic products. DDT is easily metabolized to the nontoxic (to insects) derivative DDE (p,p'-dichlorodiphenylethylene) by insects with the proper enzyme.

$$DDT \xrightarrow{\text{enzyme}} Cl-\text{C}_6\text{H}_4-\underset{\underset{CCl_2}{\|}}{C}-\text{C}_6\text{H}_4-Cl + HCl$$

DDE

The enzymatic degradation of several insecticides in resistant strains has been established. For other insecticides, the mechanism is less clear.

In any normal population of insects, there will be a few insects that will be resistant to a given insecticide and often to related insecticides. Application of a normal dose to this group kills all the susceptible insects, leaving only the resistant members of the species. The resistant strain will flourish in the absence of competition from the susceptible strain and other competitor or predator species also killed by the broad-spectrum insecticide. Under these conditions, after several generations, the insects will develop into a totally resistant population. The maintenance of the conditions under which selection occurs is favored by the use of hard insecticides or by frequent applications of nonpersistent (soft) insecticides.

The establishment of a resistant population is a perfect example of Darwinian natural selection, adaptation to the environment, or biological evolution—all being the same thing. With any species, the originally rare, but fit, resistant strain would eventually reestablish the species population, but insects, by virtue of the short time from one generation to another and the huge number of individuals in each generation, are particularly capable of doing this in a short time. It is important to realize that one individual in the population does not become resistant within its lifetime. The individual is "born" resistant; the species becomes resistant only over several generations as the susceptible members are eliminated and the resistant members take over. In the case of DDT and several other insecticides, the members of the resistant population are all descendants of the few individuals having the genetic apparatus conferring the ability to manufacture the required detoxifying enzyme.

Reproduction of Birds

Chlorinated hydrocarbons, particularly DDT, have been implicated in a number of reproductive failures of birds. These episodes include delay or failure to breed, failure to lay eggs, and the laying of eggs with thin shells that break in the nest. Most of the affected birds belong to species situated at the top of food chains. That is, they are birds of prey such as hawks, falcons, and eagles, as well as pelicans. The fat-soluble hydrocarbons are passed from lower members of the food chain to higher members. At the top of the chain, the concentration of DDT or of DDE may be tens of thousands of times higher than in the organisms at the bottom of the chain. During periods of stress, such as migration, breeding, or egg laying, the reserves of fat may be called upon for energy. Then the stored insecticides are released into the circulatory system.

An explanation for some of the reproductive failures is that DDE and other substances function as *enzyme inducers*. Their presence acts as a signal to the organism to synthesize a particular enzyme. In the case of birds, and perhaps man, the induced enzyme breaks down a sex hormone, estrogen. In birds, this hormone plays a major role in determining breeding behavior.

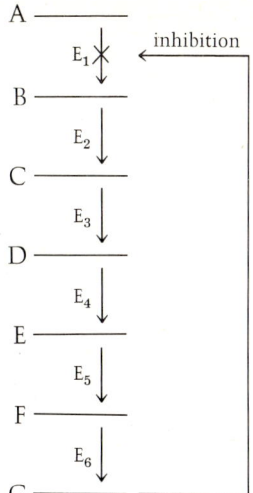

Figure 17.13. Feedback inhibition. The end product G inhibits the first enzyme in the pathway, E_1, preventing the subsequent steps from occurring. This is a common regulating mechanism in biosynthetic pathways. In addition, ATP can act as a feedback inhibitor in degradative pathways coupled to ATP production, so that the pathway is shut down when ATP has accumulated. Also, ADP and AMP, the products of ATP consumption, can act as stimulators to pathways coupled to ATP production.

Lowered estrogen levels can cause delayed breeding or failure to lay eggs. They also affect the calcium reserve laid down in the bones of birds. Because this calcium is drawn upon for the calcium carbonate of eggshells, enzyme induction by DDE can decrease the required calcium reserve and perhaps contribute to thin shells. Another proposed explanation for formation of thin shells is inhibition or precipitation of the enzyme carbonic anhydrase by DDE. This enzyme catalyzes the reaction

$$CO_2 + H_2O \rightleftharpoons H_2CO_3$$

which is the source of carbonate ions for eggshells. Because of reproductive failures attributed to increasing chlorinated hydrocarbon contamination of the environment, several species of birds are threatened with extinction.

The illustrations of resistance to chemical insecticides and delayed poisoning by them are examples of only two of the unexpected and subtle ways that chemistry and ecology come together. Although it appears likely that the hard, broad-spectrum, chlorinated hydrocarbon insecticides are going to be phased out of use (an almost complete ban has been ordered on the use of DDT in the United States), the concepts of ecological chemistry brought out in the story of DDT come up again and again as industrial man interacts with his environment. As we have seen, effects at the level of an enzyme molecule are propagated through levels of organization until they culminate in widespread disturbance in an ecosystem or the biosphere as a whole.

No attempt has been made to cover the whole insecticide story here, despite the fact that it is a perfect illustration of most of the principles of ecological chemistry. Much has been written, pro and con, about the effect of insecticides. Further information on the interplay of chemical and ecological factors can be found in the references at the end of this chapter. The book by Rudd and the article by Woodwell are recommended as starters. The article by Stokinger referenced at the end of Chapter 12 is also appropriate.

17.18 ENZYMATIC CONTROL OF METABOLIC ACTIVITY

There is no need for all the possible chemical reactions in a cell to take place at the same time or at their maximum rates. This would be extremely wasteful, and Nature is known to be economical. Only those chemical production lines are operative whose products the cell requires at any given moment. Regulatory mechanisms give the cell the ability to switch off the production line when the product is no longer needed and to turn the synthetic machinery on again when demand arises. In this way, the multitude of metabolic reactions in the cell, or in a multicellular organism, are regulated. Many of the regulation mechanisms are based on the adjustment of reaction rates, generally by enzymatic control effected either by changes in the activity of existing enzymes or by changes in the amount or kind of synthesized enzymes.

Enzyme activity is influenced by the presence of inhibitor or activator molecules called *moderators*. Figure 17.13 illustrates a type of regulation found in biosynthetic pathways. The end product of the pathway inhibits the enzyme catalyzing the first step. When the final product accumulates, it automatically turns off the chemical production line. This efficient self-regulating mechanism is variously called *feedback inhibition, end product inhibition,* or *negative feedback.* Such feedback control, where part of the output (product) of a process or system is fed back into the system as an information input, thus regulating the process, is the most common homeo-static mechanism in organisms as well as ecosystems.

Most evidence indicates that the inhibited enzymes of the first step, the regulator enzymes, have subunit or quaternary structure along the lines of the hemoglobin molecule. Furthermore, the inhibitor molecule binds at some site other than the active site of the enzyme. Somehow attachment of the inhibitory molecule(s) distorts the quaternary and tertiary structure of the enzyme so that substrate can no longer bind. This is an example of an *allosteric effect*, the change in conformation and, therefore, function of a protein due to binding of molecules (allosteric effectors) at a location away from the active site. Oxygen has an allosteric (activating) effect on the hemoglobin complex.

Another, but slower acting, control mechanism originates at the genetic, or enzyme biosynthesis, level of the cell. A cell is capable of making a great number of enzymes, but many of them are made to order on request rather than stored ready-made in the cell. In simple terms, the request for a specific enzyme is carried to the order room in the cell nucleus by a small molecule. The genetic machinery in the nucleus reads the order, which may be, "Please make some enzyme A," or perhaps, "Please turn off the synthesis of enzyme C." In the first case, the small messenger molecule (which might be a hormone or some other intermediate) is an *enzyme inducer* that activates the synthetic machinery. In the second case, the small molecule is an *enzyme repressor*. Enzyme repression and induction mechanisms regulate the kind and amount of enzymes present, not the activity of existing, ready-made enzymes. Genetic regulation will be discussed again in Section 19.15.

17.19 METABOLIC OR BIOCHEMICAL DISEASES

As a final example of the importance of enzymes in living systems, brief mention is made of the large number of diseases resulting from the absence or inactivity of enzymes in the human body. Vitamin deficiency diseases have already been pointed out as resulting from the absence of cofactors required by enzymes. Most other metabolic diseases are genetic (hereditary) in origin. The body either manufactures none of the essential enzyme, or what enzyme it does make is inactive because of genetic errors in the primary structure of the enzyme.

A simple example of a metabolic disease caused by the absence of an enzyme is phenylketonuria (PKU). The absence of an enzyme prevents the body from metabolizing the amino acid phenylalanine in the normal way. Instead, another metabolic route must be used, and this results in excesses of the alternate metabolic product, phenylpyruvic acid. This is a phenylketone and it appears in the urine. The phenylpyruvic acid metabolite impairs normal brain development in children and leads to severe mental deficiency unless the individual is restricted to a special diet. About one in 10,000 humans has an absence of the essential enzyme because of an inborn genetic defect which blocks the enzyme synthesis.

The absence of enzymes is believed by some to be responsible for the mental disease schizophrenia. Supposedly, the degradation of certain compounds is blocked, and their accumulation has hallucinogenic effects.

17.20 SUMMARY

The rate of a chemical reaction depends on the nature and concentration of the reactants, the temperature, and the presence of catalysts. The functional dependence of rate on concentration is given by the experimentally deter-mined rate equation of the reaction. The rate equation has no relation to the reaction equation except for the case of elementary, one-step reactions. Most common reactions are multistep processes consisting of a number of elementary reactions. The set of elementary steps that, when added up, gives the overall reaction, constitutes the reaction mechanism of the complex process.

There are usually several mechanisms possible for a given set of reactants and products. The mechanism that transforms the reactants to products in the fastest manner is the mechanism by which the reaction proceeds. This mechanism will affect the form of the rate equation. The rate equation may, in turn, be determined by the slowest step in the mechanism.

At equilibrium, the rates of the forward and reverse reactions in a reversi-ble process are equal. This is true whether the opposing reactions are elemen-tary or complex. It is possible to obtain the equilibrium constant by equating the experimental rate equations for the forward and reverse reactions.

Rates of most reactions increase with increasing temperature. This de-pendence on temperature is explained by the postulation of a minimum energy, the activation energy, that the reactant molecules must possess before they can transform into products. At higher temperatures, a greater fraction of the molecules have the required energy. The transition state theory of reaction rates states that reactants come together to form a transition state or activated complex, which then decomposes to products. The activation energy for the (forward) reaction is equal to the difference in potential energies of the activated complex and reactants. Catalysts speed up reactions by providing a reaction path with a different activated complex of lower energy.

Enzymes are protein molecules that act as catalysts in living systems. They are highly efficient and highly specific. Enzymes function by forming an activated complex with a reactant molecule (substrate) in which the reactant is bound to the surface of the enzyme by specific weak interactions. The catalytic activity of enzymes can be inhibited by substances that affect the higher structure of the enzyme or that block active sites on the enzyme surface. Many environmental contaminants act as enzyme inhibitors.

Enzymatic regulation of metabolism is accomplished through control of enzyme activity or enzyme synthesis. Substances that affect enzyme activity by changing the tertiary or quaternary structure of enzymes are known as allosteric effectors. Regulator substances that control enzyme synthesis at the genetic level are enzyme inducers or repressors.

Metabolic diseases are due to the absence or inactivity of appropriate enzymes in the body.

Questions and Problems

1. The reaction

$$2N_2O_5 \longrightarrow 4NO_2 + O_2$$

is first order despite the stoichiometric coefficient of N_2O_5. The initial reaction rate, $-(d[N_2O_5]/dt)$, is 1.4×10^{-3} mol/liter sec when $[N_2O_5]_0 = 2.30M$.

a. Write the rate equation for the reaction.
b. Calculate the rate constant.
c. When the concentration of N_2O_5 has dropped to $1.35M$, what is the reaction rate?

2. The stoichiometric equation for a reaction occurring in polluted air is $2NO_2 + O_3 \longrightarrow N_2O_5 + O_2$. The experimental rate equation, however, is that for a simple bimolecular process, first order with respect to both NO_2 and O_3. Suggest a mechanism for the overall reaction that involves an NO_3 intermediate.

3. The reaction $H_2 + I_2 \longrightarrow 2I$ was believed for over 50 yr to be an elementary bimolecular reaction with the experimental rate law

$$\text{rate} = \frac{d[HI]}{dt} = k[H_2][I_2]$$

It is now known that the formation of HI is a complex reaction that proceeds via a termolecular elementary reaction

$$I_2 \overset{k_2}{\rightleftharpoons} 2I \qquad \text{with } K_{eq} = \frac{[I]^2}{[I_2]}$$

$$H_2 + 2I \overset{k_3}{\rightleftharpoons} 2HI$$

a. Show that either mechanism would give the experimentally observed rate equation.
b. Show that either mechanism would give the same expression for the equilibrium constant for the reversible reaction.

4. Discuss and fully justify the statement that a catalyst has no effect on the equilibrium position of a reversible reaction.

5. The aerobic bacterial decomposition of carbonaceous organic matter in water exhibits first-order kinetic behavior. Because the process

$$\text{organic matter} + \text{bacteria} + O_2 \longrightarrow$$
$$CO_2 + H_2O$$

uses up oxygen, the deoxygenation of polluted water is also first order. Thus, $-dC/dt = k'C$, where C is the concentration of oxidizable organic matter remaining at any time t. This relationship can be converted through the use of calculus to the form

$$\log \frac{C_0}{C_t} = \frac{k't}{2.303} = kt \qquad (1)$$

The concentration of organic matter at any time is proportional to the amount of oxygen (the BOD) required to oxidize it.

a. Show that the time, $t_{1/2}$, required to oxidize one-half of the organic matter initially present is given by

$$t_{1/2} = \frac{2.303}{k'} \log 2 = \frac{0.693}{k'} = \frac{0.301}{k}$$

b. Show that the rate equation (1) can be written as

$$\log \frac{BOD_{total}}{(BOD_{total} - BOD_t)}$$

and as

$$BOD_t = BOD_{total}(1 - 10^{-kt})$$

where BOD_{total} is the ultimate BOD and BOD_t is the amount of oxygen consumed in t days. The rate constant k has the units of "per day" and values in the range 0.10–0.50/day.

c. Two different samples of polluted water were oxidized with the following results:

time, days	1	2	3	4	5	6	7	10	20

percentage of organic matter oxidized (or % of BOD exerted)
Sample 1. 21 37 50 60 68 75 80 90 99
Sample 2. 29 50 64 75 82 87 91 97 99+

What are the rate constants k for each sample?

d. For biological oxidations with rate constants $k = 0.10, 0.15,$ and 0.25 what percentage of the total or ultimate BOD is BOD_5?

e. If BOD_{total} is 300 mg/liter, what will BOD_5 be when $k = 0.10$ and 0.30?

f. Several samples have 5-day BOD's of 200 mg/liter but different ultimate BOD's. If sample 1 has $k = 0.10$/day, sample 2 has $k = 0.15$/day, and sample 3 has $k = 0.30$/day, what are their ultimate BOD's?

g. A sample has a BOD_5 of 200 mg/liter. If the rate constant k is 0.17, what is BOD_{total}?

6. The rate of bacterial deoxygenation of waste water is temperature dependent. Within the range 10–30°C, the rate constant (or the reaction rate) increases by about 1.60 for a 10°C rise in temperature, that is, $k_{20}/k_{10} = 1.60$.

a. What is the activation energy of the rate-determining step in aerobic bacterial decomposition within the temperature range 15–25°C?

b. If a waste water has a $BOD_{total} = 188$ mg/liter and a deoxygenation rate constant of 0.523/day at 20°C, what will be its BOD_5 at 30°C?

7. A waste water has a BOD_5 of 210 mg/liter with $k = 0.40$/day at 20°C. If $k_{20}/k_{10} = 1.60/10°C$ as in Problem 6, find

a. BOD_{total} at 20°C
b. k at 25°C
c. BOD_{total} at 25°C
d. BOD_3 at 25°C

8. The rates of biological reactions increase with increasing temperature within limits. Above 40–50°C, most rates reach a maximum and thereafter decrease with increasing temperature. What is the reason for this?

9. The rate of creeping of ants, chirping of crickets, and flashing of fireflies all appear to correspond to activation energies around 10,000 cal. The value would be that for the slowest step in the metabolic mechanism. A formula relating temperature t, in degrees Fahrenheit, to the chirp rate N, in chirps per minute, of a house cricket is $t = 50 + (N - 40)/4$ (S. W. Frost, "Insect Life and Insect Natural History," Dover, 1942, p 202). From the chirp rates at 68°F (20°C) and 81°F (27°C), calculate the approximate activation energy. For a higher activation energy, would the chirp rate increase faster or slower with temperature rise?

10. The rate constant of a reaction changes with temperature in the following manner:

T(°K)	k, per hr
323	1.08×10^{-4}
343	7.34×10^{-4}
363	4.54×10^{-3}
374	1.38×10^{-2}

Graphically determine the activation energy of this reaction.

11. The rates of photochemical primary (elementary) reactions are little affected by temperature. Suggest a reason for this.

12. For the Michaelis–Menten mechanism,

$$E + S \underset{k_{-1}}{\overset{k_1}{\rightleftharpoons}} ES$$

$$ES \xrightarrow{k_2} E + P$$

a. Write the rate equation for the rate of formation of P.
b. Write the overall rate equation for the production of ES.
c. Set $d[ES]/dt = 0$ in (b) and solve for [ES]. (Setting the concentration of the intermediate ES equal to a constant is known as the *steady state approximation*.)
d. Let the initial concentration of enzyme be

$[E]_0 = [E]_{total} = [E] + [ES]$, solve for [E] and substitute into the result for [ES] obtained in (c). Rearrange to obtain

$$[ES] = \frac{k_1[S][E]_{total}}{k_{-1} + k_2 + k_1[S]}$$

e. Use the results of (d) and (a) to obtain an expression for dP/dt that shows that

$$\frac{dP}{dt} = \frac{k_2 k_1 [E]_{total}[S]}{k_{-1} + k_2} = \frac{k_2}{k_M}[E]_{total}[S];$$

$$k_M = \frac{k_{-1} + k_2}{k_1}$$

when [S] is small and

$$\frac{dP}{dt} = k_2[E]$$

when [S] is large. The constant k_M is called the Michaelis-Menten constant.

13. In enzyme kinetics, the turnover number is defined as the number of substrate molecules (or moles) converted per unit time into product by a single enzyme molecule (or mole of enzyme) when the concentration of enzyme limits, that is, when all the enzyme is tied up in the enzyme–substrate complex. This condition obtains when [S] is large and the rate of reaction is a maximum. Turnover numbers up to 36 million/min are known. First convince yourself that the turnover number is given by k_2. Then calculate the turnover number, in molecules per minute, for the enzyme-catalyzed inversion of sucrose, which has a maximum rate of 0.6 mol/liter min when $[E]_{total} = 65 \ \mu g/ml$ and $[S] = 0.5M$. The molecular weight of the enzyme invertase is 120,000.

14. Discuss the wisdom of the pronouncement by an agriculture official, "If an insect can become resistant to DDT, so can I."

15. In Section 3.18, it was stated that living organisms must exhibit stability and that this overall stability was reflected in the stability of its molecular makeup. We now know that most of the complex molecules of life are thermodynamically unstable and have positive free energies of formation. What kind of stability is, therefore, inferred in Section 3.18? Could the activation energy have anything to do with the stability factor mentioned in that section?

Answers

1. b. 6.1×10^{-4}/sec
   c. 8.2×10^{-4} mol/liter sec
5. c. sample 1, $k = 0.10$/sec; sample 2, $k = 0.15$/sec
   d. 68 percent, 82 percent, 94 percent
   e. 205 mg/liter, 290 mg/liter
   f. sample 1, 293 mg/liter; sample 2, 243 mg/liter; sample 3, 207 mg/liter
   g. 233 mg/liter

6. a. 8020 cal
   b. 187 mg/liter
7. a. 212 mg/liter
   b. ~0.48/day
   d. 203 mg/liter
9. 9500 cal
13. 1×10^6 (turnover numbers up to 36 million are known).

References

REACTION RATES

Campbell, J. Arthur, "Why Do Chemical Reactions Occur?", Prentice-Hall, Englewood Cliffs, N. J., 1965.

King, Edward L., "How Chemical Reactions Occur," W. A. Benjamin, Menlo Park, Calif., 1964.

ENZYMES

Bernhard, S., "The Structure and Function of Enzymes," W. A. Benjamin, Menlo Park, Calif., 1968.

Changeux, Jean-Pierre, "The Control of Biochemical Reactions," *Scientific American*, **212**, 36–45 (April 1965).

Green, David E., and Goldberg, Robert F., "Molecular Insights into the Living Process," Academic, New York, 1967, Chapters 4, 5, 12, 14.

Koshland, D. E., Jr., "Correlation of Structure and Function in Enzyme Action," *Science*, **142**, 1533–1541 (1963).

Phillips, David C., "The Three-Dimensional Structure of an Enzyme Molecule," *Scientific American*, **215**, 78–90 (November 1966).

METAL POISONING AND TREATMENT BY CHELATING AGENTS

Chisolm, Julian, Jr., "Lead Poisoning," *Scientific American*, **224**, 15–23 (February 1971).

Schubert, Jack, "Chelation in Medicine," *Scientific American*, **214**, 40–50 (May 1966).

Walton, Harold F., "Chelation," *Scientific American*, **188**, 68–76 (June 1953).

PESTICIDES

"Diminishing Returns," *Environment*, **11**, 6–11, 36–40 (September 1969).
> A report on insect resistance.

O'Brien, R. D., "Insecticides. Action and Metabolism," Academic, New York, 1967.

Peakall, David B., "Pesticides and the Reproduction of Birds," *Scientific American*, **222**, 73–78 (April 1970).

Risebrough, Robert, with Brodine, Virginia, "More Letters in the Wind," *Environment*, **12**, 16–27 (January–February 1970).
> Environmental contamination by polychlorinated biphenyls (PCB's), industrial chemicals similar to chlorinated hydrocarbon insecticides.

Rudd, Robert, "Pesticides and the Living Landscape," University of Wisconsin Press, Madison, 1964.

Woodwell, G. M., "Toxic Substances and Ecological Cycles," *Scientific American*, **216**, 24–31 (March 1967).

18

The Green Plant

(Reprinted by permission of Publishers–Hall Syndicate.)

If the implication that elephants eat lions is ignored, this comic strip gives a good illustration of a simple food chain by describing the dining preferences of a number of organisms. In Figure 18.1, this same food chain is sketched and some descriptive terms are added. This is an example of a grazing food chain, one beginning with plants. Organisms removed from the plant base by the same number of feeding steps are said to be in the same *trophic level*. Thus, on the second trophic level in our food chain, there might be aphids, crickets, and potato bugs. On the third trophic level, we could have a big bug such as a praying mantis. The actual species on each trophic level will be determined by the particular ecosystem. In the sea, the first (producer) level would be represented by plankton, while the second (primary consumer) level would include mollusks such as clams. Also, a particular animal is not re-

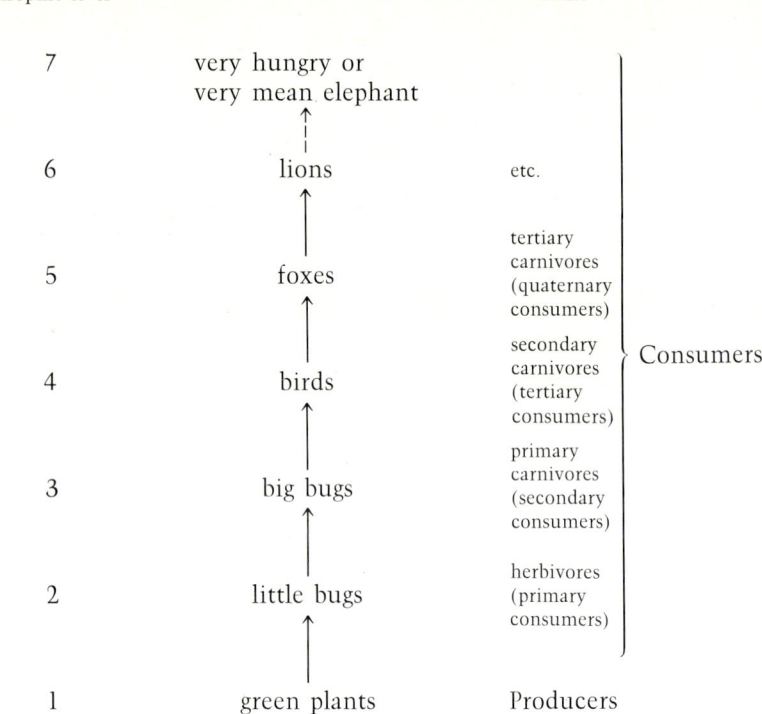

throphic level		name
7	very hungry or very mean elephant	
6	lions	etc.
5	foxes	tertiary carnivores (quaternary consumers)
4	birds	secondary carnivores (tertiary consumers)
3	big bugs	primary carnivores (secondary consumers)
2	little bugs	herbivores (primary consumers)
1	green plants	Producers

Consumers

Figure 18.1. A simple food chain according to Ira.

stricted to a given level. A robin may participate in several trophic levels, by eating plants, or by eating little bugs directly. This is typical of organisms in the higher levels, but not of those (the little bugs) in the lower levels.

In nature, trophic relationships among a community of organisms are seldom represented by this kind of linear, neatly ordered chain. Not only do several organisms in a community function at various levels, but different kinds of nongrazing food chains involving decomposer and parasitic organisms may become linked to the grazing chain of a given ecosystem. In addition, a number of food chains in different ecosystems may interact to varying extents. When the complete feeding pathways are put together, a complex food web rather than a simple linear chain results. We might show the beginnings of complexity by expanding the present food chain as shown in Figure 18.2, where the elephant, a herbivore, has been dropped (but gently). Some poetic license is claimed here because carnivorous plants such as Venus's-flytraps usually do not occur in the same community as potato bugs.

Organisms on any one trophic level obtain the energy for their living processes, their movement, maintenance of body structure, and growth from organisms on lower trophic levels. In general, only about 10 percent of the energy entering a lower level will be passed on to the next higher level. For this reason there will be little energy left to support life on the higher levels, and the steps rarely exceed four or five. It is much more efficient for the elephant to eat plants than lions. We will discuss the efficiency of energy transfer between trophic levels in more detail in Section 20.18.

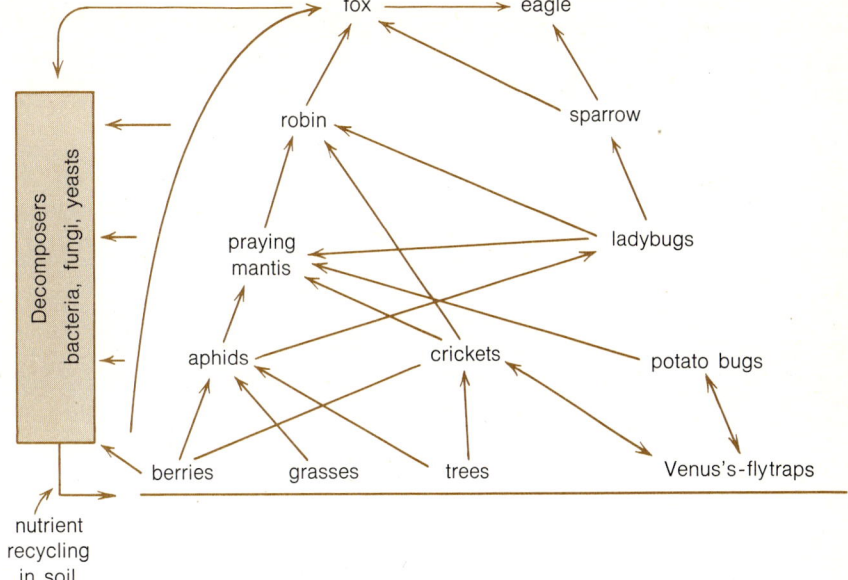

Figure 18.2. More complex feeding relations. The directions of the arrows indicate who eats what and also show the direction of energy transfer from the eaten to the eater. The decomposer chain (not shown in detail) has as its basis the remains of organisms on all trophic levels. Nutrients (but not energy) are recycled to the first trophic level by the decomposer chain. Many other feeding relationships could have been shown.

The main thing to appreciate now is that the ultimate energy basis for any food chain or web is the green plant. These plants, on the land and in the upper layers of the seas, trap the energy of sunlight and convert this energy into the chemical free energy of carbohydrates and other organic compounds used by heterotrophic organisms. As the chemical free energy is used in the biosphere, it is degraded to unavailable heat according to the second law of thermodynamics. Solar energy continually replenishes the energy store of life in the biosphere through the process of photosynthesis carried out by green plants. The photosynthetic process, as commonly understood, is the combination of carbon dioxide and water, with the help of light, to form reduced carbon compounds and oxygen. The simple overall equation, as written several times before, is

$$CO_2 + H_2O \xrightarrow{h\nu} [CH_2O] + O_2$$

with $[CH_2O]$ representing a unit of a carbohydrate molecule.

Although there are other sources of energy on earth that are unrelated to solar energy (nuclear energy) or to photosynthesis (falling water, direct conversion of solar energy into electric and chemical energy), photosynthesis is the ultimate source of all biological energy. So far as living matter is concerned, photosynthesis has rightly been called the process by which light winds up the clock of life on earth. If Ira picks too many plants he will not only spoil the whole world, he will find that he and the lion have much in common.

In the next section, we delve into the molecular mechanism of photosynthesis and see how the energy of the sun is captured and converted into the chemical fuel of life.

Photosynthesis

18.1 THE PHOTOSYNTHETIC EQUATION

Photosynthesis is a redox process, formally the reverse of respiration,

$$CO_2 + H_2O \xrightarrow{h\nu} [CH_2O] + O_2$$

where CO_2 is an electron acceptor, hydrogen atom acceptor, or oxidizing agent and H_2O is an electron donor, hydrogen donor, or reducing agent.

The overall process is a reduction of carbon(IV) in CO_2 to carbon($-$I) in carbohydrates. The elemental oxygen(0) comes exclusively from the oxygen($-$II) in water and not from the CO_2. The latter fact is shown if the overall equation is written as

$$CO_2 + 2H_2O \xrightarrow{h\nu} [CH_2O] + H_2O + O_2$$

Photosynthesis is therefore basically a splitting of the water molecule followed by the reduction of CO_2. The photosynthetic process is not limited to H_2O as the reducing agent. A more general equation is

$$CO_2 + 2H_2D \xrightarrow{light} [CH_2O] + H_2O + 2D$$

where H_2D may be H_2S, an organic compound, or even hydrogen gas (D = nothing). Certain photosynthetic bacteria use hydrogen donors other than water, and in such cases molecular oxygen is not a product. Under certain other conditions, oxidizing agents such as NO_3^-, N_2, and H^+ may be used in place of CO_2 in photosynthetic reactions.

Photosynthetic organisms use the energy of light to reduce CO_2. They should not be confused with chemosynthetic organisms, which use the energy obtained from chemical oxidation to carry out carbon reduction (Section 16.14). We shall not consider these exceptions further and shall confine our discussion to the CO_2–H_2O photosynthetic system of the higher green plants.

18.2 ENERGETICS OF PHOTOSYNTHESIS

The free energy of combustion of glucose, $C_6H_{12}O_6$, has already been given as -686 kcal/mol. Because the fixation of CO_2 by photosynthesis is exactly the reverse of the respiratory combustion of glucose, so far as thermodynamics is concerned, the photosynthetic process is endoergic by $+686$ kcal/mol of glucose formed or 114 kcal/mol of CO_2 reduced to $[CH_2O]$.

$$CO_2 + H_2O \xrightarrow{h\nu} [CH_2O] + O_2 \qquad \Delta G = +114 \text{ kcal/mol } CO_2$$

The reduction of CO_2 to $[CH_2O]$ is a nonspontaneous reaction whose occurrence is made possible only by the driving force of solar energy. The energy stored in the synthesized carbohydrate is unlocked during the spontaneous respiration reaction. Despite the favorable free energy change for this reaction we know that the oxidation of carbohydrate is not rapid, or else life would be considerably more hectic and much shorter than it is. The recombination of reduced carbon dioxide (carbohydrate) and oxygen is slowed by activation energy barriers. At the proper times and in the proper amounts, enzymes push these barriers down so that oxidation may take place with the controlled release of energy.

The endoergic nature of the photosynthetic process may be viewed from other angles. First, we might note that the stability of the products, $[CH_2O]$

and O_2, as measured by their bond energies, is less than the stability of the CO_2 and H_2O reactant molecules. The less stable a state, the higher is its potential energy. The high energy of the products is due primarily to the lower bond energy of the C—H bonds in $[CH_2O]$ as compared to the O—H bonds in H_2O and the conversion of a strong C=O bond into a weaker O=O bond as shown in the reaction

$$O{=}C{=}O \;+\; \underset{O}{\overset{H\quad H}{\diagdown\quad\diagup}} \;\longrightarrow\; O{=}O \;+\; \underset{H}{\overset{H}{\mid\; C{=}O \;\mid}}$$

The approximate bond energies are

C=O	170 kcal/mol
O—H	111 kcal/mol
O=O	118 kcal/mol
C—H	99 kcal/mol

A second way to look at the energetics of the overall photosynthetic process is in terms of its redox nature. A hydrogen atom, or an electron, is transferred from one compound to another. The reduction potentials of the substances involved give us a starting point for the discussion of electron transport in photosynthesis. We shall see that the characteristic process in photosynthesis is the excitation of an electron to a higher energy state. As the electron reverts back to its initial state, its energy is tapped off in the form of chemical free energy.

The overall equation

$$CO_2 + 2H_2O \longrightarrow [CH_2O] + H_2O + O_2$$

can be formally separated into two half-reactions,

$$2H_2O \longrightarrow O_2 + 4H^+ + 4e^- \qquad E^{0\prime} = -0.82 \text{ V}$$
$$CO_2 + 4H^+ + 4e^- \longrightarrow [CH_2O] + H_2O \qquad E^{0\prime} = -0.40 \text{ V}$$

(The single electrode potentials are for $pH = 7$, and the reduction potential for CO_2 is calculated from thermodynamic data, not measured directly.) The potential for the overall reaction is $E^{0\prime}_{total} = -1.22$ volts, and the total free energy change $\Delta G^{0\prime} = -n\mathcal{F}E^{0\prime}_{total}$, is

$$\Delta G^{0\prime} = -(4)(23.06 \text{ kcal/V})(-1.22 \text{ V})$$
$$= 113 \text{ kcal/mol of } CO_2$$

If the couples are arranged according to the reduction potentials, it is easily seen that the transferred electrons are going against the potential gradient.

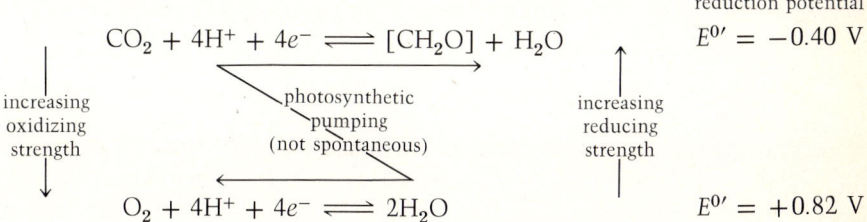

reduction potential

$$CO_2 + 4H^+ + 4e^- \rightleftharpoons [CH_2O] + H_2O \qquad\qquad E^{0\prime} = -0.40 \text{ V}$$

increasing oxidizing strength

photosynthetic pumping (not spontaneous)

increasing reducing strength

$$O_2 + 4H^+ + 4e^- \rightleftharpoons 2H_2O \qquad\qquad E^{0\prime} = +0.82 \text{ V}$$

The spontaneous (exoergic) flow of electrons (or H atoms) is from a strong reducing agent ([CH$_2$O]) to a strong oxidizing agent (O$_2$) to form CO$_2$, that is, downhill from a negative potential to a more positive potential. This occurs in the respiratory chain. In photosynthesis, on the other hand, electrons are being pumped uphill from a weak reducing agent (H$_2$O) into a weak oxidant (CO$_2$) to produce [CH$_2$O].

18.3 STAGES IN THE PHOTOSYNTHETIC PROCESS

The use of the reduction potential scheme allows us to identify three general stages in the photosynthetic reduction of CO$_2$ to a carbohydrate. These are

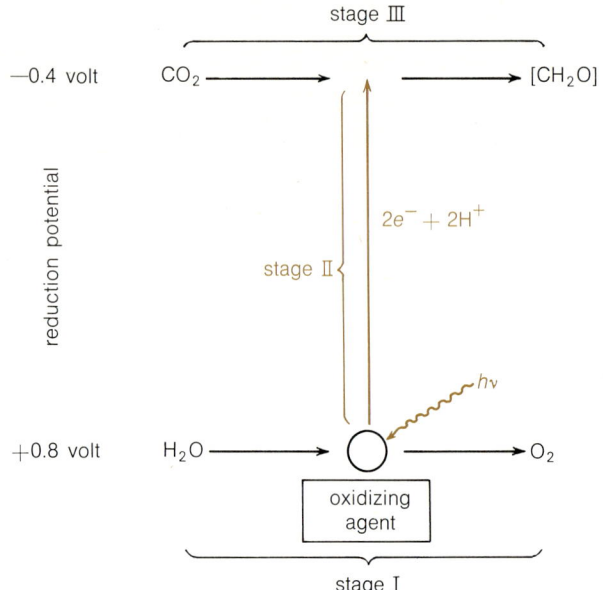

Figure 18.3. Stages in photosynthesis. Stage I is the oxidation of water to give electrons and oxygen. Stage II is the excitation of electrons by light to an energy level sufficient to carry out stage III, the reduction of CO$_2$ to [CH$_2$O].

shown in Figure 18.3. The three stages, which do not necessarily occur consecutively, are now described briefly.

Stage I: The Oxidation of Water [or O(−II)]

Little is known about the mechanism of this stage, except that it is a series of reactions catalyzed by enzymes and that one of these enzymes contains manganese. The overall half-reaction is

$$2H_2O \longrightarrow \tfrac{1}{2}O_2 + H_2O + 2H^+ + 2e^-$$

and the mechanism may involve hydroxyl free radicals,

$$2H_2O \longrightarrow 2H^+ + 2OH^-$$
$$2OH^- \longrightarrow 2 \cdot OH + 2e^-$$
$$2 \cdot OH \longrightarrow H_2O + \tfrac{1}{2}O_2$$

In any event, the electrons released by water are taken up by an as yet unidentified oxidizing agent, which passes on the electrons to stage II.

Stage II: Light-Induced Electron Excitation

In this stage, electrons from stage I are boosted to higher energy levels by the absorption of light by photosynthetic pigment molecules. The most important of these molecules is chlorophyll. The structure of chlorophyll is shown in Figure 18.4. The molecule contains a porphyrin group similar to that in the

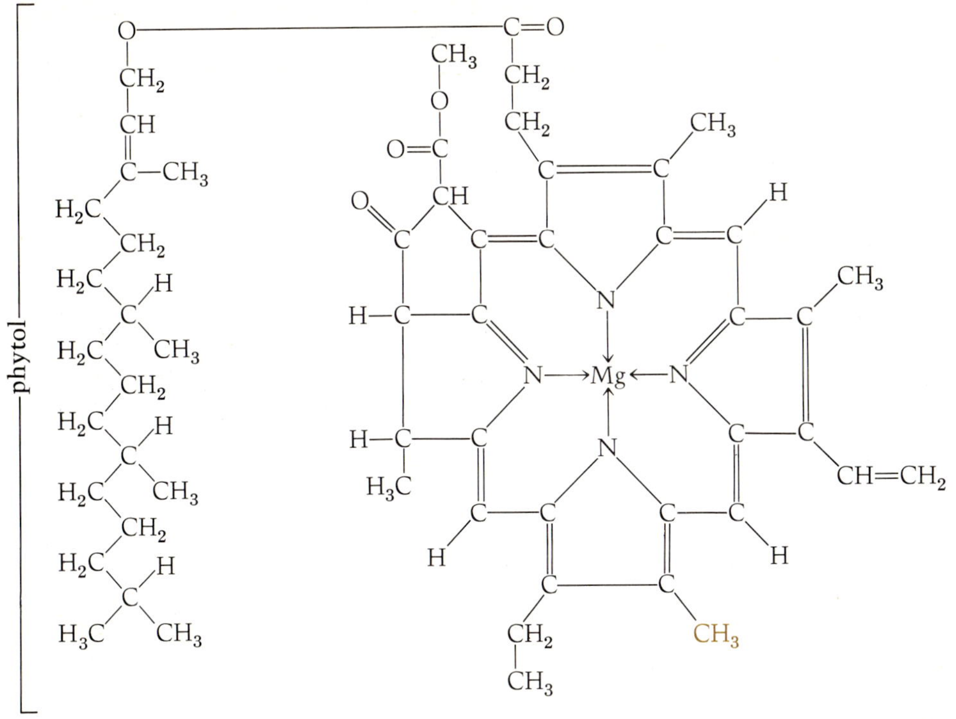

Figure 18.4. Chlorophyll *a*. In chlorophyll *b*, the colored —CH₃ group is replaced by

$$-\overset{\underset{\displaystyle |}{\displaystyle H}}{C}=O.$$

heme proteins, except that iron is replaced by magnesium. In green plants, there are two chlorophyll molecules, chlorophyll *a* and chlorophyll b, which differ very slightly in structure. The absorption spectrum of chlorophyll *a* is shown in Figure 18.5 together with the variation of the efficiency of photo-

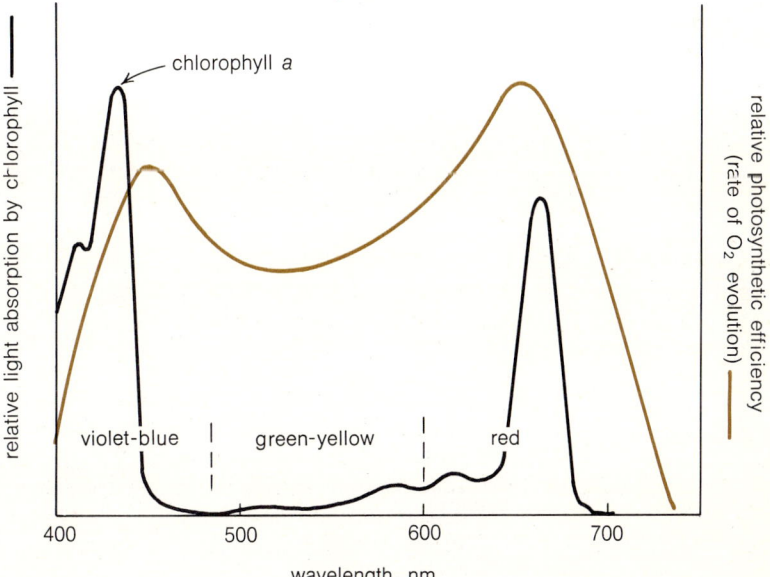

Figure 18.5. The efficiency of photosynthesis at various wavelengths compared to the absorption spectrum of the chlorophyll *a* molecule indicates the importance of chlorophyll in photosynthesis.

synthesis (rate of O_2 evolution) with wavelength. (Unless otherwise stated, the term "chlorophyll" will refer to chlorophyll a.) The correspondence of the two curves indicates that chlorophyll is the major light-absorbing substance. (Chlorophyll b has a very similar spectrum.) The absorption spectrum of chlorophyll shows strong absorption in the red, near about 670 nanometers (nm). Hence, the transmitted light is green. The chlorophyll molecule has considerable π electron delocalization because of its highly conjugated bond structure. The absorption of light in the visible region is the result of excitation of these delocalized π electrons to higher energy π^* orbitals.

In stage II, light falling on a chlorophyll molecule promotes an electron to a higher level, producing an excited chlorophyll molecule, Chl*. The excited molecule then transfers its energized electron to an electron acceptor designated as electron acceptor 2 in Figure 18.6. The chlorophyll molecule, minus

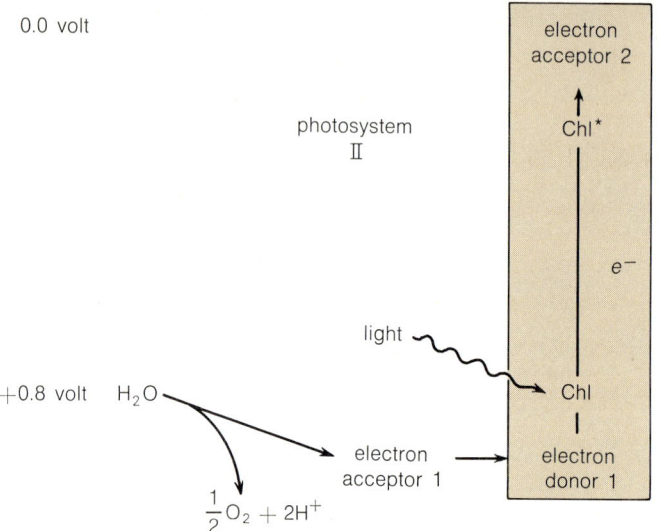

Figure 18.6. The connection between stage I and the first step of stage II in photosynthesis.

the electron, can be considered to return to the ground state, where it picks up an electron from the oxidizing agent which, as electron acceptor 1, had gained an electron from water back in stage I. The sequence of events can be represented by the equations

$$\text{Chl} + h\nu \longrightarrow \text{Chl*}$$

$$\text{Chl*} \longrightarrow \text{Chl}^+ + e^-$$
$$\text{(to electron acceptor 2)}$$

$$\text{Chl}^+ + e^- \text{ (from electron donor 1)} \longrightarrow \text{Chl}$$

Figure 18.6 diagrams the sequence so far on the redox potential scale.

Experiment has shown the existence of two (and possibly three) light-induced reactions in photosynthesis. One, which we have identified as that occurring in photosystem II, is indicated in Figure 18.6. Another light reaction takes the electron from photosystem II and boosts it to even higher energies. This extra boost occurs in photosystem I which, despite the name, follows photosystem II. Chlorophyll plays the same role in photosystem I as in photosystem II. The situation is shown in Figure 18.7. Here photosystems II and I, both driven by light captured by chlorophyll, are connected by a downhill flow of electrons from the electron acceptor–donor, X–X$^-$, to the electron acceptor–donor, Y–Y$^-$, neither of which are identified. In this exoergic fall of electrons to a slightly lower energy level, the electrons are passed along by a group of intermediate electron carriers (represented by the dots on the line from X to Y). During this process, the energy of the falling electron is tapped off and conserved in the form of the chemical bond energy of ATP, whose formation is coupled to the exoergic electron flow. You will recall that the same bucket brigade of electrons similarly coupled to ATP production occurs in the respiratory chain in mitochondria. Some of the cyclic electron carriers in photosynthesis are cytochrome molecules just as in the respiratory chain.

Figure 18.7. The chemical changes in photosynthesis, showing the complete electron transport chain from water to NADP$^+$. The points on the lines connecting the two photosystems and on the line connecting photosystem I to NADPH represent cyclic electron donor–acceptor molecules such as cytochromes. Electron flow is normally from water to NADP$^+$ via a noncyclic path through the two photosystems, but there is evidence of a cyclic flow utilizing only photosystem I.

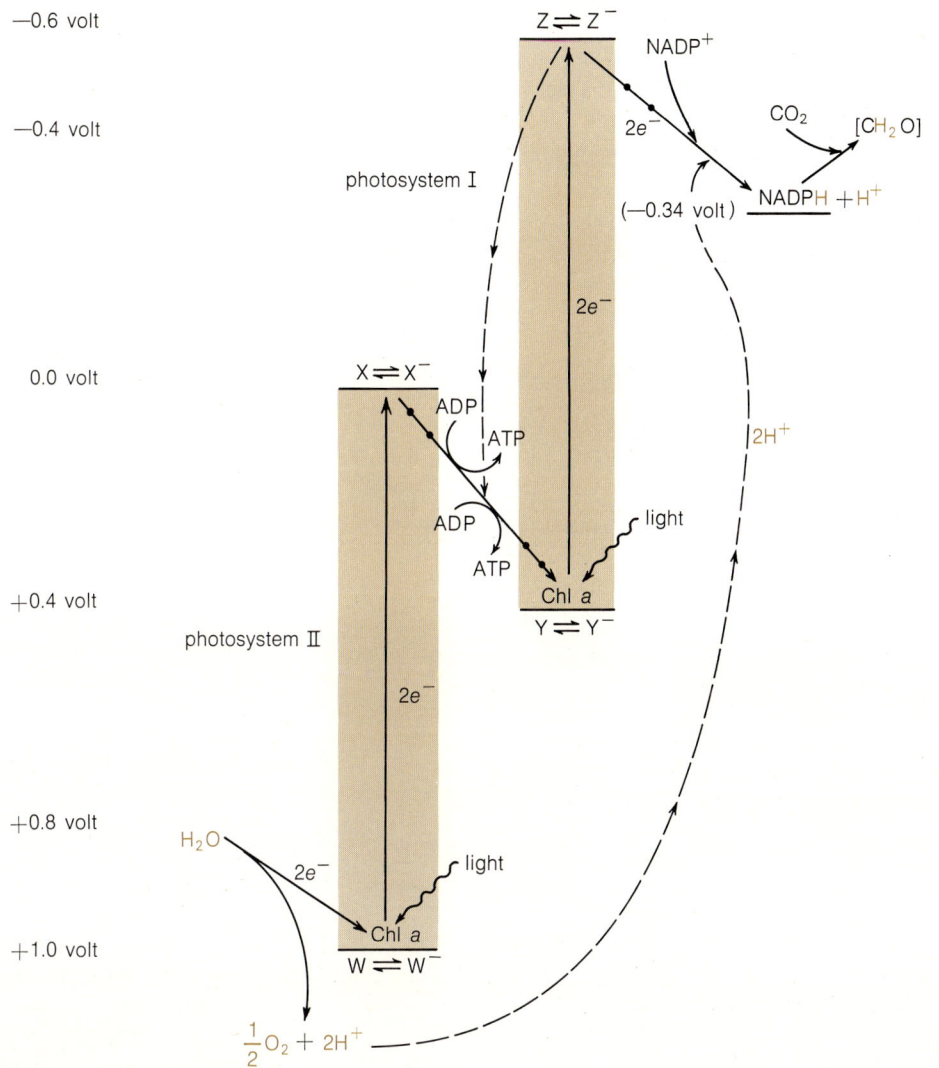

The immediate activating substance in each photosystem is the chlorophyll *a* molecule. This is the molecule that transfers its excited electron to the acceptor X or Z at the top and obtains another electron from the donors, W⁻ or Y⁻, at the bottom. In photosynthetic organisms, chlorophyll *a* is accompanied by other pigment molecules. Among these are chlorophyll *b* and the yellow-to-red carotenoid pigments in the higher plants. These accessory pigments absorb in spectral regions where chlorophyll *a* absorption is weak but where the intensity of sunlight is great. They help the organism gather the energy of sunlight and then pass on this energy to the chlorophyll *a* molecules of the photosystems. In plants, the green chlorophyll pigments mask the yellow, orange, and red colors of the accessory pigments. Only as the chlorophyll disappears when leaves turn in the fall or when fruit ripens do the underlying pigments show their colors.

After electrons have been boosted by photosystem I to another electron acceptor (Z), the electrons then slide downhill a bit via one or more electron carriers to the coenzyme electron acceptor NADP⁺, nicotinamide adenine dinucleotide phosphate (Figure 18.8). Along with the hydrogen ions from stage I, the electrons and NADP⁺ form the reducing agent NADPH, so that

Figure 18.8. Oxidized and reduced forms of the coenzyme nicotinamide adenine dinucleotide phosphate (NADP).

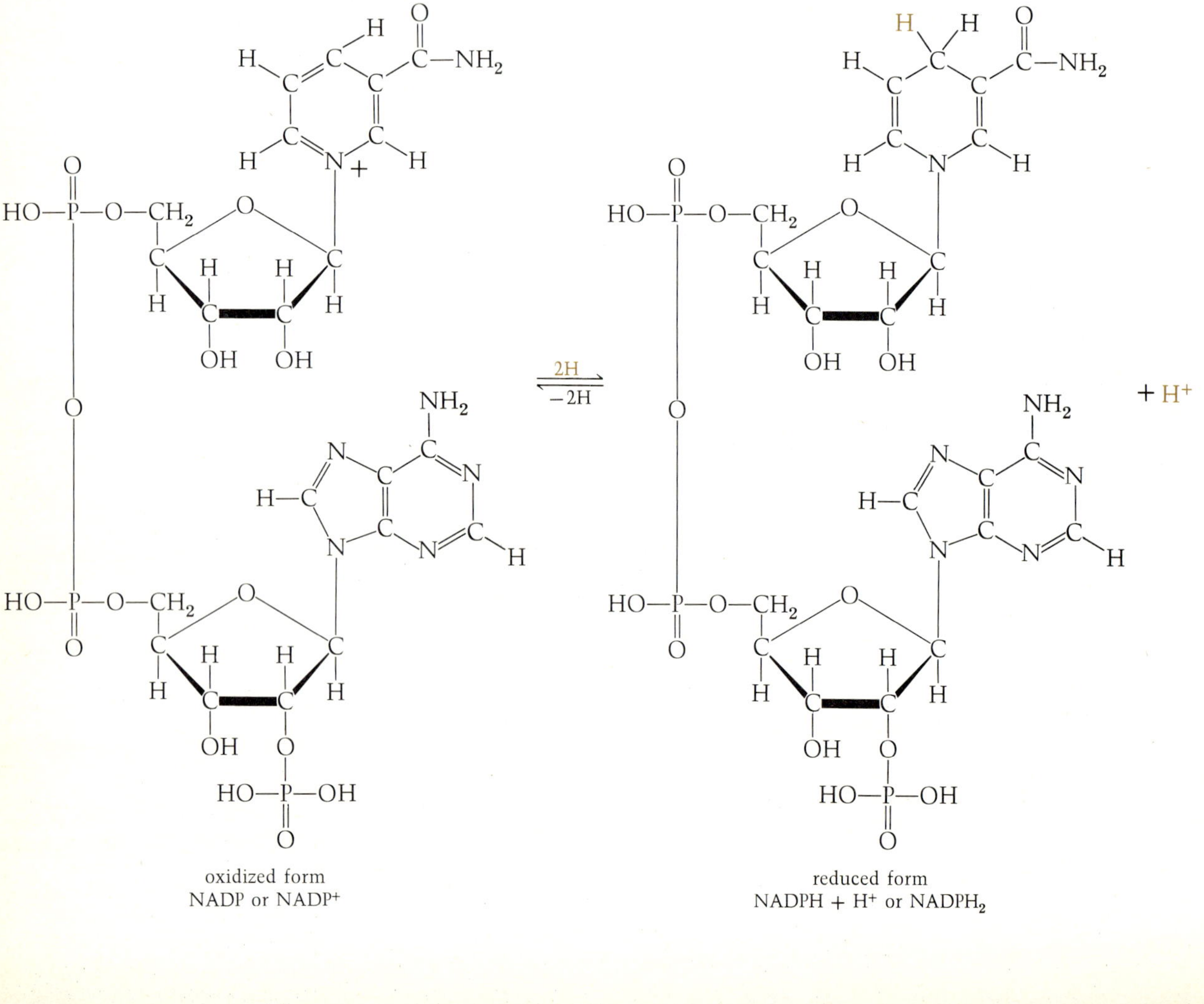

oxidized form
NADP or NADP⁺

reduced form
NADPH + H⁺ or NADPH₂

the overall process appears as a hydrogen transfer. The NADPH, with the help of energy obtained from ATP, supplies the reducing power for the reduction of CO_2 to carbohydrates in stage III.

Stages I and II are the light-induced reactions of photosynthesis. They take place entirely in the highly structured cellular organelle, the chloroplast, of photosynthetic cells (except in bacteria). In the chloroplasts, the pigment molecules, the electron donors, acceptors, and carriers, and the necessary enzymes are arranged in close physical proximity to each other to facilitate electron transfer and other reactions. This is again similar to the arrangement of molecules in the folded membranes of the mitochondria. Just as the mitochondria are the power plants of respiring cells, the chloroplasts are the power houses of photosynthetic cells. The cells of green plants contain both mitochondria and chloroplasts, because the plants must also degrade some of the high-energy molecules photosynthesized by the chloroplasts. This type of respiration most obviously occurs at night, when the plants give off CO_2 instead of oxygen. Another type of respiration known as photorespiration also takes place in the daytime, simultaneously with photosynthesis.

In terms of electron energy levels, or redox potentials, respiration is just the return of the high-energy electron in glucose or carbohydrates to its ground state in the water molecule. The wheel of life is spun by the energy given up by the falling electron.

Stage III: Biosynthesis of Carbohydrate

This final stage of photosynthesis is known as the dark reaction because it does not directly require light. Light is only necessary for the production of the reducing agent NADPH and of the ATP necessary for driving the synthesis reactions of stage III. The conversion of CO_2 to carbohydrate might be written in an oversimplification as

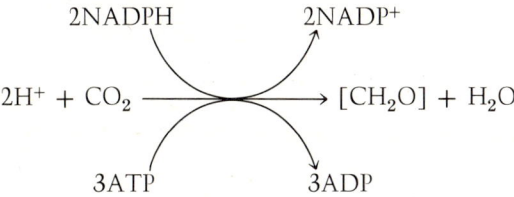

This equation, however, conceals the fact that the biosynthesis of carbohydrates, or even of a simple hexose sugar, takes place through an intricate series of reactions. The metabolic pathway for this synthesis is known in much more detail than are the pathways of stages I or II. Carbon dioxide is not reduced directly by NADPH. Instead, it is enzymatically incorporated into an organic molecule as a carboxyl group (—COOH), which is later reduced by NADPH. The process is essentially the reverse of glycolysis. In the resulting carbohydrate, the electron from stage II retains the energy acquired in the light reactions.

To conclude this section on the mechanism of photosynthesis, the overall equations for the three stages and for the complete process are given:

$$\text{light} \begin{cases} \text{stage I:} & 2H_2O \longrightarrow O_2 + 4H^+ + 4e^- \\ \text{stage II:} & 2NADP^+ + 2H^+ + 4e^- \longrightarrow 2NADPH \end{cases}$$

$$\text{dark} \quad \text{stage III:} \ 2NADPH + 2H^+ + CO_2 \longrightarrow$$
$$2NADP^+ + H_2O + [CH_2O]$$

$$\overline{\text{overall:} \qquad\qquad CO_2 + H_2O \longrightarrow [CH_2O] + O_2}$$

The emphasis on the use of photosynthesis by green plants to manufacture carbohydrates should not lead to the belief that carbohydrates are the only substances that plants synthesize. Although plants are predominantly composed of carbohydrates, they also synthesize proteins, fats, and nucleic acids because these are essential to their life. Plants also can produce all the vitamins needed by man plus many toxins, drugs, flavors, and other substances.

18.4 SOME ECOLOGICAL ASPECTS OF PHOTOSYNTHESIS

The importance of photosynthesis in the biosphere cannot be overestimated, for without photosynthesis there would be no biosphere. The photosynthetic process plays a key role in the biogeochemical cycles of oxygen and carbon and in the regulation of the oxygen and carbon dioxide content of the atmosphere. The annual amount of carbon dioxide fixed as reduced organic compounds by photosynthesis is of the order of 100 billion metric tons.

A comparison of photosynthesis and respiration provides a good illustration of the larger material-cycling and energy-degradation processes in the biosphere. So far as energy is concerned, photosynthesis takes in a "small number" of high-energy photons in the visible and ultraviolet region. The energy of these photons is eventually degraded to an equal amount of heat energy by respiration (Figure 18.9). The heat energy appears as a much larger

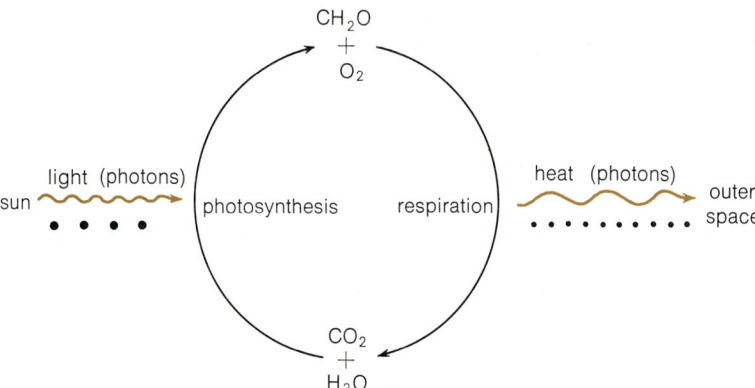

Figure 18.9. Energy absorption and degradation in the biosphere.

number of infrared photons of smaller energy radiated by the earth to outer space. The entropy of the large number of low-energy photons is greater than the entropy of the incoming solar photons. The increase in entropy corresponds to a decrease in free energy. Any decrease in entropy due to the unrespired complex molecules present at any time in plants is not sufficient to balance the entropy increase resulting from respiration and other radiation processes on the earth.

The rate of photosynthesis of green plants may be affected by a number of factors. Among these are light intensity, CO_2 pressure, and temperature. Under conditions of high light intensity, the partial pressure of CO_2 can be the limiting factor controlling the photosynthetic rate; under low light and high CO_2, the rate is limited by light intensity. When CO_2 is limiting, the rate of photosynthesis is increased by a rise in temperature because the rate-determining CO_2 reduction stage (stage III) is a temperature-sensitive nonphotochemical reaction. On the other hand, when the overall rate is limited by (low) light intensity, a temperature increase little affects the rate, because the rate-determining photochemical reactions of stages I and II are not temperature sensitive. Insensitivity to temperature is common to most photochemical processes. When CO_2 is the limiting factor, an increase in its partial pressure leads to an increase in the rate of photosynthesis. Thus, in greenhouses the concentration of CO_2 is often increased by injecting the waste gases from the fossil fuel fired heating system into the greenhouse at up to ten times the normal (0.03 percent) concentration in air. Under these conditions, the time for plant maturation is decreased and increased yields are obtained.

The metabolic pathway of photosynthesis may be blocked or altered at any number of points by chemical agents. Some weed killers or herbicides act by directly inhibiting electron flow in some step in stage II of photosynthesis, thereby preventing production of ATP or reduction of NADP. The air pollutant PAN also works this way, causing the plant to die for lack of energy. The plant literally starves to death. Other common herbicides (Section 18.5) act instead at the genetic or protein synthesis level. Ozone, a constituent, like PAN, of photochemical smog, is believed to kill plants by increasing respiration, thereby depleting the food supply.

In 1968, evidence was presented that showed inhibition by DDT of photosynthesis in marine phytoplankton. These plants are responsible for about 70 percent of the total photosynthesis on earth and are the primary producers for the marine food chains as well. The concentration of DDT in water necessary to cause significant inhibition of photosynthesis is about 10 ppb, which is about ten times the solubility of DDT in water. Thus, in the open seas, it is unlikely that the inhibitory concentration of DDT, in both dissolved and colloidal form, will be reached. It may be a different story in the highly productive estuarine areas, so the situation bears watching.

18.5 PLANT HORMONES

As with higher animals, the growth and development of a mature, multicellular plant is the result of a vast number of integrated chemical changes. The chemical changes are regulated in space and time so that certain specialized cells emerge at the proper place and time from the division of seemingly unspecialized cells. The growth and placement of the many differentiated cells is then coordinated to ensure their proper location and growth rate with respect to one another. When the normal whole plant has developed and matured, regulating mechanisms continue to control growth by participation in the timing of periodic events such as flowering, dropping of foliage, and dormancy.

The pattern of growth and development of a plant is determined by the information or instructions encoded in its genetic material and, secondarily, by environmental influences such as temperature, light intensity and duration, and moisture. When the plant must respond to its inherent genetic instructions or to the environment, it often does so through the action of plant hormones. Hormones, in general, are defined as chemical agents or messengers formed in one part of an organism that then travel to other parts of the organism where they exert their regulatory effect. The effects of plant hormones are probably achieved by altering reaction rates through an effect on enzyme synthesis or through the activation or inhibition of enzymes already present. Only very small quantities of the hormones are involved.

There are many compounds that behave as hormones. Some act as growth promoters, others as growth inhibitors. Many act in different ways at different sites in a plant. Some normally interact with one another to give a variety of synergistic or antagonistic effects depending on their relative proportions. The whole subject of plant growth and development and the effect of hormones is a large, complex, and incompletely understood field. Here we only list some major types of plant hormones and discuss two of them briefly.

There are at least five or six main classes of plant hormones—auxins, gibberellins, cytokinins, abscissins or dormins, brassins—and one specific compound—ethylene—that regulate growth through their effects on cell enlargement or cell division. Practical use has been made of artificially applied, synthetic plant hormones for many years. One of the most common applications has been the use of auxin-type compounds as herbicides (weed killers or defoliants).

The natural auxin in plants is indole-3-acetic acid,

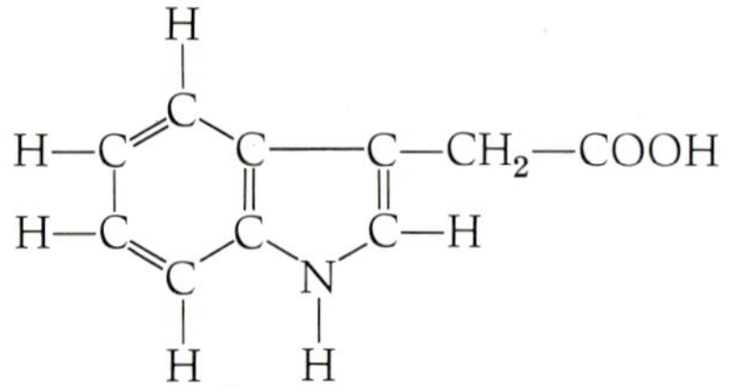

Plant growth is regulated by indole-3-acetic acid by its action in stimulating cell elongation. Two of the many synthetic compounds having hormone properties similar to the natural auxin are 2,4-dichlorophenoxyacetic acid (2,4-D) and 2,4,5-trichlorophenoxyacetic acid (2,4,5-T)or their esters:

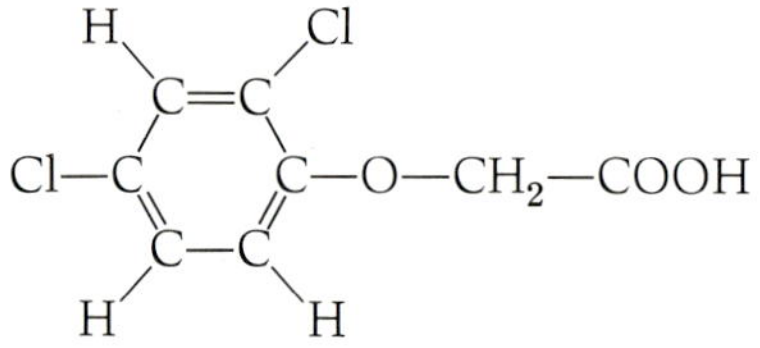

2,4-D

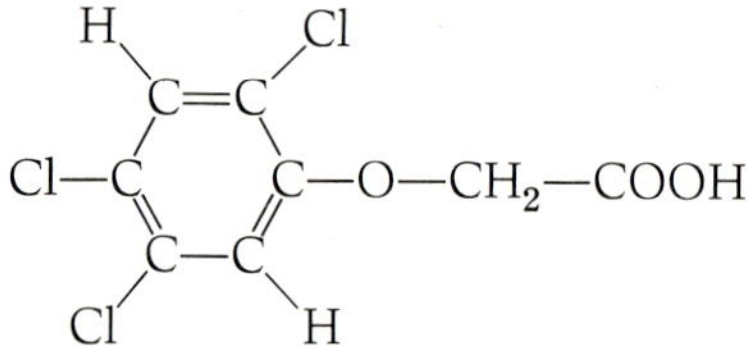

2,4,5-T

Although plants require low concentrations of auxins for their normal growth, applications of abnormally high concentrations of 2,4-D or 2,4,5-T to a plant can cause an imbalance between auxin concentration and other hormones working in conjunction with auxin. The upset in hormone balance and in plant metabolism can lead to the death of the plant. Whether the effect is due to inhibition or overstimulation of the growth of certain cells seems to be unclear. Auxin herbicides are selective in their action for broad-leaved plants. Therefore, the broad-leaved weeds in a lawn of narrow-leaved grasses can be easily eradicated with 2,4-D or 2,4,5-T. The same compounds are also efficient defoliants for trees, cotton, and other plants.

One ecological consequence of herbicide use was mentioned in Section 14.16 in connection with nutrient dumping and the resultant destruction of soil fertility and structure in tropical soils. Also of concern are any direct effects of herbicides on animals. So far 2,4,5-T has come under the most fire, because tests appear to show that it is both toxic and teratogenic at low doses. In addition, the most toxic chlorine-containing compound known, 2,3,7,8-tetrachlorodibenzo-*para*-dioxin (TCDD), is a by-product of the manufacture of 2,4,5-T and is unavoidably present in the "pure" herbicide to the extent of 0.5–1.0 ppm. This compound is also highly teratogenic in experimental animals. Restrictions have been placed on the use of 2,4,5-T pending further studies.

Ethylene, $CH_2=CH_2$, is often considered to be a gaseous plant hormone in a class by itself. Studies have shown that ethylene is a normal product of healthy plants. Other evidence indicates that high auxin concentrations induce ethylene formation in plants and that some of the effects attributed to auxins are due directly to ethylene. Among other effects, ethylene promotes premature leaf fall, fruit ripening, and the fading of flowers in some plants. This has led to difficulties in fruit- and flower-growing areas beset by air pollution, because ethylene is produced by burning hydrocarbons. The formation of ethylene accelerates as fruit ripens. The ethylene produced by one overly ripe or rotten apple in a barrel accelerates the spoilage of other apples in the barrel. The ripening of fruit and the associated production of ethylene can be slowed by storing the fruit in an atmosphere of high carbon dioxide or nitrogen concentration or by decreasing the temperature.

Food and Fiber

The most important product of the plant kingdom is food, a subject we shall take up very shortly. The common phrase "food and fiber" also indicates the importance of the indigestible (to humans) cellulose component of plants. Cellulose fibers vary in length from 350 cm in flax, jute, hemp, and sisal, to about 4 cm in cotton, to perhaps 0.2–5 mm in wood.

18.6 SOME CHEMISTRY OF WOOD PULPING

Cellulose is the principal component of all wood. Paper is a three-dimensional web of cellulose fibers made from wood pulp, which in turn is made by the separation of the cellulose component of wood from the other nonfibrous components. In addition to cellulose, wood also contains two other classes of substance. One class, the hemicelluloses, is a nonfibrous polysaccharide of the simple pentose sugar, D-xylose. Lignin is a noncarbohydrate polymer of aromatic alcohols joined by ether linkages. Its structure is not known with any certainty. Lignin acts as a cement to hold the cellulose fibers together. On a dry weight basis, wood is about 50 percent cellulose and 25 percent each hemicellulose and lignin.

In most paper-making operations, wood chips are first put through a chemical pulping process designed to dissolve and leach out the nonfibrous lignin and some or all of the hemicellulose. The separated cellulose fibers are left behind. The sulfite and the kraft processes are both used in pulping operations, but the latter process is most popular. In the *sulfite process*, the pulping agent is a buffered solution of sulfurous acid (made from SO_2 and $CaCO_3$ or CaO). The primary reaction is between bisulfite ions (HSO_3^-) and the lignin of the wood to form soluble lignin sulfonic acids or calcium salts of the acids:

$$HSO_3^- + \text{lignin—OH} \longrightarrow \text{lignin—SO}_3^- + H_2O$$

The *kraft process* uses an alkaline pulping solution of NaOH and sodium sulfide, Na_2S. (This process is also called the sulfate process because the Na_2S was originally made from sodium sulfate.) The mechanism of solubilization of lignin by this solution is complex and incompletely understood, but the eventual result is probably the hydrolysis of the ether linkages.

The digestion of wood in both the sulfite and the kraft processes results not only in degradation of lignin, but in hydrolysis of the hemicellulose carbohydrate and, inevitably, some of the cellulose. The reaction products, the sugars and organic acids, end up in the spent cooking solution (black liquor) and must be disposed of. These large-volume liquid wastes would cause a huge biochemical oxygen demand (BOD) if the effluents were released directly into natural waters. For economic reasons, recycling and recovery of chemicals is an integral part of kraft mills. In this process, most of the organic wastes are burned to CO_2 and H_2O and then harmlessly released to the air instead of to waters. However, the wood digestion reaction and the recovery operation both produce highly detectable amounts of volatile, foul-smelling sulfur compounds such as H_2S and CH_3SH. These are also released to the atmosphere and produce a not easily forgotten odor in the vicinity of kraft pulp mills. No odor-abatement procedures in use at the present time have come close to success in eliminating the stink. The kraft process has some minor water pollution problems as well. The process produces a discolored pulp, which must be bleached [usually with chlorine gas, calcium hypochlorite, $Ca(OCl)_2$, or ClO_2] for high-quality paper. Some of the effluents from the bleaching are toxic to fish and others contribute to the BOD. In sulfite mills, air pollution problems are minimal but, if the mills do not recycle and recover chemicals, water pollution is severe. Unlike kraft mills, sulfite mills have not always recycled and recovered these chemicals.

We tend to think of wood in terms of its use as a construction material or as a source of pulp for a wide variety of paper products. The latter includes new paper clothing as well as newsprint and containers. However, a number of important chemical compounds are derived from wood, either directly or as by-products of pulping operations. For instance, wood pulp provides cellulose for the production of cellulose acetate, which is used as a plastic or in acetate rayon fibers. Dissolution of native cellulose from wood pulp and its reprecipitation provides us with the regenerated cellulose of cellophane and viscose rayon fiber. Surprisingly, however, over half the wood cut worldwide is still used as fuel.

18.7 FOOD AND THE ECOSYSTEM

Over two-thirds of the population of the United States lives in or around cities. Other nations have or are striving for the same somewhat dubious hallmark of industrialization. So it is not unexpected that many people have little comprehension of the problems involved in the growth and supply of the food on their tables.

The process of food production, of course, begins with photosynthesis. About 50 percent of the solar radiation heading for the earth reaches the surface of the planet, where it is absorbed. Plants may absorb about 50 percent of the radiation incident upon them. Under average natural conditions, only about 1 percent of the solar energy absorbed by a green plant is converted into chemical energy by the photosynthetic fixation of carbon dioxide. The remaining 99 percent is immediately lost as heat or through vaporization of water. Finally, up to one-half of the small fraction of solar energy fixed by the plant is used by the plant itself for respiration. What chemical energy is left over is the basis for all other life on earth.

There is not a great deal man can do to increase the small percentage utilization of the sun's energy by photosynthesis under natural conditions. One possibility is to find some way to inhibit the respiratory process in plants so that the plant consumes a smaller portion of the energy trapped in photosynthesis. In the absence of a practical means of using solar energy more efficiently, man has attacked the problem from the other direction. He has modified the ecosystem so that he can grow more food per year or per acre, lose less of it to competitors and disease, and increase the ease of harvesting. We have already seen how the use of fertilizers and other soil nutrients and modifiers fits into this picture. What is not so obvious is the fact that the man-made agricultural ecosystem is an inherently unstable system, which tends toward a state of greater diversity—a state that does not favor man above all other organisms. To maintain his unstable agricultural ecosystem, man must inject large quantities of energy into it. The ramifications of man's actions will be examined in a little more detail later. First, we should realize why the problems of food growth are among the most serious in the history of the world.

18.8 WORLD FOOD NEEDS Ever since the publication of Thomas Robert Malthus' "An Essay on the Principle of Population As It Affects the Future Improvement of Mankind" in 1789 (revised 1803), the twin factors of man's food requirements and his population and fertility have been recognized as inextricably linked. The three basic propositions of Malthus were as follows:

1. Population is necessarily limited by the means of subsistence.

2. Population increases where the means of subsistence increase unless prevented by very powerful checks.

3. The checks to population increase and the checks that repress the superior power of population and keep its effects on a level with the means of subsistence are all resolvable into vice, misery, and moral restraint.

According to Malthus, population tends to increase at a geometric rate (1, 2, 4, 8, 16, 32, . . .), while food supply normally increases only arithmetically (1, 2, 3, 4, 5, . . .). Thus, population must inevitably exceed food supply.

The coming of the Industrial Revolution and the resultant rapid increase in food production delayed a worldwide test of Malthus' predictions except in small areas such as Ireland during the potato blight of 1845. Today the world is moving steadily toward a new test of these propositions, but on a much wider scale. World population is projected to double about every 30 yr. It is predicted that population will go from over 3.5 billion in 1970 to 7.0 billion in 2000.

Worldwide, the present average increase in agricultural productivity about matches the rate of population increase, and present levels of food supply could presumably be maintained for some time. But there are many reasons for believing that a point must eventually be reached when the world's average increase in food production will begin to fall behind its population growth.

Unless food supply is increased at a rate equal to, or greater than, population increase, hunger will be the major world crisis. Increasing food production will be only a short-term solution; population control must be the final solution to Malthus' "dismal propositions" of the approaching collision between world population and mass starvation.

Today our levels of food production are not adequate, even if the surplus production of the developed agricultural producers were spread equally around the world. At least 50 percent of the people in the world either do not get enough to eat, or do not eat food of the required quality, or both. Food quantity is generally measured by its *caloric* value, the enthalpy change when the food is oxidized to CO_2, H_2O, and such nitrogen compounds (for proteins) as would be produced in the body. The average caloric values per gram of foodstuffs as they are utilized in the human body (at 20°C) are

carbohydrate:	4.1 kcal/g
fat:	9.5 kcal/g
protein:	4.3 kcal/g

The average adult human requires a minimum daily intake of about 2500 kcal of food energy. But a body cannot exist on calories alone. Also required are the nitrogen, sulfur, vitamins, minerals, and amino acids that play important structural and functional roles. The presence of these in the diet, especially the amino acids vital to protein syntheses, establishes the quality of the diet. Human metabolism requires 20-odd amino acids. Some of these can be manufactured by the body from nitrogen and carbon compounds in the food, but 11 amino acids, the essential acids, must be taken in preformed in the protein of the diet. We have already noted how the absence of specific vitamins and minerals can lead to certain dietary diseases. Human diet requires about 20 g of high-quality protein per 1000 kcal. The absence of sufficient protein leads to several general wasting diseases such as kwashiorkor. Protein deficiency diseases are particularly damaging to the mental and physical development of children, and no amount of feeding or education in later years can repair the damage.

18.9 MODERN EFFORTS TO INCREASE FOOD SUPPLY

Throughout man's history, the search for an adequate food supply has shaped civilization. Because of the urban origins of many of us, we tend to forget this fact. But if world food supply cannot keep up with population growth, we shall be constantly reminded of our dependence on agriculture.

The first great step in the relation of man and food was probably the domestication of plants and animals and the abandoning of the nomadic life. Then came the widespread use of cereal grains grown from seed in cultivated fields. With the invention of the internal combustion engine, the energy of fossil fuels rapidly replaced the muscle power of man and work animals for the production of food. Irrigation was practiced even before mechanization, and its use has steadily grown as arid lands have been brought under cultivation. Less rapid in coming was the realization that soil conservation must be practiced to retain the soil and prevent it from wearing out. In the twentieth century, widespread use of chemical fertilizers enhanced the fertility of marginal and overfarmed soils, increasing both the yield per acre and the quality of the food grown. The protein content of a given plant depends on the nutrients that it can obtain from the soil as well as on the plant species. Soil conservation and fertility are especially important factors in the agriculture of tropical countries, where the most severe mismatches in population and food supply exist.

Beginning in the 1940's, the yield of food was again increased through the use of chemical pesticides. This new era was ushered in with the discovery of DDT. Originally used in the field of public health to control insects that transmit diseases such as malaria, typhus, and sleeping sickness to man, after World War II DDT and other insecticides were turned on the insects that normally took 10–50 percent of a crop. Other pesticides were developed to eradicate the weeds, rodents, fungi, and other organisms that interfere with high-yield agriculture.

For the most part, all of the above innovations improved crop yield. In the 1960's, more attention was turned to the equally important area of food quality. Enhancement of the quality of food has been accomplished by the application of genetic engineering to plants. The discovery that plants can be biologically designed by selective breeding to increase crop yields, protein content, resistance to disease or insects, and adaptation to various climates has led to what has been called the Green Revolution.

Rice, wheat, and corn are the major cereal grains on which the world depends for human and animal food, so they have been the subjects of most breeding experiments. Hybridization, the crossing of strains to obtain offspring with more desirable qualities than those possessed by either of the parent strains, has long been applied to corn. Hybrid breeding has now produced strong-stem dwarf wheat and rice that, when grown under good fertilization and moisture conditions, can replace the tall-growing varieties that tended to fall over and escape harvesting. The new hybrids also mature faster and are less sensitive to variation in day length, so they may be planted more often and in any season. Plant breeding has produced a corn high in lysine, one of the essential amino acids. Similar increases in essential amino acid content in the protein of wheat and rice are desirable.

Another avenue to increasing food quality, as opposed to quantity, is the fortification of foods with added amino acids. The fortifying amino acids can be obtained from a number of sources, synthetic as well as natural. One of the more promising sources is the production of high-protein concentrates from petroleum hydrocarbons, natural gas, and cellulose wastes (including garbage). Suitable microorganisms, usually yeast or bacteria, when supplied with the proper nutrients, in addition to hydrocarbons (NH_4^+, PO_4^{3-}, K^+, trace elements, vitamins) can ferment or decompose these substances and convert them into protein. The protein ends up as the protoplasm of the single-cell organisms. Their bodies, after washing and drying, result in a powder having a protein content of 50 percent or more. The product is called single-cell protein (SCP) and can be used to fortify other foods. High-protein concentrates also are obtained from fish meal, algae, leaf extracts, and the oil of cotton seeds, soybeans, and sunflowers.

Many problems of cost, distribution, and cultural and psychological acceptance remain to be solved before any of these concentrated protein products are produced and used in significant amounts. It is difficult for many people to change their dietary habits even when they are hungry. This is especially true if the concentrate has an unusual taste or if it is known that it was produced from garbage or fish heads.

It is ironic that a major problem associated with single-cell protein is its high nucleic acid content. The metabolic products of nucleic acids, urea and uric acid, can cause gout, a disease often associated with high living and rich foods.

18.10 MONOCULTURE

The techniques of modern high-capacity agriculture, namely mechanization, irrigation, fertilization, and pest control have tremendously increased food production and forestalled starvation for countless people. These practices are now essential, and the clock cannot be turned backwards. Nevertheless, our present system of agriculture has had some serious environmental side effects arising from these techniques. That modern techniques can or must be used in our agricultural system is due to another more basic agricultural practice used by modern man to increase food production. This practice, known as *monoculture*, consists of planting single crop species exclusively, usually in continuous and large land areas ranging up to thousands of acres. To understand the advantages and disadvantages of a monoculture, we should compare, say, the large wheat field ecosystem to the natural grassland or forest ecosystem replaced by the monoculture. In the natural ecosystem, there would be a variety of plant and animal species, called the community, existing together. This community would be stabilized and regulated both in kinds of species and population of any given species, by biological controls of competition and predation. The trophic structure or food chain or web in this natural system would be complex, reflecting the many possible kinds of competition between species for the same food supply and the many varieties of predator–prey relationships. The diversity and complexity of the living community confers stability, so that a shift in the population of any species is compensated by some other shift tending to return the population to its original level. A simple example would be the substitution of a different food source by a consumer if its favored food source disappeared or declined. Even in this natural ecosystem, man may be one of the species in the community.

When a grassland is plowed or a forest cleared and the land planted with a single plant species such as wheat, the original complex, stable community of organisms is replaced by a much simpler community. Wheat will be the only producer and the only food supply for the primary consumers. The number of these herbivore species will decline, but the populations of the remaining species (i.e., insects) will tend to increase massively because of less competition and predation. Likewise, the diversity of secondary consumers will decrease, but the populations of the remaining species may be larger than in the natural community. The net result is a simplification of the community and the trophic structure of the original ecosystem. Accompanying this simplification is a decrease in stability of the ecosystem, because the removal of biological controls makes excessive multiplication of a given species and hence, population instability, more likely.

Such ecological simplification in monoculture agricultural ecosystems is a normal and necessary part of food production. Having the entire crop in one place permits more efficient planting, maintenance, and harvesting procedures through mechanization. Greater productivity per acre results, because other unwanted plants are presumably not present. With plentiful harvests, man can obtain more of his food needs directly from plants; that is, he can shorten the food chain. By eating plants, man obtains ten times the amount of the energy entering an ecosystem than he obtains by eating animals that eat the plants. This is a consequence of the loss of energy at each step in a food chain.

But monoculture has disadvantages as well. The inherent instability of the simplified ecosystem, with the ever-present possibility that the entire single-species crop will be lost to drought, disease, or insects, puts man, as one of the consumers, in a precarious situation. By its nature, monoculture encourages massive population increases of plant-eating insects and other pests and discourages the presence of their natural predators. Our problems with pests and the necessity of pest control therefore follow naturally from our agricultural system.

A cultivated field is also unstable in the sense that such a simplified ecosystem will naturally revert, through a series of stages, to the more diverse ecosystem represented by the grassland or forest it replaced. This process is called *ecological succession* (see also Section 20.19). The so-called early stages of any ecosystem are characterized by simple community structure. In later stages, the community becomes more diverse until, in time, a stable ecosystem results. Much of the effort and energy required to maintain a monoculture is used to prevent its reversion to the stable, but less productive, state.

The solar energy entering an ecosystem and fixed photosynthetically as the chemical energy of organic matter over a given time period is known as the gross primary production (GP) of that ecosystem. Some of this energy R is used in the respiration of plants, animals, and microorganisms. The remaining energy, in the form of organic matter, is the net productivity (NP) of the system; $NP = GP - R$. The early successional stages of ecosystems are characterized by high net productivities and low respiratory energy losses. In later, more stable, stages R increases, and in the final stage R may equal GP, so that the net productivity is zero and no more energy is accumulated as food. Obviously, man must rely on young, simplified ecosystems for the production of his food. His agricultural practices are based on the artificial simplification of ecosystems (often by the destruction of the stable system) and the harvesting and removal of the net production. One way to look at part of the energy requirements of the monoculture system is in terms of man's efforts to keep respiratory energy losses at a minimum by continually blocking the natural tendency of R to increase. Respiratory losses will increase, and net production decrease, if organic matter is allowed to accumulate and decompose or if man's competitors (insects, rodents, birds, etc.) eat any of the crop. Man therefore spends large amounts of his energy and that of fossil fuels to harvest his crop and to wage war against his competitors.

In the more natural aged ecosystem much fertilization results from decomposition of a part of the crop, but in a monoculture this respiratory loss is undesirable and chemical fertilization is used. A chemical fertilizer may be regarded as a respiratory product imported from outside the monoculture system. Its application contributes to an artificial aging of the monoculture without loss in net productivity.

The practice of monoculture has somewhat the same effect on an ecosystem as pollution. Both simplify ecosystems and bring about large populations of certain tolerant organisms. The living community becomes unstable. Monoculture is essential for high efficiency, but its extension from small plots of earth to large spreads has necessitated the large-scale use of chemical pesticides and the leakage of these toxic agents out of the agricultural ecosystem into adjoining communities.

18.11 OTHER PEST CONTROL MEASURES

The pesticide problem is serious, because pest control is necessary for large-scale monoculture and for disease prevention. Since only 0.1 percent of the insects in a natural ecosystem are pests, most effort is directed toward the search for more target-specific biological and chemical pest control measures. Some of the more important methods are listed briefly here.

1. Insect hormones: the presence of natural or synthetic juvenile hormones at certain stages of an insect's life cycle interrupts normal development, preventing the insect from maturing and reproducing.

2. The infection of pests with viruses, bacteria, or parasites specifically lethal to the species. Also, the application of bacterial toxins to the pest.

3. Use of predatory insects such as the ladybug beetle and praying mantis.

4. Sterilization of insects by irradiation.

5. Development of chemical sex attractants (pheromones) or sex repellants.

6. Plant breeding to develop insect resistant plants.

7. Modified planting practices:
   a. Draw insects away from the crop species by planting some of their favorite food nearby.
   b. Crop rotation.

18.12 CHEMICAL ECOLOGY

Certain items in the above list (sex attractants and repellants, insect resistant plants, and juvenile hormones), if pursued further, would lead us into the field of *chemical ecology*. Many interactions between organisms involve chemical agents; the study of these interactions and of the chemicals on which they depend is the subject of chemical ecology. Among the pertinent interactions are those of defense, attack, feeding, mating, and other behavioral responses. Interactions may occur between members of the same species or between different species.

Chemicals secreted by one species to influence the physiological or behavioral response in other individuals of the same species are known as *pheromones*. Pheromones are used as alarm signals, sex attractants, recognition signals, and location or path markers. The pheromones, serving as messengers between organisms of the same species, are the principal means of communication among animals and plants. The intraspecific interactions represented by pheromones are usually of advantage to the one species involved.

Chemical interactions between different species are known as *allelochemic interactions*, and the chemical agents involved are allelochemic substances. Such interspecific chemical interactions usually confer a competitive advantage on the secreting species. Thus, many plants produce natural insecticides to discourage attack. Man uses some of these botanical insecticides (or their synthetic substitutes), such as nicotine sulfate and the pyrethrins (pyrethrum, originally from chrysanthemum flowers). As plants have developed mechanisms for discouraging insect attack, so have insects developed a proficiency for coping with natural insecticides. This same proficiency is brought to bear against man's synthetic insecticides. Several plant species contain appreciable amounts of various insect molting and juvenile hormones, or their chemical analogs, which can fatally affect the metamorphosis of feeding insects. Other examples of allelochemic effects are the defense mechanism of the bombardier beetle (see Problem 7, Chapter 16), production of antibiotics by lower organisms for their own protection, chemical defenses of animals against predators (venoms, bad odor or taste, irritants), and the inhibition of growth around certain plants by germination-inhibiting substances washed or vaporized from the leaves. References are listed at the end of the chapter for those who wish to learn more about the fascinating subject of chemical ecology.

18.13 SUMMARY   Green plants occupy the first trophic level of the grazing food chain. They capture and fix solar energy as the chemical free energy of complex reduced carbon compounds, which ultimately serve as food for all other heterotrophic organisms in the biosphere.

Green plants fix solar energy by photosynthesis. The overall photosynthetic process can be represented as $CO_2 + H_2O \longrightarrow [CH_2O] + O_2$, where H_2O is an electron donor and CO_2 is an electron acceptor. The overall process of formation of carbohydrate, $[CH_2O]$, is endoergic; the energy necessary to drive the reaction is provided by solar radiation. On the electronic level, photosynthesis involves the boosting of an electron in water to a higher energy state in $[CH_2O]$. Energy for respiratory metabolic processes is obtained from the oxidation of the reduced compounds.

Photosynthesis occurs in three broad stages. In stage I, water is dehydrogenated;

$$H_2O \longrightarrow \tfrac{1}{2}O_2 + 2H^+ + 2e^-$$

In stage II, light absorbed by pigment molecules such as chlorophyll is used to promote an electron to a higher energy level. During this stage, part of the absorbed energy is tapped off for the production of ATP and the reduction of $NADP^+$ to NADPH, the latter according to the equation

$$NADP^+ + H^+ + 2e^- \longrightarrow NADPH$$

Stage III is the reduction of CO_2 to a simple hexose sugar by NADPH,

$$2NADPH + 2H^+ + CO_2 \longrightarrow 2NADP^+ + H_2O + [CH_2O]$$

A number of substances may interfere with the photosynthetic process. Among these are certain herbicides and environmental pollutants. Other herbicides mimic plant hormones; when present in large amounts, these substances upset plant metabolism and development to the extent that the plant dies or the leaves fall.

In his attempt to increase food production, man simplifies agricultural ecosystems by the practice of monoculture. Such ecological simplification leads to instability. The tendency for monocultures to revert to more stable but less productive ecosystems requires that man invest large quantities of energy to hold the ecosystem in a simplified stage. The disruption of the ecology of populations and communities resulting from monoculture also forces man to rely heavily on pesticides for the control of his competitors.

Questions and Problems

1. At low light intensity and high CO_2 partial pressures, temperature has little effect on the rate of photosynthesis as measured by CO_2 uptake. At low CO_2 concentrations, however, an increase in temperature greatly increases the rate of photosynthesis. In another experiment, a plant in a CO_2-free atmosphere is illuminated. Then the light is turned off and CO_2 is supplied. At least for a short period in the dark, CO_2 is absorbed by the plant. Explain what both of these experiments indicate about the mechanism of photosynthesis (see also Problem 11, Chapter 17).

2. A colored film for application to greenhouse glass is said to increase growth of some species by 79 percent. What color is this product?

3. Photosynthesis proceeds by absorption of blue and red light, principally the latter.
   a. Calculate the energy of Avogadro's number of quanta (a mole or an *einstein*) of blue light (470 nm) and red light (650 nm).
   b. Does the absorption of light in photosynthesis result directly in the breaking of chemical bonds? Explain.

4. a. How many moles of quanta of red light (650 nm) are theoretically required for the fixation of 1 mol of CO_2 as $[CH_2O]$? Remember that quanta are indivisible.
   b. The actual quantum requirement of photosynthesis has been variously reported to be anywhere from 4 to 10. Taking 8 as a commonly accepted value, what is the thermodynamic efficiency of photosynthesis?

5. a. Would you expect the porphin portion of chlorophyll to be planar or puckered? Explain.
   b. In what region of the spectrum would chlorophyll most likely absorb radiation if its structure were not conjugated?

6. A tree may fix about 50 g of carbon per day at its greatest rate of photosynthesis. If the volume percentage of CO_2 in air is 0.03 percent, how many liters (STP) of air does the tree process in 1 day at this rate? How much oxygen does the tree contribute to the atmosphere per day?

7. Assume that 1×10^{11} tons of carbon are photosynthetically fixed as glucose each year on the earth's surface. If each mole of glucose requires 686 kcal for its formation and if the total solar energy falling on the earth's surface is 1×10^{21} kcal/yr, calculate the percentage of the solar energy that is absorbed annually by photosynthetic processes.

8. What is a scientific explanation for the cliche, "One rotten apple spoils the barrel"?

9. Explain, on the basis of chemical equilibrium, why storing fruit in an atmosphere with abnormally high concentrations of carbon dioxide or nitrogen will slow ripening of the fruit and permit longer storage time.

10. Discuss and relate the diversity, stability, entropy, and energy requirements of a monoculture ecosystem. Compare to a natural, or more mature, ecosystem.

Answers

3. a. blue, 60 kcal/mol; red, 44 kcal/mol
4. a. 3
   b. ~32 percent

6. ~310,000 liters; 134 g or 94 liters (STP) of O_2
7. ~0.09 percent

References

PHOTOSYNTHESIS

Bassham, J. A., The Path of Carbon in Photosynthesis, in *"The Living Cell,"* D. Kennedy, Ed., Freeman, San Francisco. Reprinted from *Scientific American*, June 1962.
  A discussion of stage III in photosynthesis.
Clayton, Roderik K., "Light and Living Matter," Vol. 2: "The Biological Part," McGraw-Hill, New York, 1971.

Levine, R. P., "The Mechanism of Photosynthesis," *Scientific American*, **221,** 58–70 (December 1969).
Rabinowitch, Eugene, and Govindjee, "Photosynthesis," Wiley, New York, 1969.
  A comprehensive book with good introductory material on the history of work on photosynthesis, thermodynamics, and other aspects of the general chemistry of photosynthesis.

Rabinowitch, Eugene, and Govindjee, The Role of Chlorophyll in Photosynthesis, in "The Molecular Basis of Life," by R. H. Haynes and P. C. Hanawalt, Freeman, San Francisco. Reprinted from *Scientific American*, July 1965.

Rosenberg, Jerome L., "Photosynthesis," Holt, Rinehart and Winston, New York, 1965.

FOOD

Brown, Lester R. and Finsterbusch, Gail W., "Man and His Environment: Food," Harper & Row, New York, 1972.

Brown, Lester R., "Human Food Production as a Process in the Biosphere," *Scientific American*, **223,** 161–170 (September 1970).

   Summarizes many aspects of the world food problem.

Champagnat, Alfred, "Protein from Petroleum," *Scientific American*, **213,** 13–17 (October 1965).

Gates, David M., "The Flow of Energy in the Biosphere," *Scientific American*, **224,** 89–100 (September 1971).

Handler, Philip, Ed., "Biology and the Future of Man," Oxford, New York, 1970, Chapter 15.

Woodwell, George M., "The Energy Cycle of the Biosphere," *Scientific American*, **223,** 64–74 (September 1970).

   How agriculture simplifies the ecosystem and how energy must be expended to maintain the agricultural ecosystem. A good reference for Chapter 20 also.

CHEMICAL ECOLOGY

Sondheimer, E., and Simeone, J. B., Eds., *"Chemical Ecology,"* Academic, New York, 1970.

Whittaker, R. H., and Feeny, P. P., "Allelochemics: Chemical Interactions Between Species," *Science*, **171,** 757–770 (1971).

MISCELLANEOUS

Epstein, Samuel S., "A Family Likeness," *Environment*, **12,** 16–25 (July–August 1970).

   The teratogenic effects of 2,4,5-T and other herbicides.

Williams, Carroll M., "Third-Generation Pesticides," *Scientific American*, **217,** 13–17 (July 1967).

The Reproduction of Life

The problem of defining life was approached in Chapter 15 by listing some of the characteristics of living organisms. Of these characteristics, the one chosen as most closely approaching a definition of life was the ability of complex living systems to reproduce their kind exactly, yet with allowance for change or adaptation through evolution. It is time now to delve into the chemical and molecular basis of this property of heredity. We shall see that the unique properties of any organism—whether it is a horse or a man and whether its hair is black or brown—depend on the nature of the protein enzymes its cells can synthesize. The availability of these proteins depends on the functional units of heredity, the genes, which have been passed on to the organism from its predecessors.

The essential components of genes, which reside in the cell nucleus, are nucleic acid molecules. The genetic instructions for protein synthesis are stored in the chemical structure of DNA. Other nucleic acid molecules serve as the go-betweens from the master instruction repository in the genes to the protein synthesis apparatus of the cell.

Proteins, then, are like the foreman in a cellular chemical factory. As enzymes, they are directly responsible for the proper operation of the cell at any given moment and, as we have seen, interference with the functioning of an enzyme may cause immediate death or injury to the organism. The nucleic acids, on the other hand, are like the master blueprints, which hold the instructions for the operation of a cell and for the construction of new cells. Subtle changes in the structure of the genetic material can lead not only to immediate effects, but to delayed effects, both good and bad, that can be propagated to future generations.

Molecular Genetics

19.1 THE STRUCTURE OF DNA

The primary structure of one of the two classes of nucleic acids, the deoxyribonucleic acids (DNA), was given in Section 15.19. DNA consists of a long polynucleotide chain with a backbone of alternating sugar (deoxyribose) and phosphate groups (Figure 15.18). Attached to each sugar is one of the four kinds of nitrogen-containing organic bases; the purines, adenine (A) and guanine (G), and the pyrimidines, thymine (T) and cytosine (C) (see Figures 15.21 and 15.23). Like the sequence of amino acids in proteins, the exact linear sequence of bases in a nucleic acid molecule determines the detailed function of that molecule. The sequence of nucleotides (—phosphate—sugar—base—) or of bases alone establishes the primary structure of DNA, which varies from species to species and individual to individual. Over a million nucleotide monomer units may be strung together in a single DNA strand to give (in *E. coli*, a coliform bacteria) a strand with a molecular weight of about 2 billion and a length of around 1 mm. Contrast this with an average protein molecule, which has roughly 300–500 amino acid units and a molecular weight of about 40,000.

The secondary structure of DNA results from the wrapping of two polynucleotide strands about a common axis. Each strand has a helical form similar to the α-helix in proteins. The two DNA strands wind about each other to form a double helix (Figure 19.1). This threadlike structure of DNA had been predicted for some time on the basis of various physical and optical properties, but the direct evidence for an ordered helical arrangement was provided by X-ray diffraction. In 1953, James D. Watson and Francis Crick showed that the X-ray patterns could be explained on the basis of a double helix in which the two DNA strands were held together by hydrogen bonds between their bases. They did this by building scale models incorporating the known bond angles, atomic radii, and bond distances of the individual DNA chains and fitting these chains together on a trial-and-error basis until the parameters provided by the X-ray pattern were reproduced. Watson, Crick, and the X-ray crystallographer, Maurice Wilkins, were awarded the 1962 Nobel prize in medicine and physiology.

To appreciate fully the shape of the double helix and its importance in genetics, we must look again at the organic bases attached to the DNA

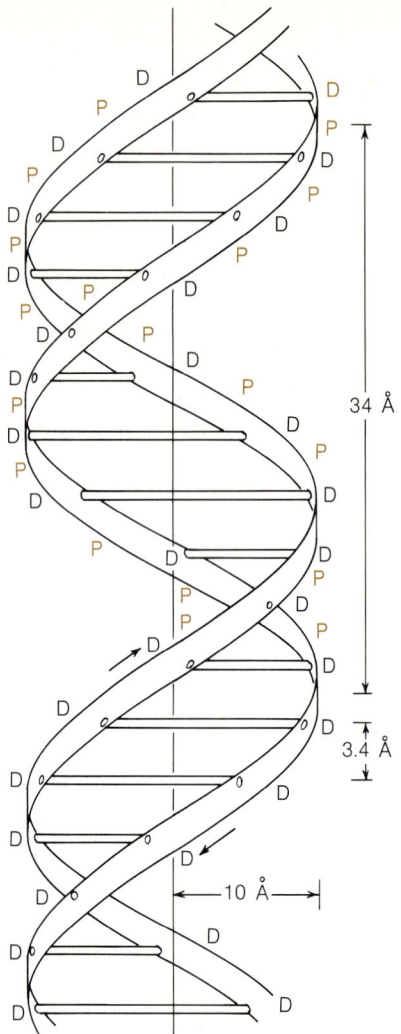

Figure 19.1. The DNA double helix. In this simplified sketch, the letters P and D represent the phosphate and deoxyribose groups along the backbone of each strand. The horizontal bars represent the pairing of base groups attached to deoxyribose units on opposing chains.

backbone. The bases are planar and point inward toward the axis of the double helix. Since the bases are hydrophobic, they tend to exclude water from the interior of the molecule. The planes of the bases are perpendicular to the axis of the double helix and parallel to each other, like the steps in a spiral stairway. A base on one strand is always faced, on the same level, by a base on the opposite chain. The paired bases on a given level are 3.4 Å above and below the planes of the paired bases on adjacent levels. The double helix makes a complete turn every 34 Å; that is, there are ten nucleotide units in each turn, each unit (or base pair) being rotated in the horizontal by 36° from the base pair next to it.

An essential feature of the Watson–Crick double helix is the restriction placed on the paired bases. Thymine (T) can pair only with adenine (A), and cytosine (C) can only be opposite guanine (G). The structural basis for these pairing restrictions can be seen from Figure 19.2. Only with A–T and

C–G pairs will the hydrogens on one base point directly at the electronegative oxygen or nitrogen on the other base so that the base pairs will be held together by the maximum number of hydrogen bonds. In addition to hydrogen bonding considerations, size factors show that interchain pairing restrictions have been

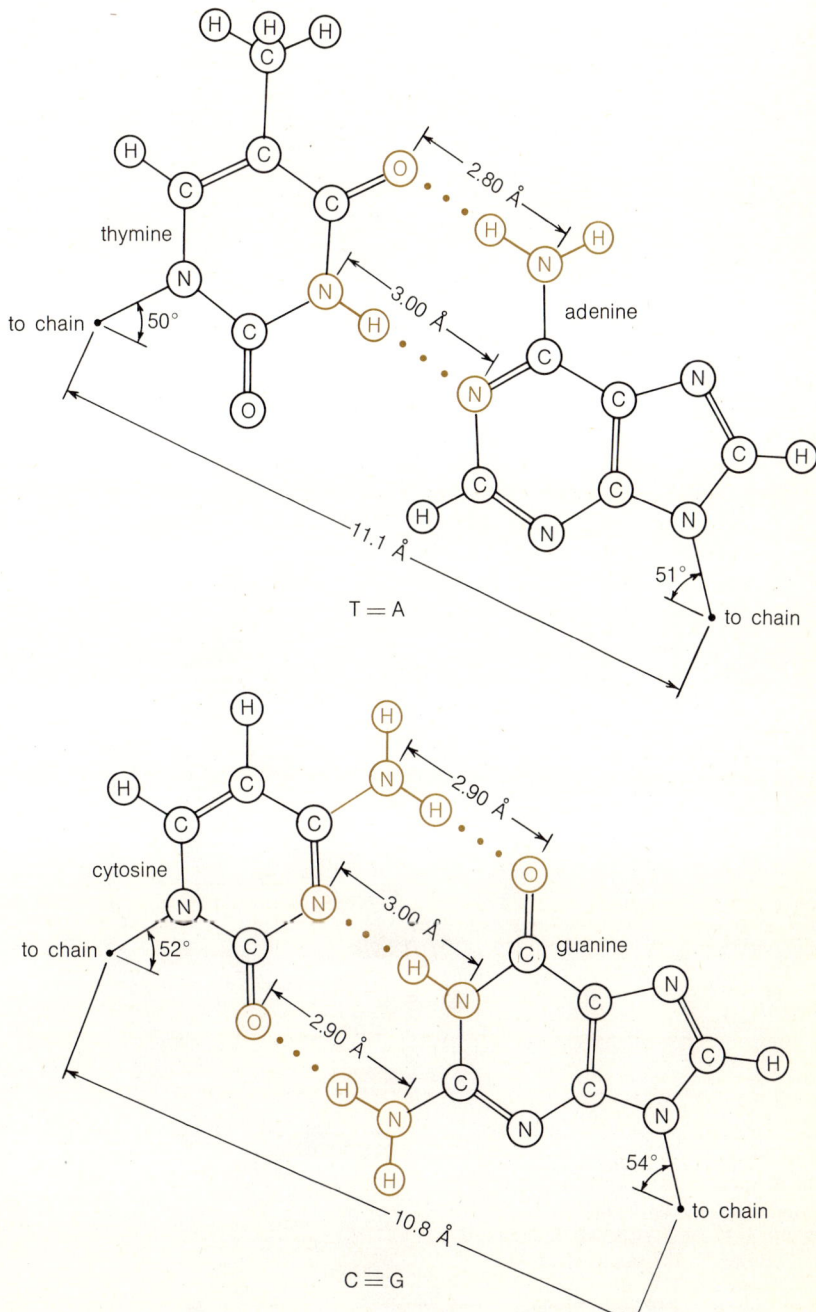

Figure 19.2. Base pairing through hydrogen bonds.

imposed. The diameter of the double helix is a constant 20 Å along its length. If large purine bases (A or G) paired with each other, leaving the two smaller pyrimidine bases (T or C) to pair up, the helix diameter would vary. Only A–T and C–G pairs would be of similar size and permit a constant helix diameter. Any other pairing scheme leads to helix distortion (A–G) or weaker

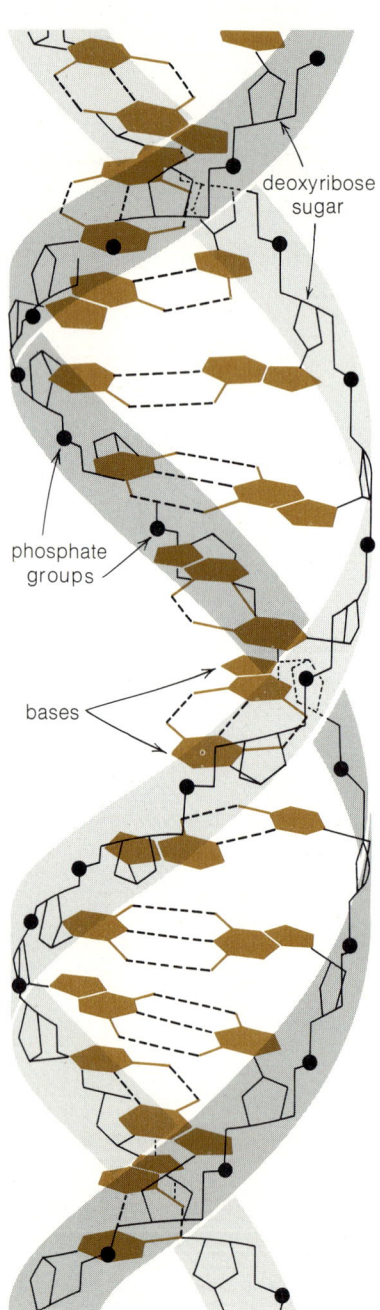

deoxyribose
sugar

phosphate
groups

bases

Figure 19.3. The DNA double helix in more detail, showing the arrangement of the hydrogen bonded base pairs. (Adapted from Philip C. Hanawalt and Robert A. Haynes, "The Repair of DNA," *Scientific American*, **216**, 38 [February 1967]). Copyright © 1967 by Scientific American, Inc. All rights reserved.

hydrogen bonds due to loss of close contact between bases (C–T). A more detailed illustration of a segment of a double-helical DNA molecule is shown in Figure 19.3. The matching of A with T and of C with G agrees with earlier chemical analyses showing a one-to-one ratio between A and T and between C and G in double-stranded DNA.

The two strands of DNA in a double helix are *complementary* by virtue of the restrictive base pairings. The sequence of bases in one chain determines the sequence of bases in the other, since A can be opposed only by T and C by G. That one strand is the complement of the other is of the greatest importance for the biological duplication, or replication, of DNA in cell division, as we shall see shortly.

In addition to secondary structure, some double-stranded DNA may exhibit tertiary and quaternary structure. For instance, in many cells and most viruses, the DNA occurs in a closed ring form, but we shall have no reason to delve into this aspect of DNA structure.

19.2 CHROMOSOMES, GENES, AND DNA

Chromosomes are rodlike structures found in the nuclei of cells. When the cell divides, the chromosomes duplicate themselves so that each of the two daughter cells gets an identical set of chromosomes. Some bacterial cells and viruses have only one chromosome; human cells contain 46 chromosomes, except for the sex or germ (sperm and egg) cells, each of which has 23 chromosomes, and mature red blood cells, which have none.

It is well established that chromosomes consist of double-stranded DNA molecules (except for those viruses whose genetic material is single-stranded DNA or RNA). In the cells of higher plants and animals, the DNA may be associated with up to 50 percent of a class of small protein molecules known as histones. Linearly arrayed along the chromosome are the genes, segments of the DNA molecule associated with some inheritable trait. A gene may be from 300 to 4500 nucleotide or base pairs long, so that one chromosomal DNA molecule may contain many genes along its length, as illustrated in Figure 19.4. The single chromosome in *E. coli* contains some 3000 genes. Each gene

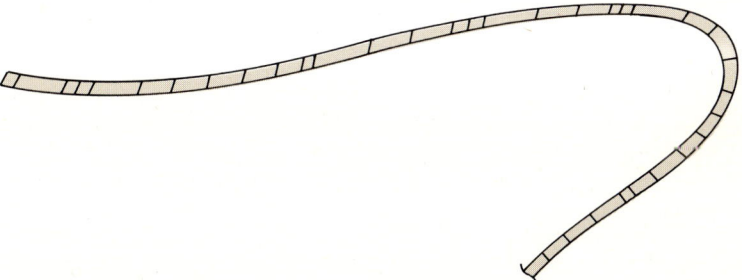

Figure 19.4. Representation of a chromosome with linear sequences of genes along its length.

directs the synthesis of a single polypeptide chain in the organism. These polypeptides are enzymes that control the chemistry of the metabolism, growth, and development of the organism.

Each cell of an organism has, in general, a complete complement of chromosomes and therefore of genes. Since the chromosomes and their subunits, the genes, are DNA molecules, there must be some mechanism to ensure that the two cells arising from the division of a parent cell each receive identical copies of the genetic material, the DNA, in the one parent cell. Thus, the paramount function of DNA is somehow to form copies of itself to pass on to new generations of cells, while a second function is to direct the chemical activities of each cell through the synthesis of enzymes.

19.3 REPLICATION OF DNA:
DNA ⟶ DNA

When a cell divides, the individual strands of the chromosome double helices separate. Each strand then acts as a template or mold for the formation of a new companion strand. The end result is the formation of two new chromosomes, each of which has one strand from the original chromosome. This process of reproduction by the synthesis of complementary strands is called semiconservative *replication*.

The two DNA strands in the double helix are intertwined in such a way that they can be separated only by unwinding the strands. This unwinding can be quite rapid, perhaps some 15,000 rev/min, considering the length of the strands and the short duplication time of some cells. As the separation of the strands proceeds by the easy breaking of the weak hydrogen bonds between the strands, free nucleotide units in the cell nucleus become attached by hydrogen bonds to their complementary bases in each strand.

As each nucleotide unit attaches to the base in one of the separating strands, it is joined to the adjacent, previously attached, nucleotide unit by an enzyme, DNA-polymerase. There are now two new but identical double strands instead of one. The identity of the two new double helices to each other and to the original DNA is ensured by the restrictions on base pairing, A with T and C with G, in the formation of the complementary strands. The unwinding, base pairing, and polymerization steps are depicted in Figure 19.5.

This mechanism of DNA replication is called *semiconservative*, because each daughter molecule conserves only one strand from the original molecule. The mechanism has been verified by replacing ^{14}N in the original molecule with the heavier isotope, ^{15}N, and determining the density of the chromosomal DNA of bacteria after successive generations. The results are shown in Figure 19.6. Only the semiconservative mechanism explains these results.

The replication of DNA explains how identical copies of the genetic information in the chromosomes are transmitted from one cell to another during division. We now turn to the way the instructions contained in the DNA molecules are encoded and how they control the synthesis of proteins within a single cell.

19.4 THE BIOSYNTHESIS
OF PROTEINS

If the DNA of the chromosomes and genes directs the synthesis of proteins, the information specifying the composition of the protein must be encoded in the DNA. This information is contained in the sequence of nucleotides (or bases) along one of the DNA strands in the double helix. Nature uses a sequence of three bases, such as ATG, to specify one, and only one, of the 20 amino acids. A gene corresponds to one polypeptide chain, and if the average protein chain is 300–500 amino acids long, the average gene will contain 900–1500 nucleotide pairs. The sequence of three nucleotide bases coding for each amino acid is known as the *genetic code*. The code will be discussed in more detail after the general features of protein synthesis are introduced.

The DNA stays in the cell nucleus and is not involved directly in the synthesis of proteins at the ribosomes (Section 16.1) outside the nucleus. Instead, copies of the information contained in selected genes (not the entire chromosome) are made by a process known as *transcription*. In transcription,

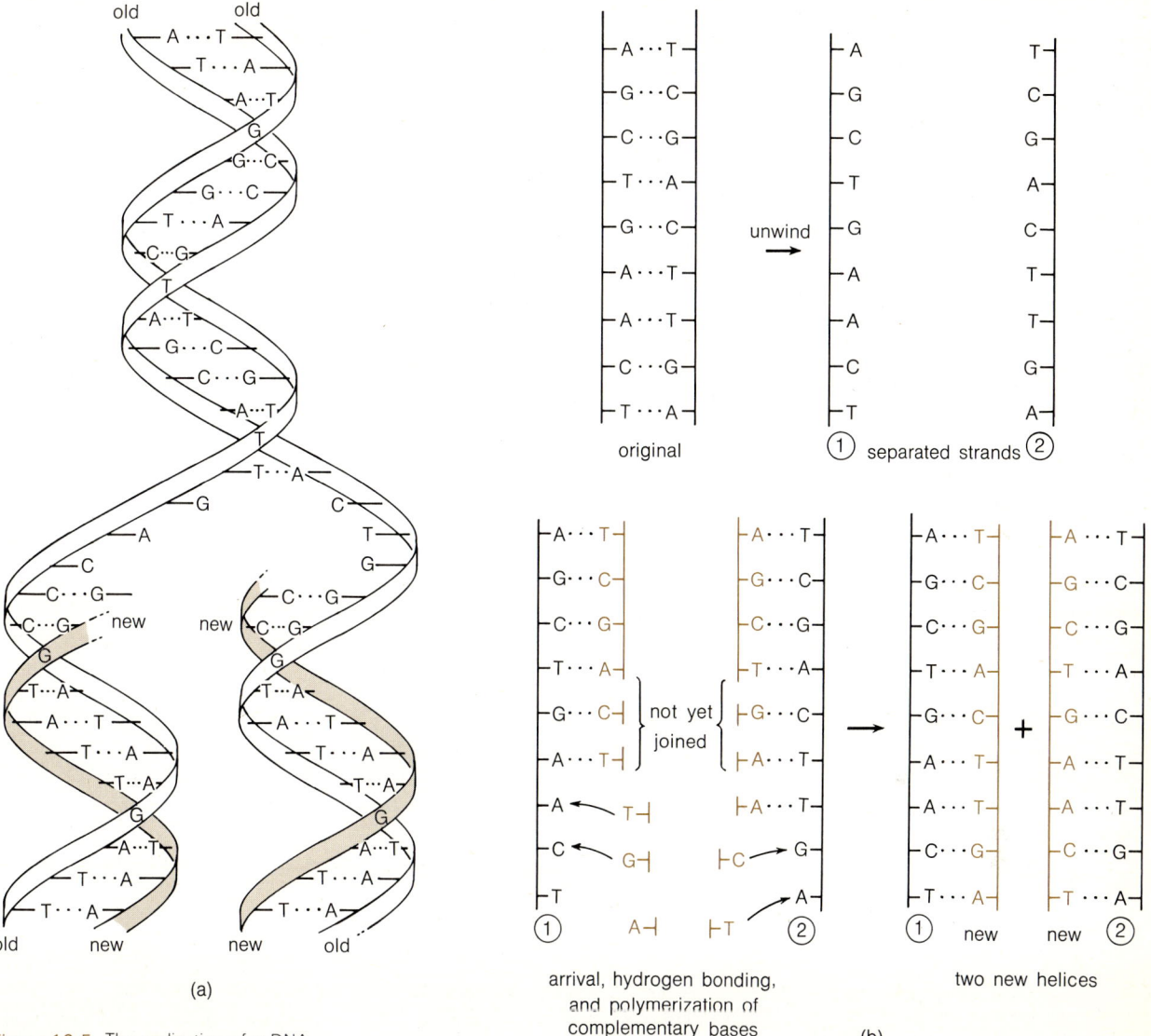

(a)

arrival, hydrogen bonding,
and polymerization of
complementary bases

two new helices

(b)

Figure 19.5. The replication of a DNA double helix begins with separation of strands. Nucleotide units in the medium attach themselves to the proper bases on the strands and are joined to form two new helices. [Some authors use the letters A, T, G, and C to represent the complete nucleotide units (adenylic acid, thymidylic acid, guanylic acid, and cytidylic acid). The chains are then written as —A—G—C—T—G————] (Part (a) from James D. Watson, "Molecular Biology of the Gene," 2d ed, W. A. Benjamin, Inc., Menlo Park, Calif., 1965. Copyright © 1965 by J. D. Watson.)

the information in the DNA is transferred to a smaller ribonucleic acid (RNA) molecule in the form of a complementary nucleotide triplet. The message carried out of the nucleus by these RNA replicas is then *translated* at the ribosome into amino acid units that are joined together to form the protein. The sequence of events for the synthesis of a protein in a cell is

$$\text{replication}\left\{\begin{array}{c}\text{daughter}\\ \text{DNA} \xrightarrow{\text{transcription}} \text{RNA} \xrightarrow{\text{translation}} \text{protein}\\ \text{parent}\end{array}\right.$$

where the replication process is also included.

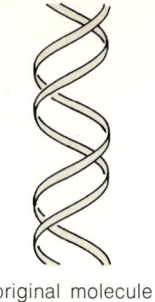

original molecule

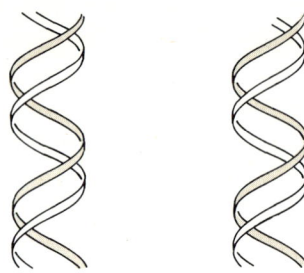

first generation molecules

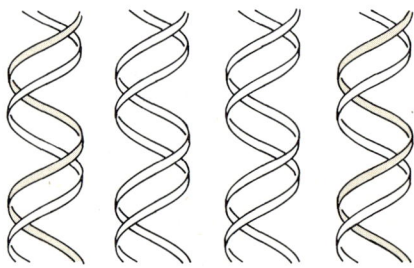

second generation molecules

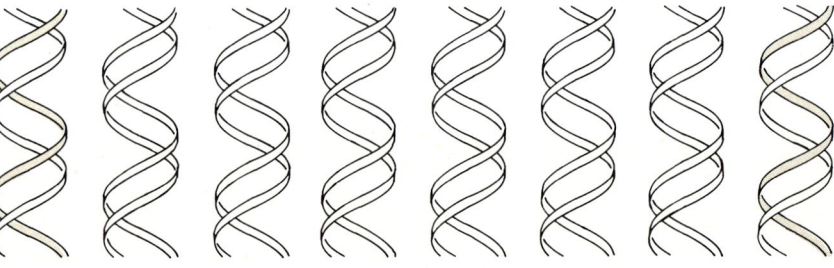

Figure 19.6. The replication of DNA by the semiconservative mechanism. The DNA in the first generation shows only one density, intermediate between that of the normal "light" and the labeled "heavy" DNA. In the second generation, both intermediate and light DNA are found. In succeeding generations, greater amounts of the light variety are produced.

third generation molecules

19.5 TRANSCRIPTION: DNA $\longrightarrow$ RNA

In the transcription process, the DNA molecule transfers its genetic information to another class of nucleic acid molecule, which serves as the template or mold for protein synthesis. These working copies of the DNA are ribonucleic acids (RNA). RNA is similar to DNA in being a linear polymer of four different nucleotides, except that the sugar in RNA is ribose instead of deoxyribose and the base thymine (T) in DNA is replaced by uracil (U) in RNA (Section 15.19). A portion of an RNA molecule is illustrated in Figure 19.7.

Transcription, then, is the synthesis of an RNA molecule complementary to one of the strands of DNA. The base pairing scheme C–G still holds, but the A–T pair is replaced by A–U, because U, also being a pyrimidine, hydrogen bonds with A just as well as T does.

Transcription is similar to DNA replication. The DNA double helix opens, and the two strands separate. Only one strand serves as the template for the synthesis of a new strand of complementary RNA (Figure 19.8). How one DNA strand is selected for copying and not the other is unclear but, presumably, one DNA strand contains a specific base sequence that is recognized as a start signal by the RNA-synthesizing enzyme. Nucleotide units in the nucleus insert themselves, by hydrogen bonds, along the template DNA strand according to the base pairing rules, with a U on a ribonucleotide unit pairing with an A on the DNA template strand. The ribonucleotide units are joined together into an RNA polymer by the enzyme RNA-polymerase. The synthesized RNA molecule peels away from the hybrid double helix it has temporarily formed with the template strand. The base sequence in the RNA is complementary to that in the DNA template strand and identical to that in the other DNA strand, except that U's substitute for the T's so that the RNA carries the equivalent information, in a slightly different alphabet, as the DNA.

The single-stranded RNA molecule so made carries the coded message for the amino acid sequence to the protein manufacturing units, the ribosomes, in the cytoplasm of the cell. This particular RNA molecule is therefore called *messenger RNA* (mRNA). There are as many of these large molecules as there are genes. The message carried by the mRNA is still coded in triplet sequences of nucleotides (or bases) and must be translated into amino acid language before it can be used for protein synthesis. Decoding and protein synthesis are carried out in the translation step.

19.6 TRANSLATION OF mRNA: RNA $\longrightarrow$ PROTEIN

The translation process seems to be more complex than transcription. Only the general features are given here. Translation again involves a template process whereby a new molecule, *transfer RNA* (tRNA), is attached to the mRNA by hydrogen bonding between complementary base pairs. Each tRNA molecule carries with it an amino acid, and the tRNA and its associated amino acid are aligned in the order directed by the base sequence code on the mRNA. The aligned amino acids are joined together by enzymes to form a protein molecule.

Translation begins with the covalent attachment of an individual amino acid in the cytoplasm to a small (70–80 nucleotides) tRNA molecule. There is a specific tRNA molecule for each amino acid, but all tRNA's have similar tertiary structures resulting from hydrogen bonding between bases in the same molecule. The distinctive folded, cloverleaf shape of a tRNA molecule is shown in Figure 19.9. In addition to the major bases, U, A, G, and C, tRNA

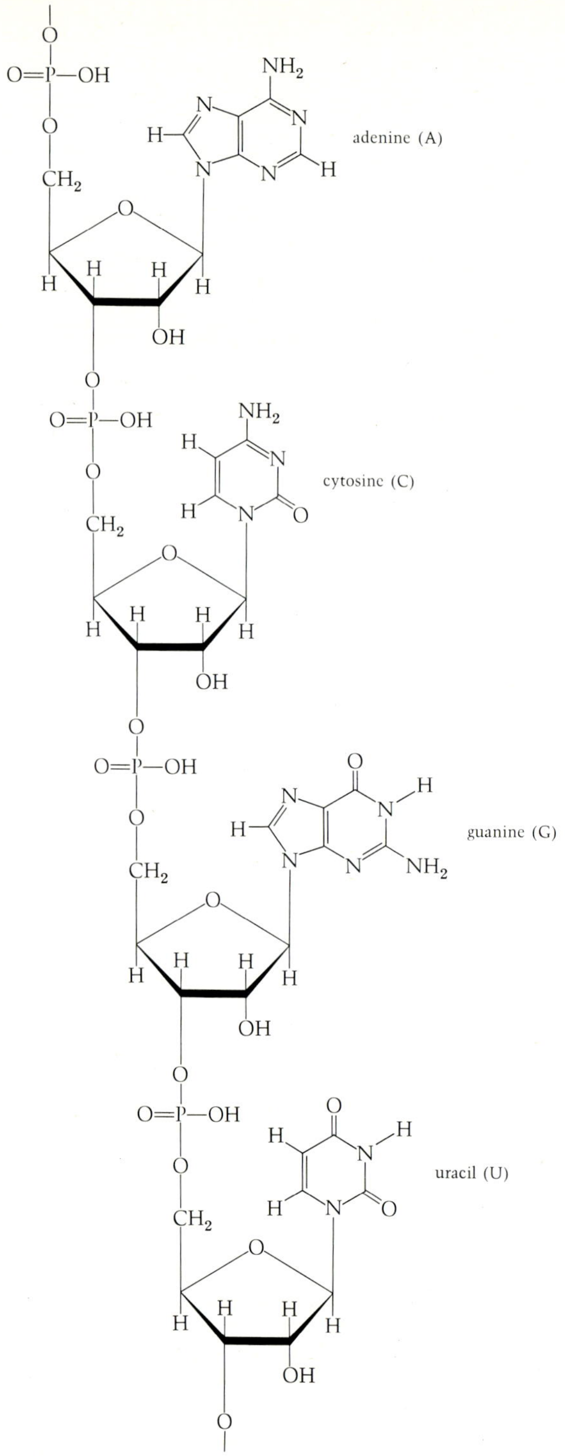

adenine (A)

cytosine (C)

guanine (G)

uracil (U)

Figure 19.7. A segment of ribonucleic acid, RNA.

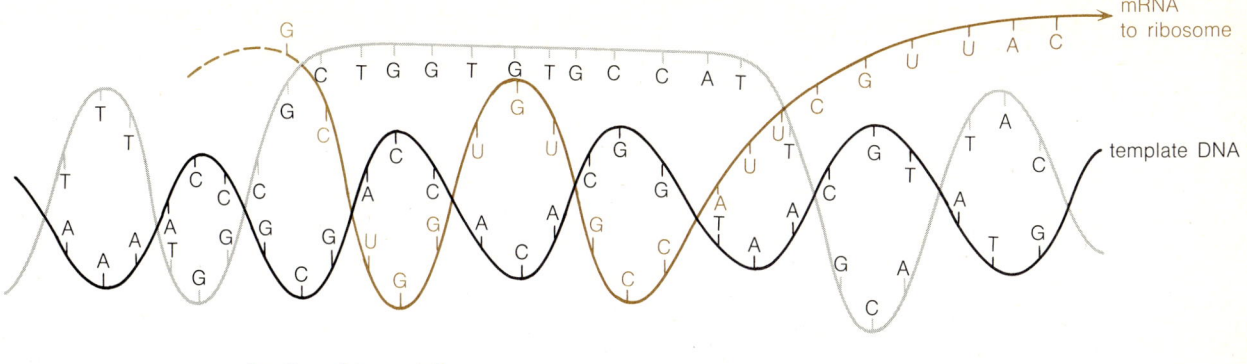

mRNA
to ribosome

template DNA

⟵——————— direction of transcription

Figure 19.8. Transcription of DNA into
complementary mRNA. In transcription,
mRNA is synthesized when the two
strands of DNA separate, and one of
them serves as a template for the assem-
bly of nucleotides into a mRNA molecule. The polymerization of mRNA is catalyzed by the enzyme RNA-polymerase.
The mRNA peels off as it is formed. One
end of the RNA may be attached to a
ribosome.

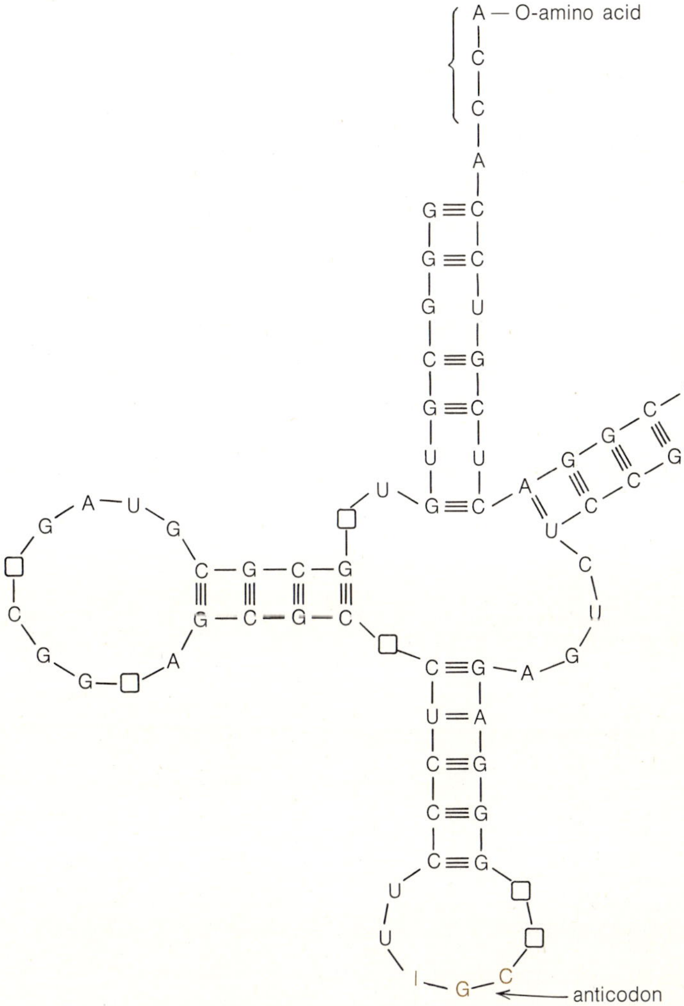

Figure 19.9. Tertiary structure of a
transfer RNA molecule. The molecule is
held in characteristic folded form by intrachain base pairing. The specific amino
acid unit is attached at the upper arm.
The base triplet at the loop of the lower
arm is the *anticodon* corresponding to a
complementary triplet *codon* on the
mRNA. A particular codon recognizes
only one specific anticodon and thus determines the nature of the amino acid at
the other end of the tRNA molecule. The
remaining two arms of the cloverleaf
are probably used for binding of the
unique enzyme needed to bind the specific amino acid to the particular tRNA
and for attachment of the tRNA to the
ribosome. Unlettered spaces represent
the unusual bases often found in tRNA,
of which I (inosine), appearing in the
anticodon here, is one example.

A — O-amino acid

anticodon

molecules also contain some less common bases. The amino acid is enzymatically bound to the end of one arm of the cloverleaf. At the loop at the end of another arm is a sequence of three bases, the *anticodon*, complementary to a base triplet, the *codon*, on the mRNA template. The tRNA, carrying a specific amino acid unit on its other end, attaches itself to the complementary or matching triplet of bases on the mRNA. Each amino acid-specific tRNA molecule has a unique anticodon with the proper base triplet permitting it to hydrogen bond with a codon on the mRNA. Thus, if the codon on mRNA is GUG, then only the tRNA with the anticodon CAC can position itself at that point of the mRNA. The codon for each amino acid is known, and this one is for the amino acid valine.

The matching of mRNA codon and tRNA anticodon and the synthesis of protein occur on the ribosomes. Ribosomes are large, two-unit particles on which tRNA and mRNA are bound and brought into contact. The mechanism is shown in Figure 19.10. The mRNA might be thought of as passing through the channel between the subunits of the ribosome, just as a tape passes between the playback heads of a tape recorder. Alternatively, the ribosome can be considered to move along the mRNA, matching tRNA molecules, with their bound amino acids, to the successive base triplet codons of the mRNA. The moving ribosome concept is used to explain Figure 19.10.

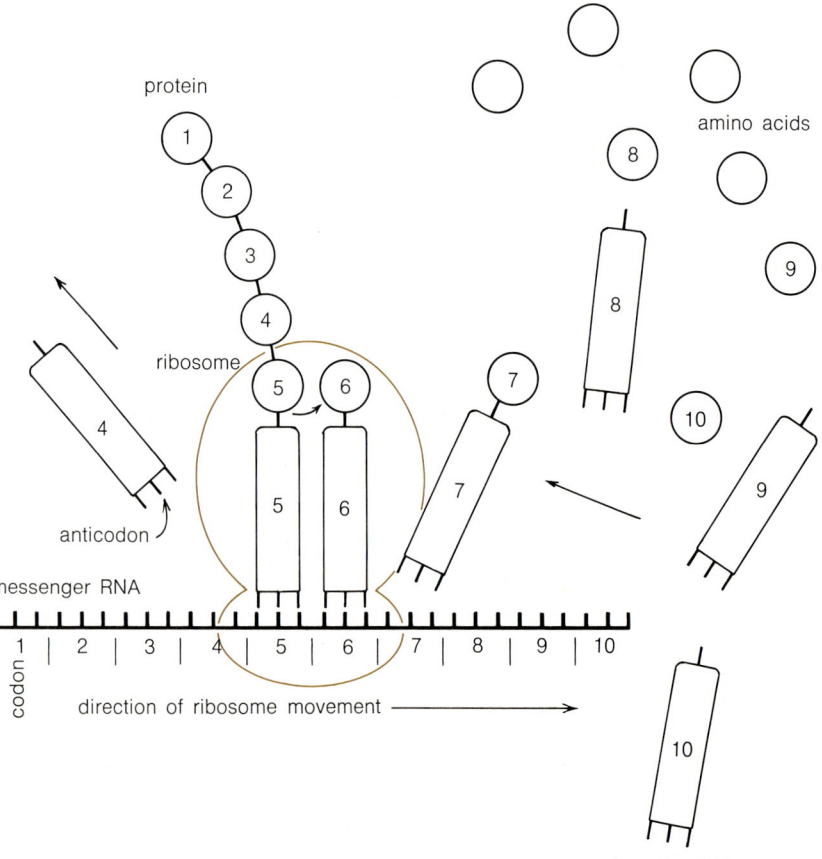

Figure 19.10. Mechanism of mRNA translation by tRNA on ribosomes. Two tRNA molecules (simplified in shape here) are attached to the larger subunit of the ribosome. The earlier tRNA (5) holds the growing polypeptide chain, while the newly arrived tRNA, having an anticodon complementary to the mRNA codon, has brought amino acid 6 into contact with the end of the growing chain. The polypeptide chain is enzymatically transferred and bonded to amino acid 6. The ribosome then moves on to cover codons 6 and 7, and the transfer process takes place again as the empty 5 tRNA moves off the mRNA. (From Masayusa Nomura, "Ribosomes," *Scientific American*, **223**, 30 [October 1969]). Copyright ©️ 1969 by Scientific American, Inc. All rights reserved.

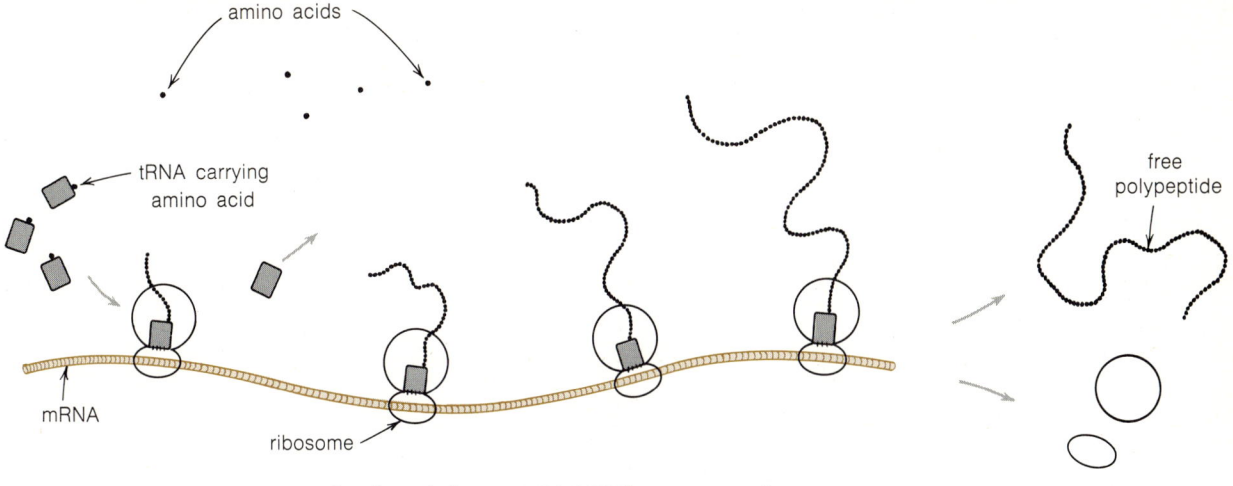

amino acids

tRNA carrying
amino acid

free
polypeptide

mRNA

ribosome

direction of ribosome movement ⟶

Figure 19.11. Several ribosomes, collectively called a polyribosome, traverse a mRNA molecule. Each ribosome reads the information (codons) from the mRNA, successively adding amino acid units onto the growing polypeptide chain. At the end of the genetic message on the mRNA, a codon signal terminates the synthesis and the polypeptide is released from the ribosome.

Usually, several ribosomes move along the mRNA simultaneously with polypeptide chains in various stages of elongation on the ribosomes, as shown in Figure 19.11. On each ribosome, the polypeptide chain grows as the ribosome moves along the mRNA, much as a long wood shaving might curl out of a wood plane as the plane is moved down a board. During its growth, the polypeptide begins folding into its functional tertiary structure as determined by the sequence of amino acids along its length. The primary structure of the protein is, of course, ultimately determined by the sequence of bases on the DNA.

19.7 THE GENETIC CODE

To complete the story of protein synthesis, it is necessary to specify not only how a code written in the four bases A, U (or T), G, and C is read, but how the code words are translated into another language of the 20 amino acids in proteins. The genetic code is the dictionary used to translate the four-letter language of nucleic acids into the 20-letter language of proteins. The four letters of the nucleic acid language are commonly those of the mRNA codons, A, U, G, and C, since these uniquely specify the base sequence in DNA.

Code words of at least three letters are needed to specify the 20 common amino acids, for words of only one letter could uniquely stand for only four amino acids; words of two letters (AU, UA, AC, CA, etc.) could specify only 16 amino acids; while words of three letters (AUU, UAU, UUA, . . .) can code for 64 amino acids. Therefore, a triplet code more than suffices.

The genetic code is given in Figure 19.12. The assignment of base triplets to amino acids was accomplished principally by the addition of synthetic mRNA of known base sequence to isolated (cell-free) ribosomes, followed by determination of the amino acid composition of the polypeptide produced or determination of the amino acid composition of the tRNA's absorbed on the ribosomes.

The code is read sequentially, without any punctuation marks between the base triplet codons. Starting at one end of the mRNA, the message is read three bases at a time until a termination codon is reached. The absence of punctuation marks has important consequences in the study of mutations, and these will be considered shortly.

second letter

		U	C	A	G	
	U	UUU UUC } phe UUA UUG } leu	UCU UCC UCA UCG } ser	UAU UAC } tyr UAA term. UAG term.	UGU UGC } cys UGA term. UGG try	U C A G
	C	CUU CUC CUA CUG } leu	CCU CCC CCA CCG } pro	CAU CAC } his CAA CAG } gln	CGU CGC CGA CGG } arg	U C A G
	A	AUU AUC } lleu AUA AUG (initiate)	ACU ACC ACA ACG } thr	AAU AAC } asn AAA AAG } lys	AGU AGC } ser AGA AGG } arg	U C A G
	G	GUU GUC GUA GUG } val	GCU GCC GCA GCG } ala	GAU GAC } asp GAA GAG } glu	GGU GGC GGA GGG } gly	U C A G

first letter · third letter

Figure 19.12. The genetic code. All but three of the triplets code for an amino acid. Some amino acids have more than one code word; the code is degenerate. The three code words UAA, UAG, and UGA act as signals for termination of polypeptide chains. All chains begin with AUG (met), although the methionine residue may be absent from the completed polypeptide chain. Thus, AUG is a chain initiation signal. If a portion of the base sequence in mRNA is AAGAGUCCAU-CACUUAAU, the corresponding amino acid sequence in the protein made from this portion will be -lys-ser-pro-ser-leu-asn-. The code is universal for all living things.

19.8 REVERSALS AND INTERFERENCE IN INFORMATION TRANSFER

As presented here, genetic information flows only from DNA to RNA to proteins by a series of irreversible steps. This unidirectional process of DNA ⟶ RNA ⟶ protein is known as the *central dogma* in the transmission of genetic information and, as such, was believed to be inviolable. There is now evidence that there is an enzyme that permits RNA to act as a template for the synthesis of DNA. In this case, the RNA carries information that can be transmitted to DNA, thus determining the base sequence in DNA. The synthesized DNA, in turn, can either replicate or serve as a template for the normal transcription of RNA and protein. Evidence that RNA serves as a template for the enzymatic synthesis of DNA (reverse transcription) has been obtained by using various tumor-producing viruses having a chromosome made of RNA instead of DNA. Such viruses, unlike other RNA viruses, which replicate their RNA directly into new copies of RNA and then to protein, can use an RNA ⟶ DNA ⟶ RNA ⟶ protein sequence. (The central dogma has been extended to include the RNA ⟶ DNA transfer as a "special transfer." The only transfer that is specifically prohibited is that beginning with protein.) So far, the evidence for reverse transcription appears to be limited to RNA viruses that cause tumors and cancerous transformations. Thus, the process may be linked to cancer, perhaps by the incorporation of the synthesized viral DNA into the DNA of animal cells. We shall not pursue this subject, but two references are listed at the end of the chapter.

The transfer of genetic information from cell to cell and from nuclear chromosome to protein structure is a complicated process that involves many template structures and enzymes. There are many places where interference with the information transfer may occur with possible lethal consequences for the cell. Some of the enzymes used to attach specific amino acids to specific tRNA's can be made to accept molecules other than the proper amino acids and to incorporate these molecules into proteins. Other competitive inhibitors have different mechanisms; that is, they do not substitute for the amino acids in the polypeptide chain.

Most of the antibiotics exert their influence somewhere in the information-transfer process. Actinomycin D binds to the DNA and interferes with the action of RNA polymerase, thereby blocking transcription (mRNA synthesis). Another antibiotic accomplishes the same thing by binding to the enzyme rather than the DNA. Some other antibiotics cause premature polypeptide chain termination at the ribosomes or interfere with the transfer of the growing polypeptide chain to the next amino acid. Streptomycin inhibits polypeptide chain initiation and causes errors in the reading of the codons on mRNA. So far as humans are concerned, all of this is fine if the affected cells belong to rapidly multiplying pathogenic bacteria. Many antibiotics, however, suffer from some of the same drawbacks as pesticides. They can kill beneficial bacteria in humans, and pathogenic bacteria can develop resistance to antibiotics.

Lethal antibiotic effects make themselves felt in a single cell generation. More insidious and longer lasting are the effects resulting from small changes in the master instructions encoded in the DNA molecules of the genes.

19.9 MUTATIONS

Compared to most of the large molecules in a living system, DNA is quite stable, as it should be if it is to carry the master instructions of the organism. Once DNA is formed in a cell it persists until the cell dies; that is, it is metabolically stable. This stability comes from the mass of hydrogen bonds holding the covalently bonded strands together, the hydrophobic (apolar) interactions between the bases on different levels and, possibly, electron delocalization (resonance) stabilization resulting from the linkage of the conjugated bases.

Just the same, the DNA molecule can be altered in composition and structure by a number of agents. The double helix can be unwound and the DNA denatured by low pH or heat. The covalent backbone of one or both of the strands in the helix may be broken by radiation, temperature, or chemicals. Or less drastic alterations may occur, such as the substitution of one base for another or the addition or loss of bases. When any of these alterations occurs in such a way that they are passed on to succeeding generations of a cell or of an organism, the alteration is a *mutation*. In general, a mutation is a change in the primary structure of DNA. The potential consequences of mutations result from changes in the genetic instructions inherent in the primary structure of the DNA molecule. We shall be interested in two broad groups of mutations. One involves gross (visible in a microscope) alterations of chromosomes, such as the breaking off of a piece of the chromosome. The other group, *point mutations*, includes alterations in only one or a small number of nucleotides (or bases) in a gene.

When a single chromosome breaks in two, the broken ends usually come together again through natural repair processes. However, in a cell with a single chromosome broken into several pieces or with a number of broken chromosomes, the pieces may rejoin in various orders as illustrated in Figure 19.13. The rearrangement or loss of large portions of the chromosomes garbles the genetic message.

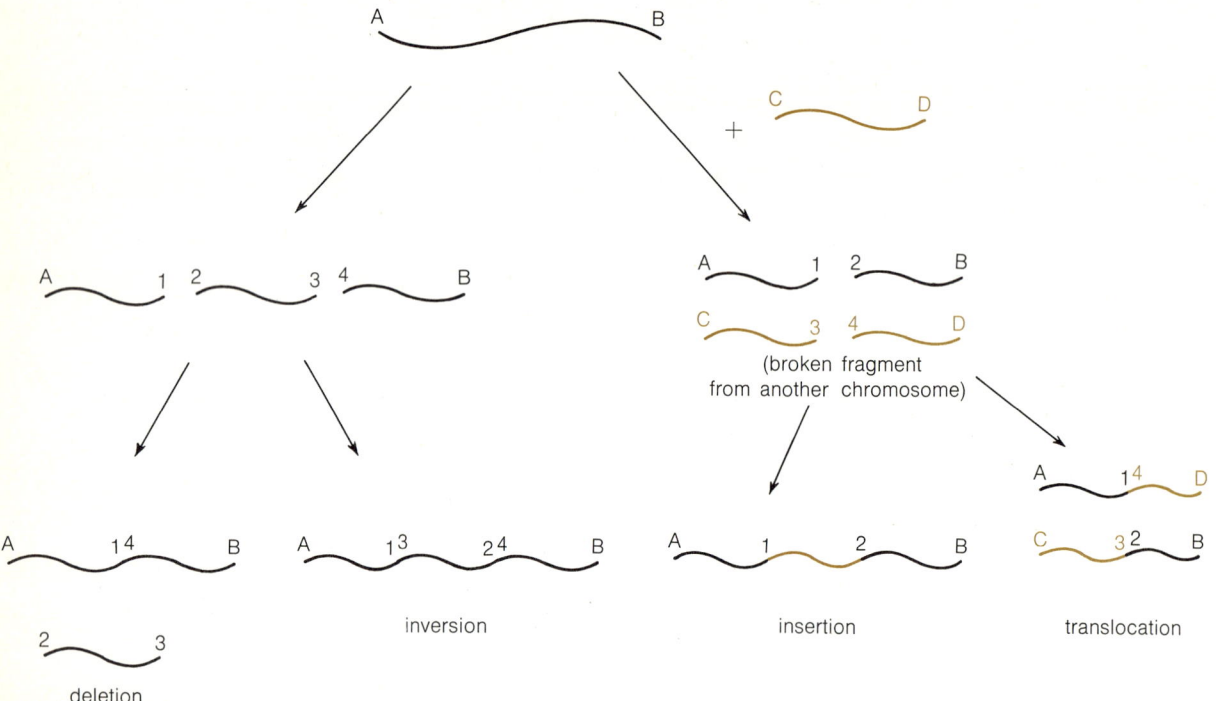

deletion

Figure 19.13. Some types of chromosome breaks.

Point mutations involving the gain, loss, or rearrangement of one or more nucleotides can be divided into four major classes, as shown in Figure 19.14. Before looking at the mechanism of formation of some of these point mutations, let's look at how they would affect protein synthesis when one DNA strand is read. In the transition mutants one purine–pyrimidine base pair is substituted for another, a purine for a purine and a pyrimidine for a pyrimidine as, for instance, an A–T pair replaced by a G–C pair. Unless the change introduces a triplet corresponding to a degenerate codon in mRNA (i.e., CCA $\longrightarrow$ CCG), there can be two results; a different amino acid will be substituted in the synthesized polypeptide or, if the new base leads to a chain termination codon, the polypeptide will be prematurely ended at that point. Transversion mutations, the replacement of a purine–pyrimidine pair by a pyrimidine–purine pair, would have effects similar to transitions.

Addition or deletion of nucleotide pairs has more serious consequences, because the entire genetic message or set of codons after the mutation is

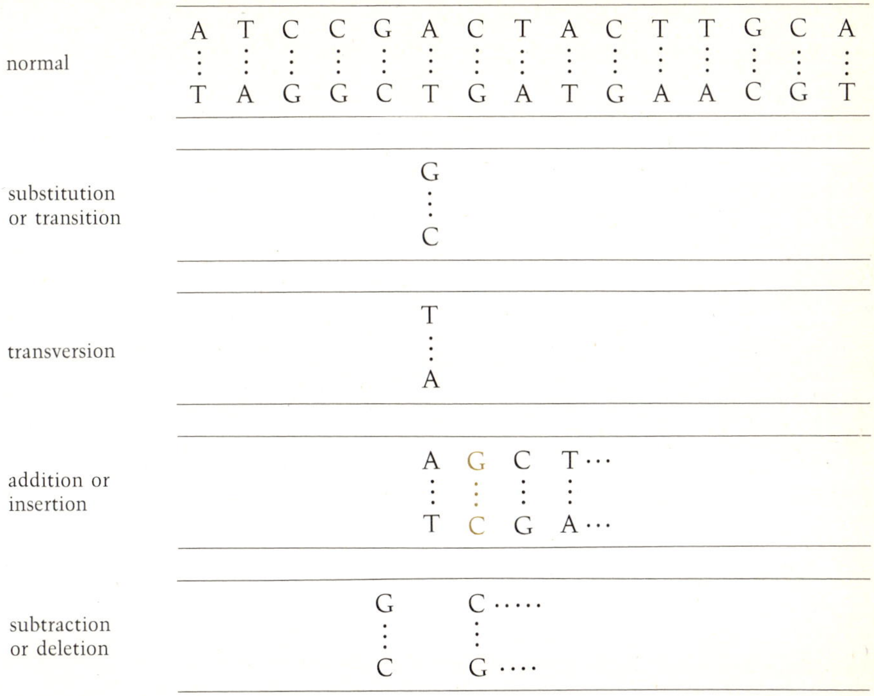

normal

A	T	C	C	G	A	C	T	A	C	T	T	G	C	A
:	:	:	:	:	:	:	:	:	:	:	:	:	:	:
T	A	G	G	C	T	G	A	T	G	A	A	C	G	T

substitution
or transition

G
:
C

transversion

T
:
A

addition or
insertion

A G C T···
: : : :
T C G A···

subtraction
or deletion

G   C·····
:   :
C   G····

Figure 19.14. Point mutations in DNA.

thrown out of kilter, and quite different proteins may be produced. This is because the code is written without punctuation marks to show where one nonoverlapping triplet (codon) ends and another begins. The code is read sequentially from the beginning of the gene to the end. These mutations are called *frame-shift* mutations, because addition shifts the codon reading frame backward by one base and deletion shifts the frame one notch forward. Examples of different situations are given in Figure 19.15, using only one strand (only one strand is read anyway) and a simplified base sequence for clarity.

19.10 MUTAGENS Many agents, both physical and chemical, can induce mutations in DNA. Among the physical agents are high-energy or ionizing radiations such as X-rays or gamma rays, and beams of elementary particles such as protons. Lower energy ultraviolet radiation is also an efficient mutagen. The chemical mutagens resist classification. Radiation and chemicals cause chromosome breaks and point mutations, but our illustrations will be limited to the point mutations.

Radiation Ionizing radiation interacts with materials in a number of ways, the simplest of which is the production of free radicals. The radiation first knocks an electron out of a molecule, creating a free energetic electron and a positive ion. The electron may combine with another molecule, which then dissociates to another ion and a highly reactive free radical. Because biological systems are 80 percent water, the primary target is water instead of DNA. Ionizing

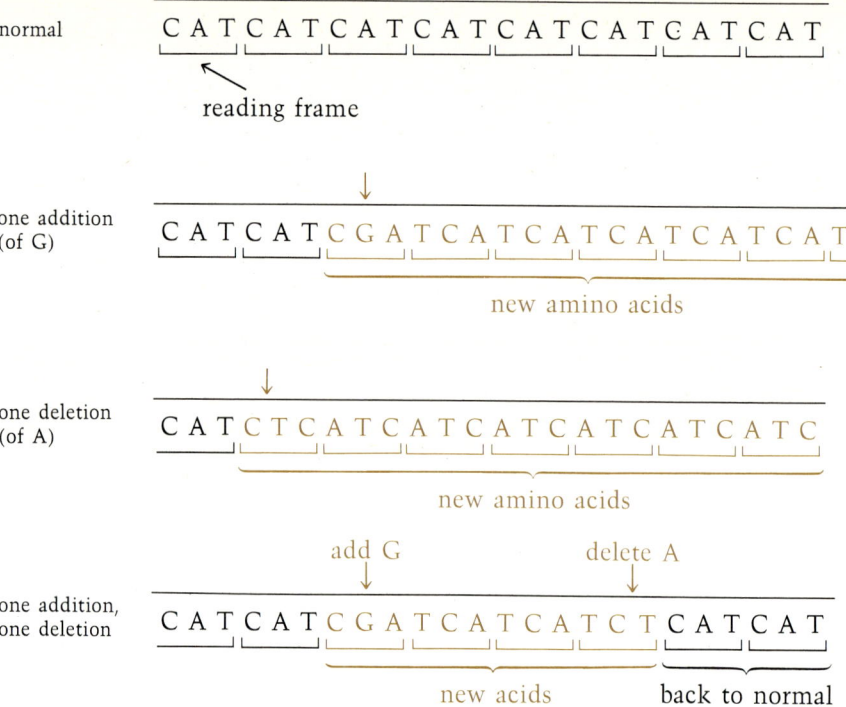

Figure 19.15. Different frame-shift mutations in DNA. Only the template strand of DNA is shown.

radiation produces free radicals from water by way of the series of reactions

$$H_2O + h\nu \longrightarrow H_2O^+ + e^-$$
$$e^- + H_2O \longrightarrow H_2O^-$$
$$H_2O^+ \longrightarrow H^+ + \cdot OH$$
$$H_2O^- \longrightarrow OH^- + H \cdot$$
$$H^+ + OH^- \longrightarrow H_2O$$
$$\overline{H_2O + h\nu \longrightarrow H \cdot + \cdot OH} \qquad \text{net reaction}$$

The reaction of the free radical $\cdot$ OH, as well as other products (H_2 from $H \cdot$, H_2O_2) with DNA leads to a number of mutations and other effects that can interfere with DNA replication and transcription.

Nonionizing ultraviolet radiation can produce many of the point mutations, but its principal and most lethal effect results from the blocking of DNA replication by the formation of covalent bonds between adjacent pyrimidine bases (principally T) in the same strand,

$$T{=}T$$

These dimers prevent replication and are responsible for the germicidal action of ultraviolet radiation. There is good evidence, however, for the existence of enzymatic repair processes in cells for the rejoining of breaks in DNA strands and for the excision of dimer defects.

Chemical Mutagens The mutagenic activity of chemicals has become a matter of great concern because of the growing exposure to more and more chemicals in the environment. Many of these chemicals have demonstrated mutagenic properties in lower organisms but not in humans. This does not mean that there is no mutagenesis in humans. The outward effect, if any, of any mutation is hard to detect in humans for a number of reasons, among which are the long generation time, the rarity of expression of the effect (see recessive character of genes, Section 9.11), and the legal and moral problems of experimenting on humans.

The potential for mutagenic effects in humans has been detected through the visible gross chromosome aberrations, in human cells as well as in bacteria and mice. Although many scientists assume that any chemical that can break chromosomes can also induce the potentially more serious point mutations, the point mutations cannot be detected at once, nor has any genetic damage resulting from such mutations been observed. Testing of mutagens is further complicated because a chemical may be mutagenic in one species but not in another.

Many chemical mutagens are known, but their exact mechanisms of action are poorly understood. Nitrous acid is one of the few mutagens whose mechanism of action is well known. It can be produced in the stomach by the action of acids on sodium nitrite, a common food perservative. Nitrous acid oxidatively deaminates the bases G, A, and C. The deamination of C, cytosine, for example, results in the replacement of the amino group by an oxygen atom and gives uracil, U. Whereas cytosine normally pairs with G, uracil pairs with A. During subsequent DNA replication a transition mutation emerges as shown in Figure 19.16. Although mutations induced by nitrous acid have been observed in lower organisms, no mutation effects are known in man; but since HNO_2 is present only at low pH, its mutagenic effects should be noted in the cells of the stomach.

$$\vdash C \cdots G \dashv \xrightarrow{HNO_2} \vdash U \quad G \dashv \xrightarrow[\text{replication}]{\text{first}} \vdash U \cdots A \dashv + \vdash C \cdots G \dashv$$

Figure 19.16. Induction of transitional mutation by nitrous acid. Deamination of C leads to a C–G to T–A transition.

$$\vdash U \cdots A \dashv \xrightarrow[\text{U–A only}]{\text{replication of}} \vdash U \cdots A \dashv + \vdash T \cdots A \dashv$$

Nitrosamines are compounds of the type $(RCH_2)_2N—N{=}O$, prepared by the action of nitrous acid on a dialkylamine,

$$\begin{array}{c} RCH_2 \\ \diagdown \\ \diagup \\ RCH_2 \end{array} NH + HNO_2 \longrightarrow \begin{array}{c} RCH_2 \\ \diagdown \\ \diagup \\ RCH_2 \end{array} N—N{=}O + H_2O$$

The dialkyl (secondary) amines are fairly common compounds found in foods and drugs, besides being natural building blocks for proteins in the body. The reaction to form nitrosamines may occur in the stomach with nitrous acid or outside the body in foods containing nitrites. Metabolic products of the nitrosamines are known to be both carcinogenic and mutagenic. It is possible that the two effects are related. Tumor (cancer) production may result from induced mutations and subsequent breakdown in the normal control of cell division. The induced point mutations due to nitrosamines are believed to be transitional mutations caused by a change (an alkylation) of the guanine base so that it can sometimes pair with thymine in DNA replication. The result is a conversion of a G–C pair to an A–T pair after replication.

19.11 SOME CONSEQUENCES OF INDUCED MUTATIONS IN HUMANS

Assuming that an induced mutation has changed instructions in some gene of a human cell, what might happen to the cell, to the human, and to the descendants of that human? To avoid defining genetic terminology, we shall confine our attention to (1) the kind of mutation and (2) the kind of cell involved.

First, there are the gross chromosome mutations resulting from breaks, rearrangements, and deletions of large segments of genes. Many chromosome breaks occur routinely during cell division, but the broken portions generally are rejoined or parts of different broken chromosomes are realigned by cellular mechanisms in such a way that no gross mutation results. However, breaks of the type shown in Figure 19.13 do lead to such large-scale errors that the cell can no longer function if the break is not properly repaired. The mutation is then lethal and the cell dies.

Point mutations may or may not be lethal. Transition and transversion mutations usually affect only one amino acid in the polypeptide chains (unless a termination codon is introduced). Unless this amino acid resides at the active center of an enzyme, the synthesized protein will continue to function, and the cell will survive. On the other hand, addition and deletion point mutations can so alter the composition of the protein (or of the mRNA or tRNA formed on the DNA template) that the mutation is lethal. Mutations in genes that encode enzymes such as DNA and RNA polymerases will result in these enzymes making errors in replication and transcription. Such mutations will be especially detrimental because their effects will be multiplied during replication.

The sex or germ cells are the ones that come together, one from each parent, to form a *somatic* cell. This single somatic cell divides to form two other somatic cells, and so on. The somatic cells are the body cells of the skin, liver, and muscle, for example, and make up the bulk of the human body. In humans, germ cells have 23 chromosomes, whereas the somatic cells derived from the union of a sperm and egg each have 46 chromosomes, 23 from one parent, 23 from the other. Each set of 23 chromosomes comprises a set of genes sufficient for the construction of a human being. Therefore, each somatic cell has a double set of genes or a double dose of genetic information. If a gene in one set directs the synthesis of a particular enzyme connected with some inheritable trait, there is a corresponding gene in the other set for that same enzyme or trait. An induced mutation in a somatic cell of an adult human either kills that one cell or affects only a comparatively small number of progeny cells. In either case, little or no harm is done to the adult organism except in the special case of a mutation that may disrupt the mechanism regulating cell division and cause cancerous growth. A mutation in a somatic cell of a fetus or embryo, however, may have serious consequences, because these cells are rapidly dividing and are the ancestors of many cells to come.

Mutations in germ cells are of special concern, because these mutations may be passed on to succeeding generations and may appear in both the germ and somatic cells. However, many generations may be required before the effect of the mutation shows up as a physical abnormality or as a genetic (biochemical) disorder in an offspring. An example is sickle-cell anemia, an inherited disease caused by a mutation in one of the genes responsible for the synthesis of two of the four polypeptide chains in hemoglobin. These two chains each have 146 amino acid units. The mutant chains differ from normal chains only in having a negatively charged glutamic acid residue in place of a neutral valine residue in one position. This subtle change causes a decrease in the solubility of the hemoglobin when it becomes deoxygenated. The precipitation of deoxyhemoglobin causes a deformation of the red blood cell to a sickle or crescent shape. As a result of the precipitation of their hemo-globin and their jagged shape, the sickled cells tend to clot in the small blood vessels. This leads to a cutoff in oxygen supply to tissues served by these vessels and inability of the sickled cells to get to the lungs where they can pick up oxygen. Sickle-cell anemia results in premature death only to those who carry two mutant genes for the polypeptide chain concerned. Sickle-cell anemia is due to a recessive gene. This means that the disease will not be apparent unless both hemoglobin genes are mutant. The recessive character apparently results because the one good gene synthesizes enough normal hemoglobin to meet the needs of the body. The hemoglobin in the blood of a carrier of only one mutant recessive hemoglobin gene will consist of one-half normal and one-half abnormal hemoglobin. Figure 19.17 shows how the effects of a recessive gene may become apparent in future generations.

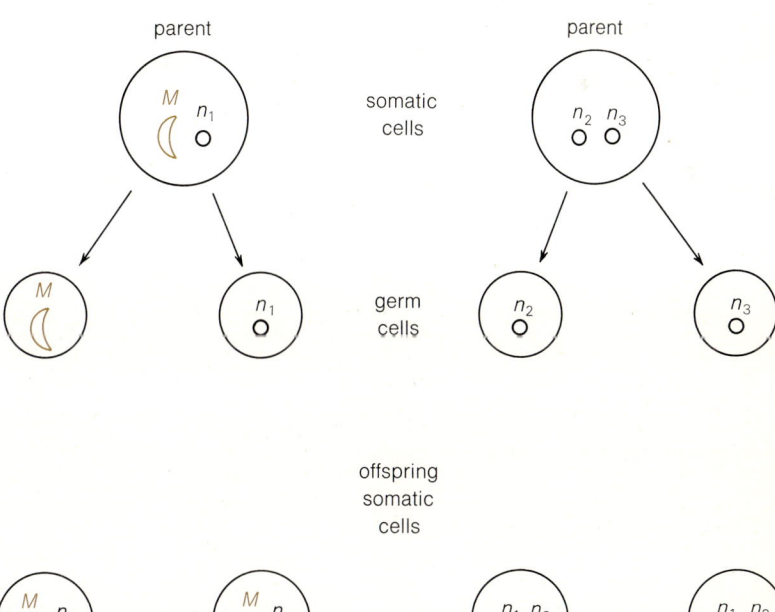

Figure 19.17. Passage of a recessive gene M from parents to offspring. If both parents should happen to carry the reces-sive gene, that is, if one of the normal genes such as $n_2 = M$, then one-fourth of the offspring may wind up with an MM pair and the genetic disorder will surface.

The mutation responsible for the abnormal hemoglobin must have arisen thousands of years ago from a spontaneous or random process caused by a stray cosmic ray or a natural error of incorporation of a base into DNA. Normally, such a lethal gene would be weeded out of the population by the process of natural selection. It appears that the sickle-cell trait had some advantage in that the abnormal red blood cells were resistant to a malarial parasite of the cells. As a result, those people with the mutant gene were relatively immune to the more lethal malaria. As malaria killed off more of the population having normal hemoglobin, those with the abnormal hemoglobin survived to reproduce and pass on the mutant gene to their offspring.

The spontaneous mutations responsible for sickle-cell anemia differ from the induced mutations due to exposure to chemicals and radiation insofar as they are of much lower frequency, are unavoidable, and are not traceable to a specific external cause. Even though most spontaneous mutations are harmful, even lethal, they have been of utmost value because the very few successful ones, over the 3 billion yr since life first arose on earth, have been responsible for the survival and diversity of species through evolution and natural selection. In the course of evolution the beneficial mutant genes have been retained and accumulated, whereas most of the damaging or useless genes have been eliminated. True, some previously beneficial genes, such as the sickle-cell mutant, have outlived their usefulness and are now considered harmful, but the present set of human genes seems to be exquisitely tailored for man's survival in the present environment. For this reason, almost every induced mutation now will contribute to a deterioration of man's genetic endowment and lead to damaging or lethal effects. For reasons discussed in the next section, these often undetected recessive mutations now accumulate over the generations, and the work of natural selection is being subverted. Deliberate radiation-induced mutations are, however, useful in plant breeding, where man assists nature in the selection process.

For a number of reasons, there is now an increasing mutational load in the human gene pool. This pool will be drawn upon by the next generation and combine with population growth to fuse what has been called a biological time bomb. Initially, mutations will be recessive and their effects indiscernible in the carriers of only one recessive gene. As more people with more recessive traits come into being, the chances of the mating of carriers of like recessive genes increases. At that time, natural selection again begins to operate with a vengeance as an increasing number of offspring carrying two of these recessive genes are produced. Presently, each human being is the carrier of four to eight recessive genes for lethal genetic diseases.

A major problem is the detection and prevention of induced mutations and their effects. This is not an easy job, because point mutations cannot be detected except through their delayed outward effects. Furthermore, mutagenesis by chemicals and radiation occurs in many genes at once, and the discernible effects are not likely to be traceable to any particular mutagen.

19.12 HUMAN BIOLOGICAL ENGINEERING

Of equal concern for the degradation of the quality of the human gene pool by induced mutations is the deterioration believed to be occurring from the perpetuation of genetic disorders and certain traits already present in man's genetic makeup. The transmission and accumulation of genetic diseases such as sickle-cell anemia from generation to generation is one example. However, other less well-defined human attributes—intelligence, temperament, overall physical vigor—are also believed to be genetically determined to some degree and would likewise be passed on to future generations.

As modern medical knowledge steadily decreases mortality rates, allowing carriers of harmful genes to survive through their reproductive years and beget more carriers, the incidence of unfavorable genes in the population accumulates. Survival is aided by the practice of *euphenics*, the modification, control, or imitation of gene action without affecting the intrinsic character of the genetic complement of the individual. Examples of euphenics are immunization against infectious diseases and administration of hormones and other substances that act directly on genes or the substances synthesized by genes. Another common example is the artificial supply of a required substance that the organism is unable to make because of a mutant gene. Specific illustrations are

1. the use of antigens in vaccines to activate genes to produce antibodies,
2. the use of urea to break a bond in abnormal hemoglobin and cause the sickle cells to assume the normal shape, and
3. the administration of insulin for diabetes.

The application of euphenics, or medical euphenics to be more exact, can be regarded as a form of biological engineering in that it shapes human development to permit survival in the face of a genetic endowment that would normally lead to nonsurvival. Modern technology and many of our humanitarian social and legal procedures team up with euphenics to allow the survival of many people who would be naturally selected out of a more primitive culture. The recognition that factors such as euphenics coupled with increasing population might speed the accumulation of unfavorable genes or genetic traits in the population led, in the 1890's, to the idea of *eugenics*. Eugenics seeks ways to improve the genetic stock of future generations of the human race by the preferential selection of genes already present in the human gene pool. Originally, this was supposed to be done by selective breeding of those judged to possess "superior" qualities in much the same way that we breed animals and plants to obtain superior strains. A type of negative eugenics or dysgenics could be used to cull the "unfit" from the population, as by sterilization. Eugenics is practiced today by the genetic counselling of prospective parents, because the presence of some recessive genes can be inferred from a person's genealogy or detected by biochemical tests. Some use is also made of artificial insemination using sperm banks from "superior" individuals. A potential eugenic method is the *cloning* of individuals. Cloning, as used here, is the asexual reproduction of an individual by replacing the nucleus of an unfertilized egg cell with the nucleus of any somatic cell from an individual. The egg develops as if it had been fertilized. Since the transplanted nucleus contained the full genetic component of the donor, an exact copy of the donor of the nucleus might then develop. Theoretically, there should be no limit to the number of copies that could be made of any individual or "parent." Amphibians have been successfully cloned, and the extension of the technique to man is probably only a matter of time.

Eugenic procedures have several disadvantages. First, they amount to inbreeding, which decreases the genetic diversity of the species and reduces its ability to adapt to new environments. Second, eugenic procedures assume a decision as to which genetic traits should be propagated and which should be eliminated. Who is the genetic engineer making this decision? Or who decides that an individual is so superior that the world needs several exact copies of that individual?

Finally, with our growing knowledge of molecular genetics, the prospect of direct manipulation of the genes has opened up. This might be done by the replacement of mutant genes with isolated or synthesized normal genes or by the selective mutation of a gene to reverse a previous random mutation. For instance, it has recently been shown that viruses may be used to transport an operative gene into a cell unable to synthesize an enzyme needed for normal metabolism. Viruses are essentially nonliving packets of genetic material—DNA or RNA—that enter (infect) cells and use the cell's metabolic machinery to make large numbers of new viruses—a situation that generally kills the infected cell. If the virus has picked up a gene from a bacteria it has previously infected, this gene could be transferred (transduced) to a human cell "lacking" this gene. Experiment has shown that the gene introduced in this way produces the missing enzyme in the human cell. It has not been proved, however, that the added gene becomes permanently incorporated into the chromosomes of the human cell and replicates itself.

Techniques such as these are variously called *genetic engineering, genetic therapy,* or *genetic intervention.* At present, the technical obstacles to direct genetic manipulation are formidable. In any case, there is little hope of using genetic therapy to modify traits such as intelligence which, unlike monogenic traits like eye color and genetic disease, depend not on one gene, but on the combined effects of a great many genes.

Population Growth and Control

19.13 THE POPULATION PROBLEM

It is now common knowledge that the world has a population problem, although many countries and many people think that it is someone else's problem. World population growth is considered by many to be the most important general problem for the immediate future of man and one that must be solved quickly. Overpopulation has long afflicted various regions of the world. Only recently has the worldwide population growth caused concern that world population will soon exceed the ability of the earth to support its inhabitants. The rise in population has resulted from a steadily decreasing death rate caused by the introduction of modern medicine and public health procedures and by more efficient production of food.

Increasing population, along with misdirected technology, affluence and consumption, and urbanization is a contributor to environmental degradation. Overpopulation is the backdrop for depletion and dissipation of natural resources and the extermination of other species. Overpopulation also leads to social, economic, and political problems. Hunger, poverty, crime, the chaos in our cities, the ineffectiveness and unresponsiveness of human institutions, social unrest, political instability leading to war and totalitarian governments, and psychological distress due to crowding are just a few of the effects attributed to unchecked population growth.

It is true the earth can support many more people than the 3.5 billion now on it if an all-out effort is made to increase food supply and distribution. But if we feel that there is more to life than physical survival at a subsistence level, then the human race has already passed the optimum population level. There has been much written on population growth, its cause, effects, and control. Our only concern here will be with control measures, in particular the oral contraceptive or birth control pill.

In any population, animal or plant, control of population growth may involve a decrease in birth rate or an increase in death rate. (Migration is also used by animals, but it is a temporary measure.) Nature utilizes both of these methods. Eventually, nature will impose a check or a reversal to the growth of the human species in the form of higher mortality. The consequences are not likely to be pleasant. Modern man is unlikely intentionally to speed the death of those already living, wars and the possibility of legal euthanasia excepted. Therefore, if he is to avoid nature's own remedy, he must decrease his birth rate. This can be done by abortion, sterilization, or contraception. Contraception now appears to be the most effective method for large-scale use in birth control in the United States, although the other methods are being used effectively in other nations.

19.14 ORAL CONTRACEPTIVES

The most effective of the contraceptive methods at this time is the hormonal birth control pill. The pill is presently composed of modifications of two classes of female sex hormones, the estrogens and the progesterones. The exact mechanism of action of such contraceptive pills is unknown, but this uncertainty is common for a large number of drugs. On the physiological level, the administration of these steroid sex hormones probably acts by a negative feedback mechanism suppressing the production of two protein hormones by the pituitary gland. The reproductive cycle of the female involves an integrated system of regulatory interactions between the pituitary body attached to the brain and the ovaries. The messengers between the two are sex hormones circulating in the blood. The pituitary generates two protein hormones that influence the ovaries to produce a mature egg (ovum) and release the egg into the Fallopian tube. The pituitary hormones also induce the production of the steroid hormones, estrogen and progesterone, by the ovaries. These hormones prepare the uterus for the implantation and nourishment of a fertilized egg.

Both of the ovarian hormones exert a feedback effect on the pituitary, checking the output of the two pituitary hormones. If pregnancy occurs, the ovaries continue to produce estrogen and progesterone at high levels. The presence of these hormones, especially estrogen, shuts down the supply of the pituitary hormones for the duration of the pregnancy, preventing further ovulation. It is believed that a synthetic estrogen or progesterone, when introduced into the body from outside, gives feedback signals that cause the pituitary to withhold the protein hormones needed for ovulation. Without ovulation, conception cannot take place. In effect, the pill simulates the ovarian hormonal conditions of pregnancy.

Both estrogen and progesterone are steroid hormones, characterized by four fused rings. Figure 19.18 shows the structures of a natural estrogen (estradiol), a natural progesterone, and a synthetic progestogen having the hormonal activity of progesterone and used in some brands of birth control pills. The natural hormones are synthesized from cholesterol in the body. Other hormonelike substances, such as the fatty acid prostaglandins, may become important agents in controlling population growth.

19.15 GENETIC REGULATION OF METABOLISM: ENZYME INDUCTION

On a molecular level, the action of oral contraceptives can possibly be explained by metabolic regulatory mechanisms operating on gene activity. Instead of affecting the activity of an existing enzyme by inhibition or activation, many regulators also operate by turning genes on or off. When the gene controls the synthesis of an enzyme, the regulator substance becomes an *enzyme inducer* if it switches a gene on, an *enzyme repressor* if it prevents the gene from operating. It is believed that the bulk of the genes in a cell are turned off most of the time, because few specialized or differentiated cells have any use at all times for the full complement of substances encoded in the complete genetic blueprints in every cell. Many genes are turned off or their activity repressed by protein molecules (histones) attached directly to the DNA double helix at the beginning of the genetic message. The presence of this protein repressor prevents the transcription of the gene into mRNA

Figure 19.18. Female sex hormones.

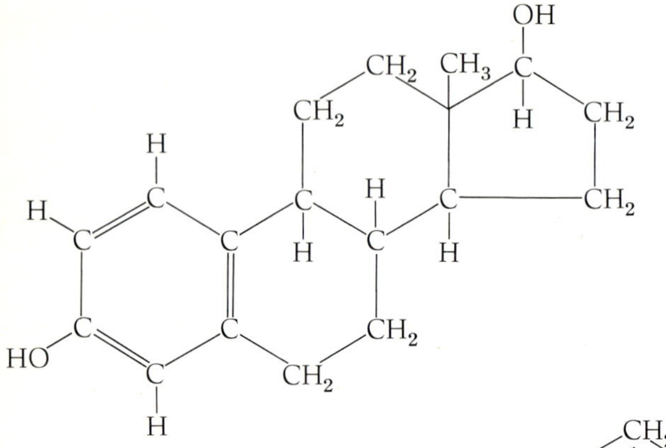

estradiol

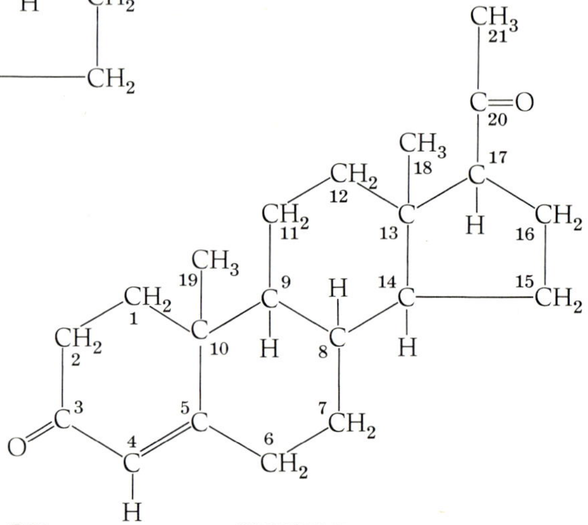

progesterone

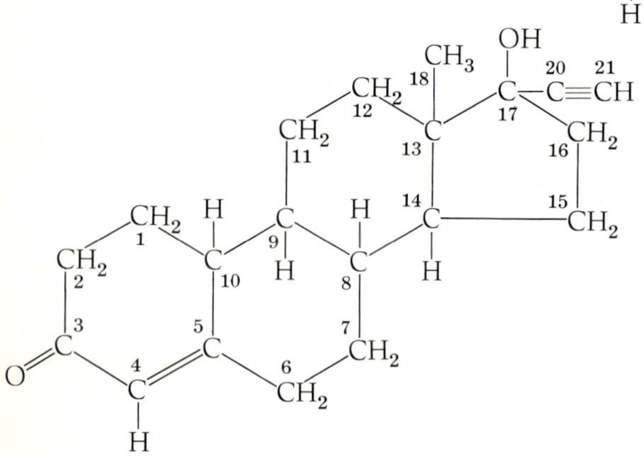

norethindrone (synthetic)

and the subsequent biosynthesis of protein. The repressor is itself the product of a separate control or regulatory gene. Each gene or set of related genes is controlled by a specific protein repressor.

A repressed gene may be turned on by the binding of a small molecule onto the large repressor molecule. Such binding may alter the shape of the repressor molecule, causing it to detach from the DNA surface and allowing the gene to function. In this case, the small molecule deactivating the repressor is the inducer. In other cases, a repressor molecule may not attach to the DNA and suppress gene activity unless it is bound to a small regulator molecule. Then the regulator is an enzyme repressor or a *corepressor*. The effects of the small regulator molecules on a repressor molecule are sketched in Figure 19.19.

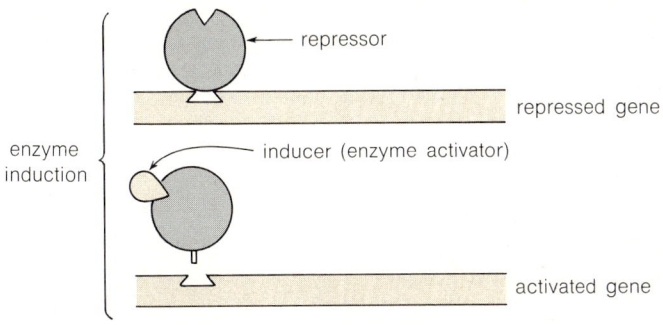

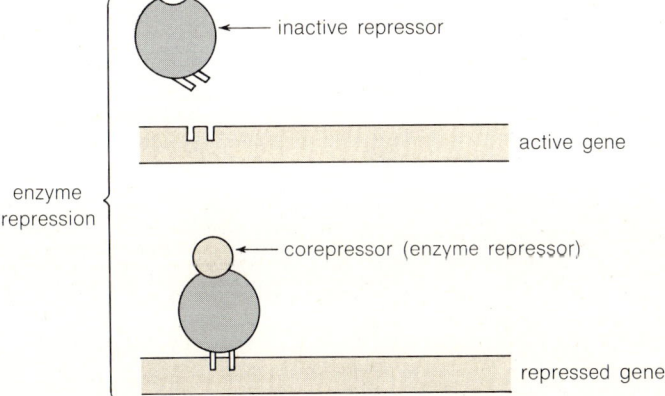

Figure 19.19. On the assumption that this gene is responsible for the synthesis of an enzyme, enzyme induction occurs when a normally active repressor molecule is removed from the gene by binding with a small inducer molecule. Enzyme repression may occur with other genes if a normally inactive repressor binds to the gene when in association with a corepressor substance.

Many hormones, or their metabolic products, apparently act at the gene level as inducers of protein synthesis. Thus, estrogen binds to and deactivates a repressor molecule on a gene controlling the synthesis of protein by cells in the uterus as the uterus prepares itself to implant an egg. Estrogen and progesterone or synthetic compounds that imitate their action might also serve as corepressors for genes responsible for the synthesis of the pituitary hormones. This could be the basis for their contraceptive action. It is now known that many other nonsteroid hormones act by causing the synthesis of an intermediate substance (cyclic AMP), which induces production or stimulates the activity of other enzymes.

The mechanism of enzyme repression briefly described here is thought to contribute to the action of certain teratogenic (monster-producing) substances. These are substances that, when taken by pregnant females, produce deformities in the newborn. A great deal of cell differentiation and organ growth and development occurs during the embryonic and fetal stages. This means that a lot of different genes are being switched on and off. If a substance ingested by the mother finds its way through the placenta into the fetus, it may block the activation of genes essential to a critical period of development. Thus, in the late 1950's, thalidomide, an apparently harmless (to adults) tranquilizer drug, caused gross deformities such as short, flipperlike arms in the newborn after their mothers had taken the drug during pregnancy. Tests had shown few harmful effects to pregnant mice and their offspring. Unfortunately, the genetic makeup and metabolic activity of mice is not the same as that of a human fetus. In any case, extreme caution should be used in taking any drug, even aspirin, during pregnancy.

19.16 SUMMARY   The genetic information of cellular genes and chromosomes is stored in the sequence of bases along the DNA molecule. In the chromosomes, DNA consists of two strands in a double-helical arrangement, with the strands held together by hydrogen bonds between opposing bases on each strand. The two strands are complementary because the base adenine (A) can pair only with thymine (T), and cytosine (C) can pair only with guanine (G).

Prior to cell division, the DNA of the chromosomes undergoes semiconservative replication (DNA $\longrightarrow$ DNA). In this process, complementary strands of DNA are enzymatically synthesized on each separated strand of the original DNA.

The DNA of the chromosomes also directs the synthesis and identity of cellular proteins. The first step in protein synthesis is the transcription (DNA $\longrightarrow$ RNA) of the genetic instructions encoded on one strand of DNA into a molecule of messenger RNA. The instructions are coded in the DNA as a series of triplet base sequences. Each set of three bases ultimately identifies an amino acid that will be inserted into the synthesized protein, and the order of the triplet bases along the DNA determines the primary structure and the biological function of the protein. The mRNA enzymatically synthesized along a portion (gene) of the DNA template has a base sequence complementary to the DNA, except that the base uracil (U) replaces thymine.

The mRNA carries the transcribed set of triplet codes to a ribosome, where the translation (RNA $\longrightarrow$ protein) step takes place. In translation, the appropriate amino acid is brought up to the mRNA template by a specialized molecule of RNA, transfer RNA. The tRNA can recognize only one triplet code on the mRNA and can carry only the one amino acid corresponding to that set of three bases. As tRNA molecules deliver their amino acids, the acids are joined together to form the protein molecule.

Errors or alterations in the instructions encoded in the DNA molecule may lead to genetic mutations, with the result that a synthesized protein will be inactive or missing. Mutations may be induced by radiation or chemicals, and they are particularly serious if they occur in the DNA of germ cells, for they will then be passed on to succeeding generations. Serious consequences can result from mutations involving only one base in a DNA molecule, because the genetic code is read sequentially, without punctuation marks. A variety of measures have been used, or are being investigated, to prevent the buildup of harmful mutations in the human gene pool. Among these methods is direct genetic intervention into the genetic machinery of the cell.

Many hormones and other metabolic regulators act at the genetic level by turning genes on and off. Certain environmental contaminants may produce biological damage by interfering with the action of genetic regulators.

Questions and Problems

1. Why are the bases A, T, G, C, and U planar?
2. How many groups of molecules of different density would result at each generation if the heavy DNA strands stayed together during replication through each generation?
3. If the average molecular weight of a nucleotide (base) pair is 600 and the average gene contains 1200 nucleotide pairs, how many genes are in a chromosome with a molecular weight of 2.4×10^6?
4. The single chromosome in *E. coli* is a double-stranded DNA molecule containing about 4.5×10^6 base pairs, which are spaced about 3.4 Å apart. What is the total length of the chromosome? Compare with the length of the *E. coli* bacteria cell, which is 20,000 Å.
5. Determine the amino acid sequences corresponding to the DNA base sequences shown in Figure 19.15(b).
6. What are the amino acid sequences of peptides formed according to the triplet codes (starting on the left) of the following mRNA's?
   a. GCAAACUCCGGUAUCUAC
   b. UAUCUAUCUAUCUAU
   c. GGUUCGCAGCUCCUGAUC
7. Write a reasonable equation for the oxidative deamination of cytosine by nitrous acid to form uracil.
8. If G $\longrightarrow$ G* by treatment with nitrosamine (or

a metabolic product) and G* can pair with T, show how an A–T base pair can be the result of a transitional mutation.
9. What would be the effect of the following?
   a. two additions (or deletions) in the same gene
   b. three additions (or deletions)
   c. one addition (or deletion) and two deletions (or additions)
10. Bromination of uracil results in 5-bromouracil (BU), which can be incorporated into DNA in place of thymine without destroying the capacity for DNA replication. One form of the 5-bromo-uracil analog of thymine can pair more readily with guanine than with adenine.
    a. What kind of mutation will result from the replacement of T by BU in an A–T base pair?
    b. Occasionally, BU may replace C; what mutation will result from replacement of C in a G–C base pair in DNA? (During replication, consider only the form of BU that pairs with A.)
11. Radioisotopes are widely used in cancer therapy. The ionizing radiation emitted by these substances kills normal somatic cells as well as the rapidly dividing cancer cells. Why would short exposures to radiation be effective and reasonably safe for treatment of cancer without unduly harming the normal cells?

Answers

2. first generation: two groups; second generation: two groups
3. 3000 genes

4. 0.15 cm or about 750 times the cell length
6. a. ala-asn-ser-gly-ileu-tyr
   b. tyr-leu-ser-ileu-tyr

References

MOLECULAR GENETICS
General
Green, David E., and Goldberger, Robert F., "Molecular Insights into the Living Process," Academic, New York, 1967, Chapter 11.
  A short, readable account of the steps from DNA to protein.
Haynes, Robert H., and Hanawalt, Philip C., Eds., "The Molecular Basis of Life," Freeman, San Francisco, 1968.
  A collection of articles from *Scientific American*. Chapters 6, 20–22 are especially pertinent to this chapter.
Smith, C. U. M., "Molecular Biology," M.I.T. Press, Cambridge, Mass., 1968, Chapters 13, 14.
  An elementary discussion of heredity through the experimental deciphering of the genetic code.
Stent, Gunther S., "Molecular Genetics: An Introductory Narrative," Freeman, San Francisco, 1971.
Watson, J. D., "Molecular Biology of the Gene," 2nd ed., W. A. Benjamin, Menlo Park, Calif., 1970, Chapters 6–13.
  A comprehensive and authoritative, but readable, account of the whole subject.
Watson, J. D., "The Double Helix," Atheneum, New York, 1968.
  A nontechnical account of how the structure of DNA was derived. Recommended reading.
Reverse (RNA ⟶ DNA) transcription
Culliton, Barbara J., "Reverse Transcription: One Year Later," *Science*, **172,** 926–928 (1971).
Temin, Howard M., "RNA-Directed DNA Synthesis," *Scientific American*, **226,** 24–33 (January 1972).
Mutations and chemical mutagens
Asimov, Isaac, and Dobzhansky, Theodosius, "The Genetic Effects of Radiation," U. S. Atomic Energy Commission, Washington, D. C., 1966.

  Good introductory sections on elementary genetics and mutations.
Auerbach, C., "The Chemical Production of Mutations," *Science*, **158,** 1141–1147 (1967).
Hollaender, Alexander, Ed., "Chemical Mutagens. Principles and Methods for Their Detection," Vol. 1, Plenum, New York, 1971, Chapters 1, 2.
Pizzarello, Donald J., and Witcofski, Richard L., "Basic Radiation Biology," Lea and Febiger, Philadelphia, 1967.
  Good background sections on basic genetics and mutations.
Human biological engineering
Davis, Bernard D., "Prospects for Genetic Intervention in Man," *Science*, **170** 1279–1283 (1970).
Friedmann, Theodore, and Roblin, Richard, "Genetic Therapy for Human Genetic Disease?" *Science*, **175,** 949–953 (1972).
Kass, Leon R., "The New Biology: What Price Relieving Man's Estate?" *Science*, **174,** 779–788 (1971).
  A discussion of the basic ethical and social problems arising from the use of biomedical technologies, including genetic engineering. Recommended reading in connection with both this chapter and Chapter 21.
Roslansky, John D., Ed., "Genetics and the Future of Man," North-Holland, Amsterdam, 1966.
Wills, Christopher, "Genetic Load," *Scientific American*, **222,** 98–107 (March 1970).

POPULATION GROWTH AND CONTROL
The population problem
Ehrlich, Paul R., and Ehrlich, Anne H., "Population, Resources, Environment," 2nd ed, Freeman, San Francisco, 1972.

Genetic regulation of metabolism

Butler, J. A. V., "Gene Control in the Living Cell," Basic Books, New York, 1968.

An easy-to-read book touching on a variety of subjects from enzymes to cancer.

Davidson, Eric H., "Hormones and Genes," *Scientific American*, **212,** 36–45 (June 1965). Also included in Haynes, *op. cit.*

Ptashne, Mark, and Gilbert, Walter, "Genetic Repressors," *Scientific American*, **222,** 36–44 (June 1970).

SIX

Energy

A moderately active adult human requires about 2500 kilocalories of energy per day to stay alive. In Chapters 16 and 18 we discussed the sources of this biological energy and its transformation and utilization by the organism. A modern, industrially based culture produces and consumes much more energy than this on a daily per capita basis. It is the availability of cheap and abundant energy over and above the biological minimum for subsistence that powers our civilization and helps maintain a high standard of living. But as it now stands, large-scale energy production is also a major contributor to the depletion of nonrenewable natural resources and to environmental pollution.

The patterns and extents of nonbiological energy consumption in societies change as energy resources are depleted and as other energy sources are brought to light by exploration or by scientific and technological advances. In turn, energy resources and technologies help determine the structure of the society using them. Today we are beginning a period of growing dependence on nuclear energy, and we will devote major attention to its scientific background and its use, present and future, in power production.

Just as the flow of energy through societies affects the structure of a society, so also does the flow of biological energy through ecosystems serve to organize, maintain, and guide the development of the ecosystem. This phenomenon of ecological succession is briefly reviewed here.

Our Energy Environment

20.1 THE HISTORY OF ENERGY USE

From the beginning, man has relied for his energy needs on the tiny fraction of solar radiation captured by photosynthesis. In primitive cultures, only the solar energy fixed in plants or passed on to the flesh of animals provided the minimum biological requirement of 2500 kcal/day. More advanced agricultural societies increased the amount of solar energy available to man's slowly increasing population, but most of the biological energy in excess of the subsistence level had to be expended in agricultural work. In time, the use of recently fixed solar energy in the forms of wood for heat, animal oils for light, and windmills and water wheels for motive power added to the total amount of energy produced and consumed. Nevertheless, population growth caused the per capita energy consumption to increase only slightly above the minimum biological level.

715

Only in the latter half of the nineteenth century did the per capita consumption of energy begin to increase rapidly in the developed nations as the use of the fossil fuels increased. In the United States from 1870 to 1900, coal supplanted wood as the most important fuel. After 1900, petroleum's contribution to the total energy supply began to increase rapidly, until today its share is greater than that from coal. In the 1940's, natural gas consumption began to rise rapidly, until it too became a more important contributor to the total energy production in the United States than coal. In addition to fossil fuels, hydroelectric power and nuclear energy contribute to total energy supply. The contribution of each is minor compared to the fossil fuels, but the use of nuclear energy is destined to increase, while water power must inevitably decrease relative to other sources.

The total energy consumption in a developed country can be divided into a number of broad categories based on the economic sector using the energy, the specific use to which the energy is put, or the source of the energy. For instance, based on the economic sector, in 1967 in the United States, industry used approximately 40 percent of the energy produced (the primary metals industry is the largest user); homes, 20 percent (mostly for space heating with natural gas); transportation, 20 percent (almost all as petroleum products for motive power); commerce, 10 percent; all other uses, including agriculture and transmission losses, 10 percent. More useful to us are the data given in Table 20.1 on energy sources in the United States in 1970. Sources are given

Table 20.1
Energy sources, United States, 1970

source	total energy,[a] %	electric energy,[b] %
coal	20.1	55.0
oil	43.0	7.3
natural gas	32.8	19.3
water	3.8	15.1
nuclear	0.3	3.3

SOURCES: [a] Commerce Clearing House, Chicago.
[b] National Air Pollution Control Administration.

for total energy production and for electric energy production. Although electric energy consumed in the United States in 1971 amounted to about 20 percent of the total energy consumed, the demand for electric energy is growing rapidly at an average rate of 10 percent per year versus an average growth rate of 3.5 percent for total energy consumption. This amounts to more than doubling the electric energy production every 10 yr. Prospects are that in another hundred years, most of the energy consumed in the United States will be, and perhaps must be, in the form of electricity.

No figures will be given for energy consumption worldwide or in other countries, developed, developing, or underdeveloped, but it is worth noting that most of the people in the world today are subsisting at a level very close to the per capita biological minimum. The United States, with 6 percent of the world's population, accounts for over one-third of the world energy consumption. The average American consumes 200,000 kcal/day or nearly 100 times the biological minimum, and this consumption is now rising at the annual rate of 2.5 percent.

20.2 FOSSIL FUELS

Table 20.1 shows the present dependence on fossil fuels for energy production in the United States. Most other countries depend even more heavily on fossil fuels. Because these fuels are the most important source of energy, we shall examine their origin and utilization in more detail.

As the name implies, the fossil fuels—coal, oil, and natural gas—were formed over geological time periods, beginning about 600 million yr ago, from the remains of living organisms that died and underwent anaerobic bacterial decomposition. The complex carbon compounds of the living organism are degraded, in the absence of oxygen, to simpler and more stable reduced organic substances. This partially decomposed organic material is found to some extent in all sedimentary rocks, but in certain places it has accumulated locally into predominantly organic sediments from which the reduced material may be easily extracted.

In order for the complex compounds of the dead organism to break down into simpler reduced substances, the organic matter must, within a reasonable period of time after death, be protected from the oxidizing nature of the present atmosphere. Otherwise, it will be oxidized and return to the atmosphere as CO_2.

Each of the fossil fuels, except for natural gas, is a complex and variable mixture of carbon compounds. The composition of each gives us a clue to its origin and to some of the likely processes of its formation from the starting materials.

Coal

Coal is formed predominantly from land vegetation composed of lignin and cellulose. Its origin is readily apparent because of the fossil remains and stratification found in beds of coal. Coal is a very complex mixture and, while its gross chemical composition is known, many of its components have not been identified. There seems no doubt that coal is the product of partial decomposition, under anaerobic conditions, of buried vegetation, followed by the action of heat and pressure.

The characteristics of coal depend on its origin and on the conditions under which it was formed. In general, coals derived mainly from cellulose materials are ranked according to the degree of increased exposure to the metamorphic conditions of heat and pressure. The overall change is essentially one of increased reduction. The resulting coal types or ranks are shown in Table 20.2. The chemical reactions in the degradation of cellulose and lignin to coal are little understood.

Table 20.2
Coal ranks and the trends in their composition and caloric value

rank	weight percent, carbon	weight percent, hydrogen	weight percent, oxygen	
wood	50	6.2	43	|
peat	55	6.3	37	increasing
lignite	73	5.2	21	caloric
bituminous	84	5.6	8.7	value
anthracite	94	2.8	2.7	↓

Most coal contains from 1 to 3 percent sulfur. About half of the sulfur content is due to the presence of iron pyrite, FeS_2. The use of coal as a fuel is responsible for the greater part of sulfur dioxide air pollution and for much of the particulate matter. If the methods presently available for the removal of pyrites from coal were more commonly used, a major part of the air pollution problem would be eliminated.

Petroleum and Natural Gas

Liquid petroleum is a complex mixture of organic compounds, principally hydrocarbons. Of these hydrocarbons, straight-chain alkanes predominate. Cycloalkanes (ring alkanes), such as cyclopentane, cyclohexane, and their derivatives, are present in lesser proportion, and there are even smaller amounts of aromatic hydrocarbons. The predominance of saturated hydrocarbons in petroleum, and the absence of alkenes, indicates that fats and possibly proteins were the likely starting points for petroleum formation. Saturated hydrocarbons can be formed by fairly simple changes from these starting materials.

Current theory is that petroleum is formed by the low-temperature anaerobic decomposition of marine organisms in the reducing environment of deep ocean bottom sediments. Marine organisms contain a greater proportion of fats and proteins than land organisms.

Before crude oil can be put to any use, the complex mixture must be refined by separating it into various fractions. Boiling point differences are used in the separation process, which is called *fractional distillation*. When a solution containing two or more volatile components is heated, the vapor phase is always richer in the components having the higher vapor pressures (Section 12.24), producing a difference in composition between the liquid and the vapor. If the vapor is removed and condensed to a liquid, the substances with the lower boiling points will tend to be concentrated in this distillate. The distillate may be redistilled to concentrate further the more volatile components in its vapor. This redistillation procedure can be carried out as many times as necessary until the pure components of the original solution are separated. In the refining of crude petroleum, the process is stopped when separation into small groups of relatively homogeneous compounds has occurred. Fractions obtained are listed in order of increasing boiling points in Table 20.3, along with the carbon chain length of the hydrocarbons commonly found in each fraction. The first five fractions serve as the primary fuels in modern technology.

Table 20.3
Fractions from the distillation of crude oil

fraction	approximate composition	approximate boiling range, °C
natural gas	C_1–C_4	-161–$+20$
petroleum ether (naphtha)	C_5–C_7	20–100
gasoline	C_6–C_{12}	50–200
kerosene	C_{12}–C_{16}	200–300
fuel oil, diesel oil, gas oil	C_{16}–C_{20}	300–500
lubricating oils, greases, petroleum jelly, paraffin wax	C_{18}–up	400–up (vacuum distilled)
asphalt, petroleum coke		residue

Natural gas is not obtained solely from the industrial distillation of crude oil. It is commonly taken directly from the earth, where it may occur as a gaseous capping over an oil pool or in a pocket not directly associated with an oil pool. The approximate composition of unprocessed natural gas is 60–85 percent methane, 5–10 percent ethane, 3–18 percent propane, and 1–14 percent higher hydrocarbons, mostly butane. Other gases present in small quantities may be nitrogen, carbon dioxide, hydrogen sulfide, and helium. The propane and butane components of natural gas are often separated by liquefaction under pressure and marketed for cooking and heating as bottled or liquefied petroleum gas (LPG).

Coal Gasification Natural gas is the cleanest fossil fuel, but it is in the shortest supply. Coal represents three-fourths of the fossil fuel reserves in the United States, but it is the dirtiest fossil fuel. To take advantage of the plentiful supply of coal and the cleanliness of gas, the conversion of coal into gas is becoming more and more popular.

There are a number of coal gasification processes, but all use the same basic set of reactions, although perhaps in different sequences. In the indirect gasification process, powdered coal is heated to drive off volatile constituents and to produce a gas rich in methane and hydrogen. The solid residue from the initial devolatizing step is called "char," with a composition we shall represent simply as C. The true gasification step involves the high-temperature reaction of the char with superheated steam to form a gas consisting mainly of carbon monoxide and hydrogen. Some of the hydrogen reacts with char to form methane. The rest of the hydrogen from the gasification step, plus some from the initial devolatizing, reacts further with carbon monoxide in a methanation reaction to form more methane. The principal reactions are summarized below.

$$\text{coal} \xrightarrow{\;600°C\;} H_2 + CH_4 + \text{char (C)} \qquad \text{(devolatizing)}$$

$$C + H_2O(g) \xrightarrow[\text{68 atm}]{\;1000°C\;} CO + H_2 \qquad \text{(gasification)}$$

$$C + 2H_2 \xrightarrow{\;800°C\;} CH_4 \qquad \text{(hydrogasification)}$$

$$CO + 3H_2 \xrightarrow{\qquad} CH_4 + H_2O \qquad \text{(methanation)}$$

and also

$$CO + H_2O \xrightarrow[\text{catalyst}]{\;300–500°C\;} CO_2 + H_2$$

The devolatizing, gasification, and $CO + H_2O$ reactions furnish additional hydrogen for the methanation step. The product gas is, after purification, virtually free from sulfur, carbon monoxide, and free hydrogen and has a heating value nearly equivalent to processed natural gas. Overall, coal gasification amounts to the addition of hydrogen to coal.

Gasoline Refining The gasoline fraction initially obtained from petroleum distillation is insufficient in quantity to meet the demand for gasoline and lacks the proper combustion characteristics with respect to antiknock or octane number properties. This *straight-run* fraction is augmented by the *cracking* of higher boiling fractions. In this process, the alkane molecules with more than 12 carbons are broken down to lower molecular weight alkenes and alkanes in the C_6–C_{12} range. The performance of gasoline in engines is rated on the octane number scale. Pure n-heptane, a poor fuel which knocks badly when burned in an internal combustion engine, has been arbitrarily assigned an octane number of zero. An octane number of 100 has been assigned to 2,2,4-trimethylpentane (isooctane), which has good antiknock properties. The octane number of a gasoline is determined by comparing its performance in a standard test engine with various mixtures of n-heptane and 2,2,4-trimethyl-pentane.

The octane numbers of the straight-run distillate range from 20 to 70 because of the presence of high percentages of straight-chain alkanes, all of which are poor fuels. Branched-chain alkanes, alkenes, and aromatics have higher octane numbers. Toluene, $C_6H_5CH_3$, for example, has an octane number of 120. The octane number of gasoline can be increased by the addition of tetraethyl or tetramethyl lead formulations or by increasing the concentration of branched-chain or aromatic hydrocarbons in the mixture. Catalytic *isomerization* converts straight-chain hydrocarbons such as n-butane to their higher octane branched-chain isomers, such as isobutane. The process of *reforming* (sometimes considered to include isomerization) results in increased amounts of high-octane aromatic hydrocarbons in the gasoline by the conversion of aliphatic hydrocarbons. Antiknock properties are especially important in high-compression engines. The imminent removal of lead antiknock compounds (Section 7.8) from gasolines means that an increase in aromatic hydrocarbons will be necessary to maintain octane ratings. The use of the aromatics in place of the lower octane alkanes is likely to intensify the photochemical smog problem. In addition, many aromatics or their residues are believed to be carcinogenic. This is another example of how the battle against environmental pollution often involves hard choices.

20.3 LIFETIMES OF THE FOSSIL FUEL RESERVES

The fossil fuels are clearly nonrenewable resources. Within a few centuries, we have used up materials whose formation required hundreds of millions of years. It is estimated that the earth's coal supplies can serve as one of the major sources of energy for another 300–400 yr, but for only 100–200 yr if coal is the main energy source. Similarly, the petroleum products, crude oil, and natural gas from all sources will be able to function as an energy supply for less than a century, with gas being the first to go. Since large-scale production and use of coal, crude oil, and natural gas only began about a century ago, this means that our fossil fuel age has been a short but eventful period in human history, an era of growth that may never be repeated.

Most of the coal produced in the United States goes into the generation of electric power. Of the petroleum consumed, 54 percent is used for motive power as gasoline, diesel oil, or jet fuel to provide the fuel for almost all of the nation's transportation. Together, oil and gas supply 90 percent of all the nation's space heating requirements. Aside from their use as energy sources, 2 percent of the oil and gas consumed provide the raw materials for the organic chemical industry that supplies our plastics, synthetic rubber, nylon, paint, and so on. In addition, coal is used not only as a fuel, but also for its reducing properties (Section 13.19).

To meet the day when fossil fuel supplies dwindle and reserves must be saved for chemical uses, new energy sources must be found. Hydroelectric power is not the answer. Dam sites are limited, and the reservoirs behind them eventually fill with sediment. There are a number of other sources of energy. Among these are geothermal energy, tidal power, solar energy, and nuclear energy. The first two are localized and inadequate. Solar energy, of course, is the only truly unlimited source of energy, and although the capture of large quantities of solar energy, aside from photosynthesis and possibly space heating, appears now to be economically and technologically unpromising, the eventual utilization of solar energy generates great interest and hope. This leaves nuclear energy as our remaining large-scale source of economically competitive energy to prevent the collapse of our energy-based culture and to contribute to a rising standard of living in the less developed parts of the world.

Nuclear Energy

Throughout this book, we have talked only about energies associated with the forces between or among atoms, ions, and electrons. The chemical energies absorbed or released in chemical reactions are of the order of 100 kcal/mol. Nuclear or atomic energies, on the other hand, are associated with the forces between the protons and neutrons in the nucleus of atoms. These nuclear forces are over a million times stronger than the ionic, covalent, and van der Waals forces with which we have been concerned. When the structure or composition of an atomic nucleus is altered, nuclear energy is either absorbed or evolved. To appreciate the origin of nuclear energy, how it can be released, and some of the environmental implications of large-scale use of nuclear energy, it is helpful to have a general knowledge of nuclear chemistry and radioactivity.

20.4 NATURAL RADIOACTIVE DECAY

The alteration of a nucleus may occur spontaneously, or it may be induced artificially. The natural radioactive disintegration of certain nuclei is a good example of a spontaneous nuclear change. Many of the elements have naturally occurring isotopes whose nuclei spontaneously transmute to the nuclei of other elements. The natural radioactive decay of such unstable isotopes is accompanied by the emission of helium nuclei, ^4_2He, or of high-speed electrons, $_{-1}^{0}e$. The subscript refers to the atomic number or nuclear charge and the superscript to the mass number. A helium nucleus is called an alpha (α) particle, and an electron is a beta (β^-) particle, so these kinds of decay are called alpha and beta emission. High-energy electromagnetic radiation, gamma (γ) rays, may be emitted along with either alpha or beta emission. All the isotopes of the elements beyond bismuth in the periodic table and also of $_{43}\text{Tc}$ and $_{61}\text{Pm}$ are radioactive, while several other elements may have both radioactive and stable, naturally occurring isotopes. Among the latter class are elements such as hydrogen and carbon, which have both stable (^1_1H, ^2_1H, $^{12}_6\text{C}$) and radioactive (^3_1H, $^{14}_6\text{C}$) isotopes in nature.

Three of the naturally occurring heavy elements, $^{238}_{92}\text{U}$, $^{235}_{92}\text{U}$, and $^{232}_{90}\text{Th}$, are the parent elements of radioactive decay series involving a succession of radioactive disintegrations culminating in the production of a stable lead isotope. The uranium-238 series is illustrated in Figure 20.1. In addition to the three radioactive series mentioned, another series, the neptunium series, is known but is not found in nature. Unlike the other three series, it ends in bismuth, not lead. The sequence of reactions in Figure 20.1 is given in

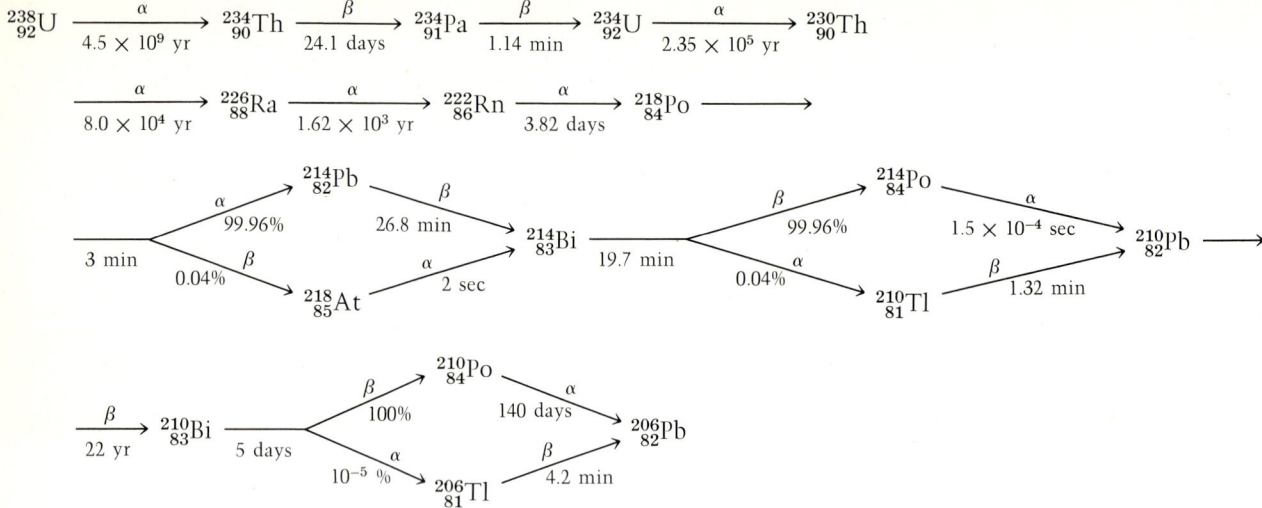

Figure 20.1. The natural uranium-238 radioactive decay series. Half-lives are given in years, days, minutes, or seconds. Some nuclei decay by different paths. The percentage disintegrating in different ways is indicated at branch points.

abbreviated notation. The first reaction in the uranium-238 series,

$$^{238}_{92}\text{U} \xrightarrow{\alpha} {}^{234}_{90}\text{Th}$$

if written out to show the ejection of the alpha particle, is

$$^{238}_{92}\text{U} \longrightarrow {}^{234}_{90}\text{Th} + {}^{4}_{2}\text{He}$$

Similarly, the beta decay of ^{234}Th in the complete notation is

$$^{234}_{90}\text{Th} \longrightarrow {}^{234}_{91}\text{Pa} + {}^{0}_{-1}e$$

These equations show how the subscripts and the superscripts must both balance in nuclear equations. The net effect of beta decay is the conversion of a neutron in the nucleus to a proton and an electron, with the electron being ejected,

$$^{1}_{0}n \longrightarrow {}^{1}_{1}\text{H} + {}^{0}_{-1}e$$

Note the balance of charges and mass numbers here.

An example of the spontaneous decay of a lighter naturally occurring isotope is

$$^{14}_{6}\text{C} \longrightarrow {}^{14}_{7}\text{N} + {}^{0}_{-1}e$$

In radioactive decay, the original unstable nucleus is called the *parent*, and the nucleus resulting from the decay of the parent is the *daughter*. Thus, ^{238}U is the parent of the daughter nucleus ^{234}Th, and ^{234}Th is, in turn, the parent of ^{234}Pa.

20.5 RATE OF RADIOACTIVE DECAY

The rate of radioactive decay for any unstable nucleus is the number of nuclei decaying per unit time, dN/dt (see instantaneous rate, Section 17.2). All radioactive disintegration processes are kinetically of the first order, meaning that the rate is proportional to N, the number of parent nuclei present at any time. In mathematical form,

$$\text{rate of decay} = -\frac{dN}{dt} = kN$$

where k is the decay (or rate) constant. Unlike chemical reactions, the value of k in radioactive decay is independent of temperature, chemical combination, or other changes in the chemical or physical environment of the nucleus. It is a characteristic only of the particular decaying nucleus. Using calculus, the above equation can be reworked to

$$\log \frac{N_0}{N_t} = \frac{kt}{2.303}$$

or

$$k = \frac{2.303}{t} \log \frac{N_0}{N_t}$$

where N_0 is the number of parent nuclei originally present and N_t is the number remaining after time t. The time taken for one-half of the initial number of atoms to decay is called the *half-life*, $t_{1/2}$, and is characteristic of the parent since if $N_t = \frac{1}{2}N_0$, then

$$t_{1/2} = \frac{2.303}{k} \log 2 = \frac{0.693}{k}$$

The half-life for ^{14}C beta decay is approximately 6000 yr. Given 1000 ^{14}C atoms, 500 atoms will be left after 6000 yr. After another 6000 yr, only 250 atoms of ^{14}C will remain. Alternatively, on a macroscopic scale, 1000 g of ^{14}C will be decreased to 500 g in 6000 yr and to 250 g in another 6000 yr.

Half-lives are customarily used instead of decay constants to describe the relative radioactivity of a nucleus, and they have been included in Figure 20.1. The more radioactive a nucleus, the more disintegrations per unit time and the shorter the half-life.

Radioactive Dating
The ages of minerals and dead organic matter can be determined if the half-lives of appropriate constituent nuclei are known and if the concentrations of parent and daughter nuclei can be measured. For instance, the age of a mineral can be calculated from the concentrations of ^{238}U and ^{206}Pb in the mineral.

EXAMPLE
Analysis of a uranium-containing mineral shows the ratio of the number of gram-atoms of ^{206}Pb to ^{238}U to be 0.052. Since the ^{206}Pb could come only from ^{238}U originally present, 0.052 g-atoms of lead can be considered to have been derived from an initial 1.052 g-atoms of ^{238}U. The $t_{1/2}$ for ^{238}U determines the decay constant k for the entire ^{238}U series, since it is the longest-lived member. Therefore,

$$k = \frac{0.693}{4.5 \times 10^9 \text{ yr}}$$

$$= 1.5 \times 10^{-10}$$

$$= \frac{2.303}{t} \log \frac{^{238}U_0}{^{238}U_t}$$

$$1.5 \times 10^{-10} = \frac{2.303}{t} \log \frac{1.052}{1.052 - 0.052}$$

$$t = 3.4 \times 10^8 \text{ yr for the age of the rock}$$

This calculation assumes that no ^{238}U and no ^{206}Pb produced by the decay has escaped from the mineral.

EXAMPLE

Another way to determine the age of a mineral is to compare the relative abundances of two natural radioactive isotopes in the mineral. If both ^{238}U and ^{235}U are assumed to have been originally formed in approximately equal amounts (Section 20.13) and to have been present in equal amounts when a mineral was formed, their present difference in abundance will be the result of more rapid decay of ^{235}U, with $^{235}t_{1/2} = 7.1 \times 10^8$ yr as compared to $^{238}t_{1/2} = 4.5 \times 10^9$ yr for ^{238}U.

If the present abundances are ^{238}U, 99.28 percent, and ^{235}U, 0.72 percent, how long has it been since both had the same abundances?

SOLUTION

$$\log {}^{238}N_0 - \log {}^{238}N_t = {}^{238}k\frac{t}{2.303}$$

$$\log {}^{235}N_0 - \log {}^{235}N_t = {}^{235}k\frac{t}{2.303}$$

But we have assumed that $^{235}N_0 = {}^{238}N_0$. Therefore,

$$\log {}^{238}N_t + {}^{238}k\frac{t}{2.303} = \log {}^{235}N_t + {}^{235}k\frac{t}{2.303}$$

$$\log {}^{238}N_t - \log {}^{235}N_t = \log \frac{{}^{238}N_t}{{}^{235}N_t}$$

$$= \frac{t}{2.303}({}^{235}k - {}^{238}k)$$

but $^{238}N_t/^{235}N_t$ can be measured:

$$\log \frac{{}^{238}N_t}{{}^{235}N_t} = \log \frac{99.28}{0.72} = \log 138$$

Therefore,

$$\log 138 = \frac{t}{2.303}({}^{235}k - {}^{238}k)$$

$$t = \frac{2.303 \log 138}{{}^{235}k - {}^{238}k}$$

$${}^{235}k = \frac{0.693}{{}^{235}t_{1/2}} = \frac{0.693}{7.1 \times 10^8} = 9.8 \times 10^{-10} \text{ per yr}$$

and

$${}^{238}k = \frac{0.693}{4.5 \times 10^9} = 1.5 \times 10^{-10} \text{ per yr}$$

Therefore,

$$t = \frac{2.303 \log 138}{(9.8 - 1.5) \times 10^{-10}}$$

$$= \frac{(2.303)(2.14)}{8.3 \times 10^{-10}}$$

$$= 5.9 \times 10^9 \text{ yr}$$

Since the abundances given are the present average earthly abundances, this time value approximates the age of the earth if it is assumed that both uranium isotopes were present in equal amounts at the time of formation of the earth.

The beta decay of the radioactive carbon-14 isotope is the basis for the determination of the age of dead organic matter such as wood. Carbon-14 is continually formed in the upper atmosphere by the bombardment of nitrogen atoms by neutrons produced by cosmic rays:

$$\ce{_0^1n} + \ce{_7^{14}N} \longrightarrow \ce{_6^{14}C} + \ce{_1^1H}$$

The carbon-14 then decays by way of beta emission to $_7^{14}N$ with a half-life of 5760 yr. The radioactive carbon atoms readily enter the natural carbon cycle in CO_2 and carbonates and are incorporated into plants and animals. Studies have shown that an equilibrium ratio of ^{14}C to ^{12}C was established long ago in the atmosphere and, as a consequence, in living organisms. The carbon of living plants and animals is in equilibrium with the atmosphere and contains the same constant ratio of ^{14}C to ^{12}C. This ratio gives 15.3 beta disintegrations per minute per gram of carbon. When the living organism dies, no further ^{14}C is brought into its tissues, and the ratio of ^{14}C to ^{12}C decreases due to the decay of the unstable ^{14}C. The disintegration rate per gram of carbon decreases correspondingly.

By comparing the equilibrium disintegration rate of 15.3 per min to the actual decay activity of 1 g of carbon from an archeological specimen, the age, from the time of death, of the specimen can be determined. The equation

measured disintegration rate (per min), $R = 15.3e^{-0.693t/5760}$

or

$$\log \frac{15.3}{R} = \frac{0.693t}{(2.303)(5760)}$$

may be used to find t in years.

20.6 INDUCED NUCLEAR REACTIONS

Natural radioactive decay is concerned with the spontaneous disintegration of naturally occurring isotopes. These isotopes may have been present at the formation of the earth, they may have been decay products of the original radioisotopes, or they may have been formed by naturally occurring nuclear reactions as ^{14}C was formed by neutron bombardment of ^{14}N. However, nuclear reactions of the $^{14}N \longrightarrow {}^{14}C$ type and others may also be induced artificially if a supply of energetic bombarding particles is available.

The most obvious way in which induced (and natural) nuclear reactions differ from radioactive decay is in the presence of two reactants instead of one. One reactant is usually a relatively stable target nucleus. The nuclear reaction is initiated by bombarding the target nucleus with high-energy radiation such as X-radiation or gamma rays, or with projectile particles. Common projectile particles are neutrons, electrons, protons, alpha particles, and deuterons ($_1^2H$). Less common are tritons ($_1^3H$) and heavier nuclei such as ^{12}C, with six positive charges.

The products of the collision between the target nucleus and a bombarding species may be an isotope of the same or a different element plus various particles or radiation. The product nucleus itself may also be radioactive. If it is radioactive, it will decay spontaneously with the emission of particles or radiation or both until some stable nucleus is formed. Man-made or artificially induced nuclear reactions, as distinct from naturally occurring nuclear reactions, often yield artificial radioisotopes that no longer exist in nature because of their short half-lives. These man-made radioisotopes are important in many areas of scientific research, where they serve as easily detected "tagged" atoms in chemical reactions. Nuclear reactions in which new elements are produced are called *transmutation reactions* and are the modern version of the alchemist's attempts to convert base metals such as lead into gold.

As examples of nuclear reactions, we shall limit ourselves to two processes, one initiated by alpha particles and the other by neutrons.

Alpha-Induced Reactions   The two most common types of reactions induced by alpha particles lead to the production of protons or neutrons. These reactions are exemplified by

$$^{14}_{7}\text{N} + ^{4}_{2}\text{He} \longrightarrow ^{17}_{8}\text{O} + ^{1}_{1}\text{H} \quad \text{or} \quad ^{14}_{7}\text{N} \ (\alpha, p) \ ^{17}_{8}\text{O} \tag{1}$$

and

$$^{9}_{4}\text{Be} + ^{4}_{2}\text{He} \longrightarrow ^{12}_{6}\text{C} + ^{1}_{0}n \quad \text{or} \quad ^{9}_{4}\text{Be} \ (\alpha, n) \ ^{12}_{6}\text{C} \tag{2}$$

Note that the mass numbers and atomic numbers balance.

At the right of each equation is an abbreviated notation often used in writing nuclear reactions. The symbol for the target nucleus is followed, in parenthesis, by the projectile and the ejected particle or radiation, then by the symbol for the product nucleus.

The alpha particles used in these reactions are obtained directly from the decay of a natural radioactive isotope. With the exception of carbon and oxygen, all elements from boron to potassium give protons when bombarded with alpha particles obtained from radioactive decay. The ejection of protons from elements of higher atomic number requires that the alpha particle be artificially accelerated before striking the target nucleus. Reaction (2) led to the discovery of the neutron and can be used as a laboratory source of neutrons. Reaction (1) was the first artificially produced nuclear transmutation, announced by Rutherford in 1919.

Another alpha-induced reaction of historical interest involves magnesium:

$$^{24}_{12}\text{Mg} + ^{4}_{2}\text{He} \longrightarrow ^{27}_{14}\text{Si} + ^{1}_{0}n$$

The product ^{27}Si is an unstable isotope of silicon, one of the first artificial radioisotopes to be produced. It decays by positron emission to ^{27}Al:

$$^{27}_{14}\text{Si} \longrightarrow ^{27}_{13}\text{Al} + ^{0}_{1}e$$

A positron is a positively charged particle with the same mass and an electron.

The Compound Nucleus   It is often helpful to think of a nuclear reaction taking place by the merger of the target nucleus and the projectile to form an activated *compound nucleus*, which breaks up quickly (about 10^{-14} sec) into the products of the reaction. In the nitrogen–helium reaction of reaction (1), this compound nucleus would have been $^{18}_{9}\text{F}$. This compound nucleus is at a much higher energy level than a normal $^{18}_{9}\text{F}$ nucleus, because it contains most of the kinetic energy of the incoming alpha particle. This energy is distributed among the protons and neutrons in the excited nucleus. At some time in the 10^{-14}-sec period, a proton obtains more than the average energy and is ejected from the nucleus in much the same way that an energetic molecule evaporates from the surface of a liquid. In other systems, a neutron or an alpha particle is ejected rather than a proton, but the mechanism is the same.

Neutron-Induced Reactions

Examples of neutron induced reactions are

$$^{14}_{7}N + ^{1}_{0}n \longrightarrow ^{14}_{6}C + ^{1}_{1}H \quad \text{or} \quad ^{14}_{7}N \, (n, \, p) \, ^{14}_{6}C$$

$$^{27}_{13}Al + ^{1}_{0}n \longrightarrow ^{24}_{11}Na + ^{4}_{2}He \quad \text{or} \quad ^{27}_{13}Al \, (n, \, \alpha) \, ^{4}_{2}He$$

$$^{23}_{11}Na + ^{1}_{0}n \longrightarrow ^{24}_{11}Na + \gamma \quad \text{or} \quad ^{23}_{11}Na \, (n, \, \gamma) \, ^{24}_{11}Na$$

All the product nuclei in these reactions happen to be radioactive and decay further by electron or positron emission. These emissions are usually accompanied by gamma radiation:

$$^{14}_{6}C \longrightarrow ^{14}_{7}N + ^{0}_{-1}e \qquad t_{1/2} = 5760 \text{ yr}$$

$$^{24}_{11}Na \longrightarrow ^{24}_{12}Mg + ^{0}_{-1}e \qquad t_{1/2} = 15 \text{ hr}$$

A special type of neutron-induced reaction is the fission reaction that occurs with heavy elements. After the nucleus of the heavy element absorbs a neutron, the activated compound nucleus breaks up into 2 nuclei of about equal mass, and 2 or more neutrons are ejected:

$$^{235}_{92}U + ^{1}_{0}n \longrightarrow ^{142}_{56}Ba + ^{91}_{36}Kr + 3 \, ^{1}_{0}n$$

Reactions of this type are the basis of the atomic fission bomb and of nuclear fission reactors for energy and neutron production. Details of fission reactions will be discussed in Section 20.12.

Potential Barriers

The products of a nuclear reaction, or the way the compound nucleus splits up, are determined by a number of factors. Among these are the nature of the target nucleus, the nature of the incident projectile, and the kinetic energy of the projectile. Charged projectiles like the alpha particle and the proton must overcome the electrostatic repulsion of the positively charged nucleus before they can enter the nucleus to form a compound nucleus. This potential barrier varies with the nuclear charge and with the charge on the particle, increasing with increasing positive charge on either reactant particle. The height of the potential barrier increases as the incident particle approaches the nucleus until the projectile is sufficiently close to be attracted to the nucleus by the attractive nuclear forces among neutrons and protons. If the projectile has sufficient kinetic energy to surmount the coulombic potential barrier, it can enter the nucleus and transfer its kinetic energy to the compound nucleus. The minimum energy required to get over the barrier is analogous to the activation energy of chemical reactions. There is no potential barrier for the collision of a neutron with a nucleus. Thus, low-energy neutrons are ideal projectiles, their only disadvantage being that, because they are uncharged, they cannot be speeded up to higher kinetic energies in particle accelerators. In addition to the potential energy barrier for the entry of a positive projectile into the target nucleus, there is also a similar barrier that must be overcome for the ejection of a positive particle from the compound nucleus.

After using some of its kinetic energy (unless it is a neutron) to enter the nucleus, some or all of the remaining kinetic energy of a projectile is transferred to the compound nucleus. The exact reaction products depend on how much energy is transferred. That is, a number of products may be formed from the same reactants, depending on the energy of the incoming projectile.

20.7 SOME APPLICATIONS
OF INDUCED NUCLEAR
REACTIONS AND THEIR
PRODUCTS
Neutron Activation Analysis

When a sample is irradiated with particles, the atoms in the sample often form radioactive products. The emissions of the products can be easily detected and measured with appropriate radiation detectors or disintegration rate counters and are, therefore, useful for analytical purposes. In neutron activation analysis, an intense beam of low-energy neutrons is absorbed by elements in the sample. The products of the capture of neutrons by nuclei emit gamma rays and, possibly, particles as well.

The simplest gamma-ray emission occurs in the small number of cases in which a bombarding neutron produces an excited state of the parent nucleus, this excited nucleus then returning to its ground state with the emission of gamma radiation:

$$^{86}_{38}\text{Sr} + ^{1}_{0}n \longrightarrow ^{87}_{38}\text{Sr*}$$
$$^{87}_{38}\text{Sr*} \longrightarrow ^{87}_{38}\text{Sr} + \gamma$$

More often, the parent nucleus is transmuted to an excited state of another element, which decays, as shown in the previous section, through electron or positron emission accompanied by gamma radiation. In any case, the energy or the wavelength of the gamma radiation has a discrete value or set of values characteristic of the decaying nucleus. The radioactive or excited nucleus is, in turn, characteristic of the original absorbing nucleus. Because the intensity of the characteristic gamma radiation is proportional to both the number of emitting and absorbing nuclei, a measurement of intensity at the proper wavelengths allows a determination of the concentration of the activated parent nucleus in the sample. Neutron activation analysis can identify and determine the concentration of some elements present in amounts of only 10^{-12} g and is a powerful and accurate method for the analysis of trace elements and environmental contaminants such as mercury and lead in foods and organisms. Neutron activation analysis can also be used for the detection of cystic fibrosis in infants, because the one infant in 2500 with this disease has a higher concentration of copper in his fingernails than does a normal infant.

Incidentally, the fact that decaying nuclei radiate gamma radiation of definite energies shows that nuclei exist in certain discrete energy levels determined by nuclear quantum numbers that govern the occupancy of neutrons and protons in the nuclear energy levels. These allowed nuclear energy levels and nuclear quantum numbers are analogous to the electronic levels and quantum numbers for an atom.

Tracer Studies

The characteristic and easily detectable emissions resulting from the decay of radioactive isotopes, either natural or man-made, are widely used to follow a substance through chemical and physical processes. In ecological studies, the path followed by radioactive isotopes through biogeochemical cycles aids in tracing the various reactions making up the cycles and in deducing the magnitude of biological magnification of substances in ecosystems. The assumptions made in tracer studies are that the chemical properties of the radioactive isotope are identical to those of the stable isotopes, that the radioactive nature of the isotopes does not change its chemical or physical properties, and that the disintegration rates, as measured by suitable detectors, are proportional to the number of unstable nuclei present. Many of the widely used radioactive tracers do not exist in nature. Artificially induced nuclear reactions have greatly increased the scope of tracer studies by providing man-made radioisotopes with appropriate chemical and decay characteristics.

20.8 STABILITY OF NUCLEI

Of over 1600 different nuclear species known, only 325 are naturally occurring species. The rest are artificial radioactive nuclei produced by induced nuclear reactions. About 275 of the naturally occurring nuclei are stable. What makes some nuclei stable and some unstable? The answer must lie in the forces holding the nulcear particles, the protons and the neutrons, together in the nucleus. There must be attractive forces between these particles. There must also be repulsive forces between like-charged protons. In radioactive nuclei, the repulsive forces apparently are greater than the attractive forces.

Little is known about the attractive forces except that they are very different in nature and magnitude from the electrostatic forces we usually encounter in chemistry. Attractive nuclear forces operate only between adjacent protons, between adjacent neutrons, and between adjacent neutrons and protons. That is, the attractive forces are very short range and are nonelectrostatic interactions. By contrast, proton–proton electrostatic repulsion forces, although weaker at short range than nuclear forces, are stronger at long range and operate at greater particle separation. Their action is not limited to adjacent protons or to proton pairs alone. In the larger nuclei of heavy atoms, all the protons are affected by the electrostatic repulsive forces, but the attractive forces are effective only between pairs of adjacent nuclear particles. Partially as a result of these factors, it is found that there are stable nuclei of the lighter elements having neutron-to-proton ratios of unity, that is, $Z = N$. With the heavier elements, above $Z = 20$, stable nuclei always must contain more neutrons than protons because more neutrons are needed to counter the increased proton–proton coulombic repulsions. As Figure 20.2 shows, there are optimum neutron-to-proton ratios for stability of nuclei. Unstable nuclei lying outside the stability range will tend to readjust their N/Z ratios to stable ratios by various radioactive decay processes.

Nuclei with high N/Z ratios generally attain stability by electron emission. As before, the electron is created in the nucleus by the conversion of a neutron into a proton and an electron. The net result is a decrease of one in the value of N and an increase of one in the value of Z, leading to a decrease in the N/Z ratio:

$$_{Z}^{A}X \longrightarrow _{Z+1}^{A}Y + _{-1}^{0}e$$

Beta emission occurs for both natural and synthetic radioisotopes of this type.

Nuclei with low N/Z ratios go to stable nuclei by increasing the N/Z value. The most common ways of doing this are positron emission, electron capture, and alpha decay. Positron emission is known only among synthetic radioisotopes. The positron is considered to be produced, in the unstable nucleus, from the conversion of a proton into a neutron,

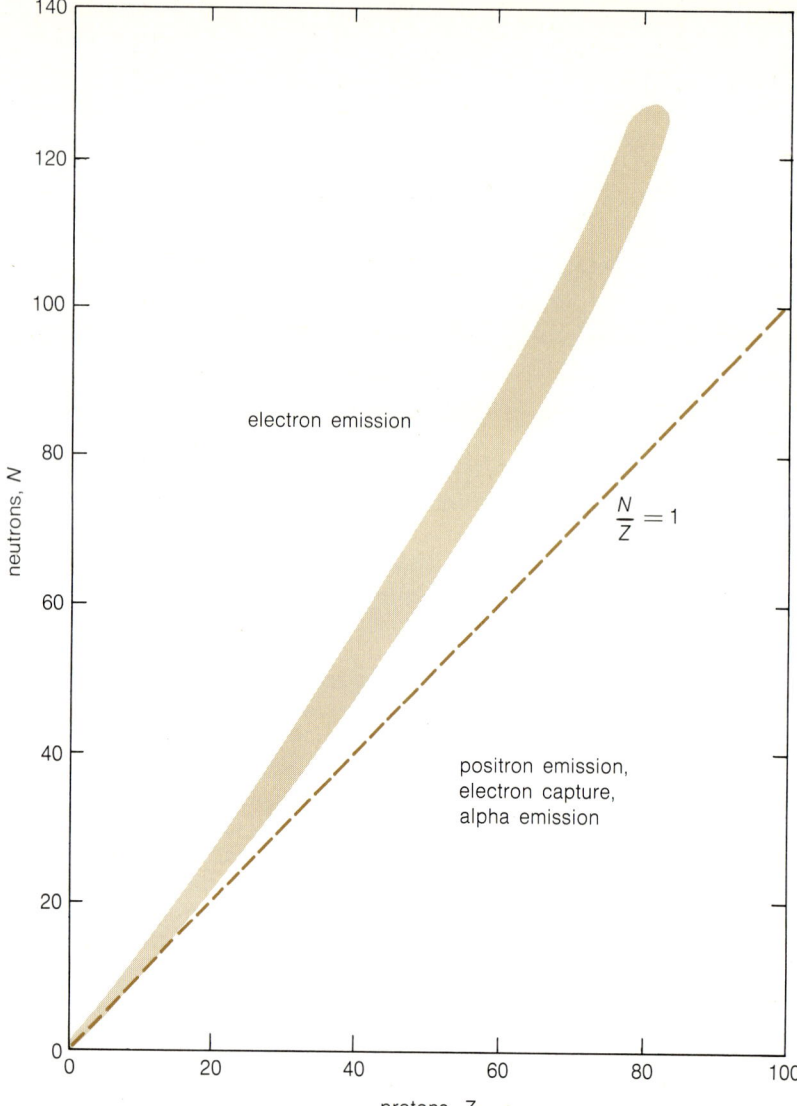

Figure 20.2. Stable nuclei exist within a narrow range of neutron-to-proton ratios. The shaded area represents the region of nuclear stability. Nuclei with larger numbers of neutrons than optimum lie above the stable nuclei and tend to attain a stable neutron-to-proton ratio by electron (beta) emission. Unstable nuclei with too small a neutron-to-proton ratio decay by positron or alpha emission or electron capture.

$$\frac{1}{1}p \longrightarrow \frac{1}{0}n + \frac{0}{1}e$$

The reaction for the element X is

$$\frac{A}{Z}X \longrightarrow \frac{A}{Z-1}Y + \frac{0}{1}e$$

Electron capture results from the capture of an inner orbital electron, usually one in the $1s$ orbital, by the nucleus, with the conversion of a proton into a neutron:

$$\frac{1}{1}p + \frac{0}{-1}e \longrightarrow \frac{1}{0}n$$

and

$$^A_ZX + ^0_{-1}e \longrightarrow ^A_{Z-1}Y$$

(The symbol X does not include the orbital electrons.) This process is usually accompanied by the emission of characteristic X-radiation as a higher energy electron drops into the vacant inner orbital.

20.9 BINDING ENERGY

Another, more quantitative, measure of nuclear stability is the binding energy of the nucleus. This is the energy required to break up a nucleus into its component protons and neutrons. The binding energy cannot be measured directly, but it can be easily calculated by using Einstein's relation for the equivalence of mass and energy, $E = mc^2$. This relation can also be written as $\Delta E = (\Delta m)c^2$, where ΔE and Δm are the energy change and mass change, respectively, in a nuclear reaction. As an example, the binding energy of the $^{23}_{11}Na$ nucleus will be calculated using the hypothetical reaction

$$^{23}_{11}Na \longrightarrow 11\ ^1_1H + 12\ ^1_0n$$

In atomic mass units, the rest masses of the $^{23}_{11}Na$ *atom* (including 11 electrons), the hydrogen *atom* (proton plus electron), and the neutron are

$$m_{^{23}Na} = 22.989773\ \text{amu}$$
$$m_{^1_1H} = 1.007825\ \text{amu}$$
$$m_{^1_0n} = 1.008665\ \text{amu}$$

The mass change in the reaction is

$$\Delta m = \text{mass of products} - \text{mass of reactants}$$
$$\Delta m = 11(1.007825) + 12(1.008665) - 22.989773$$
$$\Delta m = 11.086075 + 12.103980 - 22.989773$$
$$\Delta m = 0.200282\ \text{amu}$$

Since there is no change in the number of extranuclear electrons, the mass change is entirely nuclear in origin. For the breakup of the ^{23}Na nucleus there is an increase in mass.

From Einstein's relation, the energy equivalent of the mass difference may be calculated in various units. With $c = 3 \times 10^{10}$ cm/sec and m in grams, E (or ΔE) comes out in ergs. One atomic mass unit (amu) is $1/(6.023 \times 10^{23})$ g, so 1 amu of mass is equivalent to 14.94×10^{-4} ergs. In discussions of nuclear reactions, it is common to express energy in terms of electron volts, eV, or millions of electron volts, MeV. One MeV is the energy gained by an electron when it is accelerated through a potential difference of 10^6 volts. The energy corresponding to 1 MeV is $\Delta E = q_e V = (1.602 \times 10^{-19}$ coulomb) $\times$ $(10^6$ volts$) = 1.602 \times 10^{-13}$ joules. Using the conversion 1 joule $= 10^7$ ergs, we can get the important relation

$$1\ \text{amu} = 931\ \text{MeV}$$

(Other energy relationships that will be useful later are 1 MeV $= 3.83 \times 10^{-14}$ cal $= 4.45 \times 10^{-20}$ kW hr.) The *total binding energy* of the ^{23}Na nucleus is then $(0.20028\ \text{amu})(931\ \text{MeV/amu}) = 186\ \text{MeV}$. This is the energy required to decompose the nucleus into its constituent particles. Conversely, it is the energy that would be released if it were possible to bring the 11 protons and 12 neutrons together to form the nucleus.

For comparison with other nuclei, it is more appropriate to use the binding energy per nuclear particle (nucleon). This is obtained by dividing the total binding energy by the atomic mass number A, the sum of the protons and neutrons. The binding energy per nucleon in ^{23}Na is $186/23 = 8.1$ MeV. If the binding energies per nuclear particle for stable nuclei are graphed against mass number, the curve in Figure 20.3 is obtained. Nuclei with the higher

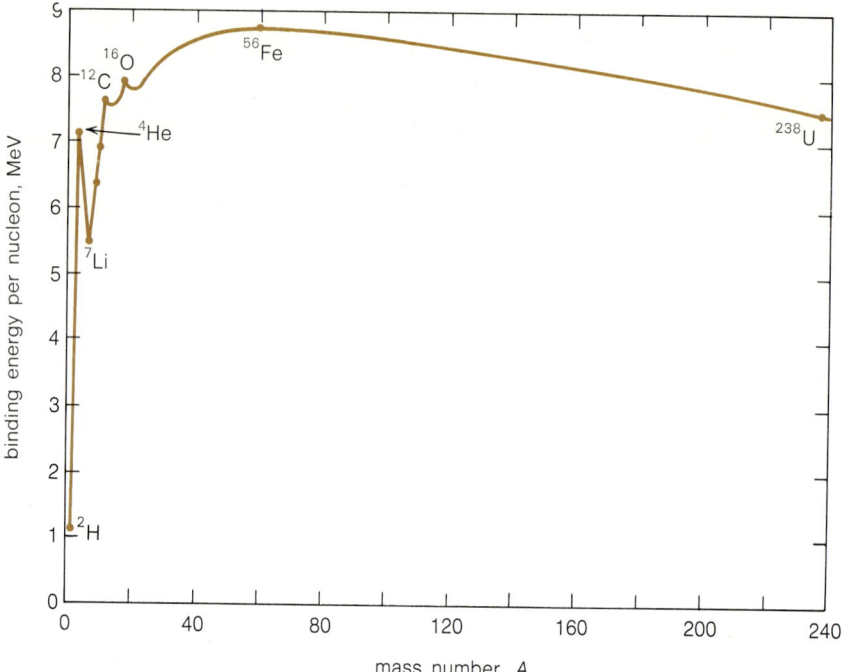

Figure 20.3. Nuclear binding energies. Binding energies per nucleon (neutrons and protons are both called nucleons) for the stable nuclei fall on a curve with a maximum near ^{56}Fe. Radioactive nuclei tend to lie in the area below the curve.

binding energies are the most stable, because it takes more energy to break up the nucleus. Nuclei with lower binding energies tend to transform, subject to potential energy barriers, to more stable nuclei.

20.10 ENERGETICS OF NUCLEAR REACTIONS

The use of the mass–energy equivalence in the calculation of the total binding energy of a nucleus shows how the energy changes in nuclear reactions can be calculated. Nuclear transformations, like chemical reactions, can be exoergic or endoergic. As with chemical reactions, spontaneous nuclear processes, such as radioactive decay, are energy evolving, the mass of the products being less than the mass of the reactants. The mass lost in the reaction reappears as the kinetic energy of the products and as radiation. Some induced nuclear reactions, although exoergic and thermodynamically spontaneous, require a certain kinetic energy for the projectile particles so that they can surmount the potential barrier to their entry to the nucleus.

Other induced nuclear reactions are endoergic, the reaction being accompanied by an increase in mass of the system. The increase in mass of the products over the reactants is supplied in the form of kinetic energy of the incident projectile particle. In addition to the reaction energy that must be supplied for an endoergic reaction, an even larger amount of projectile energy may be necessary to enable the projectile to overcome any electrostatic potential energy barrier on entering the target nucleus.

As already mentioned, the energy changes in nuclear reactions are of a much higher order of magnitude than those in ordinary chemical reactions. The energy evolved in an ordinary chemical reaction is around 100 kcal/mol. A low-energy nuclear process might evolve 1 MeV *per nucleus*. For a mole (or gram-atom) of nuclei undergoing change, the total energy evolved would be, in calories,

$$(1 \text{ MeV})(3.83 \times 10^{-14} \text{ cal/MeV})(6.02 \times 10^{23} \text{ mol}^{-1})$$
$$\simeq 23 \times 10^9 \text{ cal/mol}$$
$$= 23 \times 10^6 \text{ kcal/mol}$$

Thus, a typical nuclear reaction is some one million times more energetic than a typical chemical reaction.

EXAMPLE | For the reaction

$$^{14}_{7}\text{N} \ (\alpha, p) \ ^{17}_{8}\text{O}$$

or

$$^{14}_{7}\text{N} + ^{4}_{2}\text{He} \longrightarrow ^{1}_{1}\text{H} + ^{17}_{8}\text{O}$$

the masses of the product and reactant *atoms* are

$$m_{^{1}\text{H}} = 1.007825 \text{ amu}$$
$$m_{^{17}\text{O}} = 16.999133 \text{ amu}$$
$$m_{^{4}\text{He}} = 4.002604 \text{ amu}$$
$$m_{^{14}\text{N}} = 14.003074 \text{ amu}$$

$$\Delta m = \text{mass of products} - \text{mass of reactants}$$
$$= (1.007825 + 16.999133) - (4.002604 + 14.003074)$$
$$= 18.006958 - 18.005678$$
$$= 0.001280 \text{ amu}$$
$$\Delta E = (0.001280 \text{ amu})(931 \text{ MeV/amu}) = 1.19 \text{ MeV}$$

In this calculation, the atomic masses include the masses of the electrons, but the electron masses occur on both sides of the equation and cancel. This reaction is endoergic. The alpha particle must have 1.19 MeV in addition to the kinetic energy needed to overcome the nuclear repulsion before this reaction can take place.

20.11 NUCLEAR FISSION | A fission reaction is one in which the activated compound nucleus breaks up into two fragments of approximately equal mass. Both spontaneous and particle-induced fission reactions are known. The reactions of interest here are neutron-induced fissions of the heavy elements uranium, thorium, and plutonium, such as is illustrated by $^{235}_{92}\text{U}$:

$$^{235}_{92}\text{U} + ^{1}_{0}n \ (0.03 \text{ eV}) \longrightarrow ^{142}_{56}\text{Ba} + ^{91}_{36}\text{Kr} + 3 \ ^{1}_{0}n \ (1.8 \text{ MeV})$$

This particular reaction has been induced by *thermal* neutrons of 0.03-eV kinetic energy, the energy they would have at room temperature. By contrast, the ejected neutrons are *fast* neutrons.

This is not the only reaction that will occur when the ^{235}U nucleus absorbs a neutron. Other products may result, but the products generally have mass numbers in the range 70–166. When a heavy nucleus captures a neutron and then undergoes fission, the usual products are two fragments, one of higher mass number than the other, plus several neutrons. One of the many other possibilities is

$$^{235}_{92}U + ^{1}_{0}n \longrightarrow [^{236}_{92}U] \longrightarrow ^{133}_{51}Sb + ^{99}_{41}Nb + 4\ ^{1}_{0}n$$

Such ^{235}U fission reactions produce about 200 MeV of energy release, together with an average of 2.5 neutrons. These reactions occur with highest probability with low-energy thermal neutrons. About 85 percent of the 200 MeV released appears as kinetic energy of the fission fragments.

As with the natural isotope $^{235}_{92}U$, the artificial nuclei $^{233}_{92}U$ and $^{239}_{94}Pu$ are both fissionable by both slow and fast neutrons, but the probability of fission occurring is again much larger with the slower neutrons. The most common uranium isotope, $^{238}_{92}U$, and the common thorium isotope, $^{232}_{90}Th$, are fissionable only with fast neutrons. However, particularly with ^{238}U, a competing reaction, the radiative capture of 11 eV neutrons is more probable than fission. This reaction,

$$^{238}_{92}U + ^{1}_{0}n\ (11\ eV) \longrightarrow ^{239}_{92}U + \gamma$$

is of great importance, because the product ^{239}U decays to the slow neutron-fissionable ^{239}Pu by the steps

$$^{239}_{92}U \xrightarrow[t_{1/2}\ =\ 24\ min]{\beta^-} ^{239}_{93}Np \xrightarrow[2.3\ days]{\beta^-} ^{239}_{94}Pu$$

Thorium-232 also absorbs fast neutrons and eventually produces slow neutron-fissionable ^{233}U.

$$^{232}_{90}Th + ^{1}_{0}n \longrightarrow ^{233}_{90}Th \xrightarrow[22\ min]{\beta^-} ^{233}_{91}Pa \xrightarrow[27\ days]{\beta^-} ^{233}_{92}U$$

Chain Reactions

A neutron fission reaction produces more neutrons than are needed to initiate the reaction. If the neutrons produced are used as reactants for further fission events, chain reactions can occur, as shown in Figure 20.4. Such reactions are the basis of atomic fission bombs and of nuclear reactors for power production. A chain reaction that is allowed to proceed rapidly, in an uncontrolled manner, will release large amounts of energy in a short time (10^{-4} sec) and result in a devastating explosion. If, however, the reactions are controlled so that the energy is released gradually, then an atomic pile or nuclear reactor may be constructed for power production.

Let us first consider the uncontrolled neutron fission reaction of ^{235}U or of ^{239}Pu. Given a mass of ^{235}U, a fission reaction might be initiated by a neutron resulting from a rare spontaneous fission of ^{235}U or from cosmic rays. If three fast neutrons are generated by this initial fission, they may then cause fission of three other ^{235}U, producing 9 neutrons, which, in turn, might produce 27 neutrons, each capable of initiating 27 new fission processes or $3n$ fissions after n events. This buildup of ever-increasing numbers of ^{235}U fissions takes place in an extremely short period of time, each fission process releasing about 200 MeV of energy. The result is an explosion.

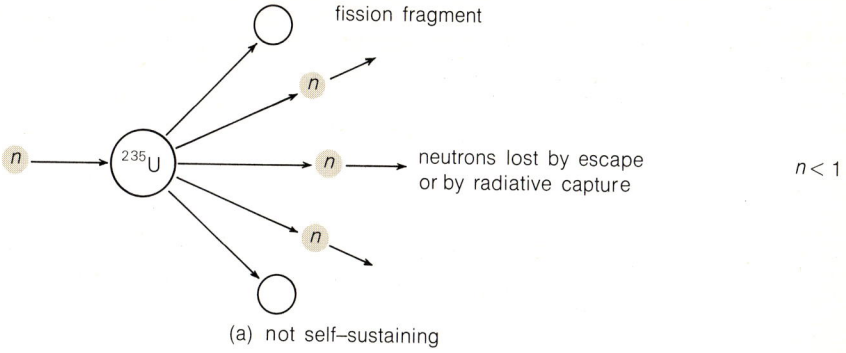

fission fragment

neutrons lost by escape
or by radiative capture

$n < 1$

(a) not self–sustaining

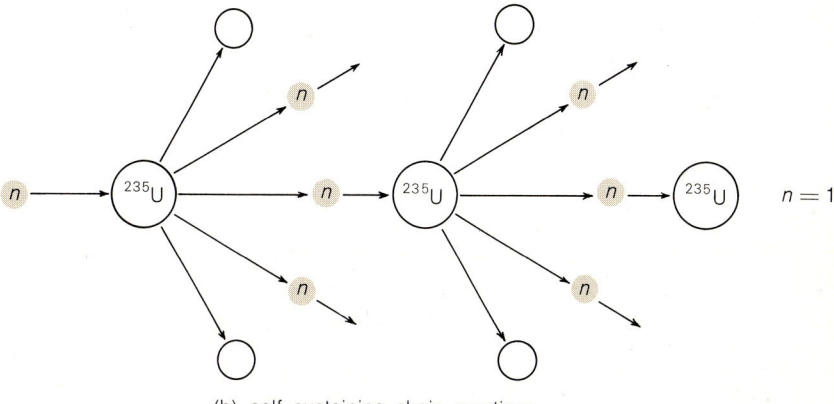

$n = 1$

(b) self–sustaining chain reaction

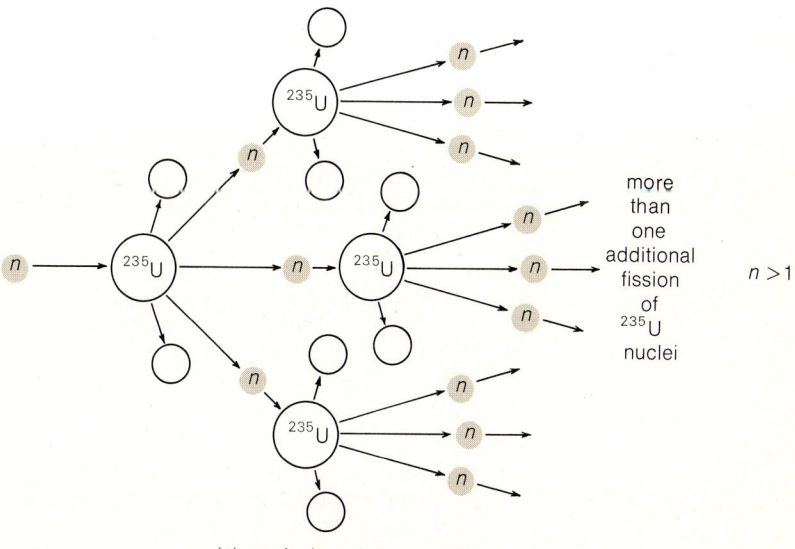

more
than
one
additional
fission
of
^{235}U
nuclei

$n > 1$

Figure 20.4. (a) If less than 1 of the
neutrons ejected by fission is effective in
causing further fission, a chain reaction
cannot be sustained. (b) If 1 neutron
from each fission event goes on to cause
another fission, a self-sustaining chain
reaction takes place. (c) If more than
1 neutron per fission is effective in producing additional fissions, an explosion
results.

(c) explosive chain reaction

The ideal conditions for such an explosive chain reaction are that (1) none of the neutrons are absorbed by impurities in nonfission processes and that (2) neutrons are not lost by escape from the reacting material. For these requirements to be fulfilled, the ^{235}U or ^{239}Pu in the mass must be extremely pure. Uranium-238, while fissionable with fast neutrons, can, if present in any quantity, cause neutron loss by the competing ^{238}U (n, γ) ^{239}U reaction. In addition, the ratio of mass to surface area of the hunk of fissionable material must be large enough so that most neutrons remain inside the mass to cause fission rather than escaping through the surface.

The size factor leads to the existence of a critical size or mass below which a chain reaction cannot take place because of the loss of neutrons through the surface of the fissionable ^{235}U or ^{239}Pu. A mass of ^{235}U, for instance, that is below the critical size will not explode, but if it is joined to another piece so that the combined size is greater than critical, an explosion will result. The critical size is that in which more than 1 neutron of the 2–3 produced in a fission reaction is available to cause another fission. For pure ^{235}U of normal density, the critical size is a sphere about 10 cm diameter.

As mentioned above, ^{238}U, ^{235}U, and ^{239}Pu are all fissionable by fast neutrons produced in fission reactions. However, ^{238}U also swallows up these neutrons through the competing ^{238}U (n, γ) ^{239}U reaction. Therefore, in producing a bomb it was necessary to use pure ^{235}U or ^{239}Pu. The ^{235}U isotope occurs to the extent of 0.72 percent in naturally occurring uranium. The rest is almost all (99.28 percent) ^{238}U. A major problem was the separation of the desired ^{235}U from the ^{238}U. Because these are isotopes of the same chemical element, there are no significant chemical differences to speak of that would allow their separation. Instead, slight differences in their physical properties due to their different masses were used to effect separation. Thus, lighter gaseous $^{235}UF_6$ diffuses faster through porous barriers than $^{238}UF_6$ (Graham's law), $^{238}UF_6$ settles faster in the high gravitational field of a centrifuge, and lighter ^{235}U ions are deflected more in a magnetic field than ^{238}U ions. The production of ^{239}Pu (also used in bombs) from ^{238}U will be discussed later.

The deadly effects of an atom bomb explosion are not confined to the physical damage done by the shock wave. The explosion of a fission bomb is accompanied by large quantities of lethal neutrons and gamma rays, which can damage the cells and molecules in living organisms, often causing genetic mutations in succeeding generations long after the explosion. The bomb also generates intense radiation, from ultraviolet to visible and infrared, which can cause eye and skin burns at a large distance from the target.

Radioactive fallout is the term given to the variety of fission products that are thrown up by the explosion. Many of these products are radioactive (mostly beta and gamma emitters) and can be carried literally around the world by currents in the atmosphere before they fall to earth. There they may be absorbed by plants and, in turn, by animals, with accompanying adverse effects. One of the most dangerous of these radioactive products is long-lived ^{90}Sr, which accumulates in the bones in place of calcium. There it affects the red blood cells and leads to leukemia or cancer of the blood.

At one time there were high hopes for the use of nuclear fission explosions for peaceful purposes such as large-scale excavation. However, the *Plowshare* program to find safe applications has so far succeeded only in using underground explosions to fracture natural gas-bearing rock formations to increase the flow of gas. Although gas production may be increased tremendously (after removing residual radioactivity), the cost of the project has so far exceeded the value of the gas produced.

20.12 CONTROLLED NUCLEAR FISSION

An uncontrolled nuclear fission explosion is not of any use for the production of energy for industrial use. If the energy released by nuclear fission is to be harnessed, some way has to be found to control the number of neutrons released so that the chain reaction does not get out of hand. This means that, on the average, only 1 of the 2 to 3 neutrons produced by each fission process must be used for a succeeding fission. If less than 1 is available, the chain dies out; if more than 1 is available, the chain grows explosively.

Thus, an important part of any nuclear reactor designed to give a self-sustaining, but controlled, chain reaction are *control rods* that can be inserted into the fissionable material to regulate the number of neutrons. These control rods are made of materials that readily capture neutrons, generally by an (n, γ) process. Cadmium, tantalum, and boron carbide are among the substances used as neutron absorbers in control rods.

The next point to consider is the optimum energy of the neutrons in the fissionable material that makes up the core of the nuclear reactor. Both ^{235}U and ^{239}Pu are fissionable with slow or fast neutrons, but the probability of fission is much greater with very slow thermal neutrons having an energy of only 0.03 eV. The neutrons produced as products of a fission reaction are quite fast. For greatest efficiency, these must be slowed down but not captured. In addition, if there is any ^{238}U mixed with the ^{235}U, most neutrons must be slowed down rapidly to the thermal range before ^{238}U can grab them while they are in the 11-eV range that causes the ^{238}U (n, γ) ^{239}U reaction. To accomplish this slowing down, *moderators* are used. A moderator must have two qualities:

1. It must be, or contain, a light element, because loss of energy by the neutron is greatest on elastic collision (not capture) with a light atom with mass not too much larger than the neutron mass.

2. It must not capture neutrons by any nuclear reaction, because then the neutrons would not be available for fission.

Common moderators are H_2O, D_2O, and graphite.

Finally, the kinetic energy of the fission products is shared by impact with other atoms in the reactor with generation of large amounts of heat. This heat represents the energy produced by the fission. The heat is carried away from the pile by circulating a coolant through the pile. The coolant carries the heat to a heat exchanger, where it is transferred to ordinary water to generate steam. Water, various gases, and liquid sodium have been used as coolant–heat exchangers. In a modern helium-cooled reactor, helium at 47 atm and 780°C circulates about the core and transfers heat to water.

A shielding material such as concrete is needed to prevent escape of neutrons and gamma radiation from the reactor. Figure 20.5 shows how all these components go together in a nuclear reactor for power production.

Present-day power reactors use natural uranium or uranium slightly enriched in the ^{235}U isotope. The fuel is usually in the form of the heat-stable oxide, U_3O_8. The fission of ^{235}U supplies fast neutrons, some of which are slowed to 11 eV and are captured by ^{238}U to form ^{239}Pu in the (n, γ) process.

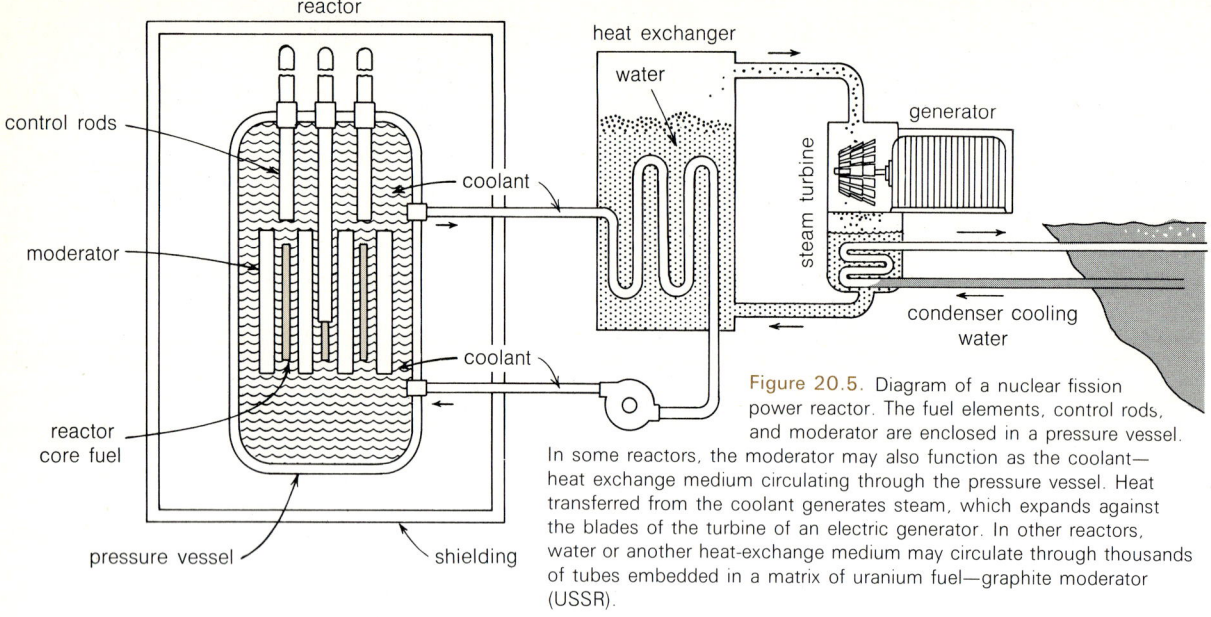

Figure 20.5. Diagram of a nuclear fission power reactor. The fuel elements, control rods, and moderator are enclosed in a pressure vessel. In some reactors, the moderator may also function as the coolant—heat exchange medium circulating through the pressure vessel. Heat transferred from the coolant generates steam, which expands against the blades of the turbine of an electric generator. In other reactors, water or another heat-exchange medium may circulate through thousands of tubes embedded in a matrix of uranium fuel—graphite moderator (USSR).

Other neutrons escape capture by ^{238}U if they have been slowed down rapidly enough. When these neutrons reach thermal energies, they are absorbed by ^{235}U nuclei in the fuel and cause energy release through fission. The amount of ^{239}Pu produced in these thermal reactors is negligible compared to the amount of ^{235}U consumed. For this reason, the days of these reactors are numbered. They would eventually consume the world's reserves of ^{235}U, the only naturally occurring isotope fissionable by slow neutrons, without producing enough of the synthetic fissionable ^{239}Pu to replace the ^{235}U.

Breeder Reactors

It is possible to design reactors that will produce more fissionable fuel (^{239}Pu or ^{233}U) than is consumed as ^{235}U. In these *breeder reactors*, one of the neutrons emitted by the fission of ^{235}U or ^{239}Pu in the reactor core is used to sustain the fission process. The remainder of the fission neutrons (1.5 on the average) are absorbed by ^{238}U or ^{232}Th surrounding the core, with the production of fissionable ^{239}Pu or ^{233}U, respectively, as shown in Figure 20.6.

There are many technological, design, and safety problems to be solved before breeder reactors become operational for large-scale electric power generation. The first commercial breeder in the United States, a liquid metal (Na) cooled fast breeder reactor (LMFBR) using the ^{238}U–^{239}Pu cycle, is expected to become operational in the early 1980's.

20.13 NUCLEAR FUSION

A nuclear reaction in which light-weight nuclei combine to form heavier and more stable nuclei is called a *fusion* reaction. As the binding energy curve indicates, such reactions are capable of releasing especially large quantities of energy. Fusion reactions occur in stars and in the hydrogen bomb. The fusion process has not yet been controlled to provide energy for peaceful purposes on earth, but it is discussed here because of the important long-range implications of fusion power.

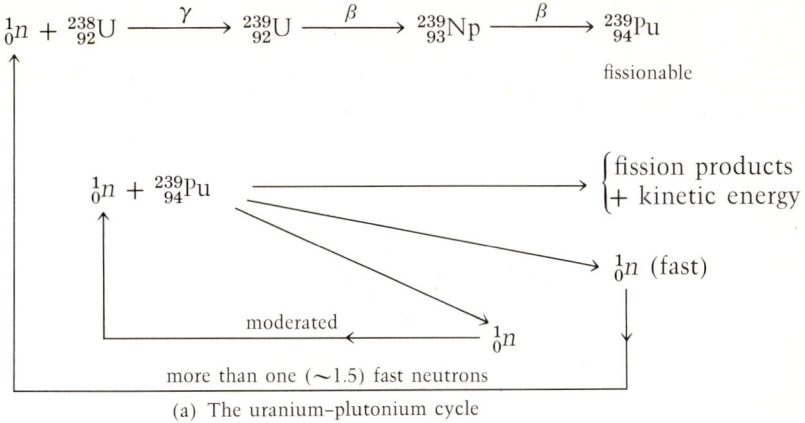

$$\begin{array}{ccccccc} {}^{1}_{0}n + {}^{238}_{92}U & \xrightarrow{\gamma} & {}^{239}_{92}U & \xrightarrow{\beta} & {}^{239}_{93}Np & \xrightarrow{\beta} & {}^{239}_{94}Pu \end{array}$$

fissionable

$${}^{1}_{0}n + {}^{239}_{94}Pu \longrightarrow \begin{cases} \text{fission products} \\ + \text{ kinetic energy} \end{cases}$$

$${}^{1}_{0}n \text{ (fast)}$$

moderated $\quad {}^{1}_{0}n$

more than one (~1.5) fast neutrons

(a) The uranium–plutonium cycle

$$\begin{array}{ccccccc} {}^{1}_{0}n + {}^{232}_{90}Th & \xrightarrow{\gamma} & {}^{233}_{90}Th & \xrightarrow{\beta} & {}^{233}_{91}Pa & \xrightarrow{\beta} & {}^{233}_{92}U \end{array}$$

fissionable

fission products + kinetic energy

$${}^{1}_{0}n + {}^{233}_{92}U \qquad \qquad {}^{1}_{0}n \text{ (fast)}$$

$${}^{1}_{0}n$$

moderated

more than one (~1.5)
slow neutrons

moderate

(b) The thorium–uranium cycle

Figure 20.6. Breeding, the production of more slow-neutron-fissionable fuel than is consumed, results when the average 1.5 neutrons not involved in sustaining the fission chain reaction are used to convert nonfissionable ^{238}U or ^{232}Th to slow-neutron-fissionable ^{239}Pu or ^{233}U. Fast neutrons work best in the U–Pu cycle, slow neutrons best in the Th–U cycle. Any slow-neutron-fissionable isotope (^{235}U, ^{239}Pu, or ^{233}U) could be used to start either cycle.

An example of a fusion reaction is

$${}^{2}_{1}H + {}^{3}_{1}H \longrightarrow {}^{4}_{2}He \text{ (3.5 MeV)} + {}^{1}_{0}n \text{ (14.1 MeV)}$$

The total energy released is 17.6 MeV per fusion event. On a mass or nucleon basis, this reaction is considerably more energetic than a fission reaction. This particular reaction has been used in the hydrogen bomb, but other fusion processes, which produce helium may also be possible, for example,

$${}^{7}_{3}Li + {}^{1}_{1}H \longrightarrow 2 {}^{4}_{2}He + 17.5 \text{ MeV}$$

Because of the high potential barrier for bringing two positive nuclei together in any fusion reaction, very high temperatures, in the neighborhood of $100 \times 10^{6}°C$ (10 keV), are required to give the reactants sufficient kinetic energy to initiate the reaction. For this reason, fusion reactions are also called *thermonuclear* processes. In the fusion bomb, the required temperature may be provided by using a fission bomb as a detonator.

The stars, including our sun, obtain their energy from thermonuclear processes. A number of nuclear reaction mechanisms involving carbon, nitrogen, oxygen, helium, and hydrogen nuclei have been proposed for the reactions occurring in stars. Whatever the mechanism, the net reaction is the fusion of 4 protons into one helium nucleus.

$$4 \, {}^1_1\text{H} \longrightarrow {}^4_2\text{He} + 2 \, {}^0_1 e + \gamma$$

The energy release is 26.7 MeV for each He produced. It is estimated that the sun converts about 600 million tons of hydrogen into helium every second. The sun has used up approximately one-half of its hydrogen in its 4.5 billion yr of existence, but it has enough hydrogen left for many billion years more.

The controlled release of energy from thermonuclear reactions requires that the temperature of the reactants be raised to the ignition point at which fusion might begin. After the fuel is ignited, the reactants must be held together long enough ($\sim 10^{-2}$ sec) for a sizable number of fusion events to take place. (The product of gas density and confinement time must be about 10^{14} sec particles/cm^3.) The confinement problem is by far the most difficult. At the required ignition temperature, all reactants exist as completely ionized gases (plasma) consisting of free electrons and bare atomic nuclei. Aside from the fact that no solid container can hold a hot gas at 100 million°C, any kinetic energy the reactant nuclei might have would be lost by collision with the walls of any material container. However, the hot fully ionized gas, or *plasma*, can be influenced by magnetic fields, and a variety of "magnetic bottles" whose walls are magnetic lines of force are being investigated in the attempt to confine the low-density plasmas needed for controlled fusion.

Among the fusion reactions considered for controlled fusion are the deuterium–tritium process used in hydrogen bombs and the deuterium–deuterium reactions

$$
{}^2_1\text{H} + {}^2_1\text{H}
\begin{cases}
\xrightarrow{50\%} {}^3_2\text{He} + {}^1_0 n + 3.2 \text{ MeV} \\
\xrightarrow{50\%} {}^3_1\text{H} + {}^1_1\text{H} + 4.0 \text{ MeV}
\end{cases}
$$

Ignition temperature is 40×10^6°C for the 17.6-MeV deuterium–tritium reaction and 400×10^6°C for the deuterium–deuterium set. As with fission reactions, most of the energy of fusion reactions appears as kinetic energy of the lighter products. In the simplest case, this kinetic energy can be transferred to coolant–heat exchangers such as liquid lithium to generate steam to run dynamos. If neutrons carry most of the energy away, this scheme could also produce more tritium fuel by the reaction

$$
{}^6_3\text{Li} + {}^1_0 n \longrightarrow {}^4_2\text{He} + {}^3_1\text{T} + 4.8 \text{ MeV}
$$

Great technical difficulties lie in the path of commercial use of controlled fusion for power generation. Nevertheless, there appear to be no scientific reasons why it should not be possible.

Origin of the Elements

Our sun, in its present and probably its past state, cannot make the medium and heavy-weight elements. Yet all these elements do exist on the sun as well as on the earth, and it is believed that these elements originated in stellar nuclear reactions that took place before the sun and earth were formed. An explanation of this situation requires a brief look at the history of a star's life. One theory of stellar evolution pictures a star as forming from the coming together of large quantities of hydrogen. As the star contracts under the influence of gravitational attraction, its interior temperature increases to the point where fusion of hydrogen nuclei to helium nuclei can occur. As the star ages, hydrogen is used up, helium is formed, and the star continues to contract. Temperatures in the core become high enough to cause the fusion of helium nuclei by the reaction

$$3\ {}^{4}_{2}He \longrightarrow {}^{12}_{6}C$$

Further contraction and temperature increase lead to the formation of heavier elements, up to ${}^{56}_{26}Fe$, by fusion reactions such as

$${}^{4}_{2}He + {}^{12}_{6}C \longrightarrow {}^{16}_{8}O$$

A star that is massive enough can condense nuclei until most of its core material is composed of nuclei in the iron region. Increased gravitational attraction, however, will lead to fission reactions of the type

$${}^{56}_{26}Fe \longrightarrow 13\ {}^{4}_{2}He + 4\ {}^{1}_{0}n$$

This one results in the absorption of 124.5 MeV. Such reactions lead to a rapid collapse of the core. The outer layers in turn collapse, with the release of large amounts of energy. The result is an explosion, a supernova, which throws stellar matter into space. Our solar system is believed to be composed of the material ejected by such an exploding star. The swarms of fast neutrons ejected in the fission reactions that led to the collapse can be absorbed by heavy nuclei in the iron region. Beta decay of the resulting neutron-rich nuclei then increases the atomic number. By successive neutron capture and beta decay, the known heavy elements above iron were formed. This matter, containing all elements, later aggregated to form the sun and the planets.

20.14 ELECTRIC POWER GENERATION

With the exception of hydroelectric power stations, all present power plants for the large-scale generation of electric power are thermal plants. This means that the heat derived from the burning of fossil fuels or the fission of nuclear fuels is used to generate steam to turn the shaft of a turbine connected to a generator of electric energy. Fusion reactors, when they become operational for commercial use, probably no earlier than 2000, will probably use the same heat $\longrightarrow$ mechanical energy (work) $\longrightarrow$ electricity sequence. As fossil fuels are used up and as nuclear energy takes over, electric power will become a more important portion of our total energy consumption. Because of this, it is well to examine some aspects of electric power production, particularly from nuclear fuels.

Electric *energy* is usually measured in terms of kilowatt hours. The following relations are helpful in converting from electrical units to thermal and nuclear energy units.

$$1\ kW\ hr = 8.6 \times 10^{5}\ cal = 2.27 \times 10^{19}\ MeV$$

The capacities of electric power plants are rated by the amount of electric energy they can deliver in 1 day. That is, they are rated in terms of *power*, the rate at which work is done or energy is delivered. The common units of power plant capacity are kilowatts ($kW = 10^3$ W) or megawatts ($MW = 10^6$ W). Therefore, the capacity of a power plant is obtained by dividing the kilowatt hours of electric energy produced each day by 24 hr.

EXAMPLE | A coal-fired power plant burns 1400 tons of coal a day. One ton of coal delivers 6.55×10^9 cal of thermal energy. Therefore, the thermal energy available from 1400 tons of coal is 9.2×10^{12} cal or 10.6×10^6 kW hr. The thermal power obtained from the coal is 10.6×10^6 kW hr/24 hr = 440,000 kW.

Because of the second law of thermodynamics, all of this thermal energy cannot be converted into mechanical energy. A modern fossil-fueled steam turbine plant has an efficiency of about 40 percent. This means that only 40 percent of the thermal energy obtained from the burning coal can be converted to electric energy; the rest is wasted. Thus, the capacity of this power plant, in terms of electric power delivered, is

$$0.40(440,000 \text{ kW}) \simeq 176,000 \text{ kW}$$
$$= 176 \text{ MW}$$

Now let us see what power capacity might be obtained from a fission reactor fueled with ^{235}U or any of the other slow-neutron-fissionable nuclei, all of which release about 200 MeV per fission.

EXAMPLE | The energy released by the complete fission of 1 kg of ^{235}U is found by multiplying the energy released per fission event by the number of atoms undergoing fission:

$$\left(200\frac{\text{MeV}}{\text{atom}}\right)\left(6.02 \times 10^{23}\frac{\text{atoms}}{\text{g-atom}}\right)\left(\frac{1000 \text{ g}}{235 \text{ g/g-atom}}\right)$$
$$\left(4.45 \times 10^{-20}\frac{\text{kW hr}}{\text{MeV}}\right) = 23 \times 10^6 \text{ kW hr of } \textit{thermal} \text{ energy.}$$

Conventional (nonbreeder) nuclear fission power plants are less efficient than fossil fuel plants in converting thermal energy into electric energy. Their efficiency is about 33 percent, and therefore, only 7.7×10^6 kW hr of electric energy would be produced. The electric power generated in 1 day by the fission of 1 kg of ^{235}U would be 7.7×10^6 kW hr/24 hr = 3.3×10^5 kW = 330,000 kW or 330 MW.

Comparison of the two examples shows that complete fission of 1 kg of pure ^{235}U is approximately equal to 3000 tons of coal in the generation of electric power. This is also equal to the heat energy in about 13,000 barrels (42 gal/bbl) of crude oil.

Modern electric power plants are generally in the range of 100–1000 megawatts (100,000–1,000,000 kW) of *electric* (not thermal) power capacity. (The abbreviations MWe and kWe are sometimes used to denote electric power delivered by a plant as opposed to thermal power produced by the fuel before conversion to electrical power.) A 350 megawatt (MWe) plant is considered to have an output sufficient for a city of 500,000 people. Large coal-fired plants with capacities of up to 5000 MWe are under construction in the southwest United States.

20.15 THE SUPPLY OF FISSION AND FUSION POWER

Nuclear fuels are often considered to be the electric power source of the future. But the nuclear fission age will be short-lived if conventional nonbreeder reactors consuming naturally occurring ^{235}U are used exclusively. These reactors will use up forever the 0.72 percent of ^{235}U in natural uranium, with little or no production of ^{239}Pu. It is estimated that the domestic reserves of inexpensive U_3O_8 will be used up by 1985. After that the price of U_3O_8 will climb, and nuclear fission power will no longer be competitive with electricity derived from fossil fuels. This situation will change only if breeder reactors become operational in the early 1980's. Then, if the use of nonbreeding reactors is discontinued, the supply of fissionable material may be doubled every 10 yr.

If the problems of controlled nuclear fusion power are solved, the world's power reserves would be almost infinite. For the deuterium–deuterium fusion reaction, the fuel would be obtained from the world's oceans. Only 1 percent of the deuterium in the oceans would release an amount of energy equal to 500,000 times the world's original supply of fossil fuel. The total energy supply from the deuterium–tritium reaction would be much less, because most tritium is produced from a nuclear reaction using lithium-6, and its supply would be limited by the smaller supplies of lithium-6.

20.16 ENVIRONMENTAL IMPACT OF POWER PRODUCTION

The earth has finite limits to its reserves of natural resources and to its ability to absorb wastes. So far as the supply of fuel resources for the production of electric power is concerned, we appear to be in good shape if breeder, and later, fusion reactors become operational. Now we turn to some pollution problems arising from large-scale generation of electric power. These problems, some of which have been discussed in previous chapters, are listed in Table 20.4. Included are effects arising from production and processing of the energy sources as well as their conversion to thermal and electric energy.

In a nuclear fission reactor, only a small part of the fissionable fuel can be consumed before the products begin to interfere with the efficiency of the fission process. The fuel elements must be removed and processed to remove the highly radioactive fission products. The purified fuel is then reused. The fission products are the ashes of the fission fire and must be somehow disposed of. Some of the fission products have long half-lives and must be isolated for thousands of years before their radiation level decreases to a biologically harmless level. Various methods have been used for storage or disposal of liquid and solid wastes. But because the 1000 commercial nuclear fission plants expected to be operating in the United States by the year 2000 may generate 3 million kg of radioactive ashes per year, the disposal problem is serious. By 2000, nuclear power will supply about 23 percent of the nation's total energy demand and 50 percent of its electric power. Burial in underground salt mines is one possibility, since the presence of the salt deposits indicates the absence of groundwater that could leach the wastes, become contaminated, and transport radioactive materials out of the mines. Another solution may be the injection of liquid wastes into very deep wells. This method has been used on a limited scale for radioactive and other wastes, and is becoming more popular. Great care must be taken to see that wastes are not injected into geological formations that will result in groundwater contamination or transport of wastes to the surface, often at great distances from the injection point.

Table 20.4
Environmental impact of electric energy production

energy source	impact on			
	land and soil	water	air	other
coal	strip mining (38% of all U. S. coal is strip mined) disposal of waste from processing and burning	acid mine drainage thermal pollution	SO_2 NO_2 particulates trace metals (Hg, etc.)	
oil	oil spills from pipelines strip mining and waste disposal for oil shales	oil spills disposal of brines brought up in drilling thermal pollution	SO_2 NO_2 CO hydrocarbons odor	
gas		thermal pollution	(cleanest of fossil fuels; also in shortest supply)	
fission	disposal of radioactive wastes from mining and spent reactor fuels	thermal pollution mining and processing effluents	volatile radioisotopes: ^{85}Kr, ^{131}I, ^3H	reactor accidents; radiation
(fusion)		thermal pollution (?)	^3H	

Disposal of nonradioactive liquid wastes into a 2-mile deep well in Denver, Colorado, has been blamed for over 1000 small earthquakes in that area in the mid-1960's, apparently the result of lubrication of faults by the wastes. Another suggestion for disposal of radioactive wastes is to load them into rockets and fire them into the sun. The problems here are the huge amounts of energy needed to accomplish this plus the disastrous consequences of an aborted flight.

Thermal pollution of water by power plants is a serious problem. A large power plant requires hundreds of millions of gallons of cooling water daily. It has been estimated that one-sixth of the total fresh water runoff in the United States will be required for coolant purposes in 1980, almost all of it being used by electric power plants. In 1970, about 10 percent of the stream flow was used for cooling purposes.

Waste heat from thermal power plants and the need for cooling water is a consequence of the second law of thermodynamics. The second law provides a quantitative measure of the efficiency of the conversion of heat into work. All thermal power plants withdraw heat from a high-temperature source and convert a fraction of it into mechanical energy (work). This requires a heat engine in which a working substance such as water absorbs heat at the high temperature and is converted to steam. The steam expands against a piston or a turbine blade. The expanded steam is then condensed into water in a condenser at a lower temperature, and the condensed steam is recycled to the boiler. The condenser is maintained at a low temperature by cooling water. The efficiency of such a steam engine is given by the ratio

$$\frac{w}{q_h} = \frac{T_h - T_c}{T_h}$$

where w is the useful work obtained when q_h units of heat are absorbed at the higher (boiler or heat exchanger) temperature T_h, and the steam is condensed at the lower temperature T_c. At T_c, heat is given up to the surroundings by transfer to the cooling water. It is this heat that causes thermal pollution. The lower the efficiency of the steam engine or turbine, the greater will be the fraction of the absorbed heat q_h that will be rejected to the surroundings. The efficiency of any power plant therefore depends on the temperature of the steam (usually superheated) that it can produce (T_h) and the temperature of the condensers. In addition to this unavoidable heat loss governed by the second law, other heat losses may occur through friction, radiation, and other equipment design problems.

A fossil-fueled power plant can operate at higher temperatures (T_h) than a nonbreeder fission reactor, so its contribution to thermal pollution is somewhat less. Thus, about 45–50 percent of the heat value of the fuel is transferred to the environment at the condensers in a fossil-fueled plant, compared to 55–65 percent heat rejection in a nonbreeder fission plant. Breeder fission and fusion nuclear plants will be approximately 50 percent efficient. The excess heat from nuclear reactors could be used for water desalination plants designed in conjunction with the power reactor, for space heating, and for crop production in greenhouses or sea farms. Our concern with thermal pollution has been restricted to localized sources and effects. In a larger sense, all the waste heat must be radiated back into space, whether it was originally released to cooling water or to the atmosphere. The heat released by increased human activity is much less than 1 percent of the total solar energy absorbed by the earth and is apparently radiated to space with no effect on global climatic conditions. Localized sources of heat such as cities do, however, experience climatic changes resulting from their comparatively large rates of heat emission.

In addition to sulfur dioxide and particulate matter emitted in stack effluents of coal-fired power stations, other more insidious pollutants arising from coal may be heavy metals. For instance, some coal has been found to have a mercury content of up to 33 ppm. Assuming only 1 ppm Hg and a yearly consumption of 3 billion tons of coal worldwide, 3000 tons/yr of mercury might be released to the environment by the burning of coal.

Finally, the subject of reactor safety must be mentioned. Fission reactors contain a supercritical amount of fuel, but there is little danger of a real explosion. The major problems are heat deterioration of the material covering the fuel elements and loss or blockage of the reactor core coolant. The latter would lead to overheating of the core and possible melting of the fuel elements. Fusion reactors would not be susceptible to such accidents because of the small amounts of fuel required and because heat buildup would automatically cause the plasma to disperse. Both fission and fusion reactors would introduce some radiation into the environment. In addition to radioactive solid wastes, fission plants release small amounts of radioactive wastes to the atmosphere and to the water. Among these is tritium, which would also be the principal radioactive waste from a deuterium–tritium fusion reactor. The radiation hazards associated with nuclear plants are currently a matter for argument.

The many considerations involved in the production and utilization of energy resources often point in different directions. It is the reconciliation of opposing factors that constitutes the major problem in energy resource management in the coming decades. What is the balance between the rising demand for energy and the depletion of energy resources? How does a higher standard of living derived from high energy consumption equate with the quality of life? Who will determine the balance between the desire for low-cost energy and the greater environmental pollution that often results from the cheaper fuels? Should power plants be located close to population centers, or should they be built far away from the cities where the power is used. Questions such as this must be answered as the current and inevitably increasing confrontation between energy needs and preservation of the environment continues.

The idea that electric power consumption will continue to double every decade owing to increasing population and a standard of living demanding such energy consumption is considered untenable by many students of the environment and natural resources. It is obvious that man's consumption of irreplaceable natural resources cannot continue at an increasing rate indefinitely, any more than his population can increase without bound. These observers contend that the time has passed when we should be planning for growth and consumption without limit. Rather, it is time now to devise the means and the philosophy to limit population and consumption of natural resources and to make an effort to lessen the already considerable impact of man on his environment.

Energy and Ecology

20.17 ENERGY FLOW IN ORGANIZED SYSTEMS

Life is a flow of energy. Energy flows into and through cells and organisms, where it is used for maintenance, growth, and work by the living system. Energy flows through ecosystems, determining the structure and guiding the development of the ecosystem. Energy flows through the biosphere, driving all the biological and physical processes on earth. The ultimate source of all this energy is the sun. In order for energy to flow, it must go somewhere. This destination or energy sink is outer space. No matter what size or kind of system is considered—cell, organism, ecosystem, human society, biosphere— the structure and function can be ultimately related to a flow of radiant energy through the system to the heat sink of outer space.

The role of the source and sink of energy is most apparent if the entire biosphere or surface of the earth is considered. The earth's surface absorbs radiant energy from the sun. The earth also must radiate, on the average, an equal amount of energy to outer space. The radiated energy is in the form of infrared or heat radiation. In general, in the course of energy flow through any system, all energy is eventually degraded to heat and reradiated to outer space. Some of the energy may be stored within the system for varying lengths of time. Of most interest is the storing of energy as the chemical energy of organic matter, living and dead.

We are interested in open systems, which exchange energy or matter with their surroundings. Only by such an exchange can the system maintain itself in an ordered nonequilibrium state. Otherwise, the system will spontaneously tend to an equilibrium state of minimum free energy and maximum entropy (disorder). We have already seen that living organisms are nonequilibrium systems, as are ecosystems, the biosphere, and the earth as a whole. A special case of a nonequilibrium system is the steady state system, where the rates of energy flow into and out of the system are equal. A steady state with respect to energy will generally also be associated with a steady state with respect to the flow of matter into and out of the system.

In another special case, there is no flow of matter into and out of the system. Instead, matter is continually recycled within the system. Such a cyclic flow of matter is a consequence of the energy flow through the system. Again the earth's surface serves as an example, first of the energy steady state and second of the cycling of matter. The hydrologic, or water, cycle illustrates how incident solar radiation is absorbed to evaporate and transport water. Subsequently, the water vapor condenses and falls back to earth, releasing thermal energy which is radiated to outer space. Also on the earth's surface, solar energy stored in organic molecules is degraded by processes of metabolism and decay. These processes produce waste heat and, in addition, waste products such as carbon dioxide and water, which remain in the system but are continually recycled between the living and the nonliving parts of the biosphere. The energy for the cycling processes is extracted from the energy flowing through the biosphere.

Morowitz (1968) has stated these and other ideas about energy flow in the biosphere. Some of the more important ones are collected below.

1. The flow of energy through a system acts to organize that system.

2. The maintenance of the order or structure of a system requires continuous work, which can be supplied only by energy flow from a source to a sink. Order can be maintained only in a nonequilibrium system, since equilibrium is equated with disorder and maximum entropy. The degree of order in a system is related to the distance from equilibrium.

The higher the temperature of the system, the greater must be the rate of energy flow to maintain the nonequilibrium state. (Higher temperatures favor disorder.)

3. Energy flow leads to the cyclic flow of matter in systems.

4. A system through which energy constantly flows tends to evolve to a condition of maximum stability, maximum order and complexity, maximum amount of stored energy, and minimum rate of energy flow through the system. This is the steady state.

These ideas are stated here with no more proof than can be found in earlier parts of this book. Some of them will serve as background for a short discussion of two areas of ecological energetics: (1) the efficiency of energy transfer between trophic levels in ecosystems and (2) ecological succession, the evolution of ecosystems with time. Both of these subjects have been mentioned very briefly before, particularly in Chapter 18.

20.18 ENERGY TRANSFER IN COMMUNITIES

The flow of energy through the living part, the community, of an ecosystem begins with the fixation of solar energy by plants. We have seen that only a very small part of the radiation absorbed by plants is actually converted into food energy by photosynthesis. Perhaps 50 percent of the energy fixed in this way is used by the plants for self-maintenance, work of biosynthesis, and other energy-requiring functions known collectively as respiration. During respiration, the chemical energy of food is converted to heat and dissipated.

The total amount of solar energy assimilated, or the total amount of organic matter manufactured by photosynthesis over a period of time, is the gross primary productivity, which was denoted by GP in Chapter 18. The amount of energy remaining and stored as organic matter after respiration is the net primary production, NP. Net primary production can also be treated as a rate, representing the rate of accumulation of plant organic matter. The units of productivity might be kilocalories per square meter per year, the square meter representing the portion of the ecosystem sampled.

The net primary production of the producer organisms represents energy available to the consumer (herbivore) organisms on the next trophic level. Some of the energy of the food available to the herbivores will be used for herbivore respiration; some may not be available or metabolized. The remainder will end up as net production (body tissue) of the herbivores. Similarly, the net production of the herbivores is available to carnivores on the next trophic level, but because of respiration and other losses on the carnivore level, only a fraction of the herbivore net production will be incorporated as carnivore net production. As a rough approximation, only 10 percent of the net production on one trophic level ends up as net production on the next highest level. This 10 percent figure applies to natural ecosystems. It is possible, for instance, for herbivores to extract higher proportions of energy from plants by harvesting the entire plant population. However, this destroys the ability of the population to reproduce and grow under natural conditions. Man, as a predator, is guilty of this sort of thing when, by overfishing or overgrazing, he attempts to increase his food yield too much. The point remains, however, that there is still a large energy loss at each step in a food chain. The efficiency of energy transfer is illustrated in Figure 20.7. Also indicated is the energy dissipated as heat by respiratory activities at each level. The right side of Figure 20.7 has been compressed in the inset to show a form of ecological pyramid. This energy pyramid shows the energy flow (the rate of food production) at each trophic level and clearly shows the decrease in available energy at higher trophic levels.

The ecological energy pyramid shows how the energy available for growth and reproduction rapidly decreases for organisms at higher trophic levels. This explains why the population of organisms on higher trophic levels is almost always less than that on lower levels. There is not enough energy remaining at the high levels to support large populations. Man gets around this by feeding at a number of levels and by importing energy from external sources into his agricultural ecosystems.

The diminution of energy at higher levels means that carnivores must eat large quantities of organisms on lower levels in order to obtain sufficient energy for growth and maintenance of their population. For instance, in order for a pelican to increase in weight by 1 kg (net production), it must consume 10 kg of fish. To produce 10 kg of fish requires 100 kg of plankton.

The concentration of any toxic material such as mercury or insecticides will be biologically magnified in the pelican by a factor of 100 over the concentration in the plankton if the toxic material is not broken down or metabolized and excreted along the food chain. Because the plankton may have already concentrated the foreign substance in their own tissue, sometimes by a factor of many thousands over the concentration in sea water, the amount of environmental contaminants in organisms near the top of food

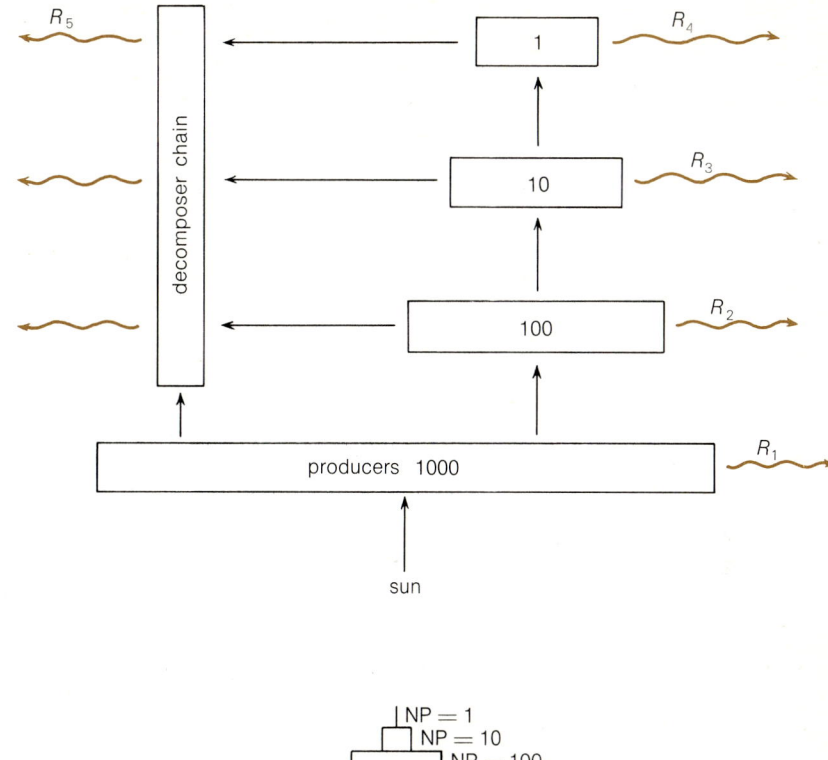

Figure 20.7. Energy flow in a community consists of transfer of chemical (food) energy between trophic levels and dissipation of respiratory energy R as heat at each level. The numbers correspond to net production at each level in arbitrary units. The wavy arrows indicate dissipation of heat to space, and the straight arrows show the direction of energy transfer. The total or community respiration R is equal to the sum of the R_i's shown here. The ecological pyramid of energy shows the rate of energy flow at each trophic level in terms of net production.

chains can build up rapidly. This is especially true in the aquatic food chains, which tend to be complex webs with many energy transfer steps. On the other hand, complex food webs with several steps may protect the dwellers on high trophic levels if the environmental contaminant is so toxic that some of the organisms on lower levels are killed before they can be eaten. In this case, complex food webs serve to protect species such as man so long as the levels of toxic wastes do not rise to the point where the complexity of the food web is broken down. Then the toxic substances may pass directly to man.

20.19 ENERGY AND ECOLOGICAL SUCCESSION

The continued flow of energy through an ecosystem leads to progressive and fairly predictable changes in the structure and function of the ecosystem as the ecosystem develops or ages over a period of time. The process of sequential change is known as *ecological succession* and may be regarded as resulting from the interaction of the organisms present in the ecosystem at any given time (the community) with the physical environment and with each other. Organisms tolerant to the new physical environment take hold and begin to modify the environment chemically and physically. They may improve the fertility and structure of the soil with their waste products or by breaking up the rock with their roots and exuded chemicals. The early organisms prepare the ground, figuratively and very often literally, for the entry of new species into the ecosystem. The new species replace all or part of the earlier community and constitute a new community of organisms. In this respect, succession is often regarded as a sequence of communities, each replacing one another in a given area.

In Chapter 18 it was noted that the overall change from a young ecosystem such as a bare field to an older, more mature ecosystem such as a forest is characterized by a movement toward increased complexity and stability of the ecosystem. Our concern in Chapter 18 was with the change in community structure of the ecosystem, in particular, the trend toward species diversity in older ecosystems. But many other changes take place during succession, and these changes interact with each other to produce the characteristic properties of the ecosystem at any point in time. Eventually, ecological succession is believed to terminate in a mature stage of the ecosystem with a terminal or *climax* community.

Table 20.5 lists trends in a number of ecosystem properties as succession proceeds from young to mature stages. Of present interest is the change in ecosystem energy flow as succession proceeds. Initially, there is an excess of available energy, and this is channeled into the rapid growth of the few autotrophic species predominant in the ecosystem. As succession goes on, organic matter accumulates and can serve as food for heterotrophic herbivores, carnivores, and decay organisms. These organisms make no contribution to gross primary production, but they do increase the net production of the community and the total respiratory loss of energy. The total amount of organic matter in the ecosystem increases continually as the system ages. As organic matter increases, nutrient elements become tied up in the organic matter but, with the help of decomposers, begin cycling between the biotic and abiotic portions of the ecosystem.

Eventually, gross primary productivity slows down and the energy flowing into the system is channeled into maintenance of the complex organization of the mature ecosystem. Growth and rate of reproduction (productivity) slow down as accumulation of excess organic matter gives way to maintenance of a constant but maximum quantity of organic matter in a highly ordered ecosystem characterized by stability and diversity, conservation of nutrients, and a steady state flow of energy. The energy flowing into the mature ecosystem (GP) equals the amount flowing out (R). Less energy is required to maintain the structure of a mature ecosystem than a younger one with less order and higher productivity.

This is as far as we shall go with a description of succession in natural ecosystems. However, a glance at Table 20.5 shows that there are many similarities between the evolution of ecosystems and the development of multicellular organisms, nations, and human societies. All generally start out with high growth rates, rapid reproduction rates, simple structures, large inputs of energy, and large accumulation of products and exploitation of natural resources with little recycling. In the United States and other young developed and developing nations, high consumption of energy has been accompanied by rapid increase in population and the rate of production of material things (gross national product). If the trends of ecological succession apply to nations and societies, then it is possible that population growth, productivity, energy consumption, and depletion of natural resources will level off as our human ecosystem reaches a mature steady state.

property	young stage	mature stage
energetics		
1. gross primary production/total community respiration, GP/R	generally greater than 1	GP/$R \simeq 1$ (steady state)
2. net production (rate) NP of entire community	high	low
3. food chains	short, simple, linear	complex, weblike
community structure		
4. total organic matter (living and dead)	small	large
5. inorganic nutrients	mostly in abiotic environment	mostly tied up in living matter
6. species diversity	low	high
biogeochemical (nutrient) cycles		
7. mineral cycles	open (acyclic)	closed
8. exchange rate of nutrients between organisms and environment	fast	slow
other qualities		
9. reproduction rate	rapid	slow
10. lifetime of individuals	short	long
11. growth rate	high	low
12. nutrient conservation	poor	good
13. stability to outside pressures	poor	good
14. organization (complexity)	low	high

SOURCE: Selected from Eugene P. Odum, *Science*, **164**, 262–270 (1969). Copyright © 1969 by the American Association for the Advancement of Science.

Many of the qualities of a mature human ecosystem or society, if there is one, would be desirable. Among these are longer life, stable populations, order, cooperation among individuals, conservation of resources, and more efficient use of energy. Unfortunately, many of these qualities are incompatible with present economic systems, all of which are based on growth. But it appears that energy flow, and not economics, exerts the controlling influence on community and population structure, and man may have no choice in the matter if energy is wasted.

20.20 SUMMARY Modern industrial cultures are based on the existence of inexpensive and abundant sources of energy. Today the fossil fuels—coal, oil, and natural gas—are the primary sources of large-scale energy, but the lifetimes of the fossil fuel reserves are finite, and we shall soon have to depend more and more on nuclear energy sources, particularly for electric power.

The attractive forces acting between protons and neutrons in atomic nuclei are about one million times larger than the typical covalent forces of chemistry. When the repulsive forces among nuclear particles exceed the attractive forces, a nucleus will undergo spontaneous radioactive decay, with the emission of alpha or beta particles and the transmutation of the element to another element. The relative stability of a nucleus is inversely proportional to its half-life, the time needed for one-half of a sample to decay.

Nuclear reactions involve the bombardment of nuclei with projectile particles or radiation. Nuclear transmutation may occur in such reactions, and the product nuclei may be stable or radioactive. Nuclear reactions, like chemical reactions, may be exoergic or endoergic, but the energies involved are of much larger magnitude than those of chemical reactions. The difference in mass between products and reactants is equivalent to the energy change. By Einstein's relation, $E = mc^2$, 1 amu is equivalent to 931 MeV = 3.56×10^{-11} cal. The stability of nuclei is measured quantitatively by the binding energy per nucleon in the nucleus.

Present-day nuclear energy sources rely on controlled neutron-induced fission of unstable uranium and plutonium nuclei in nuclear reactors. Future nuclear energy production may rely on nuclear fusion reactions, where two light nuclei condense into a heavier and more stable nucleus with the evolution of energy.

Nuclear power plants are not without environmental impact. Their air pollution potential is quite small, but the problems of thermal pollution and disposal of radioactive waste products are more serious than those connected with fossil fuel plants.

Every organized system requires that energy continually flow through it if the system is to maintain itself in an ordered nonequilibrium state capable of doing work. In communities, energy flow occurs from one trophic level to another. There is an inevitable loss of useful energy at each transfer point, resulting in the ecological pyramid of energy. Energy flow through ecosystems results in the phenomenon of ecological succession. Ecosystems evolve from highly productive young states, characterized by simple structure and low stability to mature stages exhibiting complexity, stability, low net productivity, material recycling, and minimum energy flow.

Questions and Problems

1. Draw the structure of 2,2,4-trimethylpentane (isooctane).
2. An ideal solution of two components, A and B, contains 0.75 mole fraction of A. The normal boiling point of the solution is 137°C. At this temperature, A and B have vapor pressures of 863 mm and 453 mm, respectively.

a. What is the composition of the initial distillate (i.e., the vapor in equilibrium with the liquid) from this mixture?

b. If the vapor is condensed and redistilled, how will the normal boiling point of the second distillate compare with that of the original solution in part (a)? How will the composition of the vapor obtained compare with the concentration of the initial distillate in part (a)?

3. An observer notes that a power plant emits a white stack effluent on a very cold day. He accuses this plant of air pollution, citing as evidence that another power plant emits no visible effluent under the same conditions. When the case comes to court, it is established that one plant burns natural gas and the other plant burns coke.
   a. Which plant do you think emits the visible effluent?
   b. Will the case be thrown out of court?

4. Why do you think solar energy might be unpromising for large-scale electric power generation?

5. What if all transportation ran on rechargeable batteries? Would air pollution problems connected with transportation be changed? In what way?

6. The composition of Illinois #6 coal is represented by $C_{100}H_{85}S_{2.1}N_{1.5}O_{9.5}$.
   a. To what rank does this coal belong?
   b. How much SO_2, in kilograms, would theoretically be produced from the combustion of 1 ton (2000 lb) of this coal?
   c. How much CH_4 gas, in kilograms, could theoretically be obtained from gasification of 1 ton of this coal if
      (1) hydrogen is plentiful?
      (2) only the hydrogen supplied by the coal itself is available?

7. Write equations for
   a. the alpha decay of
      1. $^{232}_{90}Th$
      2. $^{228}_{88}Ra$
   b. the beta decay of
      3. $^{3}_{1}H$
      4. $^{40}_{19}K$

8. Complete and write the equations for
   a. $^{19}_{9}F$ (α, p) _____
   b. $^{6}_{3}Li$ (___, α) $^{3}_{2}He$
   c. _____ (n, γ) $^{24}_{11}Na$

9. Carbon-14 has a half-life of 5760 yr. If live wood has a radioactivity of 15.3 counts/min and a sample of dead wood gives 9.4 counts/min for the same weight of wood, how old is the dead wood?

10. One gram of carbon obtained from an archeological specimen has a disintegration rate of 4.6 counts/min. What is the age of the specimen?

11. A piece of charcoal was found, along with some fossilized hot dogs and marshmallows, in some archeological diggings. The ratio of ^{14}C to ^{12}C in the charcoal is only 50 percent of the ^{14}C to ^{12}C ratio in charcoal made from a newly cut tree. Approximately how long ago did the wiener roast take place? What if the charcoal had a $^{14}C/^{12}C$ ratio 25 percent of that of newly made charcoal?

12. The ^{206}Pb content in a sample of uranium ore is 30 percent of the weight of ^{238}U in the ore. If all the ^{206}Pb originated from the decay of ^{238}U originally present and neither isotope has been leached from the ore over the years, what is the ore's age?

13. Calculate the energy changes in the following reactions:
   a. $^{14}_{7}N$ (n, p) $^{14}_{6}C$
   b. $^{27}_{13}Al$ (n, α) $^{24}_{11}Na$
   c. $2\ ^{2}_{1}H \longrightarrow\ ^{4}_{2}He$

   Atomic masses are $p = ^{1}_{1}H = 1.007825$ amu; $n = 1.008665$ amu; $^{2}_{1}H = 2.014102$ amu; $\alpha = ^{4}_{2}He = 4.002604$ amu; $^{14}_{6}C = 14.003242$ amu; $^{14}_{7}N = 14.003074$ amu; $^{24}_{11}Na = 23.990967$ amu; $^{27}_{13}Al = 26.981535$ amu.

14. The exact atomic masses of many fission fragments are not available, so the energy released in a fission reaction cannot always be calculated from mass differences. Calculate the approximate energy release in the hypothetical fission processes

   $$^{235}_{92}U + ^{1}_{0}n \longrightarrow 2\ ^{118}_{46}Pd$$

   using information given in Figure 10.3. Give the energy in MeV per fission event and in calories per gram-atom.

15. The heat of combustion of carbon is 94 kcal/g-atom (compare with the previous problem). What is the change in mass when 1 g-atom of carbon (graphite) is burned to CO_2?

16. The caloric value of 1 short ton of coal is 6.55×10^6 kcal. Calculate the thermal energy obtainable from the fission of 1 lb of ^{235}U and determine its equivalent in tons of coal.

17. If a deuterium–tritium fusion power plant consumes 130 g of tritium per day, what is the power capacity of the plant if the efficiency of conversion of thermal to electric energy is 40 percent?

18. In the deuterium–deuterium fusion process, one of the two equally probable fusion reactions produces a tritium atom, which then reacts with another deuterium to give $^{4}_{2}He$, $^{0}_{1}n$, and 17.6 MeV.

a. Show that the net result of the three fusion reactions involves five deuterium atoms and evolves a total 24.8 MeV of energy.

b. One cubic meter of water contains 34.4 g of deuterium. Potentially how much thermal energy, in kilowatt hours, could be obtained from the fusion of this much deuterium according to the net reaction of part (a)?

19. The total electric energy consumption in the United States in 1980 is projected to be 2810×10^9 kW hr.

    a. If all of this energy were to be generated by a ^{235}U nonbreeder fission reactor, how many kilograms and how many pounds of ^{235}U would be required?

    b. If all the energy were produced by the deuterium–tritium fusion process, how many grams of tritium would be consumed?

20. a. If the efficiency of a fossil fuel plant is 40 percent, how much heat is lost to the surroundings for each kilowatt of electric power produced?

    b. If the efficiency of a fission plant is 30 percent, how much heat is dissipated per kilowatt of net electricity production?

21. a. How would you increase the efficiency of a power plant located on the ocean—by drawing cooling water from the upper layers or from the bottom layers of the ocean?

    b. Which water, after leaving the condensers, might be more suitable for pumping into pounds for marine agriculture?

22. An 800-MWe power plant burns coal with 3 percent sulfur content. When 90 percent of the SO_2 in the stack gas is removed by the alkalized alumina process, 180 tons of sulfur per day can be recovered.

    a. How many tons of coal are burned per day?

    b. If the plant efficiency is 40 percent, what is the caloric value, per ton, of the coal used?

    c. How many tons of 80 percent sulfuric acid could be produced?

23. A 2085-MWe power plant in the southwestern United States reportedly burns 26,400 tons of coal per day. According to the U. S. Department of the Interior, the coal contains 30 ppb Hg, but the Environmental Protection Agency claims that the mercury content is 180–390 ppb. What are the emissions, in pounds per year, of mercury into the atmosphere, using 30 ppb and 285 ppb as the mercury contents of the coal?

24. Discuss the connection of the second law of thermodynamics to the ecological energy pyramid.

25. Discuss some of the correspondences and discrepancies of the course of ecological succession as applied to nations such as the United States, India, Canada, England, and Australia. You might also compare cities such as New York to smaller cities and suburban areas.

Answers

2. a. $X_A^{Vap} = 0.85$
6. C = 86 percent; H = 6 percent, O = 11 percent, bituminous
9. 4050 yr
10. ~10,000 yr
12. 2×10^9 yr
13. a. 0.625 MeV evolved
    b. 3.14 MeV absorbed
    c. 23.83 MeV evolved

14. ~200 MeV = 4.6×10^{12} cal/g-atom
15. 4.4×10^{-9} g
18. b. 2.27×10^6 kW hr
20. a. 1.5 kW of heat
    b. 2.3 kW of heat
22. a. 6700 tons
    b. 6.2×10^9 cal/ton
    c. 690 tons
23. 580 lb/yr, 5500 lb/yr

References

GENERAL

Scientific American, **224**, (September 1971).

The entire issue is devoted to "Energy and Power." Highly recommended reading.

ENERGY PRODUCTION: NATURAL RESOURCE CONSUMPTION AND ENVIRONMENTAL IMPACT

Hubbert, M. King, Energy Resources, in "Resources and Man," edited by the National Academy of Sciences-

National Research Council, Freeman, San Francisco, 1969.

Mills, G. Alex, Johnson, H. R., and Perry, H., "Fuels Management in an Environmental Age," *Environmental Science and Technology*, **5,** 30–38 (1971).

Mills, G. Alex, "Gas from Coal: Fuel of the Future," *Environmental Science and Technology*, **5,** 1178–1183 (1971).

Singer, S. Fred, "Human Energy Production As a Process in the Biosphere," *Scientific American*, **223,** 175–190 (September 1970).

NUCLEAR CHEMISTRY AND NUCLEAR ENERGY

Choppin, Gregory, "Nuclei and Radioactivity," W. A. Benjamin, Menlo Park, Calif., 1964.

Feare, Thomas E., "Fusion Scientists Optimistic over Progress," *Chemical and Engineering News*, December 20, 1971, p 29.

Gillette, Robert, "Nuclear Reactor Safety: A Skeleton at the Feast?", *Science*, **172,** 918–919 (1971).

Gough, William C., and Eastlund, B. J., "The Prospects of Fusion Power," *Scientific American*, **224,** 50–64 (February 1971).

Harvey, B. "Nuclear Chemistry," Prentice-Hall, Englewood Cliffs, N. J., 1965.

Odum, Eugene P., "Fundamentals of Ecology," 3rd ed., Saunders, Philadelphia, 1971. Chapter 17.
    A discussion of the biological and environmental effects of radiation and radioactive substances. Recommended reading to supplement the material here.

Rose, David J., "Controlled Nuclear Fusion: Status and Outlook," *Science*, **172,** 797–808 (1971).
    Rather technical, but includes a discussion of environmental aspects.

Seaborg, Glenn T., and Bloom, J. L., "Fast Breeder Reactors," *Scientific American*, **223,** 13–21 (November 1970).

Wahl, W. H., and Kramer, Henry H., "Neutron Activation Analysis," *Scientific American*, **216,** 68–82 (April 1967).

ENERGY AND ECOLOGY

Margalef, R., "On Certain Unifying Principles in Ecology," *The American Naturalist*, **97,** 357–374 (1963).
    Ecological succession.

Morowitz, H., "Energy Flow in Biology," Academic, New York, 1968, Chapters 4, 5.

Odum, Eugene P., "The Strategy of Ecosystem Development," *Science*, **164,** 262–270 (1969).

Odum, Eugene P., "Fundamentals of Ecology," 3rd ed., Saunders, 1971, Chapter 3.

Odum, Howard T., "Environment, Power and Society," Wiley, New York, 1971.

Phillipson, John, "Ecological Energetics," Edward Arnold, London, 1966.

Woodwell, George M., "The Energy Cycle of The Biosphere," *Scientific American*, **223,** 64–74 (September 1970).
    Recommended reading.

Epilogue

Before this century is done, there will be an evolution in our values and the values of human society, not because man has become more civilized but because, on a blighted earth, he will have no choice. This evolution—actually a revolution whose violence will depend on the violence with which it is met— must aim at an order of things that treats man and his habitat with respect.

Peter Matthiessen—*Sal Si Puedes*

Technology is the application of neutral scientific knowledge of man's world to that world. It is the principal link between what man knows about his environment and what he does to that environment. Of the three major factors leading to deterioration of man's total environment—population growth, affluence and increased per capita consumption, and advanced but imperfect technology—unwise use of technology is perhaps the most important. Nevertheless, all of these factors feed on one another, and their combined effect is aggravated by our love of growth and by a system that charges society as a whole for the external or hidden costs of pollution caused by the manufacture and use of specific products.

If major flaws in our technology are responsible for much of the damage to the biosphere, then an important step in environmental protection should be a more efficient and systematic way of evaluating the consequences of current and developing technologies. This process of evaluation, called technological assessment, would attempt to comprehend and make informed decisions about the implications of technological development.

Technological assessment goes on all the time, but the decision making has usually been confined to special interest groups, and most decisions have been based on short-range economic goals. The new technological assessment process would recognize that technology affects many groups in indirect, subtle, or delayed ways and that these groups should therefore be represented in a collective decision-making process. The new process would realize that technological impact is not restricted to the physical or biological environment, but also affects, and is affected by, the social, political, economic, psychological, and moral underpinnings of human culture. The proper exercise of technological assessment would be to make decisions and suggest alternatives before a new technology, with possible undesirable side effects, is put into operation and becomes entrenched.

There are many who believe that more and bigger doses of technology will raise man to ever higher material, if not spiritual, standards of living. At the other extreme are those who say that technology has become a sordid boon, an end in itself which we have allowed to distort humanity and ravage the environment. They yearn for a return to nature. Neither approach is possible or desirable in today's world. The proper application of technological assessment could help us avoid an unthinking exploitation of scientific knowledge and attain a better technology. In this way, those who now enjoy a high quality of life (not to be confused with a high material standard) and who wish the same for their children, may continue to do so, while the disadvantaged of the world may still hope for a better life.

Because the direction, application, and effects of technology are so intimately related to social and cultural factors, effective technological assessment will require a total analysis of all effects in the human ecosystem. In a larger sense, technological assessment is then a part of the study and practice of systems analysis in human ecology, the interrelations of man and his total environment. This total environment includes not only the physical and biological surroundings, but other men, social groups, and man-made institutions. Analysis of such a complex system calls for detailed scientific knowledge of nature, of man, and of man in nature. No longer can this knowledge be compartmentalized and be the sole property of specialists and experts. If a truly participatory technological assessment is to be achieved, the layman must have some basic background in a number of disciplines to help him participate in and evaluate the arguments of technological assessment. The goal should be to create a society whose members can discriminate and assess, who have enough knowledge and insight to guide them in the rejection of bad ideas and an acceptance of good ideas. This text has been an effort in this direction, at least with respect to some of the chemical background needed to generate an ecological conscience in both scientists and laymen.

No book can ever provide anyone with the true basis for deciding what is a good idea and what is a bad idea. This is given by an individual's value system and what his set of values tells him about the implications of his conduct toward his fellow men and his natural surroundings. Whether a person's values are generated within or imposed from without, as the above quotation implies, is not for us to decide here. What seems quite certain is that the values, priorities, and sensitivities of men and society must undergo some kind of change or evolution if further environmental as well as social crises are to be avoided. The important questions we must ask are first of all political, social, and moral, not scientific or technical. Their answers will lead to our survival as human beings and not merely as organisms.

References

Branscomb, Lewis M., "Taming Technology," *Science,* **171,** 972–977 (1971).

Brooks, Harvey, "Can Science Survive in the Modern Age?" *Science,* **174,** 21–30 (1971).
> Contains some ideas about the need for knowledge of basic science in technology assessment (or participatory technology).

Carroll, James D., "Participatory Technology," *Science,* **171,** 647–653 (1971).

Kash, Don E., and White, Irvin L., "Technology Assessment: Harnessing Genius," *Chemical and Engineering News,* November 29, 1971, pp. 36–41.

Katz, Milton, "Decision-Making in the Production of Power," *Scientific American,* **224,** 191–200 (September 1971).

"Technology: Processes of Assessment and Choice." Report of the National Academy of Sciences. Committee on Science and Astronautics, U. S. House of Representatives, U. S. Government Printing Office, Washington, D. C., July, 1969.
> This report is summarized in *Scientific American,* **222,** 13–20 (February 1970).

Appendix

Mathematical Operations

A.1 EXPONENTIAL NOTATION

Exponential notation is used to simplify the writing and mathematical manipulation of very large and very small numbers. An exponential number is written in the form

$$M \times 10^a$$

where M is usually a number between 1 and 10 and the exponent a is a positive or negative whole number. Examples of numbers in exponential notation are

$$3600 = 3.6 \times 10^3$$
$$2{,}000{,}000 = 2 \times 10^6$$
$$0.00016 = 1.6 \times 10^{-4}$$
$$100{,}000 = 1 \times 10^5 \text{ or } 10^5$$

In working with powers of ten, it is helpful to have in mind some of the following relations.

$$10^6 = 1{,}000{,}000$$
$$10^5 = 100{,}000$$
$$10^4 = 10{,}000$$
$$10^3 = 1{,}000 = 10 \times 10 \times 10$$
$$10^2 = 100 = 10 \times 10$$
$$10^1 = 10$$
$$10^0 = 1 \qquad \text{by definition}$$

$$10^{-1} = 0.1 = \frac{1}{10}$$

$$10^{-2} = 0.01 = \frac{1}{100} = \frac{1}{10^2}$$

$$10^{-3} = 0.001 = \frac{1}{1000} = \frac{1}{10^3}$$

In carrying out arithmetic operations, the coefficients of exponential numbers are treated as ordinary numbers. Thus, exponential numbers are multiplied by multiplying the coefficients and by adding the exponents.

$$(M \times 10^a)(N \times 10^b) = M \times N \times 10^{a+b}$$
$$10^3 \times 10^2 = 10^5$$
$$(4 \times 10^5)(2 \times 10^{-2}) = 8 \times 10^{5-2}$$
$$= 8 \times 10^3$$

Exponential numbers are divided by dividing the coefficients and subtracting the exponents.

$$\frac{M \times 10^a}{N \times 10^b} = \frac{M}{N} \times 10^{a-b}$$

$$\frac{10^5}{10^3} = 10^{5-3} = 10^2$$

$$\frac{1.5 \times 10^3}{3.0 \times 10^{-2}} = \frac{1.5}{3.0} \times 10^{3-(-2)}$$

$$= 0.50 \times 10^5$$
$$= 5.0 \times 10^4$$

To add or subtract exponential numbers, first convert them to the same power of 10, and then add or subtract the coefficients but leave the exponential terms unchanged.

$$2.0 \times 10^2 + 3.2 \times 10^3 = 0.20 \times 10^3 + 3.2 \times 10^3$$
$$= 3.4 \times 10^3$$

Alternatively, this could have been written

$$2.0 \times 10^2 + 32 \times 10^2 = 34 \times 10^2 = 3.4 \times 10^3$$

Powers or roots of exponential numbers are obtained by multiplying or dividing the exponents, respectively.

$$(M \times 10^a)^b = M^b \times 10^{a \times b}$$
$$(3 \times 10^3)^2 = 3^2 \times 10^6 = 9 \times 10^6$$
$$(M \times 10^a)^{1/b} = M^{1/b} \times 10^{a/b}$$
$$(4 \times 10^6)^{1/2} = 4^{1/2} \times 10^{6/2} = 2 \times 10^3$$
$$(1.6 \times 10^5)^{1/2} = (16 \times 10^4)^{1/2} = 4 \times 10^2$$

A.2 SIGNIFICANT FIGURES

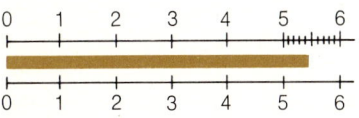

Figure A.1

All experimentally measured quantities are subject to some uncertainty in their numerical values. The uncertainty in a measured quantity depends on the measuring instrument and the skill of the experimenter in reading the instrument. As an example, consider the line segment in Figure A.1. Two centimeter scales are adjacent to the segment, one scale being more finely subdivided than the other. The reading on the fine scale is 5.45 cm; on the rough scale, 5.4 cm. In each case, the last figure has to be *estimated* by the reader, since there are no scale divisions corresponding to hundredths (0.01) or tenths (0.1) of a centimeter, respectively. Thus, the last digits in 5.45 and 5.4 are uncertain to some extent. To show this uncertainty, the values might be reported as

$$5.45 \pm 0.01 \text{ cm}$$

and

$$5.4 \pm 0.1 \text{ cm}$$

meaning that the actual length of the segment lies somewhere within the range 5.44–5.46 cm using one scale or 5.3–5.5 cm on the other.

Instead of using the $\pm$ notation, we can simply write the number 5.45 cm with the understanding that there is some uncertainty (usually ± 0.01) in the last figure. The number 5.45 is said to have three *significant figures*. Similarly, 5.4 cm has two significant figures, reflecting the use of the rough scale for its measurement.

The number of significant figures in any number is obvious except in the case of zeros. Zeros used to place the decimal point in a number are not considered to be significant figures. Several examples will illustrate these points.

number	significant figures
2.766	4
120,053.1	7
0.1066	4
2	1
2.0	2
2.00	3
46,001	5
0.00065	2
0.01065	4
0.058	2
0.0580	3

In the numbers 2.0 and 2.00, the zeros are significant figures because they are not used to locate the decimal point; the number 2 (or 2.) could have been written without them. In the number 0.0580, the last zero is significant, but the other zeros are not significant figures.

A number such as 46,000 has at least two significant figures but may have up to five significant figures. Written as 46,000, the actual precision of the number is ambiguous. The uncertainty may be lifted by writing the number as $46,000 \pm 1,000$ (two significant figures; the zeros are only needed to reach the end of the number and locate the decimal point), as $46,000 \pm 100$ (three significant figures), as $46,000 \pm 10$ (four significant figures), or as $46,000 \pm 1$ (five significant figures). A more convenient method of clarifying the number of significant figures in cases such as this is the use of exponential notation. Only the significant figures, including significant zeros, are shown in the coefficient.

number	significant figures
4.6×10^4	2
4.60×10^4	3
4.600×10^4	4
4.6000×10^4	5
2×10^3	1
6.5×10^{-4}	2
1.065×10^{-2}	4

Exact whole number may be considered to have an infinite number of significant figures.

In computations with experimentally measured quantities, the certainty of the calculated result is determined by the numbers involved. Two rules of thumb are stated here, followed by examples.

Multiplication and Division

In multiplication or division operations, the answer may contain no more significant figures than the factor with the fewest significant figures. This figure is underlined in the following examples.

$$\underline{9.6} \times 10.21 = 98$$

$$\frac{\underline{1} \times 10^4}{1.666 \times 10^{-5}} = 6 \times 10^9$$

$$\left(\frac{2}{3}\right)\left(\frac{\underline{3.2}}{32.00}\right) = 0.067 = 6.7 \times 10^{-2}$$

(the 2 and 3 are exact whole numbers here)

Addition and Subtraction

The result of an addition or subtraction operation should have no more *decimal places* (digits to the right of the decimal point) than the term with the least number of decimal places.

$$1.6 + 2.002 + \underline{12} = 15.602 = 16$$
$$24.421 - \underline{6.0} = 18.421 = 18.4$$

In the above examples, the answers have been rounded off to give the proper number of significant figures or decimal places. Rules for rounding off are as follows.

1. Determine the number of significant figures desired and consider the first (nonsignificant) digit of the group to be discarded.

2. If the digit to be dropped is less than 5, the preceding (significant) digit is left unchanged. Thus, to three significant figures, 18.736 becomes 18.7, and 0.01644 becomes 0.0164.

3. If the digit to be dropped is greater than 5, increase the preceding digit by 1. The numbers 20.58 and 0.167664 become, to three significant figures, 20.6 and 0.168.

4. If the digit dropped is 5, increase the preceding digit by one if this digit is odd but leave it alone if it is even. Examples are 21.65 and 21.756, which become 21.6 and 21.8 when rounded off to three significant figures.

A.3 LOGARITHMS

The *common* logarithm of a number is the power or exponent to which the base number 10 must be raised to give the number. That is, if the number is N and

$$N = 10^x$$

then the logarithm of N, log N, is x,

$$\log N = x$$

or

$$N = 10^{\log N}$$

(Natural logarithms, to the base $e = 2.71828$, are used in higher mathematics. The conversion from natural logs, ln, to common logs, log, is $\ln N = 2.303 \log N$.)

Logarithms of integral powers of ten are easy to find. For instance,

$$\log 1 = \log 10^0 = 0$$
$$\log 10 = \log 10^1 = 1$$
$$\log 100 = \log 10^2 = 2$$

Logarithms of other numbers are not so self-evident. For instance, $5 = 10^{0.6990}$ and $\log 5 = 0.6990$. Tables of logarithms list directly the logs of numbers between 1 and 10. Thus, to find log 6.23, we look for 62 in the left-hand column, then move to the right to the number under the 3 in the top line. The number here is 7945. Because 6.23 is between 1 and 10, its logarithm should be between 0 and 1, so the logarithm of 6.23 is 0.7945.

$$\log 6.23 = 0.7945 = 0.794$$

Similarly,

$$\log 1.01 = 0.0043$$
$$\log 8.91 = 0.9499 = 0.950$$

To find the logarithm of numbers less than 1 or greater than 10, use is made of the relation

$$\log (M \times N) = \log M + \log N$$

The log of a product is equal to the sum of the logs of the factors. Using exponential notation, any number less than 1 or greater than 10 can be written as an integral power of ten times a number between 1 and 10.

$$23.6 = 2.36 \times 10^1$$
$$0.000623 = 6.23 \times 10^{-4}$$

Logarithms	N	0	1	2	3	4	5	6	7	8	9
	10	0000	0043	0086	0128	0170	0212	0253	0294	0334	0374
	11	0414	0453	0492	0531	0569	0607	0645	0682	0719	0755
	12	0792	0828	0864	0899	0934	0969	1004	1038	1072	1106
	13	1139	1173	1206	1239	1271	1303	1335	1367	1399	1430
	14	1461	1492	1523	1553	1584	1614	1644	1673	1703	1732
	15	1761	1790	1818	1847	1875	1903	1931	1959	1987	2014
	16	2041	2068	2095	2122	2148	2175	2201	2227	2253	2279
	17	2304	2330	2355	2380	2405	2430	2455	2480	2504	2529
	18	2553	2577	2601	2625	2648	2672	2695	2718	2742	2765
	19	2788	2810	2833	2856	2878	2900	2923	2945	2967	2989
	20	3010	3032	3054	3075	3096	3118	3139	3160	3181	3201
	21	3222	3243	3263	3284	3304	3324	3345	3365	3385	3404
	22	3424	3444	3464	3483	3502	3522	3541	3560	3579	3598
	23	3617	3636	3655	3674	3692	3711	3729	3747	3766	3784
	24	3802	3820	3838	3856	3874	3892	3909	3927	3945	3962
	25	3979	3997	4014	4031	4048	4065	4082	4099	4116	4133
	26	4150	4166	4183	4200	4216	4232	4249	4265	4281	4298
	27	4314	4330	4346	4362	4378	4393	4409	4425	4440	4456
	28	4472	4487	4502	4518	4533	4548	4564	4579	4594	4609
	29	4624	4639	4654	4669	4683	4698	4713	4728	4742	4757
	30	4771	4786	4800	4814	4829	4843	4857	4871	4886	4900
	31	4914	4928	4942	4955	4969	4983	4997	5011	5024	5038
	32	5051	5065	5079	5092	5105	5119	5132	5145	5159	5172
	33	5185	5198	5211	5224	5237	5250	5263	5276	5289	5302
	34	5315	5328	5340	5353	5366	5378	5391	5403	5416	5428
	35	5441	5453	5465	5478	5490	5502	5514	5527	5539	5551
	36	5563	5575	5587	5599	5611	5623	5635	5647	5658	5670
	37	5682	5694	5705	5717	5729	5740	5752	5763	5775	5786
	38	5798	5809	5821	5832	5843	5855	5866	5877	5888	5899
	39	5911	5922	5933	5944	5955	5966	5977	5988	5999	6010
	40	6021	6031	6042	6053	6064	6075	6085	6096	6107	6117
	41	6128	6138	6149	6160	6170	6180	6191	6201	6212	6222
	42	6232	6243	6253	6263	6274	6284	6294	6304	6314	6325
	43	6335	6345	6355	6365	6375	6385	6395	6405	6415	6425
	44	6435	6444	6454	6464	6474	6484	6493	6503	6513	6522
	45	6532	6542	6551	6561	6571	6580	6590	6599	6609	6618
	46	6628	6637	6646	6656	6665	6675	6684	6693	6702	6712
	47	6721	6730	6739	6749	6758	6767	6776	6785	6794	6803
	48	6812	6821	6830	6839	6848	6857	6866	6875	6884	6893
	49	6902	6911	6920	6928	6937	6946	6955	6964	6972	6981
	50	6990	6998	7007	7016	7024	7033	7042	7050	7059	7067
	51	7076	7084	7093	7101	7110	7118	7126	7135	7143	7152
	52	7160	7168	7177	7185	7193	7202	7210	7218	7226	7235
	53	7243	7251	7259	7267	7275	7284	7292	7300	7308	7316
	54	7324	7332	7340	7348	7356	7364	7372	7380	7388	7396

Logarithms (*Continued*)

N	0	1	2	3	4	5	6	7	8	9
55	7404	7412	7419	7427	7435	7443	7451	7459	7466	7474
56	7482	7490	7497	7505	7513	7520	7528	7536	7543	7551
57	7559	7566	7574	7582	7589	7597	7604	7612	7619	7627
58	7634	7642	7649	7657	7664	7672	7679	7686	7694	7701
59	7709	7716	7723	7731	7738	7745	7752	7760	7767	7774
60	7782	7789	7796	7803	7810	7818	7825	7832	7839	7846
61	7853	7860	7868	7875	7882	7889	7896	7903	7910	7917
62	7924	7931	7938	7945	7952	7959	7966	7973	7980	7987
63	7993	8000	8007	8014	8021	8028	8035	8041	8048	8055
64	8062	8069	8075	8082	8089	8096	8102	8109	8116	8122
65	8129	8136	8142	8149	8156	8162	8169	8176	8182	8189
66	8195	8202	8209	8215	8222	8228	8235	8241	8248	8254
67	8261	8267	8274	8280	8287	8293	8299	8306	8312	8319
68	8325	8331	8338	8344	8351	8357	8363	8370	8376	8382
69	8388	8395	8401	8407	8414	8420	8426	8432	8439	8445
70	8451	8457	8463	8470	8476	8482	8488	8494	8500	8506
71	8513	8519	8525	8531	8537	8543	8549	8555	8561	8567
72	8573	8579	8585	8591	8597	8603	8609	8615	8621	8627
73	8633	8639	8645	8651	8657	8663	8669	8675	8681	8686
74	8692	8698	8704	8710	8716	8722	8727	8733	8739	8745
75	8751	8756	8762	8768	8774	8779	8785	8791	8797	8802
76	8808	8814	8820	8825	8831	8837	8842	8848	8854	8859
77	8865	8871	8876	8882	8887	8893	8899	8904	8910	8915
78	8921	8927	8932	8938	8943	8949	8954	8960	8965	8971
79	8976	8982	8987	8993	8998	9004	9009	9015	9020	9025
80	9031	9036	9042	9047	9053	9058	9063	9069	9074	9079
81	9085	9090	9096	9101	9106	9112	9117	9122	9128	9133
82	9138	9143	9149	9154	9159	9165	9170	9175	9180	9186
83	9191	9196	9201	9206	9212	9217	9222	9227	9232	9238
84	9243	9248	9253	9258	9263	9269	9274	9279	9284	9289
85	9294	9299	9304	9309	9315	9320	9325	9330	9335	9340
86	9345	9350	9355	9360	9365	9370	9375	9380	9385	9390
87	9395	9400	9405	9410	9415	9420	9425	9430	9435	9440
88	9445	9450	9455	9460	9465	9469	9474	9479	9484	9489
89	9494	9499	9504	9509	9513	9518	9523	9528	9533	9538
90	9542	9547	9552	9557	9562	9566	9571	9576	9581	9586
91	9590	9595	9600	9605	9609	9614	9619	9624	9628	9633
92	9638	9643	9647	9652	9657	9661	9666	9671	9675	9680
93	9685	9689	9694	9699	9703	9708	9713	9717	9722	9727
94	9731	9736	9741	9745	9750	9754	9759	9763	9768	9773
95	9777	9782	9786	9791	9795	9800	9805	9809	9814	9818
96	9823	9827	9832	9836	9841	9845	9850	9854	9859	9863
97	9868	9872	9877	9881	9886	9890	9894	9899	9903	9908
98	9912	9917	9921	9926	9930	9934	9939	9943	9948	9952
99	9956	9961	9965	9969	9974	9978	9983	9987	9991	9996

so that

$$\begin{aligned}
\log 23.6 &= \log (2.36 \times 10^1) \\
&= \log 2.36 + \log 10^1 \\
&= 0.3729 + 1 \\
&= 1.3729 = 1.373 \\
\log 0.000623 &= \log (6.23 \times 10^{-4}) \\
&= \log 6.23 + \log 10^{-4} \\
&= 0.7945 - 4 \\
&= -3.2055 = -3.206
\end{aligned}$$

Other relationships often used in working with logarithms are

$$\log \frac{M}{N} = \log M - \log N$$

$$\log M^a = a \log M$$

Often, the logarithm of a number is given and the number must be found. This is the case when the pH of a solution is measured and the hydrogen ion concentration must be derived. If a positive logarithm is given, the quantity to the right of the decimal point (the mantissa) is located in the body of the log table and the number corresponding to it is read from the left-hand column and topmost line. Thus, the number whose logarithm is 0.5465 is 3.520. The number whose logarithm is 5.6665 is 4.640×10^5. Only the mantissa is given in the log table. (In our examples, we use the rule that the number and the mantissa of its logarithm have the same number of significant figures.)

If a negative logarithm is given, it must first be converted to a form with a positive mantissa, since only positive values are listed in the table. As an example, the number whose logarithm is -6.8962 can be found by the following procedure:

1. $-6.8962 = 0.1038 - 7$.
2. The number whose log (mantissa) is 0.1038 is 1.270.
3. The number whose log is -7 is 10^{-7}.
4. By $\log MN = \log M + \log N$, the number whose log is $0.1038 - 7$ is 1.270×10^{-7}.

One more example:

$$\begin{aligned}
\text{pH} &= 7.45 \\
\log [\text{H}^+] &= -7.45 \\
&= 0.55 - 8 \\
[\text{H}^+] &= 3.6 \times 10^{-8} M
\end{aligned}$$

A.4 UNITS AND PROPORTION IN PROBLEM SOLVING

Numerical problems in chemistry, as in any other physical science, contain experimentally measured quantities. These quantities consist of a number giving the magnitude of the quantity plus a dimensional term giving the units used in the measurement of the quantity. Thus, a particular distance is written not merely as 3.6, but as 3.6 meters (m), or 360 centimeters (cm), or 3600 millimeters (mm), or 142 inches (in.). Most physical quantities are derived quantities expressed in terms of a few fundamental *dimensions*. The most important fundamental dimensions are length, mass, and time. These dimensions are measured in standard *units* in various systems. In this book, the metric system is used almost exclusively. In this system, the fundamental units of length, mass, and time are the meter (or centimeter), the kilogram (or gram), and the second, respectively. In the English system, the fundamental units of the corresponding dimensions are the foot, the pound mass, and the second. Within each system, a number of other units, such as the millimeter, the inch, the ounce, and the hour, can be defined in terms of the fundamental units. As examples, quantities such as 2.6 m and 142 in. have the dimensions of length. Similarly, 10 kg and 2.4 g have the dimensions of mass, and 2.6 sec, 24 hr, and 1 day have the dimensions of time. Derived quantities such as velocity and volume have dimensions of length/time and (length)^3, respectively, in whatever units they are expressed.

In any physical equation, it is essential that both the dimensions and the units match in all terms of the equation. So far as dimensions are concerned, an equation $X = Y$ or $X - Y = 0$ must have X and Y in the same dimensions, because an equation such as 2 quarts -6 miles $= 0$ is meaningless.

Factoring Units   Units can be multiplied, divided, and otherwise factored in the same way as numbers in an equation. To illustrate how this applies to the requirement that the units match in an equation, we can use the simple equation for the distance s traveled in a time t by a body moving at a constant velocity v, $s = vt$. If $v = 10$ m/sec and $t = 5$ sec, then $s = vt = (10)(5) = 50$. As written, this is numerically correct, but to check the unit-matching requirement, the solution should have been written

$$s = \left(10\frac{m}{sec}\right)(5\ sec) = 50\ m$$

Treating the units separately and factoring,

$$\left(\frac{m}{\cancel{sec}}\right)(\cancel{sec}) = m$$

shows that the product vt has the same dimensions and the same units as required for the distance s. If we had tried to write $s = v/t$, the insertion of units would have quickly shown that this was an incorrect equation.

$$s(m) \neq \frac{m/sec}{sec} = \frac{m}{sec^2}$$

Neither dimensions nor units match.

Two points follow from this simple example:
1. Know the units of your answer.
2. Insert all units in the calculation and multiply and divide them as well as the numerical factors. The units should eventually factor out to give the dimensions and the units in the desired answer.

The use of these two guidelines forms the basis of one approach, the factor-unit method, to the solution of many numerical problems. Using this method, some problems can be solved completely without the use of any defining equations. Other problems can at least be checked by inspection of the units. These examples are given here.

EXAMPLE   Convert 2.0 ft to centimeters given the conversions

1 ft $= 12$ in.
1 in. $= 2.54$ cm

These conversions are not physical equations. They are ways of writing the conversion factors

$$1 \frac{\text{ft}}{12 \text{ in.}} \quad \text{and} \quad 1 \frac{\text{in.}}{2.54 \text{ cm}}$$

or

$$12 \frac{\text{in.}}{\text{ft}} \quad \text{and} \quad 2.54 \frac{\text{cm}}{\text{in.}}$$

SOLUTION

The answer sought has units of centimeters. To get centimeters from feet, the units of the calculation would have to be arranged as follows:

$$\text{cm} = (\text{ft}) \left(\frac{\text{in.}}{\text{ft}} \right) \left(\frac{\text{cm}}{\text{in.}} \right) = \text{cm}$$

Inserting numbers,

$$\text{cm} = (2.0 \text{ ft}) \left(\frac{12 \text{ in.}}{1 \text{ ft}} \right) \left(\frac{2.54 \text{ cm}}{1 \text{ in.}} \right)$$

or

$$\text{cm} = (2.0 \text{ ft})(12 \text{ in./ft})(2.54 \text{ cm/in.})$$
$$= 61 \text{ cm}$$

EXAMPLE

How many atoms are in 2.0 liters of N_2 gas at STP?

SOLUTION

The answer must be in *atoms*. The data is in *liters*. With units only,

$$\text{atoms} = (\text{liters}) \left(\frac{\text{mol}}{\text{liters}} \right) \left(\frac{\text{molecules}}{\text{mol}} \right) \left(\frac{\text{atoms}}{\text{molecule}} \right)$$

Inserting known conversion factors,

$$\text{atoms} = (2.0 \text{ liters}) \left(\frac{1 \text{ mol}}{22.4 \text{ liters}} \right) \left(\frac{6.02 \times 10^{23} \text{ molecules}}{1 \text{ mol}} \right) \left(\frac{2 \text{ atoms}}{1 \text{ molecule}} \right)$$
$$= 1.1 \times 10^{23} \text{ atoms}$$

EXAMPLE

The volume of a gas, in liters, is reported as the result of a calculation based on the perfect gas equation with $P = 1$ atm, $n = 2$ mol, $T = 273°K$, and $R = 1.987$ cal/°K mol. Using these values, however, the volume of gas should come out in the units

$$V = \frac{nRT}{P} = \frac{(\text{mol})}{(\text{atm})} \left(\frac{\text{cal}}{\text{mol °K}} \right) (°K) = \frac{\text{cal}}{\text{atm}}$$

Although the dimensions are correct, the volume comes out in calories per atmosphere, not liters. The reported volume, in liters, must be incorrect. If the gas constant R, in the units liter atm/mol °K, had been used rather than R in cal/°K mol, the correct volume in liters would have been obtained. This shows how the factor-unit method can be used to check a calculation even if it is not used in setting up the computation.

Proportion | The factor-unit approach sidesteps the reasoning process based on proportion and proportionalities, even though the approach is based on proportion. Some people feel that the use of proportion is old fashioned and prefer to go to the factor-unit immediately. Others feel that mere juggling of units is dangerous if there is nothing else to back up the factor-unit approach. You decide for yourself after looking at the following examples of problems solved by proportion.

EXAMPLE | What distance is covered in 5 sec by a body going 10 m/sec?

REASONING | If the body travels 10 m in 1 sec, how many meters will it go in 5 sec?

The proportion

$$\frac{10 \text{ m}}{1 \text{ sec}} = \frac{s}{5 \text{ sec}} \qquad (\text{or } 10 : 1 = s : 5)$$

can be thought of as "10 m is (corresponds) to 1 sec as (in the same ratio as) s m is to 5 sec."

Solving for s,

$$s = \frac{10 \text{ m}}{1 \text{ sec}} \times 5 \text{ sec}$$

$$= 50 \text{ m}$$

EXAMPLE | If 20 liters weigh 100 g, what is the weight of 4 liters?

REASONING | One hundred grams is to 20 liters as m grams is to 4 liters, or

$$\frac{100 \text{ g}}{20 \text{ liters}} = \frac{m \text{ g}}{4 \text{ liters}}$$

$$m = \frac{100 \text{ g}}{20 \text{ liters}}(4 \text{ liters})$$

$$= 20 \text{ g}$$

Try this by the factor-unit method.

EXAMPLE | How many atoms are in 2.0 liters of N_2 gas at STP?

SOLUTION | The most important thing you should know is the volume of 1 mol of N_2 at STP. This is the conversion factor 1 mol/22.4 liters used in the factor-unit method. Without writing out the proportion equation, we have

$$\text{moles} = \frac{(2 \text{ liters})(1 \text{ mol})}{22.4 \text{ liters}} = \frac{2.0}{22.4} \text{mol}$$

Again, by proportion,

$$\text{molecules} = \left(\frac{2}{22.4} \text{ mol}\right)\left(\frac{6.02 \times 10^{23} \text{ molecules}}{\text{mol}}\right)$$

$$= \left(\frac{2}{22.4}\right)(6.02 \times 10^{23}) \text{ molecules}$$

By a third proportion (again not written out),

$$\text{atoms} = \left(\frac{2}{22.4}\right)(6.02 \times 10^{23}) \; \text{molecules} \left(\frac{2 \text{ atoms}}{1 \text{ molecule}}\right)$$

$$= 1.1 \times 10^{23} \text{ atoms}$$

In these examples units are included, but they have been used to check the calculation, not to design the computation. In the last example, the three separate reasoning steps could have been strung together in one line of calculations, but the end result is still reached by working with the figures first and the units second, whereas the factor-unit method puts all the units down first, then adds the figures.

Appendix **B**

Chemical Calculations

B.1 CALCULATIONS BASED ON CHEMICAL FORMULAS
Formula from Composition

The chemical formula of a substance, in addition to showing which elements compose the substance, indicates the *relative* proportions of the different elements in the compound. Thus, the formula, NO_2 indicates that nitrogen and oxygen are present in the *ratio* of one atom of nitrogen to two atoms of oxygen. The words *relative* and *ratio* are used here because at this point we cannot be sure whether the formula is NO_2, N_2O_4, N_3O_6, or some other formula with the same relative numbers of atoms. The formula giving only the relative numbers of atoms is called the *simplest formula*. The *true* or *molecular* formula will be some integral multiple of the simplest formula.

The symbol of a chemical element signifies either 1 atom or 1 g-atom of the element; the simplest formula thus gives the relative numbers of gram-atoms of each element in the compound. The formula NO_2 represents

1 g-atom (or 1 mol) of N *atoms* associated with 2 g-atoms (or 2 mol) of O atoms.

To find the simplest formula of a compound from its composition, the relative number of gram-atoms of each element must be calculated from the weights of the component elements present in a sample of the compound.

EXAMPLE

What is the simplest formula of an iron oxide if a 1.347-g sample of the oxide contains 0.405 g of oxygen and 0.942 g of iron?

SOLUTION

First find the number of gram-atoms of each element present:

$$\frac{0.942 \text{ g}}{55.85 \text{ g/g-atom}} = 0.0169 \text{ g-atom of Fe}$$

$$\frac{0.405 \text{ g}}{16.00 \text{ g/g-atom}} = 0.0253 \text{ g-atom of O}$$

Next find the ratio of the gram-atoms of each element by dividing the gram-atoms of each element by the lowest result, which can now be regarded as a pure number:

$$\frac{0.0169 \text{ g-atom}}{0.0169} \simeq 1 \text{ g-atom of Fe}$$

$$\frac{0.0253 \text{ g-atom}}{0.0169} \simeq 1.5 \text{ g-atom of O}$$

Multiply by 2 to obtain the set of smallest whole numbers:

$$2(1) \simeq 2 \text{ g-atom of Fe}$$
$$2(1.5) \simeq 3 \text{ g-atom O}$$

The relative number (in whole numbers) of gram-atoms of Fe to O is therefore 2 to 3, and, the simplest formula of the oxide is Fe_2O_3.

EXAMPLE

A compound contains 40.0 percent C, 6.7 percent H, and 53.3 percent O. One gram of the compound as a gas occupies 489 ml at 80°C and 755 mm pressure. Find its molecular formula.

SOLUTION

The sample weight is immaterial and, for convenience, may be considered to be 100 g so that the percentage composition gives the weight of each element. Thus, we can find the number of gram-atoms of each element present:

$$\frac{40.0 \text{ g}}{12.0 \text{ g/g-atom}} = 3.36 \text{ g-atom of C}$$

$$\frac{6.7 \text{ g}}{1.0 \text{ g/g-atom}} = 6.7 \text{ g-atom of H}$$

$$\frac{53.3 \text{ g}}{16.0 \text{ g/g-atom}} = 3.36 \text{ g-atom of O}$$

Dividing by 3.36 gives, for the relative number of gram-atoms of each element, 1 g-atom of C, 2 g-atom of H, and 1 g-atom of O. The simplest formula is CH_2O.

To determine the true formula, the molecular weight of the compound must be known. This is obtained from the *PVT* data. The volume of 1 g of the compound at 755 mm and 80°C is 489 ml. Thus, the volume of 1.0 g at STP is

$$(489 \text{ ml}) \left(\frac{273°K}{273 + 80°K} \right) \left(\frac{755 \text{ mm}}{760 \text{ mm}} \right) = 376 \text{ ml} = 0.376 \text{ liter}$$

If 0.376 liter weighs 1.0 g, then 1 mol of gas, or 22.4 liters at STP, weighs

$$\left(\frac{1.0 \text{ g}}{0.376 \text{ liter}} \right) \left(22.4 \frac{\text{liters}}{\text{mol}} \right) = 59 \text{ g/mol}$$

Because the molecular weight of CH_2O is only 30 g/mol, the true formula must be $2(CH_2O) = C_2H_4O_2$.

Composition from Formula   The percentage composition of a compound of known formula is easily derived from the molecular weight of the substance.

EXAMPLE What are the percentages of aluminum and oxygen in Al_2O_3? The molecular weight (or formula weight) of Al_2O_3 is

$$2(26.98) + 3(16.00) = 102.0 \text{ g/mol}$$

$$\text{percent aluminum} = \frac{2(26.98)}{102.0} \times 100 = 52.92\%$$

$$\text{percent oxygen} = \frac{3(16.00)}{102.0} \times 100 = 47.08\%$$

$$(\text{or } 100 - 52.92\%)$$

The quantity of an element in a specified amount of compound may be calculated in at least two ways: (1) from the fraction or percentage of the element in the compound or (2) on a mole basis.

EXAMPLE How many grams of aluminum are in 25.0 g of Al_2O_3?

SOLUTION The fraction of Al in Al_2O_3 is $2(26.98)/101.96 = 0.5292$ g of Al/g of Al_2O_3 (or 52.92 percent = 52.92 g of Al/100 g of Al_2O_3). There is the same fraction of Al in 25.0 g of Al_2O_3,

$$\left(0.5292 \frac{\text{g of Al}}{\text{g of Al}_2O_3} \right) (25.0 \text{ g of Al}_2O_3) = 13.2 \text{ g of Al}$$

On a mole basis,

$$25.0 \text{ g of Al}_2O_3 = \frac{25.0 \text{ g}}{101.96 \text{ g/mol of Al}_2O_3}$$

$$25.0 \text{ g of Al}_2O_3 = 0.245 \text{ mol of Al}_2O_3$$

Since 1 mol of Al_2O_3 contains

$$2(26.98 \text{ g}) = 53.96 \text{ g of Al}$$

0.245 mol contains

$$(0.245 \text{ mol}) \left(53.96 \frac{\text{g}}{\text{mol}} \right) = 13.2 \text{ g of Al}$$

B.2 CALCULATIONS BASED ON CHEMICAL EQUATIONS

A balanced chemical equation shows at a glance the relative numbers of moles of reactants and products involved in the reaction. From the mole relationships, weight (or volume, in the case of gases) relationships may be derived.

Weight–Weight Relationships
EXAMPLE

A reaction in the alkalized alumina process for removal of SO_2 from effluent gases is

$$4SO_2 + Na_2O + Al_2O_3 \longrightarrow Na_2SO_3 + Al_2(SO_3)_3$$

How many grams of sodium sulfite would be obtained from the complete reaction of 100 g of sulfur dioxide?

SOLUTION

The balanced equation tells us that 4 mol of SO_2 yield 1 mol of Na_2SO_3, or that 1 mol of SO_2 yields $\frac{1}{4}$ mol of Na_2SO_3. The molecular weight of SO_2 is $32.06 + 2(16.00) = 64.06$ g/mol. The number of moles of SO_2 in 100 g is therefore

$$\frac{100 \text{ g}}{64.06 \text{ g/mol}}$$

Because 1 mol of SO_2 gives $\frac{1}{4}$ mol of Na_2SO_3, (100/64.06) mol of SO_2 will give

$$\left(\frac{100}{64.06} \text{ mol of } SO_2\right)\left(\frac{1}{4}\frac{\text{mol of } Na_2SO_3}{\text{mol of } SO_2}\right) = \text{mol of } Na_2SO_3$$

The grams of Na_2SO_3 obtained are found by multiplying the moles by its molecular weight, 124.04 g/mol of Na_2SO_3.

$$\text{grams of } Na_2SO_3 = \left(\frac{100}{64.06} \text{ mol of } SO_2\right)\left(\frac{1}{4}\frac{\text{mol of } Na_2SO_3}{\text{mol of } SO_2}\right)$$
$$\left(124.04\frac{\text{g}}{\text{mol of } Na_2SO_3}\right)$$
$$= 48.4 \text{ g}$$

EXAMPLE

An impure sample of potassium chlorate was heated to produce oxygen gas according to the equation

$$2KClO_3 \longrightarrow 2KCl + 3O_2$$

If 3.2 g of oxygen were produced, how much $KClO_3$ was in the sample?

SOLUTION

As above, it is useful to think of the equation as having its coefficients divided by the coefficient of the *substance whose quantity is known*, that is, O_2.

$$\tfrac{2}{3}KClO_3 \longrightarrow \tfrac{2}{3}KCl + O_2$$

This shows that 1 mol of oxygen corresponds to $\frac{2}{3}$ mol of $KClO_3$. Now convert weight to moles.

$$\text{moles of } O_2 = \frac{3.2 \text{ g of } O_2}{32.00 \text{ (g of } O_2/\text{mol of } O_2)}$$

$$\text{moles of } KClO_3 = \left(\frac{2}{3} \frac{\text{mol of } KClO_3}{\text{mol of } O_2}\right)(\text{moles of } O_2)$$

$$= \left(\frac{2}{3} \frac{\text{mol of } KClO_3}{\text{mol of } O_2}\right)\left[\frac{3.2 \text{ g of } O_2}{32.00 \text{ (g of } O_2/\text{mol of } O_2)}\right]$$

$$\text{grams of } KClO_3 = (\text{mol wt of } KClO_3)(\text{moles of } KClO_3)$$

$$= \left(122.55 \frac{\text{g of } KClO_3}{\text{mol of } KClO_3}\right)\left(\frac{2}{3} \frac{\text{mol of } KClO_3}{\text{mol of } O_2}\right)$$

$$\left[\frac{3.2 \text{ g of } O_2}{32.00 \text{ (g of } O_2/\text{mol of } O_2)}\right]$$

$$= 8.2 \text{ g of } KClO_3$$

These two examples illustrate the mole method of calculation. As much of the calculation as possible is kept in terms of moles of substance. Weights are converted immediately to moles at the beginning of the calculation and only returned to weight units at the very end.

EXAMPLE | How many moles of CO_2 will be formed by the combustion of 2.3 g of ethanol (mol wt = 46.07) with 3.2 g of oxygen?

$$C_2H_5OH + 3O_2 \longrightarrow 2CO_2 + 3H_2O$$

SOLUTION |

$$\text{moles of } C_2H_5OH = \frac{2.3}{46.07} = 0.050 \text{ mol}$$

$$\text{moles of } O_2 = \frac{3.2}{32.00} = 0.10 \text{ mol}$$

For complete combustion of all the ethanol, $3(0.050) = 0.15$ mol of O_2 would be required. This much oxygen is not available, so only $\frac{1}{3}(0.10) = 0.033$ mol of C_2H_5OH will react. The quantity of product formed will be determined by the reactant present in least amount relative to the stoichiometric requirements of the reaction. Here that limiting reactant is oxygen. Therefore, because 1 mol of O_2 yields $\frac{2}{3}$ mol of CO_2, 0.10 mol of O_2 will yield $\frac{2}{3}(0.10)$ mol of O_2.

Weight–Volume Relationships | If volumes of gaseous reactants or products are involved, it is usually necessary to convert from volume to moles or vice versa with proper attention to conditions of temperature and pressure. In general, this means correcting volumes to or from standard temperature and pressure to permit comparison with the gram-molecular volume of 22.4 liters for gases. See Chapter 4 for gas calculations.

EXAMPLE | What volume of dry nitric oxide, measured at 20°C and 720 mm pressure, would be obtained by the complete reaction of 10.0 g of copper(II) sulfide with hot nitric acid? The balanced equation is

$$3CuS + 8H^+ + 2NO_3^- \longrightarrow 3Cu^{2+} + 3S + 2NO + 4H_2O$$
$$(\text{or } 3CuS + 8HNO_3 \longrightarrow 3Cu(NO_3)_2 + 3S + 2NO + 4H_2O)$$

SOLUTION

$$\text{molecular weight of CuS} = 95.61$$

$$\text{moles of CuS} = \frac{10.0 \text{ g}}{95.61 \text{ g/mol}}$$

$$\text{moles of NO produced} = \left(\frac{2}{3}\frac{\text{mol of NO}}{\text{mol of CuS}}\right)\left(\frac{10.0 \text{ g of CuS}}{95.61 \text{ g/mol}}\right)$$

$$\text{liters of NO produced (STP)} = \left(\frac{2}{3}\frac{\text{mol of NO}}{\text{mol of CuS}}\right)\left(\frac{10.0 \text{ g of CuS}}{95.61 \text{ g/mol}}\right)$$
$$\left(22.4\frac{\text{NO}}{\text{mol of NO}}\right)$$

liters of NO produced at 20°C and 720 mm

$$= \left(\frac{2}{3}\frac{\text{mol of NO}}{\text{mol of CuS}}\right)\left(\frac{10.0 \text{ g of CuS}}{95.61 \text{ g/mol}}\right)\left(22.4\frac{\text{liters of NO}}{\text{mol NO}}\right)$$
$$\left(\frac{293°\text{K}}{273°\text{K}}\right)\left(\frac{760 \text{ mm}}{720 \text{ mm}}\right)$$

$$= 1.77 \text{ liters of NO}$$

Volume–Volume Relationships

According to Avogadro's law, equal volumes of gases, measured at the same temperature and pressure, contain equal numbers of molecules, or equal numbers of moles. When the gas volumes in a chemical reaction are measured under the same conditions (not necessarily STP), the balanced equation gives the relative volumes of all gases directly.

EXAMPLE

How many liters of ammonia are formed from the reaction of 100 liters of nitrogen and 500 liters of hydrogen, all measured at 500°C and 1000 atm? The equation is

$$N_2 + 3H_2 \longrightarrow 2NH_3$$

SOLUTION

The amount of N_2 available is less than one-third the amount of H_2 present. Thus, N_2 is the limiting reagent here and controls the amount of ammonia produced. (Only 300 liters of H_2 are required.) The volume of ammonia produced will be

$$\frac{2 \text{ liters of NH}_3}{1 \text{ liter of N}_2}(100 \text{ liters of N}_2) = 200 \text{ liters of NH}_3$$

Appendix

Naming of Simple Inorganic Compounds

C.1 BINARY COMPOUNDS

Binary compounds contain only two different elements. Examples are NaCl and Al_2S_3. In general, binary compounds are written and named by first stating the symbol or name of the more metallic (less electronegative) element followed by a name derived from the name of the more electronegative element by adding the suffix *-ide*.

NaCl	sodium chloride
HF(g)	hydrogen fluoride
Al_2S_3	aluminum sulfide
CaC_2	calcium carbide

Where two or more binary compounds of the same elements may occur because of the existence of variable oxidation states of the more metallic element, additional information must be included in the name of the compound. There are three methods in use.

1. The Stock system denotes the oxidation state of the electropositive element by Roman numerals.

$FeCl_2$	iron(II) chloride
$FeCl_3$	iron(III) chloride
MnO_2	manganese(IV) oxide
Mn_2O_7	manganese(VII) oxide
Hg_2Cl_2	mercury(I) chloride
$HgCl_2$	mercury(II) chloride

2. The endings -ous and -ic are used to indicate lower and higher oxidation states, respectively.

$FeCl_2$	ferrous chloride
$FeCl_3$	ferric chloride
Hg_2Cl_2	mercurous chloride
$HgCl_2$	mercuric chloride

The -ous and -ic notation is being supplanted by the Stock notation, but it is still widely used where only two oxidation states commonly occur.

3. Prefix notation is widely used for predominantly covalent binary compounds and sometimes for ionic compounds.

PCl_3	phosphorus trichloride
PCl_5	phosphorus pentachloride
N_2O	dinitrogen oxide
NO_2	nitrogen dioxide
N_2O_4	dinitrogen tetroxide
N_2O_5	dinitrogen pentoxide
NO	nitrogen oxide

The prefix notation avoids the ambiguity that sometimes arises in the other kinds of notation. Thus, in prefix notation, NO_2 and N_2O_4 have different names, but in the Stock system they are both nitrogen(IV) oxide. In the -ous and -ic suffix notation, the names nitrous oxide and nitric oxide can be applied to only two oxides of nitrogen, and they are commonly applied to N_2O and NO.

4. Exceptions to the systematic names are found with some binary compounds with well-established names and formulas.

H_2O	water (not dihydrogen oxide)
NH_3	ammonia (not H_3N, nitrogen hydride, or hydrogen nitride)

C.2 RADICALS OR POLYATOMIC IONS

A number of polyatomic ions, some derived from oxygen acids (see Appendix C.3), enter as units into compounds with other elements. Common polyatomic ions are listed here.

ammonium	NH_4^+
hydroxide	OH^-
nitrite	NO_2^-
nitrate	NO_3^-

hydrogen carbonate	HCO_3^-
hydrogen sulfate	HSO_4^-
acetate	CH_3COO^- (also commonly written as $C_2H_3O_2^-$, OAc^-, or Ac^-)
permanganate	MnO_4^-
dihydrogen phosphate	$H_2PO_4^-$
cyanide	CN^-
hypochlorite	ClO^-
chlorite	ClO_2^-
chlorate	ClO_3^-
perchlorate	ClO_4^-
sulfite	SO_3^{2-}
sulfate	SO_4^{2-}
thiosulfate	$S_2O_3^-$
carbonate	CO_3^{2-}
chromate	CrO_4^{2-}
dichromate	$Cr_2O_7^{2-}$
monohydrogen phosphate	HPO_4^{2-}
phosphate	PO_4^{3-}

Salts containing these polyatomic ions are named with the positive ion first.

$NaClO_4$	sodium perchlorate
Na_2HPO_4	sodium monohydrogen phosphate

C.3 OXYGEN ACIDS Prefixes and suffixes are used to indicate different oxidation states of the central element. Anions derived from the acids end in *-ate* or *-ite* as shown in Appendix C.2. Anion names are given in parentheses.

HNO_3	nitric acid	(nitrate)
HNO_2	nitrous acid	(nitrite)
H_2SO_4	sulfuric acid	(sulfate)
H_2SO_3	sulfurous acid	(sulfite)

The ending *-ic* goes with the higher oxidation state. If there are more than two oxidation states involved, the prefixes *hypo-* and *per-* are used.

increasing oxidation state of Cl ↓			
	$HClO$	hypochlorous acid	(hypochlorite)
	$HClO_2$	chlorous acid	(chlorite)
	$HClO_3$	chloric acid	(chlorate)
	$HClO_4$	perchloric acid	(perchlorate)

Appendix 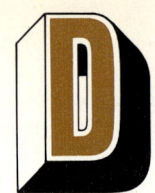 **D**

Balancing Oxidation–Reduction Reactions

Many redox reaction equations cannot be balanced easily by inspection. There are two widely used methods for balancing the more complicated reactions. Both rely on the fact that the total number of electrons gained (or total decrease in oxidation number) by the element(s) reduced must equal the total number of electrons lost (or total increase in oxidation number) by the element(s) oxidized. Both methods require a knowledge of the oxidizing and reducing agents and their products.

Balanced equations should show a balance in ionic charge as well as atoms. No electrons should appear in the final equation.

D.1 THE HALF-REACTION OR ION–ELECTRON METHOD

No oxidation numbers need be assigned when balancing by the half-reaction method unless there is difficulty in determining what elements are undergoing

780

oxidation and reduction. The procedure for the half-reaction method differs slightly for redox reactions taking place in acid or neutral or in basic solutions.

Acid or Neutral Medium
EXAMPLE

Write a balanced equation for the reaction of $Cr_2O_7^{2-}$ and Fe^{2+} to give Cr^{3+} and Fe^{3+} in an acidic or neutral medium.

SOLUTION

1. Write the skeletal half-reactions for the oxidation and reduction processes. Only the species undergoing redox and their products need be included.

$$Cr_2O_7^{2-} \longrightarrow Cr^{3+}$$
$$Fe^{2+} \longrightarrow Fe^{3+}$$

Balance each half-reaction separately as follows.

2. Balance atoms involved in redox.

$$Cr_2O_7^{2-} \longrightarrow 2Cr^{3+}$$
$$Fe^{2+} \longrightarrow Fe^{3+}$$

3. Balance oxygen atoms with H_2O.

$$Cr_2O_7^{2-} \longrightarrow 2Cr^{3+} + 7H_2O$$
$$Fe^{2+} \longrightarrow Fe^{3+}$$

4. Balance hydrogen atoms with H^+.

$$Cr_2O_7^{2-} + 14H^+ \longrightarrow 2Cr^{3+} + 7H_2O$$
$$Fe^{2+} \longrightarrow Fe^{3+}$$

5. Balance ionic charge with electrons.

$$Cr_2O_7^{2-} + 14H^+ + 6e^- \longrightarrow 2Cr^{3+} + 7H_2O$$
$$Fe^{2+} \longrightarrow Fe^{3+} + e^-$$

6. Multiply each half-reaction by the smallest integers that will equalize the number of electrons in each half-reaction.

$$Cr_2O_7^{2-} + 14H^+ + 6e^- \longrightarrow 2Cr^{3+} + 7H_2O$$
$$6(Fe^{2+} \longrightarrow Fe^{3+} + e^-)$$

7. Add the two half-reactions so that the electrons cancel. Also cancel equal amounts of H^+ and H_2O that may appear on both sides of the equation.

$$Cr_2O_7^{2-} + 6Fe^{2+} + 14H^+ \longrightarrow 2Cr^{3+} + 6Fe^{3+} + 7H_2O$$

8. Check final equation for atom and ionic charge balance. All atoms balance. There is a net ionic charge of $+24$ on each side.

Basic Medium

If a redox reaction occurs in a basic solution, hydrogen ions cannot appear in the half-reactions or in the total equation. To balance redox reactions in basic solution, follow steps 1–5 above. After completing step 5 for each half-reaction, add sufficient OH^- to *both* sides of the half-reaction to convert any H^+ present into H_2O. (Cancellation of H_2O's appearing on both sides of a half-reaction may be done here or in step 7.) Then proceed with steps 6–8 as before.

EXAMPLE | Balance the reaction for the reduction of iodine to iodide by thiosulfate ion in a basic medium.

SOLUTION | (1)
$$S_2O_3{}^{2-} \longrightarrow SO_4{}^{2-}$$
$$I_2 \longrightarrow I^-$$

(2)
$$S_2O_3{}^{2-} \longrightarrow 2SO_4{}^{2-}$$
$$I_2 \longrightarrow 2I^-$$

(3)
$$S_2O_3{}^{2-} + 5H_2O \longrightarrow 2SO_4{}^{2-}$$
$$I_2 \longrightarrow 2I^-$$

(4)
$$S_2O_3{}^{2-} + 5H_2O \longrightarrow 2SO_4{}^{2-} + 10H^+$$
$$I_2 \longrightarrow 2I^-$$

(5)
$$S_2O_3{}^{2-} + 5H_2O \longrightarrow 2SO_4{}^{2-} + 10H^+ + 8e^-$$
$$I_2 + 2e^- \longrightarrow 2I^-$$

(Add OH^-, cancel H_2O's).

$$S_2O_3{}^{2-} + 5H_2O + 10OH^- \longrightarrow 2SO_4{}^{2-} + \underbrace{(10H^+ + 10OH^-)}_{10H_2O} + 8e^-$$

or

$$S_2O_3{}^{2-} + 10OH^- \longrightarrow 2SO_4{}^{2-} + 5H_2O + 8e^-$$
$$I_2 + 2e^- \longrightarrow 2I^-$$

(6)
$$S_2O_3{}^{2-} + 10OH^- \longrightarrow 2SO_4{}^{2-} + 5H_2O + 8e^-$$
$$4(I_2 + 2e^- \longrightarrow 2I^-)$$

(7)
$$S_2O_3{}^{2-} + 4I_2 + 10OH^- \longrightarrow 2SO_4{}^{2-} + 8I^- + 5H_2O$$

(8) All atoms balance. There are 12 negative ionic charges on each side.

D.2 THE OXIDATION NUMBER METHOD

In acidic solution $Cr_2O_7{}^{2-}$ and H_2S react to form Cr^{3+} and S. Write a balanced equation for this reaction.

EXAMPLE | 1. Write a skeletal equation with the formulas of the oxidizing and reducing agents on one side and their products on the other.

$$Cr_2O_7{}^{2-} + H_2S \longrightarrow Cr^{3+} + S \quad \text{(acid solution)}$$

2. Assign oxidation numbers to the elements undergoing oxidation and reduction.

$$\overset{+6}{Cr_2}O_7{}^{2-} + \overset{-2}{H_2S} \longrightarrow \overset{+3}{Cr^{3+}} + \overset{0}{S}$$

3. Balance the atoms undergoing oxidation and reduction.

$$\overset{+6}{Cr_2}O_7{}^{2-} + \overset{-2}{H_2S} \longrightarrow 2\overset{+3}{Cr^{3+}} + \overset{0}{S}$$

4. Determine the *total* oxidation number change (or total electron loss or gain) for each element involved in redox.

$$\overset{2e^- \text{ loss}}{\overset{+6 \qquad -2\diagup \quad +3 \qquad \diagdown 0}{Cr_2O_7{}^{2-} + H_2S \longrightarrow Cr^{3+} + S}}$$

3e^- gain per Cr atom

total gain = 6e^-

5. Adjust coefficients (not subscripts) to balance the change in oxidation numbers (or to equalize the electron loss and gain). Multiply H_2S and S by $\frac{6}{2} = 3$ so the total electron loss becomes (3S atoms $\times$ 2e^- loss per S atom) = 6e^- loss.

$$Cr_2O_7{}^{2-} + 3H_2S \longrightarrow 2Cr^{3+} + 3S$$

6. Balance ionic charges with H^+ when the reaction occurs in acid or neutral solution or with OH^- in basic solution.

$$Cr_2O_7{}^{2-} + 3H_2S + 8H^+ \longrightarrow 2Cr^{3+} + 3S$$

7. Balance H atoms with H_2O.

$$Cr_2O_7{}^{2-} + 3H_2S + 8H^+ \longrightarrow 2Cr^{3+} + 3S + 7H_2O$$

8. Check the equation for atom and ionic charge balance. Check oxygen balance first. Each side has 7 O's and equal numbers of other atoms. There is a net ionic charge of +6 on each side.

If a molecular rather than a net ionic equation is desired, the compounds used to furnish the ions appearing in the net ionic equation must be known. For instance, the balanced reaction for the oxidation of oxalate ion, $C_2O_4{}^{2-}$, by permanganate ion is

$$2MnO_4{}^- + 5C_2O_4{}^{2-} + 16H^+ \longrightarrow 2Mn^{2+} + 5CO_2 + 8H_2O$$

This equation shows only the substances directly involved as reactants and products. If the reactant ions were derived from solutions of $KMnO_4$, $Na_2C_2O_4$, and HCl, then the molecular equation could be obtained as follows.

$$(2MnO_4{}^- + 2K^+) + (5C_2O_4{}^{2-} + 10Na^+) + (16H^+ + 16Cl^-) \longrightarrow$$
$$(2Mn^{2+} + 4Cl^-) + (2K^+ + 2Cl^-) + (10Na^+ + 10Cl^-) + 5CO_2 + 8H_2O$$

or

$$KMnO_4 + 5Na_2C_2O_4 + 16HCl \longrightarrow$$
$$2MnCl_2 + 2KCl + 10NaCl + 5CO_2 + 8H_2O$$

The K^+, Na^+, and Cl^- ions are added to both sides of the equation. Note how the Cl^- are apportioned on the right side.

Qualitative Solubility Rules for Simple Inorganic Compounds in Water

1. All nitrates (NO_3^-), nitrites (NO_2^-), and acetates ($C_2H_3O_2^-$) are soluble in water except silver acetate and silver nitrite, which are moderately insoluble.

2. All common lithium, sodium, potassium, and ammonium compounds are soluble.

3. All common chlorides, bromides, and iodides are soluble except those of silver, mercury(I), lead, and HgI_2. The lead halides are soluble in hot water.

4. Common sulfates (SO_4^{2-}) are soluble except those of lead, strontium, barium, and calcium. $AgSO_4$ is slightly soluble.

5. Common carbonates (CO_3^{2-}), phosphates (PO_4^{3-}), chromates (CrO_4^{2-}), and sulfites (SO_3^{2-}) are insoluble except those of sodium, potassium, and ammonium.

6. Common sulfides (S^{2-}) are insoluble except those of the alkali (group IA) metals and ammonium. Sulfides of the alkaline earth (group IIA) metals and aluminum hydrolyze in water to form insoluble hydroxides.

7. Hydroxides (OH^-) are generally insoluble. Exceptions are the very soluble hydroxides of sodium, potassium, and ammonium and the moderately soluble hydroxides of barium, strontium, and calcium.

Appendix F

Units, Conversion Factors, and Physical Constants

Metric System Prefixes

prefix and symbol	power	example
tera, T	10^{12}	
giga, G	10^{9}	
mega, M	10^{6}	megawatt, MW
kilo, k	10^{3}	kilogram, kg
hecto, h	10^{2}	
deka, da	10	
deci, d	10^{-1}	
centi, c	10^{-2}	centimeter, cm
milli, m	10^{-3}	millimeter, mm

micro, μ	10^{-6}	microgram, μg
nano, n	10^{-9}	nanometer, nm
pico, p	10^{-12}	
femto, f	10^{-15}	
atto, a	10^{-18}	

Selected Units, Symbols, and Conversion Factors

Length
1 meter (m) = 100 cm = 1000 mm
1 angstrom (Å) = 10^{-10} m = 10^{-8} cm

The preferred names for the micron and millimicron are now the micrometer, μm (10^{-6} m) and the nanometer, nm (10^{-9} m), respectively.

1 inch (in.) = 2.54 cm
1 kilometer (km) = 0.6214 mile

Mass
1 kilogram (kg) = 1000 grams (g)
$\qquad$ = 2.2046 pounds (lb)
1 atomic mass unit (amu) = 1.6604×10^{-24} g
1 metric ton = 1000 kg
1 (short) ton = 2000 lb = 907.18 kg

Volume
1 liter = 1000 milliliters (ml)
$\qquad$ = 10^{-3} cubic meters (m^3)
$\qquad$ = 10^3 cubic centimeters (cm^3)
$\qquad$ = 1.0567 quarts (qt)
$\qquad$ = 0.264 U.S. gallon (gal)

Energy
1 calorie (cal) = 4.184 joules (J)
$\qquad$ = 4.184×10^7 g cm^2/sec^2 (ergs)
1 liter atmosphere (liter atm) = 24.217 cal
1 electron volt (eV) = 1.6021×10^{-19} J
$\qquad$ = 3.83×10^{-20} cal
1 kilowatt hour (kW hr) = 8.604×10^5 cal
$\qquad$ = 2.27×10^{19} million electron volts (MeV)
1 amu is equivalent to 931 MeV.

Pressure
1 atmosphere (atm) = 1.0133×10^6 g/cm sec^2 (dyne/cm^2)
$\qquad$ = 760 mmHg
$\qquad$ = 14.696 lb/in.^{2}
1 mmHg = 1 torr

Other
1 cycle per second (cps or sec^{-1}) = 1 hertz (Hz)

Selected Physical Constants

| Speed of light in vacuum, c | 2.9979×10^{10} cm/sec |
| Avogadro's number, N_A or N | 6.0225×10^{23} molecules/mol |

Electron rest mass, m 9.1091×10^{-28} g

Electron charge, e 1.60210×10^{-19} coulomb (C)

4.80298×10^{-10} electrostatic units (esu, $g^{1/2}$ $cm^{3/2}$ sec^{-1})

Gas constant, R 0.08205 liter atm/°K mol

1.987 cal/°K mol

8.314 J/°K mol

Planck's constant, h 6.6256×10^{-27} erg sec

Boltzmann's constant, k 1.3805×10^{-16} erg/°K molecule

Faraday's constant, $\mathfrak{F}$ $96{,}487$ coulombs/equivalent

$23{,}061$ cal/volt equivalent

Index